THE ELEMENTS

3	4	5	6	7	8
				1 **H** 1.00794	2 **He** 4.00260
5 **B** 10.81	6 **C** 12.011	7 **N** 14.0067	8 **O** 15.9994	9 **F** 18.998403	10 **Ne** 20.179
13 **Al** 26.98154	14 **Si** 28.0855	15 **P** 30.97376	16 **S** 32.06	17 **Cl** 35.453	18 **Ar** 39.948

28 **Ni** 58.69	29 **Cu** 63.546	30 **Zn** 65.38	31 **Ga** 69.72	32 **Ge** 72.59	33 **As** 74.9216	34 **Se** 78.96	35 **Br** 79.904	36 **Kr** 83.80
46 **Pd** 106.42	47 **Ag** 107.8682	48 **Cd** 112.41	49 **In** 114.82	50 **Sn** 118.69	51 **Sb** 121.75	52 **Te** 127.60	53 **I** 126.9045	54 **Xe** 131.29
78 **Pt** 195.08	79 **Au** 196.9665	80 **Hg** 200.59	81 **Tl** 204.383	82 **Pb** 207.2	83 **Bi** 208.9804	84 **Po** (209)	85 **At** (210)	86 **Rn** (222)

62 **Sm** 150.36	63 **Eu** 151.96	64 **Gd** 157.25	65 **Tb** 158.9254	66 **Dy** 162.50	67 **Ho** 164.9304	68 **Er** 167.26	69 **Tm** 168.9342	70 **Yb** 173.04
94 **Pu** (244)	95 **Am** (243)	96 **Cm** (247)	97 **Bk** (247)	98 **Cf** (251)	99 **Es** (252)	100 **Fm** (257)	101 **Md** (258)	102 **No** (259)

Chemical Principles

Sixth Edition

Saunders College Publishing
Complete Package for Teaching
General Chemistry with Chemical Principles, 6th Edition

Masterton
Instructor's Manual to Accompany **Chemical Principles**, 6th Edition.

Slowinski, Wolsey & Masterton
Chemical Principles in the Laboratory, 4th Edition.

Slowinski, Wolsey & Masterton
Instructor's Manual to Accompany **Chemical Principles in the Laboratory**, 4th Edition.

Hurley
Study Guide/Workbook to Accompany **Chemical Principles**, 6th Edition.

Kroening & Edgar
Student Solutions Manual to Accompany **Chemical Principles**, 6th Edition.

Stanitski
Test Bank to Accompany **Chemical Principles**, 6th Edition.

Computerized Test Bank to Accompany **Chemical Principles**, 6th Edition.

Masterton, Slowinski & Stanitski
Overhead Transparencies to Accompany **Chemical Principles**, 6th Edition. 100

Wilkie
Saunders Computer Chemistry Package

Smith & Chabay
COMPress: **Introduction to General Chemistry**

Shakhashiri, Schreiner & Meyer
General Chemistry Audio-Tape Lessons, 2nd Edition.

Shakhashiri, Schreiner & Meyer
Workbook for General Chemistry Audio-Tape Lessons, 2nd Edition.

Masterton & Slowinski
Mathematical Preparation for General Chemistry, 2nd Edition.

Slowinski & Masterton
Qualitative Analysis and the Properties of Ions in Aqueous Solution

William L. Masterton
Professor of Chemistry
University of Connecticut
Storrs, Connecticut

Emil J. Slowinski
Professor of Chemistry
Macalester College
St. Paul, Minnesota

Conrad L. Stanitski
Professor of Chemistry
Randolph-Macon College
Ashland, Virginia

Chemical Principles

Sixth Edition

Saunders Golden Sunburst Series **Saunders College Publishing**
Philadelphia New York Chicago
San Francisco Montreal Toronto
London Sydney Tokyo Mexico City
Rio de Janeiro Madrid

Address orders to:
383 Madison Avenue
New York, NY 10017

Address editorial correspondence to:
West Washington Square
Philadelphia, PA 19105

Text Typeface: Times Roman
Compositor: General Graphic Services, Inc.
Acquisitions Editor: John Vondeling
Project Editor: Sally Kusch
Copyeditor: Ruth Melnick
Art Director: Carol Bleistine
Art/Design Assistant: Virginia A. Bollard
Text Design: Arlene Putterman
Cover Design: Lawrence R. Didona
Text Artwork: Tom Mallon and Larry Ward
Production Manager: Tim Frelick
Assistant Production Manager: Maureen Iannuzzi

Cover credit: Pouring steel. © Alvis Upitis/THE IMAGE BANK

Library of Congress Cataloging in Publication Data

Masterton, William L., 1927–
  Chemical principles.

  (Saunders golden sunburst series)
  Includes index.
   1. Chemistry.   I. Slowinski, Emil J.   II. Stanitski,
Conrad L.   III. Title.
QD31.2.M38   1985   540   84-22217

ISBN 0-03-070744-7

Chemical Principles, 6th edition   ISBN 0-03-070744-7

© 1985 by CBS College Publishing. Copyright 1981
by Saunders College Publishing.
Copyright 1966, 1969, 1973, and 1977 by W. B.
Saunders Company. All rights reserved. Printed in
the United States of America.
Library of Congress catalog card number 84-22217.

  56   32   98765432

CBS College Publishing
Saunders College Publishing
Holt, Rinehart and Winston
The Dryden Press

Preface

A preface is always the most difficult part of a textbook to write. Too often, it becomes an annotated table of contents. Even worse, a preface may resemble an advertisement for the text, promising all things to everyone. Ideally, it should tell you what the authors' objectives are and how they have tried to achieve them. Let's see if we can do that.

This edition of *Chemical Principles*, like its predecessors, is principles-oriented. We believe that, if the general chemistry course is to be meaningful, the student must master the basic ideas of stoichiometry, chemical bonding, kinetics, thermodynamics, and chemical equilibrium. We have tried to express these ideas as clearly as possible, using simple, direct language and relevant examples. However, we are under no illusion that a student will thoroughly understand these topics after reading this textbook. A considerable amount of work will be required by you, the instructor, and, most of all, by your students.

You may be interested in our rationale for the organization of the basic principles of chemistry referred to above. Specifically, we:

— introduce stoichiometry, including the concept of limiting reactant, very early in the text (Chapter 3). In later chapters this topic is reviewed by being applied to reactions in solution (precipitation reactions in Chapter 18, acid-base reactions in Chapter 19, and redox reactions in Chapter 23).

— divide chemical bonding into two chapters. Chapter 9 presents elementary concepts of ionic and covalent bonding with heavy emphasis on Lewis structures. These serve as a foundation for discussing molecular geometry and hybridization in Chapter 10. That chapter closes with a section on molecular orbitals which you may or may not wish to cover.

— cover thermochemistry (including the First Law) early, in Chapter 5. This makes it possible to refer to ΔH and ΔE throughout later chapters. We find that our students have considerable trouble with thermochemistry, perhaps because it is the first topic they encounter that was not covered in high school. We have rewritten this material, trying to make it less abstract and more intelligible.

— place thermodynamics midway through the text. Chapter 14 emphasizes the meaning, calculation, and applications of ΔS and ΔG; it concludes with a discussion of the Second Law. Chemical kinetics is covered after thermodynamics, in Chapter 16. This order seems the more logical to us; you may or may not agree.

— spread chemical equilibrium over several chapters, starting with gas-phase equilibria in Chapter 15. The solubility product constant K_{sp} is covered in Chapter 18, K_w in Chapter 19. A separate chapter (20) is devoted to acid-base equilibria (K_a, K_b), the reciprocal rule, and the rule of multiple equilibria. Formation constants (K_f) of complex ions are covered in Chapter 21. This treatment may seem disjointed, but our experience suggests that repeated exposure to equilibrium principles, each time in a somewhat different context, is very helpful.

Opinions differ considerably as to how descriptive chemistry should be covered in a general chemistry textbook. Some authors essentially ignore it; others defer it to a series of chapters at the end. The effect is about the same in the two cases; little or no descriptive chemistry actually gets taught. In our 5th edition, we tried to be innovative, interspersing descriptive chapters of a review nature throughout the text. As many people have pointed out to us, we made some mistakes in this area, but we believe that the approach is a sound one.

In this edition, we have refined this approach to descriptive chemistry. There are $6\frac{1}{2}$ chapters devoted entirely to descriptive chemistry. These can be handled in any of three ways. They can be discussed in lecture in the usual manner. Alternatively, you can assign these chapters for students to read on their own, perhaps emphasizing the review problems at the end of each chapter. Finally, you can skip one or more of these chapters, since they introduce no new principles (really, *really and truly*, they don't). The chapters that fall into this category are:

— Chapter 4 (Sources of the Elements), which reviews the mole concept, formulas, and stoichiometry.

—Sections 8.4–8.6 of Chapter 8, which deal with the properties of metals with particular emphasis on those in Groups 1 and 2 of the Periodic Table.

— Chapter 13 (Structures of the Nonmetals and Their Binary Compounds), which reviews chemical bonding, molecular structure, and the properties of condensed phases.

— Chapter 17 (The Atmosphere), reviewing thermodynamics, gaseous equilibrium, and kinetics.

— Chapter 22 (Qualitative Analysis), which ties together solution chemistry and solution equilibria.

— Chapters 25 and 26, which deal with the aqueous redox chemistry of the metals and nonmetals.

Several other chapters cover descriptive chemistry in a more traditional way, including those on coordination chemistry (Chapter 21) and organic chemistry (Chapter 28).

Within each chapter in the text are several worked examples dealing with major topics. Each example is followed by an exercise illustrating the same principle. Students should be able to work the exercise readily if they understand the principle. Scattered throughout each chapter are marginal notes. Many of these are of the "now, hear this" variety; a few make points that we forgot to emphasize in the body of the text. Some, probably fewer than we think, are humorous.

Each chapter ends with 65 ± 1 questions and problems. These are virtually all new with this edition, although they are similar to ones used earlier. Problems are arranged in matched pairs, side by side. Answers to problems 31–65 are given in Appendix 5. In the center of the book are a series of color plates, most of which were taken by Ray Boyington or S. Ruven Smith (see however Color Plate 13.1). These are designed to illustrate some of the more colorful aspects of inorganic chemistry, regardless of whether they happen to be "pretty" or not.

The changes we have made in this edition are based in large part on comments from students and their instructors. We are particularly grateful to those instructors who have provided us with written reviews. These include: Marcia Davies, Creighton University; Rick Bearden, Midway College; Robert Conley, New Jersey Institute of Technology; Joel Goldberg, University of Vermont; Robert Flanagan, Diablo Valley College; Charles Russ, University of Maine; David Adams, North Shore Community College; R.W. Ohline, New Mexico Institute of Mining and Technology; Fred Redmore, Highland Community College; Gordon Ewing, New Mexico State University; Barbara Sawrey, San Diego State University; and James Carr, University of Nebraska.

Finally, we express our appreciation to the people at Saunders who have made our task more enjoyable. These include John Vondeling, Margaret Mary Kerrigan, and Sally Kusch, who have massaged our egos and tolerated our idiosyncracies.

William L. Masterton
Emil J. Slowinski
Conrad L. Stanitski

Contents
Overview

Contents

Chapter 1
Matter
and
Measurements

You are beginning a course in chemistry, the science that deals with matter, the "stuff" of which the universe is composed. Chemistry examines the structure of matter, its properties, and the changes it undergoes. In chemistry, we make use of both experiment and theory; facts and principles complement one another.

The science of chemistry is really not very old, dating from about 1800. At that time, the first verifiable theories were proposed and confirmed by experiment. During the nineteenth century, the basic foundations of chemistry were developed and the first industrial applications were made. In the first half of the twentieth century, chemists and physicists working together established the basic structure of matter at the submicroscopic level. The last 40 years have seen the emergence of whole new areas of chemistry, most of them closely related to biology.

The most beneficial of the chemical discoveries of this century have come in the field of medicine. Chemists, working in industrial and academic laboratories, isolated the antibiotics we use to control diseases. They also synthesized the drugs used to treat such perennial health problems as high blood pressure, diabetes, arthritis, and mental depression. Most of the substances used for these purposes were unknown in 1950. Their effect on the health of people throughout the world has been dramatic.

In the past decade, a new science, molecular genetics, has developed on the borderline between chemistry and biology. This involves manipulation of the genetic material in living cells, called DNA, which determines how a living organism will grow and develop.

We must admit that not all of the effects of modern chemistry have been positive. Chemistry has contributed to the development of weapons, ranging from napalm to nuclear bombs. Although these may have prevented some wars, they have been used to kill, maim, and burn as well. Chemists, along with people in a great many other professions, must bear some of the responsibility for this destruction.

In environmental areas, chemistry has contributed both to the problems and to the control of pollution. Dioxin and many other hazardous organic materials are products of the chemical industry. On the other side of the coin, chemists carry out most of the research aimed at controlling or preventing pollution.

In this, a general course, our major goal is to introduce the principles that underlie all of chemistry. Whenever possible, we will try to point out how these principles apply to our daily lives in such areas as human health, environmental pollution, and energy conservation. We cannot emphasize too strongly that the principles of chemistry derive from experiment; *chemistry is an experimental science*. In this chapter, we will describe the results of a great many different experiments. It is important that you understand the experiments that allow us to classify substances as elements or compounds (Section 1.4), identify them on the basis of their properties (Section 1.5), and separate them from one another (Section 1.6).

All experiments involve measurements of one type or another. We begin this chapter by considering what kinds of quantities are commonly measured in the chemistry laboratory (Section 1.1). Every measurement, no matter how carefully made, is subject to experimental error. The magnitude of this error can be estimated in a simple way using *significant figures* (Section 1.2). A measured quantity such as length, volume, or mass ordinarily can be expressed in any of several different units. Using an approach involving *conversion factors* (Section 1.3), it is possible to change from one unit to another.

> Most chemists are experimentalists, not theoreticians

1.1
Measurements

Chemistry is based upon concepts that require the measurement of such quantities as length, volume, mass, and temperature. These measurements are quantitative; that is, they have numbers associated with them. In this section, we will consider some of the simple instruments used in quantitative measurements. We will also look at the units used to express these measured quantities.

Length

Most of us are familiar with a simple measuring device found in every general chemistry laboratory. This is the meter stick, which reproduces, as accurately as possible, the basic unit of length in the metric system, the **meter** (m). The length of a normal stride is about 1 m; a typical room has a height of 2 to 3 m.

A meter stick is divided into 100 equal parts, each one *centimeter* (cm) in length (1 cm = 10^{-2} m). A centimeter, in turn, is divided into ten equal parts, each one millimeter (mm) long (1 mm = 10^{-3} m). A much larger unit, familiar to runners, is the kilometer (1 km = 10^3 m). The prefixes *kilo-, centi-,* and *milli-* are used in the metric system to designate units obtained by multiplying by 1000, 0.01, and 0.001, respectively (Table 1.1). Another unit used to express the dimensions of tiny particles such as atoms is the *nanometer* (nm):

> 100 cm = 1 m = 1000 mm

$$1 \text{ nm} = 10^{-9} \text{ m}$$

Table 1.1
Common Metric Prefixes

10^6 mega-	10^{-1} deci-	10^{-3} milli-
10^3 kilo-	10^{-2} centi-	10^{-6} micro-
		10^{-9} nano-

Volume

Units of volume in the metric system are simply related to those of length. The *cubic centimeter* (cm³) represents the volume of a cube one centimeter on an edge. A larger unit is the *liter* (L), which is exactly 1000 cm³:

$$1 \text{ L} = 1000 \text{ cm}^3 \tag{1.1}$$

A softball has a volume of about $\frac{1}{2}$ L, a basketball a volume of somewhat more than 7 L. A *milliliter* (mL), 1/1000 of a liter, has the same volume as a cubic centimeter:

$$1 \text{ mL} = 1 \text{ cm}^3$$

1000 mL = 1 L

The device most commonly used to measure volumes in general chemistry is the graduated cylinder. With this, we can measure out a known volume of a liquid, accurate to perhaps 0.1 mL. When greater accuracy is required, we use a pipet or buret (Fig. 1.1). A pipet is calibrated to deliver a fixed volume of liquid (for example, 25.00 ± 0.01 mL) when filled to the mark and allowed to drain

Written: 25 mL
Spoken: "25 em ell"

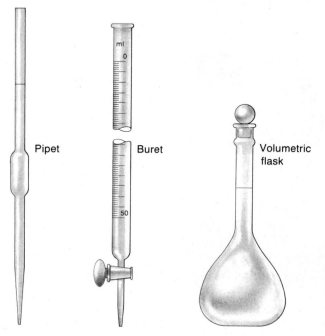

Pipet   Buret   Volumetric flask

FIGURE 1.1 Instruments used with liquids to deliver a fixed volume (pipet), deliver a variable volume (buret), or contain a fixed volume (volumetric flask).

normally. Variable volumes can be delivered with about the same accuracy from a buret. With a buret, final and initial volumes must be read carefully to calculate the volume of liquid withdrawn. A volumetric flask is shown at the right of Figure 1.1. It is designed to contain a specified volume of liquid (for example, 50, 100, . . . , 1000 mL) when filled to a level marked on the narrow neck.

Mass

The mass of a sample is a measure of the amount of matter it contains. In the metric system, mass may be expressed in grams (g), kilograms (kg), or milligrams (mg):

$$1 \text{ kg} = 10^3 \text{ g}; \qquad 1 \text{ mg} = 10^{-3} \text{ g}$$

This book weighs about 2 kg (2×10^3 g).

Chemists measure mass on a balance (Fig. 1.2). To show what is involved in weighing ("massing") an object, consider the two-pan balance shown at the left of the figure. With nothing on either pan, the balance comes to rest with the two pans at the same height. To weigh an object, we place it on the left pan. We then add pieces of metal of known mass to the right pan to restore balance, bringing the pans to the same height again. Under these conditions, the masses on the two pans are equal:*

mass sample = mass metal

On the moon your mass would be the same as on the earth, but your weight would be 1/6 as great

---

*Strictly speaking, it is the weights that are equal when the pans are at the same height. However, weight = k(mass), where k is a proportionality constant that has a fixed value at a given location. Thus, k(mass sample) = k(mass metal), and we see that the masses must be equal.

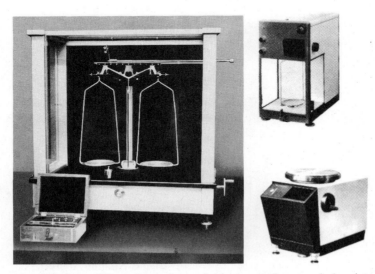

FIGURE 1.2 When the authors of this text took general chemistry, they used a double-pan balance of the type shown at the left. You don't know how lucky you are to be using a single-pan balance such as the ones shown at the right of the figure.

Nowadays, teaching and research laboratories use single-pan balances of the type shown at the right of Figure 1.2. These balances operate on the same principle as the one just described. They are a great deal less tedious to operate, however. You can weigh a sample to within ± 0.001 g in a few seconds on a single-pan balance.

Temperature

The concept of temperature is familiar to all of us. This is because our bodies are so sensitive to temperature differences. When we pick up a piece of ice, we feel cold because its temperature is lower than that of our hand. After drinking a cup of coffee, we may refer to it as "hot," "lukewarm," or "atrocious." In the first two cases, at least, we are describing the extent to which its temperature exceeds ours. From a slightly different viewpoint, temperature is the factor that determines the direction of heat flow. Anyone brave enough to swim in a Minnesota lake in January feels cold because heat is absorbed from his body. If he takes a hot shower afterward, which he certainly will, heat flows in the reverse direction. In general, whenever two objects at different temperatures touch each other, heat flows from the one at the higher to the one at the lower temperature.

Such people are rare, even in Minnesota

To measure temperature we can use a mercury-in-glass thermometer. Here, we take advantage of the fact that mercury, like other substances, expands as temperature increases. When the temperature rises, the mercury in the thermometer expands up a narrow tube. The total volume of the tube is only about 2% of that of the bulb at the base. In this way, a rather small change in volume is made readily visible.

Thermometers used in chemistry are marked in degrees *Celsius* (centigrade), named after the Swedish astronomer Anders Celsius (1701–1744). On this scale, the freezing point of water is taken to be 0°C. The boiling point of water at one atmosphere pressure is 100°C. When we place a mercury-in-glass thermometer in a beaker containing crushed ice and water, the mercury comes to rest exactly at the 0° mark. In a beaker of boiling water, the mercury rises to the 100° mark. The distance between these two marks is divided into 100 equal parts. Each of these corresponds to 1 Celsius degree. Thus, a temperature of 45°C corresponds to a mercury level 45% of the way from the 0° to the 100° mark.

The thermometer is made that way

A temperature scale in common use in the United States today is based on the work of Daniel Fahrenheit (1686–1736). He was a German instrument maker who was the first to use the mercury-in-glass thermometer. On this scale the normal freezing and boiling point of water are taken to be 32° and 212°, respectively; that is,

$$32°F = 0°C; \quad 212°F = 100°C$$

As you can see from Figure 1.3, the Fahrenheit degree is smaller than the Celsius degree. The distance between the freezing and boiling points of water is 100° on the Celsius scale and 180° on the Fahrenheit scale. The relation between temperatures expressed on the two scales is

$$°F = 1.8(°C) + 32 \tag{1.2}$$

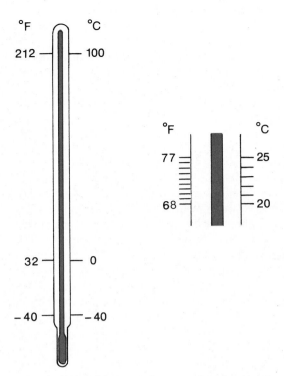

FIGURE 1.3 On the Celsius scale, the distance between the freezing and boiling points of water is 100°; on the Fahrenheit scale, it is 180°. This means that the Celsius degree is nine fifths as large as the Fahrenheit degree, as is evident from the magnified section of the thermometer at the right. Using Equation 1.2, you should be able to show that the two scales coincide at −40°.

Another temperature scale that is particularly useful for gases is the absolute or Kelvin scale. The relationship between temperatures in K and °C is

$$K = °C + 273.15 \qquad (1.3)$$

This scale is named after Lord Kelvin, an English physicist. He showed that it is impossible to reach a temperature lower than 0 K (−273.15°C).

Lord Kelvin was 10 when he entered college, was 26 when he proved you can't get below 0 K.

EXAMPLE 1.1 Express normal body temperature, 98.6°F, in °C and K.

Solution Substituting in Equation 1.2, we obtain

$$98.6 = 1.8(°C) + 32$$

Solving:

$$1.8(°C) = 98.6 − 32 = 66.6$$

$$°C = 66.6/1.8 = 37.0$$

Applying Equation 1.3:

$$K = 37.0 + 273.15 = 310.2$$

EXERCISE Convert 25°C to °F and to K. Answer: 77°F; 298 K.

1.2
Uncertainties in Measurements; Significant Figures

Every measurement we make carries with it a degree of uncertainty, or error. How large this error is depends upon the nature of the measuring device and the skill with which we use it. Suppose, for example, we try to measure out 8 mL of liquid using a 100-mL graduated cylinder. Here, the volume is likely to be in error by at least 1 mL. With such a crude measuring device, we will be fortunate to obtain a volume closer to 8 than to 7 or 9 mL. To obtain greater accuracy, we might use a narrow 10-mL cylinder, on which the divisions are much farther apart. The volume we measure now may be within 0.1 mL of the desired value of 8 mL; that is, it is likely to fall in the range 7.9 to 8.1 mL. Using a buret we can do even better. If we are very careful, we may reduce the uncertainty to 0.01 mL.

The person who makes a measurement such as this should indicate the uncertainty associated with it. Such information is vital to anyone who wants to repeat an experiment or judge its accuracy. There are many ways to do this. We might report the three volume measurements referred to above as

8 ± 1 mL   (large graduated cylinder)
8.0 ± 0.1 mL   (small graduated cylinder)
8.00 ± 0.01 mL   (buret)

In this text, we will drop the $\pm$ notation and simply write

8 mL;   8.0 mL;   8.00 mL

When we do this, it is understood that there is an *uncertainty of one unit in the last digit* (1 mL, 0.1 mL, 0.01 mL).

This method of citing the degree of confidence in a measurement is often described in terms of **significant figures.** We say that in 8.00 mL there are three significant figures. Each of the three digits in 8.00 has experimental meaning. Similarly, there are two significant figures in 8.0 mL and one significant figure in 8 mL.

The significant figures in a number contribute to its relative precision

Counting Significant Figures

Frequently, we need to know the number of significant figures in a measurement that someone else has reported. We do this by applying the following common-sense rules.

1. *All nonzero digits are significant.* There are three significant figures in 5.37 cm and four significant figures in 4.293 cm.

2. *Zeros between nonzero digits are significant.* There are three significant figures in 106 g or in 1.02 g.

3. *Zeros beyond the decimal point at the end of a number are significant.* As pointed out above, there are two significant figures in 8.0 mL and three significant figures in 8.00 mL.

4. *Zeros preceding the first nonzero digit in a number are not significant.* In a mass measurement of 0.002 g, there is only one significant figure—the "2" at the end. The zeros serve only to fix the position of the decimal point. This becomes obvious if we express the mass in exponential (scientific) notation (Appendix 4). In that case, we would write 0.002 g as

2×10^{-3} g

Now, clearly, there is only one significant figure. The uncertainty is $\pm 1 \times 10^{-3}$ g.

The 10^{-3} term does not contribute any significant figures

Sometimes the number of significant figures in a reported measurement is ambiguous. Suppose you are told that a coin weighs "50 g." You cannot be sure how many of these digits are meaningful. Perhaps the coin was weighed to the nearest gram (50 ± 1 g). If so, both the "5" and the "0" are known; there are two significant figures. Then again, the coin might have been weighed only to the nearest ten grams (50 ± 10 g). In this case, only the "5" is known accurately; there is only one significant figure. About all we can do in situations like this is to wish that the person who carried out the weighing had used exponential notation. The mass should have been reported as either

$$5.0 \times 10^1 \text{ g} \qquad \text{(2 significant figures)}$$

or

or even 5.00×10^1 g
(3 sig. fig.)

$$5 \times 10^1 \text{ g} \qquad \text{(1 significant figure)}$$

EXAMPLE 1.2 Three different students weigh the same object, using different balances. They report the following masses:
a. 15.02 g   b. 15.0 g   c. 0.01502 kg
How many significant figures are there in each value?

Solution
a. 4
b. 3. The zero after the decimal point is significant. It indicates that the object was weighed to the nearest 0.1 g.
c. 4. The zeros at the left are not significant. They are there only because the mass was expressed in kilograms rather than grams. Note that "15.02 g" and "0.01502 kg" represent the same mass.
   The same answers could have been obtained, perhaps with more confidence, by expressing the masses in exponential notation:

$$15.02 \text{ g} = 1.502 \times 10^1 \text{ g} \qquad \text{(4 significant figures)}$$
$$15.0 \text{ g} = 1.50 \times 10^1 \text{ g} \qquad \text{(3 significant figures)}$$
$$0.01502 \text{ kg} = 1.502 \times 10^{-2} \text{ kg} \qquad \text{(4 significant figures)}$$

EXERCISE Give the number of significant figures in 2.6×10^2 cm³; 2.40 $\times 10^{-3}$ cm³. Answer: 2; 3.

Significant Figures in Multiplication and Division

Most of the quantities that we measure are not end results in themselves. Instead, they are used to calculate other quantities, often by multiplication or division. We might, for example, measure mass and volume in order to determine a concentration (Example 1.3). The precision of any such derived result is limited by those of the measurements upon which it is based. **When experimental quantities are multiplied or divided, the number of significant figures in the result is the same as that in the quantity with the smallest number of significant figures.**

EXAMPLE 1.3   A chemist analyzes a sample of polluted air for sulfur trioxide, a precursor of acid rain. She finds that there is 0.0361 g of sulfur trioxide in a sample of air that has a volume of 12 L. Express the concentration of sulfur trioxide in grams per liter.

Solution

$$\text{concentration} = \frac{\text{mass sulfur trioxide}}{\text{volume of air}} = \frac{0.0361 \text{ g}}{12 \text{ L}} = 0.0030 \text{ g/L}$$
$$= 3.0 \times 10^{-3} \text{ g/L}$$

Note that only two significant figures should be retained in the answer, since there are only two significant figures in the volume (12 L). The "extra" significant figure in the measured mass is wasted.

EXERCISE   What is the area, in square meters, of a rectangular field that is measured to be 362 m long and 29.725 m wide? Answer: 1.08×10^4 m².

You can't increase the precision of a result by multiplication or division

Uncertainties in Addition and Subtraction

When measured quantities are added or subtracted, the uncertainty in the result is found in a quite different way than in multiplication or division. It is determined by the absolute *uncertainty* (rather than the number of significant figures) in the least precise measurement. Suppose, for example, we wish to calculate the total mass of a solution containing 10.21 g of instant coffee, a "pinch" (0.2 g) of sugar, and 256 g of water. The uncertainties in these masses are

	MASS (g)		UNCERTAINTY
Instant coffee	10.21	±	0.01 g
Sugar	0.2	±	0.1 g
Water	256	±	1 g
Total mass	266	±	1 g

The sum of the masses cannot be more precise than that of the water, which has the largest uncertainty, ±1 g. The total mass should be reported as 266 g rather than 266.4 g or 266.41 g. The general rule illustrated by this example is as follows:

When experimental quantities are added or subtracted, the number of digits beyond the decimal point in the result is the same as that in the quantity with the smallest number of digits beyond the decimal point.

Exact Numbers

In applying the rules that we have cited, you should keep in mind one important point: certain numbers involved in calculations are exact rather than approximate. If you were asked to express in liters a measured volume of 536 cm³, using the relation

$$1 \text{ L} = 1000 \text{ cm}^3$$

your answer should be given to three significant figures. This is the number of significant figures in the measured quantity, 536 cm³. The "1" and the "1000" in the equation are defined quantities. There are exactly 1000 cubic centimeters in exactly one liter. A similar situation applies with the equation relating Fahrenheit and Celsius temperatures:

$$°F = 1.8(°C) + 32$$

The number of students in a room is exact

The numbers 1.8 and 32 are exact. Hence, they do not affect the precision of any calculation involving a temperature conversion.

1.3
Conversion of Units

Often we need to convert measurements expressed in one unit (e.g., grams) to another unit (milligrams or kilograms). To do this, we follow what is known as a **conversion factor** approach. Suppose, for example, we want to convert a volume of 536 cm³ to liters. We know that

$$1 \text{ L} = 1000 \text{ cm}^3$$

Dividing both sides of this equation by 1000 cm³ gives a quotient equal to one:

$$\frac{1 \text{ L}}{1000 \text{ cm}^3} = \frac{1000 \text{ cm}^3}{1000 \text{ cm}^3} = 1$$

We multiply 536 cm³ by the quotient 1 L/1000 cm³, which is called a *conversion factor*. Since the conversion factor equals one, this does not change the value of the volume. However, it does accomplish the desired conversion of units:

The undesired units cancel

$$536 \text{ cm}^3 \times \frac{1 \text{ L}}{1000 \text{ cm}^3} = 0.536 \text{ L}$$

The relation 1 L = 1000 cm³ can be used equally well to convert a volume in liters, say 1.28 L, to cubic centimeters. In this case, we obtain the necessary conversion factor by dividing both sides of the equation by 1 L:

$$\frac{1000 \text{ cm}^3}{1 \text{ L}} = \frac{1 \text{ L}}{1 \text{ L}} = 1$$

Multiplying 1.28 L by the quotient 1000 cm³/1 L converts the volume from liters to cubic centimeters:

$$1.28 \text{ L} \times \frac{1000 \text{ cm}^3}{1 \text{ L}} = 1280 \text{ cm}^3 = 1.28 \times 10^3 \text{ cm}^3$$

Notice that a single relation (1 L = 1000 cm³) gives us two conversion factors:

$$\frac{1 \text{ L}}{1000 \text{ cm}^3} \quad \text{and} \quad \frac{1000 \text{ cm}^3}{1 \text{ L}}$$

Both of these are equal numerically to one. In making a conversion, we choose the factor that cancels out the unit we want to get rid of:

initial quantity × conversion factor(s) = desired quantity

Conversions between English and metric units are made in a similar way (Example 1.4). The relations required can be obtained from Table 1.2.

Table 1.2
Relations Between Length, Volume, and Mass Units

	METRIC		ENGLISH		METRIC-ENGLISH
Length					
1 km	$= 10^3$ m	1 ft	$= 12$ in	1 in	$= 2.54$ cm
1 cm	$= 10^{-2}$ m	1 yd	$= 3$ ft	1 m	$= 39.37$ in
1 mm	$= 10^{-3}$ m	1 mile	$= 5280$ ft	1 mile	$= 1.609$ km
1 nm	$= 10^{-9}$ m $= 10$ Å				
Volume					
1 m^3	$= 10^6$ $cm^3 = 10^3$ L	1 gallon	$= 4$ qt $= 8$ pt	1 ft^3	$= 28.32$ L
1 cm^3	$= 1$ mL $= 10^{-3}$ L	1 qt (Can.)	$= 69.35$ in^3	1 L	$= 0.8799$ qt (Can.)
		1 qt (U.S. liq.)	$= 57.75$ in^3	1 L	$= 1.057$ qt (U.S. liq.)
Mass					
1 kg	$= 10^3$ g	1 lb	$= 16$ oz	1 lb	$= 453.6$ g
1 mg	$= 10^{-3}$ g	1 short ton	$= 2000$ lb	1 g	$= 0.03527$ oz
1 metric ton	$= 10^3$ kg			1 metric ton	$= 1.102$ short ton

EXAMPLE 1.4 According to a highway sign, the distance from St. Louis to Chicago is 295 miles. Express this distance in kilometers.

Solution From Table 1.2, we note that the required relation is

1 mile = 1.609 km

The 1 is exact; the 1.609 is not (4 sig. fig.)

Since we want to convert from miles to kilometers, we want a conversion factor with kilometers in the numerator and miles in the denominator. The required conversion factor is 1.609 km/1 mile:

$$295 \text{ miles} \times \frac{1.609 \text{ km}}{1 \text{ mile}} = 475 \text{ km}$$

There are 3 significant figures in the answer. Why?

EXERCISE Determine the mass in grams of a hamburger that weighs 8.0 oz. Answer: 2.3×10^2 g.

Frequently, you will need to carry out more than one conversion to work a problem. This can be done by setting up successive conversion factors (Example 1.5).

EXAMPLE 1.5 A certain U.S. car has a fuel efficiency rating of 36.2 miles per gallon. Convert this to kilometers per liter.

Solution The distance conversion can be made by using the relation

$$1 \text{ mile} = 1.609 \text{ km} \quad (1)$$

Using Table 1.2, we find that the volume conversion involves using two relations:

$$1 \text{ gallon} = 4 \text{ qt} \quad (2)$$

$$1 \text{ L} = 1.057 \text{ qt} \quad (3)$$

We set up the arithmetic in a single expression, writing conversion factors in such a way as to cancel units:

$$36.2 \, \frac{\cancel{\text{miles}}}{\cancel{\text{gallon}}} \times \frac{1.609 \text{ km}}{1 \, \cancel{\text{mile}}} \times \frac{1 \, \cancel{\text{gallon}}}{4 \, \cancel{\text{qt}}} \times \frac{1.057 \, \cancel{\text{qt}}}{1 \text{ L}} = 15.4 \text{ km/L}$$

$$(1) \qquad\qquad (2) \qquad\qquad (3)$$

Notice that three consecutive conversions were required. We first converted miles to kilometers, obtaining the fuel efficiency in kilometers per gallon. Then we converted gallons to quarts, and, finally, quarts to liters. Our final answer was in the desired units, kilometers per liter.

This approach is easier than trying to set up one grand glorious conversion factor

EXERCISE Convert a density of 3.50 g/cm³ to kilograms per cubic meter. Answer: 3.50×10^3 kg/m³.

Experiments in general chemistry often involve measured quantities other than those given in Table 1.2 (length, volume, mass). You may, for example, measure the time required to carry out a reaction. This might be expressed in days (d), hours (h), minutes (min), or seconds (s).

$$1 \text{ d} = 24 \text{ h}; \quad 1 \text{ h} = 60 \text{ min}; \quad 1 \text{ min} = 60 \text{ s}$$

In experiments involving gases, it is important to measure the pressure. Many different units are used to express pressure (Table 1.3). In this text, we will most commonly use the two units

—*millimeter of mercury* (mm Hg). This is the pressure exerted by a column of mercury 1 millimeter in height.

Table 1.3
Relations Between Pressure and Energy Units

PRESSURE
1 atm = 760 mm Hg = 1.013×10^5 Pa = 14.70 lb/in²
1 torr = 1 mm Hg
1 bar = 10^5 Pa

ENERGY
1 cal = 4.184 J = 4.129×10^{-2} L·atm
1 J = 10^7 ergs

—*atmosphere* (atm). An atmosphere (760 mm Hg) is approximately the pressure of the atmosphere on a "normal" day at sea level.

Pascals (Pa) and kilopascals (1 kPa = 10^3 Pa) will be used less frequently.

$100 \text{ kPa} \cong 1 \text{ atm}$

In every reaction, there is an energy change whose magnitude can be measured. Chemists use a variety of energy units (Table 1.3). You may be most familiar with the calorie,* which is the amount of heat required to raise the temperature of 1 gram of water by 1°C. Increasingly, nowadays, scientists are using a different unit, the **joule**. This is the preferred unit of energy in the International System of Units. We will use the joule extensively throughout this text. The basic definition of this energy unit is given in Appendix 1. Notice from Table 1.3 that 1 calorie is about 4 joules; more exactly, 1 cal = 4.184 J. This means that an energy change expressed in joules will have a magnitude about four times that for the same energy change in calories. For example, it takes about 4 joules to raise the temperature of 1 gram of water 1°C.

The joule is actually a rather small energy unit. The energy released when a match burns is of the order of 2000 J. Most often, in chemical reactions, we will express energy changes in **kilojoules**:

$$1 \text{ kJ} = 10^3 \text{ J}$$

Conversions involving these units are carried out in the ordinary way, using the relations given in Table 1.3. (A more extensive list of relations between units of all types is given in Appendix 1.) Example 1.6 illustrates the conversion from calories to joules.

EXAMPLE 1.6 When one gram of gasoline burns in an automobile engine, the amount of energy given off is about 1.03×10^4 cal. Express this in joules.

Solution From Table 1.3, we see that 1 cal = 4.184 J. To convert calories to joules, we use the conversion factor 4.184 J/1 cal.

$$1.03 \times 10^4 \text{ cal} \times \frac{4.184 \text{ J}}{1 \text{ cal}} = 4.31 \times 10^4 \text{ J} \quad \text{(3 significant figures)}$$

EXERCISE Convert 836.8 J to calories. Answer: 2.000×10^2 cal.

The conversion factor approach shown in Examples 1.4 to 1.6 will be used throughout this text. If this is your first contact with it, it may seem awkward or artificial. You will find, however, that it is the best way to solve a wide variety of problems in chemistry. It is particularly useful when multiple conversions are required (Example 1.5) or when the units may be unfamiliar to you (Example 1.6).

The conversion factor method beats proportions every time

To apply this approach to convert a quantity expressed in one unit to another unit, you proceed as follows:

*The "calorie" referred to by nutritionists is actually a kilocalorie (1 kcal = 10^3 cal). On a "2000-calorie" per day diet, you eat food capable of producing 2000 kcal = 2×10^3 kcal = 2×10^6 cal of energy.

In this text we will often summarize a method as we have done here, using bold-face color type

1. Find a relationship between the two units, using a reference such as Table 1.2 or 1.3. For example, if asked to convert between pressures in atmospheres and millimeters of mercury, you first locate in Table 1.3 the relation: 1 atm = 760 mm Hg.

2. Translate this relationship into a conversion factor. The relation 1 atm = 760 mm Hg gives you two conversion factors: 1 atm/760 mm Hg and 760 mm Hg/ 1 atm.

3. Multiply the original quantity by a conversion factor that cancels the unit you want to get rid of. If asked to convert a pressure in atmospheres, let us say X atm, to millimeters of mercury, you multiply by 760 mm Hg/1 atm, so as to cancel the unit "atm."

$$X \text{ atm} \times \frac{760 \text{ mm Hg}}{1 \text{ atm}} = 760X \text{ mm Hg}$$

4. If necessary, repeat this process, using successive conversion factors, until you obtain the quantity in the desired units.

SI Units

As indicated by the large number of entries in Tables 1.2 and 1.3, many different units are often used to express a single quantity. Pressure may be expressed in atmospheres, millimeters of mercury, pascals, etc. This proliferation of units has long been of concern to scientists. In 1960, the General Conference of Weights and Measures recommended a self-consistent set of units based upon the metric system. In the International System of Units (SI), a single base unit is used for each measured quantity. An extended discussion of this system is presented in Appendix 1. Table 1.4 lists the recommended SI units for each of the quantities referred to in this chapter.

Table 1.4
SI Units

QUANTITY	UNIT
Length	meter (m)
Volume	cubic meter (m^3)
Mass	kilogram (kg)
Temperature	kelvin (K)
Time	second (s)
Pressure	pascal (Pa)
Energy	joule (J)

We will often use liters (L) to express volumes; 1 m^3 is just too big

For the most part, we will use SI units in this text. For example, we will express atomic dimensions in nanometers rather than angstroms (1 nm = 10^{-9} m = 10 Å). Energy changes will be expressed in joules rather than calories (1 cal = 4.184 J). In dealing with pressures, however, we will more commonly use the atmosphere or the millimeter of mercury as opposed to the pascal or kilopascal. If desired, non-SI units such as the millimeter of mercury can readily be converted to SI units.

1.4
Kinds of Substances

Chemists, by methods which we will consider shortly, have isolated many thousands of pure substances. These substances can be divided into two classes. Some of them, called **elements**, cannot be broken down by chemical means into two or more pure substances. All other pure substances are **compounds**. A compound, by definition, is a pure substance that can be broken down into two or more elements.

Elements and the Periodic Table

There are, at latest count, 108 known elements. Of these, 91 occur naturally. Many elements are familiar to all of us. The charcoal used in outdoor grills is nearly pure carbon. Electrical wiring, jewelry, and water pipes are often made from copper, a metallic element. Another such element, aluminum, is used in many household utensils. The shiny liquid in the thermometers you use is still another metallic element, mercury.

In chemistry, an element is identified by its symbol. This consists of one or two letters, usually derived from the name of the element. Thus the symbol for carbon is C; that for aluminum is Al. Sometimes the symbol comes from the Latin name of the element or one of its compounds. The two elements copper and mercury, which were known in ancient times, have the symbols Cu (*cuprum*) and Hg (*hydrargyrum*). Other examples include

If you don't already know the symbols for the elements, you should learn them, now

Antimony	Sb	Lead	Pb	Sodium	Na
Gold	Au	Potassium	K	Tin	Sn
Iron	Fe	Silver	Ag		

You are probably familiar with a device used to organize the properties of the elements. This is the Periodic Table, shown on the inside cover of the text. Elements that are similar chemically fall directly beneath one another in the table. In later chapters, we will consider the rationale for this and see how the table is used for a variety of purposes. At the moment, we need only be concerned with two general features of the Periodic Table.

The Periodic Table is both convenient and useful

1. The horizontal rows are referred to as **periods**. Thus, the first period consists of the two elements, hydrogen (H) and helium (He). The second period starts with lithium (Li) and ends with neon (Ne), and so on. Each period ends with a very unreactive element called a noble gas (He, Ne, Ar, Kr, Xe, Rn).

2. The vertical columns are known as **groups**. The groups at the far left and the right of the table are numbered: Groups 1 and 2 are at the left, Groups 3 to 8 at the right. Elements that fall in these groups are often referred to as *main-group* elements. The elements in the center of periods 4 through 6 are called *transition* elements (for example, Sc through Zn in period 4).

Certain main groups are given special names. The elements in Group 1 are called alkali metals; those in Group 2 are referred to as alkaline earth metals. The Group 7 elements are called halogens; the noble gases comprise Group 8. Elements in the same main group show very similar chemical reactions. For example, sodium (Na) and potassium (K) in Group 1 both react violently with water to produce hydrogen gas.

Compounds

As you might suppose, there are a great many more compounds than elements. The elements can combine with one another in many different ways. Millions of different compounds have been prepared in the laboratory or extracted from natural sources. Each year thousands of new compounds are reported.

Every compound contains two or more elements in fixed proportion by mass. Water, by far our most abundant compound, contains the two elements hydrogen and oxygen. In a 100-g sample of pure water, there are 11.19 g of hydrogen and 88.81 g of oxygen. We would say that water contains 11.19% by mass of hydrogen and 88.81% by mass of oxygen. The chemical composition of table sugar (sucrose) is somewhat more complex. This compound contains the three elements carbon (42.10%), hydrogen (6.48%), and oxygen (51.42%).

The properties of compounds are very different from those of the elements they contain. Ordinary table salt, sodium chloride, is a white, unreactive solid. As you can guess from its name, it contains the two elements sodium and chlorine. Sodium (Na) is a shiny, extremely reactive metal. Chlorine (Cl) is a poisonous, greenish-yellow gas. Clearly, when these two elements combine to form sodium chloride, a profound change takes place.

Many different methods can be used to resolve compounds into their elements. Sometimes, but not often, heat alone is sufficient. Mercury(II) oxide, a compound of mercury and oxygen, decomposes to its elements when heated to 600°C. Joseph Priestley, an English chemist, discovered oxygen 200 years ago when he carried out this reaction by exposing a sample of mercury(II) oxide to an intense beam of sunlight focused through a powerful lens. A more common method of resolving compounds into elements is called electrolysis. This involves passing an electric current through a compound, usually in the liquid state. Through the process of electrolysis, it is possible to separate water into the two elements hydrogen and oxygen.

If you electrolyze molten sodium chloride, you get metallic sodium and chlorine gas

1.5
Properties of Substances

Every pure substance has its own unique set of properties that serve to distinguish it from all other substances. A chemist most often identifies an unknown substance by measuring its properties and comparing them to the properties recorded in the chemical literature for known substances. In this section we will consider a few of these properties, some of which you will probably measure in the general chemistry laboratory.

A substance having all the properties of benzene is benzene

Elements and compounds are sometimes identified on the basis of their *chemical properties*. These are properties observed when a substance undergoes a chemical change, converting it to one or more other substances. You might, for example, show that a red solid is mercury(II) oxide by heating it in air. At about 600°C, this compound decomposes to mercury, a silvery liquid, and oxygen, a colorless gas.

More commonly, we measure the *physical properties* of elements or compounds. These are properties that can be measured without changing the chemical identity of a substance. Table 1.5 lists several physical properties of a variety of substances, both elements and compounds.

Table 1.5
Physical Properties of Substances

SUBSTANCE	DENSITY (g/cm³)*	mp (°C)	bp (°C)	SOLUBILITY (g/100 g water)*	COLOR	SPECIFIC HEAT (J/g·°C)*
Elements						
Bromine(*l*)	3.12	−7	59	3.3	red	0.448
Chlorine(*g*)	0.00292	−101	−34	0.59	green-yellow	0.478
Copper(*s*)	8.94	1083	2567	~0	red	0.382
Iron(*s*)	7.87	1535	2750	~0	gray	0.476
Magnesium(*s*)	1.74	650	1120	~0	gray	1.01
Compounds						
Benzene(*l*)	0.879	5	80	0.13	colorless	1.72
Ethyl alcohol(*l*)	0.785	−112	78	infinite	colorless	2.43
Potassium nitrate(*s*)	2.11	334	—	40	white	0.920
Sodium chloride(*s*)	2.16	808	1473	36	white	0.866
Water(*l*)	1.00	0	100	—	colorless	4.18

*At 25°C, 1 atm.

Density

The density of a substance is the ratio of its mass to its volume:

$$\text{density} = \frac{\text{mass}}{\text{volume}}$$

(1.4) This is the definition of density

Densities of liquids and solids are most often expressed in grams per cubic centimeter (g/cm³), as in Table 1.5. With gases, densities are frequently quoted in grams per liter (g/L). Thus, instead of citing the density (d) of chlorine gas at 25°C and 1 atm as 0.00292 g/cm³, we might write it as 2.92 g/L

$$d \text{ (g/L)} = 0.00292 \frac{g}{cm^3} \times \frac{1000 \text{ cm}^3}{1 \text{ L}} = 2.92 \text{ g/L}$$

Densities of liquids or gases can be found by measuring, independently, the mass and volume of a sample (Example 1.7). For solids, density is a bit more difficult to determine. A common approach is shown in Figure 1.4. The mass of the solid sample is first measured using a balance. Its volume is found indirectly by determining the volume of liquid displaced by the solid.

EXAMPLE 1.7 Mercury is one of the most dense liquids known. A student measures out 12.1 cm³ Hg from a buret and finds that this sample weighs 164.56 g. Calculate
a. the density of mercury.
b. the mass of a sample of mercury that has a volume of 2.15 cm³.
c. the volume of another sample of mercury that weighs 94.2 g.

FIGURE 1.4 One way to determine the density of a solid is to first find its mass, m. The solid is then added to a flask of known volume, V. The volume of water, V_w, required to fill the flask is determined. The density of the solid must then be $m/(V - V_w)$.

Solution

a. Since density is the ratio of mass to volume,

$$d = \frac{164.56 \text{ g}}{12.1 \text{ cm}^3} = 13.6 \text{ g/cm}^3$$

b. Perhaps the simplest approach here is to consider density as a conversion factor relating mass to volume. For any sample of mercury,

$$13.6 \text{ g Hg} = 1 \text{ cm}^3 \text{ Hg}$$

This equation provides two conversion factors

Hence, we have

$$\text{mass Hg} = 2.15 \text{ cm}^3 \text{ Hg} \times \frac{13.6 \text{ g Hg}}{1 \text{ cm}^3 \text{ Hg}} = 29.2 \text{ g Hg}$$

c. Here we convert mass to volume, so the required conversion factor is $1 \text{ cm}^3 \text{ Hg}/13.6 \text{ g Hg}$:

$$\text{volume Hg} = 94.2 \text{ g Hg} \times \frac{1 \text{ cm}^3 \text{ Hg}}{13.6 \text{ g Hg}} = 6.93 \text{ cm}^3 \text{ Hg}$$

EXERCISE Using Table 1.5, calculate the mass of 2.15 cm^3 of ethyl alcohol; the volume of 94.2 g of ethyl alcohol. Answer: 1.69 g; $1.20 \times 10^2 \text{ cm}^3$.

Melting Point and Boiling Point

The melting point is the temperature at which a substance changes from the solid to the liquid state. If the substance is pure, the temperature stays constant during melting (Fig. 1.5A). Only when all the solid has melted does heating produce a temperature increase. The melting point behavior of an impure solid is quite different. The solid ordinarily starts to melt at a temperature below the melting point of the pure substance. Moreover, the temperature rises steadily during the

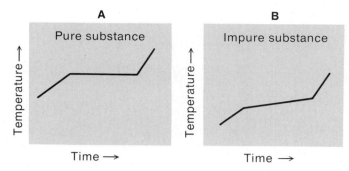

FIGURE 1.5 Time-temperature curves for the melting of a pure substance (A) and a mixture (B). A pure substance will melt at a constant temperature. If the substance contains an impurity, its melting point will increase as melting proceeds and will at all stages be lower than when pure.

melting process (Fig. 1.5B). Therefore, any evidence of an increase in temperature during melting suggests the presence of impurities.

The boiling point of a liquid is the temperature at which bubbles filled with vapor form within the liquid. For reasons to be discussed in Chapter 11, boiling point depends upon the pressure above the liquid. The normal boiling point (Table 1.5) is the temperature at which a liquid boils when the pressure above it is 1 atm (760 mm Hg). For a pure liquid, temperature remains constant during the boiling process. The heating curve for a pure liquid looks very much like that shown in Figure 1.5A. If the liquid is impure, we expect the temperature to rise steadily during boiling.

Solubility

The extent to which a substance dissolves in a particular solvent can be expressed in various ways. A common method is to state the number of grams of the substance that dissolves in 100 g of solvent at a given temperature. At 25°C, about 40 g of potassium nitrate dissolves in 100 g of water. At 100°C, the solubility of this solid is considerably greater, about 240 g/100 g of water (Fig. 1.6, p. 20).

EXAMPLE 1.8 Taking the solubility of sodium chloride, NaCl, to be 36 g/100 g of water at 25°C, calculate the mass of NaCl that dissolves in 120 g of water at this temperature.

Solution We can solve this problem by the conversion factor approach used in Examples 1.4 through 1.6. To do this we start by noting that

36 g NaCl ≏ 100 g water

where the sign ≏ means that these two quantities are equivalent as far as solubility calculations are concerned. The equivalence sign may be treated as if it were an equals sign, giving us the conversion factor

Sample A and Sample B melt at the same temperature and look alike. How might one show that they contain the same substance?

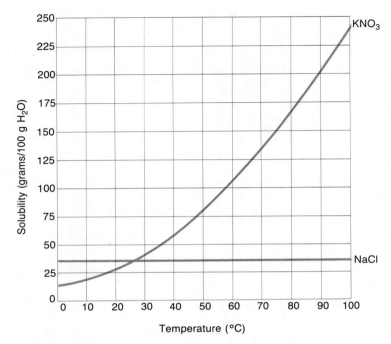

FIGURE 1.6 The solubility of potassium nitrate (KNO_3) in water increases rapidly with temperature, while that of sodium chloride (NaCl) remains nearly constant. This means that potassium nitrate can readily be purified by fractional crystallization while sodium chloride cannot.

$$\frac{36 \text{ g NaCl}}{100 \text{ g water}}$$

Now we can calculate the mass of NaCl that dissolves in 120 g of water:

$$120 \text{ g water} \times \frac{36 \text{ g NaCl}}{100 \text{ g water}} = 43 \text{ g NaCl}$$

EXERCISE   How much water is required to dissolve 52 g of NaCl at 25°C? Answer:

$$52 \text{ g NaCl} \times \frac{100 \text{ g water}}{36 \text{ g NaCl}} = 1.4 \times 10^2 \text{ g water}$$

Specific Heat

The amount of heat absorbed in raising the temperature of a substance depends upon three factors:

—the mass of the substance: the greater the mass, the more heat absorbed.

—the temperature change: the greater the change in temperature, the more heat absorbed.

—a property of a substance known as its *specific heat*, which is the amount of heat required to raise the temperature of one gram of a substance one degree Celsius. The units for specific heat are joules per gram per °C. The specific heat of water is 4.18 J/g·°C. This means that 4.18 J of heat must be absorbed to raise the temperature of 1.00 g of water by 1.00°C (e.g., from 15.00 to 16.00°C).

The amount of heat absorbed by a substance can be calculated from the equation

$$q = (\text{specific heat}) \times m \times \Delta t \tag{1.5}$$

where q is the amount of heat absorbed in joules, m is the mass of the substance in grams, and Δt is the temperature change, $t_{final} - t_{initial}$. The use of Equation 1.5 is illustrated in Example 1.9.

EXAMPLE 1.9   Taking the specific heat of copper to be 0.382 J/g·°C, calculate the amount of heat that must be absorbed to raise the temperature of a 5.00-g sample of copper from 25.00 to 32.70°C.

Solution   We substitute in Equation 1.5, with m = 5.00 g and Δt = 32.70°C − 25.00°C = 7.70°C.

$$q = 0.382 \frac{J}{g \cdot °C} \times 5.00 \text{ g} \times 7.70°C = 14.7 \text{ J}$$

EXERCISE   What mass of copper can be heated from 5.00 to 10.00°C by the absorption of 22.0 J of heat? Answer: 11.5 g.

We will have more to say about the measurement of heat flow and the significance of specific heat in Chapter 5.

Color

Some substances can be identified, at least tentatively, on the basis of their color. Gaseous nitrogen dioxide has a brown color. Bromine vapor is red, and iodine vapor is violet. A water solution of copper sulfate is blue, a solution of potassium permanganate is purple, and so on.

The colors of gases and liquids are due to the absorption of visible light. Sunlight is a mixture of radiation of various wavelengths (λ). Light falling within a certain wavelength range is associated with a particular color. For example, light in the range 400 to 450 nm appears violet; for blue light, λ = 450 to 490 nm (see Fig. 1.7, which relates wavelength to color). Bromine absorbs light in these regions and transmits light of other wavelengths. The subtraction of the violet and blue components of sunlight accounts for the red color of bromine (Table 1.6). In contrast, a water solution of potassium permanganate absorbs light in the green-yellow region, midway through the visible range. Light transmitted through such a solution is rich in the blue (shorter wavelength) and red (longer wavelength) components. Hence, it appears purple, a mixture of blue and red (see Color Plate 1.1).

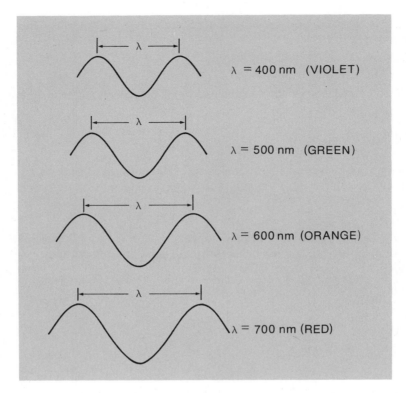

FIGURE 1.7   Relation between wavelength and color of visible light. The wavelength is the distance between successive crests of the wave. As wavelength increases, the color changes from violet (400 nm) to green (500 nm), orange (600 nm), and red (700 nm). Sunlight is a mixture of all wavelengths in the visible region as well as ultraviolet (<400 nm) and infrared (>700 nm) radiation.

Table 1.6
Colors of Substances that Absorb Light in the Visible Region

WAVELENGTH REGION	COLOR ABSORBED	COLOR TRANSMITTED
< 400 nm	ultraviolet	colorless
400–450	violet	red, orange, yellow
450–490	blue	
490–550	green	purple
550–580	yellow	
580–650	orange	blue, green
650–700	red	
> 700 nm	infrared	colorless

We can only see light in the region between 400 and 700 nm

Substances which do not absorb visible light are colorless (or white, if they are solids). These substances often absorb radiation outside the visible region. Sometimes this occurs in the ultraviolet, at wavelengths below 400 nm. Benzene, a colorless liquid, absorbs strongly around 255 nm. Ozone, a gaseous form of the element oxygen, is another substance that absorbs in the ultraviolet. Much of the harmful ultraviolet radiation of sunlight is absorbed by ozone in the upper part of the earth's atmosphere.

Absorption in the infrared region, at wavelengths greater than 700 nm, is quite common. Water and carbon dioxide are among the substances that absorb in the infrared. Their presence in the atmosphere has an insulating effect. The earth, like other warm bodies, gives off heat in the form of infrared radiation. Much of this is absorbed by water vapor and carbon dioxide, preventing excessive loss of heat to outer space.

This helps moderate the temperature changes on the earth

1.6
Separation of Mixtures

Very few elements and compounds occur in nature in the pure state. Most often, they are found in mixtures with other substances. We distinguish between two different types of mixtures.

1. A *homogeneous* (uniform) mixture, called a *solution*. In discussing solutions, we often distinguish between the components by using the words "solute" and "solvent." Most commonly, the solvent is a liquid; the solute may be a solid, liquid, or gas. Soda water is a solution of carbon dioxide (solute) in water (solvent). Seawater is a more complex solution in which there are several solid solutes, including sodium chloride; the solvent is water.

2. A *heterogeneous* (nonuniform) mixture, sometimes referred to as a coarse mixture. Most of the rocks and minerals in the earth's crust fall into this category. If you look at a piece of granite, you can distinguish several components differing from one another in color.

Chemists, in carrying out reactions, ordinarily work with pure substances. To obtain a pure substance, it is often necessary to separate it from a mixture containing impurities. Such separations are based on differences in properties between the components of a mixture. Sometimes the process is a very simple one. Iron fillings can be separated from powdered sulfur by using a magnet to extract the iron. Sand can be removed from sugar by shaking the mixture with water. The sugar goes into solution; the sand remains behind and can be filtered off.

To identify the components of a mixture, you often have to separate them out as pure substances

There are several separation techniques that are more generally useful than those just mentioned. In the remainder of this section, we will consider three of the more common methods:

1. *Distillation*, used to separate a solid from a liquid, and *fractional distillation*, by which two liquids can be separated from each other.
2. *Fractional crystallization*, a technique used routinely to purify solids.
3. *Chromatography*, a separation method that can be applied to solid, liquid, or gaseous mixtures.

Distillation and Fractional Distillation

A mixture of two substances, only one of which is volatile, can be separated by distillation. A simple distillation apparatus that can be used to separate sodium chloride from water is shown in Figure 1.8. When the solution is heated the water boils off, leaving a residue of solid sodium chloride in the distilling flask. The water may be collected by passing the vapor down a cold tube and thus condensing

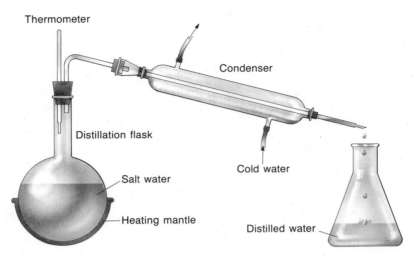

FIGURE 1.8 Pure water can be obtained from salt water by simple distillation. Since salt is not volatile, it remains in the distillation flask.

it as a liquid. In many arid areas of the world, distillation is used to obtain fresh water from seawater.

If both components of a solution are volatile liquids, simple distillation does not give a complete separation. Consider, for example, a 50-50 mixture of ether (bp = 35°C) and benzene (bp = 80°C). If this solution is heated, it begins to boil somewhat above 35°C. The vapor formed is richer than the solution in the more volatile component, ether, since ether boils out of the solution more readily than does benzene. However, the condensed vapor, called the *distillate*, still contains some benzene. The *residue* in the distilling flask becomes richer in benzene but still contains some ether. If we continue heating until half of the solution has distilled, we might find that the distillate contains 70% ether, while the residue is 70% benzene.

To improve upon this separation, we might repeat the process, using the two fractions obtained in the first distillation. This would give a distillate still richer in ether (perhaps 85%) and a residue richer in benzene. A more effective way to carry out this process of **fractional distillation** is to use a fractionating column (Fig. 1.9). The glass beads in the column provide a surface on which much of the benzene condenses and falls back into the distilling flask. The more volatile ether tends to pass through the column without condensing. If the rate of heating and the temperature of the column are carefully controlled, a good separation can be achieved. Fractional distillation is used routinely in the petroleum industry. There, the columns used are much larger and more complex than that shown in Figure 1.9. By this technique, crude oil is separated into fractions such as gasoline (bp = 40 to 200°C), kerosene (bp = 175 to 325°C), and diesel fuel (bp > 275°C).

Volatile means easily vaporized

Petroleum stills are huge, ~5 m in diameter and 100 m high

Fractional Crystallization

The reagent grade solid chemicals that you work with in the laboratory have been purified, usually by fractional crystallization. In this process, we start with an impure solid, A, which we wish to purify. It contains a relatively small amount of impurity, B. The mixture of A and B is dissolved in a minimum amount of hot

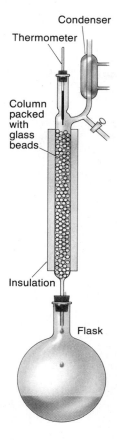

FIGURE 1.9 A simple fractionating column. Temperature decreases as one moves up the column. The less volatile component tends to condense back into the flask, while the more volatile component distills off and is recovered.

solvent, often boiling water (Fig. 1.10, p. 26). The solution is cooled, often in an ice-water bath. If all goes well, the solid that separates out on cooling will be pure A. This can be filtered off and dried. The solution that remains (the filtrate) should contain all of the impurity along with a small amount of A. This filtrate is ordinarily discarded.

To obtain a pure solid by fractional crystallization, two conditions must apply:

1. *The major component must be much more soluble at high than at low temperatures*. Otherwise, much of it will be lost; that is, it will remain in the solution upon cooling. Referring back to Figure 1.6, you can see that it would be futile to try to purify sodium chloride by fractional crystallization from water. Its solubility is nearly independent of temperature. In contrast, this process works very well with potassium nitrate, which is six times more soluble at 100°C than at 25°C.

2. *The amount of impurity, B, must be rather small*. Otherwise, it will tend to crystallize out on cooling, contaminating the desired product, A. This is likely to happen with a mixture containing 10% to 20%B. Several recrystallizations might be required to obtain pure A. With a 50-50 mixture, the technique is likely to be completely ineffective.

B will remain in solution if the amount per 100 g solvent is less than its solubility

Chromatography

Distillation, fractional distillation, and fractional crystallization have been used to separate mixtures for at least two centuries. In contrast, the process called

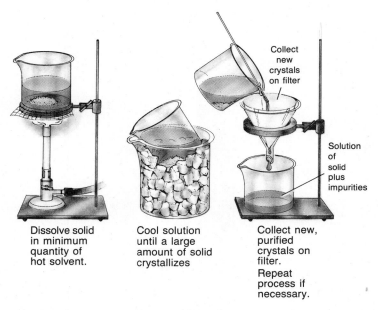

Collect
new
crystals
on filter

Solution
of
solid
plus
impurities

Dissolve solid
in minimum
quantity of
hot solvent.

Cool solution
until a large
amount of solid
crystallizes

Collect new,
purified
crystals on
filter.
Repeat
process if
necessary.

FIGURE 1.10 Steps involved in purifying a solid by fractional crystallization. Filtration is used to separate the crystals from the solution (filtrate), which contains the impurities.

chromatography is relatively new. It has become popular only within the past 25 years. The word chromatography is derived from the Greek *chroma*, meaning "color." In most of the early experiments the separated components were identified by their colors.

Chromatography has several advantages over more classical methods of separation. For one thing, it can be applied to very complex mixtures. As many as 20 amino acids can be isolated from a protein sample by this method. Moreover, it can be used for very small samples or for substances present at very low concentrations. Chromatographic techniques have been developed to detect air pollutants at concentrations of one part per million or less.

To illustrate the use of this method, let us consider a typical experiment in *paper chromatography*, which is often used in general chemistry. The sample is a liquid solution containing two or more dissolved solids. It is applied as a spot to a long strip of filter paper, near the lower edge. The spot is allowed to dry and the paper hung in a stoppered flask. Enough solvent is used to almost, but not quite, reach the spot. As the solvent rises by capillary action, the components of the sample travel along with it. A component that is very soluble tends to move farther up the paper. Another component that is less soluble, or more strongly adsorbed onto the paper, moves a smaller distance. In this way, a separation occurs. If the substances are colored, they are easily visible as colored spots on the paper (see Color Plate 1.2). If not, some process must be used to detect them and make their positions clear. If desired, the components can be recovered by cutting the paper into sections and dissolving them off the paper with a solvent.

The process just described is suitable for separating two or more solids. Many other chromatographic techniques are available. One of the most widely used is vapor phase chromatography (VPC). Here, the components are separated as va-

Chromatography offers: high resolution, high sensitivity, fast analyses, versatility

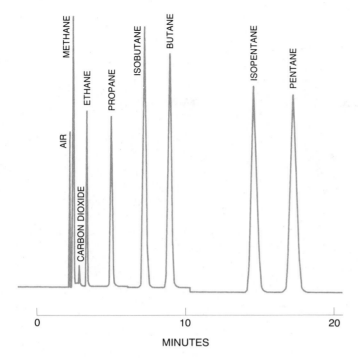

VPC is the method of choice for the separation of complex volatile mixtures

FIGURE 1.11 The components of natural gas (mostly methane) can be separated by vapor phase chromatography. With some volatile mixtures, the sample can be as small as 10^{-6} L.

pors. The mixture is injected into a heated glass tube at one end. The tube is packed with a finely divided solid, often coated with a high-boiling viscous liquid. An unreactive "carrier" gas, often helium, is passed through the tube. The components of the sample gradually separate as they vaporize into the helium or adsorb onto the packing. Usually the more volatile components move faster and come out of the column first. As successive fractions leave the column, they activate a detector and recorder. The end result is a plot such as that shown in Figure 1.11.

Summary

Chemistry is the study of matter and its composition, properties, structure, and reactions. Matter is composed of elements and compounds, which may occur as pure substances or as mixtures. A compound is a substance in which two or more elements are combined chemically. An element is unique; no two elements have the same symbol or exactly the same properties. The Periodic Table, shown inside the front cover, is a convenient tabulation of the elements. It is arranged by periods (horizontal rows) and groups (vertical columns). Elements in a group are similar chemically.

Substances are identified by their properties. These include density (Example 1.7), solubility (Example 1.8), specific heat (Example 1.9), melting point, boiling point and color. Colored substances absorb visible light (wavelength 400 to 700 nm).

Differences in properties such as boiling point or solubility permit components to be separated from a mixture. Liquids can be purified by distillation or fractional distillation, solids by fractional crystallization. Chromatography can be used to separate all types of mixtures.

Many different kinds of measurements have been discussed in this chapter. All measured quantities have an uncertainty whose magnitude depends upon the instrument used and the skill of the person using it. Significant figures indicate the degree of uncertainty in a measurement (Examples 1.2 and 1.3). Measured quantities may be expressed in various units (Tables 1.2 and 1.3). To convert a quantity from one set of units to another, we use conversion factors. Their use is illustrated in Examples 1.4 through 1.8. We will use the conversion factor approach over and over again to solve problems in future chapters, so you should become familiar with it now. Other problems will be solved by substituting into equations such as those relating temperature scales (Example 1.1) or heat absorbed to specific heat (Example 1.9).

Key Words and Concepts

These are defined in the glossary at the back of the book.

atmosphere	element	*kilo-*	period
boiling point	Fahrenheit degree	kilojoule	Periodic Table
Celsius degree	fractional crystallization	liter	physical property
centi-	fractional distillation	main group	SI unit
chemical property	gram	mass	significant figure
chromatography	group	melting point	solubility
compound	heterogeneous	meter	solution
conversion factor	homogeneous	*milli-*	specific heat
cubic centimeter	infrared	millimeter of mercury	symbol
density	joule	mixture	ultraviolet
distillation	Kelvin scale	*nano-*	wavelength

Questions and Problems

The questions and problems listed here are typical of those at the end of each chapter. Some involve discussions, others calculations. The topic emphasized in each question or problem is indicated in the headings below. Those in the "general" category may involve more than one concept. Problems indicated by an asterisk, listed at the end of each set, require extra skill and/or effort.

Most of the questions and problems are arranged in "matched pairs" next to one another (for example, 1.1 and 1.31; 1.2 and 1.32). The two problems within the pair illustrate the same concept and are of comparable difficulty. Answers are given in Appendix 5 for the member of the pair in the *right* column (for example, 1.31, 1.32).

Units and Measurements

1.1 Classify each of the following as units of mass, volume, length, density, energy, or pressure.
   a. nm   b. kg   c. J   d. m^3
   e. g/cm^3 f. atm   g. kcal

1.31 Classify each of the following as units of mass, volume, length, density, energy, or pressure.
   a. mg   b. mL   c. cm^3 d. mm
   e. kg/m^3 f. mm Hg   g. kJ

1.2 Select the larger member of each pair:
   a. 3.12 g or 3.12 mg
   b. 50 cm³ or 0.050 m³
   c. 2.69 g/L or 2.69 g/cm³
   d. 16.0 J or 0.0154 kJ

1.3 The boiling point of neon is $-246°C$. Express this in
   a. °F   b. K

1.4 The Ideal Gas Law states that PV = nRT. If P is in mm Hg, V in mL, n in mol, and T in K, what are the units of R?

1.32 Select the larger member of each pair:
   a. 502 m or 0.500 km
   b. 500 kg or 0.0500 g
   c. 150 cm³ or 150 nm³
   d. 12.0 g/cm³ or 1.20×10^3 kg/m³

1.33 The coldest nonlaboratory temperature recorded on earth was $-126.9°F$. Express this in
   a. °C   b. K

1.34 Consider the equation: $u^2 = 3RT/MM$. If u is in cm/s, MM in g/mol, and T in K, what are the units of R?

Significant Figures

1.5 How many significant figures are there in each of the following?
   a. 259.6 cm   b. 0.0142 g
   c. 3.41×10^7 nm   d. 52.3 L
   e. 1.9140 atm   f. 220 J

1.6 A student prepares a solution by dissolving 51.923 g of sugar in enough water to form 519 cm³ of solution. Calculate the number of grams of sugar per cubic centimeter of solution.

1.7 Calculate each of the following to the correct number of significant figures:
   a. $x = \dfrac{13 \text{ g}}{52.9 \text{ cm}^3}$
   b. $x = \dfrac{6.314 \text{ g}}{2.3 \text{ L}}$
   c. $x = 1.16 \text{ g} + 52 \text{ g}$
   d. $x = \dfrac{3.27 \text{ g} + 1.914 \text{ g}}{6.149 \text{ cm}^3}$

1.35 How many significant figures are there in each of the following?
   a. 1.92 cm   b. 0.560 m
   c. 6.022×10^{23} atoms d. 5.10 L
   e. 0.00291 m³   f. 5×10^3 g

1.36 Calculate the volume of a gold atom, which has a radius of 0.144 nm. Assume the atom is spherical; the volume of a sphere is given by the expression $V = 4\pi r^3/3$.

1.37 How many significant figures are there in the values of x obtained from
   a. $x = \dfrac{16.245 \text{ g}}{12.91 \text{ cm}^3}$
   b. $x = 21.24 \text{ g} - 21.22 \text{ g}$
   c. $x = (1.92 \text{ cm})(2.3 \text{ cm})$
   d. $x = \dfrac{18.1 \text{ g} - 12.2 \text{ g}}{(15.0 \text{ cm})(1.9 \text{ cm}^2)}$

Conversion Factors

1.8 Using Table 1.2, convert 2.10 ft³ to
   a. liters   b. cubic meters   c. quarts

1.9 Using Table 1.3, convert 52.0 mm Hg to
   a. atmospheres   b. pascals   c. lb/in²

1.10 In the United States, cigarettes are smoked at the rate of about 2.0×10^4 cigarettes/s. At this rate, how many are smoked in one year?

1.38 Using Table 1.2, convert 1.59 square miles to
   a. km²   b. m²   c. ft²

1.39 Using Table 1.3, convert 1.000 J to
   a. cal   b. kJ   c. liter atmospheres

1.40 The unit of land measure in the metric system is the hectare; in the English system it is the acre. A square exactly 100 m on a side has an area of one hectare; if it is 208.7 ft on a side, the volume is one acre. How many hectares are there per acre?

1.11 A world-class sprinter runs the 100.0-yard dash in 9.08 s. What is his speed in miles per hour?

1.12 When the Pharmacopoeia of London was compiled in 1618, the troy system of measure was used to prepare medicine. Among the units used were: 20 grains = 1 scruple; 3 scruples = 1 drachm; 8 drachms = 1 ounce; 12 ounces = 1 pound. Hence
a. 6.51 pounds = _____ drachms
b. 15.0 drachms = _____ pounds; = _____ ounces

1.13 A reaction requires 12.4 g of a substance X, which is only available as a mixture containing 93.2% X. How many grams of the mixture are needed to provide the amount of X required for reaction?

Elements and the Periodic Table

1.14 Referring to the table of elements inside the back cover of this text, name all the elements whose symbols are found in the names
a. New Hampshire   b. your surname
c. Lech Walesa

1.15 How many elements are there in the following periods?
a. period 1   b. period 2   c. period 3
d. period 4   e. period 5

1.16 Name the elements whose symbols are
a. Mn   b. Na   c. As   d. W   e. P

Physical and Chemical Properties

1.17 Which of the following are physical properties of bromine? Chemical properties?
a. density at 25°C, 1 atm = 3.12 g/cm³.
b. reacts with fluorine.
c. bromine vapor has an orange color.
d. normal boiling point = 58.8°C.

1.18 The mass and volume of ten glass beads are 2.62 g and 1.10 mL, respectively. What is the density of the beads?

1.19 A piece of metal weighing 16.52 g is added to a flask with a volume of 24.5 cm³. It is found that 19.6 g of water (d = 1.00 g/cm³) must be added to the metal to fill the flask. What is
a. the volume of the metal?
b. the density of the metal?

1.41 Bill Rodgers won the 1979 Boston Marathon (26.2 miles) in 2 h, 9 min, 27 s, finishing well ahead of our junior author. What was Rodgers' speed in meters per second?

1.42 During earlier times in England, land was measured in units such as fardells, nookes, yards, and hides: 2 fardells = 1 nooke; 4 nookes = 1 yard; 4 yards = 1 hide. Thus,
a. 8.00 hides = _____ fardells
b. 21 nookes = _____ hides; = _____ fardells

1.43 When a light beer is metabolized, about 58% of the "calories" are converted to useful energy. If 204 kJ of useful energy is produced from a 12-oz bottle of beer, how many nutritional calories does it contain (1 nutritional cal = 1 kcal)?

1.44 Referring to the table of elements inside the back cover of this text, name all the elements whose symbols are found in the names
a. Minnesota   b. Virginia
c. your college

1.45 How many elements are there in the following groups?
a. Group 2   b. Group 3   c. Group 8
d. the subgroup headed by Cr

1.46 Give the symbols of
a. potassium   b. copper   c. gold
d. antimony   e. lead

1.47 The following data refer to the element carbon. Classify each as a physical or chemical property.
a. reacts with oxygen to form oxides.
b. is virtually insoluble in water.
c. exists in several forms, e.g., diamond, graphite.
d. is a solid at 25°C, 1 atm.

1.48 A sample of chlorine gas with a volume of 2.50 L weighs 7.24 g. Calculate the density of chlorine in grams per cubic centimeter.

1.49 A solid weighing 12.02 g is added to a flask with a volume of 19.62 cm³. It is found that 12.02 g of methyl alcohol (d = 0.787 g/cm³) must be added to the metal to fill the flask. What is the density of the metal?

1.20 Blood plasma volume for adults is about 3.1 L. Its density is 1.020 g/cm³. About how many pounds of blood plasma are there in your body?

1.21 Air is 21% oxygen by volume; oxygen has a density of 1.31 g/L. What is the volume, in liters, of a room that contains 60.0 kg of oxygen?

1.22 The solubility of potassium chloride is 37.0 g per hundred grams of water at 30°C. Calculate
  a. the mass of potassium chloride that dissolves in 18.4 g of water.
  b. the mass of water required to dissolve 18.4 g of potassium chloride.

1.23 Use Figure 1.6 to estimate
  a. the solubility of KNO_3 at 75°C.
  b. the mass of water required to dissolve 50.0 g KNO_3 at 75°C.
  c. the mass of KNO_3 that dissolves in one liter of water (d = 1.00 g/mL) at 75°C.

1.24 The specific heat of solid aluminum is 0.902 J/g·°C. Calculate the amount of heat that must be absorbed to raise the temperature of a 3.50-g Al sample from 27.0 to 29.6°C.

1.25 It is found that 209 J of heat must be absorbed to raise the temperature of 44.5 g of nickel from 18.0 to 28.9°C. What is the specific heat of nickel?

1.26 What is the color of light having a wavelength of
  a. 300 nm   b. 500 nm
  c. 600 nm   d. 800 nm

Separation of Mixtures

1.27 Describe how you would separate each of the following from a mixture with water:
  a. ashes   b. table salt
  c. ethyl alcohol   d. fuel oil

1.28 You are given a mixture of 88 g of KNO_3 and 12 g of NaCl. You dissolve the mixture in 50.0 g of water at 100°C and cool to 10°C. Using Figure 1.6, calculate
  a. how much KNO_3 crystallizes out of solution.
  b. how much NaCl crystallizes out of solution.
  c. what percentage of the KNO_3 you started with remains in solution.

1.50 A water bed filled with water has the dimensions 8.0 ft × 7.0 ft × 0.75 ft. Taking the density of water to be 1.00 g/cm³, determine the number of kilograms of water required to fill the water bed.

1.51 A gas contains 8.00 g of nitrogen (d = 1.14 g/L) and 2.00 g of oxygen (d = 1.31 g/L). Calculate
  a. the mass percent of oxygen in the gas.
  b. the volume percent of oxygen in the gas.

1.52 At 0°C and 1 atm, 2.9×10^{-3} g of nitrogen dissolves in one hundred grams of water. Calculate
  a. the mass of nitrogen that dissolves in 1.00 mL of water (d = 1.00 g/mL)
  b. the mass of water required to dissolve 0.100 g of nitrogen.

1.53 Use Figure 1.6 to estimate
  a. the solubility of KNO_3 at 25°C.
  b. the mass of KNO_3 that dissolves in 4.8×10^3 g of water at 25°C.
  c. the mass of water needed to dissolve 1.00 g KNO_3 at 25°C.

1.54 The specific heat of solid platinum is 0.133 J/g·°C. Calculate the temperature change in °C of a 25.0-g sample of platinum when it absorbs 75.0 J of heat.

1.55 What is the specific heat of iron if 169 J must be absorbed to raise the temperature of a 65.0-g sample from 25.00 to 30.46°C?

1.56 Classify the following statements as true or false:
  a. A substance that absorbs at 600 nm is colored.
  b. Ultraviolet radiation has a wavelength above 400 nm.
  c. Infrared radiation has a wavelength above 700 nm.

1.57 Describe how you would separate each of the following from a mixture with potassium chromate, a water-soluble solid:
  a. ashes   b. table salt
  c. gasoline   d. water

1.58 Repeat the calculations called for in Problem 1.28, this time for a mixture containing 92 g of KNO_3 and 8 g of NaCl.

General

1.29 How do you distinguish
    a. an element from a compound?
    b. an element from a mixture?
    c. a solution from a heterogeneous mixture?
    d. distillation from fractional distillation?

1.30 An extensive property is one that depends upon the mass of a sample. Which of the following properties are extensive?
    a. volume   b. density   c. specific heat
    d. energy   e. melting point

1.59 How do you distinguish
    a. chemical properties from physical properties?
    b. a solute from a solution?
    c. boiling point from normal boiling point?
    d. a compound from a mixture?

1.60 An intensive property is independent of sample mass. Which of the following are intensive?
    a. solubility per 100 g of water
    b. heat required to raise the temperature 1°C
    c. boiling point

*1.61 At what point is the temperature in °C exactly twice that in °F?

*1.62 Oil spreads on water to form a film about 120 nm thick (two significant figures). How many square kilometers of ocean will be covered by the slick formed when one barrel of oil is spilled (1 barrel = 31.5 U.S. gallons)?

*1.63 A laboratory experiment requires 0.500 g of copper wire (d = 8.94 g/cm³). The diameter of the wire is 0.0179 in. Determine the length of the wire, in centimeters, to be used for this experiment. Volume of a cylinder = $\pi r^2 l$, where r = radius, l = length.

*1.64 An average human male breathes about 8.50×10^3 L of air per day. The concentration of lead (Pb) in highly polluted urban air is 7.0×10^{-6} g Pb/m³ air. Assume that 75% of the lead particles in the air are less than 1.0×10^{-6} m in diameter, and that 50% of the particles below that size are retained in the lungs. Calculate the mass of lead absorbed in this manner in one year by an average male living in this environment.

*1.65 What is the final temperature when 5.0 g of iron at 90.0°C is added to 20.0 cm³ of ethyl alcohol at 0.0°C? Use tables in this chapter for the necessary data.

In Chapter 1, we looked at some of the macroscopic properties of matter. These are properties that can be measured using samples large enough to see and weigh. They include density, melting point, boiling point, and specific heat. In this chapter, we will look at substances from a different point of view. We will examine the submicroscopic building blocks that make up elements and compounds. Our emphasis will be upon atoms, the smallest particles of elements (Section 2.1). We will also consider briefly two particles derived from atoms; molecules and ions (Section 2.3).

In this chapter, we will limit our discussion to a few basic properties of atoms. We will consider the particles of which atoms are composed: electrons, protons, and neutrons (Section 2.2). Beyond that, we will examine how atoms compare to one another in mass (Section 2.4). Finally, in Sections 2.5 and 2.6, we will consider a basic counting unit used with atoms, molecules, and ions: the mole.

2.1
Atomic Theory

The notion that matter consists of discrete particles is an old one. About 400 BC this idea appeared in the writings of Democritus, a Greek philosopher. He had been introduced to it by his teacher, Leucippus. The idea was rejected by Plato and Aristotle, who had a great deal more influence in the development of ideas than Democritus. Not until 1650 AD was the concept of atoms suggested again, this time by the French philosopher Pierre Gassendi. Sir Isaac Newton (1642–1727) supported Gassendi's arguments with these words:

> . . . it seems probable to me that God, in the Beginning, formed Matter in solid, massy, hard, impenetrable, movable Particles, of such Sizes and Figures, and with such other Properties, and in such Proportions to Space, as most conduced to the End for which he formed them. . . .

As far as we know, Newton did no experiments to test these ideas

Prior to 1800 the concept of the particulate nature of matter was based largely on intuition. Then, in 1808, an English schoolteacher, John Dalton, developed an explanation of several of the laws of chemistry. This became known as the **atomic theory**. Some of Dalton's ideas had to be discarded as chemists learned more about the structure of matter. However, the essentials of his theory have withstood the test of time. Three of the main postulates of modern atomic theory, all of which came from Dalton, are given below with examples to illustrate their meaning.

The atomic theory gave chemistry its start as a science

1. *An element is composed of tiny particles called atoms. All atoms of a given element show the same chemical properties.* The element oxygen is made up of oxygen atoms. These atoms are much too small to be seen or weighed directly. All oxygen atoms behave chemically in the same way.

2. *Atoms of different elements have different properties. In an ordinary chemical reaction, no atom of any element disappears or is changed into an atom of another element.* The chemical behavior of oxygen atoms is different from that of hydrogen atoms or any other kind of element. When hydrogen and oxygen react, all the hydrogen and oxygen atoms that react are present in the water formed. No atoms of any other element are formed.

3. *Compounds are formed when atoms of two or more elements combine. In a given compound, the relative numbers of atoms of each kind are definite and constant. In general, these relative numbers can be expressed as integers or simple fractions.* In the compound water, hydrogen atoms and oxygen atoms are combined with each other. For every oxygen atom present, there are always two hydrogen atoms.

The atomic theory explains two of the basic laws of chemistry.

1. The **Law of Conservation of Mass**. This law was first stated by the French chemist Antoine Lavoisier in 1789. In modern form, it says that *there is no detectable change in mass in an ordinary chemical reaction.* If atoms are "conserved" in a reaction (Postulate 2 above), mass will also be conserved.

2. The **Law of Constant Composition**. This tells us that *a compound always contains the same elements in the same proportions* by mass. If the atom ratio of the elements in a compound is fixed (Postulate 3), their proportions by mass must also be fixed.

It was not a simple matter to prove that pure substances exist

The validity of this law became generally recognized at about the same time that Dalton's theory appeared. Prior to 1808 many people agreed with the French chemist Berthollet. He believed that the composition of a compound could vary over wide limits, depending on how it was prepared. Joseph Proust, a Frenchman working in Madrid, refuted Berthollet by showing that the "compounds" Berthollet had cited were actually mixtures.

It is now known that in certain compounds, particularly metal oxides and sulfides, the atom ratio may vary slightly from a whole number ratio. Careful analyses of samples of nickel oxide prepared by heating nickel with oxygen at high temperatures give an atom ratio of 0.97:1.00 rather than the expected 1:1 ratio. Compounds of this type are sometimes referred to as "Berthollides" or, more frequently, *nonstoichiometric* compounds. The deviations from constant composition arise because of defects in the crystal structure.

The third postulate of the atomic theory is in many ways the most important. Among other things, it led Dalton to formulate the **Law of Multiple Proportions**. This law applies to the situation in which two elements form more than one compound. It states that, in these compounds *the masses of one element that* combine with a fixed mass of the second element are in a ratio of small whole numbers (for example 2:1). To derive the Law of Multiple Proportions, Dalton reasoned along the following lines: Suppose elements A and B form two different compounds. In one of these (AB), one atom of A might be combined with one atom of B. The second compound (AB$_2$) might contain two atoms of B per atom of A. If this is true, the mass of B combined with a fixed mass (such as one gram) of A would be twice as great in the second compound. In other words, the masses of B *per gram of A* in the two compounds would be in a 2:1 ratio. The meaning of the Law of Multiple Proportions is further illustrated in Table 2.1.

This Law is harder to state than it is to illustrate

Table 2.1
Law of Multiple Proportions Applied to the Oxides of Carbon and Sulfur

COMPOUND	ATOMS OF OXYGEN PER ATOM OF OTHER ELEMENT	GRAMS OF OXYGEN PER GRAM OF OTHER ELEMENT	WHOLE NUMBER RATIO
CO	1	1.33	2.66/1.33 = 2:1
CO$_2$	2	2.66	
SO$_2$	2	1.00	1.50/1.00 = 3:2
SO$_3$	3	1.50	

2.2
Components of the Atom

Like any useful scientific theory, the atomic theory raised more questions than it answered. Scientists wondered whether atoms, tiny as they are, could be broken down into still smaller particles. Nearly 100 years passed before the existence of subatomic particles was confirmed by experiment. Three physicists did pioneer work in this area. J. J. Thomson was an Englishman working at the Cavendish Laboratory at Cambridge. Ernest Rutherford, a native of New Zealand, carried out his research at McGill University in Montreal and at Manchester and Cambridge in England. The third of these was an American, Robert A. Millikan, who did his research at the University of Chicago.

Electrons

The first evidence for the existence of subatomic particles came from studies of the conduction of electricity through gases at low pressures. A sketch of the apparatus used is shown in Figure 2.1. When the tube is partially evacuated and connected to a spark coil, an electric current flows through it. Associated with this flow are colored rays of light, spreading out from the negative electrode (cathode). The properties of these *cathode rays* were studied during the last three decades of the nineteenth century. It was found that they were bent by both electric and magnetic fields. From a careful study of this deflection, J. J. Thomson showed in 1897 that the rays consist of a stream of negatively charged particles.

The fact that they were bent meant that they were charged

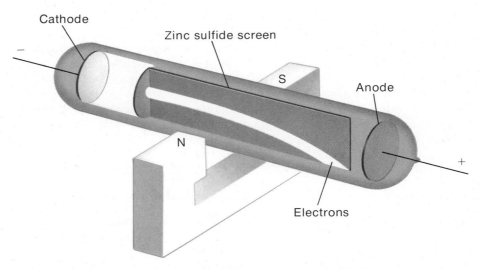

Cathode

Zinc sulfide screen

S

Anode

−

N

+

Electrons

FIGURE 2.1 Cathode ray tube. The ray, shown here as a white beam, is made up of fast-moving electrons. In an electric or magnetic field, the beam is deflected in such a way as to indicate that it carries a negative charge.

He called these particles **electrons**. Thomson went on to measure the mass-to-charge (m/e) ratio of the electron, finding that

The coulomb is a unit of electrical charge

$$m/e = 5.69 \times 10^{-9} \text{ g/coulomb}$$

The fact that this ratio is the same, regardless of what gas is in the tube, implies that the electron is a basic particle, common to all atoms.

In 1909, Millikan determined the charge on the electron, using the apparatus sketched in Figure 2.2. He measured the effect of an electric field on the rate at which charged oil drops fall between two plates. From his data, Millikan calculated the charges on the drops. He found that these charges were always an integral multiple of some smallest charge. Taking this smallest charge to be that of the electron, he arrived at a value of 1.60×10^{-19} coulomb. Combining this value with the mass-to-charge ratio from Thomson's work, we obtain the mass of the electron:

This is about as small a mass as you can find anywhere

$$m = (1.60 \times 10^{-19} \text{ coulomb}) \times (5.69 \times 10^{-9} \text{ g/coulomb})$$
$$= 9.11 \times 10^{-28} \text{ g}$$

This is roughly 1/2000 of the mass of the lightest atom, that of the element hydrogen.

Every atom contains a definite number of electrons. This number, which runs from 1 to over 100, is characteristic of a neutral atom of a particular element: all hydrogen atoms contain one electron; all atoms of the element uranium contain 92 electrons. We will have more to say in Chapter 7 about how these electrons are arranged relative to one another. At this time, we need only point out that electrons are found in the outer regions of atoms. They comprise what amounts to a cloud of negative charge about the nucleus of an atom.

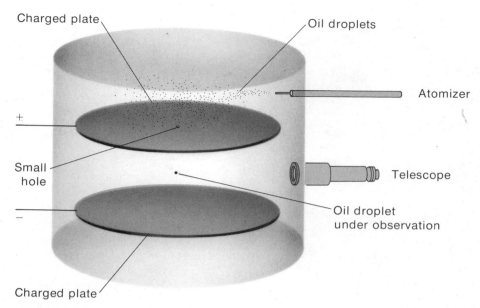

FIGURE 2.2 Oil drop experiment. At a certain voltage, the electrical and gravitational forces on a charged drop between the two plates will be exactly balanced, and the drop will not move. Knowing this voltage and the mass of the drop, it is possible to calculate the charge on the drop. This charge must be a whole-number multiple of the charge on the electron.

The Atomic Nucleus; Protons and Neutrons

A series of experiments carried out under the direction of Ernest Rutherford in 1911 shaped our ideas about the nature of the atom. The experiments were conducted by Johannes Geiger, a German physicist working with Rutherford, and Ernest Marsden, an undergraduate at Cambridge. They bombarded a piece of thin gold foil (Fig. 2.3, p. 38) with α-particles (helium atoms minus their electrons). With a fluorescent screen, they observed the extent to which the α-particles were scattered. Most of them went through the foil unchanged in direction. A few, however, were reflected back at acute angles. The relative numbers of α-particles reflected at different angles were counted. By a mathematical analysis of the forces involved, Rutherford showed that the scattering was caused by a small, positively charged center within the gold atom.

They had expected that no α-particles would be reflected at acute angles

These experiments were repeated, with similar results, using foils of many different elements. In this way, Rutherford and his co-workers showed that in all atoms there is a central nucleus which

—carries a positive charge equal in magnitude to the total negative charge of the electrons outside the nucleus.

—contains more than 99.9% of the total mass of the atom.

—has a diameter only 0.01% of that of the atom itself. If an atom could be expanded to cover this page, its nucleus would be barely visible as a tiny dot one tenth the size of the period at the end of this sentence.

The calculated density of a nucleus is enormous, about 1×10^6 tons/cm³!

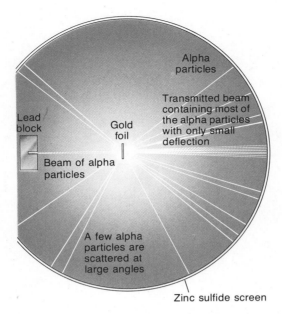

FIGURE 2.3 Rutherford scattering experiment. Most of the α-particles are essentially undeflected, but a few are scattered at large angles. In order to cause the large deflections, atoms must contain heavy, positively charged nuclei.

Since the time of Rutherford, we have learned a great deal about the properties of atomic nuclei. For our purposes in chemistry, we can consider the nucleus of an atom to consist of two different types of particles:

1. The **proton**, which has a mass nearly equal to that of the lightest atom, the hydrogen atom. Reflecting that fact, we assign the proton a *mass number* of 1. The proton carries a unit positive charge ($+1$), equal in magnitude to that of the electron (-1).
2. The **neutron**, an uncharged particle with a mass about equal to that of a proton. The neutron, like the proton, has a mass number of 1.

Atomic Number

All the atoms of a particular element have the same number of protons in the nucleus. This number is a basic property of an element, called its **atomic number**:

$$\text{atomic number} = \text{number of protons} \tag{2.1}$$

The $+$ charge on the nucleus must equal the total $-$ charge on the electrons

In a neutral atom the number of protons in the nucleus is exactly equal to the number of electrons outside the nucleus. Consider, for example, the elements hydrogen (at. no. = 1) and helium (at. no. = 2). All hydrogen atoms have one proton in the nucleus; all helium atoms have two. In a neutral hydrogen atom there is one electron outside the nucleus; in a helium atom there are two.

H atom:	1 proton,	1 electron,	atomic number = 1
He atom:	2 protons,	2 electrons,	atomic number = 2
U atom:	92 protons,	92 electrons,	atomic number = 92

Atomic numbers of the elements are given in the Periodic Table (inside front cover). They are printed in color directly above the symbol of the element. Notice that atomic number increases steadily as we move across the table. Indeed, the location of an element in the table is determined by its atomic number.

Mass Number; Isotopes

As with protons and neutrons, we can assign mass numbers to atoms. Recall that the mass number is 1 for both a proton and a neutron. Hence, we can find the mass number of an atom by adding up the number of protons and neutrons in the nucleus:

$$\text{mass number} = \text{number of protons} + \text{number of neutrons} \qquad (2.2)$$

All atoms have integral mass numbers

For an atom with 17 protons and 20 neutrons in the nucleus,

$$\text{mass number} = 17 + 20 = 37$$

As we have noted, all atoms of a given element have the same atomic number (number of protons). They may, however, differ from one another in mass and hence in mass number. This can happen because, although the number of protons in the nucleus of a given kind of atom is fixed, the number of neutrons is not. It may vary and often does. Consider the element hydrogen (at. no. = 1). There are three different kinds of hydrogen atoms. They all have one proton in the nucleus. A "light" hydrogen atom (the most common type) has no neutrons in the nucleus (mass no. = 1). Another type of hydrogen atom (deuterium) has one neutron (mass no. = 2). Still a third type (tritium) has two neutrons (mass no. = 3).

All H nuclei contain one proton

Atoms that contain the same number of protons but a different number of neutrons are called isotopes. The three kinds of hydrogen atoms just described are isotopes of that element. They have masses that are very nearly in the ratio 1:2:3. Among the isotopes of the element uranium are the following:

ISOTOPE	ATOMIC NUMBER	MASS NUMBER	NUMBER OF PROTONS	NUMBER OF NEUTRONS
Uranium-235	92	235	92	143
Uranium-238	92	238	92	146

All U nuclei contain 92 protons

The composition of a nucleus is shown by its nuclear symbol. Here, the atomic number appears as a subscript at the lower left of the symbol of the element. The mass number is written as a superscript at the upper left. The nuclear symbols of the isotopes of hydrogen are

$$^{1}_{1}\text{H}, \quad ^{2}_{1}\text{H}, \quad ^{3}_{1}\text{H}$$

For the isotopes of uranium, we write the nuclear symbols

$$^{235}_{92}\text{U}, \quad ^{238}_{92}\text{U}$$

EXAMPLE 2.1

a. Write nuclear symbols for three isotopes of oxygen (at. no. = 8) in which there are 8, 9, and 10 neutrons, respectively.

b. One of the most harmful species in nuclear fallout is a radioactive isotope of strontium, $^{90}_{38}\text{Sr}$. How many protons are there in this nucleus? how many neutrons?

Solution

a. The mass numbers must be as follows: 8 + 8 = 16; 8 + 9 = 17; 8 + 10 = 18. Thus we have

$$^{16}_{8}\text{O}, \ ^{17}_{8}\text{O}, \ ^{18}_{8}\text{O}$$

b. The number of protons is given by the atomic number, 38. To obtain the number of neutrons we subtract the number of protons from the mass number:

$$\text{number of neutrons} = 90 - 38 = 52$$

EXERCISE Write the nuclear symbol for an atom that contains 32 protons and 38 neutrons (use the Periodic Table to find the symbol of the element). Answer: $^{70}_{32}\text{Ge}$.

	Mass No.	Charge
Proton	1	+1
Neutron	1	0
Electron	0	−1

2.3
Molecules and Ions

Isolated atoms rarely occur in nature. For the most part, they are too reactive to be found by themselves. Instead, atoms tend to combine with one another in various ways. As a result, the structural units in most elements and all compounds are more complex than simple atoms. Two of the most important "building blocks" of matter are molecules and ions, both of which are formed from atoms.

Molecules

The basic structural unit in most volatile (easily vaporized) substances is the molecule. *A molecule is a group of two or more atoms held together by strong forces called chemical bonds.* Many elements consist of diatomic molecules (Fig. 2.4). The hydrogen molecule is typical; its structure may be shown as

H—H

where the dash is used to represent the bond joining the two hydrogen atoms. The molecule of the gaseous compound hydrogen chloride is also diatomic. It has the structure

The H—H molecule is much more stable than two free H atoms

H—Cl

Most often, molecular substances are represented by molecular formulas. In a molecular formula, the number of atoms of each element is indicated by a subscript written after the symbol of the element. The molecular formulas of the two substances just described are

GROUP 5	GROUP 6	GROUP 7	GROUP 8
		H_2	
N_2	O_2	F_2	
P_4	S_8	Cl_2	
		Br_2	
		I_2	

FIGURE 2.4 Molecular elements in the Periodic Table. In writing chemical equations (Chap. 3), sulfur is ordinarily represented as being monatomic; the other elements have the formulas shown.

hydrogen:	H_2	(2 H atoms per molecule)
hydrogen chloride:	HCl	(1 H atom, 1 Cl atom per molecule)

(Note that no subscript is used when there is only one atom of a particular type present.)

Most molecular substances are composed of molecules more complex than those just cited. The water molecule, for example, consists of a central oxygen atom bonded to two hydrogen atoms. In the ammonia molecule, a central nitrogen atom is bonded to three hydrogen atoms. Methane, the major component of natural gas, has as a structural unit a molecule in which a carbon atom is bonded to four hydrogen atoms. The structures and formulas of these molecules are shown in Figure 2.5.

Ions

If enough energy is applied, one or more electrons can be removed from a neutral atom. This leaves a positively charged particle somewhat smaller than the original

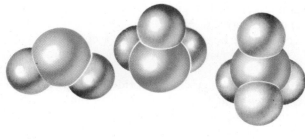

Water (H_2O)   Ammonia (NH_3)   Methane (CH_4)

FIGURE 2.5 Space-filling models of three simple molecules. The atoms shown in gray are hydrogen atoms. They are held by strong chemical bonds to a central, nonmetal atom (O, N, or C).

atom. Electrons may be added to certain atoms to form negatively charged particles larger than the original atom. Charged particles are called **ions**. An example of a positive ion (**cation**) is the Na^+ ion, formed from a sodium atom by the loss of a single electron. Another cation is Ca^{2+}, derived from a calcium atom by removing two electrons.

$$Na \text{ atom} \longrightarrow Na^+ \text{ ion} + e^-$$
$$(11\ p^+,\ 11\ e^-) \qquad (11\ p^+,\ 10\ e^-)$$

$$Ca \text{ atom} \longrightarrow Ca^{2+} \text{ ion} + 2\ e^-$$
$$(20\ p^+,\ 20\ e^-) \qquad (20\ p^+,\ 18\ e^-)$$

Two common negative ions (**anions**) are the chloride ion, Cl^-, and the oxide ion, O^{2-}. These are formed when atoms of chlorine or oxygen acquire electrons:

$$Cl \text{ atom} + e^- \longrightarrow Cl^- \text{ ion}$$
$$(17\ p^+,\ 17\ e^-) \qquad (17\ p^+,\ 18\ e^-)$$

$$O \text{ atom} + 2e^- \longrightarrow O^{2-} \text{ ion}$$
$$(8\ p^+,\ 8\ e^-) \qquad (8\ p^+,\ 10\ e^-)$$

Usually atoms gain or lose electrons by reacting with atoms of a different kind:

$$Na + Cl \rightarrow Na^+ + Cl^-$$

(in NaCl)

Notice that when an ion is formed the number of protons in the nucleus is unchanged. It is the number of electrons that increases or decreases. Negative ions contain more electrons than protons; positive ions contain fewer electrons than protons. The charge of the ion is indicated by a superscript at the upper right. The O^{2-} ion has a -2 charge; it has two more electrons than the oxygen atom. The Na^+ ion has a $+1$ charge; it has one fewer electron than the sodium atom.

EXAMPLE 2.2   Give the number of protons and electrons in the Sc^{3+} ion.

Solution   Using the Periodic Table, we find that Sc (scandium) has an atomic number of 21. Hence, the Sc^{3+} ion contains 21 protons and

$$21 - 3 = 18 \text{ electrons}$$

EXERCISE   Give the symbol of an ion that has $10\ e^-$ and $7\ p^+$; an ion that has $10\ e^-$ and $12\ p^+$. Answers: N^{3-}; Mg^{2+}.

Many compounds are made up of ions. They are called ionic compounds. Since a bulk sample of matter must be electrically neutral, ionic compounds always contain both cations ($+$ charge) and anions ($-$ charge). Ordinary table salt, sodium chloride, is made up of an equal number of Na^+ and Cl^- ions. The structure of a sodium chloride crystal is shown in Figure 2.6. Notice that there are no discrete molecules. Positive and negative ions are bonded together in a continuous network. Calcium oxide (quicklime), which is made up of Ca^{2+} and O^{2-} ions, has a similar structure.

In the ionic compounds sodium oxide and calcium chloride there are unequal numbers of cations and anions. To maintain electrical neutrality, the amount of positive charge in the compound must equal the amount of negative charge. This means that in sodium oxide there must be two Na^+ ions for every O^{2-} ion. A similar situation applies with calcium chloride: two Cl^- ions are required to balance one Ca^{2+} ion.

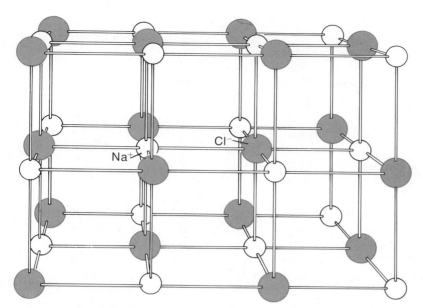

FIGURE 2.6 Ball-and-stick model of a crystal of NaCl (the Na$^+$ ions actually touch the Cl$^-$ ions). Each Cl$^-$ ion is surrounded by six Na$^+$ ions, and each Na$^+$ ion by six Cl$^-$ ions. A single crystal of NaCl contains many billions of ions arranged in this pattern.

Most alkali halides and many other ionic substances have the same structure as NaCl

The composition of an ionic compound is indicated by writing its formula. Here, subscripts are used to indicate the relative numbers of ions of each type. For the four compounds just discussed,

COMPOUND	FORMULA	INTERPRETATION
Sodium chloride	NaCl	1 Na$^+$ ion for 1 Cl$^-$ ion
Calcium oxide	CaO	1 Ca^{2+} ion for 1 O^{2-} ion
Sodium oxide	Na$_2$O	2 Na$^+$ ions for 1 O^{2-} ion
Calcium chloride	CaCl$_2$	1 Ca^{2+} ion for 2 Cl$^-$ ions

2.4
Masses of Atoms

As pointed out earlier, individual atoms are too small to be seen, let alone weighed. It is possible, however, to determine the relative masses of different atoms; that is, we can find out how heavy an atom is compared to an atom of a different element. In this section, we will consider how this is done, using quantities called **atomic masses**. Later, we will see how it is possible to calculate the masses of individual atoms, using a quantity referred to as **Avogadro's number**.

Some experiments indirectly let us "see" atoms

Atomic Masses;
The Carbon-12 Scale

Relative masses of atoms of different elements are expressed in terms of their atomic masses. **The atomic mass of an element is a number. It tells how heavy, on the average, an atom of that element is compared to an atom of another element.** Consider, for example, the two elements copper and sulfur. The atomic mass of copper, to four significant figures, is 63.55; that of sulfur is 32.06. This tells us that a copper atom is a little less than twice as heavy as a sulfur atom. More exactly,

$$\frac{\text{mass Cu atom}}{\text{mass S atom}} = \frac{\text{atomic mass Cu}}{\text{atomic mass S}} = \frac{63.55}{32.06} = 1.982$$

In the general case, for two elements Y and Z,

$$\frac{\text{mass of an atom of Y}}{\text{mass of an atom of Z}} = \frac{\text{atomic mass of Y}}{\text{atomic mass of Z}}$$

For many years oxygen was the reference element

To set up a scale of atomic masses, it is necessary to establish a standard value for one species. Over the years, several different standards have been used. In 1961 it was agreed to assign an atomic mass of exactly 12 to the most common isotope of carbon, $^{12}_{6}C$. Atomic masses used in this text are based on the carbon-12 scale. When we say that the atomic mass of sodium (Na) is 22.98977, we mean that it weighs $22.98977/12.00000 = 1.915814$ times as much as an atom of $^{12}_{6}C$.

Atomic masses are listed in the Periodic Table (inside front cover). They are given in black below the symbol of the element (H = 1.00794, He = 4.00260, and so on). Notice that, with a few exceptions, atomic mass increases in the same order as atomic number. Another table of atomic masses is given on the inside back cover of this text. Here the elements are arranged alphabetically.

Atomic Masses and Isotopic Abundances

You may have noticed that the atomic mass of carbon is a bit larger than 12, 12.011 to be exact. Recalling that $^{12}_{6}C$ has an atomic mass of exactly 12, you might wonder where the "extra" 0.011 comes from. The explanation is a simple one. The element carbon, as it occurs in nature, is a mixture of two isotopes: about 99% of all carbon atoms are of the $^{12}_{6}C$ type; about 1% represent the $^{13}_{6}C$ isotope. The presence of these heavier atoms of mass number 13 explains why the atomic mass of carbon is slightly greater than 12.

The situation just described is quite common. Most elements, as they occur in nature, consist of two or more isotopes in fixed proportions. In such cases, the atomic mass of the element is a "weighted average" of those of its isotopes. To calculate the atomic mass we must know

The atomic masses of protons and neutrons are very nearly equal to 1

—*the masses of the individual isotopes on the carbon-12 scale.* These are either equal to or very close to the mass numbers of the isotopes. For example, the atomic masses of the two isotopes $^{12}_{6}C$ and $^{13}_{6}C$ are 12.00 and 13.00, respectively.

—*the isotopic abundances (percentages).* These tell us the fraction of the total number of atoms identified with a particular isotope. In naturally occurring carbon, the abundances of $^{12}_{6}C$ and $^{13}_{6}C$ are about 99% and 1%, respectively. This means that 99 of 100 carbon atoms have an atomic mass of 12.00; 1 of 100 has an atomic mass of 13.00.

Knowing the atomic mass and abundance of each isotope, we can readily calculate the atomic mass of an element. For an element Y that consists of isotopes $Y_1, Y_2, \ldots,$

atomic mass Y = (atomic mass Y_1) × (% of Y_1)/100
+ (atomic mass Y_2) × (% of Y_2)/100
+ · · ·

(2.3)

The percentages of the several isotopes must, of course, add up to 100:

% of Y_1 + % of Y_2 + · · · = 100%

EXAMPLE 2.3   The element boron consists of two isotopes, $^{10}_{5}B$ and $^{11}_{5}B$. Their masses, on the carbon-12 scale, are 10.01 and 11.01, respectively. The abundance of $^{10}_{5}B$ is 20.0%. What is
a. the abundance of $^{11}_{5}B$?   b. the atomic mass of boron?

Solution
a. The sum of the abundances must add to 100%. Hence,

abundance $^{11}_{5}B$ = 100.0% − 20.0% = 80.0%

b. Using Equation 2.3, which relates the atomic mass of an element to the atomic masses of its isotopes

$$\text{atomic mass B} = 10.01 \times \frac{20.0}{100} + 11.01 \times \frac{80.0}{100}$$
$$= 2.00 + 8.81 = 10.81$$

Note that this is the value listed for boron in the Periodic Table. The atomic mass of boron is closer to 11 than to 10. This reflects the fact that the heavier isotope is more abundant.

EXERCISE   Taking the abundances of $^{12}_{6}C$ and $^{13}_{6}C$ to be 98.9% and 1.1% and their masses to be 12.00 and 13.00, calculate the atomic mass of carbon. Answer: 12.01.

Data of the type used in Example 2.3 can be obtained with a mass spectrometer (Section 2.7). The masses and abundances of isotopes in a sample of an element can be found very accurately, to seven or eight significant figures. The accuracy of tabulated atomic masses is limited mostly by variations in natural abundances. Sulfur is an interesting case in point. It consists largely of two isotopes, $^{32}_{16}S$ and $^{34}_{16}S$. The abundance of sulfur-34 varies from about 4.18% in sulfur deposits in Texas and Louisiana to 4.34% in volcanic sulfur from Italy. This leads to an uncertainty of 0.01 unit in the atomic mass of sulfur (32.06).

The most accurately known atomic masses are those of elements containing only one isotope

Masses of Individual Atoms; Avogadro's Number

For most purposes in chemistry, it is sufficient to know the relative masses of different atoms. We would like, however, to go one step further and calculate the mass in grams of individual atoms. We will now consider how this can be done.

To start with, it will be helpful to refer to Color Plate 2.1. This shows samples of three different elements: carbon, sulfur, and copper. One contains 12.01 g of carbon, another 32.06 g of sulfur, and the third 63.55 g of copper. The question is, what do these samples have in common? They don't have the same mass and clearly they don't look the same. They do, however, have one important property in common. Can you see what this is? (It will help to recall that the atomic masses of C, S, and Cu are 12.01, 32.06, and 63.55, respectively.) If you think about it for a moment, you should be able to answer this question. Look at it this way. A sulfur atom is 32.06/12.01 times as heavy as a carbon atom. If N carbon atoms weigh 12.01 g, then the same number, N, of sulfur atoms must weigh

$$\frac{32.06}{12.01} \times 12.01 \text{ g} = 32.06 \text{ g}$$

In other words, 32.06 g of sulfur contains the same number of atoms as does 12.01 g of carbon. By the same argument, if there are N carbon atoms in 12.01 g C, there must be the same number, N, of copper atoms in 63.55 g Cu. In summary, there are the same number of atoms in each of the three samples referred to above.

The answer to this question suggests another, more difficult question. How many atoms are there in each of these samples (12.01 g C, 32.06 g S, 63.55 g Cu)? As it happens, this problem is one that has been studied for at least a century. Several ingenious experiments have been designed to determine this number, known as **Avogadro's number** (see Problems 2.63 and 2.64). As you can imagine, it is huge. (Remember that atoms are tiny. There must be a lot of them in 12.01 g C, 32.06 g S, etc.) To four significant figures, Avogadro's number is

$$6.022 \times 10^{23}$$

To get some idea of how large this number is, suppose the entire population of the world were assigned to counting the atoms in 12.01 g of carbon. If each person counted one atom per second and worked a 48-hour week, the task would take more than ten million years.

The importance of Avogadro's number in chemistry should be clear. *It represents the number of atoms in X grams of any element, where X is the atomic mass of the element.* Thus there are

6.022×10^{23} C atoms in 12.01 g of carbon	atomic mass C = 12.01
6.022×10^{23} S atoms in 32.06 g of sulfur	atomic mass S = 32.06
6.022×10^{23} Cu atoms in 63.55 g of copper	atomic mass Cu = 63.55
6.022×10^{23} O atoms in 16.00 g of oxygen	atomic mass O = 16.00

Knowing Avogadro's number and the atomic mass of an element, it is possible to calculate the mass of an individual atom (Example 2.4a). We can also determine the number of atoms in a weighed sample of any element (Example 2.4b).

EXAMPLE 2.4 Taking Avogadro's number to be 6.022×10^{23}, calculate
a. the mass of a sulfur atom.
b. the number of sulfur atoms in a 1.000-g sample of the element.

Solution
a. We know that 6.022×10^{23} S atoms weigh 32.06 g:

$$6.022 \times 10^{23} \text{ S atoms} = 32.06 \text{ g}$$

This relation gives us a factor to "convert" a sulfur atom to grams:

If a nickel weighs twice as much as a dime, there are equal numbers of coins in 1000 g of nickels and 500 g of dimes

A real boring job

$$\text{mass S atom} = 1 \text{ S atom} \times \frac{32.06 \text{ g}}{6.022 \times 10^{23} \text{ S atoms}}$$
$$= 5.324 \times 10^{-23} \text{ g}$$

b. Here we want to go in the opposite direction, from grams to number of atoms. The appropriate conversion factor is

6.022×10^{23} atoms/32.06 g

$$\text{number of S atoms} = 1.000 \text{ g} \times \frac{6.022 \times 10^{23} \text{ S atoms}}{32.06 \text{ g}}$$
$$= 1.878 \times 10^{22} \text{ atoms}$$

EXERCISE Suppose the element referred to in the example had been copper instead of sulfur. What would have been the answers in (a) and (b)? Answers: 1.055×10^{-22} g, 9.476×10^{21} atoms.

Calculations such as those just carried out confirm what we have been saying all along: Atoms have very, very small masses ranging from 2×10^{-24} g (H atom) to 4×10^{-22} g (U atom).

2.5
The Mole

People in different professions often use special counting units. You and I eat eggs one at a time, but farmers sell them by the dozen. We spend dollar bills one at a time, but Congress distributes them by the billion. Chemists have their own counting unit—Avogadro's number. Since atoms and molecules are so small, it is convenient to talk about groups of 6.022×10^{23} of them. This counting unit is used so often in chemistry that it is given a special name—the **mole** (abbreviated as **mol**). To a chemist, **a mole means Avogadro's number of items.** Thus,

$$1 \text{ mol H atoms} = 6.022 \times 10^{23} \text{ H atoms}$$
$$1 \text{ mol O atoms} = 6.022 \times 10^{23} \text{ O atoms}$$
$$1 \text{ mol H}_2 \text{ molecules} = 6.022 \times 10^{23} \text{ H}_2 \text{ molecules}$$
$$1 \text{ mol H}_2\text{O molecules} = 6.022 \times 10^{23} \text{ H}_2\text{O molecules}$$
$$1 \text{ mol electrons} = 6.022 \times 10^{23} \text{ electrons}$$
$$1 \text{ mol pennies} = 6.022 \times 10^{23} \text{ pennies}$$

(One mole of pennies is a lot of money. It's enough to pay all the expenses of the United States for the next billion years or so.)   (If we ignore inflation)

This definition allows us to convert directly between moles and numbers of particles. For example,

$$\text{no. particles in 1.24 mol} = 1.24 \text{ mol} \times \frac{6.022 \times 10^{23} \text{ particles}}{1 \text{ mol}}$$
$$= 7.47 \times 10^{23} \text{ particles}$$

$$\text{no. moles in } 3.24 \times 10^{22} \text{ particles} = 3.24 \times 10^{22} \text{ particles} \times \frac{1 \text{ mol}}{6.022 \times 10^{23} \text{ particles}}$$
$$= 0.0538 \text{ mol}$$

Molar Masses

We can readily find the mass in grams of one mole of atoms. Recall that there are 6.022×10^{23} atoms in 1.01 g of hydrogen (atomic mass H = 1.01) or 16.00 g of oxygen (atomic mass O = 16.00). Therefore,

1 mol H weighs 1.01 g;   1 mol O weighs 16.00 g

In general, *one mole of atoms of any element weighs X grams, where X is the atomic mass of the element*. We can readily extend this idea to substances that consist of molecules or ions. For the substances represented by the formulas H_2, H_2O, and NaCl,

We set up the mole so that these relations would be true

1 mol H_2 contains 2 mol H atoms and weighs 2(1.01 g) = 2.02 g
1 mol H_2O contains 2 mol H atoms and 1 mol O atoms and weighs 2(1.01 g) + 16.00 g = 18.02 g
1 mol NaCl contains 1 mol Na atoms (22.99 g) and 1 mol Cl atoms (35.45 g) and weighs 22.99 g + 35.45 g = 58.44 g

In general, we can say that, for any substance, *one mole weighs X grams, where X is the formula mass, i.e., the sum of the atomic masses in the formula.* Thus,

FORMULA	FORMULA MASS	MOLAR MASS
H	1.01	1.01 g/mol
O	16.00	16.00 g/mol
H_2	2(1.01) = 2.02	2.02 g/mol
H_2O	2(1.01) + 16.00 = 18.02	18.02 g/mol
NaCl	22.99 + 35.45 = 58.44	58.44 g/mol

In other words, **the molar mass of a substance, in grams per mole (g/mol), is numerically equal to its formula mass.**

EXAMPLE 2.5 Calculate the molar masses (g/mol) of
a. potassium chromate, K_2CrO_4 b. sucrose, $C_{12}H_{22}O_{11}$

Solution
a. We start by calculating the formula mass of K_2CrO_4:

$$\text{formula mass } K_2CrO_4 = 2(\text{atomic mass K}) + \text{atomic mass Cr} + 4(\text{atomic mass O})$$
$$= 2(39.10) + 52.00 + 4(16.00) = 194.20$$

molar mass K_2CrO_4 = 194.20 g/mol

b. $$\text{formula mass} = 12(\text{atomic mass C}) + 22(\text{atomic mass H}) + 11(\text{atomic mass O})$$
$$= 12(12.01) + 22(1.01) + 11(16.00) = 342.34$$

molar mass $C_{12}H_{22}O_{11}$ = 342.34 g/mol

EXERCISE What is the molar mass of H_2SO_4, sulfuric acid? Answer: 98.08 g/mol.

Notice that the molar mass, which tells us the number of grams in one mole of a substance, has the units of grams per mole. In contrast, formula mass, like atomic mass, is dimensionless. The formula mass of H_2O is 18.02; its molar mass is 18.02 g/mol.

One other point should be emphasized in dealing with molar masses. In order to specify the molar mass of a substance, *we must know its formula*. It would be ambiguous, to say the least, to refer to the "molar mass of oxygen." One mole of oxygen atoms, represented by the symbol O, weighs 16.00 g; the molar mass of O is 16.00 g/mol. One mole of oxygen molecules, represented by the formula O_2, weighs 32.00 g; the molar mass of O_2 is 32.00 g/mol.

Beginning students sometimes fail to distinguish between O atoms and O_2 molecules.

Mole-Gram Conversions

Very often in chemistry we have to convert from moles to grams or vice versa. Such conversions are readily made by knowing the molar masses of the substances involved. They are illustrated in Examples 2.6 and 2.7.

EXAMPLE 2.6 Determine the number of moles in 212 g of
a. K_2CrO_4 b. $C_{12}H_{22}O_{11}$

Solution

a. Recall from Example 2.5a that the molar mass of K_2CrO_4 is 194.20 g/mol:

$$1 \text{ mol } K_2CrO_4 = 194.20 \text{ g } K_2CrO_4$$

$$\text{no. moles } K_2CrO_4 = 212 \text{ g} \times \frac{1 \text{ mol}}{194.20 \text{ g}} = 1.09 \text{ mol}$$

b. From Example 2.5b,

$$1 \text{ mol } C_{12}H_{22}O_{11} = 342.34 \text{ g } C_{12}H_{22}O_{11}$$

$$\text{no. moles } C_{12}H_{22}O_{11} = 212 \text{ g} \times \frac{1 \text{ mol}}{342.34 \text{ g}} = 0.619 \text{ mol}$$

EXERCISE How many moles are there in 212 g of H_2SO_4?
Answer: 2.16 mol.

EXAMPLE 2.7 Find the mass in grams of 1.69 mol of phosphoric acid, H_3PO_4.

Solution First, we need to know the molar mass of H_3PO_4:

formula mass = 3(atomic mass H) + atomic mass P
            + 4(atomic mass O)
            = 3(1.01) + 30.97 + 4(16.00) = 98.00

molar mass H_3PO_4 = 98.00 g/mol

Now, we can make the required conversion:

$$\text{mass } H_3PO_4 = 1.69 \text{ mol} \times \frac{98.00 \text{ g}}{1 \text{ mol}} = 166 \text{ g}$$

EXERCISE What is the mass in grams of 1.69 mol H_2O? Answer: 30.5 g.

Conversions of the type we have just carried out come up over and over again in chemistry. They will be required in nearly every chapter of this text. Clearly, you must know what is meant by a mole. Remember, a mole always represents a certain number of items, 6.022×10^{23}. Its mass, however, differs with the substance involved: A mole of H_2O, 18.02 g, weighs considerably more than a mole of H_2, 2.02 g, even though they both contain the same number of molecules. In the same way, a dozen bowling balls weigh a lot more than a dozen eggs, even though each involves the same number of items.

Moles are important because they give us a handle on the masses of substances that react chemically

We have now considered several different types of conversions involving numbers of particles, moles, and grams. Notice that:

To Convert Between	*Use the Relation*
1. Numbers of particles, moles	1 mol = 6.022×10^{23} particles
2. Moles, grams	1 mol = X grams
3. Numbers of particles, grams	6.022×10^{23} particles = X grams

where X is the formula mass (63.55 for Cu, 18.02 for H_2O, - -)

2.6
Moles in Solution; Molarity

To obtain a given amount of a pure solid in the laboratory, you would weigh it out on a balance. Suppose, however, the solid is present as a solute dissolved in a solvent such as water. In this case, you ordinarily measure out a given volume of the water solution, perhaps using a graduated cylinder. The amount of solute you obtain in this way depends not only upon the volume of solution but also upon the *concentration* of solute, i.e., the amount of solute in a given amount of solution.

Most chemical reactions occur in solution

Perhaps the most useful way to express solute concentration is in terms of **molarity (M).** The molarity of a solution represents the number of moles of solute per liter of solution. Specifically, molarity is defined by the equation

$$\text{molarity} = \frac{\text{number moles solute}}{\text{number liters solution}} \qquad (2.4)$$

Thus, a 1 M solution of NaCl would contain

1 mol (58.44 g) NaCl in one liter (1000 mL) of solution
0.5 mol (29.22 g) NaCl in one half liter (500 mL) of solution
0.1 mol (5.844 g) NaCl in one tenth liter (100 mL) of solution

and so on.

To prepare a solution to a desired molarity, we follow the procedure shown in Figure 2.7. If less accuracy is required, we might substitute a graduated cylinder or even a beaker for the volumetric flask. The calculations required to obtain the molarity of the solute are illustrated in Example 2.8.

EXAMPLE 2.8 A student adds 12.10 g NaCl to a volumetric flask and dissolves it in enough water to give a solution volume of 250.0 mL. Calculate the molarity.

1.
Take a
volumetric flask.

2.
Add carefully the
weighed amount
of solid.

3.
Add some water,
shake, and
dissolve solid.

4.
Fill flask to
1000-cm³ mark
and shake until
homogeneous
solution is
obtained.

FIGURE 2.7 To prepare a solution of a desired molarity, the calculated amount of solute is weighed out and transferred to a volumetric flask. Enough water is added so that all the solid is dissolved by shaking. More water is then added to bring the level up to the mark on the neck. The flask is then shaken repeatedly until a homogeneous solution is formed.

Solution To find the molarity, we need to know
1. *The number of moles of NaCl:* Since the molar mass of NaCl is 58.44 g/mol,

$$\text{number moles NaCl} = 12.10 \text{ g} \times \frac{1 \text{ mol}}{58.44 \text{ g}} = 0.2070 \text{ mol}$$

2. *The number of liters of solution:* Since 1 L = 1000 mL,

$$\text{no. liters solution} = 250.0 \text{ mL} \times \frac{1 \text{ L}}{1000 \text{ mL}} = 0.2500 \text{ L}$$

Now we can use the defining equation, 2.4, to calculate the molarity:

$$\text{molarity} = \frac{0.2070 \text{ mol}}{0.2500 \text{ L}} = 0.8280 \text{ mol/L}$$

With a little practice you will become comfortable with molecules, moles and molarity. We try to minimize the jargon, but we do need some

EXERCISE Suppose the solute referred to in this example had been glucose, $C_6H_{12}O_6$ (molar mass = 180.2 g/mol), instead of NaCl. What would have been the molarity? Answer: 0.2686 mol/L.

We can use the molarity of a solution to calculate

—the number of moles of solute in a given volume of solution.
—the volume of solution containing a given number of moles of solute.

Here, as in so many other cases, we use a conversion factor approach (Example 2.9).

We label the bottle 12M HCl

EXAMPLE 2.9 The bottle labeled "concentrated hydrochloric acid" in the lab contains 12.0 mol of the compound HCl per liter of solution, that is, M = 12.0 mol/L.

a. How many moles of HCl are there in 25.0 mL of this solution?

b. What volume of concentrated hydrochloric acid must be taken to contain 1.00 mol HCl?

Solution The molarity of the HCl relates the number of moles of HCl to the number of liters of solution. Since there are 12.0 mol HCl per liter, we can say that

$$12.0 \text{ mol} \simeq 1 \text{ L solution}$$

This relation gives us the conversion factors we need to go from liters of solution to moles or vice versa.

a. We first convert 25.0 mL to liters:

$$\text{number liters solution} = 25.0 \text{ mL} \times \frac{1 \text{ L}}{1000 \text{ mL}} = 0.0250 \text{ L}$$

Now we convert to moles of HCl:

$$\text{number moles HCl} = 0.0250 \text{ L} \times \frac{12.0 \text{ mol HCl}}{1 \text{ L}} = 0.300 \text{ mol HCl}$$

b. To find the volume in liters, we start with the number of moles of HCl and use the conversion factor 1 L/12.0 mol HCl:

$$\text{number liters solution} = 1.00 \text{ mol HCl} \times \frac{1 \text{ L}}{12.0 \text{ mol HCl}}$$
$$= 0.0833 \text{ L } (83.3 \text{ mL})$$

EXERCISE How many grams of HCl (molar mass = 36.46 g/mol) are there in 0.100 L of this solution? Answer: 43.8 g.

Knowing the molarity of a solution, you can readily obtain a specified amount of solute. All you have to do is to calculate the required volume as in Example 2.9b. Upon measuring out that volume, you should obtain the desired number of moles or grams of solute. Concentrations of reagents in the general chemistry laboratory are most often expressed as molarities. We will have more to say about molarity and other concentration units in Chapter 12.

2.7
Atomic Masses From Experiment

The problem of finding the relative masses of different atoms occupied the time of a great many famous chemists of the nineteenth century. We will not attempt to describe the tortuous path that led to the modern table of atomic masses. It may be helpful, however, to describe a couple of the approaches that were used. Both of these led to approximate values for atomic masses, accurate to perhaps the nearest whole number.

The first direct approach to the determination of atomic masses was proposed in 1819 by two Frenchmen, Pierre Dulong and Alexis Petit. They suggested that the amount of heat required to raise the temperature of an atom of a solid element by, let us say, 1°C, should be independent of the type of atom. In their words, "the atoms of all simple bodies have the same capacity for heat." Since one mole of every element contains the same number of atoms, a fixed amount of heat should then be required to raise the temperature of one mole of a solid element by 1°C.

The Law of Dulong and Petit is expressed most simply in terms of the property called specific heat. This is the amount of heat required to raise the temperature of one gram of a substance by 1°C. The specific heat can be measured rather easily in the laboratory. You may do this later in this or other courses. By the Law of Dulong and Petit, the product of the molar mass of a solid element multiplied by its specific heat is about 25 J/mol·°C:

$$\text{(molar mass)} \times \text{(specific heat)} \approx 25 \text{ J/mol·°C} \tag{2.5}$$

To show how this equation is used, consider the element iron. Its specific heat, as measured in the laboratory, is 0.476 J/g·°C. Substituting in Equation 2.5,

$$\text{molar mass Fe} = \frac{25 \text{ J/mol·°C}}{0.476 \text{ J/g·°C}} = 53 \text{ g/mol}$$

This leads to an atomic mass of 53 for iron, close to the true value, 55.847.

The discovery of the Periodic Table by Mendeleev in 1869 suggested another way of estimating atomic masses. Recall that atomic mass increases in a more or less regular fashion as we move across the table. This suggests that the atomic mass of an element should be close to the average of the two elements to its left and right. Consider, for example, the element scandium, which was unknown in 1869. Its neighbor on the left, calcium, has an atomic mass of about 40. To the right is titanium, with an atomic mass of 48. On this basis, the atomic mass of Sc should be about

$$\frac{40 + 48}{2} = 44$$

The true value is 44.9559, about one unit higher than this prediction.

The Mass Spectrometer

Today, classical methods of measuring atomic masses are largely of historical interest. Atomic masses can now be determined with great accuracy using a mass spectrometer (Fig. 2.8). Here, a gas such as helium, at very low pressures, is bombarded by high-energy electrons. A few helium atoms are converted to He^+ ions. These ions are focused into a narrow beam and accelerated by a voltage of 500 to 2000 V toward a magnetic field. The field deflects the ion beam out of its straight line path, toward the collector.

For ions of a given charge, let us say +1, the extent of deflection depends upon the mass of the ion. The heavier the ion, the less it will be deflected.

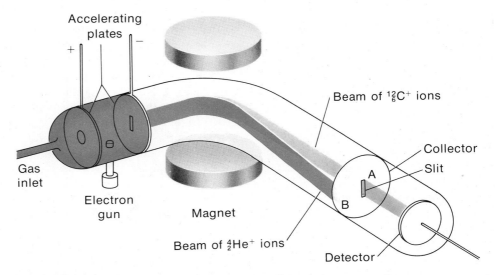

Accelerating plates

Beam of $^{12}_6C^+$ ions

Collector

Slit

A

B

Gas inlet

Electron gun

Magnet

Beam of $^4_2He^+$ ions

Detector

FIGURE 2.8 The mass spectrometer. A beam of gaseous ions is deflected in the magnetic field toward the collector plate. Light ions such as $^4_2He^+$ are deflected more than heavy ions such as $^{12}_6C^+$. By comparing the accelerating voltages required to bring the two ions to the same point, A, it is possible to determine the relative masses of the ions (Eq. 2.6).

At a given magnetic field strength and accelerating voltage (V_1), a $^{12}_6C^+$ ion might arrive at point A (Fig. 2.8). Under the same conditions, $^4_2He^+$ ions might be deflected to point B. By changing the accelerating voltage to V_2, it is possible to bring $^4_2He^+$ ions to point A instead. The accelerating voltages needed to bring the two ions to the same point are related by the equation

$$\frac{\text{mass } ^4_2He^+}{\text{mass } ^{12}_6C^+} = \frac{V_1}{V_2} \tag{2.6}$$

By carefully measuring V_1 and V_2, we can calculate the relative masses of ^4_2He and $^{12}_6C$. This way, we find that ^4_2He has a mass 0.3336 times that of $^{12}_6C$. Hence, on the carbon-12 scale, the mass of ^4_2He is

$$(0.3336)(12.00) = 4.003$$

Naturally occurring helium consists almost entirely of this isotope; therefore, the atomic mass of the element helium is about 4.003.

Finding the atomic mass of an element with more than one isotope is more difficult. Consider neon, which has three isotopes, $^{20}_{10}Ne$, $^{21}_{10}Ne$, and $^{22}_{10}Ne$. A beam of neon ions is split into three parts when it passes through the magnetic field. To obtain the atomic mass of neon, we must know the abundance as well as the mass of each isotope. The abundance can be obtained by measuring the relative peak areas produced on a recorder when the three beams reach the detector (Fig. 2.9). Knowing the mass of each isotope and its abundance, we can readily obtain the atomic mass of neon (recall Example 2.3).

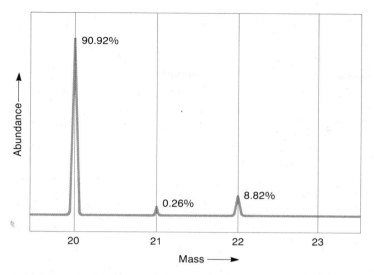

The mass spectrum
gives us both masses
and abundances

FIGURE 2.9 Mass spectrum of neon. Neon consists of three isotopes with atomic masses of 20.00 (90.92%), 21.00 (0.26%), and 22.00 (8.82%). Hence, the atomic mass of neon is 20.00(0.9092) + 21.00(0.0026) + 22.00(0.0882) = 20.18.

Summary

The three basic components of atoms are protons, neutrons, and electrons. Protons and neutrons are in the small, positively charged nucleus of an atom. Electrons, which carry a negative charge, surround the nucleus. The proton carries a positive charge equal in magnitude to that of the electron. A neutron has zero charge. In a neutral atom, there are the same number of electrons as protons. The number of protons in the nucleus (atomic number) is characteristic of a particular element. In the Periodic Table, elements are arranged in order of increasing atomic number as we move from left to right in a horizontal row (period).

The mass number of an atom is found by adding the number of neutrons and protons. Atoms of the same element (same number of protons) that differ in the number of neutrons are called isotopes (Example 2.1). The atomic mass is a number that tells us how heavy an atom is relative to a $^{12}_{6}C$ atom, which is assigned an atomic mass of exactly 12. The atomic mass of an element reflects the masses and relative abundances of its component isotopes (Example 2.3).

When an atom loses electrons, a positive ion (cation) is formed (Example 2.2). The gain of electrons by an atom leads to the formation of a negative ion (anion). Ionic compounds consist of positive and negative ions held together by strong attractive forces. In such a compound, the sum of positive charges equals the sum of negative charges. Certain elements and compounds contain discrete structural units called molecules. Molecules consist of two or more atoms joined by strong chemical bonds.

A mole represents Avogadro's number (6.022×10^{23}) of items, which may be atoms (Example 2.4), molecules, ions, etc. The molar mass of a substance can be found from its formula (Example 2.5). It is numerically equal to the sum of the atomic masses of the element(s) present. Thus, we have

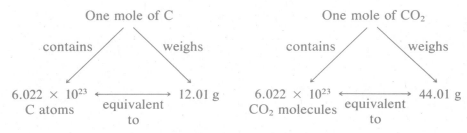

Knowing the molar mass of a substance, we can readily relate moles to grams (Examples 2.6 and 2.7).

The number of moles of a solute in solution depends upon the volume of the sample and the concentration of solute. The concentration unit, molarity, relates the number of moles of solute to the volume of solution in liters (Examples 2.8 and 2.9).

The basic concepts described above evolved from the atomic theory first proposed by John Dalton. Atomic theory leads directly to three of the basic laws of chemistry: (1) Law of Conservation of Mass; (2) Law of Constant Composition; and (3) Law of Multiple Proportions.

Key Words and Concepts

alpha particle	formula mass	molar mass
anion	ion	mole
atom	ionic compound	molecule
atomic mass	isotope	molecular formula
atomic number	Law of Conservation of Mass	neutron
Avogadro's number	Law of Constant Composition	nucleus
bond	Law of Multiple Proportions	proton
cation	mass number	solute
electron	molarity	solvent

Questions and Problems

Atomic Theory and Laws

2.1 State, in your own words, how the Law of Multiple Proportions is explained in terms of atomic theory.

2.2 Which of the three laws (if any) listed on p. 34 is illustrated by each of the following statements?
  a. An intact flash cube has the same mass before and after use.
  b. Analyses of sulfur dioxide carried out in the United States and the Soviet Union give the same value for the percent sulfur.
  c. The mass of Si combined with 1 g of oxygen in SiO_2 is a little more than twice the mass of C combined with 1 g of oxygen in CO_2.
  d. Sulfur forms two oxides with formulas SO_2 and SO_3.

2.31 State, in your own words, the three basic postulates of the atomic theory of matter.

2.32 Follow the directions for Problem 2.2 for the following:
  a. Lavoisier found that when HgO decomposed, the mass of mercury and oxygen formed equalled the mass of mercury(II) oxide decomposed.
  b. It is very unlikely that the formula for water in Virginia is $H_{2.6}O_{1.1}$.
  c. The atom ratio of oxygen to nitrogen is twice as large in one compound as it is in another compound of the two elements.
  d. Hydrogen occurs as a mixture of two isotopes, one of which is twice as heavy as the other.

2.3 Fluorine and oxygen form two compounds. In the first compound, the mass percent of F is 70.37; in the second compound it is 54.29.
  a. Calculate the mass of F combined with one gram of oxygen in each compound.
  b. Use the two values from (a) to illustrate the Law of Multiple Proportions.

Nuclear Symbols and Isotopes

2.4 Radon-222 accounts for a significant portion of the radioactivity in air. Write a nuclear symbol for this isotope of Rn.

2.5 One of the isotopes of iron can be represented as $^{57}_{26}Fe$ or Fe-57 but not as $_{26}Fe$. Explain.

2.6 The sodium isotope $^{24}_{11}Na$ is used to trace blood clots. How many
  a. protons are in its nucleus?
  b. neutrons are in its nucleus?
  c. electrons are in a sodium-24 atom?
  d. electrons and protons are in an Na^+ ion?

2.7 Complete the table below, using the Periodic Table when necessary:

SYMBOL	CHARGE	NUMBER OF PROTONS	NUMBER OF NEUTRONS	NUMBER OF ELECTRONS
——	0	10	11	——
Ba	0	——	82	——
——	+3	21	24	——
——	——	15	16	18

Atomic Masses

2.8 Calculate the mass ratio of an iron atom to an atom of
  a. Sc   b. Mo   c. Te

2.9 Suppose the atomic mass of C-12 were taken to be 1.000 instead of 12.00. On that basis, what would be the atomic mass of the following:
  a. Na   b. C   c. Ne

2.10 Copper consists of two isotopes with masses 62.96 (70.5%) and 64.96 (29.5%). Calculate the atomic mass of Cu.

2.33 Using the data in Problem 2.3,
  a. calculate the mass of oxygen combined with one gram of fluorine in both compounds.
  b. show how the data in (a) illustrate the Law of Multiple Proportions.
  c. suggest reasonable formulas for the two compounds.

2.34 A highly radioactive component of the waste from a nuclear reactor is a plutonium isotope, Pu-239. How many protons are there in an atom of this type? how many neutrons?

2.35 Explain how the two isotopes of rubidium, Rb-85 and Rb-87, differ. Write nuclear symbols for these isotopes.

2.36 An isotope of iodine used in thyroid disorders is $^{131}_{53}I$. How many
  a. protons are in its nucleus?
  b. neutrons are in its nucleus?
  c. electrons are in an I atom?
  d. neutrons and protons are in the I^- ion formed from this isotope?

2.37 Complete the table below, using the Periodic Table when necessary:

SYMBOL	CHARGE	NUMBER OF PROTONS	NUMBER OF NEUTRONS	NUMBER OF ELECTRONS
$^{98}_{42}Mo$	0	——	——	——
——	−2	34	46	——
——	+3	63	88	——
$^{207}_{82}Pb$	0	——	——	——

2.38 Arrange the following in order of increasing mass:
  a. 3 C atoms   b. 2 Si atoms
  c. S atom   d. Se atom

2.39 If the atomic mass of P were taken to be exactly 31, what would be the atomic mass of the following:
  a. N   b. Ge   c. Ag

2.40 Magnesium consists of three isotopes with masses of 23.98 (78.6%), 24.98 (10.1%), and 25.98 (11.3%). Calculate the atomic mass of Mg.

2.11 Bromine consists of two isotopes with masses of 78.92 and 80.92. Estimate the abundances of these isotopes.

2.12 Silicon (at. mass = 28.086) consists of three isotopes with masses of 27.977, 28.977, and 29.974. The abundance of the middle isotope is 4.70%. Estimate the other two abundances.

Molecules and Ions

2.13 Give the number of protons and electrons in
a. a Cr^{3+} ion   b. a Cr^{2+} ion
c. an HCl molecule   d. an H_2O molcule

2.14 Complete the table below:

SPECIES	NUMBER OF NEUTRONS	NUMBER OF PROTONS	NUMBER OF ELECTRONS
$^{45}_{21}Sc^{3+}$	___	___	___
___	16	16	18
$^{206}_{81}Tl^{+}$	___	___	___
___	34	30	28

2.15 Give the formulas of all the compounds containing no ions other than Na^+, Ca^{2+}, S^{2-}, or I^-.

Avogadro's Number

2.16 The atomic mass of molybdenum is 95.94. Calculate the
a. mass in grams of one molybdenum atom.
b. number of atoms in one milligram of molybdenum.

2.17 Determine the
a. mass of 2×10^{15} Ag atoms.
b. number of atoms in one pound of silver.

2.18 An average raindrop weighs 0.063 g. Calculate the mass in metric tons of Avogadro's number of raindrops.

2.19 How many electrons are there in
a. a Br atom?   b. a mole of Br atoms?
c. 0.0187 mol Br?   d. 0.0187 g Br?

Molar Masses and Mole-Gram Conversions

2.20 Calculate the molar masses (g/mol) of
a. Si   b. $SiCl_4$ c. $C_{12}H_{22}O_{11}$

2.21 Convert the following to moles:
a. 1.34 g H_2O b. 1.34 g Cu   c. 1.34 g N_2O

2.41 Lithium consists of two isotopes with masses of 6.015 and 7.016. Estimate the abundances of these isotopes.

2.42 Neon (at. mass = 20.18) consists of three isotopes with masses of 20.00, 21.00, and 22.00. The abundance of the middle isotope is 0.26%. Estimate the other two abundances.

2.43 Give the number of protons and electrons in
a. an F^- ion   b. an F_2 molecule
c. an HF molecule   d. an H^+ ion

2.44 Complete the following table:

SPECIES	NUMBER OF PROTONS	NUMBER OF ELECTRONS
Co^{3+}	___	___
CO	___	___
Se^{2-}	___	___
HCl	___	___

2.45 Give the formulas of compounds in which
a. the cation is K^+; the anion is O^{2-} or N^{3-}.
b. the anion is Cl^-; the cation is Fe^{2+} or Fe^{3+}.

2.46 Copper has an atomic mass of 63.55. Calculate the
a. mass of 1.5×10^{20} Cu atoms.
b. number of atoms in one gram of copper.

2.47 Repeat the calculations in Problem 2.17, substituting lead for silver.

2.48 Calculate the number of water molecules in a 50.0 m × 36.0 m swimming pool filled to a depth of 2.44 m. Take the density of H_2O to be 1.00 g/cm^3.

2.49 Give the number of protons in
a. one Cs atom   b. a mole of Cs atoms
c. 21.2 mol Cs   d. 21.2 g Cs

2.50 Calculate the molar masses of
a. K_2CrO_4 b. $C_3H_8O_2$ c. $CsNO_3$

2.51 Convert the following to moles:
a. 2.91 g C   b. 2.91 g CO   c. 2.91 g CO_2

2.22 Calculate the mass in grams of 2.42 mol of
  a. H
  b. H_2
  c. H_2O
  d. H_2O_2

2.23 Complete the following table for acetone, C_3H_6O:

NUMBER OF GRAMS	NUMBER OF MOLES	NUMBER OF MOLECULES	NUMBER OF C ATOMS
0.0880	——	——	——
——	0.00500	——	——
——	——	1.0×10^9	——
——	——	——	1.0×10^{20}

2.24 The density of ethyl alcohol, C_2H_6O, is 0.785 g/mL at 25°C. Calculate
  a. the molar mass of C_2H_6O.
  b. the number of moles in 252 mL of ethyl alcohol.
  c. the mass of 1.62 mol of ethyl alcohol.

Molarity

2.25 As a laboratory assistant, you are asked to prepare
  a. 0.240 L of 3.00 M NaOH.
  b. 0.100 L of 6.00 M KNO_3.
  Describe how you would prepare these solutions, starting with the pure solid and water.

2.26 You are given a bottle labeled "6.00 M HCl."
  a. How many moles of HCl are there in 10.2 mL of this solution?
  b. What volume of this solution contains 0.100 mol HCl?

2.27 Complete the table below for aqueous solutions:

SOLUTE	MASS OF SOLUTE	VOLUME	MOLARITY
$NaHCO_3$	2.52 g	0.125 L	——
$C_3H_8O_3$	——	0.800 L	3.50
$SrCl_2$	2.30 g	——	1.45
$Fe(NO_3)_3$	——	0.300 L	0.275

General

2.28 Which of the following statements are always true? never true?
  a. An anion contains fewer protons than the corresponding atom.
  b. One mole H_2 is heavier than 1 mol He.
  c. A -3 ion is heavier than the atom from which it is derived.
  d. The mass of 1 mol H_2O is the mass of a water molecule.

2.52 Calculate the number of moles in 3.61 g of
  a. Cl
  b. Cl_2
  c. NaCl
  d. $CaCl_2$

2.53 Complete the following table for glycerol, $C_3H_8O_3$:

NUMBER OF GRAMS	NUMBER OF MOLES	NUMBER OF MOLECULES	NUMBER OF C ATOMS
0.00450	——	——	——
——	0.0125	——	——
——	——	3.0×10^{24}	——
——	——	——	1.2×10^{15}

2.54 Liquid ammonia, NH_3, has a density of 0.691 g/mL at -40°C. Calculate
  a. the number of moles in 12.0 g NH_3.
  b. the volume of 5.62 mol NH_3.
  c. the mass of 1.62×10^{-5} mol NH_3.

2.55 How would you prepare 355 mL of 0.512 M
  a. NaCl
  b. $CaCl_2$
  c. $C_6H_{12}O_6$

2.56 On the reagent shelf in the laboratory are bottles marked 0.10 M NaI and 6.0 M acetic acid, $HC_2H_3O_2$.
  a. How many grams of solute are there in 15.0 mL of each solution?
  b. What volume of each solution must be taken to obtain 0.050 mol of solute?

2.57 Complete the table below for aqueous solutions:

SOLUTE	MASS OF SOLUTE	VOLUME	MOLARITY
$CaBr_2$	30.0 g	0.200 L	——
SO_2	——	0.500 L	0.300
NiI_2	9.36 g	——	1.25
Li_2CO_3	——	0.150 L	0.0420

2.58 Criticize each of the following statements:
  a. Atoms are heavier than ions.
  b. The number of cations in a crystal of $CaCl_2$ equals the number of anions.
  c. A C-12 atom weighs 12.0 g.
  d. There are 6.02×10^{23} atoms in one mole of NaCl.

2.29 The molecular formula of morphine, a pain-killing narcotic, is $C_{17}H_{19}NO_3$.
   a. How many atoms are there in the molecule?
   b. Which element contributes the least to the molar mass?
   c. How many C atoms are there in 10.0 mg morphine, the normal dose?

2.30 Arrange the following in order of decreasing mass:
   a. Cl_2 molecule
   b. 1.0×10^{-23} mol Cl
   c. 1.0×10^{-23} g Cl
   d. Cl atom

2.59 The hormone adrenaline has the molecular formula $C_9H_{13}NO_3$.
   a. What is the molar mass of adrenaline?
   b. What fraction of the atoms in adrenaline is accounted for by carbon?
   c. The normal concentration of adrenaline in blood plasma is 6.0×10^{-8} g/L. How many adrenaline molecules are there in one liter of plasma?

2.60 Arrange the following in order of increasing number of atoms:
   a. 1×10^{-10} mol N
   b. 1×10^{-11} mol NH_3
   c. 1×10^{-10} g NH_3
   d. 5×10^{-9} mol N_2

*2.61 Taking the mass of an electron to be 9.11×10^{-28} g, determine the mass, to six significant figures, of one mole of
   a. K^+ ions   b. F^- ions   c. KF

*2.62 Hemoglobin, the oxygen carrier in red blood cells, has four iron atoms per molecule and contains 0.340% Fe by mass. Calculate the molar mass of hemoglobin.

*2.63 One way to determine Avogadro's number is to measure the number of electrons required to plate out a known mass of a metal. It is found that 894.5 coulombs are required to form one gram of silver from Ag^+ ions. Using the known atomic mass of silver and the charge of the electron in coulombs given in this chapter, calculate the number of atoms in one mole of Ag.

*2.64 By x-ray diffraction, it is possible to determine the geometric pattern in which atoms are arranged in a crystal and the distances between atoms. In a crystal of silver, four atoms effectively occupy the volume of a cube 0.409 nm on an edge. Taking the density of silver to be 10.5 g/cm³, calculate the number of atoms in one mole of Ag.

*2.65 Suppose you arranged a mole of moles, head to tail, in a straight line stretching from the earth to the moon. Assume your average mole is 6 inches long; the distance to the moon is 250,000 miles, more or less. How many columns of moles would there be between the earth and the moon?

*2.66 Each time you inhale, you take in about 500 mL of air; each milliliter of air contains about 2.5×10^{19} molecules. It has been estimated that Abraham Lincoln, in delivering the Gettysburg Address, inhaled about 200 times.
   a. How many molecules did Lincoln take in?
   b. In the entire atmosphere, there are about 1.8×10^{20} mol of air. What fraction of the molecules in the earth's atmosphere was inhaled by Lincoln at Gettysburg?
   c. In the next breath that you take, estimate the number of molecules that were inhaled by Lincoln at Gettysburg.

Chapter 3
Chemical Formulas and Equations

In this chapter, we will consider the language of chemistry. We start by discussing chemical formulas, referred to briefly in Chapter 2. Here we will be interested in

—what formulas mean (Section 3.1).
—how they can be used to find the mass percents of the elements in a compound (Section 3.2).
—how they can be determined by experiment (Section 3.3) or by prediction (Section 3.4).

Compounds can be identified by names as well as formulas. In Section 3.5, we will see how certain simple compounds are named in a systematic way. With that background, we will go on to discuss how chemical reactions are represented by equations (Section 3.6).

Chemical equations serve many purposes. In the most general sense, they describe what happens in reactions. Here we will be particularly interested in how to use them to relate amounts of reactants and products (Sections 3.7 and 3.8). Such relationships are readily obtained by using the mole concept introduced in Chapter 2.

3.1
Types of Formulas

The **simplest** (*empirical*) formula of a compound gives the smallest whole-number ratio between the numbers of atoms of different elements in the compound. An example is the simplest formula of water, H_2O. This tells us that there are twice as many hydrogen atoms as oxygen atoms in water. In the compound potassium chlorate, there are three elements: potassium, chlorine, and oxygen. These are present in an atom ratio of 1 K:1 Cl:3 O. Hence, the simplest formula of potassium chlorate is $KClO_3$.

The **molecular** formula indicates the actual number of atoms of each type in a molecule. The molecular formula may be the same as the simplest formula. This is the case with water, H_2O: there are two atoms of hydrogen and one oxygen atom in a water molecule. In other cases, the molecular formula is a whole-number multiple of the simplest formula. Consider, for example, the compound of hydrogen and oxygen known as hydrogen peroxide. Here we write the molecular formula H_2O_2 to indicate that two hydrogen atoms are combined with two oxygen atoms in the hydrogen peroxide molecule. The simplest formula of this compound would be HO.

Molecular formulas are more informative than simplest formulas, so we use them when we can

Sometimes, to represent a compound, we go beyond the simplest or molecular formula. We write its formula in such a way as to suggest the structure of the compound. Consider, for example, the compounds called *hydrates*. These are solids, usually ionic in nature, that contain water molecules in their crystal lattice (Color Plate 3.1). An example is hydrated barium chloride. This compound contains 2 mol of water, H_2O, for every 1 mol of barium chloride, $BaCl_2$. Its formula is written as

$$BaCl_2 \cdot 2H_2O$$

(A dot is used to separate the formulas of the two compounds, $BaCl_2$ and H_2O.) We interpret the formula $CuSO_4 \cdot 5H_2O$ in a similar way. In this hydrate, there are 5 mol of water for every 1 mol of copper sulfate, $CuSO_4$.

3.2
Percent Composition from Formula

As we have seen, the formula of a compound tells us the relative number of each kind of atom present. It can also be used to determine the mass percents of the elements in the compound. These mass percents are found by obtaining the mass in grams of each element in one mole of the compound and applying the following relation:

$$\text{mass percent of element X} = \frac{\text{mass of element X}}{\text{molar mass of compound}} \times 100 \qquad \textbf{(3.1)}$$

The calculations involved are shown in Example 3.1.

EXAMPLE 3.1   Sodium hydrogen carbonate, commonly called "bicarbonate of soda," is used in many commercial products to relieve an upset stomach. It has the simplest formula $NaHCO_3$. What are the mass percents of Na, H, C, and O in sodium hydrogen carbonate (at. mass Na = 22.99, H = 1.01, C = 12.01, O = 16.00)?

Solution   In one mole of $NaHCO_3$, there are

22.99 g (1 mol) Na	1.01 g (1 mol) H
12.01 g (1 mol) C	48.00 g (3 mol) O

The mass of one mole of $NaHCO_3$ is

$$22.99 \text{ g} + 1.01 \text{ g} + 12.01 \text{ g} + 48.00 \text{ g} = 84.01 \text{ g}$$

$$\text{mass percent Na} = \frac{22.99 \text{ g}}{84.01 \text{ g}} \times 100 = 27.36$$

$$\text{mass percent H} = \frac{1.01 \text{ g}}{84.01 \text{ g}} \times 100 = 1.20$$

$$\text{mass percent C} = \frac{12.01 \text{ g}}{84.01 \text{ g}} \times 100 = 14.30$$

$$\text{mass percent O} = \frac{48.00 \text{ g}}{84.01 \text{ g}} \times 100 = 57.14$$

The percentages add up to 100, as they should:

$$27.36 + 1.20 + 14.30 + 57.14 = 100.00$$

EXERCISE What are the mass percents of carbon and oxygen in CO_2?
Answer: 27.29% C; 72.71% O.

The calculations in Example 3.1 illustrate an important characteristic of formulas. In one mole of $NaHCO_3$, there is 1 mol Na (22.99 g), 1 mol H (1.01 g), 1 mol C (12.01 g), and 3 mol O (48.00 g). In other words, the mole ratio is 1 mol Na:1 mol H:1 mol C:3 mol O. This is the same as the atom ratio in $NaHCO_3$, 1 atom Na:1 atom H:1 atom C:3 atoms O. In general, we can say that **the subscripts in a formula represent not only the atom ratio in which the different elements are combined, but also the mole ratio.** For example,

The atom ratio equals the mole ratio. Can you see why?

Formula	Atom Ratio	Mole Ratio
H_2O	2 atoms H:1 atom O	2 mol H:1 mol O
KNO_3	1 atom K:1 atom N:3 atoms O	1 mol K:1 mol N:3 mol O
$C_{12}H_{22}O_{11}$	12 atoms C:22 atoms H:11 atoms O	12 mol C:22 mol H:11 mol O

3.3
Formulas from Experiment

We saw in Section 3.2 that, given the formula of a compound, we can calculate the mass percents of the elements present. As you might suppose, this process can be reversed. If we know the mass percents of the elements from experiment, we can obtain the formula of a compound. The formula found in this way is the simplest formula. To go one step further and determine the molecular formula, it is necessary to know one other quantity, the molar mass.

Simplest Formula from Percent Composition

If we know the mass percents of the elements in a compound, we can deduce its simplest formula. To do this, we follow a three-step process:

1. Determine the number of moles of each element in a sample of the compound, for convenience, 100 g.
2. Using the results from (1), determine the simplest mole ratios for the elements present.
3. Equate the mole ratio to the atom ratio, from which the simplest formula follows directly.

To illustrate this process, consider the mineral cassiterite, a compound of tin and oxygen. Chemical analysis shows that this compound contains 78.8% tin by mass and 21.2% oxygen by mass. To determine the simplest formula, we proceed as follows:

1. In 100 g of cassiterite, there are

$$78.8\% = \frac{78.8}{100}$$

$$\frac{78.8}{100} \times 100 \text{ g} = 78.8 \text{ g Sn}; \qquad \frac{21.2}{100} \times 100 \text{ g} = 21.2 \text{ g O}$$

To find the number of moles, note that since the atomic masses of Sn and O are 118.7 and 16.00, respectively,

$$1 \text{ mol Sn} = 118.7 \text{ g Sn}; \qquad 1 \text{ mol O} = 16.00 \text{ g O}$$

Hence, in 100 g of the compound,

$$\text{no. moles Sn} = 78.8 \text{ g Sn} \times \frac{1 \text{ mol Sn}}{118.7 \text{ g Sn}} = 0.664 \text{ mol Sn}$$

$$\text{no. moles O} = 21.2 \text{ g O} \times \frac{1 \text{ mol O}}{16.00 \text{ g O}} = 1.33 \text{ mol O}$$

2. To find the simplest whole-number mole ratio, we divide by the smaller number:

2 mol O:1 mol Sn
2 atoms O:1 atom Sn
Formula is SnO_2

$$\frac{1.33 \text{ mol O}}{0.664 \text{ mol Sn}} = \frac{2.00 \text{ mol O}}{\text{mol Sn}}$$

We conclude that there are 2 mol of oxygen per mole of tin in the compound.

3. As pointed out earlier, *the mole ratio is equal to the atom ratio*. Hence, the simplest formula must be SnO_2.

The same approach can be used to determine the simplest formula of a compound containing more than two elements. The arithmetic is a bit longer but the reasoning is the same (Example 3.2).

EXAMPLE 3.2 Potassium dichromate, a red, water-soluble solid, contains the three elements potassium, chromium, and oxygen. Analysis of a sample of potassium dichromate gives the following mass percents:

K = 26.6;   Cr = 35.4;   O = 38.0

From these data, determine the simplest formula.

Solution Again, for convenience, we work with a 100-g sample. In this sample, there are 26.6 g K, 35.4 g Cr, and 38.0 g O. Noting that the atomic masses of K, Cr, and O are 39.10, 52.00, and 16.00, respectively, in the 100-g sample we have

$$\text{moles K} = 26.6 \text{ g K} \times \frac{1 \text{ mol K}}{39.10 \text{ g K}} = 0.680 \text{ mol K}$$

$$\text{moles Cr} = 35.4 \text{ g Cr} \times \frac{1 \text{ mol Cr}}{52.00 \text{ g Cr}} = 0.681 \text{ mol Cr}$$

$$\text{moles O} = 38.0 \text{ g O} \times \frac{1 \text{ mol O}}{16.00 \text{ g O}} = 2.38 \text{ mol O}$$

To find the simplest mole ratio in the compound, we divide by the smallest number, 0.680:

$$\frac{0.681 \text{ mol Cr}}{0.680 \text{ mol K}} = 1.00 \frac{\text{mol Cr}}{\text{mol K}}; \qquad \frac{2.38 \text{ mol O}}{0.680 \text{ mol K}} = 3.50 \frac{\text{mol O}}{\text{mol K}}$$

We conclude that for every mole of potassium, there is 1.00 mol of chromium and 3.50 mol of oxygen. The mole ratio and hence the atom ratio could be expressed as

1 K:1.00 Cr:3.50 O

Multiplying by two to obtain the simplest whole-number ratio, we arrive at

2 K:2 Cr:7 O

The simplest formula of potassium dichromate is $K_2Cr_2O_7$.

EXERCISE Hexane, a colorless organic liquid, contains 83.6% C and 16.4% H by mass. What is its simplest formula? Answer: C_3H_7.

The calculations in Example 3.2 are typical of those involved in determining simplest formulas from percent composition. Frequently, the mole ratio that you calculate directly involves one or more fractional numbers (for example, 3.50, 2.33). When this happens, multiply through by the smallest integer (2, 3, etc.) that will give a whole-number ratio. Thus,

Since there is some experimental error, you may need to round off to obtain integers

$$\frac{3.50}{1.00} \times \frac{2}{2} = \frac{7.00}{2.00} = 7{:}2$$

$$\frac{2.33}{1.00} \times \frac{3}{3} = \frac{6.99}{3.00} = 7{:}3$$

Determination of Percent Composition and Simplest Formula from Experiment

Once the mass percents of the elements in a compound are known, the calculation of the simplest formula is straightforward. Finding these percentages, however, is not a simple matter. They must be determined by experiment. Many different methods are possible. To illustrate the principle involved, consider the oxide of

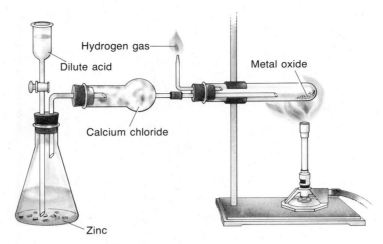

Hydrogen gas

Dilute acid

Metal oxide

Calcium chloride

Zinc

FIGURE 3.1 Reduction of oxides by hydrogen. Acid from the dropping funnel reacts with zinc to produce $H_2(g)$, which is dried by passing over $CaCl_2$. In the test tube at the right, the hydrogen reacts with the heated metal oxide. The water produced by the reaction passes off with the excess hydrogen.

tin referred to earlier. To *analyze* this compound—that is, to determine the mass percents of tin and oxygen—we might use the apparatus shown in Figure 3.1. We start with a weighed sample of the oxide, let us say 0.800 g. The sample is then heated in a stream of hydrogen gas. At high temperatures a reaction occurs: the oxygen in the compound is converted to water vapor, which escapes with the excess hydrogen. The solid residue that remains is pure tin; its mass is 0.630 g. The mass percent of tin in the oxide must then be

$$\text{mass percent Sn} = \frac{0.630 \text{ g}}{0.800 \text{ g}} \times 100 = 78.8$$

The mass percent of oxygen is found by subtracting from 100:

$$\text{mass percent O} = 100.0 - 78.8 = 21.2$$

Many simple compounds containing only two elements can be analyzed by a procedure similar to this. The general approach is to carry out a reaction in which one of the elements is produced in the pure state. In this case, an oxide of tin is reacted with hydrogen to form tin metal and water. In another case, we might cause an oxide of iron to react with carbon, forming pure iron as one of the products.

Simple organic compounds such as hexane (containing C and H only) or ethyl alcohol (containing C, H, and O) can be analyzed using the apparatus shown in Figure 3.2. A weighed sample of the compound, often only a few milligrams, is burned in oxygen. The carbon present in the sample is converted to carbon dioxide; the hydrogen present is converted to water. The amounts of CO_2 and H_2O produced are determined by measuring the increases in mass of the two absorption tubes. From these changes in mass, the masses of carbon and hydrogen can be calculated.

The amounts of CO_2 and H_2O could also be obtained by vapor phase chromatography

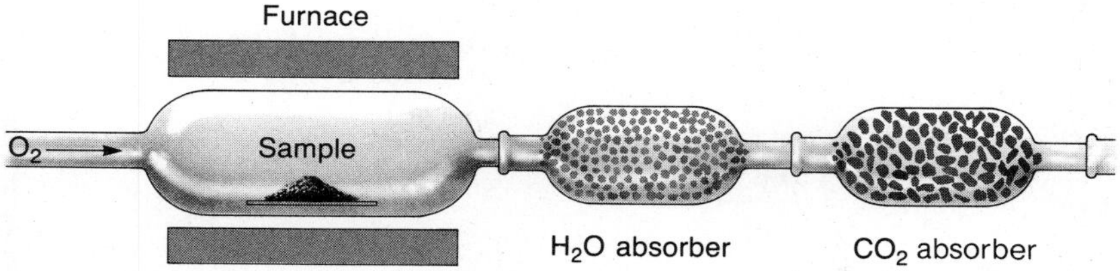

FIGURE 3.2 Combustion train used for carbon-hydrogen analysis. The absorbent for water is magnesium perchlorate, $Mg(ClO_4)_2$. Carbon dioxide is absorbed by finely divided sodium hydroxide supported on asbestos. Only a few milligrams of sample are needed for an analysis.

If oxygen was originally present in the sample, its mass is determined by difference; that is,

mass O = mass sample − (mass C + mass H)

Once the masses of the elements are known, we can calculate the percent composition of the compound and/or its simplest formula (Example 3.3).

EXAMPLE 3.3 Ethyl alcohol contains the three elements carbon, hydrogen, and oxygen. Combustion of a 5.00-g sample of ethyl alcohol gives 9.55 g CO_2 and 5.87 g H_2O. Calculate

a. the masses of C, H, and O in the 5.00-g sample, assuming all the carbon is converted to CO_2 and all the hydrogen to H_2O.
b. the percent composition of ethyl alcohol.
c. the simplest formula of ethyl alcohol.

Why do we have to make these assumptions?

Solution

a. Let us first calculate the mass of carbon in 9.55 g CO_2. Since one mole of CO_2, 44.01 g, contains one mole of C, 12.01 g, these two quantities are equivalent to one another:

$$12.01 \text{ g C} \simeq 44.01 \text{ g } CO_2$$

This relation gives us the conversion factor we need to convert from grams of CO_2 to grams of carbon:

They are chemically equivalent since they contain the same amount of carbon

$$\text{mass C} = 9.55 \text{ g } CO_2 \times \frac{12.01 \text{ g C}}{44.01 \text{ g } CO_2} = 2.61 \text{ g C}$$

One mole of H_2O (18.02 g) contains 2 mol H (2.02 g). Hence,

$$2.02 \text{ g H} \simeq 18.02 \text{ g } H_2O$$

$$\text{mass H} = 5.87 \text{ g } H_2O \times \frac{2.02 \text{ g H}}{18.02 \text{ g } H_2O} = 0.658 \text{ g H}$$

Since the total mass of the sample is 5.00 g, and it contains only C, H, and O,

$$\text{mass O} = 5.00 \text{ g} - (2.61 \text{ g} + 0.658 \text{ g}) = 1.73 \text{ g O}$$

b. $\text{mass percent C} = \dfrac{\text{mass C}}{\text{mass sample}} \times 100 = \dfrac{2.61 \text{ g}}{5.00 \text{ g}} \times 100 = 52.2$

Similarly,

$$\text{mass percent H} = \dfrac{0.658 \text{ g}}{5.00 \text{ g}} \times 100 = 13.2$$

$$\text{mass percent O} = \dfrac{1.73 \text{ g}}{5.00 \text{ g}} \times 100 = 34.6$$

c. We could now determine the simplest formula as in Example 3.2, using the percent composition data calculated in (b). Another approach, however, is to work directly with the masses of C, H, and O in the 5.00-g sample. We found in (a) that the sample contained

2.61 g C;   0.658 g H;   1.73 g O

Let us find the number of moles of each element in the sample:

$$\text{moles C} = 2.61 \text{ g C} \times \dfrac{1 \text{ mol C}}{12.01 \text{ g C}} = 0.217 \text{ mol C}$$

$$\text{moles H} = 0.658 \text{ g H} \times \dfrac{1 \text{ mol H}}{1.01 \text{ g H}} = 0.651 \text{ mol H}$$

$$\text{moles O} = 1.73 \text{ g O} \times \dfrac{1 \text{ mol O}}{16.00 \text{ g O}} = 0.108 \text{ mol O}$$

Now, we find the mole ratio:

$$\dfrac{0.217 \text{ mol C}}{0.108 \text{ mol O}} = 2.01 \dfrac{\text{mol C}}{\text{mol O}}; \qquad \dfrac{0.651 \text{ mol H}}{0.108 \text{ mol O}} = 6.03 \dfrac{\text{mol H}}{\text{mol O}}$$

Rounding off to whole numbers, we have

2 mol C:6 mol H:1 mol O

The simplest formula of ethyl alcohol is C_2H_6O.

EXERCISE Using the percentages obtained in (b), find the number of moles of each element in a 100-g sample of ethyl alcohol and its simplest formula. Answer: 4.35 mol C, 13.1 mol H, 2.16 mol O; C_2H_6O.

In a microanalysis lab, they would use about 1 mg of sample. It's faster, and safer

As we have just seen in Example 3.3, it is not necessary to work with the percent composition of a compound to find its simplest formula. All we need to know are the masses of each element in a sample of the compound. The sample may weigh 5.00 g, 100 g, or have any other mass. In any case, we first find the number of moles of each element in the sample. Using that information, we obtain the mole ratio, which is equal to the atom ratio.

To complete our discussion, it may be helpful to review the general approach used to obtain simplest formulas from experimental data.

1. **Determine the masses of each element present in a fixed mass of the compound. Sometimes, these data may be given directly. More often, you will be given:**

a. the mass percents of the elements (Example 3.2). Here, it is convenient to work with a 100-g sample of the compound so that the mass of each element in grams is given directly by its mass percent.

b. mass data for the analysis of a sample. Ordinarily, as in Example 3.3, you will know the masses of products (e.g., H_2O, CO_2) formed when a weighed sample of the compound is analyzed. In this case, you can find the mass of each element by using relations such as 2.02 g H $\simeq$ 18.02 g H_2O; 12.01 g C $\simeq$ 44.01 g CO_2.

2. Find the number of moles of each element present. To do this, you work with the masses calculated in (1) and use relations such as 1.01 g H = 1 mol H; 12.01 g C = 1 mol C.

3. Determine the simplest mole ratio between the different elements present. Working with the numbers of moles obtained in (2), you divide each number by the smallest. This gives you a ratio of the form

1 mol of element A : X mol of element B : Y mol of element C

If X and Y are whole numbers, let us say 2 and 3, the problem is solved; the formula would be AB_2C_3. If either X or Y is a fractional number, an additional step is required. Multiply through by the smallest integer that will give a whole number ratio. Thus if X were 1.5 and Y were 2.5, you would multiply by 2:

1 A : 1.5 B : 2.5 C = 2 A : 3 B : 5 C
simplest formula = $A_2B_3C_5$

Molecular Formula from Simplest Formula

As pointed out in Section 3.1, the molecular formula is a whole-number multiple of the simplest formula. That multiple may be 1 as in H_2O, 2 as in H_2O_2, or 3 as in C_3H_6. To find out what the multiple is, we need to know one more piece of data, the molar mass. By comparing the observed molar mass to that corresponding to the simplest formula, we can determine whether the multiple is 1, 2, 3, etc.

EXAMPLE 3.4 The simplest formula of vitamin C is found by analysis to be $C_3H_4O_3$. From another experiment, the molar mass is found to be about 180 g/mol. What is the molecular formula of vitamin C?

Solution Let us first calculate the sum of the atomic masses for $C_3H_4O_3$:

3(12.0) + 4(1.0) + 3(16.0) = 88.0

Thus, the molar mass of $C_3H_4O_3$ must be 88.0 g/mol. The approximate molar mass, 180 g/mol, is twice this (180/88 = 2.0). Hence, the simplest formula is multiplied by 2 to obtain the molecular formula:

molecular formula of vitamin C = 2 × $C_3H_4O_3$ = $C_6H_8O_6$

EXERCISE The simplest formula of hexane is C_3H_7. Its molar mass is about 86 g/mol. What is the molecular formula of hexane? Answer: C_6H_{14}.

You might consider this to be the recipe for finding chemical formulas

The exact molar mass of vitamin C follows from its molecular formula, and is 176.12 g

3.4
Formulas of Ionic Compounds

We have just seen how formulas can be determined by experiment. It is also possible to predict "on paper" the formulas of certain substances, notably, simple ionic compounds. This is done using the principle of **electrical neutrality**, referred to briefly in Chapter 2. Consider, for example, the ionic compound calcium chloride. The ions present are Ca^{2+} and Cl^-. It follows that, if the compound is to be electrically neutral, the formula of calcium chloride must be $CaCl_2$: two Cl^- ions are required to balance one Ca^{2+} ion.

All ionic compounds are electrically neutral

To apply this principle, you must know the charges of the ions involved. In the remainder of this section, we will consider the charges of two types of ions:

—*monatomic* ions, formed when a single atom gains or loses electrons.
—*polyatomic* ions, which are charged particles containing more than one atom.

Monatomic Ions

Figure 3.3 shows the charges of the more common monatomic ions, superimposed upon the Periodic Table. Notice that the diagonal line or stairway that runs from the upper left to the lower right of the table separates positive ions (cations) from negative ions (anions). In particular,

—*metals,* located below and to the left of this line, form *cations* by losing electrons. For example,

Na atom (11 p^+, 11 e^-) → Na^+ ion (11 p^+, 10 e^-) + e^-

Mg atom (12 p^+, 12 e^-) → Mg^{2+} ion (12 p^+, 10 e^-) + 2 e^-

—*nonmetals,* located above and to the right of the line, form *anions* by gaining electrons. For example,

F atom (9 p^+, 9 e^-) + e^- → F^- ion (9 p^+, 10 e^-)

O atom (8 p^+, 8 e^-) + 2 e^- → O^{2-} ion (8 p^+, 10 e^-)

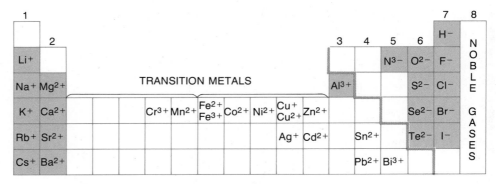

FIGURE 3.3 Charges of ions found in solid ionic compounds. The step-like, diagonal line separates metals from nonmetals and cations from anions. Ions shown in color have the same number of electrons as the neighboring noble gas atom. For example, S^{2-}, Cl^-, K^+ and Ca^{2+} all have 18 electrons, the number found in the argon atom.

Notice that each of the ions just described (Na^+, Mg^{2+}, F^-, and O^{2-}) contains the same number of electrons, 10. This is the number of electrons found in an atom of the noble gas neon (10 p^+, 10 e^-). In general, we find that **elements that are close to a noble gas (Group 8) in the Periodic Table tend to form ions that have the same number of electrons as the noble gas atom.**

Ions that have the same number of electrons as the nearest noble gas atom are shown in color in Figure 3.3. Notice that

The ions of the Group 1 metals have +1 charges.
The ions of the Group 2 metals have +2 charges.
The ions of the Group 6 nonmetals have −2 charges.
The ions of the Group 7 nonmetals have −1 charges.

It is useful to know the charges on these ions

Several metals that are farther removed from the noble gases in the Periodic Table form positive ions. The structures of these ions are not related in any direct way to those of noble gas atoms. Indeed, there is no simple way to predict their charges. Some of the more important of these ions are listed in Figure 3.3. They are derived from the **transition metals** (those in the groups near the center of the Periodic Table) and the metals of higher atomic mass in Groups 4 and 5. The most common charge among these ions, you will note, is +2. Charges of +1 (Ag^+) and +3 (Cr^{3+}, Bi^{3+}) are less common. Several of these metals form more than one cation. Thus, we have

$$Fe^{2+} \text{ and } Fe^{3+} \qquad Cu^+ \text{ and } Cu^{2+}$$

Copper forms two oxides, Cu_2O and CuO

Using Figure 3.3 and the principle of electrical neutrality, it is possible to predict the formulas of a large number of ionic compounds (Example 3.5).

EXAMPLE 3.5 Predict the formulas of the ionic compounds formed by
a. lithium and oxygen (Li and O) b. magnesium and chlorine (Mg and Cl)
c. nickel and sulfur (Ni and S) d. bismuth and fluorine (Bi and F)

Solution
a. Li_2O; 2 Li^+ ions required to balance 1 O^{2-} ion
b. $MgCl_2$; 1 Mg^{2+} ion required to balance 2 Cl^- ions
c. NiS; 1 Ni^{2+} ion required to balance 1 S^{2-} ion
d. BiF_3; 1 Bi^{3+} ion required to balance 3 F^- ions

EXERCISE Give the formulas of zinc iodide, silver sulfide, and aluminum oxide. Answer: ZnI_2; Ag_2S; Al_2O_3.

Notice that, *in writing the formula of an ionic compound, the positive ion is always placed first*. Thus, we write $CaCl_2$ not Cl_2Ca, Li_2O not OLi_2, and so on.

Polyatomic Ions

Many familiar compounds contain polyatomic ions. Sodium hydroxide (lye), $NaOH$, contains the hydroxide ion, OH^-. Calcium carbonate (limestone), $CaCO_3$, contains the carbonate ion, CO_3^{2-}. Table 3.1 lists some of the more common polyatomic ions, along with their names and charges. Notice that, with the single exception of the ammonium ion, NH_4^+, all of these polyatomic ions are anions.

Table 3.1
Some Common Polyatomic Ions

+1	−1	−2	−3
NH_4^+ (ammonium)	OH^- (hydroxide)	CO_3^{2-} (carbonate)	PO_4^{3-} (phosphate)
	NO_3^- (nitrate)	SO_4^{2-} (sulfate)	
	ClO_3^- (chlorate)	CrO_4^{2-} (chromate)	
	ClO_4^- (perchlorate)	$Cr_2O_7^{2-}$ (dichromate)	
	MnO_4^- (permanganate)		
	HCO_3^- (hydrogen carbonate)		

You should learn the names, formulas and charges of the above ions by heart

The formulas of compounds containing polyatomic ions can be predicted in much the same way as with monatomic ions. The principle of electrical neutrality is used to find the relative numbers of positive and negative ions. A minor complication arises when the number of polyatomic ions in the formula is two or greater. When this occurs, the polyatomic ion is enclosed in parentheses to avoid confusion. Thus, we have

$Ca(OH)_2$ (1 Ca^{2+} ion, 2 OH^- ions)
$(NH_4)_2S$ (2 NH_4^+ ions, 1 S^{2-} ion)

EXAMPLE 3.6 Using Figure 3.3 and Table 3.1, predict the formulas of
a. barium hydroxide   b. potassium sulfate   c. ammonium phosphate

Solution
a. $Ba(OH)_2$; 1 Ba^{2+} ion required to balance 2 OH^- ions
b. K_2SO_4; 2 K^+ ions required to balance 1 SO_4^{2-} ion
c. $(NH_4)_3PO_4$; 3 NH_4^+ ions required to balance 1 PO_4^{3-} ion

EXERCISE What is the formula of potassium hydroxide? barium phosphate? ammonium sulfate? Answer: KOH; $Ba_3(PO_4)_2$; $(NH_4)_2SO_4$.

3.5
Names of Compounds

There are several million known compounds, all of which have systematic names of one type or another. Most of these are carbon compounds, which fall into the area of organic chemistry (Chap. 28). At this time, we need only be concerned with naming inorganic compounds, a simpler task. Moreover, we will restrict our attention to two types of inorganic compounds:

—simple ionic compounds of the type described in Section 3.4. Nearly all of these contain cations derived from metals, and either monatomic or polyatomic anions.
—binary compounds formed from two different nonmetals.

Ionic Compounds

The name of an ionic compound consists of two words: the first word gives the name of the positive ion (cation), the second that of the negative ion (anion). Thus, we have

COMPOUND	CATION	ANION	NAME
$NaCl$	Na^+	Cl^-	sodium chloride
K_2SO_4	K^+	SO_4^{2-}	potassium sulfate
$Zn(NO_3)_2$	Zn^{2+}	NO_3^-	zinc nitrate

The system for naming compounds is almost, but not quite, obvious

To assign names to individual ions, note the following:

1. Monatomic positive ions take the name of the metal from which they are derived. Examples include

Na^+ *sodium* K^+ *potassium* Zn^{2+} *zinc*

There is one complication: certain metals form more than one type of positive ion. An example is iron, where we have the Fe^{2+} and Fe^{3+} ions. To distinguish between these ions, the charge must be indicated in the name. This is done by giving the charge as a Roman numeral in parentheses after the name of the metal:

Fe^{2+} *iron(II)* Fe^{3+} *iron(III)*

Similarly, for the two cations of copper, we write

Cu^+ *copper(I)* Cu^{2+} *copper(II)*

(An older system used the suffixes *-ic* for the ion of higher charge and *-ous* for the ion of lower charge. These were added to the stem of the Latin name of the metal, so that the Fe^{3+} ion was referred to as ferric, the Fe^{2+} ion as ferrous.)

2. Monatomic negative ions are named by adding the suffix *-ide* to the stem of the name of the nonmetal from which they are derived. Thus, we have

				H^-	hydride
N^{3-}	nitride	O^{2-}	oxide	F^-	fluoride
		S^{2-}	sulfide	Cl^-	chloride
		Se^{2-}	selenide	Br^-	bromide
		Te^{2-}	telluride	I^-	iodide

3. Polyatomic ions are given special names (Table 3.1, p. 72).

Following these rules, we conclude that

Old names:

—the compound copper(I) chloride contains the ions Cu^+ and Cl^-; its formula is $CuCl$.

cuprous chloride

—the compound copper(II) chloride contains the ions Cu^{2+} and Cl^-; its formula is $CuCl_2$.

cupric chloride

—the compound iron(II) hydroxide contains the ions Fe^{2+} and OH^-; its formula is $Fe(OH)_2$.

ferrous hydroxide

—the compound iron(III) hydroxide contains the ions Fe^{3+} and OH^-; its formula is $Fe(OH)_3$.

ferric hydroxide

EXAMPLE 3.7 Name the following ionic compounds:
a. CaS   b. Al(NO$_3$)$_3$ c. FeCl$_2$

Solution
a. calcium sulfide
b. aluminum nitrate
c. Recall that the chloride ion has a -1 charge (Cl$^-$). The positive ion must then be Fe^{2+}. The name is iron(II) chloride.

EXERCISE Name Al$_2$O$_3$, Fe(NO$_3$)$_3$, and Ag$_2$Se. Answer: aluminum oxide; iron(III) nitrate; silver selenide.

Binary Compounds of the Nonmetals

Which goes first in the formula is somewhat arbitrary

When a pair of nonmetals forms only one compound, that compound is named very simply. The name of the element whose symbol appears first in the formula is written first. The second part of the name is formed by adding the suffix *-ide* to the stem of the name of the second nonmetal. Examples include

HCl   hydrogen chloride
H$_2$S   hydrogen sulfide
NF$_3$ nitrogen fluoride

More often, a pair of nonmetals forms more than one compound. In those cases, the name is a bit more complex. The Greek prefixes *di* = two, *tri* = three, *tetra* = four, *penta* = five, *hexa* = six, *hepta* = seven, and *octa* = eight are used to show the number of atoms of each element. The several oxides of nitrogen have the names

There is only one compound called nitrogen oxide, but six nitrogen oxides.

N$_2$O$_5$ dinitrogen pentoxide
N$_2$O$_4$ dinitrogen tetroxide
NO$_2$ nitrogen dioxide
N$_2$O$_3$ dinitrogen trioxide
NO   nitrogen oxide
N$_2$O   dinitrogen oxide

EXAMPLE 3.8 Give the names of
a. SO$_2$ b. SO$_3$ c. PCl$_3$ d. PCl$_5$

Solution
a. sulfur dioxide
b. sulfur trioxide
c. phosphorus trichloride
d. phosphorus pentachloride

EXERCISE Give the names of SF$_4$ and SF$_6$. Answer: sulfur tetrafluoride; sulfur hexafluoride.

Many of the best known binary compounds of the nonmetals have acquired common names. These are widely and, in some cases, exclusively used. Examples include

H_2O	water	PH_3	phosphine
H_2O_2	hydrogen peroxide	AsH_3	arsine
NH_3	ammonia	NO	nitric oxide
N_2H_4	hydrazine	N_2O	nitrous oxide

Nobody calls water dihydrogen oxide

3.6
Writing and Balancing Chemical Equations

When a chemical reaction occurs, starting materials, which we call *reactants,* are converted to other substances, called *products.* A reaction can be described in words, but it is more useful to represent it by a chemical equation. Formulas of reactants appear on the left side of a chemical equation. They are separated by an arrow from the formulas of the products, written on the right side of the equation. In a balanced chemical equation, there is the same number of atoms of a given element on the two sides.

Beginning students are sometimes led to believe that writing a chemical equation is a simple, mechanical process. Nothing could be further from the truth. One point that seems obvious is often overlooked. *You cannot write an equation unless you know what happens in the reaction that it represents.* All the reactants and all the products must be identified. Moreover, you must know their formulas and physical states.

To illustrate how we arrive at a balanced equation, consider a reaction that occurs in a rocket motor. The starting materials or reactants are two liquids, hydrazine and dinitrogen tetroxide. As we saw in Section 3.5, these compounds have the molecular formulas N_2H_4 and N_2O_4, respectively. The products of the reaction are gaseous nitrogen, N_2, and liquid water, H_2O. To write a balanced equation for this reaction, we proceed as follows:

We always use molecular formulas for molecular substances

1. Write a "skeleton" equation in which the formulas of the reactants appear on the left and those of the products on the right. In this case,

$$N_2H_4 + N_2O_4 \rightarrow N_2 + H_2O$$

2. Balance the equation by taking into account the Law of Conservation of Mass. This requires that there be the same number of atoms of each element on the two sides of the equation. To accomplish this, start by writing a coefficient of 4 for H_2O, thus obtaining 4 oxygen atoms on both sides:

$$N_2H_4 + N_2O_4 \rightarrow N_2 + 4 H_2O$$

Now consider the hydrogen atoms. There are $4 \times 2 = 8$ H atoms on the right. To obtain 8 H atoms on the left, write a coefficient of 2 for N_2H_4:

$$2 N_2H_4 + N_2O_4 \rightarrow N_2 + 4 H_2O$$

Finally, consider nitrogen. There are a total of $(2 \times 2) + 2 = 6$ nitrogen atoms on the left. To balance nitrogen, write a coefficient of 3 for N_2:

$$2 N_2H_4 + N_2O_4 \rightarrow 3 N_2 + 4 H_2O$$

3. Indicate the physical state of each reactant and product, after the formula, by writing

(*g*) for a gaseous substance
(*l*) for a pure liquid
(*s*) for a solid
(*aq*) for an ion or molecule in water (aqueous) solution

In this case, the final equation is

$$2 \text{ N}_2\text{H}_4(l) \; + \; \text{N}_2\text{O}_4(l) \rightarrow 3 \text{ N}_2(g) \; + \; 4 \text{ H}_2\text{O}(l) \qquad (3.2)$$

Two points concerning the balancing process are worth emphasizing:

1. Equations are balanced by adjusting coefficients in front of formulas, never by changing subscripts within formulas. In arriving at Equation 3.2, we balanced nitrogen by writing 3 N_2, which indicates three N_2 molecules. On paper, we could have obtained six nitrogen atoms on the right by writing N_6, but that would have been absurd. Elemental nitrogen exists as diatomic molecules, N_2; there is no such thing as an N_6 molecule.

2. In balancing an equation, it is best to start with an element that appears in only one species on each side of the equation. In this case, we could have started with either oxygen or hydrogen. Nitrogen would have been a poor choice, however, since there are nitrogen atoms in both reactant molecules, N_2H_4 and N_2O_4.

Ordinarily, if you are asked to balance a chemical equation, you will be given the formulas of reactants and products. However, for one type of reaction you should be able to come up with this information on your own. We refer to the reaction of a metal with a nonmetal to form an ionic compound containing monatomic ions listed in Figure 3.3. Recall that in Example 3.5 we showed how to predict the formulas of ionic compounds of this type. Beyond that, the following information should be helpful in writing equations:

—All elements that react to form ionic compounds can be shown as monatomic solids (for example, Na(*s*), Al(*s*), S(*s*), etc.) *except*

$\text{Hg}(l)$
$\text{F}_2(g), \quad \text{Cl}_2(g), \quad \text{Br}_2(l), \quad \text{I}_2(s)$
$\text{O}_2(g), \quad \text{N}_2(g), \quad \text{P}_4(s)$

—All ionic compounds are solids.

Example 3.9 shows how this information is applied.

EXAMPLE 3.9 Write a balanced equation for the reaction between
a. lithium and oxygen   b. bismuth and fluorine

Solution

a. Recall from Example 3.5 that the formula of lithium oxide is Li_2O. The skeleton equation is

$$\text{Li} \; + \; \text{O}_2 \rightarrow \text{Li}_2\text{O}$$

(Note that elemental oxygen is diatomic.) To balance oxygen atoms, we write a coefficient of 2 for Li_2O. This requires a coefficient of 4 for Li:

6 N is also no good; the formula must be that of the species actually present

$$4 \text{ Li} + O_2 \rightarrow 2 \text{ Li}_2O$$

Inserting the proper physical states, we obtain

$$4 \text{ Li}(s) + O_2(g) \rightarrow 2 \text{ Li}_2O(s)$$

b. The formula of bismuth fluoride is BiF_3, as deduced in Example 3.5. The skeleton equation is

$$\text{Bi} + F_2 \rightarrow BiF_3$$

The final, balanced equation is

$$2 \text{ Bi}(s) + 3 F_2(g) \rightarrow 2 BiF_3(s)$$

EXERCISE Write a balanced equation for the reaction of aluminum with iodine. Answer:

$$2 \text{ Al}(s) + 3 I_2(s) \rightarrow 2 \text{ AlI}_3(s).$$

Each of the reactions referred to in Example 3.9 involves a transfer of electrons from metal to nonmetal atoms. Consider, for example, the reaction between lithium and oxygen, for which we derived the balanced equation:

$$4 \text{ Li}(s) + O_2(g) \rightarrow 2 \text{ Li}_2O(s) \tag{3.3}$$

In this reaction, lithium atoms give up electrons and are converted to Li^+ ions:

$$4 \text{ Li} \rightarrow 4 \text{ Li}^+ + 4 e^- \tag{3.3a}$$

The electrons given up by lithium atoms are taken on by oxygen molecules to form O^{2-} ions:

$$O_2 + 4 e^- \rightarrow 2 O^{2-} \tag{3.3b}$$

The product, lithium oxide, consists of Li^+ and O^{2-} ions in the 2:1 ratio required for electrical neutrality. Reaction 3.3 is, in a very real sense, the sum of the two "half-reactions," 3.3a and 3.3b.

A half-reaction, such as 3.3a, in which electrons are given up, or "lost," by a species is referred to as an **oxidation.** We would say that Li atoms, in being converted to Li^+ ions, *lose* electrons, or are *oxidized*. A half-reaction, such as 3.3b, in which electrons are taken on, or "gained," by a species is referred to as a **reduction.** We say that O_2 molecules, in being converted to O^{2-} ions, *gain* electrons, or are *reduced*. A reaction, such as 3.3, in which there is electron transfer from one species to another (i.e., from Li atoms to O_2 molecules) is referred to as an **oxidation-reduction** reaction, or *redox* reaction.

The term *oxidized* originated in reactions with oxygen, like Eqn 3.3. It is not limited to such reactions

In later chapters of this text, you will encounter many other examples of redox reactions. Frequently, the reactants will be species in water solution rather than pure substances. All redox reactions have one characteristic in common: *in the overall reaction, there is no net gain or loss of electrons.* The reaction between lithium and oxygen illustrates this principle: four electrons are lost by four Li atoms in half-reaction 3.3a; an equal number of electrons, four, are gained by an O_2 molecule in half-reaction 3.3b.

3.7
Mass Relations in Reactions

One of the main uses of balanced equations is in relating the masses of reactants and products in a reaction. We can use a balanced equation to determine, for a given amount of one reactant, how much of another reactant is required or how much product is formed. Calculations of this sort are based upon a very important principle:

The coefficients of a balanced equation represent numbers of moles of reactants and products.

To show that this statement is valid, consider the reaction between N_2H_4 and N_2O_4. Recall Equation 3.2:

An equation like this contains a lot of information

$$2\ N_2H_4(l)\ +\ N_2O_4(l) \rightarrow 3\ N_2(g)\ +\ 4\ H_2O(l)$$

The coefficients in this equation represent numbers of molecules; that is,

$$2\ \text{molecules}\ N_2H_4\ +\ 1\ \text{molecule}\ N_2O_4 \rightarrow 3\ \text{molecules}\ N_2\ +\ 4\ \text{molecules}\ H_2O$$

However, a balanced chemical equation remains valid if each coefficient is multiplied by the same number. That number might be 2, 100, or any other—in particular, *Avogadro's number, N*:

$$2N\ \text{molecules}\ N_2H_4\ +\ N\ \text{molecules}\ N_2O_4 \rightarrow 3N\ \text{molecules}\ N_2\ +\ 4N\ \text{molecules}\ H_2O$$

As we saw in Chapter 2, however, a mole represents Avogadro's number of items, N. Thus, we can write

$$2\ \text{mol}\ N_2H_4\ +\ 1\ \text{mol}\ N_2O_4 \rightarrow 3\ \text{mol}\ N_2\ +\ 4\ \text{mol}\ H_2O$$

which is the relation we set out to demonstrate. From a slightly different standpoint, we might say that, in this reaction

$$2\ \text{mol}\ N_2H_4 \simeq 1\ \text{mol}\ N_2O_4 \simeq 3\ \text{mol}\ N_2 \simeq 4\ \text{mol}\ H_2O$$

where the symbol $\simeq$ indicates that the quantities listed are chemically equivalent to one another in this reaction. The use of these relations is illustrated in Example 3.10, where we follow the conversion factor approach to relate moles of one substance to moles of another.

EXAMPLE 3.10 For the reaction
$$2\ N_2H_4(l)\ +\ N_2O_4(l) \rightarrow 3\ N_2(g)\ +\ 4\ H_2O(l)$$
determine
a. the number of moles of N_2O_4 required to react with 2.72 mol N_2H_4.
b. the number of moles of N_2 produced from 2.72 mol N_2H_4.

Solution
a. The conversion factor required follows directly from the coefficients of the balanced equation:

$$2 \text{ mol } N_2H_4 \simeq 1 \text{ mol } N_2O_4$$

$$\text{moles } N_2O_4 = 2.72 \text{ mol } N_2H_4 \times \frac{1 \text{ mol } N_2O_4}{2 \text{ mol } N_2H_4} = 1.36 \text{ mol } N_2O_4$$

b. In this case,

$$2 \text{ mol } N_2H_4 \simeq 3 \text{ mol } N_2$$

$$\text{moles } N_2 = 2.72 \text{ mol } N_2H_4 \times \frac{3 \text{ mol } N_2}{2 \text{ mol } N_2H_4} = 4.08 \text{ mol } N_2$$

How many moles of water are produced?

EXERCISE In the reaction between aluminum and iodine (Example 3.9), how many moles of AlI_3 are formed from 1.68 mol I_2? Answer: 1.12 mol AlI_3.

The approach followed in Example 3.10 is readily extended to reactions involving masses in grams of reactants and products. To do this, we convert moles to grams, knowing the molar masses of the substances involved. In this way, it is possible to use the coefficients of a balanced equation to relate

—moles of one substance to grams of another (Example 3.11a).
—grams of one substance to grams of another (Examples 3.11b and c).

EXAMPLE 3.11 The ammonia used to make fertilizers for lawns and gardens is made by reacting nitrogen of the air with hydrogen. The balanced equation for the reaction is

$$3 \text{ H}_2(g) + N_2(g) \rightarrow 2 \text{ NH}_3(g)$$

Determine
a. the mass in grams of ammonia, NH_3, formed when 1.34 mol N_2 reacts.
b. the mass in grams of N_2 required to form 1.00 kg NH_3.
c. the mass in grams of H_2 required to react with 6.00 g N_2.

Solution
a. The balanced equation tells us that

$$1 \text{ mol } N_2 \simeq 2 \text{ mol } NH_3$$

The formula mass of NH_3 is $14.01 + 3(1.01) = 17.04$. It follows that the molar mass of NH_3 is 17.04 g/mol and hence,

$$1 \text{ mol } NH_3 = 17.04 \text{ g } NH_3$$

These two relations give us the conversion factors we need to find the mass in grams of NH_3 formed from 1.34 mol N_2:

$$\text{mass } NH_3 = 1.34 \text{ mol } N_2 \times \frac{2 \text{ mol } NH_3}{1 \text{ mol } N_2} \times \frac{17.04 \text{ g } NH_3}{1 \text{ mol } NH_3}$$

$$= 45.7 \text{ g } NH_3$$

b. We carry out a three-step conversion:

1. grams $NH_3 \rightarrow$ moles NH_3 (1 mol NH_3 = 17.04 g NH_3)

2. moles $NH_3 \rightarrow$ moles N_2 (1 mol $N_2 \simeq 2$ mol NH_3)

3. moles $N_2 \rightarrow$ grams N_2 (1 mol N_2 = 28.02 g N_2)

$$\text{mass } N_2 = 1000 \text{ g NH}_3 \times \underbrace{\frac{1 \text{ mol NH}_3}{17.04 \text{ g NH}_3}}_{(1)} \times \underbrace{\frac{1 \text{ mol N}_2}{2 \text{ mol NH}_3}}_{(2)} \times \underbrace{\frac{28.02 \text{ g N}_2}{1 \text{ mol N}_2}}_{(3)}$$

$$= 822 \text{ g N}_2$$

c. As in (b), we first convert grams of nitrogen to moles (1 mol N_2 = 28.02 g N_2). Then we convert moles of N_2 to moles of H_2, using the coefficients of the balanced equation (3 mol $H_2 \simeq$ 1 mol N_2). Finally, we convert moles of H_2 to grams (1 mol H_2 = 2.02 g H_2):

$$\text{mass } H_2 = 6.00 \text{ g N}_2 \times \frac{1 \text{ mol N}_2}{28.02 \text{ g N}_2} \times \frac{3 \text{ mol H}_2}{1 \text{ mol N}_2} \times \frac{2.02 \text{ g H}_2}{1 \text{ mol H}_2}$$

$$= 1.30 \text{ g H}_2$$

EXERCISE For this reaction, calculate the mass in grams of H_2 required to form 1.0 g NH_3. Answer: 0.18 g H_2.

> Conversion factors are the only way to go in problems like this

Calculations of the types discussed in Examples 3.10 and 3.11 are common in chemistry; they will come up over and over again in subsequent chapters. You will need to relate the amount of one substance taking part in a reaction to that of another substance, given the balanced equation for the reaction. The approach you follow depends upon what two quantities you need to relate.

1. **To relate moles of one substance, A, to moles of another substance, B, you work directly with the coefficients of the balanced equation. The require conversion factor is obtained from the relation:**

   a mol A $\simeq$ b mol B

   where a and b are the coefficients of A and B, in that order, in the balanced equation.

2. **To relate grams of A to moles of B, or vice versa, two successive conversions are required. One involves the coefficients of the balanced equation as in (1). The other requires that you relate moles to grams, using the molar mass. To find the number of moles of B produced from Y grams of A:**

   $$Y \text{ grams A} \times \frac{1 \text{ mol A}}{\text{molar mass A}} \times \frac{b \text{ mol B}}{a \text{ mol A}} = \text{no. moles B}$$

> This is the most important mass relation, since we weigh substances in grams

3. **To relate grams of A to grams of B, three successive conversions are required. You first convert to moles of A, then to moles of B (using the coefficients of the balanced equation), and finally to grams of B:**

   $$Y \text{ grams A} \times \frac{1 \text{ mol A}}{\text{molar mass A}} \times \frac{b \text{ mol B}}{a \text{ mol A}} \times \frac{\text{molar mass B}}{1 \text{ mol B}} = \text{grams of B}$$

3.8
Limiting Reactant and Theoretical Yield

When the two elements aluminum and iodine are heated together, they react to form a compound, aluminum iodide. The balanced equation for this reaction was derived in Example 3.9:

$$2\ Al(s)\ +\ 3\ I_2(s) \rightarrow 2\ AlI_3(s) \tag{3.4}$$

The coefficients in this equation tell us the relative numbers of moles of reactants and products: two moles of Al (54.0 g Al) react exactly with three moles of I_2 (761.4 g I_2) to form two moles of AlI_3 (815.4 g AlI_3). If we mix aluminum and iodine in a 2:3 mole ratio, we expect both reactants to be completely consumed, forming two moles of aluminum iodide.

In a reaction, usually only one reactant is used up. The others are in excess

Ordinarily, in the laboratory, reactants are not mixed in exactly the ratio required for reactions. Instead, we use an excess of one reactant, usually the cheaper one. We might, for example, mix 3.00 mol Al with 3.00 mol I_2. In that case, the aluminum would be in excess, since only 2.00 mol Al is required to react with 3.00 mol I_2. After the reaction is over, we expect to have 1.00 mol Al remaining:

excess Al = 3.00 mol originally present − 2.00 mol consumed in reaction
= 1.00 mol

The 3.00 mol I_2 should be completely consumed in forming the 2.00 mol AlI_3:

$$\text{moles } AlI_3 \text{ formed} = 3.00 \text{ mol } I_2 \times \frac{2 \text{ mol } AlI_3}{3 \text{ mol } I_2} = 2.00 \text{ mol } AlI_3$$

After the reaction is over, the solid obtained would be a mixture of product, 2.00 mol AlI_3 (815.4 g AlI_3), together with 1.00 mol unreacted Al (27.0 g Al).

In situations such as this, we distinguish between the reactant in excess (Al) and the other reactant (I_2), called the **limiting reactant**. The amount of product formed is determined (limited) by the amount of limiting reactant. With 3.00 mol I_2, we cannot get more than 2.00 mol AlI_3, regardless of how large an excess of Al we use. The amount of product that would be formed if all the limiting reactant were consumed is referred to as the **theoretical yield** of product. If we mix 3.00 mol I_2 with excess Al, the coefficients of Equation 3.4 tell us that the theoretical yield of AlI_3 is 2.00 mol.

Often, you will be told the amounts of two different reactants and asked to determine which is the limiting reactant and calculate the theoretical yield of product. To do this, it helps to follow a systematic procedure. The one we will use involves three steps:

1. **Calculate the amount of product that would be formed if the first reactant were completely consumed.**
2. **Repeat this calculation for the second reactant; that is, calculate how much product would be formed if all of that reactant were consumed.**
3. **Choose the smaller of the two amounts calculated in (1) and (2). This is the theoretical yield of product; the reactant that produces the smaller amount is the limiting reactant. The other reactant is in excess; only part of it is consumed.**

To illustrate this procedure, it may help to cite an example far removed from chemistry. Let us suppose that a fast-food restaurant is making "double cheeseburgers" by cooking a hamburger between two pieces of cheese:

2 slices cheese + 1 hamburger patty → 1 double cheeseburger

Masterton likes double cheeseburgers.
E.J.S.

Suppose further that the restaurant has 250 slices of cheese and 150 hamburger patties. How many double cheeseburgers can they make?

If all the cheese is consumed,

$$250 \text{ slices cheese} \times \frac{1 \text{ double cheeseburger}}{2 \text{ slices cheese}} = 125 \text{ double cheeseburgers}$$

If, on the other hand, all the hamburger patties are used up,

$$150 \text{ hamburger patties} \times \frac{1 \text{ double cheeseburger}}{1 \text{ hamburger patty}}$$
$$= 150 \text{ double cheeseburgers}$$

Here the "limiting reactant" is the cheese. The "theoretical yield" of double cheeseburgers is 125. When all the cheese is consumed, there will be 150 − 125 = 25 hamburger patties left over. If we tried to make 150 double cheeseburgers, 25 of them would contain no cheese and we would have 25 unhappy customers.

Definitely bad news

EXAMPLE 3.12   Consider the reaction

$$2 \text{ Al}(s) + 3 \text{ I}_2(s) \rightarrow 2 \text{ AlI}_3(s)$$

Determine the limiting reactant and the theoretical yield of product if we start with
a. 1.20 mol Al and 2.40 mol I_2 b. 1.20 g Al and 2.40 g I_2

Solution
a. We calculate the amount of product formed from the two reactants, using conversion factors obtained directly from the coefficients of the balanced equation:

$$2 \text{ mol Al} \approx 2 \text{ mol AlI}_3; \qquad 3 \text{ mol I}_2 \approx 2 \text{ mol AlI}_3$$

1. If aluminum is the limiting reactant,

$$\text{moles AlI}_3 = 1.20 \text{ mol Al} \times \frac{2 \text{ mol AlI}_3}{2 \text{ mol Al}} = 1.20 \text{ mol AlI}_3$$

2. If iodine is the limiting reactant,

$$\text{moles AlI}_3 = 2.40 \text{ mol I}_2 \times \frac{2 \text{ mol AlI}_3}{3 \text{ mol I}_2} = 1.60 \text{ mol AlI}_3$$

3. The theoretical yield of AlI_3 is the *smaller* quantity, 1.20 mol; the limiting reactant is therefore Al, and there is an excess of I_2.

b. To calculate the amount of product, we follow the three-step path used in Example 3.11b. Note that the molar masses of Al, I_2, and AlI_3 are 27.0 g/mol, 253.8 g/mol, and 407.7 g/mol, respectively.

1. Mass AlI_3 formed if all the Al is consumed:

$$1.20 \text{ g Al} \times \frac{1 \text{ mol Al}}{27.0 \text{ g Al}} \times \frac{2 \text{ mol AlI}_3}{2 \text{ mol Al}} \times \frac{407.7 \text{ g AlI}_3}{1 \text{ mol AlI}_3} = 18.1 \text{ g AlI}_3$$

2. Mass AlI_3 formed if all the I_2 is consumed:

$$2.40 \text{ g I}_2 \times \frac{1 \text{ mol I}_2}{253.8 \text{ g I}_2} \times \frac{2 \text{ mol AlI}_3}{3 \text{ mol I}_2} \times \frac{407.7 \text{ g AlI}_3}{1 \text{ mol AlI}_3} = 2.57 \text{ g AlI}_3$$

3. The theoretical yield of AlI_3 is 2.57 g; the limiting reactant is I_2.

EXERCISE How many grams of Al are required to react with the I_2 in (b)? How many grams of Al are left over? Answer: 0.170 g; 1.03 g.

Remember that, in deciding upon the theoretical yield of product, you choose the smaller of the two calculated amounts. To see why this must be the case, let us refer back to Example 3.12b. There we started with 1.20 g Al and 2.40 g I_2. We decided that the theoretical yield of AlI_3 was 2.57 g and 1.03 g Al was left over. Thus, we have

$$1.20 \text{ g Al} + 2.40 \text{ g } I_2 \rightarrow 2.57 \text{ g } AlI_3 + 1.03 \text{ g Al}$$

This makes sense: 3.60 g of reactants yields a total of 3.60 g of "products," including the unreacted aluminum. Suppose, however, we had chosen 18.1 g AlI_3 as the theoretical yield. We would then have the nonsensical situation:

$$1.20 \text{ g Al} + 2.40 \text{ g } I_2 \rightarrow 18.1 \text{ g } AlI_3$$

This violates the Law of Conservation of Mass. There is no way we can get 18.1 g of product from 3.60 g of reactants.

The theoretical yield is the maximum amount of product that we can hope to obtain. In calculating the theoretical yield, we assume that the limiting reactant is completely (that is, 100%) converted to product. In the real world, this is unlikely to happen. Some of the limiting reactant may be consumed in competing reactions. Some of the product may be lost in separating it from the reaction mixture. For these and other reasons, the **actual yield** in a reaction is ordinarily less than the theoretical yield. Putting it another way, the **percent yield** is expected to be less than 100.

The actual yield is what you get. The theoretical yield is what you would get if everything went perfectly.

$$\text{percent yield} = \frac{\text{actual yield}}{\text{theoretical yield}} \times 100 \qquad (3.5)$$

To illustrate how percent yield is calculated, consider again the reaction described in Example 3.12b. Let us suppose that the actual yield of aluminum iodide isolated from the reaction mixture was 2.05 g. Recalling that the theoretical yield was 2.57 g, we obtain

$$\text{percent yield} = \frac{2.05 \text{ g}}{2.57 \text{ g}} \times 100 = 79.8$$

We conclude that the actual yield was a little less than 80% of the maximum yield that we could have hoped for, starting with 2.40 g I_2 and excess Al.

Summary

The formula of a compound shows its chemical composition. The subscripts in the formula give the number of moles of each element in one mole of the compound. From the formula, we can calculate the mass percents of the elements (Example 3.1). By the same token, if we know the mass percents, we can calculate the simplest formula (Example 3.2). Combustion analysis serves to determine the mass percents of carbon and hydrogen in an organic compound (Example 3.3). We can readily obtain the molecular formula from the simplest formula if we know the molar mass (Example 3.4).

We can predict the simplest formula of an ionic compound from the charges of the ions (Examples 3.5 and 3.6). Charges of monatomic ions are indicated in Figure 3.3; those of polyatomic ions are given in Table 3.1. In naming ionic compounds, as in writing their formulas, the cation comes first. If a metal forms more than one cation, a Roman numeral is used to indicate the charge of the ion (Example 3.7). In naming binary compounds of the nonmetals, Greek prefixes are used to indicate the number of atoms of each type in the formula (Example 3.8).

In order to write a chemical equation for a reaction, the formulas and physical states of reactants and products must be known (Example 3.9). The coefficients in a balanced equation represent numbers of moles. We can use them directly to relate moles of different substances taking part in a reaction (Example 3.10). By knowing the molar masses, we can go a step further and relate masses in grams of reactants and products (Example 3.11).

Ordinarily, we do not mix reactants in exactly the ratio called for by the coefficients of the balanced equation. One reactant is in excess; the other reactant is limiting. The amount of product we obtain if all the limiting reactant is consumed is referred to as the theoretical yield (Example 3.12). The actual yield is ordinarily less than the theoretical yield; in other words, the percent yield is less than 100.

Essentially everything we do in this chapter follows from the properties of chemical formulas

Key Words and Concepts

actual yield	limiting reactant	polyatomic ion
binary compound	mole	product
chemical equation	molecular formula	reactant
coefficient	oxidation	reduction
empirical formula	percent composition	simplest formula
hydrate	percent yield	theoretical yield

Questions and Problems

Percent Composition from Formula

3.1 Calculate the mass percent of each element in the hormone thyroxine, which has the formula $C_{15}H_{11}NO_4I_4$.

3.2 Garlic salt contains NaCl in addition to garlic. Analysis of a 3.50-g sample of garlic salt yields 1.15 g Cl.
a. How many grams of NaCl are in the sample?
b. What percent, by mass, of garlic salt is NaCl? (Assume there is no other Cl-containing compound in garlic salt.)

3.3 Combustion of 1.000 g of benzene, a compound of carbon and hydrogen, gives 3.383 g CO_2. What are the mass percents of C and H in benzene?

3.31 Penicillin G, a widely used antibiotic, has the formula $C_{16}H_{18}N_2O_4S$. Calculate the mass percent of each element in penicillin G.

3.32 A 500-mg sample of a commercial headache remedy contains 256 mg of aspirin, $C_9H_8O_4$.
a. What is the mass percent of aspirin in the product?
b. How many grams of carbon are there in the aspirin contained in a tablet of this product weighing 0.611 g?

3.33 Combustion of a 1.000-g sample of pentane, which contains only carbon and hydrogen, yields 1.498 g H_2O. What are the mass percents of C and H in pentane?

Simplest Formulas from Analysis

3.4 Iron reacts with sulfur to form a sulfide. If 2.561 g of iron reacts with 2.206 g of sulfur, what is the simplest formula of the sulfide?

3.5 Determine the simplest formulas of compounds with the following compositions:
a. 43.2% K, 39.1% Cl, 17.7% O
b. 62.1% C, 5.21% H, 12.1% N, 20.7% O
c. 32.3% Co, 1.66% H, 7.68% N, 58.3% Cl

3.6 When a sample of acetic acid weighing 1.540 g burns in oxygen, 2.257 g CO_2 and 0.9241 g H_2O are formed. The elements present in acetic acid are carbon, hydrogen, and oxygen.
a. What are the mass percents of C, H, and O in acetic acid?
b. What is the simplest formula of acetic acid?
c. The molar mass of acetic acid is between 50 and 70 g/mol. What is its molecular formula?

3.7 The insecticide DDD contains only C, H, and Cl. When a 3.200-g sample is burned in oxygen, 6.162 g CO_2 and 0.9008 g H_2O are formed. What is the simplest formula of DDD?

3.8 A 2.612-g sample of a copper oxide, when heated in a stream of H_2 gas, yields 0.592 g H_2O. What is the formula of the copper oxide?

3.9 A hydrate of magnesium iodide has the formula $MgI_2 \cdot x\ H_2O$. To determine a value for x, a student heats a hydrate sample until all the water is removed. A 1.628-g sample of hydrate is heated until a constant mass of 1.072 g is reached. What is x?

3.10 A certain compound has the simplest formula C_2H_4O. Its molar mass is about 90 g/mol. What is its molecular formula?

3.34 If 5.28 g of tin reacted directly with fluorine to form 8.65 g of a metal fluoride, what is the simplest formula of the fluoride?

3.35 What are the simplest formulas of compounds with the following compositions?
a. 42.10% C, 6.48% H, 51.42% O
b. 26.4% Na, 36.8% S, 36.8% O
c. 58.8% Xe, 7.17% O, 34.1% F

3.36 Phenol contains the three elements carbon, hydrogen, and oxygen. Combustion of 2.136 mg of phenol gives 5.993 mg CO_2 and 1.227 mg H_2O.
a. What is the simplest formula of phenol?
b. List three possible molar masses for phenol, based on your answer to (a).

3.37 The insecticide lindane contains only C, H, and Cl. When a 3.000-g sample is burned in oxygen, 2.724 g CO_2 and 0.5575 g H_2O are formed. What is the simplest formula of lindane?

3.38 A 4.700-g sample of a copper oxide is heated in a stream of H_2 gas and 0.592 g H_2O is formed. What is the formula of the copper oxide?

3.39 The compound $CrSO_4 \cdot x\ H_2O$ is analyzed for percent water by mass. A 1.912-g sample of the hydrate is heated to remove all the water; 1.032 g $CrSO_4$ remains after heating. What is the mass percent of water in the compound? The value of x?

3.40 A hydrocarbon has the simplest formula CH. What is its molecular formula if its formula mass is
a. 26   b. 52   c. 78

Names and Formulas of Compounds

3.11 Write the formulas of the ionic compounds below. Use the Periodic Table and, if necessary, Table 3.1 and Figure 3.3.
a. potassium dichromate
b. tin(II) phosphate
c. gold(I) sulfide
d. aluminum oxide

3.41 Follow the directions for Problem 3.11.
a. chromium(III) sulfate
b. lithium nitrate
c. lead(II) iodide
d. strontium chlorate
e. copper(I) telluride
f. manganese(II) carbonate

3.12 Complete the following table (all compounds are ionic):

NAME	FORMULA
a. _____	$LiOH$
b. potassium permanganate	_____
c. ammonium chloride	_____
d. _____	CaF_2
e. _____	$Ba(OH)_2$
f. calcium nitrate	_____
g. _____	$FeCO_3$

3.13 Complete the following table of molecular compounds:

NAME	FORMULA
a. _____	BCl_3
b. xenon hexafluoride	_____
c. nitrogen trifluoride	_____
d. _____	S_4N_4
e. silicon tetraiodide	_____
f. _____	NH_3

3.14 How many moles of ions are there in
    a. 0.200 mol of lithium oxide?
    b. 0.350 mol of barium hydroxide?
    c. 1.67 g of iron(II) phosphate?

Equation Balancing

3.15 Balance the following equations:
    a. $C_6H_{14}(l) + O_2(g) \rightarrow CO_2(g) + H_2O(l)$
    b. $Al(s) + HCl(aq) \rightarrow AlCl_3(s) + H_2(g)$
    c. $KClO_3(s) \rightarrow KClO_4(s) + KCl(s)$

3.16 Write balanced equations for the reaction of bromine with the following metals to form ionic solids:
    a. Al   b. Ba   c. K   d. Ni   e. Ag

3.17 Write a balanced equation for the
    a. reaction of magnesium with nitrogen.
    b. reaction of copper(I) oxide with oxygen to form copper(II) oxide.
    c. combustion of methyl alcohol, CH_3OH, to give carbon dioxide and water.
    d. decomposition of solid sodium azide, NaN_3, to its elements.
    e. reaction of fluorine with aluminum.

3.42 Complete the following table (all compounds are ionic):

NAME	FORMULA
a. lithium nitride	_____
b. copper(I) sulfate	_____
c. _____	Ag_2Se
d. _____	$Fe_2(SO_4)_3$
e. barium chromate	_____
f. _____	NH_4ClO_4
g. ammonium phosphate	_____

3.43 Complete the following table of molecular compounds:

NAME	FORMULA
a. _____	N_2H_4
b. diselenium diiodide	_____
c. _____	XeO_3
d. _____	IF_5
e. _____	P_4O_{10}
f. dinitrogen oxide	_____

3.44 How many moles of
    a. anions are there in 0.100 mol of calcium phosphate?
    b. cations are there in 0.750 mol of ammonium carbonate?
    c. ions are there in 0.250 g of manganese(II) chromate?

3.45 Balance the following equations:
    a. $H_2O_2(l) \rightarrow H_2O(l) + O_2(g)$
    b. $Al(s) + MgO(s) \rightarrow Mg(s) + Al_2O_3(s)$
    c. $C_2H_5OH(l) + O_2(g) \rightarrow CO_2(g) + H_2O(l)$

3.46 Write balanced equations for the reaction of lithium with the following nonmetals to form ionic solids:
    a. nitrogen   b. oxygen   c. sulfur
    d. chlorine   e. iodine

3.47 Write a balanced equation for the
    a. reaction of iron with chlorine to form iron(III) chloride.
    b. reaction of strontium with oxygen.
    c. reaction of magnesium with solid CO_2 (dry ice) to form MgO and carbon.
    d. combustion of silane gas, SiH_4, to produce water vapor and solid silicon dioxide.
    e. reaction of zinc with iodine.

Mole-Mass Relations in Reactions

3.18 The reaction of the mineral fluorapatite with sulfuric acid follows the equation

$$Ca_{10}F_2(PO_4)_6(s) + 7\ H_2SO_4(l) \rightarrow$$
$$2\ HF(g) + 3\ Ca(H_2PO_4)_2(s) + 7\ CaSO_4(s)$$

Fill in the blanks below:
a. 8.60 mol $Ca_{10}F_2(PO_4)_6$ yields ____ mol $CaSO_4$.
b. 7.25 mol $Ca(H_2PO_4)_2$ requires ____ mol H_2SO_4.
c. 0.990 mol H_2SO_4 reacts with ____ mol $Ca_{10}F_2(PO_4)_6$.
d. 3.330 mol $Ca_{10}F_2(PO_4)_6$ produces ____ mol $Ca(H_2PO_4)_2$.

3.19 Using the equation given in Problem 3.18, calculate
a. the mass of $CaSO_4$ formed from 0.660 mol $Ca_{10}F_2(PO_4)_6$.
b. the number of moles of $Ca_{10}F_2(PO_4)_6$ required to form 100.0 g HF.
c. the mass of $Ca_{10}F_2(PO_4)_6$ required to yield 1.06 g $CaSO_4$.
d. the mass of HF formed from 16.25 g of sulfuric acid.

3.20 When acetylene gas, C_2H_2, burns in air, the products are $CO_2(g)$ and $H_2O(l)$.
a. Write a balanced equation for this reaction.
b. How many moles of CO_2 are produced from 0.524 mol C_2H_2?
c. How many grams of O_2 are required to react with 2.46 mol C_2H_2?

3.21 Ethyl alcohol, C_2H_5OH, can be produced by the fermentation of grains, which contain glucose, $C_6H_{12}O_6$:

$$C_6H_{12}O_6(aq) \rightarrow 2\ C_2H_5OH(l) + 2\ CO_2(g)$$

a. How many grams of ethyl alcohol are produced from 1.00 lb of glucose?
b. Gasohol is a mixture of 10 cm³ of ethyl alcohol (d = 0.79 g/cm³) per 90 cm³ of gasoline. How many grams of glucose are required to produce the ethyl alcohol in one gallon of gasohol?

3.22 Oxygen masks for producing O_2 in emergency situations contain potassium superoxide, KO_2. It reacts with CO_2 and H_2O in exhaled air to produce oxygen:

$$4\ KO_2(s) + 2\ H_2O(g) + 4\ CO_2(g) \rightarrow$$
$$4\ KHCO_3(s) + 3\ O_2(g)$$

If a person wearing such a mask exhales 0.702 g CO_2/min, how many grams of KO_2 are consumed in five minutes?

3.48 One way to remove NO from smokestack emissions is to react it with ammonia:

$$4\ NH_3(g) + 6\ NO(g) \rightarrow 5\ N_2(g) + 6\ H_2O(l)$$

Fill in the blanks below:
a. 16.5 mol NO reacts with ____ mol NH_3.
b. 9.30 mol NO yields ____ mol N_2.
c. 0.772 mol N_2 requires ____ mol NO.
d. 22.4 mol NO produces ____ mol H_2O.

3.49 Using the equation given in Problem 3.48, calculate
a. the mass of N_2 produced from 1.25 mol NO.
b. the mass of NH_3 required to form 3.20 mol H_2O.
c. the mass of NO that yields 0.865 g H_2O.
d. the mass of NH_3 required to react with 45.0 g NO.

3.50 The combustion of butane gas, C_4H_{10}, in air yields $CO_2(g)$ and $H_2O(l)$.
a. Write a balanced equation for the reaction.
b. How many moles of C_4H_{10} are required to form 11.6 mol CO_2?
c. How many grams of H_2O are formed from 2.69 mol C_4H_{10}?

3.51 A commercial wine is about 9.7% ethyl alcohol by mass. Assume that 1.21 kg of wine is produced by the fermentation reaction given in Problem 3.21.
a. How many grams of glucose are needed to produce the ethyl alcohol in the wine?
b. What volume of $CO_2(g)$ is produced at the same time (d = 1.80 g/L)?

3.52 A crude oil burned in electrical generating plants contains about 1.2% sulfur by mass. When the oil burns, the sulfur forms sulfur dioxide gas:

$$S(s) + O_2(g) \rightarrow SO_2(g)$$

How many liters of SO_2 (d = 2.60 g/L) are produced when one kilogram of oil burns?

Limiting Reactant; Theoretical Yield

3.23 A textbook chapter contains figures, tables, and photographs. The publisher has on hand 1500 sets of figures, 500 sets of tables, and 1000 photograph sets. How many chapters can be assembled?

3.24 A gaseous mixture containing 10.0 mol H_2 and 12.0 mol Cl_2 reacts to form $HCl(g)$.
a. Write a balanced equation for the reaction.
b. Which reactant is limiting?
c. If all the limiting reactant is consumed, how many moles of HCl are formed?
d. How many moles of the excess reactant remain when the reaction is over?

3.25 Hydrogen reacts with sodium to produce solid sodium hydride, NaH. A reaction mixture contains 6.75 g Na and 3.03 g hydrogen.
a. Write a balanced equation for NaH formation.
b. Which reactant is limiting?
c. What is the theoretical yield of NaH from the above reaction mixture?
d. What is the percent yield if 4.00 g NaH is formed?

3.26 A century ago, $NaHCO_3$ was prepared from Na_2SO_4 by a three-step process:

$$Na_2SO_4(s) + 4\ C(s) \rightarrow Na_2S(s) + 4\ CO(g)$$

$$Na_2S(s) + CaCO_3(s) \rightarrow CaS(s) + Na_2CO_3(s)$$

$$Na_2CO_3(s) + H_2O(l) + CO_2(g) \rightarrow 2\ NaHCO_3(s)$$

How many kilograms of $NaHCO_3$ could be formed from a metric ton of Na_2SO_4, assuming an 85% yield in each step?

3.27 A student prepares HBr by reacting sodium bromide with phosphoric acid:

$$3\ NaBr(s) + H_3PO_4(l) \rightarrow 3\ HBr(g) + Na_3PO_4(s)$$

She needs 50.0 g HBr. If NaBr is the limiting reactant and a 40% excess of H_3PO_4 is to be used, how much of each reagent should she weigh out, assuming a yield of
a. 100%   b. 80%

3.53 A tool set contains 4 wrenches, 3 screwdrivers, and 2 pliers. The manufacturer has in stock 1000 pliers, 2000 screwdrivers, and 1500 wrenches. Can an order for 500 tool sets be filled?

3.54 Chlorine and fluorine react to form gaseous chlorine trifluoride, ClF_3. You start with 3.40 mol Cl_2 and 7.16 mol F_2.
a. Write a balanced equation for the reaction.
b. What is the limiting reactant?
c. What is the theoretical yield of ClF_3?
d. How many moles of the excess reactant remain unreacted?

3.55 Oxyacetylene torches are used for welding, reaching temperatures near 2000°C. These temperatures are due to the combustion of acetylene, C_2H_2, with oxygen:

$$C_2H_2(g) + O_2(g) \rightarrow$$
$$CO_2(g) + H_2O(g)\ \text{(unbalanced)}$$

a. Balance the equation.
b. Starting with 125 g of both C_2H_2 and O_2, which reactant is limiting?
c. What is the theoretical yield of H_2O from this reaction mixture?
d. If 22.5 g of water forms, what is the percent yield?

3.56 In the Ostwald process, nitric acid, HNO_3, is produced from NH_3 by a three-step process:

$$4\ NH_3(g) + 5\ O_2(g) \rightarrow 4\ NO(g) + 6\ H_2O(g)$$

$$2\ NO(g) + O_2(g) \rightarrow 2\ NO_2(g)$$

$$3\ NO_2(g) + H_2O(g) \rightarrow 2\ HNO_3(aq) + NO(g)$$

Assuming an 82% yield in each step, how many grams of nitric acid can be made from 1.00×10^4 g of ammonia?

3.57 A student in the inorganic laboratory wants to make 25 g of the compound $[Co(NH_3)_5SCN]Cl_2$ by the reaction

$$[Co(NH_3)_5Cl]Cl_2(s) + KSCN(s) \rightarrow$$
$$[Co(NH_3)_5SCN]Cl_2(s) + KCl(s)$$

He is told to use a 50% excess of KSCN, and that he can expect to get an 85% yield in the reaction. How many grams of each reactant should he use?

General

3.28 Which of the following statements are always true? never true?
a. A compound with the molecular formula $C_3H_6O_3$ has the same simplest formula.
b. The conversion of Br^- to Br is an oxidation.
c. The limiting reactant is the one present in the smallest amount.
d. The mass percent of carbon is greater in CO_2 than in CO.
e. Since C_2H_2 and C_6H_6 reduce to the same simplest formula, they represent the same compound.

3.29 Classify each of the following as oxidation or reduction:
a. $Ca \rightarrow Ca^{2+} + 2\,e^-$
b. $S^{2-} \rightarrow S + 2\,e^-$
c. $Cu^{2+} + e^- \rightarrow Cu^+$

3.30 Write a balanced equation for the combustion in air of a gaseous hydrocarbon containing 18.3% hydrogen. (Hydrocarbons are compounds containing the elements C and H only.)

3.58 Criticize each of the following statements:
a. In an ionic compound, the number of cations equals the number of anions.
b. The compound $Co_4(CO)_{12}$ has the same percent composition as $Co_{12}(CO)_4$.
c. In every reaction, one reactant is limiting and the other is in excess.
d. The conversion of Fe^{3+} to Fe^{2+} is an oxidation.
e. The molecular formula of calcium chloride is $CaCl_2$.

3.59 Which of the following is an oxidation half-reaction? reduction?
a. $Sr^{2+} + 2\,e^- \rightarrow Sr$
b. $Mn^{3+} + e^- \rightarrow Mn^{2+}$
c. $Zn \rightarrow Zn^{2+} + 2\,e^-$

3.60 Determine the percent composition of the compound formed when aluminum burns in air.

*3.61 Caffeine, a stimulant found in coffee, tea, and certain soft drinks, contains C, H, O, and N. Combustion of 1.000 mg of caffeine produces 1.813 mg CO_2, 0.4639 mg H_2O, and 0.2885 mg N_2. Estimate the molar mass of caffeine, which lies between 150 and 200 g/mol.

*3.62 A 1.600-g sample of magnesium is burned in air to produce a mixture of two ionic solids, magnesium oxide and magnesium nitride, Mg_3N_2. Water is added to this mixture. It reacts with the magnesium oxide to form 3.544 g of magnesium hydroxide.
a. Write balanced equations for the three reactions described above.
b. How many grams of magnesium oxide are formed by the combustion of magnesium?
c. How many grams of magnesium nitride are formed?

*3.63 A mixture of NaCl and NaBr weighing 1.234 g is heated with chlorine, which converts the mixture completely to NaCl. The total mass of NaCl is now 1.129 g. What are the mass percents of NaCl and NaBr in the original sample?

*3.64 A sample of an oxide of vanadium weighing 4.589 g was heated with hydrogen gas to form water and another oxide of vanadium weighing 3.782 g. The second oxide was treated further with hydrogen until only 2.573 g of vanadium metal remained.
a. What are the simplest formulas of the two oxides?
b. What is the total mass of water formed in the successive reactions?

*3.65 A sample of cocaine, $C_{17}H_{21}O_4N$, is diluted with sugar, $C_{12}H_{22}O_{11}$. When a 1.00-mg sample of this mixture is burned, 2.00 mg CO_2 is formed. What is the percentage of cocaine in the mixture?

3.66 The term "alum" refers to a class of compounds of general formula $MM^*(SO_4)_2 \cdot 12H_2O$, where M and M are different metals. A 20.000-g sample of a certain alum is heated to drive off the water; the anhydrous residue weighs 11.123 g. Treatment of the residue with excess NaOH precipitates all the M* as $M^*(OH)_3$, which weighs 4.388 g. Calculate the molar mass of the alum and identify the two metals, M and M*.

Chapter 4
Sources
of
the
Elements

As pointed out in Chapter 1, 91 different elements are found in nature. Relative abundances differ greatly from one element to another. Figure 4.1 gives the percentages by mass of the elements in the world around us. This includes the atmosphere, the waters of the earth, and the earth's crust to a depth of 40 km. Notice that a single element, oxygen, makes up nearly half (49.5%) of the total mass. The ten most abundant elements account for about 99%. Some of the more familiar elements are really quite rare; for example, the ten metals chromium, cobalt, copper, gold, lead, mercury, nickel, silver, tin, and zinc together account for less than 0.1%. Clearly, abundance by itself is no measure of the importance of an element.

In this chapter we will look at some of the processes used to obtain elements from natural sources. Industries that extract elements from the air, the oceans, or mineral deposits called *ores*, employ several million people in the United States alone. The products of these industries may be used directly: iron and steel in automobiles, aluminum in airplanes, chlorine as a bleach. Other elements are converted to useful compounds: nitrogen and hydrogen to ammonia, NH_3; sulfur to sulfuric acid, H_2SO_4.

Many different processes are used to obtain elements in pure form. Sometimes a simple physical separation is all that is required. This is the case with the gaseous elements found in uncombined form in the atmosphere (Section 4.1). More often, elements in nature are chemically combined with one another. This is true of many of the more reactive nonmetals (Section 4.2) and virtually all of the metals (Section 4.3). These elements must be prepared from their compounds by chemical reactions.

Our discussion of the sources and preparation of the elements is organized around the Periodic Table. We will make frequent use of the principles introduced in Chapters 1 to 3. Indeed, a major purpose of this chapter is to review such basic concepts as the mole (Chapter 2) and the use of formulas and chemical equations (Chapter 3).

A large part of the chemical industry is involved in the production of pure substances

90

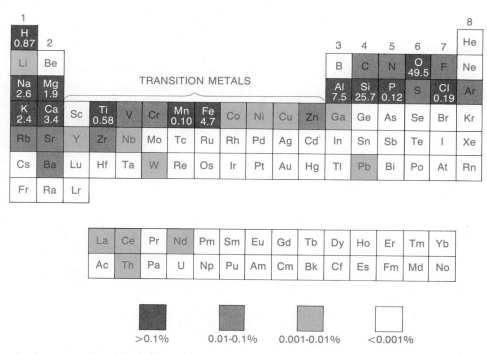

>0.1%   0.01-0.1%   0.001-0.01%   <0.001%

FIGURE 4.1 Periodic Table color coded to show the abundances (mass percents) of the elements. The ten most abundant elements are oxygen, silicon, aluminum, iron, calcium, sodium, potassium, magnesium, hydrogen, and titanium, in that order. Many of the familiar heavy elements, including silver, tin, iodine, gold, and mercury, each have abundances less than 0.0001%.

4.1
Nonmetals from the Air

A total of seven gaseous elements are found commonly in the atmosphere. These include nitrogen, oxygen, and the *noble gases* in Group 8 of the Periodic Table. All of these elements are obtained industrially from air, with one exception: helium is extracted from natural gas in certain wells in Kansas, Oklahoma, and Texas. It occurs there at concentrations up to 7%, far higher than in the atmosphere.

For the most part, air is a mixture of gaseous elements

In the separation of the components of air, it is first liquefied by cooling to about −200°C (Fig. 4.2). The liquid obtained in this way contains all the components listed in Table 4.1 except neon and helium, which remain as gases. Further cooling of the remaining gas to −250°C causes nearly pure neon to condense out as a liquid.

When liquid air at −200°C is allowed to warm up, the first substance that boils off is nitrogen, the lowest boiling species present (bp N_2 = −196°C). Careful fractionation gives nearly pure N_2 in the distillate. The residue is mostly oxygen, with about 5% argon and traces of the heavier noble gases krypton and xenon. Fractional distillation is used to separate these components. Small amounts of oxygen can be removed from the noble gases by sparking with hydrogen. The O_2 reacts with H_2 under these conditions to form water, which readily condenses out from the gas mixture.

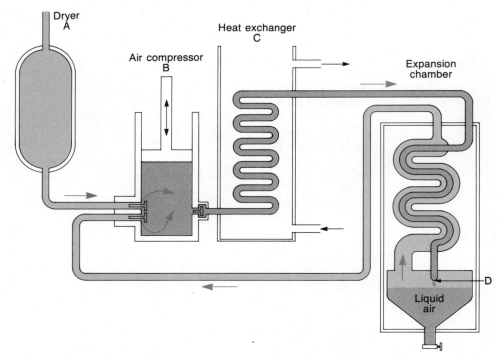

FIGURE 4.2 Liquefaction of air. Carbon dioxide and water are removed by passing the air through a drying chamber (A) packed with NaOH. The air is then compressed (B) and cooled by passing through a heat exchanger (C). Further cooling takes place upon expansion at a throttle valve (D). After going through the cycle several times the air begins to condense at D.

Table 4.1
Elements in the Atmosphere

	GROUP	bp (°C)	MOLE PERCENT IN AIR	USES
N_2	5	−196	78.08	**Synthesis of NH₃.** Packaging of foods such as instant coffee to preserve flavor; liquid N_2 used as coolant (safer than liquid air)
O_2	6	−183	20.95	**Making steel** (Section 4.3). Life support systems, waste water treatment, high temperature flames
He	8	−269	0.000524	Balloons, leak detectors
Ne	8	−246	0.00182	Neon signs, lasers
Ar	8	−186	0.934	Provides inert atmosphere in light bulbs, arc welding of Al, Mg
Kr	8	−152	0.000114	High-speed flash bulbs
Xe	8	−107	0.000009	Experimental anesthetic; ¹³³Xe used as radioactive tracer in medical diagnostic studies

About 30 × 10⁶ tons of liquid air are fractionated each year in the U.S., about 100 kg per person

The first person to isolate a noble gas was the English scientist Sir Henry Cavendish (1731–1810). He subjected a sample of atmospheric nitrogen to repeated electrical discharges in the presence of oxygen. In this way, he formed oxides of nitrogen, which he then dissolved in water. Cavendish found that a bubble of gas, about 1% of the original volume, remained undissolved.

Nothing came of this observation for over a century. In 1892, Lord Rayleigh, in the Cavendish Laboratory at Cambridge, took the next step. He found that "atmospheric" nitrogen had a density about 0.5% greater than that of pure nitrogen. Rayleigh and Sir William Ramsay, a Scotsman, showed that this difference was due to the presence of a very unreactive element in the air. In 1894, they reported the discovery of argon ("the lazy one"). Over the next 5 years, Ramsay and William Travers isolated three more noble gases from liquid air. These were neon ("the new one"), krypton ("the hidden one"), and xenon ("the stranger").

Sometimes small discrepancies have large implications

Helium was first detected in the spectrum of the sun in 1868. Ramsay, in the same year he discovered argon, separated helium from a gas given off when a uranium mineral was heated. A year later, helium was found in trace amounts in the atmosphere. The last of the noble gases to be discovered was radon, in 1898. It is produced by a natural nuclear reaction and is itself intensely radioactive. Radon is not found in detectable amounts in air.

Small amounts of oxygen and nitrogen are often prepared in the laboratory by decomposing compounds of these elements. If a little powdered manganese dioxide, MnO_2, is added to a 3% water solution of hydrogen peroxide, H_2O_2, the following reaction occurs:

$$2\ H_2O_2(aq) \rightarrow O_2(g) + 2\ H_2O \tag{4.1}$$

The MnO_2 acts as a *catalyst*: it speeds up the reaction without being consumed by it. A convenient small-scale preparation of nitrogen involves the ionic compound ammonium nitrite, NH_4NO_2. Heating a water solution of ammonium nitrite brings about the following reaction:

$$NH_4^+(aq) + NO_2^-(aq) \rightarrow N_2(g) + 2\ H_2O \tag{4.2}$$

4.2
Nonmetals from the Earth

Most elements must be extracted from the waters of the earth, its solid surface, or from underground mines. Included among these are 13 of the elements classified as nonmetals. These are listed, with their sources and principal uses, in Table 4.2, p. 94. Here, we will discuss the methods used to obtain six of these elements (S, H_2, F_2, Cl_2, Br_2, and I_2) from natural sources.

EXAMPLE 4.1 The mineral known as borax, whose formula is given in Table 4.2, is found in dried-up lake beds in the Mojave Desert of California. What mass of borax is required to give 1.00 g of boron?

The main use of borax is in Pyrex glass and glass fiber insulation

Solution The formula of borax is $Na_2B_4O_7 \cdot 10\ H_2O$. In 1 mol of borax there are

$$
\begin{array}{rl}
2(22.99\ g) = & 45.98\ g\ Na \\
4(10.81\ g) = & 43.24\ g\ B \\
7(16.00\ g) = & 112.00\ g\ O \\
10(18.02\ g) = & \underline{180.2\ \ g\ H_2O} \\
& 381.4\ \ g
\end{array}
$$

Hence, 381.4 g borax $\simeq$ 43.24 g B. This gives us the conversion factor we need:

$$\text{mass borax} = 1.00 \text{ g B} \times \frac{381.4 \text{ g borax}}{43.24 \text{ g B}} = 8.82 \text{ g borax}$$

EXERCISE   What is the mass percent of boron in borax? Answer: 11.34.

Sulfur

Large deposits of native sulfur are found in salt domes, which have areas as large as 20 km², in the states of Texas and Louisiana. The sulfur lies 60 to 600 m below the surface of the earth, mixed with calcium carbonate, $CaCO_3$, and calcium sulfate, $CaSO_4$. The sulfur is believed to have been formed from the SO_4^{2-} ions of the calcium sulfate by bacterial action.

The process used to mine sulfur is named after its inventor, Herman Frasch,

Table 4.2
Nonmetals Extracted from the Earth's Crust

	GROUP	mp (°C)	bp (°C)	PRINCIPAL SOURCE	USES
B	3	2300	2550	$Na_2B_4O_7 \cdot 10 \, H_2O$ (borax)	Alloys, O_2 scavenger
C	4	3570		Coal, petroleum, natural gas	**Reduction of iron ore (coke); adsorbent (charcoal)**
Si	4	1414	2355	SiO_2 (sand, quartz)	Transistors, solar batteries
P_4	5	44	280	$Ca_3(PO_4)_2$ (phosphate rock)	Synthesis of P_4O_{10}, H_3PO_4, for fertilizers, detergents
As	5	814		As_2S_3, other sulfides	Lead alloys (shot, batteries), semiconductors
S	6	119	444	Free element, H_2S	**Synthesis of H_2SO_4,** vulcanization of rubber
Se	6	217	685	PbSe, other selenides	Color treatment of glass
Te	6	450	990	PbTe, other tellurides	Alloys with metals
H_2		−259	−253	Natural gas, petroleum, H_2O	**Synthesis of NH_3,** conversion of vegetable oils to saturated fats
F_2	7	−220	−188	CaF_2 (fluorite), Na_3AlF_6 (cryolite)	Synthesis of UF_6, SF_6, organic fluorine compounds
Cl_2	7	−101	−34	NaCl, Cl^- in ocean	**Bleach, water treatment,** organic chlorine compounds
Br_2	7	−7	59	Br^- in salt brines	Synthesis of $C_2H_4Br_2$ (antiknock additive)
I_2	7	114	184	I^- in salt brines	Antiseptic, synthesis of NaI, AgI

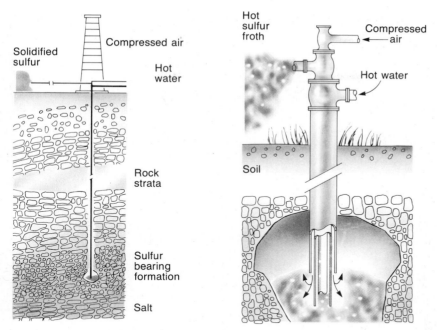

FIGURE 4.3 Frasch process for mining sulfur. Superheated water at 165°C is sent down through the outer pipe to form a pool of molten sulfur (mp = 119°C) at the base. Compressed air, pumped down the inner pipe, brings the sulfur to the surface. Sulfur deposits are often 100 m or more beneath the earth's surface, covered with quicksand and rock.

an American chemical engineer. A diagram of the Frasch process is shown in Figure 4.3. The sulfur is heated to its melting point (119°C) by pumping superheated water at 165°C down one of three concentric pipes. Compressed air is used to bring the sulfur to the surface. The air and sulfur form a frothy mixture that rises through the middle pipe. Upon cooling, the sulfur solidifies, filling huge vats that may be 0.5 km long. The sulfur obtained in this way has a purity approaching 99.9%.

Considerable amounts of sulfur are now being recovered from natural gas, which contains some hydrogen sulfide, H_2S. The hydrogen sulfide is separated and then burned in a limited amount of air. The overall reaction is

Removal of H_2S from natural gas reduces air pollution

$$2\ H_2S(g)\ +\ O_2(g) \rightarrow 2\ S(s)\ +\ 2\ H_2O(l) \tag{4.3}$$

EXAMPLE 4.2 The natural gas from a certain well in Texas contains 2.7 mole percent of H_2S. That is, in every 100 mol of gas, there is 2.7 mol H_2S. Consider a 1.00-m³ sample of this gas at 25°C and 1 atm, which contains 40.9 mol of gas. What mass of sulfur, in grams, can be recovered from this sample?

Solution The number of moles of H_2S in the sample is

$$40.9\ \text{mol gas} \times \frac{2.7\ \text{mol } H_2S}{100\ \text{mol gas}} = 1.1\ \text{mol } H_2S$$

From the coefficients of Equation 4.3,

$$2 \text{ mol } H_2S \simeq 2 \text{ mol } S$$

The atomic mass of sulfur is 32.06; the molar mass is 32.06 g/mol. Hence,

$$\text{mass S} = 1.1 \text{ mol } H_2S \times \frac{2 \text{ mol } S}{2 \text{ mol } H_2S} \times \frac{32.06 \text{ g } S}{1 \text{ mol } S} = 35 \text{ g } S$$

EXERCISE How many grams of O_2 are required to react with the H_2S in this sample? Answer: 18 g O_2.

Hydrogen

The principal method used to prepare hydrogen industrially starts with natural gas. The major component of natural gas is methane, CH_4. When this is heated with steam, a series of reactions occur. The overall reaction is a simple one:

We can make many chemicals from natural gas

$$CH_4(g) + 2 H_2O(g) \rightarrow CO_2(g) + 4 H_2(g) \tag{4.4}$$

The hydrogen formed must be separated from the carbon dioxide produced with it. The CO_2 can be removed by bubbling the gas mixture through cold water because carbon dioxide is much more soluble in water (0.034 mol/L at 25°C, 1 atm) than is hydrogen (< 0.001 mol/L).

Very pure hydrogen can be prepared by the electrolysis of water. A simple type of apparatus used for this purpose is shown in Figure 4.4. A small amount

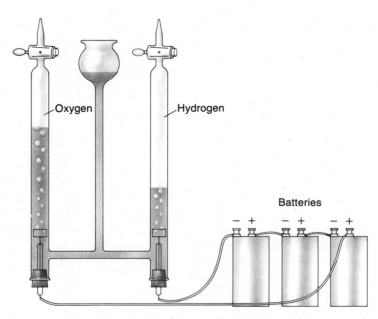

FIGURE 4.4 Electrolysis of water. Current research suggests that the energy required for electrolysis could come from sunlight. To do this, it will be necessary to add a light-absorbing material, most likely to one of the electrodes. The H_2 produced could be used as a fuel.

of a compound such as NaOH is added to the water to provide ions to carry the current. Electrical energy is absorbed to bring about the reaction:

$$2\ H_2O(l) \rightarrow 2\ H_2(g)\ +\ O_2(g) \tag{4.5}$$

The hydrogen produced in this way is more expensive than that from natural gas; the major cost is for electrical energy. About 118 kJ of energy must be supplied to form one gram of hydrogen.

Small amounts of hydrogen are often prepared in the laboratory by reacting zinc with hydrochloric acid, a water solution containing H^+ ions and Cl^- ions. The H^+ ions of the acid react with zinc atoms:

$$Zn(s)\ +\ 2\ H^+(aq) \rightarrow Zn^{2+}(aq)\ +\ H_2(g) \tag{4.6}$$

The zinc "dissolves" in the acid

Reaction 4.6 is an example of an oxidation-reduction reaction (Chap. 3), in which there is a transfer of electrons. Zinc atoms lose electrons to form Zn^{2+} ions:

$$Zn(s) \rightarrow Zn^{2+}(aq)\ +\ 2\ e^- \tag{4.6a}$$

These electrons are accepted by H^+ ions to form H_2 molecules:

$$2\ H^+(aq)\ +\ 2\ e^- \rightarrow H_2(g) \tag{4.6b}$$

Half-reaction 4.6a is an oxidation; Zn atoms are oxidized to Zn^{2+} ions when they lose electrons. Half-reaction 4.6b is a reduction; H^+ ions are reduced to H_2 molecules when they gain electrons.

EXAMPLE 4.3 Suppose 1.00 g CH_4 is mixed with 2.00 g H_2O to produce hydrogen gas by Reaction 4.4. What is the theoretical yield in grams of H_2? Which reactant is limiting?

Solution We proceed as described in Chapter 3, calculating first the theoretical yield based on CH_4, and then making a second calculation based on H_2O. Note from the coefficients of Equation 4.4 that

$$1\ mol\ CH_4 \hateq 2\ mol\ H_2O \hateq 4\ mol\ H_2$$

The molar masses of CH_4, H_2O, and H_2 are 16.04, 18.02, and 2.02 g/mol, respectively. If all the CH_4 is consumed,

$$mass\ H_2 = 1.00\ g\ CH_4 \times \frac{1\ mol\ CH_4}{16.04\ g\ CH_4} \times \frac{4\ mol\ H_2}{1\ mol\ CH_4} \times \frac{2.02\ g\ H_2}{1\ mol\ H_2}$$
$$= 0.504\ g\ H_2$$

If all the H_2O is consumed,

$$mass\ H_2 = 2.00\ g\ H_2O \times \frac{1\ mol\ H_2O}{18.02\ g\ H_2O} \times \frac{4\ mol\ H_2}{2\ mol\ H_2O} \times \frac{2.02\ g\ H_2}{1\ mol\ H_2}$$
$$= 0.448\ g\ H_2$$

We choose the smaller number: the theoretical yield of H_2 is 0.448 g; the limiting reactant is H_2O.

EXERCISE If the actual yield of H_2 is 0.376 g, what is the percent yield?
Answer: 83.9.

The Halogens:
F_2, Cl_2, Br_2, I_2

The elements in Group 7 of the Periodic Table are known as *halogens* ("salt-formers"). The first two members of the family, fluorine and chlorine, are gases at 25°C and 1 atm. Bromine is a deep red, volatile liquid boiling at 59°C. Iodine is a shiny black solid. Upon heating, iodine is readily converted to a violet vapor (see Color Plate 4.1). Astatine, the last member of the family, is radioactive, like its neighbor radon in Group 8.

The halogens are much too reactive to be found as the free elements in nature. Instead they occur as the *halide ions:* F^-, Cl^-, Br^-, and I^-. The principal source of fluorine is the water-insoluble mineral fluorite, CaF_2. Chlorine is present as the Cl^- ion in such ionic solids as NaCl, KCl, and $MgCl_2$. It is also the most abundant anion in seawater (0.53 M). Bromide and iodide ions are found in seawater, but at much lower concentrations. In certain natural brines (salt solutions), the concentrations of Br^- and I^- ions are higher.

In order to prepare the halogens, electrons must be removed from halide ions (F^-, Cl^-, Br^-, I^-); that is, the following oxidation processes must be carried out:

$$2\ F^- \rightarrow F_2 + 2\ e^-$$

$$2\ Cl^- \rightarrow Cl_2 + 2\ e^-$$

$$2\ Br^- \rightarrow Br_2 + 2\ e^-$$

$$2\ I^- \rightarrow I_2 + 2\ e^-$$

Among the halide ions, oxidation becomes easier as we move down the Periodic Table. Of the four halide ions, F^- ions are most reluctant to give up electrons to form F_2 molecules, whereas I^- ions are most readily oxidized to I_2.

The oxidation half-reactions referred to above cannot, of course, occur by themselves. Some species must accept the electrons given up by the halide ions. One species that can do this with I^- and Br^- ions is the Cl_2 molecule. This molecule has a relatively strong attraction for electrons; that is, it is readily reduced to chloride ions in the half-reaction

With every oxidation, a reduction also occurs

$$Cl_2(g) + 2\ e^- \rightarrow 2\ Cl^-(aq)$$

When chlorine gas is bubbled through a brine containing I^- ions, an oxidation-reduction reaction occurs. In this reaction, Cl_2 molecules are reduced to Cl^- ions, while I^- ions are oxidized to I_2 molecules. The equation for the reaction is

$$Cl_2(g) + 2\ I^-(aq) \rightarrow 2\ Cl^-(aq) + I_2(s) \tag{4.7}$$

A reaction similar to Reaction 4.7 occurs when a brine containing Br^- ions is treated with chlorine gas:

The color changes from yellow to red as the reaction proceeds

$$Cl_2(g) + 2\ Br^-(aq) \rightarrow 2\ Cl^-(aq) + Br_2(l) \tag{4.8}$$

In practice, an aqueous solution containing dissolved bromine and unreacted chlorine is the first product. Bromine can be separated from this solution in various ways. One method involves heating the solution to produce a gaseous mixture of bromine, chlorine, and water vapor. Upon cooling, liquid bromine separates out.

EXAMPLE 4.4   In a natural brine found in Arkansas, the concentration of Br^- ions is 5.00×10^{-3} M. For one cubic meter (1.00×10^3 L) of this brine,
a. how many moles of Br^- ion are present?
b. how many grams of Cl_2 are required to react with all the Br^- ions (Reaction 4.8)?

Solution
a. Recall (Chapter 2) that molarity, M, is defined as moles per liter. Hence,

$$\text{moles } Br^- = 5.00 \times 10^{-3} \frac{\text{mol } Br^-}{L} \times 1.00 \times 10^3 \text{ L}$$

$$= 5.00 \text{ mol } Br^-$$

b. From Reaction 4.8, 1 mol $Cl_2 \stackrel{\frown}{=} 2$ mol Br^-. The molar mass of Cl_2 is 70.90 g/mol. Hence,

$$\text{mass } Cl_2 = 5.00 \text{ mol } Br^- \times \frac{1 \text{ mol } Cl_2}{2 \text{ mol } Br^-} \times \frac{70.90 \text{ g } Cl_2}{1 \text{ mol } Cl_2} = 177 \text{ g } Cl_2$$

EXERCISE   If 210 g Cl_2 were added to one cubic meter of this brine, which would be the limiting reactant, Cl_2 or Br^-? Answer: Br^-.

Chloride ions hold on to their electrons more tightly than I^- or Br^- ions. As a result, the element chlorine is more difficult to prepare than iodine or bromine. Industrially, Cl^- ions are oxidized to Cl_2 molecules as part of an electrolysis reaction. The usual source of Cl^- ions is NaCl, either in the molten state (Section 4.3) or, more commonly, in water solution. The reaction that occurs when a water solution of sodium chloride is electrolyzed is

$$2 \text{ Cl}^-(aq) + 2 \text{ H}_2O \rightarrow Cl_2(g) + H_2(g) + 2 \text{ OH}^-(aq) \tag{4.9}$$

Most industrial Cl_2 is made by this reaction

This process will be discussed further in Chapter 23. In the laboratory, chlorine gas can be prepared by the reaction of manganese dioxide with hydrochloric acid (Fig. 4.5). The equation for the reaction is

$$MnO_2(s) + 4 \text{ H}^+(aq) + 2 \text{ Cl}^-(aq) \rightarrow Mn^{2+}(aq) + Cl_2(g) + 2 \text{ H}_2O \tag{4.10}$$

Of all the halogens, fluorine, F_2, is the most difficult to prepare. The industrial process used to make fluorine starts with the mineral fluorite, calcium fluoride, CaF_2. This is treated with sulfuric acid, H_2SO_4, and the products are hydrogen fluoride, HF, and calcium sulfate, $CaSO_4$:

$$CaF_2(s) + H_2SO_4(l) \rightarrow 2 \text{ HF}(l) + CaSO_4(s) \tag{4.11}$$

The hydrogen fluoride is mixed with potassium fluoride, KF, and electrolyzed. The electrolysis reaction can be shown most simply as

$$2 \text{ HF}(l) \rightarrow H_2(g) + F_2(g) \tag{4.12}$$

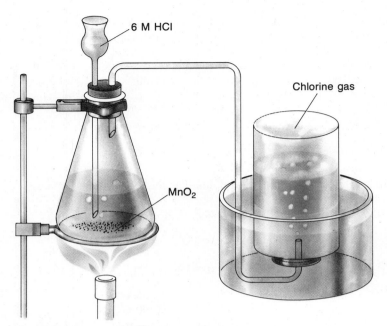

6 M HCl

Chlorine gas

MnO$_2$

FIGURE 4.5 Preparation of chlorine. If MnO$_2$ in hydrochloric acid is heated, one obtains Cl$_2$ gas, which can be collected over water and used for other reactions.

Most chemists have never seen fluorine, and don't want to

The F$_2$ molecule has such a strong attraction for electrons that the element is very dangerous to work with. It reacts violently with water, asbestos, hot glass, and most metals. Fluorine is commonly stored in containers made of monel metal, an alloy of Ni, Cu, and Fe. This is one of the few materials that is not attacked by F$_2$.

4.3
Metals

Metals occur in nature in many different chemical forms (Fig. 4.6). Most of the ores of metals fall into one of the categories described below.

Types of Ores

1. *Elements*. A few of the least reactive transition metals occur as uncombined elements. Gold is the most familiar example (Color Plate 4.2). It is sometimes found as nuggets that are almost pure gold. More often, the element occurs in veins of quartz or other rocky material.

2. *Oxides*. Here the metal is present as a cation (for example, Al^{3+}, Fe^{3+}) in an ionic compound; the anion is the oxide ion, O^{2-}. As you can see from Figure 4.6, oxide ores are quite common, particularly among the transition metals that appear toward the left of the Periodic Table. Perhaps the two best known oxide ores are those of aluminum and iron, Al$_2$O$_3$ and Fe$_2$O$_3$.

3. *Sulfides*. This type of ore is more common than any other. Most of the transition metals that appear toward the right of the Periodic Table are found as sulfides. Among these are copper(I) sulfide(Cu$_2$S), and mercury(II) sulfide(HgS).

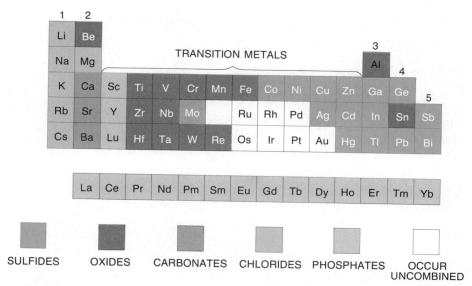

SULFIDES   OXIDES   CARBONATES   CHLORIDES   PHOSPHATES   OCCUR
UNCOMBINED

FIGURE 4.6   Sources of the metals. The best known ores, particularly common with the transition metals, are sulfides and oxides. The metals in Groups 1 and 2 usually occur as chlorides or carbonates. A few very unreactive metals, including gold and platinum, are found uncombined in nature.

Many of the heavier elements in Groups 3, 4, and 5 also occur as sulfides. Among the ores of this type are the sulfides of lead and bismuth, PbS and Bi_2S_3.

4. *Chlorides*. The Group 1 (*alkali*) metals are extracted from chloride ores (NaCl, KCl). These are found in dried-up lake beds in arid regions or in underground deposits formed by the evaporation of ancient seas. Of the Group 2 elements, magnesium occurs, in a sense, as a chloride. The principal source of this metal is seawater, where it occurs as the Mg^{2+} ion. Of the many anions in seawater, Cl^- is the most abundant.

5. *Carbonates*. The heavier Group 2 (*alkaline earth*) metals occur as carbonates. In these compounds, the metal is present as a +2 cation (Ca^{2+}, Sr^{2+}, Ba^{2+}) balanced by a carbonate ion, CO_3^{2-}. Thus, we have $CaCO_3$, $SrCO_3$, and $BaCO_3$.

6. *Phosphates*. The elements of atomic number 57 through 70, commonly called *lanthanides* or rare earths, are found as phosphates. A phosphate anion (PO_4^{3-}) is balanced by a +3 cation (for example, La^{3+}), giving compounds such as $LaPO_4$.

EXAMPLE 4.5   In the United States, most magnesium is obtained from seawater by a process to be described later in this section. However, magnesium can also be extracted from a mineral called dolomite, which contains 13.2% Mg, 21.7% Ca, 13.0% C, and 52.1% O. What is the simplest formula of this ionic compound? What ions do you think are present?

Some minerals are surprisingly pure

Solution   Proceeding as in Example 3.2, p. 64, we find the number of moles of each element in 100 g of dolomite. A sample calculation is

$$\text{moles Mg} = 13.2 \text{ g} \times \frac{1 \text{ mol}}{24.30 \text{ g}} = 0.543 \text{ mol}$$

Making the same calculation for the other elements, we find

0.543 Mg : 0.541 Ca : 1.08 C : 3.25 O

Dividing by the smallest number, 0.541,

1 Mg : 1 Ca : 2 C : 6 O

The simplest formula could be written as $MgCaC_2O_6$. A moment's reflection should convince you that the ions present are Mg^{2+}, Ca^{2+}, and $2\ CO_3^{2-}$. The formula of dolomite is often written as $MgCO_3 \cdot CaCO_3$. It is a 1:1 mixture of $MgCO_3$ and $CaCO_3$.

EXERCISE The mineral called carnallite contains 23.0% K, 14.3% Mg, and 62.7% Cl. What is its simplest formula? What ions are present? Answer: $KMgCl_3$; K^+, Mg^{2+}, $3\ Cl^-$.

Extraction of Metals from Their Ores

The processes used to obtain pure metals from their ores comprise the science of *metallurgy*. These processes ordinarily include one or more of the following steps:

1. *Concentration of the ore.* Typically, when an ore is mined, it contains a large amount of useless rocky material, such as silicon dioxide, SiO_2. In some cases, the concentration of the metal in the ore may be less than 1%. If so, the ore must be treated to remove most if not all of the impurities.

When an ore is "reduced," the volume of metal obtained is less than that of the ore

2. *Reduction of a metal cation to a metal atom.* This is required when the ore is a compound of the metal. Electrons must be supplied to bring about the reduction. Typical half-reactions include

$$Na^+ + e^- \rightarrow Na$$

$$Mg^{2+} + 2\ e^- \rightarrow Mg$$

For these reductions to occur, some other species must be oxidized. The overall reaction is one of oxidation and reduction.

3. *Refining (purification) of the metal.* Ordinarily, the metal formed by steps (1) and (2) contains at least small amounts of other metals and/or nonmetals. Depending upon the purpose for which the metal is to be used, it may or may not be necessary to remove these impurities.

In discussing the extraction of various metals, we will see examples of each of the processes just listed. Instead of describing the metallurgy of all the elements shown in Figure 4.6, we will focus upon five important and rather typical metals. These are

—Na, obtained from its chloride, NaCl
—Mg, obtained from seawater
—Al and Fe, obtained by quite different processes from their oxide ores, Al_2O_3 and Fe_2O_3
—Cu, obtained from its sulfide ore, Cu_2S.

Na from NaCl

Sodium metal is produced by the electrolysis of molten sodium chloride (Na^+ and Cl^- ions), mixed with some $CaCl_2$. The electrical cell used is shown in Figure 4.7. Electrons, from a storage battery or other source of direct electric current, enter the cell through the iron bar at the left. At this electrode,* a reduction half-reaction occurs. Sodium ions, Na^+, gain electrons and are reduced to sodium atoms:

$$2 \, Na^+ + 2 \, e^- \rightarrow 2 \, Na \tag{4.13a}$$

Electric current is carried through the cell by the movement of ions. Cations (Na^+ ions) move in one direction; anions (Cl^- ions) move in the opposite direction. At the graphite rod near the center of the cell, an oxidation half-reaction occurs: chloride ions, Cl^-, lose electrons to form Cl atoms. Pairs of Cl atoms immediately combine to form Cl_2 molecules:

$$2 \, Cl^- \rightarrow Cl_2 + 2 \, e^- \tag{4.13b}$$

The electrons given off by this half-reaction return to the storage battery.

A substantial fraction of our electrical energy is used in electrolysis reactions

---

*In an electrical cell, the electrode at which reduction occurs is commonly called the cathode; the electrode at which oxidation occurs is the anode. Electrolysis cells will be discussed in greater detail in Chapter 23.

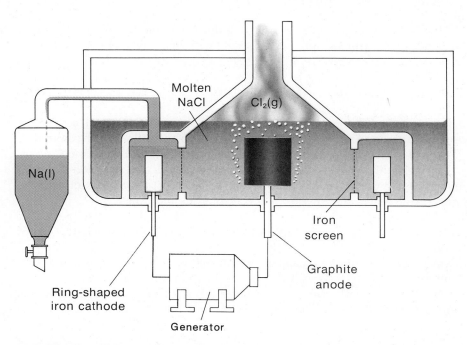

FIGURE 4.7 Electrolysis of molten sodium chloride. The iron screen is used to prevent sodium and chlorine from coming into contact with each other.

The overall reaction involved in the electrolysis of molten sodium chloride is obtained by summing Reactions 4.13a and 4.13b:

In 1807 Sir Humphrey Davy discovered sodium by electrolysis of NaOH

$$2\ NaCl(l) \rightarrow 2\ Na(l)\ +\ Cl_2(g) \tag{4.13}$$

The cell must be operated at a high temperature, about 600°C, to keep the sodium chloride melted. (Solid NaCl does not conduct a current, since the ions are not free to move.) About 14 kJ of electrical energy must be supplied for every gram of sodium produced. The sodium metal (mp = 98°C) is drawn off as a liquid. The chlorine gas formed at the other electrode is a valuable by-product. The cell is designed so that the sodium and chlorine do not come in contact with each other. If that happened, they would react by the reverse of Reaction 4.13, and we would be back where we started. (As a matter of fact, since molten sodium reacts violently with chlorine gas, we might well be worse off than when we started.)

Most of the sodium metal produced in the United States is converted to a Pb-Na alloy* which is used in the preparation of lead tetraethyl, $Pb(C_2H_5)_4$, the antiknock additive used in "leaded" gasoline. Smaller amounts of sodium are used to make inorganic compounds such as sodium peroxide (Na_2O_2) and sodium hydride (NaH), and the class of organic compounds called synthetic detergents. Perhaps the most familiar application of sodium is in the sodium vapor lamps used to illuminate highways and parking lots with a brilliant golden-white light.

Mg from Seawater

The Mg^{2+} ion is the second most abundant cation in seawater, after Na^+. Its concentration is about 0.052 mol/L. Magnesium metal is obtained from seawater by a three-step process:

1. The Mg^{2+} ions are removed from seawater by adding a solution of calcium hydroxide, $Ca(OH)_2$, which contains Ca^{2+} and OH^- ions. The OH^- ions combine with Mg^{2+} to form a precipitate of magnesium hydroxide:

$$Mg^{2+}(aq)\ +\ 2\ OH^-(aq) \rightarrow Mg(OH)_2(s) \tag{4.14}$$

This process is used by the Dow Chemical Co. in its Freeport, Texas plant. It makes about 100,000 tons of Mg per year

The magnesium hydroxide is much less soluble than $Ca(OH)_2$ and so can be recovered as a solid. The precipitate is first allowed to settle out of solution in large tanks. It is then separated from the seawater by filtration and dried.

2. The solid $Mg(OH)_2$ is treated with hydrochloric acid, a water solution containing H^+ and Cl^- ions. The H^+ ions of the hydrochloric acid react with the OH^- ions of the magnesium hydroxide to form H_2O molecules:

$$Mg(OH)_2(s)\ +\ 2\ H^+(aq)\ +\ 2\ Cl^-(aq) \rightarrow Mg^{2+}(aq)\ +\ 2\ Cl^-(aq)\ +\ 2\ H_2O \tag{4.15}$$

In this way, a water solution of magnesium chloride (Mg^{2+}, $2\ Cl^-$ ions) is formed. Evaporation of the water gives solid magnesium chloride, $MgCl_2$.

3. The magnesium chloride is electrolyzed in the molten state in a process very similar to that discussed for NaCl. The electrode half-reactions are

---

*An alloy is a solid with metallic properties that contains two or more elements, usually metals.

reduction: $Mg^{2+} + 2\,e^- \rightarrow Mg$

oxidation: $2\,Cl^- \rightarrow Cl_2 + 2\,e^-$

The overall cell process is an oxidation-reduction:

$$MgCl_2(l) \rightarrow Mg(l) + Cl_2(g)$$ (4.16)

Electrolysis of a solution of $MgCl_2$ would not produce Mg metal

EXAMPLE 4.6   Consider a 1.00-g sample of magnesium metal. How many moles of Mg are there in the sample? How many Mg atoms?

Solution   The atomic mass of Mg is 24.30; one mole of Mg weighs 24.30 g. Recall from Chapter 2 that one mole of any element contains 6.022×10^{23} atoms. Hence,

$$1 \text{ mol Mg} = 6.022 \times 10^{23} \text{ atoms} = 24.30 \text{ g}$$

$$\text{moles Mg} = 1.00 \text{ g} \times \frac{1 \text{ mol}}{24.30 \text{ g}} = 0.0412 \text{ mol}$$

$$\text{atoms Mg} = 1.00 \text{ g} \times \frac{6.022 \times 10^{23} \text{ atoms}}{24.30 \text{ g}} = 2.48 \times 10^{22} \text{ atoms}$$

EXERCISE   Consider a sample of 0.100 mol Mg. What is the mass of this sample in grams? How many atoms does it contain? Answer: 2.43 g; 6.02×10^{22} atoms.

Magnesium is used mainly as a structural metal, either pure or as an alloy with aluminum. The density of magnesium, 1.74 g/cm³, is lower than that of aluminum, 2.70 g/cm³, and much lower than that of the other common structural metal, iron (d = 7.87 g/cm³). This explains why magnesium and its alloys are used in aircraft parts, ladders, chain saws, and similar products. Small amounts of magnesium are used in flares, photographic flash ribbons, and rocket propellants. The latter applications take advantage of the fact that magnesium burns readily in air, giving off both heat and light.

Adding Al makes the alloy stronger and harder

Al from Al_2O_3

Aluminum is the third most abundant element in the earth's crust, after oxygen and silicon. In the form of Al^{3+} ions, it is a major component of such common materials as clay, feldspar, and granite. Unfortunately, aluminum cannot be extracted profitably from these materials. Instead, it is obtained from a relatively uncommon mineral called bauxite. This is a mixture of the oxides of aluminum (Al_2O_3), iron (Fe_2O_3), and silicon (SiO_2). The percentage of aluminum in bauxite is about 28.

The first step in preparing aluminum is the separation of aluminum oxide from bauxite. This is done by heating the ore with a concentrated solution of sodium hydroxide, NaOH. Of the three major components of bauxite, only Al_2O_3 dissolves. The other components are then removed by filtration. When the solution is cooled and diluted with water, aluminum hydroxide, $Al(OH)_3$, separates as a white solid. This is heated to drive off the water and form nearly pure aluminum oxide:

$$2\ Al(OH)_3(s) \rightarrow Al_2O_3(s)\ +\ 3\ H_2O(g) \qquad\qquad (4.17)$$

Aluminum metal is obtained from Al_2O_3 by electrolysis (Fig. 4.8). Cryolite, Na_3AlF_6, is added to the Al_2O_3 to produce a mixture that melts at about 1000°C. (A mixture of AlF_3, NaF, and CaF_2 may be substituted for cryolite.) The cell is heated electrically to keep the mixture molten so that ions can move through it, carrying the electric current.

> Pure Al_2O_3 melts at 2000°C, way above the MP of steel

The iron wall of the cell serves as one electrode. Here, Al^{3+} ions are reduced to form molten aluminum (mp = 660°C). The reduction half-reaction is

$$4\ Al^{3+}\ +\ 12\ e^-\ \rightarrow 4\ Al$$

The other electrode is a retractable carbon rod. There, O^{2-} ions lose electrons to form gaseous oxygen. The oxidation half-reaction is

$$6\ O^{2-} \rightarrow 3\ O_2(g)\ +\ 12\ e^-$$

The oxygen formed attacks the carbon rod, slowly consuming it to give CO and CO_2.

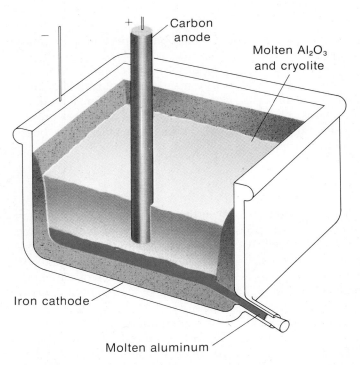

FIGURE 4.8 Electrolytic preparation of aluminum. Aluminum, being more dense than cryolite, collects at the bottom of the cell and so is protected from oxidation by the air.

The overall cell reaction is

$$2 \text{ Al}_2\text{O}_3(l) \rightarrow 4 \text{ Al}(l) + 3 \text{ O}_2(g) \tag{4.18}$$

This process consumes about 60 kJ of electrical energy per gram of aluminum formed. The high energy requirement explains in large part the interest in recycling aluminum containers. The energy cost for recovering aluminum is only about 1% of that required to extract it from aluminum oxide.

A century ago, aluminum sold for about $20/kg and the total production worldwide was about 2×10^3 kg a year. The metal was made by a costly process involving the reaction of sodium with aluminum chloride:

$$\text{AlCl}_3(s) + 3 \text{ Na}(s) \rightarrow \text{Al}(s) + 3 \text{ NaCl}(s)$$

The process for obtaining aluminum from bauxite was worked out in 1886 by Charles Hall, a graduate student at Oberlin College. The problem that Hall faced was to find a way to electrolyze Al_2O_3 at a temperature below its melting point of 2000°C. His general approach was to look for ionic compounds in which Al_2O_3 would dissolve at a reasonable temperature. After several unsuccessful attempts, Hall found that cryolite was the ideal "solvent." Curiously enough, the same electrolytic process was worked out by Heroult in France, also in 1886. Each young man (both were 22 years old) was entirely unaware of the other's work!

Some graduate students are more productive than others

After the Hall-Heroult process began to be used, the price of aluminum fell rapidly. Within 10 years it was selling for $1/kg, which is about what aluminum costs today. In the United States, the production of aluminum is greater than that of any other metal except iron (Table 4.3).

Aluminum metal is widely used in the building and construction industries. All of us are familiar with aluminum doors, windows, and siding. Most commercial aircraft are made almost entirely from aluminum and its alloys. Increasingly, aluminum is being used in automobiles to reduce mass and thereby increase gasoline mileage.

Iron and Steel

The principal high-grade ore of iron is a mineral called hematite. This consists largely of iron(III) oxide, Fe_2O_3. The main impurity is silicon dioxide, SiO_2 (sand). Smaller amounts of compounds of manganese and phosphorus are also present. Reduction of Fe^{3+} ions in Fe_2O_3 is carried out in a blast furnace (Fig. 4.9, p. 109). Typically, the furnace is about 30 m high and perhaps 10 m in diameter.

The solid "charge," admitted at the top of the blast furnace, consists of three materials. These are iron ore, limestone (CaCO_3), and coke, which is nearly pure carbon. To get the process started, a blast of compressed air or pure O_2 at 500°C is blown into the furnace through nozzles located near the bottom. Several different reactions occur, of which three are most important:

1. *Conversion of carbon to carbon monoxide.* In the lower part of the furnace coke burns to form carbon dioxide, CO_2. As the CO_2 rises through the solid mixture, it reacts further with the coke to form carbon monoxide, CO. The overall reaction is

$$2 \text{ C}(s) + \text{O}_2(g) \rightarrow 2 \text{ CO}(g) \tag{4.19}$$

Table 4.3
Production and Sources of Metals (1980)

	ANNUAL PRODUCTION (METRIC TONS)		MAJOR SOURCES OF ORES		PRICE ($/kg)
	WORLDWIDE	U.S.	WORLDWIDE	U.S.	
Fe	5.0×10^8	6.7×10^7	U.S.S.R., Brazil, Australia	Minnesota	0.2
Al	1.6×10^7	4.5×10^6	Australia, Guinea, Jamaica	Arkansas	1.7
Mn	1.2×10^7	——	U.S.S.R., So. Africa	nil	1.5
Cu	7.8×10^6	1.2×10^6	U.S., U.S.S.R., Chile	Arizona	1.9
Zn	6.1×10^6	3.7×10^5	Canada, Peru, Mexico	Tennessee	1.0
Pb	5.4×10^6	1.1×10^6	U.S., Australia	Missouri	0.8
Cr	4.5×10^6	——	So. Africa, Zimbabwe, Philippines, U.S.S.R.	nil	8.8
Ni	7.6×10^5	4.0×10^4	Canada, Australia, U.S.S.R.	nil	7.3
Mg	3.1×10^5	1.5×10^5	U.S., U.S.S.R., Norway	Texas	2.9
Na	2.3×10^5	1.6×10^5	U.S., Britain, Japan	Louisiana	1.5
Sn	2.0×10^5	3.6×10^3	Malaysia, Thailand	nil	18
Mo	1.0×10^5	6.5×10^4	U.S., Chile	Colorado	20
Ag	1.3×10^4	1.4×10^3	Mexico, Peru, Canada	Idaho	280
Hg	6.6×10^3	1.0×10^3	U.S.S.R., Spain, U.S.	Nevada	12
Au	1.4×10^3	5.0×10^1	So. Africa	So. Dakota	13000

The heat given off by this reaction maintains a high temperature within the furnace.

2. *Reduction of Fe³⁺ ions to Fe.* The CO produced by Reaction 4.19 reacts with the iron(III) oxide in the ore:

C(s) can reduce Fe₂O₃, but CO(g) does it a lot faster. Can you see why?

$$Fe_2O_3(s) + 3\ CO(g) \rightarrow 2\ Fe(l) + 3\ CO_2(g) \qquad \textbf{(4.20)}$$

Molten iron, formed at a temperature of 1600°C, collects at the bottom of the furnace. Four or five times a day, it is drawn off. The daily production of iron from a single blast furnace is about 1000 metric tons. This is enough to make 500 Cadillacs or 1000 Volkswagens.

3. *Formation of slag.* The limestone added to the furnace decomposes at about 800°C:

$$CaCO_3(s) \rightarrow CaO(s) + CO_2(g) \qquad \textbf{(4.21)}$$

The calcium oxide formed reacts with impurities in the iron ore to form a glassy material called slag. The main reaction is with SiO_2 to form calcium silicate, $CaSiO_3$:

$$CaO(s) + SiO_2(s) \rightarrow CaSiO_3(l) \qquad \textbf{(4.22)}$$

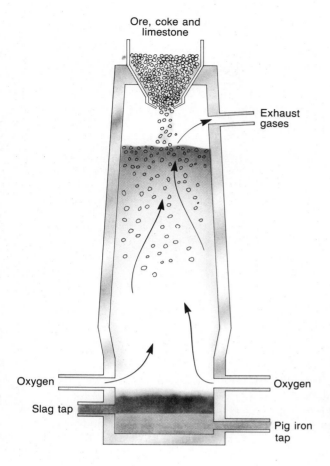

Ore, coke and limestone

Exhaust gases

Oxygen

Oxygen

Slag tap

Pig iron tap

These furnaces are big!

FIGURE 4.9 Blast furnace for production of iron. Oxygen reacting with coke furnishes the high temperatures required for reduction of the iron(III) oxide. The ore appears to be reduced by carbon monoxide, CO. Using pure oxygen instead of air greatly increases the efficiency of the process.

The slag, which is less dense than molten iron, forms a layer on the surface of the metal. This makes it possible to draw off the slag through an opening in the furnace above that used to remove the iron. The slag is used to make cement and as a base in road construction.

EXAMPLE 4.7 In operating a blast furnace, enough limestone is used to give about 2 g CaO/g SiO_2 impurity in the ore. In Reaction 4.22, what is the theoretical yield in grams of $CaSiO_3$, starting with 2.00 g CaO and 1.00 g SiO_2?

Solution From the balanced equation,

$$1 \text{ mol CaO} \simeq 1 \text{ mol } SiO_2 \simeq 1 \text{ mol } CaSiO_3$$

The molar masses of CaO, SiO_2, and $CaSiO_3$ are 56.08, 60.09, and 116.17 g/mol, respectively.

The theoretical yield of $CaSiO_3$ if all the CaO is consumed is

$$2.00 \text{ g CaO} \times \frac{1 \text{ mol CaO}}{56.08 \text{ g CaO}} \times \frac{1 \text{ mol } CaSiO_3}{1 \text{ mol CaO}} \times \frac{116.17 \text{ g } CaSiO_3}{1 \text{ mol } CaSiO_3}$$

$$= 4.14 \text{ g } CaSiO_3$$

The theoretical yield of $CaSiO_3$ if all the SiO_2 is consumed is

$$1.00 \text{ g SiO}_2 \times \frac{1 \text{ mol SiO}_2}{60.09 \text{ g SiO}_2} \times \frac{1 \text{ mol CaSiO}_3}{1 \text{ mol SiO}_2} \times \frac{116.17 \text{ g CaSiO}_3}{1 \text{ mol CaSiO}_3}$$

$$= 1.93 \text{ g CaSiO}_3$$

We want to get rid of all of the SiO_2

We conclude that the theoretical yield of $CaSiO_3$ is 1.93 g. The limiting reactant is SiO_2; there is an excess of CaO.

EXERCISE How many grams of $CaSiO_3$ could be produced starting with a mixture of 1.00 g CaO and 1.00 g SiO_2? Answer: 1.93 g.

The product that comes out of the blast furnace, called "pig iron," is highly impure. On the average, it contains about 4% carbon. Other impurities, present in smaller amounts, include silicon, manganese, and phosphorus. To make steel from pig iron, two things must be done:

1. *Remove the Si, Mn, and P.* To do this, these three elements are oxidized:

Pure O_2 is the oxidizing agent

$$Si(s) + O_2(g) \rightarrow SiO_2(s) \tag{4.23}$$

$$2 \, Mn(s) + O_2(g) \rightarrow 2 \, MnO(s) \tag{4.24}$$

$$P_4(s) + 5 \, O_2(g) \rightarrow P_4O_{10}(s) \tag{4.25}$$

The oxides formed by these reactions are removed by adding lime (CaO), which forms a slag that is separated as in the blast furnace.

2. *Lower the carbon content below 2%.* This is necessary to produce a strong ductile metal that can be heat-treated. Most of the carbon is burned to carbon dioxide:

$$C(s) + O_2(g) \rightarrow CO_2(g) \tag{4.26}$$

By controlling the amount of oxygen used, it is possible to adjust the carbon content of the steel within very narrow limits.

Most of the steel produced in the world today is made by the "basic oxygen" process. This is the only process used in Japan and Western Europe. It accounts for about 60% of steel production in the United States. A diagram of the "converter" used is shown in Figure 4.10. This is filled with a mixture of about 70% pig iron, 25% scrap iron or steel, and 5% limestone. The pig iron is brought over from the blast furnace while it is still molten. Most of the scrap comes from the steel plant itself. Some is obtained from recycling of steel products such as automobiles.

Pure oxygen under a pressure of about 10 atm enters the converter through the water-cooled "lance" at the top. When it reaches the surface of the molten metal, it causes vigorous stirring. Reactions 4.23 through 4.26 occur rapidly. Their progress is followed by an automatic, computerized system of chemical analysis. When the carbon content drops to the desired level, the supply of oxygen is cut off. (Too much oxygen would oxidize the iron, a reaction we don't want.) At this stage, the steel is ready to be poured. The whole process takes from 30 min to 1 h, and yields about 200 metric tons of steel in a single "blow."

The properties of steel depend to a large extent upon the percentage of carbon

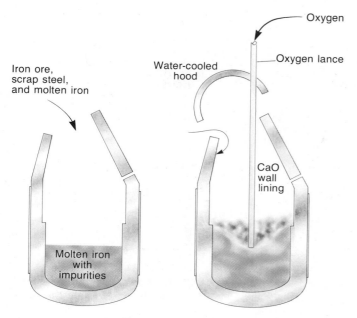

FIGURE 4.10 Converter used to make steel from pig iron by the basic oxygen process. Pure oxygen under pressure is blown into the molten metal to oxidize impurities such as Si, Mn, and P_4 to their oxides. The oxygen also reacts with carbon to form CO_2, lowering the carbon content to 2% or less.

present. "Mild" steels, containing less than 0.25% carbon, are very ductile. They are used to make steel cans and sheet for automobile bodies. Medium steels (0.25% to 0.7% C) are less ductile but much stronger. They are used for structural purposes, such as girders and bridge supports. High-carbon steels (0.7% to 1.5% C) are hard and brittle. Springs, razor blades, and surgical instruments are made from this type of steel.

> Ductile: able to be drawn into a wire

Alloy steels contain small amounts of other metals in addition to iron (and carbon). Probably the best known alloy steel is stainless steel, which contains about 15% chromium and 8% nickel. Stainless steel is very resistant to corrosion and tarnishing. Another element that is used very commonly in alloy steels is manganese. This metal gives a very hard, tough product; steel armor plate contains 10% or more of manganese.

Cu from Cu_2S

Copper is found in nature in many different forms. Most often, it occurs in chemical combination with sulfur or oxygen and, often, iron. We will consider the metallurgy of one of the simpler copper ores, called chalcocite. This contains copper(I) sulfide, Cu_2S, in highly impure form. Rocky material typically lowers the fraction of copper in the ore to 1% or less. The extraction of copper metal from chalcocite ore involves three separate steps:

1. The Cu_2S in the ore is concentrated by a process called flotation (Fig. 4.11). The finely divided ore is mixed with oil and stirred with soapy water in a large tank. Compressed air is then blown through this mixture. The oil-coated particles of copper(I) sulfide are carried to the top of the tank. There they form

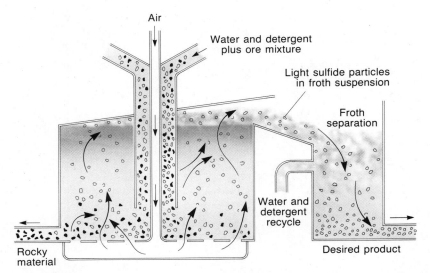

FIGURE 4.11 Low-grade sulfide ores, including Cu_2S, are often concentrated by flotation. The finely divided sulfide particles are trapped in soap bubbles, while the rocky material sinks to the bottom and is discarded.

a froth that can be skimmed off. Most of the rocky impurities, including SiO_2, settle to the bottom of the tank. In this way, the fraction of copper is raised to 20% to 40%.

2. The copper(I) sulfide is converted to the metal by blowing air through it at a high temperature. (Pure O_2 is often used instead of air.) The overall reaction that occurs is a simple one:

$$Cu_2S(s) + O_2(g) \rightarrow 2\ Cu(s) + SO_2(g) \qquad \textbf{(4.27)}$$

The solid produced is called "blister copper." It has an irregular appearance due to air bubbles that enter the copper while it is still molten. Blister copper is impure, containing small amounts of several other metals.

3. The copper is refined (purified) by electrolysis. A piece of blister copper, which serves as one electrode, is oxidized to Cu^{2+} ions that enter the water solution. These ions are reduced to copper atoms at the other electrode, a sheet of pure copper. The half-reactions are

Electrolytic copper is >99.9% pure

oxidation: $Cu(s,\ impure) \rightarrow Cu^{2+}(aq) + 2\ e^-$

reduction: $Cu^{2+}(aq) + 2\ e^- \rightarrow Cu(s,\ pure)$

The overall reaction, obtained by adding these two half-reactions, is

$$Cu(s,\ impure) \rightarrow Cu(s,\ pure) \qquad \textbf{(4.28)}$$

Thus, the net effect of electrolysis is to transfer copper metal from the impure blister copper used as one electrode to the pure copper sheet used as the other electrode.

Less active metals, including silver, gold, and platinum, which are present as impurities in the blister copper, are not oxidized. Instead, they drop off and collect at the bottom of the cell. These metals are recovered, separated from each other, and purified. Their value exceeds the cost of the energy used in the electrolysis.

EXAMPLE 4.8 Zinc, like copper, occurs as a sulfide ore. The formula of the ore, known as zinc blende, is ZnS. When ZnS is heated with oxygen, a reaction similar to Reaction 4.27 occurs. However, the solid product is the oxide, ZnO, rather than the metal.
a. Write a balanced equation for the reaction.
b. How many grams of SO_2 are formed from 1.00 kg of ore containing 22% ZnS?

Solution
a. The unbalanced equation is $ZnS(s) + O_2(g) \rightarrow ZnO(s) + SO_2(g)$. We could balance this equation by writing a coefficient of $\frac{3}{2}$ for O_2:

$$ZnS(s) + \tfrac{3}{2} O_2(g) \rightarrow ZnO(s) + SO_2(g)$$

Ordinarily, whole-number coefficients are preferred. Multiplying all the coefficients by 2 gives

$$2\ ZnS(s) + 3\ O_2(g) \rightarrow 2\ ZnO(s) + 2\ SO_2(g)$$

b. The mass of ZnS in the ore is

$$1.00 \times 10^3 \text{ g} \times 0.22 = 2.2 \times 10^2 \text{ g ZnS}$$

From the balanced equation,

$$1 \text{ mol ZnS} \simeq 1 \text{ mol SO}_2$$

Changing to grams (at. mass Zn = 65.38, S = 32.06, O = 16.00),

$$97.44 \text{ g ZnS} \simeq 64.06 \text{ g SO}_2$$

$$\text{mass SO}_2 = 2.2 \times 10^2 \text{ g ZnS} \times \frac{64.06 \text{ g SO}_2}{97.44 \text{ g ZnS}} = 1.4 \times 10^2 \text{ g SO}_2$$

The SO_2 produced in the treatment of sulfide ores such as ZnS and Cu_2S is a major contributor to air pollution.

EXERCISE How many grams of SO_2 are formed from 1.00 kg of ore containing 8.0% Cu_2S (Reaction 4.27)? Answer: 32 g.

About half of the copper produced each year is used in the electrical industry for wiring, motors, generators, and switches. Copper is one of the best electrical conductors. It is more resistant to corrosion than aluminum, its principal competitor in this area. Copper is also widely used in plumbing for pipes, valves, and fittings. There are many common alloys of copper. Perhaps the best known are brass (67% to 90% Cu, 10% to 33% Zn) and bronze (70% to 95% Cu, up to 18% Sn, and up to 25% Zn). Sterling silver is an alloy of 92.5% Ag with 7.5% Cu; pure silver is much too soft for use as tableware. The common "nickel" coin contains 25% nickel and 75% copper.

Al wiring tends to oxidize at connection points

Summary

We obtain elements from a variety of sources: the air, the oceans, and the earth. Most elements are found in combined form, as compounds, in nature. A few, including sulfur, gold, and the atmospheric gases, exist naturally as the uncombined element. In this chapter, we have discussed the sources of selected elements and the processes used to obtain them in pure form.

When an element is extracted from one of its compounds, an oxidation-reduction reaction is involved. Many nonmetals, including the halogens, are obtained from the corresponding anions by oxidation, for example,

$$2 \ Cl^- \rightarrow Cl_2 + 2 \ e^-$$

In contrast, metals are obtained from cations by reduction, for example,

$$Mg^{2+} + 2 \ e^- \rightarrow Mg$$

These processes, oxidation and reduction, are often carried out in an electrolytic cell; oxidation occurs at one electrode and reduction at the other. Alternatively, oxidation and reduction may be brought about chemically. Chlorine, Cl_2, oxidizes I^- ions to I_2; CO reduces the Fe^{3+} ions in Fe_2O_3 to iron metal.

Throughout this chapter, we have applied principles discussed in Chapters 2 and 3. Several examples were used to illustrate these principles. All the examples in this chapter involved the mole concept, introduced in Chapter 2. Beyond that,

—Examples 4.1 and 4.5 review the relationship between formula and percent composition (Chap. 3; recall Examples 3.1 and 3.2).
—Examples 4.3 and 4.7 review limiting reactant and theoretical yield (Chap. 3; recall Example 3.12).
—Examples 4.4 and 4.8 review mass relations in reactions (Chap. 3; recall Examples 3.10 and 3.11).

Key Words and Concepts

alkali metal	halide ion	noble gas
alkaline earth metal	halogen	ore
alloy	lanthanide	oxidation
catalyst	limiting reactant	reduction
electrode	metallurgy	slag
electrolysis	molarity	theoretical yield
flotation	mole	

Questions and Problems

General

4.1 Which of the following elements are extracted from the air? the oceans? the earth's crust?
a. Al   b. Ar   c. Cu   d. Fe
e. Mg   f. N_2 g. S

4.31 Which of the elements in Question 4.1 commonly occur in the elemental form in nature? Which are commonly obtained by electrolysis?

4.2 Which of the following metals commonly occur as oxides? sulfides? neither?
a. Fe   b. Cu   c. Ti
d. Na   e. Hg   f. Mg

4.3 Describe, in your own words, the processes used to obtain the following elements from their natural sources:
a. O_2 b. I_2 c. Mg   d. Cu   e. Al

4.4 Explain why
a. in the fractional distillation of liquid air, Ar comes off with O_2 rather than N_2.
b. hydrogen is ordinarily obtained from natural gas rather than water.
c. Ar occurs in nature in elemental form, but Na and Cl_2 do not.
d. cryolite is used in the electrolysis of bauxite ore.

4.5 Criticize each of the following statements:
a. The objective in steel-making is to form chemically pure iron.
b. When copper(I) sulfide is heated in air, copper(II) oxide is formed.
c. In the refining of copper, pure copper forms by oxidation.
d. Chloride ions oxidize bromine to bromide ions.

4.6 Write balanced equations for
a. the reaction of methane with steam.
b. the electrolysis of water.
c. the reaction of $Cl_2(g)$ with Br^- ions in water solution.
d. the electrolysis of molten NaCl.

4.7 Write balanced equations for
a. the reaction that occurs when Cu_2S is heated in air.
b. the reaction between Fe_2O_3 and CO.
c. the decomposition of limestone, $CaCO_3$.
d. the electrolysis of molten Al_2O_3.

4.8 Write balanced equations for
a. the oxidation of I^- ions in water solution by bromine.
b. the reaction of CaO with the P_4O_{10} produced in steel manufacture. (The product is calcium phosphate.)

4.9 Split each of the following equations into two half-equations, one involving oxidation, the other reduction:
a. $Cl_2(g) + 2Br^-(aq) \rightarrow 2Cl^-(aq) + Br_2(l)$
b. $2Na^+ + 2Cl^- \rightarrow 2Na(s) + Cl_2(g)$
c. $Mg(s) + Cl_2(g) \rightarrow Mg^{2+}(aq) + 2Cl^-(aq)$

4.32 Group 1 metals are most often found as ____ ores. Carbonate ores are common with Group ____. Metals toward the left of a transition series are usually found as ____. Metals toward the right of a transition series usually have ____ ores.

4.33 Describe, in your own words, the processes used to obtain the following elements from their natural sources:
a. N_2 b. S   c. Br_2 d. Cl_2 e. Fe

4.34 Explain why
a. Al is obtained from bauxite, rather than feldspar or granite.
b. sulfur is not obtained by strip mining.
c. Br_2, but not F_2, can be obtained by oxidation with Cl_2.
d. the sodium and chlorine produced by electrolysis of NaCl must be kept separated.

4.35 Indicate whether each of the following statements is true or false:
a. The percentage of carbon is smaller in steel than in pig iron.
b. Pure oxygen is used in steel-making because it oxidizes iron to Fe_2O_3.
c. Bauxite is the most abundant source of aluminum.
d. Iodine is found in natural brines.

4.36 Write balanced equations for
a. the reaction of chlorine gas with I^- ions in water solution.
b. the electrolysis of a water solution containing Cl^- ions.
c. the reaction of calcium fluoride with sulfuric acid.

4.37 Write balanced equations for
a. the precipitation of Mg^{2+} ions from seawater.
b. the reaction of $Mg(OH)_2$ with H^+ ions.
c. the electrolysis of molten $MgCl_2$.
d. the oxidation of phosphorus in the manufacture of steel.

4.38 Write balanced equations for
a. the reaction that occurs upon heating dolomite, $MgCO_3 \cdot CaCO_3$, which is similar to the decomposition of limestone.
b. the reaction that occurs when cinnabar, HgS, is heated in air (similar to Reaction 4.27).

4.39 Follow the directions of Question 4.9 for
a. $Cl_2(g) + 2I^-(aq) \rightarrow 2Cl^-(aq) + I_2(s)$
b. $2Al^{3+} + 3 O^{2-} \rightarrow 2Al(s) + \frac{3}{2} O_2(g)$
c. $MnO_2(s) + 4 H^+(aq) + 2 Cl^-(aq) \rightarrow Mn^{2+}(aq) + 2 H_2O + Cl_2(g)$

4.10 Define the following terms:
    a. oxidation   b. electrolysis
    c. ore   d. transition metal

4.11 List at least one use for
    a. N_2 b. S   c. Na   d. Fe   e. Cu

4.12 For each of the following metals, use Table 4.3 to calculate the fraction of the total amount that is produced in the United States:
    a. Fe   b. Al   c. Cu   d. Pb   e. Ni

Mole-Gram Relations (Chap. 2)

4.13 How many moles are there in 100.0 g of each of the following ores?
    a. Fe_2O_3 b. Al_2O_3 c. Cu_2S d. NaCl

4.14 Find the mass in grams of 1.65 mol of each of the following elemental substances:
    a. Br_2 b. Cu   c. Zn   d. S   e. P_4

4.15 Complete the following table for $CuFeS_2$, a copper ore:

mol $CuFeS_2$	g $CuFeS_2$	g Cu
0.1250	____	____
____	6.36	____
____	____	14.82

Molarity (Chap. 2)

4.16 A well-water sample contained 34.0 mg of magnesium ions per liter.
    a. What is the molarity of Mg^{2+} in the sample?
    b. What volume of this water contains 2.50 mol Mg^{2+}?

4.17 Reaction 4.11 is carried out with concentrated H_2SO_4, which is 18.0 M and has a density of 1.84 g/cm^3. In 750 cm^3 of this reagent, there is
    a. ____ mol H_2SO_4.   b. ____ g H_2SO_4.
    c. ____ g H_2O.

Percent Composition; Formula (Chap. 3)

4.18 The mineral cobaltite, CoAsS, is a source of cobalt.
    a. What is the mass percent of cobalt in this mineral?
    b. What mass of cobaltite is required to furnish 75.0 g of cobalt?

4.40 Explain what is meant by the following terms:
    a. reduction   b. limiting reactant
    c. halogen   d. fractional distillation

4.41 What are the following elements used for?
    a. O_2 b. H_2 c. Mg   d. Al   f. Hg

4.42 Using Table 4.3, find five metals for which U.S. production accounts for less than 10% of the total.

4.43 How many moles of metal can be obtained from 100.0 g of each of the ores listed in Problem 4.13?

4.44 Find the mass in grams of 0.0129 mol of
    a. Cl_2 b. Fe   c. Mn   d. O_2 e. Ar

4.45 Complete the following table for meerschaum, $Mg_2Si_3O_8 \cdot 2H_2O$:

mol meerschaum	g meerschaum	g Mg
____	455.0	____
1.65	____	____
____	____	24.8

4.46 The bromide ion concentration in seawater is about 8.1×10^{-4} M. In 5.00 L of seawater, there are about
    a. ____ mol of Br^-.
    b. ____ g Br^-.

4.47 As a laboratory assistant, you are asked to prepare 600 cm^3 of an aqueous solution 0.020 M in bromine. The density of liquid bromine is 3.12 g/cm^3. How many cubic centimeters of liquid bromine should you use to prepare this solution?

4.48 Ilmenite, $FeTiO_3$, is a commercially important source of titanium.
    a. Calculate the mass percent of titanium in ilmenite.
    b. Calculate the maximum number of grams of titanium that can be extracted from 5.00 metric tons of ilmenite.

4.19 The mineral malachite contains 57.5% Cu, 5.43% C, 36.2% O, and 0.914% H.
 a. What is the simplest formula of malachite?
 b. Which of the following formulas corresponds to the composition of malachite?

 $CuCO_3 \cdot 2H_2O$ $CuCO_3 \cdot Cu(OH)_2$
 $CuCO_3 \cdot 2Cu(OH)_2$ $Cu(OH)_2 \cdot 2CuCO_3$

4.20 Manganese occurs in nature as the mineral pyrolusite (63.2% Mn, 36.8% O). What is the simplest formula of pyrolusite?

4.21 What is the mass percent of metal in each of the following ores?
 a. Fe_2O_3
 b. $CaSO_4 \cdot 2H_2O$
 c. $Al_2O_3 \cdot 2H_2O$
 d. Zn_2SiO_4

4.49 Azurite, a deep blue mineral used as a source of copper, contains 55.3% Cu, 6.97% C, 37.1% O, and 0.585% H.
 a. What is the simplest formula of azurite?
 b. Which of the formulas in Problem 4.19b corresponds to azurite?

4.50 Bismuth occurs in nature as a sulfide ore. The mass percents of Bi and S are 81.3 and 18.7, respectively. What is the simplest formula of bismuth sulfide?

4.51 What mass of each ore referred to in Problem 4.21 is required to obtain 10.0 g of metal?

Mass Relations in Reactions (Chap. 3)

4.22 Consider the electrolysis of Al_2O_3 (Reaction 4.18).
 a. How many grams of Al are produced from 2.00 kg Al_2O_3?
 b. How many grams of O_2 are formed in (a)?
 c. About 15% of the O_2 formed in (a) reacts with the carbon electrode to form CO_2. How many moles of CO_2 are produced?

4.23 Consider the conversion of Fe_2O_3 to iron via Reaction 4.20.
 a. How many moles of carbon monoxide are needed to produce 3.50 mol of iron?
 b. How many grams of Fe_2O_3 are required to react with 0.500 mol CO?
 c. Coke (essentially pure carbon) is burned to give the CO necessary for this reaction. How many kilograms of coke must be burned to form enough CO to react with 1.00 metric ton Fe_2O_3?

4.24 Hydrofluoric acid, HF, cannot be stored in glass containers because it reacts with $CaSiO_3$ in the glass. The unbalanced equation for the reaction is

 $$CaSiO_3(s) + HF(l) \rightarrow$$
 $$CaF_2(s) + H_2O(l) + SiF_4(g)$$

 a. Balance the equation.
 b. How many moles of water are produced by 1.40 mol HF?
 c. How many grams of HF react with 10.0 g $CaSiO_3$?
 d. How many grams of CaF_2 are formed in (c), assuming 100% yield?

4.52 In the "roasting" of Cu_2S (Reaction 4.27), suppose one starts with 2.00 kg Cu_2S.
 a. How many moles of O_2 are required?
 b. What volume of O_2 is required (1 mol O_2 occupies about 24.5 L)?
 c. What volume of air is required (refer to Table 4.1; mole percent = volume percent)?

4.53 For the reaction in Problem 4.23, calculate
 a. the mass in grams of Fe_2O_3 that reacts with 1.60 kg CO.
 b. the mass in grams of CO_2 formed from 12.3 mol Fe_2O_3.
 c. the number of kilograms of coke required to form 1.00 metric ton of iron.

4.54 Small amounts of bromine can be made by the reaction of MnO_2 with hydrobromic acid. The unbalanced equation for the reaction is

 $$MnO_2(s) + Br^-(aq) + H^+(aq) \rightarrow$$
 $$Mn^{2+}(aq) + Br_2(l) + H_2O$$

 a. Balance the equation.
 b. Assuming 100% yield, how many moles of Br^- must be used to make 5.68 g of bromine?
 c. How many grams of H^+ are required in (b)?

4.25 Consider the reaction by which $Mg(OH)_2$ dissolves in hydrochloric acid (Reaction 4.15).
   a. How many moles of H^+ are required to react with 1.00 g $Mg(OH)_2$?
   b. What volume of 6.00 M HCl is required to react with 1.00 g $Mg(OH)_2$?

4.26 A sample of chalcocite containing copper(I) sulfide required 0.550 L of oxygen to convert it to blister copper (Reaction 4.27). If the sample weighed 10.0 g, what was the percent copper in the sample? Assume that copper(I) sulfide was the only source of copper. Density of oxygen = 1.31 g/L.

Limiting Reactant (Chap. 3)

4.27 Consider the reaction by which Fe_2O_3 is converted to iron (Reaction 4.20). Suppose 1.00×10^3 L of CO is available to react with 1.20 kg Fe_2O_3. Taking the density of CO to be 1.14 g/L,
   a. what is the limiting reactant?
   b. what is the theoretical yield of iron, in grams?
   c. if the actual yield of iron is 612 g, what is the percent yield?

4.28 Consider Reaction 4.2, used to produce small amounts of nitrogen. What is the limiting reactant, starting with
   a. 1.25 mol NH_4^+, 1.30 mol NO_2^-
   b. 1.00 g NH_4^+, 1.00 g NO_2^-
   c. 15.6 g NH_4^+, 0.864 mol NO_2^-

4.29 In the preparation of Br_2 from Br^- ions (Reaction 4.8), 2.40 mol $Cl_2(g)$ is added to 5.00 L of a solution 0.500 M in Br^-.
   a. What is the limiting reactant?
   b. What is the theoretical yield of Br_2, in moles?
   c. If 1.24 mol Br_2 is formed, what is the percent yield?

4.30 Pure manganese can be made by the reaction of aluminum with MnO_2:

$$3\ MnO_2(s) + 4\ Al(s) \rightarrow 3\ Mn(s) + 2\ Al_2O_3(s)$$

   a. If 5.00×10^2 g MnO_2 and 1.00×10^2 g Al are used, identify the limiting reactant.
   b. Assuming 100% yield, how much Mn is formed?
   c. How many more grams of the limiting reactant would be needed to convert all the excess reactant to products?

4.55 Consider the reaction by which I^- ions are oxidized to iodine (Reaction 4.7).
   a. How many moles of I^- react with 1.00 g of chlorine gas?
   b. If the concentration of I^- in a brine is 0.00292 M, what volume of this brine reacts with 1.00 g of chlorine?

4.56 Consider the reaction of a 1.00×10^3 g ore sample containing impure zinc blende, ZnS, with sufficient oxygen to form 85.3 L of sulfur dioxide (see Example 4.8). Assume complete conversion and that zinc blende is the only source of sulfur. Calculate the percentage of zinc blende in the ore. Density of SO_2 = 2.62 g/L.

4.57 Consider the reactions by which $CaSiO_3$ slag is formed in the blast furnace (Reactions 4.21 and 4.22). Suppose one starts with 1.00 g $CaCO_3$ and 1.00 g SiO_2:
   a. What is the limiting reactant?
   b. What is the theoretical yield of $CaSiO_3$, in grams?

4.58 Consider Reaction 4.3, used to prepare sulfur. What is the theoretical yield of sulfur if one starts with
   a. 1.60 mol H_2S, 0.715 mol O_2
   b. 1.00 g H_2S, 1.00 g O_2
   c. 50.0 g H_2S, 0.798 mol O_2

4.59 Consider the reaction between $Mg(OH)_2$ and H^+ ions (Reaction 4.15). If 2.4 L of a solution 6.0 M in H^+ is added to 2.5 mol of solid $Mg(OH)_2$.
   a. what is the limiting reactant?
   b. what is the theoretical yield of Mg^{2+}, in moles?
   c. assuming a 100% yield and a final volume of 2.5 L, what is the concentration of Mg^{2+} in the final solution?

4.60 Antimony metal can be obtained from Sb_4O_6 by reaction with carbon:

$$Sb_4O_6(s) + 6\ C(s) \rightarrow 4\ Sb(s) + 6\ CO(g)$$

   a. If 2.50×10^2 g C and 3.00×10^2 g Sb_4O_6 are used, identify the limiting reactant.
   b. Assuming 100% yield, what mass of Sb is formed?
   c. If the yield were 82.0%, what mass of Sb_4O_6 would be required to react with excess carbon to form 1.00 kg Sb?

*4.61 A certain copper ore contains 6.1% Cu_2S by mass. When the ore is concentrated by flotation, 20% of the Cu_2S is lost. Of the Cu_2S that remains, 15% is lost in the conversion to copper. How much ore is required to yield 1.00 kg of copper?

*4.62 A sample of 1.00 g $Mg(OH)_2$ is dissolved in 10.0 cm³ of 6.00 M HCl. Assuming a final volume of 10.5 cm³, what is the concentration of H^+ in the solution formed after reaction?

*4.63 Hydrogen gas can be made by a reaction similar to Reaction 4.4 except that the starting material is octane, C_8H_{18}, rather than methane. Write a balanced equation for the reaction, and determine the mass percent of H_2 in the gaseous H_2-CO_2 mixture formed.

*4.64 A sample of limestone, $CaCO_3$, is partially converted to calcium oxide, CaO, by heating (Reaction 4.21). When the residue is treated with concentrated hydrochloric acid, the following reactions occur:

$$CaCO_3(s) + 2\ H^+(aq) \rightarrow Ca^{2+}(aq) + CO_2(g) + H_2O$$

$$CaO(s) + 2\ H^+(aq) \rightarrow Ca^{2+}(aq) + H_2O$$

It is found that 2.50 L of 12.0 M HCl is required to react with 1.00 kg of the residue. What is the percentage of $CaCO_3$?

*4.65 When copper(I) sulfide is roasted in air, the atmosphere is contaminated by SO_2 (Reaction 4.27). This is eventually converted to H_2SO_4, which falls as acid rain. Suppose the sulfur dioxide formed from 1.00 g Cu_2S spreads out over 2.0×10^3 m³ of air, where it is converted to sulfuric acid. How many H_2SO_4 molecules are there in one cubic centimeter of this air?

Chapter 5
Thermochemistry

Chemical reactions such as those considered in Chapters 3 and 4 are accompanied by energy changes, which may take various forms. Consider, for example, the reactions

$$CH_4(g) + 2 O_2(g) \rightarrow CO_2(g) + 2 H_2O(l)$$

$$2 N_2H_4(l) + N_2O_4(l) \rightarrow 3 N_2(g) + 4 H_2O(l)$$

$$2 H_2O(l) \rightarrow 2 H_2(g) + O_2(g)$$

The first reaction takes place when natural gas, which is mostly methane, burns. It supplies the heat required to cook a steak on a gas range or heat water with a Bunsen burner. The reaction between hydrazine, N_2H_4, and dinitrogen tetroxide furnishes the mechanical energy to lift a rocket and its payload from the surface of the earth. In contrast, electrical energy must be supplied to decompose water to its elements, hydrogen and oxygen.

Heat is a form of energy

In this chapter, our main concern will be with one type of energy change: the heat flow associated with a chemical reaction or physical change. Studies of heat flow comprise the science of *thermochemistry*. Our discussion will concentrate upon

—the direction (or sign) of the heat flow (Section 5.1).
—the magnitude of the heat flow, expressed in kilojoules (Sections 5.2 and 5.3).
—the experimental determination of the sign and magnitude of the heat flow (Section 5.4).

In the last two sections of this chapter we will look at energy changes from a broader standpoint. We will examine the relation between heat flow and other forms of energy in Section 5.5. In Section 5.6 we will review the natural sources of energy that are available today and speculate about energy sources of the future.

5.1
Exothermic and
Endothermic
Reactions; Enthalpy
Changes

From the standpoint of heat flow, we can distinguish between two types of reactions:

1. **Exothermic** reactions, in which a reaction system evolves heat to the surroundings. Most of the reactions that you carry out in the laboratory are exothermic. A familiar example is the combustion of methane, referred to earlier:

$$CH_4(g) + 2 O_2(g) \rightarrow CO_2(g) + 2 H_2O(l) \tag{5.1}$$

This reaction gives off heat to the surroundings, which might be a beaker of water (Fig. 5.1A). The result is to increase the temperature of the water. Ordinarily,

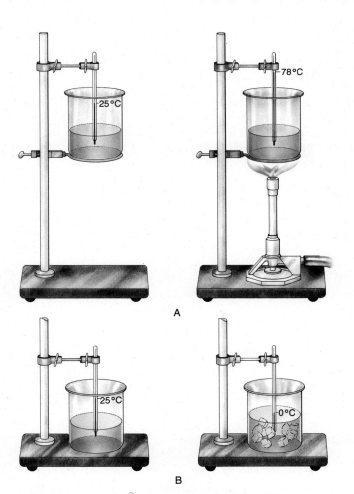

In an exothermic reaction, the products are often very hot when formed. When CH_4 burns, the CO_2 and H_2O produced are hot, and can be used to heat water or cook a steak

FIGURE 5.1 The heat given off by the exothermic reaction (A), the combustion of methane, raises the temperature of the water in the beaker. The melting of ice (B), an endothermic process, lowers the temperature of the water.

the effect of an exothermic reaction is to increase the temperature of the surroundings.

2. **Endothermic** reactions, in which a reaction system absorbs heat from the surroundings. A familiar example of an endothermic "reaction" is the melting of ice. We can represent this process by the equation

$$H_2O(s) \rightarrow H_2O(l) \tag{5.2}$$

This process absorbs heat from the surroundings, which may be a beaker of water (Fig. 5.1B). The result is to lower the temperature of the water. Ordinarily, *the effect of an endothermic reaction is to lower the temperature of the surroundings.*

As we have just seen, in exothermic and endothermic reactions there is an exchange of heat between the reaction system and the surroundings. The amount of heat that flows is related to a basic property of the reaction system. This property is a form of energy called *heat content* or, more properly, **enthalpy.** In an **exothermic** reaction, the **enthalpy** of the reaction system **decreases;** that is, the enthalpy (heat content) of the products is less than that of the reactants. This decrease in enthalpy is the source of the heat given off to the surroundings. Using the symbol H to represent enthalpy,

exothermic reaction: $\Delta H = [H_{products} - H_{reactants}] < 0$

This situation is shown in Figure 5.2A. Here the products have a lower enthalpy than the reactants. The enthalpy change, ΔH, is a negative quantity; the reaction is exothermic. Reaction 5.1, the combustion of methane, is represented by this type of enthalpy diagram. The products (1 mol CO_2 and 2 mol H_2O) have a lower enthalpy than the reactants (1 mol CH_4 and 2 mol O_2).

$$CH_4(g) + 2 O_2(g) \rightarrow CO_2(g) + 2 H_2O(l); \qquad \Delta H < 0 \tag{5.1}$$

In an **endothermic** reaction, the **enthalpy** of the reaction system **increases;** that is, the enthalpy (heat content) of the products is greater than that of the reactants. To bring about this increase in enthalpy, heat must be absorbed from the surroundings:

endothermic reaction: $\Delta H = [H_{products} - H_{reactants}] > 0$

Heat flow *into* surroundings increases their enthalpy: $\Delta H_{surr} > 0$

Heat flow *out* of reaction mixture decreases its enthalpy: $\Delta H_{reaction\ mixture} < 0$

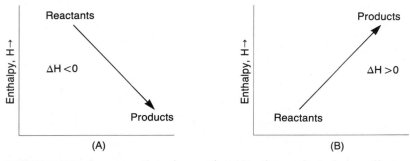

(A)   (B)

FIGURE 5.2 In an exothermic reaction (A), the products have a lower enthalpy than the reactants; thus, ΔH is negative, and heat is given off to the surroundings. In an endothermic reaction (B), the products have a higher enthalpy than the reactants; thus, ΔH is positive and heat is absorbed from the surroundings.

An enthalpy diagram for an endothermic reaction is shown in Figure 5.2B. The products have a higher enthalpy than the reactants; ΔH is a positive quantity. Reaction 5.2, the melting of ice, is represented by such a diagram. The product, liquid water, has a higher heat content than the starting material, ice:

$$H_2O(s) \rightarrow H_2O(l); \quad \Delta H > 0 \tag{5.2}$$

The development that we have just gone through is based upon the **Law of Conservation of Energy.** This law states that *energy is neither created nor destroyed in an ordinary chemical or physical change.* Here we say that the heat flow observed in the surroundings is exactly compensated for by the change in enthalpy of the reaction mixture. In an exothermic reaction such as Reaction 5.1, enthalpy is converted into heat, which flows into the surroundings. In an endothermic process such as Reaction 5.2, heat from the surroundings is converted into enthalpy of the reaction system.

For any change at T,P
$\Delta H_{surr} + \Delta H_{reaction\ mix} = 0$

> **EXAMPLE 5.1**   When 5.0 g of sodium hydroxide dissolves in 100 g of water, the temperature of the water rises from 25 to 38°C. The reaction is
>
> $$NaOH(s) \rightarrow Na^+(aq) + OH^-(aq)$$
>
> Is this reaction exothermic or endothermic? What is the sign of ΔH?
>
> **Solution**   Since the temperature of the water increased, heat must have been given off in the reaction. The reaction is exothermic; ΔH is negative.
>
> **EXERCISE**   When 5.0 g NH_4Cl dissolves in 100 g of water, the temperature of the water drops from 25 to 22°C. Write a chemical equation for the reaction involved and state the sign of ΔH. Answer: $NH_4Cl(s) \rightarrow NH_4^+(aq) + Cl^-(aq); \Delta H > 0$.

5.2 Thermochemical Equations

The development we just went through in Section 5.1 can be made quantitative. We can determine not only the sign of ΔH for a reaction but also its magnitude. To do this, we use the fact that *ΔH is equal in magnitude to the heat flow for a reaction carried out directly at constant pressure.* In other words, to determine the enthalpy change, ΔH, we need only measure the amount of heat evolved or absorbed when a reaction takes place in an open container, i.e., at constant atmospheric pressure.

$\Delta H = q_{reaction\ mixture}$

To illustrate how such a measurement might be made, consider again the combustion of methane (Reaction 5.1). Let us assume that we can arrange the apparatus shown in Figure 5.1A so that all of the heat given off by the reaction is absorbed by the water. Suppose further that burning 1.00 mol CH_4 raises the temperature of 4.00×10^3 g of water from 25.0 to 78.2°C. To calculate the amount of heat absorbed by the water, we use the relation introduced in Chapter 1:

$$q = (S.H.) \times m \times \Delta t \tag{5.3}$$

where

q = heat absorbed by the water, in joules (J)

S.H. = specific heat of water = 4.18 J/g·°C

m = mass of water = 4.00×10^3 g

$\Delta t = t_{final} - t_{initial} = 78.2°C - 25.0°C = 53.2°C$

Substituting, we obtain

$$q = 4.18\ \frac{J}{g·°C} \times 4.00 \times 10^3\ g \times 53.2°C = 8.90 \times 10^5\ J = 890\ kJ$$

$q_{water} = 890$ kJ

$\therefore q_{reaction\ mix} =$
-890 kJ $= \Delta H$

We conclude that the combustion of one mole of methane supplied 890 kJ of heat to the water. This means that 890 kJ is evolved in the reaction. That is, ΔH per mole of methane burned is -890 kJ.

The results of this calculation can be summarized quite simply by means of a thermochemical equation. This type of equation specifies the heat flow for a reaction, with the value of ΔH, in kilojoules, given to the right of the equation, following the products. The thermochemical equation for the combustion of methane is

$$CH_4(g) + 2\ O_2(g) \rightarrow CO_2(g) + 2\ H_2O(l); \qquad \Delta H = -890\ kJ \qquad \textbf{(5.1)}$$

This equation tells us that ΔH is -890 kJ (that is, 890 kJ of heat is evolved) when 1 mol CH_4 reacts with 2 mol O_2 to form 1 mol CO_2 and 2 mol H_2O. In general, a negative value of ΔH appearing in a thermochemical equation indicates an exothermic reaction.

We find that when one mole (18.02 g) of ice melts at 0°C, almost exactly six kilojoules of heat is absorbed from the surroundings. It follows that the thermochemical equation for this process is

In an equation like this, ΔH refers to the reaction mix, not the surroundings

$$H_2O(s) \rightarrow H_2O(l); \qquad \Delta H = +6.00\ kJ\ at\ 0°C \qquad \textbf{(5.2)}$$

This means that the enthalpy (heat content) of liquid water is 6.00 kJ/mol higher than that of ice; therefore, 6.00 kJ of heat must be absorbed to melt one mole of ice. In general, if ΔH in a thermochemical equation is positive, the reaction is endothermic*.

As we have just seen in connection with Reaction 5.2, the enthalpy of a substance depends upon its physical state. Hence, we must specify that state in writing a thermochemical equation. For a pure substance, we do this by writing the symbols (s), (l), or (g) after the formula, as in Equations 5.1 and 5.2. For a species in aqueous solution, we use the symbol (aq). The thermochemical equation for the process that occurs when ammonium chloride dissolves in water is

---

*In some texts, thermochemical equations are written by including the value of ΔH as a "reactant" or "product." Thus, Equations 5.1 and 5.2 might be written

$CH_4(g) + 2\ O_2(g) \rightarrow CO_2(g) + 2\ H_2O(l) + 890\ kJ$
$H_2O(s) + 6.00\ kJ \rightarrow H_2O(l)$

$$NH_4Cl(s) \rightarrow NH_4^+(aq) + Cl^-(aq); \quad \Delta H = +15.2 \text{ kJ} \qquad (5.4)$$

This equation tells us that when one mole of solid NH_4Cl dissolves, forming NH_4^+ and Cl^- ions in solution, 15.2 kJ of heat is absorbed.

The value of ΔH depends, at least slightly, upon the temperature at which a reaction is carried out. Hence, that temperature should be specified. Unless indicated otherwise (as was the case in Equation 5.2), we will take the temperature to be 25°C. Thus, we interpret Equation 5.4 to mean that ΔH is +15.2 kJ when one mole of NH_4Cl, originally at 25°C, forms ions in solution at a final temperature of 25°C.

Laws of Thermochemistry

To make effective use of thermochemical equations, we apply three basic laws of thermochemistry. These are expressed most simply in terms of ΔH:

1. *The magnitude of ΔH is directly proportional to the amount of reactant or product.* This rule tells us that the amount of heat evolved in Reaction 5.1 is directly proportional to the amount of methane burned. If 890 kJ of heat is produced when 1 mol (16.04 g) CH_4 burns, it follows that

The enthalpy H of a sample is proportional to its mass

—the combustion of two moles of CH_4 produces twice as much heat, 1780 kJ.
—the combustion of one gram of CH_4 produces about $\frac{1}{16}$ as much heat, 55.5 kJ.

This rule allows us to relate ΔH for a thermochemical equation to the amount of product or reactant. The approach followed is a simple extension of that used in Chapter 3 with ordinary chemical equations. Let us refer again to Reaction 5.1:

$$CH_4(g) + 2 O_2(g) \rightarrow CO_2(g) + 2 H_2O(l); \quad \Delta H = -890 \text{ kJ}$$

As pointed out earlier, this equation says that ΔH is −890 kJ when 1 mol CH_4 reacts with 2 mol O_2 to form 1 mol CO_2 and 2 mol of liquid H_2O. From a slightly different point of view, we can say that for this reaction,

$-890 \text{ kJ} \simeq 1 \text{ mol } CH_4$

or

$-890 \text{ kJ} \simeq 2 \text{ mol } O_2$

or

$-890 \text{ kJ} \simeq 1 \text{ mol } CO_2$

or

$-890 \text{ kJ} \simeq 2 \text{ mol } H_2O$

These "equivalences" can be used in the ordinary way to relate ΔH to the amount of reactant or product. Suppose, for example, we want to find ΔH when two moles of CH_4 burn. In that case, we use the conversion factor, −890 kJ/1 mol CH_4:

$$\Delta H = 2.00 \text{ mol CH}_4 \times \frac{-890 \text{ kJ}}{1 \text{ mol CH}_4} = -1780 \text{ kJ}$$

In another case, we might want to know how many moles of CO_2 are formed when 1.00×10^2 kJ of heat is evolved (that is, $\Delta H = -1.00 \times 10^2$ kJ). Here, the required conversion factor is 1 mol $CO_2/-890$ kJ:

$$\text{moles } CO_2 = -1.00 \times 10^2 \text{ kJ} \times \frac{1 \text{ mol } CO_2}{-890 \text{ kJ}} = 0.112 \text{ mol } CO_2$$

EXAMPLE 5.2   Ammonium nitrate, NH_4NO_3, is commonly used as an explosive. It decomposes by the following reaction:

$$NH_4NO_3(s) \rightarrow N_2O(g) + 2 \text{ } H_2O(g); \qquad \Delta H = -37.0 \text{ kJ}$$

Calculate ΔH when
a. 2.50 g of NH_4NO_3 decomposes
b. 1.00 g of H_2O is formed by this reaction

Solution
a. From the thermochemical equation, we see that

$$1 \text{ mol } NH_4NO_3 \backsimeq -37.0 \text{ kJ}$$

One mole of NH_4NO_3 weighs 80.04 g. To find ΔH, we first convert 2.50 g NH_4NO_3 to moles and then convert moles to kilojoules:

$$\Delta H = 2.50 \text{ g } NH_4NO_3 \times \frac{1 \text{ mol } NH_4NO_3}{80.04 \text{ g } NH_4NO_3} \times \frac{-37.0 \text{ kJ}}{1 \text{ mol } NH_4NO_3}$$
$$= -1.16 \text{ kJ}$$

> Someone has said that 90% of all problems in general chemistry are really conversion factor problems

We conclude that 1.16 kJ of heat is evolved when 2.50 g NH_4NO_3 decomposes.
b. Here, the appropriate relation is

$$2 \text{ mol } H_2O \backsimeq -37.0 \text{ kJ}$$

Again, a two-step conversion is required. Taking the molar mass of H_2O to be 18.02 g/mol, we have

$$\Delta H = 1.00 \text{ g } H_2O \times \frac{1 \text{ mol } H_2O}{18.02 \text{ g } H_2O} \times \frac{-37.0 \text{ kJ}}{2 \text{ mol } H_2O} = -1.03 \text{ kJ}$$

EXERCISE   What is ΔH when 0.200 mol N_2O is formed in this reaction? Answer: -7.40 kJ.

This relation between ΔH and amounts of substances is equally useful in dealing with chemical reactions or changes in physical state (Example 5.3).

EXAMPLE 5.3   Equation 5.2 tells us that when 1 mol of ice melts at 0°C, ΔH is $+6.00$ kJ. We can write a similar thermochemical equation for the vaporization of water at 100°C:

$$H_2O(l) \rightarrow H_2O(g); \qquad \Delta H = +40.7 \text{ kJ at } 100°C$$

a. Calculate ΔH when one gram of ice melts at 0°C; when 1.00 g of water boils at 100°C.

b. What mass of water, at 100°C, can be vaporized by absorbing 0.800 kJ of heat?

Solution

a. The molar mass of H_2O is 18.02 g/mol. Hence, for the melting of ice, taking ΔH to be +6.00 kJ/mol,

$$\Delta H = 1.00 \text{ g} \times \frac{1 \text{ mol}}{18.02 \text{ g}} \times \frac{6.00 \text{ kJ}}{1 \text{ mol}} = 0.333 \text{ kJ}$$

For the boiling of water, where ΔH is +40.7 kJ/mol,

$$\Delta H = 1.00 \text{ g} \times \frac{1 \text{ mol}}{18.02 \text{ g}} \times \frac{40.7 \text{ kJ}}{1 \text{ mol}} = 2.26 \text{ kJ}$$

b. Here, we convert 0.800 kJ to moles, using the conversion factor 1 mol/40.7 kJ. Then we convert moles to grams:

$$\text{mass } H_2O = 0.800 \text{ kJ} \times \frac{1 \text{ mol}}{40.7 \text{ kJ}} \times \frac{18.02 \text{ g}}{1 \text{ mol}} = 0.354 \text{ g}$$

EXERCISE Which requires the absorption of the greater amount of heat: melting 5.0 mol of ice at 0°C or boiling 0.80 mol of liquid water at 100°C? Answer: 0.80 mol of liquid water at 100°C.

The heat absorbed when a solid melts ($s \rightarrow l$) is referred to as the **heat of fusion;** that absorbed when a liquid vaporizes ($l \rightarrow g$) is called the **heat of vaporization.** Heats of fusion (ΔH_{fus}) and vaporization (ΔH_{vap}) are most often expressed in kilojoules per mole (kJ/mol). Values for several different substances are given in Table 5.1.

Table 5.1
ΔH(kJ/mol) for Phase Changes

SUBSTANCE		mp(°C)	ΔH_{fus}*	bp(°C)	ΔH_{vap}*
Benzene	C_6H_6	5	9.84	80	30.8
Bromine	Br_2	−7	10.8	59	30.0
Mercury	Hg	−39	2.33	357	59.4
Naphthalene	$C_{10}H_8$	80	19.3	218	40.5
Water	H_2O	0	6.00	100	40.7

For any substance,
$\Delta H_{fus} > 0$
$\Delta H_{vap} > 0$

*Values of ΔH_{fus} are given at the melting point, values of ΔH_{vap} at the boiling point. The heat of vaporization of water decreases from 44.9 kJ/mol at 0°C to 43.9 kJ/mol at 25°C to 40.7 kJ/mol at 100°C.

2. *ΔH for a reaction is equal in magnitude but opposite in sign to ΔH for the reverse reaction.* Another way to state this rule is to say that the amount of heat evolved in a reaction is exactly equal to the amount of heat absorbed in the reverse reaction. Using this rule, we conclude that since ΔH for the vaporization of one mole of water at 100°C is +40.7 kJ,

$$H_2O(l) \rightarrow H_2O(g); \qquad \Delta H = +40.7 \text{ kJ at } 100°C \qquad \textbf{(5.5)}$$

it follows that 40.7 kJ of heat is evolved when one mole of steam condenses (Fig. 5.3):

$$H_2O(g) \rightarrow H_2O(l); \qquad \Delta H = -40.7 \text{ kJ at } 100°C \qquad \textbf{(5.6)}$$

If you go into a sauna at 100°C, you want it to be a dry sauna

This is indeed the case. The heat evolved in the condensation of steam is the principal source of heat given off in a steam heating system. It is also the major cause of burns one gets when exposed to live steam.

EXAMPLE 5.4 For the reaction $H_2(g) + \frac{1}{2}O_2(g) \rightarrow H_2O(l)$, $\Delta H = -285.8$ kJ. Calculate ΔH for the reaction $2 H_2O(l) \rightarrow 2 H_2(g) + O_2(g)$

Solution Applying Rules 1 and 2 in succession,
1. $2 H_2(g) + O_2(g) \rightarrow 2 H_2O(l);$ $\qquad \Delta H = 2(-285.8 \text{ kJ}) = -571.6 \text{ kJ}$
2. $2 H_2O(l) \rightarrow 2 H_2(g) + O_2(g);$ $\qquad \Delta H = -(-571.6 \text{ kJ}) = +571.6 \text{ kJ}$
We conclude that 571.6 kJ must be absorbed to decompose 2 mol of liquid water to its elements.

EXERCISE Given that $2 Al_2O_3(s) \rightarrow 4 Al(s) + 3 O_2(g)$, $\Delta H = +3339.6$ kJ, obtain ΔH for the formation of one mole of Al_2O_3 from the elements. Answer: -1669.8 kJ.

3. *The value of ΔH for a reaction is the same whether it occurs directly or in a series of steps.* This means that if a thermochemical equation can be expressed as the sum of two or more other equations,

$$\text{Equation (3)} = \text{Equation (1)} + \text{Equation (2)} + \cdots$$

All of the laws of thermochemistry follow from the fact that the enthalpy H of a substance is one of its properties

then ΔH for the overall equation is the sum of the ΔH's for the individual equations:

$$\Delta H_3 = \Delta H_1 + \Delta H_2 + \cdots$$

This relationship is referred to as **Hess's Law.**

To illustrate this situation, consider the reaction between tin and chlorine:

$$Sn(s) + 2 Cl_2(g) \rightarrow SnCl_4(l) \qquad \textbf{(5.7)}$$

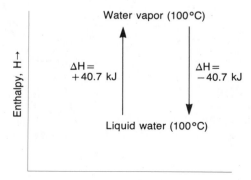

FIGURE 5.3 Water vapor (steam) at 100°C has an enthalpy 40.7 kJ/mol greater than that of liquid water at the same temperature. When 1 mol of water vaporizes at 100°C, 40.7 kJ of heat is absorbed and ΔH is $+40.7$ kJ. When 1 mol of steam condenses, 40.7 kJ of heat is evolved and ΔH is -40.7 kJ.

We can imagine that this reaction takes place in two steps. In the first step, 1 mol Sn reacts with 1 mol Cl_2 to form 1 mol $SnCl_2$. Then, in a second step, the $SnCl_2$ reacts with another mole of Cl_2 to form $SnCl_4$:

Step 1: $\quad Sn(s) + Cl_2(g) \rightarrow SnCl_2(s); \quad \Delta H_1 = -349.8$ kJ

Step 2: $\quad \underline{SnCl_2(s) + Cl_2(g) \rightarrow SnCl_4(l);} \quad \Delta H_2 = -195.4$ kJ

$\qquad\qquad Sn(s) + 2\ Cl_2(g) \rightarrow SnCl_4(l)$

$$\Delta H_1 = H_{SnCl_2} - H_{Sn} - H_{Cl_2}$$
$$\Delta H_2 = H_{SnCl_4} - H_{SnCl_2} - H_{Cl_2}$$
$$\Delta H = H_{SnCl_4} - H_{Sn} - 2H_{Cl_2}$$
$$= \Delta H_1 + \Delta H_2$$

Since these two equations add to give Equation 5.7,

$$\Delta H \text{ for Reaction } 5.7 = \Delta H_1 + \Delta H_2$$
$$= -349.8 \text{ kJ} - 195.4 \text{ kJ} = -545.2 \text{ kJ}$$

This situation is shown graphically in Figure 5.4.

We will consider several other applications of Hess's Law later in this chapter and in succeeding chapters. In some cases, such as the one just cited, we will use Hess's Law to obtain ΔH for a reaction, knowing the individual values of ΔH for each step. In other cases, we will use the law to calculate ΔH for a particular step, knowing the values of ΔH for every other step and for the overall reaction.

5.3
Heats of Formation

We have now written several thermochemical equations. In each case, we have cited the corresponding value of ΔH. Literally thousands of such equations would be needed to list the ΔH values for all the reactions that have been studied. Clearly, we need some more concise way of recording data of this sort. These data should be in a form that can easily be used to calculate ΔH for a reaction that interests us. It turns out that there is a simple way to do this, using quantities known as heats of formation. Let us first consider how these quantities are defined. Then we will show how they can be used to calculate ΔH for reactions.

The molar heat of formation of a compound, ΔH_f, is equal to the enthalpy change, ΔH, when one mole of the compound is formed from the elements in their stable forms at 25°C and 1 atm. Thus, from the equations

This is the definition of ΔH_f.

$$Ag(s) + \tfrac{1}{2} Cl_2(g) \rightarrow AgCl(s); \quad \Delta H = -127.0 \text{ kJ} \qquad \textbf{(5.8)}$$

$$\tfrac{1}{2} N_2(g) + O_2(g) \rightarrow NO_2(g); \quad \Delta H = +33.9 \text{ kJ} \qquad \textbf{(5.9)}$$

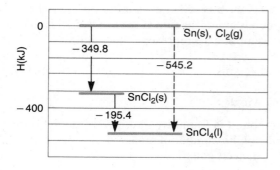

FIGURE 5.4 ΔH for the reaction $Sn(s) + 2\ Cl_2(g) \rightarrow SnCl_4(l)$ is the same whether it is carried out in one step (*dotted arrow*) or two steps (*solid arrows*). Each division on the vertical axis represents 80 kJ.

we conclude that

$$\Delta H_f \, AgCl(s) \; = \; -127.0 \text{ kJ/mol}$$

$$\Delta H_f \, NO_2(g) \; = \; +33.9 \text{ kJ/mol}$$

Heats of formation are listed for a variety of compounds in Table 5.2. Notice that, with a few exceptions, heats of formation are usually negative quantities. This means that the formation of a compound from the elements is ordinarily exothermic. Conversely, when a compound decomposes to the elements, heat usually must be absorbed.

Table 5.2
Heats of Formation (kJ/mol) at 25°C and 1 atm

$AgBr(s)$	-99.5	$C_2H_2(g)$	$+226.7$	$H_2O(l)$	-285.8	$NH_4Cl(s)$	-315.4
$AgCl(s)$	-127.0	$C_2H_4(g)$	$+52.3$	$H_2O_2(l)$	-187.6	$NH_4NO_3(s)$	-365.1
$AgI(s)$	-62.4	$C_2H_6(g)$	-84.7	$H_2S(g)$	-20.1	$NO(g)$	$+90.4$
$Ag_2O(s)$	-30.6	$C_3H_8(g)$	-103.8	$H_2SO_4(l)$	-811.3	$NO_2(g)$	$+33.9$
$Ag_2S(s)$	-31.8	$n\text{-}C_4H_{10}(g)$	-124.7	$HgO(s)$	-90.7	$NiO(s)$	-244.3
$Al_2O_3(s)$	-1669.8	$n\text{-}C_5H_{12}(l)$	-173.1	$HgS(s)$	-58.2	$PbBr_2(s)$	-277.0
$BaCl_2(s)$	-860.1	$C_2H_5OH(l)$	-277.6	$KBr(s)$	-392.2	$PbCl_2(s)$	-359.2
$BaCO_3(s)$	-1218.8	$CoO(s)$	-239.3	$KCl(s)$	-435.9	$PbO(s)$	-217.9
$BaO(s)$	-558.1	$Cr_2O_3(s)$	-1128.4	$KClO_3(s)$	-391.4	$PbO_2(s)$	-276.6
$BaSO_4(s)$	-1465.2	$CuO(s)$	-155.2	$KF(s)$	-562.6	$Pb_3O_4(s)$	-734.7
$CaCl_2(s)$	-795.0	$Cu_2O(s)$	-166.7	$MgCl_2(s)$	-641.8	$PCl_3(g)$	-306.4
$CaCO_3(s)$	-1207.0	$CuS(s)$	-48.5	$MgCO_3(s)$	-1113	$PCl_5(g)$	-398.9
$CaO(s)$	-635.5	$CuSO_4(s)$	-769.9	$MgO(s)$	-601.8	$SiO_2(s)$	-859.4
$Ca(OH)_2(s)$	-986.6	$Fe_2O_3(s)$	-822.2	$Mg(OH)_2(s)$	-924.7	$SnCl_2(s)$	-349.8
$CaSO_4(s)$	-1432.7	$Fe_3O_4(s)$	-1120.9	$MgSO_4(s)$	-1278.2	$SnCl_4(l)$	-545.2
$CCl_4(l)$	-139.5	$HBr(g)$	-36.2	$MnO(s)$	-384.9	$SnO(s)$	-286.2
$CH_4(g)$	-74.8	$HCl(g)$	-92.3	$MnO_2(s)$	-519.7	$SnO_2(s)$	-580.7
$CHCl_3(l)$	-131.8	$HF(g)$	-268.6	$NaCl(s)$	-411.0	$SO_2(g)$	-296.1
$CH_3OH(l)$	-238.6	$HI(g)$	$+25.9$	$NaF(s)$	-569.0	$SO_3(g)$	-395.2
$CO(g)$	-110.5	$HNO_3(l)$	-173.2	$NaOH(s)$	-426.7	$ZnO(s)$	-348.0
$CO_2(g)$	-393.5	$H_2O(g)$	-241.8	$NH_3(g)$	-46.2	$ZnS(s)$	-202.9

Heats of formation of compounds can be used to calculate ΔH for a reaction. To do this we apply the general rule: **ΔH for a reaction is equal to the sum of the heats of formation of the product compounds minus the sum of the heats of formation of the reactant compounds.** Using the symbol Σ to represent "the sum of,"

$$\Delta H \; = \; \Sigma \, \Delta H_f \text{ products} \; - \; \Sigma \, \Delta H_f \text{ reactants} \qquad \qquad \textbf{(5.10)}$$

In applying this relation, note the following:

1. The contribution for each compound is found by multiplying the heat of formation in kilojoules per mole by the number of moles of the compound, given by its coefficient in the balanced equation.

2. Any element *in its stable form* is omitted in taking the sums required in Equation 5.10. The heat of formation of an element is zero because of the way in which ΔH_f is defined.

EXAMPLE 5.5   Using Table 5.2, calculate ΔH for the combustion of one mole of propane, C_3H_8, according to the equation

$$C_3H_8(g) + 5\ O_2(g) \rightarrow 3\ CO_2(g) + 4\ H_2O(l)$$

Solution   Expressing ΔH in terms of heats of formation of products and reactants,

$$\Delta H = 3\ \Delta H_f\ CO_2(g) + 4\ \Delta H_f\ H_2O(l) - [\Delta H_f\ C_3H_8(g) + 5\ \Delta H_f\ O_2(g)]$$

It is useful for chemists to be able to make calculations like this for a reaction

Taking the heat of formation of $O_2(g)$ to be zero and substituting values for the other substances from Table 5.2,

$$\Delta H = 3\ mol\left(-393.5\ \frac{kJ}{mol}\right) + 4\ mol\left(-285.8\ \frac{kJ}{mol}\right)$$
$$- 1\ mol\left(-103.8\ \frac{kJ}{mol}\right) = -2219.9\ kJ$$

EXERCISE   The reaction between aluminum and iron(III) oxide, Fe_2O_3, is strongly exothermic. Referred to as the thermite reaction, it was formerly used to weld steel rails. Using Table 5.2, calculate ΔH for the reaction: $2\ Al(s) + Fe_2O_3(s) \rightarrow 2\ Fe(s) + Al_2O_3(s)$. Answer: $-847.6\ kJ$.

Equation 5.10 can also be used to calculate the heat of formation of one substance if we know ΔH for the reaction and the heats of formation of all other substances involved in the reaction. Example 5.6 illustrates the calculations involved.

EXAMPLE 5.6   The thermochemical equation for the combustion of benzene, C_6H_6, is

$$C_6H_6(l) + 15/2\ O_2(g) \rightarrow 6\ CO_2(g) + 3\ H_2O(l);\quad \Delta H = -3267.4\ kJ$$

Using Table 5.2 to obtain heats of formation for CO_2 and H_2O, calculate the heat of formation of benzene.

Solution   Taking the heat of formation of $O_2(g)$ to be zero, the relation for ΔH is

$$\Delta H = 6\ \Delta H_f\ CO_2(g) + 3\ \Delta H_f\ H_2O(l) - \Delta H_f\ C_6H_6(l)$$

Experimentally, we find ΔH_f for $C_6H_6(l)$ from its heat of combustion, as in this example

Substituting values for ΔH, $\Delta H_f\ CO_2$, and $\Delta H_f\ H_2O(l)$, we have

$$-3267.4\ kJ = 6\ mol\left(-393.5\ \frac{kJ}{mol}\right) + 3\ mol\left(-285.8\ \frac{kJ}{mol}\right)$$
$$- 1\ mol\ [\Delta H_f\ C_6H_6(l)]$$

Solving,

$$\Delta H_f\ C_6H_6(l) = +49.0\ kJ/mol$$

EXERCISE   Given that ΔH for the reaction $2\ NO_2(g) \rightarrow N_2O_4(g)$ is $-58.2\ kJ$, calculate the heat of formation of N_2O_4. Answer: $+9.6\ kJ/mol$.

The relation between ΔH and heats of formation given by Equation 5.10 is perhaps the most useful in all of thermochemistry. To show its validity, consider the reaction,

$$CO(g) + \tfrac{1}{2} O_2(g) \rightarrow CO_2(g)$$

This equation is the sum of two other chemical equations:

$$C(s) + O_2(g) \rightarrow CO_2(g) \tag{5.11a}$$
$$\underline{ CO(g) \rightarrow C(s) + \tfrac{1}{2} O_2(g)} \tag{5.11b}$$
$$CO(g) + \tfrac{1}{2} O_2(g) \rightarrow CO_2(g) \tag{5.11}$$

Applying Hess's Law,

$$\Delta H \text{ for Reaction } 5.11 = \Delta H \text{ for Reaction } 5.11a + \Delta H \text{ for Reaction } 5.11b$$

By definition, ΔH for Reaction 5.11a is the heat of formation of carbon dioxide, $\Delta H_f \, CO_2$. Since Reaction 5.11b is the reverse of that for the formation of CO from the elements, its ΔH is $- \Delta H_f \, CO$. Hence,

$$\Delta H \text{ for Reaction } 5.11 = \Delta H_f \, CO_2 - \Delta H_f \, CO$$

This is the relation we would have obtained directly by applying Equation 5.10 to this reaction.

The analysis we have just gone through can be applied to any reaction. Equation 5.10 is a special case of Hess's Law. It is simpler to apply the equation directly, as we did in Example 5.5, rather than go through a Hess's Law calculation.

Heats of Formation of Ions in Solution

It is possible to set up a table, very much like Table 5.2, for heats of formation of ions in water solution. There is, however, one problem. We cannot measure the heat of formation of an individual ion. In any reaction involving ions, at least two of them are present, as required by the principle of electrical neutrality. Consider, for example, the reaction that occurs when HCl is added to water:

$$HCl(g) \rightarrow H^+(aq) + Cl^-(aq)$$

There is no way to make only one ion from a neutral species

Here, two different ions are formed, H^+ and Cl^- ions. The same situation applies in all other reactions. We cannot carry out a reaction to form only the H^+ ion, with no other ions involved.

To get around this dilemma, the heat of formation of the H^+ ion is arbitrarily taken to be zero:

$$\Delta H_f \, H^+(aq) = 0 \tag{5.12}$$

Having done this, we can establish heats of formation for other ions. Take, for instance, the Cl^- ion. We can measure ΔH for the reaction that occurs when HCl is added to water; it is -75.1 kJ/mol; that is,

$$HCl(g) \rightarrow H^+(aq) + Cl^-(aq); \quad \Delta H = -75.1 \text{ kJ} \qquad (5.13)$$

Applying Equation 5.10,

$$-75.1 \text{ kJ} = \Delta H_f \, H^+(aq) + \Delta H_f \, Cl^-(aq) - \Delta H_f \, HCl(g)$$

Taking the heat of formation of $H^+(aq)$ to be zero and that of $HCl(g)$ to be -92.3 kJ (Table 5.2), we have

$$-75.1 \text{ kJ} = 0 + \Delta H_f \, Cl^-(aq) + 92.3 \text{ kJ}$$

or

$$\Delta H_f \, Cl^-(aq) = -75.1 \text{ kJ} - 92.3 \text{ kJ} = -167.4 \text{ kJ}$$

Heats of formation for other ions in water solution can be established in a similar way. Table 5.3 lists values for several common ions. This table can be used, much like Table 5.2, to calculate ΔH for reactions in solution. To do that, we use Equation 5.10

> Once we have ΔH_f for Cl^- ion, we can use it to get ΔH_f for many cations

$$\Delta H = \Sigma \, \Delta H_f \text{ products} - \Sigma \, \Delta H_f \text{ reactants}$$

realizing that ΔH_f for $H^+(aq)$ is zero as is ΔH_f for an element in its stable state.

Table 5.3
Heats of Formation (kJ/mol) at 25°C, 1 M

CATIONS				ANIONS			
$Ag^+(aq)$	$+105.9$	$K^+(aq)$	-251.2	$Br^-(aq)$	-120.9	$H_2PO_4^-(aq)$	-1302.5
$Al^{3+}(aq)$	-524.7	$Li^+(aq)$	-278.5	$Cl^-(aq)$	-167.4	$HPO_4^{2-}(aq)$	-1298.7
$Ba^{2+}(aq)$	-538.4	$Mg^{2+}(aq)$	-462.0	$ClO_3^-(aq)$	-98.3	$I^-(aq)$	-55.9
$Ca^{2+}(aq)$	-543.0	$Mn^{2+}(aq)$	-218.8	$ClO_4^-(aq)$	-131.4	$MnO_4^-(aq)$	-518.4
$Cd^{2+}(aq)$	-72.4	$Na^+(aq)$	-239.7	$CO_3^{2-}(aq)$	-676.3	$NO_3^-(aq)$	-206.6
$Cu^{2+}(aq)$	$+64.4$	$NH_4^+(aq)$	-132.8	$CrO_4^{2-}(aq)$	-863.2	$OH^-(aq)$	-229.9
$Fe^{2+}(aq)$	-87.9	$Ni^{2+}(aq)$	-64.0	$F^-(aq)$	-329.1	$PO_4^{3-}(aq)$	-1284.1
$Fe^{3+}(aq)$	-47.7	$Pb^{2+}(aq)$	$+1.6$	$HCO_3^-(aq)$	-691.1	$S^{2-}(aq)$	$+41.8$
$H^+(aq)$	0.0	$Sn^{2+}(aq)$	-10.0			$SO_4^{2-}(aq)$	-907.5
		$Zn^{2+}(aq)$	-152.4				

EXAMPLE 5.7 When hydrochloric acid is added to a solution of sodium carbonate, carbon dioxide gas is formed (Fig. 5.5, p. 134). The equation for the reaction is

$$2 \, H^+(aq) + CO_3^{2-}(aq) \rightarrow CO_2(g) + H_2O(l)$$

Using heats of formation from Tables 5.2 and 5.3, calculate ΔH for this reaction.

Solution Applying the general relation given above,

$$\Delta H = \Delta H_f \, CO_2(g) + \Delta H_f \, H_2O(l) - 2 \, \Delta H_f \, H^+(aq) - \Delta H_f \, CO_3^{2-}(aq)$$

Taking $\Delta H_f\ H^+(aq)$ to be zero and obtaining the other heats of formation from Tables 5.2 and 5.3,

$$\Delta H = 1\ mol\left(-393.5\ \frac{kJ}{mol}\right) + 1\ mol\left(-285.8\ \frac{kJ}{mol}\right)$$
$$- 1\ mol\left(-676.3\ \frac{kJ}{mol}\right) = -3.0\ kJ$$

EXERCISE Determine ΔH for the reaction: $Zn(s) + 2\ H^+(aq) \rightarrow Zn^{2+}(aq) + H_2(g)$. Answer: -152.4 kJ.

We have now carried out several calculations involving heats of formation (Examples 5.5–5.7). Values of ΔH_f listed in Tables 5.2 and 5.3 can be used to

1. Calculate ΔH for a reaction, knowing the heats of formation of all reactants and products. Here you apply Equation 5.10, summing the heats of formation of the products and subtracting the corresponding sum for the reactants. Note that:
   a. Since heats of formation are given in kilojoules per mole, Equation 5.10 can be used directly to calculate ΔH for a thermochemical equation, where the coefficients represent moles of products and reactants.
   b. Having found ΔH for a thermochemical equation as in (a), you can use any of the laws of thermochemistry discussed in Section 5.2 to calculate other ΔH's (e.g., ΔH per gram of reactant or product).
2. Calculate the heat of formation of a species, knowing ΔH for a thermochemical equation and heats of formation for all the other species involved in the reaction. The approach here is the same as in 1(a); the only difference is that you solve for a specified ΔH_f rather than ΔH.

Since ΔH for the reaction equals -3.0 kJ. the temperature of the reaction mixture would go up as the reaction proceeded

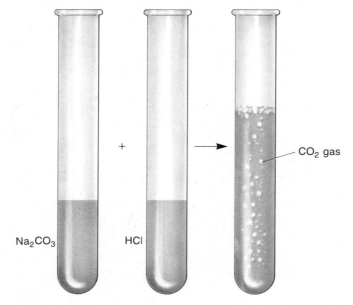

FIGURE 5.5 When water solutions of hydrochloric acid (H^+, Cl^- ions) and sodium carbonate (Na^+, CO_3^{2-} ions) are mixed, a reaction occurs. The H^+ ions of the HCl solution react with the CO_3^{2-} ions of the Na_2CO_3 solution: $2\ H^+(aq) + CO_3^{2-}(aq) \rightarrow CO_2(g) + H_2O$. The reaction is slightly exothermic, as shown in Example 5.7.

5.4
Measurement of Heat Flow; Calorimetry

In Section 5.2 we suggested how ΔH for the combustion of methane might be measured using the apparatus shown in Figure 5.1A. Actually, such a measurement would be rather inaccurate for several reasons. For one thing, some of the heat given off by the reaction would most likely be absorbed by the air of the room rather than the water in the beaker.

To make accurate thermochemical measurements, we carry out reactions within a device called a *calorimeter*. This apparatus contains water and/or other materials with a known capacity for absorbing heat. The outside walls of the calorimeter are insulated to minimize exchange of heat with the surrounding air. All of the heat evolved in the reaction is absorbed within the calorimeter.

Coffee-Cup Calorimeter

Figure 5.6 shows a simple calorimeter often used in the general chemistry laboratory. It consists of a polystyrene foam cup partially filled with water. The cup has a tightly fitting cover through which an accurate thermometer is inserted. Since polystyrene foam is a good insulator, there is very little heat flow through the walls of the cup. Essentially all the heat evolved by a reaction taking place within the calorimeter is absorbed by the water. We ordinarily neglect the small amount of heat absorbed by the polystyrene foam and the glass thermometer.

$$q_{\text{reaction mix}} = -q_{\text{water}}$$

To use this apparatus, we start by adding a weighed amount of water to the cup. We then put on the cover, making sure that the thermometer bulb is below the surface of the water. The initial temperature is recorded. The reaction under study is then carried out within the calorimeter. After the reaction is over, the final temperature is measured. Knowing the two temperatures and the mass of the water, we can calculate ΔH for the reaction. To do this we proceed as follows:

1. *Calculate the amount of heat, q, flowing into the water.* The general relation here is that introduced in Chapter 1:

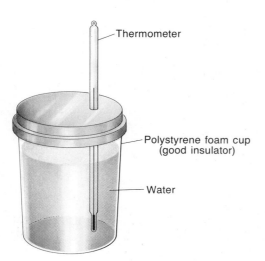

— Thermometer

— Polystyrene foam cup (good insulator)

— Water

FIGURE 5.6 Coffee-cup calorimeter. The heat given off by a reaction is absorbed by the water. Knowing the mass of the water, its specific heat, 4.18 J/(g·°C), and the temperature change as read on the thermometer, we can calculate ΔH for the reaction (Examples 5.8 and 5.9).

$$q = (\text{S.H.}) \times m \times \Delta t$$

where

S.H. = specific heat = 4.18 J/g·°C for water
m = mass of water in grams
Δt = $t_{final} - t_{initial}$ in °C

Substituting for the specific heat of water,

$$q_{water} = 4.18 \frac{J}{g \cdot °C} \times m_{water} \times \Delta t \qquad (5.14)$$

2. *Calculate ΔH for the reaction, using the relation*

$$\Delta H = -q_{water} \qquad (5.15)$$

This equation is valid because the heat flow for the reaction mixture, ΔH, must be equal in magnitude but opposite in sign to that of the water, q_{water}. We can distinguish between two types of reactions:

a. In an exothermic reaction, ΔH is negative (for example, $\Delta H = -100$ J) and q_{water} is positive (for example, $q_{water} = +100$ J). This makes sense. For an exothermic reaction, the temperature of the water rises. This means that Δt is positive; by Equation 5.14, q_{water} must also be positive.

b. In an endothermic reaction, ΔH is positive (for example, $\Delta H = +100$ J) and q_{water} is negative (for example, $q_{water} = -100$ J). This is what we would expect from Equation 5.14. In an endothermic reaction, the temperature drops; Δt is negative, so q_{water} must be negative.

Since the temperature goes up, a reaction must occur. What is that reaction?

EXAMPLE 5.8   When 0.400 g NaOH is dissolved in 100.0 g of water, the temperature rises from 25.00 to 26.03°C. Calculate

a. q_{water} b. ΔH for the solution process

Solution
a. Applying Equation 5.14,

$$q_{water} = 4.18 \frac{J}{g \cdot °C} \times 100.0 \text{ g} \times (26.03 - 25.00)°C = 430 \text{ J}$$

b. $\Delta H = -q_{water} = -430$ J

EXERCISE   Suppose that, in a different experiment with 100.0 g of water, the temperature dropped from 25.00 to 23.84°C. What would be ΔH for this process? Answer: +485 J.

The information obtained from experiments of the type described in Example 5.8 can be used to write thermochemical equations. To do this, we use the fact that ΔH for a reaction is directly proportional to amount of reactant. The calculations involved are shown in Example 5.9.

EXAMPLE 5.9 Using the information given in Example 5.8, write a thermochemical equation for the dissolving of one mole of NaOH in water; that is, find ΔH for the equation

$$NaOH(s) \rightarrow Na^+(aq) + OH^-(aq); \qquad \Delta H = ?$$

Solution We want to find ΔH when one mole of NaOH dissolves. From Example 5.8, we see that ΔH is −430 J when 0.400 g NaOH dissolves:

$$0.400 \text{ g NaOH} \approx -430 \text{ J}$$

The molar mass of NaOH is 40.0 g/mol. Hence,

$$\Delta H = 1.00 \text{ mol NaOH} \times \frac{40.0 \text{ g NaOH}}{1 \text{ mol NaOH}} \times \frac{-430 \text{ J}}{0.400 \text{ g NaOH}}$$

$$= -43,000 \text{ J} = -43.0 \text{ kJ}$$

We conclude that 43.0 kJ of heat is evolved when one mole of NaOH dissolves in water. This is 100 times the amount of heat evolved when 0.400 g (0.0100 mol) dissolves.

Putting it another way, you would have to take 43.0 kJ out of the solution to get it back to the initial temperature of the water

EXERCISE Using Tables 5.2 and 5.3, calculate ΔH for the dissolving of one mole of NaOH. Answer: −42.9 kJ.

Bomb Calorimeter

A coffee-cup calorimeter is suitable for measuring heat flows for reactions in solution. However, it cannot be used for reactions involving gases, which would escape from the cup; nor would it be appropriate for reactions in which the products reach high temperatures. The bomb calorimeter, shown in Figure 5.7, is a more versatile instrument. This type of calorimeter was used to determine most of the enthalpy changes that we have referred to in this chapter.

To use this instrument, we start by adding a weighed sample of the reactant(s)

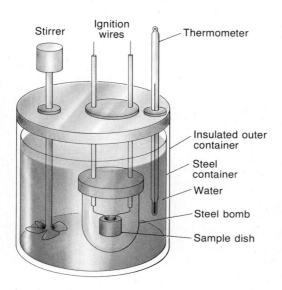

The bomb calorimeter is usually used with combustion reactions

FIGURE 5.7 Bomb calorimeter. To determine q for a reaction carried out in this apparatus, we must consider the heat absorbed by the metal parts of the calorimeter as well as by the water. To find q_{bomb}, a reaction is carried out for which q is known and Δt is measured.

to the heavy-walled steel container, called a "bomb." This is then sealed and lowered into a metal vessel that fits snugly within the insulating walls of the calorimeter. A weighed amount of water sufficient to cover the bomb is added, and the entire apparatus is closed. The initial temperature is measured precisely. The reaction is then started, perhaps by electrical ignition. In an exothermic reaction, the hot products give off heat to the walls of the bomb and to the water. The final temperature is taken to be the highest value read on the thermometer.

Analysis of the heat flow here is a bit more complex than in the case of the coffee cup. In particular, Equation 5.15 must be modified. We cannot ignore the heat flow into the metal parts of the bomb calorimeter. Our equation becomes

$$q_{reaction} = -(q_{water} + q_{bomb}) \qquad\qquad (5.16)$$

The heat evolved goes into both the water and the bomb

Here, $q_{reaction}$,* q_{water}, and q_{bomb} are the heat flows for the reaction, the water, and the bomb calorimeter, respectively. The quantity q_{water} is obtained as before:

$$q_{water} = 4.18\,\frac{J}{g \cdot {}^\circ C} \times m_{water} \times \Delta t$$

The value of q_{bomb} is obtained in a slightly different way. The bomb has a fixed mass and specific heat that remain constant for all the experiments carried out in the calorimeter. The product of the mass of the bomb times its specific heat is referred to as the "calorimeter constant" and is given the symbol C. This constant, which has the units of joules per degree Celsius (J/°C), is determined by experiment (see Problem 5.20). In terms of C,

$$q_{bomb} = C \times \Delta t$$

The use of these relations is shown in Example 5.10.

In this experiment we would burn the methane in excess O_2, at an initial pressure of about 25 atm

EXAMPLE 5.10   Methane, CH_4, is the major component of natural gas. A 1.00-g sample of methane is burned in a bomb calorimeter containing 1225 g of water. The temperature rises from 20.00 to 29.26°C. Taking C for the bomb to be 840 J/°C, calculate

a. $q_{reaction}$ for the combustion of the 1.00-g sample.
b. $q_{reaction}$ for the combustion of one mole of CH_4.

Solution

a. We start with the basic relation between the heat given off by the reaction and that absorbed by the water and bomb calorimeter:

$$\begin{aligned}
q_{reaction} &= -(q_{water} + q_{bomb}) \\
&= -\left(4.18\,\frac{J}{g \cdot {}^\circ C} \times m_{water} \times \Delta t + C\,\Delta t\right) \\
&= -\left(4.18\,\frac{J}{g \cdot {}^\circ C} \times m_{water} + C\right)\Delta t
\end{aligned}$$

---

*In a bomb calorimeter, the pressure usually changes. Hence the heat flow measured may not be exactly equal to the enthalpy change (heat flow at constant pressure). For this reason, we write "$q_{reaction}$" instead of ΔH. This point is discussed in more detail in Section 5.5.

Note that the mass of the water is 1225 g, C = 840 J/°C, Δt = 29.26°C − 20.00°C = 9.26°C. Hence,

$$q_{reaction} = -\left(4.18\frac{J}{g\cdot°C} \times 1225 \text{ g} + 840\frac{J}{°C}\right)(9.26°C)$$
$$= -5.52 \times 10^4 \text{ J} = -55.2 \text{ kJ}$$

b. From (a) we see that 55.2 kJ is evolved per gram of methane burned. One mole of methane, CH_4, weighs 16.04 g. To calculate $q_{reaction}$ for the combustion of one mole of methane, we proceed as in Example 5.9:

$$q_{reaction} = 1.00 \text{ mol } CH_4 \times \frac{16.04 \text{ g } CH_4}{1 \text{ mol } CH_4} \times \frac{-55.2 \text{ kJ}}{1.00 \text{ g } CH_4} = -885 \text{ kJ}$$

EXERCISE Repeat the calculations of Example 5.10, ignoring the heat absorbed by the bomb calorimeter. Answer: −47.4 kJ; −760 kJ.

Examples 5.8–5.10 illustrate the general approach followed in calorimetric calculations:

1. Use the data given to calculate the amount of heat, q_{cal}, absorbed by the calorimeter and its contents. In the simplest case (coffee cup calorimeter), all the heat is absorbed by water and q_{cal} is obtained by using the relation:

$$q_{cal} = 4.18\frac{J}{g\cdot°C} \times \text{mass water} \times \Delta t$$

where $\Delta t = t_{final} - t_{initial}$. In a bomb calorimeter, some heat is absorbed by the metal and:

$$q_{cal} = 4.18\frac{J}{g\cdot°C} \times \text{mass water} \times \Delta t + C \Delta t$$

where C is the calorimeter constant in joules per °C.

2. To find ΔH for the reaction taking place in the calorimeter, use the relation:

$$\Delta H = -q_{cal}$$

where q_{cal} is the quantity obtained in (1).

3. If you are asked to determine ΔH for an amount of product or reactant different from that actually used in the calorimeter reaction, use the fact that ΔH is directly proportional to amount. For example:

$$\Delta H \text{ per mole reactant} = \Delta H \text{ per gram reactant} \times \text{molar mass reactant}$$

5.5
Enthalpy, Energy, and the First Law of Thermodynamics

In this chapter, we have thus far emphasized one type of energy change—the change in enthalpy, ΔH. You will recall that ΔH represents the heat flow in a process carried out at constant pressure. For our purposes, the "process" is usually a chemical reaction and the "constant pressure" is that of the atmosphere.

It is possible to develop equations relating all types of energy changes in all types of processes. To do this, we turn to the science of *thermodynamics*. In any thermodynamic treatment, the *system* that we are studying must be carefully defined. The system may be a reaction mixture whose enthalpy change is to be determined; in another case, it might be a gas undergoing an expansion. The system is separated from its *surroundings* by a boundary, which may be real or imaginary. For practical purposes, the surroundings may be a beaker of water or the air in the laboratory. In principle, however, they include all of the universe outside the system.

The system is the sample being studied

In thermodynamics, we refer to certain quantities as being *state properties*. Their value depends only upon the "state" of the system, not upon its history. To specify the state of a system, we list

—the pressure and temperature of the system.
—the amount, chemical identity, and physical state of each substance present.

A mole of water at 25°C and 1 atm has the same enthalpy the world round

An example of a state property is volume, V. The volume of one mole of liquid water at 25°C and 1 atm has a fixed value, 18.0 cm³. This is true regardless of where the water came from or how it reached this state. Enthalpy, H, is another state property. The enthalpy of one mole $H_2O(l)$ at 25°C and 1 atm, like its volume, has a fixed value.

When a system changes from an initial to a final state, its properties change. For a state property, like enthalpy, the magnitude of the change depends only upon the nature of the two states. It does not depend upon the path by which the change occurs. Thus, for the reaction

$$Sn(s) + 2\,Cl_2(g) \rightarrow SnCl_4(l); \qquad 25°C, 1\ atm$$

ΔH is −545.2 kJ, regardless of whether the reaction occurs directly or in two steps:

$$Sn(s) + Cl_2(g) \rightarrow SnCl_2(s)$$
$$SnCl_2(s) + Cl_2(g) \rightarrow SnCl_4(l)$$

You may recall that we came to the same conclusion in Section 5.2 using Hess's Law. We now see that this law applies because H is a state property.

First Law of Thermodynamics

Thermodynamics distinguishes between two types of energy. One of these is *heat,* given the symbol q. The other is *work,* represented by the symbol w. The term work refers to any type of energy other than heat. In chemistry, we ordinarily deal with only two types of work. One of these is electrical work, such as that supplied by a storage battery. The other is mechanical work, such as that done by a gas expanding against a restraining pressure (Fig. 5.8). This is the most common type of work in the chemistry laboratory. Typically, the gas is one that is produced (or consumed) in a chemical reaction. The "restraining pressure" is simply the constant pressure of the atmosphere.

The quantities q and w have direction as well as magnitude. Heat can flow into a system, raising its temperature. In another case, heat may flow out of a system, lowering its temperature. When a gas expands, as in Figure 5.8, it does work upon the surroundings, pushing back the atmosphere. Under these condi-

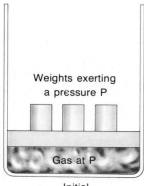
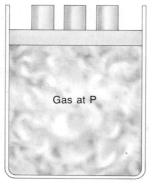

Weights exerting
a pressure P

Gas at P

Gas at P

Initial   Final

FIGURE 5.8 When a gas expands against a constant pressure P, changing its volume by ΔV, it does an amount of work equal in magnitude to $P\Delta V$. The energy of the gas decreases by this amount; $\Delta E < 0$. Conversely, if the gas is compressed by reversing the process shown here, work is done on the gas. In that case, the energy of the gas increases; $\Delta E > 0$.

tions, we would say that energy flows out of the gaseous system in the form of work. When a gas is compressed, work is done on the gas by the surroundings and energy flows into the gaseous system.

In thermodynamics, the direction of energy flow is indicated by specifying the signs of q and w. The conventions we will follow are

q is + when the system absorbs energy in the form of heat from the surroundings
q is − when the system evolves energy as heat to the surroundings
w is + when the system absorbs energy by having work done on it by the surroundings
w is − when the system evolves energy by doing work on the surroundings

q and w refer to the system, unless specified otherwise

With this background, we can state the **First Law of Thermodynamics:**
In any process, the total change in energy of the system, ΔE, is equal to the sum of the heat absorbed, q, and the work, w, done on the system.

System ← Heat flow $q > 0$

Surroundings

$$\Delta E = q + w \qquad (5.17)$$

The quantity E in this equation is called the *internal energy*.

EXAMPLE 5.11 Calculate ΔE of a gas for a process in which the gas
a. absorbs 20 J of heat and does 12 J of work by expanding.
b. evolves 30 J of heat and has 52 J of work done on it as it contracts.

Solution
a. $q = +20$ J; $w = -12$ J, since the gas does work on the surroundings

$$\Delta E = +20 \text{ J} - 12 \text{ J} = +8 \text{ J}$$

b. $q = -30$ J; $w = +52$ J, since work is done on the gas by the surroundings

$$\Delta E = -30 \text{ J} + 52 \text{ J} = +22 \text{ J}$$

EXERCISE In a certain process, a gas absorbs 25 J of heat; its volume remains constant so that no work is done. What is ΔE? Answer: +25 J.

E and ΔE

Internal energy, like enthalpy, is a state property; that is, its value is fixed when the state of the system is specified. A sample of one mole of $H_2(g)$ at 25°C and 1 atm has a certain definite internal energy. This is the sum of all the kinds of energy possessed by the hydrogen molecules. It would include such things as

—the bond energy holding the two atoms together in the H_2 molecule.
—the attractive energy between proton and electron in each H atom.
—the kinetic energy (energy of motion) of the H_2 molecule.

As you might guess, the internal energy of a system is not easy to evaluate. In thermodynamics, we make no attempt to do this. Instead, we concentrate upon the change in internal energy, ΔE. For a reaction system, there is a simple way to determine ΔE. This is to carry out the reaction in a bomb calorimeter and measure the heat flow, q. Since there is no change in volume, no mechanical work is done, and w is zero. Hence, using Equation 5.17, we conclude that for a reaction taking place in a bomb calorimeter,

$$\Delta E = q_{reaction} \qquad (5.18)$$

where $q_{reaction}$ is the heat flow for the reaction carried out at constant volume and ΔE is the difference in internal energy between products and reactants.

In a bomb calorimeter, we determine $q_{reaction}$. From that we find $\Delta E_{reaction}$, or ΔE

To show how Equation 5.18 is used, recall Example 5.10. There we showed that when 1 mol CH_4 is burned in a bomb calorimeter, 885 kJ of heat is evolved to the calorimeter and its contents. In other words,

$$CH_4(g) + 2\,O_2(g) \rightarrow CO_2(g) + 2\,H_2O(l); \qquad q_{reaction} = -885 \text{ kJ}$$

We conclude that ΔE for this reaction is -885 kJ; that is, the energy of the products (1 mol CO_2 + 2 mol H_2O) is 885 kJ less than that of the reactants (1 mol CH_4 + 2 mol O_2).

ΔH and ΔE

The enthalpy of a substance is related in a simple way to its internal energy. Thermodynamics defines enthalpy, H, to be

$$H = E + PV \qquad (5.19)$$

Here, E is internal energy, P is pressure, and V is volume. Ordinarily, the PV term in this equation is a small quantity. One mole of a gas at 25°C and 1 atm has a volume of about 24.5 L. This means that its PV product is 24.5 L·atm. Using the conversion factor 1 L·atm = 0.1013 kJ, we see that PV for one mole of a gas under these conditions is only 2.5 kJ:

$$PV = 24.5 \text{ L·atm} \times \frac{0.1013 \text{ kJ}}{1 \text{ L·atm}} = 2.5 \text{ kJ}$$

For liquids and solids, PV is even smaller (Table 5.4).

Table 5.4
Values of PV at 25°C

	L·atm/mol	kJ/mol
$CH_4(g)$	24.5	2.5
$O_2(g)$	24.5	2.5
$CO_2(g)$	24.5	2.5
$H_2O(l)$	0.018	0.0018

This means that the enthalpy of a substance is very nearly (but not quite) equal to its internal energy. More important, it means that the difference between ΔH and ΔE in a chemical reaction is small. Here,

$$\Delta H = \Delta E + \Delta(PV) \qquad (5.20)$$
$$= \Delta E + (PV)_{products} - (PV)_{reactants}$$

For the combustion of methane, using the data in Table 5.4,

$$\Delta(PV) = PV\ CO_2(g) + 2\ PV\ H_2O(l) - (PV\ CH_4(g) + 2\ PV\ O_2(g))$$
$$= -5.0\ kJ$$

Since ΔE is -885 kJ,

$$\Delta H = -885\ kJ - 5.0\ kJ = -890\ kJ$$

For practical purposes, $\Delta H = \Delta E$ for chemical reactions

As you can see, the difference between ΔH and ΔE is small, often within experimental error. This is the case with nearly all chemical reactions.

5.6
Sources of Energy

The United States today uses vast amounts of energy, about 8×10^{16} kJ/yr. With only about 6% of the world's population, it accounts for about 30% of worldwide energy usage. In the past 30 years, energy usage in the United States has more than doubled. Figure 5.9 shows where this energy comes from. As you can see, nearly 90% is obtained from the combustion of the fossil fuels petroleum, natural gas, and coal. Each of these materials consists largely of hydrocarbons, organic compounds containing the two elements carbon and hydrogen. Natural gas is mostly methane, CH_4. Petroleum is a mixture of liquid hydrocarbons with 5 to 20 carbon atoms per molecule. On the average, these molecules contain about 2.2 hydrogen atoms per carbon atom. Coal is made up of solid hydrocarbons of high molar mass. The atom ratio of hydrogen to carbon in coal is quite low, about 0.8.

We can't really "use up" energy, but, rather, we convert it from one form to another

When hydrocarbons burn in air, they release energy as heat. Typical reactions include

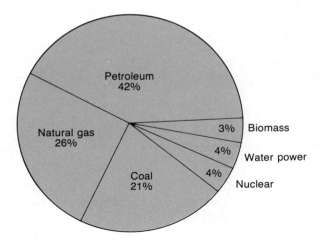

FIGURE 5.9 The major sources of energy in the United States are petroleum (42%), natural gas (26%), and coal (21%). Nuclear fission and water power make smaller contributions, although they account for nearly one fourth of electrical energy. The "biomass" contribution comes almost entirely from the combustion of wood and wood wastes. In 1860, wood supplied 90% of our energy; a century later, that figure was virtually zero. Now it has increased to 3%.

$$CH_4(g) + 2\ O_2(g) \rightarrow CO_2(g) + 2\ H_2O(l); \qquad \Delta H = -890\ kJ \qquad \textbf{(5.21)}$$

$$C_8H_{18}(l) + \tfrac{25}{2}\ O_2(g) \rightarrow 8\ CO_2(g) + 9\ H_2O(l); \qquad \Delta H = -5440\ kJ \qquad \textbf{(5.22)}$$

$$C_{10}H_8(s) + 12\ O_2(g) \rightarrow 10\ CO_2(g) + 4\ H_2O(l); \qquad \Delta H = -5155\ kJ \qquad \textbf{(5.23)}$$

The heat evolved may be used directly to warm our homes and places of business. More commonly, it is converted to mechanical or electrical energy by machines.

Today we face an "energy crisis" that no amount of wishful thinking can dissipate. The increase in price of all fuels over the past decade is but one symptom of that crisis. Its origin is simple: the United States is running out of petroleum and natural gas. Each year, known reserves of petroleum decrease; twice as much is used as is discovered. The production of oil peaked in 1970 and has decreased considerably since then.

Until about 1950, all the petroleum used in the United States came from domestic wells. Today, about one third of our petroleum is imported. For economic and political reasons, our dependence upon foreign oil is disturbing. Moreover, world reserves of petroleum are limited. Their energy equivalent is about 600×10^{16} kJ. Each year about 10×10^{16} kJ are used. Simple division gives

$$\frac{600 \times 10^{16}\ kJ}{10 \times 10^{16}\ kJ/yr} = 60\ yr$$

as a time limit for exhaustion of this fuel. Actually, this estimate may be an optimistic one. Each year, on the average, worldwide consumption of petroleum increases by 4%. If that growth rate continues, the world will run out of oil in about 30 years.

Clearly, within your lifetime, the United States will have to make some drastic adjustments to meet its energy needs. Sources that now make only a minor contribution, if any at all, will have to be used. In this section, we will look at some of the prospects in this area.

Small cars and well-insulated buildings will help a lot

Coal

Reserves of coal in the United States have an energy value 30 times that of reserves of petroleum. Worldwide, the multiple is about 10. By shifting to coal, we could satisfy our energy needs for at least a century. In some areas, such a change is quite easy to accomplish. Over the past decade, many electrical power plants have converted from oil or natural gas to coal. The cost of electrical energy generated by burning coal is only about one third of that produced by burning oil.

On the other hand, coal is hardly an ideal fuel. Its combustion has many adverse effects on both people and the environment. Most of the SO_2 that pollutes our air comes from burning coal, which typically contains 1% to 3% sulfur. Sulfur dioxide is the major factor in the formation of acid rain (Chap. 17). It is also responsible each year for the deaths of several thousand people who suffer from chronic bronchitis, asthma, or emphysema. Moreover, coal mining is a dirty, hazardous occupation. About 1 of every 1000 underground miners is killed each year by cave-ins or explosions. At least 30% of those coal miners who have worked for many years suffer from "black lung" or other lung disease.

Strip mining is safer, but also has some undesirable effects

It is difficult to use a solid fuel such as coal as an energy source for transportation. However, coal can be converted to a gaseous or liquid fuel. When oxygen and steam under pressure are blown through hot coal, reactions such as the following occur:

$$C(s) + O_2(g) \rightarrow CO_2(g); \qquad \Delta H = -393.5 \text{ kJ} \qquad \textbf{(5.24)}$$

$$C(s) + H_2O(g) \rightarrow CO(g) + H_2(g); \qquad \Delta H = +131.3 \text{ kJ} \qquad \textbf{(5.25)}$$

The product called *synthesis gas*, contains, about 40 mole percent H_2, 15% CO, 30% CO_2, and 15% CH_4. Its heating value, in kilojoules per mole, is rather low, only about one third of that for natural gas.

Instead of burning synthesis gas, it can be converted to a liquid fuel. In this process, the carbon dioxide is first removed. The remaining gas is heated to 200 to 300°C in the presence of a catalyst, usually a compound of iron or cobalt. Under these conditions, carbon monoxide and hydrogen react to form liquid hydrocarbons. The following reactions are typical:

$$9 \text{ H}_2(g) + 16 \text{ CO}(g) \rightarrow C_8H_{18}(l) + 8 \text{ CO}_2(g) \qquad \textbf{(5.26)}$$

$$17 \text{ H}_2(g) + 8 \text{ CO}(g) \rightarrow C_8H_{18}(l) + 8 \text{ H}_2O(l) \qquad \textbf{(5.27)}$$

A 60% yield of hydrocarbons in the gasoline range can be obtained. This process, known as the Fischer-Tropsch synthesis, was used in Germany during World War II to produce a low-grade gasoline. The only full-scale plant using this process today is located in South Africa, where coal is plentiful but all oil must be imported. Several small units have been built in the United States to study various modifications of the Fischer-Tropsch synthesis. At this point, the product is considerably more expensive than ordinary gasoline.

The South African plants make good quality gasoline

Nuclear Energy

At the present time, about 12% of our electrical energy comes from a process called *nuclear fission* (Chap. 27). Here, a heavy atom splits into smaller fragments when struck by a neutron. Energy is evolved as heat, which is then converted into electrical energy. In nuclear plants now in operation, the fuel used is the relatively rare isotope of uranium, ^{235}U. A typical fission reaction is

$$^{235}_{92}\text{U} + ^{1}_{0}\text{n} \rightarrow ^{90}_{38}\text{Sr} + ^{144}_{54}\text{Xe} + 2\,^{1}_{0}\text{n}; \quad \Delta H = -2 \times 10^{10} \text{ kJ} \qquad \textbf{(5.28)}$$

You can get an idea of the enormous amount of energy available from fission by calculating the amount of heat evolved per gram of uranium:

$$\frac{-2 \times 10^{10} \text{ kJ}}{235 \text{ g}} = -1 \times 10^8 \text{ (100 million) kJ/g}$$

1 g U-235 ≃ 2000 kg
petroleum

This compares to about 46 kJ/g for petroleum and natural gas.

Balanced against this advantage of nuclear fuels is the problem of storing the fission products. These products, such as strontium-90, are dangerously radioactive. They will remain so for many years to come. Moreover, there is always the possibility of a nuclear accident that could release radioactive material to the surroundings. For these reasons, among others, the development of nuclear energy has been much slower than predicted.

In many ways, the ideal energy source of the future would be *nuclear fusion* (Chap. 27). Here, light nuclei such as deuterium, $^{2}_{1}\text{H}$, and tritium, $^{3}_{1}\text{H}$, combine to form heavier nuclei. A typical fusion reaction is

$$^{2}_{1}\text{H} + ^{3}_{1}\text{H} \rightarrow ^{4}_{2}\text{He} + ^{1}_{0}\text{n}; \quad \Delta H = -1.7 \times 10^9 \text{ kJ} \qquad \textbf{(5.29)}$$

This reaction evolves about three to four times as much energy as fission, per gram of fuel:

$$\frac{-1.7 \times 10^9 \text{ kJ}}{5.0 \text{ g}} = -3.4 \times 10^8 \text{ kJ/g}$$

The products of the fusion process are not radioactive. Hence, the safety hazards associated with fission reactors are greatly reduced. Most important, the light isotopes required for fusion are quite common. There are enough of them available to supply all our energy needs for hundreds of years.

Unfortunately, no one up to now has been able to use Reaction 5.29 to generate a sustained flow of energy. For this to happen, a basic problem must be solved. There is a huge electrical repulsion between two small, positively charged hydrogen nuclei (H^+ ions). To overcome this repulsion, the nuclei must be accelerated to enormous velocities, corresponding to temperatures of millions of degrees. Maintaining such temperatures requires the development of a whole new technology. It seems unlikely that fusion reactors will contribute to meeting our energy needs before the year 2000, if then.

Solar Energy

If we could store the summer heat for winter, our heating problems would be solved

The problem of trapping and using solar energy is one that has intrigued scientists for generations (and federal granting agencies from 1974 to 1980). Each year the earth receives the enormous total of 5×10^{21} kJ of energy from the sun. A tiny fraction of this, through evaporation and condensation of water, supplies us with hydroelectric power. If this fraction could be increased to 0.0001 (1/100 of 1%), it would meet all the world's energy needs.

At present, the most practical application of solar energy is to heat water. Solar water heaters are common in Israel, Japan, and Australia. Solar energy is also being used to heat water and some homes in the United States. So-called "active" solar heating systems use a roof collector of the type shown in Figure 5.10. Here, the base of a shallow metal tray is painted black to absorb as much sunlight as possible. The glass cover exerts a "greenhouse" effect, allowing sunlight to reach the collector but preventing radiant heat from escaping. Water, passing through the collector several times, is heated to about 65°C. This water passes into a storage tank, from which heat is transferred throughout the house as in an ordinary hot water heating system.

There are several problems with solar heating systems of this type. For one thing, the components tend to break down frequently. Leaks, freeze-ups, and failures due to corrosion have been all too common. Moreover, the supply of heat is cut off in periods of cold, cloudy weather and during long winter nights. These conditions prevail during the heating season in most of the United States. Hence, an auxiliary heating system of the ordinary type must be available. This factor makes the economics of "active" solar heating systems unattractive at the present time.

Recently, emphasis has shifted to "passive" solar heating systems. These involve no machinery or circulating water. Instead, a new house is designed, or an existing house remodeled, to absorb as much of the sun's rays as possible. A common technique is to cover most of the south wall of the house with double-glazed windows. During the day, heat passing through the windows is absorbed within the house, perhaps by a concrete floor next to the windows. When the sun

Solar water heaters are practical in the summer time anywhere in the U.S.

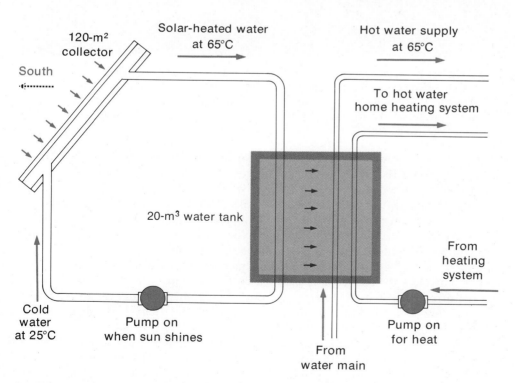

FIGURE 5.10 An "active" solar heating system. The large insulated storage tank acts as a heat exchanger, transferring heat from the roof collector to hot water faucets and radiators in the house.

FIGURE 5.11 Solar One, a joint project of the U.S. Department of Energy, Southern California Edison Company, and the Los Angeles Department of Water and Power, started operating in 1982. Sunlight is reflected by 1818 heliostats (movable mirrors) into a vessel filled with water at the top of a 100-m tower. The water is converted to steam, which generates electricity for 6000 homes in southern California.

sets, the window area is closed off by insulating drapes that prevent heat from escaping. About 100,000 passive solar homes have been built in the United States since 1980.

Within the next 20 years, solar energy may be used in areas other than heating. One area of intense research and development is the conversion of solar to electrical energy. This can be done by means of "solar cells." Such cells contain semiconductor materials made from silicon. When sunlight strikes the semiconductor, electrons are set free; their movement creates an electric current. The major drawback to solar cells is the high cost of the silicon, which must be very pure. Currently, it costs about $60/kg. Other, less expensive materials are available but they are less efficient. At present, the cost of electricity from solar cells is five to ten times that from ordinary sources.

We are making progress, slowly, with solar cells

Another way to produce electrical energy from sunlight is illustrated in Figure 5.11. This research facility, located in the Mojave Desert of California, contains more than 1800 mechanized mirrors, each with a surface area of 40 m². The mirrors can be tilted to catch the rays of the sun and reflect them into a water reservoir mounted on a tower 100 m high. Upon absorbing heat, the water is converted to steam, which operates a turbine to generate electricity. The problem here is again one of cost; this facility is not competitive with conventional power plants.

Summary

From the standpoint of thermochemistry, we can distinguish between two types of reactions. In an exothermic reaction, heat is given off to the surroundings, usually increasing their temperature. In an endothermic reaction, heat is absorbed from the surroundings, reducing their temperature. We can relate the heat flow for a reaction carried out in an open container at constant pressure to the difference in enthalpy (heat content) between products and reactants:

exothermic reaction: $\Delta H < 0$ endothermic reaction: $\Delta H > 0$ $\qquad q = \Delta H$

These basic ideas are illustrated in Example 5.1.

The magnitude of ΔH for a reaction depends upon the amounts of substances reacting or formed (Examples 5.2 and 5.3). The ΔH for a reaction in one direction is equal in magnitude but opposite in sign to ΔH for the reverse reaction (Example 5.4). Because ΔH is independent of reaction path, enthalpy changes for successive steps can be added to obtain ΔH for the overall reaction (Hess's Law).

The enthalpy change for a reaction can be calculated by using molar heats of formation of compounds or of ions in aqueous solution (Tables 5.2 and 5.3). The relation used is

$$\Delta H = \Sigma \Delta H_f \text{ products} - \Sigma \Delta H_f \text{ reactants}$$

Here it is understood that the heat of formation of an element in its stable state is zero; ΔH_f of $H^+(aq)$ is also taken to be zero (Examples 5.5 through 5.7).

Calorimetry is the study of heat flow. In a coffee-cup calorimeter, the heat evolved by a reaction is absorbed by water. The amount of heat absorbed can be calculated from the relation

$$q_{water} = 4.18 \frac{J}{g \cdot °C} \times mass_{water} \times \Delta t_{water}$$

(Examples 5.8 and 5.9). In a bomb calorimeter, part of the heat evolved in the reaction is absorbed by the metal parts of the calorimeter (Example 5.10).

Enthalpy is a state property; that is, the enthalpy of a substance depends only upon its present state, not upon its history. Hence, ΔH for a process is independent of path. The First Law of Thermodynamics relates the change in energy for a process (ΔE) to the heat flow (q) and the work term (w). It tells us that $\Delta E = q + w$ (Example 5.11). Internal energy, E, like enthalpy, is a state property. For a chemical reaction, ΔE can be determined by measuring the heat flow in a bomb calorimeter. Ordinarily, ΔH and ΔE for a reaction are nearly equal (Equation 5.20).

At present, our major sources of energy are fossil fuels, mainly petroleum and natural gas (Fig. 5.9). Within the next 20 years, this situation is very likely to change, perhaps markedly. Energy sources that may make a major contribution include coal (21% today), nuclear energy (4% today), and solar energy (0.01% today).

Key Words and Concepts

calorimeter	heat flow	mole
calorimeter constant	heat of formation (ΔH_f)	solar cell
Conservation of Energy	heat of fusion (ΔH_{fus})	specific heat
endothermic	heat of vaporization (ΔH_{vap})	state property
enthalpy (H)	Hess's Law	surroundings
exothermic	internal energy (E)	system
First Law of Thermodynamics	joule	thermochemical equation
fission	kilojoule	thermodynamics
fusion		work

Questions and Problems

Exothermic and Endothermic Reactions

5.1 Which of the following are true for an exothermic reaction?
a. The enthalpy of the system decreases.
b. ΔH has a negative sign.
c. The enthalpy of the products is higher than that of the reactants.
d. Heat is absorbed from the surroundings.

5.2 When 6.00 g of calcium chloride dissolves in 100.0 g of water, the temperature rises from 18.0 to 28.7°C. The reaction is:

$$CaCl_2(s) \rightarrow Ca^{2+}(aq) + 2\ Cl^-(aq)$$

a. Is this reaction exothermic or endothermic?
b. What is the sign of ΔH?

Laws of Thermochemistry

5.3 Consider the reaction,

$$Fe(s) + Br_2(l) \rightarrow FeBr_2(s); \Delta H = -249.8 \text{ kJ}$$

a. Is the reaction exothermic or endothermic?
b. Draw a diagram similar to Figure 5.2 for this reaction.
c. Calculate ΔH when 10.0 g $FeBr_2$ is formed.
d. How many grams of iron must react to evolve 1.00 kJ of heat?

5.4 Consider the reaction,

$$Ag^+(aq) + Cl^-(aq) \rightarrow AgCl(s); \Delta H = -65.5 \text{ kJ}$$

a. Calculate ΔH when one mole of AgCl dissolves in water.
b. What is ΔH when 1.00 g AgCl dissolves?

5.31 Which of the following are true for an endothermic reaction?
a. ΔH is positive.
b. Heat is transferred to the surroundings.
c. The enthalpy of the system increases.
d. The temperature of the surroundings decreases.

5.32 The temperature of 50.0 g of water drops from 18.0 to 9.5°C when 10.0 g KBr are dissolved in it.
a. Write a chemical equation for this reaction.
b. What is the sign of ΔH?

5.33 Nickel tetracarbonyl, $Ni(CO)_4$, decomposes upon heating:

$$Ni(CO)_4(g) \rightarrow Ni(s) + 4CO(g); \Delta H = +160.7 \text{ kJ}$$

a. Is the reaction exothermic or endothermic?
b. Draw a diagram similar to Figure 5.2 for this reaction.
c. Calculate ΔH when 1.00 g $Ni(CO)_4$ decomposes.
d. How many grams of $Ni(CO)_4$ decompose when 1.00 kJ of heat is absorbed?

5.34 Consider the reaction,

$$H^+(aq) + OH^-(aq) \rightarrow H_2O(l); \Delta H = -55.9 \text{ kJ}$$

a. Calculate ΔH when one mole of water dissociates into ions.
b. What is ΔH when 1.00 g H_2O dissociates?

5.5 Upon heating, 1.000 g $KClO_3$ decomposes to KCl and $KClO_4$, evolving 350 J of heat.
a. Write a balanced equation for the reaction.
b. Calculate ΔH for the decomposition of one mole of $KClO_3$.
c. What is ΔH for the formation of 0.250 mol $KClO_4$?

5.6 When glucose combines with O_2, the following reaction occurs:

$$C_6H_{12}O_6(s) + 6\ O_2(g) \rightarrow$$
$$6\ CO_2(g) + 6\ H_2O(l);\ \Delta H = -2820\ kJ$$

How many grams of glucose would have to be burned to heat 1.00 kg of water from 25.00 to 30.00°C (specific heat = 4.18 J/g·°C)?

5.7 Which requires the absorption of the greater amount of heat, melting 100.0 g of mercury or boiling 25.0 g of benzene (Table 5.1)?

Hess's Law

5.8 Given

$$Na(s) + \tfrac{1}{2} Cl_2(g) \rightarrow Na(g) + Cl(g);\ \Delta H = +230\ kJ$$

$$Na(g) + Cl(g) \rightarrow Na^+(g) + Cl^-(g);\ \Delta H = +147\ kJ$$

$$Na(s) + \tfrac{1}{2} Cl_2(g) \rightarrow NaCl(s);\ \Delta H = -411\ kJ$$

calculate ΔH for the reaction

$$Na^+(g) + Cl^-(g) \rightarrow NaCl(s)$$

5.9 Given

$$Fe(s) + Br_2(l) \rightarrow FeBr_2(s);\ \Delta H = -249.8\ kJ$$

$$FeBr_3(s) \rightarrow FeBr_2(s) + \tfrac{1}{2} Br_2(l);\ \Delta H = +18.4\ kJ$$

calculate the heat of formation of $FeBr_3$.

ΔH and Heats of Formation

5.10 Given

$$2\ Al_2O_3(s) \rightarrow 4\ Al(s) + 3\ O_2(g);\ \Delta H = +3339.6\ kJ$$

a. what is the heat of formation of Al_2O_3?
b. what is ΔH for the formation of 10.0 g Al_2O_3?

5.11 A first-aid hot pack uses the reaction

$$CaCl_2(s) \rightarrow Ca^{2+}(aq) + 2\ Cl^-(aq)$$

Using Tables 5.2 and 5.3, calculate the amount of heat evolved per gram of solid.

5.35 When magnesium metal and carbon dioxide gas react, 16.7 kJ is evolved per gram of Mg. The products are solid carbon and magnesium oxide.
a. Write a balanced equation for the reaction.
b. Calculate ΔH when 2.00 mol Mg reacts.
c. How much heat is evolved per gram of products formed in this reaction?

5.36 When glucose is oxidized in the body, about 40% of the energy evolved in the reaction referred to in Problem 5.6 is available for muscular activity. How much of this type of energy can be obtained from the oxidation of 1.00 g of glucose?

5.37 How much heat is evolved when 100.0 g of bromine condenses? when 100.0 g of benzene freezes (Table 5.1)?

5.38 Given

$$KCl(s) \rightarrow K^+(g) + Cl^-(g);\ \Delta H = +718\ kJ$$

$$KCl(s) \rightarrow K(s) + \tfrac{1}{2} Cl_2(g);\ \Delta H = +436\ kJ$$

$$K(s) + \tfrac{1}{2} Cl_2(g) \rightarrow K(g) + Cl(g);\ \Delta H = +211\ kJ$$

calculate ΔH for the reaction

$$K(g) + Cl(g) \rightarrow K^+(g) + Cl^-(g)$$

5.39 Given

$$V(s) + 2\ Cl_2(g) \rightarrow VCl_4(l);\ \Delta H = -569.4\ kJ$$

$$VCl_3(s) \rightarrow VCl_2(s) + \tfrac{1}{2} Cl_2(g);\ \Delta H = +128.9\ kJ$$

$$2\ VCl_3(s) \rightarrow VCl_2(s) + VCl_4(l);\ \Delta H = +140.2\ kJ$$

calculate the heat of formation of VCl_3.

5.40 Given

$$2\ CuO(s) \rightarrow 2\ Cu(s) + O_2(g);\ \Delta H = +310.4\ kJ$$

a. determine the heat of formation of CuO.
b. calculate ΔH for the formation of 50.0 g CuO.

5.41 First-aid cold packs use the reaction

$$NH_4NO_3(s) \rightarrow NH_4^+(aq) + NO_3^-(aq)$$

Using Tables 5.2 and 5.3, calculate ΔH when 1.00 g NH_4NO_3 dissolves.

5.12 Use Tables 5.2 and 5.3 to obtain ΔH for the following reactions:

a. $2 Cl^-(aq) + I_2(s) \rightarrow Cl_2(g) + 2I^-(aq)$
b. $MnO_2(s) + 4 H^+(aq) + 2 Cl^-(aq) \rightarrow$
   $Mn^{2+}(aq) + Cl_2(g) + 2 H_2O(l)$
c. $H_2SO_4(l) \rightarrow 2 H^+(aq) + SO_4{}^{2-}(aq)$

5.13 Use Table 5.2 to calculate ΔH for

a. the combustion of one mole of acetylene gas, $C_2H_2(g)$, to form $CO_2(g)$ and $H_2O(l)$.
b. the combustion of one mole of acetylene gas to form $CO_2(g)$ and $H_2O(g)$.

5.14 For the reaction

$$CaO(s) + SiO_2(s) \rightarrow CaSiO_3(s); \Delta H = -89.5 \text{ kJ}$$

use Table 5.2 to calculate ΔH_f $CaSiO_3$.

5.15 Use Tables 5.2 and 5.3 to determine the heat of formation of the $SO_3{}^{2-}$ ion, given $\Delta H = -88.0 \text{ kJ}$ for:

$$SO_3{}^{2-}(aq) + 2 H^+(aq) + 2 NO_3{}^-(aq) \rightarrow$$
$$SO_4{}^{2-}(aq) + H_2O(l) + 2 NO_2(g)$$

5.16 Ethyl alcohol, C_2H_5OH, is used in Brazil as a substitute for gasoline in internal combustion engines. Assume that $C_8H_{18}(l)$ approximates the composition of gasoline, a mixture of hydrocarbons.

a. Write balanced equations for the combustion of C_2H_5OH and of C_8H_{18} to form $CO_2(g)$ and $H_2O(l)$.
b. Use Table 5.2 to calculate the heat evolved in the combustion of one mole of ethyl alcohol; one mole of C_8H_{18} (ΔH_f $C_8H_{18} = -269.7 \text{ kJ/mol}$).
c. Which fuel gives off more heat per gram?

5.42 Use Tables 5.2 and 5.3 to calculate ΔH for the following reactions:

a. $2 Cl^-(aq) + 2 H_2O(l) \rightarrow Cl_2(g) + H_2(g) +$
   $2 OH^-(aq)$
b. $Cu(s) + 2 Fe^{3+}(aq) \rightarrow Cu^{2+}(aq) + 2 Fe^{2+}(aq)$
c. $6 I^-(aq) + 2 MnO_4{}^-(aq) + 4 H_2O(l) \rightarrow$
   $3 I_2(s) + 2 MnO_2(s) + 8 OH^-(aq)$

5.43 Use Table 5.2 to calculate

a. ΔH for the combustion of one mole of propane, $C_3H_8(g)$, to form gaseous CO_2 and liquid water.
b. ΔH for the combustion of one mole of propane to form liquid water and gaseous carbon monoxide.

5.44 For the reaction

$$NH_4NO_2(s) \rightarrow N_2(g) + 2 H_2O(l); \Delta H = -315.1 \text{ kJ}$$

Using Table 5.2, determine ΔH_f NH_4NO_2.

5.45 Use Table 5.2 to determine the heat of formation of Ca_3N_2 from the following:

$$Ca_3N_2(s) + 6 H_2O(l) \rightarrow$$
$$3 Ca(OH)_2(s) + 2 NH_3(g); \Delta H = -905.6 \text{ kJ}$$

5.46 Acetylene, C_2H_2, and butane, C_4H_{10}, are gaseous fuels.

a. Write balanced equations for the combustion of acetylene and butane, assuming the products in each case are $CO_2(g)$ and $H_2O(l)$
b. Using Table 5.2, calculate the amount of heat evolved per gram of acetylene; per gram of butane.
c. Taking the densities of $C_2H_2(g)$ and $C_4H_{10}(g)$ to be 1.07 g/L and 2.38 g/L, respectively, determine which gives off more heat per unit volume.

Calorimetry

5.17 Using the data in Problem 5.2 and assuming all the heat is absorbed by the water, calculate

a. ΔH when 6.00 g $CaCl_2$ dissolves in water.
b. ΔH for the equation written in Problem 5.2.

5.18 Consider the reaction

$$Ni(s) + Cu^{2+}(aq) \rightarrow$$
$$Ni^{2+}(aq) + Cu(s); \Delta H = -128.4 \text{ kJ}$$

This reaction is carried out in a coffee-cup calorimeter using 125 g of water. The copper formed weighs 1.00 g. Calculate the temperature change of the water (S.H. = 4.18 J/g·°C).

5.47 Using the data in Problem 5.32 and assuming all the heat is absorbed by the water, calculate

a. ΔH when 10.0 g KBr dissolves in water.
b. ΔH when one mole of KBr dissolves.

5.48 When 1.00 g $KClO_3$ is dissolved in 50.0 g of water (S.H. = 4.18 J/g·°C) in a coffee-cup calorimeter, the temperature drops from 25.00 to 23.36°C. Calculate ΔH for the process

$$KClO_3(s) \rightarrow K^+(aq) + ClO_3{}^-(aq)$$

5.19 A sample of sucrose, $C_{12}H_{22}O_{11}$, weighing 3.85 g is burned in a bomb calorimeter containing 6.00 kg of water (S.H. = 4.18 J/g·°C). The calorimeter constant is 3180 J/°C. The temperature rises from 23.40 to 25.64°C. Calculate q for the combustion of one mole of sucrose.

5.20 A 0.200-g tablet of benzoic acid gave off 5.290 kJ of heat when burned in a bomb calorimeter containing 0.5260 kg of water (S.H. = 4.184 J/g·°C). The temperature rose 2.020°C. Calculate the calorimeter constant in J/°C.

5.21 A 2.75-g sample of sucrose, $C_{12}H_{22}O_{11}$, is burned in a bomb calorimeter containing 4.80 kg of water originally at 25.00°C. The calorimeter constant is 2540 J/°C. The combustion produces 45.4 kJ of heat. Calculate the final temperature.

5.49 When 3.20 g of ethyl alcohol, $C_2H_5OH(l)$, is burned in a bomb calorimeter containing 3.50 kg of water, the temperature rises 5.52°C. The calorimeter constant is 2550 J/°C; the specific heat of water is 4.18 J/g·°C. Calculate q for the combustion of one mole of ethyl alcohol.

5.50 When 1.000 g CH_4 burns in a bomb calorimeter containing 8.060 kg of water (S.H. = 4.184 J/g·°C), the temperature rises 1.520°C. Under these conditions, 885.3 kJ of heat is evolved per mole of methane burned. Calculate the calorimeter constant in J/°C.

5.51 When 3.16 g of salicylic acid, $C_7H_6O_3$, is burned in a bomb calorimeter containing 5.00 kg of water originally at 23.00°C, 69.3 kJ of heat are evolved. The calorimeter constant is 3612 J/°C. Calculate the final temperature.

First Law

5.22 Find
  a. ΔE when a gas absorbs 60 J of heat and has 20 J of work done on it.
  b. q when 85 J of work are done by a system and its internal energy is increased by 89 J.

5.23 For the combustion of acetylene

$$C_2H_2(g) + \tfrac{5}{2} O_2(g) \rightarrow 2\ CO_2(g) + H_2O(l)$$

$\Delta(PV)$ is -3.8 kJ. When the reaction is carried out in a bomb calorimeter, 1296 kJ of heat is evolved. Calculate ΔH for the reaction.

5.52 Calculate
  a. q when a system does 65 J of work and its internal energy decreases by 90 J.
  b. ΔE for a gas that releases 50 J of heat and has 200 J of work done on it.

5.53 Consider the reaction

$$C_2H_6(g) + \tfrac{7}{2} O_2(g) \rightarrow 2\ CO_2(g) + 3\ H_2O(l)$$

  a. Using Table 5.2, calculate ΔH.
  b. For this reaction, $\Delta(PV)$ is -6.2 kJ. How much heat is evolved when this reaction is carried out in a bomb calorimeter?

Sources of Energy

5.24 The United States uses about 8.0×10^{16} kJ of energy annually. The heat of combustion of natural gas is about 43 kJ/g; its density is about 0.73 g/L. Using Figure 5.9, estimate the number of cubic meters of natural gas burned each year in the United States.

5.25 In one type of solar heating system, the sun's energy is used to bring about the reaction

$$Na_2SO_4 \cdot 10\ H_2O(s) \rightarrow$$
$$2\ Na^+(aq) + SO_4{}^{2-}(aq) + 10\ H_2O(l);$$

for which $\Delta H = +78.7$ kJ
Upon cooling, the reverse reaction occurs, giving off heat to raise the temperature of water in a storage tank. How much water can be warmed from 25.0 to 65.0°C by the heat evolved when one gram of $Na_2SO_4 \cdot 10\ H_2O$ is formed?

5.54 The heat of combustion of coal is about 32 kJ/g. Using Figure 5.9 and the information in Problem 5.24, estimate the number of metric tons of coal burned each year in the United States.

5.55 Using the information given in Problem 5.25, calculate the increase in temperature of 1.00 g of water if it absorbs the heat given off when one gram of $Na_2SO_4 \cdot 10\ H_2O$ is formed.

5.26 To conserve energy, a home owner resets the hot water heater temperature control from 64 to 49°C. The heater tank holds 150 L of water (d = 1.00 g/cm³, S.H. = 4.18 J/g·°C).
   a. How much energy can be saved per tankful by this change?
   b. If the water is heated electrically at a rate of 6.0¢ per kilowatt hour, how much money is saved per tankful (1 kW·h = 3.60 × 10³ kJ)?

5.27 Rank the following energy sources in the order you think they will contribute to meeting our energy needs in the year 2000:
   a. solar energy   b. coal
   c. nuclear fusion   d. petroleum

5.56 An average home electric dishwasher uses 50 L of hot water per load. The owner reduces the temperature of the wash water from 65 to 50°C.
   a. How much energy is saved per month (30 days) if the dishwasher is run once a day?
   b. If the water is heated electrically at 6.0¢ per kilowatt hour, how much money is saved in a month (1 kW·h = 3.60 × 10³ kJ)?

5.57 Discuss the advantages and disadvantages of concentrating research efforts on the development of the following energy sources as alternatives to petroleum:
   a. nuclear fusion   b. solar energy
   c. coal   d. nuclear fission

General

5.28 Criticize each of the following:
   a. The heat of formation of Na(l) is 0.
   b. $\Delta H = \Delta E$ for all reactions.
   c. Condensation is an endothermic process.
   d. The United States consumes about 8×10^{16} kJ of energy annually.

5.29 To raise the temperature of 27.1 mL of ethyl alcohol from 25.0 to 45.0°C requires 1.034 kJ. What is the specific heat of ethyl alcohol (d = 0.785 g/mL)?

5.30 A paperweight weighing 129 g is heated and added to 44.6 g of water in a coffee-cup calorimeter. The initial temperature of the paperweight was 95.0°C; that of the water was 25.0°C. The final temperature is 40.0°C. What is the specific heat of the paperweight? Is the paperweight brass (S.H. = 0.393 J/g·°C)?

5.58 Criticize each of the following:
   a. Hess's Law is valid because enthalpy change depends upon reaction path.
   b. Melting is an exothermic process.
   c. The use of wood as an energy source in the United States is decreasing.

5.59 The temperature of 10.0 mL of liquid mercury rises 15.0°C when it absorbs 0.2836 kJ of heat. Calculate the specific heat of mercury (d = 13.6 g/mL).

5.60 To identify an unknown metal, a student measures its specific heat. When 75.0 g of the metal at 25.0°C is placed in 50.0 g of water at 80.0°C in a coffee-cup calorimeter, the final temperature is 75.0°C. What is the specific heat of the metal? Is it likely to be lead (S.H. = 0.128 J/g·°C)?

*5.61 There are about 120 million automobiles in the United States, each driven about 2.0×10^4 km/yr on the average. Fuel economy, on the average, is about 5.5 km/L. The heat of combustion of gasoline is about 48 kJ/g and its density is 0.68 g/cm³.
   a. How much energy is used per year by automobiles?
   b. To reduce energy usage by automobiles by 0.50×10^{16} kJ/yr, what would have to be the average fuel economy, assuming the other factors stay constant?

*5.62 Walking one kilometer uses about 100 kJ of energy. This comes from oxidation of foods, which is about 30% efficient. How much energy do you "save" by walking a kilometer instead of driving a car that gets 6.0 km/L of gasoline (d of gasoline = 0.68 g/cm³; heat of combustion = 48 kJ/g)?

*5.63 On a hot day, you take a six-pack of beer on a picnic, cooling it with ice. Each (aluminum) can weighs 38.5 g and contains 12.0 oz of beer. The specific heat of aluminum is 0.902 J/g·°C; take that of beer to be 4.10 J/g·°C.
   a. How much heat must be absorbed from the six-pack to lower the temperature from 25.0 to 5.0°C?
   b. How much ice must be melted to absorb this amount of heat (ΔH_{fus} of ice is given in Table 5.1)?

*5.64 A cafeteria sets out glasses of tea at room temperature; the customer adds ice. Assuming the customer wants to have some ice left when the tea cools to 0°C, what fraction of the total volume of the glass should be left empty for adding ice? Make any reasonable assumptions needed to work this problem.

*5.65 The thermite reaction was once used to weld rails:

$$2 \text{ Al}(s) + \text{Fe}_2\text{O}_3(s) \rightarrow \text{Al}_2\text{O}_3(s) + 2 \text{ Fe}(s)$$

a. Calculate ΔH for this reaction using heat of formation data.
b. Take the specific heats of Al_2O_3 and Fe to be 0.79 and 0.48 J/g·°C, respectively. Calculate the temperature to which the products of this reaction will be raised, starting at room temperature, by the heat given off in the reaction.
c. Will the reaction produce molten iron (mp Fe = 1535°C, ΔH_{fus} = 270 J/g)?

Chapter 6
Physical
Behavior
of
Gases

In the gas phase, all substances show remarkably similar physical behavior. Consider, for example, their molar volumes. At 0°C and 1 atm, 1 mol of every gas occupies almost exactly the same volume, about 22.4 L. This is true of $O_2(g)$, $N_2(g)$, $CH_4(g)$, and any other gas you care to mention. Moreover, the volumes of different gases respond in almost exactly the same way to changes in amount, changes in pressure, or changes in temperature. We find that

This behavior is remarkably simple, all things considered

—if the amount of *any* gas is doubled from 1 mol to 2 mol, at 0°C and 1 atm, the volume doubles from 22.4 to 44.8 L.
—if the Kelvin temperature of 1 mol of *any* gas at 1 atm is doubled from 273 K (0°C) to 546 K (273°C), the volume doubles from 22.4 to 44.8 L.
—if the pressure on 1 mol of *any* gas at 0°C is cut in half, from 1 atm to $\frac{1}{2}$ atm, the volume doubles from 22.4 to 44.8 L.

It is possible to write an equation relating the volume of any gas to amount, temperature, and pressure. This equation is known as the **Ideal Gas Law;** it is central to almost everything we will have to say about gases in this chapter. After a brief review of the measurement of volume, amount, temperature, and pressure (Section 6.1), we will look at the nature and applications of the Ideal Gas Law in Sections 6.2 and 6.3. We will apply this law to gas mixtures in Section 6.4 and consider deviations from it in Section 6.5.

The fact that different gases resemble each other so closely in their physical behavior can be explained in terms of their particle structure. In all gases at ordinary temperatures and pressures, the particles (atoms or molecules) are

Atoms in He, Ar Molecules in H_2, CH_4

—*very far apart.* At 25°C and 1 atm, only about 0.1% of the volume of a gas is occupied by the particles themselves. This situation is quite different from that in liquids and solids, where particles are in contact with one another and occupy more than half of the total volume.

—moving very rapidly. As we will see in Section 6.6, which deals with the kinetic theory of gases, gas particles at 25°C are moving at speeds of the order of 200 to 2000 m/s (400 to 4000 miles/h). The particles in a gas, in contrast to those in a liquid or solid, are free to move throughout their entire container (Color Plate 6.1).

6.1
Measurements
on Gases

To completely specify the state of a gaseous substance, we cite the values of four quantities.

1. **Volume (V).** A gas expands uniformly to fill any container in which it is placed. This means that the volume of a gas is, quite simply, the volume of its container. Volumes of gases can be expressed in liters, cubic centimeters, or cubic meters:

$$1 \text{ L} = 10^3 \text{ cm}^3 = 10^{-3} \text{ m}^3 \qquad \text{(6.1)}$$

mL are OK too

2. **Amount (n).** Most commonly, the amount of matter in a gaseous sample is expressed in terms of the number of moles. In some cases, we may express amount by quoting the mass in grams of the gas. These two quantities are related by

$$n = g/MM \qquad \text{(6.2)}$$

where n is the number of moles, g is the number of grams, and MM stands for the molar mass.

3. **Temperature (T).** We ordinarily measure the temperature of a gas using a thermometer marked in degrees Celsius. However, *in any calculation involving gases, temperature must be expressed on the Kelvin scale.* To convert between °C and K, we use the relation introduced in Chapter 1:

$$T(K) = t(°C) + 273.15 \qquad \text{(6.3)}$$

Gas behavior is most easily described if we use the Kelvin scale

Typically, we express temperatures only to the nearest degree. In that case, the Kelvin temperature can be found by simply adding 273 to the Celsius temperature.

4. **Pressure (P).** In Chapter 1, we considered briefly the units used to express pressure. In this section, we will look more closely at these units and the relations between them. Before doing that, let us consider two instruments commonly used to measure pressure in the general chemistry laboratory. These are the mercury barometer and manometer.

Atmospheric Pressure, the Barometer, and the Manometer

The most familiar gas, and the only one known until about 1750, is the air above us. This gas lies over the earth in a blanket about 80 km thick. Like all earthly matter, the air is subject to the pull of gravity. The air near the earth is compressed by the mass of the air above it. For this reason, the pressure of the atmosphere has its maximum value at sea level. In Denver, Colorado, 1.6 km above sea level, the atmospheric pressure has decreased by about 20%. Above 3 km, the low pressure makes breathing uncomfortable for people not used to such altitudes. For this reason, commercial airplanes have pressurized cabins. Spacecraft, which operate at much higher altitudes, are also pressurized. At a height of 15 km, atmospheric pressure is only about 10% of that at sea level. Uncomfortable is not the word to describe how an unpressurized person would feel at that altitude.

Although the facts of atmospheric pressure are really very simple, they were not clearly understood until about 1640. Torricelli, an Italian scientist, was the first person to measure the pressure of the atmosphere accurately. The device he built to do this is still used today in chemistry and physics laboratories. It is called the mercury barometer (Fig. 6.1). This consists of a closed glass tube filled with mercury and inverted over a pool of mercury. When the tube is first inverted, mercury flows into the reservoir, leaving a nearly perfect vacuum above the mercury in the tube. After a few seconds, the mercury reaches a constant level. As shown in Figure 6.1, the pressure exerted by the mercury column exactly balances that of the atmosphere. Hence, the height of the column is a measure of the atmospheric pressure. At or near sea level, it typically varies from 740 to 760 mm, depending upon weather conditions.

The pressure exerted by a gas in a closed container can be measured using a manometer (Fig. 6.2). The principle involved here is the same as that of the

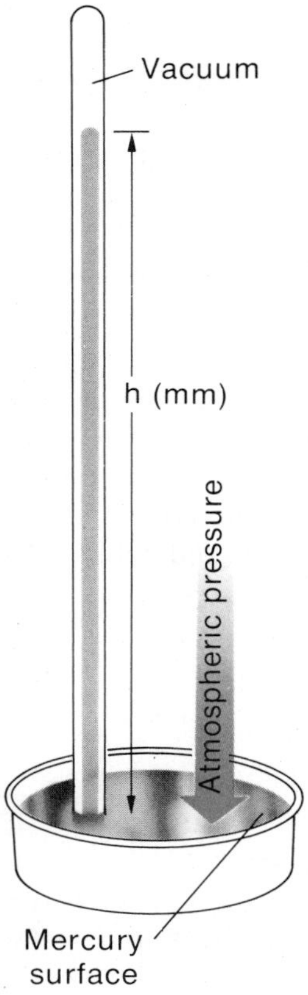

How could you prove that by an experiment?

Vacuum

h (mm)

Atmospheric pressure

Mercury surface

FIGURE 6.1   The mercury barometer. At the level of the lower mercury surface, the pressure both inside and outside the tube must be that of the atmosphere. Inside the tube the pressure is exerted by the mercury column h mm high. Hence, the atmospheric pressure must equal h mm Hg.

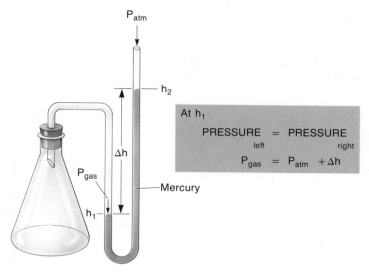

FIGURE 6.2 Manometer and the measurement of gas pressure.

barometer. A manometer consists of a U-tube partially filled with mercury. One side of the U-tube is connected to the gas container. The other side is connected to a region of known pressure, most often the atmosphere. The gas in the container exerts a force on the mercury column that tends to push it down. This force is opposed by that of the air over the other surface. The difference between the two mercury levels is a direct measure of the difference between the two gas pressures. Referring to Figure 6.2, we can say that

$$P_{gas} = P_{atm} + P \text{ due to } \Delta h \text{ mm Hg}$$

All P's must be in the same units

Knowing the atmospheric pressure as read on a barometer and the value of Δh, we can readily calculate the pressure exerted by the gas in the flask.

Pressure Units

Pressure is defined as force per unit area. However, because of the way in which pressure is measured, it is often expressed in **millimeters of mercury (mm Hg)**.* Thus, we might say that the atmospheric pressure on a certain day is 752 mm Hg. This means that the pressure of the air is equal to that exerted by a column of mercury 752 mm high.

Another unit commonly used to express gas pressure is the standard atmosphere, or simply **atmosphere (atm)**. This is the pressure exerted by a column of mercury 760 mm high with the mercury at 0°C. If we say that a gas has a pressure of 0.98 atm, we mean that the pressure is 98% of that exerted by a mercury column 760 mm high. Other pressure units include

---

*The pressure exerted by a column of mercury one millimeter high under certain specified conditions (0°C at sea level) is defined as 1 *torr*. More commonly, the unit torr, introduced to honor Torricelli, is used as a synonym for millimeter of mercury. Throughout this text, we will use millimeters of mercury rather than torr because the former has a clearer physical meaning.

—pounds per square inch. A mass of 1 lb resting on a surface 1 in² in area exerts a pressure of 1 lb/in².

—*kilopascal* (a metric pressure unit). A mass of 10 g resting on a surface 1 cm² in area exerts a pressure of approximately 1 kPa. Atmospheric pressure is ordinarily close to 100 kPa.

To convert between these pressure units, we use the relations

$$1 \text{ atm} = 760 \text{ mm Hg} = 14.70 \text{ lb/in}^2 = 101.3 \text{ kPa} \qquad \textbf{(6.4)}$$

If the pressure on 1 cm² of your body were doubled, you would definitely be aware of that fact

EXAMPLE 6.1   An announcer on a radio station in Montreal reports the atmospheric pressure to be 99.6 kPa. What is the pressure in
a. atmospheres   b. millimeters of mercury

Solution   Using the conversion factor approach,

a. $99.6 \text{ kPa} \times \dfrac{1 \text{ atm}}{101.3 \text{ kPa}} = 0.983 \text{ atm}$

b. $0.983 \text{ atm} \times \dfrac{760 \text{ mm Hg}}{1 \text{ atm}} = 747 \text{ mm Hg}$

EXERCISE   Express a pressure of 729 mm Hg in atmospheres. Answer: 0.959 atm.

6.2
The Ideal Gas Law

As pointed out at the beginning of this chapter, all gases closely resemble each other in one important aspect: the dependence of volume upon amount, temperature, and pressure. In particular,

   1. *Volume is directly proportional to amount.* Figure 6.3A shows a typical plot of volume (V) versus number of moles (n) for a gas. Notice that the graph

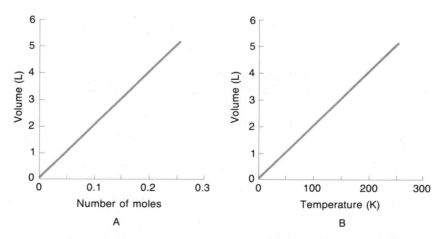

FIGURE 6.3   At constant pressure, the volume of a gas is directly proportional to the number of moles (A) and to the absolute temperature (B).

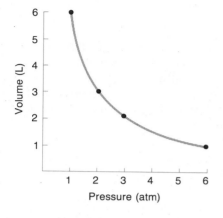

FIGURE 6.4 At constant temperature, the volume of a gas sample is inversely proportional to pressure. In this case, the volume decreases from 6 to 1 L when the pressure increases from 1 to 6 atm.

is a straight line passing through the origin. The general equation for such a plot is

$$V = k_1 n \qquad \text{(constant T, P)} \tag{6.5}$$

where k_1 is a "constant"; that is, it is independent of the individual values of V or n. This is the equation of a direct proportionality.

2. *Volume is directly proportional to absolute temperature.* The dependence of volume (V) on the Kelvin temperature (T) is shown in Figure 6.3B. Here again, the graph is a straight line through the origin. The equation of the line is

> As V increases, so does T

$$V = k_2 T \qquad \text{(constant n, P)} \tag{6.6}$$

where k_2 is a constant independent of the values of V or T. This relationship was first suggested, in a different form, by two French scientists, Charles and Gay-Lussac (see Section 6.7). It is often referred to as the *Law of Charles and Gay-Lussac.*

3. *Volume is inversely proportional to pressure.* Figure 6.4 shows a typical plot of volume (V) versus pressure (P). Notice that V decreases as P increases. The graph is a hyperbola. The general relation between the two variables is

$$V = k_3/P \qquad \text{(constant n, T)} \tag{6.7}$$

The quantity k_3, like k_1 and k_2, is a constant. This is the equation of an inverse proportionality. The fact that volume is inversely proportional to pressure was first established by Robert Boyle. Equation 6.7 is one form of *Boyle's Law.*

Equations 6.5 through 6.7 can be combined into a single equation relating volume to amount, temperature, and pressure. Since V is directly proportional to n, directly proportional to T, and inversely proportional to P, it follows that

$$V = \text{constant} \times \frac{n \times T}{P}$$

> This is probably not intuitively obvious, but true nonetheless

This equation is ordinarily written in a different form. We represent the constant by the symbol R and multiply both sides of the equation by P. This leads to the **Ideal Gas Law:**

$$PV = nRT \tag{6.8}$$

where P is the pressure, V the volume, n the number of moles, and T the Kelvin temperature. The quantity R appearing in the Ideal Gas Law is a true constant. It has the same value for all gases and is independent of P, V, n, or T.

The Ideal Gas Law incorporates, in a useful way, the relations between volume, amount, temperature, and pressure referred to earlier. To see that this is the case, we write Equation 6.8 in the form

This is one of the few relations in science that contains four variables

$$V = \frac{nRT}{P}$$

Note that

—if T and P are constant, the quotient RT/P has a constant value, which we can call k_1, and the Ideal Gas Law reduces to Equation 6.5:

$$V = k_1 n \qquad \text{(constant T, P)}$$

—if n and P are constant, the quotient nR/P has a constant value. Taking $nR/P = k_2$, we obtain Equation 6.6:

$$V = k_2 T \qquad \text{(constant n, P)}$$

—if n and T are constant, the product nRT is constant. Setting $nRT = k_3$, we obtain Equation 6.7:

$$V = k_3/P \qquad \text{(constant n, T)}$$

The Ideal Gas Law can be used to solve a wide variety of problems involving the physical behavior of gases. Later in this section, we will see how this is done. First, let us consider the value of R, the universal gas constant.

Evaluation of R, the Gas Constant

To determine the value of R, we need only establish by experiment one set of values for P, V, n, and T in Equation 6.8. This is readily done. Consider, for example, gaseous oxygen at 0°C and 1.00 atm. These conditions are often referred to as *standard temperature and pressure* (STP). We find that at STP, 32.0 g (1.00 mol) O_2 occupies a volume of 22.4 L. Solving the Ideal Gas Law for R,

$$R = \frac{PV}{nT}$$

R has the same value for all gases because any two gases at the same P, V, and T contain the same number of moles (Avogadro's Law)

Substituting P = 1.00 atm, V = 22.4 L, n = 1.00 mol, and T = 0 + 273 = 273 K,

$$R = \frac{1.00 \text{ atm} \times 22.4 \text{ L}}{1.00 \text{ mol} \times 273 \text{ K}} = 0.0821 \text{ L·atm/(mol·K)}$$

The value of R obtained under the most precise conditions, using oxygen at low pressures, is 0.082056 L·atm/(mol·K). Note that R involves the units of atmospheres, liters, moles, and K. These units must be used for pressure, volume, amount, and temperature in any problem where this value of R is employed.

In most of our work in this chapter, we will use 0.0821 L·atm/(mol·K) as the value of R. For certain purposes, however, we will need R in other units. Table 6.1 lists values of R in various sets of units.

Table 6.1
Values of R in Different Units

VALUE	WHERE USED	HOW OBTAINED
$0.0821 \; \dfrac{\text{L·atm}}{\text{mol·K}}$	Gas Law problems with V in liters, P in atm	From known values of P, V, T, n
$8.31 \; \dfrac{\text{L·kPa}}{\text{mol·K}}$	Gas Law problems with V in liters, P in kPa	1 atm = 101.3 kPa
$8.31 \; \dfrac{\text{J}}{\text{mol·K}}$	Equations involving energy in joules	1 L·atm = 101.3 J
$8.31 \times 10^7 \; \dfrac{\text{g·cm}^2}{\text{s}^2 \text{·mol·K}}$	Calculation of average speed of molecules (Section 6.6)	$1 \text{ J} = 10^7 \; \dfrac{\text{g·cm}^2}{\text{s}^2}$

Final and Initial State Problems

In a common type of problem, a gas undergoes a change from an "initial" to a "final" state. In this process, two or more of the four variables (V, n, T, P) change. You are asked to determine the effect of this change upon a particular variable, perhaps the volume V. To do this, you follow what amounts to a four-step procedure:

1. Decide, from the information given in the problem, which variables change and which quantities remain constant.
2. Write down the Ideal Gas Law for both the final and initial states. This leads to two equations, one for the final conditions, the other for the original conditions.
3. Combine these two equations so as to eliminate any quantity that remains constant. This gives a single equation relating the variables that change.
4. Solve the equation obtained in (3) for the desired variable. Then, using the information given in the problem, calculate the value of that variable.

This procedure is illustrated in Example 6.2.

EXAMPLE 6.2 An air bubble forms at the bottom of a lake, where the total pressure is 2.18 atm. At this pressure, the bubble has a volume of 3.6 cm³. What volume will it have when it rises to the surface, where the pressure is that of the atmosphere, 742 mm Hg? Assume the temperature remains constant, as does the amount of gas within the bubble.

Solution We proceed as described above.
1. From the statement of the problem, the variables are P and V; n and T remain constant.

2. Using the subscript 2 for final state and 1 for initial state:

$$\text{final state:} \quad P_2V_2 = nRT \qquad \text{initial state:} \quad P_1V_1 = nRT$$

Here, n is the (constant) number of moles and T is the (constant) temperature.

3. Since P_2V_2 and P_1V_1 are both equal to nRT, they must be equal to each other; that is,

$$P_2V_2 = P_1V_1$$

4. We are asked to find the final volume, V_2. Solving this equation for V_2, we obtain

$$V_2 = V_1 \times \frac{P_1}{P_2}$$

Keep your subscripts straight and avoid trouble

From the statement of the problem, $V_1 = 3.6$ cm³, $P_1 = 2.18$ atm, $P_2 = 742$ mm Hg. In order to calculate V_2, the pressure ratio P_1/P_2 must be dimensionless; that is, P_1 and P_2 must be expressed in the same units. To achieve this, we might convert P_2 from millimeters of mercury to atmospheres:

$$P_2 = 742 \text{ mm Hg} \times \frac{1 \text{ atm}}{760 \text{ mm Hg}} = 0.976 \text{ atm}$$

Now we can solve for V_2:

$$V_2 = 3.6 \text{ cm}^3 \times \frac{2.18 \text{ atm}}{0.976 \text{ atm}} = 8.0 \text{ cm}^3$$

Notice that the bubble expands as it rises ($V_2 > V_1$). This makes sense. Since volume is inversely proportional to pressure, it should increase when pressure decreases.

EXERCISE A gas with a volume of 20.0 cm³ at 1.00 atm expands to 50.0 cm³ at constant T. What is the final pressure? Answer: 0.400 atm.

The equation derived in Example 6.2,

$$P_2V_2 = P_1V_1 \qquad \text{(constant n, T)} \tag{6.9}$$

This Law was discovered in 1660 by Robert Boyle and is one of the oldest physical laws

is a form of Boyle's Law. It tells us that, if we hold the temperature of a sample of gas constant, the pressure-volume product remains constant. Note that in solving this equation, we must use consistent units. As we saw in Example 6.2, P_1 and P_2 must have the same units; this holds as well for V_1 and V_2.

EXAMPLE 6.3 A sealed balloon has a volume of 50.0 m³ when filled with air at 22°C and atmospheric pressure. The air in the balloon is heated. Assuming the pressure remains constant, at what temperature (°C) does the volume of the balloon become 60.0 m³?

Solution Here the variables are V and T; n and P remain constant:

$$\text{final state:} \quad PV_2 = nRT_2 \qquad \text{initial state:} \quad PV_1 = nRT_1$$

Collecting the constant terms n, R, and P on the same side of each equation, final state: $\dfrac{nR}{P} = \dfrac{V_2}{T_2}$ initial state: $\dfrac{nR}{P} = \dfrac{V_1}{T_1}$

We conclude that for this change in state:

$$\dfrac{V_2}{T_2} = \dfrac{V_1}{T_1}$$

Solving for T_2,

$$T_2 = T_1 \times \dfrac{V_2}{V_1}$$

From the statement of the problem $T_1 = 22 + 273 = 295$ K, $V_2 = 60.0$ m³, $V_1 = 50.0$ m³. Hence,

$$T_2 = 295 \text{ K} \times \dfrac{60.0 \text{ m}^3}{50.0 \text{ m}^3} = 354 \text{ K}$$

Converting to degrees Celsius, °C = K − 273 = 354 − 273 = 81°C.

EXERCISE   A stoppered flask full of air at 20°C is heated until the pressure is doubled. What is the final temperature in °C? Answer: 313°C.

The equation derived in Example 6.3,

$$\dfrac{V_2}{T_2} = \dfrac{V_1}{T_1} \qquad \text{(constant n, P)} \tag{6.10}$$

is a form of the Law of Charles and Gay-Lussac. It says that if the pressure of a sample of gas is held constant, the ratio V/T remains constant. In any calculation using this equation, the units of the two volumes must be the same. Moreover, both T_2 and T_1 *must be expressed on the Kelvin scale.*

The original form of the Law was different. See pages 183 and 184

Calculation of P, V, n, or T

In Examples 6.2 and 6.3 the value of the gas constant R was not needed, since R was canceled from the calculations. In a different type of problem, we use R to calculate one of the four quantities, P, V, n, or T, when we know the values of the other three. Example 6.4 illustrates the kind of calculation involved. Later in this chapter and in succeeding chapters we will deal with other examples of this type.

EXAMPLE 6.4   If 3.00 g SF_6 gas is introduced into an evacuated 5.00-L container at 92°C, what is the pressure in atmospheres in the container?

Solution   In this problem, only one state is involved and, for it, PV = nRT:

P = ?;   V = 5.00 L;   T = 273 + 92 = 365 K

R = 0.0821 L·atm/(mol·K)

$$n = 3.00 \text{ g } SF_6 \times \dfrac{1 \text{ mol}}{146.1 \text{ g } SF_6} = 0.0205 \text{ mol}$$

Substituting,

$$P = \frac{nRT}{V} = \frac{0.0205 \text{ mol} \times 0.0821 \dfrac{\text{L·atm}}{\text{mol·K}} \times 365 \text{ K}}{5.00 \text{ L}} = 0.123 \text{ atm}$$

Here all the quantities in the Ideal Gas Law enter the calculation directly. If we use 0.0821 L·atm/(mol·K) for R, the units for P, V, n, and T must be those that appear in R. Any quantity that is not given in these units must be converted to them before substituting in the Ideal Gas Law. Here, for example, we converted grams of SF_6 to moles by dividing by the molar mass, 146.1 g/mol.

EXERCISE   What is the pressure exerted by 2.0 mol O_2 in a 10.0-L flask at 27°C? Answer: 4.9 atm.

Calculation of Molar Mass and Density

For certain problems, it is convenient to rewrite the Ideal Gas Law in a different form. In particular, we need an equation relating P, V, and T to the mass of a gas rather than the number of moles. To obtain such a relation we note that

$$n = \frac{g}{MM}$$

where g is the mass of gas in grams. The quantity MM appearing in this relation is the **molar mass** in grams per mole. Thus, for O_2, the molar mass would be 32.0 g/mol; for N_2, MM = 28.0 g/mol, and so on. Substituting for n in the Ideal Gas Law, we have

$$PV = \frac{gRT}{MM} \tag{6.11}$$

The Ideal Gas Law in this form is useful for calculating the following:

1. The molar mass (MM) of a gas, knowing the mass (g) of a given volume (V) at a certain temperature (T) and pressure (P). Solving Equation 6.11 for MM,

   For many years this equation gave us our best way to find accurate molar masses

   $$MM = \frac{gRT}{PV} \tag{6.12}$$

2. The density (d) of a gas of known molar mass (MM) at a given temperature (T) and pressure (P). From Equation 6.11, on solving for the density, we obtain,

   $$d = \frac{g}{V} = \frac{P \times MM}{RT} \tag{6.13}$$

EXAMPLE 6.5   A sample of xenon tetrafluoride is collected in a flask with a volume of 226 mL at a pressure of 749 mm Hg and a temperature of 12°C.

The mass of the gas is found to be 1.973 g. Calculate the molar mass of xenon tetrafluoride.

Solution   We can calculate the molar mass directly, using Equation 6.12:

$$g = 1.973 \text{ g}; \quad R = 0.0821 \text{ L·atm/(mol·K)}; \quad T = 12 + 273 = 285 \text{ K}$$

$$P = 749 \text{ mm Hg} \times \frac{1 \text{ atm}}{760 \text{ mm Hg}}$$

$$= 0.986 \text{ atm}$$

$$V = 226 \text{ mL} \times \frac{1 \text{ L}}{1000 \text{ mL}}$$

$$= 0.226 \text{ L}$$

$$MM = \frac{1.973 \text{ g} \times 0.0821 \dfrac{\text{L·atm}}{\text{mol·K}} \times 285 \text{ K}}{0.986 \text{ atm} \times 0.226 \text{ L}} = 207 \text{ g/mol}$$

The calculated value compares closely with the theoretical value for XeF_4, 207.3 g/mol.

> It's rather remarkable that we can get information about the molecules in a gas from density data

EXERCISE   What is the molar mass of a gas whose density is 5.00 g/L at 25°C and 1.00 atm? Answer: 122 g/mol.

EXAMPLE 6.6   The molar mass of air can be considered to be a weighted average of those of nitrogen and oxygen, its main components. Taking the molar mass of air to be 29.0 g/mol, calculate the density of air at 27°C and 1.00 atm.

Solution   We use Equation 6.13 to calculate the density:

$$d = \frac{P \times MM}{RT} = \frac{1.00 \text{ atm} \times 29.0 \text{ g/mol}}{0.0821 \dfrac{\text{L·atm}}{\text{mol·K}} \times 300 \text{ K}} = 1.18 \text{ g/L}$$

EXERCISE   What is the density of $O_2(g)$ at 1.00 atm and 27°C? Answer: 1.30 g/L.

6.3
Volumes of Gases
Involved in Reactions

We saw in Chapter 3 that a balanced equation can be used to relate moles or masses of substances taking part in a reaction. Where gases are involved, we can extend these relations to include volumes. To do this, we use the Ideal Gas Law and the conversion factor approach described in Chapter 3. The calculations required are illustrated in Examples 6.7 and 6.8.

EXAMPLE 6.7   What mass in grams of hydrogen peroxide must be used to produce 1.00 L $O_2(g)$, measured at 25°C and 1.00 atm? The reaction is

$$2 \text{ H}_2\text{O}_2(aq) \rightarrow \text{O}_2(g) + 2 \text{ H}_2\text{O}$$

Solution   We follow a three-step procedure:

CONVERSION	REQUIRED RELATION
1. volume O_2 → moles O_2	$n = PV/RT$
2. moles O_2 → moles H_2O_2	2 mol H_2O_2 ≏ 1 mol O_2
3. moles H_2O_2 → grams H_2O_2	1 mol H_2O_2 = 34.0 g H_2O_2

Substituting into the Ideal Gas Law,

$$n\ O_2 = \frac{1.00\ \text{atm} \times 1.00\ \text{L}}{0.0821\ \text{L·atm/(mol·K)} \times 298\ \text{K}} = 0.0409\ \text{mol}\ O_2$$

The last two steps can be carried out in a single-line calculation (Chapter 3):

$$\text{grams}\ H_2O_2 = 0.0409\ \text{mol}\ O_2 \times \frac{2\ \text{mol}\ H_2O_2}{1\ \text{mol}\ O_2} \times \frac{34.0\ \text{g}\ H_2O_2}{1\ \text{mol}\ H_2O_2}$$

$$= 2.78\ \text{g}\ H_2O_2$$

We conclude that it takes somewhat less than 3 g of hydrogen peroxide to form 1 L of oxygen gas at room temperature and atmospheric pressure.

EXERCISE   What mass of H_2O is formed as a by-product in the reaction described in this example? Answer: 1.47 g H_2O.

EXAMPLE 6.8   How many liters of oxygen, measured at 740 mm Hg and 24°C, are required to burn 1.00 g of octane, $C_8H_{18}(l)$, to carbon dioxide and water?

Solution   We first write the balanced equation for the reaction, using the approach described in Chapter 3:

$$2\ C_8H_{18}(l) + 25\ O_2(g) \rightarrow 16\ CO_2(g) + 18\ H_2O(l)$$

Now we again follow a three-step approach:
1. Convert 1.00 g of octane to moles (1 mol C_8H_{18} = 114 g C_8H_{18}).
2. Convert moles of octane to moles of oxygen (2 mol C_8H_{18} ≏ 25 mol O_2).
3. Convert moles of oxygen to volume, using the Ideal Gas Law.
Combining the first two steps,

$$\text{no. moles}\ O_2 = 1.00\ \text{g}\ C_8H_{18} \times \frac{1\ \text{mol}\ C_8H_{18}}{114\ \text{g}\ C_8H_{18}} \times \frac{25\ \text{mol}\ O_2}{2\ \text{mol}\ C_8H_{18}}$$

$$= 0.110\ \text{mol}\ O_2$$

We find the volume of oxygen by substituting into the Ideal Gas Law:

$$V = \frac{nRT}{P} = \frac{0.110\ \text{mol} \times 0.0821\ \text{L·atm/(mol·K)} \times 297\ \text{K}}{(740/760)\ \text{atm}} = 2.75\ \text{L}$$

This reaction is very similar to that occurring in an automobile engine. Clearly, a large volume of air (which is only 21% O_2 by volume) must pass through the engine during combustion.

Chemistry is often just one conversion factor problem after another

$$V_{air} = V_{O_2} \times \frac{100\ \text{L air}}{21\ \text{L}\ O_2}$$
$$= 13\ \text{L}$$

EXERCISE   How many liters of $O_2(g)$ at 1.00 atm and 27°C are required to burn 1.00 g of octane? Answer: 2.71 L.

The two examples just worked illustrate what is perhaps the most important application of the gas laws to chemistry. Here your objective is to relate the volume of a gas A to the mass of another species B involved in a reaction with A.

To go from volume of A to mass of B, you follow a three-step path.

1. **Convert volume of A to moles of A, using the Ideal Gas Law**

    Volume of A = (moles of A)(RT/P)

2. **Convert moles of A to moles of B, using the coefficients of the balanced equation:**

    a moles of A $\backsimeq$ b moles of B

    where a and b are the coefficients of A and B in that order.
3. **Convert moles of B to grams of B, using its molar mass**

    moles of B = grams of B/molar mass B

To go in the reverse direction, from a known mass of B to a required volume of A, you simply go through this three-step process in reverse:

grams of B → moles of B → moles of A → Volume of A

Law of Combining Volumes

When two different gases are involved in a reaction, there is a simple relationship between their volumes, measured at the same temperature and pressure. To find this relationship, consider two gases, 1 and 2, both at T and P. Applying the Ideal Gas Law to both gases,

$$V_2 = n_2RT/P$$
$$V_1 = n_1RT/P$$

Dividing the first equation by the second, RT/P cancels and we obtain

$$V_2/V_1 = n_2/n_1 \qquad\qquad (6.14)$$

This eqn gives us a simple experimental handle on mole ratios

This tells us that the volume ratio in which two gases react at T and P is the same as the reacting mole ratio.

To see what Equation 6.14 implies, consider the reaction

$$N_2(g) + 3 H_2(g) \rightarrow 2 NH_3(g) \qquad\qquad (6.15)$$

We found in Chapter 3 that the coefficients of a balanced equation can be interpreted in terms of moles. In this case, 1 mol N_2 reacts with 3 mol H_2 to form 2 mol NH_3:

1 mol N_2 $\backsimeq$ 3 mol H_2 $\backsimeq$ 2 mol NH_3

Nitrogen   +   Hydrogen ⟶ Ammonia

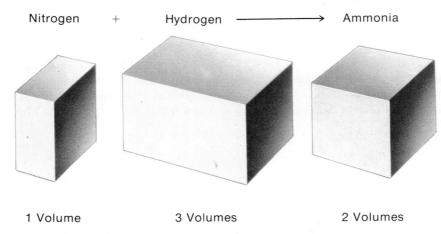

1 Volume   3 Volumes   2 Volumes

FIGURE 6.5 Gay-Lussac's Law of Combining Volumes as applied to the reaction $N_2(g)$ + 3 $H_2(g)$ → 2 $NH_3(g)$. When measured under the same conditions, the volumes of reacting gases have the same ratios as their coefficients in the equation for the reaction.

However, since the reacting volume ratio, at a given temperature and pressure, is the same as the mole ratio, 1 L N_2 must react with 3 L H_2 to form 2 L NH_3 (Fig. 6.5):

$$1 \text{ L } N_2 \approx 3 \text{ L } H_2 \approx 2 \text{ L } NH_3 \quad \text{(same T, P)}$$

We see that the reacting volume ratio is given by the coefficients of the balanced equation. More generally, we can say the following:

This Law made possible the first determination of reliable atomic masses (Cannizzaro, 1860)

   The volumes of different gases involved in a reaction, if measured at the same temperature and pressure, are in the same ratio as the coefficients in the balanced equation.

   This relation, known as the Law of Combining Volumes, was first proposed in a somewhat different form by Gay-Lussac in 1808. Its use is shown in Example 6.9.

> **EXAMPLE 6.9** In Reaction 6.15, what volume of H_2 at 22°C and 719 mm Hg is required to form 12.0 L NH_3, measured at the same temperature and pressure?
>
> **Solution** From the coefficients of the balanced equation,
>
> $$3 \text{ L } H_2 \approx 2 \text{ L } NH_3$$
>
> $$\text{volume } H_2 = 12.0 \text{ L } NH_3 \times \frac{3 \text{ L } H_2}{2 \text{ L } NH_3} = 18.0 \text{ L } H_2$$
>
> Note that it makes no difference what the temperature and pressure are, so long as they are the same for both gases.
>
> **EXERCISE** In the reaction referred to in this example, what volume of $N_2(g)$ is required? Answer: 6.0 L N_2.

6.4
Mixtures of Gases;
Dalton's Law

So far we have concentrated upon the behavior of pure gases. Frequently, however, we deal with gaseous mixtures where more than one substance is present. Here, a relation discovered by John Dalton in 1801 is very helpful. Dalton's Law states the following:

> *The total pressure of a gas mixture is the sum of the partial pressures of the components of the mixture.*

For a mixture of two gases A and B,

$$P_{tot} = P_A + P_B \tag{6.16}$$

Here, P_A is the partial pressure of gas A; P_B is the partial pressure of gas B. The partial pressure of a gas is the pressure it would exert if it were alone in the container at the same temperature as the mixture.

The properties of a gas in a mixture do not depend on the other gases present

Dalton's Law is perhaps applied most often in calculations involving gases collected over water (Fig. 6.6). Here, the gas being collected is mixed with water vapor. The only pressure that can be measured directly is the total pressure of the mixture. The partial pressure of the water vapor is readily obtained. It has a fixed value at a given temperature, shown in a table of water vapor pressures (Appendix 1). By subtracting the partial pressure of $H_2O(g)$ from the total pressure of the mixture, we obtain the partial pressure of the gas under study. This in turn can be used to determine the number of moles of that gas present (Example 6.10).

EXAMPLE 6.10   A student prepares a sample of hydrogen gas by electrolyzing water at 25°C. She collects 152 mL H_2 at a total pressure of 758 mm Hg. Using Appendix 1 to find the vapor pressure of water, calculate
a. the partial pressure of hydrogen gas.
b. the number of moles of hydrogen collected.

Solution
a. The collected gas is a mixture of hydrogen and water vapor. Using Dalton's Law,

$$P_{tot} = P_{H_2} + P_{H_2O}; \qquad P_{H_2} = P_{tot} - P_{H_2O}$$

The total pressure is given as 758 mm Hg; from Appendix 1 we find that the vapor pressure of water at 25°C is 23.76 mm Hg. Hence,

$$P_{H_2} = 758 \text{ mm Hg} - 23.76 \text{ mm Hg} = 734 \text{ mm Hg}$$

b. We use the Ideal Gas Law, where P is the partial pressure of hydrogen:

$$n = \frac{PV}{RT} = \frac{(734/760 \text{ atm})(0.152 \text{ L})}{\left(0.0821 \frac{\text{L·atm}}{\text{mol·K}}\right)(298 \text{ K})} = 0.00600 \text{ mol}$$

P must be in atm if you use this value of R

EXERCISE   Hydrogen gas is collected over water at a total pressure of 744 mm Hg at 20°C. Using the table of water vapor pressures in Appendix 1, calculate the partial pressure of the $H_2(g)$. Answer: 726 mm Hg.

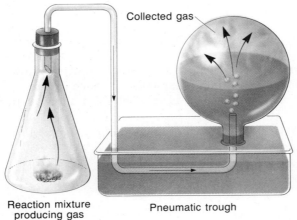

Reaction mixture
producing gas

Pneumatic trough

Collected gas

FIGURE 6.6 The collection of gases over water. With this apparatus the collected gas will contain water vapor at a partial pressure equal to the vapor pressure of water at the temperature of the system.

The validity of Dalton's Law is readily shown using the Ideal Gas Law. For a mixture of two gases A and B, we can write

$$P_{tot} = n_{tot} \times \frac{RT}{V} = (n_A + n_B) \times \frac{RT}{V}$$

Separating terms on the right side of this equation,

In the equation PV = nRT, n = n_{total}

$$P_{tot} = n_A\frac{RT}{V} + n_B\frac{RT}{V}$$

Applying the Ideal Gas Law again, this time to obtain the partial pressure of the individual gases,

$$P_A = n_A\frac{RT}{V}; \qquad P_B = n_B\frac{RT}{V}$$

Hence, $P_{tot} = P_A + P_B$, which is Equation 6.16.

Partial Pressure and Mole Fraction

As we have just seen, for a gas mixture,

$$P_A = n_A\frac{RT}{V} \qquad \text{and} \qquad P_{tot} = n_{tot}\frac{RT}{V}$$

If we divide P_A by P_{tot}, we obtain

$$\frac{P_A}{P_{tot}} = \frac{n_A}{n_{tot}} \qquad \text{or} \qquad P_A = \frac{n_A}{n_{tot}} \times P_{tot} \tag{6.17}$$

The fraction n_A/n_{tot} is referred to as the **mole fraction** of A in the mixture. It is the fraction of the total number of moles that is accounted for by gas A. Using X to represent the mole fraction,

Mole fraction is a concentration term

$$X_A = \frac{n_A}{n_{tot}} \qquad (6.18)$$

Substituting for n_A/n_{tot} in Equation 6.17, we have

$$P_A = X_A P_{tot} \qquad (6.19)$$

In other words, *the partial pressure of a gas in a mixture is equal to its mole fraction multiplied by the total pressure.*

EXAMPLE 6.11 Calculate the mole fractions of H_2 and H_2O in the gas mixture referred to in Example 6.10.

Solution In Example 6.10a, we found that the partial pressure of H_2 was 734 mm Hg; the total pressure is 758 mm Hg. Solving Equation 6.19 for mole fractions, we obtain

$$X_{H_2} = \frac{P_{H_2}}{P_{tot}} = \frac{734 \text{ mm Hg}}{758 \text{ mm Hg}} = 0.969$$

The mole fraction of water is found similarly, taking its partial pressure to be 23.76 mm Hg:

$$X_{H_2O} = \frac{P_{H_2O}}{P_{tot}} = \frac{23.76 \text{ mm Hg}}{758 \text{ mm Hg}} = 0.0313$$

In the gas mixture, $X_{H_2} + X_{H_2O} = 1$

EXERCISE The mole fraction of nitrogen in air is about 0.79. What is the partial pressure of N_2 in air at a total pressure of 740 mm Hg? Answer: 5.8×10^2 mm Hg.

6.5
Real Gases

In this chapter, we have used the Ideal Gas Law in all our calculations, assuming it applies exactly. Under ordinary conditions (and in nearly all problems in this text), this assumption is a good one; however, all real gases deviate at least slightly from the Ideal Gas Law. Ordinarily, we find that the deviations become larger at *low temperatures* and *high pressures*, where the gas particles (atoms or molecules) are relatively close to one another.

Deviations from the Ideal Gas Law arise because it neglects two factors:

1. The finite volume of gas particles.
2. Attractive forces between gas particles.

The effect of these factors depends upon the temperature and pressure.

Temperature

As the temperature is lowered, gas particles move more slowly. This makes attractive forces more important, which in turn makes the gas easier to compress. The attractive forces make the "effective pressure" operating on the gas particles greater than the measured pressure. This makes the observed molar volume, V_{obs}, less than that calculated from the Ideal Gas Law, V_{ideal}. To illustrate this effect, consider oxygen gas. At 25°C and 1 atm, the observed molar volume of O_2 is within 0.1% of that predicted by the Ideal Gas Law. Under these conditions, O_2 behaves ideally for all practical purposes. When oxygen is cooled to $-150°C$, the deviation from the Ideal Gas Law becomes ten times as great. At $-150°C$ and 1 atm, V_{obs} for O_2 is about 1% less than V_{ideal}.

Here we are speaking of gas pressures of 1 atm or lower

At the boiling point of a substance, attractive forces between particles are large enough to make a gas condense to a liquid. The closer a gas is to its boiling point, the more it will deviate from the Ideal Gas Law. Consider, for example, SO_2. Sulfur dioxide boils at $-10°C$ at 1 atm. At 25°C and 1 atm, it is close enough to the boiling point for attractive forces to be significant. The molar volume of SO_2 under these conditions is about 1% less than that calculated from the Ideal Gas Law.

Pressure

In Figure 6.7, the ratio V_{obs}/V_{ideal} for $CH_4(g)$ at 25°C is plotted as a function of pressure. As you might expect, this ratio starts off at 1.00 at zero pressure. As the pressure increases, V_{obs}/V_{ideal} drops below 1 and decreases steadily up to about 150 atm. This behavior is caused by the attraction between the molecules. The effect is in the same direction as that observed when the temperature is lowered.

The molecules in a gas take up space; that space decreases the volume through which the molecules can move

The behavior of $CH_4(g)$ at very high pressures at 25°C is shown at the right of Figure 6.7. Above about 150 atm, the ratio V_{obs}/V_{ideal} starts to increase. Here we are seeing the effect of the finite volume of the methane molecules. At high pressures their volume becomes significant compared to that of the container. The volume of the molecules contributes to the observed volume, so that V_{obs}/V_{ideal}

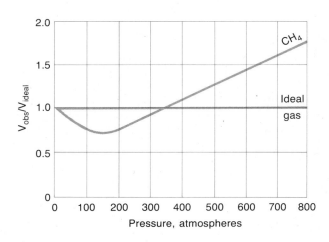

FIGURE 6.7 The pressure-volume behavior of $CH_4(g)$ at 25°C. Below about 350 atm, attractive forces between CH_4 molecules cause the observed volume to be less than that calculated from the Ideal Gas Law. At 350 atm, the effect of the attractive forces is just balanced by that of the finite volume of CH_4 molecules, and the gas appears to behave ideally. Above 350 atm, the effect of finite molecular volume predominates and $V_{obs} > V_{ideal}$.

increases. By the time that pressure reaches 350 atm, the effect of molecular volume just about cancels that of the attractive forces: V_{obs}/V_{ideal} becomes 1.00 at this point. Above 350 atm, the molecular volume factor predominates and V_{obs}/V_{ideal} becomes greater than 1.

The graph shown in Figure 6.7 for CH_4 is typical of most gases. Ordinarily,

—at low pressures, close to 1 atm, V_{obs} is nearly equal to V_{ideal}.
—at intermediate pressures, perhaps as high as several hundred atmospheres, $V_{obs} < V_{ideal}$.
—at very high pressures, $V_{obs} > V_{ideal}$.

The exact shape of the curve will depend upon the identity of the gas and the temperature.

van der Waals Equation

It is possible to write equations involving P, V, and T for gases which take into account attractions between particles and finite particle volumes. One of these is the van der Waals equation:

$$\left(P + \frac{n^2a}{V^2}\right)(V - nb) = nRT \tag{6.20}$$

In this equation, a and b are independent of P, V, n, and T. They do, however, vary from one gas to another (Table 6.2). Their values are selected to give the best possible agreement between the equation and the observed behavior of the gas. The term n^2a/V^2 reflects the attractive forces between particles, while the term nb corrects for the effect of particle volume. The constant b is approximately equal to the volume of one mole of particles. The van der Waals equation is much better than the Ideal Gas Law for predicting the behavior of gases at high pressures.

V − nb = volume in which the molecules can move

Table 6.2
van der Waals Constants

Gas	a $\left(\dfrac{L^2 \cdot atm}{mol^2}\right)$	b (L/mol)
H_2	0.244	0.027
O_2	1.360	0.032
N_2	1.390	0.039
CH_4	2.253	0.043
CO_2	3.592	0.043
SO_2	6.714	0.056
Cl_2	6.493	0.056
H_2O	5.464	0.030

6.6
Kinetic Theory of Gases

The fact that the Ideal Gas Law applies to all gases indicates that the gaseous state is a relatively simple one to treat from a theoretical point of view. Gases must have certain properties in common that cause them to follow the same natural law. Between about 1850 and 1880, Maxwell, Boltzmann, Clausius, and others developed the kinetic theory of gases to explain these similarities in the behavior of gases. They based it on the idea that all gases behave similarly insofar as particle motion is concerned. Since that time, the kinetic theory has had to be modified only slightly. In its present form it is one of the most successful of scientific theories. It ranks in stature with the atomic theory of matter.

Postulates of the Kinetic Theory

1. **Gases consist of particles (atoms or molecules) in continuous, random motion.** These particles undergo frequent collisions with one another and with the container walls. The pressure exerted by a gas is due to the forces associated with wall collisions (Fig. 6.8).

2. **Collisions between gas particles are elastic.** When two particles collide, their individual energies may change. Typically, one speeds up and the other slows down. The total energy, however, remains the same. No kinetic energy is con-

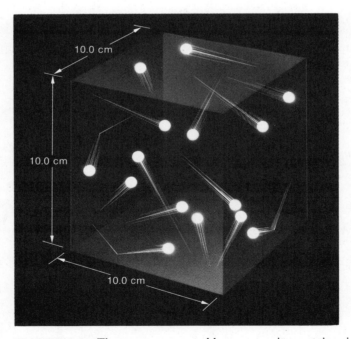

FIGURE 6.8 The pressure exerted by a gas on its container is the same in all directions and is caused by collisions of the gas molecules with the container walls. The force per collision is very small but there are a great many collisions per second on any given area.

verted to heat. As a result, the temperature of a gas insulated from its surroundings does not change.

3. **The average energy of translational motion of a gas particle is directly proportional to temperature.** At a given temperature, this average energy is the same for all gases. The energy associated with the motion of a particle from one place to another is related to its speed by the equation

$$E_t = \frac{mu^2}{2}$$

Here, E_t is the average kinetic energy of translation, m is the mass of the particle, and u is the corresponding velocity, which we will call the average speed. Kinetic theory tells us that E_t, and hence $mu^2/2$, are directly proportional to temperature:

$$E_t = \frac{mu^2}{2} = cT \tag{6.21}$$

The fact that the average energy of a gas molecule does not depend on m is the main reason gases have the same PVT behavior

In this equation, T is the absolute temperature in K. The quantity c is a constant that has the same value for all gases.

In addition to these postulates, it is assumed that the volumes of the particles are negligible as compared to container volume. Moreover, attractive forces between particles are neglected.

The postulates of the kinetic theory are easily stated. Their implications, however, are by no means obvious. The problem is that gas particles move about in a completely random manner. Their velocities are constantly changing in both magnitude and direction as they collide with each other and with the walls of their container. To treat this kind of motion in a rigorous way requires mathematics of a high level of sophistication. In this text, we will not attempt to present such a mathematical development. Instead, we will concentrate upon some of the relationships that have resulted from the kinetic theory.

Gas Pressure

As noted in Section 6.2, the pressure of a gas (P) is dependent upon container volume (V), number of moles (n), and temperature (T). We can readily understand the effect of these factors in terms of kinetic theory, which tells us that gas pressure is due to collisions with the walls of the container.

1. P increases as n increases (constant V, T) because, with more particles present, there are more collisions per unit time.
2. P increases as V decreases (constant n, T) because, in a smaller volume, gas particles strike the walls more often.
3. P increases as T increases (constant n, V) because, at a higher temperature, particles move more rapidly and collisions occur more often and with greater force.

Effusion of Gases;
Graham's Law

The flow of gas particles through a small opening or pinhole in a container is referred to as *effusion*. The rate of effusion is directly proportional to the average speed of gas particles. Thus, the rates of effusion of two gases, A and B, from the same container at the same pressure are in the ratio of their average speeds:

$$\frac{\text{rate of effusion of A}}{\text{rate of effusion of B}} = \frac{\text{average speed of A}}{\text{average speed of B}} = \frac{u_A}{u_B} \qquad (6.22)$$

By Equation 6.21, if the two gases are at the same temperature,

$$\frac{m_A u_A^2}{2} = \frac{m_B u_B^2}{2} = cT$$

or

$$\frac{u_A^2}{u_B^2} = \frac{m_B}{m_A} = \frac{MM_B}{MM_A} \qquad (6.23)$$

where MM_A and MM_B are the molar masses of A and B. If Equation 6.23 is solved for the ratio u_A/u_B, which is then substituted into 6.22, we obtain

$$\frac{\text{rate of effusion of A}}{\text{rate of effusion of B}} = \left(\frac{MM_B}{MM_A}\right)^{\frac{1}{2}} \qquad (6.24)$$

Graham stated the Law in terms of gas densities rather than MM's

This relation, in a slightly different form, was discovered experimentally by Thomas Graham in 1828. It is known as Graham's Law.

This law gives us a way of determining molar masses of gases. All we need to do is compare the rate of effusion of the gas we are interested in to that of another gas of known molar mass. Usually we measure the time required for equal amounts of the two gases to effuse under the same conditions of temperature and pressure. The measured times are *inversely* related to rates: the faster the gas effuses, the less the time required for a given amount to effuse. Hence,

$$\frac{\text{time}_B}{\text{time}_A} = \frac{\text{rate}_A}{\text{rate}_B} = \left(\frac{MM_B}{MM_A}\right)^{\frac{1}{2}} \qquad (6.25)$$

In other words, the time required for effusion increases with molar mass: heavy molecules take longer to effuse (Figure 6.9).

EXAMPLE 6.12 In an effusion experiment, it required 45 s for a certain number of moles of an unknown gas, X, to pass through a small opening into a vacuum. Under the same conditions, it took 28 s for the same number of moles of Ar to effuse. Find the molar mass of the unknown gas.

Solution We use the relation between time of effusion and molar mass, Equation 6.25:

$$\frac{\text{time for Ar}}{\text{time for X}} = \frac{28}{45} = \left(\frac{MM \text{ of Ar}}{MM \text{ of X}}\right)^{\frac{1}{2}}$$

To solve this equation, we square both sides:

$$\frac{MM \text{ of Ar}}{MM \text{ of X}} = \left(\frac{28}{45}\right)^2 = 0.39$$

Solving for the molar mass of the unknown gas,

$$\text{MM of X} = \frac{\text{MM of Ar}}{0.39} = \frac{39.9 \text{ g/mol}}{0.39} = 1.0 \times 10^2 \text{ g/mol}$$

EXERCISE A certain gas takes only one fourth as long to effuse as O_2. What is its molar mass? Answer: 2.0 g/mol.

During World War II, Graham's Law was applied to a rather complex chemical problem. It became necessary to separate $^{235}_{92}U$, which is fissionable (Chap. 5), from the more abundant isotope of uranium, $^{238}_{92}U$. Since the two isotopes have almost identical chemical properties, chemical separation was not feasible. Instead, an effusion process was worked out using uranium hexafluoride, UF_6. This compound is a gas at room temperature. Preliminary experiments indicated that $^{235}_{92}UF_6$ could indeed be separated from $^{238}_{92}UF_6$ by effusion. An enormous plant was built for this purpose in Oak Ridge, Tennessee. In this process, UF_6 effuses many thousands of times through porous barriers. The lighter fractions move on to the next stage while heavier fractions are recycled through earlier stages. Eventually, a nearly complete separation of the two isotopes is achieved.

If UF_6 were not volatile we might never have developed nuclear energy

Average Speeds of Gas Particles

We have seen that Equation 6.21 can be used in a straightforward way to calculate relative speeds of two different kinds of particles. For many purposes in chemistry, that calculation is sufficient. It is possible, however, to go one step further and calculate the average speed of a given kind of gas particle. Here again we use Equation 6.21:

$$\frac{mu^2}{2} = cT$$

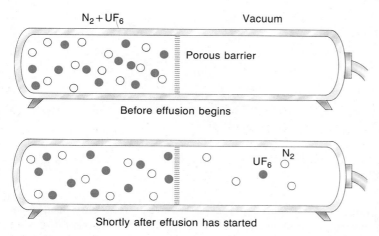

Before effusion begins

Shortly after effusion has started

FIGURE 6.9 Effusion of gases. N_2 molecules move through the barrier faster than UF_6 molecules, since they are lighter. After a short time, the gas to the right of the barrier is mostly N_2. The gas remaining in the compartment at the left is enriched in UF_6. This process can be used to separate a mixture of two gases.

This time, we need an expression for the constant c in this equation. By a development that we will not attempt to go through here, it can be shown that

$$c = \frac{3R}{2N} \tag{6.26}$$

where R is the gas constant and N is Avogadro's number. Substituting in Equation 6.21, we obtain

$$mu^2 = \frac{3RT}{N} \quad \text{or} \quad u^2 = \frac{3RT}{mN}$$

However, the product of the mass of a particle, m, times the number of particles in a mole, N, is simply the molar mass, MM. Therefore,

$$u^2 = \frac{3RT}{MM}$$

or

$$u = \left(\frac{3RT}{MM}\right)^{\frac{1}{2}} \tag{6.27}$$

We can calculate an average molecular speed by substituting into Equation 6.27. We must be careful, however, to select the proper value of R (recall Table 6.1, p. 163). If we want to obtain u in centimeters per second, we should use

$$R = 8.31 \times 10^7 \frac{g \cdot cm^2}{s^2 \cdot mol \cdot K}$$

EXAMPLE 6.13 Find the average speed of an oxygen molecule in air at room temperature (25°C).

Solution In Equation 6.27, T = 25 + 273 = 298 K; MM = 32.0 g/mol.

$$u = \left(\frac{3RT}{MM}\right)^{\frac{1}{2}} = \left(\frac{3 \times 8.31 \times 10^7 \text{ g} \cdot cm^2/(s^2 \cdot mol \cdot K) \times 298 \text{ K}}{32.0 \text{ g/mol}}\right)^{\frac{1}{2}}$$
$$= 4.82 \times 10^4 \text{ cm/s} = 482 \text{ m/s}$$

By a simple conversion (1 m/s = 2.24 miles/h) we can show that, in more familiar units, the average speed is about 1000 miles/h!

EXERCISE What is the average speed of an H_2 molecule at 25°C? Answer: 1.92×10^3 m/s.

Molecules move pretty fast in a gas, but don't go very far between collisions, $\sim 10^{-5}$ cm

According to kinetic theory, particle speeds are very high. This prediction can be tested by experiments that show that measured speeds agree quite well with those calculated. Qualitatively, the average speed of O_2 and N_2 molecules agrees with the observed speed of sound in air. Since sound is carried by molecular motion, one would expect its speed roughly to equal that of the molecules through which it passes. The speed of sound in air is about 350 m/s, comparable to the speeds of O_2 or N_2 molecules at 25°C.

Distribution of Molecular Speeds and Energies

It is possible to use Equation 6.27 to calculate the average speed of a gas particle at a particular temperature. We must remember, however, that not all particles in a gas sample will have that speed. The motion of particles in a gas is utterly chaotic. In the course of a second, a particle undergoes millions of collisions with other particles. As a result, the speed and direction of motion of a particle is constantly changing. Over a period of time, the speed will vary from almost zero to some very high value, considerably above the average.

In 1860, James Clerk Maxwell showed that different possible speeds are distributed among particles in a definite way. Indeed, he developed a mathematical expression for this distribution. His results are shown graphically in Figure 6.10 for O_2 at 25°C and at 1000°C. On the graph, we plot the relative number of molecules having a certain speed, u, against that speed. We see that at 25°C this number increases rapidly with the speed, up to a maximum of about 400 m/s. This is the most probable speed of an oxygen molecule at 25°C. Above about 400 m/s, the number of molecules moving at any particular speed decreases. A molecule is only about one fifth as likely to be moving at 800 m/s as it is to be moving at 400 m/s. For speeds in excess of about 1200 m/s, the fraction of molecules drops off to nearly zero. In general, most molecules have speeds rather close to the average value.

As temperature increases, the speed of the molecules increases. The distribution curve for molecular speeds (Fig. 6.10) shifts to the right and becomes broader. The chance of a molecule having a very high speed is much greater at 1000°C than it is at 25°C. Note, for example, that a large number of molecules have speeds greater than 1200 m/s at 1000°C.

There are almost no molecules moving at very low speeds

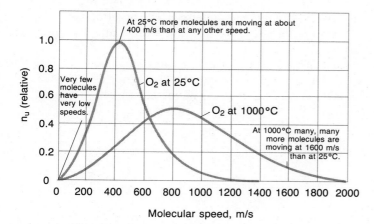

FIGURE 6.10 The fraction of O_2 molecules having a given speed at 25°C and at 1000°C. At 25°C, most of the molecules are moving at a speed close to 400 m/s. At 1000°C, the average speed is about twice as great. More important, the fraction of molecules moving at very high speeds is much greater at the higher temperature.

Since translational energy is simply related to speed (Equation 6.21), we can talk about a distribution of energies as well as speeds. A plot of the relative number of particles having a certain energy, E, against that energy would resemble Figure 6.10. Here again, *the fraction of particles having very high energies increases sharply with temperature*. As we will see in Chapter 16 it is these high-energy particles that take part in chemical reactions.

6.7 Historical Perspective on the Gas Laws

For more than 300 years, gases have been involved, in one way or another, with the evolution of chemistry and physics. Many famous scientists have made their reputations through experiments or theoretical calculations involving gases. Among these are several individuals mentioned in this or previous chapters: Boyle, Charles, Gay-Lussac, Avogadro, Cannizzaro, Joule, Lord Kelvin, and Maxwell.

Robert Boyle (1627–1691)

Boyle was born in Ireland, the fourteenth child of the Earl of Cork. He was one of the first experimental scientists. Among other things, Boyle invented a vacuum pump and used it to show that air is necessary for combustion, respiration, and the transmission of sound. His discovery, in 1660, of the law that bears his name ($P_2V_2 = P_1V_1$) grew out of his work on the properties of air.

He also used the pump to prove that the barometer measured the pressure of the atmosphere. How could he do that?

Boyle was a prolific author on a wide variety of topics ranging from religious philosophy to the structure of matter. The best known of his publications is a book entitled *The Sceptical Chemist*. Here he attacked Aristotle's theory of four elements (earth, air, fire, and water) and the three principles of Paracelsus (salt, sulfur, and mercury). Instead, Boyle proposed that matter was composed of particles of various sorts that could arrange themselves into groups. Groups of one kind constituted a chemical substance. In this sense, he used concepts of atomic and molecular theory similar to those we have today.

Jacques Charles (1746–1823) and Joseph Gay-Lussac (1778–1850)

Charles started his career as a minor bureaucrat in the government of France. When he was dismissed from that post in an economy move, he turned to the study of gases. On December 1, 1783, Charles was on the second balloon ever to lift a human being off the surface of the earth. This accomplishment so impressed Louis XVI that Charles was given a laboratory at the Sorbonne.

In 1787, Charles discovered that the temperature dependence of gas volume could be expressed by the equation

$$V = V_0 (1 + \alpha t) \tag{6.28}$$

where V_0 is the volume at 0°C, t is the Celsius temperature, and α is a constant. The quantity α is the same for all gases and is approximately equal to 1/273; that is, when the temperature increases 1°C, the volume of a gas increases by 1/273 of its value at 0°C.

Charles never published his work; Gay-Lussac found out about it by accident. He repeated Charles' work and published his results in 1802. Gay-Lussac, like Charles, was fascinated by balloons. In 1804, he ascended to a height of 7 km in a hydrogen-filled balloon. This altitude record remained unbroken for 50 years. Unlike Charles, Gay-Lussac had a keen interest in chemistry. Among his accomplishments were the isolation of the element boron, the preparation of hydrogen fluoride, and the identification of prussic acid, an extremely toxic gas, as HCN.

He was lucky to survive the experience

Gay-Lussac's most important contribution was the Law of Combining Volumes, published in 1808. This law was based on studies of several reactions involving gases, including that between hydrogen and oxygen; Gay-Lussac found that

2 volumes hydrogen + 1 volume oxygen → 2 volumes water vapor

Gay-Lussac seems not to have grasped the implications of his work. However, John Dalton saw where the Law of Combining Volumes led, and he didn't like what he saw. He realized that this simple, integral relation between volumes implied an equally simple relation between particles:

2 particles hydrogen + 1 particle oxygen → 2 particles water

At this point, Dalton, equating particles with atoms, was in trouble. One atom of oxygen could hardly yield two particles of water, both of which must contain at least one atom of oxygen (in Dalton's words, "Thou canst not split an atom"). Faced with what he took to be a direct challenge to the atomic theory, Dalton attempted to discredit Gay-Lussac's results, citing contradictory experiments of his own. His argument accomplished little except to prove that Dalton was a better theoretician than experimentalist.

Avogadro (1776–1856) and Cannizzaro (1826–1910)

In 1811, an Italian physicist at the University of Turin with the improbable name of Lorenzo Romano Amedeo Carlo Avogadro di Quarequa e di Cerreto resolved the dispute between Gay-Lussac and Dalton. He pointed out that Dalton had confused the concepts of atoms and molecules. If an oxygen molecule is *diatomic,* two molecules of water, each containing one atom of oxygen, can be formed from one oxygen molecule. Avogadro interpreted Gay-Lussac's work with hydrogen, oxygen, and water to mean that

$$2 \ H_2 \ \text{molecules} \ + \ 1 \ O_2 \ \text{molecule} \rightarrow 2 \ H_2O \ \text{molecules}$$

Going one step further, Avogadro suggested that *equal volumes of all gases at the same temperature and pressure contain the same number of molecules*.

Avogadro's Law, which we have just stated, offers a simple method of determining molar masses. All one has to do is to determine the masses of a fixed volume, say 1 L, of different gases at the same temperature and pressure. These masses must be in the same ratio as the molar masses of the different gases. Unfortunately, this argument, which seems so obvious today, made little or no impact on Avogadro's contemporaries. For one thing, scientists of that time refused to believe that an element could form a diatomic molecule.

Avogadro's ideas lay fallow for nearly half a century. They were revived in 1860 by a fellow countryman, Stanislao Cannizzaro, Professor of Chemistry at Genoa. He showed that Avogadro's Law could be used to determine not only molar masses, but also, indirectly, atomic masses. Cannizzaro had more impact than Avogadro, perhaps because he presented his ideas more clearly. He made a major contribution to the development of the atomic mass scale discussed in Chapter 2.

James Joule (1818–1889) and William Thomson (1824–1907)

The Law of Charles and Gay-Lussac, relating gas volume to temperature, was originally expressed in degrees Celsius. Half a century passed before the absolute temperature scale (K) was developed. That scale came about through the work of two remarkable British physicists. One of these was James Joule, who studied chemistry under Dalton, did physics experiments in his home, and is generally credited with discovering the Law of Conservation of Energy and the First Law of Thermodynamics. Joule seems to have been the first to recognize that if the straight line of V vs. t (Equation 6.28) is extrapolated, the volume goes to zero at $-273°C$; that is,

$$V \rightarrow 0 \qquad \text{as} \qquad t \rightarrow -273°C$$

In 1847, Joule suggested that since this relationship holds for all gases, a temperature of $-273°C$ might serve as a starting point for a new temperature scale.

Joule's suggestion was developed by William Thomson, Professor of Physics at the University of Glasgow. Thomson was something of a child prodigy; he entered college at the age of 10 and published his first scientific paper at 15. In 1848, at the age of 24, Thomson showed by a thermodynamic argument that $-273°C$ is the lowest attainable temperature with any substance in any physical state. He went on to establish the absolute temperature scale that we use today. This was only one of Thomson's accomplishments. Among other things, he played a major role in establishing the Second Law of Thermodynamics (Chap. 14). In 1892, he became Lord Kelvin, the name by which William Thomson is best known today.

He proved that there was no way to go below 0 K

James Clerk Maxwell
(1831–1879)

Maxwell was born in 1831 in Scotland. After his education at the University of Edinburgh and at Cambridge he became a professor of natural philosophy, first in Scotland and later at Cambridge. His mathematical abilities became apparent at an early age and were applied in many areas. In addition to his accomplishments with gases, which laid the foundation of the science now known as statistical mechanics, Maxwell worked extensively in thermodynamics, developing several fundamental equations that bear his name. His greatest successes, however, were in connection with the theory of light and electricity: he discovered and formulated the general equations of the electromagnetic field. He was the first to recognize that light is a form of electromagnetic radiation and anticipated the development of what we now call radio waves. Maxwell was forced to be inactive for several years of his life because of illness; he was 48 years old when he died, having completed much of his work by the time he was 30.

The number of truly outstanding theoreticians the world has known is very small, certainly numbering less than 100. James Clerk Maxwell belongs among the elite of this group. He was truly an intellectual giant, to be ranked with Newton, Einstein, and J. Willard Gibbs (Chapter 14).

Summary

The variables in gas behavior are pressure, volume, amount (number of moles), and temperature. These are related by the Ideal Gas Law:

$$PV = nRT$$

where n is the number of moles, T is the temperature in K, and R is a constant [0.0821 L·atm/(mol·K)]. This law can be used for many different kinds of calculations involving gases. It can be applied to determine relationships between two or more variables (Examples 6.2 and 6.3) or to calculate one variable by knowing the other three (Example 6.4). By using the relation n = g/MM, the Ideal Gas Law can be extended to find the molar mass of a gas (Example 6.5) or its density (Example 6.6). Going one step further, we can use the Ideal Gas Law to relate the volumes of gases involved in reactions (Examples 6.7 to 6.9). Dalton's Law enables us to obtain the partial pressures of gases in a mixture (Examples 6.10 and 6.11).

Real gases deviate at least slightly from the Ideal Gas Law because of the finite size of gas molecules and attractive forces between them. These deviations become more significant when the molecules are close together, at high pressures and low temperatures. The van der Waals equation, although more complex than the Ideal Gas Law, better represents the behavior of real gases.

The kinetic theory of gases deals with the random, continuous motion of gas molecules from place to place (translational motion). One of its basic postulates is that the average translational energy is directly proportional to the absolute temperature; the proportionality constant is the same for all gases. This relation can be used to compare the rates of effusion of different gases (Example 6.12) or to calculate average molecular speeds (Example 6.13). The distribution of molecular speeds at different temperatures is given in Figure 6.10.

Key Words and Concepts

atmosphere	Kelvin temperature scale	molar mass (MM)
Avogadro's Law	kilopascal	mole
Boyle's Law	kinetic theory	mole fraction
Dalton's Law	Law of Charles and Gay-Lussac	partial pressure
effusion	Law of Combining Volumes	torr
gas constant (R)	manometer	translational energy
Graham's Law	millimeter of mercury	van der Waals equation
Ideal Gas Law		

Questions and Problems

Pressure Measurements

6.1 Consider the manometer shown in Figure 6.2. Suppose the barometric pressure is 748 mm Hg and h_2 is 15.0 cm above h_1. What is the pressure, in atmospheres, of the gas in the flask?

6.2 Complete the following table of pressure conversions:

MM HG	ATMOSPHERES	KILOPASCALS
745	___	___
___	1.40	___
___	___	97.3

6.31 If, with the manometer shown in Figure 6.2, h_2 is 12.4 cm *below* h_1 and the atmospheric pressure is 753 mm Hg, calculate the pressure in atmospheres of the gas in the flask.

6.32 Carry out the indicated conversions between pressure units.

MM HG	ATMOSPHERES	KILOPASCALS
22.5	___	___
___	0.831	___
___	___	101

Ideal Gas Law: Final and Initial States

6.3 A sample of nitrogen gas occupies a volume of 8.50 L at 0.980 atm. If the temperature and number of moles of gas are constant, calculate the volume when the pressure is
a. increased to 1.30 atm.
b. decreased to 0.490 atm.

6.4 A helium sample at 25°C has a volume of 1.82 L. If the pressure and amount of helium are unchanged, determine what the volume will be when the temperature is
a. 50°C   b. 12°C

6.5 A 3.20-cm³ air bubble forms in a deep lake at a depth where the temperature is 8°C at a total pressure of 2.45 atm. The bubble rises to a depth where the temperature and pressure are 19°C and 1.12 atm, respectively. Assuming the amount of air in the bubble has not changed, calculate its new volume.

6.33 The pressure of a 4.14-L argon gas sample is 772 mm Hg. Assuming that the temperature and moles of argon are unchanged, calculate the new pressure when the volume becomes
a. 7.00 L   b. 3.60 L

6.34 What temperature (°C) must be reached to double the volume of a gas at constant pressure and moles of gas if the gas was at 55°C originally?

6.35 On a cold day, a person takes in a breath of 450 mL of air at 756 mm Hg and −10°C. What is the volume of this air in the lungs at 37°C and 752 mm Hg?

6.6 On a day when the pressure is 742 mm Hg and the temperature is 62°F, an open flask contains 0.100 mol of air. How many moles are present in the flask when the pressure is 1.02 atm and the temperature is 31°C?

6.7 Houses with well-water systems have ballast tanks to hold a supply of water. Typically, water flows into the tank from the well until the gauge pressure (pressure in excess of 15 lb/in²) of the air in the tank reaches 50 lb/in². As water is drawn from the tank, the air above it expands and its pressure drops. When the gauge pressure reaches 20 lb/in², the pump delivers more water to the tank. Suppose a 1.50-m³ tank is 75% full of water when the gauge pressure is 50 lb/in². How much water can be withdrawn before the pump turns on at 20 lb/in²?

6.36 A flask contains 0.0519 mol CO_2 at 25°C and 718 mm Hg. An additional 0.0100 mol CO_2 is injected into the flask and the temperature rises to 27°C. What is the new pressure in atmospheres?

6.37 Frequently, ballast tanks of the type described in Problem 6.7 lose most of their air and become "waterlogged." When this happens, the pump operates more frequently. Suppose the 1.50-m³ tank referred to in Problem 6.7 has to be 95% full of water before the gauge pressure reaches 50 lb/in². How much water can be withdrawn from the tank before the pump turns on at 20 lb/in²?

Ideal Gas Law: Calculation of One Variable

6.8 On a hot day, an amusement park balloon is filled with 44.3 g of helium. The temperature is 37°C and the pressure is 2.50 atm. Calculate the volume of the balloon.

6.9 How many moles of air are there in a 125-mL Erlenmeyer flask if the pressure is 739 mm Hg and the temperature is 18°C?

6.10 Use the Ideal Gas Law to complete the following table for helium:

PRESSURE	VOLUME	TEMP.	MOLES	GRAMS
4.00 atm	____	0°C	____	10.0
____	105 mL	25°C	____	0.800
751 mm Hg	2.50 L	100°C	____	____
202 kPa	61.5 L	____	____	20.0

6.38 A gas cylinder contains 2.50×10^3 g O_2 in a volume of 10.0 L. What pressure is exerted by the oxygen at 20°C?

6.39 A drum used to transport crude oil has a 159-L volume. How many water molecules, as steam, are required to fill the drum at one atm and 100°C? What volume of liquid water (d $H_2O(l)$ = 1.00 g/cm³) is required to produce that amount of steam?

6.40 Complete the following table for neon, assuming ideal gas behavior:

PRESSURE	VOLUME	TEMP.	MOLES	GRAMS
746 mm Hg	____	10°C	____	10.0
____	252 mL	20°C	____	15.5
1.65 atm	0.710 L	38°C	____	____
185 kPa	4.20 L	____	____	25.0

Ideal Gas Law: Density, Molar Mass

6.11 Calculate the density of gaseous butane, C_4H_{10}, at 1.00 atm and 25°C.

6.12 A sample of a volatile liquid is vaporized completely in a 240-cm³ flask at 99°C and 757 mm Hg. The condensed vapor weighs 0.564 g.
a. What is the molar mass of the liquid?
b. The liquid contains 66.6% C, 11.2% H, and 22.2% O (mass percent). What is its formula?

6.41 Calculate the densities of the following gases at 25°C and 752 mm Hg:
a. N_2 b. O_2 c. UF_6

6.42 At 99°C and 748 mm Hg, a sample of a volatile liquid is vaporized completely in a 256-cm³ flask. The condensed vapor weighs 1.097 g.
a. Calculate the molar mass of the liquid.
b. What is the formula of the liquid if it contains, in mass percent, 18.0% C, 2.26% H, and 79.7% Cl?

6.13 Exhaled air contains 74.5% N_2, 15.7% O_2, 3.6% CO_2 and 6.2% H_2O (mole percent).
a. Calculate the molar mass of exhaled air.
b. Calculate its density at 27°C and 1.00 atm and compare the value obtained to that for ordinary air (Example 6.6).

6.14 A 1.95-g sample of $X_2H_6(g)$ has a volume of 740 cm³ at 1.00 atm and 15°C.
a. What is the molar mass of X_2H_6?
b. Identify the element X.

6.43 To prevent a condition called the "bends," deep-sea divers breath a mixture containing, in mole percent, 10.0% O_2, 10.0% N_2, and 80.0% He.
a. Calculate the molar mass of this mixture.
b. What is the ratio of the density of this gas to that of air (MM = 29.0 g/mol)?

6.44 A 0.0712-g sample of $X_4H_{10}(g)$ has a volume of 30.6 cm³ at 801 mm Hg and 20°C.
a. What is the molar mass of X_4H_{10}?
b. Identify the element X.

Volumes of Gases in Reactions

6.15 For the reaction

$$2 SO_2(g) + O_2(g) \rightarrow 2 SO_3(g)$$

How many liters of sulfur trioxide would be produced from 3.0 L O_2? Assume 100% yield and that all gases are measured at the same temperature and pressure.

6.45 Consider the reaction

$$4 NH_3(g) + 5 O_2(g) \rightarrow 4 NO(g) + 6 H_2O(l)$$

Suppose 6.12 L NH_3 is mixed with 6.42 L O_2, both at 25°C and 1.00 atm.
a. Which is the limiting reactant?
b. What volume of NO is formed at 25°C and 1.00 atm?

6.16 Hydrogen gas can be made by reacting zinc with hydrochloric acid:

$$Zn(s) + 2 H^+(aq) \rightarrow Zn^{2+}(aq) + H_2(g)$$

a. What mass of zinc is required to form 1.00 L H_2 at STP (0°C, 1.00 atm)?
b. What volume of H_2 at 22°C and 729 mm Hg can be prepared from 1.00 g Zn?

6.46 Oxygen can be made by the decomposition of hydrogen peroxide:

$$2 H_2O_2(aq) \rightarrow 2 H_2O + O_2(g)$$

a. What volume of O_2 at 25°C and 1.00 atm can be made from 1.00 mL of a solution containing 3.00 mass percent H_2O_2 and having a density of 1.01 g/mL?
b. The label on the bottle of H_2O_2 claims that the solution produces ten times its own volume of oxygen gas. Does it?

6.17 Diborane, B_2H_6, is a highly explosive compound formed by the reaction:

$$3 NaBH_4(s) + 4 BF_3(g) \rightarrow$$
$$2 B_2H_6(g) + 3 NaBF_4(s)$$

What volume of diborane at 22°C and 758 mm Hg is formed from 26.0 g $NaBH_4$?

6.47 Hydrogen cyanide, HCN, is a poisonous gas allegedly purchased by Lizzie Borden in August 1892. It can be formed by the reaction

$$NaCN(s) + H^+(aq) \rightarrow HCN(g) + Na^+(aq)$$

What mass of NaCN, sodium cyanide, is required to make 2.24 L HCN at 30°C and 748 mm Hg?

6.18 Gasoline is a mixture of various hydrocarbons, of which octane, C_8H_{18}, is typical.
a. Write a balanced equation for the combustion of octane to give $CO_2(g)$ and $H_2O(l)$.
b. What volume of O_2 at 50°C and 1.00 atm is required to react with 2.00 g of octane?
c. Air is about 21% by volume of O_2. What volume of air is required in (b)?

6.48 A Volvo engine has a cylinder volume of about 500 cm³. The cylinder is full of air at 70°C and 1.00 atm.
a. How many moles of O_2 are in the cylinder (mole percent O_2 in air = 21)?
b. Assume that the hydrocarbons in gasoline have an average molar mass of 100 g/mol and react with O_2 in a 1:12 mol ratio. How many grams of gasoline should be injected into the cylinder to react with the oxygen?

Dalton's Law

6.19 Suppose exhaled air (Problem 6.13) is at a pressure of 751 mm Hg. Calculate the partial pressures of N_2, O_2, CO_2, and H_2O.

6.20 A student collects 275 cm³ O_2 gas saturated with water vapor at 23°C. The mixture exerts a total pressure of 768 mm Hg. At 23°C, the vapor pressure of $H_2O(l)$ = 21.0 mm Hg.
a. What is the partial pressure of O_2 in the sample?
b. How many grams of O_2 does the sample contain?

6.21 What volume is occupied by 1.25 g O_2 saturated with water vapor at 25°C and a total pressure of 749 mm Hg (vp H_2O at 25°C = 23.8 mm Hg).

6.49 If the gas referred to in Problem 6.43 is at a pressure of 6.18 atm, calculate the partial pressure of each gas.

6.50 To prepare a sample of hydrogen gas, a student reacts zinc with hydrochloric acid. The overall reaction is

$$Zn(s) + 2 H^+(aq) \rightarrow Zn^{2+}(aq) + H_2(g)$$

The hydrogen is collected over water at 22°C and the total pressure is 750 mm Hg (vp $H_2O(l)$ = 20 mm Hg).
a. What is the partial pressure of H_2?
b. How many grams of H_2 are there in a 2.00-L sample of wet gas?

6.51 For the sample of wet oxygen referred to in Problem 6.21, calculate
a. the number of moles of O_2.
b. the number of moles of H_2O (partial pressure = 23.8 mm Hg).
c. the total mass of the wet gas.

Real Gases

6.22 A sample of $CH_4(g)$ is at 50°C and 20 atm. Would you expect it to behave more ideally or less ideally if
a. the pressure were reduced to 1 atm?
b. the temperature were reduced to −50°C?

6.23 Calculate the pressure of one mole of H_2 in a 225-cm³ container at 0°C using the
a. Ideal Gas Law.
b. van der Waals equation (see Table 6.2).

6.52 The normal boiling points of CO and SO_2 are −192°C and −10°C, respectively.
a. At 25°C and 1 atm, which gas would you expect to have a molar volume closest to the ideal value?
b. If you wanted to reduce the deviation from ideal gas behavior, in what direction would you change the temperature? the pressure?

6.53 Using Figure 6.7, estimate the density of $CH_4(g)$ at 100 atm and 25°C and compare to the value calculated from the Ideal Gas Law.

Kinetic Theory

6.24 What is the ratio of the rate of effusion of the most dense gas known, UF_6, to that of the lightest gas, H_2?

6.25 It takes 22.0 s for a CH_4 sample to effuse down a capillary tube. Another gas at the same temperature and pressure took 44.0 s for the same amount of effusion to occur. One student claims this gas is He; another thinks it is SO_2. Which student is correct?

6.54 A gas effuses 1.05 times faster than NF_3 at the same temperature and pressure.
a. Is the gas heavier or lighter than NF_3?
b. Calculate the ratio of the molar mass of NF_3 to that of the unknown gas.

6.55 There is a tiny leak in a system containing helium gas at 22°C and 760 mm Hg. After one minute, the pressure of He drops to 752 mm Hg. The helium is replaced by hydrogen, again at 22°C and 760 mm Hg. What would you expect the pressure of H_2 to be after one minute?

6.26 Calculate the average speed of
   a. an O_2 molecule at 50°C.
   b. an N_2 molecule at -50°C.

6.27 For an SF_4 gas molecule,
   a. at what temperature will its average kinetic energy be half that at 25°C?
   b. at what temperature will it have twice the average speed it has at 25°C?

6.56 A professional tennis player can serve a tennis ball traveling at 45 m/s. At what temperature will an O_2 molecule have that average speed?

6.57 At what temperature will UF_6 molecules have the same
   a. average kinetic energy as F_2 molecules at 0°C?
   b. average speed as F_2 molecules at 0°C?

General

6.28 The pressure exerted by a column of liquid is directly proportional to its density. If you wanted to fill a barometer with water (d = 1.00 g/cm³), how long a tube should you use (d Hg = 13.6 g/cm³)?

6.29 Given that 1.00 mol of neon and 1.00 mol of hydrogen are in separate containers at the same T and P, calculate each of the following ratios:
   a. volume Ne/volume H_2.
   b. density Ne/density H_2.
   c. average translational energy Ne/average translational energy H_2.
   d. number of atoms Ne/number of atoms H.

6.30 For 1 mol of an ideal gas, sketch graphs of
   a. V vs. T at constant P, n.
   b. P vs. T at constant V, n.
   c. n vs. T at constant P, V.
   d. E_{trans} vs. T at constant P, V.

6.58 Using the information given in Problem 6.28, explain why an ordinary "suction" pump cannot pump water out of a well 50 ft deep.

6.59 A mixture of 6.4 mol of helium and 6.1 mol of oxygen occupies a 4.75-L container at 25°C. Which gas has the larger
   a. average molecular speed?
   b. average translational energy?
   c. partial pressure?
   d. mole fraction?

6.60 For an ideal gas, sketch graphs of
   a. V vs. P at constant T, n.
   b. P vs. n at constant V, T.
   c. u vs. T at constant P, V.
   d. E_{trans} vs. P at constant T, n.

*6.61 The Rankine temperature scale resembles the Kelvin scale in that 0° is taken to be the lowest attainable temperature (0°R = 0 K). However, the Rankine degree is the same size as the Fahrenheit degree, whereas the Kelvin degree is the same size as the Celsius degree. What is the value of the gas law constant in L·atm/(mol·°R)?

*6.62 A 0.100-g sample of an Al-Zn alloy reacts with HCl to form H_2:

$$Al(s) + 3 H^+(aq) \rightarrow Al^{3+}(aq) + \tfrac{3}{2} H_2(g)$$

$$Zn(s) + 2 H^+(aq) \rightarrow Zn^{2+}(aq) + H_2(g)$$

The hydrogen produced has a volume of 0.100 L at 27°C and 1.00 atm. What is the mass percent of aluminum in the alloy?

*6.63 The buoyant force on a balloon is equal to the mass of air it displaces. The gravitational force on the balloon is equal to the sum of the masses of the balloon, the gas it contains, and the balloonist. If the balloon and balloonist together weigh 150 kg, what would the diameter of a spherical hydrogen-filled balloon have to be in meters if the rig is to get off the ground at 25°C and one atm? (Take MM air = 29.0 g/mol).

*6.64 A mixture in which the mole ratio of hydrogen to oxygen is 2:1 is used to prepare water by the reaction

$$2 H_2(g) + O_2(g) \rightarrow 2 H_2O(g)$$

The total pressure in the container is 0.800 atm at 20°C before reaction. What is the final pressure in the container at 120°C after reaction, assuming an 80.0% yield of water?

*6.65 The volume percent of a gas, A, in a mixture is defined by the equation

$$\text{volume percent A} = \frac{100 \times V_a}{V}$$

where V is the total volume of the mixture and V_a is the volume that gas A would occupy alone at the same temperature and pressure. Show that, assuming ideal gas behavior, the volume percent of a gas in a mixture is the same as its mole percent. Explain why the volume percent differs from the mass percent.

Chapter 7
The
Electronic
Structure
of Atoms

In Chapter 2, we considered the structure of the atom briefly. You will recall that every atom has a tiny, positively charged nucleus, made up of protons and neutrons. The number of protons in the nucleus is characteristic of the atoms of a particular element. It is referred to as the atomic number of the element.

The nucleus is surrounded by electrons that carry a negative charge. The charge of an electron is equal in magnitude but opposite in sign to that of a proton. In a neutral atom, the number of electrons is equal to the number of protons and, hence, to the atomic number of the element.

The chemical properties of an atom are determined by its electronic structure, i.e., the number and arrangement of electrons about the nucleus. We can use the electronic structure of an atom to predict much of the chemical behavior of the corresponding element. In this chapter, we will focus upon the electron arrangements in atoms (Section 7.5). In preparation for that, it will be helpful to examine the energies of these electrons, starting with the simplest atom, hydrogen (Section 7.2), and proceeding to multielectron atoms (Sections 7.3 and 7.4). Energies of electrons in atoms are derived from the quantum theory (Section 7.1), which was developed during the early part of this century.

7.1
The Quantum Theory

The quantum theory was first proposed by Max Planck in 1900 to explain the properties of the radiation given off by hot bodies. A few years later, in 1905, it was used by Albert Einstein in describing the emission of electrons by metals exposed to light. Still later, in 1913, Niels Bohr used the quantum theory to develop a mathematical model of the hydrogen atom (Section 7.2). We now know that the quantum theory is a general one that applies to all the interactions of matter with energy. Here, we will discuss the postulates of the theory as they apply to electrons in atoms or molecules.

Postulates of the Quantum Theory

1. Atoms and molecules can only exist in certain states, characterized by definite amounts of energy. When an atom or molecule changes its state, it absorbs or emits an amount of energy just sufficient to bring it to another state.

Atoms and molecules can possess various kinds of energy. One form of energy of particular importance arises from the motion of electrons about the atomic nucleus and from the charge interactions among the electrons and between the electrons and the nucleus. This kind of energy is called *electronic energy*. Only certain values of electronic energy are allowed for an atom. The energy of systems that can exist only in discrete states is said to be *quantized*. Because the electronic energy of an atom is quantized, a change in the electronic energy level (state) of an atom involves the absorption or emission of a definite amount, or *quantum*, of energy.

The electrons in an atom can't have just any amount of energy

When an atom goes from one allowed energy state to another, it must absorb or emit just enough energy to bring its own energy to that of the final state. The lowest electronic energy state for an atom is called the **ground state**. An electronic **excited state** is any state with energy greater than that of the ground state.

2. When atoms or molecules absorb or emit light in moving from one energy state to another, the wavelength λ of the light is related to the energies of the two states by the equation

$$E_{hi} - E_{lo} = \frac{hc}{\lambda} \qquad \text{(7.1)}$$

In this equation, E_{hi} is the energy of the higher state (the one with more energy). The quantity E_{lo} is the energy of the lower state (the one with less energy). The symbols h and c represent physical constants: h is called Planck's constant, and c is the speed of light.

A ray of light can be considered to consist of photons. Each photon of wavelength λ has an energy of hc/λ. An atom or molecule can move from one electronic energy state to another by emitting a photon. If it absorbs a photon of energy hc/λ, it moves from a lower to a higher energy state and its energy increases by hc/λ; that is,

$$\Delta E = E_{final} - E_{initial} = E_{hi} - E_{lo} = hc/\lambda$$

If, on the other hand, the atom or molecule emits a photon of energy hc/λ, it moves from a higher to a lower energy state. In that case, its energy *decreases* by hc/λ:

When an atom loses energy, that energy goes into the photon that is emitted

$$\Delta E = E_{final} - E_{initial} = E_{lo} - E_{hi} = -hc/\lambda$$

3. The allowed energy states of atoms and molecules can be described by sets of numbers called quantum numbers.

It is possible to set up mathematical equations to describe the energies of electrons in atoms or molecules. Such equations usually have several solutions. All of these solutions are of the same general form and contain one or more *quantum numbers*. These numbers distinguish a particular energy state from all the others. In the usual model, quantum numbers are associated with individual electrons in an atom. Each electron is assigned a set of quantum numbers according to a set of rules (Section 7.4).

Relation Between Energy Difference and Wavelength

Figure 7.1 illustrates the meaning of Equation 7.1, the basic equation of simple quantum theory. Notice that the wavelength, λ, is inversely related to the energy difference, ΔE. If the energy states are widely separated (large ΔE), light of short wavelength, perhaps in the ultraviolet region ($\lambda < 400$ nm), is produced. Suppose, on the other hand, that the energy states are close together (small ΔE). In that case, light of longer wavelength, perhaps in the infrared region ($\lambda > 700$ nm), is emitted.

An electron moves from one energy state to another by absorbing or emitting radiation of a particular wavelength. By accurately measuring the wavelength associated with an electron transition, we can find the difference in energy between the two energy states involved. Equation 7.1 is used to obtain $E_{hi} - E_{lo}$ from λ. If we substitute

$$h = 6.626 \times 10^{-34} \frac{J \cdot s}{particle}; \qquad c = 2.998 \times 10^{8} \frac{m}{s}$$

we obtain $E_{hi} - E_{lo}$ in *joules per particle* when λ is expressed in *meters:*

$$E_{hi} - E_{lo} = \frac{hc}{\lambda} = \frac{(6.626 \times 10^{-34})(2.998 \times 10^{8})}{\lambda} \frac{J \cdot m}{particle}$$

$$= \frac{1.986 \times 10^{-25}}{\lambda} \frac{J \cdot m}{particle}$$

Ordinarily, we express energy differences in **kilojoules per mole** rather than joules per particle. Since wavelengths for electron transitions are very short, it is convenient to express them in **nanometers** rather than meters. To obtain an equation with the desired units, we use the relations

$$1 \text{ mol} = 6.022 \times 10^{23} \text{ particles}; \qquad 1 \text{ kJ} = 10^{3} \text{ J}; \qquad 1 \text{ nm} = 10^{-9} \text{ m}$$

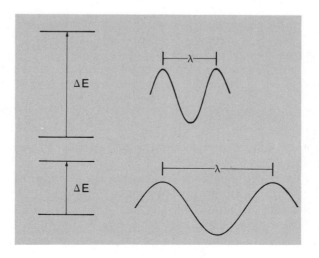

FIGURE 7.1 The wavelength of light given off when an atom emits energy is inversely proportional to the difference in energy, ΔE, between the higher and lower levels. If ΔE is large, λ is small, perhaps in the ultraviolet region. If ΔE is small, λ is large, perhaps in the infrared region. By measuring λ, it is possible to calculate the energy difference, $E_{hi} - E_{lo}$, using Equation 7.2.

These relations give us the conversion factors we need:

$$E_{hi} - E_{lo} = \frac{1.986 \times 10^{-25}}{\lambda} \frac{J \cdot m}{particle} \times \frac{6.022 \times 10^{23}\ particles}{1\ mol} \times \frac{1\ kJ}{10^3\ J} \times \frac{1\ nm}{10^{-9}\ m}$$

$$= \frac{(1.986 \times 6.022 \times 10^4)}{\lambda} \frac{kJ \cdot nm}{mol}$$

$$= \frac{1.196 \times 10^5}{\lambda} \frac{kJ \cdot nm}{mol} \tag{7.2}$$

We will find Equation 7.2 to be extremely useful; a simple application is shown in Example 7.1.

EXAMPLE 7.1   Light of wavelength 670.8 nm is given off in an electron transition between two energy states in the lithium atom. Calculate the difference in energy, ΔE, in kilojoules per mole, between these two states.

Solution   Substituting directly into Equation 7.2,

$$E_{hi} - E_{lo} = \frac{1.196 \times 10^5}{670.8\ nm} \frac{kJ \cdot nm}{mol} = 178.3\ kJ/mol$$

The energy change in transitions like this is roughly equal to that in chemical reactions

EXERCISE   For another electron transition in lithium, the energy difference is 195.9 kJ/mol. Calculate the wavelength, in nanometers, of the light given off in this transition. Answer: 610.5 nm.

Relation Between Wavelength and Frequency

Instead of specifying the wavelength, λ, of light, we may cite its *frequency, ν.* The frequency tells us how many wave cycles (successive crests or troughs) pass a given point in unit time. If we are told that, for a certain kind of wave, 10^8 cycles pass a given point in one second, we would say that*

$$frequency = 10^8\ cycles\ per\ second$$
$$= 10^8/s = 10^8\ s^{-1}$$

The velocity at which a wave moves can be found by multiplying the length of a wave cycle (λ) by the number of cycles passing a point in unit time (ν). For light waves,

ν is the frequency

$$\nu\lambda = c$$

where c, the speed of light, has a constant value, 2.998×10^8 m/s. This means that the frequency of light is related to its wavelength by the equation

---

*The unit "cycle per second" has been given a special name, hertz (Hz). Frequencies are often cited in hertz or megahertz (1 MHz = 10^6 Hz). We might say that for the radiation referred to here, $\nu = 10^8$ Hz = 10^2 MHz.

$$\nu = \frac{c}{\lambda} = \frac{2.998 \times 10^8 \text{ m/s}}{\lambda} \tag{7.3}$$

From Equation 7.3, we see that ν is inversely related to λ (see also Figure 7.2). Blue light, with a wavelength of 450 nm, has a higher frequency than red light, with a frequency of 650 nm:

blue light: $\quad \nu = \dfrac{2.998 \times 10^8 \text{ m/s}}{450 \times 10^{-9} \text{ m}} = 6.66 \times 10^{14}/\text{s}$

Both frequencies are very very high

red light: $\quad \nu = \dfrac{2.998 \times 10^8 \text{ m/s}}{650 \times 10^{-9} \text{ m}} = 4.61 \times 10^{14}/\text{s}$

Combining Equations 7.1 and 7.3, we obtain

As ν goes up, E_{photon} goes up

$$E_{hi} - E_{lo} = \frac{hc}{\lambda} = h\nu \tag{7.4}$$

This tells us that the difference in energy between two states is directly proportional to the frequency of the light absorbed or emitted when the electron transition occurs. If the energy difference is large, the light will have a relatively high

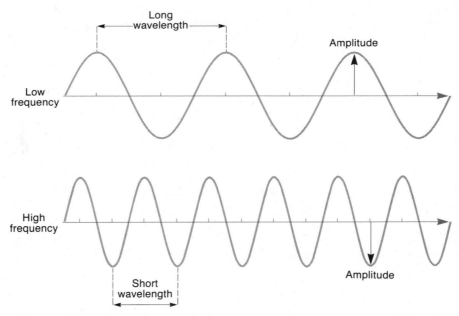

FIGURE 7.2 Three characteristics of a wave are its amplitude, wavelength, and frequency. The *amplitude* (ψ) is the height of a crest (or the depth of a trough). The two waves shown here have the same amplitude. The *wavelength* (λ) is the distance between successive crests (or successive troughs). The wavelength of the lower wave is one half that of the upper wave. The *frequency* (ν) is the number of wave cycles (successive crests or troughs) that pass a given point in unit time. If the two waves are moving from left to right at the same speed, in meters per second, the lower wave must have a frequency twice that of the upper wave.

frequency. Light of low frequency results from a transition between energy states that are close together.

EXAMPLE 7.2 In the sodium atom, there are two states that differ in energy by 203.1 kJ/mol. When an electron transition occurs from the higher of these states to the lower, energy is given off as yellow light. For this light, calculate

a. λ, using Equation 7.2. b. ν, using Equation 7.3.

Solution

a.

$$\lambda = \frac{1.196 \times 10^5 \text{ kJ·nm}}{E_{hi} - E_{lo} \text{ mol}} = \frac{1.196 \times 10^5}{203.1} \text{ nm} = 588.9 \text{ nm}$$

b.

$$\nu = \frac{2.998 \times 10^8 \text{ m/s}}{588.9 \times 10^{-9} \text{ m}} = 5.091 \times 10^{14}/\text{s}$$

This is the main wavelength in the light from sodium vapor lamps

(Frequencies of visible light are typically of the order of 10^{14} cycles per second; a lot of waves pass a given point in 1 s!)

EXERCISE An FM radio station broadcasts at a frequency of 106.5 MHz = 106.5×10^6/s. What is the wavelength in meters of these radio waves? Answer: 2.815 m.

7.2
The Atomic Spectrum of Hydrogen and the Bohr Model

By 1900, scientists had begun to speculate on the way in which positive and negative charges are arranged in atoms. Part of the problem was solved in 1911 when Rutherford showed the existence of atomic nuclei (Chap. 2). Two years later, Niels Bohr, a young Danish physicist, took the next step. He developed a mathematical model for the behavior of the electron in the hydrogen atom.

Bohr based his approach on the Rutherford atom. He also utilized the quantum theory of Planck. Most important, he had accurate information concerning energy differences between electronic states in the hydrogen atom. This came from earlier studies of the atomic spectrum of hydrogen.

Atomic Spectrum of Hydrogen

If we heat a metal in a furnace or in a flame, it gives off visible light. At 1000°C, iron looks red; at 1500°C it appears white. The emitted light can be examined by a *spectroscope,* a device that breaks up light into its component colors. When this is done, the light from the white-hot iron is found to contain nearly all colors. Under these conditions, we say that the spectrum is *continuous.*

If a sample of an element is excited electrically, as in a sodium vapor lamp or neon sign, the light given off looks quite different. We find that under these conditions, the spectrum is not continuous. Instead, the light consists of several discrete colors, which appear as lines of definite wavelength when seen in a spectroscope. The lines occurring in an atomic spectrum produced this way are characteristic of a given element and can be used to identify it (Color Plate 7.1).

Some atomic spectra are very complex

The simplest atomic spectrum is that shown by hydrogen. In a hydrogen atom, we are dealing with a single electron moving between energy states. When the atom absorbs energy, the electron moves to a higher energy level. When the electron returns to a lower level, it gives off energy as light at discrete wavelengths.

The atomic spectrum of hydrogen was first studied in the 1880s. At that time, a series of lines was discovered in the visible region. These lines comprise what is called the Balmer series. They were known to Bohr at the time he was developing his model of the hydrogen atom. Later, other series were found. One of these, called the Lyman series, is in the ultraviolet region (<200 nm). Another, the Paschen series, lies in the infrared region (>800 nm). The wavelengths of some of the more prominent lines in each of these series are listed in Table 7.1.

Table 7.1
Wavelengths (nm) of Lines in the Atomic Spectrum of Hydrogen

ULTRAVIOLET (LYMAN SERIES)	VISIBLE (BALMER SERIES)	INFRARED (PASCHEN SERIES)
121.53	656.28	1875.09
102.54	486.13	1281.80
97.23	434.05	1093.80
94.95	410.18	1004.93
93.75	397.01	
93.05		

Balmer found an equation relating the lines in his series

Bohr Model for the Hydrogen Atom

Bohr assumed that a hydrogen atom consists of a central proton about which an electron moves in a circular orbit. He related the force of attraction of the proton for the electron to the centrifugal force due to the circular motion of the electron. In this way, Bohr was able to express the energy of the atom in terms of the radius of the electron's orbit. To this point, his analysis was purely classical, based on Coulomb's Law of electrostatic attraction and Newton's laws of motion. Bohr then introduced quantum theory into his model. Boldly and arbitrarily, he imposed a condition upon a property of the electron called its angular momentum. This is given by the product mvr, where m is the electron mass, v is its speed, and r is the radius of its orbit about the nucleus. Bohr proposed that the angular momentum be given by the equation

$$mvr = nh/2\pi \tag{7.5}$$

n cannot equal zero

where h is Planck's constant and **n** is a quantum number that can have any positive integral value (1, 2, 3, . . .). According to the Bohr model of the hydrogen atom, the angular momentum of the electron is quantized. It cannot have just any value; the angular momentum mvr is restricted to values for which **n** is a positive integer (h, 2, and π are constants). Thus, the angular momentum can change only by discrete amounts, i.e., integral multiples of $h/2\pi$.

Bohr found that his quantum condition also restricted the allowed energies of the electron in the hydrogen atom. It could have only those values given by the equation

$$E = \frac{-B}{n^2} \qquad (n = 1, 2, 3, \ldots)$$

where B is a constant that can be calculated from theory. The value of B is 2.179 $\times$ 10^{-18} J/particle. Substituting,

$$E = \frac{-2.179 \times 10^{-18}}{n^2} \frac{J}{particle}$$

Here the particle is an electron in an H atom

To find E in the common units of kilojoules per mole, we carry out two successive conversions:

$$E = \frac{-2.179 \times 10^{-18}}{n^2} \frac{J}{particle} \times \frac{6.022 \times 10^{23} \text{ particles}}{1 \text{ mol}} \times \frac{1 \text{ kJ}}{10^3 \text{ J}}$$

$$= \frac{-1312}{n^2} \frac{kJ}{mol} \tag{7.6}$$

Before going further with the Bohr model, let us make three points:

1. In setting up his model, Bohr designated zero energy as the point at which the proton and electron are completely separated. Energy has to be absorbed to reach that point. This means that the electron, in all its allowed energy states within the atom, must have an energy below zero, i.e., must be negative. Hence, the minus sign in Equation 7.6.

2. In the normal hydrogen atom, the electron is in its **ground state,** for which **n** = 1. When an electron absorbs energy, it moves to a higher, **excited state.** In a hydrogen atom, the first excited state has **n** = 2; for the second excited state, **n** = 3, and so on.

The ground state is the one with the lowest possible energy

3. When an excited electron gives off energy in the form of light, it drops back to a lower energy state. Some of these transitions are shown in Figure 7.3. Notice that the electron may return to

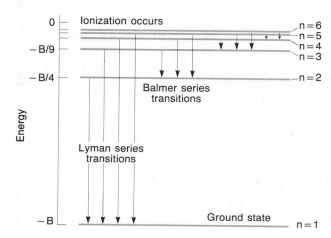

FIGURE 7.3 Some energy levels and transitions in the hydrogen atom. Lines in the Balmer series arise from transitions from upper levels (**n** > 2) to the **n** = 2 level. In the Lyman series, the lower level is **n** = 1.

—*the ground state* (**n** = 1). Electrons returning to this state produce the Lyman lines in the hydrogen spectrum. The transition between **n** = 2 and **n** = 1 yields one such line, that between **n** = 3 and **n** = 1 gives another, and so on.

Figure 7.3 shows the origin of several of the lines in the Balmer and Lyman series

—*an excited state*. Electrons returning to **n** = 2 from **n** = 3, 4, . . . are responsible for lines in the Balmer series. Transitions back to **n** = 3 give the Paschen series.

EXAMPLE 7.3 In which of the following transitions in the hydrogen atom is energy absorbed? emitted? Which transitions give lines in the Lyman series? the Balmer series?
a. **n** = 2 → **n** = 3   b. **n** = 5 → **n** = 2   c. **n** = 2 → **n** = 1

Solution
a. Energy is absorbed, since the electron moves from a lower to a higher energy state.
b. Energy is emitted; the electron moves from a higher to a lower state. This transition produces a line in the Balmer series, for which the lower state has **n** = 2.
c. Energy is emitted, as in (b). Since the electron returns to the ground state (**n** = 1), a line in the Lyman series is produced.

EXERCISE How do the wavelengths of lines in the Balmer series compare to those in the Lyman series? How do the frequencies compare? Answer: λ larger, ν smaller.

Bohr used his model to calculate the wavelengths of various lines in the spectrum of hydrogen. He found excellent agreement between theory and experiment. We can repeat his calculations by first using Equation 7.6 to obtain the energies of the individual states. Then, by subtraction, we get the energy difference between two states. Finally, using Equation 7.2, we obtain a predicted value for the wavelength.

EXAMPLE 7.4 Find the wavelength, in nanometers, of the line in the Balmer series that results from the transition from **n** = 3 to **n** = 2.

Solution Using the Bohr relation (Equation 7.6), we can calculate the energies of the two levels and hence the energy difference between the two states. Then, using the basic relation between energy difference and wavelength (Equation 7.2), we can find λ for this line:

$$E_3 = \frac{-1312}{9} \text{ kJ/mol} = -145.8 \text{ kJ/mol}$$

$$E_2 = \frac{-1312}{4} \text{ kJ/mol} = -328.0 \text{ kJ/mol}$$

$$E_{hi} - E_{lo} = E_3 - E_2 = -145.8 \text{ kJ/mol} - (-328.0 \text{ kJ/mol})$$
$$= 182.2 \text{ kJ/mol}$$

Solving Equation 7.2 for λ,

$$\lambda = \frac{1.196 \times 10^5}{E_{hi} - E_{lo}} \frac{kJ \cdot nm}{mol} = \frac{1.196 \times 10^5}{182.2} \; nm = 656.4 \; nm$$

EXERCISE Repeat this calculation for the first line in the Lyman series ($n = 2$ to $n = 1$). Answer: 121.5 nm.

The calculation we have just gone through may seem routine. You can imagine, however, how Bohr must have felt when he did it for the first time. With his value of B, Bohr predicted the wavelengths of all the known lines in the Balmer series. He obtained agreement with experiment to within 0.1% or better. Bohr also predicted the existence of the Lyman series, which was unknown in 1913. When it was discovered some years later, the wavelengths agreed almost exactly with Bohr's predictions.

Bohr's equation 7.6 is one of the most accurate in physical science

Another quantity that can be calculated from the Bohr model is the *ionization energy* of the hydrogen atom. This is the energy that must be absorbed to remove an electron from the gaseous atom, starting from the ground state:

$$H(g) \rightarrow H^+(g) + e^-; \quad \Delta E = \text{ionization energy}$$

From Equation 7.6, it is possible to calculate the ionization energy of hydrogen. To do this, we calculate the energy change, ΔE, when an electron moves from the ground state ($n = 1$, $E = -1312 \, kJ/mol$) to the state where it is completely removed from the atom ($n = \infty$, $E = 0$).

$$\Delta E = 0 - (-1312 \; kJ/mol) = 1312 \; kJ/mol$$

The value calculated in this way by Bohr was in excellent agreement with the experimental value for the ionization energy of the hydrogen atom.

7.3
The Quantum Mechanical Atom

Bohr's theory for the structure of the hydrogen atom was highly successful. Scientists of the day must have thought they were on the verge of being able to predict the allowed energy levels of all atoms. However, the extension of Bohr's ideas to atoms with two or more electrons gave, at best, only qualitative agreement with experiment. Consider, for example, what happens when Bohr's theory is applied to the helium atom. Here, the errors in calculated energies and wavelengths are of the order of 5% instead of the 0.1% error with hydrogen. There appeared to be no way the theory could be modified to make it work well with helium or other atoms. Indeed, it soon became apparent that there was a fundamental problem with the Bohr model. The idea of an electron moving about the nucleus in a well-defined orbit at a fixed distance from the nucleus finally had to be abandoned.

A lot of work was done to save the Bohr theory, all to no avail

Wave Nature of the Electron; The de Broglie Relation

Prior to 1900, it was supposed that light was wave-like in nature. However, the work of Planck and Einstein suggested that, in many processes, light behaves as

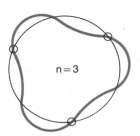
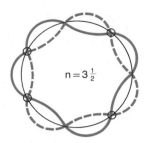

FIGURE 7.4 Imagine the electron of the hydrogen atom in the form of a wave moving in a circular path about the nucleus. In order for the wave to be stable, it must trace the same path in successive orbits. This is the case for the **n** = 3 wave, where three wavelengths bring the electron exactly back to its starting point. The condition also applies to the **n** = 4 wave; four wavelengths correspond exactly to a trip around the circle. The general condition for stability is

$$2\pi r = n\lambda$$

where $2\pi r$ is the circumference of the orbit, λ is the wavelength of the electron, and n is an integer. The **n** = $3\frac{1}{2}$ wave (B) is unstable; the electron does not trace the same path in successive orbits (the second orbit is shown by a dotted path).

if it consists of particles, called photons. Within 20 years, the dualistic, wave-particle nature of light became generally accepted. Then, in 1924, a young French physicist, Louis de Broglie, came up with a revolutionary idea about the nature of matter. Reasoning as physicists do, often with striking success, de Broglie suggested that particles might well exhibit wave properties. He showed that the wavelength, λ, associated with a particle of mass m moving at speed v would be

$$\lambda = \frac{h}{mv} \tag{7.7}$$

They diffracted electrons from a crystal, in much the same way as we diffract x-rays

where h is Planck's constant. Within a few years, Davisson and Germer, working at the Bell Telephone Laboratories, tested de Broglie's prediction and showed that a beam of electrons does indeed have wave properties. Moreover, the observed wavelength of the electron was exactly that predicted by de Broglie.

Using Equation 7.7, it is possible to show how the Bohr quantum number **n** arises in a natural way. To do this, consider Figure 7.4. Here, we imagine an electron in the form of a wave moving about the nucleus along the circumference of a circle. Under these conditions, there is a restriction on the wavelengths that the electron can have. In successive revolutions, the waves must be exactly in phase with each other; that is, they must have exactly the same height (*amplitude*) at any given point. This means that a wave must fit into the circumference of the circle, $2\pi r$, an integral number of times, **n**. In other words,

$$2\pi r = n\lambda \tag{7.8}$$

where λ is the wavelength and **n** is a whole number—that is, 1, 2, 3, . . . , but *not* 1.5, 2.1, etc. Combining Equations 7.7 and 7.8, we obtain

$$2\pi r = \frac{nh}{mv}$$

or

$$mvr = \frac{nh}{2\pi}$$

This is the condition that Bohr imposed arbitrarily on the momentum of the electron in the hydrogen atom (Equation 7.5). Using the de Broglie relation, this condition becomes physically reasonable.

more or less

Electron Cloud Diagrams

If an electron behaves like a wave, there is a fundamental problem with the Bohr atom. How does one specify the "position" of a wave at a particular instant? We can hope to determine its wavelength, its energy, and even its amplitude, but there is no obvious way to tell precisely where the electron is. Indeed, since a wave extends over space, the very idea of the position of the electron within an atom becomes nebulous, to say the least.

Scientists in the 1920s, speculating on this problem, became convinced that the Bohr model had to be abandoned. The idea of an electron revolving about the nucleus in a fixed orbit of definite radius simply did not correspond to reality. An entirely new approach was required to treat electrons in atoms and molecules. A new discipline, called *quantum mechanics,* was developed to describe the motion of small particles confined to tiny regions of space. The main concern of quantum mechanics is to find expressions for the energies of these particles.

In the quantum mechanical atom, no attempt is made to specify the position of an electron at a given instant; nor does quantum mechanics concern itself with the path that an electron takes about the nucleus. (After all, if we can't say where the electron is, we certainly don't know how it got there.) Instead, quantum mechanics deals only with the *probability* of finding a particle within a given region of space. It tells us, for example, that there is a 90% chance of finding an electron within 0.14 nm of the hydrogen nucleus. Conversely, there is a 10% chance that, in the ground state, the hydrogen electron will be farther from the nucleus than this. You may find this way of describing electron distributions disturbing. At least Albert Einstein did; he is reputed to have said that "God does not play dice with the Universe." Nevertheless, no one has ever been able to pin down an electron in an atom or molecule to a precise location. There really isn't much we can do about that.

Some think God not only plays dice, but loads them

In 1926, Erwin Schrödinger made a major contribution to quantum mechanics. Expanding upon the ideas of de Broglie, Schrödinger went one step further. He derived an equation from which one could calculate the amplitude (height) ψ, of the electron wave at various points in space. The equation is a complex one, involving the notation of the calculus, and we will not attempt to work with it. Suffice it to say that, for the electron in the hydrogen atom, the Schrödinger wave equation can be solved exactly for ψ. It turns out that there are several expressions for ψ that will satisfy this equation. Each of these solutions is associated with a set of quantum numbers (Section 7.4). Using these quantum numbers, it is possible to calculate the allowable energies for an electron in a hydrogen atom.

Once the Schrödinger equation has been solved for ψ, it becomes possible to determine the probability of finding a particle in a given region of space. It turns out that the square of the amplitude, ψ^2, is directly proportional to the probability of finding the particle at that point. For electrons, we can interpret

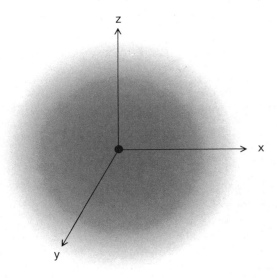

FIGURE 7.5 Electron cloud surrounding an atomic nucleus. The depth of color is proportional to the probability of finding an electron in a particular region. Notice that the probability decreases rapidly and smoothly as one moves out from the nucleus.

the value of ψ^2 as being proportional to the electric charge density at that point. "Electron cloud" diagrams, showing charge densities in atoms and molecules, are often drawn on that basis. Figure 7.5 shows the "electron cloud" surrounding the nucleus of a typical atom.

Unlike the Bohr model, the Schrödinger wave equation can be applied to atoms other than hydrogen and to molecules as well. Unfortunately, this is easier said than done. The form of the equation for multielectron systems is so complex as to defy exact solution. However, in the few cases where a satisfactory solution has been obtained, agreement with experiment is excellent. This convinces us that the approach is a correct although complex one.

We are getting better and better at solving the equation for complex systems

7.4
Quantum Numbers, Energy Levels, and Orbitals

Schrödinger found that the electron in the hydrogen atom could be described by three quantum numbers. These are now given the symbols n, ℓ, and m_ℓ. There are three such numbers because the electron requires three coordinates to describe its motion.

These might be its x, y, z coordinates

For reasons we will discuss later in this section, we find that it takes four, rather than three, quantum numbers to completely describe the state of an electron. The fourth quantum number is given the symbol m_s. Each electron in an atom has a set of four quantum numbers, n, ℓ, m_ℓ, and m_s, which fix its energy and the shape of its electron cloud. We will now discuss the quantum numbers of electrons as they are used with atoms in general.

First Quantum Number, n; Principal Energy Levels

The first quantum number, given the symbol n, is of primary importance in determining the energy of an electron. For the hydrogen atom, the energy depends only upon **n** (recall Equation 7.6). In other atoms, the energy of each electron depends mainly, but not completely, upon the value of **n**. As **n** increases, the energy of the electron increases and, on the average, it moves farther out from the nucleus. The quantum number **n** can take on only integral values, starting with 1:

$$\mathbf{n} = 1, 2, 3, 4, \ldots \tag{7.9}$$

*Electrons with the highest **n** values are farthest from the nucleus*

In an atom, the value of **n** corresponds to what we call a **principal energy level**. Thus, an electron for which **n** = 1 is said to be in the first principal level. If **n** = 2, we are dealing with the second principal level, and so on.

Second Quantum Number, ℓ; Sublevels (s, p, d, f)

Each principal energy level includes one or more **sublevels**. The sublevels are denoted by the second quantum number, ℓ. As we shall see later, the general shape of the electron cloud associated with an electron is determined by ℓ.

The quantum numbers **n** and ℓ are related. We find that ℓ can take on any integral value starting with 0 and going up to a maximum of (**n** − 1); that is,

$$\ell = 0, 1, 2, \ldots, (\mathbf{n} - 1) \tag{7.10}$$

If **n** = 1, there is only one possible value of ℓ, namely 0. This means that, in the first principal level, there is only one sublevel, for which ℓ = 0. If **n** = 2, two values of ℓ are possible, 0 and 1. In other words, there are two sublevels (ℓ = 0 and ℓ = 1) within the second principal energy level. Similarly,

*Given the value of **n**, you can predict all the possible values of ℓ*

if **n** = 3: ℓ = 0, 1, or 2   (three sublevels)
if **n** = 4: ℓ = 0, 1, 2, or 3   (four sublevels)

In general, **in the nth principal level, there are n different sublevels.**

Another method is commonly used to designate sublevels. Instead of giving the quantum number ℓ, we use a letter (s, p, d, or f*) to indicate the sublevel. A sublevel for which ℓ = 0 is referred to as an **s sublevel**. If ℓ = 1, we are dealing with a **p sublevel**. A d sublevel is one for which ℓ = 2; in an **f sublevel**, ℓ = 3:

Quantum number ℓ	0	1	2	3
Type of sublevel	s	p	d	f

Usually, in designating a sublevel, we include a number to indicate the principal level as well. Thus, we refer to a 1s sublevel (**n** = 1, ℓ = 0), a 2s sublevel (**n** = 2, ℓ = 0), and a 2p sublevel (**n** = 2, ℓ = 1). The sublevel designations for the first four principal levels are shown in Table 7.2.

---

*These letters come from the adjectives used by spectroscopists to describe spectral lines; *s*harp, *p*rincipal, *d*iffuse, and *f*undamental.

Table 7.2
Sublevel Designations for the First Four Principal Levels

n	1	2		3			4			
ℓ	0	0	1	0	1	2	0	1	2	3
Sublevel	1s	2s	2p	3s	3p	3d	4s	4p	4d	4f

This is observed experimentally, not predicted

Within a given principal level (same value of **n**), sublevels always increase in energy in the order

$$ns < np < nd < nf \tag{7.11}$$

Thus a 2p sublevel has a slightly higher energy than a 2s sublevel. By the same token, when **n** = 3, the 3s sublevel has the lowest energy, the 3p is intermediate, and the 3d has the highest energy.

Third Quantum Number, m_ℓ; Orbitals

Each sublevel contains one or more **orbitals**, designated by the third quantum number, m_ℓ. This quantum number tells us how the electron cloud surrounding the nucleus is directed in space. The value of m_ℓ is related to that of ℓ. For a given value of ℓ, m_ℓ can have any integral value, including 0, between ℓ and $-\ell$; that is,

Wheels within wheels

$$m_\ell = \ell, \ldots, +1, 0, -1, \ldots, -\ell \tag{7.12}$$

To illustrate how this rule works, consider an s sublevel ($\ell = 0$). Here, m_ℓ can have only one value, 0. This means that an s sublevel contains only one orbital. In contrast, consider a p sublevel ($\ell = 1$). Here, m_ℓ may be 1, 0, or -1. This means that within each p sublevel there are three different orbitals: one with $m_\ell = 1$, another with $m_\ell = 0$, and a third one with $m_\ell = -1$. In general, for a sublevel of quantum number ℓ, there are a total of $2\ell + 1$ orbitals. All of the orbitals in the same sublevel have essentially the same energy.

The shape of the electron cloud making up an orbital can be indicated in various ways. Consider, for example, the orbital within an s sublevel, called an s orbital. Two different ways of representing the 1s orbital are shown in Figure 7.6. The diagram in Figure 7.6A is similar to that shown in Figure 7.5. The density of shading is proportional to the probability of finding the electron at a given

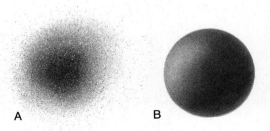

A   B

FIGURE 7.6 Two different ways of indicating the shape of the electron cloud of the 1s orbital. One diagram (A) shows how the probability of finding the electron decreases as one moves out from the nucleus. The colored sphere (B) encloses the region where the electron spends 90% of its time.

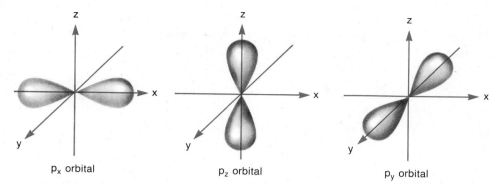

p$_x$ orbital   p$_z$ orbital   p$_y$ orbital

FIGURE 7.7 Electron clouds corresponding to the three p orbitals. The electron density in one of these orbitals is symmetrical about the x axis (p$_x$ orbital). In another orbital, it is symmetrical about the z axis (p$_z$ orbital), and in the third it is symmetrical about the y axis (p$_y$ orbital). We describe this situation by saying that the three orbitals are directed at 90° angles to each other.

n and ℓ determine the energy and shape of an orbital. m$_\ell$ fixes its orientation in space

distance from the nucleus. The diagram in Figure 7.6B is more commonly used, perhaps because it is easier to draw. The diameter of the sphere indicates the region in which there is a high probability, let us say 90%, of finding the electron.

The shapes of orbitals differ depending upon the sublevel in which they are located. Figure 7.7 shows the shape of the p orbitals. Notice that they are dumbbell shaped, in contrast to the spherical shape of s orbitals. You will recall that in a p sublevel there are three different p orbitals. As you can see from Figure 7.7, these three orbitals are oriented at right angles to one another. We may consider them to be directed along the x, y, and z axes. They are sometimes referred to as p$_x$, p$_y$, and p$_z$ orbitals.

It is possible to draw figures showing the shape and orientation of d orbitals. There are five such orbitals within a d sublevel (ℓ = 2), with m$_\ell$ = 2, 1, 0, −1, and −2. We will consider how d orbitals are represented in Chapter 21.

As you might expect, d orbitals are rather complicated

Fourth Quantum Number, m$_s$; Electron Spin

To describe an electron in an atom completely, we need to specify a fourth quantum number, m$_s$. This quantum number is associated with the spin of the electron. An electron has magnetic properties that correspond to those of a charged particle spinning on its axis (Fig. 7.8). Either of two spins are possible, clockwise or counterclockwise.

The quantum number m$_s$ was introduced to make theory consistent with experiment. In that sense, it differs from the first three quantum numbers, which came from the solution to the Schrödinger wave equation for the hydrogen atom. This quantum number is not related to n, ℓ, or m$_\ell$. It can have either of two possible values:

$$m_s = +1/2 \text{ or } -1/2 \qquad (7.13)$$

Electrons that have the same value of m$_s$ (i.e., both $+\frac{1}{2}$ or both $-\frac{1}{2}$) are said to have *parallel* spins. Electrons that have different m$_s$ values (i.e., one $+\frac{1}{2}$ and the other $-\frac{1}{2}$) are said to have *opposed* spins.

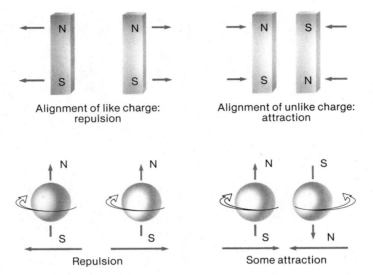

Alignment of like charge: repulsion

Alignment of unlike charge: attraction

Repulsion

Some attraction

FIGURE 7.8 In some respects, an electron behaves as if it were a spherical particle spinning about its axis. There is an analogy between the alignment of electron spins and the alignment of bar magnets (*top of figure*). Within an orbital, the more stable arrangement is the one in which the two electrons have opposed spins (*lower right*).

Pauli Exclusion Principle

Our model for electronic structure is a mix of theory and experiment

We have now considered the four quantum numbers that characterize an electron in an atom. There is an important rule, called the Pauli exclusion principle, that relates to these numbers. It tells us that *no two electrons in an atom can have the same set of four quantum numbers*. This principle was first stated by Wolfgang Pauli in 1925, again to make theory consistent with the properties of atoms.

The Pauli exclusion principle has an implication that may not be obvious at first glance. It requires that no more than two electrons can fit into an orbital. Moreover, if two electrons occupy the same orbital they must have opposed spins. To see that this is the case, consider the 2s orbital. Any electron in this orbital must have

$$n = 2, \quad \ell = 0, \quad m_\ell = 0$$

To satisfy the Pauli exclusion principle, the electrons in this orbital must have different m_s values. But there are only two possible values of m_s. Hence only two electrons can enter that orbital. If the orbital is filled, one electron must have $m_s = +\frac{1}{2}$, and the other $m_s = -\frac{1}{2}$. In other words, the two electrons must have opposed spins. Their quantum numbers are

These electrons are said to be paired

$$n = 2, \quad \ell = 0, \quad m_\ell = 0, \quad m_s = +\tfrac{1}{2}$$
$$n = 2, \quad \ell = 0, \quad m_\ell = 0, \quad m_s = -\tfrac{1}{2}$$

The same argument can be applied equally well to any other orbital.

Capacities of Principal Levels, Sublevels, and Orbitals

The rules that we have given for quantum numbers fix the capacities of principal levels, sublevels, and orbitals. In summary,

1. Each principal level of quantum number **n** contains a total of **n** sublevels.
2. Each sublevel of quantum number ℓ contains a total of $2\ell + 1$ orbitals; that is,

> an s sublevel ($\ell = 0$) contains 1 orbital
> a p sublevel ($\ell = 1$) contains 3 orbitals
> a d sublevel ($\ell = 2$) contains 5 orbitals
> an f sublevel ($\ell = 3$) contains 7 orbitals

3. Each orbital can hold two electrons, which must have opposed spins.

Applying these rules to the first three principal energy levels, we obtain Table 7.3. On the bottom line of the table, each arrow indicates an electron. In each orbital, there are two electrons with opposed spins. The number of electrons in a sublevel is found by adding up the electrons in the orbitals within that sublevel. For example, in a p sublevel ($\ell = 1$), there are six electrons, two in each of three orbitals. To find the total number of electrons in a principal level, we add the electrons in the sublevels within that principal level.

The Pauli principle limits the capacities of orbitals, sublevels, and levels

Table 7.3
Allowed Sets of Quantum Numbers for Electrons in Atoms

Level **n**	1	2			3								
Sublevel ℓ	0	0	1		0	1			2				
Orbital m_ℓ	0	0	1	0	−1	0	1	0	−1	2	1	0	−1	−2
Spin m_s ↑ $= +\frac{1}{2}$ ↓ $= -\frac{1}{2}$	⇅	⇅	⇅	⇅	⇅	⇅	⇅	⇅	⇅	⇅	⇅	⇅	⇅	⇅

EXAMPLE 7.5

a. How many electrons can fit into the principal level for which **n** = 2?
b. What is the capacity for electrons of the 3d sublevel?

Solution

a. In the **n** = 2 level, there are two sublevels, the 2s and the 2p. Of these, the 2s contains one orbital with a capacity of 2 e^-. The 2p contains three orbitals, each of which can hold 2 $e-$:

> total capacity = 1(2) + 3(2) = 8

Note that under the **n** = 2 level in Table 7.3 there are eight electrons, in four pairs.

b. Like any d sublevel, the 3d contains five orbitals. Each orbital can hold 2 e^-. This result is confirmed in Table 7.3, which shows five electron pairs (10 e^-) in the $\ell = 2$ sublevel.

EXERCISE What is the total capacity for electrons of the fourth principal energy level? Answer: 32.

Using the approach in Example 7.5, we can readily obtain the total electron capacity of any principal level or sublevel. In Table 7.4, these capacities are listed through $n = 4$. Notice that the capacity of a principal level is $2n^2$, where n is the principal quantum number. Since each orbital can hold two electrons, the maximum number of electrons in a sublevel ℓ is

$$2 \times \text{(number of orbitals)} = 2(2\ell + 1).$$

Table 7.4
Capacities of Electronic Levels and Sublevels in Atoms

LEVEL n	TOTAL NUMBER OF ELECTRONS IN LEVEL, $2n^2$	MAXIMUM NUMBER OF ELECTRONS IN SUBLEVELS, $2(2\ell + 1)$			
		s	p	d	f
1	2	2	—	—	—
2	8	2	6	—	—
3	18	2	6	10	—
4	32	2	6	10	14

7.5
Electron Arrangements in Atoms

Given the rules referred to in Section 7.4, it is possible to assign quantum numbers to each electron in an atom. Beyond that, electrons can be assigned to specific principal levels, sublevels, and orbitals. There are several ways to do this. Perhaps the simplest way to describe the arrangement of electrons in an atom is to give its **electron configuration**. This tells us the number of electrons in each principal level and sublevel. With an **orbital diagram**, we go one step further and indicate the arrangement of electrons within orbitals. Finally, we can specify the set of four quantum numbers for each electron.

Table 7.5 gives the electron configuration, orbital diagram, and quantum numbers of the electrons in hydrogen (1 e^-) and helium (2 e^-). Note that

—in electron configurations, a superscript is used to indicate the number of electrons in a given sublevel.

This notation is purely for convenience

—in an orbital diagram, arrows are used to indicate electron spins. Thus ↑ indicates an electron with $m_s = +\frac{1}{2}$; ↓ shows an electron with the opposite spin, $m_s = -\frac{1}{2}$.

—quantum numbers are specified in the order n, ℓ, m_ℓ, m_s.

Table 7.5
Electron Arrangements in the Hydrogen and Helium Atoms

	$_1$H (ONE ELECTRON)	$_2$HE (TWO ELECTRONS)
Electron configuration	$1s^1$	$1s^2$
Orbital diagram	1s	1s
	(↑)	(↑↓)
Quantum numbers (n, ℓ, m_ℓ, m_s)	$1, 0, 0, +\frac{1}{2}$	↑ : $1, 0, 0, +\frac{1}{2}$ ↓ : $1, 0, 0, -\frac{1}{2}$

Throughout the remainder of this section, we will consider the electron configurations, orbital diagrams, and electron quantum numbers in multielectron atoms. We should point out that throughout this discussion, we will be dealing with **isolated, gaseous atoms in their ground states**.

Electron Configurations

To arrive at the electron configurations of atoms, we must know the order in which different sublevels are filled. Electrons enter the available sublevels in order of their increasing energy. Usually, a sublevel is filled to capacity before the next one is entered. The relative energies of different sublevels can be obtained from experiment. Figure 7.9 is a plot of these energies for atoms through the **n** = 4 principal level.

Atoms tend to have the lowest possible energy, so electrons go into the lowest unfilled sublevels

From Figure 7.9, it is possible to predict the electron configurations of atoms of elements with atomic numbers 1 through 36. Since an s sublevel can hold only two electrons, the 1s is filled at helium ($1s^2$). With lithium (at. no. = 3), the third electron has to enter a new sublevel: this is the 2s, the lowest sublevel of the second principal energy level; lithium has one electron in this sublevel ($1s^2 2s^1$). With beryllium (at. no. = 4), the 2s sublevel is filled ($1s^2 2s^2$). The next six elements (at. no. 5 through 10) fill the 2p sublevel. Their electron configurations are

$_5$B	$1s^2 2s^2 2p^1$		$_8$O	$1s^2 2s^2 2p^4$
$_6$C	$1s^2 2s^2 2p^2$		$_9$F	$1s^2 2s^2 2p^5$
$_7$N	$1s^2 2s^2 2p^3$		$_{10}$Ne	$1s^2 2s^2 2p^6$

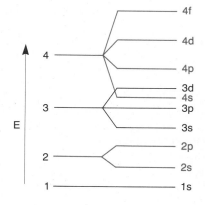

FIGURE 7.9 In general, energy increases with the principal quantum number, **n**. It is possible, however, for the lowest sublevel of **n** = 4 (that is, 4s) to be below the highest sublevel of **n** = 3 (that is, 3d). This appears to be the situation in the potassium and calcium atoms, where successive electrons enter the 4s rather than the 3d sublevel.

Beyond neon, we enter the third principal level. The 3s sublevel is filled at magnesium:

$$_{12}Mg \quad 1s^22s^22p^63s^2$$

Six more electrons are required to fill the 3p sublevel. This is filled to capacity at argon:

$$_{18}Ar \quad 1s^22s^22p^63s^23p^6$$

After argon, we observe an "overlap" of principal energy levels. The next electron enters the *lowest* sublevel of the fourth principal level (4s) instead of the *highest* sublevel of the third principal level (3d). Potassium (at. no. = 19) has one electron in the 4s sublevel; calcium (at. no. = 20) fills it with two electrons:

In atoms, 4s < 3d in energy

$$_{20}Ca \quad 1s^22s^22p^63s^23p^64s^2$$

Now, the 3d sublevel starts to fill with scandium (at. no. = 21). Recall that a d sublevel has a capacity of ten electrons. Hence the 3d sublevel becomes filled at zinc (at. no. = 30):

$$_{30}Zn \quad 1s^22s^22p^63s^23p^64s^23d^{10}$$

The next sublevel, 4p, is filled at krypton (at. no. = 36):

$$_{36}Kr \quad 1s^22s^22p^63s^23p^64s^23d^{10}4p^6$$

EXAMPLE 7.6 Find the electron configuration of the sulfur atom and the nickel atom.

Solution S atom: at. no. = 16; 16 electrons. The sublevels fill to capacity in order of increasing energy. There are two 1s electrons, two 2s, six 2p, two 3s, and four 3p electrons, making a total of 16. The configuration of the S atom is therefore $1s^22s^22p^63s^23p^4$.

The electron configuration tells us the populations of the sublevels in an atom

Ni atom: at. no. = 28; 28 electrons. Proceeding as before, this time until we have added 28 electrons, we obtain as the electron configuration of the Ni atom: $1s^22s^22p^63s^23p^64s^23d^8$. (Note that the 4s sublevel fills before the 3d.)

EXERCISE Give the electron configuration of Mn (at. no. = 25). Answer: $1s^22s^22p^63s^23p^64s^23d^5$.

We can use this general approach to find the electron configuration of any atom. The method, as we have seen, involves adding electrons one by one as atomic number increases. This is sometimes referred to as the Aufbau (building-up) process. To follow it, all we need to know is the order in which sublevels are filled. For the first 36 electrons, this is

1s, 2s, 2p, 3s, 3p, 4s, 3d, 4p

Table 7.6
Ground State Electron Configurations of Gaseous Atoms

Element	Atomic Number	1s	2s	2p	3s	3p	3d	4s	4p	4d	4f	5s
H	1	1										
He	2	2										
Li	3	2	1									
Be	4	2	2									
B	5	2	2	1								
C	6	2	2	2								
N	7	2	2	3								
O	8	2	2	4								
F	9	2	2	5								
Ne	10	2	2	6								
Na	11		Neon core		1							
Mg	12				2							
Al	13				2	1						
Si	14				2	2						
P	15				2	3						
S	16				2	4						
Cl	17				2	5						
Ar	18	2	2	6	2	6						
K	19		Argon core					1				
Ca	20							2				
Sc	21						1	2				
Ti	22						2	2				
V	23						3	2				
Cr	24						5	1				
Mn	25						5	2				
Fe	26						6	2				
Co	27						7	2				
Ni	28						8	2				
Cu	29						10	1				
Zn	30						10	2				
Ga	31						10	2	1			
Ge	32						10	2	2			
As	33						10	2	3			
Se	34						10	2	4			
Br	35						10	2	5			
Kr	36	2	2	6	2	6	10	2	6			
Rb	37		Krypton core									1
Sr	38											2
Y	39									1		2
Zr	40									2		2
Nb	41									4		1
Mo	42									5		1
Tc	43									5		1
Ru	44									7		1
Rh	45									8		1
Pd	46									10		
Ag	47									10		1
Cd	48									10		2

Table 7.6
Ground State Electron Configurations of Gaseous Atoms (*Continued*)

Element	Atomic Number		4d	4f	5s	5p	5d	5f	6s	6p	6d	7s
In	49		10		2	1						
Sn	50		10		2	2						
Sb	51		10		2	3						
Te	52		10		2	4						
I	53		10		2	5						
Xe	54		10		2	6						
Cs	55		10		2	6			1			
Ba	56		10		2	6			2			
La	57		10		2	6	1		2			
Ce	58		10	2	2	6			2			
Pr	59		10	3	2	6			2			
Nd	60		10	4	2	6			2			
Pm	61		10	5	2	6			2			
Sm	62		10	6	2	6			2			
Eu	63		10	7	2	6			2			
Gd	64		10	7	2	6	1		2			
Tb	65		10	9	2	6			2			
Dy	66		10	10	2	6			2			
Ho	67		10	11	2	6			2			
Er	68		10	12	2	6			2			
Tm	69		10	13	2	6			2			
Yb	70		10	14	2	6			2			
Lu	71		10	14	2	6	1		2			
Hf	72		10	14	2	6	2		2			
Ta	73		10	14	2	6	3		2			
W	74		10	14	2	6	4		2			
Re	75	Krypton	10	14	2	6	5		2			
Os	76	core	10	14	2	6	6		2			
Ir	77		10	14	2	6	7		2			
Pt	78		10	14	2	6	9		1			
Au	79		10	14	2	6	10		1			
Hg	80		10	14	2	6	10		2			
Tl	81		10	14	2	6	10		2	1		
Pb	82		10	14	2	6	10		2	2		
Bi	83		10	14	2	6	10		2	3		
Po	84		10	14	2	6	10		2	4		
At	85		10	14	2	6	10		2	5		
Rn	86		10	14	2	6	10		2	6		
Fr	87		10	14	2	6	10		2	6		1
Ra	88		10	14	2	6	10		2	6		2
Ac	89		10	14	2	6	10		2	6	1	2
Th	90		10	14	2	6	10		2	6	2	2
Pa	91		10	14	2	6	10	2	2	6	1	2
U	92		10	14	2	6	10	3	2	6	1	2
Np	93		10	14	2	6	10	5	2	6		2
Pu	94		10	14	2	6	10	6	2	6		2
Am	95		10	14	2	6	10	7	2	6		2
Cm	96		10	14	2	6	10	7	2	6	1	2
Bk	97		10	14	2	6	10	9	2	6		2
Cf	98		10	14	2	6	10	10	2	6		2
Es	99		10	14	2	6	10	11	2	6		2
Fm	100		10	14	2	6	10	12	2	6		2
Md	101		10	14	2	6	10	13	2	6		2
No	102		10	14	2	6	10	14	2	6		2
Lr	103		10	14	2	6	10	14	2	6	1	2
	104		10	14	2	6	10	14	2	6	2	2
	105		10	14	2	6	10	14	2	6	3	2

Beyond that, it is probably simplest to deduce the order of filling from the Periodic Table (Chap. 8).

A rule-of-thumb that can be helpful in writing electron configurations of atoms is the $n + \ell$ rule. It states that for any pair of sublevels, the one with the lower sum of $n + \ell$ fills first. Thus, a 4s sublevel ($n = 4$, $\ell = 0$; $4 + 0 = 4$) is occupied in an atom before a 3d sublevel ($n = 3$, $\ell = 2$; $3 + 2 = 5$). When the sum of $n + \ell$ is the same for a pair of sublevels, the one with the lower n value fills first; for example, a 3d sublevel ($3 + 2 = 5$) is filled before a 4p sublevel ($4 + 1 = 5$).

In Table 7.6 we list the electron configurations of elements with atomic numbers 1 through 105. They generally follow the rules we have illustrated. The major exceptions involve elements where a sublevel is close to being filled or half-filled. It appears that having a full or half-full sublevel leads to increased stability. Consider, for example, chromium (at. no. = 24). Here, a 4s electron is "promoted" to the 3d sublevel, giving chromium five 3d electrons. In copper, a 4s electron is also promoted to the 3d sublevel, thereby filling that sublevel. In both chromium and copper, there is a half-filled 4s sublevel.

We don't know why this is so, but then there are lots of things we don't know

You will note that in Table 7.6 we have saved space by using the phrases "neon core" to indicate the first ten electrons, "argon core" for the first 18 electrons, and so on. This method is often used in writing out what we call *abbreviated electron configurations*. Thus for nickel (at. no. = 28), we write

$$[_{18}Ar]4s^23d^8$$

Here the symbol $[_{18}Ar]$ indicates that the first 18 electrons have the argon configuration $1s^22s^22p^63s^23p^6$. Usually, in an abbreviated electron configuration, we start from the preceding noble gas.

Orbital Diagrams. Hund's Rule

For many purposes, electron configurations are sufficient to describe the arrangement of electrons in atoms. The energy of an electron is determined primarily by the principal level and sublevel in which it is located. Sometimes, however, we want to indicate how electrons are distributed within orbitals. To do this, we use orbital diagrams such as those shown in Table 7.5, p. 211, for hydrogen and helium.

To show how orbital diagrams are obtained from electron configurations, consider the boron atom (at. no. = 5). Its electron configuration is $1s^22s^22p^1$. We know that the pair of electrons in the 1s orbital must have opposed spins. The same is true for the two electrons in the 2s orbital. There are three orbitals in the 2p sublevel. The single 2p electron in boron could be in any one of these orbitals. Its spin could be either "up" ($m_s = +\frac{1}{2}$) or "down" ($m_s = -\frac{1}{2}$). The orbital diagram is ordinarily written

	1s	2s	2p
$_5B$	($\uparrow\downarrow$)	($\uparrow\downarrow$)	($\uparrow$)( )( )

with the first electron in an orbital arbitrarily designated by an "up" arrow, $\uparrow$.

With the next element, carbon, a complication arises. Where should we put the sixth electron? We could put it in the same orbital as the other 2p electron, in which case it would have to have the opposite spin, ↓. It could go into one of the other two orbitals, either with a parallel spin, ↑, or an opposed spin, ↓. Experiment shows that there is an energy difference between these arrangements. The most stable is the one in which the two electrons are in different orbitals with parallel spins. The orbital diagram of the carbon atom is

$$\begin{array}{cccc} & 1s & 2s & 2p \\ _6C & (\uparrow\downarrow) & (\uparrow\downarrow) & (\uparrow)(\uparrow)(\ ) \end{array}$$

This situation arises frequently with other atoms. There is a general principle that applies in all such cases. Known as Hund's rule, it can be stated as follows:

Within a given sublevel, the order of filling is such that there is the maximum number of half-filled orbitals. The single electrons in these half-filled orbitals have parallel spins.

Following this principle, we show the orbital diagrams for the elements boron through neon in Figure 7.10. Notice that

—in all filled orbitals, the two electrons have opposed spins. Such electrons are often referred to as being *paired*. There are four paired electrons in the B, C, and N atoms, six in the oxygen atom, eight in the fluorine atom, and ten in the neon atom.

—in accordance with Hund's rule, within a given sublevel there are as many half-filled orbitals as possible. Electrons in such orbitals are said to be *unpaired*. There is one unpaired electron in atoms of B and F, two unpaired electrons in C and O atoms, and three unpaired electrons in the N atom. When there are two or more unpaired electrons, as in C, N, and O, those electrons have parallel spins.

Hund's rule keeps the electrons as far apart as possible.

EXAMPLE 7.7 Construct orbital diagrams for atoms of chlorine and iron.

Solution We first need to know the electron configurations of the atoms. For a Cl atom with 17 electrons, the electron configuration is $1s^2 2s^2 2p^6 3s^2 3p^5$. For an Fe atom with 26 electrons, the configuration is $1s^2 2s^2 2p^6 3s^2 3p^6 4s^2 3d^6$.

In writing an orbital diagram, we deal first with orbitals in completed sublevels. Each such orbital is filled with two electrons of opposed spins (↑ and ↓). Then we turn our attention to partially filled sublevels. Here we add electrons one by one to the available orbitals, keeping spins parallel (↑ and ↑) as much as possible, in accordance with Hund's rule:

$$\begin{array}{ccccccc} & 1s & 2s & 2p & 3s & 3p \\ _{17}Cl & (\uparrow\downarrow) & (\uparrow\downarrow) & (\uparrow\downarrow)(\uparrow\downarrow)(\uparrow\downarrow) & (\uparrow\downarrow) & (\uparrow\downarrow)(\uparrow\downarrow)(\uparrow) \end{array}$$

$$\begin{array}{cccccccc} & 1s & 2s & 2p & 3s & 3p & 4s & 3d \\ _{26}Fe & (\uparrow\downarrow) & (\uparrow\downarrow) & (\uparrow\downarrow)(\uparrow\downarrow)(\uparrow\downarrow) & (\uparrow\downarrow) & (\uparrow\downarrow)(\uparrow\downarrow)(\uparrow\downarrow) & (\uparrow\downarrow) & (\uparrow\downarrow)(\uparrow)(\uparrow)(\uparrow)(\uparrow) \end{array}$$

Notice that there is one unpaired electron in the Cl atom; in the Fe atom, there are four unpaired electrons.

EXERCISE Show the orbital diagram for the 3d sublevel for Ni (at. no. = 28). Answer: $(\uparrow\downarrow)(\uparrow\downarrow)(\uparrow\downarrow)(\uparrow)(\uparrow)$.

Atom	Orbital diagram			Electron configuration
B	$(\uparrow\downarrow)$	$(\uparrow\downarrow)$	$(\uparrow\ )(\ \ )(\ \ )$	$1s^2 2s^2 2p^1$
C	$(\uparrow\downarrow)$	$(\uparrow\downarrow)$	$(\uparrow\ )(\uparrow\ )(\ \ )$	$1s^2 2s^2 2p^2$
N	$(\uparrow\downarrow)$	$(\uparrow\downarrow)$	$(\uparrow\ )(\uparrow\ )(\uparrow\ )$	$1s^2 2s^2 2p^3$
O	$(\uparrow\downarrow)$	$(\uparrow\downarrow)$	$(\uparrow\downarrow)(\uparrow\ )(\uparrow\ )$	$1s^2 2s^2 2p^4$
F	$(\uparrow\downarrow)$	$(\uparrow\downarrow)$	$(\uparrow\downarrow)(\uparrow\downarrow)(\uparrow\ )$	$1s^2 2s^2 2p^5$
Ne	$(\uparrow\downarrow)$	$(\uparrow\downarrow)$	$(\uparrow\downarrow)(\uparrow\downarrow)(\uparrow\downarrow)$	$1s^2 2s^2 2p^6$
	1s	2s	2p	

There are five other orbital diagrams for the C atom that are really equivalent to the one shown. Can you find them?

FIGURE 7.10 Orbital diagrams showing electron arrangements for atoms with five to ten electrons. Electrons entering orbitals of equal energy remain unpaired as long as possible.

Hund's rule, like the Pauli exclusion principle, is based upon experiment. It is possible to determine the number of unpaired electrons in an atom. With solids, this is done by studying their behavior in a magnetic field. If there are unpaired electrons present, the solid will be attracted into the field. Such a substance is said to be *paramagnetic*. If the atoms in the solid contain only paired electrons, it is slightly repelled by the field. Substances of this type are called *diamagnetic*. With gaseous atoms, the atomic spectrum can also be used to establish the presence and number of unpaired electrons.

Experimentally, we find that certain gaseous atoms are paramagnetic. Among these are H(g) and Li(g). Both of these have a single electron in a half-filled orbital. In contrast, He(g) and Be(g) are diamagnetic. In these two atoms, each orbital is completely full. Since we find that C(g) and N(g) are paramagnetic, we infer that they, like H(g) and Li(g), have half-filled orbitals. Indeed, we can go one step further and show that C(g) has two unpaired electrons, in different orbitals with parallel spins. Similarly, N(g) can be shown to have three unpaired electrons.

Quantum Numbers

Within limits, it is possible to state a complete set of quantum numbers for electrons in atoms. This was done in Table 7.5 for hydrogen and helium. Ordinarily, to specify the quantum numbers, it is simplest to start with the orbital diagram. Consider, for example, the boron atom, whose orbital diagram is given in Figure 7.10. For the first four electrons, we readily arrive at the sets of quantum numbers:

$$1, 0, 0, +\tfrac{1}{2}; \quad 1, 0, 0, -\tfrac{1}{2}; \quad 2, 0, 0, +\tfrac{1}{2}; \quad 2, 0, 0, -\tfrac{1}{2}$$

We note that the fifth electron is in a 2p orbital. Its m_ℓ value could be 1, 0, or −1. If we agree to use the highest m_ℓ value for the first orbital to be filled, the quantum numbers for the fifth electron in boron are:

$$2, 1, 1, +\tfrac{1}{2}$$

Following the same procedure, the set of quantum numbers for the sixth electron in carbon would be

$$2, 1, 0, +\tfrac{1}{2}$$

Notice that the two 2p electrons in carbon have parallel spins: $m_s = +\tfrac{1}{2}$ in both cases. For the seventh electron in nitrogen, the set of quantum numbers is:

$$2, 1, -1, +\tfrac{1}{2}$$

Summarizing the quantum numbers of each electron in the nitrogen atom,

	1s		2s		2p		
	($\uparrow$	$\downarrow$)	($\uparrow$	$\downarrow$)	($\uparrow$)	($\uparrow$)	($\uparrow$)
n	1	1	2	2	2	2	2
ℓ	0	0	0	0	1	1	1
m_ℓ	0	0	0	0	1	0	-1
m_s	$+\tfrac{1}{2}$	$-\tfrac{1}{2}$	$+\tfrac{1}{2}$	$-\tfrac{1}{2}$	$+\tfrac{1}{2}$	$+\tfrac{1}{2}$	$+\tfrac{1}{2}$

EXAMPLE 7.8 Write a complete set of quantum numbers for the four unpaired 3d electrons in an iron atom. Its orbital diagram is shown in Example 7.7.

Solution For all these electrons, $n = 3$ (third principal level), and $\ell = 2$ (d sublevel). We agreed earlier to assign the highest m_ℓ value to the first orbital filled. For a d sublevel, with $\ell = 2$, the values of m_ℓ run from 2 to -2. Thus from left to right in the orbital diagram, we have

$$(\uparrow\downarrow) \quad (\uparrow) \quad (\uparrow) \quad (\uparrow) \quad (\uparrow)$$
$$m_\ell \qquad 2 \qquad 1 \qquad 0 \qquad -1 \qquad -2$$

The unpaired electrons must all have the same spin. Since they are drawn as $\uparrow$, we arbitrarily assign them an m_s value of $+\tfrac{1}{2}$. The complete set of quantum numbers for the four unpaired electrons is

$$3, 2, 1, +\tfrac{1}{2}; \quad 3, 2, 0, +\tfrac{1}{2}; \quad 3, 2, -1, +\tfrac{1}{2}; \quad 3, 2, -2, +\tfrac{1}{2}$$

EXERCISE Give a complete set of quantum numbers for each paired 3d electron in Fe. Answer: $3, 2, 2, +\tfrac{1}{2}$; $3, 2, 2, -\tfrac{1}{2}$.

Our model for the electronic structure of atoms is an approximation, but a very useful one

In this section, we have presented a model that describes how electrons are distributed between principal levels, sublevels, and orbitals in atoms. There is one difficulty with this model that should be mentioned. It is based on the assumption that the electrons in an atom can be described by assigning them quantum numbers. This means that, in the theory, we consider the electrons to have properties as individuals. (Indeed, the quantum numbers we use are based on the wave mechanical solution of the one-electron problem and, hence, involve only the electron-nucleus interactions.) In such an atomic model, the electronic properties of the atom will simply be equal to the sum of the properties of all the electrons in the atom. Such a model is clearly incorrect. In an atom there are important electron-electron charge interactions and electron-electron spin interactions, as well as the electron-nucleus charge interaction covered by the model. So far, the theory of atomic structure has been unable to properly treat electron-electron interactions in atoms. Such interactions are usually considered to be "averaged in" when one interprets electron configurations. The qualitative structure of atoms, involving the populations of principal levels and sublevels, does not seem to be appreciably altered by such an averaging. This explains the usefulness of electron configurations. Clearly, however, the necessity for such an averaging renders impossible the use of electron configurations in quantitative calculations of atomic energy levels and atomic dimensions.

Summary

Atoms can have only certain quantized (discrete) energies; that is, their electrons can be located only in certain energy levels. When an electron moves from a higher to a lower energy level, energy is ordinarily given off in the form of light. The difference in energy between the two levels is inversely related to the wavelength of the light (Example 7.1) and directly related to its frequency (Example 7.2). The wavelengths of lines in the spectrum of the hydrogen atom can be quite accurately calculated from the Bohr model (Examples 7.3 and 7.4).

The quantum mechanical model of the atom describes electrons in terms of quantum numbers. There are four such numbers:

n designates the principal level; $n = 1, 2, 3, \ldots$
ℓ designates the sublevel (s, p, d, or f); $\ell = 0, 1, \ldots, (n - 1)$
m_ℓ designates the orbital; $m_\ell = +\ell, \ldots, +1, 0, -1, \ldots, -\ell$
m_s designates the electron spin; $m_s = +\frac{1}{2}, -\frac{1}{2}$

No two electrons in an atom can have the same set of four quantum numbers. This means, for example, that two electrons in the same orbital must have opposed spins (Tables 7.3 and 7.4).

These rules fix the capacities of principal levels, sublevels, and orbitals (Example 7.5). With one more piece of information, the order in which sublevels fill (Fig. 7.9), we can derive electron configurations of atoms (Example 7.6). Orbital diagrams can be written (Example 7.7) taking into account Hund's rule (p. 216). Finally, we can write a set of four quantum numbers for each electron in an atom (Example 7.8).

Key Words and Concepts

abbreviated electron configuration	electron spin	nanometer	Planck's constant
amplitude (ψ)	excited state	opposed spins	principal energy level
atomic spectrum	frequency (ν)	orbital	quantum mechanics
Balmer series	ground state	orbital diagram	quantum number
Bohr model	hertz	paired electron	quantum theory
de Broglie relation	Hund's rule	parallel spins	Schrödinger equation
diamagnetic	ionization energy	paramagnetic	sublevel
electron cloud	Lyman series	Pauli exclusion principle	unpaired electron
electron configuration	$n + \ell$ rule	photon	wavelength (λ)

Questions and Problems

Energy, Wavelength, and Frequency

7.1 Compare the energy of an electron in the ground state to that in an excited state. In which state does the electron have the higher energy?

7.2 A photon of green light has a wavelength of 540.0 nm. Calculate the
a. frequency in s^{-1}.
b. energy difference, $E_{hi} - E_{lo}$, in J/particle.
c. energy difference, in kilojoules/mole.

7.3 Two states differ in energy by 305 kJ/mol. When an electron moves from the higher to the lower of these states, calculate
a. λ, in nanometers.   b. ν, in (seconds)$^{-1}$.

7.4 Photons of red light from a laser have a frequency of 4.51×10^{14}/s. For this light, what is the
a. wavelength?
b. photon energy, in J/particle?
c. photon energy, in kJ/mol?

7.31 Compare the energies and frequencies of two photons, one with a short wavelength, the other with a long wavelength.

7.32 As the result of an electron transition, light is given off in the ultraviolet at 279.5 nm. For this transition, calculate
a. $E_{hi} - E_{lo}$, in joules per particle.
b. $E_{hi} - E_{lo}$, kilojoules per mole.

7.33 Two states differ in energy by 1.50×10^{-18} J/particle. When an electron moves from the higher state to the lower, calculate
a. λ, in meters.   b. λ, in nanometers.
c. ν, in seconds^{-1}.

7.34 A radio station broadcasts at a frequency of 9.65×10^{7}/s. What is the wavelength of these radio waves in
a. meters?   b. nanometers?
c. kilometers?

Bohr Model of the Hydrogen Atom

7.5 For which of the following transitions is energy absorbed? emitted?
a. $n = 1$ to $n = 4$ b. $n = 4$ to $n = 3$
c. $n = 2$ to $n = 3$ d. $n = 4$ to $n = 2$

7.6 Using Equation 7.6, calculate E for $n = 1, 2, 3$, and 4. Make a one-dimensional graph showing energy, at different values of n, increasing vertically. On this graph, indicate by vertical arrows transitions in the
a. Lyman series
b. Balmer series

7.7 The Paschen series lines in the atomic spectrum of hydrogen result from electronic transitions from $n > 3$ to $n = 3$. Calculate the wavelength, in nanometers, of a line in this series resulting from the $n = 5$ to $n = 3$ transition.

7.8 In the Pfund series $n_{lo} = 5$. Calculate the longest wavelength possible for a transition in this series.

7.9 Calculate the ionization energy (kJ/mol) for the hydrogen electron in the first excited state ($n = 2$).

7.35 Consider the transitions in Question 7.5.
a. Which ones involve the ground state?
b. Which one absorbs the most energy?
c. Which one emits the most energy?

7.36 According to the Bohr model, the radius of a circular orbit is given by the expression

$$r \text{ (in nm)} = 0.0529 \, n^2$$

Draw successive orbits for the hydrogen electron at $n = 1, 2, 3$, and 4. Indicate by arrows transitions between orbits that lead to lines in the
a. Lyman series   b. Balmer series

7.37 For the Brackett series, $n_{lo} = 4$. Calculate the wavelength in nanometers, of a transition from $n = 8$ to $n = 4$.

7.38 A line in the Balmer series occurs at 434.05 nm. Calculate the n_{hi} for the transition associated with this line.

7.39 Calculate the ionization energy (kJ/mol) for the hydrogen electron in the $n = 6$ state.

de Broglie Relation

7.10 What is the de Broglie wavelength of an electron moving at 1.0% of the speed of light (mass of electron $= 9.11 \times 10^{-31}$ kg, speed of light $= 2.998 \times 10^8$ m/s; 1 J $= 1$ kg·m²/s²)?

7.40 Calculate the de Broglie wavelength of an automobile weighing 1.68 metric tons and moving at a speed of 72 km/h (1 J $= 1$ kg·m²/s²).

Energy Levels and Sublevels

7.11 What are the possible values of m_ℓ for
  a. $\ell = 1$ b. $\ell = 2$ c. $n = 2$ (all sublevels)

7.41 What are the possible values of m_ℓ for
  a. $\ell = 0$ b. $\ell = 3$ c. $n = 3$ (all sublevels)

7.12 When $n = 4$, what are the possible values of ℓ? Identify each ℓ value with an s, p, d, or f sublevel.

7.42 When $\ell = 2$,
  a. what letter designation is used for this sublevel?
  b. what is the minimum n value?
  c. what is the maximum number of electrons in this sublevel?

7.13 State the total capacity for electrons of
  a. the principal level $n = 3$.   b. a 3d sublevel.
  c. a 3d orbital.

7.43 Give the number of orbitals in
  a. the principal level $n = 4$.
  b. a 3d sublevel.
  c. an f sublevel.

7.14 What is the
  a. minimum n value for $\ell = 3$?
  b. letter used to designate the sublevel with $\ell = 3$?
  c. number of orbitals in a sublevel with $\ell = 4$?
  d. number of different sublevels when $n = 4$?

7.44 State the relationship, if any, between the following pairs of quantum numbers:
  a. n and ℓ b. ℓ and m_ℓ c. m_ℓ and m_s

Electron Configurations

7.15 Write electron configurations for
  a. P   b. Ti   c. As   d. Al

7.45 Write electron configurations for
  a. Si   b. Fe   c. Na   d. Ga

7.16 Write abbreviated electron configurations for
  a. Cl   b. Ni   c. Zn   d. Br

7.46 Write abbreviated electron configurations for
  a. S   b. Ca   c. Co   d. Se

7.17 Give the symbol of the element of lowest atomic number that has
  a. a completed p sublevel.
  b. two 3d electrons.
  c. four 3p electrons.

7.47 Give the symbol of the element with the lowest atomic number that has
  a. a completed d sublevel.
  b. a completed 4s sublevel.
  c. two 4p electrons.

7.18 Give the symbols of all the elements that have
  a. no p electrons.
  b. from two to four d electrons.
  c. from two to four s electrons.

7.48 Give the symbol of the element that
  a. is in Group 8 but has no p electrons.
  b. has a single electron in the 3d sublevel.
  c. forms a $+1$ ion with a $1s^2 2s^2 2p^6$ electron configuration.

7.19 What fraction of the total number of electrons is in s sublevels in
  a. He   b. Ne   c. Zn

7.49 What fraction of the total number of electrons is in p sublevels in
  a. Be   b. Mg   c. Fe

7.20 Which of the following electron configurations are for atoms in the ground state? in excited states? Which are impossible?
  a. $1s^1 2s^1$
  b. $1s^2 2s^2 2p^3$
  c. $[_{10}Ne]3s^2 3p^3 4s^1$
  d. $[_{10}Ne]3s^2 3p^6 4s^3 3d^2$
  e. $[_{10}Ne]3s^2 3p^6 4f^4$
  f. $1s^2 2s^2 2p^4 3s^2$

7.50 Which of the following electron configurations are for atoms in ground states? in excited states? Which are not possible?
  a. $1s^2 2s^2$
  b. $1s^2 2s^2 3s^1$
  c. $[_{10}Ne]3s^2 3p^8 4s^1$
  d. $[_2He]2s^2 2p^6 2d^2$
  e. $[_{18}Ar]4s^2 3d^3$
  f. $[_{10}Ne]3s^2 3p^5 4s^1$

Orbital Diagrams; Hund's Rule

7.21 Give the orbital diagram of
  a. Si   b. Mn   c. Al   d. Ne

7.51 Give the orbital diagram of
  a. C   b. V   c. Ca   d. As

7.22 Give the symbols of atoms with the following orbital diagrams beyond argon:

	4s	3d	4p
a.	(↑↓)	(↑↓)(↑↓)(↑↓)(↑)(↑)	( )( )( )
b.	(↑↓)	(↑↓)(↑)(↑)(↑)(↑)	( )( )( )
c.	(↑↓)	(↑↓)(↑↓)(↑↓)(↑↓)(↑↓)	(↑)(↑)(↑)

7.52 Give the symbols of the atoms that have the following orbital diagrams

	1s	2s	2p	3s	3p
a.	(↑↓)	(↑↓)	(↑↓)(↑↓)(↑↓)	(↑)	
b.	(↑↓)	(↑↓)	(↑↓)(↑↓)(↑↓)	(↑↓)	(↑)(↑)( )
c.	(↑↓)	(↑↓)	(↑↓)(↑)(↑)		

7.23 Give the number of unpaired electrons in an atom of
  a. N   b. Mg   c. Ti

7.53 How many unpaired electrons are in an atom of
  a. Cl   b. Fe   c. Be

7.24 In what main group(s) of the Periodic Table do elements have the following numbers of unpaired electrons per atom?
  a. 0   b. 1   c. 2   d. 3

7.54 Give the symbol of the element(s) in the first transition series with the following number of unpaired electrons per atom:
  a. 0   b. 1   c. 2   d. 3   e. 4   f. 5

Quantum Numbers in Atoms

7.25 Assign a set of four quantum numbers to each electron in the carbon atom.

7.55 Assign a set of four quantum numbers to each electron in the fluorine atom.

7.26 Assign a set of four quantum numbers to
  a. the 4s electron in potassium.
  b. all the 3d electrons in Co.
  c. all the p electrons in S.

7.56 Assign a set of four quantum numbers to
  a. the 3s electrons in magnesium.
  b. all the 3d electrons in Ni.
  c. all the 3p electrons in chlorine.

7.27 For how many electrons in Ca does
  a. $n = 4$, $\ell = 0$, $m_\ell = 0$, $m_s = +\frac{1}{2}$
  b. $n = 3$, $\ell = 1$
  c. $n = 3$, $\ell = 2$, $m_\ell = 1$

7.57 For how many electrons in As does
  a. $n = 4$, $\ell = 1$, $m_\ell = 0$
  b. $n = 3$, $\ell = 2$
  c. $n = 3$, $\ell = 2$, $m_\ell = -1$

7.28 Given the following sets of electron quantum numbers, indicate those that could not occur and explain your reasoning:
  a. $2, 2, 1, +\frac{1}{2}$ b. $3, 2, 0, -\frac{1}{2}$
  c. $3, 3, 2, +\frac{1}{2}$ d. $1, 0, 0, 1$
  e. $4, 0, 2, +\frac{1}{2}$

7.58 Arrange the following sets of quantum numbers of electrons in order of increasing energy. If they have the same energy, place them together.
  a. $3, 2, -1, +\frac{1}{2}$ b. $1, 0, 0, +\frac{1}{2}$
  c. $2, 1, 1, -\frac{1}{2}$ d. $3, 2, 1, +\frac{1}{2}$
  e. $3, 1, 0, +\frac{1}{2}$ f. $2, 0, 0, +\frac{1}{2}$

General

7.29 Explain in your own words what is meant by
  a. the Pauli exclusion principle.
  b. Hund's rule.
  c. a line in an atomic spectrum.
  d. the principal quantum number.

7.30 Criticize the following statements:
  a. The energy of a photon is inversely proportional to frequency.
  b. The energy of the hydrogen electron is inversely proportional to the quantum number n.
  c. Electrons start to enter the fourth principal level as soon as the third is full.

7.59 Explain the difference between
  a. the Bohr model of the atom and the quantum mechanical model.
  b. wavelength and frequency.
  c. paramagnetism and diamagnetism.
  d. the geometries of the three different p orbitals.

7.60 Indicate whether each of the following statements is true or false. If false, correct the statement.
  a. An electron transition from $n = 3$ to $n = 1$ absorbs energy.
  b. Light emitted by an $n = 4$ to $n = 2$ transition will have a shorter wavelength than that from an $n = 5$ to $n = 2$ transition.
  c. A sublevel with $\ell = 4$ has a capacity of 18 electrons.
  d. An atom of a Group 6 element has two unpaired electrons.

*7.61 The energy of any one-electron species in its nth state is given by $E = -BZ^2/n^2$, where Z is the charge on the nucleus and B is 2.179×10^{-18} J/particle. Find the ionization energy of the He^+ ion in its first excited state in kilojoules per mole.

*7.62 In 1885, Johann Balmer, a numerologist, derived the following relation for the wavelength of lines in the visible spectrum of hydrogen:

$$\lambda = 364.6 \, n^2/(n^2 - 4)$$

where λ is in nanometers and n is an integer that can be 3, 4, 5, ... Show that this relation follows from the Bohr equation and Equation 7.1. Note that, for the Balmer series, the electron is returning to the $n = 2$ level.

*7.63 Suppose the rules for assigning quantum numbers were as follows:

$$n = 1, 2, 3, \ldots$$
$$\ell = 0, 1, 2, \ldots, n$$
$$m_\ell = 0, 1, 2, \ldots, \ell + 1$$
$$m_s = +\tfrac{1}{2} \text{ or } -\tfrac{1}{2}$$

Prepare a table similar to Table 7.3, based on these rules, for $n = 1$ and $n = 2$. Give the electron configuration for an atom with eight electrons.

*7.64 Suppose that the spin quantum number could have the values $\tfrac{1}{2}$, 0, and $-\tfrac{1}{2}$. Assuming that the rules governing the values of the other quantum numbers and the order of filling sublevels were unchanged,
  a. what would be the electron capacity of an s sublevel? a p sublevel? a d sublevel?
  b. how many electrons could fit in the $n = 3$ level?
  c. what would be the electron configuration of the element with atomic number 8? 17?

*7.65 In the photoelectric effect, electrons are ejected from a metal surface when light strikes it. A certain minimum energy, E_{min}, is required to eject an electron. Any energy absorbed beyond that minimum gives kinetic energy to the electron. It is found that when light at a wavelength of 540 nm falls on a cesium surface, an electron is ejected with a kinetic energy of 6.69×10^{-20} J. When the wavelength is 400 nm, the kinetic energy is 1.96×10^{-19} J.
  a. Calculate E_{min} for cesium, in joules.
  b. Calculate the longest wavelength, in nanometers, that will eject electrons from cesium.

Chapter 8
The
Periodic
Table
and the
Properties
of Metals

Scientists constantly seek ways to organize factual material so that similarities, differences, and trends become more apparent. The most useful device for this purpose in chemistry is the Periodic Table of the elements, referred to several times in previous chapters. The Periodic Table, you will recall, organizes elements in horizontal rows, called *periods*, and vertical columns, called *groups*. As one moves across a period or down a group, the physical properties of elements change in a smooth, regular fashion. Within a given group, the elements show very similar chemical properties.

In this chapter, we will examine the Periodic Table in some detail, with emphasis upon its use in correlating the properties of elements. After a brief historical discussion of the evolution of the Periodic Table (Section 8.1), we will consider how the electron configuration of an element is related to its position in the table (Section 8.2). With that background, we will examine how the properties of elements change as we move across or down the table (Section 8.3).

In the last three sections of this chapter we will focus on the properties of metals, which account for more than 80 of the 108 elements known today. We will relate the properties of elements to their position in the Periodic Table and their electron configurations (Section 8.4). The chemical properties of the alkali metals (Group 1) and the alkaline earth metals (Group 2) will be considered in Section 8.5. Finally, in Section 8.6, we will briefly survey the general properties of the elements in the center of the Periodic Table, those referred to as transition metals.

Throughout this chapter, we will refer frequently to the principles of electronic structure introduced in Chapter 7. Some of the more important concepts in Chapter 5 (thermochemical equations) and Chapter 6 (Ideal Gas Law) will be reviewed through examples and end-of-chapter problems.

8.1
Development of the Periodic Table

In a sense, the atomic theory of John Dalton laid the foundation for the discovery, a half-century later, of the Periodic Table. For one thing, it focused attention upon the relative masses of atoms. By the early 1800s, approximate atomic masses had been established for more than 20 elements. In 1817, Johann Dobereiner discovered a simple relationship between the atomic masses of calcium, strontium, and barium, three elements known to resemble one another chemically. He found that the atomic mass of strontium (88) was very close to the average of the atomic masses of calcium (40) and barium (137). He later discovered another such "triad" consisting of three elements that we now refer to as halogens: chlorine (at. mass = 35), bromine (at. mass = 80), and iodine (at. mass = 127):

Scientists look for relationships like this

$$80 \approx \frac{35 + 127}{2}$$

Since chemists of the time could conceive of no reason that atomic masses should be related in this way, Dobereiner's observations received almost no attention.

By 1864, the situation in chemistry had changed considerably. Many more elements had been discovered and atomic masses established more accurately. In that year, J. A. R. Newlands, an English chemist, proposed the first version of the Periodic Table. He organized his table by arranging the known elements in order of increasing atomic mass in horizontal rows seven elements long (Fig. 8.1). He pointed out that the eighth element in a sequence had chemical properties

							H
Li	Be		B	C	N	O	F
Na	Mg		Al	Si	P	S	Cl
K	Ca		Cr	Ti	Mn	Fe	Co, Ni
Cu	Zn		Y	In	As	Se	Br
Rb	Sr	La, Ce	Zr	Nb, Mo	Ru, Rh		Pd
Ag	Cd		U	Sn	Sb	Te	I
Cs	Ba, V						

Newlands (1864)

I	II	III	IV	V	VI	VII	VIII
R_2O	RO	R_2O_3	RO_2	R_2O_5	RO_3	R_2O_7	RO_4
H							
Li	Be	B	C	N	O	F	
Na	Mg	Al	Si	P	S	Cl	
K	Ca	—	Ti	V	Cr	Mn	Fe, Co, Ni
Cu	Zn	—	—	As	Se	Br	Ru, Rh, Pd
Ag	Cd	In	Sn	Sb	Te	I	
Cs	Ba						

Mendeleev (as revised, 1871)

FIGURE 8.1 Two early versions of the Periodic Table. Both are in the condensed form used by early chemists, with transition metals placed with main-group elements. The elements in color were out of place in Newlands' table.

They suggested he take up music

very similar to those of the starting one. Newlands referred to this principle as the Law of Octaves. Unfortunately, Newlands' periodic table met only with ridicule and scorn from members of the Chemical Society in London. Indeed, they refused to publish his paper.

Shortly thereafter, a much more successful proposal for a periodic table was made in Russia by Dmitri Mendeleev, Professor of Chemistry at the University of Saint Petersburg. In writing a textbook of chemistry, Mendeleev devoted separate chapters to families of elements with similar chemical properties. Early in 1869, he was writing chapters on the alkali metals (Group 1 in the modern Periodic Table), the alkaline earth metals (Group 2) and the halogens (Group 7). He listed successive members of these families in order of increasing atomic mass:

Group 7	F = 19	Cl = 35.5	Br = 80	I = 127
Group 1	Na = 23	K = 39	Rb = 85	Cs = 133
Group 2	Mg = 24	Ca = 40	Sr = 88	Ba = 137

Mendeleev was struck by the nearly constant increase in atomic mass (ΔAM) from one member of any of these families to the next one:

	ΔAM		ΔAM		ΔAM
F → Cl	16.5	Cl → Br	44.5	Br → I	47
Na → K	16	K → Rb	46	Rb → Cs	48
Mg → Ca	16	Ca → Sr	48	Sr → Ba	49

Further speculation along these lines led Mendeleev to the conclusion that the properties of elements vary *periodically* (in cycles) with their atomic masses. The Periodic Table of Mendeleev first appeared in a paper presented at a meeting of the Russian Chemical Society on March 6, 1869. The acclaim that greeted this and successive papers lifted Mendeleev from obscurity to fame. His textbook, *Principles of Chemistry*, first appeared in 1870 and was widely adopted through eight editions.

Why did the Mendeleev version meet with such success while that of Newlands failed? We can see reasons for this by comparing the two versions as shown in Figure 8.1. For the elements H through Ca, Newlands' arrangement worked well; beyond calcium he was in trouble. Although chromium and aluminum both form oxides of the formula R_2O_3 (R is the metal), placing chromium below aluminum requires that chromium (at. mass = 52) precede titanium (at. mass = 48). Additionally, iron certainly does not resemble oxygen chemically, and neither cobalt nor nickel even vaguely belong with the halogens.

Mendeleev liked to live dangerously

In contrast to Newlands, Mendeleev was creative enough to realize that the elements beyond calcium would align properly only if he left some empty spaces in the Periodic Table. He boldly suggested that new elements would be discovered to occupy the gaps he had left for them. Going one step further, Mendeleev predicted detailed physical and chemical properties for three such elements: ''ekaboron'' (scandium), ''ekaaluminum'' (gallium), and ''ekasilicon'' (germanium). Mendeleev's predictions were based on the known properties of other elements in the same group of the Periodic Table. To estimate the atomic mass of germanium, for example, he took an average of those of Si (at. mass = 28) and Sn (at. mass = 118):

predicted atomic mass Ge $\approx \dfrac{28 + 118}{2} = 73$

Since the oxides of silicon and tin were known to have the formulas SiO_2 and SnO_2, respectively, Mendeleev could predict with some confidence that the formula of the oxide of germanium would be GeO_2. Compounds of elements in the same group of the Periodic Table tend to have similar formulas.

By 1886, all of the elements predicted by Mendeleev had been isolated. They were shown to have properties very close to those he predicted (Table 8.1). The spectacular agreement between prediction and experiment removed any doubts about the validity and value of Mendeleev's Periodic Table.

Table 8.1
Predicted and Observed Properties of Germanium (Ekasilicon)

PROPERTY	PREDICTED BY MENDELEEV (1871)	OBSERVED (1886)
Atomic mass	72	72.3
Density	5.5 g/cm³	5.47 g/cm³
Specific heat	0.31 J/(g·°C)	0.32 J/(g·°C)
Melting point	very high	960°C
Formula of oxide	RO_2	GeO_2
Formula of chloride	RCl_4	$GeCl_4$
Density of oxide	4.7 g/cm³	4.70 g/cm³
Boiling point of chloride	100°C	86°C

Another person who made a major contribution to the development of the Periodic Table was the German chemist Lothar Meyer. In July 1868, in the process of revising his highly successful text, *Modern Theories of Chemistry,* Meyer compiled a Periodic Table containing 56 elements. He also prepared extensive graphs showing that such properties as molar volume were a periodic function of atomic mass. Unaware of Mendeleev's work, Meyer published his results in 1870. In 1882, the two men jointly were awarded the Davy Medal, the highest honor of the Royal Society. Five years later, the Society belatedly awarded the same medal to its own member, J. A. R. Newlands.

The Periodic Table of Mendeleev was based on atomic masses; elements were arranged in order of increasing atomic mass. If you look at a modern Periodic Table, you will find three cases in which elements are out of order insofar as atomic mass is concerned: argon (at. mass = 39.95) comes before potassium (at. mass = 39.10), cobalt (at. mass = 58.93) before nickel (at. mass = 58.69), and tellurium (at. mass = 127.60) before iodine (at. mass = 126.90). Chemically, however, the positions of these elements make sense. Argon, for example, is clearly a noble gas, not an alkali metal.

These anomalies were resolved in 1914 by a young Englishman, Henry Moseley, a student of Rutherford. Moseley was studying the properties of the radiation (x-rays) given off when elements were bombarded by high-energy electrons. He discovered that a plot of the square root of the x-ray frequency versus atomic mass gave a nearly straight line. There were, however, three pairs of

The Periodic Table was an idea whose time had come

elements that fell off the line. You guessed it: the elements that were out of order were Ar and K, Co and Ni, and Te and I. When Moseley plotted order number in the Periodic Table rather than atomic mass, these elements fell neatly in line. A graph of $\nu^{1/2}$ versus order number was a perfect straight line.

Moseley's work showed that order number in the Periodic Table had a significance that went beyond atomic mass. Moseley suggested a simple explanation:* "There is every reason to suppose that the integer that controls the x-ray spectrum is the charge on the nucleus." Today we relate position in the Periodic Table to atomic number. The Periodic Law is stated in these terms: *The properties of the chemical elements are a periodic (cyclic) function of atomic number.* By the same token, we can say the following:

The Periodic Table is an arrangement of elements in order of increasing atomic number in horizontal rows of such a length that elements with similar chemical properties fall directly beneath one another.

8.2
Electron Arrangements and the Periodic Table

If you examine the Periodic Table on the inside front cover of this text, you will note that there are seven horizontal rows, called *periods*. Each period, except the first, starts with an alkali metal (Li, Na, K, Rb, Cs, Fr). Each period, except the last, which is incomplete, ends with a noble gas (He, Ne, Ar, Kr, Xe, Rn). The length of successive periods varies from 2 to 32 elements, as shown in Table 8.2.

Table 8.2
Structure of the Periodic Table

PERIOD	NUMBER OF ELEMENTS	BEGINS WITH	ENDS WITH
1	2	$_1$H	$_2$He
2	8	$_3$Li	$_{10}$Ne
3	8	$_{11}$Na	$_{18}$Ar
4	18	$_{19}$K	$_{36}$Kr
5	18	$_{37}$Rb	$_{54}$Xe
6	32	$_{55}$Cs	$_{86}$Rn
7	32	$_{87}$Fr	at. no. 118

The numbers 2, 8, 18, and 32 should ring a bell

The vertical columns of elements in the Periodic Table are referred to as *groups*. Each group is assigned a number, written at the top of the vertical column. Unfortunately, from the time of Mendeleev, there has been no general agreement as to the numbering system to be used. We will refer to the **main-group elements,** those in the two groups at the far left and the six groups at the right, as Groups 1 through 8. The **transition elements,** in the center of the Periodic Table, are not assigned group numbers. The same is true of the **lanthanides** (at. no. = 57 to 70) and **actinides** (at. no. = 89 to 102), listed separately at the bottom of the table.

*Moseley's paper was published in April 1914. A little more than a year later, he died at the age of 27 in the senseless slaughter of British troops at Gallipoli.

The structure of the Periodic Table was established by experiment, based on the properties of elements. The fact that successive elements in a group resemble each other chemically suggests that there must be a basic similarity in the structure of their atoms. The nature of this similarity became apparent when electron configurations were established in the first two decades of this century.

To understand how position in the Periodic Table relates to electron configuration, consider the metals in the first two groups. Atoms of the Group 1 elements all have one s electron in the outermost principal energy level (Table 8.3). In each Group 2 atom, there are two s electrons in the outermost level. A similar relationship applies to the elements in any group:

The atoms of elements in a group of the Periodic Table have the same outer electron configuration.

Table 8.3
Electron Configurations of the Group 1 and 2 Elements

GROUP 1		GROUP 2	
$_3$Li	$[_2He]2s^1$	$_4$Be	$[_2He]2s^2$
$_{11}$Na	$[_{10}Ne]3s^1$	$_{12}$Mg	$[_{10}Ne]3s^2$
$_{19}$K	$[_{18}Ar]4s^1$	$_{20}$Ca	$[_{18}Ar]4s^2$
$_{37}$Rb	$[_{36}Kr]5s^1$	$_{38}$Sr	$[_{36}Kr]5s^2$
$_{55}$Cs	$[_{54}Xe]6s^1$	$_{56}$Ba	$[_{54}Xe]6s^2$

The outermost electrons in an atom determine its chemical properties

Filling of Electron Sublevels in the Periodic Table

As we have just seen, the group in which an element is located depends upon its outer electron configuration. This means that the order in which electron sublevels are filled establishes the structure of the Periodic Table. Figure 8.2 shows how this applies for all the elements. Notice the following points:

1. The elements in Groups 1 and 2 are filling an s sublevel. Thus, Li and Be in the second period fill the 2s sublevel. Na and Mg in the third period fill the 3s sublevel, and so on.

2. The elements in Groups 3 through 8 (six elements in each period) fill p sublevels, which have a capacity of six electrons. In the second period, the 2p sublevel starts to fill with B (at. no. = 5) and is completed with Ne (at. no. = 10). In the third period, the elements Al (at. no. = 13) through Ar (at. no. = 18) fill the 3p sublevel.

3. The transition metals, in the center of the Periodic Table, fill d sublevels. Remember that a d sublevel can hold 10 electrons. In the fourth period, the ten elements Sc (at. no. = 21) through Zn (at. no. = 30) fill the 3d sublevel. In the fifth period, the 4d sublevel is filled by the elements Y (at. no. = 39) through Cd (at. no. = 48). The ten transition metals in the sixth period fill the 5d sublevel. Note that transition metals fill sublevels in which the principal quantum number is 1 less than the period number, for example, 3d in the fourth period.

4. There are two sets of 14 elements each listed separately (to save space) at the bottom of the table. These elements are filling f sublevels with a principal quantum number 2 less than the period number; that is,

Group	1	2											3	4	5	6	7	8

Period

																	1 H	2 He
1																	**1s**	

	3 Li	4 Be											5 B	6 C	7 N	8 O	9 F	10 Ne
2	**2s**															**2p**		
	11 Na	12 Mg											13 Al	14 Si	15 P	16 S	17 Cl	18 Ar
3	**3s**															**3p**		
	19 K	20 Ca	21 Sc	22 Ti	23 V	24 Cr	25 Mn	26 Fe	27 Co	28 Ni	29 Cu	30 Zn	31 Ga	32 Ge	33 As	34 Se	35 Br	36 Kr
4	**4s**					**3d**										**4p**		
	37 Rb	38 Sr	39 Y	40 Zr	41 Nb	42 Mo	43 Tc	44 Ru	45 Rh	46 Pd	47 Ag	48 Cd	49 In	50 Sn	51 Sb	52 Te	53 I	54 Xe
5	**5s**					**4d**										**5p**		
	55 Cs	56 Ba	71 Lu	72 Hf	73 Ta	74 W	75 Re	76 Os	77 Ir	78 Pt	79 Au	80 Hg	81 Tl	82 Pb	83 Bi	84 Po	85 At	86 Rn
6	**6s**					**5d**										**6p**		
	87 Fr	88 Ra	103 Lr	104	105	106												
7	**7s**																	

	57 La	58 Ce	59 Pr	60 Nd	61 Pm	62 Sm	63 Eu	64 Gd	65 Tb	66 Dy	67 Ho	68 Er	69 Tm	70 Yb
6							**4f**							
	89 Ac	90 Th	91 Pa	92 U	93 Np	94 Pu	95 Am	96 Cm	97 Bk	98 Cf	99 Es	100 Fm	101 Md	102 No
7							**5f**							

FIGURE 8.2 The Periodic Table can be used to deduce the electron configurations of atoms. Elements in Groups 1 and 2 are filling an **n**s sublevel, where **n** is the number of the period. Elements in Groups 3 through 8 (except for H and He) fill an **n**p sublevel. The transition metals fill an $(\mathbf{n} - 1)$d sublevel. For example, the elements Sc through Zn in the fourth period fill the 3d sublevel. The lanthanides fill the 4f sublevel, while the actinides fill the 5f.

—14 elements in the sixth period (at. no. = 57 to 70) are filling the 4f sublevel. These elements are sometimes called rare earths or, more commonly nowadays, lanthanides, after the name of the first element in the series, lanthanum (La).

—14 elements in the seventh period (at. no. = 89 to 102) are filling the 5f sublevel. The first element in this series is actinium (Ac); collectively, these elements are referred to as actinides.

Because they have similar properties, compounds of the lanthanides are difficult to separate from one another. Until quite recently, samples of pure compounds of these elements were not available except for those of cerium, the most abundant member of the series. Chromatographic processes are now used to separate compounds of the lanthanide metals. The availability of these compounds in highly pure form has led to several commercial applications. A brilliant red phosphor now used in color TV receivers contains a small amount of europium oxide, Eu_2O_3. This is added to a base of yttrium oxide, Y_2O_3, or gadolinium oxide, Gd_2O_3. Another lanthanide oxide, Nd_2O_3, is being used as part of a liquid laser system.

Just about every element finds a use

The actinide metals are all radioactive. Only two of these elements, uranium and thorium, are found in appreciable amounts in nature. The other elements were first observed in the products of controlled nuclear reactions. In most cases, they have been produced in only very small amounts. Uranium and plutonium are used as the fuel elements in nuclear reactors and nuclear weapons (Chap. 27).

Outer Electron Configurations from the Periodic Table

We could, if we had to, use Figure 8.2 to deduce the complete electron configuration of any element. More commonly, we use the Periodic Table itself for a less ambitious purpose. Specifically, we use it to obtain the outer electron configurations of the main-group elements:

Or, we could deduce the Periodic Table from the electron configurations of the elements

Group	1	2	3	4	5	6	7	8
Outer configuration	ns^1	ns^2	ns^2np^1	ns^2np^2	ns^2np^3	ns^2np^4	ns^2np^5	ns^2np^6

where **n** *is the number of the period in which the element is located*; that is, **n** = 1 for the first period (H, He), **n** = 2 for the second period (Li → Ne), and so on. The value for **n** is also the principal quantum number of the highest occupied energy level in that atom.

EXAMPLE 8.1   Using the Periodic Table and principles of electronic structure discussed in Chapter 7, give
a. the outer electron configurations of radium (Ra) and bromine (Br).
b. the orbital diagram for the outer electrons in these two atoms.
c. a set of quantum numbers for each of the outer electrons in radium.

Solution
a. Radium (at. no. = 88) is in the seventh period, in Group 2; its outer electron configuration is $7s^2$. Bromine (at. no. = 35) is in the fourth period in Group 7. Its outer electron configuration is $4s^24p^5$.
b. The orbital diagrams are

$$
\begin{array}{llll}
 & 7s & 4s & 4p \\
\text{Ra} & (\uparrow\downarrow) & & \\
\text{Br} & & (\uparrow\downarrow) & (\uparrow\downarrow)(\uparrow\downarrow)(\uparrow)
\end{array}
$$

c. For the two 7s electrons in radium, the four quantum numbers, in the usual order, are $7, 0, 0, +\frac{1}{2}$ and $7, 0, 0, -\frac{1}{2}$.

What are the quantum numbers for the unpaired electron in Br?

EXERCISE   Give the outer electron configuration of tin, Sn. Answer: $5s^25p^2$.

8.3 Some Trends in the Periodic Table

The Periodic Table can be used for a variety of purposes. In particular, it is useful in correlating properties on an atomic scale. In this section, we will consider trends in two such properties.

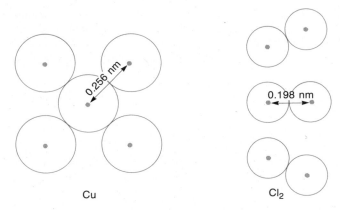

Cu Cl$_2$

$\text{Atomic radius } = \dfrac{0.256 \text{ nm}}{2} = 0.128 \text{ nm} \qquad \text{Atomic radius } = \dfrac{0.198 \text{ nm}}{2} = 0.099 \text{ nm}$

FIGURE 8.3 Atomic radii are defined by assuming that those atoms that are closest in the element are touching. The atomic radius is taken to be one half of the closest internuclear distance.

1. *Atomic radius*, which is a measure of the size of an atom. Strictly speaking, the "size" of an atom is a rather nebulous concept. The electron cloud surrounding the nucleus does not have a sharp boundary. We can, however, define and measure a quantity known as the atomic radius, assuming a spherical atom. Quite simply, the atomic radius is taken to be one half the distance between centers of touching atoms (Fig. 8.3).

2. *Ionization energy*, which is a measure of how difficult it is to remove an electron from an atom. The first ionization energy is the energy change for the removal of the outermost electron from a gaseous atom to form a $+1$ ion:

> An atom with n electrons has n ionization energies

$$M(g) \rightarrow M^{+}(g) + e^{-}; \qquad \Delta E_1 = \text{first ionization energy} \qquad \textbf{(8.1)}$$

We may also refer to second, third, etc., ionization energies. For example,

$$M^{+}(g) \rightarrow M^{2+}(g) + e^{-}; \qquad \Delta E_2 = \text{second ionization energy}$$

In general, the larger the ionization energy, the more difficult it is to remove an electron.

Atomic Radius

The atomic radii of the main-group elements are shown in Figure 8.4. Notice that, in general, atomic radius

—**decreases as we move across a period from left to right in the Periodic Table.**
—**increases as we move down a group in the Periodic Table.**

These trends can be rationalized in terms of the *effective nuclear charge* experienced by an electron at the outer edge of an atom. For any electron, the effective nuclear charge, Z_{eff}, is given by the expression

$$Z_{\text{eff}} = Z - S \qquad \textbf{(8.2)}$$

where Z is the actual nuclear charge, i.e., the atomic number, and the quantity S is the number of electrons between the outer electron and the nucleus. These electrons *shield* the outer electron from the nucleus, in effect reducing the positive charge attracting that electron toward the nucleus. For main-group elements, S is approximately equal to the number of electrons in inner, complete energy levels. Thus for the three atoms Na, Mg, and K, we have

Electron Configuration	Z	Number of Inner Electrons = S	Approx. Value Z_{eff}
$_{11}$Na $1s^2 2s^2 2p^6 3s^1$	11	$2 + 2 + 6 = 10$	$11 - 10 = 1$
$_{12}$Mg $1s^2 2s^2 2p^6 3s^2$	12	$2 + 2 + 6 = 10$	$12 - 10 = 2$
$_{19}$K $1s^2 2s^2 2p^6 3s^2 3p^6 4s^1$	19	$2 + 2 + 6 + 2 + 6 = 18$	$19 - 18 = 1$

Electrons in an outer energy level do not effectively shield one another from the nucleus. For example, a 3s electron in magnesium is not shielded by a second 3s electron that is at the same distance from the nucleus. Only those electrons in

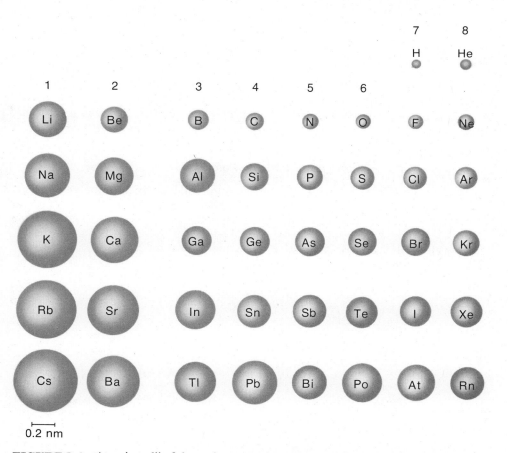

0.2 nm

FIGURE 8.4 Atomic radii of the main-group elements. Atomic radii increase as one goes down a group and in general decrease going across a row in the Periodic Table. Hydrogen has the smallest atom, cesium the largest.

the inner complete levels (1s, 2s, 2p) are effective in reducing the positive charge felt by the outer electron in magnesium.

According to the argument we have just made, effective nuclear charge should increase as we move across a period from left to right in the Periodic Table. Thus, Z_{eff} should be about 1 for Na, 2 for Mg, 3 for Al, and so on. The greater the effective nuclear charge, the stronger will be the attractive force pulling an outer electron in toward the nucleus. As this force increases, the electron is drawn in closer to the nucleus and atomic radius decreases. As a result, atoms get smaller as we move across from left to right in the Periodic Table.

As we move down in the Periodic Table, the effective nuclear charge experienced by an outer electron should remain essentially constant. For example, for the elements Li, Na, K, Rb, and Cs in Group 1, the value of Z_{eff} should be about 1. In that sense, the outer s electron in an alkali metal atom behaves like the electron in a hydrogen atom, where Z = 1. However, you will recall from our discussion of the Bohr model of the hydrogen atom that the distance of the electron from the nucleus increases with the principal quantum number. Hence, we expect atomic radius to increase as we move from Li (2s electron) to Na (3s electron), to K (4s electron), and so on. A similar argument can be applied to other groups, explaining why atoms get larger as we move down a group.

These arguments are qualitative, but work pretty well

First Ionization Energy

It takes energy to knock an electron out of an atom

As pointed out earlier, the first ionization energy is the energy that must be absorbed to remove the outermost electron from an atom. First ionization energies of the main-group elements are listed in Figure 8.5. Notice that this ionization energy

—**increases as we move across the Periodic Table from left to right**.
—**decreases as we move down in the Periodic Table**.

Comparing Figure 8.5 with 8.4, we see an inverse correlation between ionization energy and atomic radius. The smaller the atom, the more tightly its elec-

1	2		3	4	5	6	7	8
							H 1312	He 2372
Li 520	Be 900		B 801	C 1086	N 1402	O 1314	F 1681	Ne 2081
Na 496	Mg 738		Al 578	Si 786	P 1012	S 1000	Cl 1251	Ar 1520
K 419	Ca 590		Ga 579	Ge 762	As 944	Se 941	Br 1140	Kr 1351
Rb 403	Sr 550		In 558	Sn 709	Sb 832	Te 869	I 1009	Xe 1170
Cs 376	Ba 503		Tl 589	Pb 716	Bi 703	Po 812	At	Rn 1037

FIGURE 8.5 First ionization energies of the main-group elements (in kilojoules per mole). In general, ionization energy decreases as one moves down in the Periodic Table and increases as one moves across from left to right. Cesium has the smallest ionization energy and helium the largest.

trons are held to the nucleus and the more difficult they are to remove. Large atoms, such as those of the heavier Group 1 metals, have rather low ionization energies. The outer electrons are far away from the nucleus and hence are relatively easy to remove.

EXAMPLE 8.2 Consider the three elements B, C, and Al. Using only the Periodic Table, predict which of the three elements has
a. the largest atomic radius; the smallest atomic radius.
b. the largest ionization energy; the smallest ionization energy.

Solution The three elements form a block in the table:

B   C
Al

a. Since atomic radius increases as we go down in the table and decreases as we go across, Al should be the largest and C the smallest atom. The observed atomic radii are as follows: B = 0.088 nm; C = 0.077 nm; Al = 0.143 nm.
b. Since ionization energy increases as we go across in the table and decreases as we go down, C should have the highest value and Al the smallest. The observed values are as follows: B = 801 kJ/mol; C = 1086 kJ/mol; Al = 578 kJ/mol.

EXERCISE Compare the two elements Ca and Rb with regard to atomic radius and ionization energy. Answer: Rb has the larger radius and the smaller ionization energy.

In Figure 8.6, ionization energy is plotted against atomic number. This plot shows clearly that ionization energy, like so many other properties of elements, is a periodic function of atomic number; that is, it goes through successive cycles

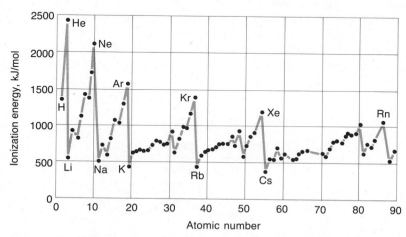

As you can see, the pattern of ionization energies is not perfectly regular

FIGURE 8.6 Ionization energy is a periodic function of atomic number. In each period, it has its minimum value with the first (alkali metal) atom and its maximum value with the last (noble gas) atom.

as atomic number increases. The minima (low points) in ionization energy occur with successive alkali metals (Li, Na, K, . . .). This reflects the fact that atoms of these metals are larger than those of elements in any other group; as a result, their ionization energies are the lowest. The maxima (high points) fall at successive noble gases (He, Ne, Ar, . . .). Atoms of these elements are the smallest and hence have the largest ionization energies.

If you look carefully at Figure 8.5 or 8.6, you will note a few exceptions to the general trends referred to above and illustrated in Example 8.2. Thus, first ionization energy actually decreases as one moves across from Be in Group 2 (900 kJ/mol) to B in Group 3 (801 kJ/mol). This is explained when we realize that the first electron removed from boron comes from the 2p sublevel, as opposed to 2s for beryllium. Since 2p is higher in energy than 2s, it is not too surprising that less energy is required to remove an electron from that sublevel.

Successive Ionization Energies

As we remove successive electrons from an atom, ionization energy increases. In the case of sodium, we have

$$Na(g) \rightarrow Na^+(g) + e^-; \qquad \Delta E_1 = 496 \text{ kJ/mol}$$
$$Na^+(g) \rightarrow Na^{2+}(g) + e^-; \qquad \Delta E_2 = 4561 \text{ kJ/mol}$$
$$Na^{2+}(g) \rightarrow Na^{3+}(g) + e^-; \qquad \Delta E_3 = 6913 \text{ kJ/mol}$$

This trend is readily explained. As the positive charge on the species increases (0 in the Na atom, +1 in the Na^+ ion, +2 in the Na^{2+} ion), electrons are attracted more and more strongly toward the nucleus. Hence, it becomes more and more difficult to remove these electrons.

Ionization energies are measured in so-called stripping experiments

Notice, however, that ionization energy does not increase smoothly as we remove successive electrons. For sodium, there is a big jump between the first and second ionization energies; ΔE_2 is nine times ΔE_1. In contrast, ΔE_3 is less than twice ΔE_2. This is readily explained when we recall the electron configuration of Na, $1s^2 2s^2 2p^6 3s^1$. The first electron to be removed is the 3s electron, which is far from the nucleus. It should leave rather easily ($\Delta E_1 = 496$ kJ/mol). To remove a second electron, we have to break into a complete 2p sublevel, which is much

Table 8.4
Successive Ionization Energies* of Third Period Elements (kJ/mol)

ELEMENT	ΔE_1	ΔE_2	ΔE_3	ΔE_4	ΔE_5	ΔE_6	ΔE_7	ΔE_8
Na	496	4561	6913	9543	13355	16606	20108	25493
Mg	738	1450	7731	10540	13623	17995	21698	25665
Al	578	1817	2745	11580	15033	18372	23292	27455
Si	786	1577	3232	4355	16083	19790	23773	29246
P	1012	1903	2910	4955	6272	21267	25405	29840
S	1000	2251	3361	4562	6996	8494	28079	31723
Cl	1251	2297	3822	5160	6544	9330	11025	33606

*Ionization energies shown in color require removing an electron from the neon core, i.e., an inner level electron.

more difficult (ΔE_2 = 4561 kJ/mol). Successive electrons come from that same sublevel, so we would expect ΔE to increase steadily but rather slowly (ΔE_3 = 6913 kJ/mol, . . .).

Similar effects are observed with other atoms. Table 8.4 shows successive ionization energies for several elements in the third period of the Periodic Table. Notice that in each case the large jump in ionization energy comes when an inner, 2p electron is removed. For sodium, this is the second electron to ionize, for magnesium, it is the third, and so on.

8.4
Metals and the
Periodic Table

Of the known elements, nearly 80% are metals. The metallic elements have several distinguishing characteristics. For one thing, they are good conductors of electricity, much better than nonmetals. For another, they tend to exist as positive ions (for example, Na^+, Mg^{2+}) in their compounds with nonmetals.

In this section we will consider how metallic character varies with position in the Periodic Table. We will then look at some of the general properties that distinguish metals from nonmetals and finally, consider the composition of some typical alloys formed by metals.

Metals, Nonmetals, and Metalloids

The diagonal line or stairway that runs from the upper left to the lower right of the Periodic Table separates metals from nonmetals. *Elements below and to the left of this line are metals.* Included among the metals are:

> We're not sure why this kind of separation exists

—all of the elements in Group 1 (except H) and Group 2.
—the heavier elements in Groups 3 (Al, Ga, In, Tl), 4 (Sn, Pb), and 5 (Bi).
—all of the transition elements.
—all of the lanthanides and actinides.

Elements above and to the right of the diagonal line are classified as nonmetals. With a few exceptions, these elements do not conduct an electric current. In their compounds with metals, nonmetals are usually present as negative ions (for example, O^{2-}, F^-).

Along the diagonal line in the Periodic Table are several elements that are difficult to classify exclusively as metals or nonmetals. They have properties in between those of elements in the two classes. In particular, their electrical conductivities are intermediate between those of metals and nonmetals. The six elements

> Some of these elements can exist in two crystalline forms, one metallic and the other nonmetallic

B	Si	Ge	As	Sb	Te
boron	silicon	germanium	arsenic	antimony	tellurium

are often called **metalloids**. Certain metalloids, notably silicon and germanium, are used in semiconductor devices such as the "chips" found in calculators, computers, and solar cells.

As this discussion implies, metallic character varies in a systematic way with position in the Periodic Table. Metallic character *decreases* as one moves *across* from left to right within a given period. For example, the third period starts with three elements that are distinctly metallic (Na, Mg, Al), moves on to a metalloid (Si), and is completed by four nonmetals (P, S, Cl, Ar). Metallic character *increases* as one moves *down* a given group. Consider, for example, Group 4. C is a nonmetal; Si and Ge are metalloids; Sn and Pb have the properties we associate with metals.

Physical Properties of Metals

You will recall (Section 8.3) that ionization energy decreases as we move down in the Periodic Table and increases as we move across from left to right. This relates directly to the trends in metallic character discussed above. Metallic elements tend to have low ionization energies. This explains why metals readily form positive ions (for example, Na^+, Mg^{2+}). It also means that in a metal there are electrons that are relatively free to move about. The physical properties associated with high electron mobility include the following:

1. *High electrical conductivity*. Metals typically have electrical conductivities several hundred times greater than those of typical nonmetals. Silver is the best electrical conductor but is too expensive for general use. Copper, with a conductivity close to that of silver, is the metal most commonly used for electrical wiring. Although a poor conductor, mercury is used in many electrical devices, such as "silent" light switches, where a liquid conductor is required.

2. *High thermal conductivity*. Of all solids, metals are the best conductors of heat. Anyone who has unwittingly picked up a hot pan can vouch for this property.

There are no molecules in metals. The outer electrons are shared by several atoms

3. *Ductility, malleability*. Most metals are ductile (capable of being drawn out into a wire) and malleable (capable of being hammered into thin sheets). Gold, for example, can be hammered into sheets so thin that they are transparent. A wire over 1 mile long can be drawn out of less than 1 g of gold. In a metal, the electrons act somewhat like a glue holding the atomic nuclei together. As a result, metal crystals can be deformed without shattering. A crystal of a nonmetal breaks into small pieces if you hammer it or try to draw it into a wire.

4. *Luster*. Polished metal surfaces reflect light. Most metals have a silvery white "metallic" color because they reflect light of all wavelengths. Gold and copper absorb some light in the blue region of the visible spectrum and so appear yellow (gold) or red (copper).

Alloys

An important characteristic of metallic elements is their ability to form alloys. An alloy is a material with metallic properties that contains two or more elements, at least one of which is a metal. Solid alloys are ordinarily prepared by melting the elements together, stirring the molten mixture until it is homogeneous, and allowing it to cool. Many alloys, notably bronze, brass, and pewter, have been made for centuries by this method.

The properties of alloys are often quite different from those of their component metals. By alloying a metal with another element, we usually

—*lower the melting point*. Solder, an alloy of lead and tin, has a melting point of 180°C as compared to 232°C for Sn and 338°C for Pb. Wood's metal has a very low melting point, about 70°C. This alloy is used in fusible plugs that melt to set off automatic sprinkler systems.

—*increase the hardness.* A small amount of copper is present in sterling silver, which is much harder than pure silver. Gold is alloyed with silver and copper to form a metal hard enough to be used in jewelry. The lead plates used in storage batteries contain small amounts of antimony to prevent them from bending under stress.

—*lower the electrical and thermal conductivity.* Copper used in electrical wiring must be extremely pure; as little as 0.03% of arsenic can lower its conductivity by 15%. Sometimes we take advantage of this effect. High-resistance nichrome wire (Ni, Cr) is used in the heating elements of hair dryers and electric toasters.

Most metallic objects are made from alloys

Table 8.5 lists the compositions and uses of some familiar alloys.

Table 8.5
Some Commercially Important Alloys

COMMON NAME	COMPOSITION, ELEMENT (MASS PERCENT)	USES
Alnico	Fe(50), Al(20), Ni(20), Co(10)	magnets
Aluminum bronze	Cu(90), Al(10)	crankcases, connecting rods
Brass	Cu(67–90), Zn(10–33)	plumbing, hardware, lighting
Bronze	Cu(70–95), Zn(1–25), Sn(1–18)	bearings, bells, medals
Cast iron	Fe(96–97), C(3–4)	castings
Coinage, US	Cu(75), Ni(25)	5¢, 10¢, 25¢, 50¢ coins
Dental amalgam	Hg(50), Ag(35), Sn(15)	dental fillings
Duriron	Fe(84), Si(14), C(1), Mn(1)	pipes, kettles, condensers
German silver	Cu(60), Zn(25), Ni(15)	tea pots, jugs, faucets
Gold, 18 carat	Au(75), Ag(10–20) Cu(5–15)	jewelry
Gold, 10 carat	Au(42), Ag(12–20), Cu(38–46)	jewelry
Gunmetal	Cu(88), Sn(10), Zn(2)	gun barrels, machine parts
Lead battery plate	Pb(94), Sb(6)	storage batteries
Lead shot	Pb(99.8), As(0.2)	shotgun shells
Magnalium	Al(70–90), Mg(10–30)	aircraft bodies
Manganese steel	Fe(86), Mn(13), C(1)	safes, armor plate, rails
Monel	Ni(60–70), Cu(25–35), Fe, Mn	instruments, machine parts
Nichrome	Ni(60), Fe(25), Cr(15)	electrical resistance wire
Pewter	Sn(70–95), Sb(5–15), Pb(0–15)	tableware
Silver solder	Ag(63), Cu(30), Zn(7)	high-melting solder
Solder	Pb(67), Sn(33)	joining metals
Stainless steel	Fe(73–79), Cr(14–18), Ni(7–9)	instruments, sinks
Steel	Fe(98–99.5), C(0.5–2)	structural metal
Sterling silver	Ag(92.5), Cu(7.5)	tableware, jewelry
Wood's metal	Bi(50), Pb(25), Sn(13), Cd(12)	automatic sprinkler systems

8.5
The Alkali and Alkaline Earth Metals

The elements in Groups 1 and 2 show to an unusual extent the properties that we associate with metals. All of these elements are excellent conductors of electricity. Compared to other elements, they have unusually large atomic radii and unusually small ionization energies. As a result, they lose electrons readily to form positive ions. This explains in part why these elements are among the most reactive of all metals.

Table 8.6
Physical Properties of the Group 1 and Group 2 Metals

PROPERTY	GROUP 1				
	$_3$Li	$_{11}$Na	$_{19}$K	$_{37}$Rb	$_{55}$Cs
Atomic mass	6.94	22.99	39.10	85.47	132.91
Outer electron configuration	$2s^1$	$3s^1$	$4s^1$	$5s^1$	$6s^1$
Atomic radius (nm)	0.152	0.186	0.231	0.244	0.262
Ionization energy (kJ/mol)	520	496	419	403	376
Density (g/cm³)	0.534	0.971	0.862	1.53	1.87
Melting point (°C)	186	98	64	39	28
Boiling point (°C)	1326	889	774	688	690
Flame color	red	yellow	violet	purple	blue
Atomic spectrum (nm)	670.8	589.6	404.7	780.0	459.3
(strong lines)	610.4	589.0	404.4	420.2	455.5

PROPERTY	GROUP 2				
	$_4$Be	$_{12}$Mg	$_{20}$Ca	$_{38}$Sr	$_{56}$Ba
Atomic mass	9.01	24.30	40.08	87.62	137.34
Outer electron configuration	$2s^2$	$3s^2$	$4s^2$	$5s^2$	$6s^2$
Atomic radius (nm)	0.111	0.160	0.197	0.215	0.217
Ionization energy (kJ/mol)	900	738	590	550	503
Density (g/cm³)	1.85	1.74	1.55	2.60	3.51
Melting point (°C)	1283	650	845	770	725
Boiling point (°C)	2970	1120	1420	1380	1640
Flame color	—	—	red	crimson	green

Table 8.6 summarizes some of the physical properties of the alkali and alkaline earth metals. Looking at the table, we note some of the trends referred to earlier in this chapter. In particular, atomic radius decreases as we move from Group 1 to Group 2 (compare Na, 0.186 nm, to Mg, 0.160 nm). As we move down a group, atomic radius increases (compare Mg, 0.160 nm, to Ca, 0.197 nm).

As we would expect, ionization energy is inversely related to atomic radius. The larger Group 1 atoms have smaller ionization energies than their neighbors in Group 2 (compare Na, 496 kJ/mol, to Mg, 738 kJ/mol). In both groups, ionization energy decreases as we move down in the Periodic Table; the larger the atom, the smaller the ionization energy.

The horizontal trend in atomic radius explains in part why the Group 2 metals have higher densities than do the Group 1 metals. An alkaline earth atom is smaller than that of an alkali metal in the same period. It is also slightly heavier (at. mass Na = 22.99, Mg = 24.30). This means that atoms of the Group 2 metals have the larger density, as do the metals themselves (d Na = 0.971 g/cm³, Mg = 1.74 g/cm³).

There are a few anomalies in the trends in density

When a compound of an alkali metal is heated in a Bunsen burner flame, an electron is excited to a higher energy state. When this electron returns to a lower state, energy is emitted as visible light. This process also occurs with compounds of the heavier metals in Group 2: Ca, Sr, and Ba. The flame colors listed in Table 8.6 and shown in Color Plate 8.1 can be used to identify many of these metals. Fireworks displays are vivid, large-scale "flame tests," using compounds of sodium (yellow), strontium (red), and barium (green). Flash powder used at rock concerts is a mixture of magnesium and potassium chlorate, $KClO_3$.

EXAMPLE 8.3   Railroad and highway flares contain $Sr(NO_3)_2$ and $SrCO_3$, which, when ignited, give off an intense red light at 640 nm. Calculate the energy difference, in kilojoules per mole, associated with this emission.

Solution   Recall from Chapter 7 the equation relating the energy difference to wavelength:

$$E_{hi} - E_{lo} = \frac{1.196 \times 10^5 \text{ kJ·nm}}{\lambda} \frac{}{\text{mol}}$$

Substituting $\lambda = 640$ nm,

$$E_{hi} - E_{lo} = \frac{1.196 \times 10^5}{640} \text{ kJ/mol} = 187 \text{ kJ/mol}$$

EXERCISE   Calculate the wavelength (nm) of the orange line in the lithium spectrum for which the energy difference is 195.9 kJ/mol. Answer: 610.5 nm.

Anyone who has worked with the alkali and alkaline earth metals is aware that they are very reactive chemically. The alkali metals and the heavier alkaline earth metals (Ca, Sr, Ba) are commonly stored under dry mineral oil or kerosene to prevent them from reacting with oxygen or water vapor in the air. Magnesium is less reactive; it is commonly available in the form of ribbon or powder. Beryllium, as one would expect from its position in the Periodic Table, is the least metallic element in these two groups. It is also the least reactive toward water, oxygen, or other nonmetals.

The chemical properties of the Group 1 and Group 2 metals are summarized in Table 8.7. In all the products of these reactions, the metals (except for Be*) are always present as positive ions. The cations of the Group 1 metals have a +1 charge: Li^+, Na^+, K^+, Rb^+, Cs^+. Those of the Group 2 metals carry a +2 charge: Mg^{2+}, Ca^{2+}, Sr^{2+}, Ba^{2+}. The properties of these ions are entirely different from those of the corresponding atoms. Sodium atoms, found in sodium metal, are violently reactive toward molecules of H_2O, O_2, or Cl_2. The Na^+ ion, found in compounds such as NaCl, is chemically inert. It is completely unreactive toward water or nonmetals. Every day, you injest about 5 g of Na^+ ions in your food. If you took in that amount of Na atoms in the form of the element, it would be a fatal dose. When a nutritionist tells you that you should take in 800 mg of "calcium" each day, it is understood that he or she is talking about Ca^{2+} ions; 800 mg of calcium atoms would send you to the emergency room of the nearest

The Na^+ ion is about as inert as an He atom

---

*Many of the compounds of beryllium are molecular, as we will see in Chapter 9.

Table 8.7
Reactions of Alkali Metals and Alkaline Earth Metals

METAL*	COMBINING SUBSTANCE	REACTION
		Group 1
All	Hydrogen	$2\ M(s) + H_2(g) \rightarrow 2\ MH(s)$
All	Halogens	$2\ M(s) + X_2 \rightarrow 2\ MX(s)$
Li	Nitrogen	$6\ M(s) + N_2(g) \rightarrow 2\ M_3N(s)$
All	Sulfur	$2\ M(s) + S(s) \rightarrow M_2S(s)$
Li	Oxygen	$4\ M(s) + O_2(g) \rightarrow 2\ M_2O(s)$
Na		$2\ M(s) + O_2(g) \rightarrow M_2O_2(s)$
K, Rb, Cs		$M(s) + O_2(g) \rightarrow MO_2(s)$
All	Water	$2\ M(s) + 2\ H_2O(l) \rightarrow 2\ M^+(aq) + 2\ OH^-(aq) + H_2(g)$
		Group 2
Ca, Sr, Ba	Hydrogen	$M(s) + H_2(g) \rightarrow MH_2(s)$
All	Halogens	$M(s) + X_2 \rightarrow MX_2(s)$
Mg, Ca, Sr, Ba	Nitrogen	$3\ M(s) + N_2(g) \rightarrow M_3N_2(s)$
Mg, Ca, Sr, Ba	Sulfur	$M(s) + S(s) \rightarrow MS(s)$
Be, Mg, Ca, Sr, Ba	Oxygen	$2\ M(s) + O_2(g) \rightarrow 2\ MO(s)$
Ba		$M(s) + O_2(g) \rightarrow MO_2(s)$
Ca, Sr, Ba	Water	$M(s) + 2\ H_2O(l) \rightarrow M^{2+}(aq) + 2\ OH^-(aq) + H_2(g)$
Mg		$M(s) + H_2O(g) \rightarrow MO(s) + H_2(g)$

*Abbreviated as "M" in reactions.

hospital. No question about it, ions such as Na^+ and Ca^{2+} must be distinguished from the corresponding atoms.

The use of Table 8.7 is illustrated in Example 8.4.

EXAMPLE 8.4 Write balanced equations for the reaction of
a. Cesium with hydrogen
b. Magnesium with nitrogen
c. Barium with oxygen
d. Lithium with bromine

Solution Using Table 8.7 as a guide and recalling that all of the nonmetals required are diatomic elements,
a. $2\ Cs(s) + H_2(g) \rightarrow 2\ CsH(s)$
b. $3\ Mg(s) + N_2(g) \rightarrow Mg_3N_2(s)$
c. $2\ Ba(s) + O_2(g) \rightarrow 2\ BaO(s)$ and $Ba(s) + O_2(g) \rightarrow BaO_2(s)$
d. $2\ Li(s) + Br_2(l) \rightarrow 2\ LiBr(s)$

EXERCISE Write the balanced equation for the reaction of lithium with
a. nitrogen; b. water. Answers:
a. $6\ Li(s) + N_2(g) \rightarrow 2\ Li_3N(s)$
b. $2\ Li(s) + 2\ H_2O(l) \rightarrow 2\ Li^+(aq) + 2\ OH^-(aq) + H_2(g)$

In the rest of this section, we will look more closely at some of the reactions listed in Table 8.7. In particular, we will examine the reactions of the alkali and alkaline earth metals with hydrogen, water, and oxygen. Then we will consider some properties of compounds of the two most familiar elements in these groups, sodium and calcium.

Reactions with Hydrogen

The metals of Groups 1 and 2 react with hydrogen at high temperatures to form ionic hydrides. For example,

$$2 \, Na(s) + H_2(g) \rightarrow 2 \, NaH(s) \tag{8.3}$$

$$Ca(s) + H_2(g) \rightarrow CaH_2(s) \tag{8.4}$$

These white, crystalline solids contain H^- ions. They are often referred to as saline hydrides because of their physical resemblance to NaCl.

    The saline hydrides react with water to produce hydrogen gas. Typical reactions are

On electrolysis of molten NaH, the H^- ions go to the + electrode

$$NaH(s) + H_2O(l) \rightarrow H_2(g) + Na^+(aq) + OH^-(aq) \tag{8.5}$$

$$CaH_2(s) + 2 \, H_2O(l) \rightarrow 2 \, H_2(g) + Ca^{2+}(aq) + 2 \, OH^-(aq) \tag{8.6}$$

In this way, saline hydrides can serve as compact, portable sources of hydrogen gas for inflating life rafts and balloons.

A tank of compressed H_2 would not be practical

EXAMPLE 8.5 What volume of hydrogen gas at 15°C and 1.00 atm can be formed by the reaction of 1.00 g of calcium hydride with water?

Solution We first calculate the number of moles of H_2 formed by Reaction 8.6. Note that the molar mass of CaH_2 is (40.08 + 2.02) g/mol = 42.10 g/mol.

$$\text{no. moles } H_2 = 1.00 \text{ g } CaH_2 \times \frac{1 \text{ mol } CaH_2}{42.10 \text{ g } CaH_2} \times \frac{2 \text{ mol } H_2}{1 \text{ mol } CaH_2}$$
$$= 0.0475 \text{ mol } H_2$$

Now we use the Ideal Gas Law to find the volume of hydrogen gas:

$$V = \frac{nRT}{P} = \frac{(0.0475 \text{ mol})\left(0.0821 \, \dfrac{L \cdot atm}{mol \cdot K}\right)(288 \text{ K})}{1.00 \text{ atm}} = 1.12 \text{ L}$$

EXERCISE What volume of hydrogen gas at 15°C and 1.00 atm can be formed by the reaction of 1.00 g NaH with water? Answer: 0.986 L.

Reactions with Water

The alkali metals react vigorously with water to liberate hydrogen. At the same time, a water solution of the alkali hydroxide is formed. The reaction of sodium is typical:

$$Na(s) + 2 \, H_2O(l) \rightarrow 2 \, Na^+(aq) + 2 \, OH^-(aq) + H_2(g); \, \Delta H = -367.6 \text{ kJ} \tag{8.7}$$

As you can see from the thermochemical equation, the reaction is strongly exothermic. Enough heat is evolved for the hydrogen to catch fire.

Among the Group 2 metals, Ca, Sr, and Ba react with water in much the same way as the alkali metals. The reaction with barium is

$$Ba(s) + 2\,H_2O(l) \rightarrow Ba^{2+}(aq) + 2\,OH^-(aq) + H_2(g);\, \Delta H = -426.6\ \text{kJ} \quad \textbf{(8.8)}$$

Beryllium does not react with water at all. Magnesium reacts very slowly with boiling water, but reacts more readily with steam at high temperatures:

$$Mg(s) + H_2O(g) \rightarrow MgO(s) + H_2(g); \quad \Delta H = -360.0\ \text{kJ} \quad \textbf{(8.9)}$$

Mg was used in the fire-bombing of London in WW II

This reaction, like Reaction 8.7 or 8.8, produces enough heat to ignite the hydrogen. Fire fighters who try to put out a magnesium fire by spraying water on it have discovered this reaction, often with tragic results. The best way to extinguish burning magnesium is to dump dry sand on it.

EXAMPLE 8.6 Consider the thermochemical equation for the reaction of magnesium with steam (Reaction 8.9).
a. Calculate ΔH when 1.00 kg of magnesium reacts.
b. If the heat evolved by the reaction of 1.00 kg of magnesium is absorbed by 50.0 kg of water (S.H. = 4.18 J/g·°C), originally at 25.0°C, what is the final temperature?

Solution
a. From the thermochemical equation, we see that 1 mol Mg $\mathrel{\hat=}$ -360.0 kJ. The atomic mass of Mg is 24.30. Hence,

$$\Delta H = 1.00 \times 10^3\ \text{g Mg} \times \frac{1\ \text{mol Mg}}{24.30\ \text{g Mg}} \times \frac{-360.0\ \text{kJ}}{1\ \text{mol Mg}}$$
$$= -1.48 \times 10^4\ \text{kJ}$$

b. The heat given off in the reaction is absorbed by the water. Hence,

$$q_{water} = +1.48 \times 10^4\ \text{kJ} = +1.48 \times 10^7\ \text{J}$$

Recall from Chapter 5 that

$$q_{water} = 4.18\ \frac{\text{J}}{\text{g·°C}} \times m_{water} \times \Delta t$$

Solving for Δt,

$$\Delta t = \frac{q_{water}}{4.18\ \dfrac{\text{J}}{\text{g·°C}} \times m_{water}} = \frac{1.48 \times 10^7\ \text{J}}{4.18\ \dfrac{\text{J}}{\text{g·°C}} \times 50.0 \times 10^3\ \text{g}} = 70.8°C$$

$$t_{final} = t_{initial} + \Delta t = 25.0°C + 70.8°C = 95.8°C$$

Hot Mg will burn as well in CO_2 as it does in steam

Note that the final temperature is very close to the boiling point of water, 100°C. This means that the reaction of 1 kg of magnesium generates enough heat to raise 50 times as much water to the boiling point! No wonder magnesium fires are difficult to extinguish.

EXERCISE In this reaction, how much heat is evolved per gram of hydrogen? Answer: 178.6 kJ

Reactions with Oxygen

As noted from Table 8.7, three different types of oxides can form when oxygen reacts with a Group 1 or 2 metal. Depending on which metal is used, the product can be a normal oxide, a peroxide, or a superoxide.

Oxides The normal oxide contains the oxide anion, O^{2-}, and the metal ion:

Group 1 oxides: M_2O (2 M^+ ions, 1 O^{2-} ion)
Group 2 oxides: MO (M^{2+} ion, O^{2-} ion)

Lithium is the only Group 1 metal that forms the normal oxide in good yield by direct reaction with oxygen. The other Group 1 oxides (Na_2O, K_2O, Rb_2O, Cs_2O) must be prepared by other means. In contrast, the Group 2 metals usually react with oxygen to give the normal oxide. Beryllium and magnesium must be heated strongly to give BeO and MgO. Calcium and strontium react more readily to give CaO and SrO. Barium, the most reactive of the Group 2 metals, catches fire when exposed to moist air. The product is a mixture of the normal oxide, BaO, and the peroxide, BaO_2.

Peroxides These compounds contain the peroxide ion, O_2^{2-}. The most important peroxides are those of sodium and barium, Na_2O_2 and BaO_2. Upon addition to water, they react vigorously to form hydrogen peroxide, H_2O_2;

Peroxides and superoxides are very strong oxidizing agents

$$Na_2O_2(s) + 2 H_2O(l) \rightarrow 2 Na^+(aq) + 2 OH^-(aq) + H_2O_2(aq) \qquad \textbf{(8.10)}$$

$$BaO_2(s) + 2 H_2O(l) \rightarrow Ba^{2+}(aq) + 2 OH^-(aq) + H_2O_2(aq) \qquad \textbf{(8.11)}$$

These reactions can be used to produce small amounts of hydrogen peroxide in the laboratory. Commercially, both Na_2O_2 and BaO_2 are used as bleaching agents.

Superoxides The superoxide ion, O_2^-, is present in these unusual compounds, the most important of which is KO_2. Virtually all the potassium metal produced today is used to make potassium superoxide, KO_2. The reaction

$$K(s) + O_2(g) \rightarrow KO_2(s) \qquad \textbf{(8.12)}$$

is carried out by spraying the metal into air. This serves not only to furnish the oxygen required but also to cool the product. Potassium superoxide is used in self-contained breathing apparatus for fire fighters and miners. It reacts with the moisture in exhaled air to generate oxygen:

KO_2 is perfect for this job

$$4 KO_2(s) + 2 H_2O(g) \rightarrow 3 O_2(g) + 4 KOH(s) \qquad \textbf{(8.13)}$$

The carbon dioxide in the exhaled air is removed by reaction with the KOH formed in Reaction 8.13:

$$KOH(s) + CO_2(g) \rightarrow KHCO_3(s) \qquad \textbf{(8.14)}$$

A person using a mask charged with KO_2 can rebreathe the same air for an extended period of time. This allows that person to enter an area where there are poisonous gases or oxygen-deficient air.

EXAMPLE 8.7 It is possible to prepare potassium peroxide, K_2O_2, by heating potassium carefully with exactly the calculated amount of oxygen. If 1.00 g K is to be converted to K_2O_2, what volume of $O_2(g)$ at 720 mm Hg and 20°C should be used?

Solution We start by writing the balanced equation for the reaction

$$2 K(s) + O_2(g) \rightarrow K_2O_2(s)$$

Noting that the atomic mass of potassium is 39.1, we have

$$78.2 \text{ g K} \doteq 1 \text{ mol } O_2$$

$$\text{no. moles } O_2 \text{ required} = 1.00 \text{ g K} \times \frac{1 \text{ mol } O_2}{78.2 \text{ g K}} = 0.0128 \text{ mol } O_2$$

Now, using the Ideal Gas Law (Chap. 6), we calculate the volume of oxygen:

$$V = \frac{nRT}{P} = \frac{(0.0128 \text{ mol})\left(0.0821 \frac{\text{L·atm}}{\text{mol·K}}\right) \times 293 \text{ K}}{(720/760 \text{ atm})} = 0.325 \text{ L}$$

EXERCISE Suppose air (21% oxygen by volume) is used instead of O_2. What volume of air at the same temperature and pressure should be used? Answer: 1.5 L.

Sodium Compounds

NaCl As pointed out in Chapter 4, NaCl is the major source of sodium metal and the compounds of that element. It occurs in impure form ("rock salt") in large underground deposits. The largest of these, more than 100 m thick, underlies parts of Oklahoma, Texas, and Kansas. The salt may be obtained by forming a water solution, which is pumped to the surface. Upon evaporation, crystals of NaCl form.

All sodium compounds are water soluble

The NaCl produced in this way is contaminated with small amounts of $MgCl_2$ and $CaCl_2$. These compounds are present in ordinary table salt. They tend to pick up water from the air, causing the salt to "cake" and refuse to pour. Small amounts of chemical drying agents are added to prevent this. About 0.01% KI is also added to table salt. This prevents a condition called goiter, an enlargement of the thyroid gland caused by iodine deficiency.

About 50 kg per person in the U.S.

NaOH About 10^7 metric tons of sodium hydroxide are produced annually. You may be familiar with NaOH in the form of lye. As a solid or concentrated solution, NaOH is used in commercial oven and drain cleaners. These products should be used with caution because concentrated NaOH is corrosive to skin and clothing. Larger amounts of NaOH are used to make soap, textiles, and paper.

Sodium hydroxide is very soluble in water, about 20 M at 25°C. The solution process is exothermic:

$$NaOH(s) \rightarrow Na^+(aq) + OH^-(aq); \quad \Delta H = -42.9 \text{ kJ} \qquad (8.15)$$

Often, enough heat is released to raise the temperature to near the boiling point. The solution is strongly basic because of the high concentration of OH^- ions. Over a period of time, concentrated NaOH solutions attack glass, becoming cloudy in the process. They also pick up CO_2 from the air:

$$2 \, OH^-(aq) + CO_2(g) \rightarrow H_2O + CO_3^{2-}(aq) \qquad (8.16)$$

This contaminates the NaOH solution with sodium carbonate, Na_2CO_3.

Na_2CO_3 and $NaHCO_3$ These two compounds, sodium carbonate and sodium hydrogen carbonate, occur as the mineral trona, a hydrate that has the composition $Na_2CO_3 \cdot NaHCO_3 \cdot 2 \, H_2O$. Large deposits of trona are found near Green River, Wyoming. They are the major source of the 7×10^6 metric tons of Na_2CO_3 and smaller amounts of $NaHCO_3$ produced annually in the United States.

Na_2CO_3 and $NaHCO_3$ are also made from NaCl and $CaCO_3$ by the Solvay process

Sodium carbonate, commonly called washing soda, is used to make glass, soap, paper, and many chemicals. Nowadays, it is also used in detergents as a substitute for phosphates. Washing soda is not a serious pollutant, since carbonates occur naturally in surface waters. However, the CO_3^{2-} ions present in Na_2CO_3 can react with Ca^{2+} ions in hard water:

$$Ca^{2+}(aq) + CO_3^{2-}(aq) \rightarrow CaCO_3(s) \qquad (8.17)$$

Calcium carbonate, which is insoluble in water, precipitates as a white or grayish solid. If this happens on a freshly laundered skirt or pair of jeans, it can be annoying, to say the least.

Sodium hydrogen carbonate, $NaHCO_3$, is commonly called bicarbonate of soda or baking soda. It is an ingredient of many commercial products used to relieve indigestion. The HCO_3^- ions react with H^+ ions, thus relieving "excess stomach acidity":

$$HCO_3^-(aq) + H^+(aq) \rightarrow CO_2(g) + H_2O \qquad (8.18)$$

A similar reaction accounts for the use of $NaHCO_3$ in baking powders. An acidic component of the baking powder provides the H^+ ions required for Reaction 8.18. The carbon dioxide is formed as tiny bubbles that expand upon warming, causing bread or pastries to "rise."

Calcium Compounds

By far the most abundant and widely used compound of calcium is the carbonate, $CaCO_3$. Calcium carbonate is found in many different forms in nature. The purest of these is the transparent mineral calcite. In less pure form, $CaCO_3$ is found as marble, limestone, and dolomite ($CaCO_3 \cdot MgCO_3$). Huge quantities of $CaCO_3$ are used in the metallurgy of iron (Chap. 4), in making other chemicals, and in the manufacture of glass.

Calcium carbonate is very insoluble in pure water. However, it dissolves to an appreciable extent in ground water, which contains dissolved carbon dioxide from the atmosphere:

$$CaCO_3(s) + H_2O(l) + CO_2(g) \rightarrow Ca^{2+}(aq) + 2\ HCO_3^-(aq) \tag{8.19}$$

This reaction is responsible for the formation of limestone caves. Such caves often contain icicle-like formations of calcium carbonate, called stalactites and stalagmites (Fig. 8.7). These are formed when Reaction 8.19 reverses within the cave.

When limestone is heated to about 800°C, it decomposes:

$$CaCO_3(s) \rightarrow CaO(s) + CO_2(g) \tag{8.20}$$

FIGURE 8.7 Stalactites (upper) and stalagmites (lower) consist of calcium carbonate. They are formed when a water solution containing Ca^{2+} and HCO_3^- ions enters a cave. Carbon dioxide is released and calcium carbonate precipitates:

$$Ca^{2+}(aq) + 2\ HCO_3^-(aq) \rightarrow CaCO_3(s) + H_2O(l) + CO_2(g)$$

(Photograph courtesy of Luray Caverns, Virginia.)

The solid product, calcium oxide, is often referred to as quicklime. It reacts with water to form calcium hydroxide:

$$CaO(s) + H_2O(l) \rightarrow Ca(OH)_2(s) \qquad (8.21)$$

This reaction is very exothermic. CaO is quick to react

Calcium hydroxide is sometimes called slaked lime or, more simply, lime. It is used to "sweeten" acidic soils and in many processes where a cheap, strong base is required. Mortar is made by mixing $Ca(OH)_2$ with sand and water. The mortar "sets" by picking up CO_2 from the air:

$$Ca(OH)_2(s) + CO_2(g) \rightarrow CaCO_3(s) + H_2O(l) \qquad (8.22)$$

The calcium carbonate formed binds the particles of sand together.

Another important calcium compound is the mineral called gypsum. This is a hydrate of calcium sulfate, $CaSO_4 \cdot 2 H_2O$. Blackboard "chalk" is nearly pure gypsum. On a larger scale, gypsum is used in making cement, wallboard, and pottery. Other uses depend upon a reaction that takes place when gypsum is heated, losing three fourths of its water of crystallization:

$$CaSO_4 \cdot 2 H_2O(s) \rightarrow CaSO_4 \cdot \tfrac{1}{2} H_2O(s) + \tfrac{3}{2} H_2O(g) \qquad (8.23)$$

The product is called plaster of Paris, or sometimes simply plaster. When it is ground to a fine powder and mixed with water, plaster is converted back to gypsum:

$$CaSO_4 \cdot \tfrac{1}{2} H_2O(s) + \tfrac{3}{2} H_2O(l) \rightarrow CaSO_4 \cdot 2 H_2O(s) \qquad (8.24)$$

Wallboard consists of a gypsum-cardboard sandwich

This process takes place with an increase in volume. Hence, the material expands to fill completely any space to which it is confined. This explains its use in making models of statues, patching holes in walls, and forming casts for broken bones.

EXAMPLE 8.8 Using Tables 5.2 and 5.3 (Chap. 5), calculate ΔH for the thermochemical equation 8.19.

Solution Recall that ΔH can be calculated from heats of formation:

$$\Delta H = \Sigma \Delta H_f \text{ products} - \Sigma \Delta H_f \text{ reactants}$$

Applying this relation to Equation 8.19,

$$\Delta H = \Delta H_f \, Ca^{2+}(aq) + 2 \Delta H_f \, HCO_3^-(aq) - \Delta H_f \, CaCO_3(s) - \Delta H_f \, H_2O(l) - \Delta H_f \, CO_2(g)$$

Substituting numbers from Tables 5.2 and 5.3,

$$\Delta H = -543.0 \text{ kJ} - 1382.2 \text{ kJ} + 1207.0 \text{ kJ} + 285.8 \text{ kJ} + 393.5 \text{ kJ}$$
$$= -38.9 \text{ kJ}$$

We conclude that the reaction is slightly exothermic.

EXERCISE Taking the heats of formation of gypsum and plaster of Paris to be -2021.1 and -1575.2 kJ/mol, respectively, calculate ΔH for Reaction 8.23; for Reaction 8.24. Answer: $+83.2$ kJ; -17.2 kJ.

8.6
Transition Metals

Table 8.8 compares some of the properties of the metals in the first transition series ($_{21}$Sc → $_{30}$Zn) with those of the two metals that precede them in the fourth period of the Periodic Table. Notice particularly the trend in atomic radius. There is a sharp decrease in the size of the atom as we move through the first few metals in the period. Then the atomic radius levels off at about 0.13 nm for the last eight transition metals. This value is only a little more than half the atomic radius of potassium (0.231 nm).

The constant radius is the result of adding electrons to the inner d sublevel

Table 8.8
Properties of Transition Versus Main-Group Metals (Fourth Period)

PROPERTY	$_{19}$K	$_{20}$Ca	$_{21}$Sc	$_{22}$Ti	$_{23}$V	$_{24}$Cr
Atomic mass	39.10	40.08	44.96	47.90	50.94	52.00
Outer electron configuration	$4s^1$	$4s^2$	$4s^23d^1$	$4s^23d^2$	$4s^23d^3$	$4s^13d^5$
Atomic radius (nm)	0.231	0.197	0.160	0.146	0.131	0.125
Ionization energy (kJ/mol)	419	590	631	658	650	653
Density (g/cm^3)	0.862	1.55	3.0	4.51	6.11	7.19
Melting point (°C)	64	845	1541	1660	1890	1857
Boiling point (°C)	774	1420	2831	3287	3380	2672

PROPERTY	$_{25}$Mn	$_{26}$Fe	$_{27}$Co	$_{28}$Ni	$_{29}$Cu	$_{30}$Zn
Atomic mass	54.94	55.85	58.93	58.69	63.55	65.38
Outer electron configuration	$4s^23d^5$	$4s^23d^6$	$4s^23d^7$	$4s^23d^8$	$4s^13d^{10}$	$4s^23d^{10}$
Atomic radius (nm)	0.129	0.126	0.125	0.124	0.128	0.133
Ionization energy (kJ/mol)	717	759	758	737	746	906
Density (g/cm^3)	7.43	7.87	8.92	9.91	8.94	7.13
Melting point (°C)	1244	1535	1495	1453	1083	420
Boiling point (°C)	1962	2750	2870	2732	2567	907

Zn is stabilized by having two filled outermost sublevels

As we would expect, there is an inverse relation between atomic radius and ionization energy. In general, the transition metals have somewhat higher first ionization energies than do the main-group elements potassium and calcium. This explains, at least in part, the difference in chemical behavior between these two different types of elements. The transition metals are generally less reactive than those in Groups 1 and 2. Consider, for example, the behavior of these metals toward water. As you will recall from Section 8.5, potassium and calcium react vigorously with water at room temperature:

$$K(s) + H_2O \rightarrow K^+(aq) + OH^-(aq) + \tfrac{1}{2} H_2(g) \qquad (8.25)$$

$$Ca(s) + 2 H_2O \rightarrow Ca^{2+}(aq) + 2 OH^-(aq) + H_2(g) \qquad (8.26)$$

Of the ten transition metals in this series, only the first one, scandium, reacts rapidly with cold water:

$$Sc(s) + 3\,H_2O \rightarrow Sc^{3+}(aq) + 3\,OH^-(aq) + \tfrac{3}{2}\,H_2(g) \qquad \textbf{(8.27)}$$

The other metals react only at high temperatures (Ti, Mn, Fe, Co, Ni, and Zn) or not at all (V, Cr, and Cu). The lower reactivity of these metals reflects the fact that they form cations less readily, as we might expect from their higher ionization energies.

As pointed out in Chapter 3, many transition metals form more than one stable cation and hence more than one set of ionic compounds. Iron forms two different chlorides, one containing the Fe^{2+} ion, the other the Fe^{3+} ion:

$FeCl_2$ iron(II) chloride $FeCl_3$ iron(III) chloride

Salts of Fe(III) and Cu(II) are the more common ones

Again, copper in its compounds can be present as the Cu^+ or Cu^{2+} ion:

$CuCl$ copper(I) chloride $CuCl_2$ copper(II) chloride

In contrast, the main-group metals calcium and potassium each form only one stable cation. Calcium in its compounds is always present as the Ca^{2+} ion, potassium as K^+:

$CaCl_2$ calcium chloride KCl potassium chloride

Table 8.9
Successive Ionization Energies* of K, Ca, Fe, and Cu

	E_1 (kJ/mol)	E_2 (kJ/mol)	E_3 (kJ/mol)
$_{19}K$	419	3051	4411
$_{20}Ca$	590	1145	4912
$_{26}Fe$	759	1561	2957
$_{29}Cu$	746	1958	3554

*Ionization energies shown in color require removing electrons from a noble-gas structure, that of argon.

This difference in behavior between transition and main-group metals is related to the difference in electron configurations. Transition metal cations are formed by the loss of successive 4s and 3d electrons,* which have energies of the same order of magnitude. Hence, it is not too difficult to go from a +1 to a +2 or a +2 to a +3 cation; there is no abrupt jump in ionization energy. With a main-group metal, the situation is quite different. Electrons are lost until an ion with a noble gas configuration (K^+, Ca^{2+}) is formed. Once that happens, further ionization is extremely unlikely. The energy requirement for removing an electron from an inner, complete energy level is simply too great. This point is illustrated by the data in Table 8.9. Notice the sharp increase between the first and second ionization energies of K and again between the second and third ionization energies

Seems reasonable

---

*Electron configurations of transition metal cations will be discussed in Chapter 9.

of Ca. Small wonder that the K^+ ion is not converted to K^{2+} or Ca^{2+} to Ca^{3+}! In contrast, successive ionization energies increase smoothly with Fe and Cu; there are none of the abrupt jumps noted with K and Ca.

The transition metals form a wide variety of colored compounds. Thus, we have

The colors come from the transition metals, not the main-group elements

Na_2CrO_4	yellow	$Fe(OH)_3$	red
$Na_2Cr_2O_7$	red	$CoCl_2 \cdot 6\ H_2O$	pink
$NaMnO_4$	purple	$NiCl_2 \cdot 6\ H_2O$	green
		$CuSO_4 \cdot 5\ H_2O$	blue

(See also Color Plate 8.2.) In contrast, compounds of main-group metals such as potassium and calcium are ordinarily white solids that dissolve in water to give colorless solutions.

Qualitatively, we can explain the colors of transition metal compounds in terms of electron configurations. In most transition metal ions there are unfilled or half-filled inner d orbitals. These differ by only small amounts of energy from orbitals occupied by electrons. The energy difference is often comparable to that of visible light. By absorbing light of a particular color, an electron can move from a lower to a higher orbital. This absorption of light accounts for the color we see when we look at a transition metal compound or its water solution. We will have more to say about the colors of transition metal compounds when we discuss complex ions in Chapter 21.

Summary

The discovery of the Periodic Table is usually credited to Mendeleev; Meyer and Moseley also made important contributions. In the table, elements are arranged in order of increasing atomic number in horizontal periods and vertical groups. Elements in the same group have the same outer electron configuration, which explains why they have similar chemical properties. Main-group elements fill s sublevels (Groups 1 and 2) or p sublevels (Groups 3 to 8). Transition metals fill inner d sublevels (3d, 4d, 5d, 6d); lanthanides fill the 4f sublevel, actinides the 5f.

As one moves across from left to right in the Periodic Table, effective nuclear charge and ionization energy increase; atomic radius and metallic character decrease. As one moves down in the table, effective nuclear charge stays constant and ionization energy decreases; atomic radius and metallic character increase. Of 108 known elements, about 84 are metals; there are 18 nonmetals and 6 metalloids.

The alkali and alkaline earth metals show trends in physical properties (Table 8.6) related to their positions in the Periodic Table. The atomic spectra of these elements yield characteristic flame colors. Chemically, the Group 1 and Group 2 metals are extremely reactive (Table 8.7). They react with hydrogen and the halogens to form binary compounds containing $+1$ cations (Group 1 metals) or $+2$ cations (Group 2 metals) and -1 anions (H^-, F^-, Cl^-, Br^-, I^-). The ionic compounds formed with oxygen include normal oxides (O^{2-} ion), peroxides (O_2^{2-} ion), and superoxides (O_2^- ion). With water, hydrogen is evolved; a solution of the Group 1 or Group 2 hydroxide is formed at the same time.

Transition metal atoms are typically smaller and more difficult to ionize (Table

8.8) than those of the alkali and alkaline earth metals. Transition metals are generally less reactive than those in Groups 1 and 2; they also form a wider variety of cations and many colored compounds. The differences in properties between transition and main-group metals can be explained, at least in part, by differences in electron configuration.

Relatively few new concepts have been introduced in this chapter; Examples 8.1 (outer electron configurations), 8.2 (trends in atomic radius and ionization energy), and 8.4 (reactions of metals) deal with these. The remaining examples review concepts introduced in previous chapters:

Chapter 5 (Thermochemistry)   Examples 8.6, 8.8
Chapter 6 (Gas Laws)   Examples 8.5, 8.7
Chapter 7 (Electronic Structure, Spectra)   Examples 8.1, 8.3

Many of the problems at the end of this chapter are also of a review nature.

Key Words and Concepts

actinide	group (Periodic Table)	nonmetal
alkali metal	ionization energy	outer electron configuration
alkaline earth metal	lanthanide	period (Periodic Table)
alloy	luster	Periodic Law
atomic radius	main-group element	Periodic Table
atomic spectra	malleable	peroxide
ductile	metallic character	shielding effect
effective nuclear charge	metalloid	superoxide
first ionization energy	noble-gas configuration	transition metal
flame test		

Questions and Problems

Periodic Table

8.1 Describe or define the following terms related to the Periodic Table:
a. group b. halogen c. alkali metal
d. actinide e. noble gas

8.2 What should be the atomic number of the element that completes the seventh period?

8.3 Give the atomic number of the element that
a. completes the 5s sublevel.
b. starts to fill the 7s sublevel.
c. has a half-filled 5p sublevel.
d. starts the lanthanide series.

8.4 State the outer electron configuration of
a. Ba b. I c. Pb d. Bi

8.31 Define or describe the following terms related to the Periodic Table:
a. period b. alkaline earth metal
c. lanthanide d. transition metal
e. metalloid

8.32 In the seventh period, what would be the atomic number of the element
a. that completes the 6d sublevel?
b. in Group 5?

8.33 Give the atomic number of the element that
a. starts to fill the 6p sublevel.
b. completes the 6p sublevel.
c. completes the 4s sublevel.
d. is the last member of the actinide series.

8.34 State the outer electron configuration of
a. Xe b. Cs c. Po d. Sb

8.5 Arrange the three elements K, Se, and Br in order of
  a. increasing atomic radius.
  b. increasing ionization energy.
  c. decreasing metallic character.

8.6 Which of the four atoms Li, Be, O, or Na has the
  a. smallest atomic radius?
  b. lowest ionization energy?
  c. most metallic character?

8.7 Classify each of the following as a metal, nonmetal, or metalloid:
  a. Mn   b. Sb   c. Br   d. La   e. No

8.8 Give the symbols of all elements
  a. with atomic numbers greater than 50 that are chemically similar to sodium.
  b. that have the outer configuration ns^2np^6.
  c. that are lanthanides with atomic number greater than 68.
  d. that are main-group metals in the sixth period.

8.9 Mendeleev predicted the properties of gallium by averaging the values for the elements above and below gallium in the Periodic Table. Use this method to estimate the density of gallium, given that the densities of Al and In are 2.70 and 7.31 g/cm^3, respectively (measured density of Ga = 5.91 g/cm^3).

Alkali and Alkaline Earth Metals

8.10 Give the formula of the compound formed when lithium reacts with
  a. N_2 b. S   c. H_2 d. H_2O e. I_2

8.11 Write a balanced equation for the reaction between
  a. potassium peroxide and water.
  b. strontium and oxygen.
  c. cesium and oxygen.
  d. strontium hydride and water.

8.12 When calcium is exposed to air, it forms an oxide. This compound reacts with water to form calcium hydroxide, which can also be made by adding calcium to water. Write a balanced equation for each reaction just described.

8.13 Which Group 1 metal
  a. is found in bicarbonate of soda?
  b. has the lowest melting point?
  c. gives a yellow flame test?
  d. has an atomic mass less than that of the preceding noble gas?

8.35 Follow the directions of Question 8.5 for the three elements Na, Si, and Cl.

8.36 Follow the directions of Question 8.6 for the four atoms Mg, Si, Cl, and K.

8.37 Classify each of the following as a metal, nonmetal, or metalloid:
  a. Ir   b. N   c. B   d. W   e. Ge

8.38 Give the symbols of all elements
  a. with the outer configuration ns^2np^2.
  b. with atomic numbers less than 30 that are chemically similar to barium.
  c. that just complete a p sublevel.
  d. that have a single outer s electron.

8.39 Predict the melting point of strontium given those of calcium (845°C) and barium (725°C). (The observed value is 770°C.)

8.40 Give the formula of the compound formed when calcium reacts with
  a. O_2 b. Br_2 c. N_2 d. H_2O

8.41 Write a balanced equation for the reaction of
  a. calcium with bromine.
  b. barium peroxide with water.
  c. potassium with sulfur.
  d. potassium with water.

8.42 When sodium is exposed to air, it forms a peroxide. This compound, upon addition to water, gives a solution of sodium hydroxide. The same solution can be obtained by adding sodium to water. Write balanced equations for the reactions just described.

8.43 Identify the Group 2 element that
  a. does not give a flame test.
  b. forms a peroxide readily.
  c. forms molecular compounds.
  d. is found in limestone.

8.14 Write a balanced equation for the reaction that occurs when
   a. calcium carbonate is heated.
   b. $CaSO_4 \cdot 2H_2O$ is heated.
   c. potassium superoxide is exposed to $H_2O(g)$.
   d. magnesium is exposed to hot steam.

Thermochemistry (Chap. 5)

8.15 A 2.50-g sample of lithium is added to water.
   a. Write a balanced equation for the reaction that occurs.
   b. Using heat of formation data from Chapter 5, calculate how much heat is evolved.

8.16 Barium peroxide decomposes upon heating to barium oxide and oxygen gas. When 1.18 g of barium peroxide decomposes, ΔH is $+0.503$ kJ. Calculate ΔH_f for barium peroxide (ΔH_f BaO $= -558$ kJ/mol).

8.17 During the making of a cast to set a broken bone, the following exothermic reaction occurs:

$$CaSO_4 \cdot \tfrac{1}{2} H_2O(s) + \tfrac{3}{2} H_2O(l) \rightarrow CaSO_4 \cdot 2 H_2O(s)$$
$$\text{plaster} \qquad\qquad\qquad\qquad \text{cast}$$

Calculate ΔH for the fabrication of a 2.50-kg cast. (ΔH_f $CaSO_4 \cdot \tfrac{1}{2} H_2O = -1575$ kJ/mol; ΔH_f $H_2O = -286$ kJ/mol; ΔH_f $CaSO_4 \cdot 2 H_2O = -2021$ kJ/mol).

8.18 Consider the reaction

$$2 Mg(s) + O_2(g) \rightarrow 2 MgO(s)$$

   a. Taking the heat of formation of MgO to be -601.8 kJ/mol, calculate ΔH when 1.00 g of MgO is formed from the elements.
   b. If the heat evolved in (a) is absorbed by 452 g of water (S.H. $= 4.18$ J/g·°C), calculate the temperature change.

Gas Laws (Chap. 6)

8.19 Above 700°C, barium peroxide decomposes:

$$2 BaO_2(s) \rightarrow 2 BaO(s) + O_2(g)$$

If the oxygen liberated by heating 12.0 g of barium peroxide is collected in a 500-cm³ flask at 20°C, what is the pressure in the flask?

8.20 A sample of 3.05 g of sodium is added to a large volume of water.
   a. Write a balanced equation for the reaction that occurs.
   b. Calculate the volume of $H_2(g)$ produced at 25°C and 742 mm Hg.

8.44 Write a balanced equation for the reaction that occurs when
   a. barium oxide is added to water.
   b. a solution containing HCO_3^- ions is made acidic.
   c. calcium hydroxide is exposed to CO_2.
   d. plaster of Paris is exposed to water.

8.45 Use heat of formation data from Chapter 5 to calculate ΔH for the reaction when 16.6 g NaCl is electrolyzed to form sodium metal and chlorine gas.

8.46 Use data from Tables 5.2 and 5.3 to calculate ΔH for the slaking of lime, which can be represented as either (a) or (b) below:
   a. $CaO(s) + H_2O(l) \rightarrow Ca(OH)_2(s)$
   b. $CaO(s) + H_2O(l) \rightarrow Ca^{2+}(aq) + 2 OH^-(aq)$

8.47 Use data from Tables 5.2 and 5.3 to
   a. calculate ΔH for the reaction given by Equation 8.17.
   b. calculate ΔH for the dissolving of 1.00 g $CaCO_3$ in water by the reverse of Reaction 8.17.

8.48 The heat of formation of BaO is -558.1 kJ/mol. Calculate
   a. ΔH when 1.00 g BaO is formed from the elements.
   b. the mass of water required to absorb the heat evolved in (a) if the temperature increases from 25.00 to 30.00°C (S.H. $H_2O = 4.18$ J/g·°C).

8.49 If 1.00 g of solid sodium is to be converted to NaCl by reaction with chlorine gas, what volume of Cl_2 at 25°C and 736 mm Hg is required?

8.50 A sample of 3.05 g of sodium, upon exposure to air, is converted to the peroxide.
   a. Write a balanced equation for the reaction.
   b. Calculate the volume of $O_2(g)$ required for the reaction at 18°C and 751 mm Hg.

8.21 A self-contained breathing apparatus contains 216 g KO_2. A fireman exhales 116 L of air at 37°C and 752 mm Hg into the apparatus. The volume percent of water in exhaled air is 6.2. Calculate
a. the number of moles of water present in the exhaled air.
b. the mass of KO_2 left after the water in (a) reacts as in Equation 8.13.

8.51 A sample of 7.55×10^3 L of air at 21°C and 741 mm Hg, containing 0.033 mole percent CO_2, is bubbled through 1.00 L of 2.50 M NaOH solution. Calculate
a. the number of moles of CO_2 in the sample.
b. the number of moles of NaOH left in solution after the CO_2 in (a) reacts as in Equation 8.16.

Electronic Structure and Atomic Spectra (Chap. 7)

8.22 Give the electron configuration and orbital diagram of the strontium atom.

8.52 Give the electron configuration and orbital diagram of the cesium atom.

8.23 Give a set of four quantum numbers for the outer electron(s) in
a. Rb   b. Ba   c. Ra

8.53 Give a set of four quantum numbers for each d electron in
a. Mn   b. Tc   c. Sr

8.24 Using the data in Table 8.6, calculate the energy difference, $E_{hi} - E_{lo}$, in kilojoules per mole, for each of the transitions responsible for lines in the rubidium spectrum.

8.54 Barium atoms impart a green color to a flame due to an electronic transition that gives off light at 554 nm. What is the energy difference, $E_{hi} - E_{lo}$, for this transition?

8.25 The yellow line in the sodium spectrum is produced by a transition between two states that differ in energy by 3.368×10^{-19} J/particle. Calculate the wavelength of the line and compare it to those listed in Table 8.6.

8.55 Magnesium does not give a flame test. The strongest line in the Mg spectrum arises from a transition between two states that differ in energy by 419.3 kJ/mol. What is the wavelength of this line? In what region of the spectrum does it fall?

General

8.26 What role did each of the following have in developing the Periodic Table?
a. Henry Moseley   b. J. A. R. Newlands
c. D. Mendeleev   d. Yuri Andropov

8.56 What role did each of the following have in developing the Periodic Table?
a. Lothar Meyer   b. John Dalton
c. J. Dobereiner   d. John Morse

8.27 The first three ionization energies for calcium are 590, 1145, and 4912 kJ/mol. The corresponding values for argon are 1520, 2666, and 3931 kJ/mol. Explain why
a. the first two ionization energies are much lower for calcium than for argon.
b. the third ionization energy for calcium is higher than that for argon.

8.57 Consider the elements in the second period. Indicate where the largest jump in ionization energy is expected to occur, following the example given:

Li: between ΔE_1 and ΔE_2

a. C: between ＿＿ and ＿＿
b. B: between ＿＿ and ＿＿
c. F: between ＿＿ and ＿＿

8.28 Describe four physical properties and one chemical property that are characteristic of metals.

8.58 Describe three ways in which alloys such as those listed in Table 8.5 can differ from the pure metals from which they are prepared.

8.29 Explain why
a. transition metals form many colored compounds.
b. atomic radius increases going down a group in the Periodic Table.
c. Group 2 metals form +2 ions.

8.59 Explain why
a. transition metals often form more than one stable cation.
b. atomic radius decreases going across a period in the Periodic Table.
c. Group 1 metals form +1 ions.

8.30 Indicate whether each of the following is true or false. If false, correct the statement.
  a. Effective nuclear charge stays about constant going across a period.
  b. Group 7 elements have an np^7 outer electron configuration.
  c. all alloys contain at least two metals.

8.60 Criticize the following statements:
  a. Energy is given off when an electron is removed from an atom.
  b. Elements are located in the Periodic Table in sequence of increasing atomic mass.
  c. The alkali metals are stored under water to protect them from air.

*8.61 Astatine is a synthetic element, first made by nuclear bombardment in 1940. On graph paper, plot the quantities listed below versus atomic number for the more common halogens. (Data can be obtained from Appendix 2.) Draw smooth curves through your points and extrapolate to estimate the atomic radius, melting point, and boiling point of astatine.
  a. atomic radius   b. melting point   c. boiling point

*8.62 A sample of 16.00 g of barium reacts with oxygen to form 19.07 g of a mixture of barium oxide and barium peroxide. Determine the composition of this mixture.

*8.63 Using data in Table 8.8, calculate the densities of (spherical) atoms of K, Ca, and Sc and compare to the listed densities of these metals.

*8.64 Using data in Table 8.5, calculate the number of atoms in a sterling silver spoon weighing 1.68 oz.

*8.65 From data given in this chapter, calculate the energy change for the processes

$$Cu(g) \rightarrow Cu^{2+}(g) + 2\,e^-$$

$$Na^{3+}(g) + 2e^- \rightarrow Na^+(g)$$

$$Al^{3+}(g) + Mg^{2+}(g) \rightarrow Al^{2+}(g) + Mg^{3+}(g)$$

*8.66 Sodium hydrogen carbonate is a home remedy for "excess stomach acid." The relieving reaction is

$$HCO_3^-(aq) + H^+(aq) \rightarrow CO_2(g) + H_2O(l)$$

To relieve 0.200 L of excess stomach acid (0.12 M HCl), how many teaspoonsful (1 teaspoonful = 7.8 g) of $NaHCO_3$ should you take? What volume of CO_2 will be produced at body temperature, 98.6°F, and 1.00 atm?

Chapter 9
Chemical
Bonding

In Chapter 8, we described the electron configurations of isolated, gaseous atoms. Ordinarily, these atoms are not chemically stable. Instead, they tend to combine chemically with other atoms. By reacting, their electron configurations change through formation of chemical bonds. In this chapter, we will be concerned with two ways in which atoms form bonds:

1. Electrons are **transferred** from one atom to another. Positive ions (cations) result from electron loss; negative ions (anions) are formed by electron gain. When sodium and chlorine atoms are brought into contact, they react vigorously to form Na^+ and Cl^- ions. These ions are the basic building blocks in ordinary table salt, NaCl. The cations and anions are held together by *ionic bonds* (Section 9.1), strong electrostatic attractions between ions of opposite charge.

2. Electrons are **shared** by atoms to form a molecule. The atoms within a molecule are held together by strong forces called *covalent bonds* (Section 9.2). A covalent bond consists of an electron pair shared between two atoms. Perhaps the simplest bond of this type is that which joins H atoms in the H_2 molecule. When two hydrogen atoms, each with one electron, come together, they form a bond. This may be shown as

H : H

where the dots represent electrons. More commonly, we show the structure of the H_2 molecule as

H—H

The understanding is that the dash represents a covalent bond, an electron pair shared by the two atoms.

Most of this chapter will be devoted to the covalent bond and the electronic structures of species containing covalent bonds. We will consider

—the distribution of valence electrons in molecules and polyatomic ions where atoms are joined by covalent bonds (Section 9.3).

—the properties of the covalent bond (Section 9.4), including bond polarity, bond distance, and bond energy.

9.1
Ionic Bonding

Ionic compounds result typically from the reaction of a metal of low ionization energy (usually a Group 1, 2, or transition metal) with a nonmetal (Group 6 or 7). Electrons are transferred from the metal to the nonmetal, forming cations (positive ions) and anions (negative ions), respectively. An example of such a reaction is that which occurs when chlorine gas passes over sodium metal:

All ionic compounds are solids at 25°C

$$\text{Na}(s) + \tfrac{1}{2}\text{Cl}_2(g) \rightarrow \text{NaCl}(s); \qquad \Delta H = -411 \text{ kJ} \qquad \textbf{(9.1)}$$

In NaCl, the Na^+ and Cl^- ions are held together by strong electrostatic forces called *ionic bonds*. These forces cause the ions to line up to form the regular crystalline structure shown in Figure 9.1.

Ionic compounds, like all stable substances, are electrically neutral. This is reflected in their formulas, where total positive charge equals total negative charge (recall the discussion in Chap. 3). The formulas of ionic compounds can be predicted using this principle of electrical neutrality (Table 9.1).

FIGURE 9.1 Sodium chloride crystals photographed through a scanning electron microscope. The crystals are magnified 50 times their actual size. Under this magnification the organized, cubic nature of the crystal is apparent.

Table 9.1
Formulas of Ionic Compounds

CATION	ANION			EXAMPLES
	X^-	X^{2-}	X^{3-}	
M^+	MX	M_2X	M_3X	NaF, Na_2O, Na_3N
M^{2+}	MX_2	MX	M_3X_2	MgF_2, MgO, Mg_3N_2
M^{3+}	MX_3	M_2X_3	MX	AlF_3, Al_2O_3, AlN

Electron Configurations of Ions

As pointed out in Chapter 3, *elements close to a noble gas in the Periodic Table form ions that have the same number of electrons as the noble gas atom.* This means that these ions have noble gas electron configurations ($1s^2$ for He; ns^2np^6 for the other noble gases). Thus the three elements preceding neon (N, O, and F) and the three elements following neon (Na, Mg, and Al) all form ions with the neon configuration, $1s^22s^22p^6$. The three nonmetal atoms achieve this structure by gaining electrons to form anions:

The ns^2np^6 outer configuration is very stable

$$_7N\ (1s^22s^22p^3)\ +\ 3\ e^-\ \rightarrow\ _7N^{3-}\ (1s^22s^22p^6)$$

$$_8O\ (1s^22s^22p^4)\ +\ 2\ e^-\ \rightarrow\ _8O^{2-}\ (1s^22s^22p^6)$$

$$_9F\ (1s^22s^22p^5)\ +\ e^-\ \rightarrow\ _9F^-\ (1s^22s^22p^6)$$

The three metal atoms acquire the neon structure by losing electrons to form cations:

$$_{11}Na\ (1s^22s^22p^63s^1)\ \rightarrow\ _{11}Na^+\ (1s^22s^22p^6)\ +\ e^-$$

$$_{12}Mg\ (1s^22s^22p^63s^2)\ \rightarrow\ _{12}Mg^{2+}\ (1s^22s^22p^6)\ +\ 2\ e^-$$

$$_{13}Al\ (1s^22s^22p^63s^23p^1)\ \rightarrow\ _{13}Al^{3+}\ (1s^22s^22p^6)\ +\ 3\ e^-$$

The species N^{3-}, O^{2-}, F^-, Ne, Na^+, Mg^{2+} and Al^{3+} are said to be *isoelectronic;* they have the same electron configuration.

There are a great many ions that have noble gas configurations, suggesting that this electron arrangement is a particularly stable one. Table 9.2 lists 24 ions of this type. Notice from the table that

Nonmetal atoms gain electrons

1. The anions are formed from neutral atoms having one, two, or three *less* electrons than the *next* noble gas. Those in Group 7 (outer configuration ns^2np^5) *gain one* electron per atom to form -1 ions. Group 6 nonmetals (outer configuration ns^2np^4) *gain two* electrons per atom to form -2 ions. Nitrogen (outer configuration $2s^22p^3$) forms -3 ions when it reacts with very reactive metals such as lithium or magnesium. Other Group 5 elements rarely, if ever, form -3 ions.

2. The cations are formed by metals whose neutral atoms have one, two, or three electrons *more* than the preceding noble gas. Atoms of Group 1 (ns^1) and Group 2 (ns^2) achieve noble gas configurations by losing one and two electrons,

respectively. Accordingly, the corresponding cations have $+1$ and $+2$ charges (Group 1 = $+1$; Group 2 = $+2$). Atoms of aluminum in Group 3 and the metals in the scandium subgroup all have three electrons more than the preceding noble gas. They form $+3$ ions: Al^{3+}, Sc^{3+}, ... (Boron, the first member of Group 3, does not form ions.)

Metal atoms lose electrons

Table 9.2
Ions with Noble Gas Electron Configurations

NOBLE GAS	CONFIGURATION			IONS				
He	$1s^2$				H^-	Li^+	Be^{2+}	
Ne	$[_2He]2s^22p^6$	N^{3-}	O^{2-}	F^-	Na^+	Mg^{2+}	Al^{3+}	
Ar	$[_{10}Ne]3s^23p^6$		S^{2-}	Cl^-	K^+	Ca^{2+}	Sc^{3+}	
Kr	$[_{18}Ar]3d^{10}4s^24p^6$		Se^{2-}	Br^-	Rb^+	Sr^{2+}	Y^{3+}	
Xe	$[_{36}Kr]4d^{10}5s^25p^6$		Te^{2-}	I^-	Cs^+	Ba^{2+}		

The transition metals to the right of the scandium subgroup do not form ions with noble gas configurations. To do so, they would have to lose four or more electrons. The energy requirement is too high for that to happen. However, as pointed out in Chapter 3, these metals do form cations with charges of $+1$, $+2$, or $+3$. It is important to note that **when transition metal atoms form positive ions, the outer s electrons are lost first.** Consider, for example, the formation of the Mn^{2+} ion from the Mn atom:

$$_{25}Mn \ (Ar \ 4s^23d^5) \rightarrow \ _{25}Mn^{2+} \ (Ar \ 3d^5) + 2 \ e^-$$

Notice that it is the 4s electrons that are lost rather than the 3d electrons. We know this is the case because the Mn^{2+} ion has been shown to have five unpaired electrons (the five 3d electrons). If two 3d electrons had been lost, the Mn^{2+} ion would have had only three unpaired electrons.

The nuclear charge is the same in the atom and ion, but the numbers of electrons differ. This shifts the relative energies of the sublevels

All the transition metals form cations by a similar process, i.e., loss of outer s electrons. Only after those electrons are lost are electrons removed from the inner d sublevel. Consider, for example, what happens with iron which, you will recall, forms two different cations. First the 4s electrons are lost to give the Fe^{2+} ion:

$$_{26}Fe \ (Ar \ 4s^23d^6) \rightarrow \ _{26}Fe^{2+} \ (Ar \ 3d^6) + 2 \ e^-$$

Then an electron is removed from the 3d level to form the Fe^{3+} ion:

$$_{26}Fe^{2+} \ (Ar \ 3d^6) \rightarrow \ _{26}Fe^{3+} \ (Ar \ 3d^5) + e^-$$

In the Fe^{2+} and Fe^{3+} ions, as in all transition metal ions, there are no outer s electrons.

The behavior just described suggests that in transition metal cations (as opposed to the atoms), the inner d sublevels are lower in energy than the outer s sublevels; that is,

Fourth period cations: 3d lower in energy than 4s
Fifth period cations: 4d lower in energy than 5s
Sixth period cations: 5d lower in energy than 6s

EXAMPLE 9.1 Give the abbreviated electron configurations of
a. an Sc atom and a Sc^{3+} ion.
b. a Zn atom and a Zn^{2+} ion.
c. a Te atom and a Te^{2-} ion.

Solution In cation formation, electrons are lost. The ion formed has a positive charge equal to the number of electrons lost. In anion formation, electron gain yields an ion with a negative charge equal to the number of electrons gained.

a. To form a +3 ion, scandium (at. no. = 21) loses three electrons. The Sc^{3+} ion has 18 electrons and the argon structure:

$$_{21}Sc \ (Ar \ 4s^2 3d^1) \rightarrow \ _{21}Sc^{3+} \ (Ar) + 3 \ e^-$$

b. A Zn atom loses two electrons to form the Zn^{2+} ion:

$$_{30}Zn \ (Ar \ 4s^2 3d^{10}) \rightarrow \ _{30}Zn^{2+} \ (Ar \ 3d^{10}) + 2 \ e^-$$

Note the loss of 4s rather than 3d electrons.

c. Two electrons are gained to give the electron configuration of the noble gas xenon (at. no. = 54):

$$_{52}Te \ (Kr \ 5s^2 4d^{10} 5p^4) + 2 \ e^- \rightarrow \ _{52}Te^{2-}(Xe)$$

EXERCISE What is the abbreviated electron configuration of $_{44}Ru^{3+}$?
Answer: $(Kr \ 4d^5)$.

Sizes of Ions

Figure 9.2 compares the radii of cations and anions to those of the corresponding atoms. Notice that

Removing or adding electrons has a large effect on the size of the particle

—*positive ions are smaller than the metal atoms from which they are formed.* The Na^+ ion has a radius, 0.095 nm, only a little more than half that of the Na atom, 0.186 nm.

—*negative ions are larger than the nonmetal atoms from which they are formed.* The radius of the Cl^- ion, 0.181 nm, is nearly twice that of the Cl atom, 0.099 nm.

These effects combine to make anions, on the whole, larger than cations. Compare, for example, the Cl^- ion (radius = 0.181 nm) to the Na^+ ion (radius = 0.095 nm). This means that in sodium chloride, and indeed in the vast majority of all ionic compounds, most of the space in the crystal lattice is taken up by anions.

The differences in radii between atoms and ions can be explained quite simply. A cation is smaller than the corresponding metal atom because the excess of protons in the ion draws the outer electrons in closer to the nucleus. In contrast, an extra electron in an anion adds to the repulsion between outer electrons. This makes a negative ion larger than the corresponding nonmetal atom.

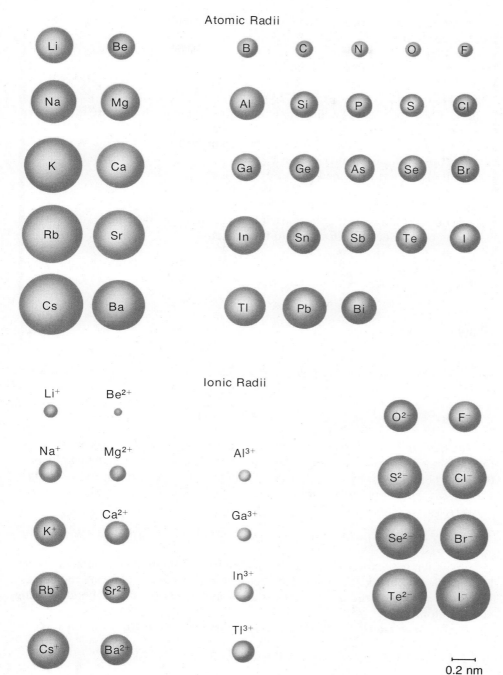

FIGURE 9.2 Sizes of atoms and ions of the main-group elements. Negative ions are always larger than the atoms from which they are derived, whereas positive ions are smaller.

EXAMPLE 9.2 For each of the following pairs, select the smaller species:
a. a Pt atom or a Pt^{2+} ion.
b. a P atom or a P^{3-} ion.
c. an Fe^{2+} ion or an Fe^{3+} ion.

Solution
a. Cations are always smaller than the atoms from which they are formed. Thus a Pt^{2+} ion is smaller than a Pt atom.
b. Anions are larger than the atoms from which they are formed. A phosphorus atom is smaller than a P^{3-} ion.
c. The Fe^{3+} ion is formed from Fe^{2+} by loss of an electron:

$$Fe^{2+} \rightarrow Fe^{3+} + e^-$$

The Fe^{3+} ion is smaller because it has a greater positive charge, drawing the outer electrons in closer to the nucleus.

EXERCISE Select the larger member of each of the following pairs: Tl atom and Tl^{3+} ion; Br atom and Br^- ion; Tl^+ ion and Tl^{3+} ion. Answer: Tl, Br^-, Tl^+.

Heats of Formation of Ionic Compounds; Lattice Energy

When an ionic compound is formed from the elements, the reaction is always exothermic. Recall, for example, Reaction 9.1:

$$Na(s) + \tfrac{1}{2} Cl_2(g) \rightarrow NaCl(s); \qquad \Delta H = -411 \text{ kJ}$$

We see from this thermochemical equation that 411 kJ of heat are given off in the formation of a mole of sodium chloride. We can gain some understanding of where this heat comes from by considering a possible step-wise path for the reaction.

1. *The elemental substances in their stable states at 25°C and 1 atm are converted to gaseous atoms.*

$$Na(s) + \tfrac{1}{2} Cl_2(g) \rightarrow Na(g) + Cl(g); \qquad \Delta H_1 = +230 \text{ kJ}$$

Na(g) has a lot more energy than Na(s)

As you might expect, this step is endothermic. It requires that enough energy be absorbed to break the bonds between chlorine atoms in chlorine molecules. Energy is also required to vaporize sodium metal and break the bonds joining sodium atoms.

2. *Electron transfer occurs between the gaseous atoms to form ions.*

$$Na(g) + Cl(g) \rightarrow Na^+(g) + Cl^-(g); \qquad \Delta H_2 = +147 \text{ kJ}$$

Interestingly, this step is also endothermic. It can be considered as the sum of two processes: (a) electron loss from a mole of sodium atoms ($\Delta H = +496$ kJ/mol) and (b) electron gain by a mole of chlorine atoms ($\Delta H = -349$ kJ/mol). Overall for step 2, $\Delta H = (+496 - 349)$ kJ/mol $= +147$ kJ/mol.

3. *The gaseous ions combine to form an ionic solid.*

$$Na^+(g) + Cl^-(g) \rightarrow NaCl(s); \qquad \Delta H_3 = -788 \text{ kJ}$$

This is a highly exothermic process because of the strong electrostatic attraction between oppositely charged ions.

The value of ΔH for the overall reaction must be, by Hess's Law, the sum of the enthalpy changes for the individual steps:

$$\Delta H = \Delta H_1 + \Delta H_2 + \Delta H_3$$
$$= +230 \text{ kJ} + 147 \text{ kJ} - 788 \text{ kJ} = -411 \text{ kJ}$$

From this analysis we see that the evolution of energy in the overall reaction is due to the exothermic nature of the final step. This enthalpy change, in which the crystal lattice is formed from gaseous ions, is referred to as the **lattice energy.**

It is possible to go through an analysis of this type for the formation of any alkali halide. The magnitudes of the several enthalpy changes vary with the particular compound. However, the signs are always the same. Steps 1 and 2 are endothermic. The lattice energy, ΔH_3, is always a large negative quantity. It makes the overall reaction exothermic and the ionic compound stable (Table 9.3).

Table 9.3
Lattice Energies (kJ/mol) for the Alkali Halides

	F^-	Cl^-	Br^-	I^-
Li^+	−1036	−857	−813	−758
Na^+	−922	−788	−752	−704
K^+	−820	−718	−688	−648
Rb^+	−790	−692	−665	−629
Cs^+	−734	−660	−636	−603

The lattice energy is a measure of the strength of the ionic bond between oppositely charged ions. The smaller the ions, the closer they approach one another and the stronger is the bond between them. This effect is evident from the data in Table 9.3. The lattice energy has its largest negative value, −1036 kJ/mol, for lithium fluoride, where the ions are the smallest (radius Li^+ = 0.060 nm, radius F^- = 0.136 nm). At the other extreme, in cesium iodide, where the ions are the largest (radius Cs^+ = 0.169 nm, radius I^- = 0.216 nm), the lattice energy is only −603 kJ/mol.

LiF is the least soluble in water of all of the alkali halides. Can you see why this is reasonable, given Table 9.3?

9.2
Nature of the Covalent Bond

Ionic bonds found in compounds such as NaCl can be interpreted in a simple way. The cations (Na^+) and anions (Cl^-) are held together by electrostatic forces between oppositely charged ions. The nature of the covalent bond in H_2 is more difficult to visualize. The structures

H : H or H—H

can be misleading if they are taken to mean that the two electrons are fixed in position between the two nuclei. A more accurate picture of the electron density

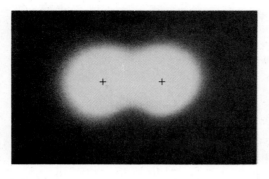

FIGURE 9.3 Electron density in H_2. The depth of shading is proportional to the probability of finding an electron in a particular region. In chemical bonds there tends to be a concentration of electronic charge between the nuclei.

in H_2 is shown in Figure 9.3. At a given instant, the two electrons may be located at any of various points about the two nuclei. However, they are more likely to be between the nuclei than at the far ends of the molecule.

A question that has long intrigued chemists is: Why should the sharing of two electrons between two nuclei result in increased stability? Why, for example, should the H_2 molecule be more stable, by about 400 kJ/mol, than two isolated H atoms? The first plausible answer to this question was put forth in 1927 by two physicists, W. A. Heitler and T. London. They used quantum mechanics to calculate the interaction energy of two hydrogen atoms as a function of the distance between them.

At large distances of separation (far right, Fig. 9.4), there is no interaction between two hydrogen atoms. As they come closer together (moving to the left in Fig. 9.4), the two atoms experience an attraction. This leads gradually to an energy minimum, at an internuclear distance of 0.074 nm. At this distance of separation of the two atoms, the molecule is in its most stable state. It takes 436 kJ of energy to separate a mole of H_2 molecules at this energy minimum into isolated atoms. It also takes energy to bring the atoms closer together than 0.074 nm, since at shorter distances the atoms repel each other. This is why the energy curve rises steeply to the left of 0.074 nm.

All stable chemical bonds have an energy minimum of this sort

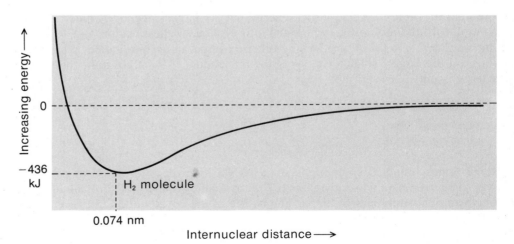

FIGURE 9.4 Energy of the H_2 molecule as a function of the distance between the two nuclei. The minimum in the curve occurs at the observed internuclear distance. Energy is compared to that of two separated hydrogen atoms.

The existence of the energy minimum shown in Figure 9.4 accounts for the stability of the H_2 molecule. The question remains, however: What causes this energy minimum? There are two major factors involved:

1. As two hydrogen atoms approach each other, the electron of one atom is attracted to the nucleus of the other. The electrostatic attraction increases steadily as the distance shortens. However, repulsion between particles of like charge (electron-electron and nucleus-nucleus) also becomes important. The attractive and repulsive forces are balanced at 0.074 nm. There the system has its lowest energy and so is most stable.

2. In the H_2 molecule, the electrons interact with both nuclei equally. In the electrostatic analysis, we treat the electrons as distinguishable particles. It turns out that we really don't know which electron is which. When we take this into account, the electrons are given a little more freedom, which leads to a lower energy. In the jargon of quantum mechanics, this effect is described in terms of "orbital overlap." We say that the H_2 molecule is stabilized by the overlap of the two 1s orbitals. Each electron, in effect, is spread over both orbitals.

The Heitler-London model explains the stability of the H_2 molecule reasonably well. Unfortunately, the equations upon which it is based cannot be solved for any but the simplest of molecules. Later, in Chapter 10, we will describe an extension of this model that explains qualitatively the stability of more complex molecules. In the remainder of this chapter, our goal is more modest. We will look at how electron pairs are distributed between two atoms joined by one or more covalent bonds.

The H_2 problem has been solved exactly by quantum mechanics

9.3
Lewis Structures;
The Octet Rule

The idea of the covalent bond was first suggested by the American physical chemist G. N. Lewis in 1916. He pointed out that the electron configuration of the noble gases appears to be a particularly stable one. Noble-gas atoms are themselves extremely unreactive. Lewis suggested that *atoms, by sharing electrons to form an electron-pair bond, can acquire a stable, noble-gas structure.* Consider, for example, two hydrogen atoms, each with one electron. The process by which they combine to form an H_2 molecule can be shown as

$$H\cdot \; + \; H\cdot \rightarrow H:H$$

using dots for electrons. In H_2, each hydrogen atom has a share in two electrons. In that sense, it has the electronic structure of the noble gas helium (at. no. = 2).

The idea is readily extended to other simple molecules. An example is the F_2 molecule, formed from two fluorine atoms. You will recall that a fluorine atom has the electron configuration $1s^2 2s^2 2p^5$. It has seven electrons in its outermost principal energy level ($n = 2$). These electrons are referred to as **valence electrons**. Showing them as dots about the symbol of the element, we can represent the fluorine atom as

$$:\!\ddot{F}\cdot$$

As with ions, noble gas structures tend to be stable in molecules

When two of these atoms combine, we have

$$:\ddot{F}\cdot \; + \; \cdot\ddot{F}: \; \rightarrow \; :\ddot{F}:\ddot{F}:$$

From our drawing of the F_2 molecule, you can see that each atom owns six valence electrons outright and shares two others. Putting it another way, each F atom is surrounded by eight valence electrons. By this model, both F atoms achieve the electron configuration $1s^2 2s^2 2p^6$, which is that of the noble gas neon. This, according to Lewis, explains why the F_2 molecule is stable and why F atoms combine to form F_2 rather than F_3, F_4, ...

The structures just written are referred to as Lewis structures, or sometimes as "flyspeck formulas." In writing the Lewis structure for a species, we include only those electrons in the outermost energy level. These so-called *valence electrons* are the ones that take part in covalent bonding. For the main-group elements, the only ones that we will be concerned with here, **the number of valence electrons is equal to the group number in the Periodic Table** (Table 9.4).

Table 9.4
Lewis Structures of Atoms

GROUP	OUTER ELECTRON CONFIGURATION	NUMBER OF VALENCE ELECTRONS	EXAMPLE
1	ns^1	1	Li·
2	ns^2	2	·Be·
3	$ns^2 np^1$	3	·Ḃ·
4	$ns^2 np^2$	4	·Ċ·
5	$ns^2 np^3$	5	·N̈·
6	$ns^2 np^4$	6	·Ö·
7	$ns^2 np^5$	7	:F̈·
8	$ns^2 np^6$	8	:N̈e:

In the Lewis structure of a molecule, a covalent bond between atoms is ordinarily shown as a straight line between bonded atoms. Unshared electron pairs, belonging entirely to one atom, are shown as dots. The Lewis structures for the molecules formed by hydrogen with C, N, O, and F are

$$CH_4 \qquad NH_3 \qquad H_2O \qquad HF$$

Notice that in each case the central atom (C, N, O, or F) is surrounded by eight valence electrons. In each of these molecules, a single electron pair is shared between two bonded atoms. These bonds are called **single bonds**. There are four single bonds in CH_4, three in NH_3, two in H_2O, and one in HF. Pairs of electrons not used in bonding are called *unshared pairs*. There are none in CH_4, one unshared pair in NH_3, two unshared pairs in H_2O, and three unshared pairs in HF.

Bonded atoms can share more than one electron pair. When this happens, we say that multiple bonds join the atoms. A **double bond** occurs when bonded atoms share two electron pairs; in a **triple bond**, three pairs of electrons are shared. In ethylene (C_2H_4) and acetylene (C_2H_2), the carbon atoms are linked by a double bond and a triple bond, respectively. Using two parallel lines to represent a double bond and three for a triple bond, we write the structures of these molecules as

$$\begin{array}{ccc} H & & H \\ \diagdown & & \diagup \\ & C = C & \\ \diagup & & \diagdown \\ H & & H \end{array} \qquad\qquad H-C\equiv C-H$$

ethylene, C_2H_4 　　　　　 acetylene, C_2H_2

There are no unshared pairs of electrons in these molecules

Note that each carbon is surrounded by eight valence electrons and each hydrogen with two. The atoms most commonly joined by multiple bonds are C, O, N, and S.

Many polyatomic ions can be assigned simple Lewis structures. For example, the OH^- and NH_4^+ ions can be shown as

$$(:\ddot{O}-H)^- \qquad \text{and} \qquad \left[\begin{array}{c} H \\ | \\ H-N-H \\ | \\ H \end{array}\right]^+$$

In both these ions, hydrogen atoms are joined by covalent bonds to nonmetal atoms (O, N). In both ions there are eight valence electrons. With the OH^- ion, this is one more than the number contributed by the neutral atoms ($6 + 1 = 7$). The extra electron is accounted for by the -1 charge of the ion. With the NH_4^+ ion, four hydrogen atoms and a nitrogen atom supply nine valence electrons ($4 + 5 = 9$). One of these is missing in the NH_4^+ ion, accounting for its $+1$ charge.

These examples illustrate the principle that atoms in covalently bonded species tend to have noble-gas structures. This rule is often referred to as the **octet rule**. Nonmetals, except for hydrogen, achieve a noble-gas structure by sharing in an "octet" of electrons (eight). Hydrogen atoms, in stable molecules or polyatomic ions, are surrounded by two electrons.

Most molecules obey the octet rule

Writing Lewis Structures

For very simple molecules, Lewis structures can often be written by inspection. Usually, though, you will save time and avoid confusion by following these steps:

1. *Draw a skeleton structure for the molecule or ion, joining atoms by single bonds.* In some cases, only one arrangement of atoms is possible; in others, experimental evidence must be used to decide between two or more alternative structures.

2. *Count the number of valence electrons.* For a molecule, we simply sum up

the valence electrons of the atoms present. For a *polyatomic anion,* electrons are *added* to take into account the negative charge. For a *polyatomic cation,* a number of electrons equal to the positive charge must be *subtracted.*

3. *Deduct two valence electrons for each single bond written in step 1. Distribute the remaining electrons as unshared pairs so as to give each atom eight electrons, if possible.*

The application of these steps and some further guiding principles are shown in Example 9.3.

EXAMPLE 9.3 Draw Lewis structures for
a. hypochlorous acid, HOCl.
b. methyl alcohol, CH_4O.
c. the silicate ion, $SiO_4{}^{4-}$.

Solution

a. Several skeletons could be drawn for HOCl. In the correct structure, both the H atom and the Cl atom are bonded to a central oxygen atom:

$$H—O—Cl$$

This skeleton is analogous to that of water, H—O—H. In this and all other structures that we will write, *hydrogen forms only one covalent bond,* since it needs only two electrons to acquire a noble-gas configuration.

To obtain the total number of valence electrons, we add those contributed by the hydrogen atom (one valence electron), oxygen atom, and chlorine atom. Oxygen is in Group 6 and so has six valence electrons; chlorine, in Group 7, has seven valence electrons.

$$H \quad O \quad Cl$$

total valence electrons $= 1 + 6 + 7 = 14$

In the skeleton, there are two bonds, accounting for four valence electrons. That leaves ten to be distributed as unshared pairs. Putting two unshared pairs ($4\ e^-$) on the oxygen atom and three unshared pairs ($6\ e^-$) on chlorine gives

$$H—\overset{..}{\underset{..}{O}}—\overset{..}{\underset{..}{Cl}}:$$

This is indeed a reasonable structure for HOCl. H is surrounded by $2\ e^-$ (the single bond to O), O is surrounded by $8\ e^-$ (two single bonds and two unshared pairs), and Cl is surrounded by $8\ e^-$ (one single bond, three unshared pairs).

b. In principle, several different skeletons could be written for the methyl alcohol molecule, CH_4O. The correct one is obtained by noting that each hydrogen atom forms only one bond and *each carbon atom typically forms four bonds with no unshared pairs.*

The only skeleton consistent with these rules is

$$\begin{array}{c} H \\ | \\ H—C—O—H \\ | \\ H \end{array}$$

The extra guiding principles are shown in italics

The number of valence electrons in the molecule is simply the sum of those from the carbon, hydrogen, and oxygen atoms:

$$\text{total valence electrons} = 4 + 4(1) + 6 = 14$$

Deducting ten electrons for the five covalent bonds in the skeleton leaves four. This is just enough to complete the octet of oxygen. The Lewis structure is

$$\begin{array}{c} \text{H} \\ | \\ \text{H}-\text{C}-\overset{..}{\underset{}{\text{O}}}-\text{H} \\ | \\ \text{H} \end{array}$$

c. In ions such as SiO_4^{4-}, SO_4^{2-}, and CO_3^{2-} we generally find that *each oxygen atom is bonded to a central, nonmetal atom.* Following this general rule, we write

$$\left[\begin{array}{c} \text{O} \\ | \\ \text{O}-\text{Si}-\text{O} \\ | \\ \text{O} \end{array}\right]^{4-}$$

When it comes to Lewis structures, a little experience helps

The number of valence electrons is found by adding the charge of the ion to the total contributed by silicon (Group 4) and oxygen (Group 6) atoms:

$$\begin{array}{ccc} \text{Si} & \text{O} & \text{Charge} \end{array}$$
$$\text{total valence electrons} = 4 + 4(6) + 4 = 32$$

Using rule 3, we deduct eight electrons for the four covalent bonds in the skeleton structure, leaving 24. Putting six electrons around each oxygen atom gives us a plausible Lewis structure for the silicate ion:

$$\left[\begin{array}{c} :\overset{..}{\underset{}{\text{O}}}: \\ | \\ :\overset{..}{\underset{..}{\text{O}}}-\text{Si}-\overset{..}{\underset{..}{\text{O}}}: \\ | \\ :\overset{..}{\underset{..}{\text{O}}}: \end{array}\right]^{4-}$$

EXERCISE Draw the Lewis structure of the arsenate ion, AsO_4^{3-}. Answer:

$$\left[\begin{array}{c} :\overset{..}{\underset{}{\text{O}}}: \\ | \\ :\overset{..}{\underset{..}{\text{O}}}-\text{As}-\overset{..}{\underset{..}{\text{O}}}: \\ | \\ :\overset{..}{\underset{..}{\text{O}}}: \end{array}\right]^{3-}$$

Sometimes, when you follow these rules, you find in the last step that there are "too few electrons to go around." That is, there are not enough electrons left to give each atom an octet. This dilemma can be avoided by making electron pairs do "double duty." By moving an unshared pair to a position between bonded atoms, it is counted in the octet of both atoms. The result of such a shift is to form a multiple bond from a single bond. *Forming a double bond by moving one electron pair corrects a deficiency of two electrons*. If you are four electrons shy, you must form a triple bond (or two double bonds). The process involved is shown in Example 9.4.

EXAMPLE 9.4 Draw Lewis structures for
a. SO_2 b. H_2CO c. N_2

Solution
a. Here, a central sulfur atom is bonded to oxygens. The skeleton is

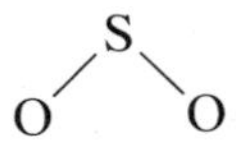

The number of valence electrons is $6 + 2(6) = 18$. Subtracting four electrons for the two covalent bonds leaves 14. These electrons could be spent by filling out the oxygen octets, which accounts for 12, and putting two around the sulfur:

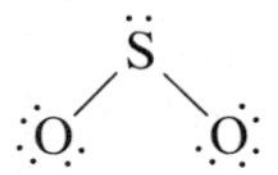

This leaves sulfur with only six valence electrons, two less than an octet. To correct this, shift an unshared pair from one of the oxygen atoms to form a double bond with sulfur. The final Lewis structure is

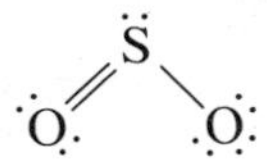

b. A reasonable skeleton, which happens to be the correct one, is

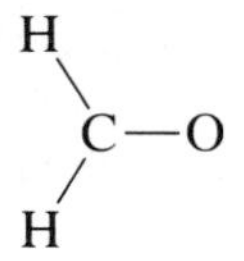

The total number of valence electrons is $2(1) + 4 + 6 = 12$. Subtracting the six electrons used in the three bonds in the skeleton leaves six valence electrons to distribute. We might put these as unshared pairs around the oxygen atom:

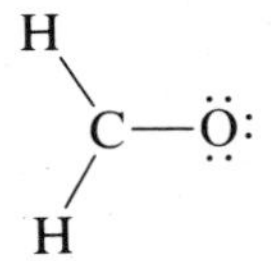

This, however, leaves carbon with only six valence electrons. To supply the two electrons to complete the octet, an unshared pair from the oxygen is used to form a double bond with carbon. The correct Lewis structure is

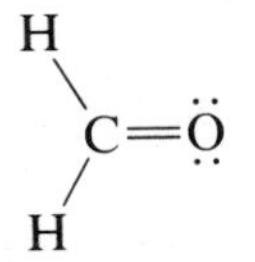

Double and triple bonds are much less common than single bonds

Note that carbon here, as in almost all cases, forms four bonds. In this case, it participates in two single bonds and one double bond.

  c. The skeleton is N—N. Since nitrogen is in Group 5, the total number of valence electrons is 2(5) = 10. The single bond in the skeleton consumes two electrons, so there are eight left. Distributing these as unshared pairs gives the structure

$$:\overset{..}{N}—\overset{..}{N}:$$

Note that both N atoms are two electrons shy of an octet. In other words, there is a deficiency of four electrons. Move two unshared pairs, one from each nitrogen, to give a triple bond:

$$:N{\equiv}N:$$

EXERCISE Draw the Lewis structure of HOCN. Answer: H—$\overset{..}{\underset{..}{O}}$—C≡N:

Resonance Forms

In certain cases, the Lewis structure does not adequately describe the properties of the ion or molecule that it represents. Consider, for example, the SO_2 structure derived in Example 9.4. This structure implies that there are two different kinds of sulfur-to-oxygen bonds in SO_2. One of these appears to be a single bond, the other a double bond. Yet experiment tells us that there is only one kind of bond in the molecule. In particular, we find that the two sulfur-to-oxygen distances are identical, 0.143 nm.

One way to explain this situation is to assume that each of the bonds in SO_2 is intermediate between a single and double bond. The fact that the observed bond distance is half-way between those expected for a single and double bond lends support to this idea. To express this concept within the framework of Lewis structures, we write two structures

with the understanding that the true structure is intermediate between them. These are referred to as *resonance forms*. The concept of resonance is invoked whenever a single Lewis structure does not adequately describe the properties of substances such as sulfur dioxide.

Another species for which it is necessary to invoke the idea of resonance is the nitrate ion. Here three equivalent structures can be written:

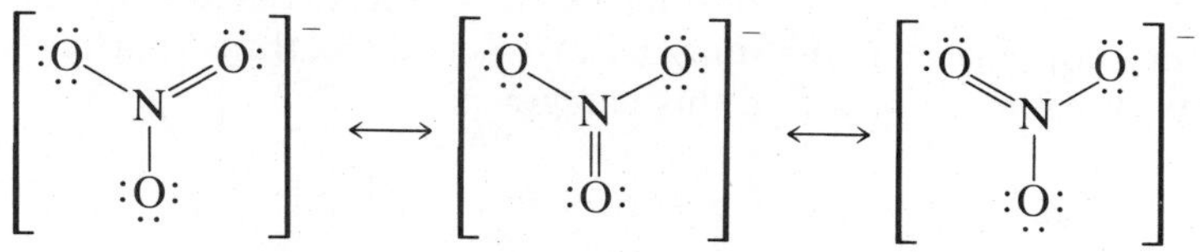

In all the resonance forms for a species, the nuclei remain fixed, and the electrons are moved around

to explain the experimental observation that the three nitrogen to oxygen bonds in the NO_3^- ion are identical in all respects.

We will encounter other examples of molecules and ions whose properties can be interpreted in terms of resonance. In this connection it may be well to point out the following:

1. Resonance forms do not imply different kinds of molecules. Sulfur dioxide is built up of only one type of molecule, whose structure is assumed to be between those of the two resonance forms.

2. Resonance can be anticipated when it is possible to write two or more Lewis structures that are about equally plausible. In the case of the nitrate ion, the three structures we have written are equivalent. One could, in principle, write many other structures, but none of them would put eight electrons around each atom, so presumably would not make a major contribution to the true structure of the nitrate ion.

Resonance structures usually obey the octet rule

3. In writing resonance forms, one can shift only electrons, not atoms. The structure

could not be a resonance form of the nitrate ion, since the atoms are arranged in a quite different way.

EXAMPLE 9.5 Write three resonance forms for SO_3.

Solution SO_3 has the same number of atoms (4) and valence electrons (24) as NO_3^-. We would expect the Lewis structures to be similar. The resonance forms are

EXERCISE One Lewis structure for CO_2 is $:\overset{..}{O}=C=\overset{..}{O}:$ Suggest other Lewis structures for CO_2 that would obey the octet rule. Answer: $:\overset{..}{\underset{..}{O}}-C\equiv O:$ and $:O\equiv C-\overset{..}{\underset{..}{O}}:$ These are ordinarily considered to be resonance forms of CO_2.

Exceptions to the Octet Rule

Although most of the molecules and polyatomic ions that we talk about in general chemistry follow the octet rule, there are some familiar species that do not. Among these are molecules containing an odd number of valence electrons. Nitric oxide, NO, and nitrogen dioxide, NO_2, fall in this category:

There is no way to satisfy the octet rule with an odd number of electrons

NO: no. valence electrons = 5 + 6 = 11

NO_2:   no. valence electrons = 5 + 6(2) = 17

For such *odd electron* species (sometimes called free radicals) it is impossible to write Lewis structures in which each atom obeys the octet rule. The NO molecule is considered to be a resonance hybrid with the two contributing structures

$$\cdot \overset{\cdot\cdot}{N}=\overset{\cdot\cdot}{O}: \longleftrightarrow :\overset{\cdot\cdot}{N}=\overset{\cdot\cdot}{O}\cdot$$

Several different resonance structures can be written for NO_2, of which the more plausible are of the type

$$\cdot N \overset{\displaystyle \overset{\cdot\cdot}{O}:}{\underset{\displaystyle :\overset{\cdot\cdot}{\underset{\cdot\cdot}{O}}:}{\Big|}}$$

Species such as NO and NO_2, in which there are unpaired electrons, are **paramagnetic**; they show a weak attraction toward a magnetic field. Elementary oxygen is also paramagnetic, which suggests that the conventional Lewis structure is incorrect (see Color Plate 9.1),

$$:\overset{\cdot\cdot}{O}=\overset{\cdot\cdot}{O}:$$

This structure is wrong

since it requires that all the electrons be paired. The paramagnetism of oxygen could be explained by the structure

$$:\overset{\cdot\cdot}{O}-\overset{\cdot\cdot}{O}:$$

in which there are two unpaired electrons. However, this structure, like the one written previously, is unsatisfactory. In the first place, it does not conform to the octet rule; much more important, it does not agree with experimental evidence. The distance between the two oxygen atoms in O_2 (0.121 nm) is considerably smaller than that ordinarily observed with an O—O single bond (0.148 nm). These properties of oxygen are difficult to explain in terms of simple Lewis structures. As we shall see in Chapter 10, the molecular orbital approach leads to a more satisfactory picture of the electron distribution in the O_2 molecule.

The bond in O_2 has to be a double bond

There are other species for which Lewis structures written to conform to the octet rule are unsatisfactory. Examples include the fluorides of beryllium and boron, which exist in the vapor as molecules of BeF_2 and BF_3, respectively. Although one could write multiple bonded structures for these molecules in accordance with the octet rule, experimental evidence suggests the structures

$$:\overset{\cdot\cdot}{F}-Be-\overset{\cdot\cdot}{F}: \qquad \text{and} \qquad \overset{\displaystyle :\overset{\cdot\cdot}{F} \quad \overset{\cdot\cdot}{F}:}{\underset{\displaystyle :\overset{\cdot\cdot}{F}:}{\underset{\displaystyle |}{B}}}$$

in which the central atom is surrounded by four and six valence electrons, respectively, rather than eight.

At the opposite extreme, certain of the halides of the heavier nonmetals have structures in which the central atom is surrounded by more than eight valence electrons. In PF_5 and PCl_5, the phosphorus atom is joined by single bonds to each of five halogen atoms and consequently must be surrounded by ten bonding electrons. An analogous structure for SF_6 requires that a sulfur atom have 12 valence electrons around it. We will have more to say about the structures of these molecules in Chapter 10.

9.4
Covalent Bond
Properties

Three important properties of a covalent bond are its polarity, length (distance), and energy of dissociation. The polarity describes how the bonding electrons are distributed between the two bonded atoms. The bond length is the distance between the centers of the two bonded atoms. The strength of the bond is reflected by its bond energy, which tells us how much energy is required to break the bond.

Polar and Nonpolar Covalent Bonds

As we might expect, the two electrons in the H_2 molecule are shared equally by the two nuclei. Stated another way, a bonding electron is as likely to be found in the vicinity of one nucleus as another. Bonds of this type are described as **nonpolar**. We find nonpolar bonds whenever the two atoms joined are identical, as in H_2 or F_2.

In the HF molecule, the distribution of the bonding electrons is somewhat different from that found in H_2 or F_2. Here, the density of the electron cloud is greatest about the fluorine atom. The bonding electrons, on the average, are shifted toward fluorine and away from the hydrogen (Fig. 9.5). Bonds in which the electron density is unsymmetrical are referred to as **polar bonds.**

Atoms of two different elements always differ at least slightly in their affinity for electrons. Hence, covalent bonds between unlike atoms are always polar. Consider, for example, the H—F bond. Since fluorine has a stronger attraction for electrons than does hydrogen, the bonding electrons are displaced toward the fluorine atom. The H—F bond is polar, with a partial negative charge at the fluorine atom and a partial positive charge at the hydrogen atom.

The ability of an atom to attract to itself the electrons in a covalent bond is described in terms of **electronegativity.** The greater the electronegativity of an atom, the greater its affinity for bonding electrons. Electronegativities can be estimated in various ways. One method, based on bond energies, leads to the scale listed in Table 9.5. Here each element is assigned a number ranging from 4.0 for the most electronegative element, fluorine, to 0.7 for cesium, the least electronegative. Among the main-group elements, electronegativity increases moving from left to right in the Periodic Table. Ordinarily, it decreases as we move down a given group.

Electro-
negativity

Gr 1 → Gr 8
↓ incr
decr

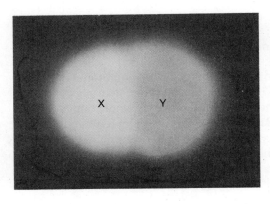

FIGURE 9.5 If two atoms, X and Y, differ in electronegativity, the bond between them is polar. The electron cloud associated with the bonding electrons is concentrated around the more electronegative atom, in this case X. One way to show this is to write:

$$\overset{(-)}{X}—\overset{(+)}{Y}$$

Where the (−) and (+) represent partial charges.

Table 9.5
Electronegativity Values

H 2.1						
Li 1.0	Be 1.5	B 2.0	C 2.5	N 3.0	O 3.5	F 4.0
Na 0.9	Mg 1.2	Al 1.5	Si 1.8	P 2.1	S 2.5	Cl 3.0
K 0.8	Ca 1.0	Sc 1.3	Ge 1.8	As 2.0	Se 2.4	Br 2.8
Rb 0.8	Sr 1.0	Y 1.2	Sn 1.8	Sb 1.9	Te 2.1	I 2.5
Cs 0.7	Ba 0.9	La–Lu 1.0–1.2	Pb 1.9	Bi 1.9	Po 2.0	At 2.2

The extent of polarity of a covalent bond is related to the difference in electronegativities (EN) of the bonded atoms. If this difference is large, as in HF (EN H = 2.1, F = 4.0), the bond will be strongly polar. Where the difference is small, as in H—C (EN H = 2.1, C = 2.5), the bond will be only slightly polar. Thus, in the carbon-hydrogen bond, the bonding electrons are only slightly displaced toward the carbon atom.

In a pure (nonpolar) covalent bond, the electrons are equally shared. In a pure ionic bond, there has been a complete transfer of electrons from one atom to another. We can think of a polar covalent bond as being intermediate between these two extremes. In this sense, we can relate bond polarity to *partial ionic character*. The greater the difference in electronegativity between two elements, the more ionic will be the bond between them. The relation between these two variables is shown in Figure 9.6, p. 278. A difference of 1.7 units corresponds to a bond with 50% ionic character. Such a bond might be described as being halfway between a pure covalent (nonpolar) and a pure ionic bond.

The electrons are displaced toward the more electronegative atom

It is clearly an oversimplification to refer to a bond between two elements as being "ionic" or "covalent." Consider, for example, the bonding in compounds formed by a Group 1 or 2 metal with a nonmetal in Group 6 or 7. The difference in electronegativity ranges from a minimum of 0.6 for Be—Te to a maximum of 3.3 for Cs—F. The percentage of ionic character varies similarly, from about 10% for BeTe to 95% for CsF.

The difference in electronegativity between oxygen or fluorine on the one hand and a Group 1 or 2 metal on the other always exceeds 1.7 units. In this sense, the bonding in the oxides and fluorides of these metals is mainly ionic. The same is true for the oxide and fluoride of aluminum in Group 3. Here, electronegativity differences of 2.0 and 2.5 units correspond to about 65% and 80% ionic character in Al_2O_3 and AlF_3, respectively. The situation is quite different with the chloride, bromide, and iodide of aluminum. In each of these compounds, elec-

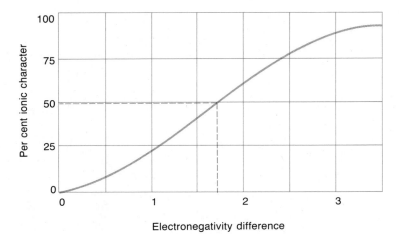

FIGURE 9.6 Relation between ionic character of a bond and the difference in electro-negativity of the bonded atoms.

tronegativity differences less than 1.7 units imply that the bonding is mainly covalent.

EXAMPLE 9.6 Using only the Periodic Table, arrange the following bonds in order of increasing polarity (ionic character): F—Cl, F—F, F—Na.

Solution The F—F bond must come first; since the atoms are identical, the bond is nonpolar. To decide which bond comes next, recall that elec-tronegativity increases as we move across in the Table and decreases as we move down. Chlorine lies below fluorine so fluorine should be somewhat more electronegative. On the other hand, sodium lies far to the left of fluorine and is below it. Hence, sodium should have a much lower elec-tronegativity than fluorine. The proper sequence is

F—F < F—Cl < F—Na

We progress from a nonpolar to a slightly polar to a strongly polar, essen-tially ionic, bond.

EXERCISE Which of the three bonds, B—C, C—N, or B—Si would you expect to be least polar? Answer: B—Si.

Bond Distances

It is possible to estimate the distance between atoms joined by a covalent bond by adding their atomic radii (Appendix 2). For a nonpolar bond, such as that in F_2, we calculate a value just twice the atomic radius:

F—F distance = 2(radius F) = 0.064 nm + 0.064 nm = 0.128 nm

This is exactly the observed internuclear distance, since the atomic radius of a nonmetal is defined as one half the distance between centers of bonded atoms in the element.

When two unlike atoms are joined by a covalent bond, the observed bond distance is usually smaller than the sum of the atomic radii. Thus, for HF, we would calculate the following, using atomic radii:

H—F distance = radius H + radius F = 0.037 nm + 0.064 nm = 0.101 nm

The observed distance is only 0.092 nm. There is a simple way to explain the bond shortening in HF and other polar molecules. It appears that the introduction of partial ionic character into a covalent bond strengthens it and, hence, tends to pull the bonded atoms closer together.

The 10% difference is a big effect

When we compare bond distance in a multiple bond to that of a single bond between the same two atoms, we find that the multiple bond distance is smaller. Compare, for example, the three types of carbon-carbon bonds:

C—C (in C_2H_6)   bond distance = 0.154 nm

C=C (in C_2H_4)   bond distance = 0.133 nm

C≡C (in C_2H_2)   bond distance = 0.120 nm

Bond Energies

The strength of the bond between two atoms can be described using a quantity called bond energy. (More properly, but less commonly, it is called bond enthalpy). **The bond energy is defined as ΔH when one mole of bonds is broken in the gaseous state.** From the equations (see also Fig. 9.7),

$$H_2(g) \rightarrow 2 H(g); \quad \Delta H = +436 \text{ kJ} \tag{9.2}$$

$$Cl_2(g) \rightarrow 2 Cl(g); \quad \Delta H = +243 \text{ kJ} \tag{9.3}$$

Chemical bonds are not easily broken

we conclude that the H—H bond energy is +436 kJ/mol while that for Cl—Cl is +243 kJ/mol. In Reaction 9.2, 1 mol of H—H bonds in H_2 is broken; in Reaction 9.3, 1 mol of Cl—Cl bonds is broken.

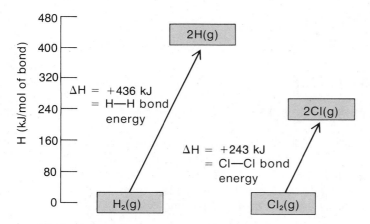

FIGURE 9.7 Bond energies in H_2 and Cl_2. The H—H bond energy is greater than the Cl—Cl bond energy (+436 kJ/mol vs. +243 kJ/mol). This means that the bond in H_2 is stronger than that in Cl_2. Bond energies are always positive quantities: heat has to be absorbed to break a bond.

Bond energies for a variety of single bonds are listed in Table 9.6. You will note that bond energy is always a positive quantity. Energy is always absorbed when chemical bonds are broken. Conversely, heat is given off when bonds are formed from gaseous atoms. Thus, we have

$$H(g) + Cl(g) \rightarrow HCl(g); \quad \Delta H = -(B.E.\ H—Cl) = -431\ kJ$$

$$H(g) + F(g) \rightarrow HF(g); \quad \Delta H = -(B.E.\ H—F) = -565\ kJ$$

Here the abbreviation B.E. is used to stand for bond energy (kJ/mol of bonds).

Table 9.6
Single Bond Energies (kJ/mol) at 25°C

	H	C	N	O	S	F	Cl	Br	I
H	436	414	389	464	339	565	431	368	297
C		347	293	351	259	485	331	276	218
N			159	222	—	272	201	243	—
O				138	—	184	205	201	201
S					226	285	255	213	—
F						153	255	255	277
Cl							243	218	209
Br								193	180
I									151

The single-bond energies listed in Table 9.6 cover a wide range, from 138 kJ/mol for the O—O bond to 565 kJ/mol for the H—F bond. One factor that contributes to bond strength is polarity. In general, a polar bond is stronger than might be expected if the bonding electrons were equally shared. To illustrate this effect, consider the H—F bond. If the bonding electrons were equally shared, the bond energy should be an average of those for the nonpolar bonds H—H and F—F:

$$\frac{436\ kJ/mol\ +\ 153\ kJ/mol}{2} = 295\ kJ/mol$$

The extra energy in the bond comes from its ionic character

The actual bond energy in H—F is nearly twice as great, 565 kJ/mol. We might say that the "extra" bond energy due to polarity is

$$565\ kJ/mol\ -\ 295\ kJ/mol\ =\ 270\ kJ/mol$$

The extent of this effect depends upon the difference in electronegativity between the bonded atoms. Consider the H—I bond, where the difference in electronegativity is only 0.4 (EN I = 2.5, H = 2.1) as compared to 1.9 for H—F (EN F = 4.0, H = 2.1). The H—I bond energy from Table 9.6 is 297 kJ/mol. The average of the H—H and I—I bond energies is

$$\frac{436\ kJ/mol\ +\ 151\ kJ/mol}{2} = 293\ kJ/mol$$

In this case, the "extra" bond energy is only 4 kJ/mol:

297 kJ/mol − 293 kJ/mol = 4 kJ/mol

In general, the greater the difference in electronegativity between bonded atoms, the more strongly polar the bond, and the greater the "extra" bond energy.

Another factor that affects bond energy is the number of electron pairs between the bonded atoms. The bond energy is larger for a multiple bond than for a single bond between the same two atoms. In other words, the multiple bond is stronger. This effect is shown by the data in Table 9.7 where the bond energies of single, double, and triple bonds are compared. It appears that the extra electron pairs strengthen the bond, making it more difficult to separate the bonded atoms from each other.

Table 9.7
Comparison of Bond Energies (kJ/mol)

SINGLE BOND	BOND ENERGY	DOUBLE BOND	BOND ENERGY	TRIPLE BOND	BOND ENERGY
C—C	347	C=C	612	C≡C	820
C—N	293	C=N	615	C≡N	890
C—O	351	C=O	715	C≡O	1075
C—S	259	C=S	477	—	—
N—N	159	N=N	418	N≡N	941
N—O	222	N=O	607	—	—
O—O	138	O=O	498	—	—
S—O	347	S=O	498	—	—

A triple bond is *roughly* three times as strong as a single bond

In Chapter 5, we showed how heats of formation can be used to calculate ΔH for a chemical reaction. In principle, a table of bond energies can be used for this same purpose, at least for reactions involving only gaseous, molecular species. To do this, we apply Hess's Law. A simple case is shown in Example 9.7.

EXAMPLE 9.7 Using Table 9.6, estimate ΔH for the reaction

$$H_2(g) + Cl_2(g) \rightarrow 2\ HCl(g)$$

Solution Let us imagine this reaction as taking place in two steps:
1. The reactant molecules, H_2 and Cl_2, break down to free atoms:

$$H_2(g) \rightarrow 2\ H(g)$$

$$Cl_2(g) \rightarrow 2\ Cl(g)$$

2. The atoms formed in (1) combine to form HCl molecules:

$$2\ H(g) + 2\ Cl(g) \rightarrow 2\ HCl(g)$$

In the first step, we *break* 1 mol of H—H bonds and 1 mol of Cl—Cl bonds

$$\Delta H_1 = \text{B.E. H—H} + \text{B.E. Cl—Cl} = +436\ \text{kJ} + 243\ \text{kJ} = +679\ \text{kJ}$$

In the second step, *2 mol* of H—Cl bonds are *formed:*

$$\Delta H_2 = -2 \text{ (B.E. H—Cl)} = -2(431 \text{ kJ}) = -862 \text{ kJ}$$

By Hess's Law:

$$\Delta H = \Delta H_1 + \Delta H_2 = +679 \text{ kJ} - 862 \text{ kJ} = -183 \text{ kJ}$$

EXERCISE Using Table 9.6, estimate ΔH for $2\,BrCl(g) \rightarrow Br_2(g) + Cl_2(g)$.
Answer: 0 kJ.

The procedure illustrated by Example 9.7 can be applied to any gas-phase reaction. To calculate ΔH from bond energies, we follow a three-step procedure:

$\Delta H_1 > 0$

1. **Decide what bonds are broken in the reaction. Calculate ΔH for bond breaking (ΔH_1) for this step by adding the appropriate bond energies.**

$\Delta H_2 < 0$

2. **Decide what bonds are formed in the reaction. Calculate ΔH for bond formation (ΔH_2). To do this, note that ΔH to form a bond is *minus* the bond energy; that is,**

$$\Delta H \text{ to form X—Y bond} = -(\text{B.E. X—Y})$$

3. **Apply Hess's Law: $\Delta H = \Delta H_1 + \Delta H_2$.**

Example 9.8 applies this method to an organic reaction.

EXAMPLE 9.8 Using bond energies, estimate ΔH for the reaction of ethylene, C_2H_4, with chlorine:

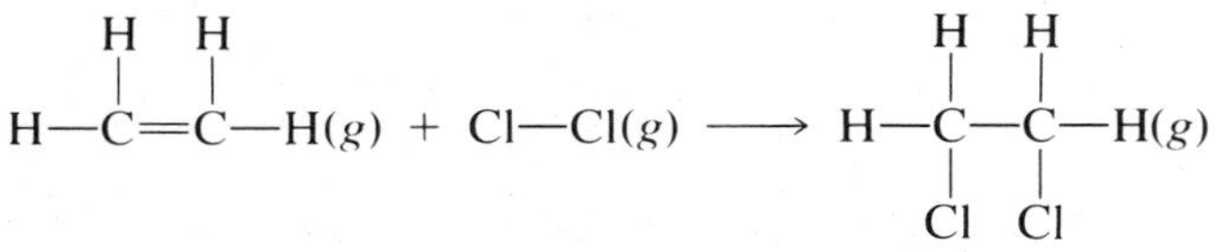

Solution We start by taking an inventory of bonds:

    reactants: 4 C—H bonds, 1 C=C bond, 1 Cl—Cl bond
    products: 4 C—H bonds, 1 C—C bond, 2 C—Cl bonds

Note that the four C—H bonds are present in both reactants and products; they are neither broken nor formed.
The bonds broken are those in the reactants, one C=C and one Cl—Cl bond.

$$\Delta H_1 = \text{B.E. C=C} + \text{B.E. Cl—Cl} = (612 + 243)\text{kJ/mol}$$
$$= +855 \text{ kJ/mol}$$

The bonds formed are those in the products, one C—C and two C—Cl bonds

$$\Delta H_2 = -(\text{B.E. C—C} + 2 \text{ B.E. C—Cl}) = -(347 + 662)\text{kJ/mol}$$
$$= -1009 \text{ kJ/mol}$$

To calculate the overall value of ΔH, we apply Hess's Law:

$$\Delta H = \Delta H_1 + \Delta H_2 = 855 \text{ kJ/mol} - 1009 \text{ kJ/mol} = -154 \text{ kJ/mol}$$

EXERCISE Calculate ΔH for the reaction of ethylene with fluorine to give $C_2H_4F_2$. Answer: -552 kJ.

We should point out one difficulty with calculating ΔH for reactions from bond energies. In most cases it is not possible to assign an exact value to a bond energy. The bond energy varies to some extent with the species in which the bond is found. Consider, for example, the enthalpy changes for the two reactions

$$H{-}O{-}H(g) \rightarrow H(g) + O{-}H(g); \qquad \Delta H = +502 \text{ kJ}$$

$$O{-}H(g) \rightarrow H(g) + O(g); \qquad \Delta H = +426 \text{ kJ}$$

The ΔH values are quite different, even though both involve breaking an O—H bond. The bond energy quoted in Table 9.6, $+464$ kJ/mol, is an average, calculated from ΔH values for many different reactions in which O—H bonds are broken.

A similar situation applies with other bond energies; the amount of energy required to break a given type of bond varies from one molecule to another. For that reason, the value of ΔH calculated from tables of bond energies is likely to be in error, often by 10 kJ or more. That is why, in Example 9.8, we asked you to "estimate" the value of ΔH. The accurate value for ΔH of this reaction, calculated from heats of formation as in Chapter 5, is -182 kJ/mol.

In Example 9.7, the result obtained should be correct. Can you see why?

Summary

An electron transfer reaction from a metal to a nonmetal atom forms an ionic compound. Such a compound contains cations and anions held together by strong electrostatic forces called ionic bonds. The formation of an ionic compound is an exothermic process because the lattice energy (Table 9.3) is a large negative quantity.

Many common cations and anions are isoelectronic with the nearest noble gas (Table 9.2). Transition metal cations, in contrast to the atoms from which they are formed, do not contain outer s electrons (Example 9.1). Cations are smaller than the parent atoms; anions are larger (Example 9.2).

In a molecule or polyatomic ion, nonmetal atoms are joined by covalent (electron pair) bonds. Two atoms may share one, two, or three electron pairs to form a single, double, or triple bond, respectively. Covalent bonds are nonpolar if the two atoms joined are the same; otherwise, they are polar (Example 9.6). Polar bonds are shorter and stronger than one would expect if they were nonpolar. Multiple bonds are shorter and stronger than single bonds. The strength of a bond is measured by the bond energy; ΔH for a gas-phase reaction can be estimated from tables of bond energies (Examples 9.7 and 9.8).

In molecules or ions, atoms of main-group elements ordinarily have octet structures; H atoms are surrounded by only one shared electron pair. Lewis structures are drawn using the octet rule to locate shared and unshared pairs. These structures are derived for molecules or polyatomic ions using the rules given on p. 269. The application of these rules is illustrated in Examples 9.3 and 9.4. Species with an odd number of valence electrons (for example, NO and NO_2) cannot follow the octet rule. In a few other molecules, including BeF_2, BF_3, PF_5, and SF_6, there are fewer than or more than four electron pairs around the central atom. For certain species, the concept of resonance is invoked when a single Lewis structure is inadequate (Example 9.5).

Key Words and Concepts

abbreviated electron configuration	electronegativity	Lewis structure	polar bond
anion	free radical	multiple bond	resonance
bond energy	Hess's Law	noble-gas structure	single bond
cation	ionic bond	nonpolar bond	triple bond
covalent bond	ionic radius	octet rule	unshared pair
double bond	isoelectronic	octet structure	valence electron
	lattice energy	paramagnetic	

Questions and Problems

Ionic Bonding

9.1 Give the formulas of
   a. the cations formed by K, Ca, and Sc.
   b. the anions formed by S and Cl.
   c. the ionic compounds containing a cation in (a) combined with an anion in (b).

9.2 Write formulas for the following ionic compounds:
   a. lithium nitride   b. barium nitrate
   c. iron(II) sulfate   d. chromium(III) oxide

9.3 Give the name of the element whose
   a. -2 ion is isoelectronic with Xe.
   b. -1 ion is isoelectronic with He.
   c. $+2$ ion has the same number of electrons as Cu^+.
   d. -3 ion has the same number of electrons as O^{2-}.

9.4 Give the formulas of four Group 1 halides in which the cation and anion have the same electron configuration.

9.5 Write the electron configuration for
   a. an Mg atom; an Mg^{2+} ion.
   b. an S atom; an S^{2-} ion.
   c. a Ti atom; a Ti^{2+} ion.
   d. a Cr^{2+} ion; a Cr^{3+} ion.

9.6 How many unpaired electrons are there in each of the following ions?
   a. Fe^{2+} b. Fe^{3+} c. Ni^{2+} d. Cl^-

9.7 Select the larger member of each pair in Question 9.5.

9.8 List the following species in order of decreasing radius: Na, K, Mg, Mg^{2+}

9.31 Give the formulas of
   a. the anions formed by O and F.
   b. the cations formed by Na, Ni, and Al.
   c. the ionic compounds containing a cation in (b) combined with an anion in (a).

9.32 Write formulas for the following ionic compounds:
   a. calcium carbonate   b. iron(II) phosphate
   c. potassium sulfide   d. magnesium iodide

9.33 Name the element whose
   a. -1 ion is isoelectronic with Kr.
   b. -2 ion is isoelectronic with Ar.
   c. $+3$ ion has the same number of d electrons as Ti^{2+}.
   d. -2 ion has the same number of 2p electrons as Na^+.

9.34 Give the formulas of four compounds containing Group 2 and Group 6 ions in which the cation and anion have the same electron configuration.

9.35 Give the electron configuration of
   a. a P atom; a P^{3-} ion.
   b. an Rb atom; an Rb^+ ion.
   c. an Sc atom; an Sc^{3+} ion.
   d. a Co^{2+} ion; a Co^{3+} ion.

9.36 Give the number of unpaired electrons in
   a. Cu   b. Cu^+ c. Cu^{2+} d. Zn^{2+} e. K^+

9.37 Select the smaller member of each pair in Question 9.35.

9.38 Arrange the members of each of the following sets in order of increasing size:
   a. I^-, Br, Br^- b. Cr, Cr^{2+}, Cr^{3+}
   c. Se, Te, Te^{2-} d. F^-, Cl^-, I^-

9.9 For CsBr,

$Cs(s) + \frac{1}{2} Br_2(l) \rightarrow Cs(g) + Br(g); \Delta H = +191$ kJ

$Cs(g) + Br(g) \rightarrow Cs^+(g) + Br^-(g); \Delta H = +50$ kJ

$Cs(s) + \frac{1}{2} Br_2(l) \rightarrow CsBr(s); \Delta H = -395$ kJ

Calculate the lattice energy of CsBr.

9.39 Using the lattice energy of KI given in Table 9.3 and the following data,

$K(s) + \frac{1}{2} I_2(s) \rightarrow K(g) + I(g); \Delta H = +197$ kJ

$K(g) + I(g) \rightarrow K^+(g) + I^-(g); \Delta H = +123$ kJ

calculate the heat of formation of KI.

Lewis Structures

9.10 Write the Lewis structure of
a. $GeBr_4$ b. ClO_2^- c. AsH_3 d. NO^+

9.11 Draw Lewis structures for the following polyatomic ions:
a. PO_4^{3-} b. NO_3^- c. SO_3^{2-} d. ClO_4^-

9.12 Draw Lewis structures for the following species (the skeleton is indicated by the way the molecule is written):
a. HO—NO_2 b. Cl—NO_2
c. FN—NF

9.13 Formation of dioxirane, H_2CO_2, has been suggested as a factor in smog formation. The molecule contains an oxygen-oxygen bond. Draw its Lewis structure.

9.14 There are two compounds with the molecular formula C_2H_6O. Draw a Lewis structure for each compound.

9.15 Give the formula of a molecule which you would expect to have the same Lewis structure as
a. ClO^- b. $H_2PO_4^-$ c. PH_4^+ d. SiO_4^{4-}

9.16 Write a Lewis structure for
a. SCN^- b. $S_2O_3^{2-}$ c. NH_2OH d. $P_2O_7^{4-}$

9.17 Write reasonable Lewis structures for the following, none of which obey the octet rule:
a. NO   b. SO_2^- c. BCl_3 d. CO^+

9.40 Write the Lewis structure of
a. $CHCl_3$ b. PCl_3 c. NO_2^- d. NH_4^+

9.41 Follow the directions of Question 9.11 for
a. PCl_4^+ b. CN^- c. ClF_2^+ d. N_3^-

9.42 Follow the directions of Question 9.12 for the following species:
a. H_2N—NF_2 b. (HO—CO_2)$^-$
c. N—N—O   d. O—N—N—O

9.43 Radioastronomers have detected the isoformyl ion, HOC^+, in outer space. Write the Lewis structure for this ion.

9.44 Two different molecules have the formula $C_2H_4Cl_2$. Draw Lewis structures for both.

9.45 Give the formula of a polyatomic ion which you would expect to have the same Lewis structure as
a. HCl   b. F_2 c. N_2 d. CCl_4

9.46 Write a Lewis structure for
a. $NFCl_2$ b. $H_2PO_4^-$ c. BCl_4^- d. HSO_3^-

9.47 Write reasonable Lewis structures for the following, none of which obey the octet rule:
a. NO_2 b. CH_3 c. BeH_2 d. CO^-

Resonance Structures

9.18 Draw resonance structures for
a. SO_3 b. NO_2^- c. SCN^-

9.48 Draw resonance structures for
a. SeO_2 b. NO_3^- c. CO_2

9.19 The Lewis structure for hydrazoic acid may be written as

$$H\diagdown N{=}N{=}\ddot{N}:$$

a. Draw two other resonance forms for this molecule.

b. Is

another resonance form of hydrazoic acid? Explain.

9.49 The oxalate ion, $C_2O_4{}^{2-}$, has the skeleton structure

$$\begin{matrix} O & & O \\ \diagdown & & \diagup \\ C & {-} & C \\ \diagup & & \diagdown \\ O & & O \end{matrix}$$

a. Complete the Lewis structure of this ion.

b. Draw resonance forms for $C_2O_4{}^{2-}$, equivalent to the Lewis structure drawn in (a).

c. Is

$$\left[\ddot{O}{=}\ddot{C}{-}\ddot{O}{-}C{-}\ddot{O}: \atop \quad\quad\quad\quad \underset{:O:}{\|}\right]^{2-}$$

another resonance form of the oxalate ion? Explain.

Bond Properties

9.20 Arrange the following bonds in order of increasing polarity, using only the Periodic Table:

P—O,   P—P,   P—C,   P—N

9.21 Which one of the following bonds would you expect to be most polar?

I—Cl,   I—Br,   I—S,   or   I—I

9.22 Toward which atom are the bonding electrons shifted in the following bonds?

a. C—Cl   b. O—S   c. H—F   d. Cl—I

9.23 Compare the lengths of the nitrogen-nitrogen bonds in

$$H{-}\underset{\underset{H}{|}}{N}{-}\underset{\underset{H}{|}}{N}{-}H, \quad N{\equiv}N, \quad N{=}N{=}O$$

In which molecule is the bond longest? shortest?

9.24 Which of the bonds referred to in Question 9.23 is the weakest? the strongest?

9.50 Using electronegativity trends in the Periodic Table, list the following bonds in order of decreasing polarity:

N—F,   N—N,   N—O,   N—S

9.51 Of the following bonds, which one is least polar?

Si—P,   Si—As,   or   P—Ge

9.52 On which atom is the partial positive charge located in the following polar bonds?

a. N—O   b. F—Br   c. H—O   d. N—C

9.53 Compare the lengths of the carbon-oxygen bonds in

$$H_2C{=}O, \quad H_3C{-}OH, \quad C{\equiv}O$$

In which molecule is the bond longest? shortest?

9.54 Which of the bonds referred to in Question 9.53 is the strongest? the weakest?

ΔH and Bond Energies

9.25 Calculate the amount of heat that must be absorbed (kJ/mol) to dissociate the following molecules into gaseous atoms:

a. H_2O b. NH_3 c. CH_4

9.55 Calculate ΔH (kJ/mol) for the formation of the following molecules from gaseous atoms:

a. N_2 b. CH_2O c. C_2H_4

9.26 Using bond energies, estimate ΔH for the reaction

$$Cl_2(g) + 3 F_2(g) \rightarrow 2 ClF_3(g)$$

9.27 Using Tables 9.6 and 9.7, estimate ΔH for the reaction

$$C_2H_2(g) + 2 Br_2(g) \rightarrow C_2H_2Br_4(g)$$

9.56 Using bond energies, estimate ΔH for the reaction

$$N_2(g) + 3 H_2(g) \rightarrow 2 NH_3(g)$$

9.57 Using Tables 9.6 and 9.7, estimate ΔH for the reaction

$$C_2H_4(g) + HBr(g) \rightarrow C_2H_5Br(g)$$

General

9.28 Give the electron configurations of monatomic ions and Lewis structures of polyatomic ions in
a. iron(II) chloride   b. iron(III) hydroxide
c. calcium phosphate

9.29 Complete the following statements:
a. Bond energy _____ as the number of bonds between two atoms increases.
b. Atoms in a double bond are _____ than those joined by a single bond.
c. Bonds between atoms in molecules consist of _____ _____.
d. Electronegativity _____ going down in the Periodic Table.

9.30 Explain why
a. negative ions are larger than the corresponding atoms.
b. lattice energy is a negative quantity.
c. there is a strong attractive force between H atoms at short distances.
d. energy is evolved in the formation of HF from the elements.

9.58 Follow the directions of Question 9.28 for
a. manganese(II) nitrate
b. magnesium carbonate
c. potassium sulfate

9.59 Criticize each of the following statements:
a. A cation is larger than the atom from which it is formed.
b. Metals typically gain electrons to form ions with noble-gas configurations.
c. Electron loss always leads to cation formation.
d. In a polar bond, the more electronegative atom carries a partial positive charge.

9.60 Explain why
a. molecules with an odd number of valence electrons do not follow the octet rule.
b. scandium, a transition metal, forms an ion with a noble-gas structure.
c. lattice energy becomes a smaller negative number as ions become larger.

*9.61 Benzyne, C_6H_4, is a cyclic molecule. Write a plausible Lewis structure for it and indicate possible resonance structures.

*9.62 Write as many Lewis structures as you can for N_2F_2, following the octet rule.

*9.63 A certain element reacts with chlorine to form a compound which is a gas at 85°C and 1.00 atm with a density of 4.66 g/L.
a. What is the molar mass of the gas?
b. Identify the element involved.
c. Draw the Lewis structure of the molecule.

*9.64 Consider the hypothetical reaction: $Na(s) + Cl_2(g) \rightarrow NaCl_2(s)$, where the product contains Na^{2+} and Cl^- ions. Using data in this chapter and Chapter 8, and assuming the lattice energy of $NaCl_2$ to be that of $MgCl_2$, -2494 kJ/mol, estimate the heat of formation of $NaCl_2$ and comment upon its stability.

*9.65 The percentage of ionic character in a polar covalent bond can be calculated by the equation

$$\text{percent ionic character} = 100[1 - e^{-(\Delta EN)^2/4}]$$

where e is the base of natural logarithms and ΔEN is the difference in electronegativity between the two atoms joined by the bond. Use this equation to calculate the percentage of ionic character when $\Delta EN = 1.7$.

Chapter 10
Molecular
Structure

In Chapter 9, we looked at two important types of bonding: ionic and covalent. Covalent bonding typically leads to the formation of molecules (for example, HF and H_2O). The arrangement of electrons in these molecules can be shown in a very simple way by drawing their Lewis structures.

In this chapter, we will build upon Lewis structures to obtain answers to important questions related to molecular structure:

1. What is the "shape" of a molecule? That is, how are the atoms located in space with respect to one another (Section 10.1)? As we will see, starting with the Lewis structure of a molecule, we can use a simple model to predict

—the angles between bonds formed by a central atom.
—the orientation of other atoms around a central atom.

A knowledge of molecular geometry is also helpful in predicting whether a molecule will be polar or nonpolar (Section 10.2).

2. How are valence electrons in a molecule distributed among electron orbitals? What are the shapes of these orbitals and in what order are they occupied? Here we will consider two different approaches:

—*the atomic orbital, or valence bond, model* (Section 10.3). In this approach, the valence electrons are distributed among orbitals characteristic of the individual atoms. The valence bond model can be used with molecules that "violate" the octet rule as well as those that "obey" it. In Section 10.4, we will apply this model to molecules such as PF_5 and SF_6 in which the central atom has an "expanded octet," i.e., more than eight electrons around it.
—*the molecular orbital model* (Section 10.5). In this model, valence electrons are distributed among orbitals characteristic of the molecule as a whole. Molecular orbital theory has been applied with striking success to a wide

variety of substances; however, our discussion will be limited to simple diatomic molecules such as N_2, O_2, and F_2.

10.1
Molecular Geometry

The geometry of a diatomic molecule such as Cl_2 or HCl can be described very simply. Since two points define a straight line, the molecule must be linear:

Cl—Cl   H—Cl

No question about it, Cl_2 and HCl are linear

With molecules containing three or more atoms, the geometry is not so obvious. Here we must be concerned with the angles between bonded atoms, called *bond angles*. Consider, for example, a molecule of the type XY_2, where X represents the central atom and Y an atom bonded to it. Two geometries are possible. The atoms might be in a straight line, giving a linear molecule with a bond angle ($\angle$ Y—X—Y) of 180° (Fig. 10.1). On the other hand, they could be arranged in a nonlinear pattern. In that case, XY_2 would be a bent molecule with a bond angle less than 180° (Fig. 10.1).

The major features of molecular geometry can be predicted on the basis of a quite simple principle—electron pair repulsion. This principle is the essence of the *v*alence-*s*hell *e*lectron-*p*air *r*epulsion (VSEPR) model, first suggested by Sidgwick and Powell in 1940. It was developed and expanded later by R. J. Gillespie of McMaster University in Canada. According to the VSEPR model: **The electron pairs surrounding an atom repel one another and are oriented as far apart as possible.**

This is a very useful rule, one of the best

In this section we will apply this model to predict the geometry of some rather simple molecules and polyatomic ions. In all of these species, a central atom is surrounded by two, three, or four electron pairs (five and six pairs of electrons are considered separately in Section 10.4). We will be interested in

—the bond angles in the molecule.
—the position of other atoms relative to the central atom.
—the general shape of the molecule, as described by such words as "linear," "bent," and so on.

We begin with the simplest case: species with only single bonds and no unshared pairs around the central atom. The discussion will then be extended to include

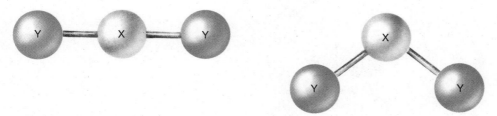

FIGURE 10.1 Two possible geometries for a molecule of general formula XY_2. Certain molecules of this type, including BeF_2 and CO_2 are linear (180° bond angle). Others, including H_2O and SO_2, are bent (bond angle <180°). Knowing the Lewis structure, it is possible to predict whether a molecule will be linear or bent (see text discussion).

species with unshared pairs on the central atom, and finally to multiply bonded species.

Species with Single Bonds and No Unshared Pairs Around the Central Atom

Figure 10.2 shows the geometries expected when two, three, or four electron pairs around a central atom are used to form single bonds. In each case there are no unshared pairs and the bonding electron pairs are oriented as far apart as possible. With two electron pairs (Be in BeF_2), this gives a 180° bond angle and, hence, a **linear** molecule. Three electron pairs around a central atom (B in BF_3) are directed toward the corners of an **equilateral triangle.** The bond angles are 120°. The molecule is planar; all four atoms, including the central atom, are in the same plane.

In the most common case, with molecules that follow the octet rule, the central atom is surrounded by four electron pairs. Suppose that all of these pairs are used to form a total of four single bonds. Here, electron pair repulsion predicts a three-dimensional structure. The four electron pairs are directed toward the corners of a regular **tetrahedron.** This is a three-dimensional structure with four faces and four corners. All the bond angles are 109.5°, the tetrahedral angle. Methane, CH_4, is a classic example of a molecule with this structure. The carbon atom is at the center of the tetrahedron. There is a hydrogen atom at each corner.

The tetrahedron is a basic unit in the structure of many organic molecules. Indeed, whenever carbon forms four single bonds, it shows tetrahedral geometry.

The CH_4 molecule is a perfect tetrahedron

Number of electron pairs	Orientation of electron pairs	Predicted bond angles		
2	Straight line	180°	BeF_2	
3	Equilateral triangle	120°	BF_3	
4	Tetrahedron	109.5°	CH_4	

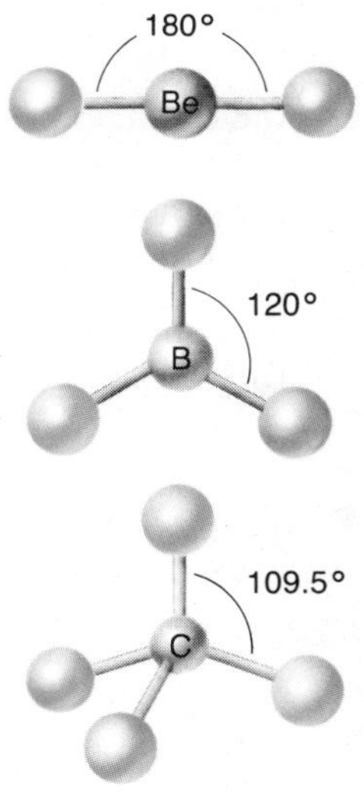

FIGURE 10.2 Geometry of electron pairs around a central atom (Be, B, C). The electron pairs orient themselves so as to be as far apart as possible. The most common situation involves four electron pairs directed toward the corners of a regular tetrahedron, as in CH_4.

Many polyatomic ions also show this structure. Consider the Lewis structures of NH_4^+ (p. 269) and SiO_4^{4-} (p. 271). Note that in both cases the central atom is surrounded by four electron pairs, all of which are used to form single bonds. In both ions, the central atom is in the center of a regular tetrahedron, symmetrically surrounded by four other atoms. Any molecule XY_4, or polyatomic ion with a central atom X surrounded by four single bonds, will have tetrahedral geometry.

Species with Unshared Pairs Around the Central Atom

In many molecules and polyatomic ions, one or more of the electron pairs around the central atom is unshared. The VSEPR model is readily extended to predict the geometries of these species. Here we should stress two points:

1. The bond angles are approximately equal to those observed when the central atom has only single bonds and no unshared pairs (180° for two pairs, 120° for three pairs, 109.5° for four pairs).

2. The words used to describe the geometry differ from those applied to molecules where the central atom has only single bonds and no unshared pairs. This is because, in describing molecular geometry, *we refer only to the positions of the bonded atoms*. These positions can be determined experimentally; positions of unshared pairs cannot be established by experiment. Hence the locations of unshared pairs are not specified in describing molecular geometry.

With these principles in mind, let us consider the NH_3 molecule. You will recall from Chapter 9 that the nitrogen atom in this molecule has an octet structure. It is surrounded by three single bonds and one unshared pair:

$$H \overset{\displaystyle \ddot{N}}{\underset{\displaystyle H}{\diagup \, | \, \diagdown}} H$$

At the left of Figure 10.3, we show the apparent orientation of the four electron pairs around the N atom in NH_3. Notice that, as in CH_4, the four pairs are directed toward the corners of a tetrahedron. At the right of Figure 10.3, we show the positions of the atoms in NH_3. The nitrogen atom is located above the center of an equilateral triangle formed by the three hydrogen atoms. We describe this

Molecular geometry refers to the positions of the nuclei, not the electrons

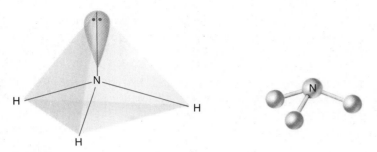

FIGURE 10.3 Two ways of showing the geometry of the NH_3 molecule. The orientation of the electron pairs, including the unshared pair (*color*), is shown at the left. The orientation of the atoms is shown at the right. The nitrogen atom is located directly above the center of the equilateral triangle formed by the three hydrogens atoms. The NH_3 molecule is commonly described as being pyramidal.

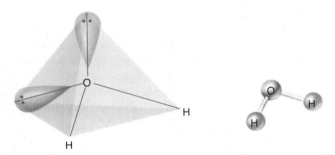

FIGURE 10.4 Two ways of showing the geometry of the H_2O molecule. At the left, the two unshared pairs are shown. As you can see from the drawing at the right, H_2O is a bent molecule. The bond angle, 105°, is a little smaller than the predicted value, 109.5°; the unshared pairs spread out over a larger volume than that occupied by the shared pairs.

NH_3 is not tetrahedral

molecule as **pyramidal.** The nitrogen atom is at the apex of the pyramid while the three hydrogen atoms form its base. The molecule is three-dimensional as the word "pyramid" implies.

The development we have just gone through for NH_3 is readily extended to the water molecule, H_2O. Here, the Lewis structure shows that the central oxygen atom is surrounded by two single bonds and two unshared pairs:

$$H \diagdown \overset{\displaystyle \ddot{O}}{} \diagup H$$

The diagram at the left of Figure 10.4 emphasizes that the four electron pairs are oriented tetrahedrally. At the right, the positions of the atoms are shown. Clearly they are not in a straight line; we describe the H_2O molecule as being **bent.**

From what we have said about the geometries of NH_3 and H_2O, you might expect the bond angles ($\angle$ H—N—H or $\angle$ H—O—H) to be the tetrahedral angle, 109.5°. Actually, in both cases the observed bond angle is slightly less than the ideal value. In NH_3 (three single bonds, one unshared pair around N), it is found to be about 107°. In H_2O (two single bonds, two unshared pairs around O), the bond angle is about 105°.

These effects are usually not very large

These effects can be explained in a rather simple way. An unshared pair is attracted by one nucleus, that of the atom to which it belongs. In contrast, a bonding pair is attracted by two nuclei, those of the two atoms it joins. Hence we might expect the electron cloud of an unshared pair to spread out over a larger volume than that of a bonding pair. In NH_3, this tends to force the bonding pairs closer to one another, thereby reducing the bond angle. Where there are two unshared pairs, as in H_2O, this effect is more pronounced. In general, the VSEPR model predicts that unshared electron pairs will occupy slightly more space than bonding pairs. This principle is sometimes expressed in a different way by establishing a "hierarchy" of electron pair repulsions. These repulsions decrease in the order

$$UP - UP > UP - BP > BP - BP$$

where UP stands for an unshared pair and BP for a bonding pair.

The argument we have just gone through for NH_3 and H_2O is readily extended to a molecule such as GeF_2,

$$:\ddot{F}\diagdown^{Ge}\diagup\ddot{F}:$$

Here, a central germanium atom is surrounded by three electron pairs, one of which is unshared. The geometry of GeF_2 can be derived from that of BF_3 (Fig. 10.2). We would predict that GeF_2 should be a bent molecule with a bond angle somewhat less than the ideal value of 120°. This is indeed the case.

EXAMPLE 10.1 Predict the geometries of the following molecules
a. BeH_2 b. H_2S c. PH_3

Solution
a. In the BeH_2 molecule, there are only four valence electrons (two from the Be atom, one from each of the two H atoms). The Lewis structure has to be

$$H-Be-H$$

There are only two electron pairs around the beryllium atom. These electron pairs are oriented to be as far apart as possible. The molecule is linear with a bond angle of 180°. Its geometry is the same as that of BeF_2 (Fig. 10.2).

b. In H_2S there are eight valence electrons (six from the S atom in Group 6, one from each of the two H atoms). The Lewis structure is

$$H\diagup^{\ddot{S}}\diagdown H$$

The H_2S molecule should be bent, like H_2O, with a bond angle somewhat less than the ideal value of 109.5°.

c. Again there are eight valence electrons (five from the P atom in Group 5, three from the hydrogen atoms). The Lewis structure is

$$H\diagup^{\ddot{P}}_{|}\diagdown H$$
$$H$$

The PH_3 molecule is pyramidal with a bond angle somewhat less than 109.5°. The geometry is very similar to that of ammonia, NH_3.

Actually, it is 93.3°

Also 93.3°

EXERCISE Describe the geometries of the HOCl molecule and the SiO_4^{4-} ion, whose Lewis structures were given in Example 9.3, Chapter 9. Answer: bent, tetrahedral.

Example 10.1 illustrates a couple of points concerning molecular geometry:
  1. You must know or be able to derive the Lewis structure of a species before you can predict its geometry.
  2. The way a Lewis structure is written does not necessarily imply its geometry. In writing the Lewis structures of H_2S and PH_3, we arranged the atoms in such a way as to suggest the proper geometry. More commonly, they are written

$$H-\ddot{S}-H \quad \text{and} \quad H-\ddot{P}-H$$
$$|$$
$$H$$

with no indication of the bond angles.

Species Containing
Multiple Bonds

The VSEPR model is readily extended to species in which double or triple bonds are present. Here a simple principle applies:

Insofar as molecular geometry is concerned, a multiple bond behaves as if it were a single electron pair.

We can readily understand why this should be the case. The four electrons in a double bond, or the six electrons in a triple bond, must be located between the two atoms, as are the two electrons in a single bond. This means that the electron pairs in a multiple bond must occupy the same region of space as those in a single bond. Hence, the "extra" electron pairs in a multiple bond have no effect upon geometry.

To illustrate this principle, consider the CO_2 molecule. Its Lewis structure is

$$:\ddot{O}=C=\ddot{O}:$$

The central atom, carbon, has two double bonds and no unshared pairs. For purposes of determining molecular geometry, we pretend that the double bonds are single bonds, ignoring the "extra" bonding pairs. The bonds are directed to be as far apart as possible, giving a 180° O—C—O bond angle. The CO_2 molecule, like BeF_2, is linear:

F—Be—F   O=C=O
    180°   180°

We can extend this idea to the acetylene molecule, C_2H_2. You will recall that here the two carbon atoms are joined by a triple bond:

H—C≡C—H

Each carbon atom behaves as if it were surrounded by two electron pairs. Both of the bond angles (H—C≡C and C≡C—H) are 180°. The molecule is linear; the four atoms are in a straight line. The two "extra" electron pairs in the triple bond do not affect the geometry of the molecule.

In ethylene, C_2H_4, there is a double bond between the two carbon atoms. The molecule has the geometry to be expected if each carbon atom had only three pairs of electrons around it.

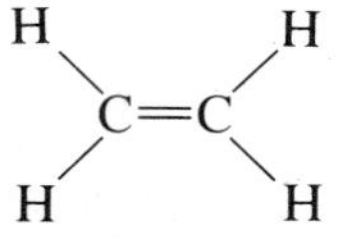

```
H              H
 \            /
  C==C
 /            \
H              H
```

The six atoms are located in a plane, with bond angles of 120°.

■ **EXAMPLE 10.2** Predict the geometry of the SO_2 molecule. ■

Solution   Consider the Lewis structure of SO_2, shown on p. 272. Around the central sulfur atom there is one double bond, one single bond, and one unshared pair of electrons. Pretending that the double bond is a single electron pair, we predict a bent molecule with a bond angle slightly less than 120°. The observed angle is 119.5°.

$$O \overset{\ddot{S}}{\diagup \diagdown} O$$

EXERCISE   Predict the geometry of the SO_3 molecule, whose Lewis structure was given in Chapter 9. Answer: The structure is like that of BF_3. The sulfur atom is at the center of an equilateral triangle with the three oxygen atoms at the corners.

The geometries of a great many molecules can be predicted by the approach we use here

Summary of Molecular Geometries: Unshared Pairs and Bonded Atoms

We have now considered the geometries of a large number of molecules in which a central atom is surrounded by two, three, or four electron pairs. It is possible to develop a general method to predict the geometry of any such species. To see what this method is, let us look at the pairs of molecules listed below, which have very similar geometries.

$O{=}C{=}O$ and $F{-}Be{-}F$ (linear, 180° bond angle)

$O \overset{\ddot{S}}{\diagup\diagdown} O$ and $F \diagdown_{}\overset{Ge}{}\diagup_{} F$ (bent, ~ 120° bond angle)

$\underset{\underset{O}{\|}}{O \diagdown\overset{S}{}\diagup O}$ and $\underset{\overset{|}{F}}{F \diagdown \overset{B}{} \diagup F}$ (equilateral triangle, 120° bond angle)

We have explained these similarities by saying that, insofar as geometry is concerned, a multiple bond behaves like a single bond.

We could express this principle in a slightly different way. Notice that the two members of each pair have

—the same number of atoms bonded to the central atom (two in CO_2 and BeF_2, two in SO_2 and GeF_2, three in SO_3 and BF_3).
—the same number of unshared pairs around the central atom (zero in CO_2 and BeF_2, one in SO_2 and GeF_2, zero in SO_3 and BF_3).

Generalizing, we might say that the geometry of a molecule containing a single central atom is determined by two factors. One of these is *the number of unshared pairs on the central atom*. The other is the *number of atoms bonded to the central atom*. It really doesn't matter whether these atoms are joined by single, double or triple bonds.

Using the principle just stated, everything we have said about molecular geometry can be summarized quite simply. This is shown in Table 10.1, where we consider all possible geometries in which a central atom (X) is surrounded by

To get this information, you need to draw the Lewis structures

Table 10.1
Geometries of Species in which a Central Atom, X, is Surrounded by Four, Three, or Two Electron Pairs

No. Atoms Bonded to X	No. Unshared Pairs	Species Type	Ideal Bond Angle	Geometry of Species	Examples
4	0	XY_4	109.5°	tetrahedral	CH_4
3	1	XY_3E	109.5°*	pyramidal	NH_3
2	2	XY_2E_2	109.5°*	bent	H_2O
3	0	XY_3	120°	equilateral triangle	BF_3, SO_3
2	1	XY_2E	120°*	bent	GeF_2, SO_2
2	0	XY_2	180°	linear	BeF_2, CO_2

* In species containing unshared pairs, the bond angle is expected to be somewhat less than the ideal value.

This table is based on the VSEPR rules. Use it or the rules to predict geometry

two to four bonded atoms (Y) or unshared electron pairs (E). With the aid of the table you can predict the molecular geometry of a species by going through the following steps:

1. **Draw the Lewis structure of the species.**
2. **Count the number of bonded atoms (Y) and unshared pairs (E) around the central atom.**
3. **Using the results of (2), classify the species in one of the categories listed in Table 10.1 (for example, XY_4, XY_3E, . . .). Then use the information in the table to predict the geometry and ideal bond angle.**

EXAMPLE 10.3 Predict the geometries of species which have the following Lewis structures:

a.
$$\left[\ddot{\text{:O}}\text{—}\overset{\displaystyle}{\underset{\displaystyle :\ddot{\text{O}}:}{\text{Cl}}}\text{—}\ddot{\text{O}}: \right]^{-}$$

b.
$$\left[\ddot{\text{:O}}\text{—}\overset{\displaystyle}{\underset{\displaystyle \overset{\|}{:\ddot{\text{O}}:}}{\text{N}}}\text{—}\ddot{\text{O}}: \right]^{-}$$

c. $:\ddot{\text{N}}\text{=}\text{N}\text{=}\ddot{\text{O}}:$

Solution
a. The central atom, chlorine, is bonded to three oxygen atoms; it has one unshared pair. The ClO_3^- ion is of the type XY_3E, similar to NH_3. It is pyramidal; the ideal bond angle is 109.5°.
b. The central atom, nitrogen, is bonded to three oxygen atoms; it has no unshared pairs. The NO_3^- ion is of the type XY_3, similar to BF_3 and SO_3. It has the geometry of an equilateral triangle, with the nitrogen atom at the center. The bond angle is 120°.
c. The central nitrogen atom is bonded to two other atoms with no unshared pairs. The type is XY_2, analogous to BeF_2 and CO_2. The molecule is linear with a bond angle of 180°.

EXERCISE Derive the Lewis structure and then predict the geometry of ONCl. Answer: :Ö=N̈—C̈l: bent.

10.2
Polarity of Molecules

A polar molecule is one that contains positive and negative poles. There is a partial positive charge (positive pole) at one point in the molecule and a partial negative charge (negative pole) of equal magnitude at another point. As shown in Figure 10.5, polar molecules orient themselves in the presence of an electric field. The positive pole in the molecule aligns with the external negative charge and the negative pole with the external positive charge. In contrast, there is no charge separation in a nonpolar molecule. In an electric field, nonpolar molecules, such as H_2, show no preferred orientation; they are oriented randomly.

If a molecule is diatomic, we can readily decide whether it is polar or nonpolar. We need only be concerned with bond polarity, as discussed in Chapter 9. If the two atoms are the same, as in H_2 or F_2, the bond is nonpolar; so is the molecule. Suppose, on the other hand, that the two atoms differ, as in HF. In this case, the bond is polar; the bonding electrons are shifted toward the more electronegative F atom. There is a negative pole at the F atom and a positive pole at the H atom. This is sometimes indicated by writing

$$H \longrightarrow F$$

The arrow points toward the negative end of the polar bond (F atom); the + sign is at the positive end (H atom). The HF bond and hence the HF molecule is a *dipole*; it contains positive and negative poles. The HF molecule, like all polar molecules, lines up in an electric field (Fig. 10.5). Any diatomic molecule composed of two different atoms behaves this way; it is polar.

Any bond between two different atoms is polar

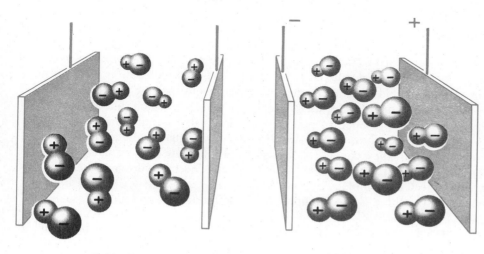

Field off   Field on

FIGURE 10.5 Orientation of polar molecules in an electric field. In the absence of an external electric field, polar molecules are randomly oriented. In an electric field, polar molecules, such as HF, line up as shown. ("Field on"). Nonpolar molecules such as H_2 do not line up.

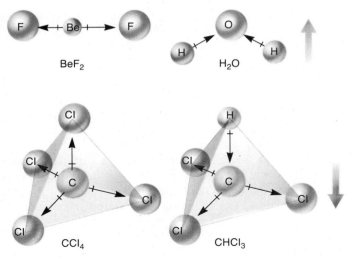

FIGURE 10.6 Polarity of molecules. In all of these molecules, the bonds are polar, as indicated by the ↔ notation. However, in BeF_2 and CCl_4, the bond dipoles cancel and the molecule is nonpolar. In H_2O and $CHCl_3$, on the other hand, there is a net dipole and the molecule is polar. (The *broad arrow* beside the molecule points to the negative pole).

If a molecule contains more than two atoms, it is not so easy to decide whether it is polar or nonpolar. In this case, we must be concerned not only with bond polarity but also with molecular geometry. To illustrate what is involved, consider the two molecules BeF_2 and H_2O, shown at the top of Figure 10.6.

Not all molecules with polar bonds are polar

1. In the linear BeF_2 molecule, the two Be ↔ F dipoles are in opposite directions and cancel one another. The molecule has no net dipole and hence is nonpolar. From a slightly different point of view, we might say that in BeF_2 the centers of positive and negative charge coincide with each other, at the Be atom. There is no way that a BeF_2 molecule can line up in an electric field.

2. In the bent H_2O molecule, the two H ↔ O dipoles do not cancel each other. Instead, they add to give the H_2O molecule a net dipole. The center of negative charge is located at the O atom; this is the negative pole of the molecule. The center of positive charge is located midway between the two H atoms; the positive pole of the molecule is at that point. The H_2O molecule is polar. It lines up in an electric field with the oxygen atom oriented toward the positive electrode.

Is CH_2Cl_2 polar?

3. Carbon tetrachloride, CCl_4, is another molecule which, like BeF_2, is nonpolar despite the presence of polar bonds. Each of its four bonds is a dipole, C ↔ Cl. However, because the four bonds are arranged symmetrically around the carbon atom, they cancel. As a result, the molecule has no net dipole; it is nonpolar. If one of the Cl atoms in CCl_4 is replaced by hydrogen, the situation changes. The $CHCl_3$ molecule is not symmetrical; the H ↔ C dipole does not cancel with the three C ↔ Cl dipoles. Hence the $CHCl_3$ molecule is polar.

EXAMPLE 10.4 Determine whether each of the following is polar or nonpolar:
   a. SO_2 b. BF_3 c. CO_2

Solution
a. In Example 10.2, we decided that the SO_2 molecule is bent, with a bond angle of 120°. We would expect it to be slightly polar. There should be

a positive pole (partial + charge) at the less electronegative sulfur atom and a negative pole (partial − charge) midway between the two oxygens.

b. BF_3, as noted in Figure 10.2, is a planar, triangular molecule. Because of this symmetrical geometry, the B ↔ F dipoles cancel.

The molecule is nonpolar.

c. As pointed out earlier, CO_2 is linear, analogous to BeF_2. The C ↔ O dipoles cancel one another; the molecule is nonpolar.

EXERCISE Is N_2O polar? Answer: Yes (see Example 10.3 for the geometry of N_2O).

10.3
Atomic Orbitals;
Hybridization

To this point, we have used Lewis structures to describe the arrangement of atoms in molecules. These can be very useful. Lewis structures allow us to predict molecular geometries and polarities. However, they do not give us information about the energies of electrons in molecules. Neither do they tell us anything about the orbitals in which the bonding electrons are located.

In the 1930's a theoretical treatment of the covalent bond was developed by Linus Pauling and J. C. Slater, among others. It is referred to as the *atomic orbital,* or **valence bond,** model. According to this model, a covalent bond consists of a pair of electrons of opposed spin within an atomic orbital. For example, a hydrogen atom forms a covalent bond by accepting an electron from another atom to complete its 1s orbital. Using orbital diagrams, we could write

Lewis structures are closely associated with the valence bond model

	1s
Isolated H atom	(↑)
H atom in a stable molecule	(↑↓)

The second electron, shown in color, is contributed by another atom. This could be another H atom in H_2, a F atom in HF, a C atom in CH_4, and so on.

This simple model is readily extended to other atoms. The fluorine atom (electron configuration $1s^2 2s^2 2p^5$) has a half-filled p orbital:

	1s	2s	2p
Isolated F atom	(↑↓)	(↑↓)	(↑↓) (↑↓) (↑)

By accepting an electron from another atom, F can complete this 2p orbital:

	1s	2s	2p
F atom in HF, F_2, . . .	(↑↓)	(↑↓)	(↑↓) (↑↓) (↑↓)

According to this model, it would seem that in order for an atom to form a covalent bond, it must have an unpaired electron. Indeed, the number of bonds formed by an atom should be determined by its number of unpaired electrons. Since hydrogen has an unpaired electron, a H atom should form one covalent bond, as indeed it does. The same holds for the F atom, which ordinarily forms only one bond. Noble gas atoms, such as He and Ne, which have no unpaired electrons, should not form bonds at all. This is the case for He and Ne.

When we try to extend this simple idea beyond hydrogen, the halogens, and the noble gases, problems arise. Consider, for example, the three atoms Be (at. no. = 4), B (at. no. = 5), and C (at. no. = 6):

	1s	2s	2p
Be atom	(↑↓)	(↑↓)	( )( )( )
B atom	(↑↓)	(↑↓)	(↑)( )( )
C atom	(↑↓)	(↑↓)	(↑)(↑)( )

Notice that the beryllium atom has no unpaired electrons, the boron atom has one, and the carbon atom two. Simple valence bond theory would predict that Be, like He, should not form covalent bonds. A boron atom should form one bond, carbon two. Experience tells us that these predictions are wrong. Beryllium forms two bonds in BeF_2; B forms three bonds in BF_3. Carbon, in all its stable compounds, forms four bonds, not two.

To explain these and other discrepancies, simple valence bond theory must be modified. It is necessary to invoke a new kind of atomic orbital, called a hybrid orbital.

Hybrid Orbitals: sp, sp², sp³

The formation of two bonds by beryllium can be explained if we assume that, prior to reaction, one of the 2s electrons is promoted to the 2p level. Thus we could write

	1s	2s	2p			1s	2s	2p
ground state Be atom	(↑↓)	(↑↓)	( )( )( )	→	excited Be atom	(↑↓)	(↑)	(↑)( )( )

With two unpaired electrons, the Be atom can form two covalent bonds, as in BeF_2:

	1s	2s	2p
Be in BeF_2	(↑↓)	(↑↓)	(↑↓)( )( )

Electron promotion allows Be to form two bonds

(The colored arrows indicate the two electrons supplied by the fluorine atoms. The horizontal lines enclose the orbitals involved in bond formation.) Forming two Be—F bonds releases enough energy to more than compensate for that absorbed in promoting the 2s electron of beryllium.

There is one basic objection to this model. It implies that two different kinds of bonds are formed. One would be an "s" bond, since the two electrons are filling an s orbital. The other would be a "p" bond, with a pair of electrons in a p orbital. Experimentally, we find that the properties of the two bonds in BeF_2 are identical. This suggests that the two orbitals used for bond formation by beryllium must be equivalent. In atomic orbital terminology, we say that an s and

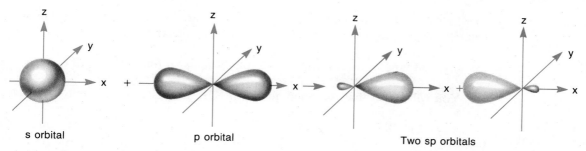

FIGURE 10.7 Formation of sp hybrid orbitals. The mixing of an s orbital with a p orbital gives two new orbitals known as sp hybrids. These two orbitals, oriented at a 180° angle, are capable of holding a total of two electron pairs; sp hybrid orbitals are found in BeF_2 and C_2H_2.

a p orbital have been mixed, or *hybridized* to give two new equivalent orbitals. These new orbitals are referred to as **sp hybrid orbitals:**

Hybridization makes the bonds in BeF_2 equivalent

one s atomic orbital + one p atomic orbital → *two* sp hybrid orbitals

Notice (Fig. 10.7) that *the number of atomic orbitals mixed equals the number of hybrid orbitals formed.* This is always true in hybridization of orbitals.

A similar argument can be used to explain why boron forms three bonds in BF_3. We assume that one of the 2s electrons in boron is promoted to the 2p level:

	1s	2s	2p
Excited B atom	($\uparrow\downarrow$)	($\uparrow$)	($\uparrow$)($\uparrow$)( )

Each fluorine atom puts an electron into a half-filled orbital to give

	1s	2s	2p
B in BF_3	($\uparrow\downarrow$)	($\underline{\uparrow\downarrow}$)	($\underline{\uparrow\downarrow}$)($\underline{\uparrow\downarrow}$)( )

The orbitals in which the bonding electrons are located, enclosed by horizontal lines, are referred to as **sp² hybrid orbitals.** They are formed by mixing an s orbital with two p orbitals:

one s atomic orbital + two p atomic orbitals → *three* sp² hybrid orbitals

sp²
↗ ↖
one s and two p orbitals

To explain why carbon forms four bonds, we again invoke promotion of an electron:

	1s	2s	2p
Excited C atom	($\uparrow\downarrow$)	($\uparrow$)	($\uparrow$)($\uparrow$)($\uparrow$)

followed by hybridization. In CH_4 or any other molecule in which carbon forms four single bonds, the carbon atom has the orbital diagram

	1s	2s	2p
C in CH_4	($\uparrow\downarrow$)	($\underline{\uparrow\downarrow}$)	($\underline{\uparrow\downarrow}$)($\underline{\uparrow\downarrow}$)($\underline{\uparrow\downarrow}$)

Carbon furnishes one electron to each of four sp³ hybrid orbitals

The four orbitals in which the bonding electrons are located are called **sp^3 hybrid orbitals.** They are formed by mixing an s orbital with three p orbitals:

one s atomic orbital + three p atomic orbitals → *four* sp^3 hybrid orbitals

You will recall that the bond angles in NH_3 and H_2O are very close to that in CH_4. This suggests that the four electron pairs surrounding the central atom in NH_3 and H_2O, like those in CH_4, occupy sp^3 hybrid orbitals. In NH_3, three of these orbitals are filled by bonding electrons, the other by the unshared pair on the nitrogen atom. In H_2O, two of the sp^3 orbitals of the oxygen atom contain bonding electron pairs; the other two contain unshared pairs. The situation in NH_3 and H_2O is not unique. In general, we find that *unshared as well as shared electron pairs can be located in hybrid orbitals.*

We should emphasize that hybrid orbitals have their own unique properties, quite different from those of the orbitals from which they are formed. Table 10.2 gives the orientation in space of hybrid orbitals. The geometries are exactly what we would predict on the basis of electron pair repulsion. In each case, the several hybrid orbitals are directed so as to be as far apart as possible.

sp³ hybrids are very common in species that obey the octet rule

Table 10.2
Hybrid Orbitals and Their Geometries

NUMBER OF BONDS	ATOMIC ORBITALS	HYBRID ORBITALS	ORIENTATION	EXAMPLE
2	s, one p	sp	linear	BeF_2
3	s, two p	sp^2	equilateral triangle	BF_3
4	s, three p	sp^3	tetrahedron	CH_4

Hybridization in Molecules Containing Multiple Bonds

In Section 10.1, we pointed out that, insofar as geometry is concerned, a multiple bond acts as if it were a single bond. In other words, the "extra" electron pairs in a double or triple bond have no effect upon the geometry of the molecule. This behavior is explained in terms of hybridization.

The extra electron pairs in a multiple bond (one pair in a double bond, two pairs in a triple bond) are not located in hybrid orbitals.

From this point of view, the geometry of a molecule is fixed by the electron pairs in hybrid orbitals about a central atom. These orbitals are directed to be as far apart as possible. The hybrid orbitals contain

—all unshared electron pairs.
—electron pairs forming single bonds.
—one and only one of the electron pairs in a double or triple bond.

To illustrate this rule, consider the ethylene (C_2H_4) and acetylene (C_2H_2) molecules. You will recall that the bond angles in these molecules are 120° for ethylene and 180° for acetylene. This implies sp^2 hybridization in C_2H_4 and sp hybridization in C_2H_2 (Table 10.2). Using colored lines to represent hybridized electron pairs,

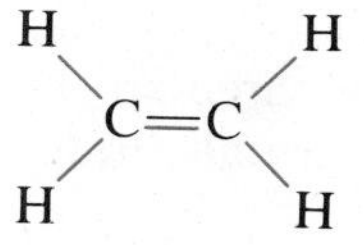

ethylene

H—C≡C—H

acetylene

The hybridization in the C atoms is

sp^2 in C_2H_4
sp in C_2H_2

In both cases, only one of the electron pairs in the multiple bond is hybridized.

EXAMPLE 10.5 Describe the hybridization of
a. carbon in CF_4.   b. nitrogen in the NO_3^- ion.   c. carbon in CO_2.

Solution   The Lewis structures are needed to apply the rules cited above.
a. For CF_4, the Lewis structure is

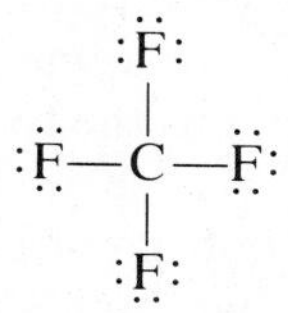

Carbon forms four single bonds, as in CH_4. The hybridization is sp^3.
b. The structure of the NO_3^- ion was given in Example 10.3:

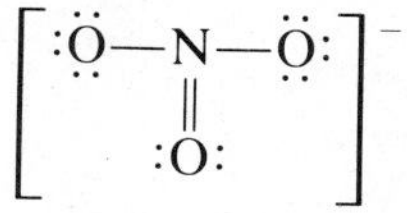

Three electron pairs are in hybrid orbitals; two single bonds and one of the pairs in the double bond. Three hybrid orbitals are required; the hybridization is sp^2.
c. The carbon atom in CO_2 forms two double bonds:

$$:\ddot{O}=C=\ddot{O}:$$

One electron pair in each double bond is hybridized. This requires two hybrid orbitals; carbon shows sp hybridization.

sp hybrids are involved if an atom forms:

1. two double bonds
2. one single and one triple bond

EXERCISE   What is the hybridization of the oxygen atoms in CO_2? Answer: sp^2.

Sigma and Pi Bonds

We have noted that the extra electron pairs in a multiple bond are not hybridized and have no effect upon molecular geometry. At this point, you may well wonder what happened to those electrons. Where are they in the molecule? To answer this question, let us consider the situation in ethylene, C_2H_4. As we have seen, each of the carbon atoms shows sp^2 hybridization. The orbital diagram of either carbon atom would be

	1s	2s	2p
C in C_2H_4	$(\uparrow\downarrow)$	$(\uparrow\downarrow)$	$(\uparrow\downarrow)\,(\uparrow\downarrow)\,(\uparrow)$

Three of the four valence electrons of carbon are located in three sp^2 hybrid orbitals. The fourth valence electron is located by itself in an unhybridized p orbital.

This situation is shown pictorially at the top of Figure 10.8. Let us focus attention on the carbon atom at the left. Each of the three sp^2 hybrid orbitals, shown in grey, contains a pair of bonding electrons. Two of these pairs bond the carbon atom to hydrogen atoms; the third pair joins the carbon atom at the left to the other carbon atom, shown at the right. The two lobes of the p orbital shown in color at the left contain the fourth valence electron. The carbon atom at the right has exactly the same structure; there are two electrons in each of three sp^2 hybrid orbitals and one electron in an unhybridized p orbital.

The orbital picture shown at the top of Figure 10.8 is unstable. The p orbitals on the two carbon atoms, each containing one electron, overlap one another. When this happens, we get the structure at the bottom of Figure 10.8. The orbital shown in color there consists of two lobes, one located above a line joining the two carbon atoms, the other below that line. This orbital contains the two electrons forming the "extra" bond between the carbon atoms in C_2H_4.

Looking at the orbital geometry of C_2H_4 shown at the bottom of Figure 10.8, we see that there is an important difference between the two bonds joining the carbon atoms.

> The p orbitals are ⊥ to the plane of the molecule

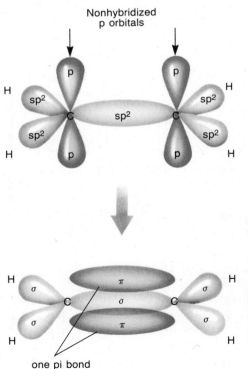

FIGURE 10.8 Bonding in ethylene. The diagram at the top shows the orbitals in the C_2H_4 molecule before pi bond formation. Each carbon atom has three electron pairs in sp^2 hybrid orbitals; it also has one electron in an unhybridized p orbital. The pi bond shown in the diagram at the bottom is formed by the overlap of p orbitals of the two carbon atoms. All of the other bonds in the molecule are sigma bonds, with a high electron density along the bond axis.

1. One bond consists of an electron pair in an orbital (grey) where the electron density is concentrated along a line joining the two carbon atoms. Bonds of this type are called **sigma (σ) bonds** to denote that the electrons are most likely to be found directly along the bond axis. Notice that the carbon-to-hydrogen bonds are also sigma bonds.

All single bonds are σ bonds

2. The second bond consists of an electron pair in an orbital (color) in which the electron density is zero along a line joining the two carbon atoms. Instead, these electrons spend most of their time in lobes above and below the bond axis where the unhybridized p orbitals overlap. Bonds of this type are referred to as **pi (π) bonds**.

From this point of view, we might say that the double bond in C_2H_4

$$\begin{matrix} H \diagdown & & \diagup H \\ & C=C \\ H \diagup & & \diagdown H \end{matrix}$$

consists of a sigma bond and a pi bond. The four single bonds joining carbon to hydrogen are all sigma bonds. The situation in acetylene, C_2H_2, is similar:

$$H-C\equiv C-H$$

The two bonds joining carbon to hydrogen atoms are sigma bonds. The triple bond between the two carbon atoms consists of a sigma bond and two pi bonds. In other words, one of the electron pairs joining the two atoms is located in an orbital concentrated along the bond axis. The other two electron pairs are in orbitals where the electron density is zero along that axis.

For any molecule, we can say that

—*each single covalent bond is a sigma bond.*
—*each double covalent bond consists of a sigma bond and a pi bond.*
—*each triple covalent bond consists of one sigma bond and two pi bonds.*

These rules describe the σ, π character of chemical bonds

EXAMPLE 10.6   Give the number of sigma and pi bonds in molecules with the following Lewis structures:

a. $H-C\equiv C-\ddot{\underset{..}{C}}l:$ $:\ddot{O}=\dot{N}-\ddot{\underset{..}{C}}l:$

Solution

a. In C_2HCl, there are three sigma bonds, the two single bonds and one of the bonds in the triple bond between the carbon atoms. There are two pi bonds, the remaining bonds in the triple bond.
b. There are three bonds in the molecule, a double bond and a single bond. Of these, two are sigma bonds (the single bond and one of the double bonds). There is one pi bond, formed by the "extra" electron pair in the double bond.

EXERCISE   Give the number of sigma and pi bonds in CO_2. Answer: two of each.

10.4
Expanded Octets

In Sections 10.1 to 10.3, we considered the molecular geometry, polarity, and hybridization of species in which a central atom is surrounded by two, three, or four electron pairs. In this section, we will illustrate the principles discussed to this point by extending them to molecules where the central atom is surrounded by five or six electron pairs. Perhaps the best known molecules of the "expanded octet" type are phosphorus pentachloride, PCl_5, and sulfur hexafluoride, SF_6. Their Lewis structures are

Molecules like PCl_5 and SF_6 are relatively rare

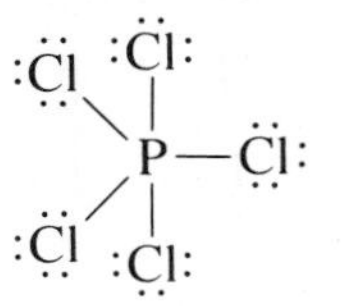

(10 e^- around P) 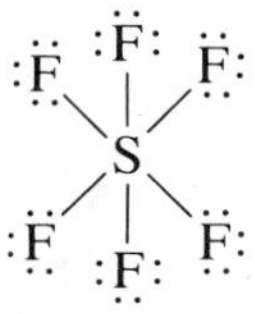   (12 e^- around S)

Some of the electrons around the central atom may be unshared pairs, as in the fluorides of xenon, XeF_2 and XeF_4:

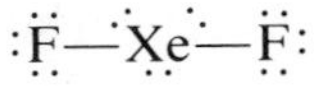

(10 e^- around Xe)

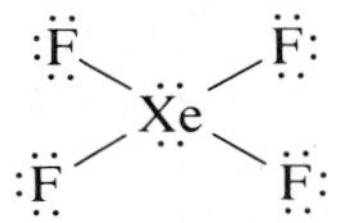

(12 e^- around Xe)

Atoms with expanded octets have more than four pairs of electrons

In these molecules, and in all the examples of expanded octets that we will consider in this chapter, *a central atom, M, is joined by single bonds to halogen atoms* (F, Cl, Br, or I). The central atom M can be any nonmetal atom in the third, fourth, or fifth periods of the Periodic Table. Most frequently, it is one of the following elements:

Third:	P	S	Cl	
Fourth:	As	Se	Br	Kr
Fifth:	Sb	Te	I	Xe

With molecules of this type, there is a simple way to determine the number of valence electrons surrounding the central atom. Almost without exception, a halogen atom shares one of its electrons with the central atom. Hence,

To find the number of valence electrons around the central atom, M, add one electron for each halogen atom bonded to M to the number contributed by the central atom itself (its group number in the Periodic Table).

Applying this rule to the four molecules whose structures are shown above, we have

no. valence e^- around P in PCl_5 = 5 + 5 = 10

no. valence e^- around S in SF_6 = 6 + 6 = 12

no. valence e^- around Xe in $XeF_2 = 8 + 2 = 10$

no. valence e^- around Xe in $XeF_4 = 8 + 4 = 12$

Hybrid Orbitals in Expanded Octets; sp^3d and sp^3d^2

The "extra" electron pairs in expanded octets are accommodated by using d orbitals. To explain the formation of five bonds by phosphorus in PCl_5, we start by writing the abbreviated orbital diagram for the P atom in the ground state:

$$\begin{array}{cccc} & & 3s & 3p \\ \text{P atom (ground state)} & [_{10}Ne] & (\uparrow\downarrow) & (\uparrow)(\uparrow)(\uparrow) \end{array}$$

We now assume that one of the 3s electrons is promoted to a 3d orbital, giving five unpaired electrons:

$$\begin{array}{ccccc} & & 3s & 3p & 3d \\ \text{P atom (excited state)} & [_{10}Ne] & (\uparrow) & (\uparrow)(\uparrow)(\uparrow) & (\uparrow)(\ )(\ )(\ )(\ ) \end{array}$$

Now, by accepting electrons from five halogen atoms (colored arrows), phosphorus forms five bonds.

$$\begin{array}{ccccc} & & 3s & 3p & 3d \\ \text{P atom in PCl}_5 & [_{10}Ne] & \underline{(\uparrow\downarrow)} & \underline{(\uparrow\downarrow)(\uparrow\downarrow)(\uparrow\downarrow)} & \underline{(\uparrow\downarrow)}(\ )(\ )(\ )(\ ) \end{array}$$

The bonding electrons in PCl_5 occupy five hybrid orbitals described as **sp^3d hybrid orbitals.** They are formed by mixing s, p, and d orbitals from the same principal level:

The P atom furnishes one electron to each of the 5 sp^3d oribtals

one s orbital + three p orbitals + one d orbital → *five* sp^3d hybrid orbitals

We can visualize a very similar process to account for the formation of six bonds by sulfur in SF_6:

$$\begin{array}{ccccc} & & 3s & 3p & 3d \\ \text{S (ground state)} & [_{10}Ne] & (\uparrow\downarrow) & (\uparrow\downarrow)(\uparrow)(\uparrow) & (\ )(\ )(\ )(\ )(\ ) \\ \text{S (excited state)} & [_{10}Ne] & (\uparrow) & (\uparrow)(\uparrow)(\uparrow) & (\uparrow)(\uparrow)(\ )(\ )(\ ) \\ \text{S atom in SF}_6 & [_{10}Ne] & \underline{(\uparrow\downarrow)} & \underline{(\uparrow\downarrow)(\uparrow\downarrow)(\uparrow\downarrow)} & \underline{(\uparrow\downarrow)(\uparrow\downarrow)}(\ )(\ )(\ ) \end{array}$$

The orbitals occupied by the bonding electrons in SF_6 are **sp^3d^2 hybrid orbitals.**

one s orbital + three p orbitals + two d orbitals → *six* sp^3d^2 hybrid orbitals

In all species where a central atom is surrounded by more than eight electrons, d orbitals are used. With 10 valence electrons, one d orbital is involved; with 12 valence electrons, two d orbitals must be used. This explains why nonmetal atoms in the second period (N, O, F) do not form expanded octets. There are no 2d orbitals; d orbitals first become available when $n = 3$; that is, in the third period.

Seems reasonable, once you think about it

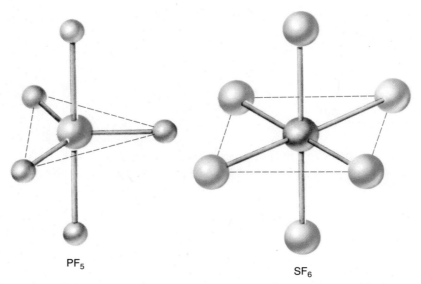

PF₅

SF₆

FIGURE 10.9 Geometry of PF₅ and SF₆. In PF₅, the phosphorus atom is at the center of a triangular bipyramid (two pyramids sharing a triangular base). In SF₆, the sulfur atom is at the center of a regular octahedron. This is a figure with eight sides, all of which are equilateral triangles, and six vertices, each equidistant from the center.

Geometry of Expanded Octets

The VSEPR model can be extended to species having more than eight electrons (four pairs). The question that must be answered here is: How do we direct five or six electron pairs so that they are as far apart as possible? The answer to this question is not obvious, but is given in Figure 10.9, which shows the geometry of PF₅ (five bonds around P) and SF₆ (six bonds around S). Note the following:

This structure mini-mizes the repulsion be-tween five electron pairs

    1. The PF₅ molecule is a **triangular bipyramid** (two pyramids with triangular faces joined through the base). The P atom is at the center of an equilateral triangle formed by three F atoms. The other two F atoms are located above and below the phosphorus, in a straight line with it. The five F atoms are equidistant from the P atom.

    2. The SF₆ molecule is an **octahedron.** We can think of an octahedron as being formed by fusing two square pyramids through the base. The S atom is at the center of a square formed by four F atoms. The other two fluorine atoms are located above and below the sulfur, in a straight line with it. The six F atoms are equidistant from the S atom.

    In molecules such as XeF₂ and XeF₄, one or more of the electron pairs around the central atom is unshared. To deduce the geometries of these molecules, we need to know what happens when successive corners are removed from an octahedron or triangular bipyramid. In many cases, more than one geometry is possible. The observed geometries are shown in Figure 10.10. These structures are ones which assign as much space as possible to unshared pairs of electrons. Consider, for example, the structure XeF₄. Two geometries are possible:

6 Electron Pairs

Bonds	Unshared Pairs	Species Type	Geometry	Example	Description
6	0	XY_6	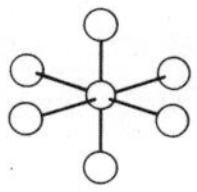	SF_6	Octahedron
5	1	XY_5E	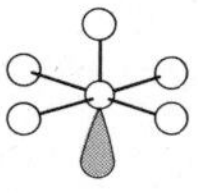	ClF_5	Square pyramid
4	2	XY_4E_2	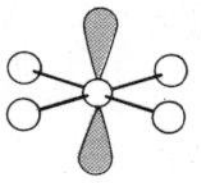	XeF_4	Square planar

XeF$_4$ and SF$_4$ are not tetrahedral molecules

5 Electron Pairs

Bonds	Unshared Pairs	Species Type	Geometry	Example	Description
5	0	XY_5	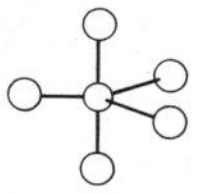	PF_5	Triangular bypyramid
4	1	XY_4E	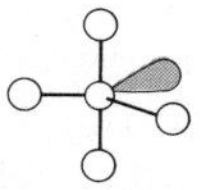	SF_4	Distorted tetrahedron
3	2	XY_3E_2	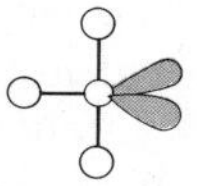	ClF_3	T-shaped
2	3	XY_2E_3	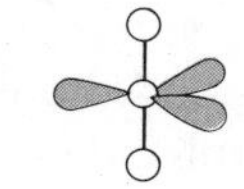	XeF_2	Linear

FIGURE 10.10 Geometries of molecules with expanded octets. The colored ovals represent unshared electron pairs. The structures of ClF$_5$ and XeF$_4$ are derived from that of SF$_6$ by removing successive corners of an octahedron.

In the structure at the left, the lone pairs are as close to one another as possible; in the structure at the right, they are far apart, at opposite ends of the molecule. VSEPR theory predicts and experiment confirms that the structure at the right is the correct one. The xenon atom is at the center of a square with fluorine atoms at each corner of the square.

EXAMPLE 10.7 Consider the IF$_3$ molecule.
a. How many valence electrons are there around the central I atom?
b. What is the hybridization of iodine?
c. Describe the geometry of the molecule.

The PhD thesis of EJS was on the geometries of IF$_5$ and IF$_7$. See if you can predict them

Solution
a. Applying the rule cited earlier,

no. valence e^- around I $= 7 + 3 = 10$

There are five electron pairs around I; the Lewis structure of IF$_3$ is

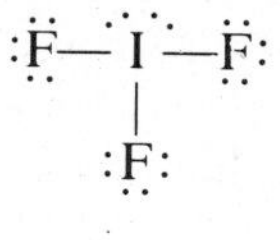

:F̈—Ï—F̈:
    |
   :F̈:

b. Five electron pairs are located in sp³d hybrid orbitals.

c. Referring to Figure 10.10, we conclude that IF₃ is T-shaped.

EXERCISE In BrF₅, how many electron pairs are there around Br? What is the hybridization of Br? Describe the geometry of BrF₅. Answer: 6, sp³d², square pyramid.

10.5
Molecular Orbitals

In Sections 10.3 and 10.4, we used valence bond theory to explain bonding in molecules. It accounts, at least qualitatively, for the stability of the covalent bond in terms of the overlap of atomic orbitals. By invoking hybridization, valence bond theory can account for the molecular geometries predicted by electron-pair repulsion. Where Lewis structures are inadequate, as in SO_2, the concept of resonance allows us to explain the observed properties.

A major weakness of valence bond theory has been its inability to predict the magnetic properties of molecules. We discussed this problem in Chapter 9 with regard to the O_2 molecule. The same problem arises with the B_2 molecule found in boron vapor at high temperatures. This molecule, like O_2, is paramagnetic but has an even number (six) of valence electrons. The octet rule, or valence bond theory, would predict that all electrons are paired in these molecules. This is inconsistent with their paramagnetism, which requires two unpaired electrons.

Valence bond theory has many strengths and a few weaknesses, like most of us

The deficiencies of valence bond theory arise from an inherent weakness. It assumes that the electrons in a molecule occupy atomic orbitals of the individual atoms. In the CH_4 molecule, the bonding is described in terms of the 1s orbitals of the H atom and the four sp³ orbitals of the C atom. Clearly, this is an approximation. Each bonding electron in CH_4 must really be in an orbital characteristic of the molecule as a whole.

Following this idea, molecular orbital theory tries to treat bonds in terms of orbitals involving an entire molecule. The molecular orbital approach involves three basic operations:

1. The atomic orbitals of atoms are combined to give a new set of molecular orbitals characteristic of the molecule as a whole. Here, *the number of molecular orbitals formed is equal to the number of atomic orbitals combined*. When two H atoms combine to form H_2, two s orbitals, one from each atom, yield two molecular orbitals. In another case, six p orbitals, three from each atom, give a total of six molecular orbitals.

2. The molecular orbitals are arranged in order of increasing energy. In principle these energies can be obtained by solving the Schrödinger equation. In practice, we cannot make precise calculations for any but the simplest of molecules. Instead, the relative energies of molecular orbitals are deduced from experimental observations. Spectra and magnetic properties of molecules are used.

We find MO structures in much the same way that we obtain electron configurations

3. The valence electrons in a molecule are distributed among the available molecular orbitals. The process followed is much like that used with electrons in atoms. In particular, we find the following:

   a. *Each molecular orbital can hold a maximum of two electrons.*

   b. *Electrons go into the lowest energy molecular orbital available.* A higher orbital starts to fill only when each orbital below it has its quota of two electrons.

c. *Hund's rule is obeyed.* When two orbitals of equal energy are available to two electrons, one electron goes into each, giving two half-filled orbitals.

To illustrate molecular orbital theory, we will apply it to the diatomic molecules of the elements in the first two periods of the Periodic Table.

Hydrogen and Helium; Combination of 1s Orbitals

Molecular orbital (MO) theory predicts that two 1s orbitals will combine to give two molecular orbitals. One of these has an energy lower than that of the atomic orbitals from which it is formed (Fig. 10.11). Placing electrons in this orbital gives a species which is more stable than the isolated atoms. For that reason the lower molecular orbital in Figure 10.11 is called a **bonding orbital.** The other molecular orbital has a higher energy than the corresponding atomic orbitals. Electrons entering it are in an unstable, higher energy state. It is referred to as an **antibonding orbital.** In energy, this orbital lies about as far above the individual atomic orbitals as the bonding orbital lies below them.

The 1s orbitals on the H atoms interact to form two MO's

The electron density in these molecular orbitals is shown at the right of Figure 10.11. Notice that the bonding orbital has a high density between the nuclei. This accounts for its stability. In the antibonding orbital, the chance of finding the electron between the nuclei is very small. The electron density is concentrated at the far ends of the "molecule." This means that the nuclei are less shielded from each other than they are in the isolated atoms. Small wonder that the antibonding orbital is higher in energy than the atomic orbitals from which it is formed.

The electron density in both molecular orbitals is symmetrical about the axis between the two nuclei. This means that both of these are sigma orbitals. In MO notation, the 1s bonding orbital is designated as σ_{1s}^{b}. The antibonding orbital is given the symbol σ_{1s}^{*}. In general, the superscript "b" refers to a bonding orbital. An asterisk is used to designate an antibonding orbital.

σ orbitals have high electron density on the internuclear axis

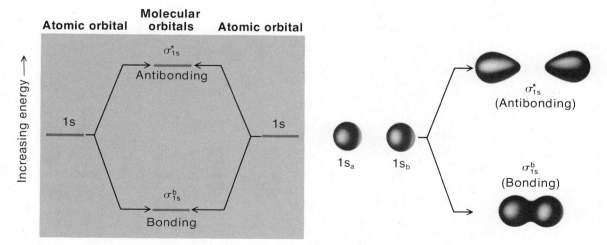

FIGURE 10.11 Molecular orbital formation. Two molecular orbitals are formed by combining two 1s atomic orbitals. In the bonding molecular orbital there is a high electron density between the atoms, which leads to a low energy. In the antibonding orbital the electron density is low between the atoms, giving a higher energy.

In the H_2 molecule, there are two 1s electrons. They fill the σ^b_{1s} orbital, giving a single bond. In the He_2 molecule, there would be four electrons to distribute, two from each atom. These would fill both the bonding and antibonding orbitals. As a result, the net number of bonds in He_2 would be zero. The general relation is

In H_2, no. bonds $=$
$\dfrac{2 - 0}{2} = 1$

$$\text{no. bonds} = \frac{B - NB}{2}$$

where B is the number of electrons in bonding orbitals and NB the number of electrons in antibonding orbitals. Here, $B = NB = 2$, so the number of bonds is zero. The stability of He_2 would be no greater than that of two isolated He atoms. We would predict that the He_2 molecule should not exist, and indeed it does not.

Second Period Elements; Combination of 2s and 2p Orbitals

Among the diatomic molecules of the second period elements are three familiar ones, N_2, O_2, and F_2. The molecules Li_2, B_2, and C_2 are less common but have been observed and studied in the gas phase. In contrast, the molecules Be_2 and Ne_2 are either highly unstable or nonexistent. Let us see what molecular orbital theory predicts about the structure and stability of these molecules. We start by considering how the atomic orbitals containing the valence electrons (2s and 2p) are used to form molecular orbitals.

Combining two 2s atomic orbitals, one from each atom, gives two molecular orbitals. These are very similar to the ones discussed above. They are designated as σ^b_{2s}(sigma, bonding, 2s) and σ^*_{2s}(sigma, antibonding, 2s).

Consider now what happens to the 2p orbitals. In an isolated atom, there are three such orbitals, oriented at right angles to each other. We designate these atomic orbitals as p_x, p_y, and p_z. When two p_x atomic orbitals, one from each

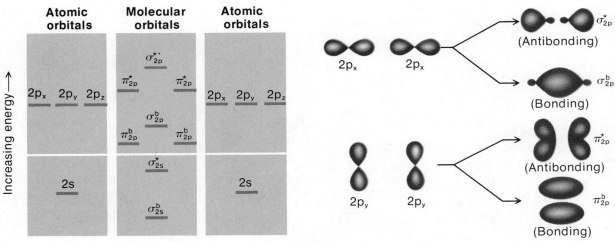

FIGURE 10.12 Molecular orbitals obtained by combining 2p atomic orbitals on two atoms. There are two π^b_{2p} orbitals and two π^*_{2p} orbitals because the $2p_y$ and $2p_z$ orbitals can participate in π orbital formation. The molecular orbitals from the 2s electrons have the structure shown in Figure 10.11.

atom, combine, they form the two molecular orbitals shown at the upper right of Figure 10.12. These are both sigma orbitals, symmetrical about the x axis. One of them, called σ_{2p}^b, is a bonding orbital. The other, σ_{2p}^*, is antibonding. These orbitals have electron cloud densities that resemble those of the sigma molecular orbitals shown in Figure 10.11.

The p_y or p_z atomic orbitals combine with one another in a different way. As you can see from Figure 10.12, two such atomic orbitals, one from each atom, combine to form two pi molecular orbitals. In these orbitals, the electron density is not symmetrical about the axis. The higher energy orbital, π_{2p}^*, where the electron density is concentrated away from the nuclei, is antibonding. The lower energy orbital, π_{2p}^b, with a high electron density between the nuclei, is a bonding orbital. There are two such sets of orbitals. One is formed by p_y orbitals. The other, not shown in the cloud density diagram at the right of Figure 10.12, is formed by p_z orbitals.

The π MO's are formed from p electrons

Summarizing the molecular orbitals available to the valence electrons of the second period elements, we have

one σ_{2s}^b and one σ_{2s}^* orbital.

one σ_{2p}^b and one σ_{2p}^* orbital.

two π_{2p}^b and two π_{2p}^* orbitals.

Table 10.3
Predicted and Observed Properties of Diatomic Molecules of Second Period Elements

	OCCUPANCY OF ORBITALS							
	σ_{2s}^b	σ_{2s}^*	π_{2p}^b	π_{2p}^b	σ_{2p}^b	π_{2p}^*	π_{2p}^*	σ_{2p}^*
Li_2	(↑↓)	( )	( )	( )	( )	( )	( )	( )
Be_2	(↑↓)	(↑↓)	( )	( )	( )	( )	( )	( )
B_2	(↑↓)	(↑↓)	(↑)	(↑)	( )	( )	( )	( )
C_2	(↑↓)	(↑↓)	(↑↓)	(↑↓)	( )	( )	( )	( )
N_2	(↑↓)	(↑↓)	(↑↓)	(↑↓)	(↑↓)	( )	( )	( )
O_2	(↑↓)	(↑↓)	(↑↓)	(↑↓)	(↑↓)	(↑)	(↑)	( )
F_2	(↑↓)	(↑↓)	(↑↓)	(↑↓)	(↑↓)	(↑↓)	(↑↓)	( )
Ne_2	(↑↓)	(↑↓)	(↑↓)	(↑↓)	(↑↓)	(↑↓)	(↑↓)	(↑↓)

	PREDICTED PROPERTIES		OBSERVED PROPERTIES	
	Number of Unpaired e^-	Number of Bonds	Number of Unpaired e^-	Bond Energy (kJ/mol)
Li_2	0	1	0	105
Be_2	0	0	0	unstable
B_2	2	1	2	289
C_2	0	2	0	628
N_2	0	3	0	941
O_2	2	2	2	494
F_2	0	1	0	153
Ne_2	0	0	0	nonexistent

The relative energies of these orbitals are shown at the left of Figure 10.12. This order applies in the molecules of the second period elements, at least through N_2.

To obtain the MO structure of the diatomic molecules of the elements in the second period, we fill the available molecular orbitals in order of increasing energy. The results are shown in Table 10.3. Note the agreement between MO theory and the properties of these molecules. In particular, the number of unpaired electrons predicted agrees with experiment. There is also a general correlation between the predicted number of bonds

$$\text{no. bonds} = \frac{B - NB}{2}$$

and the bond energy. We would expect a double bond (C_2 or O_2) to be stronger than a single bond (Li_2, B_2, F_2). A triple bond, as in N_2, should be still stronger.

A major triumph of MO theory is its ability to explain the properties of O_2. It explains how this molecule can have a double bond and, at the same time, have two unpaired electrons. You will recall that simple valence bond theory could not do this. Again, molecular orbital theory succeeds where valence bond theory fails in explaining paramagnetism in B_2. For N_2 and F_2, the two theories predict the same number of bonds (triple and single, respectively), with no unpaired electrons.

Molecular orbital theory can also be applied to predict the properties of molecules containing different kinds of atoms. Example 10.8 uses the MO approach to explain the structure of the NO molecule.

EXAMPLE 10.8 Using MO theory, predict the electronic structure, number of bonds, and number of unpaired electrons in the NO molecule.

Solution We have 11 valence electrons to account for (five from nitrogen, six from oxygen). Placing these in molecular orbitals in order of increasing energy,

$$\sigma_{2s}^b \quad \sigma_{2s}^* \quad \pi_{2p}^b \quad \pi_{2p}^b \quad \sigma_{2p}^b \quad \pi_{2p}^* \quad \pi_{2p}^* \quad \sigma_{2p}^*$$

NO (↑↓) (↑↓) (↑↓) (↑↓) (↑↓) (↑) ( ) ( )

There are eight electrons in bonding orbitals (two in σ_{2s}^b, four in π_{2p}^b, two in σ_{2p}^b) and three in antibonding orbitals (two in σ_{2s}^*, one in π_{2p}^*). Hence,

$$\text{no. bonds} = \frac{8 - 3}{2} = \frac{5}{2}$$

There is one unpaired electron (in the π_{2p}^* orbital).

EXERCISE How many bonds and unpaired electrons are there in NO^+?
Answer: 3, 0.

We would predict the NO bond to be stronger than the O_2 bond, and it is

The bonding in molecules containing more than two atoms can also be described in terms of molecular orbitals. We will not attempt to do this here; the energy level structure is considerably more complex than the one we have considered. However, one point is worth mentioning. In polyatomic species, a pi molecular orbital can be considered to be spread over the entire molecule rather than being concentrated between two atoms. The electrons in such an orbital are shared by all the atoms in the molecule rather than being "localized" between two specific atoms.

We can apply this concept of "delocalized" molecular orbitals to a species such as the sulfur trioxide molecule. You will recall that valence bond theory has to invoke resonance to explain the fact that the three bonds in SO_3 have the same length. It considers the SO_3 molecule to be a resonance hybrid of the three structures:

For big molecules, MO theory is harder to apply than VB theory

Molecular orbital theory gives a more plausible picture of the bonding in sulfur trioxide. It tells us that the two extra electrons in the double bond are found in a delocalized pi orbital associated with the entire molecule (Fig. 10.13). There are three sigma bonds in the molecule, shown simply as straight lines in Figure 10.13. This explains why the three bond distances are the same.

Summary

The geometry (shape) of a polyatomic ion or simple molecule can be predicted from its Lewis structure by considering repulsions among electron pairs about the central atom (Examples 10.1–10.3 and 10.7). The geometry depends upon the number of bonds and unshared pairs about the central atom (Table 10.1 and Fig. 10.10). Insofar as geometry is concerned, a multiple bond behaves as if it were a single bond.

A polar molecule has a net dipole. The polarity of each bond *and* the molecular geometry must be considered to predict whether a molecule is polar or nonpolar (Example 10.4).

Valence bond theory considers a covalent bond to consist of a pair of electrons of opposed spin occupying an atomic orbital. In most cases, the atomic orbitals used in bonding are hybridized. This way, atoms make maximum use of their valence electrons to form two bonds (Be), three bonds (B), four bonds (C), five bonds (P), or six bonds (S). The corresponding hybridizations are sp, sp^2, sp^3, sp^{3}d, and sp^{3}d^2, respectively (Examples 10.5 and 10.7). In a multiple bond, the

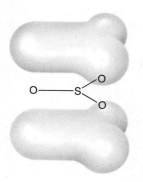

FIGURE 10.13 In the SO_3 molecule, there are four bonds. According to MO theory, three of these are sigma bonds; the fourth is a pi bond. The orbital in which the pi electrons are located is shown in color in the figure. It consists of two lobes, one above the plane of the molecule, the other below that plane.

extra electron pairs are not in hybridized orbitals; instead they form pi bonds (Example 10.6).

In the molecular orbital model, atomic orbitals are combined to form an equal number of molecular orbitals. A given MO may be bonding or antibonding (Figs. 10.11 and 10.12). The relative energies of different molecular orbitals govern the order in which they are filled (Table 10.3, Example 10.8).

Key Words and Concepts

antibonding orbital	expanded octet	polar molecule
atomic orbital model	hybrid orbital	pyramidal molecule
bent molecule	Lewis structure	resonance
bond angle	linear molecule	sigma bond
bond distance	molecular geometry	tetrahedron
bonding orbital	molecular orbital	triangular bipyramid
bond polarity	nonpolar molecule	VSEPR model
delocalized orbital	octahedron	valence bond model
dipole	pi bond	valence electron
electron pair repulsion	polar bond	

Questions and Problems

Molecular Geometry

10.1 Describe the geometry of each of the following molecules (the way the Lewis structure is written does not necessarily imply the geometry):

a.
$$\begin{array}{c} H \\ | \\ H-C-\ddot{B}r: \\ | \\ :\ddot{B}r: \end{array}$$

b. $:\ddot{S}=C=\ddot{O}:$

c. $:\ddot{O}=\ddot{N}-\ddot{C}l:$

d.
$$\begin{array}{c} :\ddot{C}l-\ddot{N}-H \\ | \\ H \end{array}$$

10.2 Describe the geometry of the following polyatomic ions:

a.
$$\left[\begin{array}{c} :\ddot{O}-C-\ddot{O}: \\ || \\ :O: \end{array}\right]^{2-}$$

b. $\left[:\ddot{O}-\ddot{N}=\ddot{O}:\right]^{-}$

c.
$$\left[\begin{array}{c} :\ddot{O}-\ddot{C}l-\ddot{O}: \\ | \\ :O: \end{array}\right]^{-}$$

d. $\left[:\ddot{S}=C=\ddot{N}:\right]^{-}$

10.31 Describe the geometry of each of the following molecules (the way the Lewis structure is written does not necessarily imply the geometry):

a. $:\ddot{C}l-Be-\ddot{C}l:$

b. $:\ddot{O}-\ddot{S}=\ddot{O}:$

c. $:\ddot{C}l-Ge-\ddot{C}l:$

d. $:N\equiv N-\ddot{O}:$

10.32 Describe the geometry of the following polyatomic ions:

a.
$$\left[\begin{array}{c} :\ddot{O}: \\ | \\ :\ddot{O}-Mn-\ddot{O}: \\ | \\ :O: \end{array}\right]^{-}$$

b.
$$\left[\begin{array}{c} :\ddot{O}-\ddot{N}-\ddot{O}: \\ || \\ :O: \end{array}\right]^{-}$$

c. $\left[:\ddot{O}-\ddot{O}-\ddot{O}:\right]^{2-}$

d.
$$\left[\begin{array}{c} H \\ | \\ H-N-H \\ | \\ H \end{array}\right]^{+}$$

10.3 Give all the bond angles (109.5°, 120°, or 180°) in the following molecules:

a.
$$H-\overset{\overset{\displaystyle H}{|}}{\underset{\underset{\displaystyle H}{|}}{C}}-C\equiv C-H$$

b.
$$H-\overset{}{\underset{\underset{\displaystyle :O:}{\|}}{C}}-\ddot{O}-H$$

c.
$$H-\overset{\overset{\displaystyle H}{|}}{C}=\overset{\overset{\displaystyle H}{|}}{C}-H$$

10.4 Draw the Lewis structure and describe the geometry of

a. NF_3 b. CH_3Cl c. N_3^- d. PCl_4^+

10.5 Draw the Lewis structure and describe the geometry of

a. $HOClO$ b. PO_3^{3-} c. $AsCl_3$ d. O_3

10.6 For the structures shown in Question 10.2, determine
a. the number of unshared pairs around the central atom.
b. the number of atoms bonded to the central atom.
c. the bond angle, using Table 10.1.

10.7 An objectionable component of smoggy air is acetylperoxide which has the skeleton structure

$$H_3C-\overset{\overset{\displaystyle O}{\|}}{C}-O-O-\overset{\overset{\displaystyle O}{\|}}{C}-CH_3$$

a. Draw the Lewis structure of this compound.
b. Indicate all the bond angles.

10.8 List all the bond angles in the following hydrocarbons:
a. C_2H_6 b. $H_3C-\overset{\overset{\displaystyle }{|}}{C}=\overset{\overset{\displaystyle }{|}}{C}-CH_3$ c. C_2H_2
  $\qquad\qquad\qquad\quad\; H \;\; H$

10.9 Consider the following molecules: SiH_4, PH_3, H_2S. In each case, a central atom is surrounded by four electron pairs. In which of these molecules would you expect the bond angle to be less than 109.5°? Explain your reasoning.

Molecular Polarity

10.10 Which of the species in Question 10.1 are dipoles?

10.11 Which species in Question 10.2 are dipoles?

10.33 Give all of the bond angles (109.5°, 120°, or 180°) in the following ions:

a. $\left[:\ddot{O}-\dot{N}=\ddot{O}\right]^-$

b. $\left[:\ddot{O}-\overset{\overset{\displaystyle }{\|}}{\underset{\underset{\displaystyle :O:}{}}{C}}-\ddot{O}:\right]^{2-}$

c. $\left[:\ddot{O}-\overset{\overset{\displaystyle :\ddot{O}:}{|}}{\underset{\underset{\displaystyle :\ddot{O}:}{|}}{Si}}-\ddot{O}:\right]^{4-}$

10.34 Draw the Lewis structure and describe the geometry of

a. $HOCl$ b. C_2HCl c. OCN^- d. HCO_2^-

10.35 Draw the Lewis structure and predict the geometry of

a. SiF_4 · b. Cl_2CO c. BrO_3^- d. NI_3

10.36 Follow the directions of Question 10.6 for the structures listed in Question 10.32.

10.37 Peroxypropionyl nitrate (PPN) is an eye irritant found in smog. Its skeleton structure is

$$H_3C-\overset{\overset{\displaystyle H}{|}}{\underset{\underset{\displaystyle H}{|}}{C}}-\overset{\overset{\displaystyle O}{\|}}{C}-O-O-\overset{\overset{\displaystyle O}{\|}}{N}-O$$

Write the Lewis structure for PPN and give all the bond angles.

10.38 List all the bond angles in the following hydrocarbons:
a. CH_4 b. $H-\overset{\overset{\displaystyle }{|}}{C}=\overset{\overset{\displaystyle }{|}}{C}-H$ c. $H_3C-C\equiv C-H$
  $\qquad\qquad\quad\; H \;\; H$

10.39 In each of the following molecules, a central atom is surrounded by a total of three atoms or unshared pairs: BCl_3, $SnCl_2$, SO_2. In which of these molecules would you expect the bond angle to be less than 120°? Explain your reasoning.

10.40 Which of the species in Question 10.31 are dipoles?

10.41 Which species in Question 10.32 are dipoles?

10.12 There are two different molecules with the formula N_2F_2:

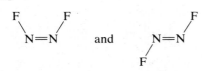

Is either molecule polar? Explain.

10.42 There are three compounds with the formula $C_2H_2Cl_2$:

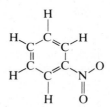

Which of these molecules are polar? nonpolar?

Hybridization

10.13 Give the hybridization of the central atom in each molecule in Question 10.1.

10.14 Give the hybridization of each atom in Question 10.2.

10.15 Give the hybridization of each C, N, and O atom in 3-pyridine carboxylic acid (unshared electron pairs are not shown):

$$
\begin{array}{c}
H-C \overset{\displaystyle H}{\underset{\displaystyle}{\overset{|}{C}}} C-C \overset{\displaystyle O}{\underset{\displaystyle}{\overset{\|}{}}} -O-H \\
H-C \underset{\displaystyle N}{\overset{\displaystyle}{}} C-H
\end{array}
$$

10.16 What is the hybridization of carbon in

   a. CH_3Cl b. $\left[O-\overset{\displaystyle}{\underset{\displaystyle O}{\overset{|}{C}}} -O \right]^{2-}$

   c. $O{=}C{=}O$ d. $H-\overset{\displaystyle}{\underset{\displaystyle O}{\overset{\|}{C}}} -OH$

10.17 Predict the hybridization of the central atom (underlined) in
   a. $\underline{O}F_2$ b. $(H_2N)_2\underline{C}O$ c. $O\underline{P}Cl_3$

10.43 Give the hybridization of the central atom in each molecule in Question 10.31.

10.44 Give the hybridization of each atom in Question 10.33.

10.45 Give the hybridization of each C, N, and O atom in nitrobenzene (unshared electron pairs are not shown):

$$
\begin{array}{c}
H \\
| \\
H \overset{\displaystyle C}{\underset{\displaystyle}{}} \overset{\displaystyle}{\underset{\displaystyle}{}} \overset{\displaystyle H}{C} \\
H-C \overset{\displaystyle}{\underset{\displaystyle}{}} C-N \overset{\displaystyle O}{\underset{\displaystyle}{}} \\
| \quad \| \\
H \quad O
\end{array}
$$

10.46 What is the hybridization of nitrogen in

   a. $\left[O-\overset{\displaystyle}{\underset{\displaystyle O}{\overset{|}{N}}} -O \right]^{-}$ b. $H-\overset{\displaystyle\cdot\cdot}{\underset{\displaystyle Cl}{\overset{}{N}}} -H$

   c. $:N{\equiv}N:$ d. $:N{\equiv}N{-}O$

10.47 What is the hybridization of the central atom (underlined) in
   a. $\underline{N}HF_2$ b. $F_2\underline{C}O$ c. $HO\underline{I}O_2$

Sigma and Pi Bonds

10.18 Give the number of sigma and pi bonds in each species listed in Question 10.16.

10.19 Give the formula of a molecule or ion in which an atom of
   a. N forms four bonds using sp^3 hybrid orbitals.
   b. N forms a pi and two sigma bonds.
   c. C forms four bonds, three in which it uses sp^2 hybrid orbitals.

10.48 Give the number of sigma and pi bonds in each species listed in Question 10.46.

10.49 Give the formula of a molecule or ion in which an atom of
   a. B forms three bonds using sp^2 hybrid orbitals.
   b. B forms four bonds using sp^3 hybrid orbitals.
   c. O forms a pi bond.
   d. C forms two pi bonds.

Expanded Octets

10.20 Consider the species
    a. SeF_6 b. KrF_2 c. $TeBr_4$ d. IF_3
How many electron pairs are there (five or six) around the central atom in each species? What is the hybridization of the central atom?

10.21 Describe the geometry of each species in Question 10.20.

10.22 Describe the geometry to be expected for species in which there are, around the central atom,
    a. six bonds.   b. five bonds and one unshared pair.
    c. four bonds and two unshared pairs.

10.23 For a polyatomic anion in which a central atom is bonded to halogen atoms, the number of valence electrons around the central atom is the sum of the group number of the central atom, the number of halogen atoms bonded to the central atom, and the charge of the anion. Determine the number of electron pairs around the central atom and the hybridization in
    a. ICl_4^- b. $SiCl_6^{2-}$ c. PCl_4^-

10.50 Consider the species
    a. RnF_4 b. $SeCl_4$ c. PCl_5 d. SF_5Cl
How many electron pairs are there (five or six) around the central atom in each species? What is the hybridization of the central atom?

10.51 Describe the geometry of each species in Question 10.50.

10.52 Describe the geometry of species in which there are, around the central atom,
    a. five bonds.   b. four bonds and one unshared pair.
    c. three bonds and two unshared pairs.

10.53 Using the information given in Question 10.23, deduce the hybridization in
    a. ClF_2^- b. XeF_3^- c. $GeCl_4^{2-}$

Molecular Orbitals

10.24 Using MO theory, list the number of electrons in each of the 2s and 2p MO's, the number of bonds, and the number of unpaired electrons in
    a. CO   b. C_2^- c. F_2^-

10.25 Consider the $+1$ ions formed by removing an electron from each of the diatomic molecules of the elements of atomic numbers 3 through 9. Using the MO approach, determine the number of bonds and unpaired electrons in each of these ions.

10.26 Using MO theory, predict the number of bonds in
    a. CN^- b. CN^+ c. F_2

10.54 Follow the directions of Problem 10.24 for
    a. O_2 b. O_2^- c. O_2^{2-}

10.55 Suppose in building up molecular orbitals the σ_{2p}^b were placed below the π_{2p}^b. Prepare a diagram similar to Table 10.3 based on this assignment. For which species would this change in relative energies affect the prediction of number of bonds? number of unpaired electrons?

10.56 Using MO theory, predict the number of bonds in
    a. OF^+ b. OF^- c. H_2^+ d. H_2^-

General

10.27 Describe how you determine
    a. whether or not a molecule is polar.
    b. the hybridization of an atom in a molecule.
    c. the number of valence electrons around the central atom in a nonmetal halide.

10.57 Describe how you determine
    a. the approximate bond angles in a molecule.
    b. the number of sigma and pi bonds in a molecule.
    c. the geometry around a central atom surrounded by six electron pairs.

10.28 Consider the $S_2O_3^{2-}$ ion, which has a skeleton similar to that of SO_4^{2-}.
  a. Draw the Lewis structure.
  b. Describe the geometry, including the approximate bond angles.
  c. Is this species a dipole?
  d. What is the hybridization for each atom?

10.29 Consider the sulfite ion, SO_3^{2-}.
  a. Draw the Lewis structure.
  b. Describe the geometry around the central sulfur atom.
  c. Is this species a dipole?
  d. What is the hybridization about the sulfur?

10.30 Complete the following table:

SPECIES	ATOMS AROUND CENTRAL ATOM X	UNSHARED PAIRS AROUND X	GEOMETRY
XY_2E_2	——	——	——
——	3	0	——
XY_4E_2	——	——	——
—	——	——	trigonal bipyramid

10.58 Follow the directions of Question 10.28 for the SF_4 molecule.

10.59 Follow the directions of Question 10.29 for the SF_6 molecule.

10.60 For each of the species in Question 10.30,
  a. give the hybridization for X.
  b. state whether the species is a dipole (assuming all Y atoms are the same).

*10.61 Draw the Lewis structure and describe the geometry of the hydrazine molecule, N_2H_4. Would you expect this molecule to be polar?

*10.62 The geometries shown in Figure 10.10 are not the only ones possible for species in which the central atom has an expanded octet and one or more pairs of unshared electrons. Suggest other possible geometries for ClF_3 and XeF_2. Can you explain, using the VSEPR model, why the geometries in Figure 10.10 are preferred?

*10.63 Consider the polyatomic ion IO_6^{5-}. How many pairs of electrons are there around the central iodine atom? What is its hybridization? Describe the geometry of the ion.

*10.64 There are two compounds with the molecular formula C_3H_6.
  a. Write the Lewis structure for each one.
  b. What is the hybridization of each carbon atom in each molecule?
  c. Describe the geometry and bond angles in the two molecules.

*10.65 The molecule XeF_6 does not have an octahedral structure even though there are six atoms bonded to xenon. Explain this and suggest a possible geometry for the molecule.

Chapter 11
Liquids
and
Solids

Chapter 6 was devoted to a study of the physical behavior of gases. We found that, at ordinary temperatures and pressures, all gases behave similarly in that they follow the Ideal Gas Law. In this chapter we will examine liquids and solids. Unfortunately, there are no simple relations analogous to the Ideal Gas Law that can be used to correlate the properties of liquids and solids. In these two states of matter, each substance has its own unique properties, quite different from those of other substances.

There is a simple explanation for this difference in behavior between gases on the one hand and liquids or solids on the other. In a liquid or solid, the particles are much closer together than they are in a gas. Indeed, the particles—which may be atoms, ions, or molecules—touch each other. As a result, attractive forces become important; they play a major role in determining the physical properties of liquids and solids. Since attractive forces between particles vary in nature and magnitude from one substance to another, different liquids and solids have quite different properties.

In studying liquids and solids, we will be concerned with two major topics:

1. The equilibria between different phases of a pure substance. We start by examining the conditions under which a liquid is in equilibrium with its vapor (Section 11.1). Then, in Section 11.2, we will consider the principles that relate to all three types of phase equilibria (liquid-vapor, solid-vapor, liquid-solid).

2. The relation between particle structure and physical properties. We will look at

—the four different structural types of solids: *ionic, molecular, network covalent,* and *metallic* (Section 11.3). Substances in each of these categories have characteristic properties that are directly related to the strength of the forces between particles.
—the nature of intermolecular forces (Section 11.4) and their effect upon the melting points and boiling points of molecular substances.

—the geometric patterns in which particles are arranged in metallic and ionic crystals (Section 11.5).

11.1
Liquid-Vapor
Equilibrium

A lake is an open container

All of us are familiar with the process of vaporization, in which a liquid is converted to a vapor. In an open container, this process continues until all the liquid is gone. Water in a beaker or a flower pot evaporates into the atmosphere, as you may have noted if you forgot to have someone water your plants when you were on vacation. In evaporation, molecular movement is mainly in one direction: molecules steadily leave the liquid and diffuse away into the air.

In a closed container, such as the flask shown in Figure 11.1, the situation is quite different. Suppose we put a small amount of a volatile liquid into the flask and stopper it. At first, the movement of molecules is primarily in one direction, from liquid to vapor. Here, however, the vapor molecules cannot escape from the container. Some of them collide with the surface and re-enter the liquid. As time passes, and the concentration of molecules in the vapor increases, so does the rate of condensation. Eventually, it becomes equal to the rate of vaporization. When this happens, we say that the liquid and vapor are in a state of *dynamic equilibrium:*

$$\text{liquid} \rightleftharpoons \text{vapor}$$

Both processes continue to occur

The double arrow implies that the forward process (vaporization) and the reverse process (condensation) are occurring at the same rate.

Vapor Pressure

Once the equilibrium shown in Figure 11.1 is reached, the concentration of molecules in the vapor does not change with time. This means that *the pressure*

FIGURE 11.1 When a liquid is placed in an evacuated flask, vaporization occurs. At first (A), molecules move in only one direction, from liquid to vapor. As the concentration of molecules in the vapor increases, condensation becomes important. The net rate of vaporization decreases (B and C). Finally (D), vaporization and condensation occur at the same rate. Equilibrium is established, and the pressure of the vapor remains constant with time.

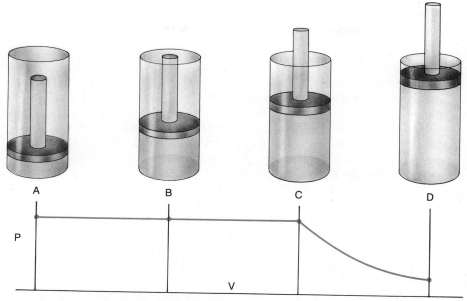

FIGURE 11.2   The pressure of a vapor in equilibrium with a liquid is independent of the volume of the container as long as there is liquid present (A and B). When all the liquid is vaporized (C), a further increase in volume (D) decreases the pressure in accordance with Boyle's Law. Since system C consists only of gas, its volume can be calculated from the Ideal Gas Law if the temperature and amount of sample are known.

exerted by the vapor over the liquid remains constant. The pressure of vapor in equilibrium with a liquid is called the **vapor pressure.** This quantity is a characteristic property of a given liquid at a particular temperature. It varies from one liquid to another, depending upon the strength of the intermolecular forces. At 25°C, the vapor pressure of water is 24 mm Hg; that of benzene at the same temperature is 92 mm Hg.

Water has stronger intermolecular forces than benzene

It is important to realize that, *as long as both liquid and vapor are present, the pressure exerted by the vapor is independent of the volume of the container.* To see what this statement means, consider Figure 11.2. This shows the vaporization of benzene at 25°C in a container whose volume can be varied by raising the piston. We start by placing a small amount of liquid benzene in the container with the piston at the level shown in Figure 11.2A. Vaporization occurs until a pressure of 92 mm Hg is established. The equilibrium is then disturbed by raising the piston to the level shown in Figure 11.2B. The pressure drops for a moment. However, liquid quickly evaporates to establish the original pressure, 92 mm Hg. This process is repeated each time the piston is raised to create a larger volume. Eventually, the container volume becomes large enough so that all the liquid vaporizes (Fig. 11.2C). From that point on, with no liquid present, the system behaves as an ideal gas and follows Boyle's Law (Fig. 11.2D).

To have liquid-vapor equilibrium, you need to have some liquid

The process just described can be reversed. If the vapor is compressed to the point where its pressure becomes equal to the vapor pressure of the liquid (point C in Fig. 11.2), liquid begins to condense. Further decrease in volume does not cause an increase in pressure. Instead, condensation continues, keeping the pressure of the vapor constant (points A and B in Fig. 11.2).

Vapor Pressure vs.
Temperature

The vapor pressure of a liquid always increases as temperature rises. Water evaporates more readily on a hot, dry day. Stoppers in bottles of volatile liquids such as ether or gasoline may pop out when the temperature rises. Unpleasant odors such as those of pyridine or butyric acid become more noticeable at higher temperatures.

We can study the effect of temperature upon vapor pressure quite simply. We confine a liquid-vapor mixture in a container of fixed volume and steadily increase the temperature. At higher temperatures, more of the molecules in the liquid have enough energy to overcome the intermolecular attractive forces and escape into the vapor. The concentration of molecules in the vapor phase increases rapidly. The pressure of vapor in equilibrium with the liquid goes up accordingly.

The vapor pressure of water, which is 24 mm Hg at 25°C, becomes 92 mm Hg at 50°C. At 100°C, the vapor pressure of water reaches 1 atm (760 mm Hg). The data for water, benzene, and ether are plotted in Figure 11.3. As you can see, the graph of vapor pressure vs. temperature is not a straight line. The slope of the curve increases steadily as we move to higher temperatures. In other words, vapor pressure increases more and more rapidly as temperature rises. This reflects the fact that there are two factors which cause vapor pressure to increase with temperature:

At 0°C the vapor pressure of water is only 4.6 mm Hg

1. As temperature increases, molecules in the vapor move more rapidly. They strike the walls more often and with greater force. As we saw in Chapter 6, this causes the pressure to increase.

2. As temperature increases, a larger fraction of molecules acquire enough energy to escape from the liquid. This increases the concentration of molecules

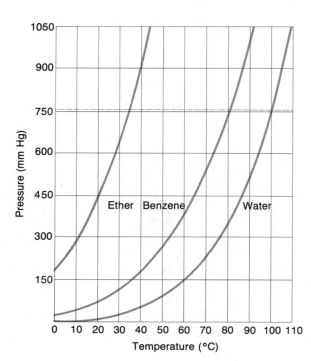

FIGURE 11.3 Effect of temperature upon the vapor pressures of ether, benzene, and water. Note that vapor pressure increases more rapidly as temperature rises. At a given temperature, vp ether > vp benzene > vp water. The normal boiling points of these three liquids (vp = 1 atm, *broken line*) are 35, 80, and 100°C.

in the vapor and hence the pressure it exerts. One can calculate that the concentration of H_2O molecules in the vapor is 25 times greater at 100°C than at 25°C. This is the reason that the vapor pressure of a liquid increases exponentially with temperature.

In working with the relationship between two variables, such as vapor pressure and temperature, scientists prefer to deal with linear (straight-line) functions. Straight-line graphs are easier to construct and to interpret. In this case, it is possible to obtain a linear function by making a simple shift in variables. Instead of plotting vapor pressure (P) vs. temperature (T), we plot the *logarithm of the vapor pressure* ($\log_{10} P$) vs. the *reciprocal of the absolute temperature* (1/T). Such a plot for water is shown in Figure 11.4; as with all other liquids, a plot of $\log_{10} P$ vs. 1/T is found to be linear.

The general equation of the straight line in Figure 11.4 is

$$\log_{10} P = A - \frac{\Delta H_{vap}}{2.30\ RT} \qquad (11.1)$$

T in this equation, as in all natural laws, is on the Kelvin scale

Here, ΔH_{vap} is the heat of vaporization of the liquid, in joules per mole, R is the gas constant and, in the units needed here, has a value of 8.31 J/mol·K. The quantity A is a constant for a particular liquid; its value need not concern us here.

For many purposes it is convenient to have a two-point equation relating the vapor pressure (P_2) at one temperature (T_2) to that (P_1) at another temperature (T_1). We obtain such an equation by applying Equation 11.1 at the two temperatures:

at T_2: $\log_{10} P_2 = A - \dfrac{\Delta H_{vap}}{(2.30)(8.31)T_2}$

at T_1: $\log_{10} P_1 = A - \dfrac{\Delta H_{vap}}{(2.30)(8.31)T_1}$

By subtracting, we eliminate the constant A:

$$\log_{10} P_2 - \log_{10} P_1 = -\frac{\Delta H_{vap}}{(2.30)(8.31)}\left[\frac{1}{T_2} - \frac{1}{T_1}\right] = \frac{\Delta H_{vap}}{(2.30)(8.31)}\left[\frac{1}{T_1} - \frac{1}{T_2}\right]$$

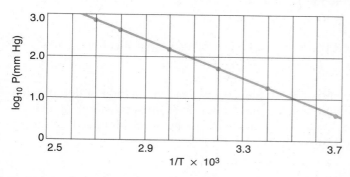

FIGURE 11.4 A plot of the logarithm of the vapor pressure of a liquid ($\log_{10} P$) vs. the reciprocal of the Kelvin temperature (1/T) is a straight line. The heat of vaporization of the liquid can be calculated from the slope:

$$\Delta H_{vap} = -2.30\ R\ \frac{\Delta \log_{10} P}{\Delta (1/T)}$$

Rearranging to a somewhat more convenient form:

This equation relates
vapor pressure at one
temperature to that at
another

$$\log_{10}\frac{P_2}{P_1} = \frac{\Delta H_{vap}}{(2.30)(8.31)}\left[\frac{T_2 - T_1}{T_2 T_1}\right]$$ (11.2)

Equation 11.2 is known as the Clausius-Clapeyron equation, honoring Rudolph Clausius and B. P. E. Clapeyron. Clausius was a prestigious nineteenth century German scientist. Clapeyron was a French engineer who first proposed a modified version of the equation in 1834. The use of the equation is illustrated in Example 11.1.

EXAMPLE 11.1 Isopropyl alcohol (rubbing alcohol) has a vapor pressure of 91 mm Hg at 40°C. Calculate its vapor pressure at 25°C, taking the heat of vaporization to be 5.31×10^4 J/mol.

Solution It is most convenient to use the subscript 2 for the *higher* temperature and pressure. With that choice, we have

$$T_2 = 40 + 273 = 313 \text{ K}; \quad T_1 = 25 + 273 = 298 \text{ K};$$
$$P_2 = 91 \text{ mm Hg}; \quad P_1 = ?$$

Substituting in the Clausius-Clapeyron equation, 11.2,

If you need to review
logarithms, see the
Appendix

$$\log_{10}\frac{P_2}{P_1} = \log_{10}\frac{91 \text{ mm Hg}}{P_1} = \frac{5.31 \times 10^4 \text{ J/mol}}{(2.30)(8.31 \text{ J/mol·K})}\left[\frac{313 \text{ K} - 298 \text{ K}}{313 \text{ K} \times 298 \text{ K}}\right]$$
$$= +0.45$$

Taking antilogs,

$$\frac{91 \text{ mm Hg}}{P_1} = 10^{+0.45} = 2.8$$

Solving for P_1,

$$P_1 = \frac{91 \text{ mm Hg}}{2.8} = 32 \text{ mm Hg}$$

This value is reasonable in the sense that lowering the temperature should reduce the vapor pressure.

EXERCISE Calculate the vapor pressure of isopropyl alcohol at 50°C. Answer: 170 mm Hg.

Equation 11.1 is the basis of a common experimental method of determining ΔH_{vap}, the heat of vaporization of a liquid. Vapor pressure is measured at a series of temperatures. The data are used to construct a straight-line plot like that shown in Figure 11.4. Compare Equation 11.1,

$$\log_{10} P = A - \frac{\Delta H_{vap}}{2.30 \text{ RT}}$$

to the general equation of a straight line,

$$y = A + Bx; \quad B = \text{slope}$$

In this case, $\log_{10} P = y$, $1/T = x$. We see that

$$\text{slope} = \frac{-\Delta H_{vap}}{2.30\,R}$$

$$\Delta H_{vap}\ (\text{J/mol}) = -(2.30)(8.31) \times \text{slope} \qquad \textbf{(11.3)}$$

$$\text{slope} = \frac{\text{change in } y}{\text{change in } x}$$

The heat of vaporization is calculated from the measured slope of the straight-line plot, using Equation 11.3. (Note that since the slope is itself negative, ΔH_{vap} is calculated to be a positive quantity, as it should be.)

Boiling Point

When we heat a liquid in an open container bubbles form, usually at the bottom, where heat is applied. The first small bubbles we see are air, driven out of solution by the increase in temperature. Eventually, at a certain temperature, large vapor bubbles form throughout the liquid. These vapor bubbles rise to the surface and break. When this happens, we say that the liquid is boiling.

No bubbles, no boiling

    The temperature at which a liquid boils depends upon the pressure above it. To understand why this is the case, consider Figure 11.5. This shows vapor bubbles rising in a boiling liquid. For a vapor bubble to form, the pressure within it, P_1, must be at least equal to the pressure above it, P_2. Since P_1 is simply the vapor pressure of the liquid, we conclude that **a liquid boils at a temperature at which its vapor pressure becomes equal to the pressure above its surface.** If this pressure is 1 atm (760 mm Hg), we refer to the temperature as the **normal boiling point.** (When the term "boiling point" is used without qualification, normal boiling point is implied.) The normal boiling point of water is 100°C; its vapor pressure is 760 mm Hg at that temperature.

    As you might expect, the boiling point of a liquid can be reduced by lowering the pressure above it. Water can be made to boil at 25°C by evacuating the space above it. When a pressure of 24 mm Hg, the equilibrium vapor pressure at 25°C, is reached, the water starts to boil. Chemists often take advantage of this effect in purifying a high-boiling compound that might decompose or oxidize at its normal

If $P_1 < P_2$, no bubbles can form

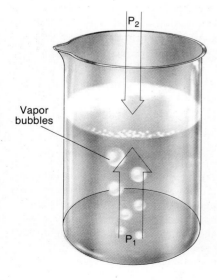

Vapor bubbles

FIGURE 11.5 A liquid boils when its vapor pressure (P_1) slightly exceeds the pressure above it (P_2).

boiling point. They boil it at a reduced temperature under vacuum and condense the vapor.

If you have been fortunate enough to camp in the high Sierras or the Rockies, you may have noticed that it takes longer to boil foods at high altitudes. The reduced pressure lowers the temperature at which water boils in an open container. This slows down the physical and chemical changes that take place when foods like potatoes or eggs are cooked. In principle, this problem can be solved by using a pressure cooker. In that device, the pressure that develops is high enough to raise the boiling point of water above 100°C. Pressure cookers are indeed used in places like Cheyenne, Wyoming (elevation 1900 m), but not by mountain climbers, who have to carry all their equipment on their backs.

It's hard enough as it is

Critical Temperature and Pressure

Consider an experiment in which liquid benzene is placed in a previously evacuated glass tube (Fig. 11.6). The tube is then sealed and heated to higher and higher temperatures. The pressure of the vapor rises steadily. It is 1 atm at 80°C, 14 atm at 200°C, and 43 atm at 280°C. Nothing spectacular happens (unless there happens to be a weak spot in the tube) until we reach 289°C, at which point the vapor pressure is 48 atm. Suddenly, as we pass this temperature, the meniscus between the liquid and vapor disappears! The tube now contains only one phase, benzene vapor.

We find that it is impossible to have liquid benzene at temperatures above 289°C, regardless of how much pressure is applied. Even at pressures as high as 1000 atm, benzene vapor does not liquefy at 290 or 300°C. This behavior is typical of all substances. There is a temperature, called the **critical temperature,** above which the liquid phase of a pure substance cannot exist. The pressure that must be applied to cause condensation at that temperature is called the *critical pressure*. Quite simply, the critical pressure is the vapor pressure of the liquid at the critical temperature.

The critical pressure is the maximum possible vapor pressure for a liquid

Table 11.1 lists the critical temperatures of several common substances. The species in the column at the left all have critical temperatures below 25°C. They are often referred to as "permanent gases." Applying pressure at room temperature will not condense a permanent gas. It must be cooled as well. The permanent gases are stored and sold under high pressures, often 150 atm or greater. When the valve on the cylinder is opened, gas escapes and the pressure drops in accordance with the Ideal Gas Law.

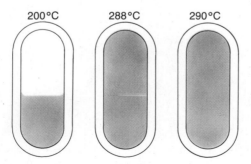

200°C   288°C   290°C

FIGURE 11.6 When a sample of benzene is heated in a sealed tube, the properties of the liquid and vapor approach one another near the critical temperature, 289°C. As one heats the sample through that temperature the meniscus disappears.

Table 11.1
Critical Temperatures (°C)

"PERMANENT GASES"		"CONDENSABLE GASES"		"LIQUIDS"	
Helium	−268	Carbon dioxide	31	Ethyl ether	194
Hydrogen	−240	Ethane	32	Ethyl alcohol	243
Nitrogen	−147	Propane	97	Benzene	289
Argon	−122	Ammonia	132	Bromine	311
Oxygen	−119	Chlorine	144	Water	374
Methane	−82	Sulfur dioxide	158		

The first liquefaction of helium was no mean feat

The gases listed in the center column of Table 11.1 have critical temperatures above 25°C. They are available commercially as liquids in high-pressure cylinders. When the valve on a cylinder of propane is opened, the gas that escapes is replaced by vaporization of liquid. The pressure quickly returns to its original value. Only when the liquid is completely vaporized does the pressure drop as gas is withdrawn (recall Fig. 11.2). This indicates that almost all of the propane is gone, and it's time to order a new tank.

11.2
Phase Diagrams

In Section 11.1, we discussed several features of the equilibrium between a liquid and its vapor. For a pure substance, at least two other types of phase equilibria need be considered: the equilibrium between a solid and its vapor, which is similar in general to liquid-vapor equilibrium, and that between solid and liquid at the melting (freezing) point. Many of the important relations in all these equilibria can be shown in a **phase diagram.** A phase diagram is a graph that shows the pressures and temperatures at which different phases are in equilibrium with each other. The phase diagram of water is shown in Figure 11.7, p. 330. The diagram at the left covers a wide temperature range; it is not drawn to scale. The graph at the right of Figure 11.7 covers a much smaller range, −5 to 30°C.

To understand what a phase diagram implies, consider first the three lines AB, AC, and AD on the graph at the right of Figure 11.7. Each of these lines tells us the pressures and temperatures at which two adjacent phases are in equilibrium:

1. Line AB is a portion of the vapor pressure-temperature curve of liquid water. At any temperature and pressure along this line, liquid water is in equilibrium with water vapor. From the curve we see that at point A, these two phases are in equilibrium at 0°C and about 5 mm Hg (more exactly, 0.01°C and 4.56 mm Hg). At B, corresponding to 25°C, the pressure exerted by the vapor in equilibrium with liquid water is about 24 mm Hg. From the diagram at the left of Figure 11.7 we see that, when line AB is extended, the equilibrium pressure of water vapor becomes 760 mm Hg at 100°C, the normal boiling point of water. The line ends at 374°C, the critical temperature of water, where the pressure is 218 atm.

Only if the state of a system lies on a line can we have two phases in equilibrium

2. Line AC represents the vapor pressure curve of ice. At any point along this line, such as −3°C and 3 mm Hg (point C) or 0°C and 5 mm Hg (point A), ice and vapor are in equilibrium with each other.

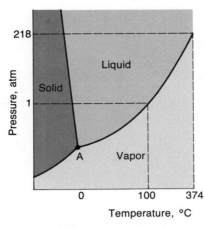
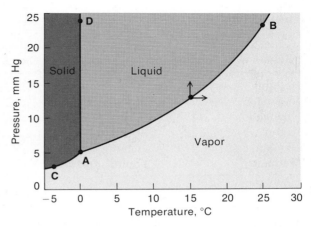

FIGURE 11.7 Phase diagram of water. An extended phase diagram (not to scale) is shown at the left. The graph at the right enlarges the portion of the phase diagram from −5 to 30°C. The triple point is at A.

3. Line AD gives the temperatures and pressures at which liquid water is in equilibrium with ice.

Point A on the phase diagram is the only one at which all three phases, liquid, solid, and vapor, are in equilibrium with each other. It is called the **triple point.** For water, the triple point temperature is 0.01°C. At this temperature liquid water and ice have the same vapor pressure, 4.56 mm Hg.

In the three areas of the phase diagram labeled "solid," "liquid," and "vapor," only one phase is present. To understand this, consider what happens to an equilibrium mixture of two phases when the pressure or temperature is changed. Suppose we start at the point on AB indicated by a filled circle. Here liquid water and vapor are in equilibrium at 15°C and 13 mm Hg. If we increase the pressure on such a mixture, condensation should occur. The phase diagram confirms this: by increasing the pressure at 15°C (*vertical arrow*), we move up into the liquid region. In another experiment, we might hold the pressure constant but increase the temperature. This change should cause the liquid to vaporize. The phase diagram tells us that this is indeed what happens: an increase in temperature (*horizontal arrow*) shifts us into the vapor region.

A system whose state lies in the Liquid region is all liquid

EXAMPLE 11.2 Consider a sample of H_2O at point D in Figure 11.7.
a. What phase(s) is (are) present?
b. If the temperature of the sample were reduced at constant pressure, what would happen?
c. How would you convert the sample to vapor without changing the temperature?

Solution
a. Point D is on the solid-liquid equilibrium line. Ice and liquid water are present.
b. This corresponds to moving horizontally to the left from point D. The sample freezes completely to ice.
c. Reduce the pressure to below the triple point value, perhaps to 4 mm Hg.

EXERCISE What phase(s) is (are) present at 15 mm Hg and 5°C? Answer: only $H_2O(l)$.

The phase diagrams of other substances resemble that of water; however, they differ in such features as the triple point temperature and the orientation of the solid-liquid line (AD in Fig. 11.7). Differences of this type cause such substances as iodine, carbon dioxide, and benzene to behave quite differently from water when they undergo phase changes.

Sublimation

The process by which a solid changes directly to vapor without passing through the liquid phase is called sublimation. A solid can sublime only at temperatures below the triple point; above that temperature it will melt to liquid (Fig. 11.7). At temperatures below the triple point, a solid can be made to sublime by reducing the pressure of the vapor above it to less than the equilibrium value. To illustrate what this means, consider the conditions under which ice sublimes. This happens on a cold, dry, winter day when the temperature is below 0°C and the pressure of water vapor in the air is less than the equilibrium value (4.56 mm Hg at 0°C). The rate of sublimation can be increased by evacuating the space above the ice. This is how foods are freeze-dried. The food is frozen, put into a vacuum chamber, and evacuated to a pressure of 1 mm Hg or less. The ice crystals formed upon freezing sublime, which leaves a product whose mass is only a fraction of that of the original food.

We lose a lot of snow in Minnesota by sublimation

Iodine sublimes more readily than ice because its triple point pressure, 90 mm Hg, is much higher. Iodine sublimes upon heating in an open test tube (Color Plate 11.1) below the triple point temperature, 114°C. If we exceed the triple point, the solid melts. No such problem arises with solid carbon dioxide (dry ice). It has a triple point pressure above 1 atm (5.2 atm at -57°C). Liquid carbon dioxide cannot be made by heating dry ice in an open container. No matter what we do, solid CO_2 passes directly to the vapor at 1 atm pressure.

How could we make liquid CO_2?

Fusion

For a pure substance, the **melting point** is identical to the **freezing point**. It represents the temperature at which solid and liquid phases are in equilibrium. Melting points are usually measured in an open container, i.e., at atmospheric pressure. For most substances, the melting point at 1 atm (the "normal" melting point) is virtually identical with the triple point temperature. For water, the difference is only 0.01°C.

MP = FP

Although the effect of pressure upon melting point is very small, we are often interested in its direction. To decide whether the melting point will be increased or decreased by compression, we apply a simple principle: **an increase in pressure favors the formation of the more dense phase.** We can distinguish between two types of behavior (Fig. 11.8, p. 332):

1. *The solid is the more dense phase* (Fig. 11.8A). The solid-liquid equilibrium line is inclined to the right, shifting away from the y-axis as it rises. At higher pressures, the solid becomes stable at temperatures above the normal melting point. In other words, the melting point is raised by an increase in pressure. This behavior is shown by most substances.

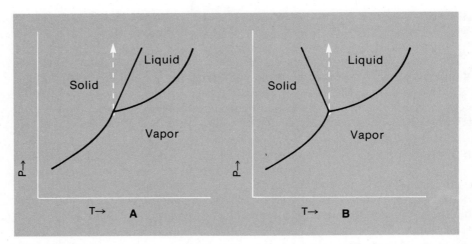

FIGURE 11.8 Effect of pressure on the melting point of a solid. (A) When the solid is the more dense phase, an increase in pressure converts liquid to solid; the melting point increases. (B) If the liquid is the more dense phase, an increase in pressure converts solid to liquid, and the melting point decreases.

2. *The liquid is the more dense phase* (Fig. 11.8B). The liquid-solid line is inclined to the left, toward the *y*-axis. An increase in pressure favors the formation of liquid; that is, the melting point is decreased by raising the pressure. Water is one of the few substances that behaves this way; ice is less dense than liquid water. The effect is exaggerated for emphasis in Figure 11.8B. Actually, an increase in pressure of 134 atm is required to lower the melting point of ice by 1°C.

This effect can easily break a water pipe if it gets too cold

11.3
Structural Types of Solids

Crystalline solids consist of a regular array of structural units, which may be atoms, molecules, or ions. The physical properties of a solid depend upon the nature of its structural units and the forces between them. There are four common types of solids, classified according to particle structure:

—*ionic* compounds (such as NaCl and KBr), built of cations and anions.
—*molecular* species (such as I_2 and H_2O), consisting of discrete molecules.
—*network covalent* materials (such as SiO_2, C), in which nonmetal atoms are connected by a continuous network of covalent bonds. These substances contain no small, discrete molecules.
—*metallic* substances (such as Na and Fe), consisting of metal cations surrounded by mobile electrons.

Ionic Solids

An ionic solid consists of cations and anions, for example, K^+, Br^-. No simple, discrete molecules are present in KBr or other ionic compounds; rather, the ions are held in a regular, repeating arrangement by strong ionic bonds, electrostatic interactions between oppositely charged ions. Because of this structure, ionic solids show the following properties:

1. Ionic solids are nonvolatile and high-melting (typically 600 to 2000°C). Ionic bonds must be broken to melt the solid, separating oppositely charged ions from each other. Only at high temperatures do the ions acquire enough kinetic energy for this to happen.

If the MP is 200°C, it's not ionic

2. Ionic solids do not conduct electricity because the charged ions are fixed in position. They become good conductors, however, when melted or dissolved in water. In both cases, the melt or solution, the ions (such as K^+ and Br^-) are free to move through the liquid and thus can conduct an electric current.

3. Many, but not all, ionic compounds (for example, $CaCl_2$ but not $CaCO_3$) are soluble in water, a polar solvent. In contrast, ionic compounds are insoluble in nonpolar solvents such as benzene (C_6H_6) or carbon tetrachloride (CCl_4). In Chapter 18, we will consider the water solubilities of ionic compounds in greater detail.

Molecular Substances

Many of the most familiar substances in the world around us and in the laboratory are molecular; that is, they consist of molecules in which a small group of atoms are joined to one another by covalent bonds. Water, our most common compound, is molecular, as are nitrogen, oxygen, iodine, carbon dioxide, and many other nonmetallic elements and compounds. Virtually all simple organic compounds, including benzene, ethyl alcohol, and sugar, consist of molecules.

Just about any substance that is a liquid or gas at 25°C is molecular

Molecular substances typically tend to be:

1. *Nonconductors of electricity when pure* (liquids or solids). Since molecules are uncharged, they cannot carry an electric current. In most cases, including iodine and ethyl alcohol, water solutions of molecular substances are nonconductors. A few polar molecules, including HCl, react with water to form ions:

$$HCl(g) \rightarrow H^+(aq) + Cl^-(aq)$$

and hence produce a conducting water solution.

2. *Insoluble in water but soluble in nonpolar solvents such as* CCl_4 *or benzene*. Iodine is only slightly soluble in water (0.0013 mol/L at 25°C); the water solubility of naphthalene is even lower (0.0002 mol/L at 25°C). Both of these molecular species are much more soluble in benzene (0.48 mol/L for iodine, 4.6 mol/L for naphthalene). A few molecular substances, including ethyl alcohol, are very soluble in water. As we will see later (Section 11.4) such substances have intermolecular forces similar to those of water.

3. *Volatile, with appreciable vapor pressures at room temperature.* This accounts for the odors we associate with such molecular substances as H_2S (rotten eggs), ethyl butyrate (oil of pineapples), and butyric acid (rancid butter and sweaty socks). It also explains why many molecular solids, notably naphthalene (mothballs) and iodine, can be purified by sublimation (Color Plate 11.1).

4. *Low melting and boiling points.* Many molecular substances are gases at 25°C and 1 atm (for example, N_2, O_2, and CO_2), which means that they have boiling points below 25°C. Others (such as H_2O and benzene) are liquids with melting (freezing) points below room temperature. Of the molecular substances that are

Molecules don't stick together very tightly

solids at ordinary temperatures, most melt around 100°C (mp I_2 = 114°C, naphthalene = 80°C). The upper limit for melting and boiling points of molecular substances is about 300°C.

The generally low melting and boiling points of molecular substances reflect the fact that the forces between molecules (intermolecular forces) are weak. *All we need do to melt or boil a molecular substance is to set the molecules free from one another.* This requires only that we supply enough energy to overcome the attractive forces between molecules. The forces within a molecule (intramolecular forces) are strong, since they are covalent bonds. These bonds remain intact when a molecular substance melts or boils.

Intermolecular forces are weaker than intramolecular forces, but intercollegiate teams are stronger than intramural teams

Network Covalent Solids

Covalent bond formation need not, and often does not, lead to small, discrete molecules. It can lead instead to structures of the type shown in Figure 11.9. Here, all the atoms are held together by a network of electron-pair bonds. Substances with this type of structure are always solids at 25°C; they are referred to as *network covalent solids,* or, sometimes, as macromolecular. The entire crystal, in effect, consists of one huge molecule.

Network covalent solids have certain characteristic properties. They are

1. *High melting, often with melting points above 1000°C.* To melt a macromolecular solid, covalent bonds between atoms must be broken. In this respect, solids of this type differ markedly from molecular solids, which have much lower melting points.

2. *Insoluble in all common solvents.* For solution to occur, covalent bonds throughout the solid would have to be broken.

3. *Poor electrical conductors.* In a typical network covalent substance, there are no charged particles to carry a current.

Table 11.2 lists the melting points of some typical network covalent substances. You will note the list includes elements (C, Si, Ge) as well as compounds. In Chapter 13, we will look at the structures of the network covalent crystals of the Group 4 nonmetals. Here, we will consider a typical network covalent compound, silicon dioxide.

Diamond is a network covalent solid

Quartz, the most common form of SiO_2, is the main component of sea sand. In quartz, each silicon atom bonds tetrahedrally to four oxygen atoms. Each oxygen atom bonds to two silicon atoms, thus linking adjacent tetrahedra to one another (Fig. 11.10). Note that no discrete SiO_2 molecules are present; the network of covalent bonds extends throughout the entire crystal. The strong Si—O bond, with a bond energy of 368 kJ/mol, must be broken to melt quartz; thus, quartz has a very high melting point, about 1700°C. Unlike most solids, quartz does not melt sharply to a liquid. Instead, it turns to a viscous mass over a wide temperature range, first softening at about 1400°C. The viscous fluid probably contains long

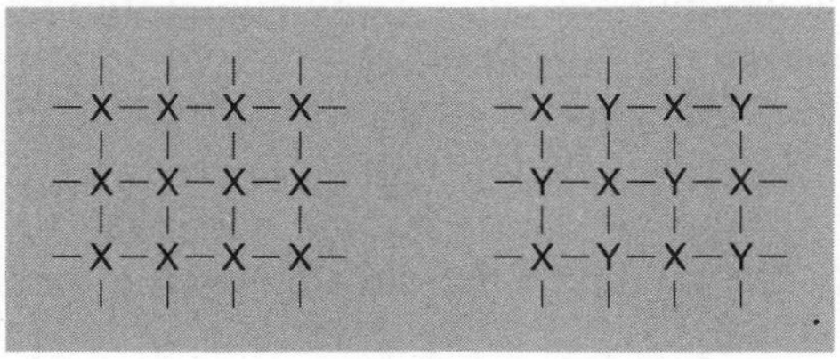

FIGURE 11.9 Two-dimensional representation of network covalent crystals of an element X or a compound XY. Because there is a continuous network of covalent bonds, there are no small, discrete molecules.

Table 11.2
Melting Points of Network Covalent Substances

	mp (°C)		mp (°C)
C	3570	SiC	>2700
Si	1414	BN	3500
Ge	937	SiO_2	1700

—Si—O—Si—O— chains, with enough bonds broken so that the material can flow.

Ordinary glass is made by heating a mixture of sand, limestone ($CaCO_3$), and soda ash (Na_2CO_3) to the melting point. Carbon dioxide is given off from the hot mass, which contains the oxides of silicon, sodium, and calcium in a mole ratio of about 7:1:1. The "soft" glass produced this way softens at about 600°C and has a wider range of high viscosity than quartz. Hard glass, called Pyrex or Kimax, is made from a melt of silicon, boron, aluminum, sodium, and potassium oxides. It is much superior to soft glass in withstanding chemical attack and thermal shock. Hard glass softens at about 800°C and is readily worked in the flame of a gas-oxygen torch. Nearly all the glassware used in chemical equipment nowadays is made of hard glass.

Small amounts of various substances are often added to glass to produce different colors. Chromium(III) oxide (Cr_2O_3) gives a green glass, CoO a blue glass, and MnO_2 a violet glass. Milky white glass contains calcium fluoride, CaF_2; SnO_2 produces an opaque glass. Eyeglasses that darken in the sunlight contain small amounts of white, finely dispersed silver chloride, AgCl. Exposure to sunlight converts some of the Ag^+ ions to metallic silver, which is black. The reaction is reversible; in a dark room, AgCl is re-formed and the glass becomes clear again.

Pyrex and photogray glass were invented by chemists at the Corning Glass Works

O   Si

FIGURE 11.10 Crystal structure of quartz SiO_2. In SiO_2 every silicon atom (*black*) is linked tetrahedrally to four oxygen atoms.

Metals; Electron-Sea Model

The unique properties of metals—malleability, ductility, and high thermal and electrical conductivities—were described in Chapter 8. These properties suggest a structure in which electrons are relatively mobile. Only if this is true can we explain why metals are good conductors.

Electron mobility can be explained by a simple model of bonding in metals, known as the **electron-sea model.** The metallic lattice is pictured as an array of positive ions (metal atoms minus their valence electrons). These are anchored in position, like buoys in a mobile "sea" of electrons. These electrons are "delo-calized"; that is, they are not attached to any particular positive ion, but rather can wander throughout the lattice. Figure 11.11 shows what a tiny portion of a metal crystal looks like according to the electron-sea model.

This simple picture of metallic bonding offers an obvious explanation of the high electrical conductivity of metals. It can also be used to explain many of the other properties of metals. High thermal conductivity is explained by assuming that heat is carried through the metal by collisions between electrons, which occur frequently. Since electrons are not restricted to a particular bond, they can absorb and re-emit light over a wide wavelength range. This explains why metal surfaces are excellent reflectors.

The harder metals have atoms with several va-lence electrons

According to the electron-sea model, the strength of the metallic bond is directly related to the charge of the positive ions in the lattice. Looking at Figure 11.11, we would expect the "lattice energy" of magnesium (+2 ions) to be greater than that of sodium (+1 ions). This agrees with the fact that magnesium has a higher melting point than sodium (650°C vs. 98°C). In general, the melting points of metals cover a wide range, from −40°C for mercury to 3380°C for tungsten. This suggests that the strength of the metallic bond varies greatly from one metal to another.

In addition to those we have mentioned, metals as a class have the following properties. They are

1. *Nonvolatile.* Mercury is the only metal whose volatility is of concern at ordinary temperatures. Its vapor pressure is very low (2×10^{-6} atm at 25°C,

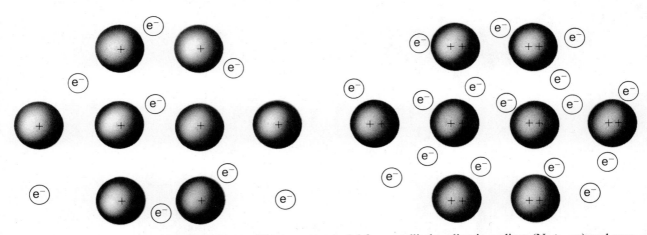

FIGURE 11.11 Electron-sea model for metallic bonding in sodium (Na^+, e^-) and magnesium (Mg^{2+}, $2e^-$). The hardness of metals increases with the number of electrons available for metallic bonding.

4×10^{-4} atm at 100°C); however, mercury vapor is toxic at the parts-per-million level. That is why your instructor will insist that you clean up any mercury spilled in the laboratory.

2. *Insoluble in water and other common solvents.* No metals "dissolve" in water in the true sense. As we saw in Chapter 8, a few very active metals react chemically with water to form hydrogen gas. Liquid mercury dissolves many metals, forming solutions called amalgams. Perhaps the most familiar amalgams are those of silver and gold. These are formed when ores of these metals are extracted with mercury. An Ag-Sn-Hg amalgam is used in filling teeth.

Much of what we have said about the properties of the four structural types of substances is summarized in Table 11.3 and Example 11.3.

Without silver amalgam, dentists would be in big trouble, and so would we

Table 11.3
Types of Substances: Properties as Related to Structure

STRUCTURAL UNITS	FORCES WITHIN UNITS	FORCES BETWEEN UNITS	PROPERTIES	EXAMPLES
Ions	—	Ionic bond	High mp. Conductors in molten state or water solution. Usually soluble in water, insoluble in organic solvents.	NaCl MgO
Molecules a. Nonpolar	Covalent bond	Dispersion (see Section 11.4)	Low mp, bp; often gas or liquid at 25°C. Nonconductors. Insoluble in water, soluble in organic solvents.	H_2 CCl_4
b. Polar	Covalent bond	Dispersion, dipole, H bond	Similar to nonpolar but generally higher mp and bp, more likely to be water soluble.	HCl NH_3
Atoms	Covalent bond	—	Hard, very high-melting solids. Nonconductors. Insoluble in common solvents.	C SiO_2
Cations, mobile electrons	—	Metallic bond	Variable mp. Good conductors in solid. Insoluble in common solvents.	Na Fe

EXAMPLE 11.3 A certain substance is a liquid at room temperature and is insoluble in water. Suggest which of the four types of basic structural units is present in the substance and list additional experiments that could be carried out to confirm your prediction.

Solution   The fact that the substance is a liquid suggests that it is probably molecular. The fact that it is insoluble in water agrees with this classification. To confirm that it is indeed molecular, we might measure the conductivity of the liquid, which should be essentially zero. It should also be soluble in most organic solvents.

EXERCISE   A certain solid is high melting (above 1000°C) and is a nonconductor. Which type of substance is it likely to be? How could you make a final decision as to the type of substance? Answer: It could be ionic or network covalent; test the conductivity of the melt.

11.4
Intermolecular Forces

As we have seen, the generally low melting and boiling points of molecular substances reflect the weakness of attractive forces between molecules. In this section, we will look at the nature of these forces and how they affect the properties of molecular substances. In general, we expect to find a correlation between the strength of intermolecular forces and the temperatures at which molecular substances melt or boil: the stronger these forces are, the higher will be the melting point or boiling point.

This should be obvious. Is it?

Trends in Melting and Boiling Points

Of the several factors that influence the physical properties of molecular substances, the most obvious is molar mass (MM). Both melting point and boiling point tend to increase with molar mass. Compare, for example, the four halogens (Group 7 elements) listed in Table 11.4. There is a steady increase in both melting point and boiling point as we go from F_2 (MM = 38 g/mol) to I_2 (MM = 254

Table 11.4
Effect of Molar Mass on Melting Point and Boiling Point of Molecular Substances

	HALOGENS			
	F_2	Cl_2	Br_2	I_2
MM (g/mol)	38	71	160	254
mp (°C)	−220	−101	−7	114
bp (°C)	−188	−34	59	184

	HYDROCARBONS (STRAIGHT CHAIN)								
	CH_4	C_2H_6	C_3H_8	C_4H_{10}	C_5H_{12}	C_6H_{14}	C_8H_{18}	$C_{16}H_{34}$	$C_{20}H_{42}$
MM (g/mol)	16	30	44	58	72	86	114	226	282
mp (°C)	−184	−172	−188	−135	−130	−94	−56	20	38
bp (°C)	−162	−88	−42	0	36	69	126	288	345

Table 11.5
Boiling Points of Nonpolar vs. Polar Substances

NONPOLAR			POLAR		
Formula	MM (g/mol)	bp (°C)	Formula	MM (g/mol)	bp (°C)
N_2	28	−196	CO	28	−192
SiH_4	32	−112	PH_3	34	−85
GeH_4	77	−90	AsH_3	78	−55
Br_2	160	59	ICl	162	97

g/mol). The same trend is evident with the hydrocarbons at the bottom of the table. As molar mass increases, we go from gases at 25°C and 1 atm (CH_4 through C_4H_{10}) to liquids (C_5H_{12} through $C_{16}H_{34}$), to low-melting, paraffin-like solids.

Another factor that affects the volatility of molecular substances is their polarity. Polar species melt and boil at slightly higher temperatures than nonpolar substances of comparable molar mass (Table 11.5).

The effect of polarity is usually small enough to be obscured by differences in molar mass. For example, in the series HCl → HBr → HI, boiling point increases steadily with molar mass. This happens despite the fact that polarity is decreasing as we move from HCl (most polar) to HI (least polar). We find, however, that when hydrogen is bonded to a small, highly electronegative atom (N, O, or F), polarity has a much greater effect on boiling point. Hydrogen fluoride, HF, despite

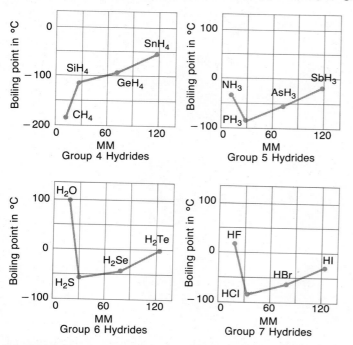

FIGURE 11.12 Boiling points of hydrogen compounds of elements in Groups 4 through 7. Among the Group 4 hydrides, boiling point increases steadily with molar mass. In Groups 5, 6, and 7, the first member (NH_3, H_2O, and HF) has an abnormally high boiling point.

its low molar mass (20 g/mol), has the highest boiling point of all the hydrogen halides. Water (MM = 18 g/mol) and ammonia (MM = 17 g/mol) also have abnormally high boiling points (Fig. 11.12). In these cases, the effect of polarity reverses the normal trend expected from molar mass alone.

Dipole Forces

As pointed out in Chapter 10, polar molecules have a net dipole, i.e., a center of positive charge separated from a center of negative charge. The effect of polarity on physical properties can be explained in terms of the way in which these molecular dipoles align. Adjacent polar molecules line up so that the negative end of the dipole on one molecule is as close as possible to the positive end of its neighbor (Fig. 11.13). Under these conditions, there is an electrostatic attraction between adjacent molecules. This attractive force between polar molecules is called a *dipole force*.

	BP
ICl	97°C
NaCl	1400°C

The dipole forces between ICl molecules are similar in origin to those between Na^+ and Cl^- ions in NaCl; however, the dipole forces in ICl are much weaker than the ionic bonds in NaCl. The unequal electronegativities of iodine (2.5) and chlorine (3.0) produce only partial plus and minus charges in ICl. In NaCl, a complete transfer of electrons generates ions with full plus and minus charges.

When iodine chloride is heated to 27°C, the rather weak dipole forces are unable to keep the molecules rigidly aligned and the solid melts. Dipole forces are still important in the liquid state, since the polar molecules remain close to one another. Only in the gas, where the molecules are far apart, do dipole forces become negligible. Hence, boiling points as well as melting points of polar compounds such as ICl are somewhat higher than those of nonpolar substances of comparable molar mass.

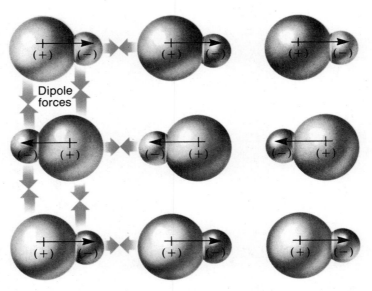

FIGURE 11.13 Dipole forces in the ICl crystal. The (+) and (−) indicate fractional charges on the I and Cl atoms in the polar molecules. The existence of these partial charges causes the molecules to line up in the pattern shown. Adjacent molecules are attracted to each other by dipole forces between the (+) pole of one molecule and the (−) pole of the other.

Hydrogen Bonds

The unusually high boiling points of HF, H_2O, and NH_3 (Fig. 11.12) result from an unusually strong type of dipole force called a *hydrogen bond*. The hydrogen bond is a force exerted between an H atom attached to an F, O, or N atom in one molecule and an unshared electron pair on the F, O, or N atom of another molecule:

X—H -- - - - :X—H   X = N, O, or F

    hydrogen bond

Hydrogen bonds are much stronger than ordinary dipole forces

There are two reasons that hydrogen bonds are stronger than ordinary dipole forces:

1. The difference in electronegativity between hydrogen (2.1) and fluorine (4.0), oxygen (3.5), or nitrogen (3.0) is quite large. It causes the bonding electrons in HF, H_2O, and NH_3 to be markedly displaced from the hydrogen atom. The hydrogen atom, insofar as its interaction with a neighboring molecule is concerned, behaves almost like a bare proton. The hydrogen bond is strongest in HF, where the difference in electronegativity is greatest. It is weakest in NH_3, where the difference in electronegativity is relatively small.

In HF(l), the molecules are hooked together in chains by H bonds

2. The small size of hydrogen allows the unshared electron pair of an F, O, or N atom of one molecule to approach the H atom in another very closely. It is significant that hydrogen bonding occurs only with these three elements. All of them have small atomic radii. The larger chlorine and sulfur atoms, with electronegativities (3.0, 2.8) similar to nitrogen, do not form hydrogen bonds in such compounds as HCl and H_2S.

Hydrogen bonds can exist in many molecules other than HF, H_2O, and NH_3. The basic requirement is simply that hydrogen be bonded to fluorine, oxygen, or nitrogen (Example 11.4).

EXAMPLE 11.4 Show, using a broken line (- - - -), how hydrogen bonds can form between adjacent molecules of

a. $CH_3—CH_2—\ddot{O}$, ethyl alcohol   b. $CH_3—\ddot{N}$, methyl amine
              H

Solution

a. The hydrogen bond joins the oxygen atom of one molecule to the hydrogen that is bonded to oxygen in the second molecule:

   $CH_3—CH_2—\ddot{O}$
                 H
         H - - - - :$\ddot{O}$:
            $CH_2—CH_3$

b. Proceeding as in (a),

        H
  $CH_3—\ddot{N}$
           H
     H - - - - :N—CH_3
           H

EXERCISE Would you predict hydrogen bonding in acetic acid,

$$CH_3—C—OH? \text{ in acetone, } CH_3—C—CH_3? \text{ Answer: In acetic acid but not}$$

in acetone.

Water has many unusual properties in addition to its high boiling point. Its heat of vaporization, 2257 J/g at 100°C, is greater than that of any other common liquid. Here again, hydrogen bonding is responsible. A large amount of energy has to be absorbed to separate the closely packed molecules in the liquid and break hydrogen bonds between them.

Perhaps the most unusual property of water is its decrease in density upon freezing. Ice is one of the very few solids that has a density less than that of the liquid from which it is formed (d ice at 0°C = 0.917 g/cm³, d water at 0°C = 1.000 g/cm³). This behavior is an indirect result of hydrogen bonding. When water freezes to ice, an open hexagonal pattern of molecules results (Fig. 11.14). Each oxygen atom in an ice crystal is bonded to four hydrogens. Two of these are attached by ordinary covalent bonds at a distance of 0.099 nm. The other two involve hydrogen bonds 0.177 nm in length. The large proportion of "empty space" in the ice structure explains why ice is less dense than water. Indeed, water starts to decrease in density if cooled below 4°C. This implies that the formation of an open structure occurs over a temperature range rather than taking place solely at the freezing point. It appears that even in water at room temperature some of the molecules are in an open, ice-like pattern. More and more molecules assume this pattern as the temperature drops. Below 4°C, the transition to the open structure outweighs the normal contraction upon cooling. Thus, water expands as its temperature drops below 4°C.

The fact that water has its greatest density at 4°C has profound effects upon our environment. In winter, at the bottoms of lakes and rivers, there is a layer of water at this temperature in which fish and other marine life find food and oxygen. Ice collects at the surface. If ice were more dense than water, it would sink to the bottom. This would make it easier for lakes or rivers to freeze completely, thus destroying marine life.

The hydrogen bonds in H₂O are about twice as long as the O—H bonds

All in all, it's best that ice floats

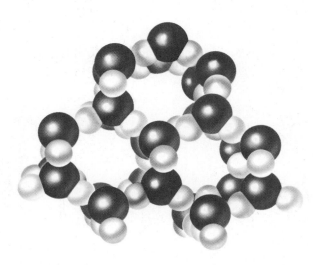

FIGURE 11.14 Crystal structure of ice. Each oxygen atom (*large sphere*) of an H₂O molecule forms hydrogen bonds with H atoms (*small spheres*) in two adjacent molecules. This leads to a hexagonal pattern with a large amount of empty space, which accounts for the low density of ice.

Dispersion Forces*

The fact that boiling point ordinarily increases with molar mass implies a type of intermolecular force, common to all substances, whose magnitude increases with molar mass. This is known as a *dispersion force*. Its origin is more difficult to visualize than that of a dipole force or hydrogen bond. Like them, it is basically electrical in nature. Dispersion forces, however, are due to what we might call temporary rather than permanent dipoles.

All molecules are attracted to one another by dispersion forces

On the average, electrons in a nonpolar molecule such as H_2 are as close to one nucleus as the other. However, at a given instant, the electron cloud may be concentrated at one end of the molecule (position 1A in Fig. 11.15). A fraction of a second later it may be at the opposite end of the molecule (position 1B). The situation is similar to that of a person watching a tennis match from a position directly in line with the net. At one instant his eyes are focused on the player to his left. A moment later, they shift to the player on his right. Over a period of time, he looks to one side as often as the other. The "average" position of focus of his eyes is straight ahead.

The momentary concentration of the electron cloud on one side or the other sets up a temporary dipole in the H_2 molecule. This in turn induces a similar dipole in an adjacent molecule. When the electron cloud in the first molecule is at 1A, the electrons in the second molecule are attracted to 2A. As the first electron cloud shifts to 1B, the electrons of the second molecule are pulled back to 2B. These temporary dipoles, both in the same direction, lead to an attractive force between the molecules. This is the dispersion force.

Relative strengths of intermolecular forces: H bonds > dispersion forces > dipole forces

The strength of dispersion forces depends upon how readily electrons in a molecule can be moved about, or *polarized*. As one might expect, the ease of polarization depends upon molecular size. Large molecules in which the electrons are far removed from the nuclei, are relatively easy to polarize. Small, compact

---

*These forces are sometimes referred to as van der Waals forces. Properly speaking, van der Waals forces include all the intermolecular forces referred to in this section, as well as one other, the force between a dipole in one molecule and an induced dipole in an adjacent molecule. It is the sum of all these forces that determines the magnitude of the deviation of real gases from ideal behavior (cf. van der Waals equation, Chap. 6).

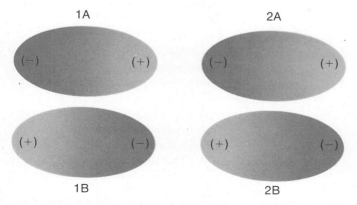

FIGURE 11.15 Temporary dipoles in H_2 molecules create an attractive force between adjacent molecules. This dispersion force is present in all molecules, but is the only intermolecular force with nonpolar molecules such as H_2.

molecules are difficult to polarize. In general, molecular size and molar mass parallel each other. This explains why dispersion forces increase in strength with molar mass.

EXAMPLE 11.5 What types of intermolecular forces are present in H_2? CCl_4? OCS? NH_3?

Solution The H_2 and CCl_4 molecules are nonpolar (Chap. 10). Thus, only dispersion forces are present among neighboring molecules. Both OCS and NH_3 are polar molecules:

There are dipole forces as well as dispersion forces with OCS molecules. In NH_3, there are hydrogen bonds and dispersion forces.

EXERCISE In which one of the above substances would you expect dispersion forces to be weakest? strongest? Answer: H_2; CCl_4.

11.5
Crystal Structures;
Unit Cells

Occasionally in nature one sees large planar rock faces that imply that the rock is one large crystal

Solids tend to crystallize* in definite geometric forms (Color Plate 11.2) that can often be seen by the naked eye. In ordinary table salt, we can distinguish small, cubic crystals of NaCl. Large, beautifully formed crystals of such minerals as fluorite, CaF_2, are found in nature. It is possible to observe distinct crystal forms of many metals under a microscope.

Crystals have definite geometric forms because the particles that form them are arranged in a definite three-dimensional pattern. These particles may be all of the same type, as is the case with atoms in a metal. In an ionic compound, two different kinds of particles are involved, cations and anions. In this section, we will look at some of the simpler ways in which atoms or ions can be arranged in a crystal.

Metals

The crystal structures of metals can be described most simply in terms of what is known as a **unit cell.** The unit cell is the smallest structural unit that, repeated over and over again in three dimensions, generates the crystal. There are many different types of unit cells. We will consider three of the simpler cells (Fig. 11.16):

1. **Simple cubic cell.** This is a cube that consists of eight atoms whose centers are located at the corners of the cell. Although this cell is the easiest to visualize, it is seldom found in nature. It is unstable because it contains a relatively large amount of empty space. Only about one half of the total volume (52.36%) is

---

*Some solids, such as glass and tar, are amorphous rather than crystalline. They show no distinct geometric form, even under the microscope. This is because there is no long-range order in the solid; that is, structural units are not arranged in a regular pattern that persists throughout the crystal.

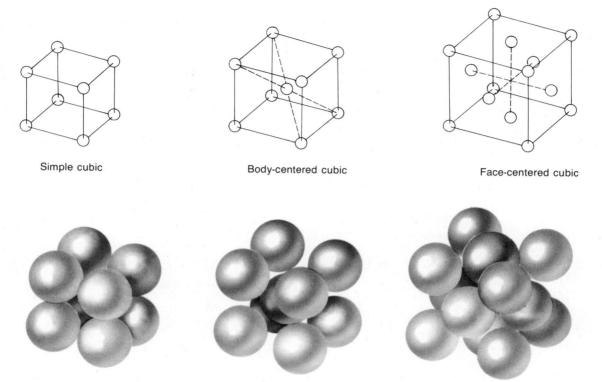

Simple cubic   Body-centered cubic   Face-centered cubic

FIGURE 11.16 Three types of cubic lattices. In the simple cubic lattice, there are 8 atoms at the corners of the cube. In the body-centered lattice, in addition to the eight atoms, there is an atom at the center of the cube. In the face-centered lattice, in addition to the eight atoms, there is an atom at the center of each of the six faces of the cube.

occupied by the atoms themselves. The rest of the space (47.64%) is accounted for by the "holes" between the atoms when they are in contact with each other.

2. **Body-centered cubic cell.** This is a cube with atoms at each corner and one in the center of the cube. Here, the atoms are packed more efficiently than in the simple cubic cell. About two thirds of the total volume (68.02%) is occupied by the atoms. All the Group 1 metals and barium in Group 2 crystallize in this pattern.

3. **Face-centered cubic cell.** This is a cube with an atom at each corner and in the center of each face of the cube. In a face-centered cubic structure, the atoms are packed as efficiently as possible. Nearly three fourths of the total volume (74.04%) is occupied by the atoms. Calcium and strontium in Group 2 have this type of unit cell, as do many of the transition metals.

This structure is called cubic close-packed

By experiment, it is possible to determine the nature and dimensions of the unit cell in a metal crystal. This information can be used to obtain the atomic radius of the metal. To see how this is done, let us consider the geometries of the three unit cells depicted in Figure 11.16.

1. In a simple cubic cell, atoms at the corners of the cube touch each other. From the diagram on p. 346, it should be clear that, for a simple cubic cell, the relation between atomic radius **r** and the length of one edge of the cube **s** is

$$2\mathbf{r} = \mathbf{s} \tag{11.4}$$

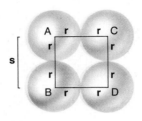

$$AB = BD = CD = AC = 2r = s$$

2. In a face-centered cubic cell, atoms at the corners of the cube do not touch each other. Instead, atom contact is made along a face diagonal as shown in the diagram below. The length of the face diagonal, BC, is $s\sqrt{2}$, where s is the length of one edge of the cell (CD or BD). Since there are four radii along the face diagonal, it follows that, for a face-centered cubic unit cell,

$$4r = s\sqrt{2} \tag{11.5}$$

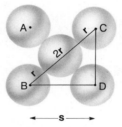

In all of these structures it is assumed that the closest atoms are in contact

3. In a body-centered cubic cell, atom contact is along a body diagonal; that is, the atom at the center of the cube touches atoms at opposite corners of the cube. The length of a body diagonal of a cube is $s\sqrt{3}$. Hence, for a body-centered cubic unit cell,

$$4r = s\sqrt{3} \tag{11.6}$$

EXAMPLE 11.6 Silver crystallizes in a structure with a face-centered cubic (FCC) unit cell 0.407 nm on an edge. Calculate the atomic radius of silver.

This is how we find the radii of metal atoms

Solution The appropriate relation for a face-centered cubic cell is

$$4r = s\sqrt{2}$$

Substituting $s = 0.407$ nm and solving for r:

$$r = \frac{0.407 \text{ nm} \times \sqrt{2}}{4} = \frac{0.407 \text{ nm} \times 1.41}{4} = 0.144 \text{ nm}$$

EXERCISE Vanadium crystallizes in a body-centered cubic (BCC) unit cell 0.303 nm on an edge. What is the atomic radius of vanadium? Answer: 0.131 nm.

LiCl   NaCl   CsCl

FIGURE 11.17 Three types of ionic crystals. In LiCl, the Cl^- ions are in contact with each other, forming a face-centered cubic lattice. In NaCl, the Cl^- ions are forced slightly apart by the larger Na^+ ions. In CsCl, the structure is quite different: the large Cs^+ ion at the center touches Cl^- ions at each corner of the cube.

Ionic Crystals

The geometry of ionic crystals, where there are two different kinds of ions, is more difficult to describe than that of metals. However, in many cases we can visualize the packing in terms of the unit cells discussed above. Lithium chloride, LiCl, is a case in point. Here, the larger Cl^- ions form a face-centered cubic lattice (Fig. 11.17). The smaller Li^+ ions fit into "holes" between the Cl^- ions. This puts an Li^+ ion at the center of each edge of the cube.

The Li^+ ions "rattle" in the Cl^- lattice

In the sodium chloride crystal, the Na^+ ion is slightly too large to fit into holes in a face-centered lattice of Cl^- ions (Fig. 11.17). As a result, the Cl^- ions are pushed slightly apart so that they are no longer touching and only Na^+ ions are in contact with Cl^- ions. However, the relative positions of positive and negative ions remain the same as in LiCl: each anion is surrounded by six cations and each cation by six anions.

NaCl is face-centered in both Na^+ and Cl^- ions

The structures of LiCl and NaCl are typical of all the alkali halides (Group 1 cation, Group 7 anion) except those of cesium. In CsCl, the Cs^+ ion is much too large to fit into a face-centered cubic array of Cl^- ions. We find that CsCl crystallizes in a quite different structure (Fig. 11.17). Here, each Cs^+ ion is located at the center of a cube outlined by Cl^- ions. The Cs^+ ion at the center touches all the Cl^- ions at the corners; the Cl^- ions do not touch each other. As you can see, each Cs^+ ion is surrounded by eight Cl^- ions. It is also true that each Cl^- ion is surrounded by eight Cs^+ ions.

CsCl is simple cubic in both Cs^+ and Cl^- ions

Summary

One way to describe the equilibrium between a liquid and its vapor is to cite the vapor pressure at a particular temperature. For an equilibrium system, vapor pressure is independent of container volume (Fig. 11.2) but increases exponentially with temperature (Fig. 11.3). There is a linear relation between the logarithm of the vapor pressure and the reciprocal of the absolute temperature (Fig. 11.4 and Example 11.1). The vapor pressure reaches 1 atm at the normal boiling point. It has its maximum value, called the critical pressure, at the critical temperature. Above that temperature, the liquid cannot exist.

The temperature/pressure conditions for phase changes are summarized in a phase diagram (Fig. 11.7 and Example 11.2). A solid sublimes below the triple point if the vapor pressure is reduced below the equilibrium value. In solid-liquid equilibria, a pressure increase results in formation of the more dense phase (Fig. 11.8).

A solid can be classified into one of four types (ionic, molecular, network covalent, or metallic) depending upon its basic structural units. The properties of a solid depend upon its structure (Table 11.3 and Example 11.3). Molecular substances are held together by relatively weak intermolecular forces. These include dipole forces, hydrogen bonds (Example 11.4), and dispersion forces (Example 11.5).

Crystals have a repeating pattern called a unit cell. Three types of unit cells are simple cubic, face-centered cubic, and body-centered cubic (Fig. 11.16 and Example 11.6). Ions in ionic solids also are arranged in repeating patterns of this type (Fig. 11.17).

Key Words and Concepts

body-centered cubic cell	freezing point	normal boiling point
Clausius-Clapeyron equation	heat of vaporization	phase diagram
critical pressure	hydrogen bond	polarization
critical temperature	intermolecular force	simple cubic cell
dipole force	intramolecular force	sublimation
dispersion force	ionic bond	triple point
electron-sea model	macromolecular	unit cell
equilibrium	melting point	vapor pressure
face-centered cubic cell	network covalent	volatile

Questions and Problems

Vapor Pressure

11.1 A sample of water vapor in a flask of constant volume exerts a pressure of 390 mm Hg at 100°C. The flask is slowly cooled.
   a. Assuming no condensation, use the Ideal Gas Law to calculate the pressure of the vapor at 90°C; at 80°C.
   b. Compare your answers in (a) to the equilibrium vapor pressures of water: 526 mm Hg at 90°C; 355 mm Hg at 80°C. Will condensation occur at 90°C; 80°C?
   c. On the basis of your answers to (a) and (b), predict the pressure exerted by the water vapor at 90°C; at 80°C.

11.31 A sample of benzene vapor in a flask of constant volume exerts a pressure of 250 mm Hg at 80°C.
   a. Assuming no condensation, use the Ideal Gas Law to construct a graph of P (mm Hg) vs. t (°C) between 80 and 20°C.
   b. Comparing the graph in (a) to the vapor pressure curve for benzene in Figure 11.3, estimate the temperature at which condensation first occurs upon cooling.

11.2 A humidifier is used in a bedroom that is at 20°C and has a volume of 3.5×10^4 L. Assume the air is originally dry and no moisture leaves the room while the humidifier is operating.
a. How many grams of water must enter the air of the room to saturate it with water vapor (vp water at 20°C = 18 mm Hg)?
b. What will be the final pressure of water vapor in the room if 800 g of water is initially in the humidifier? 400 g?

11.3 Referring to Figure 11.3, give the temperature at which water boils when the external pressure is
a. 900 mm Hg   b. 600 mm Hg   c. 200 mm Hg

11.4 The vapor pressure of bromine at 9°C is 113 mm Hg; at 20°C, it is 184 mm Hg. Using Equation 11.2, estimate the
a. heat of vaporization of bromine.
b. vapor pressure of bromine at 50°C.

11.5 The vapor pressure of mercury is 17.3 mm Hg at 200°C; its heat of vaporization is 59.4 kJ/mol. Use the Clausius-Clapeyron equation to estimate
a. the vapor pressure of mercury at 340°C.
b. the normal boiling point of mercury.

11.6 Consider the following data for the vapor pressure of acetic acid:

vp (mm Hg)	6.0	11.6	21.3	37.3
t (°C)	10	20	30	40

Plot $\log_{10}$ vp vs. 1/T (Equation 11.1) and use your graph to estimate the heat of vaporization of acetic acid.

11.7 The normal boiling point of SO_2 is −10°C. At 32°C its vapor pressure is 5 atm. Which of the following statements concerning sulfur dioxide must be true?
a. A tank of SO_2 at 32°C that has a pressure of 4 atm must contain liquid SO_2.
b. A tank of SO_2 at 32°C that has a pressure of 1 atm cannot contain liquid SO_2.
c. The critical temperature of SO_2 must be greater than 30°C.

Phase Diagrams

11.8 Referring to Figure 11.7, state what phase(s) is (are) present at
a. 15°C, 5 mm Hg   b. 20°C, 18 mm Hg
c. 10°C, 20 mm Hg

11.32 Carbon disulfide, CS_2, has a vapor pressure of 298 mm Hg at 20°C. A sample of 6.00 g CS_2 is put into a stoppered flask at that temperature.
a. What is the maximum volume the flask can have if equilibrium is to be established between liquid and vapor?
b. If the flask has a volume of 3.0 L, what will be the pressure of $CS_2(g)$?
c. If the flask has a volume of 6.0 L, what will be the pressure of $CS_2(g)$?

11.33 Work Question 11.3, substituting benzene for water.

11.34 Carbon disulfide has a vapor pressure of 41.4 mm Hg at −22°C and 100 mm Hg at −5°C. Use the Clausius-Clapeyron equation to estimate the
a. heat of vaporization of carbon disulfide.
b. vapor pressure of CS_2 at 28°C.

11.35 Liquid ammonia has a vapor pressure of 109 mm Hg at −66°C; its heat of vaporization is 2.46×10^4 J/mol. Estimate
a. the normal boiling point of ammonia.
b. the ratio of the vapor pressure at −66°C to that at −45°C.

11.36 The data below are for the vapor pressure of cyclohexane:

vp (mm Hg)	28	78	186	389
t (°C)	0	20	40	60

Follow the instructions of Problem 11.6 to find the heat of vaporization of cyclohexane.

11.37 The critical point of CO is −139°C, 35 atm. Liquid CO has a vapor pressure of 6 atm at −171°C. Which of the following statements must be true?
a. CO is a gas at −171°C and 1 atm.
b. A tank of CO at 20°C can have a pressure of 35 atm.
c. CO gas cooled to −145°C and 40 atm pressure will condense.
d. The normal boiling point of CO lies above −171°C.

11.38 Referring to Figure 11.7, state what phase(s) is (are) present at
a. −3°C, 20 mm Hg   b. 10°C, 1 mm Hg
c. 15°C, 13 mm Hg

11.9 The triple point of iodine is 114°C, 90 mm Hg. What phase(s) is (are) present at
a. 120°C, 90 mm Hg   b. 115°C, 90 mm Hg
c. 115°C, 80 mm Hg

11.10 A pure substance X has the following properties: mp = 90°C, increasing slightly as pressure increases; normal bp = 120°C; vp = 65 mm Hg at 100°C, 20 mm Hg at the triple point.
a. Draw a phase diagram for X.
b. Label solid, liquid, and vapor regions of the diagram.
c. What changes occur if, at a constant pressure of 100 mm Hg, the temperature is raised from 100 to 150°C?

11.39 The density of solid benzene is greater than that of the liquid. Which of the following is correct? The melting point of benzene at 20 atm is
a. greater than that at 1 atm, 5°C.
b. less than that at 1 atm.
c. the same as that at 1 atm.

11.40 A pure substance A has a vapor pressure of 320 mm Hg at 125°C, 800 mm Hg at 150°C, and 60 mm Hg at the triple point, 85°C. The melting point of A decreases slightly as pressure increases.
a. Sketch a phase diagram for A.
b. From the phase diagram, estimate the normal boiling point.
c. What changes occur when, at a constant pressure of 320 mm Hg, the temperature drops from 150 to 100°C?

Types of Solids

11.11 Classify as metallic, molecular, ionic, or network covalent a solid that
a. is a nonconductor but conducts when melted.
b. dissolves in water to give a nonconducting solution.
c. melts below 100°C and reacts violently with water.

11.12 Of the four general types of solids, which one(s)
a. are generally insoluble in water?
b. are very high melting?
c. conduct electricity as solids?

11.13 Give the formula of a solid containing carbon that is
a. molecular   b. ionic
c. network covalent   d. metallic

11.14 Classify the following species as being ionic, molecular, network covalent, or metallic at 25°C, 1 atm:
a. PCl_3 b. SiC   c. bromine
d. MgO   e. chromium

11.15 Describe the nature of the structural units in
a. Fe   b. $FeCl_2$ c. Cl_2 d. C

11.41 Classify as metallic, network covalent, ionic, or molecular a solid that
a. melts below 100°C to give a nonconducting liquid.
b. conducts electricity as a solid.
c. dissolves in water to give a conducting solution.

11.42 Of the four general types of solids, which one(s)
a. are generally low boiling?
b. are ductile and malleable?
c. are generally nonvolatile?

11.43 Classify each of the following solids as molecular, network covalent, ionic, or metallic:
a. Na   b. SiO_2 c. $C_{10}H_8$
d. NaCl   e. $MgSO_4$

11.44 Apply the instructions for Question 11.14 to
a. tungsten   b. IBr   c. quartz
d. iron(III) nitrate   e. carbon dioxide

11.45 Describe the nature of the structural units in
a. K_2O b. O_2 c. K   d. BN

Intermolecular Forces

11.16 Arrange the following in order of decreasing boiling point:
a. I_2 b. F_2 c. Cl_2 d. Br_2

11.46 Arrange the following in order of increasing boiling point:
a. Ar   b. He   c. Ne   d. Xe

11.17 Which of the following would show dispersion forces? dipole forces?
 a. CH_4 b. CH_3Cl c. CH_2Cl_2
 d. $CHCl_3$ e. CCl_4

11.18 Which of the following would show hydrogen bonding?
 a. CH_3F b. CH_3-OH
 c. CH_3-O-CH_3 d. $HO-OH$

11.19 Explain in terms of intermolecular forces why
 a. C_2H_6 has a higher boiling point than CH_4.
 b. C_2H_5OH is higher boiling than C_2H_6.
 c. C_2H_5OH is lower boiling than NaF.
 d. O_2 is higher boiling than N_2.
 e. CO is slightly higher boiling than N_2.

11.20 In which of the following processes is it necessary to break covalent bonds as opposed to simply overcoming intermolecular forces?
 a. dissolving I_2 in water.
 b. boiling water.
 c. subliming dry ice, CO_2.
 d. decomposing N_2O_4 to NO_2.

11.21 For each of the following pairs, choose the lower boiling member. Explain your reason in each case.
 a. F_2 or Cl_2 b. ClO_2 or SiO_2
 c. LiCl or C_3H_8 d. HCl or HBr

11.22 What are the strongest attractive forces that must be overcome to
 a. melt ice? b. melt iodine?
 c. vaporize $CaCl_2$? d. boil CCl_4?

Crystal Structures

11.23 Barium (at. radius = 0.217 nm) crystallizes with a body-centered cubic unit cell. What is the length of a side of the cell?

11.24 Platinum forms a face-centered cubic unit cell 0.390 nm on an edge. What is the atomic radius of platinum?

11.25 In the LiCl structure shown in Figure 11.17, the chloride ions form a face-centered cubic unit cell 0.513 nm on an edge. The ionic radius of Cl^- is 0.181 nm.
 a. Along a cell edge, how much space is there between the Cl^- ions?
 b. Would an Na^+ ion (r = 0.095 nm) fit into this space? a K^+ ion (r = 0.133 nm)?

11.47 Which of the following would you expect to show dipole forces?
 a. CO b. CO_2 (linear)
 c. F_2 d. H_2S (bent)

11.48 Which of the following would show hydrogen bonding?
 a. CH_4 b. NH_3 c. H—N—N—H
       | |
       H   H
 d. $[H-F-H]^+$

11.49 Explain in terms of intermolecular forces why
 a. H_2O is higher boiling than H_2Te.
 b. H_2O_2 is higher boiling than C_2H_6.
 c. I_2 is lower melting than NaI.
 d. ICl is higher melting than Br_2.

11.50 In which of the following processes are covalent bonds broken?
 a. melting benzene.
 b. melting quartz.
 c. electrolysis of water.
 d. boiling C_2H_5OH.

11.51 Follow directions for Question 11.21 for the following substances:
 a. H_2 or O_2 b. NH_3 or PH_3
 c. PH_3 or AsH_3 d. SO_2 or SiO_2

11.52 What are the strongest attractive forces that must be overcome to
 a. melt benzene, C_6H_6? b. dissolve Br_2 in CCl_4?
 c. melt NaCl? d. boil SiH_4?

11.53 Rubidium crystallizes in a body-centered cubic unit cell 0.564 nm on a side. Calculate the atomic radius of rubidium.

11.54 The unit cell in tungsten is face-centered cubic. The atomic radius of tungsten is 0.137 nm. Calculate the volume of the unit cell.

11.55 Potassium iodide has a unit cell similar to that of NaCl (Fig. 11.17). The ionic radii of K^+ and I^- are 0.133 and 0.216 nm, in that order. What is the length of
 a. one side of the cube?
 b. the face diagonal of the cube?

11.26 For a cell of the CsCl type (Fig. 11.17), how is the length of one side of the cell, **s,** related to the sum of the radii of the ions, $r_{cation} + r_{anion}$?

11.56 Consider the CsCl cell (Fig. 11.17). The ionic radii of Cs^+ and Cl^- are 0.169 and 0.181 nm, respectively. What is the length of
a. the body diagonal?
b. the side of the cell?

General

11.27 The density of liquid mercury at 20°C is 13.6 g/cm³; its vapor pressure is 1.2×10^{-3} mm Hg.
a. What volume (cm³) is occupied by one mole Hg(*l*) at 20°C?
b. What volume (cm³) is occupied by one mole Hg(*g*) at 20°C and the equilibrium vapor pressure?
c. The atomic radius of Hg is 0.155 nm. Calculate the volume (cm³) of one mole Hg atoms (V = $4\pi r^3/3$).
d. From your answers to (a), (b), and (c), calculate the percentage of the total volume occupied by the atoms in Hg(*l*) and Hg(*g*) at 20°C and 1.2×10^{-3} mm Hg.

11.57 The density of iron at 25°C is 7.87 g/cm³.
a. What is the volume of one mole of solid iron at 25°C?
b. Iron has an atomic radius of 0.126 nm. Calculate the volume of one mole of iron atoms (V = $4\pi r^3/3$).
c. From your answers to (a) and (b), calculate the fraction of "empty space" in an iron crystal at 25°C.

11.28 Which of the following statements are true?
a. The critical pressure is the highest vapor pressure that a liquid can have.
b. To sublime a solid, it must be heated above the triple point.
c. NaF melts higher than F_2 because its molar mass is larger.
d. One metal crystallizes in a BCC cell, another in a face-centered cubic cell of the same size. The two atomic radii must be equal.

11.58 Criticize each of the following statements.
a. Vapor pressure is inversely proportional to volume.
b. Vapor pressure is directly proportional to temperature.
c. Boiling point increases with molar mass.
d. The melting point of a molecular substance depends upon the strength of the covalent bonds within the molecule.

11.29 Differentiate between
a. intramolecular forces and intermolecular forces.
b. a molecular solid and a network covalent solid.
c. a dipole force and a hydrogen bond.
d. an ionic solid and a molecular solid.

11.59 Explain the difference between
a. a face-centered cubic and a body-centered cubic unit cell.
b. melting and sublimation.
c. a boiling point and a normal boiling point.
d. a phase diagram and a vapor pressure curve.

11.30 How would you explain to your parents
a. how freeze-dried foods are made?
b. why a cook does not "cry" as much when cutting a cold onion as when cutting a warm one?
c. why atmospheric pressure decreases with altitude?
d. why mothballs in a clothes closet "disappear" without forming a puddle?

11.60 How would you explain to a high school chemistry student why
a. water cools upon standing in slightly porous clay pots?
b. a sealed tin can filled with steam at 100°C collapses when it is cooled?
c. a pressure cooker is used to prepare foods at high altitudes?
d. on a hot summer day, road tar softens but beach sand (SiO_2) does not?

*11.61 The structure of a portion of a quartz crystal is given in Figure 11.10. Explain in your own words how this structure leads to the formula SiO_2 for quartz.

*11.62 It has been suggested that the pressure exerted on a skate blade is sufficient to melt the ice beneath it and form a thin film of water, which makes it easier for the blade to slide over the ice. Assume that a skater weighs 120 lb and the blade has an area of 0.10 in². Calculate the pressure exerted on the blade (1 atm = 15 lb/in²). From information in the text, calculate the decrease in melting point at this pressure. Comment on the plausibility of this explanation, and suggest another mechanism by which the water film might be formed.

*11.63 As shown in Figure 11.17, Li^+ ions fit into a closely packed array of Cl^- ions, but Na^+ ions do not. What is the value of the r_{cation}/r_{anion} ratio at which a cation just fits into a structure of this type?

*11.64 In the text it was stated that in a simple cubic cell, 52.36% of the total volume is occupied by the atoms themselves. Show where this number comes from. (Note that in a simple cubic cell, one atom effectively occupies each cell.)

*11.65 Consider liquid water in equilibrium with its vapor at 100°C. Estimate the number of molecules per cubic centimeter in
   a. the liquid (d = 0.958 g/cm³ at 100°C).   b. the vapor.

*11.66 The density of hydrogen fluoride vapor at 28°C and 1.00 atm is 2.30 g/L. What does this tell you about the intermolecular forces in hydrogen fluoride gas?

Chapter 12
Solutions

A solution is a homogeneous mixture of a *solute* (substance being dissolved) distributed through a *solvent* (substance doing the dissolving). Solutions exist in any of the three physical states: gas, liquid, or solid. Air, the most common gaseous solution, is a mixture of nitrogen, oxygen, and lesser amounts of other gases. Many metal alloys are solid solutions, as in the U.S. "nickel" coin (25% Ni, 75% Cu). The most familiar solutions are those in the liquid state, especially ones in which water is the solvent. Aqueous solutions are most important for our purposes in chemistry and will be emphasized in this chapter.

We will consider several aspects of solutions. These include:

—methods of expressing solution concentrations by specifying the relative amounts of solute and solvent (Section 12.2).
—factors affecting solubility, including the nature of solute and solvent, the temperature, and the pressure (Section 12.3).
—effect of solutes upon such solvent properties as vapor pressure, freezing point, and boiling point (Section 12.4).

To start with, it will be helpful to review some of the terms used to describe the nature and concentrations of the components of a solution (Section 12.1).

12.1
Solution Terminology

Several different adjectives can be used to indicate the relative amounts of solute and solvent in a solution. We may describe a solution containing a small amount of solute as being "dilute." Another solution containing more solute in the same amount of solvent might be called "concentrated." In a few cases, these terms, through tradition, have taken on a quantitative meaning. Solutions of certain common reagents, labeled dilute or concentrated, have the compositions specified in Table 12.1.

Table 12.1
Concentrations of Laboratory Acid and Base Solutions

		SOLUTE	MOLES SOLUTE PER LITER	MASS PERCENT SOLUTE	DENSITY (g/cm³)
Hydrochloric acid	conc.	HCl	12	36	1.18
	dilute		6	20	1.10
Nitric acid	conc.	HNO₃	16	72	1.42
	dilute		6	32	1.19
Sulfuric acid	conc.	H₂SO₄	18	96	1.84
	dilute		3	25	1.18
Ammonia	conc.	NH₃*	15	28	0.90
	dilute		6	11	0.96

*Often labeled "NH₄OH."

The terms "dilute" and "concentrated" in this context are gradually fading from use

Relative concentrations of solutions are often expressed in a different way by using the terms "saturated," "unsaturated," and "supersaturated." A **saturated** solution is one which is in equilibrium with undissolved solute. An **unsaturated** solution contains a lower concentration of solute than the saturated solution. The unsaturated solution is not at equilibrium. If solute is added, it dissolves until saturation is reached. A **supersaturated** solution contains more than the equilibrium concentration of solute. It is unstable in the presence of excess solute.

To illustrate the meaning of these terms, consider solid sodium acetate, $NaC_2H_3O_2$ (Fig. 12.1). A saturated aqueous solution of sodium acetate at 20°C contains 46.5 g of $NaC_2H_3O_2$ in 100 g of water. Any solution in which the concentration of sodium acetate at 20°C is less than this value is unsaturated. If solid sodium acetate is added to such a solution, it will dissolve until the concentration reaches the saturation value. A supersaturated solution contains more than

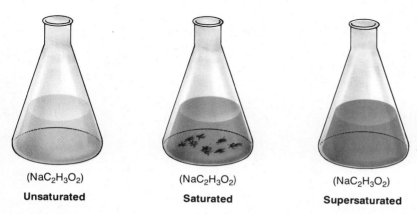

$(NaC_2H_3O_2)$
Unsaturated

$(NaC_2H_3O_2)$
Saturated

$(NaC_2H_3O_2)$
Supersaturated

FIGURE 12.1 Unsaturated, saturated, and supersaturated solutions of sodium acetate. Only the saturated solution is in equilibrium with undissolved solute. If the supersaturated solution is disturbed, say by the addition of a crystal of sodium acetate, crystallization of excess solute will occur rapidly until equilibrium between solid and solution is established.

46.5 g of $NaC_2H_3O_2$ per 100 g of water at 20°C. To prepare such a solution, we take advantage of the fact that the solubility of sodium acetate increases with temperature. At 50°C we can dissolve 80 g of $NaC_2H_3O_2$ in 100 g of water. If this solution is cooled carefully to 20°C, without shaking or stirring, the excess solute stays in solution. This produces a supersaturated solution. If a small seed crystal of sodium acetate is now added, crystallization quickly takes place. The excess solute (80 g − 46.5 g) comes out of solution, establishing equilibrium with the saturated solution.

The temperature of the mix goes up. Can you suggest why?

Nonelectrolytes vs Electrolytes

Pure water does not conduct an electric current (Fig. 12.2A). Solutes in water can be classified according to the conductivity of the solutions they form. We can distinguish between two types of solutes.

1. **Nonelectrolytes** form water solutions that do not conduct an electric current (Fig. 12.2B). Typically, these substances are molecular when pure and dissolve as molecules. Since molecules are neutral, they cannot migrate in an electric field. Hence, they do not conduct an electric current. The processes by which methyl alcohol, CH_3OH, and sugar, $C_{12}H_{22}O_{11}$, dissolve in water can be represented by the simple equations:

Most nonelectrolytes are organic

$$CH_3OH(l) \rightarrow CH_3OH(aq)$$

$$C_{12}H_{22}O_{11}(s) \rightarrow C_{12}H_{22}O_{11}(aq)$$

2. **Electrolytes** are solutes whose water solutions conduct an electric current (Fig. 12.2C). These substances exist as ions in solution. The charged ions migrate in an electric field, thereby carrying a current. Sodium chloride is a familiar

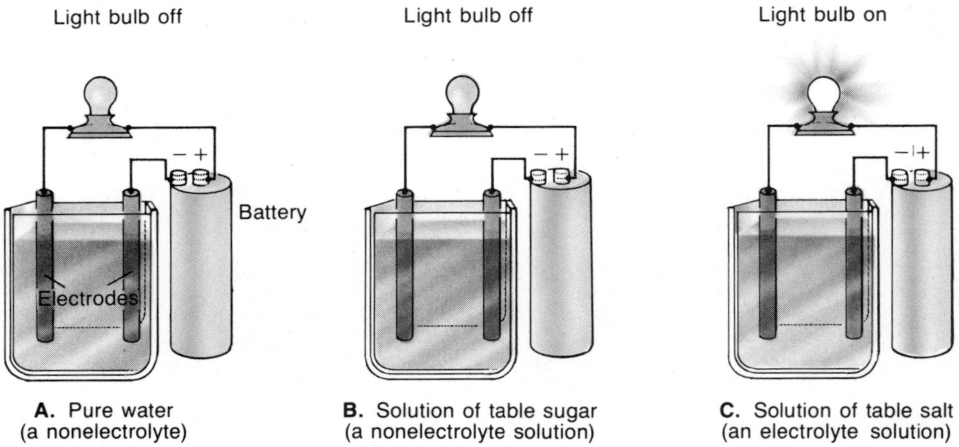

Light bulb off   Light bulb off   Light bulb on

Battery

Electrodes

A. Pure water (a nonelectrolyte)

B. Solution of table sugar (a nonelectrolyte solution)

C. Solution of table salt (an electrolyte solution)

FIGURE 12.2 An apparatus for testing electrical conductivity. For an electric current to flow, the solution must contain ions, which carry electric charge. (A) Pure water which contains very few ions, is not a conductor. (B) Dissolving sugar in water produces only sugar molecules in the solution, so it does not become a conductor. (C) If NaCl is dissolved in water, it forms ions. These conduct the current and the light bulb glows.

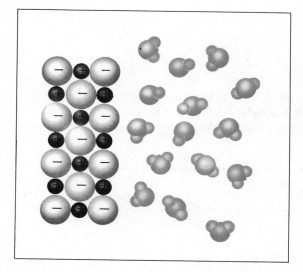 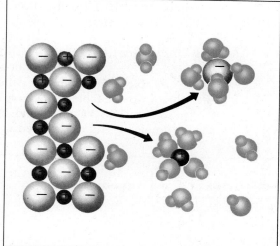

FIGURE 12.3 Dissolving NaCl in water. The attraction between the oxygen atoms in H_2O molecules and Na^+ ions in NaCl tends to bring the Na^+ ions into solution. At the same time, Cl^- ions, attracted to the hydrogen atoms of H_2O molecules, also enter the solution.

example of an electrolyte. In the solid, it consists of Na^+ and Cl^- ions. Upon dissolving in water, these ions are set free (Fig. 12.3). The solution process can be represented as

Most electrolytes are ionic compounds

$$NaCl(s) \rightarrow Na^+(aq) + Cl^-(aq)$$

Other ionic solutes behave in a similar manner. You can assume that when any ionic solid dissolves in water, it breaks up completely into ions (Example 12.1).

EXAMPLE 12.1 Write equations for the dissociation in water of each of the following ionic solids:
a. KI   b. Li_2CO_3 c. $Fe(NO_3)_3$ d. $Ce_2(SO_4)_3$

Solution In each case, the "reactant" is the ionic solid. The "products" are the ions released into water.
a. $KI(s) \rightarrow K^+(aq) + I^-(aq)$
b. $Li_2CO_3(s) \rightarrow 2\ Li^+(aq) + CO_3^{2-}(aq)$
c. $Fe(NO_3)_3(s) \rightarrow Fe^{3+}(aq) + 3\ NO_3^-(aq)$
d. $Ce_2(SO_4)_3(s) \rightarrow 2\ Ce^{3+}(aq) + 3\ SO_4^{2-}(aq)$

There are no KI molecules in a solution of KI in water

EXERCISE One mole of Na_2SO_4 and one mole of KCl are dissolved separately in water. Which solution contains the greater number of moles of ions? Answer: Na_2SO_4; 3 mol.

As noted in Chapter 11, hydrogen chloride, HCl, is typical of a certain class of molecular compounds that, *in water solution*, act as electrolytes. Pure hydrogen chloride as a solid, liquid, or gas has the properties of a molecular substance. For example, liquid HCl does not conduct an electric current. However, when put into water, HCl dissolves to form a conducting solution typical of an electrolyte. A reaction occurs with water to produce aqueous H^+ ions and Cl^- ions:

$$HCl(aq) \rightarrow H^+(aq) + Cl^-(aq)$$

Table 12.2 summarizes the solution behavior of different kinds of solutes. Note from the table that a mole of a nonelectrolyte produces *one mole* of particles in solution. On the other hand, electrolytes provide *more than one mole* of particles per mole of solute dissolved.

Table 12.2
Solution Behavior of Some Aqueous Solutes

SOLUTE	TYPE	SOLUTION EQUATION	MOLES OF SOLUTE PARTICLES PER MOLE SOLUTE
$I_2(s)$	Nonelectrolyte	$I_2(s) \rightarrow I_2(aq)$	1
$C_6H_{12}O_6(s)$ (glucose)	Nonelectrolyte	$C_6H_{12}O_6(s) \rightarrow C_6H_{12}O_6(aq)$	1
$C_{12}H_{22}O_{11}(s)$ (sucrose)	Nonelectrolyte	$C_{12}H_{22}O_{11}(s) \rightarrow C_{12}H_{22}O_{11}(aq)$	1
$HCl(g)$	Electrolyte	$HCl(g) \rightarrow H^+(aq) + Cl^-(aq)$	2 (1 H^+ and 1 Cl^-)
$LiBr(s)$	Electrolyte	$LiBr(s) \rightarrow Li^+(aq) + Br^-(aq)$	2 (1 Li^+ and 1 Br^-)
$K_2SO_4(s)$	Electrolyte	$K_2SO_4(s) \rightarrow 2 K^+(aq) + SO_4^{2-}(aq)$	3 (2 K^+ and 1 SO_4^{2-})
$LaBr_3(s)$	Electrolyte	$LaBr_3(s) \rightarrow La^{3+}(aq) + 3 Br^-(aq)$	4 (1 La^{3+} and 3 Br^-)

12.2
Concentration Units

The properties of a solution depend strongly upon the relative amounts of solute and solvent present. These are described by citing the *concentration* of solute, which tells us how much solute is present for a given amount of solvent or solution. Concentrations can be expressed in various ways. Sometimes we state the mass percent of solute:

$$\text{mass percent solute} = \frac{\text{mass solute}}{\text{total mass solution}} \times 100 \tag{12.1}$$

Usage is sometimes loose. A 5% NaCl solution means 5 mass percent NaCl

In a solution prepared by dissolving 24 g of NaCl in 152 g of water:

$$\text{mass percent NaCl} = \frac{24 \text{ g}}{24 \text{ g} + 152 \text{ g}} \times 100 = \frac{24}{176} \times 100 = 14$$

More commonly, in expressing concentrations, we work with the number of moles of solute. In the remainder of this section, we will consider three concentration units of this type. Two of these, *mole fraction* and *molarity*, were referred to in previous chapters (Chapters 2 and 6). The third, *molality*, is discussed here for the first time.

Mole Fraction

You may recall from Chapter 6 that the mole fraction (X) of a component, A, in a solution is given by the relation:

$$X_A = \frac{\text{no. moles A}}{\text{total no. moles all components}} \qquad (12.2)$$

If 1.20 mol of methyl alcohol, CH_3OH, is dissolved in 16.8 mol of water, we have

$$X_{CH_3OH} = \frac{1.20 \text{ mol}}{1.20 \text{ mol} + 16.8 \text{ mol}} = \frac{1.20}{18.0} = 0.0667$$

Total moles = moles CH_3OH + moles H_2O

The sum of the mole fractions of all the components of a solution must be 1. That is:

$$X_A + X_B + \cdots = 1 \qquad (12.3)$$

Why?

For the methyl alcohol solution just referred to, there are only two components, CH_3OH and water. The mole fraction of water is readily found.

$$X_{H_2O} = 1 - X_{CH_3OH} = 1 - 0.0667 = 0.9333$$

EXAMPLE 12.2   What are the mole fractions of CH_3OH and H_2O in a solution prepared by dissolving 1.20 g of methyl alcohol in 16.8 g of water?

Solution   We first find the numbers of moles of CH_3OH and H_2O. Their molar masses are 32.0 and 18.0 g/mol, in that order:

$$\text{no. moles } CH_3OH = 1.20 \text{ g} \times \frac{1 \text{ mol}}{32.0 \text{ g}} = 0.0375 \text{ mol } CH_3OH$$

$$\text{no. moles } H_2O = 16.8 \text{ g} \times \frac{1 \text{ mol}}{18.0 \text{ g}} = 0.933 \text{ mol } H_2O$$

To find the mole fraction of CH_3OH, we apply the defining relation, Equation 12.2:

$$X_{CH_3OH} = \frac{\text{no. moles } CH_3OH}{\text{no. moles } CH_3OH + \text{no. moles } H_2O} = \frac{0.0375}{0.0375 + 0.933}$$
$$= 0.0386$$

We could carry out a similar calculation to obtain the mole fraction of water, but it is simpler to use Equation 12.3:

$$X_{H_2O} = 1 - X_{CH_3OH} = 1 - 0.0386 = 0.9614$$

In the solution about 96% of the molecules are H_2O and 4% are CH_3OH

EXERCISE   Calculate the mole fraction of ethyl alcohol, C_2H_5OH, in a solution prepared by dissolving 1.20 g of C_2H_5OH in 16.8 g of water. Answer: 0.0272.

Molality (m)

The concentration unit molality, given the symbol m, is defined as the number of moles of solute per kilogram (1000 g) of *solvent*.

Molality is used mainly in connection with colligative properties. Sec. 12.4

$$\text{molality (m)} = \frac{\text{no. moles solute}}{\text{no. kilograms solvent}} \qquad (12.4)$$

The molality of a solution is readily calculated if the masses of solute and solvent are known (Example 12.3).

> **EXAMPLE 12.3**   A solution contains 12.0 g of glucose, $C_6H_{12}O_6$, in 95.0 g of water. Calculate the molality of glucose.
>
> **Solution**   To obtain the number of moles of solute, note that the molar mass of glucose is:
>
> $$6(12.0 \text{ g/mol}) + 12(1.0 \text{ g/mol}) + 6(16.0 \text{ g/mol}) = 180.0 \text{ g/mol}$$
>
> $$\text{no. moles solute} = 12.0 \text{ g} \times \frac{1 \text{ mol}}{180.0 \text{ g}} = 0.0667 \text{ mol}$$
>
> $$\text{no. kg solvent} = 95.0 \text{ g water} \times \frac{1 \text{ kg water}}{1000 \text{ g water}} = 0.0950 \text{ kg water}$$
>
> $$\text{molality} = \frac{0.0667 \text{ mol solute}}{0.0950 \text{ kg solvent}} = 0.702 \text{ m}$$
>
> **EXERCISE**   A solution contains 23.0 g of ethyl alcohol, C_2H_5OH, dissolved in 30.0 g of water. Calculate the molality of ethyl alcohol in this solution. Answer: 16.7 m.

Molarity (M)

In laboratory work, we are most often interested in the number of moles of solute in a given volume of solution. Here the most common concentration unit is molarity, which is the number of moles of solute per liter of solution.

Molarity: M
Molality: m

$$\text{molarity (M)} = \frac{\text{no. moles solute}}{\text{no. liters solution}} \qquad (12.5)$$

In Chapter 2, we used this defining equation for molarity to calculate

—the molarity of a solution when the amount of solute and volume of solution are known (Example 2.8, p. 51).
—the number of moles of solute in a given volume of a solution of known molarity (Example 2.9a, p. 52).
—the volume of a solution of known molarity required to contain a given number of moles of solute (Example 2.9b).

In the laboratory you will frequently need to prepare a certain volume of a solution of a specified molarity. To do this, you will most often start with pure solute. In that case, you:

—calculate the mass of solute required, using the defining equation for molarity, Equation 12.5, and the molar mass of the solute.
—weigh out the calculated mass of solute and dissolve in enough solvent to give the desired volume of solution.

EXAMPLE 12.4   How would you prepare 0.150 L of a 0.500 M NaOH solution, starting with solid sodium hydroxide and water?

Solution   We first calculate the number of moles of NaOH required. To do that, we treat molarity as a conversion factor:

0.500 mol NaOH ≏ 1 L solution

$$\text{no. moles NaOH required} = 0.150 \text{ L} \times \frac{0.500 \text{ mol NaOH}}{1 \text{ L}}$$
$$= 0.0750 \text{ mol NaOH}$$

To find the mass of NaOH required, note that the molar mass is 40.0 g/mol:

$$\text{mass NaOH required} = 0.0750 \text{ mol} \times \frac{40.0 \text{ g}}{1 \text{ mol}} = 3.00 \text{ g}$$

You should weigh out 3.00 g of NaOH and dissolve in enough water to form 0.150 L (150 mL) of solution.

EXERCISE   What mass of NaOH would be obtained by evaporating to dryness 1.00 mL of this solution? Answer: 0.0200 g.

Another way to prepare a known volume of a solution of a desired molarity is to start with a concentrated solution rather than pure solute. In this case you

—calculate the volume of the concentrated solution required.
—measure out that volume and add enough solvent to give the desired volume of the more dilute solution.

The calculations required with this method are readily made if you keep a simple point in mind. Adding solvent cannot change the number of moles of solute. In other words, the number of moles of solute is the same before and after dilution.

Dilution means adding solvent, not solute

no. moles solute in concentrated solution = no. moles solute in dilute solution

In both solutions, the number of moles of solute can be found by multiplying the molarity (M) by the volume in liters (V). Hence:

$$M_c V_c = M_d V_d \tag{12.6}$$

$M_c V_c$ = no. moles solute

where the subscripts c and d stand for concentrated and dilute solutions, respectively. The use of Equation 12.6 is illustrated in Example 12.5.

EXAMPLE 12.5   How would you prepare 0.150 L of 0.500 M NaOH, starting with a 6.00 M solution of NaOH?

Solution   This question might be restated as: What volume of 6.00 M NaOH should be diluted with water to yield 0.150 L of 0.500 M NaOH?

That quantity can readily be calculated using Equation 12.6. We need to know the volume of concentrated solution, V_c. Solving Equation 12.6 for V_c,

$$V_c = \frac{M_d \times V_d}{M_c}$$

Here

M_d = molarity of dilute solution = 0.500 mol/L
V_d = volume of dilute solution = 0.150 L
M_c = molarity of concentrated solution = 6.00 mol/L

$$V_c = \frac{0.500 \times 0.150 \text{ L}}{6.00} = 0.0125 \text{ L} \ (12.5 \text{ mL})$$

You should measure out 12.5 mL of 6.00 M NaOH and dilute with water to give 0.150 L of solution.

EXERCISE What mass of NaOH is present in 0.0125 L of 6.00 M NaOH? in 0.150 L of 0.500 M NaOH? Answer: 3.00 g.

Many water solutions in the general chemistry laboratory contain ionic solutes. As we saw in Section 12.1, these solutes are completely dissociated into ions in solution. Frequently, you will need to relate the molarity of an ionic solute to that of its ions in solution. To do this, it is convenient to start by writing dissociation equations of the type given in Example 12.1. Thus, from the equations

$$KI(s) \rightarrow K^+(aq) + I^-(aq)$$

$$Li_2CO_3(s) \rightarrow 2 \ Li^+(aq) + CO_3{}^{2-}(aq)$$

we see that

$$1 \text{ mol KI} \rightarrow 1 \text{ mol K}^+ + 1 \text{ mol I}^-$$

We label the bottle 1 M KI. In the solution, conc. K^+ = 1 M, conc. I^- = 1 M, and conc. KI molecules = 0

so that a 1 M KI solution is 1 M in K^+ and 1 M in I^-

$$1 \text{ mol Li}_2CO_3 \rightarrow 2 \text{ mol Li}^+ + 1 \text{ mol CO}_3{}^{2-}$$

so a 1 M Li_2CO_3 solution is 2 M in Li^+ and 1 M in $CO_3{}^{2-}$.

EXAMPLE 12.6 What is the concentration, in moles per liter, of each ion in
a. 0.080 M K_2SO_4?   b. 0.40 M $LaBr_3$?

Solution
a. $$K_2SO_4(s) \rightarrow 2 \ K^+(aq) + SO_4{}^{2-}(aq)$$

$$1 \text{ mol K}_2SO_4 \rightarrow 2 \text{ mol K}^+ + 1 \text{ mol SO}_4{}^{2-}$$

The concentrations of the ions are

$$\text{conc. K}^+ = \frac{0.080 \text{ mol K}_2SO_4}{1 \text{ L}} \times \frac{2 \text{ mol K}^+}{1 \text{ mol K}_2SO_4} = 0.16 \text{ M}$$

$$\text{conc. } SO_4^{2-} = \frac{0.080 \text{ mol } K_2SO_4}{1 \text{ L}} \times \frac{1 \text{ mol } SO_4^{2-}}{1 \text{ mol } K_2SO_4} = 0.080 \text{ M}$$

b. $LaBr_3(s) \rightarrow La^{3+}(aq) + 3 Br^-(aq)$

$1 \text{ mol } LaBr_3 \rightarrow 1 \text{ mol } La^{3+} + 3 \text{ mol } Br^-$

conc. La^{3+} = conc. $LaBr_3$ = 0.40 M

conc. Br^- = 3(conc. $LaBr_3$) = 3(0.40 M) = 1.2 M

conc. Br^- = 3 × conc. La^{3+}

EXERCISE   Calculate the number of moles of K^+ in 1.5 L of 0.080 M K_2SO_4. Answer: 0.24 mol.

Conversions Between Concentration Units

Sometimes you may need to convert the concentration of a solution from one unit to another. For example, you might be given the molarity of a solute in water solution and asked to determine its molality. Such conversions are best carried out by a systematic approach. In general, you should:

1. Select a particular quantity (mass or volume) of solution to work with. The quantity you choose will depend upon the concentration unit given. It is perhaps most convenient to work with

—100 g of solution if the concentration is given in mass percent.
—1 mol of solution if the concentration is given in mole fraction.
—1 L of solution if the concentration is given in molarity.
—1000 g of solvent if the concentration is given in molality.

You will find that these choices are the most convenient

2. Calculate the amount (mass or number of moles) of both solute and solvent in the quantity of solution chosen in (1).
3. Using the data obtained in (2), calculate the concentration in the desired unit.

EXAMPLE 12.7   From Table 12.1 we see that the mass percent of HCl in concentrated hydrochloric acid is 36.0. For this solution, calculate
a. the mole fraction of HCl   b. the molality of HCl

Solution
a. Let us work with 100.0 g of solution, which must contain 36.0 g HCl and 64.0 g H_2O. In order to calculate the mole fraction of HCl, we will need to know the numbers of moles of both HCl and H_2O, whose molar masses are 36.5 and 18.0 g/mol, in that order.

$$\text{no. moles HCl} = 36.0 \text{ g HCl} \times \frac{1 \text{ mol HCl}}{36.5 \text{ g HCl}} = 0.986 \text{ mol HCl}$$

$$\text{no. moles } H_2O = 64.0 \text{ g } H_2O \times \frac{1 \text{ mol } H_2O}{18.0 \text{ g } H_2O} = 3.56 \text{ mol } H_2O$$

Now we have all the information we need to calculate the mole fraction of HCl.

$$X_{HCl} = \frac{\text{no. moles HCl}}{\text{no. moles HCl + no. moles } H_2O} = \frac{0.986}{0.986 + 3.56}$$
$$= 0.217$$

b. To calculate the molality, we need to know the number of moles of HCl and the number of kilograms of water. The number of moles of HCl in 100.0 g of solution was calculated in (a) to be 0.986. The number of kilograms of water is readily calculated.

In dilute aqueous solution, $m_A \cong M_A$

$$\text{no. kilograms water} = 64.0 \text{ g water} \times \frac{1 \text{ kg}}{1000 \text{ g}}$$
$$= 0.0640 \text{ kg water}$$

$$\text{molality} = \frac{0.986 \text{ mol HCl}}{0.0640 \text{ kg water}} = 15.4 \text{ m}$$

EXERCISE The mole fraction of CH_3OH in a certain water solution is 0.200. What is the molality of CH_3OH?
Answer: 13.9 m.

When one of the units involved in a concentration conversion is molarity, you must know or be able to calculate the density of the solution. Beyond that, the approach used is the same as in any other conversion of this type (Example 12.8).

EXAMPLE 12.8 Taking the molality of HCl in concentrated hydrochloric acid to be 15.4 m, calculate its molarity. The density of the solution is given in Table 12.1 as 1.18 g/cm³ (or 1180 g/L).

Of all the concentration units, chemists most often use molarity

Solution We choose to work with an amount of solution containing one kilogram of water. That amount of solution contains

15.4 mol HCl, 1000 g water

To find the molarity of the solution, we need to know its volume in liters. That can be found, using the density, if we know the total mass of solution. To obtain that quantity, we must know the mass in grams of HCl:

$$\text{mass HCl} = 15.4 \text{ mol HCl} \times \frac{36.5 \text{ g HCl}}{1 \text{ mol HCl}} = 562 \text{ g HCl}$$

$$\text{total mass solution} = 562 \text{ g} + 1000 \text{ g} = 1562 \text{ g}$$

$$\text{volume solution} = 1562 \text{ g} \times \frac{1 \text{ L}}{1180 \text{ g}} = 1.32 \text{ L}$$

We label the bottle 12 M HCl

$$\text{molarity HCl} = \frac{\text{no. moles HCl}}{\text{no. liters solution}} = \frac{15.4 \text{ mol}}{1.32 \text{ L}} = 11.7 \text{ M}$$

EXERCISE In dilute hydrochloric acid, the molarity is 6.0 mol/L and the density is 1.10 g/cm³. How many grams of water are there in one liter of this solution? What is the molality of HCl? Answer: 1100 g − 219 g = 880 g; 6.8 M.

12.3
Principles of Solubility

The extent to which a solute dissolves in a particular solvent depends upon several factors. The most important of these are:

—the nature of solvent and solute particles and the interactions between them.
—the temperature at which the solution is formed.
—the pressure of a gaseous solute.

In this section we will consider in turn the effect of each of these factors upon solubility.

Solute-Solvent Interactions

In discussing solubility, it is sometimes stated that "like dissolves like." A more meaningful way to express this idea is to say that two substances with intermolecular forces of about the same type and magnitude are likely to be very soluble in one another. To illustrate, consider the hydrocarbons pentane, C_5H_{12}, and hexane, C_6H_{14}, which are completely miscible with each other. Molecules of these nonpolar substances are held together by dispersion forces of about the same magnitude. A pentane molecule experiences little or no change in intermolecular forces when it goes into solution in hexane.

Most nonpolar solutes are soluble in nonpolar solvents

Most nonpolar substances have very small water solubilities. Petroleum, a mixture of hydrocarbons, spreads out in a thin film on the surface of a body of water rather than dissolving. The mole fraction of pentane, C_5H_{12}, in a saturated water solution is only 0.00003. These low solubilities are readily understood in terms of the structure of liquid water. To dissolve appreciable amounts of pentane in water, it would be necessary to break the hydrogen bonds holding H_2O molecules together. There is no attractive force between C_5H_{12} and H_2O to supply the energy required to break into the water structure.

The hydrogen bonds in H_2O are fairly strong

Of the relatively few organic compounds that dissolve readily in water, most contain —OH groups. Three familiar examples are methyl alcohol, ethyl alcohol, and ethylene glycol, all of which are soluble in water in all proportions.

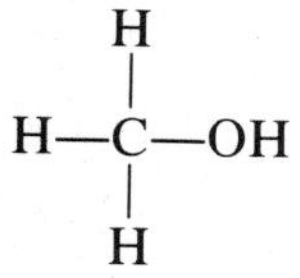

methyl alcohol 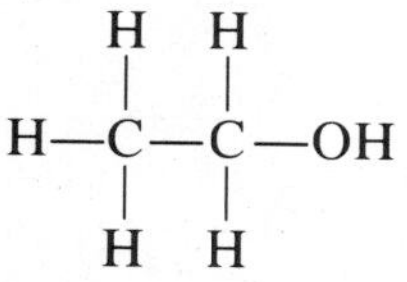   ethyl alcohol 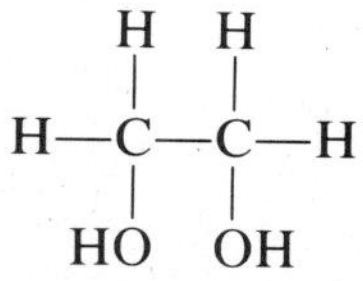   ethylene glycol

In these compounds, as in water, the principal intermolecular forces are hydrogen bonds. When a substance like methyl alcohol dissolves in water, it forms hydrogen bonds with H_2O molecules (Fig. 12.4). These hydrogen bonds, joining a CH_3OH molecule to an H_2O molecule, are about as strong as those in the pure substances.

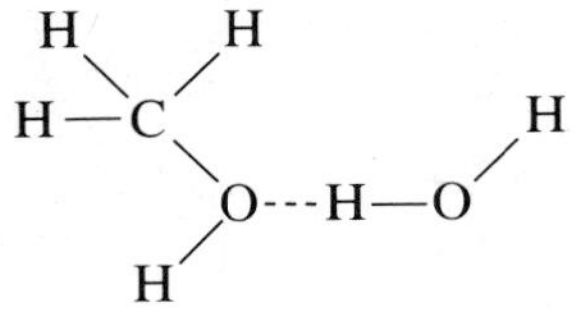

H₂C—OH structure

FIGURE 12.4 Methyl alcohol readily forms hydrogen bonds (---) with water. This explains why CH_3OH is infinitely soluble in water.

Effect of Temperature Upon Solubility

When an excess of a solid such as sodium chloride is shaken with water, it forms a saturated solution. An equilibrium is established between the solid and its ions in solution.

$$NaCl(s) \rightleftharpoons Na^+(aq) + Cl^-(aq)$$

A similar type of equilibrium is established when a gas such as carbon dioxide is bubbled through water:

$$CO_2(g) \rightleftharpoons CO_2(aq)$$

We can predict the effect of a temperature change on solubility equilibria such as these by applying a simple principle. **An increase in temperature always favors an endothermic process.** This means that if the solution process absorbs heat ($\Delta H > 0$), an increase in temperature increases the solubility. In other words, more solute goes into solution at higher temperatures. Conversely, if the solution process is exothermic ($\Delta H < 0$), an increase in temperature will decrease the solubility; heating will drive the solute out of solution.

Dissolving a solid in a liquid is usually an endothermic process; heat must be absorbed to break down the crystal lattice.

In the solution the solid, in a sense, has been melted. In the ideal case, $\Delta H = \Delta H_{fus} > 0$

$$solid + liquid \rightleftharpoons solution; \qquad \Delta H > 0 \qquad \textbf{(12.7)}$$

Applying the principle referred to above, we expect the solubilities of solids to increase as the temperature rises. Experience confirms this prediction. More sugar

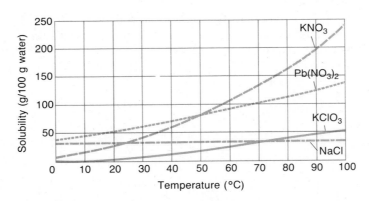

FIGURE 12.5 The water solubility of these four ionic compounds, and indeed that of most solids, increases with temperature. This reflects the fact that ΔH of solution of a solid is usually a positive quantity. With NaCl, where ΔH is very small ($\Delta H = +3.9$ kJ/mol), solubility increases only slightly with temperature.

dissolves in hot coffee than in cold coffee. Cooling a saturated solution usually causes a solid to crystallize out of solution, indicating that it is less soluble at the lower temperature. Figure 12.5 shows the effect of temperature upon the water solubility of several electrolytes; note that in each case solubility increases with temperature.

Dissolving a gas in a liquid usually evolves heat ($\Delta H < 0$):

$$\text{gas} + \text{liquid} \rightleftharpoons \text{solution}; \quad \Delta H < 0$$

This means that the reverse process (gas coming out of solution) is endothermic. Hence, it is favored by an increase in temperature; gases become less soluble as the temperature rises. This rule is followed by all gases in water. You have probably noticed this effect when heating water in an open pan or beaker. Bubbles of air are driven out of the water by an increase in temperature. The reduced solubility of oxygen in water at high temperatures (Figure 12.6A) is a major factor in the "thermal pollution" of water supplies. In the warm water discharged by large power plants, the low concentration of oxygen makes it difficult for fish and other aquatic life to survive.

Here the gas has, in a sense, been condensed. In the ideal case, $\Delta H = -\Delta H_{vap} < 0$

Effect of Pressure upon Solubility

Pressure has a major effect on solubility only for gas-liquid systems. At a given temperature, a rise in pressure increases the solubility of a gas. Indeed, at low to moderate pressures, gas solubility is directly proportional to pressure (Henry's Law; Figure 12.6B).

The solubility of KNO_3 in water does not depend appreciably on pressure

$$C_g = kP_g \tag{12.8}$$

where P_g is the partial pressure of the gas over the solution, C_g is its molarity, and k is a constant characteristic of the particular gas-liquid system. This effect arises because increasing the pressure raises the concentration of molecules in the gas phase. To balance this change and maintain equilibrium, more gas molecules enter the solution, increasing the solubility.

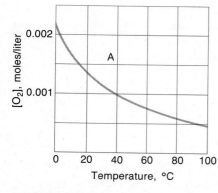
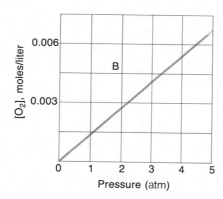

FIGURE 12.6 The solubility of $O_2(g)$ in water decreases as temperature rises (A) and increases as pressure increases (B). In the diagram at the left, the pressure is held constant at 1 atm; in (B), the temperature is held constant at 25°C.

EXAMPLE 12.9 The solubility of pure oxygen in water at 20°C and 1.00 atm pressure is 1.38×10^{-3} mol/L. Calculate the concentration of O_2 (mol/L) at 20°C and a partial pressure of 0.21 atm.

Solution One way to solve this problem is to use the solubility data for pure oxygen to calculate k in Equation 12.8. Knowing k and the partial pressure of oxygen, we can obtain the required concentration of oxygen.

k is called the Henry's Law constant

$$k = \frac{\text{conc. } O_2}{\text{pressure } O_2} = \frac{1.38 \times 10^{-3} \text{ mol/L}}{1.00 \text{ atm}} = 1.38 \times 10^{-3} \frac{\text{mol}}{\text{L·atm}}$$

Knowing k, we can now calculate the concentration of dissolved oxygen when the partial pressure is 0.21 atm.

$$\text{conc. } O_2 = \left(1.38 \times 10^{-3} \frac{\text{mol}}{\text{L·atm}} \right) (0.21 \text{ atm}) = 2.9 \times 10^{-4} \text{ mol/L}$$

The quantity just calculated is the equilibrium concentration of oxygen in water saturated with air at 20°C. In air, the partial pressure of oxygen is about 0.21 atm.

EXERCISE The partial pressure of oxygen in the lungs, at 37°C, is 101 mm Hg. Calculate the solubility of oxygen (mol/L), taking k to be 1.06×10^{-3} mol/L·atm. Answer: 0.000141 M.

The influence of partial pressure on gas solubility is used in bottling carbonated beverages such as beer, sparkling wines, and many soft drinks. These beverages are bottled under pressures of CO_2 as high as 4 atm. When the bottle or can is opened, the pressure above the liquid drops to 1 atm and the carbon dioxide bubbles rapidly out of solution. Pressurized containers for shaving cream, whipped cream, and cheese spreads work on a similar principle. Pressing a valve reduces the pressure on dissolved gas, causing it to rush from solution, carrying liquid with it as a foam.

Another consequence of the effect of pressure on gas solubility is the painful, sometimes fatal, affliction known as the "bends." This occurs when a person goes rapidly from deep water (high pressure) to the surface (lower pressure). The rapid decompression causes air, dissolved in blood and other body fluids, to bubble out of solution. These bubbles impair blood circulation and affect nerve impulses. To minimize these effects, deep-sea divers and aquanauts breathe a helium-oxygen mixture rather than compressed air (nitrogen-oxygen). Helium is only about one third as soluble as nitrogen. Hence, much less gas comes out of solution upon decompression.

SCUBA divers breathe compressed air, so have to be careful to avoid the bends

12.4
Colligative Properties
of Solutions

The properties of a solution differ considerably from those of the pure solvent. Those solution properties which depend primarily on the *concentration of solute particles* rather than their nature are called **colligative properties**. These properties include vapor pressure lowering, osmotic pressure, boiling point elevation, and freezing point depression. The fact that these properties are colligative means that all water solutions containing 0.1 mol of solute particles per liter should have about the same vapor pressure, osmotic pressure, boiling point, or freezing point. This should be true regardless of the type of solute particles present. They may be molecules, like glucose ($C_6H_{12}O_6$) or sucrose ($C_{12}H_{22}O_{11}$), or ions, as with sodium chloride (Na^+, Cl^-). In this section we will emphasize the colligative properties of nonelectrolytes, such as glucose and sucrose. Colligative properties of electrolytes will be discussed briefly at the end of this section.

An Na^+ ion and a $C_{12}H_{22}O_{11}$ molecule would have the same effect on colligative properties

The relationships among colligative properties and solute concentration are best regarded as limiting laws. They are approached most closely when the solution is very dilute. In practice, the relationships we will discuss are valid, for nonelectrolytes, to within a few per cent at concentrations as high as 1 M.

Vapor Pressure Lowering (Nonelectrolytes)

The rate at which water molecules escape from the surface is reduced in the presence of a nonvolatile solute. Concentrated aqueous solutions of nonelectrolytes such as glucose or sucrose evaporate more slowly than pure water. This reflects the fact that the vapor pressure of water in the solution is less than that of pure water. This decrease in vapor pressure is a true colligative property; that is, it is independent of the nature of the solute but directly proportional to its concentration. We find, for example, that the vapor pressure of water above a 0.10 M solution of either glucose or sucrose at 0°C is the same, about 0.008 mm Hg less than that of pure water. In 0.30 M solution, the vapor pressure lowering is almost exactly three times as great, 0.025 mm Hg.

The relationship between solvent vapor pressure and concentration is ordinarily expressed in the form of Raoult's Law:

$$P_1 = X_1 P_1^0 \tag{12.9}$$

In this equation, P_1 is the vapor pressure of solvent over the solution, P_1^0 is the vapor pressure of the pure solvent at the same temperature, and X_1 is the mole fraction of solvent. Note that since X_1 in a solution must be less than 1, P_1 must be less than P_1^0.

We can obtain a direct expression for the vapor pressure lowering by making the substitution $X_2 = 1 - X_1$, where X_2 is the mole fraction of solute in the two-component system.

$$P_1 = (1 - X_2) P_1^0$$

Rearranging,

$$P_1^0 - P_1 = X_2 P_1^0$$

The quantity $(P_1^0 - P_1)$ is the vapor pressure lowering (VPL). It is the difference between the solvent vapor pressure in the pure solvent and in solution.

$$VPL = X_2 P_1^0 \tag{12.10}$$

VPL depends on the total mole fraction of solute particles, but not on their nature

Equation 12.10 can be used to calculate the vapor pressure lowering in a solution (Example 12.10).

EXAMPLE 12.10 A solution contains 102 g of sugar, $C_{12}H_{22}O_{11}$, in 375 g of water. Calculate

a. the mole fraction of sugar.

b. the vapor pressure lowering at 25°C (vp pure water = 23.76 mm Hg).

Solution

a. The molar mass of $C_{12}H_{22}O_{11}$ is 342 g/mol; that of H_2O is 18.0 g/mol; thus,

$$\text{no. moles } C_{12}H_{22}O_{11} = 102 \text{ g} \times \frac{1 \text{ mol}}{342 \text{ g}} = 0.298 \text{ mol}$$

$$\text{no. moles } H_2O = 375 \text{ g} \times \frac{1 \text{ mol}}{18.0 \text{ g}} = 20.8 \text{ mol}$$

$$X_{\text{sugar}} = \frac{0.298}{0.298 + 20.8} = \frac{0.298}{21.1} = 0.0141$$

b. Applying Equation 12.10,

$$VPL = X_{\text{sugar}} \times P^0_{H_2O} = 0.0141 \times 23.76 \text{ mm Hg} = 0.335 \text{ mm Hg}$$

We conclude that the vapor pressure of water over this solution is:

$$23.76 \text{ mm Hg} - 0.335 \text{ mm Hg} = 23.42 \text{ mm Hg}$$

VPL tends to be quite small

EXERCISE What is the vapor pressure of water over this solution at 100°C? Answer: 749.3 mm Hg.

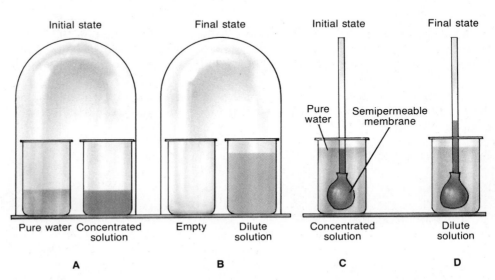

Evaporation and condensation — **Osmosis**

Initial state — Final state — Initial state — Final state

Pure water — Semipermeable membrane

Pure water / Concentrated solution — Empty / Dilute solution — Concentrated solution — Dilute solution

A — B — C — D

FIGURE 12.7 Water tends to move spontaneously from a region where its vapor pressure is high to a region where it is low. In (A) → (B), movement of water molecules occurs through the air trapped under the bell jar. In (C) → (D), water molecules move by osmosis through a semipermeable membrane. The driving force is the same in the two cases, although the mechanism differs.

One interesting effect of vapor pressure lowering is shown at the left of Figure 12.7. Here, we start with two beakers, one containing pure water and the other containing a sugar solution. These are placed next to each other, under a bell jar (Figure 12.7A). As time passes, the liquid level in the beaker containing the solution rises. The level of pure water in the other beaker falls. Eventually, by evaporation and condensation, all the water is transferred to the solution (Fig. 12.7B). At the end of the experiment, the beaker that contained pure water is empty. The driving force behind this process is the difference in vapor pressure of water in the two beakers. *Water moves from a region in which its vapor pressure is high* (pure water) *to one in which its vapor pressure is low* (sugar solution). This is a general tendency, followed by all liquids, and is responsible for a variety of natural processes, including osmosis.

It would take quite a while

Osmotic Pressure (Nonelectrolytes)

The apparatus shown in Figure 12.7C and D can be used to achieve a result similar to that found in the bell jar experiment. Here, a sugar solution is separated from water by a "semipermeable" membrane. This may be an animal bladder, a slice of vegetable tissue, or a piece of parchment. The membrane, by a mechanism that is not well understood, allows water molecules to pass through it, but not sugar molecules. Here, as before, water moves from a region where its vapor pressure is high (pure water) to a region where it is low (sugar solution). This process, taking place through a membrane permeable only to the solvent, is called **osmosis**. As a result of osmosis, the water level rises in the tube and drops in the beaker (Fig. 12.7D).

During the osmosis, only water goes through the membrane

The passage of solvent molecules through an osmotic membrane can be prevented by applying pressure to the solution (Fig. 12.8). The external pressure,

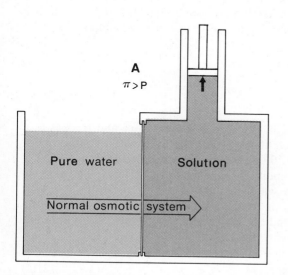
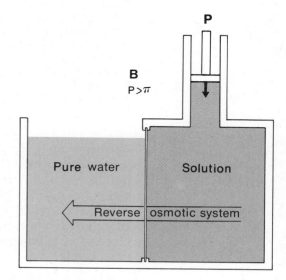

FIGURE 12.8 Osmosis can be prevented by applying to the solution a pressure P that just balances the osmotic pressure, π. If P < π, normal osmosis occurs. If P > π, water flows in the opposite direction. This process, called reverse osmosis, can be used to obtain fresh water from seawater.

P, just sufficient to prevent osmosis is equal to the **osmotic pressure** of the solution, π. If P is less than π, osmosis takes place in the normal way and water moves through the membrane into the solution (Fig. 12.8A). By making the external pressure large enough, it is possible to reverse this process (Fig. 12.8B). When $P > \pi$, water molecules move through the membrane from the solution to pure water. This process, called *reverse osmosis*, is used to obtain fresh water from seawater in arid regions of the world.

Brackish water works better than sea water

Osmotic pressure, like vapor pressure lowering, is a colligative property. For a nonelectrolyte, π is directly proportional to molarity, M. The equation relating the two quantities is very similar to the Ideal Gas Law:

$$\pi = \frac{nRT}{V} = MRT \tag{12.11}$$

n = no. moles solute
V = volume of solution

where R is the gas law constant, 0.0821 L·atm/(mol·K), and T is the Kelvin temperature. Even in dilute solution, the osmotic pressure is quite large. Suppose, for example, we are dealing with a 0.10 M solution at 25°C:

$$\pi = (0.10)(0.0821)(298) \text{ atm} = 2.4 \text{ atm}$$

A pressure of 2.4 atm is equivalent to a column of water 25 m (more than 80 ft) high. This pressure is great enough to rupture the coating of a wet, germinating seed and allow the young plant to push through the soil.

A striking example of a natural osmotic process can be seen under a microscope when red blood cells are placed in pure water. Water flows through the cell membrane to dilute the solution inside the cell. The cell swells and eventually bursts. If the red blood cells are placed in a concentrated sugar solution instead, water moves in the reverse direction. It flows from the cells, and they shrink and shrivel. To avoid these effects, solutions used in intravenous feeding are made up to have the same osmotic pressure as blood (about 7.7 atm). A 0.31 M glucose solution has this osmotic pressure and is said to be *isotonic* with blood.

It's wise to avoid these effects

Boiling Point Elevation and Freezing Point Depression (Nonelectrolytes)

A solution of a nonvolatile solute boils at a *higher* temperature and freezes at a *lower* temperature than the pure solvent. Consider, for example, a solution containing 18.0 g of glucose, $C_6H_{12}O_6$, in 100 g of water. This solution boils at 100.52°C at 1 atm pressure; the normal boiling point of pure water is 100.00°C. The glucose solution freezes at −1.86°C (fp pure water = 0.00°C).

In discussing the boiling points or freezing points of solutions, we use the terms boiling point elevation, ΔT_b, and freezing point depression, ΔT_f. These are defined so as to be positive quantities. Thus,

$$\Delta T_b = \text{bp solution} - \text{bp pure solvent} \tag{12.12}$$

$$\Delta T_f = \text{fp pure solvent} - \text{fp solution} \tag{12.13}$$

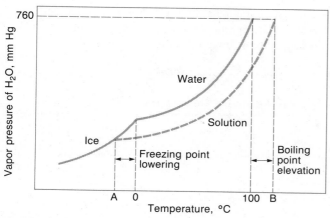

FIGURE 12.9 Since a nonvolatile solute lowers the vapor pressure of a solvent, the boiling point of a solution will be higher and the freezing point lower than the corresponding points for the pure solvent. Water solutions freeze *below* 0°C at point *A*, and boil *above* 100°C at point *B*.

For the glucose solution just referred to:

$$\Delta T_b = 100.52°C - 100.00°C = 0.52°C$$

$$\Delta T_f = 0.00°C - (-1.86°C) = 1.86°C$$

Boiling point elevation is a direct result of vapor pressure lowering. At any given temperature, a solution has a vapor pressure *lower* than that of the pure solvent. Hence, a *higher* temperature must be reached before the solution boils—that is, before its vapor pressure becomes equal to the external pressure. Figure 12.9 illustrates this reasoning graphically.

The freezing point depression, like the boiling point elevation, is a direct result of the lowering of the solvent vapor pressure by the solute. Notice from Figure 12.9 that the freezing point of the solution is the temperature at which the solvent in solution has the same vapor pressure as the pure solid solvent. This implies that it is pure solvent (e.g., ice) which separates when the solution freezes.

Boiling point elevation and freezing point depression, like vapor pressure lowering, are colligative properties. They are directly proportional to solute concentration, generally expressed as molality, m.

The equations relating boiling point elevation and freezing point depression to molality are

$$\Delta T_b = k_b \times m \qquad\qquad\qquad (12.14)$$

$$\Delta T_f = k_f \times m \qquad\qquad\qquad (12.15)$$

The proportionality constants in these equations, k_b and k_f, are called the *molal boiling point constant* and the *molal freezing point constant*. Their magnitudes depend upon the nature of the solvent (Table 12.3). Note that when the solvent is water

$$k_b = 0.52°C; \qquad k_f = 1.86°C$$

When an aqueous solution freezes, pure H_2O, ice, freezes out

Here we assume that the solute is nonvolatile

Molalities are used here, mainly for historic reasons

This tells us that a 1.00 m solution of a nonelectrolyte in water will boil at 100.52°C (at 1 atm) and freeze at −1.86°C. The organic solvents listed in Table 12.3 have values of k_f and k_b which are larger than those of water.

Table 12.3
Molal Freezing Point and Boiling Point Constants

SOLVENT	fp (°C)	k_f	bp (°C)	k_b
Water	0	1.86	100	0.52
Acetic acid	17	3.90	118	3.07
Benzene	5.50	5.10	80.0	2.53
Cyclohexane	6.5	20.2	81	2.79
Camphor	178	40.0	208	5.95
p-Dichlorobenzene	53	7.1	—	—

The use of Equations 12.14 and 12.15 is illustrated in Example 12.11.

EXAMPLE 12.11 An antifreeze solution is prepared containing 50.0 cm^3 of ethylene glycol, $C_2H_6O_2$ (d = 1.12 g/cm^3), in 50.0 g of water. Calculate the freezing point of this "50-50" mixture.

That's what we use in Minnesota

Solution We first calculate the mass of ethylene glycol in the solution, then the number of moles of ethylene glycol (molar mass $C_2H_6O_2$ = 62.0 g/mol), and then the molality. Finally, we use Equation 12.15 to calculate the freezing point lowering (k_f = 1.86°C).

$$\text{mass } C_2H_6O_2 = 50.0 \text{ cm}^3 \times 1.12 \frac{g}{cm^3} = 56.0 \text{ g}$$

$$\text{no. moles } C_2H_6O_2 = 56.0 \text{ g} \times \frac{1 \text{ mol}}{62.0 \text{ g}} = 0.903 \text{ mol}$$

$$\text{molality} = \frac{0.903 \text{ mol } C_2H_6O_2}{0.0500 \text{ kg water}} = 18.1 \text{ m}$$

$$\Delta T_f = k_f \times m = 1.86°C \times 18.1 = 33.7°C$$

Last winter, Dec. '83, we got to −35°F, only once, thank goodness   EJS

We conclude that the freezing point of the solution should be 33.7°C below that of pure water (0°C). Hence, the solution should freeze at −33.7°C. Actually, the freezing point is somewhat lower, about −37°C or −35°F. The deviation occurs because Equation 12.15 is a limiting law, strictly valid only at low concentrations.

EXERCISE Estimate the boiling point of this solution at 1 atm. Answer: 109.4°C.

As you know, we take advantage of the freezing point depression when we add antifreeze to automobile radiators in winter. Ethylene glycol is the solute commonly used in so-called "permanent" antifreezes. It has a high boiling point

(197°C), is virtually nonvolatile at 100°C and, as we have seen, raises the boiling point of water. Hence, antifreeze containing ethylene glycol does not boil away in summer driving. Earlier antifreezes contained methyl alcohol. It is cheaper than ethylene glycol but has disadvantages in summer driving. It is volatile (bp = 65°C), lowers the boiling point of water,* and tends to boil out of solution. Ethyl alcohol behaves similarly.

Determination of Molar Masses of Nonelectrolytes from Colligative Properties

The molar mass of a gas or volatile liquid can be obtained from gas density measurements, as described in Chapter 6. This method is useless for nonvolatile solids or for substances that decompose on heating. Such substances do not ordinarily exist as vapors. It turns out that colligative properties, particularly freezing point depression, can be used to determine molar masses of a wide variety of nonelectrolytes. The approach used is illustrated in Example 12.12.

EXAMPLE 12.12   A student dissolves 1.50 g of a newly prepared compound in 75.0 g of cyclohexane. She measures the freezing point of the solution to be 2.70°C; that of pure cyclohexane is 6.50°C. Cyclohexane has a k_f of 20.2°C. Using these data, calculate the molar mass of the compound.

Solution   We follow a three-step procedure. First, we obtain ΔT_f, using Equation 12.13. Then we use Equation 12.15 to calculate the molality. Finally, we obtain the molar mass, using the defining equation for molality (Equation 12.4):

1. ΔT_f = fp pure cyclohexane − fp solution = 6.50°C − 2.70°C = 3.80°C

2. $\Delta T_f = k_f \times m$; $m = \dfrac{\Delta T_f}{k_f} = \dfrac{3.80°C}{20.2°C} = 0.188$

3. $m = \dfrac{\text{no. moles solute}}{\text{no. kilograms solvent}} = \dfrac{\text{no. grams solute/MM}}{\text{no. kilograms solvent}}$

   where MM is the molar mass (g/mol). Solving the above equation for MM,

   $$MM = \frac{\text{no. grams solute}}{(m)(\text{no. kilograms solvent})}$$

   All the quantities on the right side of this equation are known. We know that 1.50 g of solute is dissolved in 75.0 g of solvent (0.0750 kg); we calculated m to be 0.188. Hence:

   $$MM = \frac{1.50}{(0.188)(0.0750)} \frac{g}{mol} = 106 \text{ g/mol}$$

For good results in this experiment, you need a sensitive thermometer (± 0.01°C), and good technique

---

*Volatile solutes ordinarily lower the boiling point because they contribute to the total vapor pressure of the solution.

EXERCISE A solution of 5.00 g of a compound X in 60.0 g of water freezes at −1.00°C. What is the molar mass of X? Answer: 155 g/mol.

In general, to determine the molar mass from freezing point lowering, you follow a three-step procedure:

1. From the observed freezing point, calculate ΔT_f:

$$\Delta T_f = \text{fp solvent} - \text{fp solution}$$

The freezing point of pure water is 0°C; freezing points of other solvents are listed in Table 12.3.

2. Knowing k_f (Table 12.3) and having calculated ΔT_f in (1), determine the molality of the solution:

$$\Delta T_f = k_f \times m; \qquad m = \Delta T_f / k_f$$

3. Using the defining equation for molality (Equation 12.4), calculate the molar mass, MM:

$$\text{molality} = \frac{\text{no. moles solute}}{\text{no. kilograms solvent}} = \frac{\text{no. grams solute/MM}}{\text{no. kilograms solvent}}$$

$$\text{MM} = \frac{\text{no. grams solute}}{(\text{molality})(\text{no. kilograms solvent})}$$

In the old days we used camphor as the solvent in this procedure (Rast method)

In carrying out a molar mass determination by freezing point depression, we must choose a solvent in which the solute is readily soluble. Usually, several such solvents are available. Of these, we tend to pick one that has a large k_f. This makes ΔT_f large and so reduces the per cent error in the freezing point measurement. From this point of view, cyclohexane or other organic solvents are better choices than water, since their k_f values are larger.

Molar masses can also be determined using other colligative properties. Osmotic pressure measurements are often used, particularly for solutes of high molar mass where the concentration is likely to be quite low. The advantage of using osmotic pressure is that the effect is relatively large, as pointed out on p. 372.

EXAMPLE 12.13 A solution contains 1.0 g of hemoglobin dissolved in enough water to form 100 cm³ of solution. The osmotic pressure at 20°C is found to be 2.75 mm Hg. Calculate:
a. the molarity of hemoglobin.
b. the molar mass, MM, of hemoglobin.

Solution

a. Rearranging Equation 12.11 to solve for the molarity, M,

$$\text{molarity} = \pi/RT$$

But, $\pi = (2.75/760)$ atm; $R = 0.0821$ L·atm/(mol·K); $T = 293$ K. Hence:

$$\text{molarity} = \frac{2.75/760}{(0.0821)(293)} \text{ mol/L} = 1.50 \times 10^{-4} \text{ mol/L}$$

b. From the defining equation for molarity:

$$\text{molarity} = \frac{\text{no. moles solute}}{\text{no. liters solution}} = \frac{\text{no. grams solute/MM}}{\text{no. liters solution}}$$

Solving for the molar mass, MM,

$$MM = \frac{\text{no. grams solute}}{(\text{molarity})(\text{no. liters solution})}$$

Recall that there is 1.0 g of hemoglobin in 100 cm^3 of solution. Hence:

$$MM = \frac{1.0 \text{ g}}{(1.50 \times 10^{-4} \text{ mol/L})(0.100 \text{ L})} = 6.7 \times 10^4 \text{ g/mol}$$

Osmotic pressure offers one of the best methods for obtaining the molar masses of proteins and industrial polymers

EXERCISE What would be the osmotic pressure at 20°C of a solution containing 1.0 g of hemoglobin (MM = 6.7 × 10^4 g/mol) per liter? Answer: 0.28 mm Hg, one tenth of that above.

Colligative Properties of Electrolytes

As noted earlier, colligative properties of dilute solutions are directly proportional to the concentration of solute *particles*. On this basis, we would predict that, at a given concentration, an electrolyte would have a greater effect upon these properties than a nonelectrolyte. When one mole of a nonelectrolyte such as glucose dissolves in water, 1 mol of solute molecules is obtained. On the other hand, one mole of the electrolyte NaCl yields 2 mol of ions. With calcium chloride, $CaCl_2$, 3 mol of ions are produced per mole of solute.

This reasoning is confirmed experimentally. Suppose, for example, we compare the vapor pressure of 1 M solutions of glucose, sodium chloride, and calcium chloride. We find that the vapor pressure lowering is smallest for glucose and largest for calcium chloride. In other words,

vp pure water > vp glucose solution > vp NaCl solution > vp CaCl$_2$ solution

Many electrolytes form saturated aqueous solutions whose vapor pressures are so low that the solids pick up water (*deliquesce*) when exposed to moist air. This occurs with calcium chloride, whose saturated solution has a vapor pressure only 20% that of pure water. If dry $CaCl_2$ is exposed to air in which the relative humidity is greater than 20%, it absorbs water and forms a saturated solution. Deliquescence continues until the vapor pressure of the solution becomes equal to that of the water in the air.

The freezing points of electrolyte solutions, like their vapor pressures, are lower than those of nonelectrolytes at the same concentration. Sodium chloride and calcium chloride are used to melt ice on highways; their aqueous solutions can have freezing points as low as −21°C and −55°C, in that order. The equation for the freezing point lowering of an electrolyte is similar to that for nonelectrolytes, except for the introduction of a multiplier, i. For aqueous solution of electrolytes

Those salt solutions unfortunately also cause rusting of car bodies

$$\Delta T_f = 1.86°C \times m \times i \tag{12.16}$$

If we assume that the ions of an electrolyte behave independently, we would predict that i should be equal to *the number of moles of ions per mole of electrolyte*. Thus i should be 2 for NaCl and $MgSO_4$, 3 for $CaCl_2$ and Na_2SO_4, and so on.

EXAMPLE 12.14 Estimate the freezing points of 0.20 m solutions of
a. KNO_3 b. $MgSO_4$ c. $Cr(NO_3)_3$
Assume that i in Equation 12.16 is the number of moles of ions formed per mole of electrolyte.

Solution
a. One mole of KNO_3 forms two moles of ions:

$$KNO_3(s) \rightarrow K^+(aq) + NO_3^-(aq)$$

Hence, i should be 2 and we have

$$\Delta T_f = (1.86°C)(0.20)(2) = 0.74°C; \qquad T_f = -0.74°C$$

b. $MgSO_4$ behaves like KNO_3, that is, i = 2:

$$MgSO_4(s) \rightarrow Mg^{2+}(aq) + SO_4^{2-}(aq)$$

$$\Delta T_f = (1.86°C)(0.20)(2) = 0.74°C; \qquad T_f = -0.74°C$$

c. For $Cr(NO_3)_3$, i = 4:

$$Cr(NO_3)_3(s) \rightarrow Cr^{3+}(aq) + 3 NO_3^-(aq)$$

$$\Delta T_f = (1.86°C)(0.20)(4) = 1.5°C; \qquad T_f = -1.5°C$$

EXERCISE What is the estimated freezing point of 0.20 m Na_2CO_3? Answer: $-1.1°C$.

These calculations are approximate. See Table 12.4

Looking at the data in Table 12.4, we see that the situation is not so simple as our discussion might imply. The observed freezing point lowerings of NaCl and $MgSO_4$ are smaller than we would predict from Equation 12.16 with i = 2. In other words, at any finite concentration, the multiplier i is less than 2. It approaches a limiting value of 2 as the solution becomes more and more dilute. This behavior is generally typical of electrolytes. Interactions occur between

Table 12.4
Freezing Point Lowerings of Solutions

	ΔT_f OBSERVED (°C)		i (CALC. FROM EQ. 12.16)	
MOLALITY	NaCl	$MgSO_4$	NaCl	$MgSO_4$
0.005	0.0182	0.0160	1.96	1.72
0.01	0.0360	0.0285	1.94	1.53
0.02	0.0714	0.0534	1.92	1.44
0.05	0.176	0.121	1.89	1.30
0.10	0.348	0.225	1.87	1.21
0.20	0.685	0.418	1.84	1.12
0.50	1.68	0.495	1.81	1.07

The deviation of i from integral values is larger for a +2: −2 electrolyte than for a +1: −1

positive and negative ions in solution. As a result, they do not behave as completely independent particles. The effect is ordinarily to make the observed freezing point depression or other colligative property less than we would expect from the number of ions in solution.

Summary

A solution consists of solute(s) and solvent. In this chapter we have dealt mainly with water solutions. Solutes can be classified as electrolytes or nonelectrolytes, depending upon their dissociation behavior in water (Table 12.2 and Example 12.1). Nonelectrolytes tend to be insoluble in water unless they are capable of forming hydrogen bonds.

Solubilities are affected by temperature. If the solution process is endothermic, as is most often the case with solids, the solute becomes more soluble as temperature increases. If the solution process is exothermic, as with gases dissolving in water, solubility decreases with increasing temperature. The solubility of gaseous solutes is increased by increasing the pressure of the gas over the solution (Example 12.9).

Solutes lower the vapor pressure of a solvent, decrease the freezing point, and increase the boiling point (Fig. 12.9 and Examples 12.10 and 12.11). These properties, along with the osmotic pressure (Figs. 12.7 and 12.8), are colligative. The extent to which vapor pressure is lowered depends primarily upon the concentration rather than the type of solute particle. The same is true of freezing point depression and boiling point elevation. Colligative properties, particularly freezing point lowering and osmotic pressure, can be used to determine the molar mass of a nonelectrolyte (Examples 12.12 and 12.13). Since electrolytes dissociate in water, they have a greater effect on colligative properties than do nonelectrolytes (Equation 12.16 and Example 12.14).

In this chapter, we have referred to four different concentration units: mass percent, mole fraction, molality, and molarity. To compare these units, consider a solution containing 15.0 g of sugar (MM = 342 g/mol) dissolved in 110 g of water. This solution is found to have a density of 1.05 g/cm^3 and hence a volume of

$$125 \text{ g} \times \frac{1 \text{ cm}^3}{1.05 \text{ g}} = 119 \text{ cm}^3$$

Thus, we have

$$\text{molarity} = \frac{\text{no. moles solute}}{\text{no. liters solution}} = \frac{15.0/342 \text{ mol}}{0.119 \text{ L}} = 0.369 \text{ mol/L}$$

$$\text{molality} = \frac{\text{no. moles solute}}{\text{no. kilograms solvent}} = \frac{15.0/342 \text{ mol}}{0.110 \text{ kg } H_2O} = 0.399 \text{ mol/kg } H_2O$$

$$X_{sugar} = \frac{\text{no. moles sugar}}{\text{no. moles sugar + no. moles water}} = \frac{15.0/342}{15.0/342 + 110/18.0}$$
$$= 0.00714$$

$$X_{water} = 1 - 0.00714 = 0.99286$$

$$\text{mass percent sugar} = \frac{\text{mass sugar}}{\text{total mass solution}} \times 100$$

$$= \frac{15.0 \text{ g}}{15.0 \text{ g} + 110 \text{ g}} \times 100 = 12.0\%$$

The use of these concentration units is further illustrated in Examples 12.2 and 12.3. Ways to prepare a solution to a desired molarity are considered in Example 12.4, where we start with pure solid, and Example 12.5, where we start with a concentrated solution. The molarity of a solute is readily converted to molarities of the corresponding ions (Example 12.6). Conversions between different concentration units are considered in Examples 12.7 and 12.8.

Key Words and Concepts

boiling point elevation	molality	semipermeable membrane
colligative property	molarity	solubility
concentrated solution	mole fraction	solute
dilute solution	nonelectrolyte	solution
electrolyte	osmosis	solvent
freezing point depression	osmotic pressure	supersaturated solution
Henry's Law	Raoult's Law	unsaturated solution
hydrogen bond	reverse osmosis	vapor pressure lowering
mass percent	saturated solution	

Questions and Problems

Solution Terminology

12.1 You are given a clear water solution containing KNO_3. How would you determine experimentally whether the solution is unsaturated, saturated, or supersaturated?

12.2 Write equations for the dissociation in water of each of the following electrolytes:
a. MgI_2 b. $KClO_4$ c. $RbHCO_3$ d. $Sc_2(SO_4)_3$

12.3 How many moles of ions are present in water solutions prepared by dissolving 0.10 mol of
a. K_2SO_4 b. $Fe(NO_3)_3$ c. $Al_2(SO_4)_3$ d. $NiSO_4$

12.31 How would you prepare a saturated solution of $CO_2(g)$ in water? a supersaturated solution?

12.32 Follow the directions of Question 12.2 for
a. $CaCl_2$ b. $LiNO_3$ c. $Ni(ClO_3)_2$ d. $Ca_3(PO_4)_2$

12.33 How many moles of each of the following electrolytes must be dissolved in water to form 0.12 mol of ions?
a. $CaCl_2$ b. $Cr(NO_3)_3$ c. $MgSO_4$ d. Na_2CO_3

Concentrations of Solutions

12.4 A solution is prepared by dissolving 1.25 g of K_2CrO_4 in 11.6 g of water. Calculate
a. the mass percent of K_2CrO_4.
b. the mass percent of H_2O.
c. the molality of K_2CrO_4.

12.34 A solution is made by dissolving 1.25 g of C_2H_5OH in 11.6 g of water. Calculate
a. the mass percent of C_2H_5OH.
b. the mass percent of H_2O.
c. the molality of C_2H_5OH.

12.5 Formalin is a preservative solution used in biology laboratories. It contains 40 cm³ of formaldehyde, CH_2O (d = 0.82 g/cm³), per 100 cm³ of water. What is the molality of formaldehyde in formalin?

12.6 A solution contains 50.0 g of carbon disulfide (CS_2) and 50.0 g of chloroform ($CHCl_3$). Calculate the mole fraction of each component.

12.7 Complete the following table for water solutions of glucose, $C_6H_{12}O_6$:

Mass Solute	Moles Solute	V Solution	M
12.5 g	——	219 mL	——
——	1.08	——	0.519
——	——	1.62 L	1.08

12.8 Describe in some detail how you would prepare 220 mL of 0.500 M KOH starting with
a. solid KOH   b. 1.25 M KOH

12.9 A solution is made by diluting 150 mL of 0.210 M $Ca(NO_3)_2$ solution with water to a final volume of 450 mL. Calculate
a. the molarity of $Ca(NO_3)_2$ in the diluted solution.
b. the molarities of Ca^{2+} and NO_3^- in the diluted solution.
c. the number of moles of NO_3^- ion in 1.00×10^2 mL of the original solution; the diluted solution.

12.10 As a lab assistant, you are asked to make 1.50 L of 0.250 M HNO_3 by diluting concentrated HNO_3, 16.0 M.
a. What volume of the concentrated acid is required?
b. Assuming the volumes are additive, what volume of water should be used in dilution?

12.11 The molality of sugar, $C_{12}H_{22}O_{11}$, in a water solution is 1.62. Calculate
a. the mole fractions of sugar and water.
b. the mass percents of sugar and water.

12.12 Complete the following table for NaOH solutions:

d (g/cm³)	Molarity	Molality	Mass % NaOH
1.05	1.32	——	——
1.22	——	——	20.0
1.35	——	11.8	——

12.35 The "proof" of an alcoholic beverage is twice the percentage by volume of ethyl alcohol, C_2H_5OH. For a 120-proof bourbon whiskey, what is the molality of ethyl alcohol (d C_2H_5OH = 0.80 g/cm³; d H_2O = 1.00 g/cm³)?

12.36 A 100.0-g solution contains 20.0 g of methyl alcohol, CH_3OH; the remainder is ethyl alcohol, C_2H_5OH. Calculate the mole fraction of each component.

12.37 Complete the following table for water solutions of sodium sulfate, Na_2SO_4:

Mass Solute	Moles Solute	V Solution	M
18.8 g	——	352 mL	——
——	0.0291	——	0.100
——	——	2.92 L	0.246

12.38 How would you prepare 6.42 L of 0.100 M Na_2CO_3 starting with
a. 0.200 M Na_2CO_3 b. 0.543 M Na_2CO_3

12.39 A solution is made by diluting 235 mL of 0.120 M $Fe(NO_3)_3$ solution with water to a final volume of 0.500 L. Calculate
a. the molarities of $Fe(NO_3)_3$, Fe^{3+}, and NO_3^- in the diluted solution.
b. the molarities of Fe^{3+} and NO_3^- in the original solution.

12.40 A student made 750 mL H_2SO_4 solution by diluting 20.0 mL of a more concentrated solution. He was not sure whether he had used concentrated H_2SO_4 (18.0 M) or dilute H_2SO_4 (3.0 M). What is the molarity of the diluted solution if
a. the concentrated acid was used?
b. the dilute acid was used?

12.41 The mole fraction of ethyl alcohol, C_2H_5OH, in a water solution is 0.0532. Calculate
a. the mole fraction of water.
b. the molality of ethyl alcohol.
c. the mass percents of both components.

12.42 Complete the following table for solutions of H_2SO_4:

d (g/cm³)	Molarity	Molality
1.14	2.33	——
1.30	——	6.80

Solubilities

12.13 Choose the member of each set that you would expect to be the most soluble in water. Explain your reasoning.
a. NH_3, CH_4, or H_2S
b. ethane or ethylene glycol
c. H_2, Br_2, or $NaOH$

12.14 A certain gaseous solute dissolves in water, evolving 4.8 kJ/mol of heat. Its solubility at 25°C and 2.00 atm is 0.010 M. Would you expect the solubility to be greater or less than 0.010 M at
a. 0°C and 5 atm? b. 50°C and 1 atm?
c. 15°C and 2 atm? d. 25°C and 1 atm?

12.15 Using Figure 12.6, estimate the solubility of O_2 in water at
a. 20°C, 1 atm b. 25°C, 1 atm
c. 25°C, 10 atm

12.43 For each of the following pairs of solutes, state which one you would expect to be the more soluble in water. Explain your reasoning.
a. $NaCl$ or CCl_4
b. CH_3—OH or CH_3—O—CH_3
c. C_6H_6 or HO—OH
d. CO_2 or SiO_2

12.44 Consider the process by which ammonium chloride dissolves in water:

$$NH_4Cl(s) \rightarrow NH_4^+(aq) + Cl^-(aq)$$

a. Using data from tables in Chapter 5, calculate ΔH for this reaction.
b. Would you expect the solubility of NH_4Cl to increase or decrease if the temperature is lowered?

12.45 A soft drink is bottled under a carbon dioxide pressure of 2.2 atm. What fraction of the dissolved CO_2 comes out of solution when the bottle is opened and P CO_2 drops to
a. 1.0 atm b. 0.010 atm

Colligative Properties

12.16 Calculate the vapor pressure lowering in water solution at 25°C (vp pure water = 23.76 mm Hg) when the mole fraction of nonelectrolyte is
a. 0.0100 b. 0.100 c. 0.200
What is the vapor pressure of water over each of these solutions?

12.17 The vapor pressure of pure CCl_4 at 25°C is 114 mm Hg. What mass of iodine, I_2, must be dissolved in 1.00 L of CCl_4 (d = 1.60 g/cm³) to lower the vapor pressure by 1.00 mm Hg?

12.18 Calculate the osmotic pressure at 25°C in solutions of sugar, $C_{12}H_{22}O_{11}$, containing the following masses of solute per liter of solution:
a. 10.0 g b. 25.0 g c. 155 g

12.19 Calculate the freezing point and normal boiling point of each of the following solutions:
a. 14.9 g glucose, $C_6H_{12}O_6$, in 100 g of water.
b. 6.00 g of urea, $(NH_2)_2CO$, in 80.0 g of water.

12.46 Repeat the calculations called for in Problem 12.16 at 65°C, where the vapor pressure of water is 188 mm Hg.

12.47 A solution contains 10.0 g of benzene (C_6H_6) and 10.0 g of toluene (C_7H_8). The vapor pressure of benzene over the solution at 20°C is 41 mm Hg. What is the vapor pressure of pure benzene at 20°C?

12.48 Calculate the osmotic pressure at 25°C of solutions containing 10.0 g of the following solutes per liter:
a. $C_6H_{12}O_6$ b. $C_2H_5O_2N$ c. $CO(NH_2)_2$

12.49 How many grams of the following nonelectrolyte solutes would have to be dissolved in 50.0 g of water to give a solution freezing at −1.40°C? What would be the normal boiling point of each solution?
a. sucrose, $C_{12}H_{22}O_{11}$
b. glycerol, $C_3H_8O_3$

12.20 An automobile radiator is filled with an antifreeze mixture containing 4.00 L of ethylene glycol, $C_2H_6O_2$ (d = 1.12 g/cm^3), with 6.00 L of water (d = 1.00 g/cm^3). Calculate
a. the molality of ethylene glycol.
b. the freezing point of the solution, assuming Equation 12.15 is valid.

12.21 Using Table 12.3, calculate the freezing point lowering and the freezing point of 0.20 m solutions of a nonelectrolyte in
a. benzene   b. cyclohexane   c. camphor

12.22 Vitamin C contains 40.9% C, 4.58% H, and 54.5% O. A solution of 6.70 g of vitamin C in 50.0 g of water freezes at −1.42°C. What is the molecular formula of vitamin C?

12.23 A student dissolves 0.180 g of a nonelectrolyte in 50.0 g of benzene and determines the freezing point to be 5.15°C. Calculate the molar mass of the solute (see Table 12.3).

12.24 The molar mass of a type of hemoglobin was determined by osmotic pressure measurement. A student measured an osmotic pressure of 4.60 mm Hg for a solution at 20°C containing 3.27 g of hemoglobin in 0.200 L of solution. What is the molar mass of the hemoglobin?

12.25 Estimate the freezing points of 0.10 m solutions of
a. K_2SO_4 b. $CsNO_3$ c. $Al(NO_3)_3$

12.26 The freezing point of 0.20 m HF is −0.38°C. Is HF primarily nonionized in this solution (HF molecules), or is it dissociated to H^+ and F^- ions?

General

12.27 A water solution containing 338 g of sugar, $C_{12}H_{22}O_{11}$, per liter has a density of 1.127 g/cm^3 at 20°C. For this solution, calculate the
a. molarity   b. molality
c. osmotic presure d. freezing point

12.50 What volume of ethylene glycol must be added to the mixture in Problem 12.20 to produce a solution freezing at −30.0°C?

12.51 Calculate the freezing point and boiling point of a solution containing 12.0 g of naphthalene, $C_{10}H_8$, in 50.0 g of benzene.

12.52 Nicotine has the empirical formula C_5H_7N. A solution of 1.6 g of nicotine in 12.0 g of water boils at 100.42°C at 760 mm Hg. What is the molecular formula of nicotine?

12.53 The freezing point of p-dichlorobenzene is 53.1°C; its k_f value is 7.10°C. A solution of 1.52 g of sulfanilamide (a sulfa drug) in 10.0 g of p-dichlorobenzene freezes at 46.7°C. What is the molar mass of sulfanilamide?

12.54 A biochemist isolated a new protein and determined its molar mass by osmotic pressure measurements. She used 0.270 g of the protein in 50.0 mL of solution and observed an osmotic pressure of 3.86 mm Hg for this solution at 25°C. What should she report as the molar mass of the new protein?

12.55 Arrange 0.10 m solutions of the following solutes in order of decreasing freezing point:
a. $C_2H_6O_2$ b. $CrCl_3$ c. $Al_2(SO_4)_3$ d. Na_2CO_3

12.56 The freezing point of 0.10 m $KHSO_3$ is −0.38°C. Which of the following equations best represents what happens when $KHSO_3$ dissolves in water?
a. $KHSO_3(s) \rightarrow KHSO_3(aq)$
b. $KHSO_3(s) \rightarrow K^+(aq) + HSO_3^-(aq)$
c. $KHSO_3(aq) \rightarrow K^+(aq) + H^+(aq) + SO_3^{2-}(aq)$

12.57 A water solution containing 0.200 kg of glycerol, $C_3H_8O_3$, in 0.800 kg of water has a density of 1.047 g/cm^3 at 20°C. For this solution, calculate the
a. molality   b. molarity
c. freezing point   d. osmotic pressure

12.28 Explain why
  a. a water solution of HCl conducts an electric current but HCl(l) does not.
  b. a water solution of NaCl conducts an electric current but NaCl(s) does not.
  c. molality and molarity are nearly the same in dilute water solution.
  d. pressure must be applied to cause reverse osmosis to occur.

12.29 Criticize the following statements:
  a. A saturated solution is always a concentrated solution.
  b. The water solubility of a solid always decreases with a drop in temperature.
  c. For aqueous solutions, molarity and molality are equal.
  d. The freezing point depression of a 0.10 m $CaCl_2$ solution is twice that of a 0.10 m KCl solution.
  e. A 0.10 m sucrose solution and a 0.10 m NaCl solution have the same osmotic pressure.

12.30 In your own words, explain why
  a. salt is added to ice in an ice cream maker to freeze the ice cream.
  b. seawater has a lower freezing point than fresh water.
  c. more ethyl alcohol, C_2H_5OH, dissolves in 100 g of water than in 100 g of octane, C_8H_{18}, at the same temperature.
  d. we believe vapor pressure lowering is a colligative property.

12.58 Explain why
  a. the freezing point of 0.10 m $CaCl_2$ is lower than that of 0.10 m $MgSO_4$.
  b. 0.10 M $CaCl_2$ has a higher electrical conductivity than 0.10 M NaCl.
  c. the solubility of solids in water usually increases as temperature increases.
  d. the solubility of gases in water decreases as temperature increases.

12.59 In your own words, explain
  a. why the concentrations of solutions used for intravenous feeding must be controlled carefully.
  b. why, when making fudge (a supersaturated sugar mixture), one must be careful to prevent it from getting "grainy."
  c. why fish in a lake seek deep, shaded places during summer afternoons.
  d. what causes the "bends" in divers.
  e. why champagne "fizzes" in a glass.

12.60 Explain, in your own words,
  a. how to determine experimentally whether a pure substance is an electrolyte or nonelectrolyte.
  b. why a cold glass of beer goes "flat" upon warming.
  c. the differences between molarity and molality.
  d. why the boiling point is raised by the presence of a solute.

*12.61 The water-soluble nonelectrolyte X has a molar mass of 410 g/mol. A 0.100-g mixture containing this substance and sugar (MM = 342 g/mol) is added to 1.00 g of water to give a solution freezing at $-0.500°C$. Estimate the mass percent of X in the mixture.

*12.62 A martini, weighing about 5.0 oz (142 g), contains 30% by mass of alcohol. About 15% of the alcohol in the martini passes directly into the blood stream (7.0 L for an adult). Estimate the concentration of alcohol in the blood (g/cm^3) of a person who drinks two martinis before dinner. (A concentration of 0.0030 g/cm^3 or more is frequently considered indicative of intoxication in a "normal" adult.)

*12.63 When water is added to a mixture of aluminum metal and sodium hydroxide, hydrogen gas is produced; this reaction is used in commercial drain cleaners:

$$2 \, Al(s) + 6 \, H_2O(l) + 2 \, OH^-(aq) \rightarrow 2 \, Al(OH)_4^-(aq) + 3 \, H_2(g)$$

A sufficient amount of water is added to 49.92 g of NaOH to make 0.600 L of solution; 41.28 g Al is added to this solution and hydrogen gas is formed.
  a. Calculate the molarity of the initial NaOH solution.
  b. How many moles of hydrogen were formed?
  c. The hydrogen was collected at 25°C and 758.6 mm Hg. The vapor pressure of water at this temperature is 23.8 mm Hg. What volume of hydrogen was generated?

*12.64 It is found experimentally that the volume of a gas that dissolves in a given amount of water is independent of the pressure of the gas; that is, if 5 cm³ of a gas dissolves in 100 g of water at 1 atm pressure, 5 cm³ will dissolve at a pressure of 2 atm, 5 atm, 10 atm, ... Show that this relationship follows logically from Henry's Law and the Ideal Gas Law.

*12.65 If osmosis were responsible for sap rising in a tree, calculate the approximate height to which the sap could rise if it were 0.13 M in sugar and the water outside the tree contained dissolved solids equivalent to a 0.02 M solution. (Note that the pressure exerted by a column of liquid is directly proportional to its density; the density of water is 0.0735 times that of mercury.)

Chapter 13
Structures
of the
Nonmetals
and Their
Binary
Compounds

Of 108 known elements, only about 20 can be classified as nonmetals or metalloids. These include

- —the noble gases in Group 8.
- —the halogens (and hydrogen) in Group 7.
- —the Group 6 elements, of which only two, oxygen and sulfur, are common.
- —the four elements nitrogen, phosphorus, arsenic, and antimony in Group 5.
- —carbon, silicon, and germanium in Group 4.

We will look at the molecular structures and some of the properties of these elements in Section 13.1.

Even though there are relatively few nonmetals, they play a major role in chemistry. Nearly all compounds contain at least one nonmetallic element. There are a great many binary compounds containing two different nonmetals. We will discuss several classes of such compounds in this chapter. These include

Over 90% of all compounds contain only nonmetallic atoms

- —the hydrogen compounds of the nonmetals in Groups 5, 6 and 7 (Section 13.2).
- —the oxygen compounds of the nonmetals (Section 13.3).
- —compounds formed by the halogens with nonmetals, including other halogens and the noble gases (Section 13.4).

In the last section of this chapter (Section 13.5) we will consider the most abundant type of binary compound of the nonmetals, the hydrocarbons, which contain the two elements hydrogen and carbon.

Throughout this chapter, we will frequently refer back to material covered in the last four chapters. In-text examples and end-of-chapter problems will review

- —Lewis structures, resonance forms, and bond energy (Chap. 9).
- —molecular geometry, polarity, hybridization, and expanded octets (Chap. 10).

—phase equilibria, solid structures, and intermolecular forces (Chap. 11).
—concentrations of solutions (Chap. 12).

13.1
The Nonmetallic Elements

Table 13.1 lists the melting and boiling points of the nonmetals and indicates their particle structures. Each of these elements has one of two structures.

Table 13.1
Structures and Properties of the Nonmetals

	GROUP 4	GROUP 5	GROUP 6	GROUP 7	GROUP 8
Structure mp, bp (°C)				Hydrogen H_2 −259, −253	Helium He −272, −269
Structure mp, bp (°C)	Carbon Net. Cov. 3570, subl.	Nitrogen N_2 −210, −196	Oxygen O_2 −218, −183	Fluorine F_2 −220, −188	Neon Ne −249, −246
Structure mp, bp (°C)	Silicon Net. Cov. 1414, 2355	Phosphorus P_4* 44, 280	Sulfur S_8 119, 444	Chlorine Cl_2 −101, −34	Argon Ar −189, −186
Structure mp, bp (°C)	Germanium Net. Cov. 937, 2830	Arsenic As_4* 814, subl.	Selenium Se_8* 217, 685	Bromine Br_2 −7, 59	Krypton Kr −157, −152
Structure mp, bp (°C)		Antimony Sb_4* 631, 1380	Tellurium Net. Cov. 450, 990	Iodine I_2 114, 184	Xenon Xe −112, −107

*These elements also form network covalent solids.

1. *Network covalent* structures, in which nonmetal atoms are bonded to one another in a continuous pattern. In a sense, a crystal of such an element consists of one huge molecule. Included in this category are the Group 4 elements (C, Si, Ge), and the heavier elements in Groups 5 and 6 (P, As, Sb; Se, Te). As pointed out in Chapter 11, elements with this structure typically have high melting and boiling points. These reflect the large amount of energy that must be absorbed to break covalent bonds so as to free nonmetal atoms from one another. The melting point of carbon, 3570°C, is perhaps the highest of all known substances.

How might we measure such a temperature?

2. *Molecular* structures, in which the basic building block is a small, discrete molecule. In the simplest case, with the noble gases in Group 8, the "molecule" is a single atom. Many nonmetals form diatomic molecules (N_2, O_2, H_2, and the halogens). The heavier elements in Groups 5 and 6 form molecules containing four atoms (for example, P_4) or eight atoms (for example, S_8).

As pointed out in Chapter 11, molecular species typically have low melting and boiling points. These ordinarily increase with molar mass, reflecting an increase in the strength of dispersion forces. Among the halogens, we start with a "permanent" gas (F_2, bp = $-188°C$), move to an easily condensable gas (Cl_2, bp = $-34°C$), then to a liquid (Br_2, bp = $59°C$), and end with a solid (I_2, mp = $114°C$).

The Lewis structures of molecules of several nonmetals were discussed in Chapter 9. You will recall that

—the noble gases (except for He) have simple octet structures:

$$:\overset{..}{\underset{..}{X}}: \qquad (X = Ne, Ar, Kr, Xe)$$

Lewis structures are valence bond structures

—the halogens consist of diatomic molecules that follow the octet rule with a single bond between the atoms:

$$:\overset{..}{\underset{..}{X}}—\overset{..}{\underset{..}{X}}: \qquad (X = F, Cl, Br, I)$$

—the nitrogen molecule contains a triple bond:

$$:N \equiv N:$$

—the oxygen molecule, O_2, cannot be described satisfactorily by a conventional Lewis structure. There is a double bond between the two atoms, but two unpaired electrons. The molecular orbital model (Chap. 10) gives a better picture of the O_2 molecule; it places two pairs of electrons in bonding orbitals and two unpaired electrons in separate nonbonding orbitals.

Ozone; Allotropy

You can sometimes smell ozone near large electric motors

Gaseous oxygen also exists in another form called *ozone*, molecular formula O_3. Ozone can be prepared in the laboratory by passing $O_2(g)$ through an electric discharge. The O_3 molecule is extremely reactive and is sometimes used in water treatment to kill bacteria or as a bleach for fabrics and paper. Small amounts of ozone are found in the atmosphere, where it can have both beneficial and harmful effects. We will have more to say about atmospheric ozone in Chapter 17.

Ozone can be represented by a Lewis structure in which there is one single bond and one double bond:

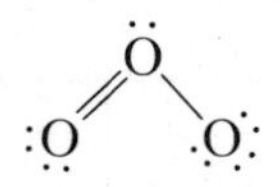

The measured bond angle (117°) is in good agreement with that predicted by the VSEPR model (120°). The bond distances, however, are identical (0.128 nm). We can rationalize this behavior by considering resonance forms for ozone:

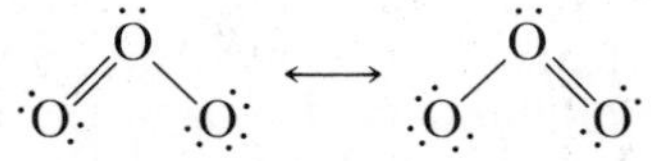

The understanding is that the true structure of ozone cannot be represented by a single Lewis structure but is a resonance hybrid of those shown above.

$O_2(l)$ and $O_2(g)$ are not allotropes

The two species O_2 and O_3 are referred to as **allotropes** of the element oxygen. The term allotropy is used to describe the situation where an element exists in two or more forms in the same physical state. Many nonmetallic elements in addition to oxygen show allotropy, as we will see in the remainder of this section.

Rhombic sulfur
a

Monoclinic sulfur
b

c

FIGURE 13.1 The two allotropes of solid sulfur, rhombic (a) and monoclinic (b), differ only in the way in which S_8 molecules are packed in the crystal. These molecules consist of eight-membered, puckered rings (c).

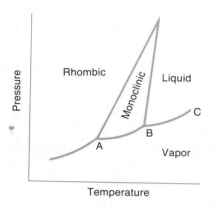

FIGURE 13.2 Phase diagram for sulfur. At point A (96°C, 0.0043 mm Hg), rhombic sulfur is in equilibrium with monoclinic sulfur and vapor. At point B (119°C, 0.027 mm Hg), monoclinic sulfur is in equilibrium with liquid and vapor.

Sulfur

Sulfur, known to the ancients as brimstone, occurs as a yellow solid that can take on several different allotropic forms. The most common of these are **rhombic** and **monoclinic** sulfur. Both allotropes are built of S_8 molecules with the structure shown in Figure 13.1. They differ only in the way that the molecules are packed in the solid, reflected in different crystal structures.

The phase diagram of sulfur is shown in Figure 13.2. This resembles the phase diagrams shown in Chapter 11, except for the introduction of a second solid phase. Rhombic sulfur, if heated slowly, is converted to monoclinic sulfur at the equilibrium temperature, 96°C (point A in Fig. 13.2). Monoclinic sulfur melts at 119°C (point B in Fig. 13.2). When liquid sulfur is cooled, along line BC, it freezes at 119°C; monoclinic crystals separate from the melt. These crystals can be kept for some time if the solid is cooled quickly to room temperature. At 25°C, the change to the more stable rhombic form is quite slow.

EXAMPLE 13.1 Using information given in Figure 13.2, determine
a. whether rhombic or monoclinic sulfur is the more dense phase.
b. the heat of sublimation of monoclinic sulfur, using the Clausius-Clapeyron equation in the form

$$\log_{10} \frac{P_2}{P_1} = \frac{\Delta H_{subl} (T_2 - T_1)}{(2.30)(8.31)(T_2 T_1)}$$

where P_2 and P_1 are the vapor pressures of the solid at temperatures T_2 and T_1, respectively, and ΔH_{subl} is the heat of sublimation in joules per mole.

Solution

a. Since the line dividing rhombic and monoclinic sulfur inclines to the right at higher pressures, the transition temperature increases as the pressure is raised. This means that an increase in pressure favors the rhombic form, raising the temperature at which it is stable. We deduce that rhombic sulfur must be more dense, as indeed it is (d rhombic = 2.07 g/cm³; d monoclinic = 1.96 g/cm³).

b. Taking the two points to be 96 and 119°C,

$$P_2 = 0.027 \text{ mm Hg}; \qquad P_1 = 0.0043 \text{ mm Hg}$$

$$T_2 = 119 + 273 = 392 \text{ K}; \qquad T_1 = 96 + 273 = 369 \text{ K}$$

$$\log_{10} \frac{0.027}{0.0043} = \frac{\Delta H_{subl}\,(392 - 369)}{(2.30)(8.31)(392)(369)}$$

Solving: $\Delta H_{subl} = 9.6 \times 10^4$ J/mol

EXERCISE Calculate the vapor pressure of monoclinic sulfur at 100°C. Answer: 0.0060 mm Hg.

The free-flowing, pale yellow liquid formed when sulfur melts contains S_8 molecules. However, upon heating to 160°C, a striking change occurs. The liquid becomes so viscous that it cannot be poured readily. At the same time its color changes to a deep reddish-brown. These effects reflect a change in molecular structure. The S_8 rings break apart and then link to one another to form long chains such as

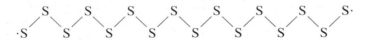

At higher temperatures the chains break down. At the BP, 444°C, S_2 molecules vaporize

Liquid sulfur between 160 and 250°C contains a high proportion of such chains. They vary in length from eight to several thousand atoms. The chains become tangled, producing a highly viscous liquid. The deep color is due to the absorption of light by the unpaired electrons at the ends of the chains.

If liquid sulfur at 200°C is quickly poured into water, a rubbery mass results (see color plate 13.1). This is referred to as "plastic sulfur." It consists of long-chain molecules that did not have time to rearrange to the S_8 molecules stable at room temperature. Within a few hours, the plastic sulfur loses its elasticity as it converts to rhombic crystals.

Phosphorus

Phosphorus, unlike nitrogen, does not form multiple bonds readily. Solid phosphorus has several allotropes. The two most common are

1. *White phosphorus*, which consists of P_4 molecules with the structure shown in Figure 13.3. It is a soft, waxy substance with a low melting point (44°C) and

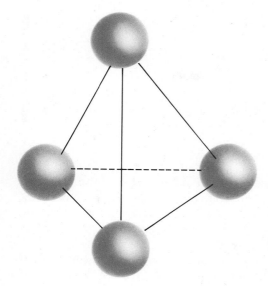

The bond angle in P_4 must be highly strained, since theory would predict it to be 109°

FIGURE 13.3 In the P_4 molecule, each phosphorus atom is at the corner of a tetrahedron, bonded to three other phosphorus atoms.

boiling point (280°C). Like most molecular substances, white phosphorus is readily soluble in such nonpolar solvents as CCl_4. The chemical reactivity of white phosphorus is so great that it is stored under water to protect it from O_2. A piece of P_4 exposed to air in a dark room glows because of the light given off upon oxidation. White phosphorus is extremely toxic. As little as 0.1 g taken internally can be fatal. Direct contact with the skin produces painful burns.

Don't fool around with white phosphorus

2. *Red phosphorus*, which is the form in which the element is usually found in the laboratory. This allotrope has properties quite different from those of white phosphorus. It is much higher melting (mp = 590°C at 43 atm) and is insoluble in common solvents. The low volatility of red phosphorus makes it much less toxic than the white form. It is also less reactive and must be heated to 250°C to burn in air. These properties are consistent with the structure of red phosphorus, which is known to be network covalent.

Before the undesirable properties of white phosphorus were known, it was used in matches. Today, two different kinds of matches are available, neither of which contains white phosphorus. The heads of "strike-anywhere" matches contain a mixture of a sulfide of phosphorus, P_4S_3, potassium chlorate, $KClO_3$, and powdered glass. When struck against a rough surface, the mixture ignites. Safety matches contain sulfur and potassium chlorate; the special surface against which they are struck contains red phosphorus and powdered glass. Friction sets off a reaction between red phosphorus and $KClO_3$.

EXAMPLE 13.2 With the aid of Figure 13.3, and following the rules given in Chapter 9, write the Lewis structure of P_4.

Solution Figure 13.3 gives the skeleton of P_4. Note that there are six single bonds in the structure, accounting for 12 valence electrons. Since phosphorus is in Group 5, each atom has 5 valence electrons. The total number of valence electrons available is 4(5) = 20. Subtracting 12 leaves 8 valence electrons to distribute. Putting two of these as an unshared pair on each P atom gives a Lewis structure in which each atom has an octet:

$$:P \overset{\overset{\ddot P}{|}}{\underset{\underset{\ddot P}{}}{|}} P:$$

EXERCISE Referring to Figure 13.3, what is the bond angle in P_4? Answer: 60°.

Carbon

Graphite crucibles never melt

Carbon exists in two different crystalline forms, diamond and graphite. The two allotropes differ in the way the atoms are bonded to one another. Both are network covalent. This explains why both diamond and graphite have very high melting points, above 3500°C. However, as you can see from Figure 13.4, the bonding patterns in the two crystals are quite different.

The graphite crystal is planar, with the carbon atoms arranged in a hexagonal pattern. Each carbon atom is bonded to three others, forming one double bond and two single bonds. The forces between the layers in graphite are of the dispersion type and are quite weak. Thus, the layers can readily slide past one another so that graphite is soft and slippery to the touch.* When you write with a "lead" pencil, which is really made of graphite, thin layers of graphite rub off onto the paper (Fig. 13.5).

---

*Studies show that the lubricating properties of graphite disappear under high vacuum. Apparently these properties require the presence of adsorbed H_2O or O_2 molecules, which act much like a film of oil on a metal surface, allowing graphite layers to slide readily past one another.

Graphite layer   Diamond crystal

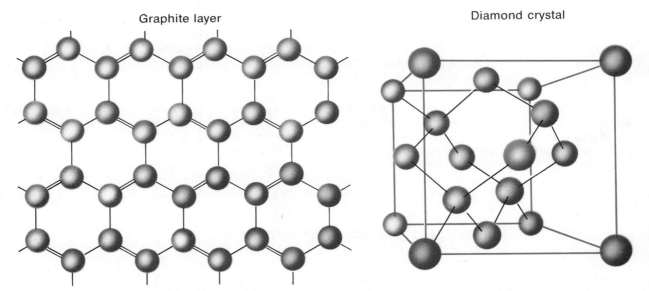

FIGURE 13.4 In diamond, each carbon atom is at the center of a tetrahedron, on each corner of which is another carbon atom. In graphite, the carbon atoms are linked together in planes of hexagons. Within a layer, a carbon atom is bonded to three other carbon atoms; one third of the bonds are double bonds. The layers are held to one another by weak dispersion forces.

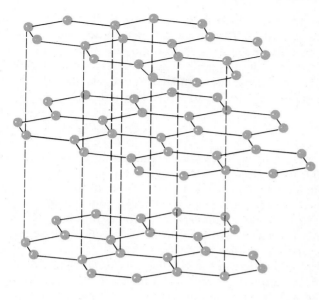

FIGURE 13.5 In graphite, successive layers of carbon atoms are stacked on top of one another. There are only weak dispersion forces between atoms in adjacent layers. As a result, one layer is easily separated from another.

In diamond, each carbon atom forms single bonds with four other carbon atoms arranged tetrahedrally around it. The bonds are strong enough (bond energy C—C = 347 kJ/mol) to produce a rugged, three-dimensional lattice. Diamond is one of the hardest of substances, and is used in industry in cutting tools and quality grindstones.

Diamonds make the best abrasives

At room temperature and atmospheric pressure, graphite is the stable form of carbon. Diamond, in principle, should slowly transform to graphite under ordinary conditions. Fortunately for the owners of diamond rings this transition occurs at zero rate unless the diamond is heated to about 1500°C, at which temperature the conversion occurs rapidly. For understandable reasons, no one has ever become very excited over the commercial possibilities of this process. The more difficult task of converting graphite to diamond has aroused much greater enthusiasm.

Since diamond has a higher density than graphite (3.51 vs. 2.26 g/cm³), its formation should be favored by high pressures. Theoretically, at 25°C and 15,000 atm graphite should turn to diamond. However, under those conditions the reaction has a negligible rate. At higher temperatures it goes faster, but the required pressure goes up too; at 2000°C, a pressure of about 100,000 atm is needed. In 1954, scientists at the General Electric laboratories were able to achieve these high temperatures and pressures and converted graphitic carbon to diamond for the first time. The synthetic diamonds produced by this process were at first quite small, but their size and quality have been improved so that one-carat diamonds are now occasionally formed. At present, most of our industrial diamonds are synthetic.

This project was a success for both the theoreticians and the business managers

Several amorphous (noncrystalline) forms of carbon are also known. Two of these are charcoal and carbon black. Charcoal can be made by heating wood or other high-carbon materials to a high temperature in the absence of air. This drives off water and other volatile substances. Carbon black is made by burning natural gas in limited air and collecting the soot on cold metal plates. This material is blended with rubber for automobile tires and the soles of running shoes, where it increases wear resistance. A form of charcoal with a very large surface area, called activated carbon, is produced in much the same way as charcoal. This material is very porous and absorbent. Activated carbon is used in industry to decolorize sugar solutions. It has also been applied to removing odors in inner soles, refrigerators, and, on a larger scale, in public water supply systems. In all the amorphous forms of carbon, the atoms are arranged in irregular hexagonal patterns. These are similar in many ways to the structure of graphite.

Silicon and Germanium

Both silicon and germanium have only one crystalline form, that of diamond. Extremely pure crystals of the two elements are nonconductors. The valence electrons are used in the four covalent bonds that each atom forms with its neighbors. The conductivity increases dramatically when small amounts of arsenic (Group 5) or boron (Group 3) are introduced into the crystal. As little as 0.0001 mole percent of these elements present in Si or Ge helps to produce the semiconductor devices used in transistors or solar cells.

1 atom As per 10^6 atoms Si

To understand the effect of impurities on the conductivity of silicon or germanium, consider Figure 13.6. An atom of arsenic, with five valence electrons, can fit into the crystal lattice of Si or Ge. Its atomic radius (0.121 nm) is close to that of Si (0.117 nm) or Ge (0.122 nm). To do so, however, an arsenic atom must

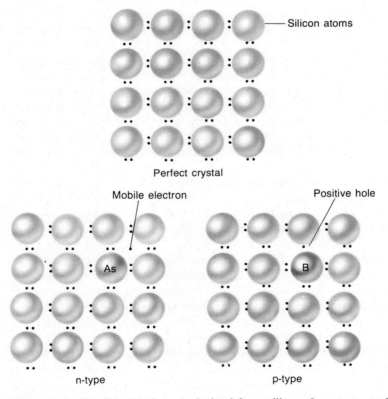

FIGURE 13.6 Semiconductors derived from silicon. In n-type semiconductors the impurity atoms furnish mobile electrons to the crystal. In p-type semiconductors there is a deficiency of electrons, since the impurity atoms have three rather than four valence electrons.

give up its fifth valence electron. This electron can move through the crystal under the influence of an electrical field. This gives an **n-type semiconductor** (current carried by the flow of negative charge). If an atom of boron or another element with three valence electrons is introduced into the lattice, a different situation arises. An electron deficiency is created at the site occupied by the foreign atom. It is surrounded by seven valence electrons rather than eight. In this sense, there is a "positive hole" in the lattice. In an electrical field, an electron moves from a neighboring atom to fill that hole. By so doing, it creates an electron deficiency around the atom which it leaves. The result is a **p-type semiconductor**. In effect, positive holes move through the lattice.

The two heavier members of Group 4, tin and lead, are not included in this chapter. They are metals rather than nonmetals or metalloids. However, tin does have an allotropic form (grey tin), which has the diamond structure. When tin metal is kept at temperatures below 13°C for long periods of time, grey tin forms as a powder. Hence articles made of tin, notably organ pipes, sometimes crumble in very cold weather. The formation of grey tin seems to spread, like an infection, from a single point. This problem was common in the cold cathedrals of northern Europe in the nineteenth century. The organ pipes were said to suffer from "tin disease," for which at that time there was no known "cure."

Heating the place would have prevented the disease

13.2
Nonmetal Hydrides

Table 13.2 gives the molecular formulas and boiling points of the hydrogen compounds of the nonmetals in Groups 5, 6, and 7. The noble gases (Group 8) do not form compounds with hydrogen. In contrast, the Group 4 elements form a great many hydrogen compounds. The most important of these are the hydrocarbons, organic compounds containing the two elements hydrogen and carbon. These will be discussed separately in Section 13.5.

Table 13.2
Hydrogen Compounds of the Nonmetals (Boiling Points in Parentheses)

GROUP 5	GROUP 6	GROUP 7
$NH_3(-33°C)$, $N_2H_4(114°C)$, $HN_3(37°C)$	$H_2O(100°C)$, $H_2O_2(151°C)$	$HF(20°C)$
$PH_3(-88°C)$, $P_2H_4(52°C)$	$H_2S(-61°C)$	$HCl(-85°C)$
$AsH_3(-55°C)$	$H_2Se(-42°C)$	$HBr(-67°C)$
$SbH_3(-18°C)$	$H_2Te(-2°C)$	$HI(-36°C)$

The boiling points and other properties of these compounds, all of which are molecular, can be related to the strength of intermolecular forces (Example 13.3).

EXAMPLE 13.3 Consider the compounds in Table 13.2.
a. Which ones show hydrogen bonding?
b. Which hydrogen halide (HF, HCl, HBr, or HI) has the strongest dispersion forces?
c. Why does HCl have a lower boiling point than either HF or HBr?

Solution

a. Hydrogen bonds occur between molecules in which hydrogen is co-valently bonded to N, O, or F. This is the case with NH_3, N_2H_4, HN_3, H_2O, H_2O_2, and HF.
b. Dispersion forces increase with molar mass; they are strongest with HI.
c. Hydrogen bonding gives HF a higher boiling point than HCl. Stronger dispersion forces explain why HBr is higher boiling than HCl.

EXERCISE Of the four hydrogen halides, which is the least polar? Answer: HI.

Of the compounds listed in Table 13.2, several are relatively rare. You are not likely to come in contact with PH_3, P_2H_4, AsH_3, SbH_3, H_2Se, H_2Te, and HN_3. On the whole, that is probably just as well: HN_3 detonates on contact, P_2H_4 bursts into flame on exposure to air, and the others are extremely poisonous. The preparation and properties of two more familiar nonmetal hydrides, NH_3 and H_2S, will be described in later chapters. In this section, we will concentrate upon

—hydrogen peroxide (H_2O_2) and hydrazine (N_2H_4).
—the hydrogen halides (HF, HCl, HBr, HI).

Hydrogen Peroxide and Hydrazine

These two compounds have several properties in common. For one thing, they are both colorless liquids at room temperature. Furthermore, both decompose spontaneously with the evolution of energy:

$$2\ H_2O_2(l) \rightarrow 2\ H_2O(l) + O_2(g); \qquad \Delta H = -196.4\ kJ \qquad \textbf{(13.1)}$$

$$N_2H_4(l) \rightarrow N_2(g) + 2\ H_2(g); \qquad \Delta H = -50.4\ kJ \qquad \textbf{(13.2)}$$

Because of these reactions, the pure liquids tend to be unstable and are dangerous to work with. Both hydrogen peroxide and hydrazine are soluble in water in all proportions. Solubility is promoted by the formation of hydrogen bonds between solute molecules (H_2O_2 or N_2H_4) and solvent (H_2O) molecules.

The aqueous solutions are reasonably stable

Hydrogen peroxide and hydrazine have quite similar Lewis structures (Example 13.4). In both molecules, two central nonmetal atoms are bonded to hydrogen.

EXAMPLE 13.4 Consider hydrogen peroxide, H_2O_2.
a. Write a reasonable Lewis structure for H_2O_2.
b. Predict the bond angle.

Solution

a. Hydrogen atoms can form only one bond. The most reasonable skeleton would be a symmetrical one in which a hydrogen is bonded to each oxygen atom:

$$\begin{array}{ccc} & & H \\ & O\!-\!O \\ H & & \end{array}$$

We start with a total of $2(1) + 2(6) = 14$ valence electrons. With three single bonds in the skeleton (six valence electrons), that leaves eight valence electrons to be distributed. Putting two unshaired pairs on each oxygen, we arrive at the correct Lewis structure:

$$\ddot{O}-\ddot{O}$$

b. The four electron pairs around each oxygen should be directed toward the corners of a regular tetrahedron. The predicted bond angle is 109.5°. Experimentally, it is found to be 105°.

EXERCISE   The skeleton of hydrazine, N_2H_4, is similar to that of hydrogen peroxide. Write a reasonable Lewis structure for N_2H_4. Answer:

$$N-N$$

In H_2O_2 there would be fast rotation around the O—O bond

You are most likely to come across hydrogen peroxide as its water solution. Two concentrations are available; one of these, containing 3 mass percent H_2O_2, is sold in drugstores. The other solution contains 30 mass percent H_2O_2. Both solutions contain stabilizers to prevent Reaction 13.1 from taking place during storage. Hydrogen peroxide is used as a disinfectant (cuts, sore throats) or as a bleach (cloth, paper, hair, etc.).

It's powerful but safe to use

Hydrazine is a common rocket fuel. The "oxidizer" used with hydrazine may be liquid oxygen, dinitrogen tetroxide (N_2O_4), or hydrogen peroxide. With all three reagents, an exothermic reaction occurs to produce a large volume of gas. With hydrogen peroxide the reaction is

$$N_2H_4(l) + 2\ H_2O_2(l) \rightarrow N_2(g) + 4\ H_2O(g); \qquad \Delta H = -642.4\ \text{kJ} \qquad \textbf{(13.3)}$$

Hydrogen Halides

These compounds can be prepared by a variety of methods. The simplest approach involves the direct reaction between the elements. This method is satisfactory for the preparation of HCl and HBr:

$$H_2(g) + Cl_2(g) \rightarrow 2\ HCl(g) \qquad \textbf{(13.4)}$$

$$H_2(g) + Br_2(l) \rightarrow 2\ HBr(g) \qquad \textbf{(13.5)}$$

Hydrogen bromide is made commercially by Reaction 13.5. The reaction between hydrogen and iodine is too slow to be of practical importance. In contrast, the reaction between hydrogen and fluorine occurs so rapidly and unpredictably that it cannot be controlled.

Small amounts of the hydrogen halides can be made in the laboratory by reacting a metal halide with a nonvolatile acid. Since the hydrogen halides produced

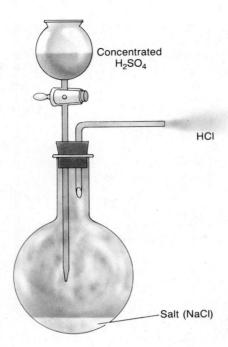

Concentrated H_2SO_4

HCl

Salt (NaCl)

FIGURE 13.7 Preparation of HCl from NaCl. To generate HBr or HI, one uses phosphoric acid, H_3PO_4. Concentrated H_2SO_4 will produce some Br_2 or I_2 if dropped on NaBr or NaI.

are gaseous, they readily separate from the reaction mixture (Fig. 13.7). Hydrogen fluoride is made commercially by the action of sulfuric acid, H_2SO_4, on the mineral fluorite, CaF_2:

$$CaF_2(s) + H_2SO_4(l) \rightarrow 2\ HF(g) + CaSO_4(s) \tag{13.6}$$

Hydrogen chloride can be made by a similar method from sodium chloride:

This reaction provides an easy way to make a little HCl(g)

$$NaCl(s) + H_2SO_4(l) \rightarrow HCl(g) + NaHSO_4(s) \tag{13.7}$$

Sulfuric acid cannot be used to prepare HBr or HI; it reacts with these compounds to form the free halogens. Phosphoric acid is used instead:

$$NaBr(s) + H_3PO_4(l) \rightarrow HBr(g) + NaH_2PO_4(s) \tag{13.8}$$

$$NaI(s) + H_3PO_4(l) \rightarrow HI(g) + NaH_2PO_4(s) \tag{13.9}$$

The only hydrogen halide that you are likely to come across in the general chemistry laboratory is HCl. This is available as its water solution, called hydrochloric acid. Hydrogen chloride, hydrogen bromide, and hydrogen iodide act as electrolytes in water. In dilute solution they are completely dissociated to H^+ ions and halide anions:

$$HX(aq) \rightarrow H^+(aq) + X^-(aq); \quad X = Cl,\ Br,\ I \tag{13.10}$$

Hydrogen fluoride behaves somewhat differently. In water solution, it forms an equilibrium mixture of H^+ ions, F^- ions, and HF molecules:

$$HF(aq) \rightleftharpoons H^+(aq) + F^-(aq)$$

Because of this behavior, hydrogen fluoride is often described as a "weak" electrolyte; HCl, HBr, and HI are "strong" electrolytes. We will have more to say about the acid properties of the hydrogen halides when we discuss acids and bases in Chapter 19.

Concentrated hydrofluoric acid reacts with glass, which we can consider to be a mixture of SiO_2 and ionic silicates such as calcium silicate, $CaSiO_3$:

$$SiO_2(s) + 4\ HF(aq) \rightarrow SiF_4(g) + 2\ H_2O \tag{13.11}$$

$$CaSiO_3(s) + 6\ HF(aq) \rightarrow SiF_4(g) + CaF_2(s) + 3\ H_2O \tag{13.12}$$

Concentrated HF is a dangerous reagent

As you might guess, HF solutions are never stored in glass bottles; plastic is used instead. Reactions 13.11 and 13.12 are sometimes used to etch glass. The glass object is first covered with a thin protective coating of wax or plastic. Then the coating is removed from the area to be etched and the glass is exposed to the HF solution. Thermometer stems and burets can be etched or light bulbs frosted in this way.

EXAMPLE 13.5 Concentrated hydrofluoric acid contains 45.0 mass percent HF and has a density of 1.14 g/cm³. Calculate the
a. molality of HF.   b. molarity of HF.

Solution

a. In 100.0 g of solution, there are 45.0 g HF and 55.0 g of water. Recall (Chap. 12) that molality is defined as the number of moles of solute per kilogram of solvent. In this case we have

$$\text{no. moles HF} = 45.0\ \text{g HF} \times \frac{1\ \text{mol HF}}{20.0\ \text{g HF}} = 2.25\ \text{mol HF}$$

$$\text{no. kilograms water} = 55.0\ \text{g water} \times \frac{1\ \text{kg}}{1000\ \text{g}} = 0.0550\ \text{kg water}$$

$$\text{molality} = \frac{2.25\ \text{mol}}{0.0550\ \text{kg water}} = 40.9\ m$$

b. The volume of 100.0 g of solution is readily calculated, knowing the density:

$$\text{no. liters solution} = 100.0\ \text{g} \times \frac{1\ \text{cm}^3}{1.14\ \text{g}} \times \frac{1\ \text{L}}{1000\ \text{cm}^3} = 0.0877\ \text{L}$$

Knowing the volume of the solution (0.0877 L) and the number of moles of HF (2.25 mol), we can readily calculate the molarity:

$$\text{molarity} = \frac{2.25\ \text{mol}}{0.0877\ \text{L}} = 25.7\ \text{M}$$

EXERCISE What volume of this solution is required to react with one mole of SiO_2 by Reaction 13.11? Answer: 156 mL.

13.3
Nonmetal Oxides

Table 13.3 lists the nonmetal oxides that are stable enough to be isolated in more or less pure form. Several other oxygen compounds (for example, SO) have been reported, usually as unstable intermediates in reactions. With some exceptions, the formulas given in the table represent molecules. The oxides of selenium (SeO_2, SeO_3), tellurium (TeO_2, TeO_3), silicon (SiO_2), and germanium (GeO_2) are known to be network covalent. The structures of As_2O_5 and Sb_2O_5 are unknown; these are simplest formulas.

Table 13.3
Oxygen Compounds of the Nonmetals*

GROUP 4	GROUP 5	GROUP 6	GROUP 7	GROUP 8
$CO_2(g)$, $CO(g)$	$N_2O_5(s)$, $N_2O_4(g)$, $N_2O_3(d)$, $N_2O_2(g)$, $N_2O(g)$, $NO_2(g)$, $NO(g)$		$F_2O(g)$, $F_2O_2(g)$	
$SiO_2(s)$	$P_4O_{10}(s)$, $P_4O_6(l)$	$SO_3(l)$, $SO_2(g)$	$Cl_2O_7(l)$, $Cl_2O_6(l)$, $Cl_2O(g)$, $ClO_2(g)$	
$GeO_2(s)$	$As_2O_5(s)$, $As_4O_6(s)$	$SeO_3(s)$, $SeO_2(s)$	$Br_3O_8(d)$, $BrO_2(d)$, $Br_2O(d)$	
	$Sb_2O_5(s)$, $Sb_4O_6(s)$	$TeO_3(s)$, $TeO_2(s)$	$I_4O_9(d)$, $I_2O_5(s)$, $I_2O_4(s)$	$XeO_4(d)$, $XeO_3(d)$

*Physical states at 25°C and 1 atm are indicated by (*s*), (*l*), and (*g*). Compounds that decompose below 25°C are listed as (*d*).

As you can see from Table 13.3, most nonmetals form more than one compound with oxygen. Typically, the lower oxide (lesser amount of oxygen) is formed at high temperatures in a limited supply of oxygen or air. Cooling with excess O_2 or air leads to further reaction. The final product is a higher oxide, containing more oxygen. As an example, white phosphorus burns in a limited supply of oxygen to form P_4O_6:

Military smoke screens are made by burning phosphorus

$$P_4(s) + 3\ O_2(g) \rightarrow P_4O_6(l) \tag{13.13}$$

In excess oxygen, the higher oxide, P_4O_{10}, is produced:

$$P_4O_6(l) + 2\ O_2(g) \rightarrow P_4O_{10}(s) \tag{13.14}$$

We will see further examples of this principle when we examine the structures and methods of preparation of the oxides of sulfur, nitrogen, and carbon.

Oxides of Sulfur

When elemental sulfur or a metal sulfide reacts with oxygen, sulfur dioxide is formed:

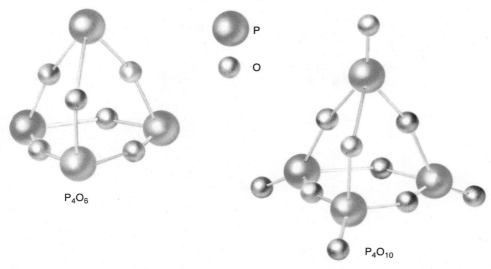

FIGURE 13.8 Molecular structures of the oxides of phosphorus, P_4O_6 and P_4O_{10}. In P_4O_6 each phosphorus atom is bonded to three oxygens; there is an unshared electron pair on each phosphorus. In P_4O_{10}, each phosphorus atom is at the center of a tetrahedron, bonded to four oxygen atoms.

$$S(s) + O_2(g) \rightarrow SO_2(g) \tag{13.15}$$

$$2\ ZnS(s) + 3\ O_2(g) \rightarrow 2\ ZnO(s) + 2\ SO_2(g) \tag{13.16}$$

This is the fire and brimstone reaction

Equation 13.16 is typical of the reactions that occur when a sulfide ore is roasted as the first step in the extraction of a metal. The Lewis structure of SO_2 was derived in Chapter 9; its resonance forms are shown on p. 273.

Sulfur dioxide in the air is slowly oxidized to the higher oxide, SO_3. This reaction is catalyzed by various solids, as will be discussed in Chapter 17:

$$2\ SO_2(g) + O_2(g) \rightarrow 2\ SO_3(g) \tag{13.17}$$

In the gaseous state, sulfur trioxide consists of discrete molecules with three resonance forms:

SO_3 reacts with water, producing sulfuric acid

As predicted by electron-pair repulsion, SO_3 is a planar molecule. The sulfur atom is at the center of an equilateral triangle formed by the three oxygen atoms. The bond angles are 120°, giving a symmetrical, nonpolar molecule.

Oxides of Nitrogen

The Lewis structures of the various oxides of nitrogen, seven in all, are shown in Figure 13.9. For each oxide, it is possible to draw resonance forms (Example 13.6).

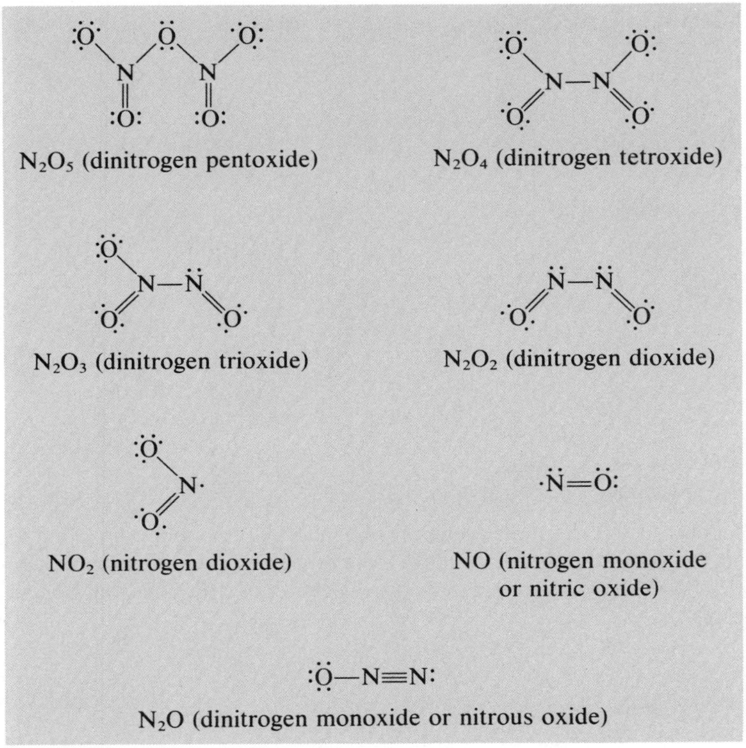

FIGURE 13.9 Lewis structures of the oxides of nitrogen. Many other resonance forms are possible.

EXAMPLE 13.6 Consider the N_2O molecule shown in Figure 13.9.
a. Draw another resonance form of N_2O.
b. What is the bond angle in N_2O?
c. Is the N_2O molecule polar or nonpolar?

Solution

a. $:\ddot{O}=N=\ddot{N}:$ or $:O\equiv N-\ddot{N}:$

b. In any of the resonance forms, the central nitrogen atom, insofar as geometry is concerned, would behave as if it were surrounded by two electron pairs. The bond angle is 180°; the molecule is linear, like BeF_2.

c. Polar (unsymmetrical).

EXERCISE What is the hybridization about the central nitrogen atom in N_2O? Answer: sp.

Of the several oxides of nitrogen, NO, NO_2, and N_2O are the most common. Nitrogen monoxide (nitric oxide) forms when the elements combine at high temperatures, above 1000°C.

$$N_2(g) + O_2(g) \rightarrow 2 \, NO(g) \qquad (13.18)$$

:$\ddot{N}=O=\ddot{N}$: is not a resonance form. Why?

This reaction occurs in an automobile engine; the exhaust gases contain small but significant amounts of NO. Upon cooling in air, NO is converted to nitrogen dioxide, NO_2:

$$2 \, NO(g) + O_2(g) \rightarrow 2 \, NO_2(g) \tag{13.19}$$

Nitrogen dioxide is a major culprit in smog formation. These two oxides can also be formed in the laboratory from nitric acid, HNO_3, by processes to be discussed in Chapter 26.

Dinitrogen oxide (nitrous oxide) is often referred to as "laughing gas." It was one of the first anesthetics and is still used for that purpose, particularly in dentistry. It was also the first aerosol propellant, used in whipped cream dispensers. Dinitrogen oxide can be made by heating ammonium nitrate *carefully*:

In the early 19th century they used to have laughing gas parties

$$NH_4NO_3(s) \rightarrow N_2O(g) + 2 \, H_2O(g) \tag{13.20}$$

At high temperatures, in an enclosed space, NH_4NO_3 can detonate; this is not a reaction to be carried out casually.

As pointed out in Chapter 9, the NO and NO_2 molecules both contain an odd number of valence electrons (11 in NO, 17 in NO_2). Hence they cannot "obey" the octet rule; each molecule contains a single unpaired electron (Fig. 13.9). These molecules can, however, combine with one another to form dimers (N_2O_4, N_2O_3, N_2O_2) in which each atom has an octet of electrons:

$$NO_2(g) + NO_2(g) \rightarrow N_2O_4(g) \tag{13.21}$$

$$NO_2(g) + NO(g) \rightarrow N_2O_3(g) \tag{13.22}$$

$$NO(g) + NO(g) \rightarrow N_2O_2(g) \tag{13.23}$$

Species like NO and NO_2, with unpaired electrons, are called free radicals

Notice from Figure 13.9 that in each of these molecules (N_2O_4, N_2O_3, N_2O_2), there is a single bond between the two nitrogen atoms. This is formed by the pairing of the single electrons on the nitrogen atoms in NO and NO_2.

Reactions 13.21 through 13.23 are successively more difficult to carry out. Cooling readily converts NO_2 to dinitrogen tetroxide via Reaction 13.21. At 25°C and 1 atm, about 75% of the NO_2 has dimerized to N_2O_4. Under the same conditions, an equimolar mixture of NO_2 and NO is only about 10% converted to dinitrogen trioxide via Reaction 13.22. Further cooling to -20°C causes N_2O_3 to condense out as a blue liquid, mixed with some NO and NO_2. Reaction 13.23 is very difficult to carry out. Even at -150°C, only small amounts of dinitrogen dioxide are observed.

EXAMPLE 13.7 Using bond energies from Table 9.6, Chapter 9, calculate ΔH for Reactions 13.21 through 13.23.

Solution Consider Reaction 13.21. Let us start by taking an inventory of bonds. In two NO_2 molecules there are (Fig. 13.9) two N—O bonds and two N=O bonds. In the N_2O_4 molecule, we see from Figure 13.9 that there are two N—O bonds, two N=O bonds, and one N—N bond. We conclude that the net effect in Reaction 13.21 is simply the *formation* of an N—N single bond. Hence,

$$\Delta H = -\text{ B.E. N—N} = -159 \text{ kJ}$$

A similar analysis for Reaction 13.22 or 13.23 leads to the same conclusion. The net effect in each case is simply the formation of an N—N bond. Hence in each case ΔH is estimated from bond energies to be -159 kJ. Experimentally, we find that the value of ΔH is somewhat different from that calculated from bond energies. However, the sign of ΔH is correct; Reactions 13.21 through 13.23 are all exothermic. This explains why lowering the temperature causes these reactions to occur. As we saw in Chapter 12, exothermic processes are favored by a decrease in temperature.

EXERCISE Using bond energies, estimate ΔH for $NO_2(g) \rightarrow NO(g) + O(g)$. Answer: $\Delta H = +222$ kJ.

Oxides of Carbon

When carbon or any of its compounds is heated to a high temperature in a limited amount of air, carbon monoxide, CO, is formed:

$$2 \text{ C}(s) + O_2(g) \rightarrow 2 \text{ CO}(g) \tag{13.24}$$

$$2 \text{ C}_8\text{H}_{18}(l) + 17 \text{ O}_2(g) \rightarrow 16 \text{ CO}(g) + 18 \text{ H}_2\text{O}(l) \tag{13.25}$$

Reaction 13.24 occurs just below the surface of a bed of burning charcoal or coal. Reaction 13.25 takes place when an automobile engine is left running in a closed garage (octane, C_8H_{18}, is typical of the hydrocarbons found in gasoline). Carbon monoxide is extremely poisonous. As little as 0.02 mole percent can cause unconsciousness. At the 0.1 mole percent level, CO is fatal if inhaled for only a few minutes.

CO is colorless and odorless

In principle, CO should react with oxygen of the air to form carbon dioxide, CO_2:

$$2 \text{ CO}(g) + O_2(g) \rightarrow 2 \text{ CO}_2(g) \tag{13.26}$$

In practice, this reaction occurs very slowly under ordinary conditions. Carbon dioxide is formed directly when hydrocarbons such as octane are burned in an excess of air:

$$2 \text{ C}_8\text{H}_{18}(l) + 25 \text{ O}_2(g) \rightarrow 16 \text{ CO}_2(g) + 18 \text{ H}_2\text{O}(l) \tag{13.27}$$

(Compare the coefficients of O_2 in Reactions 13.25 and 13.27).

The Lewis structures of CO and CO_2 are quite simple. Carbon dioxide is a resonance hybrid of three structures:

So is CO_2

$$:O\!\!\equiv\!\!C\!\!-\!\!\ddot{O}: \longleftrightarrow :\ddot{O}\!\!=\!\!C\!\!=\!\!\ddot{O}: \longleftrightarrow :\ddot{O}\!\!-\!\!C\!\!\equiv\!\!O:$$

It is a linear, nonpolar molecule. Carbon monoxide has the structure:

:C≡O:

The molecule is, of necessity, linear. It is also slightly polar, since oxygen has a stronger attraction for electrons than carbon.

EXAMPLE 13.8   How many pi bonds are there in the CO molecule? in CO_2?

Solution   In CO, the triple bond consists of one sigma bond and two pi bonds. With CO_2, whichever resonance form you look at, there are two pi bonds.

CO has the same structure as N_2, but is not nearly as inert

EXERCISE   Referring to the N_2O_4 molecule in Figure 13.9, how many pi bonds are there? how many sigma bonds? Answer: Two; five.

13.4
Nonmetal Halides

Due to their reactivity, none of the halogens (F, Cl, Br, I) occur as the free elements. Their most common compounds are those with the alkali or alkaline earth metals (for example, NaCl, CaF_2), where the bonding is ionic. The halogens

Table 13.4
Halogen Compounds of the Nonmetals

		Group 4							
Carbon	CF_4	CCl_4	CBr_4	CI_4					
		C_2Cl_4	C_2Br_4						
Silicon	SiF_4	$SiCl_4$	$SiBr_4$	SiI_4		SiF_2			
	Si_2F_6	Si_2Cl_6	Si_2Br_6	Si_2I_6					
Germanium	GeF_4	$GeCl_4$	$GeBr_4$	GeI_4		GeF_2	$GeCl_2$	$GeBr_2$	GeI_2
		Group 5							
Nitrogen	NF_3	NCl_3	NBr_3	NI_3					
Phosphorus	PF_3	PCl_3	PBr_3	PI_3		PF_5	PCl_5	PBr_5	
Arsenic	AsF_3	$AsCl_3$	$AsBr_3$	AsI_3		AsF_5	$AsCl_5$		
Antimony	SbF_3	$SbCl_3$	$SbBr_3$	SbI_3		SbF_5	$SbCl_5$		
		Group 6							
Oxygen		See Table 13.3							
Sulfur		SCl_2				SF_4	SCl_4		
	S_2F_2	S_2Cl_2	S_2Br_2	S_2I_2		SF_6			
Selenium		$SeCl_2$	$SeBr_2$			SeF_4	$SeCl_4$	$SeBr_4$	
		Se_2Cl_2	Se_2Br_2			SeF_6			
Tellurium		$TeCl_2$	$TeBr_2$			TeF_4	$TeCl_4$	$TeBr_4$	TeI_4
						TeF_6			
		Group 7							
		See Table 13.5							
		Group 8							
		See Table 13.6							

also form more than 100 binary compounds with other nonmetals. Most of the more important nonmetal halides are listed in Table 13.4. Those shown in the columns at the left of the table can be assigned noble gas structures, at least in principle.* In contrast, the species to the right of the vertical rule do not have octet structures. In molecules such as GeF_2, SF_4, and SF_6, the central atom is surrounded by 6, 10, or 12 valence electrons, respectively.

EXAMPLE 13.9 Draw Lewis structures for

a. GeF_2 b. PF_3 c. SF_4

Solution In each case it is helpful to start by determining the number of valence electrons around the central atom. For nonmetal halides this can be done by applying the rule cited in Chapter 10:

no. valence e^- = group no. of central atom
+ no. halogen atoms bonded to central atom

a. Applying this rule, we decide that there are three electron pairs around the germanium atom:

no. valence e^- = 4 + 2 = 6;   3 electron pairs

One of these pairs is unshared; the others form bonds with the two fluorine atoms. The Lewis structure is

$:\ddot{F}—\ddot{Ge}—\ddot{F}:$

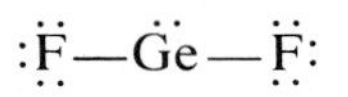

b. Here we conclude that, for the central phosphorus atom,

no. valence e^- = 5 + 3 = 8;   4 electron pairs

Three of these electron pairs are used in bond formation; the fourth pair is unshared. Each atom in PF_3 has an octet structure:

$:\ddot{F}—\ddot{P}—\ddot{F}:$
        |
       $:\ddot{F}:$

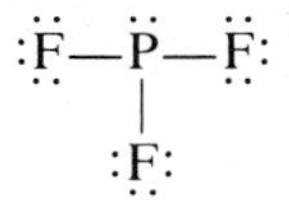

c. For the sulfur atom in SF_4:

no. valence e^- = 6 + 4 = 10;   5 electron pairs

Four of these pairs are used to form bonds with fluorine atoms; the fifth pair is unshared. The sulfur atom in SF_4 has an expanded octet:

$:\ddot{F}:$
    \\
     $\ddot{S}—\ddot{F}:$
   / |
$:\ddot{F}:$ $:\ddot{F}:$

Margin notes:

A bent molecule: sp^2 hybrid

A trigonal pyramid: sp^3 hybrid

Asymmetric: dsp^3 hybrid

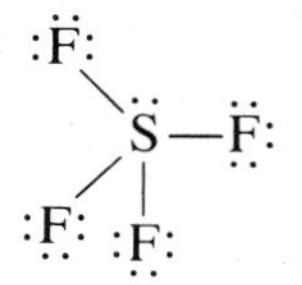

---

*Several of the compounds listed in Table 13.4 are not molecular in the solid. Some are network covalent; in at least one case, the bonding is ionic. Solid PCl_5 has been shown to consist of PCl_4^+ and PCl_6^- ions.

EXERCISE What is the total number of valence electrons in GeF_2? in PF_3? in SF_4? Answer: 18, 26, 34.

Two different halogens can react with one another to give a compound called an *interhalogen*. If chlorine and fluorine are heated together at 220 to 250°C, they react to form chlorine fluoride, a colorless gas with a choking odor:

$$Cl_2(g) + F_2(g) \rightarrow 2\ ClF(g) \tag{13.28}$$

At higher temperatures the main product is chlorine trifluoride:

$$Cl_2(g) + 3\ F_2(g) \rightarrow 2\ ClF_3(g) \tag{13.29}$$

The boiling points of ClF ($-100°C$) and ClF_3 ($12°C$) differ widely, so the two compounds are readily separated by fractional distillation.

The compositions of the known interhalogens are listed in Table 13.5. Notice that they are of two types:

1. Diatomic, polar molecules involving two different halogens (for example, ClF). These molecules follow the octet rule, with a single bond between the two atoms.
2. Polyatomic molecules in which a central atom is surrounded by three, five, or seven smaller halogen atoms (for example, ClF_3, BrF_5, IF_7). In these molecules, the central atom has an expanded octet.

Table 13.5
Interhalogens

NUMBER OF ELECTRON PAIRS AROUND CENTRAL ATOM

4	5	6	7
ClF(g)	$ClF_3(g)$	$ClF_5(g)$	$IF_7(l)$
BrF(g)	$BrF_3(l)$	$BrF_5(l)$	
BrCl(g)	$IF_3(d)$	$IF_5(l)$	
ICl(s)	$ICl_3(s)$		
IBr(s)			

Among the most interesting of the nonmetal halides are those formed with the noble gases. Until about 20 years ago, the noble gases were referred to as "inert" gases. They were believed to be completely unreactive toward other elements and compounds. The first noble-gas compound was discovered at the University of British Columbia by Neil Bartlett, a 29-year-old chemist. In the course of his research on platinum-fluorine compounds, he isolated a reddish solid which he showed to be $O_2^+(PtF_6)^-$. Bartlett realized that the ionization energy of Xe (1170 kJ/mol) is virtually identical with that of O_2 (1165 kJ/mol). This encouraged him to attempt to make the analogous compound $Xe^+(PtF_6)^-$. His success opened up a new era in noble-gas chemistry.

They never react
Never?
Well, almost never

Most of the noble-gas compounds known today involve covalent bonds between xenon and fluorine atoms. Among the few exceptions are $XeCl_2$ and KrF_2, both of which are relatively unstable. In contrast, the xenon fluorides XeF_2, XeF_4,

Table 13.6
Compounds of Xenon:
NUMBER OF ELECTRON PAIRS AROUND THE Xe ATOM

4	5	6	7
XeO_3	XeF_2	XeF_4	XeF_6
XeO_4	$XeOF_2$	$XeOF_4$	
	XeO_2F_2	XeO_2F_4	
	$XeCl_2$		

XeF_6 is not octahedral

and XeF_6 are quite stable, with negative heats of formation. Xenon tetrafluoride, a low melting solid (mp = 90°C), is readily prepared by heating the elements together at about 400°C for one hour:

$$Xe(g) + 2\ F_2(g) \rightarrow XeF_4(s); \qquad \Delta H = -260\ kJ \qquad \textbf{(13.30)}$$

As you can see from Table 13.6, xenon, in most of its compounds with fluorine or oxygen, has an expanded octet (five or six pairs of electrons). The structure of XeF_4 was discussed in Chapter 10. You may recall that it has a square planar structure with two unshared pairs on the xenon atom. This structure is shown at the right of Figure 13.10. The structures of $XeOF_4$ and XeO_2F_4 are readily derived from that of XeF_4. Oxygen atoms, each with six valence electrons, form bonds with unshared pairs on the central xenon atom. The $XeOF_4$ molecule is a square pyramid with the oxygen atom at the apex; XeO_2F_4 is octahedral. These structures are shown at the center and left of Figure 13.10.

13.5
Hydrocarbons

Organic chemistry, the study of carbon compounds, will be discussed in a systematic way in Chapter 28. Here we will consider only the molecular structures of hydrocarbons, the simplest type of organic compound. Even though these compounds contain only the two elements hydrogen and carbon, there are thousands of different hydrocarbons, each with its own unique properties. Hydrocarbons are commonly divided into classes depending upon the nature of the carbon-carbon bonds present. We distinguish between saturated, unsaturated, and aromatic hydrocarbons.

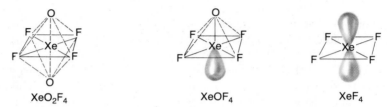

XeO_2F_4 $XeOF_4$ XeF_4

FIGURE 13.10 The molecule XeO_2F_4 has an octahedral structure. In $XeOF_4$, one corner of the octahedron is missing, giving a square pyramid. By the time we reach XeF_4, two corners of the octahedron are gone. The four F atoms in XeF_4 are at the corners of a square with the Xe atom in the middle.

Saturated Hydrocarbons: Alkanes

One large and structurally simple class of hydrocarbons includes those substances in which all the carbon-carbon bonds are single bonds. These are called *saturated* hydrocarbons or **alkanes**. In the alkanes the carbon atoms are bonded to each other in chains, which may be long or short, straight or branched.

The simplest alkanes are methane (CH_4), ethane (C_2H_6), and propane (C_3H_8):

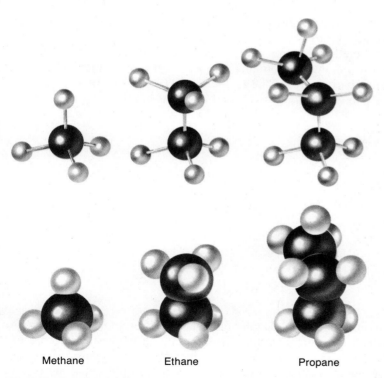

Around the carbon atoms in these molecules and indeed in any saturated hydrocarbon, there are four single bonds involving sp^3 hybrid orbitals. As we would expect from VSEPR theory, these bonds are directed toward the corners of a regular tetrahedron. The bond angles are 109.5°, the tetrahedral angle. This means that in propane (C_3H_8) and in the higher alkanes, the carbon atoms are arranged in a "zigzag" pattern. The three-dimensional structures of CH_4, C_2H_6,

The outside surfaces of these molecules consist mainly of H atoms

Methane   Ethane   Propane

FIGURE 13.11 Models showing the molecular structures of methane, ethane and propane. The bond angles in all of these compounds are equal to 109.5°, the tetrahedral angle.

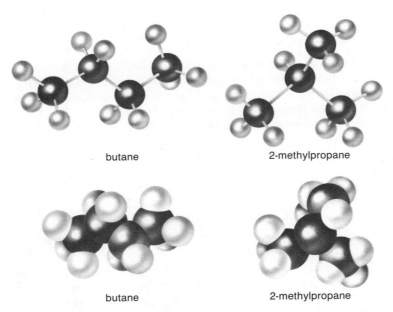

butane   2-methylpropane

butane   2-methylpropane

FIGURE 13.12 Models showing the structures of butane and 2-methylpropane. These are the two isomers of C_4H_{10}.

and C_3H_8 are indicated in Figure 13.11, using both "ball-and-stick" and "space-filling" models.

Two different alkanes are known with the molecular formula C_4H_{10}. In one of these, called butane, the four carbon atoms are linked in a "straight chain." In the other, called 2-methylpropane, there is a "branched chain." The longest continuous chain in the molecule contains three carbon atoms; there is a CH_3 branch from the central carbon atom. The geometries of these molecules are shown in Figure 13.12; the two-dimensional structural formulas are

$$—CH_3 \text{ means } —\overset{\displaystyle H}{\underset{\displaystyle H}{\overset{|}{\underset{|}{C}}}}—H$$

$$\begin{array}{c}
H \\
| \\
H—C—H \\
| \\
H—C—H \\
| \\
H—C—H \\
| \\
H—C—H \\
| \\
H
\end{array}$$

butane

$$\begin{array}{c}
H \\
| \\
H—C—H \\
| \\
H—C—CH_3 \\
| \\
H—C—H \\
| \\
H
\end{array}$$

2-methylpropane

Compounds having the same molecular formula but different molecular structures are called **structural isomers**. Butane and 2-methylpropane are referred to as structural isomers of C_4H_{10}. They are two distinct compounds with their own characteristic physical and chemical properties. Isomerism of this and other types is common among hydrocarbons and indeed among organic compounds in general. For small molecules, it is quite easy to identify the various isomers.

EXAMPLE 13.10   Draw structural formulas for the isomers of C_5H_{12}.

Solution   First sketch the straight-chain structure

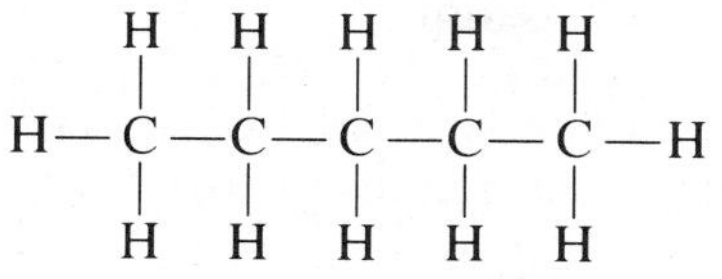

I

Counting hydrogens, we find there are 12, as required. Since each carbon atom has four bonds and each hydrogen one, this is a correct Lewis structure.

Having found one correct structure, we need to determine all the *nonequivalent* alternate structures. Two such structures are shown below (only the carbon skeletons are shown):

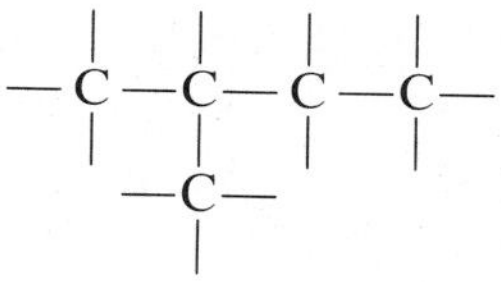

II 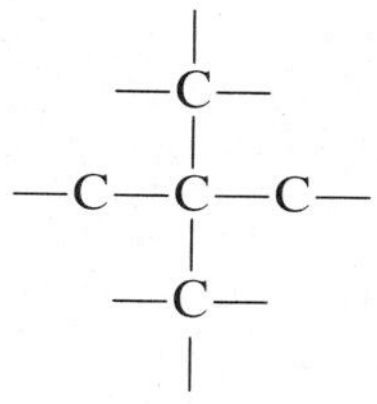   III

In structure II the longest carbon chain consists of only four atoms, so that II and I are clearly different. Similarly, in structure III, there are only three carbon atoms in the longest carbon chain, so it differs from both I and II. Looking at structures I, II, and III you will notice another important difference that distinguishes one from another. In I, no carbon atom is attached to more than two other carbon atoms. In II there is one carbon atom that is bonded to three other carbons. Finally, in III, the carbon atom in the center is bonded to four other carbon atoms.

At this point, working only with pencil and paper, you might be tempted to draw other structures, such as

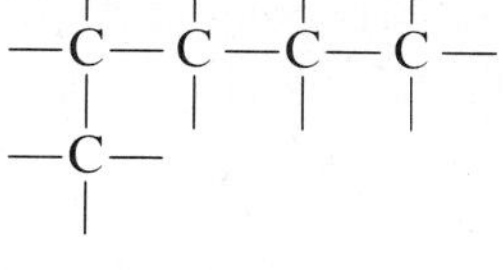

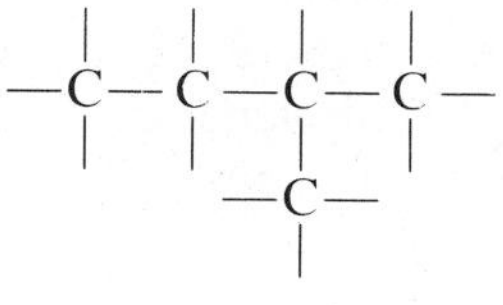

However, a few moments' reflection (or access to a molecular model kit) should convince you that these are in fact equivalent to structures written previously. In particular, the first one, like I, has a five-carbon chain in which no carbon atom is attached to more than two other carbons. The second structure, like II, has a four-carbon chain with one carbon atom bonded to three other carbons. Structures I, II, and III represent the three possible isomers of C_5H_{12}; there are no others.

Organic molecules obey the octet rule

Longest chain means longest continuous chain

There is free rotation around C—C single bonds

EXERCISE   How many isomers are there of C_6H_{14}? Answer: Five (one has a six-carbon chain, two have five-carbon chains, and two have four-carbon chains).

The intermolecular forces in alkanes are primarily of the dispersion type; the polarity of these molecules is very small and there is no hydrogen bonding. As we might expect then, the boiling points of alkanes increase with molar mass:

ALKANE	MM (g/mol)	bp (°C)
Methane, CH_4	16	−161
Ethane, C_2H_6	30	−88
Propane, C_3H_8	44	−42
Butane, C_4H_{10}	58	−1

Ordinarily, branched-chain alkanes have lower boiling points than straight-chain isomers of the same molar mass. Thus 2-methylpropane boils at −10°C as compared to −1°C for butane. This reflects the fact that dispersion forces are weaker for compact molecules. Presumably, this is because the electrons in such molecules have less room to move about and so are more difficult to polarize.

EXAMPLE 13.11   Of the three isomers of C_5H_{12} shown in Example 13.10, which would you expect to have the lowest boiling point? the highest?

The isomer with the longest chain will have the highest BP

Solution   Isomer III is the most compact molecule and should have the lowest boiling point. Isomer I is the least compact and should have the highest boiling point. The observed values are: I (36°C), II (28°C), III (10°C)

EXERCISE   Of the five isomers of C_6H_{14}, which one would you expect to be highest boiling? Answer: The straight-chain isomer, hexane (bp = 69°C).

Unsaturated Hydrocarbons: Alkenes and Alkynes

In an *unsaturated* hydrocarbon, at least one of the carbon-carbon bonds in the molecule is a multiple bond. In this section, we will discuss only two types of unsaturated hydrocarbons:

   1. **Alkenes**, in which there is one carbon-carbon double bond in the molecule. The simplest alkene is ethylene, C_2H_4. Its structural formula is

The atoms attached to a C=C bond all lie in a plane

ethylene

You may recall that we discussed the bonding in ethylene in Chapter 10. The double bond in ethylene and other alkenes consists of a sigma bond and a pi bond.

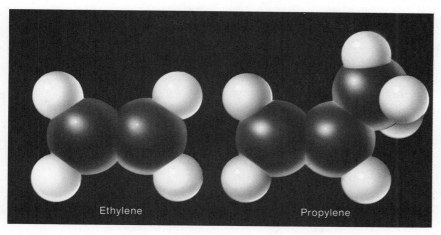

FIGURE 13.13 Space-filling models for ethylene and propylene. The ethylene molecule is planar, since the six nuclei lie in a plane. Propylene contains three carbon atoms, two of which are attached by a double bond.

The hybridization about the carbon atoms joined by the double bond is sp^2; the bond angles around these carbon atoms are 120°. Figure 13.13 shows the molecular geometry of ethylene (C_2H_4) and the next member of the alkene series, propylene (C_3H_6).

2. **Alkynes**, in which there is one carbon-carbon triple bond in the molecule. The simplest alkyne is acetylene, C_2H_2. Its structural formula is

$$H-C\equiv C-H$$

  acetylene

C_2H_2 is used in oxy-acetylene welding

The triple bond in acetylene and other alkynes consists of a sigma bond and two pi bonds. The hybridization about the triple-bonded carbon atoms is sp; the bond angles around these carbon atoms are 180° (Fig. 13.14).

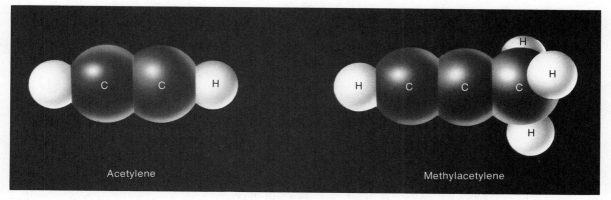

FIGURE 13.14 In acetylene and its simplest derivative, methylacetylene, two carbon atoms are linked by a triple bond. Both molecules contain four atoms on a straight line.

Aromatic Hydrocarbons

Hydrocarbons of this type can be considered to be derived from benzene, C_6H_6. Benzene, a liquid with a boiling point of 80°C, was discovered by Michael Faraday in 1825. Its formula, C_6H_6, indicates a high degree of unsaturation; yet its properties are quite different from those of alkenes or alkynes.

Kekulé, a German chemist, was one of the first to study the chemical properties of benzene. He suggested a ring structure for the benzene molecule, with alternating single and double bonds. To explain the fact that the carbon-carbon bonds are all of the same length, we might regard benzene as a resonance hybrid of the two structures

C_6H_6 molecules are perfect regular hexagons

This model is consistent with many of the properties of benzene. The molecule is a planar hexagon with bond angles of 120°. The hybridization of each carbon atom is sp^2. However, this structure is misleading in one respect. Chemically, benzene does not behave as if there were double bonds present.

It's chemically more stable than would be expected

A more satisfactory model of the electron distribution in the benzene molecule is suggested by molecular orbital theory (Chap. 10). Here we consider that:

—each carbon atom forms three sigma bonds, one to a hydrogen atom and two to adjacent carbon atoms.

—the three electron pairs remaining are not tied down (*localized*) to form three double bonds, as in the Kekulé structure. Instead, the six electrons are spread symmetrically over the entire molecule to form "delocalized" pi bonds. The orbitals occupied by these electrons are indicated in Figure 13.15.

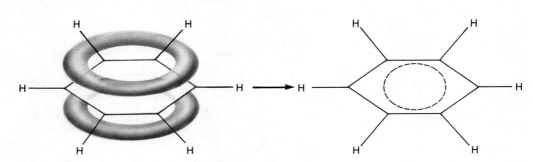

FIGURE 13.15 In benzene, three electron pairs are not localized on a particular carbon atom. Instead, they are spread out over two electron clouds of the shape shown, one above the plane of the benzene ring and the other below it.

This model is often represented by the structure

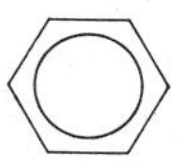

Here, it is understood that

—there is a carbon atom at each corner of the hexagon.
—there is a H atom bonded to each carbon atom.
—the circle in the center of the molecule represents the six delocalized electrons.

A few typical aromatic hydrocarbons are shown in Figure 13.16. Notice that some aromatic hydrocarbons, such as naphthalene, contain two or more benzene rings fused together. Certain compounds of this type are potent carcinogens. One of the most dangerous is 3,4-benzpyrene, which has been detected in cigarette smoke.

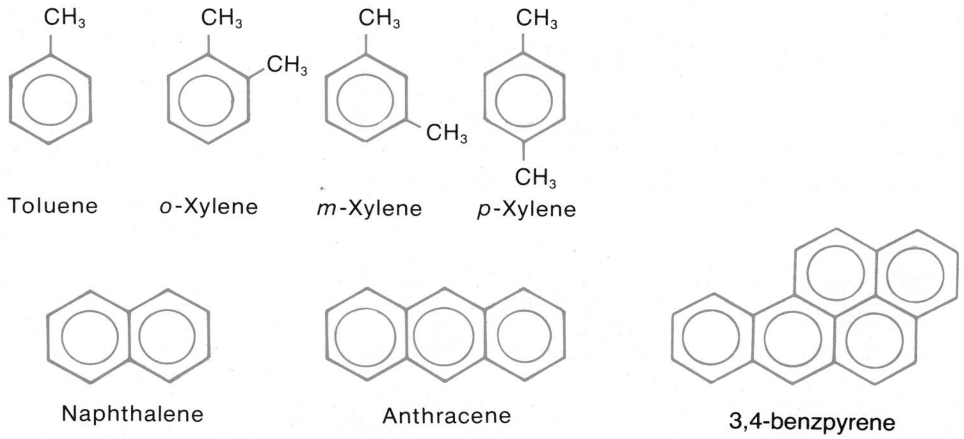

CH₃   CH₃   CH₃   CH₃

CH₃   CH₃   CH₃

Toluene   *o*-Xylene   *m*-Xylene   *p*-Xylene

Naphthalene   Anthracene   3,4-benzpyrene

FIGURE 13.16 Aromatic hydrocarbons derived from benzene. The molecular formulas are C_7H_8 (toluene), C_8H_{10} (xylene), $C_{10}H_8$ (naphthalene), $C_{14}H_{10}$ (anthracene), and $C_{20}H_{12}$ (benzpyrene).

Summary

The nonmetals comprise the elements in Groups 6, 7, and 8 of the Periodic Table along with the lighter elements in Groups 4 and 5. Atoms of many of these elements combine with each other to form molecules (H_2, F_2, Cl_2, Br_2, I_2; O_2, S_8; N_2, P_4). Other nonmetallic elements have network covalent structures (Table 13.1). Several nonmetals show allotropy; included among these are oxygen, sulfur, phosphorus, and carbon, discussed in Section 13.1.

Sections 13.2 to 13.4 dealt with binary compounds involving two nonmetals. Emphasis was placed upon

—the hydrogen compounds of the nonmetals (Table 13.2).
—the oxides of sulfur (SO_2, SO_3), carbon (CO, CO_2), and nitrogen (Fig 13.9).
—the halogen compounds of the nonmetals (Tables 13.4 to 13.6).

Throughout the first four sections of this chapter, we reviewed principles first introduced in Chapters 9 to 12. Examples 13.1 to 13.9 illustrate these principles.

Section 13.5 surveyed hydrocarbons. We distinguished between alkanes (all single bonds), alkenes (one double bond), alkynes (one triple bond), and aromatic hydrocarbons. Hydrocarbons show structural isomerism. Frequently there are two or more compounds with the same molecular formula (Example 13.10), but quite different properties (Example 13.11).

Key Words and Concepts

alkane	hydrocarbon	pi bond
alkene	hydrogen bond	polarity
alkyne	interhalogen	resonance
allotropy	isomer	saturated hydrocarbon
aromatic hydrocarbon	Lewis structure	sigma bond
bond energy	network covalent	square pyramid
branched-chain alkane	n-type semiconductor	straight-chain alkane
dipole force	octahedron	structural isomer
dispersion force	octet	unsaturated hydrocarbon
expanded octet	p-type semiconductor	VSEPR model
halogen	phase diagram	weak electrolyte
hybrid orbital		

Questions and Problems

General

13.1 Explain how the allotropic forms of each of the following elements differ from each other:
  a. oxygen and ozone.
  b. rhombic and monoclinic sulfur.
  c. diamond and graphite.
  d. white and red phosphorus.

13.2 Write balanced equations for
  a. the decomposition of hydrogen peroxide.
  b. the reaction of hydrogen peroxide with hydrazine.
  c. the preparation of HF from CaF_2.
  d. the preparation of HI from NaI.

13.3 Describe how you would prepare
  a. NO_2 from NO   b. N_2O c. SO_2 d. P_4O_6

13.4 Give the formulas of
  a. three different fluorides of xenon.
  b. two different fluorides of sulfur.
  c. three binary compounds containing carbon and chlorine.
  d. two interhalogens that follow the octet rule.

13.31 Describe the structure of
  a. liquid sulfur between 160 and 250°C.
  b. plastic sulfur.
  c. amorphous carbon.
  d. a p-type semiconductor derived from silicon.

13.32 Write balanced equations for
  a. the decomposition of hydrazine.
  b. the formation of HBr from the elements.
  c. the preparation of HCl from NaCl.
  d. the reaction of H_3PO_4 with NaBr.

13.33 Write balanced equations for the reaction of oxygen with
  a. ZnS   b. CO   c. P_4O_6 d. SO_2 e. N_2

13.34 Give the formulas of
  a. two different oxides of xenon.
  b. two different fluorides of phosphorus.
  c. two binary compounds containing chlorine and fluorine.
  d. three interhalogens that have expanded octets.

13.5 Give the symbol of the element that
    a. forms a molecule containing eight atoms.
    b. has two gaseous allotropes.
    c. is a red liquid at 25°C and 1 atm.

13.6 Classify the following statements as true or false. Correct any false statements to make them true.
    a. Allotropic forms of an element have different properties.
    b. Oxygen gas is diamagnetic.
    c. Diamond is converted to graphite at high pressures.
    d. An n-type semiconductor can be made by adding a trace of Ge to Si.
    e. Two substances with the same molecular formula have the same boiling point.

13.7 Explain the difference between
    a. allotropy and isomerism.
    b. alkenes and alkynes.
    c. the structures of graphite and diamond.
    d. the two isomeric butanes.

13.35 Write the symbol of
    a. an element that is used in jewelry and automobile tires.
    b. an element that forms an oxide referred to as "laughing gas."
    c. a solid nonmetal that reacts vigorously with O_2 at 25°C.

13.36 Criticize each of the following statements:
    a. $Br_2(l)$ and $Br_2(g)$ are allotropes.
    b. Nitrogen and phosphorus in Group 5 form oxides of similar formulas.
    c. A high reaction temperature tends to favor the formation of a higher oxide (for example, NO_2 as opposed to NO).
    d. The resonance forms of O_3 have different properties.
    e. The xenon atom in XeF_4 is surrounded by four electron pairs.

13.37 Explain the difference between
    a. alkanes and alkenes.
    b. resonance and isomerism.
    c. straight- and branched-chain hydrocarbons.
    d. ethylene and propylene.

Isomerism

13.8 Draw the structural isomers of the alkane C_6H_{14}.

13.9 Draw the structural isomers of C_4H_9Cl in which one hydrogen atom of a C_4H_{10} molecule has been replaced by chlorine.

13.10 There are three compounds with the formula $C_6H_4Cl_2$ in which two of the hydrogen atoms of the benzene molecule have been replaced by chlorine atoms. Draw structural formulas for these compounds.

13.38 Draw the structural isomers of the alkene C_4H_8.

13.39 Draw the structural isomers of $C_3H_6Cl_2$ in which two of the hydrogen atoms of C_3H_8 have been replaced by chlorine atoms.

13.40 There are three compounds with the formula $C_6H_3Cl_3$ in which three of the hydrogen atoms of the benzene molecule have been replaced by chlorine atoms. Draw structural formulas for these compounds.

Lewis Structures, Resonance, and Bond Energies (Chap. 9)

13.11 Draw the Lewis structures of
    a. H_2O_2 b. PH_3 c. $SeBr_2$ d. Cl_2O e. XeO_4

13.12 Draw Lewis structures for
    a. CO_2 b. SO_2 c. N_2O d. N_2O_4

13.13 Draw Lewis structures for each of the following "expanded octet" molecules:
    a. $XeCl_2$ b. PF_5 c. ClF_3 d. IF_5

13.14 Draw resonance forms for each of the molecules in Question 13.12.

13.41 Draw the Lewis structures of
    a. N_2H_4 b. H_2Se c. NO_2 d. PF_3 e. XeO_3

13.42 Draw Lewis structures for
    a. SO_3 b. O_3 c. HN_3 d. N_2O_5

13.43 Draw Lewis structures for each of the following "expanded octet" molecules:
    a. XeF_4 b. SF_4 c. SF_6 d. $AsCl_5$

13.44 Draw resonance forms for each of the molecules in Question 13.42.

13.15 The heat of formation of $XeF_4(g)$ is -215 kJ/mol. Calculate the bond energy for the Xe—F bond, taking that of the F—F bond to be 153 kJ/mol.

13.16 Using bond energies from Tables 9.6 and 9.7, Chapter 9, estimate ΔH for the reaction:

$$2\ CO(g)\ +\ O_2(g) \rightarrow 2\ CO_2(g)$$

13.45 The heat of formation of $ClF_3(g)$ is -163 kJ/mol. Calculate the bond energy for the Cl—F bond; those for Cl—Cl and F—F are 243 and 153 kJ/mol, respectively.

13.46 Using bond energies from Tables 9.6 and 9.7, Chapter 9, estimate ΔH for

$$N_2(g)\ +\ 3\ Cl_2(g) \rightarrow 2\ NCl_3(g)$$

Molecular Geometry, Polarity, and Hybridization (Chap. 10)

13.17 Predict the bond angles for the molecules referred to in Question 13.11.

13.18 Describe the geometry of each molecule referred to in Question 13.13, using Figure 10.10, Chapter 10, if necessary.

13.19 Classify each of the molecules in Question 13.11 as polar or nonpolar.

13.20 Give the hybridization about the central atom in each of the molecules in Question 13.13.

13.21 Consider the oxides of nitrogen shown in Figure 13.9. Give the hybridization of each N atom in each molecule.

13.22 Consider the N_2O_3 molecule shown in Figure 13.9. Give the bond angles and predict whether the molecule is polar.

13.23 Consider the SF_2 and SF_6 molecules. Write the Lewis structures for these molecules, describe the geometries, state whether they are polar, and give the hybridization about the central atom.

13.24 A plausible Lewis structure for P_4 is

:P＝P:
| |
:P＝P:

How would the properties of this molecule differ from those of the P_4 molecule shown in Example 13.2?

13.47 Predict the bond angles for the molecules referred to in Question 13.41.

13.48 Describe the geometry of each molecule referred to in Question 13.43, using Figure 10.10, Chapter 10, if necessary.

13.49 Classify each of the molecules in Question 13.12 as polar or nonpolar.

13.50 Give the hybridization about the central atom in each of the molecules in Question 13.43.

13.51 Give the number of pi and sigma bonds in each of the molecules shown in Figure 13.9.

13.52 Follow the directions of Question 13.22 for the N_2O_2 molecule shown in Figure 13.9.

13.53 Follow the directions of Question 13.23 for the ICl and IF_5 molecules.

13.54 At one time, it was believed that the O_3 molecule was a three-membered ring with the Lewis structure

How would the properties of such a molecule differ from those of the O_3 molecule shown in Section 13.1?

Phase Equilibria, Intermolecular Forces (Chap. 11)

13.25 Refer to Figure 13.2:
  a. Does the melting point of monoclinic sulfur increase or decrease as the pressure increases?
  b. How would you sublime monoclinic sulfur?
  c. What phase(s) is (are) present at 119°C and 0.027 mm Hg? 119°C and 0.010 mm Hg? 119°C and 0.050 mm Hg?

13.26 The vapor pressure of white phosphorus at 20°C is 0.0254 mm Hg; at 40°C, it is 0.133 mm Hg. Estimate the heat of sublimation of white phosphorus.

13.27 What is the major type of attractive force that must be overcome to
  a. vaporize Br_2?   b. melt ice?
  c. melt diamond?   d. vaporize XeO_4?

13.28 Sulfur hexafluoride (SF_6) and decane, $CH_3-(CH_2)_8-CH_3$, have nearly the same molar mass. Yet, decane boils at 174°C while SF_6 sublimes at −64°C. Suggest an explanation for this difference.

13.55 Refer to Figure 13.2:
  a. Is liquid sulfur more or less dense than monoclinic sulfur?
  b. Does the transition from rhombic to monoclinic sulfur absorb or evolve heat?
  c. What phase(s) is (are) present at 96°C and 0.0043 mm Hg? 98°C and 0.0043 mm Hg? 94°C and 0.0043 mm Hg?

13.56 The heat of sublimation of dry ice, $CO_2(s)$, is 27.4 kJ/mol. Dry ice sublimes at one atm and −78°C. Estimate the vapor pressure of $CO_2(s)$ at the triple point, where the temperature is −57°C.

13.57 What is the major type of attractive force that must be overcome to:
  a. sublime CO_2?   b. melt Si?
  c. convert S_8 rings to S_8 chains?   d. vaporize $NH_3(l)$?

13.58 Which one of the following compounds would you expect to have the higher boiling point? Explain your reasoning.

$$CH_3-CH_2-CH_2-CH_2Cl \quad \text{or} \quad CH_3-\overset{\overset{\displaystyle H}{|}}{\underset{\underset{\displaystyle CH_3}{|}}{C}}-CH_2Cl$$

Concentrations of Solutions (Chap. 12)

13.29 A water solution containing 10.0 mass percent H_2O_2 has a density of 1.035 g/cm³. Calculate the molality and molarity of the hydrogen peroxide solution.

13.30 A certain solution is 2.50 M in HBr.
  a. What volume of this solution must be diluted with water to obtain 0.350 L of 1.00 M HBr?
  b. How many moles of ions are there in 225 cm³ of this solution?

13.59 A water solution that is 3.15 M in hydrazine (N_2H_4) has a density of 1.01 g/cm³. Calculate the mass percent and molality of hydrazine in this solution.

13.60 Concentrated phosphoric acid is 14.7 M in H_3PO_4. What volume of this solution is required to react with 1.00 g of NaI by Equation 13.9?

*13.61 Write plausible Lewis structures for I_2O_5 and Cl_2O_7.

*13.62 In principle, there are several structural isomers of N_2O_3 in addition to the one shown in Figure 13.9. Draw Lewis structures for six of these isomers.

*13.63 A yellow-brown liquid is a compound of sulfur and chlorine with a mass percent of sulfur of 47.49.
   a. Identify the compound, referring to Table 13.4.
   b. This compound reacts with chlorine to form a solid that contains 18.4 mass percent sulfur. Determine the formula of this solid and write a balanced equation for the reaction.

*13.64 In the IF_7 molecule, the iodine atom is bonded to seven fluorine atoms. Extrapolating from the known geometries of PF_5 and SF_6, suggest a reasonable geometry for IF_7.

*13.65 Of the molecules in Table 13.4, which one contains the highest mass percent of halogen? (You should be able to answer this question without using a calculator.)

A basic goal of chemistry is to predict whether or not a reaction will occur when reactants are brought together. A reaction that occurs "by itself" without the exertion of any outside force is said to be **spontaneous**. All of us are familiar with certain spontaneous processes. For example,

—an ice cube melts when added to a glass of water at room temperature.
—a mixture of hydrogen and oxygen burns when we set a match to it.
—an iron (steel) tool exposed to moist air rusts.

In other words, the following three reactions are spontaneous at 25°C and 1 atm:

$$H_2O(s) \rightarrow H_2O(l) \tag{14.1}$$

$$2 H_2(g) + O_2(g) \rightarrow 2 H_2O(l) \tag{14.2}$$

$$2 Fe(s) + \tfrac{3}{2} O_2(g) + 3 H_2O(l) \rightarrow 2 Fe(OH)_3(s) \tag{14.3}$$

The word "spontaneous" does not imply anything about how rapidly a reaction occurs. Some spontaneous reactions, notably the rusting of iron by Reaction 14.3, are quite slow. Often a reaction that is potentially spontaneous does not occur without some sort of stimulus. A mixture of hydrogen and oxygen shows no sign of reaction in the absence of a spark or match. Once started, though, a spontaneous reaction continues by itself without further input of energy from the outside.

A spontaneous reaction can do work on its surroundings

If a reaction is spontaneous under a given set of conditions, the reverse reaction is nonspontaneous under the same conditions. Water in a beaker at room temperature does not spontaneously freeze by the reverse of Reaction 14.1; neither does it decompose to the elements by the reverse of Reaction 14.2. However, it is often possible to bring about a nonspontaneous reaction by supplying energy in the form of work. Electrolysis can be used to bring about the reaction

$$2 \, H_2O(l) \rightarrow 2 \, H_2(g) \, + \, O_2(g) \qquad \text{(14.4)}$$

To do this, electrical energy must be furnished, perhaps from a storage battery. As with all other nonspontaneous processes, Reaction 14.4 stops immediately if the source of energy is cut off.

In this chapter we will develop a general approach to predict whether a reaction is spontaneous. To do this, we return to the subject of the energy flow in chemical reactions, first mentioned in Chapter 5. We will start with a discussion of the effect of the enthalpy change, ΔH, on reaction spontaneity (Section 14.1), and continue by considering the effect upon spontaneity of the entropy change, ΔS (Section 14.2). By combining ΔH and ΔS in the proper way, it is possible to calculate a quantity called the free energy change, ΔG. The sign of ΔG is the ultimate criterion of spontaneity. As we will see in Section 14.3, a reaction for which the free energy change is negative must be spontaneous at constant temperature and pressure.

The arguments used and the equations developed throughout this chapter come from chemical thermodynamics. You may recall that we touched upon this subject in Chapter 5, in discussing the First Law of Thermodynamics. In Section 14.4, we will consider the Second Law of Thermodynamics. Here, we will be particularly interested in what the Second Law has to tell us about the spontaneity of processes taking place in the world around us.

14.1
The Enthalpy Change, ΔH

$\Delta H = q$ when P and T are constant

You will recall from Chapter 5 that, for a reaction carried out at constant temperature and pressure, the heat flow is equal to the difference in enthalpy between products and reactants, ΔH. Insofar as ΔH is concerned, we distinguish between two types of reactions. An *exothermic* reaction is one for which ΔH is a negative quantity ($\Delta H < 0$). The enthalpy of the products is less than that of the reactants; heat is given off by the reaction system to the surroundings. In contrast, for an *endothermic* reaction, ΔH is a positive quantity ($\Delta H > 0$). The enthalpy of the products is greater than that of the reactants; heat must be absorbed from the surroundings for the reaction to take place.

The enthalpy change for a reaction can be calculated from molar heats of formation, ΔH_f. The general relation, given in Chapter 5, is

$$\Delta H = \Sigma \, \Delta H_{f \text{ products}} - \Sigma \, \Delta H_{f \text{ reactants}} \qquad \text{(14.5)}$$

Recall from Chapter 5 that

—the molar heat of formation of a compound is the enthalpy change when one mole of the compound is formed from the elements in their stable forms.
—the molar heat of formation of an element is zero.
—molar heats of formation of ions in solution are established by setting ΔH_f $H^+(aq) = 0$.

Tables of heats of formation of compounds and ions in aqueous solution were given in Chapter 5. They are repeated here (Tables 14.1 and 14.2) for convenience.

Note that heats of formation (in kilojoules per mole) are given for a specific temperature (25°C), pressure (1 atm), and concentration (1 M). However, for our purposes, we can assume that *values of ΔH for reactions calculated from these tables are independent of temperature, pressure, or concentration.*

This is a good approximation

A hundred years ago many chemists felt that they had a general criterion for predicting reaction spontaneity. The prevailing idea, put forth by P. M. Berthelot in Paris and Julius Thomsen in Copenhagen, was that all spontaneous reactions

Table 14.1
Heats of Formation (kJ/mol) at 25°C and 1 atm

$AgBr(s)$	-99.5	$C_2H_2(g)$	$+226.7$	$H_2O(l)$	-285.8	$NH_4Cl(s)$	-315.4
$AgCl(s)$	-127.0	$C_2H_4(g)$	$+52.3$	$H_2O_2(l)$	-187.6	$NH_4NO_3(s)$	-365.1
$AgI(s)$	-62.4	$C_2H_6(g)$	-84.7	$H_2S(g)$	-20.1	$NO(g)$	$+90.4$
$Ag_2O(s)$	-30.6	$C_3H_8(g)$	-103.8	$H_2SO_4(l)$	-811.3	$NO_2(g)$	$+33.9$
$Ag_2S(s)$	-31.8	$n\text{-}C_4H_{10}(g)$	-124.7	$HgO(s)$	-90.7	$NiO(s)$	-244.3
$Al_2O_3(s)$	-1669.8	$n\text{-}C_5H_{12}(l)$	-173.1	$HgS(s)$	-58.2	$PbBr_2(s)$	-277.0
$BaCl_2(s)$	-860.1	$C_2H_5OH(l)$	-277.6	$KBr(s)$	-392.2	$PbCl_2(s)$	-359.2
$BaCO_3(s)$	-1218.8	$CoO(s)$	-239.3	$KCl(s)$	-435.9	$PbO(s)$	-217.9
$BaO(s)$	-558.1	$Cr_2O_3(s)$	-1128.4	$KClO_3(s)$	-391.4	$PbO_2(s)$	-276.6
$BaSO_4(s)$	-1465.2	$CuO(s)$	-155.2	$KF(s)$	-562.6	$Pb_3O_4(s)$	-734.7
$CaCl_2(s)$	-795.0	$Cu_2O(s)$	-166.7	$MgCl_2(s)$	-641.8	$PCl_3(g)$	-306.4
$CaCO_3(s)$	-1207.0	$CuS(s)$	-48.5	$MgCO_3(s)$	-1113	$PCl_5(g)$	-398.9
$CaO(s)$	-635.5	$CuSO_4(s)$	-769.9	$MgO(s)$	-601.8	$SiO_2(s)$	-859.4
$Ca(OH)_2(s)$	-986.6	$Fe_2O_3(s)$	-822.2	$Mg(OH)_2(s)$	-924.7	$SnCl_2(s)$	-349.8
$CaSO_4(s)$	-1432.7	$Fe_3O_4(s)$	-1120.9	$MgSO_4(s)$	-1278.2	$SnCl_4(l)$	-545.2
$CCl_4(l)$	-139.5	$HBr(g)$	-36.2	$MnO(s)$	-384.9	$SnO(s)$	-286.2
$CH_4(g)$	-74.8	$HCl(g)$	-92.3	$MnO_2(s)$	-519.7	$SnO_2(s)$	-580.7
$CHCl_3(l)$	-131.8	$HF(g)$	-268.6	$NaCl(s)$	-411.0	$SO_2(g)$	-296.1
$CH_3OH(l)$	-238.6	$HI(g)$	$+25.9$	$NaF(s)$	-569.0	$SO_3(g)$	-395.2
$CO(g)$	-110.5	$HNO_3(l)$	-173.2	$NaOH(s)$	-426.7	$ZnO(s)$	-348.0
$CO_2(g)$	-393.5	$H_2O(g)$	-241.8	$NH_3(g)$	-46.2	$ZnS(s)$	-202.9

Table 14.2
Heats of Formation (kJ/mol) of Aqueous Ions at 25°C, 1 M

CATIONS				ANIONS			
$Ag^+(aq)$	$+105.9$	$K^+(aq)$	-251.2	$Br^-(aq)$	-120.9	$H_2PO_4^-(aq)$	-1302.5
$Al^{3+}(aq)$	-524.7	$Li^+(aq)$	-278.5	$Cl^-(aq)$	-167.4	$HPO_4^{2-}(aq)$	-1298.7
$Ba^{2+}(aq)$	-538.4	$Mg^{2+}(aq)$	-462.0	$ClO_3^-(aq)$	-98.3	$I^-(aq)$	-55.9
$Ca^{2+}(aq)$	-543.0	$Mn^{2+}(aq)$	-218.8	$ClO_4^-(aq)$	-131.4	$MnO_4^-(aq)$	-518.4
$Cd^{2+}(aq)$	-72.4	$Na^+(aq)$	-239.7	$CO_3^{2-}(aq)$	-676.3	$NO_3^-(aq)$	-206.6
$Cu^{2+}(aq)$	$+64.4$	$NH_4^+(aq)$	-132.8	$CrO_4^{2-}(aq)$	-863.2	$OH^-(aq)$	-229.9
$Fe^{2+}(aq)$	-87.9	$Ni^{2+}(aq)$	-64.0	$F^-(aq)$	-329.1	$PO_4^{3-}(aq)$	-1284.1
$Fe^{3+}(aq)$	-47.7	$Pb^{2+}(aq)$	$+1.6$	$HCO_3^-(aq)$	-691.1	$S^{2-}(aq)$	$+41.8$
$H^+(aq)$	0.0	$Sn^{2+}(aq)$	-10.0			$SO_4^{2-}(aq)$	-907.5
		$Zn^{2+}(aq)$	-152.4				

are exothermic. If this were true, all we would have to do to predict reaction spontaneity would be to calculate the enthalpy change, ΔH, and look at its sign. If ΔH turned out to be negative, we could assume that the reaction must be spontaneous; if ΔH were positive, the reaction could not occur by itself.

It turns out that almost all exothermic chemical reactions are spontaneous at 25°C and 1 atm. Consider, for example, the formation of water from the elements and the rusting of iron:

$$2 \, H_2(g) + O_2(g) \rightarrow 2 \, H_2O(l); \qquad \Delta H = -571.6 \text{ kJ}$$

$$2 \, Fe(s) + \tfrac{3}{2} \, O_2(g) + 3 \, H_2O(l) \rightarrow 2 \, Fe(OH)_3(s); \qquad \Delta H = -791 \text{ kJ}$$

For both of these spontaneous reactions, ΔH is a negative quantity.

On the other hand, this simple rule fails for many familiar phase changes. An example is the melting of ice. This takes place spontaneously at 1 atm above 0°C, even though it is endothermic:

$$H_2O(s) \rightarrow H_2O(l); \qquad \Delta H = +6.0 \text{ kJ} \tag{14.6}$$

In another case we find that, above 100°C, liquid water vaporizes to steam at 1 atm. This process, like Reaction 14.6, absorbs heat:

$$H_2O(l) \rightarrow H_2O(g); \qquad \Delta H = +40.7 \text{ kJ} \tag{14.7}$$

There is still another basic objection to using the sign of ΔH as a general criterion for spontaneity. Endothermic reactions that are nonspontaneous at room temperature often become spontaneous when the temperature is raised. Consider, for example, the decomposition of limestone:

Spontaneity is temperature-dependent. ΔH is not

$$CaCO_3(s) \rightarrow CaO(s) + CO_2(g); \qquad \Delta H = +178.0 \text{ kJ} \tag{14.8}$$

At 25°C and 1 atm, this reaction is nonspontaneous. Witness the existence of the white cliffs of Dover and other limestone deposits over eons of time. However, if the temperature is raised to about 1100 K, the limestone decomposes to give off carbon dioxide gas at 1 atm. In other words, this endothermic reaction becomes spontaneous at high temperatures. This is true despite the fact that ΔH remains at +178.0 kJ, nearly independent of temperature.

14.2
The Entropy Change, ΔS

To decide whether a given reaction will be spontaneous at a given temperature and pressure, we must consider another factor in addition to ΔH. This is the *entropy* change for the reaction, ΔS:

$$\Delta S = S_{products} - S_{reactants} \tag{14.9}$$

The entropy of a substance, like its enthalpy, is one of its characteristic properties. Entropy is a measure of disorder or randomness. Substances that are highly disordered have high entropies. Low entropy is associated with strongly ordered substances. By "order" we mean the extent to which particles of a substance are confined to a given region of space. In a crystal, where the atoms, molecules, or ions are fixed in position, the entropy is relatively low. When a solid melts, the particles are free to move through the entire liquid volume; the entropy of the substance increases. Upon vaporization the particles acquire still greater freedom to move about, and there is a large increase in entropy.

From what we have just said, it should be clear that for Reactions 14.6 and 14.7,

$$H_2O(s) \rightarrow H_2O(l)$$

$$H_2O(l) \rightarrow H_2O(g)$$

the entropy change, ΔS, is a positive quantity. Liquid water has a higher entropy than ice; steam has a higher entropy than liquid water. Indeed, the increase in entropy is the driving force behind these processes. We might say that ice melts above 0°C and water vaporizes above 100°C *because* of the increase in randomness. This factor more than compensates for the fact that the processes are endothermic.

Many other spontaneous processes take place with an increase in entropy. Consider, for example, what happens when a tank of compressed helium is used to fill a balloon. Helium rushes out of the high-pressure tank into the balloon, which is at atmospheric pressure. We might explain this behavior by reasoning that the drop in pressure causes the helium atoms to spread out over a larger volume, thereby increasing their entropy. From a slightly different point of view, we could say that the gas molecules move spontaneously from a region of high concentration (high-pressure tank) to a region of low concentration (the balloon). Such a process is always accompanied by an increase in entropy. This is true whether we deal with a compressed gas escaping from a cylinder, or the natural process of osmosis described in Chapter 12. In that case, solvent molecules move from a region where their concentration is high (pure solvent) to a region where their concentration is lower (solution).

Spontaneity is favored by an increase in disorder, and hence by an increase in entropy

If we think about the matter carefully, we realize that there is an inherent tendency, common to inanimate substances (and probably to students and professors as well) to become disorganized. A beaker shatters when we drop it on a stone lab bench. Rivers become polluted and wilderness trails get cluttered with litter. All these processes involve an increase in entropy. They are all spontaneous, as anyone who tries to reverse them soon discovers.

Standard Molar Entropies of Elements and Compounds

Considering the rather vague way we have described entropy, you may be surprised to learn that we can determine absolute values for the molar entropies of pure substances. Actually, entropy can be defined very precisely and measured accurately. The entropy of a substance has a fixed value at a particular temperature and pressure; that is, entropy is a state function that depends only upon the state of a substance and not upon its history.

It turns out that for any pure crystal:

$$S \rightarrow 0 \text{ as } T \rightarrow 0 \text{ K}$$

Can you see why this might be so?

The details of how entropy is measured are beyond the level of this text. The results, however, are useful for a variety of purposes. They are given in Table 14.3 in the form of *standard molar entropies* (S⁰) at 25°C. These are the entropies per mole, at 1 atm, in the units of **joules per kelvin (J/K).**

Table 14.3
Standard Molar Entropies (J/K) of Elements and Compounds at 25°C and 1 atm

ELEMENTS

Ag(s)	42.7	Co(s)	28.5	I$_2$(s)	116.7	O$_2$(g)	205.0
Al(s)	28.3	Cr(s)	23.8	K(s)	63.6	Pb(s)	64.9
Ba(s)	67	Cu(s)	33.3	Mg(s)	32.5	P$_4$(s)	177.4
Br$_2$(l)	152.3	F$_2$(g)	203.3	Mn(s)	31.8	S(s)	31.9
C(s)	5.7	Fe(s)	27.2	N$_2$(g)	191.5	Si(s)	18.7
Ca(s)	41.6	H$_2$(g)	130.6	Na(s)	51.0	Sn(s)	51.5
Cl$_2$(g)	222.9	Hg(l)	77.4	Ni(s)	30.1	Zn(s)	41.6

COMPOUNDS

AgBr(s)	107.1	C$_2$H$_2$(g)	200.8	H$_2$O(l)	69.9	NH$_4$Cl(s)	94.6
AgCl(s)	96.1	C$_2$H$_4$(g)	219.5	H$_2$O$_2$(l)	88.6	NH$_4$NO$_3$(s)	151.0
AgI(s)	114.2	C$_2$H$_6$(g)	229.5	H$_2$S(g)	205.6	NO(g)	210.6
Ag$_2$O(s)	121.7	C$_3$H$_8$(g)	269.9	H$_2$SO$_4$(l)	156.9	NO$_2$(g)	240.5
Ag$_2$S(s)	145.6	n-C$_4$H$_{10}$(g)	310.0	HgO(s)	72.0	NiO(s)	38.6
Al$_2$O$_3$(s)	51.0	n-C$_5$H$_{12}$(l)	262.8	HgS(s)	77.8	PbBr$_2$(s)	161.5
BaCl$_2$(s)	126	C$_2$H$_5$OH(l)	160.7	KBr(s)	96.4	PbCl$_2$(s)	136.4
BaCO$_3$(s)	112.1	CoO(s)	43.9	KCl(s)	82.7	PbO(s)	69.5
BaO(s)	70.3	Cr$_2$O$_3$(s)	81.2	KClO$_3$(s)	143.0	PbO$_2$(s)	76.6
BaSO$_4$(s)	132.2	CuO(s)	43.5	KF(s)	66.6	Pb$_3$O$_4$(s)	211.3
CaCl$_2$(s)	113.8	Cu$_2$O(s)	100.8	MgCl$_2$(s)	89.5	PCl$_3$(g)	311.7
CaCO$_3$(s)	92.9	CuS(s)	66.5	MgCO$_3$(s)	65.7	PCl$_5$(g)	352.7
CaO(s)	39.7	CuSO$_4$(s)	113.4	MgO(s)	26.8	SiO$_2$(s)	41.8
Ca(OH)$_2$(s)	76.1	Fe$_2$O$_3$(s)	90.0	Mg(OH)$_2$(s)	63.1	SnCl$_2$(s)	122.6
CaSO$_4$(s)	106.7	Fe$_3$O$_4$(s)	146.4	MgSO$_4$(s)	91.6	SnCl$_4$(l)	258.6
CCl$_4$(l)	214.4	HBr(g)	198.5	MnO(s)	60.2	SnO(s)	56.5
CH$_4$(g)	186.2	HCl(g)	186.7	MnO$_2$(s)	53.1	SnO$_2$(s)	52.3
CHCl$_3$(l)	202.9	HF(g)	173.5	NaCl(s)	72.4	SO$_2$(g)	248.5
CH$_3$OH(l)	126.8	HI(g)	206.3	NaF(s)	58.6	SO$_3$(g)	256.2
CO(g)	197.9	HNO$_3$(l)	155.6	NaOH(s)	52.3	ZnO(s)	43.9
CO$_2$(g)	213.6	H$_2$O(g)	188.7	NH$_3$(g)	192.5	ZnS(s)	57.7

Notice from Table 14.3 that

—standard molar entropies of substances are always positive quantities ($S^0 > 0$).

All substances, elements and compounds, have some disorder at 25°C

—*elements as well as compounds have nonzero standard entropies.* This is in contrast to the situation with heats of formation, where $\Delta H_f = 0$ for elements in their stable states.

—solids as a group have lower entropies than liquids, which in turn have lower entropies than gases. Compare, for example,

	S^0 (J/K)
Si(s)	18.7
H$_2$O(l)	69.9
SO$_2$(g)	248.5

Standard Molar Entropies of Ions in Solution

It is possible to assign standard molar entropies to ions in water solution. To do this, we adopt a convention similar to that followed with heats of formation. The standard molar entropy of the H^+ ion in water solution is taken to be zero:

$$S^0\ H^+(aq) = 0$$

This is an arbitrary assumption

Standard entropies of ions based upon this convention are listed in Table 14.4. They are given in joules per kelvin at 25°C and a concentration of 1 M.*

Table 14.4
Standard Molar Entropies (J/K) of Aqueous Ions at 25°C, 1 M

CATIONS				ANIONS			
$Ag^+(aq)$	73.9	$K^+(aq)$	102.5	$Br^-(aq)$	80.7	$H_2PO_4^-(aq)$	89.1
$Al^{3+}(aq)$	−313.4	$Li^+(aq)$	14.2	$Cl^-(aq)$	55.1	$HPO_4^{2-}(aq)$	−36.0
$Ba^{2+}(aq)$	13	$Mg^{2+}(aq)$	−118.0	$ClO_3^-(aq)$	162.3	$I^-(aq)$	109.4
$Ca^{2+}(aq)$	−55.2	$Mn^{2+}(aq)$	−73.6	$ClO_4^-(aq)$	182.0	$MnO_4^-(aq)$	190.0
$Cd^{2+}(aq)$	−61.1	$Na^+(aq)$	60.2	$CO_3^{2-}(aq)$	−53.1	$NO_3^-(aq)$	146.4
$Cu^{2+}(aq)$	−98.7	$NH_4^+(aq)$	112.8	$CrO_4^{2-}(aq)$	38.5	$OH^-(aq)$	−10.5
$Fe^{2+}(aq)$	−113.4	$Ni^{2+}(aq)$	−159.4	$F^-(aq)$	−9.6	$PO_4^{3-}(aq)$	−218
$Fe^{3+}(aq)$	−293.3	$Pb^{2+}(aq)$	21.3	$HCO_3^-(aq)$	95.0	$S^{2-}(aq)$	22.2
$H^+(aq)$	0.0	$Sn^{2+}(aq)$	−24.7			$SO_4^{2-}(aq)$	17.2
		$Zn^{2+}(aq)$	−106.5				

ΔS^0 for Reactions

Using Tables 14.3 and 14.4, it is possible to calculate the standard entropy change, ΔS^0, for a variety of reactions. The appropriate relation is

$$\Delta S^0 = \Sigma\ S^0_{products} - \Sigma\ S^0_{reactants}$$

(14.10)

Use this equation to calculate entropy changes

As an example, consider the reaction

$$CaCO_3(s) \rightarrow CaO(s) + CO_2(g)$$

$$\Delta S^0 = S^0\ CaO(s) + S^0\ CO_2(g) - S^0\ CaCO_3(s)$$

$$= 39.7\ J/K + 213.6\ J/K - 92.9\ J/K = +160.4\ J/K$$

---

*Notice that several ions have negative entropies. This happens because ion entropies are calculated from experimental values of ΔS^0, using Equation 14.10, and the arbitrary convention that $S^0\ H^+(aq) = 0$. Thus, given that

$$HF(g) \rightarrow H^+(aq) + F^-(aq);\qquad \Delta S^0 = -183.1\ J/K$$

it follows that

$$-183.1\ J/K = S^0\ F^-(aq) - S^0\ HF(g)$$

$$S^0\ F^-(aq) = -183.1\ J/K + 173.5\ J/K = -9.6\ J/K$$

Values of ΔS^0 calculated in this way will, of course, have the units joules per kelvin (J/K), since these are the units listed for S^0. If we need to convert ΔS^0 to kilojoules per kelvin (kJ/K), all we need do is divide by 1000. For the decomposition of calcium carbonate,

$$\Delta S^0 = +160.4 \text{ J} \times \frac{1 \text{ kJ}}{1000 \text{ J}} = +0.1604 \text{ kJ/K}$$

You will notice that ΔS^0 for the decomposition of calcium carbonate is a positive quantity. This is reasonable since the gas formed, CO_2, has a much higher molar entropy than either of the solids, CaO or $CaCO_3$. As a matter of fact, we almost always find that *a reaction that results in an increase in the number of moles of gas is accompanied by an increase in entropy. Conversely, if the number of moles of gas decreases, we expect ΔS^0 to be a negative quantity.* Consider, for example, the reaction

$$2 \text{ H}_2(g) + \text{O}_2(g) \rightarrow 2 \text{ H}_2\text{O}(l)$$
$$\Delta S^0 = 2 \text{ S}^0 \text{ H}_2\text{O}(l) - 2 \text{ S}^0 \text{ H}_2(g) - \text{S}^0 \text{ O}_2(g)$$
$$= 139.8 \text{ J/K} - 261.2 \text{ J/K} - 205.0 \text{ J/K} = -326.4 \text{ J/K}$$

EXAMPLE 14.1 Calculate ΔS^0 for the reaction

$$\text{HCl}(g) \rightarrow \text{H}^+(aq) + \text{Cl}^-(aq)$$

Solution Using values of S^0 from Tables 14.3 and 14.4,

$$\Delta S^0 = \text{S}^0 \text{ H}^+(aq) + \text{S}^0 \text{ Cl}^-(aq) - \text{S}^0 \text{ HCl}(g)$$
$$= 0 + 55.1 \text{ J/K} - 186.7 \text{ J/K} = -131.6 \text{ J/K}$$

EXERCISE Calculate ΔS^0 for the process $\text{NaCl}(s) \rightarrow \text{Na}^+(aq) + \text{Cl}^-(aq)$. Answer: +42.9 J/K.

Strictly speaking, calculations such as those just made are valid only at 25°C, the temperature at which standard entropies are recorded in Tables 14.3 and 14.4. In practice, though, ΔS^0 for a reaction is ordinarily nearly independent of temperature. Consider, for example, the reaction

$$\text{CaCO}_3(s) \rightarrow \text{CaO}(s) + \text{CO}_2(g)$$

where ΔS^0 is $+160.4$ J/K at 25°C. One can calculate that at 100°C, ΔS^0 for this reaction is about $+160.0$ J/K, only slightly less than the value at 25°C. The entropies of all three substances involved in the reaction (CaO, CO_2, and $CaCO_3$) increase with temperature, but the difference,

$$\text{S}^0 \text{ CO}_2(g) + \text{S}^0 \text{ CaO}(s) - \text{S}^0 \text{ CaCO}_3(s)$$

ΔS^0, like ΔH is not sensitive to temperature changes

remains nearly constant. In all calculations, we will *take ΔS^0 to be independent of temperature;* that is, we will assume that the value of ΔS^0 calculated from Tables 14.3 and 14.4 remains valid at temperatures other than 25°C.

On the other hand, the entropy change for a reaction often changes appreciably with the pressure of a gas or the concentration of an ion in solution. Consider, for example, the reaction

$$HCl(g) \rightarrow H^+(aq) + Cl^-(aq)$$

We saw in Example 14.1 that the standard entropy change for this reaction, ΔS^0, is -131.6 J/K. This is the entropy change when the HCl is at a pressure of 1 atm and the H^+ and Cl^- ions are at a concentration of 1 M. To emphasize this, we might write

$$HCl(g, 1 \text{ atm}) \rightarrow H^+(aq, 1 \text{ M}) + Cl^-(aq, 1 \text{ M}); \quad \Delta S^0 = -131.6 \text{ J/K}$$

The value of ΔS changes considerably if we change the pressure of the HCl gas:

$$HCl(g, 10 \text{ atm}) \rightarrow H^+(aq, 1 \text{ M}) + Cl^-(aq, 1 \text{ M}); \quad \Delta S = -112.5 \text{ J/K}$$

or the concentrations of the ions in solution:

$$HCl(g, 1 \text{ atm}) \rightarrow H^+(aq, 0.1 \text{ M}) + Cl^-(aq, 0.1 \text{ M}); \quad \Delta S = -93.4 \text{ J/K}$$

In all our calculations in this chapter, we will deal with the standard entropy change (1 atm, 1 M). This is given the special symbol ΔS^0 to distinguish it from the entropy change under other conditions, ΔS, which may be quite different.

14.3
The Free Energy Change, ΔG

We stated earlier that there are two thermodynamic quantities that affect the spontaneity of a reaction. One of these is the enthalpy, H. The other is the entropy, S. The problem is to put these two quantities together in such a way as to arrive at a single function that can be used to determine whether a reaction is spontaneous. This problem was first solved a century ago by J. Willard Gibbs, Professor of Mathematical Physics at Yale. He introduced a new quantity into thermodynamics, now called the *Gibbs free energy* and given the symbol G. The free energy of a substance, like its enthalpy or entropy, is a state function. Its value is determined only by the state of a substance, not by how it reached the state.

The free energy G is a function of H and S

For a reaction, the change in free energy, ΔG, is the difference between the free energies of products and reactants:

$$\Delta G = G_{products} - G_{reactants} \tag{14.11}$$

Gibbs was able to show that the sign of ΔG can be used to determine whether or not a reaction is spontaneous. For a reaction carried out at constant temperature and pressure:

1. **If ΔG is negative, the reaction is spontaneous.**
2. **If ΔG is positive, the reaction will not occur spontaneously.** Instead, the reverse reaction will occur.
3. **If ΔG is 0, the reaction system is at equilibrium;** there is no tendency for it to occur in either direction.

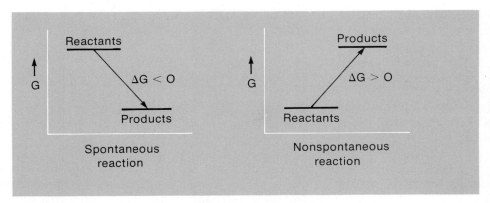

FIGURE 14.1 For a spontaneous reaction, the free energy of the products is less than that of the reactants: $\Delta G < 0$. For a nonspontaneous reaction, the reverse is true: $\Delta G > 0$.

Putting it another way, we can think of ΔG as a measure of the driving force of a reaction. **Reactions, at constant pressure and temperature, go in such a direction as to decrease the free energy of the system.** This means that the direction in which a reaction takes place is determined by the relative free energies of products and reactants. If the products have a lower free energy than the reactants ($G_{products} < G_{reactants}$), the forward reaction will occur (Fig. 14.1). If the reverse is true ($G_{reactants} < G_{products}$), the reverse reaction is spontaneous. Finally, if $G_{products} = G_{reactants}$, there is no driving force to make the reaction go in either direction.

Spontaneous processes can do work. G is a measure of the capacity of a substance to do work

Relation Among ΔG, ΔH, and ΔS

The free energy change for a reaction, ΔG, is related in a simple way to the enthalpy change, ΔH, and the entropy change, ΔS. The equation relating these quantities, called the *Gibbs-Helmholtz equation*, is

For our purposes, this equation defines ΔG

$$\Delta G = \Delta H - T\Delta S \tag{14.12}$$

where T is the temperature in K. This equation tells us that the driving force for a reaction, ΔG, depends upon two quantities. One of these is the enthalpy change due to the making and breaking of bonds, ΔH. The other is the product of the change in randomness, ΔS, times the absolute temperature, T.

The two factors that tend to make ΔG negative and hence lead to a spontaneous reaction are

1. **A negative value of ΔH.** Exothermic reactions ($\Delta H < 0$) tend to be spontaneous inasmuch as they contribute to a negative value of ΔG. On the molecular level, this means that there will be a tendency to form "strong" bonds at the expense of "weak" ones.

2. **A positive value of ΔS.** If the entropy change is positive ($\Delta S > 0$), the term $-T\Delta S$ will make a negative contribution to ΔG. Hence, there will be a tendency for a reaction to be spontaneous if the products are less ordered than the reactants.

In many physical processes, the entropy increase is the major driving force. This is the case with osmosis, where ΔH is zero and ΔS is a positive quantity. A similar situation applies when two liquids with similar intermolecular forces, such as benzene and toluene, are mixed. There is no change in enthalpy, but the entropy increases because the pure substances become diluted when they form a solution. In this and many other solution processes, it is the entropy change rather than the enthalpy change that accounts for spontaneity.

In certain reactions, ΔS is nearly zero and ΔH is the only important component of the driving force for spontaneity. An example is the synthesis of hydrogen fluoride from the elements:

$$\tfrac{1}{2} H_2(g) + \tfrac{1}{2} F_2(g) \rightarrow HF(g)$$

For this reaction, ΔH is a large negative number, -268.6 kJ. This reflects the fact that the bonds in HF are stronger than those in the H_2 and F_2 molecules. As we might expect for a gaseous reaction in which there is no change in the number of moles, ΔS is very small, about 0.0065 kJ/K. The free energy change, ΔG, at 1 atm is -270.5 kJ at 25°C, almost identical to ΔH. Even at very high temperatures, the difference between ΔG and ΔH is small, amounting to only about 13 kJ at 2000 K.

The more common case is one in which both ΔH and ΔS make significant contributions to ΔG. To determine the sign of ΔG, we must consider the values of both ΔH and ΔS as well as the temperature. The approach we will follow is described below.

The ΔH and $T\Delta S$ effects typically tend to drive a reaction in opposite directions

Standard Free Energy Change

The Gibbs-Helmholtz equation is valid under all conditions of temperature, pressure, and concentration. However, all our calculations using the equation will be restricted to *standard conditions* (1 atm pressure for gases, 1 M concentration for ions in solution). In other words, we will use the equation in the form:

$$\Delta G^0 = \Delta H - T\Delta S^0 \qquad \text{(14.13)}$$

This equation is the one we use to establish reaction spontaneity at any temperature

where ΔG^0 is the standard free energy change (1 atm, 1 M), ΔH is the enthalpy change, calculated from Tables 14.1 and 14.2,* and ΔS^0 is the standard entropy change, calculated from Tables 14.3 and 14.4.

To illustrate the use of Equation 14.13, let us apply it first at 25°C (T = 298 K). Consider the reaction

$$2 H_2(g) + O_2(g) \rightarrow 2 H_2O(l)$$

Earlier in this chapter, we showed that ΔH for this reaction is -571.6 kJ and that ΔS^0 is -326.4 J/K. Hence, at 25°C,

$$\Delta G^0 = -571.6 \text{ kJ} - 298 \text{ K} \times (-0.3264 \text{ kJ/K})$$
$$= -571.6 \text{ kJ} + 97.3 \text{ kJ} = -474.3 \text{ kJ}$$

A very spontaneous reaction, explosively so

---

*We do not ordinarily use the superscript zero with the enthalpy because ΔH, unlike both ΔG and ΔS, is essentially independent of pressure or concentration. In effect, $\Delta H = \Delta H^0 =$ standard enthalpy change (1 atm, 1 M).

Notice that

—in making a calculation of this type, the units must be consistent. If, as is ordinarily the case, we want ΔG^0 in kilojoules, ΔS^0 must be expressed in **kilojoules per kelvin.** ($1 \text{ J/K} = 10^{-3} \text{ kJ/K}$)

—the quantity ΔG^0 for this reaction is negative. This is consistent with the fact that $H_2(g)$ and $O_2(g)$, both at 1 atm, react spontaneously at 25°C to form liquid water.

EXAMPLE 14.2 For the reaction $CaSO_4(s) \rightarrow Ca^{2+}(aq) + SO_4^{2-}(aq)$, calculate, using Tables 14.1 to 14.4,

a. ΔH b. ΔS^0 c. ΔG^0 at 25°C

Solution

a. $\Delta H = \Delta H_f \ Ca^{2+}(aq) + \Delta H_f \ SO_4^{2-}(aq) - \Delta H_f \ CaSO_4(s)$
 $= -543.0 \text{ kJ} -907.5 \text{ kJ} - (-1432.7 \text{ kJ}) = -17.8 \text{ kJ}$

ΔS^0 *must* be in kJ/K if ΔG^0 is to be in kJ

b. $\Delta S^0 = S^0 \ Ca^{2+}(aq) + S^0 \ SO_4^{2-}(aq) - S^0 \ CaSO_4(s)$
 $= -55.2 \text{ J/K} + 17.2 \text{ J/K} - 106.7 \text{ J/K} = -144.7 \text{ J/K}$
 $= -0.1447 \text{ kJ/K}$

c. $\Delta G^0 = -17.8 \text{ kJ} - 298 \text{ K} \times (-0.1447 \text{ kJ/K}) = +25.3 \text{ kJ}$

We conclude that this process is not spontaneous at standard conditions at 25°C. In other words, calcium sulfate does not dissolve in water to produce a 1 M solution. This is indeed the case. The solubility of $CaSO_4$ at 25°C is considerably less than 1 mol/L, only about 0.01 mol/L.

EXERCISE For the process $NaCl(s) \rightarrow Na^+(aq) + Cl^-(aq)$, $\Delta H = +3.9$ kJ, $\Delta S^0 = +42.9$ J/K. Calculate ΔG^0 at 25°C. Answer: -8.9 kJ (NaCl dissolves spontaneously at 25°C to give a 1 M solution; its solubility at 25°C is about 5.4 mol/L).

The Gibbs-Helmholtz equation can be used to calculate the *standard free energy of formation* of a compound. This quantity, ΔG_f^0, is analogous to the heat of formation, ΔH_f. It is defined as the free energy change per mole when a compound is formed from the elements in their stable states.

EXAMPLE 14.3 Calculate the standard free energy of formation, ΔG_f^0, at 25°C, for $CH_4(g)$, using data in Tables 14.1 and 14.3.

Solution By definition, ΔG_f^0 for CH_4 is ΔG^0 for the reaction

$$C(s) + 2 H_2(g) \rightarrow CH_4(g)$$

For this reaction,

$\Delta H = \Delta H_f \ CH_4(g) = -74.8 \text{ kJ}$

$\Delta S^0 = S^0 \ CH_4(g) - S^0 \ C(s) - 2 \ S^0 \ H_2(g)$
 $= 186.2 \text{ J/K} - 5.7 \text{ J/K} - 2(130.6 \text{ J/K}) = -80.7 \text{ J/K}$
 $= -0.0807 \text{ kJ/K}$

At 25°C, that is, 298 K,

$$\Delta G^0 = \Delta H - 298\ \Delta S^0 = -74.8\ \text{kJ} - 298\ \text{K} \times (-0.0807\ \text{kJ/K})$$
$$= -50.8\ \text{kJ}$$

We conclude that ΔG_f^0 $CH_4(g)$ at 25°C is -50.8 kJ/mol.

EXERCISE What is the standard free energy of formation of liquid water at 25°C? Answer: -237.2 kJ/mol.

Tables of standard free energies of formation of compounds and ions in solution at 25°C are given in Appendix 1. If desired, these tables can be used to calculate ΔG^0 for a reaction at 25°C:

$$\Delta G^0 \text{ at } 25°C = \Sigma\ \Delta G_f^0 \text{ products at } 25°C - \Sigma\ \Delta G_f^0 \text{ reactants at } 25°C$$

This Appendix is useful only at 25°C

However, we will not calculate ΔG^0 this way. Instead, we will continue to use the Gibbs-Helmholtz equation with Tables 14.1 to 14.4.

We can also use Equation 14.13 to calculate ΔG^0 at temperatures other than 25°C. To do this, we neglect the variations of ΔH and ΔS^0 with temperature, which are ordinarily small.* In other words, we take the values of ΔH and ΔS^0 from Tables 14.1 through 14.4 and simply insert them in Equation 14.13, using the appropriate value of T. Suppose, for example, we want to know ΔG^0 at 1000°C for the reaction

$$CaCO_3(s) \rightarrow CaO(s) + CO_2(g)$$

For this reaction, we showed earlier that $\Delta H = +178.0$ kJ, $\Delta S^0 = +160.4$ J/K. Substituting T = 1000 + 273 = 1273 K, we have

$$\Delta G^0 = +178.0\ \text{kJ} - 1273\ \text{K} \times (0.1604\ \text{kJ/K})$$
$$= 178.0\ \text{kJ} - 204.2\ \text{kJ} = -26.2\ \text{kJ}$$

We conclude that this reaction is spontaneous at 1000°C and 1 atm, since ΔG^0 has a negative sign (see Fig. 14.2).

At 1000°C, $CaCO_3$ will decompose in an open container

---

*As far as Equation 14.13 is concerned, there is another reason for ignoring the temperature dependence of ΔH and ΔS. These two quantities always change in the same direction as the temperature changes (that is, if ΔH becomes more positive, so does ΔS). Hence, the two effects tend to cancel each other. The true value of ΔG at any temperature is about the same as the one we calculate taking ΔH and ΔS to be constant.

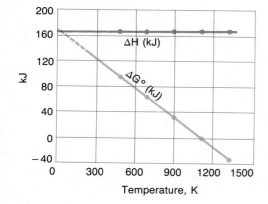

FIGURE 14.2 ΔH and ΔG^0 as a function of temperature for the reaction $CaCO_3(s) \rightarrow CaO(s) + CO_2(g)$. Below about 1100K, ΔG^0 is positive and the reaction is nonspontaneous. Above 1100K, ΔG^0 is negative and $CaCO_3$ decomposes spontaneously to CaO and CO_2 at 1 atm.

EXAMPLE 14.4 Calculate ΔG^0 for the reaction $Cu(s) + H_2O(g) \rightarrow CuO(s) + H_2(g)$ at 500 K.

Solution From Table 14.1,

$$\Delta H = \Delta H_f \, CuO(s) - \Delta H_f \, H_2O(g) = -155.2 \text{ kJ} + 241.8 \text{ kJ}$$
$$= +86.6 \text{ kJ}$$

From Table 14.3,

$$\Delta S^0 = S^0 \, CuO(s) + S^0 \, H_2(g) - S^0 \, Cu(s) - S^0 \, H_2O(g)$$
$$= 43.5 \text{ J/K} + 130.6 \text{ J/K} - 33.3 \text{ J/K} - 188.7 \text{ J/K}$$
$$= -47.9 \text{ J/K} = -0.0479 \text{ kJ/K}$$

Hence, $\Delta G^0 = +86.6 \text{ kJ} - 500 \text{ K} \times (-0.0479 \text{ kJ/K}) = +110.6 \text{ kJ}$

This example illustrates the power of this method

We conclude that this reaction is not spontaneous, since ΔG^0 is positive. Instead, the reverse reaction occurs at 1 atm and 500 K (227°C). Under these conditions, CuO is reduced to copper by being heated in a stream of H_2 gas.

EXERCISE Calculate ΔG^0 for this reaction at 1000 K. Answer: +134.5 kJ (still nonspontaneous).

From Example 14.4 and the preceding discussion, it should be clear that ΔG^0, unlike ΔH and ΔS^0, is strongly dependent upon temperature. This comes about, of course, because of the T in the Gibbs-Helmholtz equation:

$$\Delta G^0 = \Delta H - T\Delta S^0$$

Look at Figure 14.2, and think about it for a while. It will be well worth the effort

From this equation we see that ΔG^0 is a linear function of T. A plot of ΔG^0 vs. T will be a straight line with a slope of $-\Delta S^0$.

Effect of T upon Reaction Spontaneity

When the temperature of a reaction system is increased, the direction in which the reaction proceeds spontaneously may or may not change. Whether it does or not depends upon the relative signs of ΔH and ΔS^0. The four possible situations, deduced from the Gibbs-Helmholtz equation, are summarized in Table 14.5.

If ΔH and ΔS^0 have opposite signs (Table 14.5, I and II) it is impossible to reverse the direction of spontaneity by a change in temperature alone. The two terms ΔH and $-T\Delta S^0$ reinforce one another. Hence, ΔG^0 has the same sign at all temperatures. Reactions of this type are rather uncommon. One such reaction is that discussed in Example 14.4:

$$Cu(s) + H_2O(g) \rightarrow CuO(s) + H_2(g) \tag{14.14}$$

Here, ΔH is +86.6 kJ and ΔS^0 is -0.0479 kJ/K. Hence,

$$\Delta G^0 = \Delta H - T\Delta S^0$$
$$= +86.6 \text{ kJ} + T(0.0479 \text{ kJ/K})$$

Clearly, ΔG^0 is positive at all temperatures. The reaction cannot take place spontaneously at 1 atm regardless of temperature.

Table 14.5
Effect of Temperature on Reaction Spontaneity

	ΔH	ΔS^0	$\Delta G^0 = \Delta H - T\Delta S^0$	REMARKS
I	−	+	always −	spontaneous at all T; reverse reaction always nonspontaneous
II	+	−	always +	nonspontaneous at all T; reverse reaction occurs
III	+	+	+ at low T − at high T	nonspontaneous at low T; becomes spontaneous as T is raised
IV	−	−	− at low T + at high T	spontaneous at low T; at high T, reverse reaction becomes spontaneous

It is more common to find that ΔH and ΔS^0 have the same sign (Table 14.5, III and IV). When this happens, the enthalpy and entropy factors oppose each other. ΔG changes sign as temperature increases, and the direction of spontaneity reverses. At low temperatures, ΔH predominates and the exothermic reaction occurs. As the temperature rises, the quantity $T\Delta S^0$ increases in magnitude and eventually exceeds ΔH. At high temperatures, the reaction that leads to an increase in entropy occurs. In most cases, 25°C is a "low" temperature, at least at a pressure of 1 atm. This explains why exothermic reactions are usually spontaneous at room temperature and atmospheric pressure.

At low T, reactions go to the species with the strongest chemical bonds

An example of a reaction for which ΔH and ΔS have the same sign is the decomposition of calcium carbonate:

$$CaCO_3(s) \rightarrow CaO(s) + CO_2(g)$$

Here, as pointed out previously, $\Delta H = +178.0$ kJ and $\Delta S^0 = +160.4$ J/K. Hence,

$$\Delta G^0 = +178.0 \text{ kJ} - T(0.1604 \text{ kJ/K})$$

Figure 14.2, p. 433, shows a plot of ΔG^0 vs. T. Notice that

At high T, reactions go to the species with the highest disorder

—below about 1100 K (830°C), ΔH predominates, ΔG^0 is positive, and the reaction is nonspontaneous at 1 atm.
—above about 1100 K, $T\Delta S^0$ predominates, and the reaction is spontaneous at 1 atm.

EXAMPLE 14.5 At what temperature does ΔG^0 become zero for the reaction $CaCO_3(s) \rightarrow CaO(s) + CO_2(g)$?

Solution In general, $\Delta G^0 = \Delta H - T\Delta S^0$

When ΔG^0 is zero, $\Delta H = T\Delta S^0$

$$T = \frac{\Delta H}{\Delta S^0}$$

Here, $\Delta H = +178.0$ kJ, $\Delta S^0 = 0.1604$ kJ/K, so that

$$T = \frac{178.0 \text{ kJ}}{0.1604 \text{ kJ/K}} = 1110 \text{ K}$$

At this temperature, the reaction is at equilibrium at 1 atm pressure. If we put some $CaCO_3$ in a container and heat it to 1110 K, the pressure of CO_2 developed will be 1 atm.

EXERCISE Calculate the temperature at which ΔG^0 becomes zero for Reaction 14.14. Answer: The calculated value is -1810 K, which is nonsensical. There is no temperature at which ΔG^0 is zero for this reaction, because ΔH and ΔS^0 have opposite signs.

The development we went through in Example 14.5 is a general one. The temperature, T, at which a system is at equilibrium at *one atmosphere*, is given by the expression

$\Delta G^0 = 0$ here

$$T = \Delta H / \Delta S^0$$

This relationship is particularly useful for phase changes. We can relate the normal boiling point, T_b, to the heat and entropy of vaporization, ΔH_{vap} and ΔS^0_{vap}. At the normal boiling point, liquid and vapor are at equilibrium at 1 atm, so that

This is an interesting equation

$$T_b = \frac{\Delta H_{vap}}{\Delta S^0_{vap}} \tag{14.15}$$

To illustrate the use of Equation 14.15, let us apply it to water. Taking data from Tables 14.1 and 14.3, we calculate

$$\Delta H_{vap} = \Delta H_f \ H_2O(g) - \Delta H_f \ H_2O(l) = +44.0 \text{ kJ}$$
$$\Delta S^0_{vap} = S^0 \ H_2O(g) - S^0 \ H_2O(l) = +118.8 \text{ J/K}$$

From these data, we can calculate the normal boiling point of water:

$$T_b = \frac{+44.0 \text{ kJ}}{0.1188 \text{ kJ/K}} = 370 \text{ K}$$

Not bad, all things considered

This calculated boiling point is very close to the observed value, 373 K (100°C).

Pressure, Concentration, and Spontaneity

There is an important restriction upon all the calculations we have made in this section concerning reaction spontaneity. This restriction is implied when we say that if ΔG^0 is negative, the reaction is spontaneous at *standard pressure* (1 atm for gases) and *standard concentration* (1 M for ions in solution). The free energy change for a reaction varies with the pressure of a gas or the concentration of an ion. This means that the conclusions about spontaneity reached on the basis of the sign of ΔG^0 may not apply at other than standard conditions. A case in point involves the reaction

$$CaCO_3(s) \rightarrow CaO(s) + CO_2(g, 1 \text{ atm}); \qquad \Delta G^0 = -26.2 \text{ kJ at 1000°C}$$

Since ΔG^0 is negative at 1000°C, we conclude that the reaction at this temperature is spontaneous at 1 atm. However, one can calculate that when the pressure of CO_2 is 20 atm, ΔG^0 is $+5.4$ kJ, and the reaction is not spontaneous. As another example, you will recall that we showed in Example 14.2 that

$$CaSO_4(s) \rightarrow Ca^{2+}(aq, 1 \text{ M}) + SO_4{}^{2-}(aq, 1 \text{ M}); \quad \Delta G^0 \text{ at } 25°C = +25.3 \text{ kJ}$$

The positive sign of ΔG^0 indicates that $CaSO_4$ will not dissolve to form a 1 M solution at 25°C. On the other hand, one can calculate that ΔG for the formation of a 0.001 M solution of $CaSO_4$ is -8.9 kJ. Calcium sulfate should, and does, form a 0.001 M solution; its solubility at 25°C is about ten times that value.

In Example 14.5 we showed that for the decomposition of $CaCO_3$, $\Delta G^0 = 0$ at 1110 K. We interpreted this to mean that the reaction is at equilibrium at 1110 K when the pressure of CO_2 is 1 atm. You should realize that if the pressure changes, ΔG will no longer be zero, and the equilibrium temperature will change. Indeed, as we will see in Chapter 15, a reaction system such as this one will reach a position of equilibrium at any given temperature.

In the same way, a liquid-vapor system such as

$$H_2O(l) \rightleftharpoons H_2O(g)$$

can reach equilibrium over a wide range of temperatures. Depending on the pressure, a liquid can boil at any temperature between the triple point and the critical point. The normal boiling point is unique in only one respect; it is the temperature at which ΔG^0 is zero and the system is at equilibrium at *one atmosphere pressure*.

14.4
The Second Law of Thermodynamics

You may recall from Chapter 5 that the First Law of Thermodynamics has the form

$$\Delta E = q + w$$

Here, we distinguish between heat (q) and all other forms of energy, referred to collectively as work (w). You may wonder why we make the distinction between these two types of energy. There are both practical and theoretical reasons for doing so. Although all other forms of energy can be converted to heat, there is a restriction on the conversion of heat into work. The Second Law of Thermodynamics in one of its many forms tells us that

It is impossible to construct an engine, operating in cycles, that absorbs heat at a constant temperature and completely converts it into work.

In this section, we will look at some of the consequences of this and other statements of the Second Law of Thermodynamics.

There are three laws of thermodynamics. The 3rd Law is the lower marginal note on page 425

Maximum Efficiency of a Heat Engine

In a steam engine or the internal combustion engine of an automobile, there is a flow of energy that might be described as follows:

1. Heat is absorbed by the engine at a relatively high temperature, T_2.
2. Part of that heat is converted to useful, mechanical work.
3. The rest of the heat is discharged by the engine to the surroundings at a lower temperature, T_1.

From a practical standpoint, it is important to know what fraction of the heat absorbed at T_2 can be converted to work. This problem occupied the attention of a great many scientists and engineers during the first half of the nineteenth century. Two men made major contributions to its solution and, in effect, discovered the Second Law in the process. One of these was Sadi Carnot (1796–1832), an obscure artilleryman in the French army. The other was the great English physicist, Lord Kelvin (1824–1907). They showed that the maximum efficiency of a heat engine (i.e., the largest fraction of absorbed heat that can be converted to work) is

max efficiency

$= \dfrac{\text{work done}}{\text{heat absorbed at } T_2}$

$$\text{maximum efficiency} = \frac{T_2 - T_1}{T_2} \qquad \textbf{(14.16)}$$

where T_2 is the high temperature and T_1 the low temperature, both expressed on the Kelvin scale. The implications of Equation 14.16 are suggested in Example 14.6.

EXAMPLE 14.6 Consider a steam engine operating between the normal boiling point of water, 100°C, and room temperature, 25°C. What is its maximum efficiency, ignoring friction, heat leaks, etc.?

Solution

$$T_2 = 373 \text{ K}, \qquad T_1 = 298 \text{ K}$$
$$\text{maximum efficiency} = \frac{373 - 298}{373} = 0.20 = 20\%$$

To see what this means, suppose we burn enough coal or other fuel to supply 100 kJ of heat to the steam engine. Only about 20 kJ of that heat can be converted to useful work. The other 80 kJ must inevitably be wasted as heat discharged to the surroundings at 25°C.

EXERCISE Suppose T_1 were 0 K. What would be the maximum efficiency of a heat engine under that condition? Answer: 100%. (This is sometimes interpreted to mean that absolute zero is unattainable. If we could reach 0 K, heat could be completely converted into work, which would violate the Second Law.)

In an engine, it's easier to increase T_2 than it is to decrease T_1

One can readily show from Equation 14.16 that the theoretical efficiency of a steam engine goes up when T_2 is increased. This explains why superheated steam is ordinarily used in steam engines and why automobile engines are adjusted to run at as high a temperature as possible. However, the actual efficiency of a heat engine is always considerably less than the maximum allowed by the Second Law. Anyone who can build an engine and power transmission system that converts appreciably more than 30% of the absorbed heat into work is doing very well indeed.

The Entropy of the Universe

Between 1850 and 1865, the German mathematical physicist Rudolf Clausius (1822–1888) published a series of papers dealing with the Second Law of Thermodynamics. He showed that the Second Law led to a new state property which he called entropy (S). Going a step further, Clausius found that the change in entropy for a process is a criterion of its spontaneity. Specifically,

The total change in entropy, considering both system and surroundings, is positive for a spontaneous process.

We can express this version of the Second Law in the form of a simple equation:

spontaneous process: $\Delta S_{sys} + \Delta S_{surr} > 0$ **(14.17)**

Notice that it is the total entropy that increases. The entropy of a reaction system may well decrease if the entropy of the surroundings increases by a greater amount. Consider what happens when water is allowed to stand outside on a cold winter night when the temperature is −10°C (14°F). The water freezes spontaneously:

If $\Delta S_{surr} = 0$, as it would be for a reaction in a closed insulated container, the entropy of the reaction mix must increase

$$H_2O(l) \rightarrow H_2O(s); \qquad t = -10°C$$

The water (the "system" in this case) decreases in entropy; ice, with a more ordered structure than liquid water, has a lower entropy. However, this is not the only entropy change that is taking place. As the water freezes, it gives off heat to the air (the "surroundings"). The air warms up slightly; its entropy increases. It turns out that the increase in entropy of the air outweighs the decrease in entropy of the water. The freezing of water at −10°C, like all spontaneous processes, occurs with an overall increase in entropy.

The statement leading to Equation 14.17 can be phrased in a somewhat more imposing way. All natural processes are spontaneous. The sum of a system and its surroundings comprises, by definition, the universe. Hence, according to the Second Law,

The entropy of the universe is steadily increasing.

This is a rather disturbing statement. It suggests that the universe is like a clock that is slowly running down; ultimately, order in the universe must give way to chaos. However, there is at least one consolation: the universe is a very large place. For centuries, human beings have struggled, with considerable success, to decrease the entropy of their immediate surroundings. The price paid for this effort is an increase in entropy of the wider environment, much of which appears in the form of pollution. In that sense, pollution is a spontaneous process. It can be reversed but it costs us energy (and money) to do that.

This is one thing that is not worth worrying about

Free Energy and Maximum Useful Work

In principle, Equation 14.17 allows us to predict whether or not a given process will be spontaneous. All we need do is to calculate the entropy change for both the system and surroundings. If the sum of the two entropy changes is positive, the process should occur spontaneously. If the calculated value of $\Delta S_{sys} + \Delta S_{surr}$

is negative, the process is nonspontaneous. In practice, this approach is not quite as simple as it sounds. The difficulty lies in calculating the entropy change of the surroundings. In many cases, it is by no means obvious how this is to be done.

J. Willard Gibbs (1839–1903), reflecting on this and other shortcomings of nineteenth century thermodynamics, extended the Second Law to include the free energy function, G. He showed that, for a process taking place at constant temperature and pressure, the sign of ΔG for the system is a simple criterion of spontaneity. As we saw in Section 14.3,

With this approach we avoid having to deal with the surroundings

if $\Delta G_{sys} < 0$, the process is spontaneous
if $\Delta G_{sys} > 0$, the process is nonspontaneous

Moreover, the free energy change can be calculated very easily from ΔH and ΔS:

$$\Delta G = \Delta H - T\Delta S$$

(ΔG, ΔH, and ΔS in this equation all refer to changes for the system.)

Gibbs may have developed his equations by considering work as fundamental

Gibbs pointed out another important property of the free energy function. *For a process taking place at constant temperature and pressure, ΔG is a measure of the maximum amount of useful work that can be obtained.* To illustrate what this statement means, consider the combustion of methane. Using Tables 14.1 and 14.3, we can show that ΔG^0 for this reaction at 25°C is -818 kJ per mole of methane burned; that is,

$$CH_4(g) + 2\ O_2(g) \rightarrow CO_2(g) + 2\ H_2O(l); \qquad \Delta G^0 \text{ at } 25°C = -818 \text{ kJ}$$

The maximum amount of work that can be obtained from this reaction at 25°C and 1 atm is 818 kJ (per mole of methane burned). We may get much less work than that. When methane simply burns in an open flame, we get no work at all. If the combustion of methane is used to drive a gas turbine, we may get 200 to 300 kJ of electrical energy. In an electrochemical cell (Chap. 23), we can do much better, perhaps producing as much as 700 to 800 kJ of electrical work. No matter what we do, however, there is no way we can get more than 818 kJ of work from the combustion of a mole of methane. The value of ΔG sets an upper limit to the amount of work that can be obtained from the reaction.

This relation applies equally well to nonspontaneous reactions. Consider the decomposition of water:

$$H_2O(l) \rightarrow H_2(g) + \tfrac{1}{2} O_2(g); \qquad \Delta G^0 \text{ at } 25°C = +237 \text{ kJ}$$

Since ΔG is positive, work must be supplied to make this reaction go. Ordinarily, this takes the form of electrical energy, absorbed during the electrolysis of water. We must take at least 237 kJ of electrical energy from a storage battery or other source to electrolyze one mole of water.

If you think about it for a while, the argument we have just gone through leads to a very simple definition of spontaneity:

A process taking place at constant temperature and pressure is spontaneous if it is capable of doing useful work (ΔG negative). If work must be done on the system to make the process occur (ΔG positive), it is nonspontaneous.

For our purposes in chemistry, this is perhaps the most useful statement of the Second Law of Thermodynamics.

14.5 Historical Perspective: J. Willard Gibbs (1839–1903)

A century ago chemistry was primarily an empirical science. The outstanding chemists of that era were experimentalists who isolated and characterized new substances. The principles of chemistry were descriptive or correlative in nature, as illustrated by the atomic theory of Dalton and the Periodic Table of Mendeleev. Two theoreticians working in the latter half of the nineteenth century changed the very nature of chemistry by deriving the mathematical laws that govern the behavior of matter undergoing physical or chemical change. One of these was James Clerk Maxwell, whose contributions to kinetic theory were discussed in Chapter 6. The other was J. Willard Gibbs, Professor of Mathematical Physics at Yale from 1871 until his death in 1903.

In 1876 Gibbs published the first portion of a remarkable paper in the *Transactions of the Connecticut Academy of Sciences* entitled "On the Equilibrium of Heterogeneous Substances." When the paper was completed in 1878 (it was 323 pages long), the foundation was laid for the science of chemical thermodynamics. Here, for the first time, the concept of free energy appeared. Included as well were the basic principles of chemical equilibrium (Chap. 15), phase equilibrium (Chap. 11), and the relations governing energy changes in electrical cells (Chap. 23).

If Gibbs had never published another paper, this single contribution would have placed him among the greatest theoreticians in the history of science. Generations of experimental scientists have established their reputations by demonstrating in the laboratory the validity of the relationships that Gibbs derived at his desk. Many of these relationships were rediscovered by others; an example is the Gibbs-Helmholtz equation developed in 1882 by Helmholtz, who was completely unaware of Gibbs' work.

Gibbs is probably the best theoretician this country has ever produced

In the 25 years that remained to him, Gibbs made substantial contributions in chemistry, astronomy, and mathematics. Among these were two papers published in 1881 and 1884 that established the discipline known today as vector analysis. His last work, published in 1901, was a book entitled *Elementary Principles in Statistical Mechanics*. Here Gibbs used the statistical principles that govern the behavior of systems to develop thermodynamic equations that he had derived from an entirely different point of view at the beginning of his career. Here, too, we find the "randomness" interpretation of entropy that has received so much attention in the social as well as the natural sciences.

Most of us would not think that book was elementary

J. Willard Gibbs is often cited as an example of the "prophet without honor in his own country." His colleagues in New Haven and elsewhere in the United States seem not to have realized the significance of his work until late in his life. During his first 10 years as a professor at Yale he received no salary. In 1920, when he was first proposed for the Hall of Fame of Distinguished Americans at New York University, he received nine votes out of a possible 100. Not until 1950 was he elected to that body. Even today the name of J. Willard Gibbs is generally unknown among educated Americans outside of those interested in the natural sciences.

Admittedly, Gibbs himself was largely responsible for the fact that for many years his work did not attract the attention it deserved. He made little effort to publicize it; the *Transactions of the Connecticut Academy of Sciences* was hardly the leading scientific journal of its day. Gibbs was one of those rare individuals who seem to have no inner need for recognition by contemporaries. His satisfaction came from solving a problem in his mind; having done so, he was ready to pass on to other problems without being concerned whether people understood what he had done. His papers are not easy to read; he seldom cites examples to illustrate his abstract reasoning. Frequently, the implications of the laws that he derives are left for the readers to grasp on their own. One of his colleagues at Yale confessed many years later that none of the members of the Connecticut Academy of Sciences understood his paper on thermodynamics; as he put it, "We knew Gibbs and took his contributions on faith."

Gibbs achieved recognition in Europe long before his work was generally appreciated in this country. Maxwell read Gibbs' paper on thermodynamics, saw its significance, and referred to it repeatedly in his own publications. Wilhelm Ostwald, who said of Gibbs, "To physical chemistry, he gave form and content for a hundred years," translated the paper into German in 1892. Seven years later, Le Châtelier translated it into French.

Summary

Two new state functions, entropy (S) and free energy (G), were introduced in this chapter. We also made use of the enthalpy function (H), discussed in Chapter 5. Our interest is in how these quantities change in a reaction.

CHANGE	NATURE OF REACTION	EXAMPLE
$\Delta H = \Sigma \Delta H_{f \text{ products}} - \Sigma \Delta H_{f \text{ reactants}}$	$\Delta H < 0$; exothermic $\Delta H > 0$; endothermic	14.2
$\Delta S = \Sigma S_{products} - \Sigma S_{reactants}$	$\Delta S > 0$; randomness increases $\Delta S < 0$; order increases	14.1, 14.2
$\Delta G = \Delta H - T\Delta S$	$\Delta G < 0$; spontaneous $\Delta G > 0$; nonspontaneous $\Delta G = 0$; equilibrium	14.2, 14.3, 14.4

Both ΔS and ΔG depend upon the pressures or concentrations of species taking part in the reaction. We dealt only with the standard entropy change, ΔS^0, and the standard free energy change, ΔG^0 (1 atm, 1 M). Both ΔH and ΔS^0 are taken to be independent of temperature; ΔG^0 is a linear function of T (Fig. 14.2). The fact that ΔG^0 becomes zero at equilibrium allows us to calculate the temperature at which a reaction is at equilibrium at standard pressures and concentrations (Example 14.5).

The Second Law of Thermodynamics deals with the spontaneity of physical and chemical changes. It tells us that a spontaneous change is one for which

$$\Delta S_{sys} + \Delta S_{surr} > 0$$

$$\Delta G_{sys} < 0$$

The Second Law also places an upper limit on the extent to which heat can be converted to work (Equation 14.16 and Example 14.6).

Key Words and Concepts

chemical thermodynamics	free energy, G	standard free energy change, ΔG^0
endothermic	free energy change, ΔG	standard molar entropy, S^0
enthalpy, H	free energy of formation, ΔG_f	surroundings
enthalpy change, ΔH	Gibbs-Helmholtz equation	system
entropy, S	nonspontaneous reaction	work, w
entropy change, ΔS	Second Law of Thermodynamics	
exothermic	spontaneous reaction	

Questions and Problems

Enthalpy Change, ΔH

14.1 Using Tables 14.1 and 14.2, calculate ΔH for each of the following reactions.
  a. $2 H_2S(g) + 3 O_2(g) \rightarrow 2 H_2O(g) + 2 SO_2(g)$
  b. $4 NH_3(g) + 5 O_2(g) \rightarrow 4 NO(g) + 6 H_2O(l)$
  c. $2 NH_4^+(aq) + 2 OH^-(aq) \rightarrow 2 NH_3(g) + 2 H_2O(l)$

14.2 Using Tables 14.1 and 14.2, calculate ΔH for
  a. $Ca(OH)_2(s) \rightarrow CaO(s) + H_2O(g)$
  b. $Ag(s) + 2 H^+(aq) + NO_3^-(aq) \rightarrow$
     $Ag^+(aq) + H_2O(l) + NO_2(g)$
  c. $3 Ag(s) + 4 H^+(aq) + NO_3^-(aq) \rightarrow$
     $3 Ag^+(aq) + 2 H_2O(l) + NO(g)$

14.3 Using Tables 14.1 and 14.2, calculate ΔH for the reaction involved when one mole of
  a. ammonium chloride dissolves in water.
  b. silver chloride crystallizes from water.
  c. ethane, C_2H_6, burns in air to form CO_2 and liquid water.

14.4 Consider the reaction

$$Mg_3N_2(s) + 6 H_2O(l) \rightarrow 3 Mg(OH)_2(s) + 2 NH_3(g)$$
$$\Delta H = -691 \text{ kJ}$$

  a. Calculate ΔH when 75.0 g Mg_3N_2 reacts with water.
  b. Using Table 14.1, calculate the heat of formation of Mg_3N_2.

14.31 Follow the directions for Problem 14.1 for
  a. $2 NO_2(g) + 7 H_2(g) \rightarrow 4 H_2O(g) + 2 NH_3(g)$
  b. $2 HNO_3(l) + 3 H_2S(g) \rightarrow$
     $4 H_2O(l) + 2 NO(g) + 3 S(s)$
  c. $PCl_5(g) + 4 H_2O(l) \rightarrow$
     $6 H^+(aq) + 5 Cl^-(aq) + H_2PO_4^-(aq)$

14.32 Using Tables 14.1 and 14.2 calculate ΔH for
  a. $CaSO_4(s) \rightarrow CaO(s) + SO_3(g)$
  b. $H_2PO_4^-(aq) + 2 OH^-(aq) \rightarrow$
     $PO_4^{3-}(aq) + 2 H_2O(l)$
  c. $Cu(s) + 4 H^+(aq) + 2 NO_3^-(aq) \rightarrow$
     $Cu^{2+}(aq) + 2 H_2O(l) + 2 NO_2(g)$

14.33 Using Tables 14.1 and 14.2, calculate ΔH for the reaction involved when one mole of
  a. calcium sulfate dissolves in water.
  b. magnesium carbonate precipitates from water solution.
  c. methyl alcohol, CH_3OH, burns in air to form CO_2 and water vapor.

14.34 Consider the reaction

$$2 PbO(s) + 2 SO_2(g) \rightarrow 2 PbS(s) + 3 O_2(g)$$
$$\Delta H = +839.4 \text{ kJ}$$

  a. Calculate ΔH when 10.0 g PbS is formed.
  b. Using Table 14.1, calculate the heat of formation of PbS.

Entropy Change, ΔS

14.5 Predict the sign of ΔS for
 a. a candle burning.
 b. ammonia vapor condensing.
 c. butter melting.
 d. tea dissolving in water.

14.6 Predict the sign of ΔS^0 for each of the following reactions:
 a. $CuSO_4 \cdot 5\ H_2O(s) \rightarrow CuSO_4(s) + 5\ H_2O(g)$
 b. $2\ Cl(g) \rightarrow Cl_2(g)$
 c. $2\ H_2(g) + O_2(g) \rightarrow 2\ H_2O(l)$

14.7 Use Table 14.3 to calculate ΔS^0 for each of the following:
 a. $P_4(s) + 6\ Cl_2(g) \rightarrow 4\ PCl_3(g)$
 b. $Fe_2O_3(s) + 3\ H_2(g) \rightarrow 2\ Fe(s) + 3\ H_2O(l)$
 c. $4\ Al(s) + 3\ MnO_2(s) \rightarrow 3\ Mn(s) + 2\ Al_2O_3(s)$

14.8 Use Tables 14.3 and 14.4 to calculate ΔS^0 for
 a. $Zn(s) + 2\ H^+(aq) \rightarrow Zn^{2+}(aq) + H_2(g)$
 b. $H^+(aq) + OH^-(aq) \rightarrow H_2O(l)$
 c. $NH_3(g) + H_2O(l) \rightarrow NH_4^+(aq) + OH^-(aq)$

14.9 Use Tables 14.3 and 14.4 to calculate ΔS^0 for the reactions in Problem 14.1.

14.10 Calculate ΔS^0 for each of the reactions in Problem 14.3.

14.35 Predict the sign of ΔS for
 a. the freezing of water.
 b. evaporation of a seawater sample to dryness.
 c. weeding a garden.
 d. separating air into its components.

14.36 Predict the sign of ΔS^0 for each of the following reactions:
 a. $N_2(g) + 3\ H_2(g) \rightarrow 2\ NH_3(g)$
 b. $H_2(g) + Cu^{2+}(aq) \rightarrow 2\ H^+(aq) + Cu(s)$
 c. $CaCl_2(s) + 6\ H_2O(g) \rightarrow CaCl_2 \cdot 6\ H_2O(s)$

14.37 Use Table 14.3 to calculate ΔS^0 for each of the following:
 a. $N_2(g) + 3\ H_2(g) \rightarrow 2\ NH_3(g)$
 b. $2\ CuS(s) + 3\ O_2(g) \rightarrow 2\ CuO(s) + 2\ SO_2(g)$
 c. $Fe_2O_3(s) + 3\ C(s) \rightarrow 2\ Fe(s) + 3\ CO(g)$

14.38 Use Tables 14.3 and 14.4 to calculate ΔS^0 for
 a. $Zn(s) + 2\ Ag^+(aq) \rightarrow Zn^{2+}(aq) + 2\ Ag(s)$
 b. $Ag^+(aq) + Cl^-(aq) \rightarrow AgCl(s)$
 c. $HNO_3(l) \rightarrow H^+(aq) + NO_3^-(aq)$

14.39 Use Tables 14.3 and 14.4 to calculate ΔS^0 for the reactions in Problem 14.31.

14.40 Calculate ΔS^0 for each of the reactions in Problem 14.33.

ΔG^0 and the Gibbs-Helmholtz Equation

14.11 Calculate ΔG^0 at 25°C for reactions for which
 a. $\Delta H = -109.0$ kJ; $\Delta S^0 = +27.8$ J/K
 b. $\Delta H = +842$ kJ; $\Delta S^0 = -116.2$ J/K
 c. $\Delta H = +8.29 \times 10^4$ J; $\Delta S^0 = 0.115$ kJ/K

14.12 Calculate ΔG^0 at 500°C for each of the reactions in Problem 14.11.

14.13 Using the Gibbs-Helmholtz equation, calculate ΔG^0 at 25°C for each of the reactions in Problem 14.1.

14.14 Using the Gibbs-Helmholtz equation, calculate ΔG^0 at 0 K for each of the reactions in Problem 14.3.

14.15 Using data from Tables 14.1 and 14.3, calculate the standard free energy of formation, ΔG_f^0, at 25°C for
 a. $NaCl(s)$ b. $CuSO_4(s)$ c. $Fe_2O_3(s)$

14.41 Calculate ΔG^0 at 25°C for reactions for which
 a. $\Delta H = +93.5$ kJ; $\Delta S^0 = +291.6$ J/K
 b. $\Delta H = -93.5$ kJ; $\Delta S^0 = +291.6$ J/K
 c. $\Delta H = -81.2$ kJ; $\Delta S^0 = -0.239$ kJ/K

14.42 Calculate ΔG^0 at 500 K for each of the reactions in Problem 14.41.

14.43 Using the Gibbs-Helmholtz equation, calculate ΔG^0 at 127°C for each of the reactions in Problem 14.31.

14.44 Using the Gibbs-Helmholtz equation, calculate ΔG^0 at 0°C for each of the reactions in Problem 14.33.

14.45 Follow the directions of Problem 14.15 to obtain ΔG_f^0 at 227°C for
 a. $PbO_2(s)$ b. $PCl_5(g)$ c. $NH_4Cl(s)$

14.16 Oxygen can be made in the laboratory by reacting sodium peroxide with water:

$$2 \text{ Na}_2\text{O}_2(s) + 2 \text{ H}_2\text{O}(l) \rightarrow 4 \text{ NaOH}(s) + \text{O}_2(g)$$
$$\Delta H = -126.0 \text{ kJ}; \Delta G^0 = -173.8 \text{ kJ at } 25°\text{C}$$

a. Calculate ΔS^0 for this reaction. Is the sign reasonable?
b. Using Table 14.3, obtain S^0 for $\text{Na}_2\text{O}_2(s)$.
c. Using Table 14.1, obtain ΔH_f for $\text{Na}_2\text{O}_2(s)$.

14.17 Consider the reaction

$$2 \text{ CuCl}(s) + 2 \text{ OH}^-(aq) \rightarrow$$
$$\text{Cu}_2\text{O}(s) + 2 \text{ Cl}^-(aq) + \text{H}_2\text{O}(l)$$
$$\Delta H = -55.6 \text{ kJ}; \Delta S^0 = +133.1 \text{ J/K}$$

a. Calculate ΔG^0 for this reaction at 25°C.
b. Determine ΔH_f for $\text{CuCl}(s)$ (see Tables 14.1 and 14.2).
c. Calculate S^0 for $\text{CuCl}(s)$ (see Tables 14.3 and 14.4).

14.46 Sodium carbonate, also called "washing soda," can be made by heating sodium hydrogen carbonate:

$$2 \text{ NaHCO}_3(s) \rightarrow \text{Na}_2\text{CO}_3(s) + \text{CO}_2(g) + \text{H}_2\text{O}(g)$$
$$\Delta H = +128.9 \text{ kJ}; \Delta G^0 = +33.1 \text{ kJ at } 25°\text{C}$$

a. Calculate ΔS^0 for this reaction. Is the sign reasonable?
b. Calculate ΔG^0 at 0 K; at 1000 K.

14.47 Consider the reaction

$$3 \text{ MnO}_4^{2-}(aq) + 2 \text{ H}_2\text{O}(l) \rightarrow$$
$$2 \text{ MnO}_4^-(aq) + 4 \text{ OH}^-(aq) + \text{MnO}_2(s)$$
$$\Delta H = +54.5 \text{ kJ}; \Delta S^0 = +74.3 \text{ J/K}$$

a. Calculate ΔG^0 for this reaction at 25°C.
b. Determine ΔH_f of $\text{MnO}_4^{2-}(aq)$ (see Tables 14.1 and 14.2).
c. What is S^0 for $\text{MnO}_4^{2-}(aq)$ (see Tables 14.3 and 14.4)?

Temperature Dependence of Reaction Spontaneity

14.18 Discuss the effect of temperature change upon the spontaneity of the following reactions at 1 atm:
a. $2 \text{ PbO}(s) + 2 \text{ SO}_2(g) \rightarrow 2 \text{ PbS}(s) + 3 \text{ O}_2(g)$
$\Delta H = +839.4 \text{ kJ}; \Delta S^0 = +0.203 \text{ kJ/K}$
b. $\text{N}_2\text{H}_4(l) \rightarrow \text{N}_2(g) + 2 \text{ H}_2(g)$
$\Delta H = -50.4 \text{ kJ}; \Delta S^0 = +0.330 \text{ kJ/K}$
c. $2 \text{ As}(s) + 3 \text{ F}_2(g) \rightarrow 2 \text{ AsF}_3(l)$
$\Delta H = -1897.9 \text{ kJ}; \Delta S^0 = -0.318 \text{ kJ/K}$

14.19 At what temperature does ΔG^0 become zero for each of the reactions in Problem 14.18? Explain the significance of your answers.

14.20 For the reaction

$$2 \text{ NO}(g) + \text{O}_2(g) \rightarrow 2 \text{ NO}_2(g)$$

calculate
a. ΔH b. ΔS^0
c. the temperature at which $\Delta G^0 = 0$

14.21 For the decomposition of Ag_2O,

$$2 \text{ Ag}_2\text{O}(s) \rightarrow 4 \text{ Ag}(s) + \text{O}_2(g)$$

a. Using Tables 14.1 and 14.3, obtain an expression for ΔG^0 as a function of temperature. Use it to prepare a table of ΔG^0 values at 100 K intervals between 100 K and 500 K.
b. Calculate the temperature at which ΔG^0 becomes zero.

14.48 Discuss the effect of temperature upon the spontaneity of the following reactions at 1 atm:
a. $\text{Al}_2\text{O}_3(s) + 2 \text{ Fe}(s) \rightarrow 2 \text{ Al}(s) + \text{Fe}_2\text{O}_3(s)$
$\Delta H = +847.6 \text{ kJ}; \Delta S^0 = +41.2 \text{ J/K}$
b. $\text{CO}(g) \rightarrow \text{C}(s) + \frac{1}{2} \text{O}_2(g)$
$\Delta H = +110.5 \text{ kJ}; \Delta S^0 = -89.7 \text{ J/K}$
c. $\text{SO}_3(g) \rightarrow \text{SO}_2(g) + \frac{1}{2} \text{O}_2(g)$
$\Delta H = +99.1 \text{ kJ}; \Delta S^0 = +94.8 \text{ J/K}$

14.49 At what temperature does ΔG^0 become zero for each of the reactions in Problem 14.48? Explain the significance of your answers.

14.50 For the reaction

$$2 \text{ H}_2\text{S}(g) + 3 \text{ O}_2(g) \rightarrow 2 \text{ H}_2\text{O}(g) + 2 \text{ SO}_2(g)$$

calculate
a. ΔH b. ΔS^0
c. the temperature at which $\Delta G^0 = 0$

14.51 Earlier civilizations smelted iron from ore by heating it with charcoal from a wood fire:

$$2 \text{ Fe}_2\text{O}_3(s) + 3 \text{ C}(s) \rightarrow 4 \text{ Fe}(s) + 3 \text{ CO}_2(g)$$

a. Obtain an expression for ΔG^0 as a function of temperature. Prepare a table of ΔG^0 values at 200 K intervals between 200 and 1000 K.
b. Calculate the lowest temperature at which the smelting could be carried out.

14.22 The normal boiling point of $CHCl_3$ is 61°C; its heat of vaporization is 31.3 kJ/mol. The standard entropy of the liquid is 202.9 J/mol·K. Determine
a. ΔG^0 at 61°C.
b. S^0 of $CHCl_3(g)$.

14.23 Red phosphorus is formed by heating white phosphorus. Calculate the temperature at which the two forms are at equilibrium given

white P: $\Delta H_f = 0.00$ kJ/mol; $S^0 = 41.09$ J/mol·K
red P: $\Delta H_f = -17.6$ kJ/mol; $S^0 = 22.80$ J/mol·K

Second Law of Thermodynamics

14.24 A steam engine operating at a theoretical efficiency of 18% discharges liquid water to the surroundings at 35°C. What is the temperature of the steam in the engine?

14.25 Calculate the maximum amount of useful work that can be obtained from the reaction

$$2 H_2(g) + O_2(g) \rightarrow 2 H_2O(g)$$

a. at 25°C.   b. at 1000°C.

General

14.26 Which of the following are spontaneous? nonspontaneous?
a. tea dissolving in water
b. removing seeds from a watermelon
c. recharging a battery
d. burning gasoline in an automobile engine

14.27 Which of the following quantities can be taken to be independent of pressure?
a. ΔH for a reaction   b. ΔS for a reaction
c. S for a substance   d. ΔG for a reaction

14.28 Criticize each of the following statements:
a. When ΔG^0 is positive, the reaction cannot occur.
b. All endothermic reactions are nonspontaneous.
c. When ΔH and ΔS^0 are both positive, the reaction is nonspontaneous at all temperatures.
d. A process for which ΔG^0 is -500 kJ produces 500 kJ of work.

14.52 The heat of formation of $Br_2(g)$ is 30.7 kJ/mol; its standard entropy is 245.3 J/mol·K. Using Table 14.3 and Equation 14.15, estimate the normal boiling point of bromine.

14.53 Tin organ pipes in unheated churches develop tin "disease," in which white tin is converted to gray tin. Given

white Sn: $\Delta H_f = 0.00$ kJ/mol; $S^0 = 51.55$ J/mol·K
gray Sn: $\Delta H_f = -2.09$ kJ/mol; $S^0 = 44.14$ J/mol·K

calculate the temperature for the transition

$$Sn_{white}(s) \rightarrow Sn_{gray}(s)$$

14.54 Calculate the discharge temperature for a steam engine that operates at 100°C and has a theoretical efficiency of 22%.

14.55 What is the maximum amount of useful work that can be obtained from the reaction

$$C(s) + O_2(g) \rightarrow CO_2(g)$$

a. at 0 K?   b. at 500 K?

14.56 a. Under what conditions is the vaporization of water at 25°C spontaneous? nonspontaneous? neither?
b. Will a small amount of KNO_3 added to a saturated solution of KNO_3 dissolve spontaneously? Suggest a way to make the added KNO_3 dissolve.

14.57 Which of the quantities in Question 14.27 can be taken to be independent of temperature?

14.58 Comment on the *general* validity of each of the following statements:
a. An exothermic reaction is spontaneous.
b. A reaction for which ΔS is positive is spontaneous.
c. ΔS is positive for a reaction in which there is an increase in the number of moles.
d. If ΔH and ΔS are both positive, ΔG will decrease when the temperature rises.

14.29 In your own words explain why
  a. not all exothermic reactions are spontaneous.
  b. a solid has a lower entropy than the corresponding liquid.
  c. ΔG^0 is more temperature dependent than ΔH or ΔS^0.

14.30 Fill in the blanks.
  a. ΔH and ΔG^0 become equal at ____ K.
  b. At equilibrium, ΔG is ____.

14.59 In your own words explain
  a. why ΔS^0 is negative for a reaction in which the number of moles of gas decreases.
  b. how total entropy can increase even when the entropy of a reaction system decreases.
  c. why we use the superscript ^{0} for ΔG^0 and ΔS^0 but not for ΔH.

14.60 Complete the following statements:
  a. In a spontaneous process, ΔG is ____.
  b. S^0 for ice is ____ than S^0 for liquid water.

*14.61 The heat of fusion of ice is 333 J/g. For the process $H_2O(s) \rightarrow H_2O(l)$, determine

  a. ΔH b. ΔG^0 at 0°C   c. ΔS^0 d. ΔG^0 at $-10°C$ e. ΔG^0 at 10°C

*14.62 The electrolysis of Al_2O_3 to form the elements is carried out at about 1000°C. Calculate ΔG^0 for the electrolysis of one kilogram of Al_2O_3 under these conditions. Calculate the cost of the electrolysis at 6.0¢ per kilowatt hour; 1 kW·h = 3600 kJ. (The actual cost is greater because much of the electrical energy is dissipated as heat.)

*14.63 The overall reaction that occurs when sugar is metabolized is

$$C_{12}H_{22}O_{11}(s) + 12\ O_2(g) \rightarrow 12\ CO_2(g) + 11\ H_2O(l)$$

For this reaction, ΔH is -5650 kJ and ΔG^0 is -5790 kJ at 25°C.
  a. If 30% of the free energy change is actually converted to useful work, how many kilojoules of work could be obtained when one gram of sugar is metabolized at body temperature, 37°C?
  b. How many grams of sugar would you have to eat to get the energy to climb a mountain 1610 meters high? ($w = 9.79 \times 10^{-3}$ mh, where w = work in kilojoules, m is body mass in kilograms, and h is height in meters.)

*14.64 Hydrogen has been suggested as the fuel of the future. One way to store it is to convert it to a compound that can then be heated to release the hydrogen. One such compound is calcium hydride, CaH_2. This compound has a heat of formation of -188.7 kJ/mol and a standard entropy of 42.0 J/mol·K. What is the minimum temperature to which calcium hydride would have to be heated to produce hydrogen at one atmosphere pressure?

*14.65 When a copper wire is exposed to air at room temperature, it becomes coated with a black oxide, CuO. If the wire is heated above a certain temperature, the black oxide is converted to a red oxide, Cu_2O. At a still higher temperature, the oxide coating disappears. Explain these observations in terms of the thermodynamics of the reactions

$$2\ CuO(s) \rightarrow Cu_2O(s) + \tfrac{1}{2}\ O_2(g)$$

$$Cu_2O(s) \rightarrow 2\ Cu(s) + \tfrac{1}{2}\ O_2(g)$$

and estimate the temperatures at which the changes occur.

Chapter 15
Chemical
Equilibrium
in the
Gas Phase

In Chapter 11, we described the equilibrium between liquid and gaseous water. When a sample of liquid water is placed in a closed container at constant temperture, part of it vaporizes. After a few minutes, a dynamic equilibrium is established:

$$H_2O(l) \rightleftharpoons H_2O(g)$$

The double arrow implies that the two processes, vaporization ($\rightarrow$) and condensation ($\leftarrow$), are occurring at the same rate. Once equilibrium is established, the relative amounts of liquid and vapor do not change with time.

The state of this equilibrium system at a given temperature can be described in a simple way. We can cite either the equilibrium concentration of the water vapor (0.0327 mol/L at 100°C) or its equilibrium pressure (1.00 atm at 100°C). The concentration or pressure of vapor in equilibrium with liquid is independent of the volume of the container or the amount of water we started with. It depends only upon the temperature of the system.

Chemical reactions carried out in closed containers resemble in many ways the system just discussed. Ordinarily, the reactants are not completely consumed. Instead, we obtain an equilibrium mixture containing both products and reactants. At equilibrium, forward and reverse reactions take place at the same rate. As a result, the concentrations of all species at equilibrium remain constant as time passes.

The equilibrium is dynamic, not static

In this chapter, we will examine the properties of chemical systems that reach equilibrium in the gas state. A typical example is

$$N_2O_4(g) \rightleftharpoons 2\ NO_2(g)$$

The state of this equilibrium system cannot be described as simply as that of the $H_2O(l)$–$H_2O(g)$ system. In particular, the concentration of NO_2 at equilibrium is not fixed; it can take on any of a number of values, depending upon the concentration of N_2O_4. However, as we will see in Section 15.1, there is a simple relationship between the equilibrium concentrations of these two gases. This relationship is expressed in terms of a quantity called the **equilibrium constant** and given the symbol K_c. The equilibrium constant for a reaction is one of its characteristic properties. It has a fixed value at a given temperature, independent of such factors as initial concentration, container volume, or pressure. Our discussion throughout this chapter will focus upon the properties of this equilibrium constant. We will see how, among other things, one can

—write the expression for K_c corresponding to any chemical equilibrium involving gases (Section 15.2).

—use the value of K_c to predict the extent to which a reaction will take place (Section 15.3).

—use K_c along with Le Châtelier's principle to predict what will happen when an equilibrium system is disturbed in some way (Section 15.4).

—relate the equilibrium constant to the standard free energy change for a reaction (Section 15.6).

15.1
The N_2O_4–NO_2
Equilibrium System

Consider what happens when a sample of N_2O_4, a colorless gas, is placed in a closed, evacuated container at 100°C. Instantly, a reddish-brown color develops. This color is due to nitrogen dioxide, NO_2, formed by decomposition of part of the N_2O_4:

$$N_2O_4(g) \rightarrow 2\ NO_2(g) \tag{15.1a}$$

At first, this is the only reaction taking place. As soon as some NO_2 is formed, however, the reverse reaction can occur:

$$2\ NO_2(g) \rightarrow N_2O_4(g) \tag{15.1b}$$

As time passes, Reaction 15.1a slows down and 15.1b speeds up. Soon, their rates become equal. A dynamic equilibrium has been established:

NO_2 is formed at the same rate it reacts

$$N_2O_4(g) \rightleftharpoons 2\ NO_2(g) \tag{15.1}$$

At equilibrium, appreciable amounts of both gases are present. From that point on their concentrations are constant, as long as the volume of the container and the temperature remain unchanged.

The approach to equilibrium in this system is illustrated by the data in Table 15.1, plotted in Figure 15.1. We start with 0.100 mol N_2O_4 in a 1.00-L container at 100°C. The original concentration of N_2O_4 is thus 0.100 mol/L. Since there is no NO_2 present at the beginning of the experiment, its original concentration is zero. As equilibrium is approached, the *net reaction* taking place is that in the forward direction (15.1a). The concentration of N_2O_4 drops rapidly at first and then more slowly. The concentration of NO_2 increases. Finally, both concentrations level off and become constant. This indicates that we are at equilibrium. From Table 15.1 or Figure 15.1,

$$[N_2O_4] = 0.040 \text{ mol/L}$$

$$[NO_2] = 0.120 \text{ mol/L}$$

The square brackets, here and elsewhere throughout this text, represent equilibrium concentrations in moles per liter. It is important that these be distinguished from original or other nonequilibrium concentrations.

Table 15.1
Establishment of Equilibrium at 100°C in the System
$N_2O_4(g) \rightleftharpoons 2\ NO_2(g)$

As conc N_2O_4 goes down, conc NO_2 goes up

Time (s)	0	20	40	60	80	100
Conc. N_2O_4 (mol/L)	0.100	0.070	0.050	**0.040***	**0.040**	**0.040**
Conc. NO_2 (mol/L)	0.000	0.060	0.100	**0.120***	**0.120**	**0.120**

*Equilibrium concentrations are in bold type.

One feature of the data in Table 15.1 deserves further comment. Notice that, over any time interval, the increase in concentration of NO_2 is exactly *twice* the decrease in concentration of N_2O_4. Consider, for example, what happens in the 60 s required to reach equilibrium. The concentration of N_2O_4 decreases by 0.060 mol/L:

The negative sign indicates that conc N_2O_4 decreases

$$\Delta \text{ conc. } N_2O_4 = \text{final conc. } N_2O_4 - \text{orig. conc. } N_2O_4$$
$$= 0.040 \text{ mol/L} - 0.100 \text{ mol/L} = -0.060 \text{ mol/L}$$

The concentration of NO_2 increases by twice this amount, 0.120 mol/L:

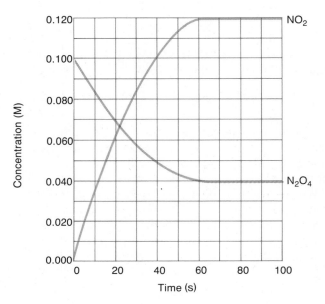

FIGURE 15.1 Approach to equilibrium in the N_2O_4–NO_2 system. The concentration of N_2O_4 starts off at 0.100 M, drops sharply at first, and finally levels off at its equilibrium value, 0.040 M. Meanwhile, the concentration of NO_2 rises rapidly at first, then increases more slowly, and finally becomes constant at the equilibrium value, 0.120 M.

Δ conc. NO_2 = 0.120 mol/L − 0.000 mol/L = 0.120 mol/L

This relationship is explained quite simply. As equilibrium is approached, N_2O_4 decomposes:

$$N_2O_4(g) \rightarrow 2\ NO_2(g)$$

From the coefficients of the balanced equation, we see that two moles of NO_2 are formed for every mole of N_2O_4 that decomposes. Hence, if x moles of N_2O_4 breaks down, 2x moles of NO_2 must form. If the concentration of N_2O_4 decreases by x mol/L, that of NO_2 must increase by twice that amount, 2x mol/L.

The total mass of the system remains constant

    There are many ways to approach equilibrium in the N_2O_4–NO_2 system. Table 15.2 gives data for three experiments in which the original concentrations are quite different. Experiment 1 is that just described, in which we start with pure N_2O_4. In Experiment 2, we start from the other side of the equilibrium system, with pure NO_2 at a concentration of 0.100 mol/L. As we approach equilibrium, some of the NO_2 combines to form N_2O_4:

$$2\ NO_2(g) \rightarrow N_2O_4(g)$$

Finally, in Experiment 3, we start with a mixture of N_2O_4 and NO_2, both at a concentration of 0.100 mol/L. Some of the N_2O_4 decomposes in this experiment. At equilibrium, there is less N_2O_4 and more NO_2 than there was originally.

Table 15.2
Equilibrium Measurements in the N_2O_4–NO_2 System at 100°C

		ORIGINAL CONC. (mol/L)	EQUILIBRIUM CONC. (mol/L)
Expt. 1	N_2O_4	0.100	0.040
	NO_2	0.000	0.120
Expt. 2	N_2O_4	0.000	0.014
	NO_2	0.100	0.072
Expt. 3	N_2O_4	0.100	0.070
	NO_2	0.100	0.160

    Looking at the data in Table 15.2, you might wonder whether these three experiments have anything in common. Specifically, is there any relationship between the equilibrium concentrations of NO_2 and N_2O_4 that is valid for all the experiments? It turns out that there is, although it is not an obvious one. The quotient $[NO_2]^2/[N_2O_4]$ is the same, about 0.36, in each case:

Expt. 1 $\dfrac{[NO_2]^2}{[N_2O_4]} = \dfrac{(0.120)^2}{0.040} = 0.36$

Expt. 2 $\dfrac{[NO_2]^2}{[N_2O_4]} = \dfrac{(0.072)^2}{0.014} = 0.37$

Rather surprising, but true

Expt. 3 $\dfrac{[NO_2]^2}{[N_2O_4]} = \dfrac{(0.160)^2}{0.070} = 0.36$

This relationship holds for any equilibrium mixture containing NO_2 and N_2O_4 at 100°C. Regardless of where we start, we find that at equilibrium,

$$\frac{[NO_2]^2}{[N_2O_4]} = 0.36 \text{ at } 100°C$$

Further experiments with this system at various temperatures lead to the following general conclusion:

At any given temperature, the quantity

$$\frac{[NO_2]^2}{[N_2O_4]}$$

is a constant, independent of the amounts of N_2O_4 and NO_2 that we start with, the volume of the container, or the total pressure. This constant is referred to as the **equilibrium constant, K_c,** for the reaction

$$N_2O_4(g) \rightleftharpoons 2 NO_2(g)$$

At 100°C, reaction will occur until

$$\frac{[NO_2]^2}{[N_2O_4]} = 0.36$$

At 100°C, K_c for this reaction is 0.36; at 150°C it has a different value, about 3.2. Any mixture of NO_2 and N_2O_4 at 100°C will react to form more NO_2 or more N_2O_4, until the ratio (conc. NO_2)²/(conc. N_2O_4) becomes equal to 0.36. At 150°C, reaction will occur until this ratio becomes equal to 3.2.

15.2
General Expression for K_c

We have seen that for the system

$$N_2O_4(g) \rightleftharpoons 2 NO_2(g); \qquad K_c = \frac{[NO_2]^2}{[N_2O_4]}$$

For every gaseous system, a similiar expression can be written. The form of that expression can be obtained readily. For the general gas-phase system,

$$a A(g) + b B(g) \rightleftharpoons c C(g) + d D(g)$$

where A, B, C, D represent different substances and a, b, c, d are their coefficients in the balanced equation

We can prove this relation by using thermodynamics

$$K_c = \frac{[C]^c \times [D]^d}{[A]^a \times [B]^b} \tag{15.2}$$

Notice that in the expression for K_c,

—the equilibrium concentrations of **products** (right side of equation) appear in the **numerator**.
—the equilibrium concentrations of **reactants** (left side of equation) appear in the **denominator**.

To illustrate the application of this law, consider the systems

$$2\ HI(g) \rightleftharpoons H_2(g) + I_2(g)$$ **(15.3)**

$$N_2(g) + 3\ H_2(g) \rightleftharpoons 2\ NH_3(g)$$ **(15.4)**

For these reactions we have

$$K_c = \frac{[H_2] \times [I_2]}{[HI]^2} \qquad \text{(for Reaction 15.3)}$$

$$K_c = \frac{[NH_3]^2}{[N_2] \times [H_2]^3} \qquad \text{(for Reaction 15.4)}$$

The equilibrium constants referred to above have the same general properties as K_c for the N_2O_4–NO_2 system:

K_c has a constant value at a given temperature, independent of original concentrations, container volume, or pressure.

It is important to realize that *the expression for K_c depends upon the form of the chemical equation written to describe the equilibrium system.* To illustrate what this statement means, consider the N_2O_4–NO_2 system, for which we wrote

$$N_2O_4(g) \rightleftharpoons 2\ NO_2(g); \qquad K_c = \frac{[NO_2]^2}{[N_2O_4]} = 0.36 \text{ at } 100°C$$

There are many other equations that could be used to describe this system. We could have written

$$\tfrac{1}{2}\ N_2O_4(g) \rightleftharpoons NO_2(g)$$

In this case the expression for the equilibrium constant would have been

$$K_c' = \frac{[NO_2]}{[N_2O_4]^{1/2}}$$

For a value of K_c to be useful, you need to know the equation for the associated reaction

We also might have written

$$2\ NO_2(g) \rightleftharpoons N_2O_4(g)$$

and arrived at

$$K_c'' = \frac{[N_2O_4]}{[NO_2]^2}$$

Since all of these equations describe the same chemical system, K_c, K_c', and K_c'' must be related in a simple way. If you examine the concentration terms in these equations, it should be apparent that

$$K_c' = (K_c)^{1/2} = (0.36)^{1/2} = 0.60 \text{ at } 100°C$$

$$K_c'' = 1/K_c = 1/0.36 = 2.8 \text{ at } 100°C$$

Each of these equations is a valid way of describing the N_2O_4–NO_2 equilibrium system at 100°C. It would be ambiguous at the very least to say that, for this system at 100°C, "the equilibrium constant is 0.36." The numerical value of K_c has meaning only in association with a particular chemical equation.

Reactions Involving Liquids or Solids as Well as Gases

In certain gas-phase reactions, one or more of the substances involved is present as a pure liquid or solid. Examples of such systems include

$$CO_2(g) + H_2(g) \rightleftharpoons CO(g) + H_2O(l) \qquad \textbf{(15.5)}$$

$$I_2(s) \rightleftharpoons 2\ I(g) \qquad \textbf{(15.6)}$$

In such cases we find that **there is no term for the liquid or solid in the expression for K_c.** Thus, we have

$$K_c = \frac{[CO]}{[CO_2] \times [H_2]} \qquad \text{(for Reaction 15.5)}$$

$$K_c = [I]^2 \qquad \text{(for Reaction 15.6)}$$

The terms for liquids (H_2O) and solids (I_2) are omitted.

To see why this simplification is possible, consider Reaction 15.5 taking place at, say, 100°C. The equation tells us that liquid water is present. There must also be water vapor present, in equilibrium with the liquid. Indeed, it is actually $H_2O(g)$ that takes part in the gas-phase reaction. The pressure of the vapor is 1.00 atm, the vapor pressure of liquid water at 100°C. The concentration of water vapor, calculated from the Ideal Gas Law, is 0.0327 mol/L. This concentration remains constant as long as there is any liquid water present, regardless of what [CO], [H_2], or [CO_2] may be. Hence, it can be incorporated into the expression for K_c, and we do not need a term for the concentration of H_2O.

A similar argument can be used to justify dropping the term [I_2] in the equilibrium constant expression for Reaction 15.6. As long as there is any solid present, the pressure of I_2 in the vapor and hence its concentration is fixed:

> For equilibrium, liquid H_2O must be present

conc. $I_2(g)$ = a constant

Hence, the term for I_2 can be absorbed into the K_c expression.

EXAMPLE 15.1 Write expressions for K_c for
a. $2\ SO_3(g) \rightleftharpoons 2\ SO_2(g) + O_2(g)$
b. $CuO(s) + H_2(g) \rightleftharpoons Cu(s) + H_2O(g)$

Solution

a. $K_c = \dfrac{[SO_2]^2 \times [O_2]}{[SO_3]^2}$

b. $K_c = \dfrac{[H_2O]}{[H_2]}$

Notice that in (b) there are no terms for copper(II) oxide or for copper metal, both of which are solids.

EXERCISE Suppose the reaction in (b) is carried out at a temperature low enough so that liquid water is present. What is the expression for K_c? Answer: $1/[H_2]$.

Determination of K_c

Numerical values of equilibrium constants are determined in much the same way as we described for the N_2O_4–NO_2 system. Typical calculations are shown in Examples 15.2 and 15.3.

EXAMPLE 15.2 When ammonium chloride is heated, it reaches equilibrium with ammonia and hydrogen chloride:

$$NH_4Cl(s) \rightleftharpoons NH_3(g) + HCl(g)$$

It is found that at equilibrium at 500°C in a 5.0-L container, there is 2.0 mol NH_3, 2.0 mol HCl, and 1.0 mol NH_4Cl. Calculate K_c for this system at 500°C.

Solution The expression for K_c is

$$K_c = [NH_3] \times [HCl]$$

(There is no term for solid NH_4Cl.) The equilibrium concentrations are found by dividing the number of moles by the volume in liters:

To evaluate K_c we need to know the equilibrium concentrations of all gaseous reactants and products

$$[NH_3] = \frac{2.0 \text{ mol}}{5.0 \text{ L}} = 0.40 \text{ mol/L}; \qquad [HCl] = \frac{2.0 \text{ mol}}{5.0 \text{ L}} = 0.40 \text{ mol/L}$$

$$K_c = 0.40 \times 0.40 = 0.16$$

EXERCISE Suppose that for the system referred to in Example 15.1a, $[SO_2] = [O_2] = 0.10$ mol/L and $[SO_3] = 0.20$ mol/L. Calculate K_c. Answer: 0.025.

EXAMPLE 15.3 Consider the equilibrium represented by Equation 15.3. Suppose we start with pure HI at a concentration of 0.100 mol/L at 520°C. The equilibrium concentration of H_2 is found to be 0.010 mol/L. Calculate
a. $[I_2]$ b. $[HI]$ c. K_c for Reaction 15.3

Solution
a. According to Equation 15.3, one mole of I_2 is formed for every mole of H_2. The concentrations of H_2 and I_2 must therefore increase by the same amount. Since they both start at zero, the two concentrations must be equal at equilibrium. The equilibrium concentration of I_2, like that of H_2, is 0.010 mol/L.
b. Two moles of HI are required to form one mole of H_2. The decrease in HI concentration must be twice the increase in the concentration of H_2:

decrease in HI conc. $= 2(0.010 \text{ mol/L}) = 0.020$ mol/L

The original concentration of HI is 0.100 mol/L; Hence,

$$\text{equilibrium conc. HI} = 0.100 \text{ mol/L} - 0.020 \text{ mol/L}$$
$$= 0.080 \text{ mol/L}$$

It may be helpful to summarize the reasoning we have gone through by means of a table.

	ORIG. CONC. (mol/L)	CHANGE IN CONC. (mol/L)	EQUIL. CONC. (mol/L)
HI	0.100	−0.020	0.080
H_2	0.000	+0.010	0.010
I_2	0.000	+0.010	0.010

c. $K_c = \dfrac{[H_2] \times [I_2]}{[HI]^2} = \dfrac{0.010 \times 0.010}{(0.080)^2} = 0.016$

K_c varies with T, but not with concentration

EXERCISE Suppose that, at a different temperature, we start with pure HI at 0.100 mol/L and find that its equilibrium concentration is 0.074 mol/L. What is the value of K_c at this temperature? Answer: $K_c = 0.031$.

15.3
Applications of K_c

A knowledge of the equilibrium constant for a reaction tells us a great deal, qualitatively and quantitatively, about the extent to which a reaction will occur. Often, knowing only the magnitude of K_c, we can predict whether a reaction is likely to be feasible. Consider, for example, a possible method of "fixing" atmospheric nitrogen—converting it to a compound—by reaction with O_2:

$$N_2(g) + O_2(g) \rightleftharpoons 2 \text{ NO}(g); \quad K_c = \frac{[NO]^2}{[N_2] \times [O_2]} = 1 \times 10^{-30} \text{ at } 25°C \quad \textbf{(15.7)}$$

We see that K_c for Reaction 15.7 is a very small number. This means that the equilibrium concentration of NO, which appears in the numerator of K_c, must be very small relative to those of N_2 and O_2, which are in the denominator. This tells us that a mixture of nitrogen and oxygen will react to only a very small extent to produce NO. Clearly, this would not be a suitable way to fix nitrogen, at least at 25°C.

An alternative approach to nitrogen fixation involves reacting it with hydrogen:

$$N_2(g) + 3 \text{ H}_2(g) \rightleftharpoons 2 \text{ NH}_3(g); \quad K_c = \frac{[NH_3]^2}{[N_2] \times [H_2]^3} = 5 \times 10^8 \text{ at } 25°C$$

Here the situation is quite different from that discussed above. Since K_c is large, the equilibrium system must contain mostly NH_3, which appears in the numerator. We expect a mixture of N_2 and H_2 to be almost completely converted to NH_3 at equilibrium. (Unfortunately, it takes essentially forever for equilibrium to be reached in this system at 25°C, but that's a problem we'll worry about in Chapter 16).

In general, we can say that if K_c is very large, the forward reaction will proceed far to the right. The equilibrium system will contain mostly products (right side of the chemical equation) with very little unreacted starting materials. Conversely, if K_c is very small, virtually no reaction will occur in the forward direction. The equilibrium system will consist almost entirely of unreacted starting materials, with very little in the way of products. Finally, if K_c is neither "very large" nor "very small," we expect the equilibrium system to contain appreciable amounts of both products and reactants.

If K_c is very small, [Reactants] = original conc

The type of prediction we have just made is qualitative. Knowing the magnitude of K_c, we can make a ballpark estimate of how far a reaction will go. The equilibrium constant can also be used to make quantitative predictions. We can use the value of K_c to predict either

—the direction of reaction, or
—the extent of reaction.

Direction of Reaction

We have seen (Section 15.2) that, for the general gas-phase reaction

$$a\ A(g) + b\ B(g) \rightleftharpoons c\ C(g) + d\ D(g)$$

the expression for the equilibrium constant K_c is

$$K_c = \frac{[C]^c \times [D]^d}{[A]^a \times [B]^b}$$

The square brackets are used to designate equilibrium concentrations in moles per liter. When we carry out a reaction in the laboratory, the original concentration quotient Q, expressed as

$$Q = \frac{(\text{orig. conc. C})^c \times (\text{orig. conc. D})^d}{(\text{orig. conc. A})^a \times (\text{orig. conc. B})^b}$$

Q is both useful and easy to calculate

will seldom be equal numerically to K_c. If it is not, reaction will occur in one direction or the other so as to bring the concentrations of products and reactants to the ratio required at equilibrium. We can distinguish two possibilities:

1. If $\quad Q < K_c$

the reaction will proceed from left to right, i.e.,

$$a\ A(g) + b\ B(g) \rightarrow c\ C(g) + d\ D(g)$$

In this way, the concentrations of products increase and those of reactants decrease. As this happens, the concentration quotient, Q, increases until it becomes equal to K_c. When we reach that point, we are at equilibrium and there is no further change.

$Q \rightarrow K_c$ as the reaction goes to equilibrium

2. If $\quad Q > K_c$

we conclude that the concentrations of products are "too high" and those of the reactants "too low" to meet the equilibrium condition. Reaction must proceed in the reverse direction, i.e.,

$$c \ C(g) + d \ D(g) \rightarrow a \ A(g) + b \ B(g)$$

increasing the concentrations of A and B while reducing those of C and D. This lowers the concentration quotient, Q, to its equilibrium value given by K_c.

EXAMPLE 15.4 Consider the following system at 100°C:

$$N_2O_4(g) \rightleftharpoons 2 \ NO_2(g); \qquad K_c = 0.36$$

Predict the direction in which the system will move to reach equilibrium if we start with
a. 0.20 mol N_2O_4 in a 4.0-L container.
b. 0.20 mol N_2O_4 and 0.20 mol NO_2 in a 4.0-L container.

Solution In each case, we first calculate the original concentrations of N_2O_4 and NO_2. We then compare the quotient,

$$Q = \frac{(\text{orig. conc. } NO_2)^2}{(\text{orig. conc. } N_2O_4)}$$

to K_c, 0.36, to decide which way reaction will proceed.

a. orig. conc. N_2O_4 = 0.20 mol/4.0L = 0.050 mol/L
    orig. conc. NO_2 = 0   (since there is no NO_2 present)

This approach can be applied to any reaction for which we know K_c

Hence, the original concentration quotient, Q, is:

$$Q = \frac{0^2}{(0.050)} = 0 < K_c$$

The reaction must proceed to the right to produce NO_2. This would be true regardless of how much N_2O_4 we started with. As long as there is no NO_2 present originally, some of it must be formed to establish equilibrium.

b. orig. conc. N_2O_4 = 0.20 mol/4.0 L = 0.050 mol/L
    orig. conc. NO_2 = 0.20 mol/4.0 L = 0.050 mol/L

$$Q = \frac{(\text{orig. conc. } NO_2)^2}{(\text{orig. conc. } N_2O_4)} = \frac{(0.050)^2}{0.050} = 0.050$$

Since 0.050 < K_c = 0.36, reaction must proceed in the forward direction to produce NO_2. The concentration of NO_2 increases, that of N_2O_4 decreases, and eventually the concentration quotient becomes equal to K_c.

EXERCISE Suppose orig. conc. N_2O_4 = orig. conc. NO_2 = 1.0 M. Which way does reaction occur to establish equilibrium? Answer: Reverse direction (right to left).

Equilibrium Concentrations

The equilibrium constant for a chemical system can be used to calculate the concentrations of the species present at equilibrium. In the simplest case, we can use K_c to obtain the equilibrium concentration of one species when we know those of all the other species (Example 15.5).

EXAMPLE 15.5   We pointed out earlier that for the equilibrium system

$$N_2(g) + O_2(g) \rightleftharpoons 2 NO(g)$$

K_c is 1×10^{-30} at 25°C. Suppose that in a mixture at 25°C the concentrations of N_2 and O_2 are 0.040 and 0.010 mol/L, respectively. Calculate the equilibrium concentration of NO.

Solution   The expression for K_c is

$$K_c = \frac{[NO]^2}{[N_2] \times [O_2]} = 1 \times 10^{-30}$$

Substituting for the concentrations of N_2 and O_2, we have

$$\frac{[NO]^2}{(0.040)(0.010)} = 1 \times 10^{-30}$$

$$[NO]^2 = (4 \times 10^{-2})(1 \times 10^{-2})(1 \times 10^{-30}) = 4 \times 10^{-34}$$

$$[NO] = 2 \times 10^{-17} \text{ mol/L}$$

The concentration of NO at equilibrium is extremely small relative to those of N_2 and O_2. This confirms the qualitative prediction that we made at the beginning of this section.

Essentially no reaction occurs

EXERCISE   At 2000°C, K_c for this system is much larger, about 0.10. Calculate [NO] at 2000°C if [N_2] = 0.040 M, [O_2] = 0.010 M. Answer: 0.0063 M.

More often, we use K_c to determine the extent of a particular reaction. Here we are given the *original* concentrations of reactants. We are asked to calculate the *equilibrium* concentrations of both reactants and products. The reasoning here is somewhat more complex than that in Example 15.5. In general, we follow a three-step process:

1. Express the equilibrium concentrations of all species in terms of a single unknown, x. As we will see, there are many different ways in which you can choose the variable x. Once you have chosen x, all the equilibrium concentrations have to be related to x in a way that is consistent with the coefficients of the balanced equation. To do this, it helps to set up an equilibrium table of the sort used in Example 15.3.

2. Substitute the concentration terms into the expression for K_c. This gives an algebraic equation involving x. Simplify this equation if possible, then solve it for x.

3. Having found x, calculate the equilibrium concentrations of all species.

EXAMPLE 15.6 For the system

$$CO_2(g) + H_2(g) \rightleftharpoons CO(g) + H_2O(g)$$

K_c is 0.64 at 900 K. Suppose we start with CO_2 and H_2, both at a concentration of 0.100 mol/L. When the system reaches equilibrium, what are the concentrations of products and reactants?

Solution

1. We must first express all the equilibrium concentrations in terms of a single unknown, x. Let us take

   x = no. moles per liter CO formed

   All the coefficients in the balanced equation are the same, 1. It follows that

   no. moles per liter H_2O formed = no. moles per liter CO formed = x

   no. moles per liter CO_2 consumed = no. moles per liter CO formed = x

   no. moles per liter H_2 consumed = no. moles per liter CO formed = x

   We conclude that, in reaching equilibrium, the concentrations of CO and H_2O *increase* by x, while those of CO_2 and H_2 *decrease* by the same amount, x. Putting this reasoning in the form of a table,

	ORIG. CONC. (mol/L)	CHANGE IN CONC. (mol/L)	EQUIL. CONC. (mol/L)
CO_2	0.100	$-x$	$0.100 - x$
H_2	0.100	$-x$	$0.100 - x$
CO	0.000	$+x$	x
H_2O	0.000	$+x$	x

You need to think about this for a while

2. We are now ready to substitute into the expression for K_c:

   $$K_c = 0.64 = \frac{[CO] \times [H_2O]}{[CO_2] \times [H_2]} = \frac{x^2}{(0.100 - x)^2}$$

   This is a second-order equation in x. Such equations can always be solved using the quadratic formula (see Example 15.7). Here, however, we can simplify the arithmetic by noting that the right side of the equation is a perfect square. Taking the square root of both sides, we have

Keep the math simple when you can

   $$(0.64)^{1/2} = 0.80 = \frac{x}{0.100 - x}$$

   Solving for x: $x = 0.80(0.100 - x) = 0.080 - 0.80x$

   $1.80x = 0.080;$ $x = 0.044$

3. Referring back to the equilibrium table,

   $$[CO] = [H_2O] = x = 0.044 \text{ mol/L}$$

   $$[CO_2] = [H_2] = 0.100 - x = 0.056 \text{ mol/L}$$

EXERCISE Suppose the original concentrations of CO_2 and H_2 were 0.100 M and 0.200 M. In that case, what would be the algebraic equation obtained in step 2? Answer:

$$0.64 = \frac{x^2}{(0.100 - x)(0.200 - x)}$$

The arithmetic involved in equilibrium calculations can be relatively simple, as it was in Example 15.6. Sometimes it is more complex (Example 15.7). The reasoning involved, however, is the same. It is always helpful to set up an equilibrium table to summarize the analysis of the problem.

EXAMPLE 15.7 Consider the system

$$N_2O_4(g) \rightleftharpoons 2\ NO_2(g); \qquad K_c = 0.36 \text{ at } 100°C$$

Suppose we start with pure N_2O_4 at a concentration of 0.100 mol/L. What are the equilibrium concentrations of NO_2 and N_2O_4?

Solution

1. Some of the N_2O_4 will decompose to achieve equilibrium. Let x be the number of moles per liter of N_2O_4 that decomposes. The balanced equation tells us that two moles of NO_2 are formed for every mole of N_2O_4 that decomposes. Hence, if the concentration of N_2O_4 decreases by x, that of NO_2 must increase by 2x. We set up an equilibrium table,

	ORIG. CONC. (mol/L)	CHANGE IN CONC. (mol/L)	EQUIL. CONC. (mol/L)
N_2O_4	0.100	$-x$	$0.100 - x$
NO_2	0.000	$+2x$	$2x$

2. $$K_c = 0.36 = \frac{(2x)^2}{0.100 - x} = \frac{4x^2}{0.100 - x}$$

This time, we cannot solve for x as simply as in Example 15.6. The denominator of the right side is not a perfect square. Instead, we use the general method of solving a quadratic equation. This involves rearranging to the form

$$ax^2 + bx + c = 0$$

and applying the quadratic formula

$$x = \frac{-b \pm \sqrt{b^2 - 4ac}}{2a}$$

To convert our expression to the desired form, we proceed as follows:

$$4x^2 = 0.36(0.100 - x) = 0.036 - 0.36x$$

$$4x^2 + 0.36x - 0.036 = 0$$

We can simplify a bit by dividing both sides of the equation by 4:

$$x^2 + 0.090x - 0.0090 = 0$$

You can also solve for x by trial and error, using your hand calculator. Try it, it's easy.

Thus, a = 1, b = 0.090, and c = −0.0090.

$$x = \frac{-0.090 \pm \sqrt{(0.090)^2 + 4(0.0090)}}{2} = \frac{-0.090 \pm \sqrt{0.0441}}{2}$$

$$= \frac{-0.090 \pm 0.21}{2} = \frac{0.12}{2} \text{ or } \frac{-0.30}{2} \text{ (that is, 0.060 or } -0.15)$$

A negative conc would be less than zero, indeed impossible

Of the two answers, only 0.060 is plausible. A value of − 0.15 for x would give a negative concentration for NO_2, which is impossible.

3. $[NO_2] = 2x = 2(0.060 \text{ mol/L}) = 0.120 \text{ mol/L}$

$[N_2O_4] = 0.100 - x = 0.040 \text{ mol/L}$

Compare these values with those listed in Experiment 1, Table 15.2. The calculations we have just carried out refer to that experiment.

EXERCISE Suppose we had chosen our unknown x to be the number of moles per liter of NO_2 formed rather than the number of moles per liter of N_2O_4 that decompose. What would be the algebraic equation obtained in step 2? Answer: $0.36 = x^2/(0.100 - \frac{1}{2}x)$.

Example 15.7 and the exercise that follows illustrate an important point concerning equilibrium problems of this type. There are many different ways in which the unknown can be chosen. It doesn't matter what choice you make, provided you are consistent in relating equilibrium concentrations. If you solve the algebraic equation cited in the exercise:

$$\frac{x^2}{0.100 - \frac{1}{2}x} = 0.36$$

you should find that x = 0.120, so that $[NO_2] = 0.120$ mol/L, $[N_2O_4] = 0.100 - \frac{1}{2}x = 0.100 - 0.060 = 0.040$ mol/L. These are, of course, the same values obtained with a different choice of unknown in Example 15.7.

15.4
Effect of Changes in Conditions upon an Equilibrium System

Once a system has attained equilibrium, it is possible to change the ratio of products to reactants by changing the external conditions. We will consider three ways in which a chemical equilibrium can be disturbed:

1. Adding or removing a gaseous reactant or product.
2. Changing the volume of the system.
3. Changing the temperature.

We can deduce the direction in which an equilibrium will shift when one of these changes is made by applying Le Châtelier's Principle. This states the following:

If a system at equilibrium is disturbed by some change, the system will shift so as to partially counteract the effect of the change.

PLATE 1.1 *Color of Potassium Permanganate.* The purple color of the solution is caused by strong absorption in the green region of the spectrum, as shown at the bottom of this plate.

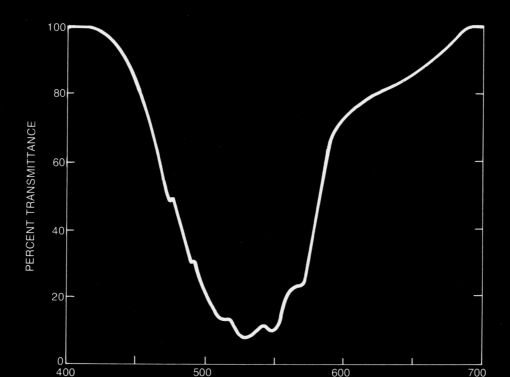

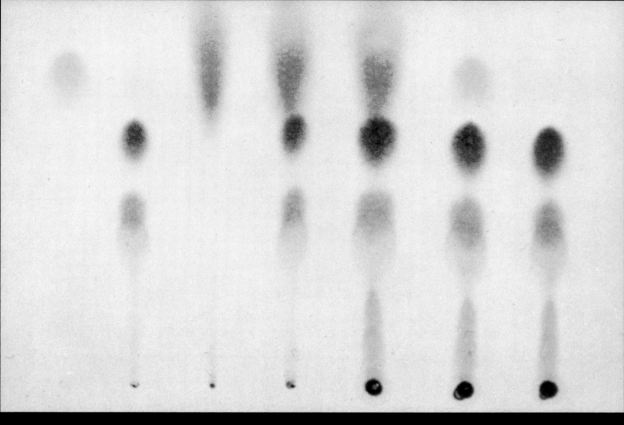

PLATE 1.2 *Paper Chromatography.* Separation of colored dyes. The original spot is shown near the bottom of the plate; the solvent has advanced to the top of the plate.

PLATE 2.1 One mole of carbon (12.01 g), sulfur (32.06 g), and copper (63.55 g). Each sample contains the same number of atoms, 6.022×10^{23}.

PLATE 3.1 $CuSO_4 \cdot 5H_2O$ (upper left) is blue; the anhydrous salt (lower left) is white. $CoCl_2 \cdot 6H_2O$ (upper right) is pink; lower hydrates such as $CoCl_2 \cdot 4H_2O$ are purple (lower right) or blue.

PLATE 4.1   *The Halogens.* Flasks containing Cl_2, Br_2, and I_2 show a gradation in color from greenish-yellow through deep red to violet. The colors shown for bromine and iodine are those of the vapors in equilibrium with $Br_2(l)$ and $I_2(s)$.

PLATE 6.1 At the left, liquid bromine slowly diffuses upward into water. At right, gaseous bromine rapidly diffuses into air.

PLATE 7.1 *Emission Spectra.* The continuous spectrum at the top contains all visible wavelengths and is typical of what is obtained when a solid such as iron is heated. The atomic spectra of gaseous Na, H, Ca, Hg and Ne are quite different. They consist of discrete lines at certain definite wavelengths. (10Å = 1 nm). *From Keenan, C.W., Wood, J.H., and Kleinfelter, D.C., General College Chemistry, 5th ed., New York, Harper & Row.*

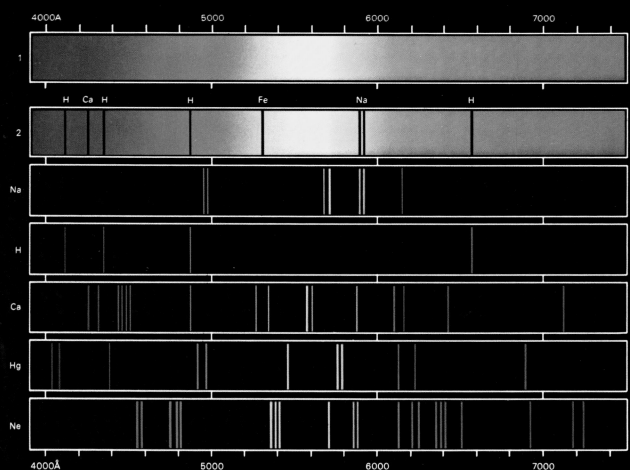

PLATE 8.1 Flame tests for the alkali metals: Li (red), Na (yellow), K (violet).

PLATE 8.2 *Nitrates of Metals of the 4th Period.* Ions having no d electrons (Ca^{2+} at far left) or a full d sublevel (Zn^{2+} at far right) are colorless. In contrast, salts of transition metals with a partially filled

PLATE 9.1 *Physical Properties of Liquid Oxygen.* Oxygen is attracted into a magnetic field; one consequence is that liquid oxygen can be suspended between the poles of an electromagnet. Both the paramagnetism and the blue color are due to the unpaired electrons in the O_2 molecule.

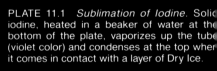

PLATE 11.1 *Sublimation of Iodine.* Solid iodine, heated in a beaker of water at the bottom of the plate, vaporizes up the tube (violet color) and condenses at the top where it comes in contact with a layer of Dry Ice.

PLATE 13.1   When liquid sulfur at 200°C is poured into cold water, a rubbery material called "plastic sulfur" forms. Here, S. Ruven Smith shows some of its (and his) properties.

PLATE 15.1 *Effect of Temperature on the N_2O_4-NO_2 Equilibrium.* At 0°C (tube in ice bath at left), N_2O_4, which is colorless, predominates. At 50°C (tube in water bath at right), some of the N_2O_4 has dissociated to give the deep brown color of NO_2.

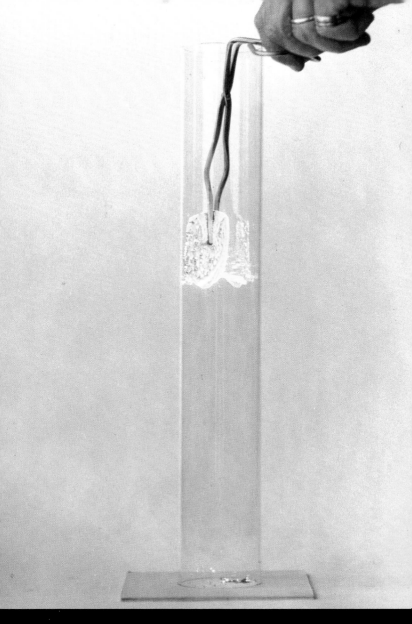

PLATE 16.1 A piece of steel wool, which has a very large surface area, burns spectacularly when heated to redness and immersed in pure oxygen.

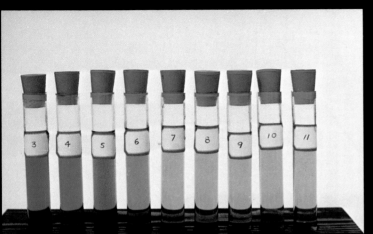

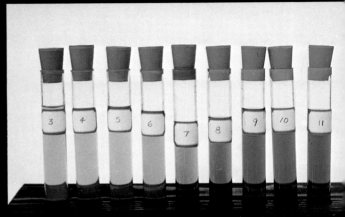

PLATE 19.1 *Color of Indicators As a Function of pH.* The photographs show the colors of methyl red (upper left), bromthymol blue (upper right), and phenolphthalein (below) in solutions ranging in pH from 3 to 11. (The labels were attached by W.L.M. just before an 8 A.M. class.) Note that each indicator is useful only in a narrow pH range of 1 to 2 units.

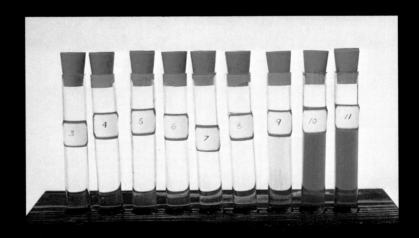

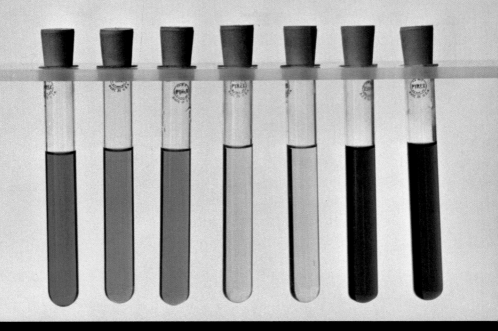

PLATE 19.2   Color of salt solutions with universal indicator, which is red in strong acid (far left) and purple in strong base (far right). NH_4I and $Zn(NO_3)_2$ are weakly acidic at about pH 5 (orange); $KClO_4$, like pure water, has a pH of 7 (yellow-green); Na_3PO_4 is basic at about pH 12 (purple).

PLATE 20.1   The three tubes at the left show the effect of adding a few drops of strong acid or strong base to pure water. The pH changes drastically, giving a pronounced color change with universal indicator. In the three tubes at the right, this experiment is repeated, using a buffer of pH 7 instead of water. This time the pH changes only very slightly, and there is no change in the color of the indicator.

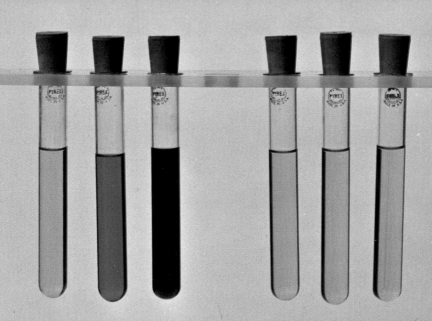

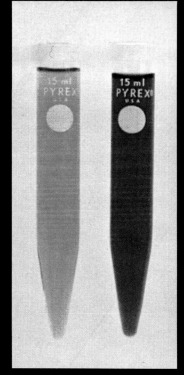

PLATE 21.1   A solution containing the Cu^{2+} Ion (light blue) turns to a deep blue when ammonia is added because of the formation of the $Cu(NH_3)_4^{2+}$ complex.

PLATE 21.2   *Color of Complex Ions of* Co^{3+}. The five compounds at the top form a spectrochemical series (Table 21.5), in which NH_3 molecules are replaced by ligands of lower field strength ($NH_3 > NCS^- > H_2O > Cl^-$). The absorption shifts to higher wavelengths with a corresponding change in color.

PLATE 21.3 *Rate of Ligand Substitution.* These solutions were prepared by dissolving *trans*-[Co(en)$_2$Cl$_2$]Cl in water at approximately 10-minute intervals. The tube at the far left shows the characteristic green color of the *trans*-Co(en)$_2$Cl$_2$$^+$ cation. As time passes, this is replaced by the red color of aquo complexes such as Co(en)$_2$H$_2$OCl^{2+} and Co(en)$_2$(H$_2$O)$_2$$^{3+}$. The overall reaction may be represented as:

$$Co(en)_2Cl_2^+ \ (aq) \ + \ 2 \ H_2O \longrightarrow Co(en)_2(H_2O)_2^{3+} \ (aq) \ + \ 2 \ Cl \ (aq)$$
$$\text{green} \hspace{4cm} \text{red}$$

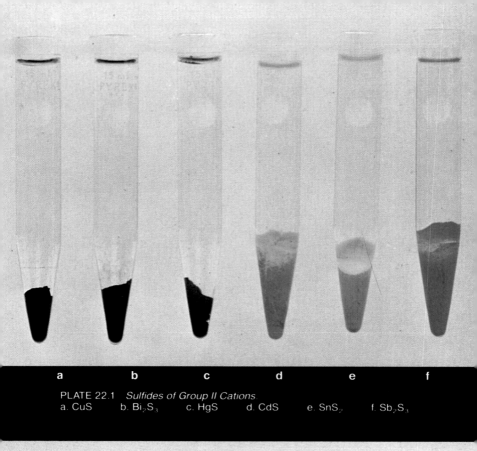

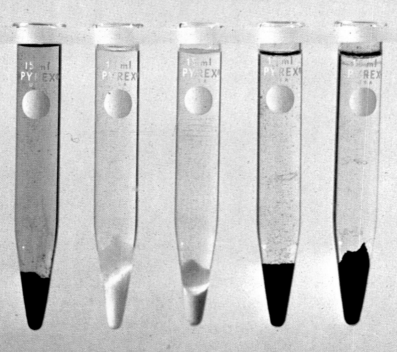

PLATE 22.1   *Sulfides of Group II Cations*
a. CuS   b. Bi$_2$S$_3$ c. HgS   d. CdS   e. SnS$_2$ f. Sb$_2$S$_3$

PLATE 23.1   When a strip of zinc is placed in a solution containing Cu^{2+} ions (left), an oxidation-reduction reaction occurs. The final result is shown at the right. Copper metal plates out and the blue color due to Cu^{2+} fades.

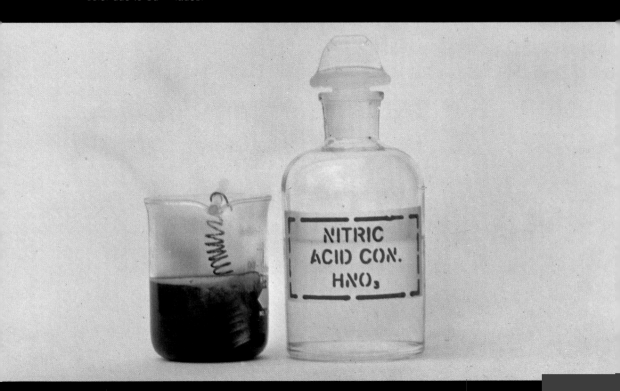

wn gas formed

PLATE 25.1 *Oxidation States of Chromium.* In the $+6$ state, chromium forms two oxyanions, $Cr_2O_7^{2-}$ (red) and CrO_4^{2-} (yellow). Compounds of Cr^{3+} are most often purple due to the presence of ions such as $Cr(H_2O)_6^{3+}$. Solutions of $CrCl_3$ are an exception; species such as $Cr(H_2O)_5Cl^{2+}$ and $Cr(H_2O)_4Cl_2^+$ shift the color to green. Compounds of Cr^{2+} (far right) are blue when freshly prepared but are rapidly oxidized to Cr^{3+} by contact with air.

PLATE 26.1 *Oxidation of Br and I by Cl$_2$.* The tube at the far left contains a colorless solution of KBr in water. Addition of chlorine brings about the oxidation-reduction reaction:

$$Cl_2(aq) + 2\,Br^-(aq) \longrightarrow Br_2(aq) + 2\,Cl^-(aq)$$

giving rise to the light color of dissolved bromine seen in the second tube. This color is intensified (third tube) by shaking with a little CCl_4, in which Br_2 is more soluble.

Tubes 4 to 6 show a similar sequence, starting with KI solution (colorless), oxidizing with Cl_2 forming iodine

$$Cl_2(aq) + 2\,I^-(aq) \longrightarrow I_2(aq) + 2\,Cl^-(aq)$$

(pale red) and concentrating the color by adding CCl_4 (deep violet).

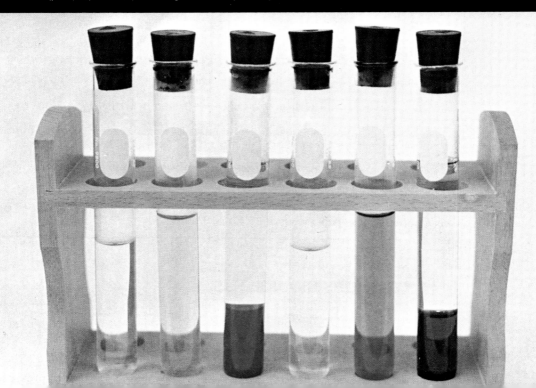

We can determine the extent of the shift by working with the equilibrium constant, K_c. In the discussion that follows, we will illustrate both of these approaches.

Adding or Removing a Gaseous Species

According to Le Châtelier's Principle, if we disturb a chemical system at equilibrium by adding a gaseous species (reactant or product), the reaction will proceed in such a direction as to consume part of the added species. Conversely, if we remove a gaseous species, the system will shift so as to restore part of that species.

Let us apply this general rule to the equilibrium system

$$N_2O_4(g) \rightleftharpoons 2\ NO_2(g)$$

Suppose this system has reached equilibrium at a certain temperature. We might disturb the equilibrium by

—*adding N_2O_4*. If we do this, reaction will occur in the forward direction (left to right). In this way, part of the N_2O_4 will be consumed. →

—*adding NO_2*, which causes the reverse reaction (right to left) to occur, using up part of the NO_2 added. ←

—*removing N_2O_4*. Here, reaction occurs in the reverse direction to restore part of the N_2O_4. ←

—*removing NO_2*, which causes the forward reaction to occur, restoring part of the NO_2 removed. →

It is possible to use K_c to calculate the extent to which reaction occurs when an equilibrium is disturbed by adding or removing a product or reactant. To show how this is done, we consider the effect of adding hydrogen iodide to the HI–H_2–I_2 system (Example 15.8):

$$2\ HI(g) \rightleftharpoons H_2(g) + I_2(g); \qquad K_c = \frac{[H_2] \times [I_2]}{[HI]^2} = 0.016 \text{ at } 520°C$$

EXAMPLE 15.8 In Example 15.3 we saw that this system is at equilibrium at 520°C when $[HI] = 0.080$ mol/L and $[H_2] = [I_2] = 0.010$ mol/L. Suppose that to this mixture we add enough HI to raise its concentration temporarily to 0.096 mol/L. When equilibrium is restored, what will $[HI]$, $[H_2]$, and $[I_2]$ be?

Solution

1. Here, the "original concentrations" are those that prevail immediately after the equilibrium is disturbed:

   orig. conc. HI = 0.096 mol/L;
   orig. conc. H_2 = orig. conc. I_2 = 0.010 mol/L

Since we added HI, some of it must decompose to bring the system back to equilibrium. When that happens, some H_2 and some I_2 are formed. Let x be the number of moles per liter of H_2 that is produced by this process. Since H_2 and I_2 are formed in a 1:1 mole ratio, the concentration of I_2 must also increase by x. Since two moles of HI are required to form one mole of H_2, the concentration of HI must decrease by 2x. The equilibrium table is

	ORIG. CONC. (mol/L)	CHANGE IN CONC. (mol/L)	EQUIL. CONC. (mol/L)
H_2	0.010	+x	0.010 + x
I_2	0.010	+x	0.010 + x
HI	0.096	−2x	0.096 − 2x

2. Substituting into the expression for K_c,

$$0.016 = \frac{(0.010 + x)(0.010 + x)}{(0.096 - 2x)^2}$$

To solve this equation, we first take the square root of both sides. Since $(0.016)^{1/2} = 0.13$, we have

$$0.13 = \frac{0.010 + x}{0.096 - 2x}$$

which solves to give x = 0.002.

3. Thus,

$$[H_2] = 0.010 + 0.002 = 0.012 \text{ mol/L}$$
$$[I_2] = 0.010 + 0.002 = 0.012 \text{ mol/L}$$
$$[HI] = 0.096 - 0.004 = 0.092 \text{ mol/L}$$

When HI was added, reaction occurred to use some of it up

Note that the equilibrium concentration of HI, 0.092 mol/L, is greater than it was originally (0.080 mol/L), but less than it was immediately after equilibrium was disturbed (0.096 mol/L).

EXERCISE Suppose that, to the original equilibrium mixture, we add enough HI to raise its concentration temporarily to 0.200 mol/L. Which of the following would be a reasonable value for the final concentration of HI: 0.060 M, 0.080 M, 0.175 M, 0.200 M, or 0.225 M? Answer: 0.175 M. (The other answers are impossible. Why?)

Notice that throughout this discussion, we have referred to the effect of adding or removing a gaseous species. If we add a pure solid or liquid to an equilibrium system, nothing happens! Consider, for example, the system

$$CaCO_3(s) \rightleftharpoons CaO(s) + CO_2(g)$$

We might establish equilibrium at 840°C with 0.100 mol $CaCO_3$, 0.100 mol CaO, and 0.011 mol CO_2 in a one-liter container. The equilibrium constant at this temperature is

$$K_c = [CO_2] = 0.011$$

Suppose now that we add some $CaCO_3$ to this equilibrium system. None of it reacts. If it did, it would produce CO_2, changing the concentration of that product.

That is impossible; the equilibrium concentration of CO_2 at 840°C is fixed at 0.011 mol/L. The same argument can be used to show that the position of this equilibrium would not be affected by adding CaO, removing $CaCO_3$, or removing CaO. In general,

> *Adding or removing a species disturbs an equilibrium system only if the concentration of that species appears in the expression for the equilibrium constant.*

Adding N_2 or He wouldn't have any effect on the $CaCO_3$—CaO—CO_2 equil system

Changes in Volume

To understand how a change in container volume can change the position of an equilibrium, consider again the N_2O_4–NO_2 system

$$N_2O_4(g) \rightleftharpoons 2\ NO_2(g)$$

Suppose we decrease the volume of this system, as shown in Figure 15.2. The immediate effect is to increase the number of molecules per unit volume. According to Le Châtelier's Principle, the system will shift so as to partially counteract this change. There is a simple way that this can happen. Some of the NO_2 molecules combine with each other to form N_2O_4 (diagram at right of Fig. 15.2). That is, reaction occurs in the reverse direction. Since two molecules of NO_2 form only one molecule of N_2O_4, this reduces the number of molecules.

You can also approach this problem by evaluating Q

It is possible, using the value of K_c, to calculate the extent to which NO_2 is converted to N_2O_4 when the volume is decreased. The results of such calculations are given in Table 15.3. As the volume is decreased from 10.0 to 1.0 L, more and more of the NO_2 is converted to N_2O_4. Notice that the total number of moles of gas decreases steadily as a result of this conversion.

The analysis we have gone through for the N_2O_4–NO_2 system can be applied

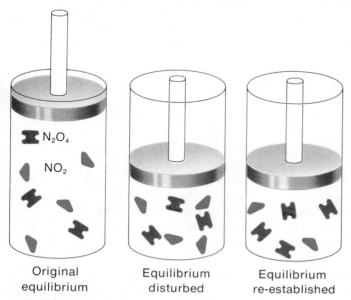

| Original equilibrium | Equilibrium disturbed | Equilibrium re-established |

FIGURE 15.2 Effect of a decrease in volume upon the $N_2O_4(g) \rightleftharpoons 2\ NO_2(g)$ system at equilibrium. The immediate effect (*middle cylinder*) is to crowd the same number of moles of gas into a smaller volume and so increase the total pressure. This is partially compensated for by the conversion of some of the NO_2 to N_2O_4, thereby reducing the total number of moles of gas present.

Table 15.3
Effect of Change in Volume on the Equilibrium System:
$N_2O_4(g) \rightleftharpoons 2\ NO_2(g)$; $K_c = 0.36$ at 100°C

V (L)	MOLES NO_2	MOLES N_2O_4	MOLES TOTAL
10.0	1.20	0.40	1.60
5.0	0.96	0.52	1.48
2.0	0.68	0.66	1.34
1.0	0.52	0.74	1.26

to any chemical equilibrium involving gases. **When the volume of an equilibrium system is decreased, reaction takes place in the direction that decreases the total number of moles of gas. When the volume is increased, the reaction that increases the total number of moles of gas takes place.**

The application of this principle to several different systems is shown in Table 15.4. In System 2, the number of moles of gas decreases, from $\frac{3}{2}$ to 1, as the reaction goes to the right. Hence, decreasing the volume causes the forward reaction to occur; an increase in volume has the reverse effect. Notice that it is the change in the number of moles of *gas* that determines which way the equilibrium shifts (System 4). When there is no change in the number of moles of gas (System 5), a change in volume has no effect on the position of the equilibrium.

Table 15.4
Effect of a Change in Volume upon the Position of Gaseous Equilibria

	SYSTEM	V INCREASES	V DECREASES
Δn_{gas}			
+1	1. $N_2O_4(g) \rightleftharpoons 2\ NO_2(g)$	→	←
$-\frac{1}{2}$	2. $SO_2(g) + \frac{1}{2} O_2(g) \rightleftharpoons SO_3(g)$	←	→
-2	3. $N_2(g) + 3\ H_2(g) \rightleftharpoons 2\ NH_3(g)$	←	→
+1	4. $C(s) + H_2O(g) \rightleftharpoons CO(g) + H_2(g)$	→	←
0	5. $N_2(g) + O_2(g) \rightleftharpoons 2\ NO(g)$	0	0

Changes in volume of gaseous systems at equilibrium ordinarily result in changes in pressure. For example, if we cut the volume of the N_2O_4–NO_2 system in half, we momentarily increase the pressure by a factor of two. Instead of saying that the shift in the position of the equilibrium comes about because of a change in volume, we might ascribe the shift to the pressure change that accompanies the volume decrease. Specifically, we could say that an *increase in pressure shifts the position of the equilibrium in such a way as to decrease the number of moles of gas* ($2\ NO_2 \rightarrow N_2O_4$).

When in doubt, evaluate Q and compare it to K_c

In discussing the effect of pressure on an equilibrium system, we must be careful to specify how the pressure change comes about. There are many different ways in which we could change the total pressure in the N_2O_4–NO_2 system without changing the volume. We might, for example, add helium at constant volume. This increases the total number of moles and hence the total pressure. It has no effect, however, upon the position of the equilibrium. In general, we do not shift the position of an equilibrium by adding an unreactive gas, since that does not change the concentrations of reactants or products.

EXAMPLE 15.9   The pressure on each of the following systems is reduced so that the volume increases from 2 L to 10 L. Which way does the equilibrium shift?

a. $2 CO_2(g) \rightleftharpoons 2 CO(g) + O_2(g)$
b. $H_2(g) + I_2(g) \rightleftharpoons 2 HI(g)$
c. $H_2(g) + I_2(s) \rightleftharpoons 2 HI(g)$

Solution

a. → (2 mol gas → 3 mol gas). The immediate effect of the volume increase is to decrease the number of molecules per unit volume. This is partially compensated for by increasing the number of molecules.
b. no effect: same number of moles of gas on both sides.
c. → (1 mol gas → 2 mol gas). Note that it is the number of moles of gas that is important; solids or liquids don't count.

EXERCISE   To increase the yield of NO_2 in the reaction

$$NO(g) + \tfrac{1}{2} O_2(g) \rightleftharpoons NO_2(g)$$

would you compress the system, expand the system, or add an inert gas? Answer: Compress it.

Changes in Temperature

Let us suppose that we increase the temperature of an equilibrium mixture of N_2O_4 and NO_2. According to Le Châtelier's Principle, the system will shift so as to partially counteract the temperature increase. This can be achieved if the forward reaction occurs:

$$N_2O_4(g) \rightarrow 2 NO_2(g); \quad \Delta H = +58.2 \text{ kJ}$$

Since the reaction is endothermic, it absorbs some of the heat used to raise the temperature. The result is to increase the concentration of NO_2 at the expense of N_2O_4. This effect brings about a color change. The reddish brown color of NO_2 is much deeper at the higher temperature (Color Plate 15.1).

In general, we can say that, for a system at equilibrium,

—**an increase in temperature causes the endothermic reaction to occur; e.g.,**

$$N_2O_4(g) \rightarrow 2 NO_2(g); \quad \Delta H = +58.2 \text{ kJ}$$

→ soaks up heat if $\Delta H > 0$

—**a decrease in temperature causes the exothermic reaction to occur; e.g.,**

$$2 NO_2(g) \rightarrow N_2O_4(g); \quad \Delta H = -58.2 \text{ kJ}$$

→ evolves heat if $\Delta H < 0$

You will recall from our earlier discussion that the equilibrium constant of a system changes with temperature. The effect of a change in temperature upon an equilibrium system is often expressed in terms of its effect upon K_c:

If the forward reaction (left to right) is endothermic, K_c becomes larger as *the temperature increases. If the forward reaction is exothermic, K_c becomes smaller as the temperature increases.*

The effect of temperature upon the magnitude of K_c is often very large. For the synthesis of ammonia from the elements, K_c changes by a factor of 50,000 when the temperature is raised from 200 to 600°C (Table 15.5).

Table 15.5
Effect of Temperature upon K_c

1. $N_2O_4(g) \rightleftharpoons 2\ NO_2(g)$; $\Delta H = +58.2$ kJ

t (°C)	0	50	100
K_c	0.0005	0.022	0.36

2. $N_2(g) + 3\ H_2(g) \rightleftharpoons 2\ NH_3(g)$; $\Delta H = -92.4$ kJ

t (°C)	200	400	600
K_c	650	0.50	0.014

EXAMPLE 15.10   Consider the system $I_2(g) \rightleftharpoons 2\ I(g)$; $\Delta H = +151$ kJ. Suppose the system is at equilibrium at 1000°C. In which direction will reaction occur if

a. I atoms are added?
b. the system is compressed?
c. the temperature is increased?

Solution
a. $2\ I(g) \rightarrow I_2(g)$
b. $2\ I(g) \rightarrow I_2(g)$ This decreases the number of particles per unit volume.
c. $I_2(g) \rightarrow 2\ I(g)$ This reaction absorbs heat.

EXERCISE   What effect, if any, will each of these changes (a, b, c) have on the magnitude of K_c? Answer: Only (c) changes K_c, making it larger.

Since decomposition reactions usually have $\Delta H > 0$, high temperatures make big molecules break up into smaller ones

15.5
K_c Versus K_p

Throughout this chapter, we have dealt with the type of equilibrium constant designated as K_c. As we have pointed out repeatedly, in the expression for K_c, equilibrium concentrations are expressed in moles per liter (molarity). There is another type of equilibrium constant in common use with gases. This constant is given the symbol K_p. In the expression for K_p, equilibrium concentrations appear as partial pressures.

These two types of equilibrium constants have many properties in common. In particular, we find the following:

1. The mathematical expression for K_p for a reaction is identical to that for K_c except that partial pressures replace molarities. Thus, we have

$$N_2O_4(g) \rightleftharpoons 2\ NO_2(g); \qquad K_c = \frac{[NO_2]^2}{[N_2O_4]}; \qquad K_p = \frac{(P_{NO_2})^2}{P_{N_2O_4}}$$

$$N_2(g) + 3\ H_2(g) \rightleftharpoons 2\ NH_3(g); \quad K_c = \frac{[NH_3]^2}{[N_2] \times [H_2]^3}; \quad K_p = \frac{(P_{NH_3})^2}{P_{N_2} \times (P_{H_2})^3}$$

Here, as always, the square brackets represent equilibrium concentrations in moles per liter. The symbol P in the expression for K_p represents an *equilibrium partial pressure in atmospheres*.

2. Terms for pure liquids or solids do not appear in either K_p or K_c:

$$CaCO_3(s) \rightleftharpoons CaO(s) + CO_2(g); \quad K_c = [CO_2]; \quad K_p = P_{CO_2}$$

3. K_p, like K_c, is independent of the starting amounts of reactants or products, the volume of the container, or the total pressure.

4. K_p, like K_c, varies with temperature. Earlier, we pointed out that for the system

$$N_2O_4(g) \rightleftharpoons 2\ NO_2(g)$$

$K_c = 0.36$ at 100°C and 3.2 at 150°C. The equilibrium constant K_p for this system is 11 at 100°C and 110 at 150°C.

As you might expect, K_p and K_c have different numerical values at a given temperature. To relate K_p to K_c, we start by relating partial pressure, P, to concentration in moles per liter, n/V. For a gas A, we can express the partial pressure using the Ideal Gas Law:

K_p and K_c must be related, since either can be used to describe an equilibrium system

$$P_A = \frac{n_A RT}{V}$$

where R is the gas constant, 0.0821 L·atm/(mol·K), and T is the Kelvin temperature:

$$P_A = \frac{n_A \times 0.0821T}{V}$$

At equilibrium, n_A/V becomes, by definition, [A], and we have

$$P_A = [A] \times 0.0821T \tag{15.8}$$

To show how Equation 15.8 can be used to relate K_p to K_c, consider the N_2O_4–NO_2 system:

$$N_2O_4(g) \rightleftharpoons 2\ NO_2(g); \quad K_p = \frac{(P_{NO_2})^2}{P_{N_2O_4}}$$

Substituting for partial pressures, using Equation 15.8,

$$K_p = \frac{[NO_2]^2 \times (0.0821T)^2}{[N_2O_4] \times 0.0821T} = \frac{[NO_2]^2 \times 0.0821T}{[N_2O_4]}$$

However, $K_c = [NO_2]^2/[N_2O_4]$, so that for this system,

$$K_p = K_c \times 0.0821T$$

We could go through a similar argument to relate K_p to K_c for any gaseous system. It is simpler, however, to work with the following general equation, valid for all systems:

$$K_p = K_c \times (0.0821T)^{\Delta n_g} \qquad (15.9)$$

The exponent Δn_g in this equation is the change in the number of moles of gas in the equation written to represent the equilibrium system. Thus, we have

$$N_2O_4(g) \rightleftharpoons 2\ NO_2(g) \qquad\qquad \Delta n_g = +1$$

$$N_2(g) + 3\ H_2(g) \rightleftharpoons 2\ NH_3(g) \qquad \Delta n_g = -2$$

$$2\ HI(g) \rightleftharpoons H_2(g) + I_2(g) \qquad\qquad \Delta n_g = 0$$

$$CaCO_3(s) \rightleftharpoons CaO(s) + CO_2(g) \qquad \Delta n_g = +1$$

EXAMPLE 15.11   At 300°C, K_c for the system

$$N_2(g) + 3\ H_2(g) \rightleftharpoons 2\ NH_3(g)$$

is 9.5. Calculate K_p for this system at the same temperature.

Solution $\Delta n_g = -2;$ $T = 573$ K;

$$K_p = K_c \times (0.0821 \times 573)^{-2} = \frac{9.5}{(0.0821 \times 573)^2} = 4.3 \times 10^{-3}$$

EXERCISE   At 520°C, K_c for the system $2\ HI(g) \rightleftharpoons H_2(g) + I_2(g)$ is 0.016. What is the value of K_p at 520°C? Answer: 0.016.

15.6
Relation Between ΔG^0 and the Equilibrium Constant

In Chapter 14 we pointed out that the free energy change, ΔG, is the basic criterion of spontaneity. A reaction occurs spontaneously if ΔG is a negative quantity. In this chapter, we discussed the extent to which a reaction takes place in terms of the equilibrium constant, K. As you might expect, these quantities are intimately related. By a thermodynamic argument that we will not go through here, it can be shown that

ln K is the natural log of K

$$\Delta G^0 = -RT \ln K$$

where ΔG^0 is the standard free energy change (1 atm for gases, 1 M for ions in solution), R is our old friend the gas constant, T is the absolute temperature, and K is the equilibrium constant. Substituting $R = 8.31 \times 10^{-3}$ (kilojoules per kelvin) and $\ln K = 2.30 \log_{10} K$,

$$\Delta G^0 \text{ (in kilojoules)} = -2.30(8.31 \times 10^{-3})T \log_{10} K$$
$$= -0.0191\ T \log_{10} K \qquad (15.10)$$

Looking at Equation 15.10, we can distinguish three possible situations:

1. If ΔG^0 is negative, $\log_{10} K$ must be positive. This means that K is greater than one. The reaction proceeds spontaneously in the forward direction when all species are at unit concentrations.

2. If ΔG^0 is positive, $\log_{10} K$ must be negative. Hence K is less than one and the reverse reaction is spontaneous when all species are at unit concentrations.

3. If, perchance, $\Delta G^0 = 0$, $\log_{10} K = 0$, $K = 1$. The reaction is at equilibrium when all species are at unit concentrations.

Equation 15.10 also allows us to give meaning to the magnitude of the free energy change. If ΔG^0 is a large negative number, K will be much greater than one. This means that the forward reaction will go virtually to completion. Conversely, if ΔG^0 is a large positive number, K will be a tiny fraction, much less than one, and the reverse reaction will go nearly to completion.

> If $\Delta G^0 \ll 0$, K is very big
>
> If $\Delta G^0 \gg 0$, $K \cong 0$

EXAMPLE 15.12 Calculate K for reactions that, at 25°C, have ΔG^0 values of

a. -40.0 kJ
b. $+40.0$ kJ

Solution We substitute in Equation 15.10 with T = 298 K:

a. $-40.0 = -0.0191(298) \log_{10} K$

$$\log_{10} K = \frac{40.0}{0.0191 \times 298} = 7.03; \qquad K = 1.1 \times 10^7$$

b. $\log_{10} K = \dfrac{-40.0}{0.0191 \times 298} = -7.03; \qquad K = 9 \times 10^{-8}$

EXERCISE What must be the value of ΔG^0 for K to be 1.0×10^{10} at 100°C? 1.0×10^{-10} at 100°C? Answer: -71 kJ; $+71$ kJ.

Throughout this discussion we have used the general symbol "K" to stand for the equilibrium constant. We did this for a simple reason. In succeeding chapters we will refer to several different equilibrium constants (for example, K_w, K_a, K_b, K_{sp}, K_f) used for different types of equilibria involving ions in aqueous solution. Equation 15.10 applies to each one of these K values, regardless of subscript.

> These are all forms of K

As we saw in Section 15.5, however, two different equilibrium constants, K_c and K_p, can be used to describe a given gas-phase equilibrium. We find that Equation 15.10 applies to K_p; that is,

gas-phase reactions: $\qquad \Delta G^0 = -0.0191 \, T \log_{10} K_p \qquad\qquad$ **(15.11)**

In this equation, ΔG^0 is the free energy change when all gases are at a partial pressure of one atmosphere and K_p is the equilibrium constant involving partial pressures.

To illustrate the use of Equation 15.11 in a simple case, consider the reaction

$$CaCO_3(s) \rightleftharpoons CaO(s) + CO_2(g)$$

In Chapter 14, we showed that ΔG^0 for this reaction is zero at 1110 K. It follows that

Put another way, $\Delta G^0 = -RT \ln K$, always, if concentrations of solutes are molarities and concentrations of gases are in atm

$$\log_{10} K_p = \frac{-\Delta G^0}{0.0191 \times 1110} = 0$$

$$K_p = 1 = P_{CO_2}; \qquad P_{CO_2} = 1 \text{ atm}$$

This means that the equilibrium pressure of CO_2 in this system is 1 atm at 1110 K. If we wanted to calculate K_c for this reaction at 1110 K, we could do so using Equation 15.9:

$$K_p = K_c(0.0821T)^{\Delta n_g} = K_c(0.0821T)^1$$

$$K_c = \frac{1}{0.0821 \times 1110} = 1.1 \times 10^{-2} = [CO_2]$$

In other words, the equilibrium concentration of CO_2 at 1110 K is about 0.011 mol/L.

Summary

When a reaction system is confined, it reaches a dynamic equilibrium where forward and reverse reactions occur at the same rate. From that point on, there is no net change in the concentrations of products or reactants. The equilibrium system can be described in terms of a ratio known as the equilibrium constant, K_c. The expression for K_c follows directly from the equation written to represent the equilibrium system (Section 15.2 and Example 15.1). The magnitude of K_c governs the extent to which forward and reverse reactions will take place. If K_c is very large, the forward reaction goes nearly to completion, while the reverse reaction does not occur to an appreciable extent. If K_c is very small, the opposite is true; the reverse reaction dominates.

To decide the direction in which a reaction system will move to establish equilibrium, we compare the original concentration quotient to the equilibrium constant, K_c. If the original concentration quotient, Q, is less than K_c, reaction occurs in the forward direction (Example 15.4). If this quotient is greater than K_c, the reverse reaction must occur to establish equilibrium.

Several calculations involving equilibrium systems were considered in this chapter. Generally, they are of three types, depending upon what is known and what is to be calculated.

KNOWN	TO BE CALCULATED	EXAMPLE
1. Equilibrium concentrations of all species	K_c	15.2
2. Initial concentrations of all species, equilibrium concentration of one species	K_c	15.3

KNOWN	TO BE CALCULATED	EXAMPLE
3. K_c and all but one equilibrium concentration	equilibrium concentration of a species	15.5
4. K_c and initial concentrations	equilibrium concentrations of all species	15.6, 15.7, 15.8

If a system at equilibrium is disturbed in some way, either the forward or the reverse reaction will occur to restore equilibrium. We can use Le Châtelier's Principle to predict which way the equilibrium will shift under stress (Examples 15.9 and 15.10). In general, we find that

—a decrease in volume (compression) causes the reaction to occur that decreases the number of moles of gas (Tables 15.3 and 15.4).
—an increase in temperature causes the reaction that is endothermic to occur; in this case, the magnitude of K_c changes (Table 15.5).

The equilibrium constant K_c is related in a simple way to the equilibrium constant K_p (Equation 15.9 and Example 15.11). The quantity K_p in turn is related to the standard free energy change, ΔG^0, for the reaction by Equation 15.11.

Key Words and Concepts

[ ] equilibrium concentration, mol/L	Le Châtelier's Principle	quadratic formula
equilibrium constant, K_c	mole	standard free energy change, ΔG^0
equilibrium constant, K_p	partial pressure	

Questions and Problems

Establishment of Equilibrium

15.1 The following data are for the system

$$A(g) \rightleftharpoons 2 B(g)$$

Time (s)	0	30	60	90	120
Conc. A (M)	0.200	0.150	0.120	0.100	0.100
Conc. B (M)	0.000	0.100	0.160	0.200	0.200

At which time(s) is the system at equilibrium?

15.2 Complete the table below for the reaction

$$A(g) + B(g) \rightleftharpoons 2 C(g)$$

Time (s)	0	20	40	60	80	100
Conc. A	0.080	0.050	___	___	0.015	___
Conc. B	0.100	___	___	___	0.040	___
Conc. C	0.000	___	0.100	___	___	0.130

At what time is equilibrium established?

15.31 For the system in Question 15.1,
a. What is the concentration of A after 3 min?
b. How does the rate of the forward reaction compare to that of the reverse reaction at 60 s? 120 s?

15.32 The following data apply to the *unbalanced* equation

$$A(g) \rightleftharpoons B(g)$$

Time (min)	0	2.0	4.0	6.0	8.0
Conc. A	0.360	0.270	0.210	0.170	0.140
Conc. B	0.000	0.180	0.300	0.380	0.440

a. Based on these data, balance the equation.
b. Has this system reached equilibrium? Explain.

Expressions for K_c

15.3 Write equilibrium constant expressions for
a. $2 SO_2(g) + O_2(g) \rightleftharpoons 2 SO_3(g)$
b. $3 O_2(g) + 2 H_2S(g) \rightleftharpoons 2 H_2O(g) + 2 SO_2(g)$
c. $2 NO_2(g) + 7 H_2(g) \rightleftharpoons 2 NH_3(g) + 4 H_2O(g)$

15.4 Write the K_c expression for
a. $Fe(s) + 5 CO(g) \rightleftharpoons Fe(CO)_5(g)$
b. $6 F_2(g) + 2 NH_3(g) \rightleftharpoons 2 NF_3(g) + 6 HF(g)$
c. $Mg_3N_2(s) + 6 H_2O(g) \rightleftharpoons 3 Mg(OH)_2(s) + 2 NH_3(g)$
d. $3 Mg(OH)_2(s) + 2 NH_3(g) \rightleftharpoons Mg_3N_2(s) + 6 H_2O(l)$

15.5 Write a chemical equation for an equilibrium system that would lead to each of the following expressions for K_c
a. $\dfrac{[H_2]^2 \times [S_2]}{[H_2S]^2}$ b. $\dfrac{[HI]^2}{[H_2] \times [I_2]}$
c. $\dfrac{[HI]^2}{[H_2]}$ d. $\dfrac{[CH_4] \times [H_2S]^2}{[CS_2] \times [H_2]^4}$

15.6 Given

$2 CO_2(g) \rightleftharpoons 2 CO(g) + O_2(g); K_c = 26$

$N_2(g) + 3 H_2(g) \rightleftharpoons 2 NH_3(g); K_c = 2 \times 10^{-5}$

determine the equilibrium constant at the same temperature for
a. $CO_2(g) \rightleftharpoons CO(g) + \frac{1}{2} O_2(g)$
b. $2 NH_3(g) \rightleftharpoons N_2(g) + 3 H_2(g)$

15.33 Write the expression for K_c for
a. $2 NO(g) + SO_2(g) \rightleftharpoons N_2O(g) + SO_3(g)$
b. $SO_2(g) + \frac{1}{2} O_2(g) \rightleftharpoons SO_3(g)$
c. $O_2(g) + 4 HCl(g) \rightleftharpoons 2 H_2O(g) + 2 Cl_2(g)$

15.34 Write the K_c expression for
a. $BaCO_3(s) \rightleftharpoons BaO(s) + CO_2(g)$
b. $CO(g) + 3 H_2(g) \rightleftharpoons CH_4(g) + H_2O(g)$
c. $Sn(s) + 2 Cl_2(g) \rightleftharpoons SnCl_4(g)$
d. $CO(g) + 3 H_2(g) \rightleftharpoons CH_4(g) + H_2O(l)$

15.35 Follow the directions for Question 15.5 for
a. $\dfrac{[Br_2] \times [Cl_2]}{[BrCl]^2}$ b. $\dfrac{[NO] \times [O_2]^{1/2}}{[NO_2]}$
c. $\dfrac{[N_2] \times [O_2] \times [Br_2]}{[NOBr]^2}$ d. $\dfrac{[H_2O]^2 \times [Cl_2]^2}{[HCl]^4 \times [O_2]}$

15.36 Given

$NOBr(g) \rightleftharpoons NO(g) + \frac{1}{2} Br_2(g); K_c = 0.71$

$2 N_2O(g) + 3 O_2(g) \rightleftharpoons 2 N_2O_4(g); K_c = 1 \times 10^6$

calculate K_c for
a. $2 NOBr(g) \rightleftharpoons 2 NO(g) + Br_2(g)$
b. $N_2O(g) + \frac{3}{2} O_2(g) \rightleftharpoons N_2O_4(g)$

Calculation of K_c

15.7 Calculate K_c at 1500°C for

$2 NO(g) \rightleftharpoons N_2(g) + O_2(g)$

given that the equilibrium concentrations of N_2, O_2, and NO are 0.040 M, 0.040 M, and 0.00035 M, respectively.

15.8 For the reaction in Problem 15.7 at a different temperature, if one starts with pure NO at 4.00 mol/L, the equilibrium concentrations of N_2 and O_2 are each 1.28 mol/L. Calculate K_c at this temperature.

15.9 When 1.36 mol H_2 and 0.78 mol CO are put in a one-liter sealed container at 160°C, the following equilibrium is established:

$CO(g) + 2 H_2(g) \rightleftharpoons CH_3OH(g)$

The equilibrium concentration of H_2 is 0.120 mol/L. Calculate
a. [CO]   b. $[CH_3OH]$ c. K_c

15.37 For the system

$2 NO(g) + Br_2(g) \rightleftharpoons 2 NOBr(g)$

analysis shows that at equilibrium at 350°C, the concentrations of NO, Br_2, and NOBr are 0.24 M, 0.11 M, and 0.037 M, respectively. Calculate K_c.

15.38 The equilibrium in Problem 15.37 is studied at a different temperature. Starting with NO and Br_2 each at 0.50 mol/L, the equilibrium concentration of NOBr is found to be 0.14 mol/L. Calculate K_c at this temperature.

15.39 Starting with 0.388 mol/L of NOCl at 220°C, the reaction

$2 NOCl(g) \rightleftharpoons 2 NO(g) + Cl_2(g)$

produces 0.0200 mol/L of Cl_2 at equilibrium. Calculate
a. [NO]   b. [NOCl]   c. K_c

K_c; Direction and Extent of Reaction

15.10 K_c is 0.56 at 300°C for the system

$$PCl_5(g) \rightleftharpoons PCl_3(g) + Cl_2(g)$$

In a 5.0-L flask, a gaseous mixture consists of 0.45 mol Cl_2, 0.90 mol PCl_3, and 0.12 mol PCl_5.
a. Is the mixture at equilibrium? Explain.
b. If not at equilibrium, which way will the system shift to establish equilibrium?

15.11 At a certain temperature, K_c is 16 for the reaction

$$2 SO_2(g) + O_2(g) \rightleftharpoons 2 SO_3(g)$$

Predict in which direction the system will move to reach equilibrium if we start with
a. 0.850 mol SO_3 in a 3.0-L container.
b. a gaseous mixture of 0.24 mol SO_2, 0.40 mol O_2, and 0.60 mol SO_3 in a 4.0-L container.

15.12 For the reaction

$$Br_2(g) \rightleftharpoons 2 Br(g)$$

$K_c = 4 \times 10^{-18}$ at 200°C. If one starts with 1.0 mol Br_2 in a 2.0-L container at 200°C, will the equilibrium system contain:
a. mostly Br_2?   b. mostly Br atoms?
c. about equal amounts of Br_2 and Br?

K_c; Equilibrium Concentrations

15.13 At 350 K, K_c is 0.14 for the reaction

$$2 BrCl(g) \rightleftharpoons Br_2(g) + Cl_2(g)$$

An equilibrium mixture at this temperature contains equal concentrations of bromine and chlorine, 0.0250 mol/L. What is the equilibrium concentration of BrCl?

15.14 For the system

$$PCl_5(g) \rightleftharpoons PCl_3(g) + Cl_2(g)$$

$K_c = 0.050$ at 250°C. In a certain equilibrium mixture at 250°C, $[PCl_3] = 3.0 \times [PCl_5]$. What is the equilibrium concentration of Cl_2?

15.15 K_c for the reaction

$$2 ICl(g) \rightleftharpoons I_2(g) + Cl_2(g)$$

is 0.11 at a certain temperature. Suppose the initial concentrations (mol/L) of ICl, I_2, and Cl_2 are 0.20, 0.00, and 0.00, respectively. Some of the ICl decomposes and the system reaches equilibrium. What is the equilibrium concentration of each species?

15.40 A gaseous reaction mixture contains 0.30 mol SO_2, 0.16 mol Cl_2, and 0.50 mol SO_2Cl_2 in a 2.0-L container; $K_c = 0.011$ for $SO_2Cl_2(g) \rightleftharpoons SO_2(g) + Cl_2(g)$
a. Is the system at equilibrium? Explain.
b. If it is not at equilibrium, in which direction will the system move to reach equilibrium?

15.41 The commercial preparation of methanol, CH_3OH, is done at elevated temperatures with the reaction

$$CO(g) + 2 H_2(g) \rightleftharpoons CH_3OH(g)$$

At a certain temperature, the K_c value is 7.3. In which direction will the system move to achieve equilibrium when the starting mixture contains
a. 0.80 M CO and 1.5 M H_2?
b. a gaseous mixture of 0.90 mol CH_3OH, 0.45 mol CO, and 0.45 mol H_2 in a 3.0-L container?

15.42 At 1500°C, K_c for the reaction in Problem 15.12 is 3×10^{-3}. At this temperature, will the equilibrium system contain mostly Br_2, mostly Br, or appreciable amounts of both species?

15.43 K_c is 2.6×10^8 at 825 K for the reaction

$$2 H_2(g) + S_2(g) \rightleftharpoons 2 H_2S(g)$$

What is the equilibrium concentration of H_2S if those of H_2 and S_2 are 0.0020 M and 0.0010 M, respectively?

15.44 For the system

$$2 HI(g) \rightleftharpoons H_2(g) + I_2(g)$$

$K_c = 0.016$ at 800 K. If, at 800 K, $[HI] = 0.20$ M and $[H_2] = [I_2]$, calculate the equilibrium concentration of H_2.

15.45 For the equilibrium in Problem 15.44, 1.00 mol HI is placed in a 4.00-L flask at 800 K. What are the equilibrium concentrations of H_2, I_2, and HI?

15.16 The reaction

$$CO(g) + H_2O(g) \rightleftharpoons CO_2(g) + H_2(g)$$

has a K_c value of 4.0 at 500°C. Calculate the concentration of all species at equilibrium starting with
a. conc. CO = conc. H_2O = 0.100 mol/L.
b. conc. CO = conc. H_2O = conc. CO_2 = conc. H_2 = 0.040 mol/L.

15.17 For the system

$$PCl_5(g) \rightleftharpoons PCl_3(g) + Cl_2(g)$$

K_c = 0.050 at 250°C. If 0.30 mol PCl_5 is placed in a 1.0-L container at this temperature, what are the equilibrium concentrations of all species?

15.18 K_c is 0.40 at a certain temperature for the system

$$SO_2Cl_2(g) \rightleftharpoons SO_2(g) + Cl_2(g)$$

Beginning with 0.200 mol SO_2Cl_2 in a 2.00-L container, calculate the equilibrium concentrations of all species.

15.19 When chlorine gas is heated, it decomposes:

$$Cl_2(g) \rightleftharpoons 2 Cl(g)$$

K_c for this reaction is 4.2×10^{-7} at 1000°C and 3.6×10^{-2} at 2000°C. Starting with a concentration of Cl_2 of 0.10 M, what will be the equilbrium concentration of Cl at
a. 2000°C?
b. 1000°C? Note that K_c at 1000°C is so small that $[Cl_2]$ will be very nearly equal to its original concentration.

15.20 For the system

$$SnO_2(s) + 2 H_2(g) \rightleftharpoons Sn(s) + 2 H_2O(g)$$

at 500°C, $[H_2O] = [H_2] = 0.10$ M.
a. Calculate K_c at this temperature.
b. Enough H_2 is added to raise its concentration temporarily to 0.18 M. When equilibrium is restored, what are the concentrations of H_2O and H_2?

15.46 For the reaction

$$2 IBr(g) \rightleftharpoons I_2(g) + Br_2(g)$$

K_c is 2.5×10^{-3} at 25°C. Calculate the equilibrium concentration of each species in a 4.0-L vessel starting with
a. 0.60 mol IBr.
b. 0.30 mol I_2, 0.30 mol Br_2.
c. 0.30 mol I_2, 0.30 mol Br_2, 0.30 mol IBr.

15.47 For the system

$$CO(g) + Cl_2(g) \rightleftharpoons COCl_2(g)$$

K_c = 3.0. If 1.5 mol CO and 1.0 mol Cl_2 are put in a 5.0-L container, what are the equilibrium concentrations of all species?

15.48 For the system

$$PBr_3(g) + Br_2(g) \rightleftharpoons PBr_5(g); \quad K_c = 0.250$$

A starting mixture of 1.00 mol PBr_3 and 3.00 mol Br_2 is used in a 1.00-L container. What are the concentrations of all species at equilibrium?

15.49 For the system

$$N_2(g) + O_2(g) \rightleftharpoons 2 NO(g)$$

$K_c = 1 \times 10^{-30}$ at 25°C and 0.10 at 2000°C. Starting with 0.030 mol/L of N_2 and 0.010 mol/L of O_2, calculate the equilibrium concentration of NO
a. at 2000°C.
b. at 25°C, making the simplification indicated in Problem 15.19b.

15.50 For the system

$$CO(g) + H_2O(g) \rightleftharpoons CO_2(g) + H_2(g)$$

equilibrium is established at a certain temperature when the concentrations of CO, H_2O, CO_2, and H_2 are 1.0×10^{-2}, 2.0×10^{-2}, 1.2×10^{-2}, and 1.2×10^{-2} M, respectively.
a. Calculate K_c.
b. If enough CO is added to raise its concentration temporarily to 1.8×10^{-2} M, what will be the final concentrations of all species?

Le Châtelier's Principle

15.21 Consider the system

$$2 H_2S(g) + 3 O_2(g) \rightleftharpoons 2 H_2O(g) + 2 SO_2(g)$$

ΔH for the forward reaction is -1036 kJ. Predict whether the forward or reverse reaction will occur when the equilibrium is disturbed by
a. expanding the container at constant temperature.
b. removing SO_2.
c. raising the temperature.
d. absorbing the water vapor.

15.22 Predict the direction in which each of the following equilibria will shift if the pressure on the system is increased by compression:
a. $ClF_5(g) \rightleftharpoons ClF_3(g) + F_2(g)$
b. $H_2O(g) + C(s) \rightleftharpoons CO(g) + H_2(g)$
c. $HBr(g) \rightleftharpoons \frac{1}{2} H_2(g) + \frac{1}{2} Br_2(g)$

15.23 For the system

$$2 P(s) + 3 H_2(g) \rightleftharpoons 2 PH_3(g)$$

K_c is 3.6×10^{-4} at 227°C and 2.3×10^{-3} at 477°C. Is the forward reaction exothermic or endothermic? Explain.

15.24 The system

$$2 X(g) + Z(g) \rightleftharpoons Q(g)$$

is at equilibrium when the concentration of X is 0.60 M. Sufficient X is added to raise its concentration temporarily to 1.0 M. When equilibrium is re-established, the concentration of X could be which of the following?
a. 0.80 M   b. 0.60 M   c. 0.90 M
d. 1.0 M   e. 1.2 M

15.51 For the system

$$N_2O_3(g) \rightleftharpoons NO(g) + NO_2(g)$$

ΔH is $+39.7$ kJ. Predict what effect each of the following changes will have on the position of the equilibrium:
a. decreasing the container size at constant temperature.
b. adding NO.
c. lowering the temperature.
d. adding helium gas.

15.52 Predict the direction in which each of the following equilibria will shift if the pressure on the system is reduced by expansion:
a. $SbCl_5(g) \rightleftharpoons SbCl_3(g) + Cl_2(g)$
b. $Ni(s) + 4 CO(g) \rightleftharpoons Ni(CO)_4(g)$
c. $CO(g) + H_2O(g) \rightleftharpoons CO_2(g) + H_2(g)$

15.53 For the system

$$CH_4(g) + Cl_2(g) \rightleftharpoons CH_3Cl(g) + HCl(g)$$

$\Delta H = -99$ kJ for the forward reaction; K_c is 1×10^{18} at 25°C. Would you expect K_c to increase or decrease when the temperature rises? Explain.

15.54 For the system in Problem 15.24, the initial equilibrium concentration of Z was 0.40 M. When equilibrium is re-established, the concentration of Z could be
a. 0.40 M   b. 0.50 M   c. 0.35 M
d. 0.30 M   e. 0.00 M

K_c and K_p

15.25 Using data in Table 15.5, calculate K_p for the system

$$N_2O_4(g) \rightleftharpoons 2 NO_2(g)$$

at each temperature listed.

15.26 Given the following K_p values at 25°C,

$$NH_4Cl(s) \rightleftharpoons NH_3(g) + HCl(g);$$
$$K_p = 7.7 \times 10^{-17}$$

$$2 NO_2(g) \rightleftharpoons 2 NO(g) + O_2(g);$$
$$K_p = 6.1 \times 10^{-13}$$

$$H_2(g) + I_2(g) \rightleftharpoons 2 HI(g);$$
$$K_p = 8.7 \times 10^2$$

calculate the corresponding values of K_c.

15.55 Using data in Table 15.5, calculate K_p for

$$N_2(g) + 3 H_2(g) \rightleftharpoons 2 NH_3(g)$$

at each temperature listed.

15.56 Consider the reaction

$$H_2O(l) \rightleftharpoons H_2O(g)$$

a. Using the table of vapor pressures of water in Appendix 1, calculate K_p for pressures in atmospheres at 25°C, 100°C, and 110°C.
b. Calculate K_c at each of these temperatures.

$\Delta G°$ and the Equilibrium Constant

15.27 Consider the reaction

$$H_2O(l) \rightleftharpoons H^+(aq) + OH^-(aq)$$

  a. Using Tables 14.1 to 14.4, calculate $\Delta G°$ at 25°C.
  b. Calculate the equilibrium constant at 25°C.

15.28 Calculate $\Delta G°$ at 25°C for each of the reactions referred to in Problem 15.26.

15.57 Consider the reaction

$$CaCO_3(s) \rightleftharpoons CaO(s) + CO_2(g)$$

  a. Using Tables 14.1 and 14.3, calculate $\Delta G°$ at 500°C.
  b. Calculate the equilibrium constant K_p at 500°C.

15.58 For the reaction

$$Cl_2(g) \rightleftharpoons 2\ Cl(g)$$

K_p is 1.0×10^{-37} at 25°C and 4.2×10^{-5} at 1000°C. Calculate $\Delta G°$ at each of these temperatures.

General

15.29 An examination question asked students to calculate K_c for the reaction

$$CS_2(g) + 4\ H_2(g) \rightleftharpoons CH_4(g) + 2\ H_2S(g)$$

given that, at equilibrium in a 5.0-L vessel, there are 5.5 mol CH_4, 1.25 mol H_2S, 1.5 mol CS_2, and 1.5 mol H_2. Four students gave the following answers, all wrong. Explain the error(s) in their reasoning.

  a. $K_c = \dfrac{(5.5)(1.25)^2}{(1.5)^5}$

  b. $K_c = \dfrac{(0.30)(0.30)^4}{(1.1)(0.25)^2}$

  c. $K_c = \dfrac{(1.1) \times (0.50)^2}{(0.30)(1.2)^4}$

  d. $K_c = \dfrac{(1.1) + (0.25)^2}{(0.30) + (0.30)^4}$

15.59 In another question related to the equilibrium given in Problem 15.29, students were asked to set up an expression to solve for $[CH_4]$ starting with 1.00 mol CS_2 and 2.00 mol H_2 in a 1.00-L container. Analyze each of the following incorrect answers for the error in the set-up, where $x = [CH_4]$:

  a. $K_c = \dfrac{(x)(x)^2}{(1.00-x)(2.00-4x)^4}$

  b. $K_c = \dfrac{(x)(2x)^2}{(1.00-x)(2.00)^4}$

  c. $K_c = \dfrac{(x)(2x)}{(1.00-x)(2.00-4x)}$

  d. $K_c = \dfrac{(x)(0.5x)^2}{(1.00-x)(2.00-2x)^4}$

15.30 Suppose that, for the system in Problem 15.29, all the equilibrium concentrations are 1.0 M. Which of the following are possible values of the original concentrations?

	CS$_2$	H$_2$	CH$_4$	H$_2$S
a.	1.0	1.0	0.0	0.0
b.	2.0	1.5	0.0	1.5
c.	1.2	1.8	0.80	0.60
d.	0.75	0.0	2.0	4.0

15.60 Consider again the system referred to in Problem 15.29. If the original concentrations are all 0.60 M, which of the following are possible equilibrium concentrations?

	[CS$_2$]	[H$_2$]	[CH$_4$]	[H$_2$S]
a.	0.40	0.40	0.80	1.00
b.	0.50	0.20	0.70	0.80
c.	0.70	1.00	0.50	0.40
d.	0.70	0.70	0.50	0.50
e.	0.50	0.50	0.50	0.50

*15.61 For the system $SO_3(g) \rightleftharpoons SO_2(g) + \frac{1}{2} O_2(g)$ at 1000 K, $K_c = 0.050$. Sulfur trioxide, originally at 1.00 atm pressure, partially dissociates to SO_2 and O_2 at 1000 K. What is the original concentration of SO_3 in moles per liter? What is its final concentration?

*15.62 At a certain temperature the reaction $Xe(g) + 2 F_2(g) \rightleftharpoons XeF_4(g)$ gives a 50% yield of XeF_4, starting with 0.40 mol Xe and 0.80 mol F_2 in a 2.0-L vessel. Calculate K_c at this temperature. How many additional moles of F_2 must be added to increase the yield of XeF_4 from Xe to 75%?

*15.63 A student studying the equilibrium $I_2(g) \rightleftharpoons 2 I(g)$ at a high temperature puts 0.10 mol I_2 in a 1.0-L container. He finds that the total pressure at equilibrium is 40% greater than it was originally. What is K_c for this reaction?

*15.64 At 308 K and a total pressure of 1.00 atm, the mole fraction of NO_2 in equilibrium with N_2O_4 is 0.39. Calculate K_p and ΔG^0 for the reaction $N_2O_4(g) \rightleftharpoons 2 NO_2(g)$.

*15.65 Using data given in Chapter 14, calculate K_c at 500°C for the reaction

$$CO(g) + H_2O(g) \rightleftharpoons CO_2(g) + H_2(g)$$

In automobile exhaust at 500°C, the concentrations of CO_2 and H_2O are about equal. What is the ratio of the equilibrium concentration of H_2 to that of CO?

Chapter 16
Rate of
Reaction

We saw in Chapter 15 that reactions for which the equilibrium constant is large tend to go virtually to completion. Consider, for example,

$$CO(g) + NO(g) \rightarrow CO_2(g) + \tfrac{1}{2} N_2(g) \tag{16.1}$$

At 25°C, K_c for this reaction is a huge number, about 10^{60}. On this basis, it would seem that Reaction 16.1 offers a way to remove the toxic gases CO and NO from automobile exhausts. Unfortunately, there is a practical difficulty. Under ordinary conditions, Reaction 16.1 occurs at what amounts to an extremely slow rate. We would have to wait essentially forever for CO and NO to be converted completely to CO_2 and N_2.

The situation just described is quite common. Many reactions that should in principle go to completion occur very, very slowly. Often, this is to our benefit. A case in point involves the fossil fuels—coal, petroleum, and natural gas. From equilibrium considerations, these fuels should be converted to CO_2 and H_2O upon exposure to air.

Actually, so should we mortals

We conclude that there is no correlation between the rate of a reaction and its equilibrium constant. To predict how rapidly a reaction will occur, we must become familiar with the principles of *chemical kinetics*. These principles are the subject of this chapter. Here, we will apply them mainly to reactions involving gases. All the reactions we will consider have very large equilibrium constants. Hence, we need only be concerned with the rate of the forward reaction and can ignore the reverse reaction.

The main emphasis in this chapter will be upon the factors that influence reaction rate. These include

—the nature of the reaction, as reflected in the activation energy (Section 16.4).

—the concentrations of reactants (Sections 16.2, 16.3).

—the presence of a catalyst (Section 16.5).
—the temperature (Section 16.6).

Before discussing the effect of these factors, it is important to state exactly what we mean by reaction rate (Section 16.1).

16.1
Meaning of
Reaction Rate

To discuss reaction rate in a meaningful way, we must first define precisely what this term means. *The rate of reaction is a positive quantity that tells us how the concentration of a reactant or product changes with time.* Most often, reaction rate is expressed in terms of reactant concentrations. Consider, for example, the decomposition of dinitrogen pentoxide, N_2O_5:

$$2\ N_2O_5(g) \rightarrow 4\ NO_2(g) + O_2(g) \tag{16.2}$$

We ordinarily take the rate of this reaction to be

$$\text{rate} = \frac{-\Delta\ \text{conc.}\ N_2O_5}{\Delta t} \tag{16.3}$$

where

Δ conc. N_2O_5 = final conc. N_2O_5 − orig. conc. N_2O_5

Δt = final time − orig. time

The minus sign in Equation 16.3 is required to make the rate a positive quantity; the concentration of N_2O_5 decreases with time.

Reaction rate has the units of concentration divided by time. We will always express concentration in moles per liter. On the other hand, time may be given in seconds, minutes, hours, days, or years. Thus, the units of reaction rate may be

$$\frac{\text{mol}}{L \cdot s},\quad \frac{\text{mol}}{L \cdot \text{min}},\quad \frac{\text{mol}}{L \cdot h},\ \ldots$$

The magnitude of the reaction rate will depend upon the time unit used. The rate with time in hours will be 60 times the rate with time in minutes. Thus, a rate of 0.05 mol/L·min would be 3 mol/L·h:

$$0.05\ \frac{\text{mol}}{L \cdot \text{min}} \times \frac{60\ \text{min}}{1\ h} = 3\ \frac{\text{mol}}{L \cdot h}$$

Reaction rate is always positive

Same rate, different units

Measurement of Rate

In order to measure the rate of a reaction, we must find how concentration changes with time. To be specific, let us consider how the rate of Reaction 16.2 might be

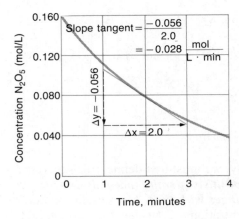

FIGURE 16.1 The rate of reaction can be determined by measuring concentration of reactant as a function of time and graphing the data. To determine the rate at a particular point, we draw a tangent to the curve and find its slope. This gives us the ratio Δ conc./Δt, which is numerically equal to the reaction rate.

measured. We start by putting 0.160 mol N_2O_5 in a one-liter container at 67°C. At one-minute intervals, we withdraw small samples of the reaction mixture and analyze them for N_2O_5. In this way, we obtain the concentration-time data listed in Table 16.1 and plotted in Figure 16.1.

Table 16.1
Rate of Decomposition of N_2O_5 at 67°C

Time (min)	0	1	2	3	4
Conc. N_2O_5 (mol/L)	0.160	0.113	0.080	0.056	0.040
Inst. rate (mol/L·min)	0.056	0.039	0.028	0.020	0.014

It's an average because the rate changes during the time interval

Using the data in Table 16.1, we can readily find the *average rate* of Reaction 16.2 over any desired time interval. Example 16.1 illustrates how this is done.

EXAMPLE 16.1 Calculate the average rate of the N_2O_5 decomposition (Equation 16.3) between

a. t = 0 and t = 1 min
b. t = 1 min and t = 2 min

Solution

a. rate $= \dfrac{-\Delta \text{ conc. } N_2O_5}{\Delta t} = \dfrac{-(0.113 \text{ mol/L} - 0.160 \text{ mol/L})}{1 \text{ min} - 0}$

$= 0.047 \dfrac{\text{mol}}{\text{L·min}}$

b. rate $= \dfrac{-(0.080 \text{ mol/L} - 0.113 \text{ mol/L})}{2 \text{ min} - 1 \text{ min}} = 0.033 \dfrac{\text{mol}}{\text{L·min}}$

Note that the rate decreases from an average value of 0.047 mol/L·min over the first minute to 0.033 mol/L·min during the second minute.

EXERCISE During the first minute, what is the change in concentration of NO_2 in this reaction? If the rate were defined as Δ conc. $NO_2/\Delta t$, what would be the average rate during the first minute? Answer: 0.094 mol/L; 0.094 mol/L·min.

Here no negative sign is used. The conc. NO_2 increases with time

Ordinarily we are interested, not in the average rate, but in the **instantaneous rate** at a given time or concentration. To find the instantaneous rate of Reaction 16.2, we use Figure 16.1. If we draw a tangent to the curve at any point, its slope will equal Δ conc. $N_2O_5/\Delta t$. From Equation 16.3, we see that the rate of reaction is equal to $-\Delta$ conc. $N_2O_5/\Delta t$. Putting these relations together,

instantaneous rate $= -$(slope of tangent to conc. vs. time curve)

From Figure 16.1 we see that the slope of the tangent at t = 2 min is -0.028 mol/L·min. Hence, the rate at that point is

$$-\left(-0.028\ \frac{\text{mol}}{\text{L·min}}\right) = 0.028\ \frac{\text{mol}}{\text{L·min}}$$

In the calculus notation, rate $=$
$$-\ \frac{d(\text{conc. }N_2O_5)}{dt}$$

Instantaneous rates at other points are obtained in a similar way. These are listed at the bottom of Table 16.1.

16.2
Reaction Rate and Concentration

In discussing the decomposition of N_2O_5, we pointed out that the rate of reaction decreases with time. From a slightly different point of view, the rate decreases as the concentration of N_2O_5 decreases. This behavior is typical of most reactions. We ordinarily find that reactions proceed more slowly as the concentration of reactant decreases. To increase the rate, we start with a higher concentration of reactant (Color Plate 16.1).

Just about all reactions slow down as time passes

It turns out that there is a simple relation between concentration and rate for the decomposition of N_2O_5. You may be able to deduce this relation from the data in Table 16.1. Notice what happens when the concentration decreases by a factor of two (from 0.160 to 0.080 or from 0.080 to 0.040 mol/L). The instantaneous rate is cut in half (from 0.056 to 0.028 or from 0.028 to 0.014 mol/L·min). This suggests that the rate of this reaction is directly proportional to the concentration of N_2O_5. Indeed, this is true, as you can see from Figure 16.2. A plot of rate vs. concentration is a straight line. If extrapolated, it would pass through the origin (rate = 0 when conc. N_2O_5 = 0).

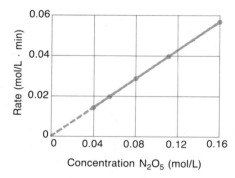

FIGURE 16.2 For the decomposition of N_2O_5, a plot of rate vs. conc. N_2O_5 is a straight line. The line, if extrapolated, passes through the origin. This means that rate is directly proportional to concentration; that is, rate = k(conc. N_2O_5). This must then be a first-order reaction.

Rate Expression and Rate Constant

Since the rate of decomposition of N_2O_5 is directly proportional to its concentration, it follows that

Double conc. N_2O_5 and you double the rate

$$\text{rate} = k(\text{conc. } N_2O_5) \quad \textbf{(16.4)}$$

Equation 16.4 is referred to as the **rate expression** for the decomposition of N_2O_5. It tells us how the rate of the reaction

$$2\ N_2O_5(g) \rightarrow 4\ NO_2(g) + O_2(g)$$

depends upon the concentration of reactant. The proportionality constant k in Equation 16.4 is called a **rate constant**. It is independent of the other quantities in the equation. However, k does depend upon the following:

—*the nature of the reaction*. "Fast" reactions typically have large rate constants. If k is very small, reaction occurs slowly at ordinary concentrations.

—*the temperature*. Ordinarily, k increases with temperature.

We can calculate the value of k in Equation 16.4 when we know the rate of reaction at any given concentration of N_2O_5. Solving this equation for k, we have

$$k = \frac{\text{rate}}{\text{conc. } N_2O_5}$$

Note from Table 16.1 that rate = 0.056 mol/(L·min) when conc. N_2O_5 = 0.160 mol/L. Hence,

Rate depends on conc., but the rate constant k does not

$$k = \frac{0.056 \text{ mol/L·min}}{0.160 \text{ mol/L}} = 0.35/\text{min}$$

The same result would be obtained, within experimental error, by using any of the other data points in Table 16.1.

Order of Reaction

Rate expressions have been established by experiment for a large number of reactions. For a process involving a single reactant

$$a\ A(g) \rightarrow \text{products}$$

the rate expression has the general form

$$\text{rate} = k(\text{conc. A})^m \quad \textbf{(16.5)}$$

The power to which the concentration of reactant A is raised in the rate expression describes the **order** of the reaction. If m in Equation 16.5 is 0, we say that the reaction is "zero order." If m = 1, the reaction is "first order"; if m = 2, it is "second order"; and so on. Note the following relationships:

—For a zero-order reaction (m = 0), the rate is independent of the concentration of reactant. Doubling the concentration leaves the rate unchanged.

—For a first-order reaction (m = 1), the rate is directly proportional to the concentration of reactant. Doubling the concentration increases the rate by a factor of two.

—For a second-order reaction (m = 2), the rate is proportional to the *square* of the concentration of reactant. Doubling the concentration increases the rate by a factor of four.

> The most common orders for reactions are 1, 2, and 0

One way to determine the order of a reaction is to obtain the initial rate (i.e., the rate at t = 0) as a function of concentration of reactant. The reasoning involved is indicated in Example 16.2.

EXAMPLE 16.2 The initial rate of decomposition of acetaldehyde, CH_3CHO,

$$CH_3CHO(g) \rightarrow CH_4(g) + CO(g)$$

was measured at a series of different concentrations with the following results:

Conc. CH_3CHO (mol/L)	0.10	0.20	0.30	0.40
Rate (mol/L·s)	0.085	0.34	0.76	1.4

Using these data, determine the order of the reaction; that is, determine the value of m in the equation

$$rate = k(conc. \ CH_3CHO)^m$$

Solution Let us write down the rate expression at two different concentrations:

$$rate_2 = k(conc._2)^m$$
$$rate_1 = k(conc._1)^m$$

Dividing the first equation by the second,

$$\frac{rate_2}{rate_1} = \left(\frac{conc._2}{conc._1}\right)^m$$

Now let us substitute data, taking $conc._2 = 0.20$ M, $conc._1 = 0.10$ M:

$$\frac{0.34}{0.085} = \left(\frac{0.20}{0.10}\right)^m$$

> This is the easiest way to find the order of a reaction

Simplifying,

$$4 = 2^m$$

Clearly, m = 2; that is, the reaction is second order.

EXERCISE Repeat the calculation, using the data at 0.40 and 0.30 M. Answer: $1.4/0.76 = (0.40/0.30)^m$. Solving, m = 2 (approximately).

Once the order of a reaction has been determined, the same rate data can be used to find the rate constant and then the rate at a new concentration (Example 16.3).

EXAMPLE 16.3 Consider the rate data for the decomposition of CH_3CHO given in Example 16.2. Knowing that the reaction is second order, determine
a. the rate constant, k.
b. the rate of reaction when conc. CH_3CHO = 0.50 mol/L.

Solution
a. Solving for k and substituting data at the first concentration listed,

$$k = \frac{rate}{(conc.\ CH_3CHO)^2} = \frac{0.085\ mol/L \cdot s}{(0.10\ mol/L)^2} = 8.5\ \frac{L}{mol \cdot s}$$

b. $rate = 8.5\ \frac{L}{mol \cdot s} \times \left(0.50\ \frac{mol}{L}\right)^2 = 2.1\ \frac{mol}{L \cdot s}$

EXERCISE At what concentration of CH_3CHO is the rate of reaction 0.20 mol/L·s? Answer: 0.15 mol/L.

Our discussion to this point illustrates the fact that *the order of a reaction must be determined experimentally. It cannot, in general, be deduced from the coefficients of the balanced equation.* The decomposition of acetaldehyde is second order even though the coefficient of CH_3CHO in the balanced equation is 1:

$$CH_3CHO(g) \rightarrow CH_4(g) + CO(g); \quad rate = k(conc.\ CH_3CHO)^2$$

Again, the decomposition of N_2O_5 is first order, even though the coefficient of N_2O_5 in the balanced equation is 2:

$$2\ N_2O_5(g) \rightarrow 4\ NO_2(g) + O_2(g); \quad rate = k(conc.\ N_2O_5)^1$$

With one reactant, we usually have a decomposition reaction

Many reactions (indeed, most reactions) involve more than one reactant. For a reaction between A and B,

$$a\ A(g) + b\ B(g) \rightarrow products$$

the general form of the rate expression is

$$rate = k(conc.\ A)^m \times (conc.\ B)^n \qquad (16.6)$$

Here we refer to m as "the order of the reaction with respect to A." Similarly, n is the order of the reaction with respect to B. The **overall order** of the reaction is the sum of the exponents m + n. Thus, for the reaction

$$CO(g) + NO_2(g) \rightarrow CO_2(g) + NO(g) \qquad (16.7)$$

the experimentally determined rate expression above 600 K is

$$rate = k(conc.\ CO) \times (conc.\ NO_2)$$

Hence, we say that this reaction is

—first order with respect to CO (m = 1).
—first order with respect to NO_2 (n = 1).
—second order overall (m + n = 2).

Note the terminology

When more than one reactant is involved in a reaction, the order is somewhat more difficult to determine experimentally. One rather straightforward approach involves holding the initial concentration of one reactant constant while varying that of the other reactant. From the rates measured under these conditions, we can find the order of the reaction with respect to the reactant whose initial concentration is changing.

To illustrate this approach, let us consider the data in Table 16.2 for the reaction

$$2 H_2(g) + 2 NO(g) \rightarrow N_2(g) + 2 H_2O(g)$$

In the first series of experiments, we hold the initial concentration of NO constant at 0.10 mol/L and vary the initial concentration of H_2. If you look at the data in Series 1, it should be clear that the rate is directly proportional to the concentration of H_2. For example, when the concentration of H_2 is doubled (from 0.10 to 0.20 mol/L), the rate doubles (from 0.10 to 0.20 mol/L·s). This means that in the general rate expression

This allows us to find the order for H_2

$$\text{rate} = k(\text{conc. } H_2)^m \times (\text{conc. NO})^n$$

m = 1. To find the value of n, we examine the data in the second series of experiments. Here, the initial concentration of H_2 is held constant at 0.10 mol/L while that of NO varies. It should be apparent that the rate is proportional to the square of the NO concentration. When the NO concentration is doubled (from 0.10 to 0.20 mol/L), the rate increases by a factor of $2^2 = 4$ (from 0.10 to 0.40 mol/L·s). When the concentration of NO increases by a factor of three (from 0.10 to 0.30 mol/L), the rate goes up by a factor of $3^2 = 9$ (from 0.10 to 0.90 mol/L·s). We conclude that in the rate expression above, n = 2. Putting the two series of experiments together, we find that

Then we find the order for NO

$$\text{rate} = k(\text{conc. } H_2) \times (\text{conc. NO})^2$$

Table 16.2
Initial Rates of Reaction (mol/L·s) at 800°C for:
$$2 H_2(g) + 2 NO(g) \rightarrow N_2(g) + 2 H_2O(g)$$

| | SERIES 1 | | | SERIES 2 | | |
	Conc. H_2 (mol/L)	Conc. NO (mol/L)	Rate	Conc. H_2 (mol/L)	Conc. NO (mol/L)	Rate
Expt. 1	0.10	0.10	0.10	0.10	0.10	0.10
Expt. 2	0.20	0.10	0.20	0.10	0.20	0.40
Expt. 3	0.30	0.10	0.30	0.10	0.30	0.90
Expt. 4	0.40	0.10	0.40	0.10	0.40	1.6

The development we have just gone through to deduce the order of the H_2–NO reaction is similar to that used in Example 16.2 to find the order of the CH_3CHO decomposition reaction. In general, given rate as a function of concentration, we can determine reaction order in one of two ways:

1. If only one reactant is involved, find the ratio of the rates at two different concentrations and apply the relation

$$\frac{rate_2}{rate_1} = \left(\frac{conc._2}{conc._1}\right)^m$$

where m is the reaction order. Ordinarily, m can be found by inspection, as in Example 16.2. It is most often a whole number (0, 1, 2, 3), although it can be a fraction ($\frac{1}{2}$, $\frac{3}{2}$, . . .).

2. If two reactants, A and B, are involved,

a. Find the ratio of the rates at two points for which conc. B is constant but conc. A differs. Using the equation above, find the order with respect to A.

b. Repeat (a), this time choosing two points at which conc. A is the same but conc. B differs. This leads to the order with respect to B.

Can you see why this works? If not, write Eqn. 16.6 for the two points, and take the ratio of the rates

16.3
Reactant
Concentration
and Time

A rate expression such as Equation 16.4 tells us how the rate of a reaction changes with concentration. However, from a practical standpoint, we are usually more interested in the relation between concentration and time. Suppose, for example, you are studying the decomposition of N_2O_5. Most likely, you would want to know how much N_2O_5 is left after 5 min, 1 h, or several days. Equation 16.4 is not of much help for that purpose.

rate = f(conc)
conc. = g(time)

These are two different, related, equations

It is possible to develop algebraic equations relating reactant concentration to time. The form of the equation may be very simple or quite complex, depending upon the order of the reaction and the number of reactants. Here, we will discuss in detail only one type of concentration-time relation—that for a first order reaction involving a single reactant:

$$a\ A(g) \rightarrow products;\qquad rate = k(conc.\ A)$$

First-Order Reactions

As we have seen, the reaction

$$2\ N_2O_5(g) \rightarrow 4\ NO_2(g)\ +\ O_2(g)$$

is first order; rate = k(conc. N_2O_5).

We would like to obtain an equation relating the concentration of N_2O_5 to time. It can be shown by using calculus that the relevant equation is

$$\log_{10}(conc.\ N_2O_5) = \log_{10}(conc.\ N_2O_5)_0 - kt/2.30$$

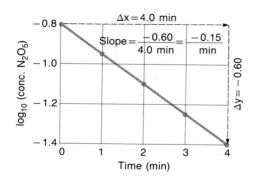

FIGURE 16.3 For a first-order reaction, a plot of log conc. vs. time is a straight line. The rate constant can be determined from the slope of the line, k = −2.30 (slope). In this case, k = −2.30(−0.15/min) = 0.35/min.

Here, k is the rate constant and t is time. The subscript $_0$ is used to indicate the original concentration of N_2O_5, at t = 0.

In Figure 16.3, the base 10 logarithm of conc. N_2O_5 vs. time is plotted using the data in Table 16.1. As you can see, we obtain a straight line. The general equation of a straight line is

$$y = a + bx$$

In this case, y is $\log_{10}$ (conc. N_2O_5) and x is time, t. The intercept, a, is $\log_{10}$ (conc. $N_2O_5)_0$. The slope, b, is $−k/2.30$. From Figure 16.3, we see that y *decreases* by 0.60 units while x increases by 4.0 min. Hence,

Take your time working this through. It's not too hard

$$\text{slope} = \frac{\Delta y}{\Delta x} = \frac{-0.60}{4.0 \text{ min}} = -0.15/\text{min}$$

Thus, we have

$$\frac{-k}{2.30} = \frac{-0.15}{\text{min}}; \quad k = \frac{(2.30)(0.15)}{\text{min}} = 0.35/\text{min}$$

$$\text{slope} = -\frac{k}{2.30}$$

This is the same value of k found earlier (p. 484).

The relation between concentration and time that we have "discovered" for the decomposition of N_2O_5 applies to any first-order reaction. It is usually written in the form

$$\log_{10} \frac{X_0}{X} = \frac{kt}{2.30} \tag{16.8}$$

where X_0 is the original concentration of reactant, X is the concentration of reactant at time t, and k is the first-order rate constant.

Equation 16.8 is a very useful one. For first-order reactions, we can use it to calculate

—the concentration of reactant remaining after a given time (Example 16.4a).
—the time required for reactant concentration to drop to a certain level (Example 16.4b and c).

EXAMPLE 16.4 Taking the first-order rate constant for the decomposition of N_2O_5 to be 0.35/min, calculate

a. the concentration of N_2O_5 after 4.0 min, starting with a concentration of 0.160 mol/L.
b. the time required for the concentration to drop from 0.160 to 0.100 mol/L.
c. the time required for half of a sample of N_2O_5 to decompose.

Solution

a. Using the first-order rate law, Equation 16.8,

$$\log_{10} \frac{0.160}{X} = \frac{0.35 \times 4.0}{2.30} = 0.60$$

To solve this example you need to use Eqn. 16.8, not Eqn. 16.4

Hence,

$$\frac{0.160}{X} = 10^{0.60} = 4.0; \qquad X = \frac{0.160}{4.0} = 0.040 \text{ mol/L}$$

Referring back to Table 16.1, we see that, starting with a concentration of N_2O_5 of 0.160 mol/L, the concentration after 4.0 min is indeed 0.040 mol/L.

b. Solving Equation 16.8 for t, we have

$$t = \frac{2.30}{k} \log_{10} \frac{X_0}{X} = \frac{2.30}{0.35} \log_{10} \frac{0.160}{0.100} = 6.6 \log_{10} 1.60 = 1.3 \text{ min}$$

c. When half of the sample has decomposed,

$$X = X_0/2; \qquad X_0 = 2X; \qquad X_0/X = 2$$

Using the equation obtained in (b),

$$t = \frac{2.30}{k} \log_{10} 2 = \frac{2.30 \times 0.301}{k} = \frac{0.693}{k}$$

With k = 0.35/min, we have

$$t = 0.693/0.35 = 2.0 \text{ min}$$

EXERCISE Looking back at the data in Table 16.1, how long does it take for the concentration of N_2O_5 to drop from 0.160 to 0.080 M? from 0.080 to 0.040 M? Answer: 2 min.

The analysis of Example 16.4c and the exercise that follows reveals an important feature of a first order reaction: *The time required for one half of a reactant to decompose via a first-order reaction has a fixed value, independent of concentration.* This quantity, called the half-life, is given by the expression

$$t_{1/2} = \frac{0.693}{k} \qquad\qquad (16.9)$$

where k is the rate constant for the first-order reaction. For the decomposition of N_2O_5, where k = 0.35/min, $t_{1/2}$ = 2.0 min; thus, we have the following data:

t (min)	NUMBER OF HALF-LIVES	CONC. N_2O_5 (mol/L)	FRACTION OF N_2O_5	
			Unreacted	Reacted
0	0	0.160	1	0
2.0	1	0.080	1/2	1/2
4.0	2	0.040	1/4	3/4
6.0	3	0.020	1/8	7/8
8.0	4	0.010	1/16	15/16

Notice from Equation 16.9 that the half-life, $t_{1/2}$, is *inversely* proportional to the rate constant, k. A "fast" reaction, for which k is large, will have a short half-life. Conversely, a "slow" reaction (small value of k) will have a relatively long half life.

Knowing the half-life, say from Eqn. 16.9, you can make "quick and dirty" calculations of conc. reactant at different times

Reactions of Other Integral Orders

Among gas-phase reactions, second-order processes are perhaps more common than those of first order. Zero-order reactions, in which the rate is independent of reactant concentrations, are much less common. One such reaction is the thermal decomposition of hydrogen iodide on a gold surface:

$$2\ HI(g) \overset{Au}{\to} H_2(g) + I_2(g); \qquad rate = k(conc.\ HI)^0 = k$$

This reaction occurs at a constant rate independent of the concentration of HI. Third-order reactions are very rare. One of the few known examples is the reaction between hydrogen and nitric oxide referred to earlier:

Table 16.3
Characteristics of Zero-, First-, and Second-Order Reactions of the Form

a A(g) → products
X, X_0 = conc. A at t and t = 0, respectively

ORDER	RATE EXPRESSION	CONC.-TIME RELATION*	HALF-LIFE	LINEAR PLOT
0	rate = k	$X_0 - X = kt$	$X_0/2k$	X vs. t
1	rate = kX	$\log_{10} \dfrac{X_0}{X} = \dfrac{kt}{2.30}$	0.693/k	$\log_{10} X$ vs. t
2	rate = kX^2	$\dfrac{1}{X} - \dfrac{1}{X_0} = kt$	$1/kX_0$	$\dfrac{1}{X}$ vs. t

Zero order reaction: Half life = 30 sec How much is gone in 30 sec? in 60 sec?

*These equations are obtained by using integral calculus. For example, for a first-order reaction, $-dX/dt = kX$. Integrating from X_0 to X gives $-\ln (X/X_0) = kt$, which is equivalent to $\log_{10} (X_0/X) = kt/2.30$.

$$2 H_2(g) + 2 NO(g) \rightarrow N_2(g) + 2 H_2O(g);$$

$$\text{rate} = k(\text{conc. } H_2) \times (\text{conc. } NO)^2$$

Rather than discuss each of these reaction types in detail, let's consider some of the properties of zero-, first-, and second-order reactions, listed in Table 16.3. All the relations apply to the decomposition of a single reactant:

a $A(g) \rightarrow$ products

In the last column of Table 16.3 are the functions (X, $\log_{10}$ X, and 1/X) that, when plotted against t, give a straight line for zero-, first-, and second-order reactions, respectively. These relations offer a simple method of deciding upon the order of a reaction for which concentration-time data are available.

EXAMPLE 16.5 The following data were obtained for the gas-phase decomposition of hydrogen iodide:

No gold surface here

t (h)	0	2	4	6
Conc. HI (M)	1.00	0.50	0.33	0.25

Is this reaction zero, first, or second order in HI?

Solution It will be useful to prepare a table in which we list X, $\log_{10}$ X, and 1/X as a function of time, letting X = conc. HI.

t (h)	X	$\log_{10}$ X	1/X
0	1.00	0.00	1.0
2	0.50	−0.30	2.0
4	0.33	−0.48	3.0
6	0.25	−0.60	4.0

For a 2nd order reaction, 1/X vs time gives a straight line

If it is not obvious from the table above that the only linear plot will be that of 1/X vs. t, that point should be clear from Figure 16.4. We conclude that we are dealing with a second-order reaction.

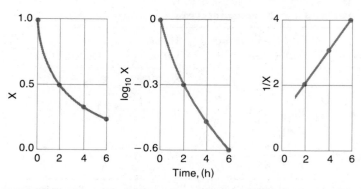

FIGURE 16.4 Decomposition of HI (Example 16.5). Here, X is the concentration of HI at time t. If the reaction were zero order, a plot of X vs. t would be linear. If it were first order, a plot of $\log_{10}$ X vs. t would be linear. For a second-order reaction such as this one, a plot of 1/X vs. t is a straight line.

> **EXERCISE** The half-life of a certain reaction is directly proportional to the original concentration of reactant. Using Table 16.3, determine the order of the reaction. Answer: Zero order.

16.4
Activation Energy

Chemical reactions ordinarily occur as a result of collisions between reacting particles. This simple statement explains many of the characteristics of reaction rates. To illustrate the usefulness of this collision model, let us apply it to the reaction between carbon monoxide and nitrogen dioxide:

$$CO(g) + NO_2(g) \rightarrow CO_2(g) + NO(g)$$

We believe that this reaction takes place as the result of collisions between CO and NO_2 molecules. This is consistent with the rate expression

$$\text{rate} = k(\text{conc. CO}) \times (\text{conc. } NO_2)$$

If we double the concentration of CO, holding that of NO_2 constant, the number of collisions in a given time doubles (Fig. 16.5). Doubling the concentration of NO_2 (holding that of CO constant) has the same effect. In general, the number of collisions per unit time is directly proportional to the concentration of CO or NO_2. The fact that the rate is also directly proportional to these concentrations implies that reaction occurs as a direct result of collisions between CO and NO_2 molecules.

There is one restriction on this simple model for the $CO–NO_2$ reaction. We can easily show that *not every collision leads to reaction*. From the kinetic theory of gases, it is possible to calculate the rate at which molecules collide with each other. For a mixture of CO and NO_2 at 700 K and concentrations of 0.10 mol/L, it turns out that every molecule should collide with about a billion other molecules in one second. If every collision were effective, the reaction between CO and NO_2 should be over in a fraction of a second. By experiment, we find this is not the case. Under the conditions specified, the half-life of the reaction is about 20 s.

If you think about it for a moment, you can see why only a small fraction of molecular collisions should be effective. The reactant molecules are held together by strong chemical bonds. For reaction to occur, these bonds must be weakened to the point of breaking. This can happen only if the molecules collide with considerable force. Slow-moving molecules do not have enough kinetic energy to react when they collide. They bounce off one another and retain their identity.

This is a fundamental postulate of rate theory

That would make for an explosion

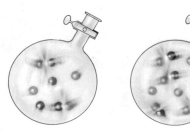

FIGURE 16.5 Effect of concentration upon reaction rate for the process $CO(g) + NO_2(g) \rightarrow CO_2(g) + NO(g)$. When the concentration of CO molecules (*open circles*) is doubled, collisions occur twice as often. More generally, the number of collisions per unit time, and hence the reaction rate, is directly proportional to the concentration of CO or NO_2.

Only molecules moving at very high speeds have enough kinetic energy for collision to result in reaction.

The high energy types get the work done

For every reaction, there is a certain minimum energy required to bring about reaction. This is referred to as the **activation energy**. It has the symbol E_a and is expressed in kilojoules. For the reaction between 1 mol CO and 1 mol NO_2, E_a is 134 kJ. The colliding molecules (CO and NO_2) must have a total kinetic energy of at least 134 kJ/mol if they are to react.

We find that the activation energy for a reaction

—*is a positive quantity* ($E_a > 0$).
—*depends upon the nature of the reaction.* Other factors being equal, we expect "fast" reactions to have a small activation energy. A reaction with a large activation energy takes place slowly under ordinary conditions. The larger the value of E_a, the smaller will be the fraction of molecules having enough kinetic energy to react when they collide.
—*is independent of temperature or concentration.*

Activation Energy Diagrams

Figure 16.6 is an energy diagram for the $CO–NO_2$ reaction. Reactants, CO and NO_2, are shown at the left. Products, CO_2 and NO, are at the right. They have an energy 226 kJ less than that of the reactants; ΔH for the reaction is -226 kJ. In the center of the figure is an intermediate called an *activated complex*. This is an unstable, high-energy species that must be formed before the reaction can occur. It has an energy 134 kJ greater than the reactants and 360 kJ greater than the products. The activation energy, 134 kJ, is absorbed in converting the reactants to the activated complex. The exact nature of this species is difficult to determine. For this reaction, the activated complex might be a "pseudomolecule" made up of CO and NO_2 molecules in close contact. The path of the reaction might be more or less as follows:

The activation energy has about the same size as a bond energy

$$O{\equiv}C + O{-}N{\overset{\displaystyle\|}{}}_O \quad \rightarrow \quad O{\equiv}C \cdots O \cdots N{\overset{\displaystyle\|}{}}_O \quad \rightarrow \quad O{=}C{=}O + N{=}O$$

reactants   activated complex   products

The dotted lines stand for "partial bonds" in the activated complex. The N—O bond in the NO_2 molecule has been partially broken. A new bond between carbon and oxygen has started to form.

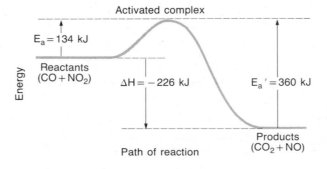

FIGURE 16.6 Concept of activation energy. During the reaction step, about 134 kJ, the activation energy E_a, must be furnished to the reactants for each mole of CO that reacts. This energy activates each $CO–NO_2$ complex to the point where reaction can proceed.

From Figure 16.6, we see that

$$CO(g) + NO_2(g) \rightarrow CO_2(g) + NO(g); \qquad E_a = 134 \text{ kJ}$$
$$CO_2(g) + NO(g) \rightarrow CO(g) + NO_2(g); \qquad E_a' = 360 \text{ kJ}$$

where E_a and E_a' are the activation energies for the forward and reverse reactions. The difference between these two quantities is equal to the enthalpy change for the forward reaction

$$CO(g) + NO_2(g) \rightarrow CO_2(g) + NO(g); \qquad \Delta H = -226 \text{ kJ}$$

In general, for any process,

$$\Delta H = E_a - E_a'; \qquad E_a = E_a' + \Delta H \qquad \qquad \text{(16.10)}$$

If the forward reaction is exothermic ($\Delta H < 0$), E_a is smaller than E_a'. This is the case with the CO–NO$_2$ reaction, where

In most reactions whose rates we study, $\Delta H < 0$

$$\Delta H = -226 \text{ kJ}; \qquad E_a = E_a' - 226 \text{ kJ}$$

Suppose, on the other hand, that the forward reaction is endothermic ($\Delta H > 0$). In that case, the activation energy for the forward reaction, E_a, is larger than that for the reverse reaction, E_a'. For example, if ΔH were $+200$ kJ, then

$$\Delta H = +200 \text{ kJ}; \qquad E_a = E_a' + 200 \text{ kJ}$$

16.5
Catalysis

Certain substances, called *catalysts*, can increase the rate of a reaction without being consumed by it. One reaction that is subject to catalysis is the decomposition of hydrogen peroxide:

$$2 H_2O_2(aq) \rightarrow 2 H_2O + O_2(g) \qquad \qquad \text{(16.11)}$$

This occurs rather slowly under ordinary conditions. However, if a solution of sodium iodide is added, reaction occurs immediately; bubbles of oxygen form. After the reaction is over, all the sodium iodide can be recovered, indicating that it is a true catalyst. This is an example of *homogeneous* catalysis: the reaction takes place within a single phase, in this case an aqueous solution.

Another reaction that is subject to catalysis is the decomposition of nitrous oxide:

$$2 N_2O(g) \rightarrow 2 N_2(g) + O_2(g) \qquad \qquad \text{(16.12)}$$

In the gas phase at ordinary temperatures, N_2O decomposes very slowly. The reaction speeds up when the N_2O is brought into contact with a metal such as gold. The gold acts as a *heterogeneous* catalyst. In its presence, Reaction 16.12 occurs at the boundary between two phases, solid and gas.

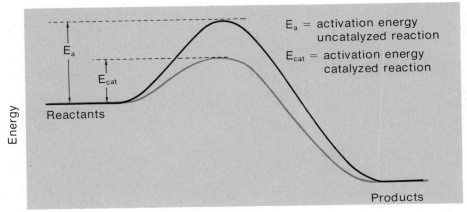

FIGURE 16.7 By changing the path by which a reaction occurs, a catalyst can lower the activation energy that is required, and so speed up the reaction.

A catalyst operates by lowering the activation energy required for reaction (Fig. 16.7). Consider, for example, Reaction 16.12. For the uncatalyzed reaction, E_a is 250 kJ. For the catalyzed reaction on a gold surface, E_a is only 120 kJ. The reduction in activation energy comes about because the catalyst provides an alternate pathway of lower energy for the reaction. In the decomposition of N_2O on gold, the gas is chemically adsorbed on the metal surface. A bond is formed between the oxygen of the N_2O molecule and a gold atom. This weakens the bond joining nitrogen to oxygen, making it easier for the N_2O molecule to break apart.

Most industrial reactions involve catalysts

The homogeneous catalysis of Reaction 16.11 is believed to take place by a two-step process:

Step 1: $H_2O_2(aq) + I^-(aq) \rightarrow H_2O + IO^-(aq)$
Step 2: $\underline{H_2O_2(aq) + IO^-(aq) \rightarrow H_2O + O_2(g) + I^-(aq)}$
$2\ H_2O_2(aq) \rightarrow 2\ H_2O + O_2(g)$

Notice that the end result is the same as in the direct reaction (Equation 16.11). The I^- ions are not consumed in the reaction. For every I^- ion used up in the first step, one is produced in the second step. The activation energy for this two-step process is much smaller than for the uncatalyzed reaction.

The catalyst, I^- ion, is regenerated

We see from Figure 16.7 that a catalyst does not affect the relative energies of reactants and products; nor does it change the equilibrium constant, K_c. Adding a catalyst does not affect the position of an equilibrium. It neither increases nor decreases the yield of product. However, by speeding up the reaction, a catalyst allows it to reach equilibrium more quickly.

Many reactions that take place slowly under ordinary conditions occur readily in the body in the presence of catalysts called *enzymes*. Enzymes are protein molecules of high molar mass. An example of an enzyme-catalyzed reaction is that of sugar (sucrose) with oxygen:

$$C_{12}H_{22}O_{11}(s) + 12\ O_2(g) \rightarrow 12\ CO_2(g) + 11\ H_2O(l)$$

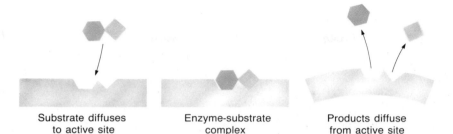

Substrate diffuses to active site   Enzyme-substrate complex   Products diffuse from active site

FIGURE 16.8 In enzyme catalysis, the substrate appears to fit on the enzyme in a "lock and key" arrangement. After absorption, the enzyme configuration often changes, which assists in cleaving the crucial substrate bond, and thereby increases the rate of reaction.

This reaction is difficult to bring about directly, such as by heating a sample of sugar in a test tube. In the body, sugar is metabolized at 37°C (98.6°F) in a series of biochemical reactions. The end products are carbon dioxide and water. Each step in the sequence is catalyzed by a particular enzyme adapted for that purpose.

Catalysis by enzymes can be interpreted in terms of the "lock and key" analogy shown in Figure 16.8. The reactant ("substrate") fits into a specific site on an enzyme surface. There, it is held in position by intermolecular forces. The substrate-enzyme complex can then react with another species such as a water molecule.

Enzyme activity is diminished in the presence of certain substances known as inhibitors. One way in which an inhibitor can operate is to occupy sites on an enzyme molecule that are supposed to be reserved for the substrate. Frequently, inhibitors have geometries closely resembling those of the substrates they replace. For example, the metabolism of citric acid is inhibited by its close relative, fluorocitric acid, which can presumably fit into the same slot in an enzyme:

Enzymes tend to be reaction-specific

$$
\begin{array}{cc}
\begin{array}{c}
\text{H} \\
| \\
\text{H}-\text{C}-\text{COOH} \\
| \\
\text{HO}-\text{C}-\text{COOH} \\
| \\
\text{H}-\text{C}-\text{COOH} \\
| \\
\text{H}
\end{array}
&
\begin{array}{c}
\text{F} \\
| \\
\text{H}-\text{C}-\text{COOH} \\
| \\
\text{HO}-\text{C}-\text{COOH} \\
| \\
\text{H}-\text{C}-\text{COOH} \\
| \\
\text{H}
\end{array}
\\[1em]
\text{citric acid} & \text{fluorocitric acid}
\end{array}
$$

Until quite recently we knew almost nothing about the molecular structure of enzymes. The lock and key model was little more than a convenient way to rationalize the kinetics of enzyme reactions. Within the past two decades the structures of some of the simpler enzymes have been established by x-ray crystallography. In some cases, the structures of substrate-enzyme complexes have been determined. Research in this area led in 1969 to the first laboratory synthesis of an enzyme, ribonuclease. For this work, Drs. Stein and Moore of Rockefeller University and Dr. Anfinsen of NIH won the 1972 Nobel Prize in chemistry.

A typical enzyme contains more than a thousand atoms

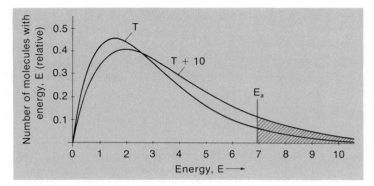

FIGURE 16.9 When the temperature is increased, the fraction of molecules with very high energies increases sharply. Hence, many more molecules possess the activation energy, E_a, and reaction occurs more rapidly. If E_a is of the order of 50 kJ, an increase in temperature of 10°C approximately doubles the number of molecules having energy E_a or greater, and thus doubles the reaction rate.

16.6
Reaction Rate
and Temperature

The rates of most reactions increase as the temperature rises. A person in a hurry to prepare dinner applies this principle in using a pressure cooker to raise the temperature for cooking potatoes, apples, or a pot roast (not all at the same time, we trust). By storing the leftovers in a refrigerator, we slow down the chemical reactions responsible for food spoilage. As a general and very approximate rule, it is often stated that an increase in temperature of 10°C doubles the reaction rate. If this rule holds, foods should cook twice as fast in a pressure cooker at 110°C as in an open saucepan, and deteriorate four times as rapidly at room temperature (25°C) as they do in a refrigerator at 5°C.

The effect of temperature on reaction rate can be explained in terms of the kinetic theory of gases. Recall from Chapter 6 that raising the temperature greatly increases the fraction of molecules having very high kinetic energies. These are the molecules that are most likely to react when they collide. The higher the temperature, the larger the fraction of molecules that can provide the activation energy required for reaction. This effect is illustrated in Figure 16.9, where the distribution of kinetic energies among gas molecules is shown at two different temperatures. The shaded areas include those molecules having a kinetic energy equal to or greater than E_a. Note that this area is considerably larger at the higher temperature. This means that, at the higher temperature, a larger fraction of the molecules will have sufficient energy to react when they collide. Hence, the

The fraction is always small, but doubling it doubles the rate

Table 16.4
Temperature Dependence of the Rate Constant for the Reaction
$$CO(g) + NO_2(g) \rightarrow CO_2(g) + NO(g)$$

T (K)	600	650	700	750	800
k (L/mol·s)	0.028	0.22	1.3	6.0	23

reaction will go faster. Putting it another way, the rate constant, k, becomes larger as the temperature increases. Table 16.4 shows this effect for the CO–NO$_2$ reaction. Notice that the rate constant increases by a factor of nearly a thousand when the temperature rises from 600 to 800 K.

$$\frac{23}{0.028} = 820$$

Relation Between k and T

The argument we have just gone through can be made quantitative. Kinetic theory tells us that the fraction, f, of molecules having an energy equal to or greater than E$_a$ is

$$f = e^{-E_a/RT} \tag{16.13}$$

If E_a = 50 kJ, at 25°C
$f \cong 2 \times 10^{-9}$

where e is the base of natural logarithms, R is the gas constant, E$_a$ is the activation energy, and T is the absolute temperature in K. If we assume that the rate constant, k, is directly proportional to f (which is approximately true),

$$k = cf = ce^{-E_a/RT}$$

where c is a proportionality constant. Taking the natural logarithm of both sides of this equation,

$$\ln k = \ln c - E_a/RT$$

Converting to base 10 logarithms,

$$\log_{10} k = \log_{10} c - \frac{E_a}{2.30\ RT}$$

Substituting R = 8.31 J/(mol·K) and setting $\log_{10} c$ = A, we obtain

$$\log_{10} k = A - \frac{E_a}{(2.30)(8.31)T} \tag{16.14}$$

Most reactions obey this equation very well

where k is the rate constant for the reaction, A is a constant independent of the other quantities in the equation, E$_a$ is the activation energy for the reaction in *joules*, and T is the temperature in K. This equation was first shown to be valid by the Swedish physical chemist Svandte Arrhenius in 1889. It is referred to as the Arrhenius equation.

Equation 16.14 is of the form

$$y = a + bx$$

where y is $\log_{10} k$ and x is 1/T. A plot of $\log_{10} k$ vs. 1/T should be a straight line. In Figure 16.10, this plot is shown for the CO–NO$_2$ reaction, with the data from Table 16.4. As you can see, it is indeed a straight line. The activation energy for the reaction can be obtained from the slope of the line:

$$slope = -E_a/(2.30 \times 8.31) \tag{16.15}$$

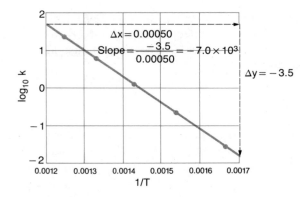

FIGURE 16.10 A plot of $\log_{10} k$ vs. $1/T$ (k = rate constant, T = Kelvin temperature) is ordinarily a straight line. From the slope of this line, the activation energy can be determined: $E_a = -2.30(8.31)(\text{slope})$. For the $CO–NO_2$ reaction shown here, $E_a = -2.30(8.31)(-7.0 \times 10^3)$ J $= 1.34 \times 10^5$ J $= 134$ kJ.

For the $CO–NO_2$ reaction, we see from Figure 16.10 that the slope is about -7.0×10^3. Hence,

This is how we find activation energies

$$E_a = -(2.30)(8.31)(-7.0 \times 10^3) \text{ J} = 134 \text{ kJ}$$

This tells us that the activation energy for this reaction is 134 kJ.

"Two-Point" Equation Relating k and T

We can use Equation 16.14 to obtain a relation between rate constants, k_2 and k_1, at two different temperatures, T_2 and T_1. We follow the same procedure used with the Clausius-Clapeyron equation (Chapter 11). At the two temperatures,

$$\log_{10} k_2 = A - \frac{E_a}{(2.30)(8.31)\, T_2}$$

$$\log_{10} k_1 = A - \frac{E_a}{(2.30)(8.31)\, T_1}$$

Subtracting the second equation from the first,

$$\log_{10} k_2 - \log_{10} k_1 = \log_{10} \frac{k_2}{k_1} = \frac{E_a}{(2.30)(8.31)} \left(\frac{1}{T_1} - \frac{1}{T_2} \right)$$

or

$$\log_{10} \frac{k_2}{k_1} = \frac{E_a}{(2.30)(8.31)} \left(\frac{T_2 - T_1}{T_2 T_1} \right) \tag{16.16}$$

In using Equation 16.16, we ordinarily take T_2 to be the higher temperature to avoid the use of negative logarithms. Remember that E_a must be expressed in joules; T_2 and T_1 are in K.

EXAMPLE 16.6

a. The activation energy of a certain reaction is 9.32×10^4 J. At 27°C, $k = 1.25 \times 10^{-2}$ L/mol·s. Calculate k at 127°C.

b. For the reaction referred to in (a), at what temperature is $k = 2.50 \times 10^{-2}$ L/mol·s?

c. What must be the value of E_a for a reaction if the rate constant is to double when the temperature increases from 15 to 25°C?

Solution

a. Using the "two-point" form of the Arrhenius equation, 16.16, we have

$$T_2 = 127 + 273 = 400 \text{ K}; \quad T_1 = 27 + 273 = 300 \text{ K}$$

$$\log_{10} \frac{k_2}{k_1} = \frac{9.32 \times 10^4}{(2.30)(8.31)}\left(\frac{400 - 300}{400 \times 300}\right) = 4.06$$

This means that

$$\frac{k_2}{k_1} = 10^{4.06} = 1.15 \times 10^4$$

Hence,

$$k_2 = 1.15 \times 10^4 k_1 = (1.15 \times 10^4)(1.25 \times 10^{-2} \text{ L/mol·s})$$
$$= 1.44 \times 10^2 \text{ L/mol·s}$$

b. Here we know k_2, k_1, E_a, and T_1; we need to calculate T_2:

$$k_2 = 2.50 \times 10^{-2} \text{ L/mol·s}; \quad k_1 = 1.25 \times 10^{-2} \text{ L/mol·s}$$

$$T_1 = 300 \text{ K}; \quad E_a = 9.32 \times 10^4 \text{ J}$$

$$\log_{10} \frac{2.50 \times 10^{-2}}{1.25 \times 10^{-2}} = 0.301 = \frac{9.32 \times 10^4 (T_2 - 300)}{(2.30)(8.31)(300)T_2}$$

Simplifying,

$$\frac{T_2 - 300}{T_2} = \frac{0.301 \times 2.30 \times 8.31 \times 300}{9.32 \times 10^4} = 0.0185$$

Solving,

$$0.9815 T_2 = 300 \text{ K}; \quad T_2 = 306 \text{ K}$$

c. If $k_2/k_1 = 2$, then $\log_{10} k_2/k_1 = \log_{10} 2 = 0.301$. Substituting in Equation 16.16,

$$0.301 = \frac{E_a(298 - 288)}{(2.30)(8.31)(298)(288)}$$

Solving,

$$E_a = 4.94 \times 10^4 \text{ J} = 49.4 \text{ kJ}$$

Note that if E_a were appreciably greater than 50 kJ, k would more than double for a 10°C rise in temperature; if E_a were smaller than 50 kJ, k would increase by less than a factor of two. Clearly, the empirical rule that a temperature increase of 10°C doubles the reaction rate is at best a crude approximation.

Most E_a values are larger than 50 kJ

EXERCISE For the reaction referred to in parts (a) and (b), how large a temperature increase is required to double the rate? Answer: About 6°C.

16.7
Reaction Mechanisms

A reaction mechanism is a description of the path, or sequence of steps, by which a reaction occurs. In the simplest case, only a single step is involved. This is a collision between two reactant molecules. This is the "mechanism" for the reaction of CO with NO_2 at high temperatures, above about 600 K:

$$CO(g) + NO_2(g) \rightarrow NO(g) + CO_2(g)$$

Some reactions involve many steps

In practice, most reactions occur in more than one step. Consider, for example, the reaction between carbon monoxide and nitrogen dioxide. Research indicates that, at low temperatures, below about 500 K, the reaction occurs in two steps:

Step 1: $NO_2(g) + NO_2(g) \rightarrow NO_3(g) + NO(g)$

Step 2: $\underline{NO_3(g) + CO(g) \rightarrow NO_2(g) + CO_2(g)}$

$\ \ CO(g) + NO_2(g) \ \rightarrow NO(g) + CO_2(g)$

The first step involves a collision between two NO_2 molecules. The NO_3 molecule formed in that step later collides, in Step 2, with a CO molecule. Notice that the overall reaction, obtained by summing Steps 1 and 2, is the same as that for the single-step mechanism referred to earlier. However, as we will see shortly, the rate expressions for the two mechanisms are quite different.

Rate Expression Determination from Reaction Mechanism

As we have seen, rate expressions for reactions must be determined experimentally. Once this has been done, it is possible to deduce a plausible mechanism leading to the observed rate expression. This, however, is a rather complex process and we will not attempt it here. Instead, we will consider the reverse process, which is much more straightforward. *Given the mechanism for a several-step reaction, how do we obtain the rate expression for the reaction?* To do this, we follow three general rules.

Each step is assumed to occur via collisions

1. **For any step, the order with respect to each reactant is its coefficient in the chemical equation for that step.** Thus, for the low temperature reaction of CO with NO_2, we have

Step 1: $NO_2(g) + NO_2(g) \rightarrow NO_3(g) + NO(g)$
$\ \ $ rate $= k_1$(conc. NO_2) $\times$ (conc. NO_2) $= k_1$(conc. NO_2)^2

Step 2: $NO_3(g) + CO(g) \rightarrow NO_2(g) + CO_2(g)$
$\ \ $ rate $= k_2$(conc. NO_3) $\times$ (conc. CO)

(The rate constants k_1 and k_2 are those for the individual steps.)

2. Often, one step in a mechanism is much slower than any other. In such cases, **the slow step is rate determining**; that is, the rate of the overall reaction is

determined by that of the slow step. The situation is analogous to that in a relay race where there is one slow runner (WLM) and two fast runners (EJS and CLS). The time required to reach the finish line will be determined mainly by WLM, provided he is much slower than EJS or CLS (as indeed he is).

In the two-step process referred to above, the first step is much slower than the second:

Step 1: $NO_2(g) + NO_2(g) \rightarrow NO_3(g) + NO(g)$ (slow)

Step 2: $NO_3(g) + CO(g) \rightarrow NO_2(g) + CO_2(g)$ (fast)

It follows that the rate of the overall reaction is that of the first step:

rate $= k_1$ (conc. NO_2)^2

Experimentally, we find that the low-temperature reaction between CO and NO_2 is indeed second order in NO_2 and zero order in CO, as this analysis predicts. This is in contrast to the situation at high temperatures with this reaction. As pointed out earlier, above about 600 K, reaction occurs in a single step involving a collision between a CO molecule and an NO_2 molecule. The high-temperature reaction, as we have emphasized throughout this chapter, is first order in NO_2 and first order in CO.

3. Frequently, reactive intermediates are present as reactants in the rate-determining step. Their concentrations are not known experimentally. Hence, they must be eliminated from the rate expression if it is to be compared with experiment. **The final rate expression must include only those species that appear in the balanced equation for the overall reaction.**

A reaction in which this situation arises is that between nitric oxide and chlorine. This reaction is believed to proceed by a two-step mechanism:

Step 1: $NO(g) + Cl_2(g) \rightleftharpoons NOCl_2(g)$ (fast)

Step 2: $\underline{NOCl_2(g) + NO(g) \rightarrow 2\ NOCl(g)}$ (slow)

$2\ NO(g) + Cl_2(g) \rightarrow 2\ NOCl(g)$ **(16.17)**

In the first step, the unstable intermediate $NOCl_2$ is formed, in equilibrium with NO and Cl_2. An $NOCl_2$ molecule then reacts with another molecule of NO in a slow, rate-determining second step.

Applying Rule 2, we find that the rate expression for Reaction 16.17 should be that for the second step:

rate $= k_2$(conc. $NOCl_2$) $\times$ (conc. NO)   **(16.18)**

The term (conc. $NOCl_2$) must be eliminated from the rate expression because the $NOCl_2$ molecule is an intermediate whose concentration cannot be determined experimentally. To eliminate (conc. $NOCl_2$), we take advantage of the fact that the first step in this reaction involves an equilibrium. Writing the equilibrium constant expression for that step, we have:

$$K_c = \frac{(\text{conc. } NOCl_2)}{(\text{conc. } NO) \times (\text{conc. } Cl_2)}$$

WLM is a better trout fisherman than we are

EJS

CLS

Every NO_3 that forms immediately reacts with CO

Frequently, intermediates exist at equilibrium concentrations

Solving for (conc. $NOCl_2$) and substituting in Equation 16.18,

$$\text{conc. } NOCl_2 = K_c \times (\text{conc. } NO) \times (\text{conc. } Cl_2)$$

$$\text{rate} = k_2 \times K_c \times (\text{conc. } NO)^2 \times (\text{conc. } Cl_2)$$

The product of the two constants, $k_2 \times K_c$, is simply the observed rate constant for Reaction 16.17. Calling that k, we predict the rate expression:

$$\text{rate} = k(\text{conc. } NO)^2 \times (\text{conc. } Cl_2) \tag{16.19}$$

Experimentally, the reaction between NO and Cl_2 is found to be second order in NO and first order in Cl_2, as Equation 16.19 predicts.

EXAMPLE 16.7 The decomposition of ozone, O_3, to diatomic oxygen, O_2, is believed to occur by a two-step mechanism:

Step 1: $O_3(g) \rightleftharpoons O_2(g) + O(g)$ (fast)
Step 2: $\underline{O_3(g) + O(g) \rightarrow 2\, O_2(g)}$ (slow)
$2\, O_3(g) \rightarrow 3\, O_2(g)$

Obtain the rate expression corresponding to this mechanism.

Solution According to Rule 2,

rate of reaction = rate of Step 2

According to Rule 1:

$$\text{rate} = k_2(\text{conc. } O_3) \times (\text{conc. } O)$$

where k_2 is the rate constant for the second step. To obtain a valid rate expression, we must eliminate (conc. O), since atomic oxygen is an unstable intermediate (Rule 3). To eliminate (conc. O), we start by writing the equilibrium constant expression for Step 1:

$$K_c = \frac{(\text{conc. } O_2) \times (\text{conc. } O)}{(\text{conc. } O_3)}$$

We now solve for (conc. O) and substitute:

$$\text{conc. } O = K_c \times \frac{(\text{conc. } O_3)}{(\text{conc. } O_2)}$$

$$\text{rate} = k_2 \times K_c \times \frac{(\text{conc. } O_3)^2}{(\text{conc. } O_2)} = k\frac{(\text{conc. } O_3)^2}{(\text{conc. } O_2)}$$

where k is the observed rate constant for the overall reaction. Notice that the concentration of O_2, a product in this reaction, appears in the denominator of the rate expression. This means that the rate is inversely proportional to the oxygen concentration. The more O_2 we have, the slower the reaction.

O_2 would be called an inhibitor of the reaction

EXERCISE What happens to the rate of this reaction if the concentration of O_3 is doubled? if the concentration of O_2 is doubled? Answer: Rate increases by a factor of four; rate is cut in half.

In summary, to find the rate expression corresponding to a given mechanism, do the following:

1. Locate the slow step in the mechanism. The rate of the overall reaction will be the rate of that step.

2. Write the rate expression for the slow step. To do this, note that the exponent of a reactant in the rate expression for a step is its coefficient in the equation for that step.

3. If the rate expression obtained in (2) contains an unstable intermediate, the term for that species must be eliminated. Frequently, this can be done by working with the equilibrium constant for a fast step in the mechanism, as in Example 16.7.

We should take note of one of the limitations of mechanism studies. Often, two different mechanisms lead to the same rate expression. When this happens, we cannot be sure which is correct. Other evidence must be considered to make a choice. A classic example of this situation is the reaction between hydrogen and iodine:

$$H_2(g) + I_2(g) \rightarrow 2\ HI(g)$$

Just because a mechanism gives the right rate expression doesn't mean it is the right mechanism

for which the observed rate expression is

$$\text{rate} = k\ (\text{conc. } H_2) \times (\text{conc. } I_2) \tag{16.20}$$

For many years, it was assumed that the H_2–I_2 reaction occurs in a single step, a collision between an H_2 molecule and an I_2 molecule. That would, of course, lead to the rate expression given by Equation 16.20. However, there is now evidence to indicate that a two-step mechanism is involved:

Step 1:	$I_2(g) \rightleftharpoons 2\ I(g)$	(fast)
Step 2:	$\underline{H_2(g) + I(g) + I(g) \rightarrow 2\ HI(g)}$	(slow)
	$H_2(g) + I_2(g) \rightarrow 2\ HI(g)$	

The first step involves the dissociation of an I_2 molecule into atoms. The second step requires a simultaneous collision between three species, a hydrogen molecule and two iodine atoms. We would expect such a collision to be a rare event, which explains why the second step is the slow one in the mechanism.

This two-step mechanism leads to the same rate expression (Equation 16.20) as would a simple collision between H_2 and I_2 molecules. To show that this is the case, we proceed in the usual way:

$$\text{rate of reaction} = \text{rate of Step 2} = k_2(\text{conc. } H_2) \times (\text{conc. } I)^2$$

From the equilibrium in Step 1,

$$K_c = \frac{(\text{conc. } I)^2}{(\text{conc. } I_2)}$$

Solving for $(\text{conc. } I)^2$ and substituting,

$$(\text{conc. } I)^2 = K_c \times (\text{conc. } I_2)$$

$$\text{rate} = k_2 \times K_c \times (\text{conc. } H_2) \times (\text{conc. } I_2) = k(\text{conc. } H_2) \times (\text{conc. } I_2)$$

In other words, the two-step mechanism and the single-step collision mechanism lead to the same rate expression.

The reaction between hydrogen and chlorine has a quite different mechanism from that of the H_2–I_2 reaction. The H_2–Cl_2 reaction is an example of a **chain reaction**. Such a reaction is started by the formation, at very low concentrations, of an extremely reactive species (Cl atoms in the case of the H_2–Cl_2 reaction). A chain reaction typically occurs very rapidly after a short induction period to allow for the formation of the reactive species.

A mixture of hydrogen and chlorine stored at room temperature in the dark shows no evidence of reaction over long periods of time. However, if the mixture is exposed to ultraviolet light or heated to 200°C, a vigorous reaction occurs. The first step in this reaction (*chain initiation*) is the reversible dissociation of a Cl_2 molecule into atoms:

$$Cl_2(g) \rightleftharpoons 2\ Cl(g) \tag{16.21a}$$

The chlorine atoms formed are extremely reactive toward hydrogen molecules:

$$Cl(g) + H_2(g) \rightarrow HCl(g) + H(g) \tag{16.21b}$$

This reaction forms a reactive hydrogen atom, which attacks a Cl_2 molecule:

$$H(g) + Cl_2(g) \rightarrow HCl(g) + Cl(g) \tag{16.21c}$$

About 10^6 HCl molecules are made for every Cl atom formed in the initiation. The mixture explodes

In this way, the chlorine atoms are regenerated and can react with more H_2 molecules. The *chain propagation*, represented by Equations 16.21b and 16.21c, occurs over and over again until all the H_2 and Cl_2 molecules are converted to HCl.

The hydrogen and chlorine atoms can be consumed by reaction with each other:

$$H(g) + Cl(g) \rightarrow HCl(g)$$
$$H(g) + H(g) \rightarrow H_2(g)$$
$$Cl(g) + Cl(g) \rightarrow Cl_2(g)$$

These processes represent *chain termination*, since they break the chain mechanism.

Summary

The rate of a reaction indicates how the concentration of a reactant or product changes with time (Example 16.1). A rate expression gives the relation between concentration of reactant and rate of reaction. The rate expression includes a rate constant, k, and concentration terms. The exponent of a concentration term indicates the order of the reaction with respect to that species. Reaction order can be determined by observing how rate varies as the initial concentration of a species changes (Example 16.2 and Table 16.2). Once the rate expression has been established, data relating rate to concentration can be used to calculate the rate constant k (Example 16.3).

As a reaction proceeds, the reactant concentration decreases with time. The relationships between concentration and time are summarized for zero-, first-, and second-order reactions in Table 16.3 and Figure 16.4. These relationships can be used to determine a concentration at a given time (Example 16.4) or to find the order of a reaction (Example 16.5).

A certain minimum energy, the activation energy, is necessary for a reaction to occur when two molecules collide (Fig. 16.6). In general, reaction rate is inversely related to activation energy; a fast reaction implies a low activation energy. A catalyst increases reaction rate by lowering the activation energy.

As temperature increases, a greater fraction of molecules have energies that exceed the activation energy (Fig. 16.9). Thus, we expect an increase in temperature to increase the rate of reaction (Table 16.4). Equation 16.16 relates temperatures, rate constants, and activation energy (Example 16.6).

A reaction mechanism shows the individual steps by which reaction occurs. Using certain rules (p. 505) a rate expression may be derived from the mechanism (Example 16.7).

Key Words and Concepts

activated complex	half-life	overall order
activation energy	heterogeneous catalysis	rate constant, k
Arrhenius equation	homogeneous catalysis	rate-determining step
catalyst	initial rate	rate expression
chain reaction	kinetics	rate of reaction
enzyme	mechanism of reaction	second-order reaction
first-order reaction	order of reaction	zero-order reaction

Questions and Problems

Meaning of Reaction Rate

16.1 For the reaction $A(g) \rightarrow 2 B(g)$, the concentration of A drops from 4.0 to 2.5 mol/L in 10.0 s. Calculate the average rate in terms of
  a. $-\Delta$ conc. $A/\Delta t$ in mol/L·min.
  b. Δ conc. $B/\Delta t$ in mol/L·s.

16.2 Express the rate of the reaction

$$2 \text{ HI}(g) \rightarrow H_2(g) + I_2(g)$$

  a. in terms of Δ conc. H_2.
  b. in terms of Δ conc. HI, if you want the rate to be the same as in (a).

16.3 Consider the following data for the decomposition of N_2O_5 at 57°C:

t (min)	0	2	4	6	8
Conc. N_2O_5	0.160	0.126	0.099	0.078	0.061

  a. Calculate the average rate between 2 and 6 min.
  b. Plot this data, as in Figure 16.1. Draw a tangent to the curve to find the rate at 4 min.

16.31 In the reaction: $2 X(g) \rightarrow Z(g)$, the concentration of X drops from 0.50 to 0.10 mol/L in 45 min. Calculate the average rate in terms of
  a. the decrease in concentration of X in mol/L·s.
  b. the increase in concentration of Z in mol/L·min.

16.32 For the reaction

$$4 \text{ HBr}(g) + O_2(g) \rightarrow 2 Br_2(g) + 2 H_2O(g)$$

the rate expression is most often written as rate $= -\Delta$ conc. $O_2/\Delta t$. Write expressions equivalent to this in terms of
  a. conc. HBr b. conc. Br_2 c. conc. H_2O

16.33 Using the data in Problem 16.3, find
  a. the average rate between 4 and 8 min.
  b. the rate at 6 min, using the plot referred to in Problem 16.3.

Rate Expressions

16.4 Complete the following table for the first-order reaction $D(g) \rightarrow$ products.

CONC. D (mol/L)	k (min^{-1})	RATE (mol/L·min)
0.60	5.0×10^{-2}	——
0.040	——	2.8
——	0.17	0.085

16.5 At 600 K the decomposition of NO_2 is second order with a rate of 2.0×10^{-3} mol/L·s when the NO_2 concentration is 0.080 M.
a. Write the rate expression.
b. Calculate k, the rate constant.
c. What are the units of k?
d. What is the rate when conc. NO_2 = 0.020 M?

16.6 The reaction $CO(g) + NO_2(g) \rightarrow CO_2(g) + NO(g)$ at 400°C is first order in both CO and NO_2. The rate constant is 0.50 L/mol·s. At what concentration of CO is the rate 0.10 mol/L·s when the concentration of NO_2 is
a. 0.40 mol/L?   b. equal to that of CO?

16.7 The reaction $E(g) \rightarrow$ products is second order. The rate is 0.080 mol/L·s when the concentration of E is 0.25 mol/L.
a. Write the rate expression.
b. Calculate k, the rate constant.
c. To double the rate, what must the concentration of E become?

16.8 The greatest increase in rate for the reaction between X and Z where rate = k(conc. X)(conc. Z)^2 will be caused by
a. doubling conc. Z.   b. doubling conc. X.
c. tripling conc. X.   d. lowering the temperature.

16.34 Complete the following table for the reaction $2 A(g) + Z(g) \rightarrow$ products, which is first order in both reactants.

CONC. A (mol/L)	CONC. Z (mol/L)	k (L/mol·s)	RATE (mol/L·s)
0.450	0.300	4.0	——
——	0.053	0.32	0.018
0.75	0.80	——	0.010

16.35 The decomposition of ammonia on tungsten is zero order with a rate of 2.5×10^{-4} mol/L·min at 1100 K when conc. NH_3 = 0.040 M.
a. Write the rate expression.
b. Calculate k, the rate constant.
c. What are the units of k?
d. What is the rate when conc. NH_3 = 0.015 M?

16.36 The reaction $O_3(g) + O(g) \rightarrow 2 O_2(g)$ is first order in both reactants with a rate constant of 7.8×10^5 L/mol·s at 25°C. At what concentration of O_3 is
a. the rate 2.0×10^{-8} mol/L·s when conc. O = 1.0×10^{-8} mol/L?
b. the rate 3.6×10^{-10} mol/L·s when conc. O = 2.2×10^{-7} mol/L?

16.37 The second-order reaction $J(g) \rightarrow$ products has a rate of 2.5 mol/L·min when conc. J = 0.15 M.
a. Write the rate expression.
b. Determine the rate constant k.
c. If the rate constant were half as large, what would conc. J have to be for the rate to remain 2.5 mol/L·min?

16.38 The greatest decrease in reaction rate for the reaction between A and D where rate = k × (conc. A)^2 (conc. D) is caused by
a. halving conc. D.   b. halving conc. A.
c. doubling conc. D.   d. halving conc. A and conc. D.

Determination of Reaction Order

16.9 In a certain reaction, $A(g) \rightarrow B(g)$, the rate is measured when conc. A is 0.10 M and again when it is 0.040 M. What is the order of the reaction if the ratio of the new rate to the original rate is
a. 0.40?   b. 1.0?   c. 0.16?

16.39 For the reaction referred to in Problem 16.9, the rate is measured again when the concentration of A is 0.010 M. What is the order if the ratio of this rate to the original rate is
a. 0.10?   b. 0.010?   c. 0.0010?

16.10 For a reaction involving a single reactant A, the following data are obtained:

Rate (mol/L·min)	0.020	0.016	0.013	0.010
Conc. A (M)	0.100	0.090	0.080	0.070

Determine the order of the reaction.

16.11 The following data refer to the reaction of A with D:

CONC. A (M)	CONC. D (M)	INITIAL RATE (mol/L·s)
0.10	0.10	4.0×10^{-4}
0.20	0.20	1.6×10^{-3}
0.50	0.10	1.0×10^{-2}
0.50	0.50	1.0×10^{-2}

a. What is the order with respect to A?
b. What is the order with respect to D?
c. Calculate k for the reaction.
d. When the concentrations of A and D are both 0.25 M, what is the rate?

16.12 In solution at constant H^+ concentration, I^- reacts with H_2O_2 to produce I_2:

$$2\,H^+(aq) + 2\,I^-(aq) + H_2O_2(aq) \rightarrow$$
$$I_2(aq) + 2\,H_2O$$

The reaction rate can be followed by monitoring iodine production. The following data apply:

CONC. I$^-$ (M)	CONC. H$_2$O$_2$ (M)	INITIAL RATE (mol/L·s)
0.020	0.020	3.3×10^{-5}
0.040	0.020	6.6×10^{-5}
0.060	0.020	9.0×10^{-5}
0.040	0.040	1.3×10^{-4}

a. What is the order with respect to I^-?
b. What is the order with respect to H_2O_2?
c. What is the rate when conc. $I^- = 0.010$ M, conc. $H_2O_2 = 0.030$ M?

16.13 The following data are obtained for the decomposition of acetaldehyde, CH_3CHO, at 700 K. Following the procedure used in Example 16.5, find the reaction order.

t (s)	0	50	100	150	200
Conc. CH$_3$CHO	0.400	0.333	0.286	0.250	0.222

16.40 For another reaction involving a single reactant, the following rate data are obtained:

Rate (mol/L·min)	0.020	0.020	0.019	0.021
Conc. reactant (M)	0.100	0.090	0.080	0.070

Determine the order of the reaction.

16.41 The data below are for the reaction of NO with Cl_2 to form NOCl at 295 K:

CONC. Cl$_2$ (M)	CONC. NO (M)	INITIAL RATE (mol/L·s)
0.050	0.050	1.0×10^{-3}
0.150	0.050	3.0×10^{-3}
0.050	0.150	9.0×10^{-3}

a. What is the order with respect to NO? with respect to Cl_2?
b. Write the rate expression.
c. Calculate k, the rate constant.
d. Determine the reaction rate when the concentrations of Cl_2 and NO are 0.20 M and 0.40 M, respectively.

16.42 The reaction $2\,X(g) + Z(g) \rightarrow$ products has the following rate data:

CONC. X (M)	CONC. Z (M)	INITIAL RATE (mol/L·h)
0.100	0.100	4.6×10^{-4}
0.200	0.100	9.1×10^{-4}
0.300	0.100	1.3×10^{-3}
0.100	0.200	1.8×10^{-3}

a. Evaluate the order of the reaction with respect to each reactant and the overall order.
b. Write the rate expression and calculate k.
c. Determine the initial rate when conc. X = 0.300 M, conc. Z = 0.150 M.

16.43 The following data are obtained for the decomposition of CH_3NO_2 at 500 K. Following the procedure used in Example 16.5, find the reaction order.

t (s)	0	300	600	900	1200
Conc. CH$_3$NO$_2$	0.200	0.145	0.105	0.076	0.055

First-Order Reactions

16.14 The following data are for the gas-phase decomposition of ethyl chloride, C_2H_5Cl, at 740 K:

t (min)	CONC. (M)	t (min)	CONC. (M)
0	0.200	4	0.187
1	0.197	8	0.175
2	0.193	16	0.153
3	0.190		

    a. By plotting the data, show that the reaction is first order.
    b. From the graph, determine k.
    c. Using k, find the time for the concentration to drop to one fourth of the original concentration.

16.15 The half-life of ethyl bromide, C_2H_5Br, at 720 K is 650 s. For this first-order reaction,
    a. find the rate constant.
    b. determine the time required for conc. C_2H_5Br to drop from 0.050 to 0.0125 M.
    c. find the concentration of C_2H_5Br one hour after the time elapsed in (b).

16.16 The first-order rate constant for the decomposition of ethyl chloride, C_2H_5Cl, at 700 K is 2.50×10^{-3} min^{-1}. Starting with a concentration of 0.200 mol/L, calculate
    a. conc. C_2H_5Cl after one hour; one day.
    b. the time required for conc. C_2H_5Cl to fall to one half its original value.

16.17 In the first-order decomposition of N_2H_4 at 740 K, it is found that 30.0% of a sample has decomposed in 13.2 min. How long will it take 50.0% of the sample to decompose?

16.44 The following data apply to the first-order decomposition of dimethyl ether, $(CH_3)_2O$:

t (s)	CONC. (M)
0	0.01000
200	0.00916
400	0.00839
600	0.00768
800	0.00703

    a. From a plot of conc. vs. time, estimate the rate at t = 800 s.
    b. From a plot of $\log_{10}$ conc. vs. time, determine k, the rate constant. Use k to obtain the rate at t = 800 s.
    c. Which of the rates just calculated in (a) or (b) do you think is more accurate? Explain.

16.45 The half-life for the first-order decomposition of acetone, $(CH_3)_2CO$, is 5.8 s at 650°C. Calculate
    a. the rate constant, k.
    b. how long it will take for the acetone concentration to change from 0.020 to 0.0020 mol/L.
    c. the acetone concentration 8.0 s after an initial concentration of 6.0 mol/L.

16.46 The first-order decomposition of C_2H_5Cl has a rate constant of 0.0170/min at 470°C. Starting with a concentration of 0.400 mol/L of C_2H_5Cl, calculate
    a. the concentration of C_2H_5Cl after 15.0 min.
    b. how long it will take for the concentration to drop from 0.400 to 0.100 mol/L
    c. the half-life.

16.47 The first-order decomposition of diazomethane, CH_2N_2, has a half-life of 17.3 min at 873 K. The concentration of CH_2N_2 is 0.058 mol/L after 10.0 min.
    a. What was the original concentration?
    b. How long will it take for 60% of the original sample to decompose?

Activation Energy, Catalysis, and ΔH

16.18 Consider the data for several systems for the conversion of reactants (R) to products (P) at the same temperature.

SYSTEM	E_a (kJ)	E_a' (kJ)
1	50	70
2	85	25
3	12	40

a. Which system has the highest forward rate?
b. What is ΔH of reaction for System 1?
c. For which system(s) is the forward reaction endothermic?

16.19 For a certain reaction, E_a is 50 kJ and ΔH is +10 kJ. In the presence of a catalyst, the activation energy is lowered to 30 kJ. Draw a diagram similar to Figure 16.7 for this reaction.

16.20 For the reaction

$$H_2(g) + I_2(g) \rightarrow 2\ HI(g)$$

E_a is 163 kJ. Taking ΔH_f HI(g) = +26 kJ, and ΔH_f $I_2(g)$ = +62 kJ, calculate the activation energy for the reverse reaction. Prepare a diagram similar to Figure 16.6 for this reaction.

16.48 Three reactions have the following activation energies:

Reaction	A	B	C
E_a (kJ)	145	210	48

Which reaction would you expect to be the fastest? the slowest? Explain.

16.49 The activation energy of a certain reaction is 70 kJ. In the presence of a catalyst, E_a is reduced to 45 kJ. For the reaction, ΔH is −60 kJ. Draw a diagram similar to Figure 16.7 for this reaction.

16.50 The activation energy for the decomposition of ammonia to the elements is about 300 kJ:

$$NH_3(g) \rightarrow \tfrac{1}{2} N_2(g) + \tfrac{3}{2} H_2(g)$$

Using Table 14.1, obtain the activation energy for the reverse reaction and draw a digram for this reaction similar to Figure 16.6.

Reaction Rate and Temperature

16.21 The following data are for the gas-phase decomposition of acetaldehyde:

$k \left(\dfrac{L}{mol \cdot s} \right)$	0.0105	0.101	0.60	2.92
T (K)	700	750	800	850

Plot these data and determine the activation energy for the reaction.

16.22 For the gas-phase decomposition of N_2O_5, k is 1.4/s at 400 K and 43/s at 450 K. What is the activation energy for this reaction?

16.23 For a certain reaction, E_a = 82 kJ. The rate constant k is 1.2×10^{-2} L/mol·s at 300 K. What is k at 500 K?

16.51 The following data were obtained for the reaction

$$SiH_4(g) \rightarrow Si(s) + 2\ H_2(g)$$

k (s^{-1})	0.048	2.3	49	590
t (°C)	500	600	700	800

Plot these data ($\log_{10}$ k vs. 1/T) and find the activation energy for the reaction.

16.52 The rate constants, in L/mol·s, are 1.1 and 3.0 at 823 and 873 K, respectively, for the gas-phase reaction

$$CH_4(g) + 2\ S_2(g) \rightarrow CS_2(g) + 2\ H_2S(g)$$

Calculate the activation energy of the reaction.

16.53 For the reaction in Problem 16.23, at what temperature does k = 1.2×10^{-3} L/mol · s?

16.24 Cold-blooded animals decrease their body temperature in cold weather to match that of their environment. The activation energy of a certain enzyme-catalyzed reaction in a cold-blooded animal is 65 kJ. By what percentage is the rate of this reaction decreased if the body temperature of the animal drops from 35 to 25°C?

16.25 Using Equation 16.13, calculate the fraction of molecules at 300 K with an energy of
a. 0.0 kJ   b. 10.0 kJ   c. 20.0 kJ

16.54 The chirping rate of a cricket, X, in chirps per minute, near room temperature is given by

$$X = 7.2\,t - 32$$

where t is the temperature in °C. Calculate the chirping rates at 25 and 35°C, and use them to estimate the activation energy for this reaction.

16.55 Using Equation 16.13, calculate the fraction of molecules with an energy of 50.0 kJ at
a. 300 K   b. 400 K   c. 500 K

Reaction Mechanisms

16.26 For the reaction

$$H_2(g) + Br_2(g) \rightarrow 2\,HBr(g)$$

a proposed mechanism is

Step 1:	$Br_2 \rightleftharpoons 2\,Br$	(fast)
Step 2:	$Br + H_2 \rightarrow HBr + H$	(slow)
Step 3:	$H + Br_2 \rightarrow HBr + Br$	(fast)

Using this mechanism, determine the rate expression in terms of the concentrations of H_2 and Br_2.

16.27 At low temperatures, the rate law for the reaction $CO(g) + NO_2(g) \rightarrow CO_2(g) + NO(g)$ is as follows: rate = constant × (conc. NO_2)^2. Which of the following mechanisms is consistent with this rate law?
a. $CO + NO_2 \rightarrow CO_2 + NO$
b. $2\,NO_2 \rightleftharpoons N_2O_4$ (fast)
   $N_2O_4 + 2\,CO \rightarrow 2\,CO_2 + 2\,NO$ (slow)
c. $2\,NO_2 \rightarrow NO_3 + NO$ (slow)
   $NO_3 + CO \rightarrow NO_2 + CO_2$ (fast)
d. $2\,NO_2 \rightarrow 2\,NO + O_2$ (slow)
   $2\,CO + O_2 \rightarrow 2\,CO_2$ (fast)

16.28 For the reaction

$$2\,NO(g) + 2\,H_2(g) \rightarrow N_2(g) + 2\,H_2O(g)$$

a proposed mechanism is

Step 1:	$2\,NO \rightleftharpoons N_2O_2$
Step 2:	$N_2O_2 + H_2 \rightarrow N_2O + H_2O$
Step 3:	$N_2O + H_2 \rightarrow N_2 + H_2O$

The observed rate expression is as follows: rate = k(conc. NO)^2(conc. H_2). If the above mechanism is correct, which is the rate-determining step? Explain your reasoning.

16.56 A possible mechanism for the reaction between H_2 and I_2 is

$I_2 \rightleftharpoons 2\,I$	(fast)
$H_2 + I \rightarrow HI + H$	(slow)
$H + I \rightarrow HI$	(fast)

Is this mechanism consistent with the rate expression, Equation 16.20?

16.57 Two mechanisms are proposed for the reaction

$$2\,NO(g) + O_2(g) \rightarrow 2\,NO_2(g)$$

Mechanism 1: $NO + O_2 \rightleftharpoons NO_3$ (fast)
              $NO_3 + NO \rightarrow 2\,NO_2$ (slow)

Mechanism 2: $NO + NO \rightleftharpoons N_2O_2$ (fast)
              $N_2O_2 + O_2 \rightarrow 2\,NO_2$ (slow)

Show that each of these mechanisms is consistent with the observed rate law, rate = k(conc. NO)^2 × (conc. O_2).

16.58 For the reaction $CO(g) + Cl_2(g) \rightarrow COCl_2(g)$, two mechanisms are proposed:

(1) $Cl_2 + CO \rightarrow CCl_2 + O$ (slow)
     $O + Cl_2 \rightarrow Cl_2O$ (fast)
     $Cl_2O + CCl_2 \rightarrow COCl_2 + Cl_2$ (fast)

(2) $Cl_2 \rightleftharpoons 2\,Cl$ (fast)
     $Cl + CO \rightleftharpoons COCl$ (fast)
     $Cl_2 + COCl \rightarrow COCl_2 + Cl$ (slow)

Which mechanism is consistent with the observed rate expression, rate = k(conc. CO)(conc. Cl_2)$^{3/2}$?

General

16.29 How does each of the following affect reaction rate?
a. The passage of time.
b. Decreasing container size for a gas phase reaction.
c. Adding a catalyst.

16.30 In your own words explain why
a. a decrease in temperature slows the rate of a reaction.
b. doubling the concentration of a reactant does not always double the rate of the reaction.
c. a flame lights a cigarette but the cigarette continues to burn after the flame is removed.
d. a catalyst does not change the equilibrium constant, K_c.

16.59 Explain why each of the following typically increases reaction rate.
a. Increasing the temperature.
b. Increasing reactant concentration.
c. Using an enzyme.

16.60 In terms of the concepts of this chapter, explain in your own words why
a. diamond is not converted to graphite even though the process has a large K_c value.
b. the rate constant for a zero order reaction equals the reaction rate.
c. an inhibitor reduces the rate of an enzyme-catalyzed reaction.
d. the slow step in a reaction determines the overall rate of the reaction.

*16.61 Using calculus, derive the equation for
a. the concentration-time relation for a second order reaction (see Table 16.3).
b. the concentration-time relation for a third order reaction, $3\ A \rightarrow$ products.

*16.62 The following data apply to the reaction

$$A(g) + 3\ B(g) + 2\ C(g) \rightarrow \text{products}$$

(concentrations are given in mol/L):

CONC. A	CONC. B	CONC. C	RATE
0.20	0.40	0.10	X
0.40	0.40	0.20	8X
0.20	0.20	0.20	X
0.40	0.40	0.10	4X

Determine the rate law for the reaction.

*16.63 In a first-order reaction, let us suppose that a quantity, X, of reactant is added at regular intervals of time, Δt. At first the amount of reactant in the system builds up; eventually, however, it levels off at a "saturation value" given by the expression

$$\text{saturation value} = \frac{X}{1 - 10^{-a}} \qquad \text{where a} = 0.30\ \frac{\Delta t}{t_{1/2}}$$

This analysis applies to the intake of mercury into the body, where one takes in a certain amount each day. The half-life for elimination of mercury appears to be about 70 days. Suppose that a person eats 50 g of fish containing 0.50 part per million of mercury each day. Using the above equation, calculate the number of grams of Hg in this body at "saturation," and compare to the value at which symptoms of mercury poisoning appear, about 0.014 g. How many grams of fish would he need to eat per day to reach the toxic limit?

*16.64 The reaction $Cl_2(g) + CH_4(g) \rightarrow CH_3Cl(g) + HCl(g)$ occurs through a chain mechanism in which Cl atoms and CH_3 radicals are chain propagators. No monatomic H atoms are believed to be involved. Write a sequence of reactions for the chain mechanism. Identify particular reactions as chain-initiating, chain-propagating, or chain-terminating steps.

The decomposition of N_2O_5 is believed to occur by the following mechanism:

$$2 \ N_2O_5(g) \rightleftharpoons N_2O_5^*(g) + N_2O_5(g)$$

$$N_2O_5^*(g) \rightleftharpoons NO_2(g) + NO_3(g)$$

$$NO_2(g) + NO_3(g) \rightarrow NO_2(g) + O_2(g) + NO(g)$$

$$NO(g) + NO_3(g) \rightarrow 2 \ NO_2(g)$$

where $N_2O_5^*$ represents an activated molecule. Show that the rate of formation of O_2 is directly proportional to the concentration of N_2O_5.

Life on this planet depends upon the relatively thin layer of air that surrounds it. The atmosphere accounts for only about 0.0001% of the total mass of the earth. Yet it is the reservoir from which we draw oxygen for metabolism, carbon dioxide for photosynthesis, and nitrogen, whose compounds are essential to plant growth. Our climate is governed by the movement of water vapor from the earth's surface into the atmosphere and back again.

Even trace components of the atmosphere can affect the delicate balance of life. Small amounts of ozone at a height of about 30 km absorb most of the harmful ultraviolet radiation of the sun. On the other hand, as little as 0.2 part per million of ozone near the earth's surface promotes smog formation.

Table 17.1 gives the mole fractions of gases in the atmosphere. Two species are omitted. One is water vapor, whose mole fraction may vary from 0.06 in the tropics to 0.0005 in polar regions. The other comprises suspended particles (e.g.,

Table 17.1
Composition of Clean, Dry Air at Sea Level

COMPONENT	MOLE FRACTION	COMPONENT	MOLE FRACTION	COMPONENT	MOLE FRACTION
N_2	0.7808	Ne	1.82×10^{-5}	SO_2	$<1 \times 10^{-6}$
O_2	0.2095	He	5.24×10^{-6}	O_3	$<1 \times 10^{-7}$
Ar	0.00934	CH_4	2×10^{-6}	NO_2	$<2 \times 10^{-8}$
CO_2	0.00034	Kr	1.14×10^{-6}	I_2	$<1 \times 10^{-8}$
		H_2	5×10^{-7}	NH_3	$<1 \times 10^{-8}$
		N_2O	5×10^{-7}	CO	$<1 \times 10^{-8}$
		Xe	8.7×10^{-8}	NO	$<1 \times 10^{-8}$

99% of the molecules in air are either N_2 or O_2

dust, smoke), which vary in both concentration and chemical composition. There are five major components of air (N_2, O_2, Ar, H_2O, and CO_2). Together, their mole fractions total about 0.99997.

In previous chapters (4 and 13) we looked at the physical and chemical properties of several components of the atmosphere (N_2, O_2, and the noble gases). With that background, we will consider in this chapter what might be called "selected topics" in atmospheric chemistry. These include

—the preparation of three important chemicals (NH_3, HNO_3, H_2SO_4) by reactions involving the two major components of the atmosphere, nitrogen and oxygen (Section 17.1).
—the effect of water vapor and carbon dioxide on our weather and climate (Section 17.2).
—the chemistry of the upper atmosphere (Section 17.3).
—air pollution (Section 17.4).

Throughout this chapter, we will review and expand upon the principles introduced in the past three chapters. In the text discussion and in examples, we will apply the concepts of chemical thermodynamics (Chapter 14), chemical equilibrium (Chapter 15), and reaction rates (Chapter 16).

17.1
Industrial Chemicals from Atmospheric Gases: NH₃, HNO₃, and H₂SO₄

The two major components of air, N_2 and O_2, are used in the manufacture of a variety of chemical products. In this section we will look at three large-scale industrial processes in which one or both of these gases are reactants. These processes are

—the Haber process for making ammonia from nitrogen and hydrogen.
—the Ostwald process for making nitric acid from ammonia, oxygen, and water.
—the contact process for making sulfuric acid from sulfur, oxygen, and water.

Nitrogen Fixation and the Haber Process for Making Ammonia

Combined ("fixed") nitrogen in the form of protein is essential to all forms of life. There is more than enough elementary nitrogen in the air, about 5×10^{18} kg, to meet all our needs. The problem is to convert the element to compounds that can be used by plants to make proteins. At room temperature and atmospheric pressure, N_2 does not react with any other element. Its inertness is due to the strength of the triple bond holding the N_2 molecule together:

The N≡N bond is one of the strongest known

$$:N{\equiv}N:(g) \rightarrow 2 \cdot\ddot{N}\cdot(g); \quad \Delta H = 941 \text{ kJ}$$

The high stability of the bond implies that the activation energy for any reaction of N_2 is likely to be high. Consequently, we expect the rate of reaction to be slow.

For thousands of years, nitrogen compounds have been added to the soil to increase the yield of food crops. Until about 100 years ago, the only way to do this was to add "organic nitrogen" (i.e., manure). Late in the nineteenth century, it became common practice in the United States and Western Europe to use sodium nitrate, $NaNO_3$, imported from Chile. Then, in 1908, Fritz Haber in Germany showed that atmospheric nitrogen could be fixed by reacting it with hydrogen to form ammonia. The reaction that Haber used was

$$N_2(g) + 3 H_2(g) \rightleftharpoons 2 NH_3(g); \qquad \Delta H = -92.4 \text{ kJ} \qquad \textbf{(17.1)}$$

It was called Chile saltpeter

His research was supported by German industrialists, who wanted to convert the ammonia to nitric acid, a starting material for making explosives. In 1913, the first large-scale ammonia plant went into production. During World War I, the Haber process produced enough ammonia and nitric acid to make Germany independent of foreign supplies of sodium nitrate, which were cut off by the British blockade. As has happened so many times, a scientific development was used first for military purposes and only later to meet social needs.

The Haber process is now the main synthetic source of fixed nitrogen in the world. Its feasibility depends upon choosing conditions under which nitrogen and hydrogen will react rapidly to give a high yield of ammonia. At room temperature and atmospheric pressure, the position of the equilibrium favors the formation of ammonia ($K_c = 5 \times 10^8$). However, the rate of reaction is virtually zero. Equilibrium can be reached more rapidly by increasing the temperature. However, since Reaction 17.1 is exothermic, high temperatures reduce K_c and hence the yield of ammonia. High pressures, on the other hand, have a favorable effect on both the rate of reaction and the position of the equilibrium (Table 17.2). An increase in pressure brings the gas molecules closer together. They collide more frequently, so equilibrium is reached more rapidly. High pressure also increases the relative amount of ammonia at equilibrium, since Reaction 17.1 results in a decrease in the number of moles of gas (4 mol → 2 mol).

Quite a few factors are involved in designing a process like this one

Much of Haber's research involved finding a catalyst to make Reaction 17.1 take place at a reasonable rate without going to very high temperatures. Nowadays, the catalyst used is a special mixture of iron, potassium oxide, and aluminum oxide. The reaction is carried out at 400 to 450°C and pressures of 200 to 600 atm. Ammonia (bp = −33°C) is condensed out as a liquid from the gaseous mixture. Unreacted hydrogen and nitrogen are recycled to raise the yield of ammonia.

Table 17.2
Effect of Temperature and Pressure Upon the Yield of Ammonia in the Haber Process ($[H_2] = 3[N_2]$)

°C	K_c	MOLE PERCENT NH_3 IN EQUILIBRIUM MIXTURE				
		10 atm	50 atm	100 atm	300 atm	1000 atm
200	650	51	74	82	90	98
300	9.5	15	39	52	71	93
400	0.5	4	15	25	47	80
500	0.08	1	6	11	26	57
600	0.014	0.5	2	5	14	31

K_c changes by a factor of 50,000 over the temperature range shown

EXAMPLE 17.1 Using tables in Chapter 14, calculate, for the reaction

$$N_2(g) + 3 H_2(g) \rightarrow 2 NH_3(g)$$

a. ΔH b. ΔS^0 c. the temperature at which $\Delta G^0 = 0$

Solution

a. $\Delta H = 2 \Delta H_f NH_3(g) = -92.4$ kJ

b. $\Delta S^0 = 2 S^0 NH_3(g) - S^0 N_2(g) - 3 S^0 H_2(g) = -198.3$ J/K
$$= -0.1983 \text{ kJ/K}$$

c. $\Delta G^0 = \Delta H - T\Delta S^0$

when $\Delta G^0 = 0$; $T = \dfrac{\Delta H}{\Delta S^0} = \dfrac{-92.4 \text{ kJ}}{-0.1983 \text{ kJ/K}} = 466$ K $= 193°C$

This is the temperature at which N_2, H_2, and NH_3, all at 1 atm pressure, are in equilibrium with each other.

EXERCISE Is this reaction spontaneous at 1 atm pressure and 200°C?
Answer: No; $\Delta G^0 = +1.4$ kJ.

NH₃ is the #2 chemical produced in the U.S., about 20×10^6 tons per year

Most of the ammonia produced today is used to make fertilizers. The pure liquid under pressure can be used directly. More commonly, it is converted to compounds containing the NH_4^+ ion. This is done by adding the ammonia to an acidic water solution:

$$NH_3(aq) + H^+(aq) \rightarrow NH_4^+(aq) \tag{17.2}$$

If the acid used is nitric acid, the final product is ammonium nitrate, NH_4NO_3. When sulfuric acid is used, ammonium sulfate, $(NH_4)_2SO_4$, is formed.

Smaller amounts of ammonia are used as a refrigerant in large-size ice-making machines. Household ammonia, a concentrated water solution of NH_3, is used for a variety of cleaning purposes.

The hydrogen used as a reactant in the Haber process accounts for virtually the entire cost of the ammonia formed. As pointed out in Chapter 4, most of our hydrogen comes from the reaction of natural gas with steam. Until quite recently, the price of hydrogen had been stable for decades at about 2 cents a kilogram. A sharp rise in the price of natural gas changed this situation in the early 1970s. It raised the price of hydrogen, the ammonia made from hydrogen, and the fertilizers made from ammonia. The effect has been particularly severe for many under-developed countries. They must have an abundant supply of cheap fertilizer to avert mass starvation. Two approaches are possible. One is to find a cheaper way to make hydrogen. The other is to discover a process for fixing nitrogen that does not involve hydrogen as a reactant.

Genetic engineering may furnish us corn and wheat that make their own fertilizer, like alfalfa does

Certain bacteria found in the soil or on the roots of legumes (peas, beans, clover, alfalfa) "fix" nitrogen on a large scale. It is estimated that these bacteria produce 10^8 metric tons of NH_3 annually; this is five times the amount formed by the Haber process. The bacterial conversion is catalyzed by enzymes described collectively as "nitrogenase." The mechanism of the process is not completely understood. Two different enzymes seem to be involved. Both enzymes contain iron and sulfur atoms; one of them contains molybdenum as well. Molecular nitrogen forms a weak complex with the metal atoms of the enzymes. This is then converted to a more stable ammonia complex. This process has been simulated in the laboratory, using model compounds containing another transition metal, titanium.

The Ostwald Process for Making Nitric Acid

About 15% of the ammonia made by the Haber process is converted to nitric acid. The process by which this is done was developed by a German chemist, Wilhelm Ostwald. The reaction takes place in three steps. All of these are carried out at a relatively low pressure, 1 to 10 atm.

1. Ammonia is burned in air at about 1000°C, using a platinum catalyst. Under these conditions, more than 95% of the ammonia is converted to nitric oxide, NO:

$$4 \, NH_3(g) + 5 \, O_2(g) \rightarrow 4 \, NO(g) + 6 \, H_2O(g) \tag{17.3}$$

With no catalyst you get N_2 and H_2O

2. The gaseous mixture produced is mixed with more air. This lowers the temperature and brings about the reaction

$$2 \, NO(g) + O_2(g) \rightleftharpoons 2 \, NO_2(g) \tag{17.4}$$

3. The nitrogen dioxide produced in the second step is passed through water. In this way a solution of nitric acid is formed:

$$3 \, NO_2(g) + H_2O(l) \rightarrow NO(g) + 2 \, HNO_3(aq) \tag{17.5}$$

The NO formed as a by-product is recycled in Reaction 17.4. The aqueous solution formed by Reaction 17.5 contains about 60 mass percent HNO_3. To obtain the anhydrous acid, H_2SO_4 is added and the mixture is distilled. Nearly pure nitric acid (bp = 86°C) boils off and is condensed.

EXAMPLE 17.2 Consider Reaction 17.4: $2 \, NO(g) + O_2(g) \rightleftharpoons 2 \, NO_2(g)$; $\Delta H = -113.0$ kJ. Applying the principles discussed in Chapters 15 and 16, what would you expect to happen to the rate and the yield of NO_2 at equilibrium if
a. the pressure were increased?
b. the temperature were increased?
c. a catalyst were used?

Optimizing the yield for a reaction like this is of economic importance

Solution
a. An increase in pressure should increase the rate (higher concentration of reactants). It should also increase the yield of NO_2 (3 mol gas → 2 mol gas).
b. One would expect an increase in temperature to increase the rate; k usually increases with T (see, however, Problems 17.29 and 17.59). An increase in temperature would decrease the yield of NO_2 since the forward reaction is exothermic.

c. A suitable catalyst would speed up the reaction but would have no effect upon the yield of NO_2.

EXERCISE Suppose pure O_2 were used in this reaction instead of air (same total pressure). What effect would you expect this to have on the rate and yield of NO_2? Answer: Should increase both.

We found out it was an explosive when a boat-load blew up in Texas City in 1947

The major use of nitric acid today is in the manufacture of ammonium nitrate, NH_4NO_3. This compound is an important component of fertilizers and many explosives. Upon heating to 200°C or detonation, ammonium nitrate decomposes rapidly to gaseous products:

$$NH_4NO_3(s) \rightarrow N_2O(g) + 2\ H_2O(g); \qquad \Delta H = -37.0\ \text{kJ} \qquad \textbf{(17.6)}$$

Nitric acid is also used to make other explosives, including nitroglycerine, trinitrotoluene (TNT), and cellulose nitrate, which is the main ingredient in smokeless powder.

The Contact Process for Making Sulfuric Acid

Sulfuric acid is made in larger quantities than any other chemical. Virtually all of it is produced by a three-step process, which we will now consider.

1. Elemental sulfur is burned in air to form sulfur dioxide:

$$S(s) + O_2(g) \rightarrow SO_2(g) \qquad \textbf{(17.7)}$$

2. Sulfur dioxide is converted to sulfur trioxide by bringing it into "contact" with oxygen on the surface of a solid catalyst:

$$SO_2(g) + \tfrac{1}{2}\ O_2(g) \rightleftharpoons SO_3(g); \qquad \Delta H = -99.1\ \text{kJ} \qquad \textbf{(17.8)}$$

The impurities occupy active sites on the surface of the catalyst

This is the key step in the process and the most difficult to carry out. The catalyst used today is an oxide of vanadium, V_2O_5. Platinum is equally effective but is more expensive and more easily "poisoned" (made ineffective) by impurities in the gases.

Since Reaction 17.8 is exothermic, we expect the equilibrium constant to decrease as the temperature rises. This is indeed the case (Table 17.3). Hence, according to equilibrium principles, a low temperature is preferred. Too low a temperature, however, reduces the rate to the point where Reaction 17.8 becomes impractical. The temperature actually used in the contact process represents a

Table 17.3
Equilibrium Constant for the Reaction
$SO_2(g) + 1/2\ O_2(g) \rightleftharpoons SO_3(g); \qquad \Delta H = -99.1\ \text{kJ}$

t (°C)	25	200	400	500	600	700	800
K_c	9.2×10^{12}	5.0×10^6	2300	400	70	20	7

compromise between equilibrium and rate considerations. A mixture of SO_2 and O_2 is first passed over the catalyst at 600°C; about 80% of the SO_2 is quickly converted to SO_3. The sulfur trioxide is removed from the gas mixture and the remaining SO_2 and O_2 are recycled over a second catalyst bed. This time the temperature is lower, about 450°C. It takes a while under these conditions, but eventually the yield of SO_3 is raised to above 99%.

3. The sulfur trioxide formed by Reaction 17.8 is converted to sulfuric acid by reaction with water:

$$SO_3(g) + H_2O \rightarrow H_2SO_4(aq) \tag{17.9}$$

SO_3 is a choking gas

This reaction cannot be carried out directly. If SO_3 is bubbled through water, H_2SO_4 is formed as a fog of tiny particles that are difficult to condense. Instead, SO_3 is absorbed in concentrated sulfuric acid to form an intermediate product, $H_2S_2O_7$, called pyrosulfuric acid. Subsequent addition of water to $H_2S_2O_7$ forms H_2SO_4:

$$SO_3(g) + H_2SO_4(l) \rightarrow H_2S_2O_7(l)$$
$$\underline{H_2S_2O_7(l) + H_2O \rightarrow 2\ H_2SO_4(aq)}$$
$$SO_3(g) + H_2O \rightarrow H_2SO_4(aq) \tag{17.9}$$

EXAMPLE 17.3 At 600° C, K_c for Reaction 17.8 is 70 (Table 17.3). Suppose that, at this temperature, $[SO_2] = 0.020$ mol/L and $[O_2] = 0.010$ mol/L. Calculate the equilibrium concentration of SO_3 under these conditions.

Solution The expression for K_c is

$$K_c = \frac{[SO_3]}{[SO_2] \times [O_2]^{1/2}} = 70$$

Solving for $[SO_3]$,

$$[SO_3] = 70 \times [SO_2] \times [O_2]^{1/2}$$

Substituting for $[SO_2]$ and $[O_2]$,

$$[SO_3] = 70 \times 0.020 \times (0.010)^{1/2} = 0.14 \text{ mol/L}$$

EXERCISE Assume the original mixture contained only SO_2 and O_2. What must the original concentration of SO_2 have been? What percentage of the SO_2 was converted to SO_3? Answer: 0.16 M; 88%.

Sulfuric acid has a variety of uses. About 70% of the sulfuric acid produced is now used to make fertilizers of one type or another. One of these, ammonium sulfate, $(NH_4)_2SO_4$, was mentioned earlier. Another is calcium dihydrogen phosphate, $Ca(H_2PO_4)_2$. This is made by treating "phosphate rock," which is mostly calcium phosphate, with sulfuric acid:

H_2SO_4 is the work-horse among industrial acids

$$Ca_3(PO_4)_2(s) + 2\ H_2SO_4(l) \rightarrow Ca(H_2PO_4)_2(s) + 2\ CaSO_4(s) \tag{17.10}$$

17.2
Water Vapor and
Carbon Dioxide;
Weather and Climate

The properties and composition of the atmosphere determine our weather over the short term and our climate over the years. Many factors affect both weather and climate. Among these are the presence in air of two gases: water vapor and carbon dioxide.

Relative Humidity

The concentration of water vapor in the air is often expressed in terms of *relative humidity*:

$$R.H. = \frac{P_{H_2O}}{P^0_{H_2O}} \times 100\% \qquad (17.11)$$

where P_{H_2O} is the partial pressure of water vapor in the air and $P^0_{H_2O}$ is the equilibrium vapor pressure of water at the same temperature. On a day when the temperature is 25°C ($P^0_{H_2O}$ = 23.8 mm Hg) and the partial pressure of water vapor in the air is 20.0 mm Hg,

> At 84% R.H., the air contains 84% of the water it could hold at that temperature

$$R.H. = \frac{20.0}{23.8} \times 100\% = 84.0\%$$

Our comfort depends upon relative humidity as well as temperature. At relative humidities below about 30%, evaporation of water from body surfaces is extensive and rapid. Mouth and nasal membranes dry out, allowing viruses to enter the lungs more readily. This may be why colds are so common in the winter months, when the relative humidity indoors is often quite low. Most of us are familiar with the uncomfortable effects of high relative humidities, above 80%. Perspiration fails to evaporate, leaving us feeling clammy and hot.

Changes in temperature can bring about marked variations in relative humidity. From Equation 17.11, we note that relative humidity is inversely related to the equilibrium vapor pressure of water, $P^0_{H_2O}$. Since $P^0_{H_2O}$ increases with temperature, this means that relative humidity tends to drop when the temperature rises (Example 17.4).

EXAMPLE 17.4 On a cold winter day, when the outside temperature is 0°C (32°F), the partial pressure of water in the air is 3.0 mm Hg. Calculate the relative humidity of the outside air and the relative humidity of the same air when it is brought into a house and warmed to 20°C (68°F).

Solution The equilibrium vapor pressure of water is 4.6 mm Hg at 0°C and 17.5 mm Hg at 20°C (Appendix 1). Consequently,

$$R.H. \text{ outside} = \frac{3.0}{4.6} \times 100\% = 65\%$$

$$\text{R.H. inside} = \frac{3.0}{17.5} \times 100\% = 17\%$$

EXERCISE At what temperature would the relative humidity become 10%?
Answer: 29°C (84°F), where vp H_2O is 30.0 mm Hg.

Cloud Formation and Weather Modification

If a warm air mass is suddenly cooled, $P^0_{H_2O}$ in Equation 17.11 drops and the relative humidity rises. When it reaches 100%, liquid water condenses. This is precisely the way in which clouds are formed in the atmosphere. Clouds consist of many billions of tiny droplets of liquid water, on the average perhaps 0.01 mm in diameter. Droplets of this size are too small to fall to the earth's surface as rain. The growth of small water droplets at temperatures above 0°C is ordinarily a very slow process. The rate of growth is increased in the presence of dust particles, which act as nuclei upon which small droplets can condense. This explains why volcanic eruptions are often followed by rainstorms.

Why don't they fall?

More frequently, ice crystals formed in the colder upper regions of clouds act as nuclei for precipitation. In principle, ice crystals should form at 0°C; in practice, they seldom develop unless the temperature drops to at least −15°C. To stimulate the formation of ice crystals, clouds are sometimes seeded with dry ice (Fig. 17.1). The sublimation of solid carbon dioxide absorbs enough heat from the cloud to reduce the temperature below that required for ice crystal formation. Another substance that is frequently used is finely divided silver iodide, which

FIGURE 17.1 Cloud seeding with Dry Ice. The particles of dry ice sublime rapidly in the cloud, cooling the fine water droplets to the point where they freeze and act as nuclei for condensation of other droplets in the cloud.

has a crystal structure similar to that of ice. The presence of silver iodide tends to prevent supercooling and, hence, allows ice crystals to form at temperatures close to 0°C.

Silver iodide has also been used with some success in seeding hurricanes. The objective here is to add so much silver iodide that an enormous number of tiny ice crystals form. These crystals are too small to bring about precipitation, but the heat evolved in their formation tends to dissipate the storm clouds at the eye of the hurricane. One difficulty is that hurricane seeding may simply shift the storm off course without weakening it appreciably. For this reason, seeding experiments are limited to storms that are far removed from populated areas.

Another difficulty is that AgI costs a dollar a gram

The Greenhouse Effect

The mean global temperature at the earth's surface is about 15°C (59°F). This temperature is determined by a delicate balance between:

—the energy that is absorbed from the sun. This energy covers a broad spectrum of wavelengths from the ultraviolet (< 400 nm) through the visible (400 to 800 nm) into the infrared (> 800 nm).

—the energy emitted back into space by the earth. This consists of infrared radiation, mostly at wavelengths between 5000 and 25,000 nm.

Any change in the amount of energy absorbed or emitted by the earth could upset this balance, affecting our climate. In this connection, consider what happens to the infrared radiation given off by the earth. Part of it is absorbed by the atmosphere rather than being radiated to outer space. Two gases in the air, H_2O and CO_2, absorb infrared radiation (Fig. 17.2). In this way, they act as an insulating blanket to prevent heat from escaping; this is often referred to as a "greenhouse effect." Were it not for this effect, the earth's temperature would be much lower, of the order of −25°C. On Mars, which has a very thin atmosphere, the surface temperature is −50°C. In contrast, Venus, which has a dense atmosphere consisting mostly of CO_2, has a temperature of 400°C. This difference in temperature is much greater than could be explained by the fact that Venus is closer to the sun; in large part it is due to the greenhouse effect.

The atmosphere has many roles besides providing us O_2

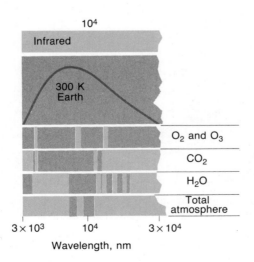

FIGURE 17.2 The earth gives off infrared radiation in the wavelength range 3000 to 30,000 nm, with a maximum emission close to 10,000 nm. A small amount of this is absorbed by O_2 and O_3. Larger amounts are absorbed by CO_2 and H_2O, as indicated by the wide absorption bands (*black*) for these two species.

Of the two gases, water vapor absorbs more infrared radiation than carbon dioxide because its concentration is higher. This property of water vapor accounts for the fact that the temperature drops less on nights when there is a heavy cloud cover. In desert regions, where there is very little water vapor, large variations between day and night temperatures are common.

The concentration of CO_2 in the earth's atmosphere is low, about 340 ppm*, but appears to be increasing exponentially. A century ago the CO_2 concentration was less than 300 ppm. Projections indicate that by early in the next century, the amount of CO_2 in the air may be twice what it is today. The increase in CO_2 content has been caused by man's activities. Increased consumption of fossil fuels is mainly responsible. Extensive land clearing, which reduces the amount of CO_2 consumed in photosynthesis, is also a factor.

A doubling of CO_2 concentration in the atmosphere could increase the mean global temperature by 2 or 3°C. Such an increase would affect the world's climate. It is quite possible that the polar ice caps would melt, raising the level of the oceans by 5 to 6 m. That would flood many coastal areas, including half the state of Florida. On a more optimistic note, an increase in CO_2 concentration is known to promote photosynthesis. Doubling the amount of carbon dioxide in the air could increase the world's food supply by raising yields of such crops as rice, wheat, and soybeans.

Nuclear weapons are a more likely source of big trouble

17.3
The Upper Atmosphere

The species in air near the surface of the earth are mostly stable atoms and molecules such as N_2, O_2, Ar, CO_2, and H_2O. Above about 30 km, the situation is quite different. Exploration of the upper atmosphere by rockets and satellites reveals additional species such as those listed in Table 17.4. Some of these are

Table 17.4
Formation of High-Energy Species in the Upper Atmosphere

MOLECULAR FRAGMENTS			CATIONS		
Reaction	ΔE (kJ/mol)	λ_{max}* (nm)	Reaction	ΔE (kJ/mol)	λ_{max}* (nm)
$NO_2 \rightarrow NO + O$	+305	392	$NO \rightarrow NO^+ + e^-$	+892	134
$O_2 \rightarrow O + O$	+494	242	$O_2 \rightarrow O_2^+ + e^-$	+1165	103
$H_2O \rightarrow H + OH$	+502	238	$O \rightarrow O^+ + e^-$	+1314	91
$NO \rightarrow N + O$	+632	189	$H \rightarrow H^+ + e^-$	+1312	91
$N_2 \rightarrow N + N$	+941	127	$N_2 \rightarrow N_2^+ + e^-$	+1510	79

*Longest wavelength of light that can supply the energy indicated.

---

*For a gaseous species,

parts per million (ppm) = 10^6 X

where X is the mole fraction. Recall from Table 17.1 that the mole fraction of CO_2 in air is 0.00034. Hence, the concentration of CO_2 in parts per million is 0.00034 × 10^6 = 340 ppm.

neutral particles (oxygen, nitrogen, and hydrogen atoms, and OH radicals). Others are positive ions formed from diatomic molecules (NO^+, O_2^+, N_2^+) or atoms (O^+, H^+).

All the reactions listed in Table 17.4 are endothermic. They require either the breaking of a chemical bond or the removal of an electron. In the upper atmosphere, sunlight is the source of energy for such reactions. The wavelengths of light given in Table 17.4 are calculated from the Einstein equation (Chap. 7) in the form

$$\lambda_{max} \text{ (nm)} = \frac{1.196 \times 10^5}{\Delta E \text{ (kJ/mol)}} \qquad (17.12)$$

Notice that all of the wavelengths listed are in the ultraviolet region, below 400 nm. Very little of this high-energy radiation reaches the surface of the earth, so the reactions in Table 17.4 are not very likely to occur in the air around us. Above 30 km, significant amounts of radiation in the near ultraviolet (200 to 400 nm) are available, and species such as O, H, and OH begin to appear. At still higher altitudes, sunlight contains radiation in the far ultraviolet (<200 nm). This can bring about highly endothermic processes, forming such species as N, O_2^+, and N_2^+.

The high energy species are formed by photochemical reactions

The concentrations of species such as those listed in Table 17.4 are much higher in the upper atmosphere than we would expect on the basis of equilibrium considerations. Consider, for example, the system

$$O_2(g) \rightleftharpoons 2 \, O(g)$$

We can calculate (Example 17.5) that the equilibrium constant, K_c, for this reaction is a very small number, about 4×10^{-83}. On that basis, it would seem that the forward reaction should occur to only a tiny extent. This is indeed the case at the earth's surface, where virtually all elemental oxygen is in the form of O_2 rather than oxygen atoms. However, at 120 km, there are about as many O atoms as O_2 molecules. The concentration of atomic oxygen is far higher than we would expect from the value of K_c.

One way to explain this situation is to consider the distance between particles in the atmosphere. At the earth's surface, atoms and molecules in the air are close together. The *mean free path* (average distance traveled by a particle between collisions) is only 7×10^{-8} m. Collisions are frequent and equilibrium is rapidly established. At high altitudes, the situation is quite different. Particles are much farther apart; the mean free path at 120 km is 4 m. Once an oxygen atom is formed at this altitude, it may be a long time before it collides with another one to form an O_2 molecule. Moreover, at this altitude, oxygen atoms have very high kinetic energies. Even if they collide, two atoms may have too much energy to stick together. As a result, thermodynamic equilibrium is never established. The concentration of O atoms relative to O_2 molecules is much higher than it should be from equilibrium considerations.

To have chemical equilibrium you need lots of collisions

EXAMPLE 17.5 For the reaction $O_2(g) \rightarrow 2 \, O(g)$, ΔG^0 is $+460$ kJ at 25°C. At this temperature, calculate
a. K_p b. K_c

Solution

a. We use Equation 15.11, p. 471: $\Delta G^0 = -0.0191T \log_{10} K_p$

$$\log_{10} K_p = \frac{-\Delta G^0}{0.0191T} = \frac{-460}{0.0191 \times 298} = -81.0$$

$$K_p = 1 \times 10^{-81}$$

b. Recall Equation 15.9, p. 470: $K_p = K_c(0.0821T)^{\Delta n_g}$. Here, $\Delta n_g = 2 - 1 = 1$:

$$K_c = \frac{K_p}{0.0821T} = \frac{1 \times 10^{-81}}{0.0821 \times 298} = 4 \times 10^{-83}$$

EXERCISE ΔH for this reaction is $+495$ kJ. Taking ΔG^0 to be $+460$ kJ at 298 K, calculate ΔS^0. Answer: $+1.2 \times 10^2$ J/K.

Ozone

One component of the upper atmosphere has received more attention than any other in recent years. This is ozone, an allotropic form of oxygen, whose electronic structure was discussed in Chapter 13. Interest in O_3 centers upon its ability to absorb ultraviolet light. From 95% to 99% of sunlight in the wavelength range of 200 to 300 nm is absorbed by ozone in the upper atmosphere. If this radiation were to reach the surface of the earth, it could have several adverse effects. A decrease in O_3 concentration of only 5% could increase the incidence of skin cancer by 25%.

O_3 is a good UV absorber, much better than O_2 at 250 nm

The concentration of ozone in the earth's atmosphere passes through a maximum of about 10 ppm at an altitude of 30 km. The ozone in this region is formed by a two-step process. The first step involves the dissociation of an O_2 molecule:

$$O_2(g) \rightarrow 2\,O(g) \tag{17.13}$$

This is followed by a collision between an oxygen atom and an O_2 molecule:

$$O_2(g) + O(g) \rightarrow O_3(g) \tag{17.14}$$

Ozone molecules formed by Reaction 17.14 decompose by several mechanisms. One of the most important is

$$O_3(g) + O(g) \rightarrow 2\,O_2(g) \tag{17.15}$$

This reaction takes place at a rather slow rate by direct collision between an O_3 molecule and an O atom. It can occur more rapidly by a two-step process in which a trace component of the upper atmosphere acts as a catalyst. Two such catalysts that have received a great deal of attention are listed in Table 17.5.

Since NO acts as a catalyst for ozone decomposition, an increase in nitric oxide concentration in the upper atmosphere could cause the depletion of ozone. This was one factor that influenced the United States to abandon the development of the supersonic transport (SST). These planes burn jet fuel with air at high

Table 17.5
Mechanism for the Catalytic Decomposition of Ozone

	CATALYST	
	NO Molecule	**Cl Atom**
Mechanism	$NO + O_3 \rightarrow NO_2 + O_2$	$Cl + O_3 \rightarrow ClO + O_2$
	$NO_2 + O \rightarrow NO + O_2$	$ClO + O \rightarrow Cl + O_2$
Overall reaction	$O_3 + O \rightarrow 2\,O_2$	$O_3 + O \rightarrow 2\,O_2$

temperatures. It was suggested that combustion might produce significant amounts of NO by the reaction

$$N_2(g) + O_2(g) \rightarrow 2\,NO(g) \tag{17.16}$$

Most of the NO in the upper atmosphere today is formed from N_2O by the reaction

$$N_2O(g) + O(g) \rightarrow 2\,NO(g) \tag{17.17}$$

The N_2O is formed by the decomposition of nitrogen-containing fertilizers (recall Equation 17.6).

In recent years, concern has focused on the Cl-catalyzed decomposition of ozone. At the time this mechanism was discovered, in 1973, it was believed to be unimportant. There was no known source of Cl atoms in the upper atmosphere. Less than a year later it was suggested that two organic compounds, $CFCl_3$ and CF_2Cl_2, might be a source of Cl. These compounds are widely used as refrigerants and aerosol propellants. They decompose to form chlorine atoms when exposed to ultraviolet radiation at 200 nm:

$$CFCl_3(g) \rightarrow CFCl_2(g) + Cl(g) \tag{17.18}$$

$$CF_2Cl_2(g) \rightarrow CF_2Cl(g) + Cl(g) \tag{17.19}$$

Light at this wavelength is readily available at a height of 40 km. Significant amounts of $CFCl_3$, CF_2Cl_2, and Cl atoms have been detected at this altitude. So far, the evidence is mixed regarding the lowering of O_3 concentration by this mechanism. Calculations suggest that the Cl-catalyzed reaction could eventually reduce the amount of ozone in the upper atmosphere by 5% to 9%. To ensure that this does not happen, the use of $CFCl_3$ and CF_2Cl_2 as aerosol propellants has been phased out in the United States, Canada, and Sweden. They are being replaced by other gases, including CO_2, C_3H_8, and C_4H_{10}. The latter two compounds, propane and butane, have the disadvantage of being flammable.

EXAMPLE 17.6 Consider the reaction mechanisms

$$O_3(g) + O(g) \rightarrow 2\,O_2(g); \qquad k_1 = 5 \times 10^6 \text{ L/mol·s}$$

$$O_3(g) + NO(g) \rightarrow NO_2(g) + O_2(g); \qquad k_2 = 1 \times 10^7 \text{ L/mol·s}$$

a. Write the rate expressions for the two mechanisms.

Another reason we dropped the SST is that supersonic flight is energy-inefficient

Don't light up while you're applying shaving cream

b. Calculate the ratio of the two rates (rate$_2$/rate$_1$) at an altitude of 40 km. Take the concentrations of O and NO to be 2×10^{-12} and 3×10^{-12} mol/L, respectively.

Solution

a. rate$_1$ = k$_1$(conc. O$_3$) × (conc. O)
   rate$_2$ = k$_2$(conc. O$_3$) × (conc. NO)

b. Dividing,

$$\frac{\text{rate}_2}{\text{rate}_1} = \frac{k_2}{k_1} \times \frac{\text{conc. NO}}{\text{conc. O}} = \frac{1 \times 10^7}{5 \times 10^6} \times \frac{3 \times 10^{-12}}{2 \times 10^{-12}} = 3$$

This calculation suggests that, under these conditions, O$_3$ is reacting with NO three times as rapidly as with O atoms.

EXERCISE Suppose the concentration of O were twice that of NO. How would the two rates compare in that case? Answer: Equal.

17.4
Air Pollution

Since the discovery of fire, mankind has polluted the atmosphere with noxious gases and soot. When coal began to be used as a fuel in the fourteenth century, the problem became one of public concern. Increased fuel consumption by industry, concentration of population in urban areas, and the advent of motor vehicles have, over the years, made the problem worse. Today a major cause of pollution in our atmosphere is the gasoline engine.

We're slowly making progress on this problem

Any substance whose addition to the atmosphere produces a measurable adverse effect on human beings or the environment can be called a pollutant. A host of materials fit this broad definition. Suspended particles (soot, dust, smoke) qualify, as do radioactive species produced by fallout from nuclear testing. In this chapter we will limit our attention to a few major pollutants:

—the oxides of sulfur (SO$_2$, SO$_3$), formed when sulfur or sulfur compounds burn in air, and sulfuric acid (H$_2$SO$_4$), formed when sulfur trioxide reacts with water.
—carbon monoxide (CO), formed by the incomplete combustion of hydrocarbon fuels.
—the oxides of nitrogen (NO, NO$_2$), formed in high-temperature combustion processes.

Sulfur Compounds
(SO$_2$, SO$_3$, H$_2$SO$_4$);
Acid Rain

Most of the coal burned in heating and power plants in the United States contains from 1% to 3% sulfur, much of which is in the form of minerals such as pyrite, FeS$_2$. Combustion converts the sulfur to sulfur dioxide:

$$4 \text{ FeS}_2(s) + 11 \text{ O}_2(g) \rightarrow 2 \text{ Fe}_2\text{O}_3(s) + 8 \text{ SO}_2(g) \qquad \textbf{(17.20)}$$

About two thirds of the SO_2 that enters the atmosphere comes from the combustion of coal. Smaller amounts are formed from heating oil and from metallurgical processes including the "roasting" of sulfide ores (Chapter 4).

Sulfur dioxide at concentrations as low as 0.3 ppm can cause acute injury to plants. Most healthy adults can tolerate considerably higher SO_2 levels without apparent adverse effects. However, individuals who suffer from chronic respiratory diseases such as bronchitis or asthma are much more sensitive. Increasing the concentration of SO_2 from 0.1 to 0.2 ppm can cause them to start coughing and experience severe difficulties in breathing.

You can smell SO_2 at very low concentrations

Much of the sulfur dioxide in polluted air is converted to sulfur trioxide by reactions such as

$$SO_2(g) + \tfrac{1}{2} O_2(g) \rightarrow SO_3(g) \tag{17.8}$$

$$SO_2(g) + 2\, OH(g) \rightarrow SO_3(g) + H_2O(g) \tag{17.21}$$

Reaction 17.8 is, you will recall, one of the steps involved in the commercial preparation of sulfuric acid. It is catalyzed by suspended solids in the atmosphere such as those found in coal smoke. Reaction 17.21 involves the OH radical

$$H\!-\!\ddot{\underset{..}{O}}\cdot$$

produced by photochemical reactions of the type discussed in Section 17.3.

EXAMPLE 17.7 The major source of the OH radicals involved in Reaction 17.21 is the reaction between oxygen atoms and water molecules:

$$O(g) + H_2O(g) \rightarrow 2\, OH(g)$$

The rate constant for this second-order reaction is 2.8×10^{-3} L/mol·s at 27°C. If the activation energy is 77 kJ, calculate k for this reaction at an altitude of 3 km, where the temperature is -4°C.

Solution We use the two-point form of the Arrhenius equation (Chapter 16):

$$\log_{10} \frac{k_2}{k_1} = \frac{E_a(T_2 - T_1)}{(2.30)(8.31)T_2T_1}$$

Here,

$$k_2 = 2.8 \times 10^{-3} \text{ L/mol·s}; \qquad E_a = 7.7 \times 10^4 \text{ J}$$

$$T_2 = 27 + 273 = 300 \text{ K}; \qquad T_1 = -4 + 273 = 269 \text{ K}$$

We substitute these values and solve for k_1:

$$\log_{10} \frac{k_2}{k_1} = \frac{7.7 \times 10^4}{(2.30)(8.31)} \left(\frac{300 - 269}{300 \times 269} \right) = 1.55$$

$10^{1.55} = 35$

$$\frac{k_2}{k_1} = 35; \qquad k_1 = \frac{k_2}{35} = \frac{2.8 \times 10^{-3}}{35} = 8.0 \times 10^{-5} \text{ L/mol·s}$$

EXERCISE   At the lower edge of the stratosphere (altitude 11 km), the temperature is $-55°C$. Calculate k for this reaction at that temperature. Answer: 2.5×10^{-8} L/mol·s.

Most of the adverse effects of sulfur oxides in the atmosphere are caused by the sulfuric acid produced when SO_3 reacts with water:

$$SO_3(g) + H_2O \rightarrow H_2SO_4(aq)$$

The sulfuric acid forms as tiny droplets high in the atmosphere. These may be carried by prevailing winds as far as 1000 miles before falling to the earth as *acid rain*. Coal-burning utilities in the Midwest appear to be largely responsible for the high acidity of rainfall in the northeastern United States and Canada (Fig. 17.3). The concentration of H^+ ions in acid rain is 5 to 100 times that in distilled water.

The effects of acid rain are particularly severe in areas where the bedrock is granite or other materials incapable of neutralizing H^+ ions. This situation prevails in the Adirondacks and the mountains of northern New England. Lakes in these regions are exposed annually to as much as 4 metric tons of sulfuric acid per square kilometer. As the concentration of acid builds up, marine life, from algae to brook trout, dies. The end product is a crystal clear, totally sterile lake. There are over 200 such dead lakes in the Adirondacks alone, all of them attributable to acid rain.

$4000 \text{ kg/km}^2 = 4 \text{ g/m}^2$

There is increasing evidence that acid rain has an adverse effect on trees as well as marine life (Fig. 17.4). Concern over this effect convinced West Germany and Switzerland to commit themselves to a policy of reducing SO_2 emissions by

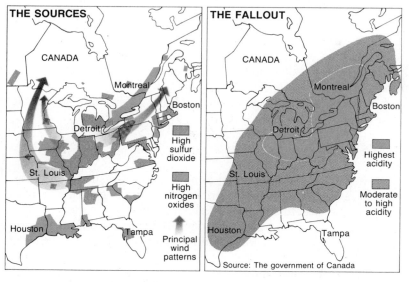

As you might expect, the Canadians strongly object to our polluting their lakes with acid rains

FIGURE 17.3   Acid rain is caused by dissolved H_2SO_4 and, to a lesser extent, HNO_3. Most of the sulfur oxides that are largely responsible for acid rain come from midwestern states such as Illinois, Indiana, and Ohio. Prevailing winds carry the acid droplets to the northeast, as far as the maritime provinces of Canada.

FIGURE 17.4 Since 1965, nearly half the spruces have died on Camels Hump in Vermont's Green Mountains. Mounting evidence suggests that acid rain may be the cause. (Photo courtesy of David Like, Botany Department, University of Vermont.)

50%. Canada has proposed a joint program with the United States to achieve the same goal. So far, the response from Washington has been to "study the problem."

It appears that much of the damage caused by acid rain is due to the leaching of toxic metal cations from the soil. For example, H^+ ions in acid rain can react with insoluble aluminum compounds in the soil, bringing Al^{3+} ions into solution. The following reaction is typical:

$$Al(OH)_3(s) + 3\ H^+(aq) \rightarrow Al^{3+}(aq) + 3\ H_2O \qquad \text{(17.22)}$$

Al^{3+} ions in water solution are taken up by the marine life and the roots of trees, where they can have a lethal effect.

Sulfuric acid also attacks building materials such as limestone or marble (calcium carbonate):

$$CaCO_3(s) + H_2SO_4(aq) \rightarrow CaSO_4(s) + CO_2(g) + H_2O \qquad \text{(17.23)}$$

The calcium sulfate formed is soluble enough to be gradually washed away (Fig. 17.5). This process is responsible for the deterioration of the Greek ruins on the Acropolis in Athens. These structures have suffered more damage in the twentieth century than in the preceding 2000 years. Another effect that is due to sulfuric acid is the deterioration of the paper in books and documents. Manuscripts printed before 1750 are almost immune to sulfur oxides. At about that time modern methods of papermaking were introduced. These leave traces of metal oxides, which catalyze the conversion of SO_2 to SO_3 and hence to sulfuric acid.

To reduce "acid rain," it is most important to lower sulfur dioxide emissions from coal-burning power plants. One way to solve the problem is to remove sulfur compounds from the coal before it is burned. Pyrite, FeS_2, can be separated from coal by "washing" with a concentrated solution of calcium chloride (d = 1.35

FIGURE 17.5 Marble statues sometimes become unrecognizable as a result of attack by oxides of sulfur. (From Wagner, R. H., *Environment and Man,* W. W. Norton & Co., New York, 1971.)

g/cm³). The coal (d = 1.2 g/cm³) floats in this solution, while pyrite (d = 4.9 g/cm³) sinks to the bottom. To use this method, the coal must be pulverized. Moreover, the method does not work with organic sulfur compounds, which are common in coal.

This method is expensive

A different approach involves adding a chemical to react with sulfur dioxide after it is formed. One possibility is to add limestone, $CaCO_3$, to the furnace where coal is burning. The limestone decomposes:

$$CaCO_3(s) \rightarrow CaO(s) + CO_2(g)$$

The calcium oxide formed reacts with sulfur dioxide and oxygen:

$$CaO(s) + SO_2(g) + \tfrac{1}{2} O_2(g) \rightarrow CaSO_4(s) \tag{17.24}$$

It is not easy to clean up smokestack gas

Alternatively, a "scrubber" charged with a water solution of calcium hydroxide can be inserted directly into the smoke stack. A reaction very similar to Reaction 17.24 occurs, this time in aqueous solution:

$$Ca^{2+}(aq) + 2\,OH^-(aq) + SO_2(g) + \tfrac{1}{2} O_2(g) \rightarrow CaSO_4(s) + H_2O \tag{17.25}$$

Both of these approaches produce large amounts of calcium sulfate that must be disposed of. Partly for this reason, they add to the cost of the electrical energy produced by the combustion of coal.

Carbon Monoxide (CO)

The incomplete combustion of hydrocarbon fuels produces significant amounts of carbon monoxide. The principal culprit here is the automobile. If enough oxygen is present, all the carbon atoms in the fuel are converted to CO_2:

$$C_8H_{18}(l) + 12.5\ O_2(g) \rightarrow 8\ CO_2(g) + 9\ H_2O(l) \tag{17.26}$$

However, if the fuel mixture is too "rich," that is, contains too much fuel and too little air, considerable amounts of CO may be formed:

$$C_8H_{18}(l) + 11.5\ O_2(g) \rightarrow 6\ CO_2(g) + 2\ CO(g) + 9\ H_2O(l) \tag{17.27}$$

In principle, the carbon monoxide should be converted to CO_2 in the atmosphere:

$$CO(g) + \tfrac{1}{2}\ O_2(g) \rightarrow CO_2(g); \qquad K_c = 3 \times 10^{45} \text{ at } 25°C \tag{17.28}$$

but this reaction, like the conversion of SO_2 to SO_3, is ordinarily quite slow.

Carbon monoxide taken into the lungs reduces the ability of the blood to transport oxygen through the body. It does this by forming a complex with the hemoglobin of the blood that is more stable than that formed by oxygen (Chap. 21):

Young people who are parking are too often victims of CO

$$CO(g) + Hem \cdot O_2(aq) \rightleftharpoons O_2(g) + Hem \cdot CO(aq) \tag{17.29}$$

$$K_c = \frac{[Hem \cdot CO] \times [O_2]}{[Hem \cdot O_2] \times [CO]} = 210$$

The equilibrium constant for Reaction 17.29 is large enough so that low levels of carbon monoxide can convert significant amounts of hemoglobin to the CO complex. Symptoms of carbon monoxide poisoning show up when 10% of the hemoglobin is tied up by CO. When the fraction rises to 20%, death can result unless the victim is removed from the poisonous atmosphere.

EXAMPLE 17.8 Carbon monoxide is present in cigarette smoke. It is estimated that a heavy smoker is exposed to a CO concentration of 2.1×10^{-6} mol/L (50 ppm). Taking the concentration of O_2 to be 8.8×10^{-3} mol/L, calculate
a. the ratio $[CO]/[O_2]$.
b. the ratio $[Hem \cdot CO]/[Hem \cdot O_2]$.

Solution

a. $\dfrac{[CO]}{[O_2]} = \dfrac{2.1 \times 10^{-6} \text{ mol/L}}{8.8 \times 10^{-3} \text{ mol/L}} = 2.4 \times 10^{-4}$

b. $K_c = \dfrac{[Hem \cdot CO] \times [O_2]}{[Hem \cdot O_2] \times [CO]} = 210$

Solving for the ratio $[Hem \cdot CO]/[Hem \cdot O_2]$,

$$\frac{[\text{Hem} \cdot \text{CO}]}{[\text{Hem} \cdot \text{O}_2]} = 210 \times \frac{[\text{CO}]}{[\text{O}_2]} = 210 \times 2.4 \times 10^{-4} = 0.050$$

This calculation suggests that in the blood stream of a heavy smoker about 5% of the hemoglobin is tied up as the CO complex. As a result, smokers are more susceptible to carbon monoxide poisoning than nonsmokers.

EXERCISE The equilibrium constant of Reaction 17.29 differs from one mammal to another. In a rabbit, K_c is about 100. What is the ratio $[\text{Hem} \cdot \text{CO}]/[\text{Hem} \cdot \text{O}_2]$ in a rabbit (Fig. 17.6) who smokes two packs of cigarettes a day? Answer: 0.024.

Some rabbit!

At least three fourths of the carbon monoxide in polluted air comes from automobiles. Since 1975, the principal method of reducing emissions of CO (and unburned hydrocarbons) has been to install catalytic converters in cars (Fig. 17.7). Typically, a catalytic converter contains from 1 to 3 g of platinum mixed with other heavy metals (Rh, Pd) embedded in a base of aluminum oxide, Al_2O_3. In the presence of platinum, CO and hydrocarbons in the exhaust are converted to carbon dioxide and water.

Cars equipped with catalytic converters cannot use "leaded" gasoline, which contains a mixture of $\text{Pb}(\text{C}_2\text{H}_5)_4$ and $\text{C}_2\text{H}_4\text{Br}_2$ to prevent knocking. As little as two tankfuls of gasoline containing this additive can "poison" the converter, making it ineffective. Unleaded gasoline is somewhat more expensive to make. It also contains relatively large amounts of aromatic hydrocarbons, some of which are carcinogenic.

FIGURE 17.6 This rabbit was taught to smoke up to nine cigarettes a day in an experiment conducted at the Tbilisi Institute of Oncology in the U.S.S.R. The rabbits in this experiment learned to enjoy smoking, but by the end of the fourth year they had all developed chronic pneumonia and emphysema. Photograph by G. Kikvadze (Fotokhronika TASS). Reprinted by permission of Sovfoto.

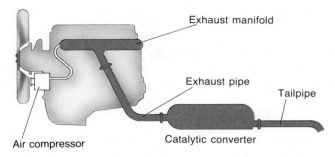

FIGURE 17.7 Catalytic converters on late-model cars contain a "three-way" catalyst. This is designed to convert CO to CO_2, unburned hydrocarbons to CO_2 and H_2O, and NO to N_2. The active components of the catalyst are the precious metals platinum and rhodium; palladium is sometimes used as well.

Nitrogen Oxides (NO, NO_2)

The high-temperature combustion of fuels is the principal source of nitrogen oxide air pollutants. Detectable amounts of nitric oxide are produced by the reaction

$$N_2(g) + O_2(g) \rightarrow 2\ NO(g); \qquad \Delta H = +180.8\ kJ$$

In urban air, NO is converted to NO_2 by a mechanism that is poorly understood. Direct reaction with O_2 (Reaction 17.4) is too slow to account for the rapid build-up of NO_2 (see Example 17.9).

> It may be cold in St. Paul, but there is essentially no smog

The oxides of nitrogen play a key role in the formation of *photochemical smog*. This type of air pollution was first noted in Los Angeles, but it is now common in cities from Honolulu to Washington, D.C. Typically, it develops on bright, sunny mornings when the concentration of NO_2 is relatively high (Fig. 17.8). The key step in smog formation is the dissociation of nitrogen dioxide:

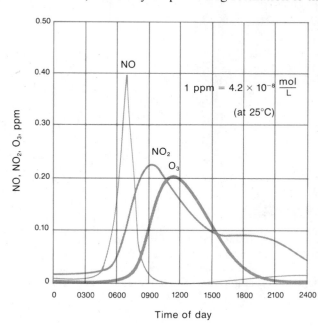

FIGURE 17.8 Average concentrations of the pollutants NO, NO_2, and O_3 on a smoggy day in Los Angeles. The concentration of NO from automobile exhaust builds up in the early morning rush hours. Later, NO_2 and O_3 are produced. The peak NO concentration of 0.40 ppm corresponds to about 1.7×10^{-8} mol/L at 25°C.

$$NO_2(g) \rightarrow NO(g) + O(g) \tag{17.30}$$

This occurs on exposure to light at the edge of the visible region (392 nm; Table 17.4). The oxygen atoms produced react with O_2 molecules:

$$O_2(g) + O(g) \rightarrow O_3(g)$$

The product, ozone, is a major component of photochemical smog. Ozone molecules, oxygen atoms, and nitric oxide molecules attack organic compounds in the air. Unsaturated hydrocarbons, containing multiple carbon-carbon bonds, such as ethylene and propylene, are particularly reactive:

O_3 is fine in the upper atmosphere, but not down here

$$
\begin{array}{cc}
\text{H} \quad \text{H} & \text{H} \quad \text{H} \\
| \quad | & | \quad | \\
\text{H}-\text{C}=\text{C}-\text{H} & \text{H}-\text{C}=\text{C}-\text{CH}_3 \\
\text{ethylene} & \text{propylene}
\end{array}
$$

Many different products are formed. Among these are acrolein and peroxyacetyl nitrate (PAN), which cause eye irritation:

$$
\begin{array}{cc}
\text{H}-\text{C}=\text{C}-\text{C}=\text{O} & \text{CH}_3-\text{C}-\text{O}-\text{O}-\text{N}-\text{O} \\
| \quad | \quad | & \quad\quad || \quad\quad\quad || \\
\text{H} \quad \text{H} \quad \text{H} & \quad\quad \text{O} \quad\quad\quad \text{O} \\
\text{acrolein} & \text{PAN}
\end{array}
$$

EXAMPLE 17.9 Consider the reaction $2\,NO(g) + O_2(g) \rightarrow 2\,NO_2(g)$, for which the rate expression is

$$\text{rate} = k(\text{conc. NO})^2 \times (\text{conc. O}_2), \quad \text{where } k = 2.5 \times 10^7 \text{ L}^2/\text{mol}^2 \cdot \text{h}$$

a. What is the rate of this reaction at the peak NO concentration in smog formation (Fig. 17.8), 1.7×10^{-8} mol/L? The concentration of O_2 in the air is 8.8×10^{-3} mol/L.

b. At this rate, how long would it take for the concentration of NO_2 to build up to its peak value, 1.0×10^{-8} mol/L?

Solution

a. $\text{rate} = 2.5 \times 10^7 \dfrac{\text{L}^2}{\text{mol}^2 \cdot \text{h}} \times (1.7 \times 10^{-8}\,\text{mol/L})^2 \times (8.8 \times 10^{-3}\,\text{mol/L})$

$= 6.4 \times 10^{-11} \dfrac{\text{mol}}{\text{L} \cdot \text{h}}$

b. $\text{time} = \dfrac{\text{conc. NO}_2}{\text{rate}} = \dfrac{1.0 \times 10^{-8}\,\text{mol/L}}{6.4 \times 10^{-11}\,\text{mol/L} \cdot \text{h}} = 160 \text{ h}$

Clearly, this cannot be the reaction by which NO_2 is produced from NO in smog formation. It is much too slow. From Figure 17.8, about 3 h is required for the concentration of NO_2 to build up to its peak value.

It has to come from someplace

| **EXERCISE** Using the calculated rate, determine the concentration of NO_2 produced by this reaction in 3.0 h. Answer: 1.9×10^{-10} mol/L.

Oxides of nitrogen also contribute to the formation of acid rain. Nitrogen dioxide reacts with water vapor in the atmosphere (recall Equation 17.5) to form nitric acid, HNO_3. Acid rainfall in the western United States appears to be due mainly to HNO_3 rather than H_2SO_4.

Of the various air pollutants, oxides of nitrogen are perhaps the most difficult to clean up. Until quite recently, the only method used to reduce NO emissions from automobiles was to recirculate from 10% to 20% of the exhaust gases. This lowers the combustion temperature and hence cuts down on the formation of NO from the elements, an endothermic reaction. Such a system has an inherent problem. Unless the amount of exhaust gas recirculated is controlled very precisely, there are adverse effects on both performance and fuel economy.

Starting with 1981 model cars, a different method has been used to lower NO emissions. This involves modifying the catalytic converter by introducing a small amount of rhodium, a rare and expensive transition metal. In the presence of a Pt–Rh catalyst, the reactions

Leaded gasoline poisons the catalyst

$$CO(g) + NO(g) \rightarrow CO_2(g) + \tfrac{1}{2} N_2(g) \qquad \textbf{(17.31)}$$

and, to a lesser extent

$$H_2(g) + NO(g) \rightarrow H_2O(g) + \tfrac{1}{2} N_2(g) \qquad \textbf{(17.32)}$$

occur rapidly, drastically reducing the NO concentration.

| **EXAMPLE 17.10** For Reaction 17.31, ΔH is -373.4 kJ and ΔS^0 is -99.1 J/K. Over what temperature range is this reaction spontaneous at 1 atm pressure?

Solution The equation for ΔG^0 is

$$\Delta G^0 = -373.4 \text{ kJ} + T(0.0991 \text{ kJ/K})$$

At "low" temperatures, the reaction is spontaneous; at some high temperature the second term dominates and ΔG^0 becomes positive. To find that temperature, we set ΔG^0 equal to zero. This gives us the equilibrium temperature, T_e:

$$T_e = \frac{373.4 \text{ kJ}}{0.0991 \text{ kJ/K}} = 3770 \text{ K} = 3500°C$$

You might say that Reaction 17.31 is always spontaneous

We conclude that Reaction 17.31 should occur spontaneously at 1 atm at any temperature below 3500°C. In effect, this means that in designing a catalytic converter to remove NO by this reaction, one does not have to worry about reaction spontaneity; 3500°C is a very high temperature.

EXERCISE Calculate ΔG^0 at 1000°C. Answer: -247.2 kJ.

17.5
Two Historical Perspectives

Fritz Haber (1868–1934)

The scientific career of Fritz Haber was characterized by theoretical studies in areas of emerging practical importance. Training as an organic chemist and self-development as a physical chemist permitted Haber to relate chemistry to engineering. His most important achievement was the synthesis of ammonia from nitrogen and hydrogen.

Haber's approach to the synthesis of ammonia was not original. The French chemist Le Châtelier (Chap. 15) first studied this reaction, but an explosion led him to abandon the project. The German chemist Nernst, a contemporary of Haber, was the first to accomplish the high-pressure synthesis. Disagreement with Nernst's data led Haber to seek the optimum temperature, pressure, and catalyst. After obtaining about 8% ammonia at 600°C and 200 atm, Haber turned the process over to engineers. In 1918 he received the Nobel Prize for his work.

During World War I, Haber directed chemical warfare in Germany, introducing the use of chlorine and mustard gas into warfare. After the war, he attempted unsuccessfully to recover gold from the oceans. In this way, he hoped to repay Germany's war debts (which were never paid).

In 1911, Haber became director of the new Kaiser Wilhelm Institute for Chemistry. Under his influence, the Institute became world famous, attracting outstanding students and professors. In 1933 Haber, who was Jewish, was removed from his directorship of the Institute by the Nazis. In his resignation letter he wrote, "For more than 40 years I have selected my collaborators on the basis of their intelligence and their character, not on the basis of their grandmothers. I am unwilling for the rest of my life to change this method which I have found so good." After a brief stay in England, Haber died enroute to a research directorship in Israel.

Scientists vary greatly in the ways they look at morality

Friedrich Wilhelm Ostwald (1853–1932)

The conversion of ammonia to nitric acid was one of many contributions from this multifaceted individual. His early interest in chemistry, physics, painting, literature, music, and philosophy became life-long pursuits. With Van't Hoff, Arrhenius, and Gibbs, Ostwald helped to establish the fledgling discipline of physical chemistry. He speculated that he became a physical chemist because his chemical education was carried out in Russia rather than Germany, where the emphasis was on organic chemistry. Early work on catalysis and reaction rates (1887) earned him the 1909 Nobel Prize in chemistry.

At 53, Ostwald retired as director of the University of Leipzig Physical Chemistry Institute and spent his remaining years in research and writing. From 1890 on, his work and personal philosophy were organized about his science of "energetics" (he even named his house *Energie*). This point of view made him a vigorous opponent of the concept of atoms and molecules! Matter, he believed, was a combination of energies occurring simultaneously at one place: "In fact, energy is the unique, real entity in the world and matter is not a bearer but a manifestation of energy." Eventually (1909), in a preface to his general chemistry text, Ostwald grudgingly accepted the atomic theory, in part because of the work of Perrin in kinetic molecular theory and discoveries by Thomson and Rutherford.

"Be not the first by whom the new is tried, nor yet the last to cast the old aside"

A pacifist and an advocate of the conservation of natural energy sources, Ostwald was a visionary. He believed in international harmony and composed Ido, an international language. His love of painting led him to spend the last 20 years of his life developing the science of color. He created color standards, dyes, and a quantitative theory of color. An inspiring teacher, his textbooks for general and analytical chemistry revolutionized the teaching of these subjects. When Ostwald died, a former student wrote, "He was loved and followed by more people than any chemist of our time."

Summary

This chapter describes selected topics in atmospheric chemistry, including

- —the fixation of nitrogen to form ammonia (Haber process) and the conversion of ammonia to nitric acid (Ostwald process).
- —the effect of water and carbon dioxide upon our weather and climate (relative humidity, cloud formation, greenhouse effect).
- —photochemical reactions taking place in the upper atmosphere (Tables 17.4 and 17.5).
- —the sources, effects, and treatment of air pollution by sulfur oxides, sulfuric acid, carbon monoxide, and nitrogen oxides.

Several concepts from Chapters 14, 15, and 16 were reviewed in this chapter. These include

- —calculations involving ΔH, ΔS^0, and ΔG^0 (Examples 17.1 and 17.10).
- —determination of the equilibrium constant from ΔG^0 (Example 17.5).
- —equilibrium calculations (Examples 17.3 and 17.8).
- —effect of changes in conditions upon the position of an equilibrium and the rate at which it is reached (Example 17.2).
- —reaction mechanisms (Example 17.6).
- —rate constant and its temperature dependence (Examples 17.7 and 17.9).

Key Words and Concepts

acid rain	mean free path	parts per million (ppm)
catalytic converter	mole fraction	relative humidity
contact process	nitrogen fixation	smog
greenhouse effect	Ostwald process	wavelength
Haber process		

Questions and Problems

General

17.1 Write balanced equations (one or more) to represent the preparation of
a. ammonia by the Haber process.
b. sulfuric acid by the contact process.
c. ammonium sulfate.

17.2 Discuss the effect of changes in pressure and temperature upon the rate and yield of NH_3 in the Haber process.

17.3 The relative humidity is 64% on a hot summer day when the temperature is 30°C.
a. What is the pressure of water vapor in the air at this temperature (see Appendix 1 for vapor pressures of water)?
b. If the air is cooled to 25°C, what will the relative humidity become?

17.4 A 30.0-L sample of air contains 0.540 g $H_2O(g)$ at 30°C.
a. Using the Ideal Gas Law, calculate the partial pressure of water vapor in the air.
b. What is the relative humidity (vp H_2O at 30°C = 31.8 mm Hg)?

17.5 The bond energy of the C—Cl bond is 331 kJ/mol. Use Equation 17.12 to calculate the maximum wavelength of radiation that can break this bond and hence lead to depletion of the ozone layer by Reactions 17.18 and 17.19.

17.6 In the atmosphere, O_2 molecules are converted to oxygen atoms by ultraviolet radiation below 242 nm. Use Equation 17.12 to calculate the bond energy of O_2.

17.7 Classify the following as true or false. If the statement is false, correct it.
a. Fertilizer prices have risen in recent years because the price of N_2 has skyrocketed.
b. If the temperature of an air mass drops, relative humidity increases.
c. CO_2 in the atmosphere has a warming effect because it absorbs infrared radiation from the sun.

17.8 Describe the possible effects of
a. an increase in Cl concentration in the upper atmosphere.
b. an increase in NO concentration in the morning rush hour.
c. an increase in suspended particles in air.
d. continued use of leaded gasoline.

17.31 Follow directions for Question 17.1 for the preparation of
a. nitric acid by the Ostwald process.
b. sulfur trioxide.
c. ammonium nitrate.

17.32 What effect would increases in pressure and temperature have upon the rate of formation and yield of SO_3 by Reaction 17.8?

17.33 On a winter day when the temperature is 5°C, the vapor pressure of water in the air is 2.3 mm Hg.
a. What is the relative humidity of the outside air (use Appendix 1 to find the vapor pressure of water)?
b. What does the relative humidity become if the air is warmed to 20°C?

17.34 The dew point is the temperature at which water vapor in the air condenses upon cooling at constant pressure. Calculate the dew point on a day when the temperature is 20°C and the relative humidity is 40%. Use the table of water vapor pressures in Appendix 1.

17.35 The dissociation energy of ozone is 105.3 kJ. Calculate the maximum wavelength of a photon (Equation 17.12) that could cause ozone to dissociate.

17.36 Radiation below 192 nm can dissociate SO_2. Use Equation 17.12 to calculate the dissociation energy of SO_2.

17.37 Criticize the following statements:
a. In the Haber process, a high pressure is used primarily to increase reaction rate.
b. Ice crystals form in clouds when the temperature drops below 0°C.
c. Ozone is found only in the upper atmosphere.
d. Sulfur dioxide is a more serious air pollutant than SO_3.

17.38 Describe the effects that may result from
a. an increase in atmospheric CO_2 concentration.
b. a decrease in the concentration of O_3 in the upper atmosphere.
c. an increase in SO_3 concentration in the atmosphere.

17.9 Explain
   a. the greenhouse effect.
   b. why chemical reactions that occur in the upper atmosphere (above 30 km) are not common near the surface of the earth.
   c. the relation between SO_3 production in air and the contact process.

17.10 During the day, typical concentrations (ppm) of air pollutants in a metropolitan area are

TIME	NO	NO$_2$	O$_3$	HYDRO-CARBONS
4 A.M.	0.07	0.06	0.02	0.28
6 A.M.	0.14	0.09	0.02	0.31
7 A.M.	0.14	0.16	0.02	0.36
8 A.M.	0.04	0.18	0.03	0.43
10 A.M.	0.01	0.06	0.08	0.33
Noon	0.01	0.04	0.19	0.25
2 P.M.	0.01	0.04	0.19	0.20
4 P.M.	0.02	0.05	0.12	0.15

Discuss these trends in light of the reactions mentioned in this chapter.

ΔH, ΔS^0, ΔG^0 (Chap. 14)

17.11 In the upper atmosphere, OH radicals are formed by the reaction

$$O(g) + H_2O(g) \rightarrow 2\ OH(g)$$

For this reaction, ΔH is $+78.5$ kJ and ΔS^0 is 17.6 J/K. Calculate ΔG^0 at
   a. 0 K   b. 300 K   c. 1000 K

17.12 Predict the sign of ΔS^0 for each of the following reactions. Confirm your predictions using Table 14.3.
   a. $N_2(g) + 3\ H_2(g) \rightarrow 2\ NH_3(g)$
   b. $NH_3(g) + HCl(g) \rightarrow NH_4Cl(s)$
   c. $SO_2(g) + \frac{1}{2} O_2(g) \rightarrow SO_3(g)$
   d. $2\ NO_2(g) \rightarrow 2\ NO(g) + O_2(g)$

17.13 Ammonium nitrate is made by reacting ammonia with nitric acid:

$$NH_3(g) +\ HNO_3(l) \rightarrow NH_4NO_3(s)$$

   a. Using Table 14.1, calculate ΔH for this reaction.
   b. Using Table 14.3, calculate ΔS^0.
   c. Calculate ΔG^0 at 25°C.

17.39 Explain why
   a. a catalytic converter is used in an automobile exhaust system.
   b. catalytic converters are expensive.
   c. even a low-level concentration of CO may be dangerous.

17.40 The 1980 United States emission standard for CO was 4.4 g CO/km (7.0 g/mi). In a large metropolitan area there are 100,000 automobiles driven an average of 32 km/day (20 mi/day).
   a. Assuming that all the cars meet this standard, how many kilograms of CO are emitted per day?
   b. Assuming an air volume of 5000 km^3 for the metropolitan area and a maximum allowable CO concentration of 4×10^{-7} mol/L, does the daily CO output exceed this level?

17.41 For the formation of SO_3 by Reaction 17.21, ΔH is -425.1 kJ and ΔS^0 is -170.8 J/K. Calculate ΔG^0 at each temperature referred to in Problem 17.11.

17.42 Follow the directions of Problem 17.12 for
   a. $SO_3(g) + H_2O(l) \rightarrow H_2SO_4(l)$
   b. $C_3H_8(g) + 5\ O_2(g) \rightarrow 3\ CO_2(g) + 4\ H_2O(g)$
   c. $C_3H_8(g) + 5\ O_2(g) \rightarrow 3\ CO_2(g) + 4\ H_2O(l)$
   d. $C_3H_8(g) + \frac{7}{2} O_2(g) \rightarrow 3\ CO(g) + 4\ H_2O(g)$

17.43 Ammonia can be converted to urea, $(NH_2)_2CO$, a fertilizer, by the reaction

$$2\ NH_3(g) + CO_2(g) \rightarrow (NH_2)_2CO(s) + H_2O(l)$$

   a. Calculate ΔH for this reaction, taking ΔH_f urea $= -333.2$ kJ/mol and using Table 14.1.
   b. Calculate ΔS^0 (S^0 urea $= 104.6$ J/mol·K).
   c. Calculate ΔG^0 at 298 K.

17.14 One way to reduce NO emissions from smoke stacks is to react it with methane:

$$CH_4(g) + 4\,NO(g) \rightarrow 2\,N_2(g) + CO_2(g) + 2\,H_2O(g)$$

   a. Calculate ΔH for this reaction.
   b. Calculate ΔS^0 for this reaction.
   c. Over what temperature range is this reaction spontaneous at 1 atm?

17.15 Removal of SO_2 from stack gases by reacting it with MgO has been proposed:

$$MgO(s) + SO_2(g) + \tfrac{1}{2}\,O_2(g) \rightarrow MgSO_4(s)$$

Using data from Tables 14.1 and 14.3, calculate the maximum temperature at which this reaction is spontaneous.

Equilibrium Calculations (Chap. 15)

17.16 Discuss the effects of changes in temperature and applied pressure on the position of each of the following equilibria:
   a. $NO(g) + O_3(g) \rightleftharpoons NO_2(g) + O_2(g)$; $\Delta H < 0$
   b. $CH_4(g) + H_2O(g) \rightleftharpoons CO(g) + 3\,H_2(g)$; $\Delta H > 0$
   c. $NO(g) + NO_2(g) \rightleftharpoons N_2O_3(g)$; $\Delta H < 0$
   d. $N_2(g) + O_2(g) \rightleftharpoons 2\,NO(g)$; $\Delta H > 0$

17.17 Consider the equilibrium

$$CO(g) + Hem \cdot O_2(aq) \rightleftharpoons O_2(g) + Hem \cdot CO(aq)$$

for which $K_c = 210$. Calculate the ratio $[Hem \cdot CO]/[Hem \cdot O_2]$ when
   a. conc. CO = conc. O_2.
   b. conc. CO = 0.15 conc. O_2

17.18 Consider the data in Table 17.2
   a. Determine the number of moles of NH_3, N_2, and H_2 in an equilibrium mixture containing one mole of gas at 300 atm and 400°C. Note that $[H_2] = 3\,[N_2]$.
   b. Use the Ideal Gas Law to determine the volume of one mole of gas at these conditions.
   c. Combine your answers from (a) and (b) to calculate the equilibrium concentrations (mol/L) of NH_3, N_2, and H_2.
   d. From your answers in (c), calculate K_c at 400°C. Compare your value with that given in the table.

17.19 Use data from Table 17.3 for the reaction at 700°C to calculate
   a. K_p b. ΔG^0

17.44 Catalysts used to lower NO emissions in automobiles also catalyze undesirable side reactions such as

$$2\,NO(g) + 5\,H_2(g) \rightarrow 2\,H_2O(g) + 2\,NH_3(g)$$

   a. Using tables in Chapter 14, obtain an expression for ΔG^0 for this reaction as a function of temperature.
   b. Over what temperature range is this reaction spontaneous at 1 atm?

17.45 What would the temperature be in Problem 17.15 if CaO had been used in place of MgO?

17.46 How would you change the temperature and applied pressure to obtain the maximum yield of products in each of the following systems?
   a. $4NH_3(g) + 5\,O_2(g) \rightleftharpoons 4NO(g) + 6H_2O(g)$; $\Delta H < 0$
   b. $2\,NO(g) + O_2(g) \rightleftharpoons 2\,NO_2(g)$; $\Delta H < 0$
   c. $CO(g) + NO(g) \rightleftharpoons CO_2(g) + \tfrac{1}{2}\,N_2(g)$; $\Delta H < 0$
   d. $CO(g) + \tfrac{1}{2}\,O_2(g) \rightleftharpoons CO_2(g)$; $\Delta H < 0$

17.47 For the equilibrium in Problem 17.17, calculate the ratio $[CO]/[O_2]$ when 5.0% of the hemoglobin is in the form of the CO complex.

17.48 Consider the data in Table 17.2,
   a. Calculate the mole percent of all three gases at 500°C and 100 atm.
   b. Use the Ideal Gas Law to calculate the number of moles of gas in a 10.0-L container at 100 atm and 500°C.
   c. Use the results from (a) and (b) to calculate the equilibrium concentrations of all three gases in a 10.0-L vessel at 100 atm and 500°C.
   d. Use your answers in (c) to calculate K_c at 500°C. Compare your value with that in the table.

17.49 For the reaction given in Table 17.3, K_c is 4.3 at 850°C. Calculate the values, at this temperature, of
   a. K_p b. ΔG^0 c. ΔS^0

17.20 Ammonia produced from NO in automobile exhaust can be reconverted to NO:

$$2 NH_3(g) + O_2(g) \rightarrow 2 NO(g) + 3 H_2(g)$$

At 298 K, ΔG^0 is 206.7 kJ. Calculate
a. K_p
b. K_c

17.50 Hydrogen is produced in automobile exhaust by the reaction of carbon monoxide with water:

$$CO(g) + H_2O(g) \rightarrow CO_2(g) + H_2(g)$$

At 298 K, ΔG^0 for this reaction is -28.5 kJ. At this temperature, calculate
a. K_p
b. K_c

Reaction Rates (Chap. 16)

17.21 The reaction

$$NO(g) + N_2O(g) \rightarrow NO_2(g) + N_2(g)$$

is first order in each reactant. The rate is 1.92×10^{-15} mol/L·s when the NO and N_2O concentrations are each 5.0×10^{-8} mol/L.
a. Calculate k for the forward reaction.
b. At conc. NO = 2.0×10^{-8} mol/L, the rate is measured as 7.4×10^{-20} mol/L·s. What is conc. N_2O?

17.51 The reaction

$$2 NO_2(g) \rightarrow 2 NO(g) + O_2(g)$$

is a second-order reaction. The rate constant is 0.498 L/mol·s at 592 K.
a. What is the reaction rate at initial conc. NO_2 = 0.010 mol/L?
b. What concentration of NO_2 is present when the reaction rate is reduced to half of that determined in (a)?

17.22 The following reaction is involved in smog formation:

$$O_3(g) + NO(g) \rightarrow O_2(g) + NO_2(g)$$

This reaction has been shown to be first order in both ozone and NO with a rate constant of 1.2×10^7 L/mol·s. Calculate the concentration of NO_2 formed per second in polluted air, where O_3 and NO concentrations are both 2×10^{-8} mol/L. From the magnitude of your answer, would you expect the conversion of NO to be rapid or slow?

17.52 Formaldehyde, CH_2O, is a major eye irritant in smog. It may be formed by the reaction $O_3(g) + C_2H_4(g) \rightarrow 2 CH_2O(g) + O(g)$; Δ conc. $CH_2O/\Delta t$ is first order in both O_3 and C_2H_4, with k = 2×10^3 L/mol·s. The concentrations of O_3 and C_2H_4 in heavily polluted air are estimated to be about 5×10^{-8} and 1×10^{-8} mol/L, respectively. What is the rate of production of formaldehyde in mol/L·s? How long will it take to build up a formaldehyde concentration of 1×10^{-8} mol/L? This is the threshold above which eye irritation becomes noticeable. (Assume constant concentrations of O_3 and C_2H_4.)

17.23 The rate constant for the decomposition of ozone by the mechanism $O_3(g) + O(g) \rightarrow 2 O_2(g)$ is 5.0×10^6 L/mol·s. Calculate the rate of ozone decomposition when the concentrations of O_3 and O are 3.0×10^{-8} and 1.2×10^{-14} mol/L, respectively.

17.53 If ozone is being produced by the mechanism $O_2(g) + O(g) \rightarrow O_3(g)$ at the same rate it is decomposing by the reaction in Problem 17.23, and if the concentrations of O_2 and O are 3.9×10^{-4} and 1.2×10^{-14} mol/L, respectively, what is the rate constant for the above reaction?

17.24 The rate constant for the NO-catalyzed decomposition of O_3 (Table 17.5) is 5.4×10^9 L/mol·s, as compared to 5.0×10^6 L/mol·s for the direct decomposition (Problem 17.23). The second step in the catalyzed decomposition is rate determining. If the concentration of NO_2 in the upper atmosphere is 1/2000 that of O_3, what is the ratio of the rate of the catalyzed reaction to that of the direct decomposition?

17.54 It is estimated that the amount of O_3 in the upper atmosphere decomposing by the NO-catalyzed reaction is four times that decomposing by the direct reaction. Using the rate constants given in Problem 17.24, estimate the ratio of the concentrations of NO_2 and O_3.

17.25 For the first-order thermal decomposition of ozone

$$O_3(g) \rightarrow O_2(g) + O(g)$$

$k = 3 \times 10^{-26}$ s^{-1} at 25°C. What is the half-life for this reaction in years? Comment on the likelihood that this reaction contributes to the depletion of the ozone layer.

17.26 One of the reactions that destroys NO in the catalytic converter of an automobile is

$$CO(g) + NO(g) \rightarrow CO_2(g) + \tfrac{1}{2} N_2(g)$$

From the following data, determine the order of reaction with respect to both CO and NO.

RATE (mol/L·min)	CONC. CO (M)	CONC. NO (M)
3.2×10^{-9}	4.0×10^{-6}	4.0×10^{-8}
1.6×10^{-9}	2.0×10^{-6}	4.0×10^{-8}
0.90×10^{-9}	2.0×10^{-6}	3.0×10^{-8}
0.20×10^{-9}	1.0×10^{-6}	2.0×10^{-8}

Calculate the rate constant for the reaction.

17.27 For the upper atmosphere reaction

$$OH(g) + H(g) \rightarrow H_2(g) + O(g)$$

E_a is 29 kJ. The heats of formation of $O(g)$, $H(g)$, and $OH(g)$ are 248 kJ, 218 kJ, and 40 kJ, respectively. What is the activation energy for the reverse reaction?

17.28 For the reaction in Example 17.7, calculate k at $-25°C$.

17.29 A possible mechanism for the reaction

$$2 NO(g) + O_2(g) \rightarrow 2 NO_2(g)$$

is

$$NO(g) + NO(g) \rightleftharpoons N_2O_2(g) \qquad \text{(fast)}$$

$$N_2O_2(g) + O_2(g) \rightarrow NO_2(g) + NO_2(g) \qquad \text{(slow)}$$

Derive the rate expression for this reaction, in terms of the concentrations of NO and O_2.

17.30 Consider the following suggested mechanism:

$$NO_2(g) + NO_2(g) \rightleftharpoons N_2O_4(g) \qquad \text{(fast)}$$

$$N_2O_4(g) + F_2(g) \rightarrow 2 NO_2F(g) \qquad \text{(slow)}$$

a. Write the equation for the overall reaction.
b. Write the rate law for the overall reaction in terms of the concentration of NO_2 and F_2.

17.55 For the reaction in Problem 17.25, how long would it take to lower the ozone concentration by 10.0%?

17.56 Another reaction that lowers NO emissions with a catalytic converter is

$$H_2(g) + NO(g) \rightarrow H_2O(g) + \tfrac{1}{2} N_2(g)$$

Use the following data to find the order of reaction with respect to both H_2 and NO.

RATE (mol/L·min)	CONC. H$_2$ (M)	CONC. NO (M)
2.4×10^{-9}	4.0×10^{-6}	4.0×10^{-8}
1.8×10^{-9}	3.0×10^{-6}	4.0×10^{-8}
1.0×10^{-9}	3.0×10^{-6}	3.0×10^{-8}
0.30×10^{-9}	2.0×10^{-6}	2.0×10^{-8}

Calculate the rate constant for the reaction.

17.57 In the upper atmosphere, water molecules react with oxygen atoms:

$$H_2O(g) + O(g) \rightarrow 2 OH(g)$$

For this reaction, E_a is 77 kJ. For the reverse reaction, E_a' is 4 kJ.
a. What is ΔH for the forward reaction?
b. Given $\Delta H_f \ H_2O(g) = -242$ kJ, and $\Delta H_f \ O(g) = 248$ kJ, calculate $\Delta H_f \ OH(g)$.

17.58 For the reaction in Example 17.7, calculate the temperature at which k is one half of its value at 27°C.

17.59 Experimentally, it is found that the rate constant for the formation of NO_2 from NO *decreases* as temperature increases. Using the rate expression derived in Problem 17.29, explain how this can happen. (ΔH for the formation of N_2O_2 from NO is a negative quantity.)

17.60 The following mechanism has been proposed for the uncatalyzed decomposition of O_3:

$$O_3(g) \rightleftharpoons O_2(g) + O(g) \qquad \text{(fast)}$$
$$O(g) + O_3(g) \rightarrow 2 O_2(g) \qquad \text{(slow)}$$

Find the rate law associated with this mechanism. Express it in terms of the concentrations of O_3 and O_2.

*17.61 Two moles of SO_2 and one mole of O_2 are placed in a 1.00-L container at 900°C. At this temperature, K_c = 2.4 for the system $SO_2(g) + \frac{1}{2} O_2(g) \rightleftharpoons SO_3(g)$. Calculate the concentration of SO_3 at equilibrium.

*17.62 Most of the ozone in the upper atmosphere exists in a layer at an altitude between 20 and 40 km.
   a. Calculate the total volume in liters of this layer, taking the radius of the earth to be 6.4×10^3 km and using the following relation for the difference in volume between two concentric spheres of radii r_1 and r_2: $\Delta V \approx 4 \pi r_1^2(r_2 - r_1)$.
   b. Taking the average pressure of air in this layer to be 0.013 atm and the temperature to be $-20°C$, calculate the total number of moles of gas in the layer.
   c. Assuming the average concentration of ozone in the layer to be 7 ppm, what is the total mass in grams of the ozone in the upper atmosphere?

*17.63 The ozone present in polluted air is formed by the two-step mechanism

$$NO_2(g) \xrightarrow{k_1} NO(g) + O(g) \quad \text{(first order)}$$
$$O(g) + O_2(g) \xrightarrow{k_2} O_3(g) \quad \text{(second order)}$$

It is known that $k_1 = 6.0 \times 10^{-3}$ s^{-1} and $k_2 = 1.0 \times 10^6$ L/mol·s. The concentrations of NO_2 and O_2 in polluted air are about 3.0×10^{-9} and 1.0×10^{-2} mol/L, respectively. One can assume that the concentration of atomic oxygen reaches a "steady state," a low constant concentration at which point it is being consumed in the second reaction at the same rate as it is being produced in the first.
   a. Calculate the steady rate concentration of $O(g)$ in polluted air.
   b. Calculate the rate of formation of O_3 in polluted air.
   c. If the rate of formation of O_3 remains constant, how long would it take for its concentration to build up to one part per million in air at 25°C and 1 atm? Under such conditions air contains about 0.041 mol/L.

*17.64 Suppose that a molecule of O_2 absorbs a photon at a wavelength of 220 nm and subsequently decomposes to 2 oxygen atoms. Taking ΔH for this reaction to be +495 kJ, how much "extra energy" in kilojoules will the oxygen atoms have? If all this energy is absorbed in raising the kinetic energy of $O(g)$ and, hence, its temperature, what will the final "apparent" temperature be? (The heat capacity of $O(g)$ is 22 J/mol·K.)

*17.65 A certain type of coal contains 2.0% by mass of sulfur.
   a. What mass of sulfur dioxide will be produced by burning 100 g of this coal?
   b. What volume of SO_2 at 25°C and one atm will be produced?
   c. If 1000 liters of air are used to burn the coal, what will be the concentration of SO_2 in the stack gas in ppm?

Most of the reactions we have considered to this point have involved pure substances as reactants. As important as such reactions are, reactions taking place in water solution are more common, both in the general chemistry laboratory and in the world around us. Such reactions will occupy our attention throughout most of the remainder of this text. We start in this chapter with what is perhaps the simplest type of reaction in water, that of precipitation.

When water solutions of two different electrolytes are mixed, we often find that an insoluble solid precipitates out of solution. To identify the solid, we must know which ionic compounds are soluble in water and which are insoluble. Information on solubilities can be summarized by a set of solubility rules (Section 18.1). With the help of these rules, we can write net ionic equations (Section 18.2) to represent precipitation reactions. These equations can be used in the usual way to relate amounts of reactants and products.

When a solid precipitates from solution, equilibrium is established with its ions in solution. This type of equilibrium is described by a quantity called the solubility product constant, K_{sp}. In Section 18.3, we will consider how the expression for K_{sp} is written and how it is used in practical calculations involving precipitation reactions. This chapter concludes (Section 18.4) with a brief discussion of water softening, a process used to remove Ca^{2+} ions from water before they have a chance to form undesirable precipitates.

18.1
Solubilities of
Ionic Solids

When an ionic solid dissolves in water, there is a strong interaction between polar H_2O molecules and the charged ions that make up the solid. The extent to which solution occurs depends upon a balance between two forces, both electrical in nature:

1. The force of attraction between H_2O molecules and the ions of the solid, which tends to bring the solid into solution. If this factor predominates, we expect the compound to be very soluble in water, as is the case with NaCl, NaOH, and many other ionic solids.

2. The force of attraction between oppositely charged ions, which tends to keep them in the solid state. If this is the major factor, we expect the water solubility to be very low. The fact that $CaCO_3$ and $BaSO_4$ are almost insoluble in water implies that interionic attractive forces predominate with these ionic solids.

Unfortunately, we cannot estimate from first principles the relative strengths of these two forces for a given solid. For this reason, among others, we cannot predict in advance the water solubilities of electrolytes. Ionic solids cover an enormous range of solubilities. At one extreme we have lithium chlorate, $LiClO_3$, which dissolves to the extent of 35 mol/L at room temperature. Mercury(II) sulfide, HgS, is at the other extreme. Its calculated solubility at 25°C is 10^{-26} mol/L. This means that, in principle at least, about 200 L of a saturated solution of HgS would be required to contain a single pair of Hg^{2+} and S^{2-} ions.

It would be hard to find them

Data on the solubility of common ionic solids are given in Table 18.1 in the form of solubility rules. These rules are quite simple to interpret. For example, the following facts should be evident from the table:

$Ni(NO_3)_2$ is soluble (all nitrates are soluble).
$BaCl_2$ is soluble ($BaCl_2$ is not one of the three insoluble chlorides listed).
PbS is insoluble (Pb is not a Group 1 or Group 2 element).

Table 18.1
Solubility Rules*

NO_3^-	All nitrates are soluble.
Cl^-	All chlorides are soluble except AgCl, Hg_2Cl_2, and $PbCl_2$.
SO_4^{2-}	Most sulfates are soluble; exceptions include $SrSO_4$, $BaSO_4$, and $PbSO_4$.
CO_3^{2-}	All carbonates are insoluble except those of the Group 1 elements and NH_4^+.
OH^-	All hydroxides are insoluble except those of the Group 1 elements, $Sr(OH)_2$, and $Ba(OH)_2$. ($Ca(OH)_2$ is slightly soluble.)
S^{2-}	All sulfides except those of Group 1 and 2 elements and NH_4^+ are insoluble.

*Insoluble compounds are those that precipitate when we mix equal volumes of 0.1 M solutions of the corresponding ions.

18.2 Precipitation Reactions

When solutions of two different electrolytes are mixed, we sometimes observe that an insoluble solid comes out of solution. This solid is referred to as a *precipitate*. The chemical reaction by which it is formed is called a *precipitation reaction*. As an example, consider what happens when solutions of barium chloride

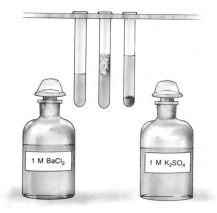

FIGURE 18.1 When solutions of $BaCl_2$ and K_2SO_4 are mixed, a white precipitate forms (center test tube). This solid, which slowly settles out (test tube at right), can be identified as $BaSO_4$, which is very insoluble in water. It is formed when Ba^{2+} ions from the $BaCl_2$ solution combine with SO_4^{2-} ions from the K_2SO_4 solution.

and potassium sulfate are mixed. We observe that a white precipitate forms (Fig. 18.1). Using the solubility rules given in Table 18.1, it is possible to identify the precipitate.

We start by noting that in solutions of these two ionic solids the following ions are available:

$BaCl_2$ solution:	Ba^{2+} and Cl^-
K_2SO_4 solution:	K^+ and SO_4^{2-}

In solution ionic solids exist as ions

The precipitate must result from an interaction between two ions. We expect the cations to repel each other and the anions to do likewise. The interaction of Ba^{2+} with Cl^- ions or K^+ with SO_4^{2-} ions would simply produce the *soluble* starting materials. Two other ion pairings are possible:

K^+ and Cl^-
Ba^{2+} and SO_4^{2-}

Potassium chloride, KCl, is soluble in water, so we deduce that K^+ and Cl^- ions do not combine with each other. On the other hand, barium sulfate, $BaSO_4$, is insoluble. This means that Ba^{2+} and SO_4^{2-} ions cannot be present together in solution at appreciable concentrations. We conclude that Ba^{2+} and SO_4^{2-} ions must combine to form a precipitate of $BaSO_4$. The white, insoluble solid shown in Figure 18.1 is barium sulfate.

If no precipitate formed, $BaSO_4$ would have to be soluble

Equations for Precipitation Reactions

The precipitation reaction that occurs when solutions of $BaCl_2$ and K_2SO_4 are mixed can be represented by a simple equation. We have seen that the product is solid $BaSO_4$. This is formed by the reaction between Ba^{2+} and SO_4^{2-} ions in aqueous solution. The equation is

$$Ba^{2+}(aq) + SO_4^{2-}(aq) \rightarrow BaSO_4(s) \tag{18.1}$$

Notice that we include in this equation only those ions that participate in the reaction. To be specific, K^+ and Cl^- ions do not appear in Equation 18.1. They are "spectator" ions, which are present in the solution before and after the precipitation of barium sulfate. Equations such as this, which involve ions and exclude any species that do not take part in the reaction, are often referred to as *net ionic equations*. They do not differ in any essential way from the equations we have been writing all along. Other examples of net ionic equations for precipitation reactions include

—the precipitation of Ag_2CrO_4 when solutions of $AgNO_3$ and K_2CrO_4 are mixed:

$$2\ Ag^+(aq)\ +\ CrO_4{}^{2-}(aq) \rightarrow Ag_2CrO_4(s) \qquad\qquad \textbf{(18.2)}$$

The spectator ions do not precipitate

(spectator ions: $NO_3{}^-$ and K^+)

—the precipitation of $Fe(OH)_3$ when solutions of $FeCl_3$ and $NaOH$ are mixed:

$$Fe^{3+}(aq)\ +\ 3\ OH^-(aq) \rightarrow Fe(OH)_3(s) \qquad\qquad \textbf{(18.3)}$$

(spectator ions: Cl^- and Na^+)

Using Table 18.1, you should be able to identify any precipitate formed when two solutions are mixed, and write an equation for its formation. To do this, you follow what amounts to a three-step procedure:

1. Decide what ions are present in the two solutions. On that basis, determine the two possible products. In each of these, a cation from one solution is combined with an anion from the other solution.

2. Using the solubility rules, decide whether either or both of the products in (1) will actually precipitate.

3. Having decided what precipitate forms, write a net ionic equation to describe the reaction. The "product" is the insoluble solid; the "reactants" are the corresponding ions in water solution.

EXAMPLE 18.1 Write a net ionic equation for any precipitation reaction that occurs when 0.1 M solutions of the following ionic compounds are mixed:
a. NaOH and $Cu(NO_3)_2$ b. $Ba(OH)_2$ and $NiSO_4$
c. RbCl and LiOH

Solution In each case we first decide what the two possible products are. Then, using the solubility rules, we decide what compound, if any, precipitates. Finally, we write an equation for the precipitation reaction.
a. *Possible products:* $NaNO_3$ (Na^+ ions from NaOH; $NO_3{}^-$ ions from the $Cu(NO_3)_2$ solution); $Cu(OH)_2$ (Cu^{2+} ions from $Cu(NO_3)_2$, OH^- ions from the NaOH solution).
Identity of precipitate: The solubility rules indicate that $NaNO_3$ is soluble and $Cu(OH)_2$ is insoluble. We conclude that a precipitate of $Cu(OH)_2$ will form.
Equation: $Cu^{2+}(aq)\ +\ 2\ OH^-(aq) \rightarrow Cu(OH)_2(s)$

b. The possible products are barium sulfate, $BaSO_4$, and nickel hydroxide, $Ni(OH)_2$. $BaSO_4$ is listed in Table 18.1 as being insoluble. Since Ni is not a Group 1 element, we deduce that $Ni(OH)_2$ is insoluble. Thus, we predict that both of these compounds will precipitate. We write two equations, since two entirely different precipitation reactions take place:

$$Ba^{2+}(aq) + SO_4^{2-}(aq) \rightarrow BaSO_4(s)$$

$$Ni^{2+}(aq) + 2\ OH^-(aq) \rightarrow Ni(OH)_2(s)$$

Experiment confirms our deductions. If we look at the precipitate under a microscope, we see white $BaSO_4$ crystals and green particles of $Ni(OH)_2$.

c. Both possible products, RbOH and LiCl, are soluble. No precipitation reaction occurs, so no equation can be written.

Usually we get only one precipitate, if that

EXERCISE Write net ionic equations for the precipitation reactions that occur when solutions of $SrCl_2$ and Na_2SO_4 are mixed; NH_4Cl and $Pb(NO_3)_2$. Answer:

$$Sr^{2+}(aq) + SO_4^{2-}(aq) \rightarrow SrSO_4(s)$$

$$Pb^{2+}(aq) + 2\ Cl^-(aq) \rightarrow PbCl_2(s)$$

Mole Relations in Precipitation Reactions

The coefficients in the net ionic equation for a precipitation reaction can be used in the usual way (Chap. 3) to relate moles of reactants and products. The number of moles of an ion in a given volume of solution can be deduced from solute concentrations, as discussed in Chapter 12. Examples 18.2 and 18.3 are typical of the calculations involved.

EXAMPLE 18.2 Consider the precipitation of Ag_2CrO_4 (Equation 18.2). What volume of 0.125 M $AgNO_3$ solution is required to react completely with 50.0 cm^3 of 0.100 M K_2CrO_4 solution?

Solution A reasonable procedure is the following:
1. Determine the number of moles of CrO_4^{2-} in the K_2CrO_4 solution.
2. Determine the number of moles of Ag^+ required, using the balanced equation for the precipitation reaction.
3. Determine the volume of $AgNO_3$ solution required to give the number of moles of Ag^+ calculated in (2).

The spectator ions, NO_3^- and K^+, are not involved in this kind of calculation

$$\text{no. moles } CrO_4^{2-} = 0.0500\ L \times \frac{0.100\ mol\ K_2CrO_4}{1\ L} \times \frac{1\ mol\ CrO_4^{2-}}{1\ mol\ K_2CrO_4}$$

$$= 0.00500\ mol\ CrO_4^{2-}$$

$$\text{no. moles } Ag^+ = 0.00500\ mol\ CrO_4^{2-} \times \frac{2\ mol\ Ag^+}{1\ mol\ CrO_4^{2-}}$$

$$= 0.0100\ mol\ Ag^+$$

$$V\ AgNO_3 = 0.0100\ mol\ Ag^+ \times \frac{1\ mol\ AgNO_3}{1\ mol\ Ag^+} \times \frac{1\ L}{0.125\ mol\ AgNO_3}$$

$$= 0.0800\ L = 80.0\ cm^3$$

EXERCISE If 20.0 cm^3 of 0.0500 M $Ba(NO_3)_2$ is required to react completely with 25.0 cm^3 Na_2SO_4 solution, what is the molarity of the Na_2SO_4? Answer: 0.0400 M.

EXAMPLE 18.3 When 0.200 L of 0.100 M $NiCl_2$ solution is added to 0.300 L of 0.200 M NaOH, the following reaction occurs:

$$Ni^{2+}(aq) + 2\ OH^-(aq) \rightarrow Ni(OH)_2(s)$$

Calculate
a. the number of moles of Ni^{2+} and OH^- originally present.
b. the number of moles of $Ni(OH)_2$ formed, assuming a 100% yield.

Solution

a. $\quad$ no. moles $Ni^{2+} = 0.200\ L \times \dfrac{0.100\ mol\ NiCl_2}{1\ L} \times \dfrac{1\ mol\ Ni^{2+}}{1\ mol\ NiCl_2}$

$$= 0.0200\ mol\ Ni^{2+}$$

no. moles $OH^- = 0.300\ L \times \dfrac{0.200\ mol\ NaOH}{1\ L} \times \dfrac{1\ mol\ OH^-}{1\ mol\ NaOH}$

$$= 0.0600\ mol\ OH^-$$

b. Let us follow the approach used in "limiting reactant" problems in Chapter 3. We calculate the theoretical yield of $Ni(OH)_2$:
   1. If all the Ni^{2+} is consumed,

$$\text{no. moles } Ni(OH)_2 = 0.0200\ mol\ Ni^{2+} \times \dfrac{1\ mol\ Ni(OH)_2}{1\ mol\ Ni^{2+}}$$

$$= 0.0200\ mol\ Ni(OH)_2$$

   2. If all the OH^- is consumed,

$$\text{no. moles } Ni(OH)_2 = 0.0600\ mol\ OH^- \times \dfrac{1\ mol\ Ni(OH)_2}{2\ mol\ OH^-}$$

$$= 0.0300\ mol\ Ni(OH)_2$$

We choose the *smaller* amount: 0.0200 mol $Ni(OH)_2$ is formed.

In most precipitation reactions, one of the ions is in excess

EXERCISE After the reaction is over, how many moles of OH^- are left?
Answer: 0.0200 mol.

Precipitation Titrations

Precipitation reactions can be used to determine the concentration of species in solution or in a solid mixture. A common approach involves a laboratory technique referred to as *precipitation titration*. We start with a water solution containing the ion to be analyzed for. That ion is precipitated by adding from a buret a "standard solution" of an appropriate reagent. The concentration of the standard solution is accurately known and the volume of solution required is carefully measured. The calculations involved are illustrated in Example 18.4; the experimental set-up is shown in Figure 18.2.

The standard solution has a known concentration

EXAMPLE 18.4 To determine the percentage of silver in an alloy, a chemist dissolves a 0.750-g sample of the alloy in nitric acid. This brings the silver into solution as the Ag^+ ion. The solution is diluted with water and titrated with a 0.150 M solution of potassium thiocyanate, KSCN. A precipitate forms:

$$Ag^+(aq) + SCN^-(aq) \rightarrow AgSCN(s)$$

She finds that 41.7 cm³ of the KSCN solution is required for the titration. What is the mass percent of silver in the alloy?

Solution We start by calculating the number of moles of SCN^- used in the titration:

$$\text{no. moles } SCN^- = 0.0417 \text{ L} \times \frac{0.150 \text{ mol KSCN}}{1 \text{ L}} \times \frac{1 \text{ mol } SCN^-}{1 \text{ mol KSCN}}$$
$$= 0.00626 \text{ mol } SCN^-$$

Since one mole of Ag^+ reacts with one mole of SCN^-,

$$\text{no. moles } Ag^+ = 0.00626 \text{ mol } SCN^- \times \frac{1 \text{ mol } Ag^+}{1 \text{ mol } SCN^-}$$
$$= 0.00626 \text{ mol } Ag^+$$

This kind of approach is often used in quantitative analysis

We can now calculate the mass of silver (1 mol Ag = 107.87 g), and finally the mass percent of silver in the 0.750-g sample:

$$\text{mass Ag} = 0.00626 \text{ mol} \times \frac{107.87 \text{ g}}{1 \text{ mol}} = 0.675 \text{ g}$$

$$\text{mass percent Ag} = \frac{0.675 \text{ g}}{0.750 \text{ g}} \times 100 = 90.0$$

EXERCISE What volume of 0.150 M K_2CrO_4 would have been required if it had been used instead of 0.150 M KSCN, precipitating Ag_2CrO_4 instead of AgSCN? Answer: 20.8 cm³.

In carrying out this titration, or any titration, we must know when the reaction is over so as to stop adding reagent from the buret. To do this, we add an *indicator* to the original solution. This material shows, usually by a change in color, when the species being analyzed for has been consumed. In the titration of Ag^+ with

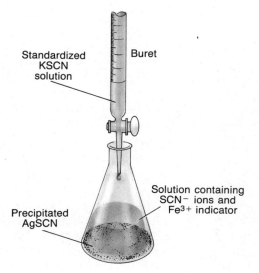

Standardized KSCN solution

Buret

Solution containing SCN^- ions and Fe^{3+} indicator

Precipitated AgSCN

FIGURE 18.2 Titration of Ag^+ with Ag^+. When all the Ag^+ has been precipitated, Fe^{3+} ions react with SCN^- ions to give a blood-red color.

There are many useful indicators, like Fe^{3+}

SCN^- ions, a few drops of $Fe(NO_3)_3$ solution serve as an indicator. When all of the Ag^+ ions have been precipitated, the Fe^{3+} ions react with excess SCN^- ions to form a blood-red complex usually assigned the formula $FeSCN^{2+}$:

$$Fe^{3+}(aq) + SCN^-(aq) \rightarrow FeSCN^{2+}(aq)$$
$$\text{red}$$

Formation of a red color tells us that the reaction we are interested in, that between Ag^+ and SCN^- ions, is over and we had better stop adding KSCN solution.

18.3
Solubility Equilibria

It's valid for many insoluble compounds

In discussing precipitation reactions, we assumed, in effect, that they went to completion. For the kind of calculations carried out in Section 18.2, that assumption is valid. We must keep in mind, however, that precipitation reactions, like all reactions, reach a position of equilibrium. Putting it another way, even the most "insoluble" electrolyte dissolves to at least a slight extent, thereby establishing equilibrium with its ions in solution.

There are many ways in which equilibrium can be established between a slightly soluble solid and its ions in solution. Consider, for example, the equilibrium between $SrCrO_4(s)$ and an aqueous solution containing Sr^{2+} and CrO_4^{2-} ions:

$$SrCrO_4(s) \rightleftharpoons Sr^{2+}(aq) + CrO_4^{2-}(aq) \qquad \textbf{(18.4)}$$

One way to establish this equilibrium is to shake a sample of solid strontium chromate with water. When we do this, Reaction 18.4 occurs in the forward direction. We can follow the progress of this reaction by observing, either visually or with an instrument (Fig. 18.3), the intensity of the yellow color imparted to the solution by the CrO_4^{2-} ion. When the intensity (or the percentage of light absorbed) becomes constant, we conclude that equilibrium has been reached. In this case, the Sr^{2+} and CrO_4^{2-} ions in equilibrium with solid $SrCrO_4$ are at equal concentrations:

They both come from the $SrCrO_4$

$$[Sr^{2+}] = [CrO_4^{2-}]$$

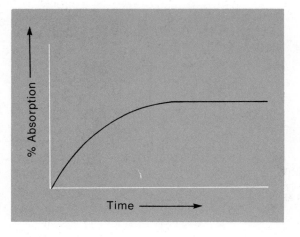

FIGURE 18.3 Establishment of equilibrium in the system

$$Sr\,CrO_4(s) \rightarrow$$
$$Sr^{2+}(aq) + CrO_4^{2-}(aq).$$

As $SrCrO_4$ dissolves, yellow chromate ions enter the solution. These ions absorb visible light at 370 nm; the percentage absorption increases with the concentration of CrO_4^{2-}. Eventually, this concentration levels off at its equilibrium value, about 6×10^{-3} M.

This must be true because, since $SrCrO_4$ was shaken with pure water, these are the only ions that enter the solution, and that solution has to be electrically neutral.

There is a quite different way to establish the equilibrium represented by Equation 18.4. We might mix solutions of $Sr(NO_3)_2$ and K_2CrO_4. In that case, Sr^{2+} and CrO_4^{2-} ions react with one another to precipitate $SrCrO_4$. In other words, equilibrium is established by Reaction 18.4 occurring in the reverse direction, from right to left. Under these conditions, we ordinarily find that the concentrations of Sr^{2+} and CrO_4^{2-} ions in solution are not equal. If we add an excess of $Sr(NO_3)_2$ solution, we expect to find that

$$[Sr^{2+}] > [CrO_4^{2-}]$$

If an excess of K_2CrO_4 solution is used,

$$[CrO_4^{2-}] > [Sr^{2+}]$$

We conclude that the concentrations of Sr^{2+} and CrO_4^{2-} ions in equilibrium with $SrCrO_4$ are not necessarily equal to each other. It turns out, however, that there is a simple relationship between these two concentrations, regardless of how equilibrium is established. That relationship can be applied to any slightly soluble ionic solid. It involves a type of equilibrium constant given the symbol K_{sp}.

This is the usual way we encounter this kind of equilibrium

The Solubility Product Constant, K_{sp}

Consider once again the equilibrium between solid $SrCrO_4$ and its ions in solution. Ordinarily, we write the equation for such an equilibrium system with the solid on the left and aqueous ions on the right:

$$SrCrO_4(s) \rightleftharpoons Sr^{2+}(aq) + CrO_4^{2-}(aq)$$

The equilibrium constant expression for this system can be written by applying the rules cited in Chapter 15. It has the form

$$[Sr^{2+}] \times [CrO_4^{2-}] = K_{sp}$$

The solid must be present, but does not appear in the equation for K_{sp}

The symbol K_{sp} represents a particular type of equilibrium constant called a **solubility product constant.** Like all equilibrium constants, K_{sp} has a fixed value for a given system at a particular temperature. At 25°C, K_{sp} for $SrCrO_4$ is about 4×10^{-5}; that is,

$$[Sr^{2+}] \times [CrO_4^{2-}] = 4 \times 10^{-5}$$

This relation tells us that the product of the two ion concentrations at equilibrium must equal 4×10^{-5}, regardless of how equilibrium is established.

For other slightly soluble ionic solids, we can write K_{sp} expressions analogous to that for strontium chromate. For $PbCl_2$ and Ag_2CrO_4,

$$PbCl_2(s) \rightleftharpoons Pb^{2+}(aq) + 2\ Cl^-(aq)$$
$$[Pb^{2+}] \times [Cl^-]^2 = K_{sp}\ PbCl_2 = 1.7 \times 10^{-5} \tag{18.5}$$

$$Ag_2CrO_4(s) \rightleftharpoons 2\ Ag^+(aq) + CrO_4{}^{2-}(aq)$$
$$[Ag^+]^2 \times [CrO_4{}^{2-}] = K_{sp}\ Ag_2CrO_4 = 2 \times 10^{-12} \qquad\qquad \textbf{(18.6)}$$

More generally, we can express the solubility product principle as follows:

In any water solution in equilibrium with a slightly soluble ionic compound, the product of the concentrations of its ions, each raised to a power equal to its coefficient in the solubility equation, is a constant. This constant, K_{sp}, has a fixed value at a given temperature, independent of the concentrations of the individual ions. Values of K_{sp} for a variety of slightly soluble ionic solids are given in Table 18.2.

Notice that in the expression for K_{sp}, as with K_c in Chapter 15, *there is no term for the concentration of solid*. A small amount of solid must be present for equilibrium to be established but, beyond that, it doesn't matter how much solid is used. Adding or removing solid does not change the equilibrium concentrations of ions in solution. In that sense, the solid has no effect upon the position of the

---

**Table 18.2
Solubility Product Constants at 25°C**

Acetates	$AgC_2H_3O_2$	2×10^{-3}	Iodides	AgI	1×10^{-16}
				PbI_2	1×10^{-8}
Bromides	$AgBr$	5×10^{-13}	Phosphates	Ag_3PO_4	1×10^{-15}
	$PbBr_2$	5×10^{-6}		$Ca_3(PO_4)_2$	1×10^{-33}
				$Pb_3(PO_4)_2$	1×10^{-54}
Carbonates	$BaCO_3$	2×10^{-9}			
	$CaCO_3$	5×10^{-9}	Sulfates	$BaSO_4$	1.4×10^{-9}
	$MgCO_3$	2×10^{-8}		$CaSO_4$	3×10^{-5}
	$SrCO_3$	7×10^{-10}		$PbSO_4$	1×10^{-8}
Chlorides	$AgCl$	1.6×10^{-10}	Sulfides	Ag_2S	1×10^{-49}
	Hg_2Cl_2	1×10^{-18}		CdS	1×10^{-26}
	$PbCl_2$	1.7×10^{-5}		CoS	1×10^{-20}
				CuS	1×10^{-35}
Chromates	Ag_2CrO_4	2×10^{-12}		FeS	1×10^{-17}
	$BaCrO_4$	2×10^{-10}		HgS	1×10^{-52}
	$PbCrO_4$	1×10^{-16}		MnS	1×10^{-15}
	$SrCrO_4$	4×10^{-5}		NiS	1×10^{-19}
				PbS	1×10^{-27}
Fluorides	BaF_2	2×10^{-6}		ZnS	1×10^{-20}
	CaF_2	2×10^{-10}			
	PbF_2	4×10^{-8}			
Hydroxides	$Al(OH)_3$	5×10^{-33}			
	$Cr(OH)_3$	4×10^{-38}			
	$Fe(OH)_2$	1×10^{-15}			
	$Fe(OH)_3$	5×10^{-38}			
	$Mg(OH)_2$	1×10^{-11}			
	$Zn(OH)_2$	5×10^{-17}			

solubility equilibrium. Hence, it does not belong in the expression for the equilibrium constant.

> **EXAMPLE 18.5** Write K_{sp} expressions for each of the following slightly soluble ionic solids:
>
> a. CoS b. CaF_2 c. As_2S_3
>
> **Solution** The dissociation equations are
>
> $CoS(s) \rightleftharpoons Co^{2+}(aq) + S^{2-}(aq)$
>
> $CaF_2(s) \rightleftharpoons Ca^{2+}(aq) + 2 F^{-}(aq)$
>
> $As_2S_3(s) \rightleftharpoons 2 As^{3+}(aq) + 3 S^{2-}(aq)$
>
> The equilibrium constant expressions follow from the dissociation equations and the rule cited earlier:
>
> a. K_{sp} CoS $= [Co^{2+}] \times [S^{2-}]$
>
> b. K_{sp} CaF_2 $= [Ca^{2+}] \times [F^{-}]^2$
>
> c. K_{sp} As_2S_3 $= [As^{3+}]^2 \times [S^{2-}]^3$
>
> **EXERCISE** Write K_{sp} expressions for $Fe(OH)_2$ and $Ca_3(PO_4)_2$. Answer:
>
> K_{sp} $Fe(OH)_2$ $= [Fe^{2+}] \times [OH^{-}]^2$
> K_{sp} $Ca_3(PO_4)_2$ $= [Ca^{2+}]^3 \times [PO_4^{3-}]^2$

You need to know the charges on the ions

Solubility product constants such as those listed in Table 18.2 can be used for a variety of purposes. In particular, we can use K_{sp} for a slightly soluble ionic solid to calculate

—the concentration of one ion in equilibrium with the solid when we know that of the other ion.
—whether or not the solid will precipitate when two solutions are mixed.
—the water solubility of the solid and the effect of a "common ion" upon that solubility.

K_{sp} and the Equilibrium Concentrations of Ions

We can use the relation

$$[Sr^{2+}] \times [CrO_4^{2-}] = K_{sp} \ SrCrO_4 = 4 \times 10^{-5}$$

to calculate the equilibrium concentration of one ion when we know that of the other. Suppose, for example, we find that when $SrCrO_4$ is precipitated out of solution, the concentration of CrO_4^{2-} in equilibrium with it is 2×10^{-3} M. It follows that

$$[Sr^{2+}] = \frac{K_{sp} \ SrCrO_4}{[CrO_4^{2-}]} = \frac{4 \times 10^{-5}}{2 \times 10^{-3}} = 2 \times 10^{-2} \ M$$

If in another case we know that $[Sr^{2+}] = 1 \times 10^{-4}$ M,

$$[CrO_4^{2-}] = \frac{K_{sp} \, SrCrO_4}{[Sr^{2+}]} = \frac{4 \times 10^{-5}}{1 \times 10^{-4}} = 4 \times 10^{-1} \text{ M}$$

Example 18.6 illustrates the same kind of calculation for a different type of electrolyte.

EXAMPLE 18.6 A precipitate of $PbCl_2$ is formed by mixing solutions containing Pb^{2+} and Cl^- ions. Use the K_{sp} value for $PbCl_2$ given in Table 18.2 to calculate
a. the concentration of Pb^{2+} in equilibrium with $PbCl_2$ if $[Cl^-] = 0.10$ M.
b. the concentration of Cl^- in equilibrium with $PbCl_2$ if $[Pb^{2+}] = 0.10$ M.

Solution The expression for K_{sp} is

$$[Pb^{2+}] \times [Cl^-]^2 = K_{sp} \, PbCl_2 = 1.7 \times 10^{-5}$$

a. Solving for Pb^{2+},

$$[Pb^{2+}] = \frac{K_{sp} \, PbCl_2}{[Cl^-]^2} = \frac{1.7 \times 10^{-5}}{(1.0 \times 10^{-1})^2} = 1.7 \times 10^{-3} \text{ M}$$

b. Solving for Cl^-,

$$[Cl^-]^2 = \frac{K_{sp} \, PbCl_2}{[Pb^{2+}]} = \frac{1.7 \times 10^{-5}}{1.0 \times 10^{-1}} = 1.7 \times 10^{-4}$$

$$[Cl^-] = (1.7 \times 10^{-4})^{1/2} = 1.3 \times 10^{-2} \text{ M}$$

EXERCISE Using Table 18.2, calculate $[Mg^{2+}]$ in contact with $Mg(OH)_2$ if $[OH^-] = 1 \times 10^{-4}$ M. Answer: 1×10^{-3} M.

K_{sp} and Precipitate Formation

In Section 18.2, we used the solubility rules (Table 18.1) to predict whether or not a precipitate will form when two solutions are mixed. That type of prediction is limited to the situation where the ions involved are at a concentration of 0.1 M or greater. If the ion concentrations are appreciably less than 0.1 M, a precipitate may not form even though the solid is listed as being "insoluble" in water.

We can use K_{sp} values such as those listed in Table 18.2 to make a more general prediction concerning precipitate formation, regardless of the concentrations of the ions involved. To do this, we compare a quantity called the *ion product*, P, to the solubility product constant, K_{sp}. The form of the expression for P is the same as that for K_{sp}. However, P differs from K_{sp} in that it involves original rather than equilibrium concentrations. Thus, for $SrCrO_4$, we have

P is analogous to Q in Chapter 15

$$P = (\text{orig. conc. } Sr^{2+}) \times (\text{orig. conc. } CrO_4^{2-}); \quad K_{sp} = [Sr^{2+}] \times [CrO_4^{2-}]$$

The "original concentrations" are those in the solution before precipitation occurs. In contrast, the terms $[Sr^{2+}]$ and $[CrO_4^{2-}]$ refer to equilibrium concentrations, those that are established after precipitation. For $PbCl_2$, we would write

$$P = (\text{orig. conc. } Pb^{2+}) \times (\text{orig. conc. } Cl^-)^2; \quad K_{sp} = [Pb^{2+}] \times [Cl^-]^2$$

We can distinguish three cases:

1. If $P > K_{sp}$, the solution contains a higher concentration of ions that it can hold at equilibrium. In other words, the solution is supersaturated. **A precipitate forms,** decreasing the concentrations until the ion product becomes equal to K_{sp} and equilibrium is established.

2. If $P < K_{sp}$, the solution contains a lower concentration of ions than is required for equilibrium with the solid. The solution is unsaturated. **No precipitate forms;** equilibrium is not established.

3. If $P = K_{sp}$, the solution is just saturated with ions and is at the point of precipitation.

EXAMPLE 18.7 Chromate ions are added to a solution in which the original concentration of Sr^{2+} is 1×10^{-3} M. Assuming the concentration of Sr^{2+} stays constant,

a. will a precipitate of $SrCrO_4$ ($K_{sp} = 4 \times 10^{-5}$) form when conc. CrO_4^{2-} $= 3 \times 10^{-2}$ M?

b. will a precipitate of $SrCrO_4$ form when the concentration of CrO_4^{2-} is 5×10^{-2} M?

c. at what concentration of CrO_4^{2-} does a precipitate just start to form?

Solution

a. $P = (\text{orig. conc. } Sr^{2+}) \times (\text{orig. conc. } CrO_4^{2-}) = (1 \times 10^{-3}) \times (3 \times 10^{-2}) = 3 \times 10^{-5}$

Since P is less than K_{sp}, 4×10^{-5}, no precipitate forms.

b. $P = (1 \times 10^{-3}) \times (5 \times 10^{-2}) = 5 \times 10^{-5}$; $\quad P > K_{sp}$.

A precipitate forms, reducing the concentrations of Sr^{2+} and CrO_4^{2-} ions until the ion product becomes equal to 4×10^{-5}.

c. The precipitate starts to form when the ion product is equal to K_{sp}:

$$\text{conc. } CrO_4^{2-} = \frac{K_{sp} \ SrCrO_4}{\text{conc. } Sr^{2+}} = \frac{4 \times 10^{-5}}{1 \times 10^{-3}} = 4 \times 10^{-2} \text{ M}$$

When the concentration of CrO_4^{2-} is less than 4×10^{-2} M, as in (a), no precipitate forms. When the concentration of CrO_4^{2-} exceeds 4×10^{-2} M, as in (b), a precipitate forms.

If $P < K_{sp}$, there is no equilibrium between the ions and the solid

EXERCISE Will PbI_2 ($K_{sp} = 1 \times 10^{-8}$) precipitate from a solution in which the concentrations of Pb^{2+} and I^- are both 1×10^{-3} M? Answer: No.

The calculations in Example 18.7 were simplified by assuming that the concentration of one ion (Sr^{2+}) stayed constant while the other ion (CrO_4^{2-}) was added. In reality, this assumption is unlikely to be valid. Mixing two solutions decreases the concentrations of the ions present because the volume increases. We have to take this effect into account before we can decide whether or not a precipitate forms. In particular, we have to calculate the concentrations of the ions after mixing but before precipitation. These are the "original concentrations" to be used in the expression for the ion product, P. The calculations involved are shown in Example 18.8.

One solution dilutes the other when they are mixed

EXAMPLE 18.8 A student mixes 0.200 L of 0.0060 M $Sr(NO_3)_2$ solution with 0.100 L of 0.015 M K_2CrO_4 solution to give a final volume of 0.300 L. Determine

a. the concentration of Sr^{2+} after mixing.
b. the concentration of CrO_4^{2-} after mixing.
c. whether or not a precipitate of $SrCrO_4$ ($K_{sp} = 4 \times 10^{-5}$) will form under these conditions.

Solution

a. To find the concentration of Sr^{2+} after mixing, we need to know the number of moles of that ion present:

$$\text{no. moles } Sr^{2+} = 0.200 \text{ L} \times \frac{0.0060 \text{ mol } Sr(NO_3)_2}{1 \text{ L}} \times \frac{1 \text{ mol } Sr^{2+}}{1 \text{ mol } Sr(NO_3)_2}$$

$$= 1.2 \times 10^{-3} \text{ mol } Sr^{2+}$$

The volume after mixing is 0.300 L. Hence,

The "original concentration" is that after mixing

$$\text{conc. } Sr^{2+} = \frac{1.2 \times 10^{-3} \text{ mol}}{0.300 \text{ L}} = 4.0 \times 10^{-3} \text{ mol/L}$$

b. Proceeding in the same way,

$$\text{no. moles } CrO_4^{2-} = 0.100 \text{ L} \times \frac{0.015 \text{ mol } K_2CrO_4}{1 \text{ L}} \times \frac{1 \text{ mol } CrO_4^{2-}}{1 \text{ mol } K_2CrO_4}$$

$$= 1.5 \times 10^{-3} \text{ mol } CrO_4^{2-}$$

$$\text{conc. } CrO_4^{2-} = \frac{1.5 \times 10^{-3} \text{ mol}}{0.300 \text{ L}} = 5.0 \times 10^{-3} \text{ M}$$

c. $P = (\text{orig. conc. } Sr^{2+}) \times (\text{orig. conc. } CrO_4^{2-})$
$= (4.0 \times 10^{-3}) \times (5.0 \times 10^{-3}) = 2 \times 10^{-5}$

Since P is less than K_{sp} (4×10^{-5}), we conclude that no precipitate should form.

EXERCISE Suppose the student ignored the effect of mixing the solutions upon ion concentrations and took orig. conc. $Sr^{2+} = 0.0060$ M, orig. conc. $CrO_4^{2-} = 0.015$ M. Would he conclude that a precipitate should form under these conditions? Answer: Yes.

K_{sp} and Water Solubility

To express the water solubility of a slightly soluble ionic solid, we most often cite the number of moles of the solid that dissolves in one liter of water. The solubility of an ionic compound and its solubility product constant are related but they are not equal. For example,

Think about this until you get the idea

COMPOUND	WATER SOLUBILITY (mol/L)	SOLUBILITY PRODUCT CONSTANT, K_{sp}
$SrCrO_4$	6×10^{-3}	4×10^{-5}
$CaCO_3$	7×10^{-5}	5×10^{-9}
BaF_2	8×10^{-3}	2×10^{-6}

Knowing one of these two quantities, we can always, in principle at least,* calculate the other one, as shown in Examples 18.9 and 18.10.

EXAMPLE 18.9

a. When excess solid $SrCrO_4$ is shaken with water at 25°C, it is found that 6×10^{-3} mol dissolves per liter. Use this information to calculate K_{sp} for $SrCrO_4$.

b. The K_{sp} of $CaCO_3$ is 5×10^{-9}; what is its water solubility in moles per liter?

Solution

a. Strontium chromate dissolves in water according to the equation

$$SrCrO_4(s) \rightleftharpoons Sr^{2+}(aq) + CrO_4^{2-}(aq)$$

For every mole of $SrCrO_4$ that dissolves, one mole of Sr^{2+} and one mole of CrO_4^{2-} enter the solution. Hence, when 6×10^{-3} mol/L of solid dissolves, the equilibrium concentrations of Sr^{2+} and CrO_4^{2-} must both be 6×10^{-3} M:

$$K_{sp} = [Sr^{2+}] \times [CrO_4^{2-}] = (6 \times 10^{-3}) \times (6 \times 10^{-3})$$
$$= 4 \times 10^{-5} \quad \text{(1 sig. fig.)}$$

b. Calcium carbonate dissolves in water according to the equation

$$CaCO_3(s) \rightleftharpoons Ca^{2+}(aq) + CO_3^{2-}(aq)$$

For every mole of $CaCO_3$ that dissolves, one mole of Ca^{2+} and one mole of CO_3^{2-} are formed. Hence, if we let s be the solubility of $CaCO_3$ (that is, s = no. moles $CaCO_3$ dissolving per liter), then $[Ca^{2+}] = [CO_3^{2-}] = s$.

Hence,

$$K_{sp} = [Ca^{2+}] \times [CO_3^{2-}] = (s) \times (s) = s^2$$

Solving for s,

$$s = (K_{sp})^{1/2} = (5 \times 10^{-9})^{1/2} = (50 \times 10^{-10})^{1/2} = 7 \times 10^{-5} \text{ M}$$

The solubility is related to, but not equal to, the solubility product, K_{sp}

EXERCISE When TlI is shaken with one liter of water, about 0.07 g dissolves. Calculate K_{sp} of TlI. Answer: 4×10^{-8}.

As Example 18.9 indicates, there is a simple relationship between K_{sp} and the water solubility, s, for an electrolyte like $SrCrO_4$ or $CaCO_3$, which dissolves to form equal numbers of cations and anions. In that case, $K_{sp} = s^2$. With other types of electrolytes, the relationship is quite different (Example 18.10).

---

*Experimentally, we usually find that the solubility is slightly greater than that predicted from K_{sp}. For example, the measured solubility of PbI_2 in water at 25°C is 1.7×10^{-3} M. This compares to a value of 1.3×10^{-3} M, calculated from the K_{sp} of PbI_2. The reason for this is that some of the lead in PbI_2 goes into solution in the form of species other than Pb^{2+}. For example, we can detect ions such as $Pb(OH)^+$ and PbI^+ in a water solution of lead iodide.

EXAMPLE 18.10 Taking K_{sp} of BaF_2 to be 2×10^{-6}, calculate its water solubility in moles per liter.

Solution When barium fluoride dissolves in water, the following equilibrium is established:

$$BaF_2(s) \rightleftharpoons Ba^{2+}(aq) + 2 F^-(aq)$$

When one mole of BaF_2 dissolves, one mole of Ba^{2+} ions and two moles of F^- ions are formed. It follows that if s moles of BaF_2 dissolves per liter, then s moles per liter of Ba^{2+} ions and 2s moles per liter of F^- ions are formed. In other words,

$$[Ba^{2+}] = s; \qquad [F^-] = 2s$$

where s is the water solubility in moles per liter of BaF_2.

To evaluate s, we substitute into the K_{sp} expression:

$$[Ba^{2+}] \times [F^-]^2 = K_{sp} \, BaF_2 = 2 \times 10^{-6}$$

$$(s) \times (2s)^2 = 4s^3 = K_{sp} \, BaF_2 = 2 \times 10^{-6}$$

$$s = \left(\frac{2 \times 10^{-6}}{4}\right)^{1/3} = 8 \times 10^{-3} \, M$$

> Here the ion concentrations are not both equal to s

EXERCISE Find the relationship between K_{sp} and s for Ag_2CrO_4; for Ag_3PO_4. Answer: $K_{sp} = 4s^3$; $K_{sp} = 27s^4$.

K_{sp} and the Common Ion Effect

In our calculations of solubilities of ionic compounds, as in Examples 18.9 and 18.10, it was understood that the solid was dissolving in pure water. Sometimes, however, we dissolve an ionic solid in a solution that already contains ions. In particular, the solution may contain an ion in common with the solid. We might, for example, dissolve $CaCO_3$ in 0.1 M Na_2CO_3 solution or in 0.1 M $CaCl_2$. In the first case, the "common ion" is CO_3^{2-}; in the second case, it is Ca^{2+}.

> There is always more than one source for the common ion

We can use K_{sp} for a slightly soluble electrolyte to estimate its solubility in a solution containing a common ion. Example 18.11 shows how this is done.

EXAMPLE 18.11 Taking K_{sp} of $CaCO_3$ to be 5×10^{-9}, estimate its solubility (moles per liter) in 0.1 M Na_2CO_3 solution.

Solution Here, as in Example 18.9, we establish the equilibrium:

$$CaCO_3(s) \rightleftharpoons Ca^{2+}(aq) + CO_3^{2-}(aq)$$

For every mole of $CaCO_3$ that dissolves, one mole of Ca^{2+} is formed. If the solubility of $CaCO_3$ in moles per liter is taken to be s, then

$$[Ca^{2+}] = s$$

The equilibrium concentration of Ca^{2+} can be calculated from the K_{sp} expression:

$$[Ca^{2+}] = s = \frac{K_{sp} \, CaCO_3}{[CO_3{}^{2-}]}$$

To solve for s, we take $[CO_3{}^{2-}]$ to be 0.1 M, the concentration of $CO_3{}^{2-}$ ion in the Na_2CO_3 solution. (In principle, $[CO_3{}^{2-}]$ is slightly greater than 0.1 M because of the $CaCO_3$ that dissolves. In practice, we can safely ignore this effect since $CaCO_3$ is only very slightly soluble.)

$$s = \frac{5 \times 10^{-9}}{0.1} = 5 \times 10^{-8} \, M$$

$$\begin{aligned}[CO_3{}^{2-}] &= (0.1 + s)M \\ &= (0.1 + 5 \\ &\quad \times 10^{-8})M \\ &= 0.1 \, M \end{aligned}$$

EXERCISE Estimate the solubility of $CaCO_3$ in 0.1 M $CaCl_2$ solution. Answer: 5×10^{-8} M.

Comparing the results of Examples 18.9 and 18.11,

SOLVENT	SOLUBILITY OF $CaCO_3$ (mol/L)
Pure water	7×10^{-5}
0.1 M Na_2CO_3	5×10^{-8}
0.1 M $CaCl_2$	5×10^{-8}

we see that $CaCO_3$ is much less soluble in 0.1 M Na_2CO_3 or $CaCl_2$ solution than it is in pure water. The effect is a general one. *An ionic solid is always less soluble in a solution containing a common ion than it is in pure water.*

Qualitatively, we can explain this effect in terms of Le Châtelier's Principle. Suppose we establish the equilibrium

$$CaCO_3(s) \rightleftharpoons Ca^{2+}(aq) + CO_3{}^{2-}(aq)$$

and then add Ca^{2+} ions (0.1 M $CaCl_2$) or $CO_3{}^{2-}$ ions (0.1 M Na_2CO_3) to the solution. We expect the equilibrium to shift to the left, consuming part of the added ions and precipitating $CaCO_3$. This is just another way of saying that the solubility is reduced by adding a common ion. The more soluble the compound is, the greater the amount precipitated by adding a common ion (Fig. 18.4).

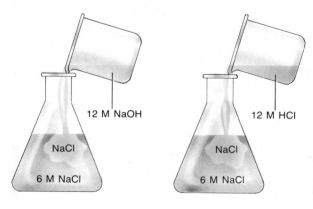

FIGURE 18.4 Sodium chloride can be precipitated from its saturated solution (6 M) by adding a solution containing either Na^+ or Cl^- ions at a concentration greater than 6 M. Both 12 M NaOH and 12 M HCl will bring about this reaction, illustrating the "common ion" effect.

18.4
Water Softening

Precipitation reactions are not confined to the laboratory. They can and do occur in the world around us, often when we don't want them to. "Hard" water contains Ca^{2+} ions and sometimes other cations (Mg^{2+} and Fe^{2+}) that form precipitates with a variety of anions. If hard water containing SO_4^{2-} anions is heated or evaporated, calcium sulfate may precipitate:

$$Ca^{2+}(aq) + SO_4^{2-}(aq) \rightarrow CaSO_4(s) \tag{18.7}$$

Calcium sulfate is one of the few ionic solids that is less soluble at high temperatures. Hard water containing HCO_3^- ions precipitates calcium carbonate upon heating as the result of the two-step reaction

$$2\ HCO_3^-(aq) \rightarrow CO_3^{2-}(aq) + CO_2(g) + H_2O$$
$$\underline{Ca^{2+}(aq) + CO_3^{2-}(aq) \rightarrow CaCO_3(s)}$$
$$Ca^{2+}(aq) + 2\ HCO_3^-(aq) \rightarrow CaCO_3(s) + CO_2(g) + H_2O \tag{18.8}$$

This reaction goes on all this time in my hot water heater

EJS

The $CaSO_4$ or $CaCO_3$ deposits, often as a tightly adherent scale, on steam irons, teakettles, boilers, and steam pipes. This lowers the heat conductivity and, with pipes, may even block the flow of water.

EXAMPLE 18.12 Typically, hard water contains 50 mg Ca^{2+} and 100 mg HCO_3^- per liter. Assuming Reaction 18.8 goes to completion, calculate the mass in grams of precipitate formed when 20 L of hard water is boiled.

Solution Let us first calculate the numbers of moles of Ca^{2+} and HCO_3^- present (1 mol Ca^{2+} = 40.1 g; 1 mol HCO_3^- = 61.0 g):

$$\text{no. moles } Ca^{2+} = \frac{50 \times 10^{-3}\ g}{1\ L} \times \frac{1\ mol}{40.1\ g} \times 20\ L = 2.5 \times 10^{-2}\ mol$$

$$\text{no. moles } HCO_3^- = \frac{100 \times 10^{-3}\ g}{1\ L} \times \frac{1\ mol}{61.0\ g} \times 20\ L$$
$$= 3.3 \times 10^{-2}\ mol$$

The molar mass of $CaCO_3$ is 100.0 g/mol. If all the Ca^{2+} ions are consumed,

$$\text{yield } CaCO_3 = 2.5 \times 10^{-2}\ mol\ Ca^{2+} \times \frac{1\ mol\ CaCO_3}{1\ mol\ Ca^{2+}} \times \frac{100.0\ g\ CaCO_3}{1\ mol\ CaCO_3}$$
$$= 2.5\ g\ CaCO_3$$

If all the HCO_3^- ions are consumed,

$$\text{yield } CaCO_3 = 3.3 \times 10^{-2}\ mol\ HCO_3^- \times \frac{1\ mol\ CaCO_3}{2\ mol\ HCO_3^-} \times \frac{100.0\ g\ CaCO_3}{1\ mol\ CaCO_3}$$
$$= 1.7\ g\ CaCO_3$$

Choosing the smaller number, we conclude that 1.7 g of $CaCO_3$ is formed from 20 L of hard water.

EXERCISE Taking K_{sp} of $CaCO_3$ to be 5×10^{-9}, calculate the concentration of $CO_3{}^{2-}$ required to precipitate $CaCO_3$ from hard water containing 50 mg Ca^{2+} per liter. Answer: 4×10^{-6} M.

In the home, hard water is most easily detected by its interaction with soap. Most soaps are sodium salts of organic acids. A typical component of soap, sodium stearate, has the formula $NaC_{18}H_{35}O_2$ and consists of Na^+ cations and $C_{18}H_{35}O_2{}^-$ anions. The calcium salts of soaps, such as calcium stearate, $Ca(C_{18}H_{35}O_2)_2$, are insoluble. Hence, when soap is added to hard water, a precipitate forms:

$$Ca^{2+}(aq) + 2\ C_{18}H_{35}O_2{}^-(aq) \rightarrow Ca(C_{18}H_{35}O_2)_2(s) \qquad \textbf{(18.9)}$$

Reactions of this type are undesirable for a couple of reasons. In the first place, they waste soap. Moreover, the calcium salt precipitates as a gray deposit upon laundry ("ring around the collar") or as a greasy film on plumbing fixtures ("bathtub ring").

There are many ways to remove Ca^{2+} ions from hard water, thereby softening it. One approach is to add sodium carbonate:

$$Ca^{2+}(aq) + CO_3{}^{2-}(aq) \rightarrow CaCO_3(s)$$

It's hard to tell how much Na_2CO_3 to add

The $CaCO_3$ can be filtered off, leaving water in which the concentration of Ca^{2+} is very low (recall Example 18.11). A more common method of softening water in the home uses a process called *cation exchange*. This exchange takes place when hard water is passed through a column containing a mineral called a *zeolite*. In appearance, zeolite crystals look very much like grains of sand. The structure of such a crystal is shown in Figure 18.5, p. 566. Zeolites differ from simple ionic compounds in one important way. The negative ion is network covalent (Chap. 11). It consists of Al, Si, and O atoms bonded together in a network structure that forms the backbone of the crystal. Trapped in holes within this anionic lattice are Na^+ ions. There are just enough Na^+ ions to cancel the negative charge of the network covalent anion.

When pure water passes through the zeolite, nothing much happens. The Na^+ ions cannot leave the crystal; that would create an imbalance of charge. However, if hard water comes in contact with the zeolite, the situation changes. Now, Na^+ ions migrate out of the lattice, to be replaced by Ca^{2+} ions. The process can be represented by the equation

Hard water is a problem in many parts of this country

$$Ca^{2+}(aq) + 2\ NaZ(s) \rightarrow CaZ_2(s) + 2\ Na^+(aq) \qquad \textbf{(18.10)}$$

Here, Z represents a small portion of the zeolite anion. The result of this reaction is to exchange a Ca^{2+} ion for two Na^+ ions. Ordinarily, Na^+ ions are much less objectionable than Ca^{2+} ions. (The reverse may be true, however, for people on low-sodium diets; they need a different kind of water softener.)

After a zeolite has been used for some time, most of the holes in the lattice become filled with Ca^{2+} ions. When that happens, exchange of cations with hard water stops. However, the zeolite is readily returned to its original state. To do this, the column is flushed with a concentrated solution of NaCl. Reaction 18.10 is reversed, and the water softener is ready for reuse.

At present most cation exchangers use synthetic organic resins rather than

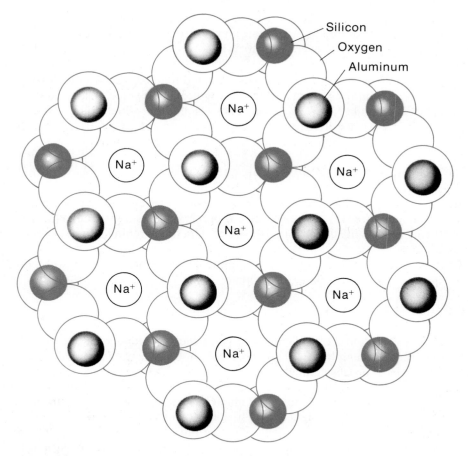

FIGURE 18.5 Structure of a natural zeolite with the simplest formula $NaAlSiO_4$. Na^+ ions are held loosely in the anionic lattice, which consists of Al, Si, and O atoms. When water containing Ca^{2+} ions is passed through the zeolite, Ca^{2+} ions enter the lattice while Na^+ ions take their place in solution.

zeolites. They operate much like the natural zeolites but have a greater capacity for cation exchange. They are also more expensive.

Summary

Ionic solutes differ widely in their water solubility (Table 18.1). The solubility rules are useful in predicting what precipitate if any will form when two solutions are mixed. Precipitation reactions can be represented by net ionic equations (Example 18.1). The coefficients in the net ionic equation for a precipitation reaction can be used to relate numbers of moles of reactants and products (Examples 18.2, 18.3, and 18.12). Precipitation titrations are useful to analyze for a component in a mixture (Example 18.4).

A slightly soluble ionic solute reaches equilibrium with its ions in aqueous solution (Fig. 18.3). A solubility product expression can be written for such an equilibrium (Example 18.5). The solubility product constant, K_{sp},

$$M_xN_y(s) \rightleftharpoons x\ M^{y+}(aq) + y\ N^{x-}(aq); \qquad K_{sp} = [M^{y+}]^x \times [N^{x-}]^y$$

can be used to calculate

> —the concentration of one ion in solution in equilibrium with the solid when we know that of the other ion (Example 18.6).
> —whether or not a precipitate will form (Examples 18.7, 18.8).
> —the solubility of the solid in pure water (Examples 18.9, 18.10) or in a solution containing a common ion (Example 18.11).

Key Words and Concepts

cation exchange	limiting reactant	solubility rules
common ion effect	net ionic equation	spectator ion
hard water	precipitate	titration
indicator	solubility	water softening
ion product, P	solubility product constant, K_{sp}	zeolite
Le Châtelier's Principle		

Questions and Problems

Solubilities of Ionic Solids

18.1 Using Table 18.1, determine which of the following are water soluble:
  a. $BaSO_4$ b. K_2CO_3
  c. $Fe(OH)_2$ d. $NiSO_4$

18.2 Describe how you could prepare
  a. $Fe(OH)_2(s)$ from a solution of $FeBr_2$.
  b. $BaSO_4(s)$ from sulfuric acid.
  c. $AgCl(s)$ from a solution of silver nitrate.

18.31 Use the solubility rules to determine which of the following are water soluble:
  a. $CaCl_2$ b. $Mg(OH)_2$
  c. $Ce(NO_3)_3$ d. $(NH_4)_2S$

18.32 What reagent would you add to a solution of $CuSO_4$ to precipitate
  a. $Cu(OH)_2$ b. CuS c. $CuCO_3$

Precipitation Reactions; Net Ionic Equations

18.3 Write net ionic equations to explain the formation of
  a. a blue precipitate when solutions of $CuCl_2$ and sodium hydroxide are mixed.
  b. two different white precipitates when solutions of zinc sulfate and barium sulfide are mixed.

18.4 Using Table 18.1, decide whether a precipitate will form when the following solutions are mixed. If a precipitate forms, write a net ionic equation for the reaction.
  a. $Hg_2(NO_3)_2$ and HCl
  b. KNO_3 and $NaOH$
  c. $Fe_2(SO_4)_3$ and $Ba(OH)_2$
  d. $AsCl_3$ and Na_2S
  e. $Ni(NO_3)_2$ and $(NH_4)_2CO_3$

18.33 Write net ionic equations for the formation of
  a. a yellow precipitate when solutions of $BaCl_2$ and K_2CrO_4 are mixed.
  b. a white precipitate (that is soluble in acid) when solutions of KOH and $AlCl_3$ are mixed.

18.34 Follow the directions of Question 18.4 for solutions of
  a. $Ba(NO_3)_2$ and $Cr_2(SO_4)_3$
  b. $FeCl_3$ and KOH
  c. $Al_2(SO_4)_3$ and $NaCl$
  d. $CuSO_4$ and K_2S
  e. $Pb(NO_3)_2$ and NH_4Cl

18.5 Write a net ionic equation for any precipitation reaction that occurs when 0.1 M solutions of the following are mixed:
a. $CaCl_2$ and $AgNO_3$
b. $(NH_4)_2S$ and $Pb(NO_3)_2$
c. SrS and KOH
d. $Fe(NO_3)_3$ and NaOH

18.35 Follow the instructions for Question 18.5 for the following pairs of solutions:
a. Na_2CO_3 and $BaCl_2$
b. $ScCl_3$ and $Ni(NO_3)_2$
c. $ZnSO_4$ and $(NH_4)_2S$
d. Na_2CO_3 and $CaCl_2$

Precipitation Reactions; Mole Relations, Titrations

18.6 A precipitate forms when solutions of silver nitrate and scandium chloride are mixed:

$$Ag^+(aq) + Cl^-(aq) \rightarrow AgCl(s)$$

a. What volume of 0.0250 M $ScCl_3$ is needed to react completely with 10.0 mL of 0.105 M $AgNO_3$?
b. How many grams of AgCl form in (a)?

18.36 Mixing solutions of $Na_2C_2O_4$ and $LaCl_3$ precipitates lanthanum oxalate:

$$2\ La^{3+}(aq) + 3\ C_2O_4{}^{2-}(aq) \rightarrow La_2(C_2O_4)_3(s)$$

a. What is the molarity of an $LaCl_3$ solution if 20.0 mL is required to react with 14.1 mL of 0.0226 M $Na_2C_2O_4$?
b. How many grams of lanthanum oxalate form in (a)?

18.7 What volume of 0.100 M $AgNO_3$ is required to react completely with
a. 50.0 mL of 0.0814 M Na_2S
b. 62.6 mL of 0.197 M HCl
c. 12.7 mL of 0.240 M K_2CrO_4

Write a balanced net ionic equation for each reaction.

18.37 What volume of 0.0800 M $CuSO_4$ is required to react completely with
a. 15.8 mL of 0.156 M NaOH
b. 36.1 mL of 0.251 M $Ba(NO_3)_2$
c. 52.4 mL of 0.182 M $(NH_4)_2S$

Write a balanced net ionic equation for each reaction.

18.8 When 0.0400 L of 0.100 M KI is added to 0.0350 L of 0.0350 M $Hg_2(NO_3)_2$, a precipitate of Hg_2I_2 forms:

$$Hg_2{}^{2+}(aq) + 2\ I^-(aq) \rightarrow Hg_2I_2(s)$$

Calculate
a. the numbers of moles of $Hg_2{}^{2+}$ and I^- present originally.
b. the moles of Hg_2I_2 formed (100% yield).
c. the number of moles of the ion in excess remaining in solution after precipitation.

18.38 When 0.0250 L of 0.300 M LiF is mixed with 0.0150 L of 0.110 M $Th(NO_3)_4$, a precipitate of ThF_4 forms:

$$Th^{4+}(aq) + 4\ F^-(aq) \rightarrow ThF_4(s)$$

Calculate
a. the numbers of moles of Th^{4+} and F^- present originally.
b. the moles of ThF_4 formed (100% yield).
c. the volume of 0.110 M $Th(NO_3)_4$ required to precipitate the remaining F^- ions.

18.9 A chemist analyzes an alloy for silver as in Example 18.4. An alloy sample weighing 0.249 g requires 40.2 mL of 0.0502 M KSCN for titration. What is the mass percent of silver in the alloy?

18.39 A student determines the mass percent of barium in an alloy by reaction with sulfuric acid, which precipitates barium sulfate. An alloy sample weighing 2.016 g requires 20.0 mL of 0.200 M H_2SO_4 for titration. What is the mass percent of barium in the alloy?

18.10 A student determines the mass percent of Cl^- in a mixture by weighing out 0.400 g, dissolving in 20 mL of water, and titrating with 15.6 mL of 0.120 M $AgNO_3$ solution. What is the mass percent of Cl^- in the mixture?

18.40 What is the mass percent of NaCl in a solid mixture if 21.2 mL of 0.100 M $AgNO_3$ is required to react completely with 0.200 g of the solid?

18.11 A 1.250-g fertilizer sample was analyzed for phosphorus by precipitating it as $MgNH_4PO_4$ and heating this to $Mg_2P_2O_7$, which weighed 0.3233 g. Calculate the mass percent of phosphorus in the fertilizer.

18.41 If 0.260 g As_2S_3 is precipitated from 1.200 g of an insecticide, what is the mass percent of arsenic in the insecticide?

Expression for K_{sp}

18.12 Write the dissociation equation and the K_{sp} expression for each of the following slightly soluble electrolytes:
a. TlI
b. $Eu(OH)_3$
c. $Pb_3(PO_4)_2$
d. $Zn(OH)_2$

18.42 Write the dissociation equation and the K_{sp} expression for each of the following slightly soluble electrolytes:
a. $Ca(IO_3)_2$
b. Ag_2SO_3
c. $Mn_3(AsO_4)_2$
d. PbC_2O_4

18.13 Write the dissociation equations on which the following K_{sp} expressions are based:
a. $[Cu^{2+}]^2 \times [P_2O_7^{4-}]$
b. $[Ni^{2+}]^3 \times [AsO_4^{3-}]^2$
c. $[Fe^{3+}] \times [OH^-]^3$
d. $[Mg^{2+}] \times [NbO_3^-]^2$

18.43 Write the dissociation equations affiliated with the following K_{sp} expressions:
a. $[Zr^{4+}] \times [OH^-]^4$
b. $[Pb^{2+}] \times [OH^-] \times [Br^-]$
c. $[K^+]^2 \times [SiF_6^{2-}]$
d. $[Bi^{3+}]^2 \times [S^{2-}]^3$

K_{sp}: Equilibrium Concentrations of Ions

18.14 Complete the following table (K_{sp} MgF_2 = 1×10^{-7}):

$[Mg^{2+}]$	$[F^-]$
	1×10^{-3} M
5×10^{-3} M	
	3×10^{-3} M
2×10^{-4} M	

18.44 Complete the following table (K_{sp} Ag_2CrO_4 = 2×10^{-12}):

$[Ag^+]$	$[CrO_4^{2-}]$
1×10^{-4} M	
	4×10^{-3} M
4×10^{-5} M	
	1×10^{-2} M

18.15 Using Table 18.2, calculate the concentration of each of the following ions in equilibrium with 1×10^{-3} M S^{2-}:
a. Cu^{2+}
b. Zn^{2+}
c. Ag^+

18.45 Using Table 18.2, calculate the concentration of each of the following ions in equilibrium with 1×10^{-4} M OH^-:
a. Al^{3+}
b. Fe^{2+}
c. Fe^{3+}

18.16 Sodium chloride is added to a solution containing 1.0×10^{-4} M Ag^+ (K_{sp} $AgCl = 1.6 \times 10^{-10}$).
a. At what concentration of Cl^- does a precipitate start to form?
b. Enough NaCl is added to make $[Cl^-] = 2.0 \times 10^{-2}$ M. What is $[Ag^+]$? What percentage of the Ag^+ originally present remains in solution?

18.46 A solution contains 0.01 M Ba^{2+}. Potassium chromate is added to precipitate $BaCrO_4$ ($K_{sp} = 2 \times 10^{-10}$).
a. At what concentration of CrO_4^{2-} does a precipitate start to form?
b. When $[CrO_4^{2-}] = 1 \times 10^{-4}$ M, what is $[Ba^{2+}]$? What percentage of the Ba^{2+} originally present remains in solution?

K_{sp}: Precipitate Formation

18.17 The Pb^{2+} concentration is 4.1×10^{-6} M in a stream near a lead-mining operation. Will $PbSO_4$ ($K_{sp} = 1 \times 10^{-8}$) precipitate if the concentration of SO_4^{2-} is 0.010 M? 0.0010 M?

18.47 A ground-water source contains 2.5 mg of F^- ion per liter.
a. What is conc. F^- in moles per liter?
b. Will CaF_2 ($K_{sp} = 2 \times 10^{-10}$) precipitate when conc. $Ca^{2+} = 5.0 \times 10^{-3}$ M?

18.18 A solution contains 0.010 M Ba^{2+} and 0.010 M Sr^{2+}. Sufficient CrO_4^{2-} is added to make the CrO_4^{2-} concentration 1.0×10^{-4} M. Will a precipitate form and, if so, what is it?

18.19 A student mixes 0.10 L of 0.00045 M $AgNO_3$ solution with 0.25 L of 0.00075 M K_2CrO_4 solution to give a final volume of 0.35 L. Determine
  a. the concentration of Ag^+ after mixing.
  b. the concentration of CrO_4^{2-} after mixing.
  c. whether a precipitate of Ag_2CrO_4 ($K_{sp} = 2 \times 10^{-12}$) forms.

18.20 Will a precipitate form when 0.20 L of 0.10 M $Pb(NO_3)_2$ is mixed with 0.30 L of 0.010 M $MgBr_2$? (Assume the volumes are additive.)

18.48 A solution is 0.01 M in CO_3^{2-} and 1.0×10^{-5} M in OH^-. Enough Mg^{2+} is added to make its concentration 0.01 M. Will a precipitate form under these conditions? Show calculations.

18.49 A student mixes 250 mL of 0.010 M $Pb(NO_3)_2$ with 250 mL of 0.020 M HCl to give a total volume of 500 mL. Determine
  a. the concentration of Pb^{2+} after mixing.
  b. the concentration of Cl^- after mixing.
  c. whether a precipitate of $PbCl_2$ ($K_{sp} = 1.7 \times 10^{-5}$) forms.

18.50 Will a precipitate form when 0.15 L of 1.0×10^{-3} M $MgCl_2$ is mixed with 0.25 L of 1.0×10^{-4} M NaOH? (Assume the volumes are additive.)

K_{sp}: Water Solubility

18.21 The Tl^+ concentration is 1.4×10^{-2} M in a saturated TlCl solution. Calculate the K_{sp} of TlCl.

18.22 Using data from Table 18.2, calculate the water solubility of
  a. Ag_2CrO_4 in moles per liter.
  b. $MgCO_3$ in grams per liter.

18.23 Calculate the mass in grams of $BaSO_4$ ($K_{sp} = 1.4 \times 10^{-9}$) in 0.150 L of a saturated solution.

18.24 Derive a relationship between s, the molar solubility in water, and K_{sp} for each of the solutes in Question 18.13.

18.51 A saturated solution of BaF_2 contains 1.38 g/L. Calculate
  a. the solubility in moles per liter.
  b. K_{sp} of BaF_2.

18.52 Using K_{sp} values in Table 18.2, calculate the water solubility of
  a. AgCl in grams per liter.
  b. PbF_2 in moles per liter.

18.53 What volume of saturated $PbCl_2$ solution ($K_{sp} = 1.7 \times 10^{-5}$) contains 1.0 g $PbCl_2$?

18.54 Obtain a relationship between s, the molar solubility in water, and K_{sp} for each of the solutes in Question 18.43.

K_{sp}: Common Ion Effect

18.25 Predict what effect each of the following has on the position of the equilibrium

$$PbCl_2(s) \rightleftharpoons Pb^{2+}(aq) + 2\,Cl^-(aq); \Delta H = +26.0 \text{ kJ}$$

  a. addition of $Pb(NO_3)_2$ solution.
  b. increase in temperature.
  c. addition of Ag^+, forming AgCl.
  d. addition of hydrochloric acid.

18.26 Calculate the molar solubility of PbI_2 ($K_{sp} = 1 \times 10^{-8}$) in
  a. pure water
  b. 0.010 M KI
  c. 0.020 M $Pb(NO_3)_2$

18.55 Consider the equilibrium

$$MgCO_3(s) \rightleftharpoons Mg^{2+}(aq) + CO_3^{2-}(aq); \Delta H = -25.3 \text{ kJ}$$

What effect does each of the following have on the position of this equilibrium?
  a. addition of $MgCl_2$ solution.
  b. addition of OH^-, precipitating $Mg(OH)_2$.
  c. decrease in temperature.

18.56 Calculate the molar solubility of Hg_2Cl_2 ($K_{sp} = 1 \times 10^{-18}$) in
  a. pure water
  b. 0.0050 M $Hg_2(NO_3)_2$
  c. 0.10 M HCl

Water Softening

18.27 Explain how a concentrated NaCl solution regenerates a home water-softening system.

18.28 A certain household uses 1.20 m³ of soft water a day. The water softener contains an automatic timer that shuts off the system every seven days for recharging. If the water supply contains 65 ppm Ca^{2+} (65 mg Ca^{2+} per liter), how many kilograms of NaCl are required to recharge the system?

18.57 Explain why a precipitate forms when a solution containing Ca^{2+} and SO_4^{2-} ions is heated.

18.58 How many kilograms of washing soda, Na_2CO_3, must be added to soften 200 L of hard water containing 60 ppm Ca^{2+} (60 mg Ca^{2+} per liter)?

General

18.29 In your own words, explain
  a. why hard water containing Ca^{2+} and HCO_3^- ions can cause problems in hot water heaters.
  b. what a "bathtub ring" consists of in areas where the water is hard.
  c. why it is not completely accurate to describe $BaSO_4$ as "insoluble" in water.

18.30 Criticize the following statements.
  a. In a water solution saturated with AgCl, conc. Ag^+ = conc. Cl^-.
  b. Since K_{sp} of CaC_2O_4 is larger than K_{sp} of $Ag_2C_2O_4$, calcium oxalate must be more soluble in water than silver oxalate.
  c. The increased solubility of AgCl in KNO_3 solution is an example of the common ion effect.

18.59 Complete the following statements.
  a. Hard water is caused by ____ ions.
  b. The solubility of CaC_2O_4 is ____ in water than in aqueous $CaCl_2$.
  c. Adding enough OH^- to double the OH^- concentration in equilibrium with $Ca(OH)_2$ causes the Ca^{2+} concentration to drop to ____ of its original value.
  d. ____ is an insoluble Group 2 hydroxide.

18.60 Correct each of the following false statements.
  a. A zeolite softens water by exchanging Al^{3+} ions for Ca^{2+} ions.
  b. The presence of a common ion always increases the solubility of an ionic solute.
  c. Adding sufficient Ba^{2+} will cause all of the SO_4^{2-} to precipitate from a sodium sulfate solution.

*18.61 The inside of a teakettle is coated with 10.0 g of $CaCO_3$. If the teakettle is washed with one liter of pure water, what fraction of the precipitate is removed? How many successive one-liter portions of water would have to be used to remove half of the $CaCO_3$?

*18.62 The concentrations of various cations in sea water, in moles per liter, are

Ion	Na^+	Mg^{2+}	Ca^{2+}	Al^{3+}	Fe^{3+}
Molarity (M)	0.46	0.050	0.01	4×10^{-7}	2×10^{-7}

  a. At what conc. of OH^- does $Mg(OH)_2$ start to precipitate?
  b. At this concentration, will any of the other ions precipitate?
  c. If enough OH^- is added to precipitate 50% of the Mg^{2+}, what percentage of each of the other ions will be precipitated?
  d. Under the conditions in (c), what mass of precipitate will be obtained from one liter of sea water?

*18.63 A solution is 0.10 M in Fe^{2+} and 0.10 M in Co^{2+}.
  a. When H_2S is added slowly, what precipitate first forms?
  b. What is the concentration of the first cation when the second cation starts to precipitate?

*18.64 An alloy of copper and silver is dissolved in nitric acid and treated with H_2S to precipitate both CuS and Ag_2S. It is found that a 0.500-g sample gives 0.730 g of sulfide. What is the mass percent of silver in the mixture?

*18.65 Determine the number of moles of $PbCl_2$ (K_{sp} = 1.7 × 10^{-5}) that dissolves in one liter of 0.010 M $Pb(NO_3)_2$ solution. (In this case, a considerable amount of $PbCl_2$ dissolves, so the concentration of Pb^{2+} at equilibrium is significantly greater than 0.01 M.)

Chapter 19
Acids
and
Bases

In previous chapters, we have referred from time to time to water solutions of acids and bases. Such solutions are among the most useful laboratory reagents. You are probably familiar with solutions of hydrochloric acid (HCl), nitric acid (HNO_3), and sulfuric acid (H_2SO_4). Two basic solutions commonly used in the general chemistry laboratory are made by dissolving sodium hydroxide (NaOH) and ammonia (NH_3) in water. Acidic and basic solutions are also common in the home. Vinegar, cranberry juice, and battery fluid are, to varying degrees, acidic. Oven and drain cleaners, whitewash (like that Tom Sawyer used), and antacid remedies are basic.

This chapter focuses upon the properties of acidic and basic water solutions. We start with the following working definitions:

1. An acid is a substance that, when added to water, produces **hydrogen ions (protons)**, H^+.*
2. A base is a substance that, when added to water, produces **hydroxide ions**, OH^-.

Acidic water solutions have certain properties in common. They react with active metals, such as zinc, to evolve hydrogen gas:

$$Zn(s) + 2\,H^+(aq) \rightarrow Zn^{2+}(aq) + H_2(g) \tag{19.1}$$

---

*In water solution, the H^+ ion is hydrated and is often written as the hydronium ion, H_3O^+, which has the Lewis structure

$$\left[\begin{array}{c} H \\ | \\ H-\underset{\cdot\cdot}{O}-H \end{array} \right]^+$$

For simplicity, we will ordinarily describe the properties of acidic solutions in terms of the H^+ ion.

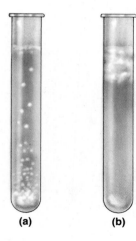

(a)   **(b)**

FIGURE 19.1 The H^+ ions present in a water solution of an acid react with calcium carbonate to form carbon dioxide: $CaCO_3(s) + 2 H^+(aq) \rightarrow Ca^{2+}(aq) + CO_2(g) + H_2O$ (tube a). The OH^- ions present in a solution of a base precipitate $Mg(OH)_2$ from a solution containing Mg^{2+} ions: $Mg^{2+}(aq) + 2 OH^-(aq) \rightarrow Mg(OH)_2(s)$ (tube b).

and with metal carbonates, such as $CaCO_3$, to form carbon dioxide gas (Fig. 19.1a):

$$CaCO_3(s) + 2 H^+(aq) \rightarrow Ca^{2+}(aq) + CO_2(g) + H_2O \qquad (19.2)$$

Acidic solutions also affect the color of certain organic dyes. For example, litmus turns from blue to red in acidic solution.

Water solutions of bases also have identifying properties. They feel slippery and turn the color of indicators. Litmus turns from red to blue in basic solution. Basic solutions react with several cations, including Mg^{2+}, to form precipitates (Fig. 19.1b):

It's easy to find out whether a solution is acidic or basic

$$Mg^{2+}(aq) + 2 OH^-(aq) \rightarrow Mg(OH)_2(s) \qquad (19.3)$$

As you might suppose, acidic and basic solutions differ in concentrations of H^+ or OH^- ions. In Section 19.1 we will look at the relationship between $[H^+]$ and $[OH^-]$ in all types of water solutions. Then, in Section 19.2, we will see how the acidity or basicity of a solution is described by a quantity called pH. Sections 19.3 to 19.5 consider the nature of the substances that act as acids or bases, the extent to which they produce H^+ or OH^- ions, and the chemical equations written to show how these ions are formed.

The last three sections of this chapter cover the general topic of acid-base reactions. In Section 19.6 we will consider how such reactions are represented by net ionic equations. In Section 19.7 the use of acid-base titrations in chemical analysis is discussed. Finally, in Section 19.8, we will look at two more general models of acids and bases that can be applied to acid-base reactions taking place in solvents other than water.

19.1
Water Dissociation; Acidic, Neutral, and Basic Solutions

The acidic and basic properties of aqueous solutions are dependent upon an equilibrium that involves the solvent, water. Water, when pure or as a solvent, tends to dissociate to some extent into hydrogen ions and hydroxide ions:

$$H_2O \rightleftharpoons H^+(aq) + OH^-(aq) \tag{19.4}$$

The forward reaction proceeds only slightly before equilibrium is reached. Only a small fraction of the total number of water molecules is dissociated (about 1 in 500 million).

Applying the general rules from Chapter 15 for equilibrium systems, we can write the equilibrium constant expression for Reaction 19.4 as

$$K_c = \frac{[H^+] \times [OH^-]}{[H_2O]}$$

In all aqueous solutions, the concentration of H_2O molecules is essentially the same, about 1000/18.0 or 55 mol/L. Hence, the term $[H_2O]$ can be combined with K_c to give a new constant, K_w, called the dissociation constant of water:

$$K_c \times [H_2O] = K_w = [H^+] \times [OH^-] = 1.0 \times 10^{-14} \tag{19.5}$$

At 25°C, K_w is 1.0×10^{-14}. This small value reflects the slight dissociation of water into H^+ and OH^- ions. In any aqueous solution or in pure water, the product of $[H^+]$ times $[OH^-]$ at 25°C is always 1.0×10^{-14}.

The value is small, but important

We can readily calculate $[H^+]$ and $[OH^-]$ in pure water. From Equation 19.4, we see that equal amounts of these two ions form when water dissociates. Applying Equation 19.5 to pure water,

$$[H^+] = [OH^-]; \quad [H^+] \times [OH^-] = [H^+]^2 = 1.0 \times 10^{-14}$$

$$[H^+] = 1.0 \times 10^{-7} M = [OH^-]$$

Any aqueous solution in which $[H^+]$ *equals* $[OH^-]$ is called a *neutral* solution. It has an $[H^+]$ equal to 1.0×10^{-7} M at 25°C.

Ordinarily, the concentrations of H^+ and OH^- in a solution are not equal. Note that Equation 19.5 represents an inverse relation. As the concentration of one of the ions, for example, $[H^+]$, increases, that of the other, $[OH^-]$, must decrease. In this way the product $[H^+] \times [OH^-]$ remains constant at 1.0×10^{-14}. The inverse relationship between these two quantities is shown in Figure 19.2. Example 19.1 illustrates how we can calculate the concentration of one of these ions when we know that of the other one.

If we know $[H^+]$, we can always find $[OH^-]$

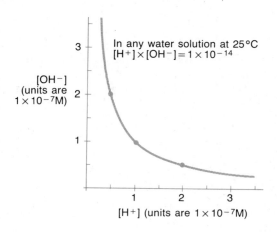

In any water solution at 25°C
$[H^+] \times [OH^-] = 1 \times 10^{-14}$

$[OH^-]$
(units are
$1 \times 10^{-7}M$)

$[H^+]$ (units are $1 \times 10^{-7}M$)

FIGURE 19.2 A graph of OH^- versus H^+ looks very much like a graph of gas volume versus pressure. In both cases, the two variables are inversely proportional to one another. When H^+ gets larger, OH^- gets smaller.

EXAMPLE 19.1 In a certain tap-water sample, $[H^+] = 3.0 \times 10^{-7}$ M. What is the concentration of $[OH^-]$?

Solution From Equation 19.5,

$$[OH^-] = \frac{K_w}{[H^+]} = \frac{1.0 \times 10^{-14}}{3.0 \times 10^{-7}} = 3.3 \times 10^{-8} \text{ M}$$

EXERCISE What is $[H^+]$ in a sea water sample if $[OH^-] = 2.0 \times 10^{-6}$ M? Answer: 5.0×10^{-9} M.

An aqueous solution where $[H^+]$ is greater than $[OH^-]$ is termed acidic. An aqueous solution in which $[OH^-]$ is greater than $[H^+]$ is basic (alkaline). Therefore,

That's how we define acidic and basic

if $[H^+] > 1.0 \times 10^{-7}$ M, $[OH^-] < 1.0 \times 10^{-7}$ M, **solution is acidic**

if $[OH^-] > 1.0 \times 10^{-7}$ M, $[H^+] < 1.0 \times 10^{-7}$ M, **solution is basic**

Table 19.1 indicates some possible combinations of concentrations of these ions. The acidity (concentration of H^+) progressively decreases in solutions 1 through 9. Solutions 1 to 4 are decreasingly acidic; solution 5 is neutral, and solutions 6 to 9 are increasingly basic.

Table 19.1
Relations Between $[H^+]$, $[OH^-]$, and pH in Aqueous Solutions

$[H^+]$ and $[OH^-]$ can vary over a very wide range

	SOLUTION								
	No. 1	**No. 2**	**No. 3**	**No. 4**	**No. 5**	**No. 6**	**No. 7**	**No. 8**	**No. 9**
$[H^+]$	10^0	10^{-2}	10^{-4}	10^{-6}	10^{-7}	10^{-8}	10^{-10}	10^{-12}	10^{-14}
$[OH^-]$	10^{-14}	10^{-12}	10^{-10}	10^{-8}	10^{-7}	10^{-6}	10^{-4}	10^{-2}	10^0
pH	0	2	4	6	7	8	10	12	14
	Acidic				Neutral		Basic		

19.2
pH

As we have seen, the acidity or basicity of a solution can be described in terms of its H^+ concentration. Sorenson, in 1909, proposed an alternative method of accomplishing this. He suggested using a term called pH (for "power of the hydrogen ion"), defined as

$$pH = -\log_{10}[H^+] = \log_{10} 1/[H^+] \tag{19.6}$$

Thus, we have

$$[H^+] = 1 \times 10^{-4} \text{ M}; \qquad pH = -\log_{10}(1 \times 10^{-4}) = -(-4.0) = 4.0$$

$$[H^+] = 1 \times 10^{-7} \text{ M}; \quad \text{pH} = -\log_{10}(1 \times 10^{-7}) = -(-7.0) = 7.0$$

$$[H^+] = 1 \times 10^{-10} \text{ M}; \quad \text{pH} = -\log_{10}(1 \times 10^{-10}) = -(-10.0) = 10.0$$

Most aqueous solutions have hydrogen ion concentrations between 1 M and 10^{-14} M. By Equation 19.6, such solutions have pH's lying between 0 and 14. In this case, it is perhaps more convenient to express acidity in terms of pH rather than $[H^+]$. This avoids using small fractions or negative exponents.

It's easier to say that pH = 2 than it is to say $[H^+]$ = 0.01 M

Looking at the pH values in Table 19.1, we see that $[H^+]$ and pH are inversely related. As $[H^+]$ decreases from 10^{-2} to 10^{-4} M, pH increases from 2 to 4. In general, **the higher the pH, the less acidic (more basic) the solution.** A solution of pH 4 has a lower concentration of H^+ and a higher concentration of OH^- than does a solution of pH 2. Notice also that when pH increases by one unit, $[H^+]$ decreases by a factor of 10. A solution of pH 3 has a hydrogen ion concentration one tenth that of a solution of pH 2 and ten times that of a solution of pH 4.

Previously, we used concentration of H^+ or OH^- to differentiate acidic, neutral, and basic solutions; pH can also be used for this purpose:

if pH < 7.0,   solution is acidic

if pH = 7.0,   solution is neutral

The larger the pH, the smaller the $[H^+]$

if pH > 7.0,   solution is basic

Table 19.2 shows the pH of some common solutions. Example 19.2 shows how Equation 19.6 can be used to calculate pH from $[H^+]$, or vice versa.

Table 19.2
pH of Some Common Liquids

Lemon juice	2.2–2.4	Urine, human	4.8–8.4
Wine	2.8–3.8	Cow's milk	6.3–6.6
Vinegar	3.0	Saliva, human	6.5–7.5
Tomato juice	3.5	Drinking water	6.5–8.0
Beer	4–5	Blood, human	7.3–7.5
Cheese	4.8–6.4	Seawater	8.3

EXAMPLE 19.2   Calculate

a. the pH of a lemon juice solution in which $[H^+]$ is 5.0×10^{-3} M.

b. the $[H^+]$ of human blood at pH 7.4.

c. the pH of a solution of baking soda, $NaHCO_3$, with an $[OH^-]$ of 2.5×10^{-6} M.

Solution   Problems of this type are most simply worked using a scientific calculator. The use of calculators and operations involving logarithms is discussed in Appendix 4.

a. You should find on your calculator that

$$\log_{10}(5.0 \times 10^{-3}) = -2.30$$

Since pH = $-\log_{10}[H^+]$, then pH = $-(-2.30) = 2.30$.

b. If the pH is 7.4, then

$$\log_{10} [H^+] = -7.4; \qquad [H^+] = 10^{-7.4}$$

To find $[H^+]$, enter -7.4 on your calculator. Then either

—punch the 10^x key, if you have one, or
—punch the INV and then the LOG key.

Either way, you should find that $[H^+] = 4 \times 10^{-8}$ M.
c. First let us calculate $[H^+]$ using the K_w expression:

$$[H^+] = \frac{1.0 \times 10^{-14}}{2.5 \times 10^{-6}} = 4.0 \times 10^{-9} \text{ M};$$
$$pH = -\log_{10}(4.0 \times 10^{-9}) = 8.40$$

EXERCISE Calculate $[H^+]$ and $[OH^-]$ of a beer with pH 4.7. Answer: $[H^+] = 2 \times 10^{-5}$ M, $[OH^-] = 5 \times 10^{-10}$ M.

> You should think about why punching those buttons gives you the answer you want

Several experimental methods can be used to determine the pH of an aqueous solution. Most often, a so-called "pH meter" is used (Fig. 19.3). This instrument translates the H^+ concentration of a solution into an electrical signal that is converted into a digital display or a deflection on a meter that reads pH directly.

A more colorful (but time-consuming) way to determine pH is to use acid-base indicators. These substances undergo a color change within a narrow pH range, usually about 1 to 2 pH units (Color Plate 19.1). By using two or more of the indicators whose colors in acidic and basic solution are listed in Table 19.3, you can estimate quite accurately the pH of a solution. Suppose, for example, that a solution turns red when a drop of phenolphthalein is added. This means that its pH is 10 or greater. If you find that the same solution gives a yellow color with alizarin yellow, its pH must be 10 or less. Putting these two observations together, you can conclude that the solution must have a pH of just about 10.

> Seems reasonable, once you think about it

FIGURE 19.3 The simplest way to determine the pH of a solution is to measure it with an electrical device called a pH meter. Care must be taken in making such a measurement; the electrodes are easily broken.

Table 19.3
Typical Acid-Base Indicators

INDICATOR	pH INTERVAL	ACID COLOR (LOWER pH)	BASE COLOR (HIGHER pH)
Methyl violet	0.0–1.6	yellow	violet
Methyl yellow	2.9–4.0	red	yellow
Methyl orange	3.1–4.4	red	yellow
Methyl red	4.8–6.2	red	yellow
Bromthymol blue	6.0–8.0	yellow	blue
Thymol blue	8.0–9.6	yellow	blue
Phenolphthalein	8.2–10.0	colorless	pink
Alizarin yellow	10.1–12.0	yellow	red

Strips of "pH paper," which is coated with a mixture of indicators, are used widely to test the pH of biological fluids, soil, ground water, and foods. Depending upon the indicators used, a test strip can measure pH over a wide or narrow range.

19.3
Strong and
Weak Acids

For a species to act as an acid, it must supply H^+ ions to water. The acid, which may be a molecule or ion, contains hydrogen atoms. The H^+ ions are formed by the dissociation of the acid in water.

Strong acids dissociate completely in water, forming H^+ ions and anions. A typical strong acid is HCl. It undergoes the following reaction upon addition to water:

Strong acids are strongly dissociated

$$HCl(aq) \rightarrow H^+(aq) + Cl^-(aq) \qquad \textbf{(19.7)}$$

This reaction goes to completion. In a dilute water solution of hydrochloric acid there are no HCl molecules, only H^+ ions and Cl^- ions. Consider, for example, a 0.10 M solution of HCl, prepared by adding 0.10 mol HCl to water to form one liter of solution. The concentration of HCl molecules is virtually zero. The concentrations of H^+ and Cl^- are 0.10 M. The pH of the solution is 1.00. Any way you look at it, the HCl is completely dissociated into ions.

Instead of 0.1 M HCl, it might be better to label the bottle 0.1 M H^+, 0.1 M Cl^-

In contrast, a weak acid is only partially dissociated in water. As an example of a weak acid, consider HF. When hydrogen fluoride is added to water, the following *reversible* reaction occurs:

$$HF(aq) \rightleftharpoons H^+(aq) + F^-(aq) \qquad \textbf{(19.8)}$$

In a solution of hydrofluoric acid, there are undissociated HF molecules as well as H^+ and F^- ions. In 0.10 M HF, prepared by adding 0.10 mol HF to enough water to give one liter of solution, more than 90% of the HF molecules remain undissociated. The concentrations of H^+ and F^- ions are less than 0.01 M. The pH of the solution is slightly greater than 2.

Experimentally, there are several ways to distinguish between strong and weak acids. One is to measure the electrical conductivities of their water solutions. A 0.10 M HF solution has a low conductivity, reflecting the fact that there are only a few H^+ and F^- ions present. In contrast, since 0.10 M HCl is completely dissociated into ions, it has a high conductivity, more than ten times that of 0.10 M HF. In its colligative properties, hydrogen fluoride is intermediate between a nonelectrolyte such as a glucose and a strong electrolyte such as HCl. The freezing point of 0.10 M HF is $-0.21°C$, as compared to $-0.19°C$ for 0.10 M glucose and $-0.37°C$ for 0.10 M HCl.

You could also mea-sure the pH

The Strong Acids

There are very few strong acids. For our purposes, we need consider only the six species listed in Table 19.4. You should learn the names and molecular formulas of these six strong acids. They will be referred to again and again, in this and following chapters.

Notice that all the substances that act as strong acids are molecular species when pure. In dilute water solution, they dissociate completely to form an H^+ ion and an anion. The dissociation reactions are similar to Reaction 19.7 for HCl. Examples include

$$HNO_3(aq) \rightarrow H^+(aq) + NO_3^-(aq) \tag{19.9}$$

$$H_2SO_4(aq) \rightarrow H^+(aq) + HSO_4^-(aq) \tag{19.10}$$

Table 19.4
Common Strong Acids

STRONG ACID	MOLECULAR FORMULA	MOLECULAR STRUCTURE
Hydrochloric acid	HCl	H—Cl
Hydrobromic acid	HBr	H—Br
Hydriodic acid	HI	H—I
Nitric acid	HNO_3	$\begin{array}{c} \text{H—O—N—O} \\ \parallel \\ \text{O} \end{array}$
Sulfuric acid	H_2SO_4	$\begin{array}{c} \text{O} \\ \mid \\ \text{H—O—S—O—H} \\ \mid \\ \text{O} \end{array}$
Perchloric acid	$HClO_4$	$\begin{array}{c} \text{O} \\ \mid \\ \text{H—O—Cl—O} \\ \mid \\ \text{O} \end{array}$

HCl, HNO_3 and H_2SO_4 rank among the most important industrial chemicals

We see from Table 19.4 that in all the strong acids, a hydrogen atom is covalently bonded to a highly electronegative nonmetal atom (Cl, Br, I, or O). Moreover, in all the acids containing oxygen (HNO_3, H_2SO_4, and $HClO_4$), the hydrogen atom is bonded to oxygen. When the molecule dissociates, the O—H bond breaks. In terms of Lewis structures, we can show the dissociation of nitric acid as follows:

$$\text{H}-\ddot{\text{O}}-\text{N}-\ddot{\text{O}}: \;\rightarrow\; \text{H}^+ \;+\; \left(:\ddot{\text{O}}-\text{N}-\ddot{\text{O}}:\right)^-$$

Species That Act as Weak Acids

A wide variety of solutes behave as weak acids in water. For convenience, they can be classified into three categories: molecules, anions and cations.

1. Molecules containing an ionizable hydrogen atom There are literally thousands of molecular weak acids, most of them organic in nature. Many organic acids, including acetic, lactic, and pyruvic acids, are physiologically important. A few common weak acids are listed in Table 19.5, p. 582. When added to water, these weak acids form an equilibrium mixture. This consists mostly of undissociated molecules, mixed with a few H^+ ions and anions. (We will have more to say about the position of this equilibrium in Chapter 20). The equations for the dissociation of these weak acids are entirely analogous to that for HF (Equation 19.8):

Weak acids are weakly dissociated

$$HClO(aq) \rightleftharpoons H^+(aq) + ClO^-(aq) \tag{19.11}$$

$$HC_2H_3O_2(aq) \rightleftharpoons H^+(aq) + C_2H_3O_2{}^-(aq) \tag{19.12}$$

The properties of 0.10 M solutions of HClO or $HC_2H_3O_2$ resemble those of 0.10 M HF. All three species are weak electrolytes with small electrical conductivities. Each solution has a pH greater than 1 (Table 19.5).

You will recall that in all strong acids containing oxygen, the hydrogen atom that dissociates is bonded to oxygen. Looking at Table 19.5, you can see that this is also true for weak acids. In H_3PO_4, hydrogen is bonded to oxygen rather than phosphorus; in HClO, hydrogen is bonded to oxygen rather than chlorine, and so on. Moreover, in the organic acids listed in Table 19.5 (acetic, pyruvic, and lactic acids), the hydrogen that ionizes is part of a —COOH group. We might represent their dissociation as

Most acids contain O as well as H atoms

$$\text{R}-\text{C}-\ddot{\text{O}}-\text{H} \;\rightarrow\; \left(\text{R}-\text{C}-\ddot{\text{O}}:\right)^- \;+\; \text{H}^+$$

(R = CH_3 for acetic acid, CH_3—CHOH for lactic acid, CH_3CO for pyruvic acid).

Acid-base indicators, referred to previously, are weak acids. We might represent the dissociation of an indicator by the equation

$$HIn(aq) \rightleftharpoons H^+(aq) + In^-(aq)$$

Table 19.5
Some Common Molecular Weak Acids

WEAK ACID	MOLECULAR FORMULA	MOLECULAR STRUCTURE	CONC. H$^+$ IN 0.10 M SOLN.	pH (0.10 M SOLN.)
Phosphoric acid	H_3PO_4	H—O—P—O—H (with O above and O—H below P)	0.024	1.62
Pyruvic acid	$HC_3H_3O_3$	CH$_3$—C—C—O—H (with O, O below)	0.018	1.75
Hydrofluoric acid	HF	H—F	0.0080	2.10
Nitrous acid	HNO_2	H—O—N (with O below)	0.0065	2.19
Lactic acid	$HC_3H_5O_3$	CH$_3$—CHOH—C—O—H (with O below)	0.0037	2.43
Acetic acid	$HC_2H_3O_2$	CH$_3$—C—O—H (with O below)	0.0013	2.87
Carbonic acid	H_2CO_3	H—O—C—O—H (with O below)	0.00021	3.69
Hydrogen sulfide	H_2S	H—S—H	0.00010	4.00
Hypochlorous acid	HClO	H—O—Cl	0.000056	4.25
Hydrogen cyanide	HCN	H—C≡N	0.0000063	5.20

The formula HIn stands for the undissociated weak acid molecule, *which has a color different from that of the In$^-$ ion.* In the case of bromthymol blue (Table 19.3), the weak acid molecule is colored yellow, whereas the anion is blue.

The position of the above equilibrium is sensitive to the concentration of H$^+$ ions. If [H$^+$] is "high," the equilibrium lies far to the left and the principal species present is the HIn molecule. We see its color (yellow with bromthymol blue) when we look at a solution containing a few drops of the indicator. When [H$^+$] is "low," the equilibrium shifts to the right, forming In$^-$ ions. The solution takes on the color of In$^-$ (blue with bromthymol blue).

We can drive the reaction right or left by changing the pH

2. Anions containing an ionizable hydrogen atom Let us refer back for a moment to Equation 19.10 for the dissociation of H_2SO_4. Notice that the anion formed, HSO_4^-, contains a hydrogen atom. In water, the HSO_4^- ion undergoes further dissociation, producing an H$^+$ ion and an SO_4^{2-} ion. The dissociation reaction is reversible, so we classify HSO_4^- as a weak acid:

$$HSO_4^-(aq) \rightleftharpoons H^+(aq) + SO_4^{2-}(aq) \qquad \text{(19.13)}$$

In practice, very few anions give acidic solutions when added to water. The only other anion of this type that we need be concerned with is the $H_2PO_4^-$ ion (see Example 19.3 below).

3. Cations   The ammonium ion, NH_4^+, behaves as a weak acid in water because of the following reversible reaction:

$$NH_4^+(aq) \rightleftharpoons H^+(aq) + NH_3(aq) \qquad \text{(19.14)}$$

> NH_4^+ is the most common ionic weak acid

The products are an ammonia molecule, NH_3, and an H^+ ion that makes the solution acidic. Note that the behavior of the NH_4^+ ion in water is very similar to that of the HF molecule (Equation 19.8) or the HSO_4^- ion (Equation 19.13). All three species contain hydrogen atoms that are converted to H^+ ions when the weak acid dissociates.

You may be surprised to learn that most metal cations, except those of Groups 1 and 2, are weak acids. At first, it is not at all obvious how a cation such as Zn^{2+} can make a water solution acidic. To understand how this is possible, we must realize that this cation and others like it are *hydrated* in water solution. When $ZnCl_2$ or $Zn(NO_3)_2$ is added to water, the cation formed is $Zn(H_2O)_4^{2+}$. Here, a Zn^{2+} ion is bonded to four water molecules. This complex cation is slightly dissociated in water, according to the following equation:

$$Zn(H_2O)_4^{2+}(aq) \rightleftharpoons H^+(aq) + Zn(H_2O)_3(OH)^+(aq) \qquad \text{(19.15)}$$

The H^+ ion, which makes the solution acidic, comes from the ionization of one of the H_2O molecules bonded to Zn^{2+}. The OH^- ion formed at the same time remains bonded to Zn^{2+} and so does not directly affect the pH of the solution. (It does, however, affect the charge of the complex cation, reducing it from $+2$ to $+1$.)

> Remember what makes a solution acidic or basic

EXAMPLE 19.3   Write equations for the dissociation of the following weak acids in water:
a. HNO_2 b. $H_2PO_4^-$ c. $Fe(H_2O)_6^{3+}$

Solution   In each case, a proton (H^+) is formed to make the solution acidic. The other product is the residue from the weak acid after removal of the proton. All the reactions go to equilibrium, as indicated by a double arrow.
a. $HNO_2(aq) \rightleftharpoons H^+(aq) + NO_2^-(aq)$
b. $H_2PO_4^-(aq) \rightleftharpoons H^+(aq) + HPO_4^{2-}(aq)$
c. $Fe(H_2O)_6^{3+}(aq) \rightleftharpoons H^+(aq) + Fe(H_2O)_5(OH)^{2+}(aq)$

EXERCISE HPO_4^{2-} and $Fe(H_2O)_5(OH)^{2+}$ can each dissociate to form an H^+ ion and another product. Give the formulas of the "other products."
Answer: PO_4^{3-}, $Fe(H_2O)_4(OH)_2^+$.

Polyprotic Acids

As our discussion of strong and weak acids has implied, certain species contain more than one ionizable hydrogen atom. Such species, called polyprotic (many-proton) acids, always dissociate in distinct steps; that is, they lose one hydrogen atom at a time. The most common polyprotic acid is sulfuric acid, H_2SO_4:

$$H_2SO_4(aq) \rightarrow H^+(aq) + HSO_4^-(aq)$$

$$HSO_4^-(aq) \rightleftharpoons H^+(aq) + SO_4^{2-}(aq)$$

As pointed out earlier, the first dissociation is complete, so that H_2SO_4 qualifies as a strong acid. The second dissociation goes to a position of equilibrium, making HSO_4^- a "weak" acid.

Several of the weak acids listed in Table 19.5 are polyprotic. An example is carbonic acid, H_2CO_3, which is formed when CO_2 dissolves in water. Carbonic acid dissociates in two steps;

$$H_2CO_3(aq) \rightleftharpoons H^+(aq) + HCO_3^-(aq) \tag{19.16}$$

$$HCO_3^-(aq) \rightleftharpoons H^+(aq) + CO_3^{2-}(aq) \tag{19.17}$$

The pH of any polyprotic acid is fixed by the first dissociation

Reaction 19.16 occurs to a much greater extent than 19.17. This reflects the fact that it is easier to remove a positively charged ion, H^+, from a neutral molecule, H_2CO_3, than from a negative ion, HCO_3^-. This is generally true for all polyprotic acids; essentially all of the H^+ ions come from the first dissociation.

19.4
Strong and
Weak Bases

For a species to act as a base, it must form OH^- ions in water solution. Bases, like acids, are classified as "strong" or "weak." As with acids, there are only a few strong bases but a great many weak bases.

Strong Bases

A strong base dissociates completely in water to release OH^- ions. Sodium hydroxide, NaOH, is the most common strong base. It dissolves readily in water to give a solution containing Na^+ and OH^- ions:

$$NaOH(s) \rightarrow Na^+(aq) + OH^-(aq) \tag{19.18}$$

As with all strong bases, this reaction goes to completion. In a 0.10 M NaOH solution, prepared by dissolving 0.10 mol NaOH in enough water to give one liter of solution, the concentration of *undissociated* NaOH is virtually zero. In this solution, the concentrations of Na^+ and OH^- are 0.10 M. The pH is 13.00.

Strong bases are limited to

These are the only strong bases

1. **The hydroxides of the Group 1 metals** (LiOH, NaOH, KOH, RbOH, CsOH).
2. **The hydroxides of the heavier Group 2 metals,** $Ca(OH)_2$, $Sr(OH)_2$, and $Ba(OH)_2$. With these compounds, two moles of OH^- are produced for every mole of solid that dissociates:

$$Ca(OH)_2(s) \rightarrow Ca^{2+}(aq) + 2\ OH^-(aq) \tag{19.19}$$

Of the strong bases, only NaOH and, to a much lesser extent, KOH, are commonly used in the chemistry laboratory. All compounds of lithium, rubidium, and cesium, including the hydroxides, are expensive. The Group 2 hydroxides have limited solubilities. Calcium hydroxide is sometimes used in industry when a strong base is needed and high solubility is not critical.

NaOH is the work-horse strong base

Weak Bases

Weak bases do not furnish OH^- ions directly by dissociation. Rather, the OH^- ions are generated by the reaction of the weak base with water. Ammonia, NH_3, is a common weak base that reacts with water as follows:

$$NH_3(aq) + H_2O \rightleftharpoons NH_4^+(aq) + OH^-(aq) \qquad (19.20)$$

The forward reaction occurs to only a slight extent. In a 0.10 M solution of NH_3, prepared by adding 0.10 mol NH_3 to enough water to form one liter of solution, nearly 99% of the NH_3 molecules remain unreacted. The concentration of NH_3 is about 0.099 M. In contrast, the concentrations of NH_4^+ and OH^- are only about 0.001 M. The pH of the solution is about 11.

The reaction goes about 1% to the right

Certain organic compounds, called amines, behave like ammonia. Methylamine, CH_3NH_2, is typical:

$$CH_3NH_2(aq) + H_2O \rightleftharpoons CH_3NH_3^+(aq) + OH^-(aq) \qquad (19.21)$$

Note the similarity between this reaction and 19.20. Indeed, the CH_3NH_2 molecule is very similar to ammonia, except that a CH_3 group is substituted for an H atom:

$$
\begin{array}{ccc}
& \overset{\displaystyle H}{\underset{\displaystyle |}{}} & \\
H-\overset{\displaystyle |}{\underset{\displaystyle |}{C}}-\overset{\displaystyle \cdot\cdot}{N}-H & & H-\overset{\displaystyle \cdot\cdot}{N}-H \\
& \overset{\displaystyle |}{H}\ \ \overset{\displaystyle |}{H} & \overset{\displaystyle |}{H}
\end{array}
$$

methyl amine   ammonia

Most weak bases are anions. A typical example is the fluoride ion, F^-. It undergoes the following reversible reaction with water:

$$F^-(aq) + H_2O \rightleftharpoons HF(aq) + OH^-(aq) \qquad (19.22)$$

As with NH_3, the forward reaction occurs only to a slight extent. In a solution in which the F^- concentration is 0.10 M, the concentration of OH^- at equilibrium is only about 10^{-6} M. As small as this is, it is enough to make the solution basic, with a pH of about 8. A reaction similar to reaction 19.22 occurs with the anion of any weak acid. With the acetate ion, $C_2H_3O_2^-$, the reaction is

The anion X^- of any weak acid HX is itself a weak base

$$C_2H_3O_2^-(aq) + H_2O \rightleftharpoons HC_2H_3O_2(aq) + OH^-(aq) \qquad (19.23)$$

Looking at Equations 19.20 through 19.23, we see that they resemble each other very closely. In each case, a weak base (NH_3, CH_3NH_2, F^-, $C_2H_3O_2^-$) picks up a proton (H^+ ion) from a water molecule. The products are

—a weak acid (NH_4^+, $CH_3NH_3^+$, HF, $HC_2H_3O_2$).

—an OH^- ion, which makes the solution basic.

The general reaction is

The weak base competes with H_2O for H^+ ions, and gets a few

$$\text{weak base}(aq) + H_2O \rightleftharpoons \text{weak acid}(aq) + OH^-(aq) \qquad \textbf{(19.24)}$$

Since the reaction does not go to completion, relatively few OH^- ions are formed. This is why we refer to species such as NH_3, CH_3NH_2, F^-, and $C_2H_3O_2^-$ as *weak* bases.

> **EXAMPLE 19.4**   Write an equation to explain why each of the following species produces a basic water solution:
> a. NO_2^- b. CO_3^{2-} c. HCO_3^-
>
> **Solution**   In each case, the weak base reacts reversibly with a water molecule, picking up a proton from it. Two species are formed. One is an OH^- ion, which makes the solution basic. The other product is a molecule or ion formed by adding H^+ to the weak base.
>
> a. $NO_2^-(aq) + H_2O \rightleftharpoons HNO_2(aq) + OH^-(aq)$
>
> b. $CO_3^{2-}(aq) + H_2O \rightleftharpoons HCO_3^-(aq) + OH^-(aq)$
>
> c. $HCO_3^-(aq) + H_2O \rightleftharpoons H_2CO_3(aq) + OH^-(aq)$
>
> This reaction occurs for the fact that $NaHCO_3$ ("bicarbonate of soda") is slightly basic. It occurs to a greater extent than the acid dissociation of HCO_3^- (Reaction 19.17).
>
> **EXERCISE**   Write an equation for the reaction with water of the weak base $C_2H_5NH_2$. Answer:
>
> $$C_2H_5NH_2(aq) + H_2O \rightleftharpoons C_2H_5NH_3^+(aq) + OH^-(aq)$$

19.5
Acid-Base Properties of Salt Solutions

After completing Sections 19.3 and 19.4, you should be able to predict correctly that an aqueous solution of HI or H_2SO_4 is acidic while a solution of NaOH or NH_3 is basic. Solutions of $NaNO_2$ or NH_4I might be more difficult for you to classify. These two compounds, and many others, such as NaCl, $Zn(NO_3)_2$, and $CuSO_4$, are **salts. A salt is an ionic compound containing a cation other than H^+ and an anion other than OH^- or O^{2-}.**

This is what we mean by a salt

In dilute water solution, a salt is completely dissociated into ions. A water solution labeled "$NaNO_2$" actually contains Na^+ and NO_2^- ions:

$$NaNO_2(s) \rightarrow Na^+(aq) + NO_2^-(aq) \qquad \textbf{(19.25)}$$

A solution prepared by dissolving ammonium iodide in water contains NH_4^+ and I^- ions:

$$NH_4I(s) \rightarrow NH_4^+(aq) + I^-(aq) \qquad \textbf{(19.26)}$$

It follows that the acid-base properties of a salt such as $NaNO_2$ or NH_4I are determined by the behavior of its ions. To decide whether a water solution of $NaNO_2$ is acidic, basic, or neutral, we must consider the effect of Na^+ and NO_2^- ions on the pH of water. The acid-base behavior of the NH_4^+ and I^- ions will determine whether a solution of NH_4I has a pH less than, equal to, or greater than 7.

Essentially all the properties of a salt solution are those of its ions

Some ions have no effect upon the pH or $[H^+]$ of water. We describe such ions as being neutral. Other ions are acidic; they increase $[H^+]$ to greater than 10^{-7} M. For the most part, these are acidic cations of the type discussed in Section 19.3. Finally, as we saw in Section 19.4, there are many anions derived from weak acids that are basic. They react with water to form OH^- ions, thereby lowering $[H^+]$ below 10^{-7} M. Table 19.6 summarizes the acid-base behavior of ions commonly present in water solution. Note that the basic anions and acidic cations listed are only typical examples of the large numbers of ions that fall into these categories.

Table 19.6
Acid-Base Properties of Some Common Ions in Water Solution

	NEUTRAL		BASIC		ACIDIC	
Anion	Cl^- Br^- I^-	NO_3^- ClO_4^- SO_4^{2-}	$C_2H_3O_2^-$ F^- CO_3^{2-} S^{2-} PO_4^{3-}	CN^- NO_2^- HCO_3^- HS^- HPO_4^{2-}	HSO_4^- $H_2PO_4^-$	
Cation	Li^+ Na^+ K^+	Ca^{2+} Ba^{2+}	none		Mg^{2+} Al^{3+} NH_4^+ transition metal ions	

Neutral Ions

A neutral ion does not react with water to produce H^+ or OH^- ions. Hence, it does not affect the pH. There are relatively few neutral ions. We see from Table 19.6 that

—**the neutral anions are those derived from strong acids.**
—**the neutral cations are those derived from strong bases.**

A typical neutral anion is the chloride ion, produced by the dissociation of hydrochloric acid:

$$HCl(aq) \rightarrow H^+(aq) + Cl^-(aq)$$

You will recall from Section 19.3 that HCl is completely dissociated in water; it is a strong acid. This means that there is no tendency for the reverse of the above reaction to occur. Chloride ions, wherever they come from, do not combine with H^+ ions. In particular, Cl^- ions from a salt such as NaCl do not pick up H^+ ions from water. As a result, Cl^- ions and the other neutral anions listed in Table 19.6 do not change the $[H^+]$ or pH of water.

Explain in your own words why the NO_3^- ion is neutral

A similar argument applies to cations such as Na^+, produced by the dissociation of strong bases such as NaOH. Dissociation is complete:

$$NaOH(s) \rightarrow Na^+(aq) + OH^-(aq)$$

NaOH molecules cannot exist in water solution

Hence, there is no tendency for the reverse reaction to occur; that is, Na^+ ions, regardless of their source, do not combine with OH^- ions in water. As a result, Na^+ and other cations derived from strong bases are neutral.

Basic Anions

Recall from Section 19.4 that any anion derived from a weak acid acts as a weak base in water solution. There is a small army of such anions. Those listed in Table 19.6 are typical examples. In contrast, there are no common basic cations.

Acidic Ions

Acidic ions include

—all cations except those of the alkali metals and the heavier alkaline earths.
—the HSO_4^- and $H_2PO_4^-$ anions (recall the discussion on p. 582).

EXAMPLE 19.5 Consider the four salts

$$NH_4I, \quad Zn(NO_3)_2, \quad KClO_4, \quad Na_3PO_4$$

For each salt,
a. indicate the ions present.
b. classify both anion and cation as acidic, basic, or neutral.
c. state whether the salt solution will be acidic, basic, or neutral.
d. write net ionic equations to explain acidity or basicity of the salt.

Solution
a. NH_4^+, I^- ions; Zn^{2+}, NO_3^- ions; K^+, ClO_4^- ions; Na^+, PO_4^{3-} ions.

b. and c. It is convenient to prepare a table classifying both ions of the salt and then the salt itself as acidic, basic, or neutral.

acidic + neutral
→ acidic

basic + neutral
→ basic

SALT	CATION	ANION	SOLUTION OF SALT
NH_4I	NH_4^+(acidic)	I^-(neutral)	acidic
$Zn(NO_3)_2$	Zn^{2+}(acidic)	NO_3^-(neutral)	acidic
$KClO_4$	K^+(neutral)	ClO_4^-(neutral)	neutral
Na_3PO_4	Na^+(neutral)	PO_4^{3-}(basic)	basic

These predictions are confirmed by experiment; see Color Plate 19.2.

d. for NH_4I; $NH_4^+(aq) \rightleftharpoons H^+(aq) + NH_3(aq)$
for $Zn(NO_3)_2$; $Zn(H_2O)_4^{2+}(aq) \rightleftharpoons H^+(aq) + Zn(H_2O)_3(OH)^+(aq)$
for Na_3PO_4; $PO_4^{3-}(aq) + H_2O \rightleftharpoons HPO_4^{2-}(aq) + OH^-(aq)$

EXERCISE   Write net ionic equations to explain why a solution of NaH_2PO_4 is acidic while a solution of Na_2HPO_4 is basic. Answer:

$$H_2PO_4^-(aq) \rightleftharpoons H^+(aq) + HPO_4^{2-}(aq);$$
$$HPO_4^{2-}(aq) + H_2O \rightleftharpoons H_2PO_4^-(aq) + OH^-(aq)$$

The procedure outlined in Example 19.5 is a general one. If you are asked to predict whether a given salt solution is acidic, basic, or neutral and write equations to explain your predictions, it helps to follow a systematic process.

1. *Determine what ions are present* (cation and anion).

2. *Decide whether each ion is acidic, basic, or neutral.* Table 19.6 is useful here. Note, however, that most of the information in that table can be deduced if you learn the strong acids and bases.

3. *Using your answers in (2), decide whether the salt will be acidic, basic, or neutral.* If both cation and anion are neutral, the salt must be neutral. If one ion is acidic and the other neutral, the salt will be acidic. If one ion is basic and the other neutral, the salt will be basic.*

4. *Write net ionic equations for the reaction of each acidic or basic ion.* Follow the examples in Sections 19.3 and 19.4. Note that neutral ions are "spectators": since they have no effect upon acidity or basicity, they are not included in the equation.

19.6
Acid-Base Reactions

When an acidic water solution is mixed with a solution containing a base, a reaction occurs. The nature of this reaction, and the equation we write for it, depend upon whether the acid and base are strong or weak. In this section, we will look at several different types of acid-base reactions. All those considered have large equilibrium constants and, for all practical purposes, *go to completion; the reactants are essentially all converted to products.*

The limiting reactant is, anyway

Reactions of Strong Acids with Strong Bases

Consider what happens when we add a solution of a strong acid such as HNO_3 to a solution of a strong base such as NaOH. Both HNO_3 and NaOH are completely dissociated into ions:

solution of HNO_3: H^+, NO_3^- ions

solution of NaOH: Na^+, OH^- ions

The acid-base reaction that occurs involves the H^+ ion of the HNO_3 solution and the OH^- ion of the NaOH solution. The equation for the reaction is simply

$$H^+(aq) + OH^-(aq) \rightarrow H_2O \qquad (19.27)$$

$K_c = 10^{14}$ for this reaction

This is the net reaction that occurs when *any strong acid reacts with any strong base*. Note that we do not include in the equation spectator ions such as Na^+ or NO_3^-, which do not take part in the reaction.

---

*In principle, there is a fourth possibility: one ion might be acidic, the other basic. In this case, all bets are off. Without further information, it is impossible to predict whether the salt will be acidic, basic, or neutral. An example of such a salt is $(NH_4)_3PO_4$, which happens to be basic.

The reaction between a strong acid and a strong base is called **neutralization.** If just enough base is added to react with all the acid, the resulting solution is neutral. For example, if equal volumes of 0.10 M HNO_3 and 0.10 M NaOH are used, the final solution will contain only Na^+ and NO_3^- ions. Since both these ions are neutral, the solution will have a pH of 7.

Robert Boyle, of gas law fame, was probably the first to recognize that when an acid reacts with a base, they neutralize each other's properties. Nearly 150 years passed before the nature of this reaction was determined. The delay came because Lavoisier (1787) insisted that oxygen was the fundamental component of all acids. In 1811, Humphry Davy showed that hydrochloric acid contained no oxygen. Shortly thereafter (1814), Gay-Lussac concluded that it is the hydrogen in acids that neutralizes bases.

Reactions of Weak Acids with Strong Bases

When a strong base such as NaOH is added to a weak acid, a reaction similar in many ways to Reaction 19.27 occurs. The equation for the reaction, however, is different. To see why this is the case, let us consider the nature of the **principal species** present in the two solutions. The strong base NaOH is completely dissociated to Na^+ and OH^- ions. In a solution of a weak acid, HX, the situation is quite different. Here there are very few H^+ and X^- ions; the principal species is the undissociated HX molecule:

solution of NaOH: Na^+, OH^- ions

solution of weak acid HX: HX molecule

The acid-base reaction involves the HX molecule and the OH^- ion as reactants. The products are an H_2O molecule and an X^- ion in solution. The equation for the reaction is

$$HX(aq) + OH^-(aq) \rightarrow H_2O + X^-(aq) \tag{19.28}$$

where HX stands for any weak acid, such as HF, $HC_2H_3O_2$, etc. By the same token, X^- is the anion derived from that acid (F^-, $C_2H_3O_2^-$, etc.).

You will recall that anions derived from weak acids are themselves weak bases. This means that a solution prepared by reacting equal amounts of HX and OH^- will be slightly basic. Consider, for example, what happens when we mix equal volumes of 0.10 M acetic acid ($HC_2H_3O_2$) and 0.10 M NaOH. The final solution contains the Na^+ ion, which is neutral, and the acetate ion, $C_2H_3O_2^-$, which is basic. The pH of this solution is greater than 7 (about 9).

Reactions of Strong Acids with Weak Bases

As an example of this type of reaction, consider what happens when a strong acid such as HCl is added to a water solution of ammonia, NH_3. Here, the acid is completely dissociated; HCl is a strong acid. In contrast, since ammonia is a weak base, the principal species present in its solution is the NH_3 molecule:

solution of HCl: H^+, Cl^- ions

solution of NH_3: NH_3 molecule

The acid-base reaction involves the H^+ ion of the HCl solution and the NH_3 molecule of the ammonia solution. The product is the NH_4^+ ion:

$$H^+(aq) + NH_3(aq) \rightarrow NH_4^+(aq) \tag{19.29}$$

Recall that the ammonium ion, NH_4^+, is a weak acid. Hence, if equal amounts of H^+ and NH_3 are used, the final solution is slightly acidic, with a pH less than 7 (about 5).

As we saw in Section 19.4, many anions act as weak bases. When a strong acid is added to a solution containing a basic anion, an acid-base reaction occurs. Consider, for example, what happens when hydrochloric acid is added to a solution of sodium fluoride. HCl is a strong acid, completely dissociated into H^+ and Cl^- ions. NaF, like all salts, is completely dissociated. Hence, the principal species are

In any acid-base reaction, an acidic species reacts with a basic species

HCl solution: H^+, Cl^- ions

NaF solution: Na^+, F^- ions

The H^+ ion of the HCl solution reacts with the F^- ion of the NaF solution to form the weak acid, HF:

$$H^+(aq) + F^-(aq) \rightarrow HF(aq) \tag{19.30}$$

Note the similarity between this reaction and Reaction 19.29. In both cases, a weak base (NH_3 or F^-) reacts with an H^+ ion to form a weak acid (NH_4^+ or HF).

EXAMPLE 19.6   Write a net ionic equation for each of the following reactions in water solution:
a. nitrous acid, HNO_2, with NaOH.
b. potassium acetate, $KC_2H_3O_2$, with hydrochloric acid, HCl.
c. hydrobromic acid, HBr, with potassium hydroxide, KOH.

Solution
a. Nitrous acid is a weak acid; sodium hydroxide is a strong base. Hence the principal species present are

solution of HNO_2: HNO_2 molecule

solution of NaOH: Na^+, OH^- ions

Writing down the principal species helps

The acid-base reaction is analogous to 19.28:

$$HNO_2(aq) + OH^-(aq) \rightarrow H_2O + NO_2^-(aq)$$

b. Potassium acetate, a salt, and HCl, a strong acid, are both completely dissociated into ions:

solution of $KC_2H_3O_2$: K^+, $C_2H_3O_2^-$ ions

solution of HCl: H^+, Cl^- ions

The reaction is that between the H^+ ion and the weak base, $C_2H_3O_2^-$. The equation is entirely analogous to Equation 19.30:

$$H^+(aq) + C_2H_3O_2^-(aq) \rightarrow HC_2H_3O_2(aq)$$

c. Here we have a strong acid, HBr, and a strong base, KOH. The principal species are

solution of HBr: H^+, Br^- ions

solution of KOH: K^+, OH^- ions

The reaction is a simple neutralization:

$$H^+(aq) + OH^-(aq) \rightarrow H_2O$$

EXERCISE Write an equation for the reaction that occurs when solutions of Na_2CO_3 and HCl are mixed; solutions of $NaHCO_3$ and HCl. Answer:

$$H^+(aq) + CO_3^{2-}(aq) \rightarrow HCO_3^-(aq)$$

$$H^+(aq) + HCO_3^-(aq) \rightarrow H_2CO_3(aq)$$

The procedure followed in Example 19.6 is a general one. If you are asked to write an equation for an acid-base reaction between two solutions, you follow what amounts to a three-step procedure.

1. *Decide upon the nature of the principal species present in both solutions.* These may be molecules (HF, NH_3) or ions (H^+, Cl^- in HCl; Na^+, OH^- in NaOH; Na^+, F^- in NaF).

2. *Decide what species take part in the acid-base reaction.* Again, these may be molecules (HF, NH_3) or ions (H^+ in HCl; OH^- in NaOH; F^- in NaF).

3. *Write a balanced net ionic equation for the acid-base reaction.* There are three possibilities:*

a. strong acid–strong base: $H^+(aq) + OH^-(aq) \rightarrow H_2O$
b. weak acid–strong base: e.g., $HF(aq) + OH^-(aq) \rightarrow H_2O + F^-(aq)$
c. strong acid–weak base: e.g., $H^+(aq) + NH_3(aq) \rightarrow NH_4^+(aq)$
  $H^+(aq) + F^-(aq) \rightarrow HF(aq)$

19.7
Acid-Base Titrations

You may recall that in Chapter 18 we described a type of experiment known as a titration. There we dealt with precipitation reactions. Acid-base titrations are perhaps more common. Here one of the reactants is an acidic water solution, the other a basic water solution. The acid-base reaction that occurs may be any one of the three types discussed in Section 19.6: strong acid–strong base, weak acid–strong base, or strong acid–weak base.

---

*In principle, we could have a fourth type of reaction, that between a weak acid and a weak base. Such reactions do not go to completion; they produce an equilibrium mixture of reactants and products. For that reason, among others, we will not consider such reactions in this chapter.

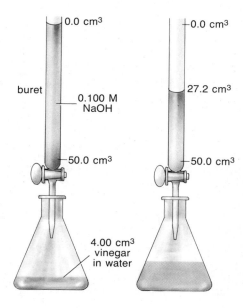

0.0 cm³

buret

0.100 M
NaOH

50.0 cm³

4.00 cm³
vinegar
in water

0.0 cm³

27.2 cm³

50.0 cm³

FIGURE 19.4 Determination of the percentage of acetic acid in a vinegar sample. A specified volume of vinegar is added to the flask along with a few drops of an acid-base indicator. Standard NaOH solution is added from a buret until the indicator changes color. Knowing the volumes of the two solutions and the concentration of the NaOH, we can calculate the concentration of $HC_2H_3O_2$ in the vinegar.

Most often, an acid-base titration is used to determine the concentration of a species in solution. If that species is a base, we titrate with a "standard solution" of a strong acid, usually HCl. That is, we determine the volume of an HCl solution of known concentration that is required to react completely with a sample of the basic solution. Conversely, to find the concentration of an acidic species, we titrate with a solution of a strong base, usually NaOH. An example of such a titration is shown in Figure 19.4. Here, we want to determine the concentration of acetic acid in vinegar. To do this, we titrate with an NaOH solution of known concentration, 0.100 M. We carefully measure the volume of this solution required to react with a known volume of vinegar. From the data obtained, it is possible to determine the concentration of $HC_2H_3O_2$ in the vinegar (Example 19.7).

The reaction goes essentially to completion

EXAMPLE 19.7 By law, vinegar must contain at least 4.0% acetic acid, which corresponds to 0.67 M acetic acid. A 4.00-mL sample of Brand X vinegar is titrated with NaOH. It is found that 27.2 mL of 0.100 M NaOH is required for complete reaction. Does the concentration of acetic acid in the vinegar meet the legal standard?

Solution The equation for the acid-base reaction is

$$HC_2H_3O_2(aq) + OH^-(aq) \rightarrow C_2H_3O_2^-(aq) + H_2O$$

To find the molarity of acetic acid, we follow a three-step procedure:
1. Calculate the number of moles of OH^- ion used in the titration, using the volume (27.2 mL) and the molarity (0.100 M) of NaOH.
2. Calculate the number of moles of $HC_2H_3O_2$ consumed in the reaction.
3. Determine the molarity of $HC_2H_3O_2$ using the volume of the vinegar sample, 4.00 mL.

$$\text{no. moles } OH^- = 0.0272 \text{ L} \times \frac{0.100 \text{ mol NaOH}}{1 \text{ L}} \times \frac{1 \text{ mol } OH^-}{1 \text{ mol NaOH}}$$

$$= 0.00272 \text{ mol } OH^-$$

$$\text{no. moles HC}_2\text{H}_3\text{O}_2 = 0.00272 \text{ mol OH}^- \times \frac{1 \text{ mol HC}_2\text{H}_3\text{O}_2}{1 \text{ mol OH}^-}$$

$$= 0.00272 \text{ mol HC}_2\text{H}_3\text{O}_2$$

$$\text{conc. HC}_2\text{H}_3\text{O}_2 = \frac{0.00272 \text{ mol}}{0.00400 \text{ L}} = 0.680 \text{ mol/L}$$

To be honest about it, this example was rigged

Brand X vinegar does comply with the law.

EXERCISE   How many milliliters of 0.0977 M HCl are required to titrate 30.0 mL of 0.103 M NaOH? Answer: 31.6 mL.

Acid-base titrations can be used for a variety of purposes. We have just seen how they can be applied to find the concentration of a species in solution. They can also be used to find the percentage of an acidic or basic component of a solid mixture (Example 19.8).

EXAMPLE 19.8   A research chemist isolates a sample of nicotinic acid, $HC_6H_4NO_2$ (molar mass = 123 g/mol). To determine its purity, she titrates 0.450 g of the sample with 0.100 M NaOH. She finds that 36.2 mL NaOH is required. The reaction is

Nicotinic acid is monoprotic

$$HC_6H_4NO_2(aq) + OH^-(aq) \rightarrow H_2O + C_6H_4NO_2^-(aq)$$

Assuming any impurity present does not react with the base, calculate
a. the mass in grams of nicotinic acid in the sample.
b. the mass percent of nicotinic acid in the sample.

Solution
a. We first find the number of moles of NaOH, equate that to the number of moles of acid, and finally calculate the mass in grams of nicotinic acid:

$$\text{no. moles NaOH} = 0.100 \frac{\text{mol}}{\text{L}} \times 0.0362 \text{ L} = 3.62 \times 10^{-3} \text{ mol}$$

$$\text{no. moles acid} = 3.62 \times 10^{-3} \text{ mol}$$

$$\text{mass acid} = 3.62 \times 10^{-3} \text{ mol} \times 123 \text{ g/mol} = 0.445 \text{ g}$$

b. $\text{mass percent acid} = \dfrac{0.445 \text{ g}}{0.450 \text{ g}} \times 100 = 98.9\%$

EXERCISE   Suppose that 35.8 mL NaOH had been required, instead of 36.2 mL. What percentage would you then calculate for nicotinic acid? Answer: 97.9%.

Indicators in Acid-Base Titrations

In an acid-base titration, we must know when to stop adding reagent. In other words, we must be able to tell at what point the reaction is complete. This is accomplished by adding an acid-base indicator. The indicator should change color when the reaction is complete—that is, when equivalent quantities of acid and base have been used. The point in the titration at which this occurs is called the *equivalence point*. If the indicator changes color before this point is reached, too little reagent will be added and the reaction will not be complete. If it changes too late, after the equivalence point has been passed, too much reagent will be added.

This is sometimes called the stoichiometric endpoint

To choose the proper indicator for a particular titration, it is helpful to refer to what is known as a titration curve. Data for such a curve are obtained by determining the pH of the solution as a function of the volume of titrant added. Two such curves are shown in Figure 19.5. At the left, we show how pH changes when a sample of 1.00 M acetic acid, $HC_2H_3O_2$, is titrated with 1.00 M NaOH. The equation for the reaction is

$$HC_2H_3O_2(aq) + OH^-(aq) \rightarrow H_2O + C_2H_3O_2^-(aq)$$

At the right, we show what happens to pH when HCl is added to a water solution of ammonia, NH_3. Here the reaction is

$$NH_3(aq) + H^+(aq) \rightarrow NH_4^+(aq)$$

Looking first at Figure 19.5a, we see that the pH starts off at about 2.3. This is the pH of 1.00 M $HC_2H_3O_2$, a weak acid. As OH^- ions are added the pH increases, rather slowly in the early stages of the titration. Near the equivalance point, the pH climbs more steeply, and then nearly levels off as excess NaOH is added. At the equivalence point, we have a solution of $NaC_2H_3O_2$. As noted earlier, this solution is basic, with a pH of about 9. Phenolphthalein, which changes from colorless to red at about pH 9, would be an excellent indicator for this titration. If we used methyl red (color change at pH 5), we would stop the titration much too early, when reaction is only about 65% complete. This situation is common in titrations of *weak acids* with *strong bases*. For such a titration we choose an indicator, such as phenolphthalein, that *changes color above pH 7.*

A good acid-base indicator changes color at the equivalence point in the titration

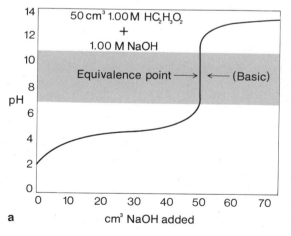

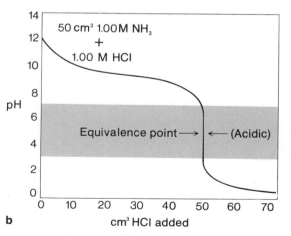

a   cm³ NaOH added   b   cm³ HCl added

FIGURE 19.5 The curve at the left shows how pH changes as a solution of the weak acid $HC_2H_3O_2$ is titrated with a strong base, NaOH. The indicator used must change color close to pH 9, which is the pH of the solution of $NaC_2H_3O_2$ formed when the reaction is complete. At the right is the curve obtained when the weak base NH_3 is titrated with the strong acid HCl. The solution of NH_4Cl formed has a pH of 5, so we must use an indicator that changes color close to pH 5.

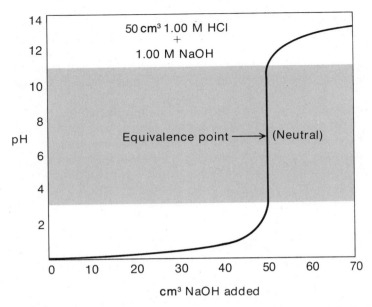

FIGURE 19.6 In the titration of HCl with NaOH, the pH is 7 at the equivalence point. However, the pH changes very rapidly near this point. Addition of a single drop of NaOH can change the pH by six units. As a result, almost any indicator is suitable for this titration.

From Figure 19.5b, we see that in the titration of NH_3 with HCl, the pH starts off at about 11.7. This is the pH of 1.00 M NH_3, a weak base. As HCl is added the pH drops. At the equivalence point, we have a solution of NH_4Cl, with a pH on the acid side, about 5. Methyl red, which changes from yellow to red at pH 5, would be an ideal indicator for this titration. Phenolphthalein (color change at pH 9) would not be suitable. It would turn from pink to colorless too soon, before all the NH_3 had reacted. In general, in any titration of a *weak base* with a *strong acid,* we should choose an indicator that *changes color below pH 7.*

In Figure 19.6, the curve is shown for the titration of a strong acid (1.00 M HCl) with a strong base (1.00 M NaOH). As we would expect, the pH at the equivalence point is 7, corresponding to a neutral solution of NaCl. This would suggest the use of an indicator such as bromthymol blue, which changes color at pH 7. Looking more closely at the shape of this curve, however, we see that the

Table 19.7
Indicators in Acid-Base Titrations

REACTION TYPE	EXAMPLE	EQUIVALENCE POINT		INDICATOR
		SPECIES	pH	
Weak acid–strong base	$HC_2H_3O_2$–NaOH	Na^+, $C_2H_3O_2^-$	>7	phenolphthalein (pH 9)
Strong acid–weak base	HCl–NH_3	NH_4^+, Cl^-	<7	methyl red (pH 5)
Strong acid–strong base	HCl–NaOH	Na^+, Cl^-	7	bromthymol blue (pH 7), or phenolphthalein (pH 9), or methyl red (pH 5)

choice of indicator is not as critical here as in the two cases just discussed. Notice that the steep portion of the curve is much longer than it is in the curves shown in Figure 19.5. Indeed, a single drop of 1.00 M NaOH (about 0.02 cm^3) changes the pH from 4 to 10 (see Example 19.9). Methyl red (color change at pH 5) or phenolphthalein (color changes at pH 9) could be used. In general, for the titration of a strong acid with a strong base, we have a much wider choice of indicators than in other types of acid-base titrations (Table 19.7). This is because the pH changes so rapidly near the equivalence point.

EXAMPLE 19.9 If 50.00 mL of 1.000 M HCl is titrated with 1.000 M NaOH, find the pH of the solution after the following volumes of 1.000 M NaOH have been added:

a. 49.99 mL   b. 50.00 mL   c. 50.01 mL

Solution During the titration the following reaction occurs:

$$H^+(aq) + OH^-(aq) \rightarrow H_2O$$

a. At this point, we have

$$(50.00 - 49.99) \text{ mL} = 0.01 \text{ mL} = 1 \times 10^{-5} \text{ L}$$

of unneutralized 1.000 M HCl in a total volume of almost exactly 100 mL (0.100 L) of solution. Consequently,

$$[H^+] = \frac{1 \times 10^{-5} \text{ L} \times 1.000 \text{ mol/L}}{0.100 \text{ L}} = 1 \times 10^{-4} \text{ M}; \qquad pH = 4.0$$

b. The 50.00 mL of 1 M NaOH exactly neutralizes the 50.00 mL of 1 M HCl, to give a solution of NaCl. Since both Na$^+$ and Cl$^-$ are neutral species, the pH is that of pure water, 7. This is the equivalence point in the titration. Note that only 0.01 mL of base is required at the equivalence point to move the pH from 4 to 7. This volume is much less than that of 1 drop of reagent.

c. We are now past the equivalence point and OH$^-$ is in excess. Specifically, we have

$$(50.01 - 50.00) \text{ mL} = 0.01 \text{ mL} = 1 \times 10^{-5} \text{ L}$$

of unneutralized 1.000 M NaOH in a total volume just slightly greater than 100 mL (0.100 L) of solution:

$$[OH^-] = \frac{1 \times 10^{-5} \text{ L} \times 1.000 \text{ mol/L}}{0.100 \text{ L}} = 1 \times 10^{-4} \text{ M}$$

$$[H^+] = \frac{1 \times 10^{-14}}{1 \times 10^{-4}} = 1 \times 10^{-10} \text{ M}; \qquad pH = 10.0$$

Clearly, a small excess of base drives the pH from 7 to 10. Since the pH around the equivalence point is so sensitive to added base, any indicator with an end point between 4 and 10 would be satisfactory. Phenolphthalein is often used.

EXERCISE Calculate [H$^+$] and the pH when 25.00 mL HCl has been added. Answer:

You can use the method in this example to find the pH of any mixture of a strong acid and a strong base

$$[H^+] = \frac{25.00 \times 10^{-3} \text{ L} \times 1.000 \text{ mol/L}}{75.00 \times 10^{-3} \text{ L}} = 0.3333 \text{ mol/L}; \quad pH = 0.48$$

19.8
General Models of Acids, Bases, and Acid-Base Reactions

Thus far in this chapter, we have considered an acid to be a substance that produces an excess of H^+ ions in water solution. A base was similarly defined to be a substance that, directly or indirectly, forms excess OH^- ions in water solution. This approach, first proposed by the Swedish chemist Svante Arrhenius in 1884, is a very practical one. In particular, it allows us to predict whether such substances as HCl, $Ca(OH)_2$, NH_4Cl, $NaCl$, and K_2CO_3 will form acidic, basic, or neutral solutions.

The Arrhenius model does, however, have one disadvantage. It greatly restricts the number of reactions that can be considered to be of the acid-base type. Over the years, many other, more general models of acids and bases have been proposed. In this section, we will discuss two such models. Curiously enough, they were both suggested in the same year, 1923. One of them was proposed independently by Brönsted in Denmark and Lowry in England. The other came from a man we have heard of before (Chap. 9), the American physical chemist, G. N. Lewis.

Brönsted-Lowry Concept

According to this model, an acid-base reaction is one in which there is a *proton transfer* from one species to another. The species that gives up or **donates the proton** (H^+ ion) is referred to as an **acid**. The molecule or ion that **accepts the proton** (H^+ ion) is a **base**.

A simple example of a Brönsted-Lowry acid-base reaction is that between acetic acid and hydroxide ions:

$$\underset{\text{acid}}{HC_2H_3O_2(aq)} + \underset{\text{base}}{OH^-(aq)} \rightarrow C_2H_3O_2^-(aq) + H_2O \tag{19.31}$$

Here, $HC_2H_3O_2$ gives up a proton to an OH^- ion. Hence, $HC_2H_3O_2$ is acting as an acid (proton donor), while OH^- is a base (proton acceptor). This conclusion is not particularly startling. We used almost the same words earlier in describing this reaction by the Arrhenius model.

Consider, however, the reverse of Reaction 19.31:

Rather surprisingly, H_2O acts here as an acid

$$\underset{\text{base}}{C_2H_3O_2^-(aq)} + \underset{\text{acid}}{H_2O} \rightarrow HC_2H_3O_2(aq) + OH^-(aq) \tag{19.32}$$

This is the equation written earlier to explain why a solution containing acetate ions is basic. At the time, we did not refer to it as an acid-base reaction. According to the Brönsted-Lowry model, it is. The $C_2H_3O_2^-$ anion accepts a proton from a water molecule and hence acts as a base. The H_2O molecule donates a proton to the acetate ion and so acts as a Brönsted-Lowry acid.

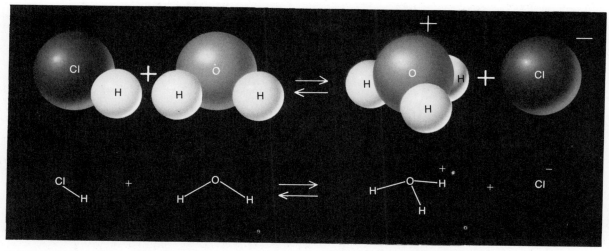

FIGURE 19.7 When HCl is added to water, there is a proton transfer from an HCl to an H_2O molecule, forming a Cl^- ion and an H_3O^+ ion. In this reaction, HCl acts as a Brönsted-Lowry acid, H_2O as a Brönsted-Lowry base.

The Brönsted-Lowry model can be extended still further. Consider the equation written earlier for the dissociation of HCl in water:

$$HCl(aq) \rightarrow H^+(aq) + Cl^-(aq)$$

Here, although HCl appears to be the proton donor, there is not any obvious proton acceptor. According to Brönsted and Lowry, the proton acceptor is really a water molecule. They would rewrite this equation as

$$\underset{\text{acid}}{HCl(aq)} + \underset{\text{base}}{H_2O} \rightarrow Cl^-(aq) + H_3O^+(aq) \tag{19.33}$$

They would if they were alive

In this equation, HCl is donating a proton to H_2O (Fig. 19.7). Thus, HCl is acting as an acid and H_2O as a base. Equation 19.33 offers a plausible explanation as to why HCl dissociates in water. It does so by reacting with water to produce two more stable species, the Cl^- and H_3O^+ ions. In the Arrhenius model, the reason for the dissociation of HCl is not apparent.

You will note from this discussion that the H_2O molecule can act as either a Brönsted-Lowry acid (Equation 19.32) or base (Equation 19.33). When it acts as an acid, it donates a proton to another species and is converted to an OH^- ion. When water acts as a base, it accepts a proton, forming the hydronium ion, H_3O^+. The dissociation of water can be expressed in these terms. Here, water serves as both the acid and base:

$$\underset{\text{base}}{H_2O} + \underset{\text{acid}}{H_2O} \rightarrow H_3O^+(aq) + OH^-(aq) \tag{19.34}$$

We can consider the H_3O^+ ion to be a hydrated proton

The hydronium ion is formed by the acceptance of an H^+ ion by an H_2O molecule.

Several other molecules and ions can behave as both acids and bases in the Brönsted-Lowry sense. Among these is the HCO_3^- ion (Fig. 19.8):

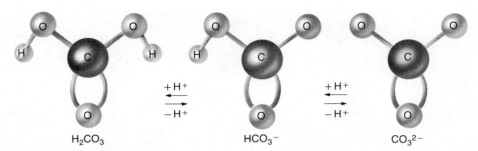

H_2CO_3 $\qquad$ HCO_3^- $\qquad$ CO_3^{2-}

FIGURE 19.8 The HCO_3^- ion can act as a Brönsted-Lowry acid, giving up a proton to form the CO_3^{2-} ion. In a different reaction, it can act as a Brönsted-Lowry base, accepting a proton to form H_2CO_3 (or $CO_2 + H_2O$).

$$HCO_3^-(aq) + H_2O \rightarrow H_2CO_3(aq) + OH^-(aq) \qquad (19.35)$$
$$\text{base} \qquad\quad \text{acid}$$

$$HCO_3^-(aq) + H_2O \rightarrow CO_3^{2-}(aq) + H_3O^+(aq) \qquad (19.36)$$
$$\text{acid} \qquad\quad \text{base}$$

In discussing reactions such as 19.31 through 19.36, we often use a special terminology. A species formed from an acid by the loss of a proton is referred to as the *conjugate base* of that acid. Thus, we refer to the $C_2H_3O_2^-$ ion as the conjugate base of $HC_2H_3O_2$; the OH^- ion is the conjugate base of H_2O, and so on. A species formed from a base by gaining a proton is called the *conjugate acid* of that base. The conjugate acid of H_2O is the H_3O^+ ion; that of the HCO_3^- ion is H_2CO_3, and so on:

$$\text{conjugate acid} \underset{\text{gain of } H^+}{\overset{\text{loss of } H^+}{\rightleftharpoons}} \text{conjugate base}$$

$$\lfloor\text{conjugate acid-base pair}\rfloor$$

EXAMPLE 19.10 Consider the reaction

$$H_2PO_4^-(aq) + C_2H_3O_2^-(aq) \rightleftharpoons HPO_4^{2-}(aq) + HC_2H_3O_2(aq)$$

Classify each of the four species involved as a Brönsted-Lowry acid or base.

Acid + base ⇌ conjugate base + conjugate acid

Solution In the forward reaction, the $H_2PO_4^-$ ion donates a proton to the $C_2H_3O_2^-$ ion. In the reverse reaction, $HC_2H_3O_2$ donates a proton to HPO_4^{2-}. Hence, $H_2PO_4^-$ and $HC_2H_3O_2$ are Brönsted-Lowry acids; $C_2H_3O_2^-$ and HPO_4^{2-} are Brönsted-Lowry bases.

EXERCISE What is the conjugate base of NH_4^+? the conjugate acid of N_2H_4? Answers: NH_3; $N_2H_5^+$.

The Lewis Concept

We have seen that the Brönsted-Lowry model extends the Arrhenius picture of acid-base reactions considerably. However, the Brönsted-Lowry model is restricted in one important respect. It can be applied only to reactions involving a proton transfer. For a species to act as a Brönsted-Lowry acid, it must contain an ionizable hydrogen atom.

The Lewis acid-base model removes this restriction. A **Lewis acid** is a species that in an acid-base reaction, **accepts** an **electron pair**. In this reaction, a **Lewis base donates** the **electron pair.**

From a structural point of view, the Lewis concept of a base does not differ in any essential way from the Brönsted-Lowry concept. In order for a species to accept a proton and thereby act as a Brönsted-Lowry base it must possess an unshared pair of electrons. Consider, for example, the NH_3 molecule, the H_2O molecule, and the F^- ion, all of which can act as Brönsted-Lowry bases:

$$H-\overset{\displaystyle .}{\underset{\displaystyle |}{N}}-H \qquad H-\overset{\displaystyle ..}{O}-H \qquad (:\overset{..}{\underset{..}{F}}:)^-$$
$$H$$

In the Lewis model, bases have unshared electron pairs

Each of these species contains an unshared pair of electrons that is utilized in accepting a proton to form the NH_4^+ ion, the H_3O^+ ion, or the HF molecule:

$$\left[\begin{array}{c}H\\H:\overset{\displaystyle ..}{N}:H\\H\end{array}\right]^+ \qquad \left[\begin{array}{c}H:\overset{\displaystyle ..}{O}:H\\H\end{array}\right]^+ \qquad H:\overset{..}{\underset{..}{F}}:$$

Clearly, NH_3, H_2O, and F^- can also be Lewis bases, since they possess an unshared electron pair that can be donated to an acid. We see then that the Lewis concept does not significantly change the number of species that can behave as bases.

On the other hand, the Lewis concept greatly increases the number of species that can be considered to be acids. The substance that accepts an electron pair and therefore acts as a Lewis acid can be a proton:

$$\underset{\text{acid}}{H^+(aq)} + \underset{\text{base}}{H_2O} \rightarrow H_3O^+(aq)$$

$$\underset{\text{acid}}{H^+(aq)} + \underset{\text{base}}{NH_3(aq)} \rightarrow NH_4^+(aq)$$

The H^+ ion is a possible reactant in this model

It can equally well be a cation, such as Zn^{2+}, which can accept electron pairs from a Lewis base:

$$\underset{\text{acid}}{Zn^{2+}(aq)} + \underset{\text{base}}{4\ H_2O} \rightarrow Zn(H_2O)_4^{2+}(aq)$$

$$\underset{\text{acid}}{Zn^{2+}(aq)} + \underset{\text{base}}{4\ NH_3} \rightarrow Zn(NH_3)_4^{2+}(aq)$$

We will discuss reactions of this type in greater detail in Chapter 21.

Another important class of Lewis acids comprises molecules containing an incomplete octet of electrons. A classic example is boron trifluoride, BF_3, which reacts readily with ammonia, accepting a pair of electrons:

$$F-B + :N-H \rightarrow F-B-N-H$$

with $F-B$ bearing F above and F below on the boron, H above and H below on the nitrogen, and the product $F-B-N-H$ bearing F above and F below on boron, H above and H below on nitrogen.

acid   base

The Lewis model is commonly used in organic chemistry to consider the catalytic behavior of such Lewis acids as $ZnCl_2$ and BF_3. In general, when proton transfer reactions are involved, most chemists use the Arrhenius or Brönsted-Lowry concepts. Table 19.8 summarizes the acid-base models we have discussed. As you might guess, other models have been proposed since 1923. The three listed, however, will suffice for our purposes.

In one model, acids are categorized as hard or soft

Table 19.8
Alternative Definitions of Acids and Bases

MODEL	ACID	BASE
Arrhenius	supplies H^+ to water	supplies OH^- to water
Brönsted-Lowry	H^+ donor	H^+ acceptor
Lewis	electron pair acceptor	electron pair donor

19.9
Historical
Perspective

Gilbert Newton Lewis
(1875–1946)

The Lewis concept of acids and bases, like the Lewis structures discussed in Chapter 9, was the product of the American physical chemist, G. N. Lewis. Born in Massachusetts, Lewis grew up in Nebraska, then came back East to obtain his B.S. (1896) and Ph.D. (1899) at Harvard. Although he stayed on for a few years as an instructor, Lewis seems never to have been happy at Harvard. A precocious student and an intellectual rebel, he was repelled by the highly traditional atmosphere that prevailed in the chemistry department there in his time. Many years later, he refused an honorary degree from his alma mater.

After leaving Harvard, Lewis made his reputation at M.I.T., where he was promoted to full professor in only four years. In 1912, he moved across the country to the University of California at Berkeley as Dean of the College of Chemistry and department chairman. He remained there for the rest of his life. Under his guidance, the chemistry department at Berkeley became perhaps the most prestigious in the country. Among the faculty and graduate students that he attracted were five future Nobel Prize winners: Harold Urey in 1934, William Giauque in 1949, Glenn Seaborg in 1951, Willard Libby in 1960, and Melvin Calvin in 1961.

In administering the chemistry department at Berkeley, Lewis demanded excellence in both research and teaching. Virtually the entire staff was involved in the general chemistry program; at one time eight full professors carried freshman sections. Several department members became leaders of chemical education in America. Among them was Joel Hildebrand (1881–1983), who came to California in 1913 and was still active in teaching and research 60 years later.

Like so many physical chemists, G. N. Lewis maintained throughout his life a fascination with chemical thermodynamics. His Ph.D. thesis was in this area, as were all his early publications. Many of the standard electrode potentials given in Table 24.1, Chapter 24, are based on data obtained by Lewis and his students. In 1923 he published with Merle Randall a text entitled *Thermodynamics and the Free Energy of Chemical Substances*. Sixty years later, a revised edition of that text is still widely used in graduate courses in chemistry.

Lewis' interest in chemical bonding and structure dates from 1902. In attempting to explain "valence" to a class at Harvard, he devised an atomic model to rationalize the octet rule. His model was deficient in many respects; for one thing, Lewis visualized cubic atoms with electrons located at the corners. Perhaps this explains why his ideas of atomic structure were not published until 1916. In that year, Lewis conceived of the electron-pair bond. This concept and its implications were elaborated upon in a book that he published in 1923, *Valence and the Structure of Atoms and Molecules* (recently reprinted by Dover Publications). Here, in Lewis' characteristically lucid style, we find many of the basic principles of covalent bonding that are accepted today. Here too is the Lewis definition of acids and bases as electron-pair acceptors and donors. Curiously enough, this general approach to acid-base reactions seems to have been virtually ignored for 15 years until revived in a paper published by Lewis in 1938.

The years from 1923 to 1938 were relatively unproductive for G. N. Lewis insofar as his own research was concerned. The applications of the electron-pair bond came largely in the areas of organic and quantum chemistry; in neither of these fields did Lewis feel at home. In the early 1930s, he published a series of relatively minor papers dealing with the properties of deuterium. Then, in 1939, he began to publish in the field of photochemistry. Of approximately 20 papers in this area, several were of fundamental importance, comparable in quality to the best work of his early years. Retired officially in 1945, Lewis died a year later while carrying out an experiment on fluorescence.

Summary

The acidity or basicity of a water solution can be expressed in terms of $[H^+]$, $[OH^-]$, or pH. These quantities are related by the equations

$[H^+] \times [OH^-] = K_w = 1.0 \times 10^{-14}$ (Example 19.1)

$pH = -\log_{10} [H^+]$ (Example 19.2)

acidic solution: $[H^+] > [OH^-]$; $[H^+] > 10^{-7}$ M   pH < 7 basic solution: $[OH^-] > [H^+]$; $[OH^-] > 10^{-7}$ M   pH $>$ 7

neutral solution: $[H^+] = [OH^-] = 10^{-7}$ M   pH $=$ 7

Strong acids are completely dissociated in dilute water solution. There are only six common strong acids (Table 19.4): HCl, HBr, HI, HNO_3, $HClO_4$, and H_2SO_4. All other acids are weak; that is, they are partially dissociated to H^+ ions in water. Species that act as weak acids may be molecules (H_2S), cations (NH_4^+, Al^{3+}), or in rare cases, anions ($H_2PO_4^-$) (Example 19.3). The only strong bases, completely dissociated to OH^- ions in water, are the hydroxides of the metals in Groups 1 and 2. Weak bases include ammonia (NH_3) and anions, such as CN^- and CO_3^{2-}, that are derived from weak acids (Example 19.4). These species furnish OH^- ions by reacting reversibly with water:

$$\text{weak base} + H_2O \rightleftharpoons \text{weak acid} + OH^-$$

An aqueous solution of a salt may be acidic, basic, or neutral, depending upon the nature of the cation and anion (Table 19.6 and Example 19.5).

Acidic and basic water solutions react with each other. The nature of the reaction and the equation written to represent it depend upon the strength of both the acid and the base (Example 19.6). The solution formed at the equivalence point may be neutral (strong acid, strong base), acidic (strong acid, weak base), or basic (weak acid, strong base). The pH at the equivalence point determines the choice of indicator (Example 19.9). The concentration of an acidic or basic species can be determined by an acid-base titration (Examples 19.7 and 19.8).

The model of acid-base reactions described above is that of Arrhenius. Other, more general models have been proposed (Table 19.8). One of the most useful of these is the Brönsted-Lowry model (Example 19.10).

Key Words and Concepts

acid	conjugate acid	pH
acid-base indicator	conjugate base	polyprotic acid
acid-base titration	equivalence point	salt
acidic solution	hydronium ion	strong acid
Arrhenius acid, base	K_w	strong base
base	Lewis acid, base	weak acid
basic solution	neutral solution	weak base
Brönsted-Lowry acid, base	neutralization	

Questions and Problems

$[H^+]$, $[OH^-]$, and pH

19.1 Calculate $[OH^-]$ in solutions that have the following values of $[H^+]$:
a. 2.0×10^{-4} M   b. 6.0 M
c. 3.2×10^{-9} M   d. 5×10^{-6} M

19.31 Calculate $[H^+]$ in solutions in which $[OH^-]$ is
a. 0.010 M   b. 2.8×10^{-3} M
c. 7.3×10^{-11} M   d. 3.0 M

19.2 Calculate the pH of solutions with the following $[H^+]$:
a. 1×10^{-4} M b. 12 M
c. 2.7×10^{-6} M d. 7.6×10^{-9} M

Classify each of these solutions as acidic or basic.

19.3 Determine $[H^+]$ and $[OH^-]$ in solutions with the following pH values:
a. 6.0 b. 7.2 c. 0.00 d. -0.78

Which of these solutions are acidic? basic?

19.4 Solution 1 has a pH of 3.8; solution 2 has $[H^+] = 2 \times 10^{-7}$ M. Which solution is more acidic? Which has the higher pH?

19.5 One solution has a pH of 3.7, another a pH of 4.7. What is the ratio of the H^+ ion concentrations in the two solutions? the OH^- ion concentrations?

19.6 The pH of human urine can be as high as 8.50. What is its $[H^+]$? $[OH^-]$?

Strong and Weak Acids

19.7 Classify each of the following as a weak or strong acid and write an equation for its dissociation in water:
a. nitric acid b. nitrous acid
c. hydrobromic acid d. acetic acid
e. phosphoric acid

19.8 Write an equation for the dissociation in water of each of the following weak acids:
a. $Al(H_2O)_6^{3+}$ b. $Zn(H_2O)_3(OH)^+$
c. $H_2PO_4^-$ d. $Mn(H_2O)_6^{2+}$

19.9 Find $[H^+]$, $[OH^-]$, and the pH of the following solutions of strong acids and bases:
a. 0.40 M HI.
b. 0.33 M RbOH.
c. a solution made by dissolving 12.0 g KOH in water to make 200 cm³ of solution.
d. a solution made by diluting 20.0 cm³ of 12 M HCl with water to make 200 cm³ of solution.

19.10 Which of the following are true regarding a 0.10 M solution of a weak acid, HA?
a. The A^- conc. is 0.10 M.
b. The $[H^+] \approx [A^-]$.
c. The $[HA] \gg [A^-]$.
d. The pH is 1.

19.32 Obtain the pH of a solution in which $[H^+]$ is
a. 6×10^{-2} M b. 1.5 M
c. 1.2×10^{-8} M d. 4.4×10^{-4} M
Classify each of these solutions as acidic or basic.

19.33 Find $[H^+]$ and $[OH^-]$ in solutions where the pH is
a. 9.0 b. -1.10 c. 5.2 d. 6.90
Which of these solutions are acidic? basic?

19.34 Solution A has $[OH^-] = 1.6 \times 10^{-5}$ M; solution B has $[H^+] = 2.0 \times 10^{-8}$ M. Which solution is more basic? Which has the lower pH?

19.35 One solution has a pH of 4.5. What must be the pH of another solution in which $[H^+]$ is five times as large? one fifth as large?

19.36 The most acidic rainfall ever measured occurred in 1974 in Scotland. The pH of the rain was 2.4.
a. Calculate its $[H^+]$.
b. Approximately how many times greater was its $[H^+]$ than its $[OH^-]$?

19.37 Follow the directions of Question 19.7 for
a. $HClO_4$ b. H_2SO_4 c. $HClO$ d. HI e. HF

19.38 Follow the directions of Question 19.8 for
a. $Ni(H_2O)_5(OH)^+$ b. $Fe(H_2O)_6^{2+}$
c. HSO_4^- d. $Cr(H_2O)_5(OH)^{2+}$

19.39 Determine $[H^+]$, $[OH^-]$, and the pH of the following solutions of strong acids and bases:
a. 0.050 M CsOH.
b. 0.12 M $HClO_4$.
c. a solution made by dissolving 10.2 g LiOH in water to make 500 cm³ of solution.
d. a solution made by diluting 4.0 cm³ of 3.0 M NaOH with water to a volume of 250 cm³.

19.40 Which of the following are true regarding a 1 M aqueous solution of a strong acid, HX?
a. The X^- concentration is 1 M.
b. The HX concentration is 1 M.
c. The sum of $[H^+]$ and $[X^-]$ is 2 M.
d. The pH is zero.

Strong and Weak Bases

19.11 Classify each of the following as a strong or weak base:
a. LiOH   b. $CH_3CH_2NH_2$ c. $Ca(OH)_2$ d. F^-

19.12 Write a balanced equation to indicate how each of the species in Question 19.11 acts as a base in water.

19.13 Write an equation to account for the basicity of water solutions containing
a. BO_3^{3-} b. CH_3NH_2 c. CN^-

Acid-Base Properties of Salt Solutions

19.14 Use the term acidic, basic, or neutral to describe 1 M solutions of the following salts:
a. NH_4I b. NaCN   c. $CsNO_3$
d. $AlCl_3$ e. KNO_2

19.15 Explain, using balanced net ionic equations where appropriate, your answers to Question 19.14.

19.16 Classify solutions of each of the following salts as acidic, basic, or neutral and, where appropriate, write net ionic equations to explain your answers:
a. Na_2SO_4 b. K_2CO_3
c. $Al(NO_3)_3$ d. $NaHSO_4$

19.17 Write formulas for four salts that
a. contain K^+ and are basic.
b. contain K^+ and are neutral.
c. contain Br^- and are neutral.
d. contain Br^- and are acidic.

Acid-Base Reactions

19.18 Write a net ionic equation for each of the following acid-base reactions in water:
a. Propionic acid, $HC_3H_5O_2$, with KOH.
b. Methyl amine, CH_3NH_2, with HBr.
c. HBr and $Ba(OH)_2$.

19.19 Write a balanced net ionic equation for the reaction of each of the following solutions with a strong acid:
a. NaF   b. $Ca(OH)_2$ c. NH_3

19.20 Consider the reactions in Questions 19.18 and 19.19. In each case, state whether the solution formed at the equivalence point will be neutral, acidic, or basic.

19.41 Classify each of the following as a strong or weak base:
a. CO_3^{2-} b. CsOH   c. $Sr(OH)_2$ d. NO_2^-

19.42 Follow the directions for Question 19.12 for the species in Question 19.41.

19.43 Write an equation to explain why water solutions of the following are basic:
a. HCO_3^- b. SO_3^{2-} c. HS^-

19.44 Indicate whether 0.6 M solutions of the following salts have a pH less than 7, equal to 7, or greater than 7:
a. KCl   b. $ZnBr_2$ c. $Ba(C_2H_3O_2)_2$
d. LiI   e. Na_2S

19.45 For each salt in Question 19.44 that is acidic or basic, write a net ionic equation to show where the H^+ or OH^- ions come from.

19.46 Follow the directions of Question 19.16 for
a. $Ca(ClO_4)_2$ b. $(NH_4)_2SO_4$
c. ZnI_2 d. $NaC_2H_3O_2$

19.47 Write the formulas of four salts containing Fe^{3+} that are acidic; four salts containing PO_4^{3-} that are basic.

19.48 Write a balanced net ionic equation for the reaction that occurs between solutions of
a. HI and Na_2S.
b. $Ca(OH)_2$ and $NaHCO_3$.
c. $HClO_4$ and $Sr(OH)_2$.

19.49 Write a net ionic equation for the reaction that occurs when a strong acid is added to a solution of
a. LiOH   b. Na_3PO_4 c. CH_3NH_2

19.50 For the reactions in Questions 19.48 and 19.49, state whether the solution at the equivalence point will have a pH less than 7, equal to 7, or greater than 7.

Acid-Base Titrations

19.21 A student finds that 22.6 mL of 0.108 M HCl is required to neutralize a 20.0-mL sample of KOH. What is the molarity of the KOH?

19.22 Calculate the volume of 0.208 M NaOH required to titrate
a. 12.6 mL of 0.280 M HF.
b. 15.0 mL of 0.144 M $HClO_4$.
c. 7.50 g of concentrated $HC_2H_3O_2$, which is 99.7% pure.

19.23 A vitamin C capsule is analyzed by titrating with 0.125 M NaOH. It is found that 22.2 mL of base is required to react with a capsule weighing 0.508 g. What is the percentage of vitamin C, $C_6H_8O_6$, in the capsule? (One mole of vitamin C reacts with one mole of OH^-.)

19.24 Which of the acid-base indicators in Table 19.7 would be useful in the following titrations? Explain.
a. KOH with $HClO_4$ b. HF with KOH
c. KCN with HCl   d. $HC_2H_3O_2$ with KOH

19.25 Fifty ml of 0.2000 M NaOH is titrated with 0.2000 M HCl. What is the pH of the solution after the following volumes of HCl have been added:
a. 0.00 ml   b. 25.00 ml   c. 49.99 ml
d. 50.00 ml   e. 50.10 ml   f. 100.00 ml
Use your data to plot a titration curve similar to Figure 19.6.

19.51 What volume of 0.245 M $Ba(OH)_2$ is required to neutralize 29.4 mL of 0.196 M HCl?

19.52 Determine the volume of 0.125 M HCl required to titrate
a. 25.0 mL of 0.188 M LiOH.
b. 12.0 mL of 6.13 M NH_3.
c. 13.5 mL of butylamine, $C_4H_9NH_2$, which has a density of 0.740 g/cm³.

19.53 The percentage of $NaHCO_3$ in a powder used for stomach upsets is found by titrating with 0.106 M HCl. If 14.9 mL of the HCl is required to react with 0.302 g of the powder, what is the percentage of $NaHCO_3$?

19.54 Three acid-base indicators and their pH color changes are methyl red (5), bromthymol blue (7), and phenolphthalein (9). Which should be used for the following titrations? Explain.
a. HNO_3 with KOH   b. NH_3 with HBr
c. HNO_2 with NaOH   d. $NaC_2H_3O_2$ with HCl

19.55 Fifty ml of 0.1000 M HCl is titrated with 0.1000 M NaOH. What is the pH of the solution after the following volumes of NaOH have been added:
a. 0.00 ml   b. 25.00 ml   c. 49.90 ml
d. 50.00 ml   e. 50.01 ml   f. 75.00 ml
Use your data to plot a titration curve similar to Figure 19.6.

Acid-Base Models

19.26 What is the conjugate
a. acid of HSO_4^- b. acid of CO_3^{2-}
c. base of HNO_2 d. base of NH_4^+

19.27 For each of the following reactions, indicate the Brönsted-Lowry acids and bases:
a. $HNO_2(aq) + H_2O \rightarrow H_3O^+(aq) + NO_2^-(aq)$
b. $H_2O + S^{2-}(aq) \rightarrow HS^-(aq) + OH^-(aq)$
c. $CN^-(aq) + HC_2H_3O_2(aq) \rightarrow$
   $C_2H_3O_2^-(aq) + HCN(aq)$
What are the conjugate acid-base pairs?

19.56 What is the conjugate
a. base of $Ni(H_2O)_6^{2+}$?
b. acid of $Zn(H_2O)_3(OH)^+$?
c. base of $HC_2H_3O_2$?   d. acid of HPO_4^{2-}?

19.57 For each of the following reactions, indicate the Brönsted-Lowry acids and bases:
a. $H_2O + HF(aq) \rightarrow H_3O^+(aq) + F^-(aq)$
b. $NH_3(aq) + H_2O \rightarrow NH_4^+(aq) + OH^-(aq)$
c. $OH^-(aq) + NH_4^+(aq) \rightarrow H_2O + NH_3(aq)$
What are the conjugate acid-base pairs?

19.28 Which of the following species can act as Brönsted-Lowry acids? Brönsted-Lowry bases? Lewis acids? Lewis bases?

a. H—Ö—H   b. Cu^{2+}

c. $\left(H-\ddot{O}-C-\ddot{O}: \atop \parallel \atop :\ddot{O}: \right)^{-}$

19.58 Which of the following species can act as Brönsted-Lowry acids? Brönsted-Lowry bases? Lewis acids? Lewis bases?

a. $\left(H-\underset{\underset{H}{|}}{\ddot{O}}-H \right)^{+}$ b. Ni^{2+}

c. $H-\underset{\underset{H}{|}}{\ddot{N}}-H$

General

19.29 A sample of 25.00 mL of 0.150 M HNO_3 is added to 15.00 mL of 0.100 M KOH. For the solution formed, calculate
a. $[H^+]$ b. pH   c. OH^- d. K^+ conc.

(Assume the two volumes are additive)

19.59 A solution is made by mixing 22.00 mL of 0.200 M NaOH with 18.00 mL of 0.165 M HBr. Calculate, assuming the volumes are additive,
a. $[H^+]$ b. pH   c. Br^- conc.

19.30 Give an example of
a. a weak acid that does not contain oxygen atoms.
b. a salt containing C atoms that gives a basic solution.
c. an indicator that changes color in acidic solution.
d. a conjugate acid-base pair, both members of which give basic water solutions.

19.60 Give an example of
a. a strong acid containing two ionizable protons.
b. an indicator that changes color at pH 7.
c. a salt containing a cation with a +1 charge that gives an acidic water solution.
d. a neutral molecule that is a Lewis base.

*19.61 A chemistry student needs an aqueous solution of pH 8.0. To prepare it, he decides to dilute 1.0 M HCl with water until $[H^+]$ becomes 1.0×10^{-8} M. Will this work? Explain. Would dilution of 1.0 M NaOH work?

*19.62 Silver hydroxide, AgOH, is insoluble in water. Describe a simple qualitative experiment that would enable you to determine whether AgOH is a strong or weak base.

*19.63 Stomach acid is approximately 0.020 M HCl. What volume of this acid is neutralized by an antacid tablet that weighs 330 mg and contains 41.0% $Mg(OH)_2$, 36.2% $NaHCO_3$, and 22.8% NaCl?

*19.64 In a recent accident, 20,000 gallons of concentrated nitric acid spilled from a tank car in a Denver railyard. Sodium carbonate was spread on the acid to react with it to form $CO_2(g)$. Using Table 12.1 in Chapter 12 and Table 1.2 in Chapter 1, calculate the mass in grams of sodium carbonate required.

*19.65 Using the data in Table 19.5, estimate the freezing point of 0.10 M H_3PO_4.

*19.66 Using Tables 14.1 and 14.2 in Chapter 14, calculate ΔH for the reaction of
a. 1.00 L of 0.100 M NaOH with 1.00 L of 0.100 M HCl.
b. 1.00 L of 0.100 M NaOH with 1.00 L of 0.100 M HF, taking the heat of formation of HF(aq) to be −320.1 kJ/mol.

Chapter 20
Acid-Base
Equilibria

As we saw in Chapter 19, most acids and bases are weak. When a weak acid or base is dissolved in water, dissociation proceeds to only a small extent. An equilibrium is established between products and reactants. In this chapter we will discuss acid-base equilibria in a quantitative way. The approach followed will be similar to that used in Chapter 15 with gaseous equilibria and in Chapter 18 with solubility equilibria.

A fundamental property of a weak acid is its dissociation constant, K_a (Section 20.1). Using K_a, we can calculate the H^+ ion concentration and pH in a solution prepared by dissolving a weak acid in water (Section 20.2). We can also use K_a to explain and predict the properties of buffers, solutions in which we have large amounts of both a weak acid and its conjugate base (Section 20.3).

The equilibrium constant for the reaction of a weak base with water is given the symbol K_b. As we will see in Section 20.4, K_b for a weak base can be related to K_a for its conjugate weak acid:

$$K_b \times K_a = 1.0 \times 10^{-14}$$

Relationships such as this one between different equilibrium constants are very useful in solution chemistry. In Section 20.5, we will look at two general relations that can be applied to all equilibrium systems. These are known as the *reciprocal rule* and the rule of *multiple equilibria*.

20.1
The Equilibrium Constant for Dissociation of a Weak Acid, K_a

The dissociation in water solution of a weak acid, HB, is expressed by the general equation

$$HB(aq) \rightleftharpoons H^+(aq) + B^-(aq) \qquad (20.1)$$

The products are a proton, H^+, and the conjugate base, B^-, of the weak acid.

Following the rules given in Chapter 15, we can write an expression for the equilibrium constant for Reaction 20.1:

$$K_a = \frac{[H^+] \times [B^-]}{[HB]} \qquad (20.2)$$

The equilibrium constant, K_a, is called the *ionization constant* or *acid dissociation constant* of the weak acid HB.

To illustrate the form taken by K_a for various weak acids, consider the three species HF, NH_4^+, and HSO_3^-:

We will use the Arrhenius model in our equilibrium calculations

WEAK ACID		CONJUGATE BASE	K_a
$HF(aq)$	$\rightleftharpoons$	$H^+(aq) + F^-(aq)$	$\dfrac{[H^+] \times [F^-]}{[HF]}$
$NH_4^+(aq)$	$\rightleftharpoons$	$H^+(aq) + NH_3(aq)$	$\dfrac{[H^+] \times [NH_3]}{[NH_4^+]}$
$HSO_3^-(aq)$	$\rightleftharpoons$	$H^+(aq) + SO_3^{2-}(aq)$	$\dfrac{[H^+] \times [SO_3^{2-}]}{[HSO_3^-]}$

The concentrations that appear in the expression for K_a are, as always, *equilibrium concentrations in moles per liter.*

The K_a values of some common weak acids are given in Table 20.1. These constants are a measure of the extent to which the acid dissociates in water solution. **The smaller the dissociation constant, the weaker the acid.** Notice that in Table 20.1 the acids are arranged in decreasing order of K_a. For example, acetic acid, $HC_2H_3O_2$ ($K_a = 1.8 \times 10^{-5}$) lies below hydrofluoric acid, HF ($K_a = 7.0 \times 10^{-4}$). This means that at a given concentration, let us say 0.10 M,

All weak acids are weak, but some are weaker than others

—$[H^+]$ is lower for $HC_2H_3O_2$ than for HF (1.3×10^{-3} vs. 8.0×10^{-3} M).
—pH is higher for $HC_2H_3O_2$ than for HF (2.87 vs. 2.10).

Both of these statements reflect the fact that acetic acid is weaker than hydrofluoric acid.

In comparing the extents of dissociation of weak acids, we sometimes use the term pK_a, which is analogous to pH:

$$pK_a = -\log_{10} K_a \qquad (20.3)$$

Thus, we have

$$HF: \quad pK_a = -\log_{10}(7.0 \times 10^{-4}) = -(-3.15) = 3.15$$

$$HC_2H_3O_2: \quad pK_a = -\log_{10}(1.8 \times 10^{-5}) = -(-4.74) = 4.74$$

In general, *the larger the value of pK_a, the weaker the acid.*

Table 20.1
Dissociation Constants of Weak Acids and Bases

	ACID		K_a	BASE		K_b
	Sulfurous acid	H_2SO_3	1.7×10^{-2}	HSO_3^-		5.9×10^{-13}
	Hydrogen sulfate ion	HSO_4^-	1.2×10^{-2}	SO_4^{2-}		8.3×10^{-13}
	Phosphoric acid	H_3PO_4	7.5×10^{-3}	$H_2PO_4^-$		1.3×10^{-12}
	Hydrofluoric acid	HF	7.0×10^{-4}	F^-		1.4×10^{-11}
	Nitrous acid	HNO_2	4.5×10^{-4}	NO_2^-		2.2×10^{-11}
	Formic acid	$HCHO_2$	1.8×10^{-4}	CHO_2^-		5.6×10^{-11}
	Benzoic acid	$HC_7H_5O_2$	6.6×10^{-5}	$C_7H_5O_2^-$		1.5×10^{-10}
	Acetic acid	$HC_2H_3O_2$	1.8×10^{-5}	$C_2H_3O_2^-$		5.6×10^{-10}
	Propionic acid	$HC_3H_5O_2$	1.4×10^{-5}	$C_3H_5O_2^-$		7.1×10^{-10}
	Carbonic acid	H_2CO_3	4.2×10^{-7}	HCO_3^-		2.4×10^{-8}
	Hydrogen sulfide	H_2S	1×10^{-7}	HS^-		1×10^{-7}
	Dihydrogen phosphate ion	$H_2PO_4^-$	6.2×10^{-8}	HPO_4^{2-}		1.6×10^{-7}
	Hydrogen sulfite ion	HSO_3^-	5.6×10^{-8}	SO_3^{2-}		1.8×10^{-7}
	Hypochlorous acid	HClO	3.2×10^{-8}	ClO^-		3.1×10^{-7}
	Boric acid	H_3BO_3	5.8×10^{-10}	$H_2BO_3^-$		1.7×10^{-5}
	Ammonium ion	NH_4^+	5.6×10^{-10}	NH_3		1.8×10^{-5}
	Hydrocyanic acid	HCN	4.0×10^{-10}	CN^-		2.5×10^{-5}
	Hydrogen carbonate ion	HCO_3^-	4.8×10^{-11}	CO_3^{2-}		2.1×10^{-4}
	Hydrogen phosphate ion	HPO_4^{2-}	1.7×10^{-12}	PO_4^{3-}		5.9×10^{-3}
	Hydrogen sulfide ion	HS^-	1×10^{-13}	S^{2-}		1×10^{-1}

(left margin, vertical: Decreasing Acid Strength ↓ ; right margin, vertical: Increasing Base Strength ↓)

For the acids: $HB(aq) \rightleftharpoons H^+(aq) + B^-(aq)$ $\qquad K_a = \dfrac{[H^+] \times [B^-]}{[HB]}$

For the bases: $B^-(aq) + H_2O \rightleftharpoons HB(aq) + OH^-(aq)$ $\qquad K_b = \dfrac{[HB] \times [OH^-]}{[B^-]}$

The value of K_a for a weak acid must be found experimentally. There are several ways to do this. Perhaps the most common involves measuring $[H^+]$ or pH in a solution prepared by dissolving a known amount of the weak acid to form a given volume of solution. The calculation is illustrated in Example 20.1.

EXAMPLE 20.1 Acetylsalicyclic acid, more commonly known as aspirin, is a weak organic acid whose formula we will represent as HAsp. A water solution is prepared by dissolving 0.1000 mol HAsp per liter. The concentration of H^+ in this solution is found to be 0.0057 mol/L. Calculate K_a for aspirin.

Solution The dissociation reaction for aspirin is

$$HAsp(aq) \rightleftharpoons H^+(aq) + Asp^-(aq); \qquad K_a = \frac{[H^+] \times [Asp^-]}{[HAsp]}$$

To calculate K_a, we need to know the *equilibrium* concentrations of H^+ ions, Asp^- ions, and HAsp molecules. The equilibrium concentration of

H^+ is given as 0.0057 mol/L. From the dissociation reaction (left to right), we note that one mole of Asp^- is produced with every mole of H^+. Therefore, in this solution,

$$[H^+] = [Asp^-] = 0.0057 \text{ mol/L}$$

For every mole of H^+ produced, a mole of HAsp must dissociate. This means that 0.0057 mol/L of HAsp must dissociate. Since the original concentration of HAsp was 0.1000 mol/L,

$$[HAsp] = 0.1000 \text{ mol/L} - 0.0057 \text{ mol/L} = 0.0943 \text{ mol/L}$$

Summarizing this reasoning in the form of a table,

	ORIG. CONC. (mol/L)	CHANGE IN CONC. (mol/L)	EQUIL. CONC. (mol/L)
HAsp	**0.1000**	−0.0057	0.0943
Asp^-	**0.0000**	+0.0057	0.0057
H^+	**0.0000**	+0.0057	**0.0057**

+ means increase
− means decrease

(Boldface numbers are those given or implied in the statement of the problem; the other numbers are deduced using the dissociation equation.)

We now have all the information we need to calculate K_a:

$$K_a = \frac{[H^+] \times [Asp^-]}{[HAsp]} = \frac{(0.0057)^2}{0.0943} = 3.4 \times 10^{-4}$$

EXERCISE In a solution prepared by dissolving 0.100 mol of lactic acid per liter, $[H^+] = 3.7 \times 10^{-3}$ M. Calculate K_a for lactic acid. Answer: 1.4×10^{-4}.

Sometimes we use the term *percent dissociation* to describe the acidity of a solution prepared by dissolving a weak acid in water. Percent dissociation refers quite simply to the percentage of the weak acid molecules originally present that dissociate to form H^+ ions; that is,

Note the distinction between original and equilibrium concentrations

$$\text{percent dissociation HB} = \frac{[H^+]}{\text{orig. conc. HB}} \times 100 \qquad (20.4)$$

For the solutions referred to in Example 20.1,

$$\text{percent dissociation 0.10 M aspirin} = \frac{0.0057}{0.1000} \times 100 = 5.7$$

$$\text{percent dissociation 0.10 M lactic acid} = \frac{3.7 \times 10^{-3}}{0.100} \times 100 = 3.7$$

As you might guess, percent dissociation at a given concentration is directly related to K_a. The larger the value of K_a, the greater the percent dissociation (Table 20.2). Percent dissociation also depends upon the concentration of weak acid, increasing as concentration decreases.

Table 20.2
Percent Dissociation of Weak Acids

| WEAK ACID | K_a | PERCENT DISSOCIATION | |
		ORIG. CONC. 1.0 M	ORIG. CONC. 0.10 M
Aspirin	3.4×10^{-4}	1.8	5.7
Lactic acid	1.4×10^{-4}	1.2	3.7
Acetic acid	1.8×10^{-5}	0.42	1.3
Carbonic acid	4.2×10^{-7}	0.065	0.20
Hypochlorous acid	3.2×10^{-8}	0.018	0.057
Ammonium ion	5.6×10^{-10}	0.0024	0.0075
Hydrocyanic acid	4.0×10^{-10}	0.0020	0.0063

As original conc. goes down, % diss. goes up

20.2
Determination of [H⁺] in Solutions of Weak Acids

The dissociation constant of a weak acid can be used for a variety of purposes. Among these is the determination of $[H^+]$ or pH in a weak acid solution. As pointed out in Example 20.1,

In a solution prepared by dissolving a weak acid, HB, in water,

$$[H^+] = [B^-]$$
$$[HB] = \text{orig. conc. HB} - [H^+]$$

Example 20.2 illustrates how these relations are used in a relatively simple case.

EXAMPLE 20.2 Nicotinic acid, $C_6H_5O_2N$ ($K_a = 1.4 \times 10^{-5}$), is another name for niacin, an important vitamin. Determine $[H^+]$ in a solution prepared by dissolving 0.10 mol of nicotinic acid, HNic, to form one liter of solution.

Solution We start by writing the expression for the dissociation of this weak acid:

$$\text{HNic}(aq) \rightleftharpoons H^+(aq) + \text{Nic}^-(aq); \qquad K_a = \frac{[H^+] \times [\text{Nic}^-]}{[\text{HNic}]}$$
$$= 1.4 \times 10^{-5}$$

Originally, before dissociation occurs, the concentration of HNic is 0.10 M. At that point, there are no Nic^- ions and virtually no H^+ ions (neglecting the few produced by the dissociation of H_2O). When dissociation occurs, this situation changes. We will let x be the number of moles of H^+ formed per liter. According to the equation for dissociation, 1 mol of Nic^- is formed and 1 mol of HNic is consumed for every mole of H^+ formed. Summarizing this reasoning in the form of a table,

	ORIG. CONC. (M)	CHANGE IN CONC. (M)	EQUILIBRIUM CONC. (M)
H$^+$	0.00	$+x$	x
Nic$^-$	0.00	$+x$	x
HNic	0.10	$-x$	$0.10 - x$

This kind of table helps you keep things straight

Substituting into the expression for K$_a$,

$$\frac{(x)(x)}{0.10 - x} = 1.4 \times 10^{-5}$$

This is a quadratic equation. It could be rearranged to the form $ax^2 + bx + c = 0$ and solved for x, using the quadratic formula. Such a procedure is time-consuming and, in this case, unnecessary. Remember that nicotinic acid is a weak acid, only slightly dissociated in water. The equilibrium concentration of HNic, $0.10 - x$, is probably only very slightly less than its original concentration, 0.10 M. We therefore make the approximation $0.10 - x \approx 0.10$. This simplifies the equation considerably:

Think about this until you understand it

$$\frac{(x)(x)}{0.10} = 1.4 \times 10^{-5}; \quad x^2 = 1.4 \times 10^{-6}$$

Taking square roots, we have

$$x = 1.2 \times 10^{-3} \text{ M} = [\text{H}^+] = [\text{Nic}^-]$$

$$[\text{HNic}] = 0.10 - x = 0.10 - 0.0012 = 0.10 \text{ M (two sig. fig.)}$$

The fact that [H$^+$], 0.0012 M, is so much less than the original concentration of nicotinic acid, 0.10 M, justifies the approximation that $0.10 - x \approx 0.10$.

EXERCISE Calculate the percent dissociation of the nicotinic acid in this example. Answer: 1.2%.

In general, the value of K$_a$ is seldom known to better than ±5%. Hence, in the expression

$$K_a = \frac{x^2}{a - x}$$

You can't neglect x in the numerator, since that's what you're trying to find

where $x = [\text{H}^+]$ and $a = $ *original* concentration of weak acid, you can neglect the x in the denominator if doing so does not introduce an error of more than 5%. In other words,

$$\text{if} \quad \frac{x}{a} = \frac{\text{percent dissociation}}{100} \leq 0.05$$

then we can take $\quad a - x = a$

In most of the problems you will work, this condition holds. Looking back at Table 20.2, you can see that the dissociation is usually less than 5%. When this is the case, the approximation $a - x \approx a$ is valid and you can calculate [H$^+$] quite simply.

Sometimes, however, you will find that the $[H^+]$ you calculate is greater than 5% of the original concentration of weak acid. If this happens, you can follow either of the approaches suggested in Example 20.3.

EXAMPLE 20.3 Calculate $[H^+]$ in a 0.100 M solution of nitrous acid, HNO_2, for which $K_a = 4.5 \times 10^{-4}$.

Solution Proceeding as in Example 20.2, we arrive at the equation

$$\frac{x^2}{0.100 - x} = 4.5 \times 10^{-4}$$

Making the same approximation as before, $0.100 - x \approx 0.100$,

$$x^2 = 0.100 \times 4.5 \times 10^{-4} = 4.5 \times 10^{-5}$$

$$x = 6.7 \times 10^{-3} \approx [H^+]$$

In this case, x is more than 5% of a; that is, the calculated $[H^+]$ is more than 5% of the original concentration of undissociated acid:

$$\frac{6.7 \times 10^{-3}}{0.100} \times 100 = 6.7\%$$

$\dfrac{\% \text{ diss.}}{100} > 0.05$

In other words, about 6.7% of the HNO_2 originally present has dissociated. The approximation, $a - x \approx a$, is not valid in this case.

To obtain a better value for x, we have a choice of two approaches. One of these is called the method of successive approximations; the other uses the quadratic formula.

1. *The method of successive approximations.* We know that

$$[HNO_2] = 0.100 - x$$

Our first approximation was to take $x = 0$. We can do better by making a second approximation. Here, we use the value just calculated for x, 6.7×10^{-3}, or 0.0067. Now,

$$[HNO_2] = 0.100 - 0.0067 = 0.093 \text{ M}$$

Substituting in the expression for K_a,

$$\frac{x^2}{0.093} = 4.5 \times 10^{-4}; \qquad x^2 = 4.2 \times 10^{-5}$$

$$x = 6.5 \times 10^{-3} \text{ M} \approx [H^+]$$

A third method is trial and error, which can always be used

This value is closer to the true $[H^+]$, since 0.093 M is a better approximation for $[HNO_2]$ than was 0.100 M. If you're still not satisfied, you can go one step further. Using 6.5×10^{-3} for x instead of 6.7×10^{-3}, you can recalculate $[HNO_2]$ and solve again for x. If you do you will find that your answer does not change. In other words, "you have gone about as far as you can go." This is generally true in calculations involving $[H^+]$ in a solution of a weak acid. Usually, the first approximation is sufficient; almost never do you have to go beyond a second approximation.

2. *The quadratic formula.* This gives an exact solution for x, but is more time-consuming. Here, we would rewrite the equation

$$\frac{x^2}{0.100 - x} = 4.5 \times 10^{-4}$$

This is the general quadratic equation

in the form $ax^2 + bx + c = 0$. Doing this, we obtain

$$x^2 + (4.5 \times 10^{-4} x) - (4.5 \times 10^{-5}) = 0$$

Thus, $a = 1; b = 4.5 \times 10^{-4}; c = -4.5 \times 10^{-5}$. Applying the quadratic formula,

$$x = \frac{-b \pm \sqrt{b^2 - 4ac}}{2a}$$

$$= \frac{-4.5 \times 10^{-4} \pm \sqrt{(4.5 \times 10^{-4})^2 + (18.0 \times 10^{-5})}}{2}$$

It's easy to make math errors with the quadratic formula

If you carry out the arithmetic properly, you should get two answers for x:

$$x = 6.5 \times 10^{-3} \quad \text{and} \quad -6.9 \times 10^{-3}$$

The second answer is physically ridiculous; the concentration of H^+ cannot be a negative quantity. The first answer is the same one we obtained by the method of successive approximations.

EXERCISE Suppose the original concentration of HNO_2 in this example had been 1.00 M instead of 0.100 M. Making the approximation $[HNO_2] = 1.00$ M, what would you calculate for $[H^+]$? Would you need to go beyond this first approximation? Answer: 0.021 M; no.

20.3
Determination of [H⁺] in Buffer Solutions

There are many different ways in which we can establish an equilibrium in solution between a weak acid, HB, and its conjugate base, B^-:

$$HB(aq) \rightleftharpoons H^+(aq) + B^-(aq)$$

We might, as discussed in Section 20.2, dissolve the weak acid in water. In that case, some of the HB molecules dissociate, producing H^+ and B^- ions.

Another common way of establishing this equilibrium is to add to water both the weak acid, HB, and its conjugate base, B^-. Thus, to establish the equilibrium

$$HC_2H_3O_2(aq) \rightleftharpoons H^+(aq) + C_2H_3O_2^-(aq)$$

Sodium acetate is a common salt

we might dissolve both acetic acid, ($HC_2H_3O_2$ molecules) and sodium acetate ($Na^+, C_2H_3O_2^-$ ions) in water. This is the kind of system we will be talking about in this section.

A solution prepared by adding roughly equal amounts of a weak acid and its conjugate base to water acts as a buffer. That is, its pH is resistant to change (Color Plate 20.1) upon the addition of strong acid (H^+ ions) or strong base (OH^- ions). Later in this section we will see why this is the case. First though, let's see how one calculates $[H^+]$ or pH of a buffer prepared in the manner just described.

[H⁺] in Solutions Made by
Adding HB and B⁻ to Water

We can readily calculate the concentration of H^+ in a solution prepared by adding both a weak acid, HB, and its conjugate base, B^-, to water. To do this, we assume that equilibrium is established without consuming appreciable amounts of either HB or B^-. In other words,

[HB] = orig. conc. HB

[B⁻] = orig. conc. B⁻

These are very good approximations

The concentration of H^+ in such a system is obtained by solving the K_a expression:

$$K_a = \frac{[H^+] \times [B^-]}{[HB]}$$

$$[H^+] = K_a \times \frac{[HB]}{[B^-]} \tag{20.5}$$

EXAMPLE 20.4 A buffer is prepared by dissolving 1.00 mol of lactic acid, HLac ($K_a = 1.4 \times 10^{-4}$), and 1.00 mol of sodium lactate, NaLac, in enough water to form one liter of solution. Calculate $[H^+]$ and the pH of this buffer.

Solution We determine [HLac] and [Lac⁻] and then use Equation 20.5 to find $[H^+]$:

[HLac] = orig. conc. HLac = 1.00 mol/L

Sodium lactate, like all salts, is completely dissociated in water. Hence,

[Lac⁻] = orig. conc. NaLac = 1.00 mol/L

Thus,

$$[H^+] = K_a \times \frac{[HLac]}{[Lac^-]} = 1.4 \times 10^{-4} \times \frac{1.00}{1.00} = 1.4 \times 10^{-4} \text{ M}$$

$$pH = -\log_{10}(1.4 \times 10^{-4}) = 3.85$$

Very little HLac is used up in establishing the H^+ ion concentration

EXERCISE Suppose a buffer were prepared by dissolving 1.00 mol of acetic acid ($K_a = 1.8 \times 10^{-5}$) and 1.00 mol of sodium acetate in enough water to form a liter of solution. What would be the pH? Answer: 4.74.

In a buffer system such as that referred to in Example 20.4, it is best if the concentrations of the weak acid, HB, and its conjugate base, B^-, are roughly equal. Looking at Equation 20.5, we see that this means that, in a buffer, the concentration of H^+ will be roughly equal to K_a of the weak acid. Putting it another way, the pH of a buffer will be rather close to the pK_a of the weak acid.

This principle must be taken into account in preparing buffer systems. To make up a buffer to a given pH, we choose a weak acid-weak base pair where pK_a of the acid is close to the desired pH (Table 20.3). Working with that system, we then determine the relative amounts of weak acid and weak base that should be used. This procedure is illustrated in Example 20.5.

Table 20.3
Buffer Systems at Different pH Values

	BUFFER SYSTEM			
DESIRED pH	WEAK ACID	WEAK BASE	K_a OF WEAK ACID	pK_a
4	HLac	Lac$^-$	1.4×10^{-4}	3.85
5	$HC_2H_3O_2$	$C_2H_3O_2{}^-$	1.8×10^{-5}	4.74
6	H_2CO_3	$HCO_3{}^-$	4.2×10^{-7}	6.38
7	$H_2PO_4{}^-$	$HPO_4{}^{2-}$	6.2×10^{-8}	7.21
8	HClO	ClO$^-$	3.2×10^{-8}	7.50
9	$NH_4{}^+$	NH_3	5.6×10^{-10}	9.25
10	$HCO_3{}^-$	$CO_3{}^{2-}$	4.8×10^{-11}	10.32

EXAMPLE 20.5 Suppose you want to prepare an $HC_2H_3O_2$–$C_2H_3O_2{}^-$ buffer with a pH of 5.00.
a. What is $[H^+]$ in this buffer?
b. What is the ratio $[HC_2H_3O_2]/[C_2H_3O_2{}^-]$ in this buffer?
c. How many moles of sodium acetate, $NaC_2H_3O_2$, should be added to one liter of 0.100 M acetic acid to prepare this buffer?

Solution
a. $[H^+] = 10^{-5.00} = 1.0 \times 10^{-5}$ M

b. Recall (Equation 20.5) that

$$[H^+] = K_a \times \frac{[HB]}{[B^-]}; \qquad \frac{[H^+]}{K_a} = \frac{[HB]}{[B^-]}$$

For acetic acid,

$$\frac{[HC_2H_3O_2]}{[C_2H_3O_2{}^-]} = \frac{[H^+]}{K_a} = \frac{1.0 \times 10^{-5}}{1.8 \times 10^{-5}} = 0.56$$

c. Solving for the concentration of acetate ion,

$$[C_2H_3O_2{}^-] = \frac{[HC_2H_3O_2]}{0.56} = \frac{0.100}{0.56} = 0.18 \text{ M}$$

You should add 0.18 mol $NaC_2H_3O_2$ to one liter of 0.100 M acetic acid.

This is probably the easiest way to make a buffer at a given pH

EXERCISE How many grams of $HC_2H_3O_2$ and $NaC_2H_3O_2$ should be used to prepare one liter of this buffer? Answer: 6.0 g, 15 g.

Effect of Added H^+ or OH^- on Buffer Systems

A buffer resists pH changes because it contains one species (HB molecules) that reacts with OH^- ions and another species (B$^-$ ions) that reacts with H^+ ions. Consider, for example, a buffer made by adding acetic acid ($HC_2H_3O_2$ molecules) and sodium acetate (Na$^+$, $C_2H_3O_2{}^-$ ions) to water. When a strong base such as NaOH is added, the OH^- ions react with acetic acid molecules:

$$HC_2H_3O_2(aq) + OH^-(aq) \rightarrow C_2H_3O_2^-(aq) + H_2O$$

Addition of a strong acid such as HCl results in reaction of the H^+ ions of the acid with the acetate ions of the buffer:

$$H^+(aq) + C_2H_3O_2^-(aq) \rightarrow HC_2H_3O_2(aq)$$

Acids react with bases, but not with their conjugate bases

As pointed out in Chapter 19, both of these reactions go virtually to completion. Hence, the added OH^- or H^+ ions are consumed. They are unable to cause the drastic pH change that would occur if they were added to water or an unbuffered solution.

The pH of a buffer changes slightly if a strong acid or strong base is added to it. By adding base, we convert a small amount of the weak acid HB to its conjugate base, B^-. Addition of strong acid converts a small amount of B^- to HB. In both cases, the ratio $[HB]/[B^-]$ changes; this in turn changes the $[H^+]$ or pH of the buffer. The effect is ordinarily small, as indicated by Example 20.6.

EXAMPLE 20.6 Consider the buffer described in Example 20.4, where $[HLac] = [Lac^-] = 1.00$ M (K_a HLac $= 1.4 \times 10^{-4}$). You will recall that in this buffer the pH is 3.85. Suppose now that we add a strong acid or strong base to this buffer. Calculate
a. $[HLac]$, $[Lac^-]$, $[H^+]$, and pH after addition of 0.10 mol H^+ per liter.
b. $[HLac]$, $[Lac^-]$, $[H^+]$, and pH after addition of 0.10 mol OH^- per liter.

Solution
a. When H^+ ions are added, the following reaction occurs:

$$H^+(aq) + Lac^-(aq) \rightarrow HLac(aq)$$

The addition of 0.10 mol H^+ produces 0.10 mol HLac and consumes 0.10 mol Lac^-. Originally, we had 1.00 mol of both HLac and Lac^-:

$$[HLac] = 1.00 \text{ M} + 0.10 \text{ M} = 1.10 \text{ M}$$

$$[Lac^-] = 1.00 \text{ M} - 0.10 \text{ M} = 0.90 \text{ M}$$

$$[H^+] = K_a \times \frac{[HLac]}{[Lac^-]} = 1.4 \times 10^{-4} \times \frac{1.10}{0.90} = 1.7 \times 10^{-4};$$

$$pH = 3.77$$

Note that the pH decreases by only 0.08 unit. Addition of the same amount of H^+ to a liter of pure water would decrease the pH by 6 units, from 7 to 1.

There is a big difference in behavior

b. This time, the reaction is

$$HLac(aq) + OH^-(aq) \rightarrow Lac^-(aq) + H_2O$$

Here, the concentration of HLac *decreases* by 0.10 M; that of Lac^- *increases* by 0.10 M:

$$[HLac] = 1.00 \text{ M} - 0.10 \text{ M} = 0.90 \text{ M}$$

$$[Lac^-] = 1.00 \text{ M} + 0.10 \text{ M} = 1.10 \text{ M}$$

$$[H^+] = K_a \times \frac{[HLac]}{[Lac^-]} = 1.4 \times 10^{-4} \times \frac{0.90}{1.10} = 1.2 \times 10^{-4};$$

$$pH = 3.92$$

The pH again changes by less than 0.1 unit.

EXERCISE   Suppose that, in (a), we had added 0.50 mol H^+ instead of 0.10 mol. What would be the final pH? Answer: 3.38.

Example 20.6 and the exercise that follows illustrate an important limitation of buffer systems. A buffer has a limited capacity to absorb H^+ or OH^- ions. Notice from Example 20.6 that when 0.50 mol H^+ was added to the buffer, it produced a rather large change in pH, decreasing it from 3.85 to 3.38. Addition of 1.00 mol or more of H^+ ion would have an even more drastic effect. Indeed, it would consume all the Lac^- ions, thereby destroying the buffer (Fig. 20.1).

Calculations such as those made in Example 20.6 are typical of "buffer problems." To determine the pH of a buffer after addition of strong acid or base, use the following procedure:

1. Determine the concentrations or relative numbers of moles of the weak acid HB and its conjugate base B^- after addition of H^+ or OH^-. Note that

Any buffer problem will yield to this approach

    a. Addition of strong acid increases [HB] and decreases [B^-]

$$[HB] = \text{orig. conc. HB} + x; \qquad [B^-] = \text{orig. conc. } B^- - x$$

where x is the number of moles per liter of H^+ added.

    b. Addition of strong base increases [B^-] and decreases [HB]

$$[B^-] = \text{orig. conc. } B^- + y; \qquad [HB] = \text{orig. conc. HB} - y$$

where y is the number of moles per liter of OH^- added.

2. Having determined [HB] and [B^-], calculate [H^+]:

$$[H^+] = K_a \times \frac{[HB]}{[B^-]}$$

Then calculate the pH $= -\log_{10} [H^+]$.

Many natural systems are buffered. Among these is blood, which has a pH of 7.4, corresponding to an H^+ ion concentration of 4×10^{-8} M. There are several

The buffer works best when [HLac] = [Lac⁻]

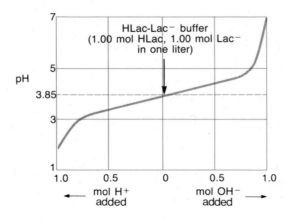

FIGURE 20.1   Effect of adding either acid or base to a lactic acid–sodium lactate buffer. The buffer itself has a pH of 3.85. Addition of OH^- ions produces a small increase in pH until most of the lactic acid is gone. Then the pH rises sharply (upper right of curve). The buffer behaves similarly toward addition of H^+ ions (lower left). The pH decreases slowly until nearly all the lactate ions are gone, then drops sharply.

different buffers that maintain a constant pH in blood. The most important of these is the H_2CO_3–HCO_3^- system:

$$CO_2(g) + H_2O \rightleftharpoons H_2CO_3(aq) \rightleftharpoons H^+(aq) + HCO_3^-(aq) \qquad (20.6)$$

In blood, there is one H_2CO_3 molecule for every ten HCO_3^- ions:

$$\frac{[H_2CO_3]}{[HCO_3^-]} = \frac{[H^+]}{K_a\ H_2CO_3} = \frac{4 \times 10^{-8}}{4.2 \times 10^{-7}} = 0.1$$

The large amount of HCO_3^- gives blood a large capacity for absorbing strong acid without appreciable change in pH. This is essential because many cellular processes produce acidic waste products.

Blood is buffered mainly against addition of acid

If the pH of blood drops significantly below 7.4, a condition called *acidosis* is created. The nervous system is depressed; fainting and even coma can result. Acidosis can be caused by breathing disorders such as emphysema, which allow the CO_2 concentration to build up, or metabolic diseases such as diabetes, which produce abnormally high concentrations of organic acids. *Alkalosis*, abnormally high pH, is much less common. Hyperventilation (rapid, heavy breathing) occasionally causes alkalosis by reducing the CO_2 concentration in the blood.

Buffers are widely used in biology as well as chemistry to maintain constant pH. Biologists and biochemists often calculate the pH of buffers by using a relation known as the Henderson-Hasselbach equation. This is readily obtained from Equation 20.5. Taking the base 10 log of both sides gives

$$\log_{10} [H^+] = \log_{10} K_a + \log_{10} \frac{[HB]}{[B^-]}$$

Multiplying both sides by -1 and remembering that $-\log x = \log 1/x$, we obtain the equation

$$pH = -\log_{10} K_a + \log_{10} \frac{[B^-]}{[HB]} = pK_a + \log_{10} \frac{[B^-]}{[HB]} \qquad (20.7)$$

This is just Eqn. 20.5 in different form. If you prefer it, use it

20.4
The Equilibrium Constant for the Reaction of a Weak Base with Water, K_b

As we noted in Chapter 19, a weak base reacts with water, acquiring a proton from an H_2O molecule and leaving an OH^- ion behind. Perhaps the most common weak base is the NH_3 molecule:

$$NH_3(aq) + H_2O \rightleftharpoons NH_4^+(aq) + OH^-(aq) \qquad (20.8)$$

We can write an equilibrium constant expression for Reaction 20.8 in the usual way:

$$K_b = \frac{[NH_4^+] \times [OH^-]}{[NH_3]}$$

The concentration of water, which remains virtually constant at about 55 M, is incorporated into the equilibrium constant. The constant K_b is often called the *base dissociation constant* of NH_3.

This is not a good name, but we are stuck with it

Most weak bases are anions, which we may represent as B^-:

$$B^-(aq) + H_2O \rightleftharpoons HB(aq) + OH^-(aq)$$

$$K_b = \frac{[HB] \times [OH^-]}{[B^-]} \qquad (20.9)$$

Applying this general relation to the acetate ion, $C_2H_3O_2^-$,

$$C_2H_3O_2^-(aq) + H_2O \rightleftharpoons HC_2H_3O_2(aq) + OH^-(aq)$$

$$K_b = \frac{[HC_2H_3O_2] \times [OH^-]}{[C_2H_3O_2^-]} \qquad (20.10)$$

Values of K_b for NH_3, $C_2H_3O_2^-$, and many other weak bases are listed in Table 20.1, p. 611, opposite K_a values for the conjugate weak acids. From Table 20.1 we note that K_b for NH_3 (1.8×10^{-5}) is considerably larger than K_b for $C_2H_3O_2^-$ (5.6×10^{-10}). This means that ammonia is a stronger base than the acetate ion. The $[OH^-]$ and pH of 0.10 M NH_3 are higher than those of 0.10 M $NaC_2H_3O_2$.

Relation Between K_a and K_b

Looking at Table 20.1, we see that there is an inverse relation between K_a of a weak acid and K_b of its conjugate base. As acid strength decreases (smaller K_a), base strength increases (larger K_b). We can derive a simple equation relating these two quantities. To do so, we start by writing general expressions for K_a and K_b. Using HB to represent the weak acid and B^- for the weak base,

In solution at equilibrium, both equations are obeyed

$$HB(aq) \rightleftharpoons H^+(aq) + B^-(aq); \qquad K_a = \frac{[H^+] \times [B^-]}{[HB]}$$

$$B^-(aq) + H_2O \rightleftharpoons HB(aq) + OH^-(aq); \qquad K_b = \frac{[HB] \times [OH^-]}{[B^-]}$$

Let us now multiply the two expressions:

$$K_a \times K_b = \frac{[H^+] \times [B^-]}{[HB]} \times \frac{[HB] \times [OH^-]}{[B^-]}$$

Canceling terms and simplifying, we obtain

$$K_a \times K_b = [H^+] \times [OH^-]$$

As we saw in Chapter 19, the product $[H^+] \times [OH^-]$ is a constant (K_w) with a value of 1.0×10^{-14} at 25°C. Hence,

$$K_a \times K_b = K_w = 1.0 \times 10^{-14} \qquad (20.11)$$

Applying Equation 20.11 to several species from Table 20.1, we obtain

ACID	BASE	K_a	K_b	$K_a \times K_b$
$HC_2H_3O_2$	$C_2H_3O_2^-$	1.8×10^{-5}	5.6×10^{-10}	1.0×10^{-14}
HCN	CN^-	4.0×10^{-10}	2.5×10^{-5}	1.0×10^{-14}
HCO_3^-	CO_3^{2-}	4.8×10^{-11}	2.1×10^{-4}	1.0×10^{-14}

As K_a goes down, K_b goes up

Equation 20.11 allows us to calculate K_b for a weak base if K_a is known, or vice versa (Example 20.7).

EXAMPLE 20.7 Butyric acid, HBut, has a K_a of 2.0×10^{-5}. Calculate K_b for the butyrate ion, But^-.

Solution   Applying Equation 20.11,

$$K_b \, But^- = \frac{1.0 \times 10^{-14}}{K_a \, HBut} = \frac{1.0 \times 10^{-14}}{2.0 \times 10^{-5}} = 5.0 \times 10^{-10}$$

EXERCISE Pyruvic acid, HPy, has a K_a of 3.3×10^{-3}. Which is the stronger acid, HBut or HPy? Which is the stronger base, But^- or Py^-? Answer: HPy; But^-.

Determination of $[OH^-]$ in a Solution of a Weak Base

We saw in Section 20.2 that the K_a of a weak acid can be used to calculate $[H^+]$ in a solution of that acid. In a very similar way, we can use K_b to obtain $[OH^-]$ in a solution of a weak base. If desired, we can go a step further and calculate $[H^+]$ or pH (Example 20.8).

EXAMPLE 20.8   For the butyrate ion, But^-, K_b is 5.0×10^{-10}. For a 1.0 M solution of sodium butyrate, NaBut, calculate
a. $[OH^-]$ b. $[H^+]$ c. pH

Solution   When sodium butyrate dissolves in water, it dissociates completely into Na^+ and But^- ions. The Na^+ ion does not react with water, but the butyrate ion does:

Salts in general are completely ionized. Since But^- is a conjugate base, its solution should be basic

$$But^-(aq) + H_2O \rightleftharpoons HBut(aq) + OH^-(aq);$$

$$K_b = \frac{[HBut] \times [OH^-]}{[But^-]} = 5.0 \times 10^{-10}$$

a. Let us represent $[OH^-]$ by x. From the chemical equation, we see that when one mole of OH^- is formed, one mole of HBut is also formed. At the same time, one mole of But^- is consumed. Setting up a table, as in Example 20.2,

	ORIG. CONC. (mol/L)	CHANGE IN CONC. (mol/L)	EQUIL. CONC. (mol/L)
HBut	0.00	$+x$	x
OH$^-$	0.00	$+x$	x
But$^-$	1.0	$-x$	$1.0 - x$

Substituting into the expression for K_b,

$$\frac{x^2}{1.0 - x} = 5.0 \times 10^{-10}$$

Since K_b is so small, it seems safe to make the approximation that $1.0 - x \approx 1.0$. In this case, $x^2 = 5.0 \times 10^{-10}$; $x = [OH^-] = 2.2 \times 10^{-5}$ M. (Note that x is far less than 5% of the original concentration of But$^-$, so the approximation is valid.)

b. Recall from Chapter 19 that $[H^+] \times [OH^-] = 1.0 \times 10^{-14}$; thus,

$$[H^+] = \frac{1.0 \times 10^{-14}}{2.2 \times 10^{-5}} = 4.5 \times 10^{-10} \text{ M}$$

The solution is indeed basic

c. $pH = -\log_{10} [H^+] = -\log_{10} (4.5 \times 10^{-10}) = 9.35$

EXERCISE What is $[OH^-]$ in a 0.10 M solution of a weak base that has a K_b of 1.0×10^{-9}? Answer: 1.0×10^{-5} M.

20.5
Relations Between
Equilibrium Constants

In Section 20.4 we showed that for a conjugate acid-base pair

$$K_a \times K_b = 1.0 \times 10^{-14}$$

This is a very useful relationship. Knowing one of the quantities K_a or K_b, we can calculate the other directly (Example 20.7).

This kind of situation arises quite frequently in solution chemistry. Often we find that the equilibrium constant we need for a reaction can be determined from the known equilibrium constants for other, closely related reactions. In this section, we will look at two general relations that allow us to make calculations of this type.

Forward and
Reverse Reactions

We can calculate the equilibrium constant, K, for a reaction if we know the equilibrium constant, K', for the reverse reaction. To do this, we use the general principle that K is the reciprocal of K':

$$K = \frac{1}{K'} \tag{20.12}$$

where K' is the equilibrium constant for the reaction written in the reverse direction.

To illustrate the use of this **reciprocal rule,** suppose we want to know K for the neutralization reaction

$$H^+(aq) + OH^-(aq) \rightleftharpoons H_2O; \quad K = ? \tag{20.13}$$

Recall that for the reverse reaction, the ionization of water,

$$H_2O \rightleftharpoons H^+(aq) + OH^-(aq); \quad K_w = 1.0 \times 10^{-14} \tag{20.14}$$

Hence, the equilibrium constant for Reaction 20.13 should be the reciprocal of that for Reaction 20.14; that is,

$$K = \frac{1}{1.0 \times 10^{-14}} = 1.0 \times 10^{14}$$

The validity of the reciprocal rule can be demonstrated by writing the expressions for K and K_w and multiplying:

$$K = \frac{1}{[H^+] \times [OH^-]} \qquad\qquad K_w = [H^+] \times [OH^-]$$

$$K \times K_w = \frac{[H^+] \times [OH^-]}{[H^+] \times [OH^-]} = 1; \quad K = 1/K_w$$

The reciprocal rule always works

The reciprocal rule is particularly helpful in calculating equilibrium constants for acid-base reactions. Indeed, we have just calculated one such equilibrium constant. You will recognize Equation 20.13 as a simple neutralization. The equilibrium constant that we calculated, 1.0×10^{14}, is that for the reaction of a strong acid such as HCl with a strong base such as NaOH. The fact that K is so large supports what we noted in Chapter 19: this type of reaction goes to completion.

Consider now the reaction of a weak acid, HB, with a strong base, OH^- ion. The equation for the reaction is

$$HB(aq) + OH^-(aq) \rightleftharpoons B^-(aq) + H_2O \tag{20.15}$$

If you compare this equation to Equation 20.9, p. 622, you will note that this acid-base reaction is simply the reverse of the reaction of the weak base, B^-, with water. For the latter reaction, the equilibrium constant is K_b. Hence, for Reaction 20.15,

$$K = \frac{1}{K_b \text{ of } B^-}$$

It's very useful here

Thus for the reaction of acetic acid with NaOH, we have

$$HC_2H_3O_2(aq) + OH^-(aq) \rightleftharpoons C_2H_3O_2^-(aq) + H_2O$$

$$K = \frac{1}{K_b\, C_2H_3O_2^-} = \frac{1}{5.6 \times 10^{-10}} = 1.8 \times 10^9$$

We see that K is a very large number. This is true for all acid-base reactions of this type, since $K_b \ll 1$. Hence we conclude that the reaction of a weak acid with a strong base should go to completion, as indeed it does.

Just about, anyhow

EXAMPLE 20.9 For the general reaction of a weak base with a strong acid,

$$B^-(aq) + H^+(aq) \rightleftharpoons HB(aq)$$

a. relate the equilibrium constant, K, to K_a of HB.
b. calculate K when $B^- = C_2H_3O_2^-$.

Solution
a. The equation is the reverse of that for the dissociation of HB, that is, $HB(aq) \rightarrow H^+(aq) + B^-(aq)$. Applying the reciprocal rule,

$$K = \frac{1}{K_a \text{ of HB}}$$

This is K for the re-action of HCl and $NaC_2H_3O_2$ in solution

b. For acetic acid, we see from Table 20.1 that $K_a = 1.8 \times 10^{-5}$. Hence,

$$K = 1/(1.8 \times 10^{-5}) = 5.6 \times 10^4$$

We see that here, as with the other types of acid-base reactions we have discussed, K is a large number (Table 20.4).

EXERCISE What is the value of K for the reaction of the weak base NH_3 with H^+ ions? Answer: $1/(5.6 \times 10^{-10}) = 1.8 \times 10^9$.

Table 20.4
Equilibrium Constants for Acid-Base Reactions

TYPE OF REACTION	EXPRESSION FOR K	EXAMPLE	K
Strong acid–strong base	$1/K_w$	HCl–NaOH	1.0×10^{14}
Weak acid–strong base	$1/K_b$	$HC_2H_3O_2$–NaOH	1.8×10^9
Strong acid–weak base	$1/K_a$	HCl–NH_3	1.8×10^9

Multiple Equilibria

The following relation is very useful in dealing with solution equilibria:

If a reaction can be expressed as the sum of two other reactions, K for the overall reaction is the product of the equilibrium constants for the individual reactions; that is, if   Reaction 3 = Reaction 1 + Reaction 2

then $K_3 = K_1 \times K_2$ **(20.16)**

To illustrate the use of this principle, often referred to as the **rule of multiple equilibria,** consider the equations for the stepwise dissociation of the weak polyprotic acid H_2S:

$$H_2S(aq) \rightleftharpoons H^+(aq) + HS^-(aq); \qquad K_1 = 1 \times 10^{-7}$$

$$HS^-(aq) \rightleftharpoons H^+(aq) + S^{2-}(aq); \qquad K_2 = 1 \times 10^{-13}$$

If we add these two equations, the HS^- cancels and we obtain

$$H_2S(aq) \rightleftharpoons 2\,H^+(aq) + S^{2-}(aq); \qquad K = ?$$ **(20.17)**

Using the rule of multiple equilibria, we can calculate the equilibrium constant for Reaction 20.17:

$$K = K_1 \times K_2 = 1 \times 10^{-20}$$ **(20.18)**

The validity of this rule can be shown by working with the expressions for K_1, K_2, and K:

$$K_1 = \frac{[H^+] \times [HS^-]}{[H_2S]}; \quad K_2 = \frac{[H^+] \times [S^{2-}]}{[HS^-]}; \quad K = \frac{[H^+]^2 \times [S^{2-}]}{[H_2S]}$$

The rule follows directly from the law of chemical equilibrium

Multiplying K_1 by K_2, we obtain the expression just written for K:

$$K_1 \times K_2 = \frac{[H^+] \times [HS^-]}{[H_2S]} \times \frac{[H^+] \times [S^{2-}]}{[HS^-]} = \frac{[H^+]^2 \times [S^{2-}]}{[H_2S]} = K$$

EXAMPLE 20.10   Given that the first and second dissociation constants of H_2CO_3 are 4.2×10^{-7} and 4.8×10^{-11}, calculate K for the reaction

$$H_2CO_3(aq) \rightleftharpoons 2\,H^+(aq) + CO_3^{2-}(aq)$$

Solution   The expression for the first and second dissociations are

$$H_2CO_3(aq) \rightleftharpoons H^+(aq) + HCO_3^-(aq); \qquad K_1 = 4.2 \times 10^{-7}$$

$$HCO_3^-(aq) \rightleftharpoons H^+(aq) + CO_3^{2-}(aq); \qquad K_2 = 4.8 \times 10^{-11}$$

If we add these two equations, the HCO_3^- ions cancel and we obtain

$$H_2CO_3(aq) \rightleftharpoons 2\,H^+(aq) + CO_3^{2-}(aq)$$

The equilibrium constant K for this reaction must then be the product of $K_1 \times K_2$:

$$K = K_1 \times K_2 = 2.0 \times 10^{-17}$$

EXERCISE Taking K for Reaction 20.17 to be 1×10^{-20}, calculate $[S^{2-}]$ in a solution in which $[H^+] = 0.3$ M, $[H_2S] = 0.1$ M.
Answer: 1×10^{-20} M.

The rule of multiple equilibria can also be applied to many other types of reactions. We can use it, for example, in conjunction with the reciprocal rule to calculate the equilibrium constant for the reaction of a weak acid with a weak base (Example 20.11).

EXAMPLE 20.11 Consider the reaction between the weak acid $HC_2H_3O_2$ and the weak base F^-:

$$HC_2H_3O_2(aq) + F^-(aq) \rightleftharpoons HF(aq) + C_2H_3O_2^-(aq)$$

a. Express this equation as the sum of two other equations.
b. Determine the equilibrium constants for the two equations in (a).
c. Using the rule of multiple equilibria, calculate K for the equation given above.

Solution
a. The two equations are

We write these equations to bring in K_a, K_b, K_w, or K_{sp}

$$1. \quad HC_2H_3O_2(aq) \rightleftharpoons H^+(aq) + C_2H_3O_2^-(aq)$$
$$2. \quad \underline{F^-(aq) + H^+(aq) \rightleftharpoons HF(aq)}$$
$$HC_2H_3O_2(aq) + F^-(aq) \rightleftharpoons C_2H_3O_2^-(aq) + HF(aq)$$

b. Equation 1 represents the dissociation of acetic acid:

$$K_1 = K_a \, HC_2H_3O_2 = 1.8 \times 10^{-5}$$

Equation 2 is the *reverse* of the equation for the dissociation of HF:

$$K_2 = 1/K_a \, HF = 1/(7.0 \times 10^{-4}) = 1.4 \times 10^3$$

c. $K = K_1 \times K_2 = (1.8 \times 10^{-5}) \times (1.4 \times 10^3) = 0.025$
Notice that K for this reaction is a small number, which indicates that the reaction does not go to completion. This is generally true for most weak acid–weak base reactions, which explains why such reactions are never used in acid-base titrations (Chap. 19).

EXERCISE Calculate K for the reaction $HF(aq) + C_2H_3O_2^-(aq) \rightleftharpoons HC_2H_3O_2(aq) + F^-(aq)$. Answer: 40.

We can summarize the results of this section by suggesting a general procedure for finding the equilibrium constant for a reaction.
1. Check to see if the equilibrium constant you want is directly available in a table such as Table 20.1 (K_a of weak acids, K_b of weak bases) or Table 18.2 (K_{sp} values).
2. If the equilibrium constant is not listed for the reaction you are interested in, examine the equation for the reaction closely. Is it the reverse of an equation for which you can find the value of K? If so, apply the reciprocal rule to obtain the equilibrium constant.
3. If neither (1) nor (2) applies, you will probably need to use the rule of multiple equilibria. To do this, express the given equation as the sum of two or more simpler

equations. Find the equilibrium constants for each of these equations, using the reciprocal rule if necessary. Finally, use the rule of multiple equilibria to find K for the overall reaction.

Summary

In this chapter, we discussed two different types of equilibrium constants.
1. **K_a,** the dissociation constant of a weak acid:

$$HB(aq) \rightleftharpoons H^+(aq) + B^-(aq); \qquad K_a = \frac{[H^+] \times [B^-]}{[HB]}$$

K_a can be determined if you know $[H^+]$ in a solution prepared by dissolving the weak acid HB in water (Example 20.1). It can be used to calculate the following:

a. The $[H^+]$, pH, or percent dissociation in a solution prepared by dissolving a weak acid in water. Here, $[H^+]^2 = K_a \times [HB]$; percent dissociation $= 100 \times [H^+]/(\text{orig. conc. HB})$ (Examples 20.2 and 20.3).

b. For a buffer, one of the two quantities $[H^+]$ or the ratio $[HB]/[B^-]$ if you know the value of the other quantity. Here, $[H^+] = K_a \times [HB]/[B^-]$ (Examples 20.4 and 20.5).

c. The $[H^+]$ or pH of a buffer after addition of a small amount of a strong acid or strong base (Example 20.6).

2. **K_b,** the dissociation constant of a weak base:

$$B^-(aq) + H_2O \rightleftharpoons HB(aq) + OH^-(aq)$$

K_b can be calculated from K_a of the conjugate weak acid by using the relation $K_a \times K_b = 1.0 \times 10^{-14}$ (Example 20.7). It can be used to obtain $[OH^-]$ in a solution of a weak base. Here, $[OH^-]^2 = K_b \times [B^-]$ (Example 20.8).

The equilibrium constant for a reaction is the reciprocal of the equilibrium constant for the reverse reaction: $K = 1/K'$ (Example 20.9). If Reaction 1 + Reaction 2 = Reaction 3, then $K_3 = K_1 \times K_2$ (Examples 20.10 and 20.11).

Key Words and Concepts

acid dissociation constant, K_a
base dissociation constant, K_b
buffer
5% rule
pH

pK_a
percent dissociation
quadratic formula
reciprocal rule
rule of multiple equilibria

strong acid
strong base
successive approximations,
  method of
weak acid
weak base

Questions and Problems

Acid Dissociation Constant, K_a

20.1 Write the dissociation equation and the K_a expression for each of the following acids:
  a. HNO_2 b. $H_2PO_4^-$ c. H_2SO_3

20.31 Write the dissociation equation and the K_a expression for each of the following acids:
  a. HCN b. PH_4^+ c. HS^-

20.2 Referring to Table 20.1, give the pK_a value for
a. HNO_2 b. H_2CO_3 c. $H_2PO_4^-$

20.3 Consider these acids:

Acid	A	B	C	D
K_a	2×10^{-6}	4×10^{-4}	5×10^{-3}	1×10^{-2}

a. Arrange the acids in order of decreasing acid strength.
b. Which acid has the highest pK_a value?

20.4 A solution prepared by dissolving 0.100 mol HN_3, hydrazoic acid, to form one liter of solution has $[H^+] = 1.4 \times 10^{-3}$ M. Calculate K_a of hydrazoic acid.

20.5 The pH of a solution prepared by dissolving 0.150 mol of formic acid, $HCHO_2$, to form 0.500 L of solution is 2.14. Calculate K_a of formic acid.

20.6 The percent dissociation of a certain weak acid in 0.10 M solution is 0.025%. Calculate K_a of the weak acid.

[H⁺] in Weak Acid Solutions

20.7 For hypobromous acid ($K_a = 2.1 \times 10^{-9}$), calculate the $[H^+]$ in solutions prepared by adding the following number of moles of HBrO to water to make one liter of solution:

a. 1.0 b. 0.20

20.8 Calculate the pH of solutions of acetic acid ($K_a = 1.8 \times 10^{-5}$) prepared by adding the following numbers of moles of acetic acid to 2.50 L of water:
a. 3.75 b. 0.700

20.9 Lactic acid ($K_a = 1.4 \times 10^{-4}$) is present in sore muscles after vigorous exercise. For a 1.5 M solution of lactic acid, calculate
a. $[H^+]$ b. $[OH^-]$
c. pH d. percent dissociation

20.10 For a 0.10 M solution of benzoic acid ($K_a = 6.6 \times 10^{-5}$), calculate
a. $[H^+]$ b. $[OH^-]$ c. pH
d. percent dissociation

20.32 Calculate K_a for weak acids that have the following pK_a values:
a. 2.0 b. 6.5 c. 3.81

20.33 Consider these acids:

Acid	A	B	C	D
pK_a	6.8	4.4	1.9	11.2

a. List the acids in order of increasing acid strength.
b. Which acid has the largest K_a value?

20.34 When 0.200 mol HClO is dissolved to form one liter of solution, it is found that $[ClO^-] = 8.0 \times 10^{-5}$ M. Calculate K_a of hypochlorous acid.

20.35 Benzoic acid, $HC_7H_5O_2$, is present in many berries. A benzoic acid solution prepared by dissolving 3.31 g of acid per liter has a pH of 2.89. What is the K_a of benzoic acid?

20.36 Citric acid is 8.6% dissociated in a solution prepared by dissolving 0.100 mol of acid to form one liter. Calculate K_a of citric acid.

20.37 Follow the directions of Problem 20.7 for hypochlorous acid, HClO ($K_a = 3.2 \times 10^{-8}$).

20.38 Follow the directions of Problem 20.8 for the $H_2PO_4^-$ ion ($K_a = 6.2 \times 10^{-8}$).

20.39 Follow the instructions for Problem 20.9 for a 0.80 M solution of NH_4Cl ($K_a\ NH_4^+ = 5.6 \times 10^{-10}$).

20.40 For a 0.20 M solution of propionic acid ($K_a = 1.4 \times 10^{-5}$), calculate
a. $[H^+]$ b. $[OH^-]$ c. pH
d. percent dissociation

20.11 Chloroacetic acid, $ClCH_2COOH$, has a K_a of 1.4×10^{-3}. Calculate $[H^+]$ in a 0.120 M solution of this acid using
a. the quadratic formula.
b. the method of successive approximations.

20.41 Dichloroacetic acid, $Cl_2CHCOOH$, has a K_a of 3.3×10^{-2}. Using the method of successive approximations, find $[H^+]$ in a 1.2 M solution of this acid.

Buffers

20.12 Calculate $[H^+]$ and pH in a solution in which $[HNO_2]$ is 0.10 M and $[NO_2^-]$ is
a. 0.20 M b. 0.10 M c. 0.050 M d. 0.010 M

20.42 Calculate $[H^+]$ and pH in a solution in which $[F^-]$ = 0.10 M and $[HF]$ is
a. 0.20 M b. 0.10 M c. 0.050 M d. 0.010 M

20.13 Calculate $[H^+]$ and pH in a buffer made by adding 0.050 mol $H_2PO_4^-$ to 2.00 L of a solution 0.025 M in HPO_4^{2-}.

20.43 Calculate $[H^+]$ and pH in a buffer made by adding 0.030 mol $H_2PO_4^-$ to 3.00 L of a solution 0.020 M in HPO_4^{2-}.

20.14 Consider the weak acids and their conjugate bases listed in Table 20.1. Which acid-base pair would be best for a buffer at a pH of
a. 3.4 b. 9.3 c. 6.4

20.44 Follow the instructions for Problem 20.14 for a pH of
a. 7.3 b. 3.2 c. 2.0

20.15 To make a buffer using HF and F^- in which the pH is 3.00,
a. what must be the ratio $[HF]/[F^-]$?
b. how many moles of HF must be added to a liter of 0.100 M NaF to give this pH?
c. how many moles of NaF must be added to a liter of 0.100 M HF to give this pH?

20.45 An NH_4Cl–NH_3 buffer is to be prepared with a pH of 9.00.
a. What must be the ratio $[NH_4^+]/[NH_3]$?
b. What volume of 1.00 M NH_4Cl should be added to a liter of 1.00 M NH_3 to form this buffer?

20.16 Consider a buffer made by adding 0.050 mol $NaC_2H_3O_2$ and 0.040 mol $HC_2H_3O_2$ to a liter of water.
a. What is the pH of this buffer?
b. What is the pH after addition of 0.010 mol NaOH?
c. What is the pH after addition of 0.010 mol HCl?

20.46 For the buffer referred to in Problem 20.16, what is the pH after adding
a. 0.020 mol HCl
b. 0.030 mol NaOH
c. 0.010 mol $Ca(OH)_2$

20.17 Blood is buffered mainly by the HCO_3^-–H_2CO_3 system (K_a $H_2CO_3 = 4.2 \times 10^{-7}$). The normal pH of blood is 7.40.
a. What is the ratio $[H_2CO_3]/[HCO_3^-]$?
b. What does the pH become if 10% of the HCO_3^- ions are converted to H_2CO_3?
c. What does the pH become if 10% of the H_2CO_3 molecules are converted to HCO_3^-?

20.47 A buffer is prepared in which the ratio $[HCO_3^-]/[CO_3^{2-}]$ is 4.0.
a. What is the pH of this buffer (K_a $HCO_3^- = 4.8 \times 10^{-11}$)?
b. Enough strong acid is added to make the pH of the buffer 9.40. What is the ratio $[HCO_3^-]/[CO_3^{2-}]$ at this point?

20.18 Which of the following would form a buffer if added to one liter of 0.20 M $HC_2H_3O_2$?
a. 0.10 mol $NaC_2H_3O_2$
b. 0.10 mol NaOH
c. 0.30 mol NaOH
d. 0.10 mol HCl
Explain your answers.

20.48 Which of the following would form a buffer if added to one liter of 0.20 M NaOH?
a. 0.10 mol $HC_2H_3O_2$
b. 0.30 mol $HC_2H_3O_2$
c. 0.10 mol $NaC_2H_3O_2$
d. 0.20 mol HCl
Explain your reasoning in each case.

The Base Dissociation Constant, K_b

20.19 Calculate $[OH^-]$ and pH in a solution prepared by dissolving 0.20 mol NH_3 ($K_b = 1.8 \times 10^{-5}$) per liter.

20.20 For hydrazoic acid, HN_3, the K_a is 1.9×10^{-5}.
a. Calculate the K_b of azide ion, N_3^-.
b. Determine the $[OH^-]$ and pH of 0.050 M NaN_3, sodium azide.

20.21 The K_a of anilinium ion, $C_6H_7NH^+$, is 2.5×10^{-5}; that of pyridinium ion, $C_5H_5NH^+$, is 2.4×10^{-7}.
a. Which is the stronger acid? Explain.
b. Calculate the K_b of aniline, C_6H_7N, and of pyridine, C_5H_5N.
c. Which is the stronger base, aniline or pyridine?

20.22 Calculate the pH of 0.25 M KCN (see Table 20.1).

20.23 Using Table 20.1, determine the pH of a 0.12 M solution of Na_3PO_4 (use successive approximations).

20.49 Estimate the pH of "concentrated NH_3" (Table 12.1, Chap. 12), taking K_b of NH_3 to be 1.8×10^{-5}.

20.50 Sodium lactate, $NaC_3H_5O_3$, is given intravenously to replenish body electrolytes. For lactate ion, $C_3H_5O_3^-$, $K_b = 7.3 \times 10^{-11}$.
a. What is the K_a of lactic acid?
b. Calculate the $[OH^-]$ and pH of 0.15 M sodium lactate.

20.51 The pK_a of formic acid is 3.74; that of acetic acid is 4.74.
a. Which is the stronger acid?
b. Determine the K_b of the formate ion; the acetate ion.
c. Which is the stronger base, the formate ion or the acetate ion?

20.52 Determine the pH of 0.10 M NaClO (see Table 20.1).

20.53 Using Table 20.1, determine the pH of a 0.0120 M solution of Na_2CO_3 (use successive approximations).

Relations Between K's

20.24 Using Table 20.1, calculate K for
a. $H^+(aq) + ClO^-(aq) \rightleftharpoons HClO(aq)$
b. $H_2PO_4^-(aq) + OH^-(aq) \rightleftharpoons HPO_4^{2-}(aq) + H_2O$
c. $H_2PO_4^-(aq) + H^+(aq) \rightleftharpoons H_3PO_4(aq)$

20.25 Using Table 20.1, obtain values of K for the following reactions:
a. $HNO_2(aq) + F^-(aq) \rightleftharpoons HF(aq) + NO_2^-(aq)$
b. $HNO_2(aq) + OH^-(aq) \rightleftharpoons H_2O + NO_2^-(aq)$
c. $H^+(aq) + NO_2^-(aq) \rightleftharpoons HNO_2(aq)$

20.26 K_a for maleic acid, $H_2C_4H_2O_4$, is 1.2×10^{-2}; K_a for the $HC_4H_2O_4^-$ ion is 6.0×10^{-7}. Calculate K for
a. $C_4H_2O_4^{2-}(aq) + H^+(aq) \rightleftharpoons HC_4H_2O_4^-(aq)$
b. $C_4H_2O_4^{2-}(aq) + 2 H^+(aq) \rightleftharpoons H_2C_4H_2O_4(aq)$
c. $H_2C_4H_2O_4(aq) + OH^-(aq) \rightleftharpoons HC_4H_2O_4^-(aq) + H_2O$

20.54 Using Table 20.1, calculate K for
a. $HPO_4^{2-}(aq) + H_2O \rightleftharpoons H_2PO_4^-(aq) + OH^-(aq)$
b. $HPO_4^{2-}(aq) + OH^-(aq) \rightleftharpoons PO_4^{3-}(aq) + H_2O$
c. $HPO_4^{2-}(aq) + H^+(aq) \rightleftharpoons H_2PO_4^-(aq)$

20.55 Using Table 20.1, obtain values of K for the following reactions ($K_w = 1.0 \times 10^{-14}$):
a. $H^+(aq) + C_2H_3O_2^-(aq) \rightleftharpoons HC_2H_3O_2(aq)$
b. $HC_2H_3O_2(aq) + F^-(aq) \rightleftharpoons HF(aq) + C_2H_3O_2^-(aq)$
c. $HC_2H_3O_2(aq) + OH^-(aq) \rightleftharpoons H_2O + C_2H_3O_2^-(aq)$

20.56 Citric acid, $H_3C_6H_5O_7$, is a triprotic acid found in lemon juice. Its acid dissociation constants are $K_1 = 7.5 \times 10^{-4}$; $K_2 = 1.7 \times 10^{-5}$; $K_3 = 4.0 \times 10^{-7}$. Calculate
a. K for the reaction

$$H_3C_6H_5O_7(aq) \rightleftharpoons 3 H^+(aq) + C_6H_5O_7^{3-}(aq)$$

b. the concentration of $C_6H_5O_7^{3-}$ ion in a solution of pH 5.00 where conc. citric acid = 0.10 M.

General

20.27 Explain why
  a. the pH increases when sodium benzoate is added to a benzoic acid solution.
  b. the pH of 0.10 M HNO_2 is greater than 1.0.
  c. a buffer resists changes in pH caused by addition of H^+ or OH^-.

20.28 Using Table 20.1, arrange 0.10 M solutions of the following compounds in order of increasing pH:
  a. HCl   b. NaOH   c. $NaC_2H_3O_2$
  d. NH_4Cl e. $HC_2H_3O_2$ f. NH_3

20.29 For each solution given below, indicate the main solute species present (0.10 M or greater) and calculate the pH of the solution:
  a. 0.20 M $HC_2H_3O_2$
  b. 0.30 M KCN
  c. 0.40 M KOH

20.30 Cinnamic acid (HCin), the active component in cinnamon oil, has a K_a of 3.5×10^{-5}. What is the pH of a solution that is 0.20 M in
  a. HCin   b. Cin^- c. both HCin and Cin^-

20.57 Indicate whether each of the following statements is true or false; if it is false, correct it.
  a. The acetate concentration in 0.10 M $HC_2H_3O_2$ is the same as in 0.10 M $NaC_2H_3O_2$.
  b. A buffer can be destroyed by adding too much strong acid.
  c. If $K_a = 3.1 \times 10^{-5}$, $K_b = 3.2 \times 10^{-9}$.

20.58 Using Table 20.1, arrange 0.10 M solutions of the following compounds in order of decreasing pH:
  a. Na_2CO_3 b. NaCl   c. NaCN
  d. H_2CO_3 e. HCl   f. HCN

20.59 Follow the directions of Problem 20.29 for the following solutions:
  a. 0.20 M NH_3
  b. 0.30 M HCl
  c. 0.40 M NH_4Cl

20.60 Ascorbic acid (HAsc) is better known as vitamin C. Its K_a is 5.0×10^{-5}. Calculate the pH of a solution that is 0.60 M in
  a. HAsc   b. Asc^- c. HAsc and Asc^-

*20.61 Show by calculation that when the concentration of a weak acid decreases by a factor of 10, its percent dissociation increases by a factor of $10^{1/2}$.

*20.62 Sulfuric acid, H_2SO_4, dissociates in two steps. The first dissociation constant is very large; the dissociation constant for the HSO_4^- ion is 1.2×10^{-2}. Estimate the pH of 0.10 M H_2SO_4.

*20.63 It is found that 0.20 M solutions of the three salts NaX, NaY, and NaZ have pH's of 7.0, 8.5, and 10.0, respectively. Arrange the acids HX, HY, and HZ in order of increasing acid strength. Where you can, find K_a for the acids.

*20.64 If 50.00 cm^3 of 1.000 M $HC_2H_3O_2$ ($K_a = 1.8 \times 10^{-5}$) is titrated with 1.000 M NaOH, what is the pH of the solution after the following volumes of NaOH have been added:
  a. 0.00 cm^3 b. 25.00 cm^3 c. 49.90 cm^3 d. 50.00 cm^3 e. 50.10 cm^3 f. 100.00 cm^3
Use your data to construct a plot similar to that shown in Figure 19.5 (pH vs. volume NaOH added).

*20.65 Calculate K for the reaction by which $Fe(OH)_3$ dissolves in strong acid:

$$Fe(OH)_3(s) + 3\ H^+(aq) \rightleftharpoons Fe^{3+}(aq) + 3\ H_2O$$

*20.66 When CO_2 dissolves in water, it forms the weak acid H_2CO_3 ($K_a = 4.2 \times 10^{-7}$). When the solution is made basic, H_2CO_3 is converted first to HCO_3^- ($K_a = 4.8 \times 10^{-11}$) and then to CO_3^{2-}. Of these three species, H_2CO_3, HCO_3^-, and CO_3^{2-}, show by calculation which one is present at the highest concentration at pH
  a. 4.00   b. 6.00   c. 8.00   d. 10.00   e. 12.00

Chapter 21
Complex Ions;
Coordination
Compounds

In previous chapters we have referred from time to time to compounds of the transition metals. Many of these have relatively simple formulas such as $CuSO_4$, $CrCl_3$, and $Fe(NO_3)_3$. These compounds are ionic: the transition metal is present as a simple cation (Cu^{2+}, Cr^{3+}, Fe^{3+}). In that sense, they resemble the ionic compounds formed by the main-group metals, such as $CaSO_4$ and $Al(NO_3)_3$.

It has been known for more than a century, however, that transition metals also form a variety of ionic compounds with more complex formulas such as

$$[Cu(NH_3)_4]SO_4 \qquad [Cr(NH_3)_6]Cl_3 \qquad K_3[Fe(CN)_6]$$

They are also called coordination complexes

In these so-called *coordination compounds*, the transition metal is present as a *complex ion*. A complex ion is a species in which a central metal ion, usually derived from a transition metal, is bonded to molecules or anions called *ligands*. Thus, we have

$Cu(NH_3)_4^{2+}$: Cu^{2+} ion is bonded to four NH_3 molecules.
$Cr(NH_3)_6^{3+}$: Cr^{3+} ion is bonded to six NH_3 molecules
$Fe(CN)_6^{3-}$: Fe^{3+} ion is bonded to six CN^- ions.

This chapter is devoted to complex ions and the important role they play in inorganic chemistry. We will consider in turn

—the compositions of complex ions and the coordination compounds they form (Section 21.1).
—the geometries of complex ions (Section 21.2).
—the electronic structures of the central metal ions in complexes (Section 21.3).
—the reaction rates and equilibrium constants for complex ion formation (Section 21.4).
—the uses of coordination compounds in chemical analysis, in industry, and in medicine (Section 21.5).

21.1
Compositions of Complex Ions and Coordination Compounds

A water solution containing the Cu^{2+} ion has a pale blue color. When aqueous ammonia, NH_3, is added, the color changes to a deep, almost opaque, blue (Color Plate 21.1). The color change is due to a chemical reaction in which four NH_3 molecules combine with a Cu^{2+} ion:

$$Cu^{2+}(aq) + 4\ NH_3(aq) \rightarrow Cu(NH_3)_4{}^{2+}(aq) \qquad (21.1)$$

light blue   deep blue

In electron-dot notation, we can represent this reaction as

The nitrogen atom of each NH_3 molecule contributes a pair of unshared electrons to form a covalent bond with the Cu^{2+} ion. This bond and others like it, where both electrons are contributed by the same atom, is referred to as a **coordinate** covalent bond. There are four such bonds in the $Cu(NH_3)_4{}^{2+}$ ion.

The $Cu(NH_3)_4{}^{2+}$ ion is commonly referred to as a **complex ion.** We use the term complex ion to indicate a charged species in which *a metal ion is joined by coordinate covalent bonds to neutral molecules and/or negative ions.* Species such as $Al(H_2O)_6{}^{3+}$ and $Zn(H_2O)_3(OH)^+$, found in previous chapters, are further examples of complex ions. The metals that show the greatest tendency to form complex ions are those toward the right of the transition series (in the first transition series, $_{24}Cr$ through $_{30}Zn$). Nontransition metals, including Al, Sn, and Pb form a more limited number of stable complex ions.

The metal cation in a complex is called the *central ion.* The molecules or anions bonded directly to it are called **ligands.** The number of bonds formed by the central ion is called its **coordination number.** In the $Cu(NH_3)_4{}^{2+}$ ion, the central ion is Cu^{2+}. The ligands are NH_3 molecules. Since the Cu^{2+} ion forms a total of four bonds, it has a coordination number of 4.

An ion such as $Cu(NH_3)_4{}^{2+}$ cannot exist by itself in the solid state. The $+2$ charge of this ion must be balanced by anions with a total charge of -2. A typical compound containing the $Cu(NH_3)_4{}^{2+}$ ion is

[Cu(NH$_3$)$_4$]Cl$_2$:   1 Cu(NH$_3$)$_4{}^{2+}$ ion, 2 Cl$^-$ ions

Compounds such as this, which contain a complex ion, are referred to as **coordination compounds.** The formula of the complex ion is set off by brackets, [ ], to make the structure of the compound clear.

In aqueous solution, most transition metal ions are present in complexes

Table 21.1 gives the formulas of a series of coordination compounds formed by the Pt^{2+} ion, which has a coordination number of 4. Notice the following points:

1. Compounds 1 and 2 contain complex cations with charges of $+2$ and $+1$, respectively. These coordination compounds are analogous to the simple ionic compounds $CaCl_2$ and KCl. In both $[Pt(NH_3)_4]Cl_2$ and $CaCl_2$, two Cl^- ions are balanced by a $+2$ ion, $Pt(NH_3)_4^{2+}$ in one case, Ca^{2+} in the other.

2. Compounds 4 and 5 contain complex anions with charges of -1 and -2, respectively. In the solid, these are balanced by K^+ ions (1 K^+ per $Pt(NH_3)Cl_3^-$ ion, 2 K^+ per $PtCl_4^{2-}$ ion).

Uncharged complexes are relatively uncommon

3. Compound 3 is a neutral complex, with zero charge. There are no ions present.

Table 21.1
Coordination Compounds Containing Complexes of Pt^{2+}

COORDINATION COMPOUND	COMPLEX	CHARGE OF COMPLEX	ANALOGOUS SIMPLE IONIC COMPOUND
1. $[Pt(NH_3)_4]Cl_2$	$Pt(NH_3)_4^{2+}$	$+2$	$CaCl_2$
2. $[Pt(NH_3)_3Cl]Cl$	$Pt(NH_3)_3Cl^+$	$+1$	KCl
3. $[Pt(NH_3)_2Cl_2]$	$Pt(NH_3)_2Cl_2$	0	
4. $K[Pt(NH_3)Cl_3]$	$Pt(NH_3)Cl_3^-$	-1	KNO_3
5. $K_2[PtCl_4]$	$PtCl_4^{2-}$	-2	K_2SO_4

As we have pointed out, a coordination compound, like all compounds, must have a net charge of zero. The charge of a complex ion within such a compound is readily determined. It is the algebraic sum of the charges of the bare metal ion and the ligands. Applying this principle to the compounds in Table 21.1, where the species within the complexes are

Pt^{2+} (charge $= +2$); NH_3 molecules (charge $= 0$); Cl^- ions (charge $= -1$)

we obtain, for the charges of the complex ions,

Complex 1: $+2 + 4(0) = +2$
Complex 2: $+2 + 3(0) + 1(-1) = +1$
Complex 3: $+2 + 2(0) + 2(-1) = 0$
Complex 4: $+2 + 1(0) + 3(-1) = -1$
Complex 5: $+2 + 4(-1) = -2$

EXAMPLE 21.1 Determine the charge of the central metal ion in each of the following:
a. $Zn(H_2O)_3(OH)^+$ b. $Pt(NH_3)_3Cl_3^-$ c. $Cr(CN)_6^{3-}$

Solution The first step in each case is to identify the ligands and their charges. Then we use the fact that the charge of the complex is the sum of that of the central ion and those of the ligands.

a. There are three H_2O molecules (0 charge) and an OH^- ion (-1 charge). The overall charge of the complex is $+1$. If we let x be the charge on the zinc,

$$+1 = x + 3(0) + 1(-1)$$

Solving for x, $x = +2$; Zn^{2+} ion

This is just simple algebra

b. The ligands are one NH_3 molecule (0 charge) and three Cl^- ions (-1 charge). The charge of the complex ion itself is -1:

$$-1 = x + 1(0) + 3(-1) ;   x = +2;   Pt^{2+} \text{ ion}$$

c. Here, the complex ion is an anion with a -3 charge. Each cyanide ion, CN^-, has a -1 charge:

$$-3 = x + 6(-1);   x = +3;   Cr^{3+} \text{ ion}$$

EXERCISE   In a certain complex ion, a central Cr^{3+} ion is bonded to two ammonia molecules, three water molecules, and a hydroxide ion. Give the formula and charge of the complex ion. Answer: $Cr(NH_3)_2(H_2O)_3 (OH)^{2+}$.

The specialized nomenclature of coordination compounds is discussed in Appendix 3.

Ligands; Chelating Agents

In principle, any molecule or anion with an unshared pair of electrons can donate them to a metal ion to form a coordinate covalent bond. We expect species such as the ammonia molecule, the water molecule, the hydroxide ion, and the chloride ion to act as ligands:

There are few, if any, cationic ligands. Can you suggest why?

In practice, a ligand usually contains an atom of one of the more electronegative elements (C, N, O, S, F, Cl, Br, I). Several hundred different ligands are known. Those most commonly encountered in general chemistry are NH_3 and H_2O molecules and Cl^- and OH^- ions.

Ligands are classified according to the number of electron pairs donated. *Monodentate* ligands provide one electron pair per ligand molecule or ion. The species NH_3, H_2O, Cl^-, and OH^- are of this type. *Bidentate* (two-toothed) ligands furnish two electron pairs per molecule or ion, a tridentate ligand furnishes three pairs, and so on. The general term *polydentate* is used for any ligand that supplies more than one pair of electrons. The structures of three important polydentate ligands are shown in Figure 21.1. These are the oxalate ion (ox), the ethylenediamine molecule (en), and the ethylenediaminetetraacetate anion (EDTA). Note that ethylenediamine and the oxalate ion are both bidentate; EDTA is hexadentate, forming six bonds per ligand.

EDTA usually forms 1:1 complexes with cations

The complexes formed by polydentate ligands are often called **chelates** (Greek *chela,* meaning crab's claw). The structures of the chelates formed by Cu^{2+} with

FIGURE 21.1 ■ Structures of three chelating agents, often abbreviated as "ox," "en," and EDTA. The numbers indicate the location of the unshared electron pairs that form coordinate covalent bonds with metal ions.

EDTA forms many complexes of importance in analytical chemistry

ethylenediamine and with the oxalate ion are shown in Figure 21.2. Notice that in these complexes, as in the $Cu(NH_3)_4^{2+}$ ion (p. 635), Cu^{2+} has a coordination number of 4. When we write these complexes as $Cu(en)_2^{2+}$ and $Cu(ox)_2^{2-}$, we must remember that each ligand is forming two bonds; thus, the coordination number of Cu^{2+} is $2 \times 2 = 4$.

EXAMPLE 21.2 Give the coordination number of the central metal ion in
a. $Cu(en)_2(NH_3)_2^{2+}$ b. $Fe(en)(ox)Cl_2^{-}$

Solution The coordination number is the number of bonds formed by the central metal ion.
a. Ammonia is a monodentate ligand; each NH_3 forms one bond. Each bidentate ethylenediamine (en) forms two bonds. Thus, there are six bonds to Cu^{2+}; its coordination number is 6.
b. Ethylenediamine (en) and oxalate (ox) are both bidentate (Fig. 21.1); each chloride is monodentate. Hence, Fe^{3+} has a coordination number of 6 in this case.

EXERCISE What is the coordination number of Pt in $Pt(en)_2^{2+}$?
Answer: 4.

For a ligand to act as a chelating agent, it must have at least two pairs of unshared electrons. Moreover, these electron pairs must be far enough removed from one another to give a chelate ring with a stable geometry. In chelates formed by ethylenediamine or the oxalate ion, such as those shown in Figure 21.2, the ring contains five atoms (four from the ligand plus the metal atom). Four-membered rings are less stable; three-membered rings do not occur. If we compare the three molecules,

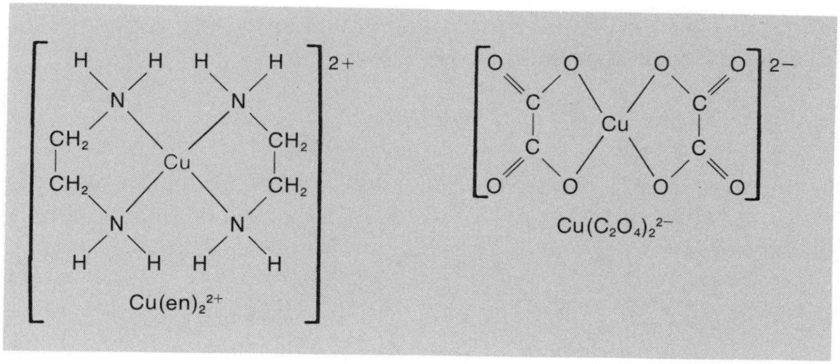

FIGURE 21.2 Structures of the chelates formed by Cu^{2+} with the ethylenediamine molecule (en) and the oxalate ion $(C_2O_4^{2-})$. In both complexes, Cu^{2+} has a coordination number of 4. The Cu^{2+} ion is in the center of a square; the four atoms to which it is bonded outline the corners of the square.

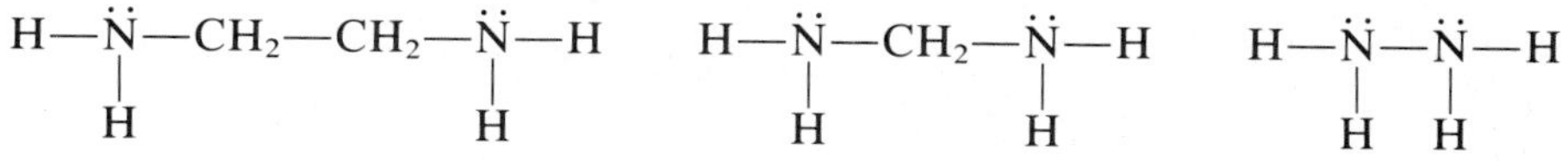

ethylenediamine   methylenediamine   hydrazine

we find that

—methylenediamine is a less effective chelating agent than ethylenediamine, since it would have to form four-membered as opposed to five-membered rings.

—hydrazine does not act as a chelating agent. To do so, it would have to form a three-membered ring with a metal atom. This would require a highly unstable 60° bond angle.

Hydrazine can form complexes, even though it doesn't exist in chelates

Coordination Number

As shown in Table 21.2, the most common coordination number is 6. A coordination number of 4 is less common. A value of 2 is restricted largely to Cu^+, Ag^+, and Au^+. Odd coordination numbers (1, 3, 5, 7, 9) are very rare.

Table 21.2
Coordination Number and Geometry of Complex Ions*

METAL ION	COORDINATION NUMBER	GEOMETRY	EXAMPLE
Ag$^+$, Au$^+$, Cu$^+$	2	linear	Ag(NH$_3$)$_2^+$
Cu^{2+}, Ni^{2+}, **Pd^{2+}**, **Pt^{2+}**	4	square planar	Pt(NH$_3$)$_4^{2+}$
Al^{3+}, Au$^+$, **Cd^{2+}**, Co^{2+}, Cu$^+$, Ni^{2+}, **Zn^{2+}**	4	tetrahedral	Zn(NH$_3$)$_4^{2+}$
Al^{3+}, Co^{2+}, **Co^{3+}**, **Cr^{3+}**, Cu^{2+}, **Fe^{2+}**, **Fe^{3+}**, Ni^{2+}, **Pt^{4+}**	6	octahedral	Co(NH$_3$)$_6^{3+}$

*Most common coordination number indicated by bold type.

A few metal ions show only one coordination number in their complex ions. Thus, Co^{3+} always shows a coordination number of 6, as in

$$Co(NH_3)_6^{3+}, \qquad Co(NH_3)_4Cl_2^+, \qquad Co(en)_3^{3+}$$

Other metal ions, such as Al^{3+} and Ni^{2+}, have variable coordination numbers, depending upon the nature of the ligand. With molecules or very small anions as ligands, we usually observe the higher coordination number, as in

More small ligands can fit around the metal ion

$$Al(H_2O)_6^{3+}, \qquad AlF_6^{3-}, \qquad Ni(H_2O)_6^{2+}, \qquad Ni(NH_3)_6^{2+}$$

With larger anions, the lower coordination number is more common:

$$AlCl_4^-, \qquad Al(OH)_4^-, \qquad Ni(CN)_4^{2-}$$

21.2 Geometry of Complex Ions

Coordination Number = 2

Complex ions in which the central metal ion forms only two bonds to ligands are *linear;* that is, the two bonds are directed at 180° angles. The structures of $CuCl_2^-$, $Ag(NH_3)_2^+$, and $Au(CN)_2^-$ may be represented as

$$(Cl-Cu-Cl)^- \qquad \left[\begin{array}{c} H \\ H-N-Ag-N-H \\ H \end{array} \begin{array}{c} H \\ \\ H \end{array} \right]^+ \qquad (N\equiv C-Au-C\equiv N)^-$$

Coordination Number = 4

Four-coordinate metal complexes may have either of two different geometries (Fig. 21.3). The four bonds from the central metal ion may be directed toward the corners of a regular tetrahedron. Two common *tetrahedral* complexes are $Zn(NH_3)_4^{2+}$ and $CoCl_4^{2-}$. In the other geometry, known as *square planar,* the four bonds are directed toward the corners of a square. The $Pt(NH_3)_4^{2+}$ and $Ni(CN)_4^{2-}$ ions are of this type.

So is $Cu(NH_3)_4^{2+}$

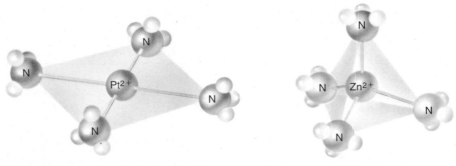

FIGURE 21.3   Structure of $Pt(NH_3)_4^{2+}$ (square planar) and $Zn(NH_3)_4^{2+}$ (tetrahedral). Geometrical isomers can exist for square planar complexes but not for tetrahedral complexes.

Certain square planar complexes occur in two different forms with quite different properties. The complex $[Pt(NH_3)_2Cl_2]$ has two forms differing in absorption spectrum, water solubility, melting point, and chemical reactivity. In one form, made by reacting NH_3 with the $PtCl_4^{2-}$ ion, the two NH_3 ligands are at adjacent corners of the square. In the other form, prepared by reacting $Pt(NH_3)_4^{2+}$ with HCl, the NH_3 molecules are at opposite corners of the square. These two forms are referred to as **geometric isomers.** Their structures differ only in the spatial arrangement of ligands about the central metal ion. The form in which like ligands are as close as possible is called the **cis** isomer. The **trans** isomer has these groups as far apart as possible.

Cl   NH₃   Cl   NH₃
  \   /   \   /
   Pt   Pt
  /   \   /   \
Cl   NH₃   NH₃   Cl

   cis   *trans*

Geometric isomerism can occur with any square planar complex of the type Ma_2b_2 or Ma_2bc, where M refers to the central metal and a, b, c are different ligands. Because all positions in a tetrahedral complex are equivalent, geometric isomerism cannot occur with such complexes.

If you make a sketch, you can see why

Coordination Number = 6

Octahedral geometry is characteristic of this coordination number. The six ligands surrounding the central metal ion in complexes such as $Fe(CN)_6^{3-}$ and $Co(NH_3)_6^{3+}$ are located at the corners of a regular octahedron. An octahedron is a geometric figure with six corners and eight faces, each of which is an equilateral triangle. The metal ion is located at the center of the octahedron (Fig. 21.4). The six ligands at the corners are *equidistant from the metal ion at the center*. The skeleton structure at the right of Figure 21.4 is easier to draw and more commonly used. From this skeleton, we see that an octahedral complex can be regarded as a derivative of a square planar complex. The two extra ligands are located above and below the square, on a line perpendicular to the square at its center.

The six ligand sites in an octahedral complex are equivalent

Geometric isomerism can occur in octahedral complexes. To see how this is possible, consider Figure 21.4. For any given position of a ligand, four equivalent positions are equidistant while a fifth is at a greater distance. Suppose, for example, we choose position 1 as a point of reference. Groups at positions 2, 3, 4, and 5 will be equidistant from 1; 6 is farther away. We may refer to positions 1 and 2,

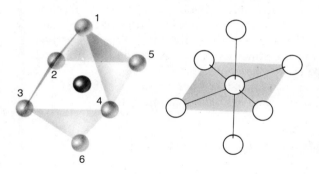

FIGURE 21.4 The drawing at the left shows six ligands at the corners of an octahedron with a metal ion at the center. A simpler way to represent an octahedral complex is shown at the right.

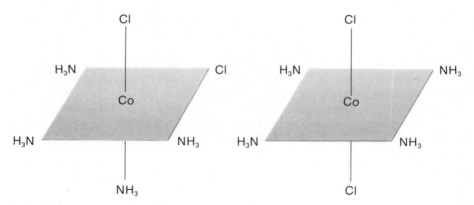

FIGURE 21.5 *Cis* and *trans* isomers of the $[Co(NH_3)_4Cl_2]^+$ complex ion. Note that the two Cl atoms are closer in the *cis* isomer (left) than in the *trans* isomer (right).

1 and 3, 1 and 4, or 1 and 5 as being *cis* to each other. On the other hand, positions 1 and 6 are *trans*. Hence, a complex ion like $Co(NH_3)_4Cl_2^+$ can exist in two isomeric forms. In the *cis* isomer, the two Cl^- ions are close together. In the *trans* form, they are far apart (Fig. 21.5). These two isomers differ in their physical and chemical properties. The most striking difference is color. Compounds of Co^{3+} containing the *cis* complex ion tend to be violet, while those containing the *trans* isomer are often green (Color Plate 21.2).

EXAMPLE 21.3   How many isomers are possible for the neutral complex $[Co(NH_3)_3Cl_3]$?

Solution   It is best to approach this problem systematically. We might start by putting two NH_3 molecules in *trans* positions, perhaps at the "top" and "bottom" of the octahedron (Fig. 21.6a). We then ask ourselves: In how many spatially different positions can we place the third NH_3 molecule? A moment's reflection should convince you that there is really only one choice. All four of the remaining positions are equivalent in that they are *cis* to the two groups that we have already located. Choosing one of these positions arbitrarily, we get our first isomer (Fig. 21.6b).

To see if there are other isomers we start again, this time locating two NH_3 molecules *cis* to each other (Fig. 21.6c). If we were to place the third NH_3 molecule at one of the other corners of the square, we would simply reproduce the first isomer. (Remember that the symmetry of a regular octahedron requires that the distance across the diagonal of the square be the same as that from "top" to "bottom.") We are left with two equivalent positions. Placing the third NH_3 molecule arbitrarily at the "top," we arrive at a second isomer (Fig. 21.6d), distinctly different from the first because all three NH_3 molecules are *cis* to one another.

We have now exhausted in a logical manner all the possibilities for geometric isomerism, finding two isomers. There are no others.

EXERCISE   How many geometric isomers are there for $Cr(NH_3)_2Cl_4^-$? for the square planar complex $[Pt(NH_3)_2ClBr]$? Answer: Two; two.

This is another example that requires a bit of thought

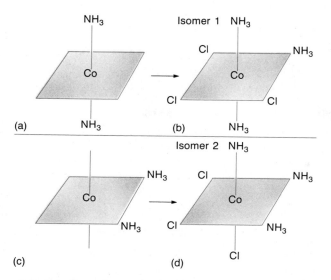

Two NH$_3$ ligands are *trans*

All NH$_3$ ligands are *cis*

FIGURE 21.6 Isomers of Co(NH$_3$)$_3$Cl$_3$ (Example 21.3). You may be tempted to write down additional structures, but you will find that they are equivalent to one of the two isomers shown here.

Geometric isomerism can occur in chelated octahedral complexes (Fig. 21.7). In the *cis* isomer of Co(en)$_2$Cl$_2^+$, the two Cl$^-$ ions are close together; in the *trans* isomer, they are farther apart.

21.3
Electronic Structures
of Complex Ions

To develop the electron distribution in a complex ion, a logical starting point is the electronic structure of the central metal ion. You may recall from Chapter 9 that in a transition metal cation, the inner d sublevel (for example, 3d) is lower

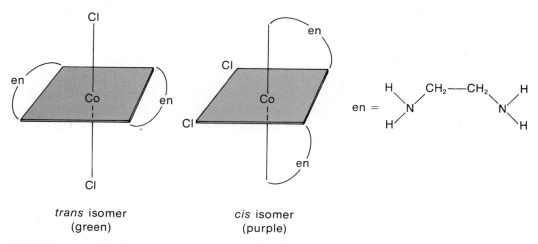

FIGURE 21.7 There are two isomers of the complex ion Co(en)$_2$Cl$_2^+$, which differ from one another in color and in certain chemical properties. For example, the *cis* isomer reacts more readily with C$_2$O$_4^{2-}$ ions than does the *trans* isomer. (Why?)

in energy than the outer s sublevel (4s). From a slightly different point of view, we might say that, in forming a transition metal cation such as Cr^{3+} or Cu^{2+}, the outer s electrons are lost first. Any way you look at it, the following statement is valid:

There are no outer s electrons in a transition metal cation. Electrons beyond the preceding noble gas are located in an inner d sublevel.

This principle is illustrated in Table 21.3. If these electron configurations do not seem familiar to you, you may want to go back to Chapter 9 and read again Section 9.1, where the structures of these ions were developed.

Table 21.3
Electron Configurations of Transition Metal Atoms and Ions

ATOMIC NUMBER	CONFIGURATION OF ATOM		CONFIGURATION OF ION	
24	Cr	$[_{18}Ar]3d^5 4s^1$	Cr^{3+}	$[_{18}Ar]3d^3$
25	Mn	$[_{18}Ar]3d^5 4s^2$	Mn^{2+}	$[_{18}Ar]3d^5$
26	Fe	$[_{18}Ar]3d^6 4s^2$	Fe^{3+}	$[_{18}Ar]3d^5$
			Fe^{2+}	$[_{18}Ar]3d^6$
27	Co	$[_{18}Ar]3d^7 4s^2$	Co^{3+}	$[_{18}Ar]3d^6$
			Co^{2+}	$[_{18}Ar]3d^7$
28	Ni	$[_{18}Ar]3d^8 4s^2$	Ni^{2+}	$[_{18}Ar]3d^8$
29	Cu	$[_{18}Ar]3d^{10} 4s^1$	Cu^{2+}	$[_{18}Ar]3d^9$
			Cu^+	$[_{18}Ar]3d^{10}$
30	Zn	$[_{18}Ar]3d^{10} 4s^2$	Zn^{2+}	$[_{18}Ar]3d^{10}$

In the ions, the 3d sublevels are more stable than 4s

With this background, we will look at two of the simpler models of the electronic structure of complex ions. One of these is the valence bond (atomic orbital) model, discussed in Chapter 10. The other is referred to as the crystal-field model.

Valence Bond (Atomic Orbital) Model

This model, for complex ions, starts with a simple assumption. It considers that electron pairs donated by ligands enter hybrid orbitals of the central metal ion. The particular orbitals that these electron pairs enter depend upon the coordination number and geometry of the complex (Table 21.4). Note that the total number of hybrid orbitals occupied by ligand electrons is equal to the coordination number.

Of the types of hybrid orbitals listed in Table 21.4, the sp and sp^3 sets were discussed earlier (Chap. 10). In square planar complexes, the four pairs of bonding electrons are accommodated by hybridizing a d orbital, an s orbital, and two p orbitals—hence, "dsp^2." For complex ions formed by metals in the first transition series, these would be one 3d, one 4s, and two 4p orbitals. In octahedral complexes, the six orbitals needed are obtained by hybridizing two d orbitals, an s orbital, and three p orbitals (d^2sp^3). In the first transition series, these are two 3d, one 4s, and three 4p orbitals.

The geometries of the hybrids are the same as in regular molecules

Orbital diagrams for the four complex ions listed in Table 21.4 are shown in Figure 21.8. Here, the hybrid orbitals are enclosed by horizontal lines. The bonding

Table 21.4
Valence Bond Model Applied to Complex Ions

COORDINATION NUMBER	GEOMETRY	HYBRID ORBITALS OCCUPIED BY LIGAND ELECTRONS	EXAMPLE
2	linear	sp	$Cu(NH_3)_2^+$
4	tetrahedral	sp^3	$Zn(NH_3)_4^{2+}$
4	square planar	dsp^2	$Ni(CN)_4^{2-}$
6	octahedral	d^2sp^3	$Cr(NH_3)_6^{3+}$

electrons within these orbitals, all of which are contributed by the ligands, are shown as colored arrows. The electrons shown in black are those contributed by the metal ion itself; none of these are involved in bonding. To save space, only those electrons beyond the argon core are shown; the inner 18 electrons of each metal ion are omitted.

The orbital diagrams shown in Figure 21.8 can be derived in a series of simple steps:

1. Determine the coordination number. This is the number of electron pairs (two, four, or six) contributed by the ligands. Locate these electron pairs in the hybrid orbitals, one pair per orbital, using Table 21.4 as a guide. If the coordination number is 4, you must know (or be given enough information to deduce) the geometry. Only then can you decide whether the hybridization is dsp^2 or sp^3. For an octahedral complex such as $Cr(NH_3)_6^{3+}$, the six pairs of electrons are located in d^2sp^3 hybrid orbitals.

These electrons form coordinate covalent bonds

2. Determine the charge of the central metal ion, using the discussion in Section 21.1, if necessary. Now find the number of electrons beyond the argon shell in that ion. To do this, start with the atomic number of the metal, deduct the number of electrons lost in forming the ion, and finally, subtract 18. Thus, for Cr^{3+},

$$\text{no. } e^- \text{ beyond Ar} = 24 - 3 - 18 = 3$$

3. Locate the electrons from step 2 in the lowest available orbitals, following Hund's rule. For Cr^{3+}, we have three 3d orbitals left vacant after locating the

Metal	Ion	Complex Ion	Orbital Diagram 3d	4s	4p	Hybridization
Cu^+	$3d^{10}$	$Cu(NH_3)_2^+$	(1↓)(1↓)(1↓)(1↓)(1↓)	(1↓)	(1↓)( )( )	sp
Zn^{2+}	$3d^{10}$	$Zn(NH_3)_4^{2+}$	(1↓)(1↓)(1↓)(1↓)(1↓)	(1↓)	(1↓)(1↓)(1↓)	sp^3
Ni^{2+}	$3d^8$	$Ni(CN)_4^{2-}$	(1↓)(1↓)(1↓)(1↓)(1↓)	(1↓)	(1↓)(1↓)( )	dsp^2
Cr^{3+}	$3d^3$	$Cr(NH_3)_6^{3+}$	(1)(1)(1)(1↓)(1↓)	(1↓)	(1↓)(1↓)(1↓)	d^2sp^3

FIGURE 21.8 Orbital diagrams derived from the valence bond model for complexes of Cu^+ (linear), Zn^{2+} (tetrahedral), Ni^{2+} (square planar), and Cr^{3+} (octahedral). Electron pairs furnished by the ligands are shown in color.

bonding electrons in step 1. We put an electron into each of these orbitals, giving each electron the same spin. This gives the orbital diagram shown at the bottom of Figure 21.8 for $Cr(NH_3)_6^{3+}$.

EXAMPLE 21.4 Write an orbital diagram for $Fe(CN)_6^{4-}$.

Solution Since the coordination number is 6, we expect the bonding electrons from the CN^- ligands to enter d^2sp^3 hybrid orbitals. We put them there:

Octahedral complexes always involve d^2sp^3 hybrids

$$\begin{array}{cccc} 3d & 4s & 4p \\ (\ )(\ )(\ )(\uparrow\downarrow)(\uparrow\downarrow) & (\uparrow\downarrow) & (\uparrow\downarrow)(\uparrow\downarrow)(\uparrow\downarrow) \end{array}$$

We must now decide how many electrons are contributed by the central metal ion. To do this, we must first deduce its charge. Since the complex ion has a charge of -4 and each CN^- ion has a charge of -1, we must be dealing with a $+2$ ion of iron; that is,

$$Fe^{2+} + 6\ CN^- \rightarrow Fe(CN)_6^{4-}$$

In the $+2$ ion of Fe (at. no. $= 26$), there must be 24 e^-; 6 of these are located beyond the argon core:

total no. e^- in $Fe^{2+} = 26 - 2 = 24$
no. e^- beyond Ar in $Fe^{2+} = 24 - 18 = 6$

We have just enough room in the empty 3d orbitals to accommodate these 6 electrons, one pair to an orbital. Hence the completed orbital diagram is

$$Fe(CN)_6^{4-} \quad [_{18}Ar]\ \begin{array}{cccc} 3d & 4s & 4p \\ (\uparrow\downarrow)(\uparrow\downarrow)(\uparrow\downarrow)(\uparrow\downarrow)(\uparrow\downarrow) & (\uparrow\downarrow) & (\uparrow\downarrow)(\uparrow\downarrow)(\uparrow\downarrow) \end{array}$$

EXERCISE Write the orbital diagram for the $Co(NH_3)_6^{3+}$ ion. Answer: Same as $Fe(CN)_6^{4-}$. (Why?)

The valence bond model has been quite successful in explaining certain of the properties of complex ions, notably their magnetic behavior. The structures shown in Figure 21.8 imply that the complex ions $Cu(NH_3)_2^+$, $Zn(NH_3)_4^{2+}$, and $Ni(CN)_4^{2-}$ should all be *diamagnetic;* they have no unpaired electrons. In contrast, the $Cr(NH_3)_6^{3+}$ ion, with three unpaired electrons, should be *paramagnetic*. These predictions are confirmed by experiment. The first three complexes are weakly repelled by a magnetic field. On the other hand, $Cr(NH_3)_6^{3+}$ is attracted into the field with a force that corresponds to three unpaired electrons.

Complexes often contain unpaired electrons

With certain complex ions, however, the observed magnetic behavior does not agree with the structure predicted by valence bond theory. An example is the $Fe(H_2O)_6^{2+}$ ion. We might expect its orbital diagram to be identical with that of the $Fe(CN)_6^{4-}$ ion shown in Example 21.4. Hence, it should be diamagnetic, with no unpaired electrons. By experiment, the $Fe(H_2O)_6^{2+}$ ion is found to be paramagnetic, with four unpaired electrons. It is possible to rationalize this behavior within the framework of valence bond theory.* However, it is by no means obvious why the CN⁻ ion should form one type of complex with Fe^{2+} and the H_2O molecule a quite different type.

In another area, valence bond theory has been notably deficient. It does not explain the most striking property of coordination compounds, their brilliant colors. A quite different approach, known as the crystal-field model, has been much more successful in dealing with this matter.

Nobody's perfect

Crystal-Field Model

The crystal-field model interprets the bonding between ligands and a metal ion as primarily electrostatic in nature. In this approach, ligand electron pairs are *not* donated into orbitals of the central metal ion, as in atomic orbital theory. Instead, it is assumed that the only effect of the ligands is to create an electrostatic field around the d orbitals of the metal ion. *This field changes the relative energies of different d orbitals.* In the isolated metal ion, before interaction occurs, all the d orbitals in a sublevel, such as 3d, have the same energy. After interaction, the d orbitals are split into two groups with different energies.

The metal ion remains intact in this theory

To illustrate the crystal-field model, let us consider the formation of the octahedral $Fe(CN)_6^{4-}$ ion:

$$Fe^{2+} + 6\ CN^- \rightarrow Fe(CN)_6^{4-}$$

In the free ($3d^6$) Fe^{2+} ion, all the 3d orbitals are equal in energy. Hence, the six 3d electrons are distributed in accordance with Hund's rule:

$$3d$$

$$Fe^{2+} \qquad (\uparrow\downarrow)(\uparrow)(\uparrow)(\uparrow)(\uparrow)$$

To form the octahedral complex, the six CN⁻ ions must approach along the x, y, and z axes (Fig. 21.9a, p. 648). This causes a splitting of the five 3d orbitals into two sets:

1. A higher energy pair, referred to as $d_{x^2-y^2}$ and d_{z^2} orbitals.
2. A lower energy trio, called the d_{xy}, d_{yz}, and d_{xz} orbitals (Fig. 21.9b).

To understand why the splitting occurs, we look at the orientation of the electron density clouds associated with the five 3d orbitals (Fig. 21.10). Notice that the two orbitals called $d_{x^2-y^2}$ and d_{z^2} have their maximum electron density directly along the x, y, and z axes, respectively. In contrast, the other d orbitals have their electron densities concentrated between the axes rather than along

---

*One "explanation" is that the d orbitals occupied by ligand electrons in $Fe(H_2O)_6^{2+}$ are those in the fourth principal energy level rather than the third. The electrons in the Fe^{2+} ion (six beyond Ar) could then spread over all the 3d orbitals, giving the structure

		3d	4s	4p	4d
$Fe(H_2O)_6^{2+}$	[₁₈Ar]	$(\uparrow\downarrow)(\uparrow)(\uparrow)(\uparrow)(\uparrow)$	$(\uparrow\downarrow)$	$(\uparrow\downarrow)(\uparrow\downarrow)(\uparrow\downarrow)$	$(\uparrow\downarrow)(\uparrow\downarrow)(\ )(\ )(\ )$

which does indeed have four unpaired electrons.

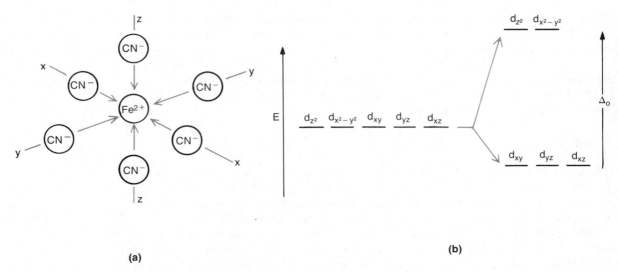

(a)

(b)

FIGURE 21.9 When six ligands (CN^- ions) approach a central metal ion (Fe^{2+}) along the x, y, and z axes the five orbitals are split into two groups. Two of these are raised in energy, while the other three are lowered. The difference in energy between the two groups is the crystal-field splitting energy, Δ_o.

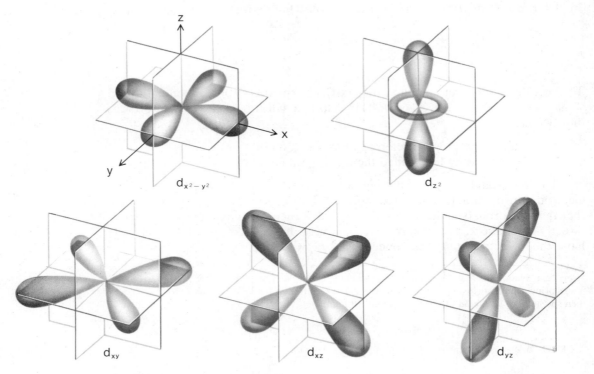

FIGURE 21.10 Spatial orientation of d orbitals. Note that the $d_{x^2-y^2}$ and d_{z^2} orbitals are oriented toward ligands approaching the corners of an octahedron.

them. Hence, as CN^- ligands approach along the x, y, and z axes, their electrostatic field is felt more strongly by electrons in $d_{x^2-y^2}$ and d_{z^2} orbitals than by those in the d_{xy}, d_{yz}, and d_{xz} orbitals. The effect is to split the 3d orbitals into two sets of differing energy. The energy difference is known as the *crystal-field splitting energy, Δ_0.*

With $Fe(CN)_6^{4-}$, Δ_0 is *large enough* to cause the six 3d electrons of Fe^{2+} to *pair up* in the three lower energy orbitals (Fig. 21.11a). In the $Fe(H_2O)_6^{2+}$ ion, the situation is quite different. The splitting energy, Δ_0, is *smaller*. Indeed, it is *not* large enough to overcome the tendency for electrons to remain unpaired. In the $Fe(H_2O)_6^{2+}$ ion, as in the Fe^{2+} ion itself, the six 3d electrons are spread out over all orbitals (Fig. 21.11b). The value of Δ_0 depends on the ligand

Looking at Figure 21.11, we see that the crystal-field model explains why $Fe(CN)_6^{4-}$ is diamagnetic while $Fe(H_2O)_6^{2+}$ is paramagnetic. In the first complex there are *no* unpaired electrons; in the second there are *four* unpaired electrons. Such pairs of complexes are known for many other transition metal cations:

1. The **high-spin** complex contains the larger number of unpaired electrons. The electrons in the complex are distributed in the same way as in the simple metal ion; Hund's rule is followed. The number of unpaired electrons in a high-spin complex such as $Fe(H_2O)_6^{2+}$ is exactly the same as in the bare Fe^{2+} ion. High-spin complexes are expected where the crystal-field splitting energy, Δ_0, is small. High spin: max no. unpaired e^-

2. The **low-spin** complex contains the smaller number of unpaired electrons. Electrons pair up in the lower energy orbitals. These orbitals fill completely before any electrons enter the higher energy orbitals. Low-spin complexes are expected when Δ_0 is large; there is a large energy difference between the two sets of d orbitals. Low-spin: min no. unpaired e^-

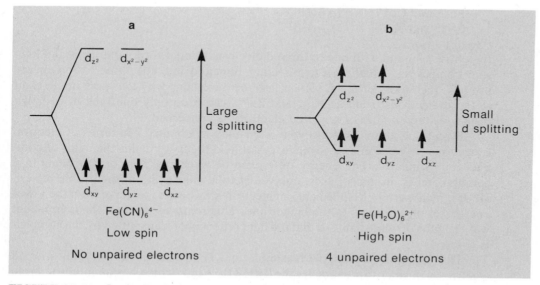

FIGURE 21.11 In the $Fe(CN)_6^{4-}$ ion, the energy difference between d orbitals, Δ_0, is large. Hence, the six 3d electrons of the Fe^{2+} ion pair up in the three orbitals of lower energy. In the $Fe(H_2O)_6^{2+}$ ion, Δ_0 is small, and the electrons spread out among the five orbitals with a maximum number (four) of unpaired electrons.

EXAMPLE 21.5 Using the crystal-field model, derive the structure of the Co^{2+} ion in low-spin and high-spin octahedral complexes.

Solution We first determine the number of d electrons available. Since the atomic number of cobalt is 27, there are $27 - 2 = 25$ electrons in Co^{2+}. Of these, 18 are accounted for by the argon core, leaving 7 in the 3d orbitals. In a low-spin complex, formed with strongly interacting ligands, 6 of these electrons are crowded into the lower three energy levels, leaving only 1 unpaired electron. In a high-spin complex, where the splitting is small, Hund's rule is followed and there are 3 unpaired electrons:

$$(\uparrow)(\quad) \qquad\qquad (\uparrow)(\uparrow)$$
$$(\uparrow\downarrow)(\uparrow\downarrow)(\uparrow\downarrow) \qquad (\uparrow\downarrow)(\uparrow\downarrow)(\uparrow)$$

low-spin   high-spin

Both types of complexes are known. The $Co(CN)_6^{4-}$ ion is of the low-spin type; $Co(H_2O)_6^{2+}$ is a high-spin complex.

EXERCISE Consider a transition metal ion with four 3d electrons, such as Cr^{2+}. How many unpaired electrons are there in a high-spin complex of this ion? a low-spin complex? Answer: Four; two.

High-spin and low-spin complexes are possible for transition metal ions containing *between four and seven inner d electrons*. Thus we expect the following ions to form two different types of complexes:

Cr^{2+} and Mn^{3+}: $[_{18}Ar]3d^4$
Mn^{2+} and Fe^{3+}: $[_{18}Ar]3d^5$
Fe^{2+} and Co^{3+}: $[_{18}Ar]3d^6$
Co^{2+} and Ni^{3+}: $[_{18}Ar]3d^7$

In contrast, an ion with one to three d electrons cannot form both high- and low-spin complexes. All the electrons enter unpaired into the three lower energy orbitals, regardless of Δ_0. A similar argument explains why ions containing eight to ten d electrons, such as Ni^{2+} and Zn^{2+}, can form only one type of complex insofar as the number of unpaired electrons is concerned.

Bright line atomic spectra were explained in Chapter 7 in terms of electron transitions between energy levels. Crystal-field theory explains the color of complex ions similarly. The energy difference between two sets of d orbitals in a complex ion is, in many cases, equivalent to a wavelength in the visible region. Hence, by absorbing visible light, an electron may be able to move from the lower energy set of d orbitals to the higher one. This removes some of the component wavelengths of white light, so that the light reflected or transmitted by the complex is colored.

The situation is particularly simple in $_{22}Ti^{3+}$, where there is only one 3d electron. Consider, for example, the $Ti(H_2O)_6^{3+}$ ion, which has an intense purple color. This ion absorbs at 510 nm, in the green region. The purple (red-blue) color that we see when we look at a solution of $Ti(H_2O)_6^{3+}$ is what is left when the green component is subtracted from the visible spectrum. Using Equation 7.2,

This concept is easy, once you get the hang of it

Color in complexes

Ground   Excited
state   state

$\Delta E = h\nu = \Delta.$

Chapter 7, we can calculate the energy difference, ΔE, corresponding to a wavelength of 510 nm:

$$\Delta E = \frac{1.196 \times 10^5 \text{ kJ} \cdot \text{nm}}{\lambda} \frac{}{\text{mol}} = \frac{1.196 \times 10^5}{5.10 \times 10^2} \frac{\text{kJ}}{\text{mol}} = 234 \frac{\text{kJ}}{\text{mol}}$$

We conclude that this is the energy absorbed to raise the 3d electron from a lower to a higher orbital. In other words, in the $Ti(H_2O)_6^{3+}$ ion, the two sets of d orbitals are separated by this amount of energy. The splitting energy, Δ_0, is 234 kJ/mol.

The magnitude of the splitting energy, Δ_0, determines the wavelength of the light absorbed by a complex and hence its color. This effect is shown in Table 21.5 and Color Plate 21.2. Notice that when we substitute for NH_3 such ligands as NCS^-, H_2O, or Cl^-, which produce smaller values of Δ_0, the light absorbed shifts to longer wavelengths (lower energies). On the basis of these and other observations, we can arrange various ligands in order of decreasing tendency to split the d orbitals. A short version of such a *spectrochemical* series is

If Δ_0 is large, the sample absorbs at the violet end of the spectrum

$$CN^- > NO_2^- > en > NH_3 > NCS^- > H_2O > F^- > Cl^-$$

strong field   decreasing Δ_0 weak field

As we move from left to right in the series, Δ_0 decreases, the wavelength of light absorbed increases, and there is an increasing tendency to form high-spin complexes.

Table 21.5
Colors of Complex Ions of Co^{3+}

COMPLEX	COLOR OBSERVED	COLOR ABSORBED	APPROXIMATE WAVELENGTH (NM) ABSORBED
$Co(NH_3)_6^{3+}$	yellow	violet	430
$Co(NH_3)_5NCS^{2+}$	orange	blue	470
$Co(NH_3)_5H_2O^{3+}$	red	blue-green	500
$Co(NH_3)_5Cl^{2+}$	purple	yellow-green	530
trans-$Co(NH_3)_4Cl_2^+$	green	red	680

Certain transition metal ions do not form colored complexes (Color Plate 8.2). Among these is Sc^{3+}, which has no d electrons. Another such ion is Zn^{2+}, in which the d orbitals are completely filled:

$$Zn^{2+} \quad [_{18}Ar]3d^{10}$$

Excitation to the 4s sublevel would require ultraviolet light

Here, there is no possibility of an electron moving from one d orbital to another, so that visible light is not absorbed. As a result, complexes such as $Zn(NH_3)_4^{2+}$ or $Zn(OH)_4^{2-}$ are colorless.

Lest you suppose that crystal-field theory can explain all the properties of complex ions, we should point out one of its weaknesses. If the bonding in complex ions is primarily electrostatic, it is hard to see why certain molecules that are only slightly polar, such as CO, can act as ligands. To explain this, it is necessary to modify crystal-field theory to take into account covalent as well as ionic bonding. This is done in a more sophisticated approach known as *ligand-field theory,* which you may study in later chemistry courses.

21.4
Complex Ions in Water; Rate and Equilibrium Considerations

For reactions involving complex ions in water solution (as indeed with all reactions), we must be concerned with two factors:

1. *The rate of reaction.* Here we will consider only one type of reaction, that in which there is an exchange of ligands. An example is

$$Cu(H_2O)_4{}^{2+}(aq) + 4\ NH_3(aq) \rightarrow Cu(NH_3)_4{}^{2+}(aq) + 4\ H_2O \tag{21.2}$$

Here, water molecules in the complex ion are replaced by NH_3 molecules.

2. *The position of the equilibrium involved.* Here we will consider the equilibrium constant for the formation of the complex ion. This is commonly called the **formation constant** (or the stability constant) and is given the symbol K_f. A typical example is

Complexes are involved in many chemical equilibria

$$Cu^{2+}(aq) + 4\ NH_3(aq) \rightleftharpoons Cu(NH_3)_4{}^{2+}(aq); \qquad K_f = \frac{[Cu(NH_3)_4{}^{2+}]}{[Cu^{2+}] \times [NH_3]^4} \tag{21.3}$$

Rate of Ligand Exchange

For reactions such as Reaction 21.2, we classify complex ions, more or less arbitrarily, into two types:

1. Complex ions that exchange ligands almost instantaneously are referred to as *labile.* Typically, they exchange ligands in water solution with a half-life of a minute or less. The $Cu(H_2O)_4{}^{2+}$ ion is of this type, as is $Ni(H_2O)_6{}^{2+}$. With the former, exchange occurs too rapidly to be measured. In the case of $Ni(H_2O)_6{}^{2+}$, the half-life for substitution by NH_3 or other ligands is of the order of 0.02 s at 25°C.

Most complexes are labile

2. Complex ions that undergo ligand substitution slowly are called nonlabile or *inert* (Color Plate 21.3). A typical example of an inert complex is $Cr(H_2O)_6{}^{3+}$. Here, the half-life for substitution by NH_3 is about 40 h at 25°C. This means that, in the time required for one $Cr(H_2O)_6{}^{3+}$ ion to react with NH_3, several million labile $Ni(H_2O)_6{}^{2+}$ ions will have undergone reaction.

An interesting illustration of the difference between labile and inert species involves the complexes of iron(II) and iron(III) with the cyanide ion. The complex of Fe^{2+}, $Fe(CN)_6{}^{4-}$, is inert and nonpoisonous. In contrast, the complex of Fe^{3+}, $Fe(CN)_6{}^{3-}$, is labile. It is poisonous because the CN^- ions released in water solution are toxic.

Only a few transition metal ions form inert complexes consistently. Among these are Cr^{3+}, Co^{3+}, Pt^{2+}, and Pt^{4+}. Complexes of these ions were among the first to be studied in the laboratory. Once formed, they are relatively easy to separate from solution in the form of pure compounds. Moreover, inert complexes retain their identity in solution long enough for their physical and chemical properties to be measured. Much of our knowledge of the structures and properties of complex ions is based upon the behavior of inert complexes in water solution.

You can't isolate isomers in labile complexes

Formation Constants

Table 21.6 lists formation constants for several complex ions. In each case, K_f applies to the formation of the complex by a reaction similar to Reaction 21.2. Note that for each complex ion listed, K_f is a large number, 10^5 or greater. This means that equilibrium considerations strongly favor complex formation. Consider, for example, the system

$$Ag^+(aq) + 2\,NH_3(aq) \rightleftharpoons Ag(NH_3)_2{}^+(aq);$$

$$K_f = \frac{[Ag(NH_3)_2{}^+]}{[Ag^+] \times [NH_3]^2} = 2 \times 10^7$$

We interpret the large K_f value to mean that the forward reaction goes virtually to completion. Addition of ammonia to a solution of $AgNO_3$ will convert nearly all of the Ag^+ ions to the $Ag(NH_3)_2{}^+$ complex. From a slightly different point of view, we might say that, since K_f is large, there will be very little tendency for the reverse reaction to occur. The $Ag(NH_3)_2{}^+$ ion is very stable toward decomposition to Ag^+ ions and NH_3 molecules.

AgCl, insoluble in water, dissolves in 6 M NH_3. Can you suggest why?

Table 21.6
Formation Constants of Complex Ions

MC₂		MC₄		MC₆	
$AgCl_2{}^-$	1×10^5	$Al(OH)_4{}^-$	1×10^{33}	$Co(NH_3)_6{}^{2+}$	1×10^5
$Ag(NH_3)_2{}^+$	2×10^7	$Cd(NH_3)_4{}^{2+}$	1×10^7	$Co(NH_3)_6{}^{3+}$	1×10^{35}
$Ag(S_2O_3)_2{}^{3-}$	1×10^{13}	$Cu(NH_3)_4{}^{2+}$	5×10^{12}	$Fe(CN)_6{}^{4-}$	1×10^{35}
$Ag(CN)_2{}^-$	1×10^{21}	$Zn(CN)_4{}^{2-}$	1×10^{17}	$Fe(CN)_6{}^{3-}$	1×10^{42}
$CuCl_2{}^-$	1×10^5	$Zn(NH_3)_4{}^{2+}$	1×10^9	$Ni(NH_3)_6{}^{2+}$	5×10^8
$Cu(NH_3)_2{}^+$	5×10^{10}	$Zn(OH)_4{}^{2-}$	3×10^{15}		

The stabilities of different complexes of the same central metal ion are directly related to their formation constants: the larger the K_f value, the more stable the complex. For the reaction

$$Ag^+(aq) + 2\,S_2O_3{}^{2-}(aq) \rightleftharpoons Ag(S_2O_3)_2{}^{3-}(aq);$$

$$K_f = \frac{[Ag(S_2O_3)_2{}^{3-}]}{[Ag^+] \times [S_2O_3{}^{2-}]^2} = 1 \times 10^{13}$$

This K_f is larger than that for $Ag(NH_3)_2^+$ ($K_f = 2 \times 10^7$). Hence, the $Ag(S_2O_3)_2^{3-}$ complex ion is more stable than $Ag(NH_3)_2^+$.

EXAMPLE 21.6   Determine the ratios
a. $[Ag(NH_3)_2^+]/[Ag^+]$ in 0.1 M NH_3
b. $[Ag(S_2O_3)_2^{3-}]/[Ag^+]$ in 0.1 M $S_2O_3^{2-}$

Solution   In each case, we write down the expression for K_f and solve for the desired ratio,

a. $K_f = 2 \times 10^7 = \dfrac{[Ag(NH_3)_2^+]}{[Ag^+] \times [NH_3]^2}$

$\dfrac{[Ag(NH_3)_2^+]}{[Ag^+]} = K_f \times [NH_3]^2 = 2 \times 10^7(0.1)^2 = 2 \times 10^5$

This means that in 0.1 M NH_3 there are 200,000 $Ag(NH_3)_2^+$ complex ions for every Ag^+ ion.

b. Proceeding as in (a),

$\dfrac{[Ag(S_2O_3)_2^{3-}]}{[Ag^+]} = K_f \times [S_2O_3^{2-}]^2 = 1 \times 10^{13}(0.1)^2 = 1 \times 10^{11}$

Note that this ratio is much higher than the comparable ratio for the $Ag(NH_3)_2^+$ complex. This is really what we mean when we say that the $Ag(S_2O_3)_2^{3-}$ complex is "more stable" than $Ag(NH_3)_2^+$.

EXERCISE   At what concentration of NH_3 is $[Ag(NH_3)_2^+] = [Ag^+]$? Answer: 2×10^{-4} M.

In a solution containing complexing ligands, like NH_3, the free cation conc. is very low

21.5
Uses of Coordination Compounds

Coordination compounds are used for a variety of purposes inside and outside the laboratory. Perhaps their most important application is in bringing water-insoluble species into solution. In discussing the metallurgy of aluminum in Chapter 4, we pointed out that Al_2O_3 is separated from Fe_2O_3 and other impurities by heating with concentrated sodium hydroxide, NaOH. The reaction that takes place involves the formation of the stable $Al(OH)_4^-$ complex ion:

Al_2O_3 is insoluble in water

$$Al_2O_3(s) + 3 H_2O + 2 OH^-(aq) \rightarrow 2 Al(OH)_4^-(aq) \qquad (21.4)$$

The stability of the $Al(OH)_4^-$ ion is great enough to make aluminum metal dissolve in a solution of a strong base. The reaction is

$$2 Al(s) + 2 OH^-(aq) + 6 H_2O \rightarrow 2 Al(OH)_4^-(aq) + 3 H_2(g) \qquad (21.5)$$

This reaction has recently found practical application in the addition of finely divided aluminum to drain cleaners containing lye, NaOH. The bubbles of hydrogen formed are supposed to loosen deposits of grease or dirt.

In many natural products, a transition metal ion is tied up in a coordination complex. Compounds of this type include hemoglobin (Fe^{2+}), chlorophyll (Mg^{2+}), and vitamin B_{12} (Co^{2+}). Other coordination compounds, made in the laboratory, are used in medicine. In the remainder of this section, we will look at a few of the biological uses of coordination compounds.

Hemoglobin and Oxygen Transport

The structure of heme, the pigment responsible for the color of blood, is shown in Figure 21.12. Notice that there is an Fe^{2+} ion at the center bonded to four nitrogen atoms at the corners of a square. Actually, the Fe^{2+} ion is part of an octahedral complex. A fifth ligand, not shown in the figure, is a large organic molecule called globin. In combination with heme, this gives the protein we refer to as hemoglobin. The sixth ligand, filling out the octahedron, is an H_2O molecule.

The globin has a molar mass of about 15000 g

The water molecule in hemoglobin can be replaced reversibly by an O_2 molecule. The product formed is called oxyhemoglobin; it has a bright red color. In contrast, hemoglobin itself is blue. This explains why arterial blood is red (high concentration of O_2), while that in the veins is bluish (low concentration of O_2).

$$\text{hemoglobin·}H_2O(aq) + O_2(aq) \rightleftharpoons \text{hemoglobin·}O_2(aq) + H_2O$$
$$\underset{\text{blue}}{} \qquad\qquad\qquad \underset{\text{red}}{} \qquad\qquad\qquad \textbf{(21.6)}$$

The equilibrium in Reaction 21.6 is sensitive to the concentration of oxygen. In the lungs, where there is a large amount of O_2 available, the equilibrium shifts to the right. Oxygen in the form of the hemoglobin complex is taken up by red blood cells and carried to the tissues. There, where the concentration of dissolved O_2 is low, the reverse reaction occurs. This supplies the oxygen required for metabolism.

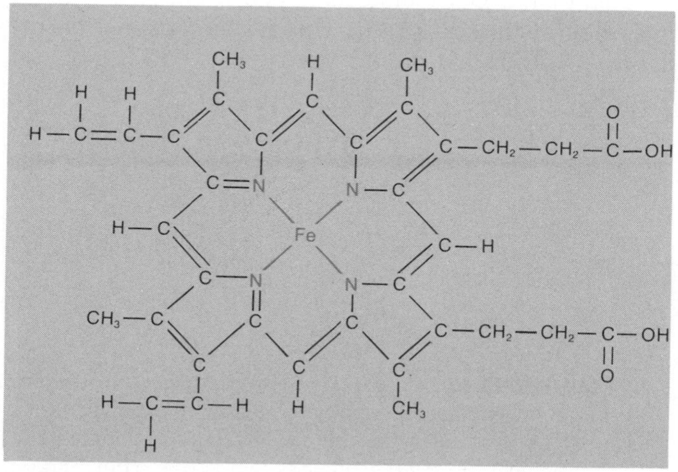

FIGURE 21.12 Structure of heme · Fe^{2+} ion is at the center of an octahedron, surrounded by four nitrogen atoms, a globin molecule, and a water molecule.

Certain molecules and anions form complexes with hemoglobin that are more stable than that formed by O_2. One such species is the CO molecule. As pointed out in Chapter 17, carbon monoxide poisoning results from the formation of the hemoglobin·CO complex in preference to hemoglobin·O_2. Another species that behaves this way is the cyanide ion, CN^-, which has the same Lewis structure as CO:

$$:C\equiv O: \qquad (:C\equiv N:)^-$$

It seems strange that N_2, $:N\equiv N:$, does not behave at all like CO and CN^-

EDTA and Lead Poisoning

Lead compounds were used widely at one time as paint pigments. Children can develop lead poisoning by swallowing paint from an old crib or a windowsill in an old building. The Pb^{2+} ion interferes with hemoglobin production by deactivating an enzyme needed to make heme.

EDTA, whose structure was shown on page 638, is a very effective chelating agent. It forms complexes with a large number of cations, including those of some of the main-group metals. The complex formed by calcium with EDTA is used to treat lead poisoning. When a solution containing the Ca–EDTA complex is given by injection, the calcium is displaced by lead. The more stable Pb–EDTA complex is eliminated in the urine. EDTA has also been used to remove radioactive isotopes of metals, notably plutonium, from body tissues.

Coordination Compounds in the Treatment of Cancer

The neutral complex cis-[Pt(NH$_3$)$_2$Cl$_2$] is an effective antitumor agent in cancer treatment (chemotherapy). The *trans* isomer is ineffective. The activity of the *cis* isomer is believed to reflect the ability of the two Cl atoms to interact with DNA, a molecule responsible for cell reproduction. A reaction occurs in which the Cl atoms are replaced by N atoms of the DNA molecule. The product is a Pt^{2+} complex that inhibits further cell growth (Fig. 21.13). This reaction cannot occur with the *trans* isomer, since the Cl atoms are too far apart.

Hopefully it doesn't also knock out normal cells

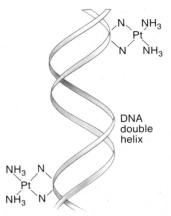

FIGURE 21.13 *cis*-Pt(NH$_3$)$_2$Cl$_2$ is effective in chemotherapy because of its ability to react with nitrogen atoms on DNA molecules, apparently causing the double helix to unwind. The *trans* isomer is ineffective. (For a more detailed discussion of this phenomenon, see *Chemical and Engineering News*, April 19, 1982, p. 36.)

21.6
Historical
Perspective

Alfred Werner and
Sophus Mads Jorgensen

The basic ideas concerning the structure and geometry of complex ions presented in this chapter were developed by one of the most gifted individuals in the history of inorganic chemistry, Alfred Werner. His theory of coordination chemistry was published in 1893 when Werner was 26 years old, holding the equivalent of an associate professorship at the University of Zurich. In his paper Werner made the revolutionary suggestion that metal ions such as Co^{3+} could show two different kinds of valences. For the compound $Co(NH_3)_6Cl_3$, Werner postulated a central Co^{3+} ion joined by "primary valences" (ionic bonds) to three Cl^- ions and by "secondary valences" (coordinate covalent bonds) to six NH_3 molecules. Moreover, he made the inspired guess that the six secondary valences were directed toward the corners of a regular octahedron. In Table 21.7 we list the familiar Werner structures for the series of compounds $Co(NH_3)_xCl_3$, where x = 6, 5, 4, or 3. All these compounds were known at the time, and many of their properties had been established. In particular, it was known from conductivity studies and by precipitation with $AgNO_3$ that the first three members of the series yielded 3, 2, and 1 mol Cl^-, respectively, when dissolved in water. This evidence was, of course, in complete agreement with Werner's theory.

Werner's structures, which seem so obvious to us today, aroused little enthusiasm among his contemporaries. The opposition was led by Sophus Mads Jorgensen, a 56-year-old professor of chemistry at the University of Copenhagen. Jorgensen was convinced that Co^{3+} could form no more than three bonds. To rationalize the existence of "addition compounds" of $CoCl_3$ containing six, five, four, or three NH_3 molecules, he invoked the chain structures shown in Table 21.7, in which NH_3 molecules are linked together much like CH_2 groups in hydrocarbons. The differing extents of ionization of these compounds in water were explained by assuming that only those chlorine atoms bonded to NH_3 groups could ionize. Chlorines attached directly to cobalt were supposed to be held so tightly that they could not ionize in water.

From the data in Table 21.7 it would appear that the controversy between Werner and Jorgensen could have been settled quite simply by studying the behavior in water solution of the compound $Co(NH_3)_3Cl_3$. Werner's structure required that this species be a nonelectrolyte with no ionizable chlorine. In contrast, the chain structure of Jorgensen implied one ionizable chlorine, i.e., a 1:1 electrolyte similar to NaCl. Unfortunately, the evidence was ambiguous: at 25°C, a water solution of $Co(NH_3)_3Cl_3$ has a conductivity intermediate between that of a nonelectrolyte and a 1:1 salt.

Werner's assumption of octahedral coordination around the Co^{3+} ion offered further opportunities for testing his ideas against those of Jorgensen. If Werner were correct, there should be two isomeric forms (*cis* and *trans*) of the compound $Co(NH_3)_4Cl_3$. At the time, only one compound of this formula was known. All of Werner's early attempts to prepare a second isomer failed, thereby weakening his position.

As the years passed, the weight of evidence began to shift toward Werner's structures. Studies at 0°C gave a very low value for the conductivity of $Co(NH_3)_3Cl_3$, which tended to support Werner's contention that the anomalous conductivity at 25°C was due to the reaction

$$Co(NH_3)_3Cl_3(s) + H_2O \rightarrow [Co(NH_3)_3(H_2O)Cl_2]^+(aq) + Cl^-(aq)$$

Moreover, Werner showed that the compound $Co(NH_3)_3(NO_2)_3$, which is entirely analogous to $Co(NH_3)_3Cl_3$, behaves as a true nonelectrolyte even at 25°C.

In 1907 Werner, after years of effort, finally succeeded in preparing a second isomer of the compound $Co(NH_3)_4Cl_3$. Jorgensen graciously accepted this new evidence as conclusive proof of Werner's structures, and the chain theory of coordination chemistry faded away. Six years later, in 1913, Alfred Werner received the Nobel Prize in chemistry.

Table 21.7
Structure and Properties of the Compounds $Co(NH_3)_xCl_3$

	STRUCTURE		MOLES Cl⁻ PER MOLE COMPOUND		
x	Werner	Jorgensen	Werner	Jorgensen	Observed
6	$[Co(NH_3)_6]^{3+}$, 3 Cl⁻	$Co\begin{smallmatrix}NH_3—Cl\\NH_3—NH_3—NH_3—NH_3—Cl\\NH_3—Cl\end{smallmatrix}$	3	3	3
5	$[Co(NH_3)_5Cl]^{2+}$, 2 Cl⁻	$Co\begin{smallmatrix}Cl\\NH_3—NH_3—NH_3—NH_3—Cl\\NH_3—Cl\end{smallmatrix}$	2	2	2
4	$[Co(NH_3)_4Cl_2]^+$, Cl⁻	$Co\begin{smallmatrix}Cl\\NH_3—NH_3—NH_3—NH_3—Cl\\Cl\end{smallmatrix}$	1	1	1
3	$[Co(NH_3)_3Cl_3]^0$	$Co\begin{smallmatrix}Cl\\NH_3—NH_3—NH_3—Cl\\Cl\end{smallmatrix}$	0	1	?

Summary

A complex ion consists of a central metal cation surrounded by ligands that may be molecules or anions. A ligand may form one or more bonds with the metal ion; if it forms more than one bond, it is called a chelating agent. The charge of the complex is the sum of the charges of the metal ion and the ligands (Example 21.1).

The coordination number of the metal ion is the number of bonds that it forms (Example 21.2). Coordination numbers of 6, 4, and 2, in that order, are most common (Table 21.2). Complexes with a coordination number of 2 are linear. Those with a coordination number of 4 may be square planar or tetrahedral. A coordination number of 6 gives an octahedral complex (Fig. 21.4). Square planar and octahedral complexes can show *cis-trans* isomerism (Example 21.3).

Two different models of bonding in complex ions were presented in this chapter. Both start with the electron configuration of the central cation (Table 21.3).

In the valence bond model, electron pairs from the ligands are fed into hybrid orbitals of the metal ion (sp, sp^3, dsp^2, or d^2sp^3). With this model, it is possible to derive orbital diagrams for complex ions (Example 21.4). In the crystal-field model, the only effect of the ligands is to change the relative energies of the d orbitals of the metal ion. In an octahedral complex, two of these orbitals are raised in energy as compared to the other three. This leads to the possibility of "high-spin" and low-spin complexes, differing in the number of unpaired electrons (Example 21.5).

In considering the reactivity of complex ions, we have to take into account two factors. One is the lability of the complex, which describes the rate at which it undergoes ligand exchange. The other is the formation constant of the complex. The larger the formation complex, the more stable the complex is and the larger the ratio of complex ion to free metal ion concentration (Example 21.6).

Key Words and Concepts

bidentate ligand	formation constant, K_f	octahedral
chelating agent	geometric *(cis, trans)* isomers	orbital diagram
complex ion	high-spin complex	paramagnetic
coordinate covalent bond	hybrid orbital	polydentate ligand
coordination compound	inert complex	spectrochemical series
coordination number	labile complex	square planar
crystal-field model	ligand	tetrahedral
crystal-field splitting energy	low-spin complex	valence bond model
diamagnetic	monodentate ligand	

Questions and Problems

Composition of Complex Ions, Coordination Compounds

21.1 Consider the complex ion $Cr(NH_3)_2(H_2O)_2Br_2^+$.
   a. Identify the ligands and their charges.
   b. What is the charge on the central metal ion?
   c. What is the formula of the sulfate salt of this cation?

21.31 Consider the complex ion $Co(C_2O_4)_2Cl_2^{3-}$.
   a. Identify the ligands and their charges.
   b. What is the charge on the central metal ion?
   c. What would be the formula and charge of the complex if the $C_2O_4^{2-}$ ions were replaced by NH_3 molecules?

21.2 Pt^{2+} forms many complexes, among them those with the following ligands. Give the formula and charge of each Pt^{2+} complex below:
a. two ammonia molecules and two chloride ions.
b. one ethylenediamine molecule and two nitrite ions, NO_2^-.
c. one chloride ion, one bromide ion, and two ammonia molecules.

21.3 If 1.00 mol of each of the following compounds is dissolved in water, how many moles of ions are present in solution?
a. $[Pt(en)Cl_2]$ b. $Na[Cr(en)_2(SO_4)_2]$
c. $K_3[Au(CN)_4]$ d. $[Ni(H_2O)_2(NH_3)_4]Cl_2$

21.4 What is the coordination number of the central metal ion in the following complexes?
a. $Ni(NH_3)_2Br_2$ b. $Fe(CN)_6^{3-}$
c. $Ti(H_2O)Cl_5^{2-}$ d. $Mn(C_2O_4)_3^{4-}$

21.5 Refer to Table 21.2 to predict the formula of the complex formed by
a. Ag^+ with en.   b. Fe^{2+} with H_2O.
c. Zn^{2+} with CN^-.   d. Fe^{3+} with $C_2O_4^{2-}$.

21.6 Classify the following ligands as monodentate, bidentate, etc.:
a. $(CH_3)_3P$

b.

c. $H_2N-(CH_2)_2-NH-(CH_2)_2-NH_2$
d. H_2O

Geometry of Complex Ions

21.7 Sketch the geometry of
a. cis-$Cu(H_2O)_2Br_4^{2-}$ b. $Co(H_2O)_2Cl_2$
c. $trans$-$Ni(NH_3)_2(en)_2^{2+}$ (tetrahedral)
d. cis-$Pt(en)_2Br_2^{2+}$ e. $trans$-$Ni(H_2O)_2Cl_2$

21.8 The compound 1,2-diaminocyclohexane,

(abbreviated "dech") is a ligand in the promising anticancer complex cis-$Pd(H_2O)_2(dech)^{2+}$. Sketch the geometry of this complex.

21.32 Give the charge on the central metal ion in each of the following:
a. VCl_6^{4-} b. $Sc(H_2O)_3Cl_3$
c. $Mn(NO)(CN)_5^{3-}$ d. $Cu(en)_2(NH_3)_2^{2+}$

21.33 For each of the compounds in Question 21.3, state which of the following it would most closely resemble in colligative properties and conductivity: $CO(NH_2)_2$ (urea), KCl, K_2SO_4, or K_3PO_4.

21.34 What is the coordination number of the central metal ion in the following complexes?
a. $Ni(en)_2Cl_2$ b. $Mo(CO)_4Br_2$
c. $Cd(CN)_4^{2-}$ d. $Co(CN)_5(OH)^{3-}$

21.35 Follow the directions for Question 21.5 for complexes formed by
a. Co^{3+} with NH_3.   b. Ag^+ with CN^-.
c. Zn^{2+} with OH^-.   d. Pt^{4+} with en.

21.36 Which of the following would you expect to be effective chelating agents?
a. CH_3CH_2OH
b. $H_2N-(CH_2)_3-NH_2$

c.
d. PH_3

21.37 Sketch the geometry of
a. $Ag(CN)_2^-$ b. $Zn(en)Cl_2$
c. cis-$Ni(H_2O)_2Cl_2$ (tetrahedral)
d. $trans$-$Cr(H_2O)_4Cl_2^+$ e. $Cu(C_2O_4)_3^{4-}$

21.38 The acetylacetonate ion ($acac^-$),

forms complexes with many metal ions. Sketch the geometry of $Fe(acac)_3$.

21.9 Which of the following octahedral complexes show geometric isomerism? If isomers are possible, draw their structures.
a. $Fe(CN)_6^{4-}$
b. $Fe(H_2O)_2Cl_3Br^-$
c. $Fe(H_2O)_2Cl_2Br_2^{2-}$

21.10 How many different octahedral complexes of Co^{3+} can you write using only ethylenediamine and/or Cl^- as ligands?

21.39 Follow the directions of Question 21.9 for
a. $Cr(NH_3)_5(H_2O)^{3+}$
b. $Cr(H_2O)_4Cl_2^+$
c. $Cr(H_2O)_3Cl_3$
d. $Cr(H_2O)_2Cl_4^-$

21.40 Draw as many structural formulas as possible for octahedral complexes in compounds of the formula $Co(NH_3)_4Cl_2Br$.

Electronic Structure of Metal Ions

21.11 Give the electron configuration of
a. Ti^{3+}
b. V^{2+}
c. Ni^{3+}
d. Cu^+
e. Ru^{4+}

21.12 Write an orbital diagram and determine the number of unpaired electrons in each species in Question 21.11.

21.41 Give the electron configuration of
a. Fe^{2+}
b. Cr^{2+}
c. Zn^{2+}
d. Co^{2+}
e. Nb^{2+}

21.42 Write an orbital diagram and determine the number of unpaired electrons for each species in Question 21.41.

Valence Bond Model

21.13 Using the valence bond model, draw orbital diagrams to indicate the electron distribution around the central metal ion in
a. $Fe(H_2O)_6^{3+}$
b. square planar $Ni(CN)_4^{2-}$
c. $Zn(OH)_4^{2-}$ (tetrahedral)
d. $Cr(en)_3^{3+}$

21.14 Use the valence bond model to show why the $CuCl_2^-$ ion is diamagnetic while the $CuCl_4^{2-}$ ion is paramagnetic.

21.15 State the number of unpaired electrons you would expect to find in the tetrahedral complexes of the following ions, using the valence bond model:
a. V^{3+}
b. Ni^{2+}
c. Cu^+

21.16 Using valence bond theory, draw orbital diagrams showing the electron distribution about the central metal ion in each square planar complex listed in Table 21.2.

21.43 Follow the directions of Question 21.13 for the following complexes:
a. $Co(H_2O)_4^{2+}$ (tetrahedral)
b. $Fe(en)Cl_4^-$
c. $Ni(NH_3)_6^{2+}$
d. $Mn(H_2O)_4^{2+}$ (square planar)

21.44 Use the valence bond model and orbital diagrams to account for the fact that square planar Ni^{2+} complexes are diamagnetic while the tetrahedral complexes are paramagnetic.

21.45 State the number of unpaired electrons you would expect to find in the octahedral complexes of the following ions, using the valence bond model:
a. Co^{3+}
b. Cr^{3+}
c. Mn^{2+}
d. Fe^{2+}

21.46 Follow the directions of Question 21.16 for each tetrahedral complex listed in Table 21.2.

Crystal-Field Model

21.17 Using the crystal-field model, give the electron distribution in the low-spin and high-spin complexes of
a. Fe^{2+}
b. Mn^{2+}

21.47 Follow the directions of Question 21.17 for
a. Co^{3+}
b. Mn^{3+}

21.18 V^{3+} does not form high- and low-spin octahedral complexes. Use crystal-field theory to explain why this is so.

21.19 Using the crystal-field model, account for the fact that $Co(NH_3)_6^{3+}$ is diamagnetic while CoF_6^{3-} is paramagnetic.

21.20 Give the number of unpaired electrons in octahedral complexes with strong-field ligands for
a. Mn^{3+} b. Co^{3+} c. Rh^{3+} d. Ti^{2+} e. Mo^{2+}

21.21 $Ti(NH_3)_6^{3+}$ has a d orbital electron transition where ΔE is 300 kJ/mol. Calculate the wavelength (nm) for this transition.

21.48 Using the crystal-field model, explain why Mn^{3+} forms high-spin and low-spin octahedral complexes but Mn^{4+} does not.

21.49 $Cr(CN)_6^{4-}$ is less paramagnetic than $Cr(H_2O)_6^{2+}$. Account for this using electron distribution and crystal-field theory.

21.50 For the species in Question 21.20, indicate the number of unpaired electrons with weak-field ligands.

21.51 The 460-nm absorption in MnF_6^{2-} corresponds to its crystal-field splitting energy, Δ_0. Find Δ_0 in kJ/mol.

Lability, Formation Constants of Complex Ions

21.22 The *cis* and *trans* isomers of $Co(en)_2(H_2O)(NCS)^{2+}$ differ in color. Suggest an experimental technique that might be useful to measure the rate of *cis* to *trans* isomerization.

21.23 For the reaction

$$Co(NH_3)_5Cl^{2+}(aq) + H_2O \rightarrow$$
$$Co(NH_3)_5(H_2O)^{3+}(aq) + Cl^-(aq)$$

the rate expression is

$$\text{rate} = k(\text{conc. } Co(NH_3)_5Cl^{2+})$$

where $k = 2.2 \times 10^{-6}$/s at 25°C. What fraction of the complex has reacted after one day?

21.24 K_f for $Mn(C_2O_4)_2^{2-}$ is 6.3×10^5. Calculate the ratio $[Mn(C_2O_4)_2^{2-}]/[Mn^{2+}]$ in a solution in which $[C_2O_4^{2-}]$ is
a. 0.10 M   b. 1.0×10^{-4} M

21.25 When NH_3 is added to a 0.10 M solution of $AgNO_3$, nearly all of the Ag^+ ions are converted to the $Ag(NH_3)_2^+$ complex ($K_f = 2 \times 10^7$). What do you expect $[Ag(NH_3)_2^+]$ to be in this solution? If $[NH_3]$ is 2.0 M, calculate $[Ag^+]$.

21.26 Use Table 21.6, the reciprocal rule, and the rule of multiple equilibria to calculate K for the reaction

$$CuCl_2^-(aq) + 2\,NH_3(aq) \rightleftharpoons$$
$$Cu(NH_3)_2^+(aq) + 2\,Cl^-(aq)$$

21.52 Describe a simple experiment to demonstrate the lability of $Ni(H_2O)_6^{2+}$ compared to the inertness of $Cr(H_2O)_6^{3+}$.

21.53 Using the information in Question 21.23, calculate the time required for 25% of the complex to react.

21.54 K_f for $Cu(en)_2^+$ is 6.3×10^{10}. Calculate the concentration of en for a $[Cu(en)_2^+]/[Cu^+]$ ratio of
a. 1.0×10^4 b. 5.0×10^6

21.55 When NaCN is added to 0.10 M $AgNO_3$, nearly all of the Ag^+ ions are converted to the extremely stable $Ag(CN)_2^-$ complex ($K_f = 1 \times 10^{21}$). If the equilibrium concentration of CN^- is 1.0×10^{-4} M, estimate $[Ag^+]$.

21.56 Cyanide ions, CN^-, will displace thiosulfate ions, $S_2O_3^{2-}$, from the $Ag(S_2O_3)_2^{3-}$ complex ion. Show that this is true by calculating K for the reaction

$$Ag(S_2O_3)_2^{3-}(aq) + 2\,CN^-(aq) \rightleftharpoons$$
$$Ag(CN)_2^-(aq) + 2\,S_2O_3^{2-}(aq)$$

General

21.27 In your own words, explain why
a. $H_2N(CH_2)_3NH_2$ is a bidentate ligand.
b. AgCl dissolves in NH_3.
c. there are no geometric isomers of tetrahedral complexes.

21.28 Explain in your own words what each of the following terms means when applied to a complex ion:
a. labile   b. stable   c. square planar
d. high spin   e. low spin

21.29 Indicate each of the following statements as true or false. If false, correct the statement to make it true.
a. In $Pt(NH_3)_4Cl_4$, platinum has a +4 charge and a coordination number of 6.
b. Complexes of Cu^{2+} are brightly colored while those of Zn^{2+} are colorless.
c. The K_f value for the octahedral complexes A and B are 1×10^{31} and 1×10^{24}, respectively. This means that complex B is more stable than complex A.

21.30 Analysis of a coordination compound gives the following results: 22.0% Co, 31.4% N, 6.78% H, and 39.8% Cl. One mole of the compound dissociates in water to form four moles of ions.
a. What is the formula of the compound?
b. Write an equation for its dissociation in water.

21.57 Determine whether each of the following is true or false. If the statement is false, correct it.
a. The coordination number of Fe^{3+} in $Fe(H_2O)_4(C_2O_4)^+$ is 5.
b. Cu^+ has two unpaired electrons.
c. $Co(CN)_6^{3-}$ is expected to absorb at a longer wavelength than $Co(NH_3)_6^{3+}$.

21.58 Give two examples of each of the types of complexes referred to in Question 21.28.

21.59 Explain why
a. oxalic acid solution removes rust stains.
b. EDTA is used in plant food.
c. a pale green solution of nickel turns blue when NH_3 is added.
d. the properties of Co^{3+} complexes are easier to study than those of Co^{2+} complexes.

21.60 A chemist synthesizes two coordination compounds. One compound decomposes at 210°C, the other at 240°C. Analysis of the compounds gives the same mass percent data: 52.6% Pt, 7.6% N, 1.63% H, and 38.2% Cl. Both compounds contain a +4 central ion.
a. What is the simplest formula of the compounds?
b. Draw structural formulas for the complexes present.

*21.61 A certain coordination compound has the simplest formula $PtN_2H_6Cl_2$. It has a molar mass of about 600 g/mol and contains both a complex cation and a complex anion. What is its structure?

*21.62 A child eats 10.0 g of paint containing 5.0% Pb. How many grams of the sodium salt of EDTA should he receive to bring the lead into solution?

*21.63 Two coordination compounds decompose at different temperatures but have the same mass percent analysis data: 20.25% Cu, 15.29% C, 7.07% H, 26.86% N, 10.23% S, and 20.39% O. Each contains Cu^{2+}.
a. Determine the simplest formula of the compounds.
b. Draw the structural formulas of the complex ion in each case.

*21.64 Using data in Table 21.6, calculate the standard free energy change at 25°C for the reaction
$Zn(NH_3)_4^{2+}(aq) + 4 OH^-(aq) \rightarrow Zn(OH)_4^{2-}(aq) + 4 NH_3(aq)$.

*21.65 When solid $Al(OH)_3$ is stirred with a concentrated solution of NaOH, it dissolves.
a. Write an equation for the reaction involved.
b. Calculate the equilibrium constant for the reaction.

Chapter 22
Qualitative
Analysis

In previous chapters, we have seen how chemical reactions of different types can serve as a basis for *quantitative analysis*. There we were interested in determining quantitatively how much of a species is present in a solution or solid mixture. Titrations involving precipitation reactions and acid-base reactions are particularly useful in quantitative analysis.

The subject of this chapter is *qualitative analysis*. Here our main goal is to identify each species in a mixture without too much concern for the relative amounts of different species. In particular, we will look at the qualitative analysis of cations in water solution.

Many physical methods can be used to detect ions in solution. Several such methods have been mentioned in previous chapters, including chromatography (Chap. 1) and emission spectroscopy (Chap. 7). The approach discussed in this chapter is based upon chemical reactions. These reactions are used in a standard scheme of analysis designed to test for about 20 common cations in water solution. An advantage of this approach is that it illustrates and reinforces the principles of solution chemistry discussed in the last several chapters. Indeed, a major purpose of this chapter is to review and extend concepts introduced in Chapters 18 through 21.

In Section 22.1, we will see how cations can be separated chemically into four different groups, each containing between three and seven ions. Then (Section 22.2) we will consider in some detail the analysis of one group (Ag^+, Pb^{2+}, and Hg_2^{2+} ions). In Section 22.3 we will look at the different kinds of reactions used in qualitative analysis and the equations written to represent them. This material should be familiar since the reactions are ones discussed in Chapters 18 through 21. In Section 22.4, we will examine solution equilibria in qualitative analysis. Here again, the equilibrium constants used (K_a, K_{sp}, etc.) are ones introduced in previous chapters.

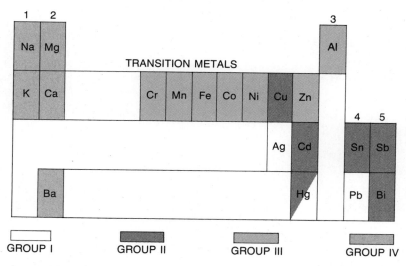

FIGURE 22.1 Metals whose cations are in the standard qualitative analysis scheme (the two cations of mercury, Hg^{2+} and Hg_2^{2+}, are in different groups).

22.1
Analysis of Cations:
An Overview

The metals whose cations are ordinarily included in the qualitative analysis scheme are indicated in Figure 22.1. Notice that they include the more common

—main-group metals (Na, K; Mg, Ca, Ba; Al).
—transition metals (Cr, Mn, Fe, Co, Ni, Cu, Zn; Ag, Cd; Hg).
—post-transition metals (Sn, Pb; Sb, Bi).

This scheme includes 20 cations, all of the common ones

Other metals are ordinarily omitted either because they are too toxic (As, Tl), very expensive (Au, Pt), or relatively rare (Li, Rb, Sr, Ga, and a host of others).

The analysis for cations in a mixture requires a systematic approach. Generally, the procedure is to remove successive groups of cations by precipitation from solution. The standard scheme of qualitative analysis separates cations into four groups (Table 22.1). Concentrations of reagents and solution pH are adjusted so that only one group is affected by a given precipitating agent. The group precipitate is removed from the solution by centrifuging the mixture and pouring the remaining solution from the solid. Within a given group, cations are separated and identified by selective chemical reactions.

Many of the metals listed in Figure 22.1 form cations other than those listed in Table 22.1. For example, tin forms the Sn^{2+} ion as well as Sn^{4+}; iron forms Fe^{3+} as well as Fe^{2+}. The ion listed in Table 22.1 is the one present when the group is separated. Ordinarily, in qualitative analysis we do not attempt to distinguish between different ions of a given metal. For example, we do not try to distinguish between Fe^{2+} and Fe^{3+}.

Table 22.1
Cation Groups of Qualitative Analysis

GROUP	CATIONS	PRECIPITATING REAGENT/CONDITIONS
I	Ag^+, Pb^{2+}, Hg_2^{2+}	6 M HCl
II	Cu^{2+}, Bi^{3+}, Hg^{2+}, Cd^{2+}, Sn^{4+}, Sb^{3+}	0.1 M H_2S at a pH of 0.5
III	Al^{3+}, Cr^{3+}, Co^{2+}, Fe^{2+}, Mn^{2+}, Ni^{2+}, Zn^{2+}	0.1 M H_2S at a pH of 9
IV	Ba^{2+}, Ca^{2+}, Mg^{2+}; Na^+, K^+, NH_4^+	0.2 M $(NH_4)_2CO_3$ at a pH of 9.5 No precipitates with Na^+, K, NH_4^+; separate tests for identification

Group I: Ag^+, Pb^{2+}, Hg_2^{2+}

These three ions are colorless. The Ag^+ and Pb^{2+} ions are the only common cations of silver and lead. The Hg_2^{2+} ion is one of two cations formed by mercury. It has an unusual structure: there is a covalent bond between the two mercury atoms

$$(Hg : Hg)^{2+}$$

Salts containing this cation have formulas such as $Hg_2(NO_3)_2$ and Hg_2Cl_2.

The three ions in Group I are unique among the common cations in one respect: they form insoluble chlorides. Hence, they can be separated from other cations by adding 6 M HCl. The Group I cations precipitate as white chlorides: $AgCl$, $PbCl_2$, and Hg_2Cl_2. These solids can be centrifuged off, leaving a solution that may contain cations in Groups II, III, and IV.

These are the only common insoluble chlorides

Group II: Cu^{2+}, Bi^{3+}, Hg^{2+}, Cd^{2+}, Sn^{4+}, Sb^{3+}

Of these cations, only Cu^{2+} is colored. Dilute solutions of copper(II) salts are most often a pale blue, the color of the $Cu(H_2O)_4^{2+}$ complex ion. Of the other cations in the group, none have partially filled d orbitals. Thus, it is not too surprising that Bi^{3+}, Hg^{2+}, Cd^{2+}, Sn^{4+}, and Sb^{3+} tend to form white salts that dissolve to give colorless solutions.

To separate the ions in Group II from those in Groups III and IV, the H^+ ion concentration is first adjusted to about 0.3 M (pH = 0.5). The solution is then saturated with hydrogen sulfide, a toxic, foul-smelling gas. At the rather high H^+ ion concentration, the equilibrium

$$H_2S(aq) \rightleftharpoons 2\,H^+(aq) + S^{2-}(aq)$$

The Group II sulfides are extremely insoluble

lies far to the left. The concentration of S^{2-} is very low, about 10^{-20} M. However, this is sufficient to precipitate the very insoluble sulfides of the Group II cations, such as CuS ($K_{sp} = 1 \times 10^{-35}$). Several of these sulfides have characteristic colors (Color Plate 22.1). By observing the color of the Group II precipitate and that of the original solution, you can often deduce which ions are likely to be present.

Group III: Al^{3+}, Cr^{3+}, Co^{2+}, Fe^{2+}, Mn^{2+}, Ni^{2+}, Zn^{2+}

Of the cations in this group, only Al^{3+} and Zn^{2+} are colorless. The other five are transition metal ions with incomplete 3d sublevels. Solutions containing Ni^{2+}, Cr^{3+}, and Co^{2+} are most often brightly colored. The $Ni(H_2O)_6^{2+}$ ion imparts a characteristic green color to solutions of nickel(II) salts. Solutions containing chromium(III) and cobalt(II) can have different colors depending upon the nature of the complex ion present:

$Cr(H_2O)_6^{3+}$ $Cr(H_2O)_5Cl^{2+}$ $Co(H_2O)_6^{2+}$ $CoCl_4^{2-}$

   purple   green   red   blue

Qual analysis is often a colorful activity

The Group III cations form sulfides that are more soluble than those of Group II. Hence, they do not precipitate at the very low S^{2-} concentration (10^{-20} M) present at $[H^+] = 0.3$ M. To bring the Group III cations out of solution, $[H^+]$ is reduced to about 10^{-9} M and the solution is saturated with H_2S. Under these conditions, the equilibrium

$$H_2S(aq) \rightleftharpoons 2\,H^+(aq) + S^{2-}(aq)$$

shifts to the right. The concentration of S^{2-} becomes high enough to precipitate Co^{2+}, Fe^{2+}, Mn^{2+}, Ni^{2+}, and Zn^{2+} as the sulfides. Of these sulfides, CoS, FeS, and NiS are black (Color Plate 22.2); MnS, like the Mn^{2+} ion itself, is a pale pink. Zinc sulfide, ZnS, is the only white sulfide in either Group II or III.

Two ions in Group III, Al^{3+} and Cr^{3+}, precipitate as hydroxides rather than sulfides. Aluminum hydroxide, $Al(OH)_3$, and chromium hydroxide, $Cr(OH)_3$, both come down as gelatinous solids; $Al(OH)_3$ is white, while $Cr(OH)_3$ is green.

The sulfides are less stable than the hydroxides under the prevailing conditions

EXAMPLE 22.1 How would you analyze a solution that might contain Ag^+, Sn^{4+}, and Zn^{2+} but no other cations?

Solution Ag^+ is in Group I, Sn^{4+} in Group II, and Zn^{2+} in Group III. First add 6 M HCl; if a precipitate forms, it must be AgCl (white). Then adjust $[H^+]$ to 0.3 M and saturate with H_2S. If Sn^{4+} is present, it should precipitate as SnS_2 (yellow). Finally, adjust the pH to 9 and saturate again with H_2S; Zn^{2+} will precipitate as ZnS (white).

EXERCISE What is the color of the original solution? Answer: Colorless.

Group IV: Ba^{2+}, Ca^{2+}, Mg^{2+}, Na^+, K^+, NH_4^+

All of these cations are colorless in solution. Their salts are typically white and generally more soluble than those of the transition metal ions. The Group IV cations do not form precipitates with Cl^- or H_2S. The alkaline earth cations—

Ba^{2+}, Ca^{2+}, and, to a lesser extent, Mg^{2+}—are precipitated as carbonates in the qualitative analysis scheme. The precipitating solution is buffered at pH 9.5 with NH_4^+ ions and NH_3 molecules.

The three remaining cations in Group IV—Na^+, K^+, and NH_4^+—stay in solution, since their carbonates are soluble. These ions are tested for individually. The tests are carried out on the original solution, rather than one from which Groups I, II, and III have been removed. One reason for this is that reagents added in the separation scheme commonly contain Na^+ and NH_4^+ ions.

The two alkali metal ions, Na^+ and K^+, are determined by flame tests (recall Color Plate 8.1). The Na^+ ion gives an intense yellow color. The test for K^+ (violet) is much weaker. It is often observed through a cobalt-blue glass plate to minimize interference by Na^+.

To test for NH_4^+, a small sample of the original solution is heated with 6 M NaOH. An acid-base reaction occurs between the OH^- ion and the weak acid NH_4^+:

$$NH_4^+(aq) + OH^-(aq) \rightarrow NH_3(aq) + H_2O \tag{22.1}$$

Ammonia gas can be detected by its odor or its basic properties. It is the only common gas that turns moist red litmus blue.

It's not easy to precipitate salts of Na^+, K^+ and NH_4^+

In qual analysis the Groups are removed sequentially from solution

EXAMPLE 22.2 A solution may contain any of the ions in Groups I through IV. Addition of 6 M HCl gives no precipitate. The H^+ concentration is adjusted to 0.3 M and the solution is saturated with H_2S; no precipitate forms. However, when $[H^+]$ is reduced to 10^{-9} M, and the solution is again saturated with H_2S, a precipitate forms. This precipitate is later shown to be a Group III hydroxide. In the Group IV analysis, a precipitate is formed with $(NH_4)_2CO_3$. In this solution, what ions *may* be present from
a. Group I?   b. Group II?   c. Group III?   d. Group IV?

Solution
a. No Group I ions (would precipitate as chlorides).
b. No Group II ions (would precipitate as sulfides).
c. Al^{3+} or Cr^{3+}. These are the only ions in Group III that precipitate as hydroxides in the group separation.
d. Any Group IV ion may be present. The carbonate precipitate could be $BaCO_3$, $CaCO_3$, or $MgCO_3$.

EXERCISE How would you test the original solution for Na^+? NH_4^+? Answer: Flame test for Na^+ (yellow); heat with strong base and test for NH_3 to detect NH_4^+.

22.2
Analysis for Group I

We will not attempt to describe the procedures used to separate and identify all the individual cations listed in Table 22.1. It will, however, be useful to illustrate the general approach followed by considering in some detail the analysis for Group I. Here, only three ions are involved: Ag^+, Pb^{2+}, and Hg_2^{2+}.

A flow sheet for the analysis of the Group I cations is shown in Figure 22.2.

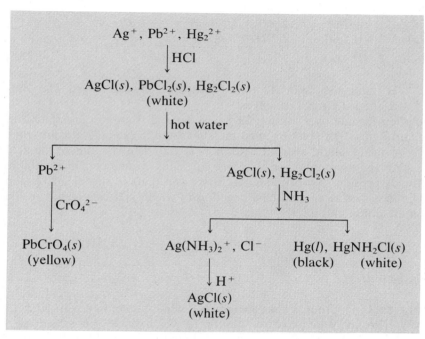

FIGURE 22.2 Flow sheet for the analysis of the Group I cations.

After precipitation with HCl, the next step in the analysis involves heating the solids with water. Lead chloride, $PbCl_2$, is considerably more soluble than either AgCl or Hg_2Cl_2. Its solubility increases with temperature to the extent that it dissolves fairly readily at 100°C:

We keep the procedure simple when we can

$$PbCl_2(s) \rightarrow Pb^{2+}(aq) + 2\ Cl^-(aq) \qquad (22.2)$$

The hot solution is quickly centrifuged to separate it from AgCl or Hg_2Cl_2. The presence of Pb^{2+} in the solution is detected by adding a solution of potassium chromate, K_2CrO_4. A bright yellow precipitate of lead chromate forms if lead is present:

$$Pb^{2+}(aq) + CrO_4{}^{2-}(aq) \rightarrow PbCrO_4(s) \qquad (22.3)$$
$$\text{yellow}$$

Any precipitate remaining after the removal of Pb^{2+} is treated with aqueous ammonia. If Hg_2Cl_2 is present, it undergoes the following reaction with NH_3:

$$Hg_2Cl_2(s) + 2\ NH_3(aq) \rightarrow HgNH_2Cl(s) + Hg(l) + NH_4{}^+(aq) + Cl^-(aq)$$
$$\text{white} \qquad\quad \text{black} \qquad\qquad\qquad (22.4)$$

The gray precipitate formed is a mixture of finely divided mercury, which appears black, and $HgNH_2Cl$, which is white.

The precipitate is often black

The addition of aqueous ammonia brings AgCl into solution. It does this by forming the stable $Ag(NH_3)_2{}^+$ complex ion. We might consider that this reaction occurs in two steps:

$$AgCl(s) \rightarrow Ag^+(s) + Cl^-(aq)$$
$$\underline{Ag^+(aq) + 2\ NH_3(aq) \rightarrow Ag(NH_3)_2{}^+(aq)}$$
$$AgCl(s) + 2\ NH_3(aq) \rightarrow Ag(NH_3)_2{}^+(aq) + Cl^-(aq) \qquad \textbf{(22.5)}$$

In a sense, NH_3 molecules and Cl^- ions "compete" for Ag^+ ions. The NH_3 molecules "win" and the AgCl dissolves.

The solution formed by Reaction 22.5 is separated from any precipitate of $Hg_2{}^{2+}$ by centrifuging. The solution must then be tested to establish the presence of silver. To do this, we add a strong acid, 6 M HNO_3. The H^+ ions of the acid destroy the $Ag(NH_3)_2{}^+$ complex ion. The free Ag^+ ions formed combine with the Cl^- ions already present to precipitate AgCl. We can think of this as a two-step reaction. The first step is an acid-base reaction in which NH_3 molecules in the complex ion are converted to $NH_4{}^+$ ions:

You can often destroy a complex by destroying the ligands in solution

$$Ag(NH_3)_2{}^+(aq) + 2\ H^+(aq) \rightarrow Ag^+(aq) + 2\ NH_4{}^+(aq)$$
$$\underline{Ag^+(aq) + Cl^-(aq) \rightarrow AgCl(s)}$$
$$Ag(NH_3)_2{}^+(aq) + 2\ H^+(aq) + Cl^-(aq) \rightarrow AgCl(s) + 2\ NH_4{}^+(aq) \qquad \textbf{(22.6)}$$

EXAMPLE 22.3   A Group I unknown gives a white precipitate with HCl. Treatment with hot water gives a solution to which K_2CrO_4 is added; a yellow precipitate forms. The white precipitate remaining from the hot water treatment dissolves completely in NH_3. What ions are present? absent?

Solution Pb^{2+} is present ($PbCl_2$ dissolves in hot water and then precipitates $PbCrO_4$). $Hg_2{}^{2+}$ is absent (Hg_2Cl_2 would not dissolve in NH_3). Ag^+ is present (AgCl dissolves in NH_3).

EXERCISE   If HNO_3 is added to the solution formed with NH_3, a white precipitate forms. Write the equation for the reaction involved. Answer: Equation 22.6.

Example 22.3 is typical of "paper unknowns" used to test your understanding of procedures in qualitative analysis. You should realize that, in framing questions of this sort, confirmatory tests are ordinarily not included. Consider, for example, the unknown referred to in Example 22.3. In the laboratory you would ordinarily confirm the presence of silver by adding nitric acid to obtain a white precipitate of AgCl. This information is not included in the body of the example. In principle at least, the unknown must contain Ag^+; otherwise, the white precipitate would not dissolve in NH_3.

22.3
Reactions in
Qualitative Analysis

A wide variety of different reactions are carried out in qualitative analysis. You may add an acid or base to adjust the pH of a solution before separating one group of cations from another; you may oxidize or reduce one cation to separate it from another. Most often, however, you will be carrying out reactions in which

—a complex ion is formed or decomposed.
—a precipitate is formed or dissolved.

Formation of Complex Ions

In one type of reaction in qualitative analysis, a cation in water solution is converted to a complex ion. Most often, the complexing agent (ligand) is an NH_3 molecule or OH^- ion. Typical reactions include

$$Cu^{2+}(aq) + 4\ NH_3(aq) \rightarrow Cu(NH_3)_4^{2+}(aq) \tag{22.7}$$

$$Al^{3+}(aq) + 4\ OH^-(aq) \rightarrow Al(OH)_4^-(aq) \tag{22.8}$$

Cations in the qualitative analysis scheme that commonly form complexes with NH_3 or OH^- are listed in Table 22.2.

Table 22.2
Complexes of Cations with NH_3 and OH^-

CATION	NH_3 COMPLEX	OH^- COMPLEX
Ag^+	$Ag(NH_3)_2^+$	
Pb^{2+}		$*Pb(OH)_3^-$
Cu^{2+}	$Cu(NH_3)_4^{2+}$ (blue)	
Cd^{2+}	$Cd(NH_3)_4^{2+}$	
Sn^{4+}		$Sn(OH)_6^{2-}$
Sb^{3+}		$Sb(OH)_4^-$
Al^{3+}		$Al(OH)_4^-$
Ni^{2+}	$Ni(NH_3)_6^{2+}$ (blue)	
Zn^{2+}	$Zn(NH_3)_4^{2+}$	$Zn(OH)_4^{2-}$

*Most likely composition: $Pb(H_2O)_3(OH)_3^-$

Quite a few cations form NH_3 or OH^- complexes

EXAMPLE 22.4 Write a net ionic equation for the formation of a complex ion when a solution containing Zn^{2+} is treated with a solution of
a. NH_3 b. NaOH

Solution
a. Referring to Table 22.2, we note that the complex ion has the formula $Zn(NH_3)_4^{2+}$. The reactants are a Zn^{2+} ion and four NH_3 molecules, all in water solution:

$$Zn^{2+}(aq) + 4\ NH_3(aq) \rightarrow Zn(NH_3)_4^{2+}(aq)$$

b. $Zn^{2+}(aq) + 4\ OH^-(aq) \rightarrow Zn(OH)_4^{2-}(aq)$
Note that the Na^+ ions take no part in the reaction and therefore are not included in the equation.

EXERCISE Write a net ionic equation for the reaction that occurs when a solution of silver nitrate is treated with ammonia. Answer: $Ag^+(aq) + 2\ NH_3(aq) \rightarrow Ag(NH_3)_2^+(aq)$.

If a complex ion is brightly colored, its formation may serve to identify a particular cation. In Group III, Fe^{3+} can be identified by adding SCN^- ions to form the blood-red complex $Fe(SCN)^{2+}$. The presence of Ni^{2+} is shown by adding

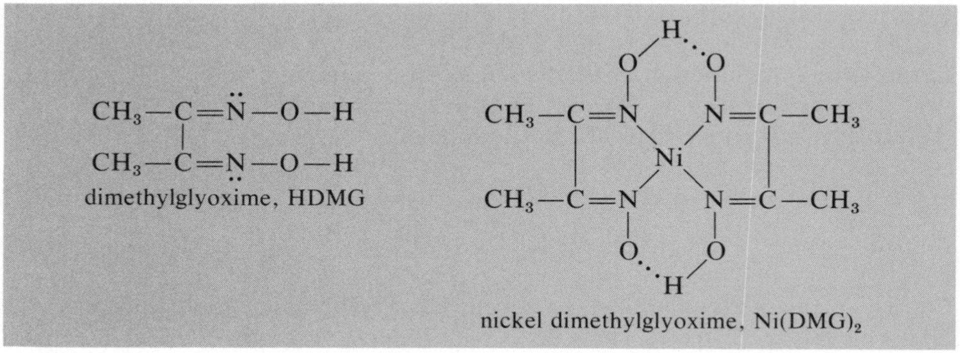

FIGURE 22.3 Structure of dimethylglyoxime (DMG) and its complex with Ni^{2+}.

the chelating agent dimethylglyoxime, which forms a complex (Fig. 22.3) that comes out of solution as a rose-red precipitate.

Decomposition of Complex Ions

Often in qualitative analysis, we want to destroy a complex ion formed in an earlier step. For example, in one step of the analysis for Group III, Al^{3+} is converted to $Al(OH)_4{}^-$. This separates aluminum from other ions in this group. To test for aluminum later on, the $Al(OH)_4{}^-$ complex ion must be converted back to Al^{3+}.

This is one of the reasons we use OH^- or NH_3 complexes

Complexes containing OH^- ligands can be decomposed back to the free cation by adding a strong acid. The H^+ ions of the strong acid react with the OH^- ions, converting them to water molecules. The behavior of the $Al(OH)_4{}^-$ ion is typical:

$$Al(OH)_4{}^-(aq) + 4\ H^+(aq) \rightarrow Al^{3+}(aq) + 4\ H_2O \qquad (22.9)$$

Complexes containing NH_3 ligands are also unstable in strong acid. Thus, we have

$$Ag(NH_3)_2{}^+(aq) + 2\ H^+(aq) \rightarrow Ag^+(aq) + 2\ NH_4{}^+(aq) \qquad (22.10)$$

In general, **complexes containing NH_3 molecules or OH^- ions as ligands are stable in basic solution but decompose in strong acid.**

EXAMPLE 22.5 Write balanced equations for the reactions of H^+ ions with the complexes of Zn^{2+} given in Table 22.2.

Solution With the OH^- complex, reaction with H^+ forms H_2O and the free Zn^{2+} ion:

$$Zn(OH)_4{}^{2-}(aq) + 4\ H^+(aq) \rightarrow Zn^{2+}(aq) + 4\ H_2O$$

The reaction with the ammonia complex is similar, except that $NH_4{}^+$ ions are formed rather than H_2O molecules:

$$Zn(NH_3)_4^{2+}(aq) + 4 H^+(aq) \rightarrow Zn^{2+}(aq) + 4 NH_4^+(aq)$$

EXERCISE Write a balanced equation for the reaction of the hydroxide complex of Sn^{4+} with H^+. Answer: $Sn(OH)_6^{2-}(aq) + 6 H^+(aq) \rightarrow Sn^{4+}(aq) + 6 H_2O$.

Formation of Precipitates

In qualitative analysis, precipitation reactions are more common than any other type. As we have seen, all the group separations of cations involve precipitations of one type or another. For example, the Group I cations, such as Pb^{2+}, are precipitated by adding Cl^- ions:

$$Pb^{2+}(aq) + 2 Cl^-(aq) \rightarrow PbCl_2(s) \tag{22.11}$$

Selective precipitation is a simple way to separate cations

The alkaline earth cations in Group IV are precipitated by adding CO_3^{2-} ions in solution. The reaction with Ba^{2+} is typical:

$$Ba^{2+}(aq) + CO_3^{2-}(aq) \rightarrow BaCO_3(s) \tag{22.12}$$

EXAMPLE 22.6 Write net ionic equations for the precipitation reactions that occur when

a. a solution of Na_2S is added to a qualitative analysis unknown containing Bi^{3+} ions.
b. a solution of NaOH is added to a qualitative analysis unknown containing Fe^{3+} ions.

Solution The products formed are bismuth(III) sulfide (Bi^{3+}, S^{2-} ions) and iron(III) hydroxide (Fe^{3+}, OH^- ions). The principle of electrical neutrality requires that the formulas of these compounds be Bi_2S_3 and $Fe(OH)_3$. The net ionic equations are

a. $2 Bi^{3+}(aq) + 3 S^{2-}(aq) \rightarrow Bi_2S_3(s)$

Just a little review here

b. $Fe^{3+}(aq) + 3 OH^-(aq) \rightarrow Fe(OH)_3(s)$

Note that the Na^+ ions in the Na_2S and NaOH solutions are simply spectator ions and therefore are not included in the equations.

EXERCISE Write balanced equations for the precipitation of Cd^{2+} by OH^-; by S^{2-}. Answer:

$$Cd^{2+}(aq) + 2 OH^-(aq) \rightarrow Cd(OH)_2(s);$$
$$Cd^{2+}(aq) + S^{2-}(aq) \rightarrow CdS(s)$$

In qualitative analysis you often form precipitates in a somewhat more subtle way than that shown in Example 22.6. Consider, for example, the analysis for Cd^{2+} in Group II. Here you precipitate the sulfide, CdS, by adding hydrogen sulfide, H_2S, rather than S^{2-} ions. In this case, two different processes are involved:

dissociation of H_2S: $\qquad H_2S(aq) \rightleftharpoons 2 H^+(aq) + S^{2-}(aq) \tag{22.13a}$

precipitation of CdS: $\qquad Cd^{2+}(aq) + S^{2-}(aq) \rightarrow CdS(s) \tag{22.13b}$

To obtain the net ionic equation for the overall reaction, we add Equations 22.13a and 22.13b to obtain

$$Cd^{2+}(aq) + H_2S(aq) \rightarrow CdS(s) + 2\,H^+(aq) \tag{22.13}$$

A similar situation applies when a water solution of ammonia, NH_3, is used as a source of OH^- ions. Recall (Chap. 19) that ammonia is a weak base. It undergoes the following reaction with water:

$$NH_3(aq) + H_2O \rightleftharpoons NH_4^+(aq) + OH^-(aq) \tag{22.14a}$$

The OH^- ions formed can then precipitate a cation such as Pb^{2+}:

When you get a precipitate with NH_3 it's usually a hydroxide

$$Pb^{2+}(aq) + 2\,OH^-(aq) \rightarrow Pb(OH)_2(s) \tag{22.14b}$$

To obtain the net ionic equation for the reaction of Pb^{2+} with a water solution of ammonia, we multiply Equation 22.14a by 2 and add it to Equation 24.14b:

$$2\,NH_3(aq) + 2\,H_2O \rightleftharpoons 2\,NH_4^+(aq) + 2\,OH^-(aq)$$
$$\underline{Pb^{2+}(aq) + 2\,OH^-(aq) \rightarrow Pb(OH)_2(s)}$$
$$Pb^{2+}(aq) + 2\,NH_3(aq) + 2\,H_2O \rightarrow Pb(OH)_2(s) + 2\,NH_4^+(aq) \tag{22.14}$$

In practice, it is not necessary to add two different equations in precipitation reactions involving H_2S or NH_3 (Example 22.7).

EXAMPLE 22.7 Write net ionic equations for the precipitation reactions that occur when
a. H_2S is added to a solution containing Bi^{3+}.
b. NH_3 is added to a solution containing Fe^{3+}.

Solution The precipitates are the same as those obtained in Example 22.6: Bi_2S_3 and $Fe(OH)_3$. The equations, however, are somewhat more complex.
a. To form one mole of Bi_2S_3, we need two moles of Bi^{3+} and three moles of H_2S. The balanced equation is

$$2\,Bi^{3+}(aq) + 3\,H_2S(aq) \rightarrow Bi_2S_3(s) + 6\,H^+(aq)$$

Note that when H_2S is used to precipitate a cation, H^+ ions appear as a by-product; compare Equation 22.13.
b. To form one mole of $Fe(OH)_3$, we need one mole of Fe^{3+}. Three moles of NH_3 are required to furnish the three moles of OH^- ions needed. The equation is

$$Fe^{3+}(aq) + 3\,NH_3(aq) + 3\,H_2O \rightarrow Fe(OH)_3(s) + 3\,NH_4^+(aq)$$

When NH_3 is used to precipitate cations, NH_4^+ forms as by-product.

EXERCISE Write balanced equations for the precipitation of Tl^+ by H_2S and NH_3. Answer:

$$2\,Tl^+(aq) + H_2S(aq) \rightarrow Tl_2S(s) + 2\,H^+(aq);$$

$$Tl^+(aq) + NH_3(aq) + H_2O \rightarrow TlOH(s) + NH_4^+(aq)$$

It is of interest to compare the equations written in Example 22.6 to those involved in Example 22.7. The cations are the same (Bi^{3+}, Fe^{3+}) in the two cases, as are the precipitates (Bi_2S_3, $Fe(OH)_3$). The sources of the anions are different. In Example 22.6 we worked with electrolyte solutions where S^{2-} and OH^- ions were principal species. Hence, those ions appeared directly in the equations we wrote for the precipitation reactions. In Example 22.7, we used the nonelectrolytes H_2S and NH_3 as a source of the precipitating anions. The principal species in these solutions are molecules, H_2S and NH_3, which appear in the equations as such. *In writing any net ionic equation, we always work with "principal species," i.e., ions or molecules present at high concentrations.*

Dissolving Precipitates

Very often in qualitative analysis, you will need to dissolve a precipitate. You may want to bring an ion into solution so that it can be tested for more readily. Many different methods can be used to do this. In the simplest case, it may be possible to bring a solid into solution by heating with water. This works with $PbCl_2$, which has a solubility of about 0.03 M at 100°C.

Two methods are used more often than any others to dissolve precipitates in qualitative analysis. One of these involves treating the solid with a strong acid, most often with 6 M HCl. The other uses a complexing agent, most often 6 M NH_3 or 6 M NaOH. We will now consider the reactions involved in these methods.

1. **Precipitates containing a basic anion can often be brought into solution with 6 M HCl.** This works with all metal hydroxides in qualitative analysis. The H^+ ions of the HCl combine with the OH^- ions of the solid to form H_2O molecules. The cation of the hydroxide passes into solution. The reaction with zinc hydroxide is typical:

$$Zn(OH)_2(s) + 2\ H^+(aq) \rightarrow Zn^{2+}(aq) + 2\ H_2O \qquad \textbf{(22.15)}$$

K_c for this reaction is very large. Can you evaluate it?

Notice that this equation is very similar to that for the neutralization of a strong base by a strong acid: $H^+(aq) + OH^-(aq) \rightarrow H_2O$. The difference is that the OH^- ions required for the reaction are tied up in the insoluble zinc hydroxide. The principal species is $Zn(OH)_2(s)$, so that is what we put in the equation.

Hydrochloric acid can also be used to dissolve many water-insoluble salts in which the anion is a weak base. In particular, we can use 6 M HCl to dissolve

—*all the carbonates* (CO_3^{2-}) in the qualitative analysis scheme. Here, the product is the weak acid H_2CO_3, which then decomposes into CO_2 and H_2O. The equation for the reaction of H^+ ions with $ZnCO_3$ may be written

$$ZnCO_3(s) + 2\ H^+(aq) \rightarrow Zn^{2+}(aq) + H_2CO_3(aq) \qquad \textbf{(22.16)}$$

CO_2 effervesces in this reaction

—*many sulfides* (S^{2-}). The "driving force" behind this reaction is the formation of the weak acid H_2S. The reaction in the case of zinc sulfide is

$$ZnS(s) + 2\ H^+(aq) \rightarrow Zn^{2+}(aq) + H_2S(aq) \qquad \textbf{(22.17)}$$

This reaction produces a bad odor

The Group III sulfides, including ZnS, dissolve in 6 M HCl, at least when freshly precipitated. In contrast, the Group II sulfides are much less soluble in strong acid.

Reactions 22.16 and 22.17 are strong acid–weak base reactions quite similar to those discussed in Chapter 19. Addition of strong acid to solutions containing the weak bases CO_3^{2-} or S^{2-} would give the reactions

$$CO_3^{2-}(aq) + 2 H^+(aq) \rightarrow H_2CO_3(aq)$$

$$S^{2-}(aq) + 2 H^+(aq) \rightarrow H_2S(aq)$$

Equations 22.16 and 22.17 resemble those just written except that the source of the basic anion is an insoluble solid, $ZnCO_3$ or ZnS, rather than a water solution.

EXAMPLE 22.8 Write net ionic equations to explain why each of the following precipitates dissolves in 6 M HCl:
a. $Al(OH)_3$　　　　　　b. $BaCO_3$　　　　　　c. MnS

Solution

a. The OH^- ions in $Al(OH)_3$ are converted to H_2O molecules by reacting with H^+ ions. Three moles of H^+ ions are required to react with one mole of $Al(OH)_3$:

$$Al(OH)_3(s) + 3 H^+(aq) \rightarrow Al^{3+}(aq) + 3 H_2O$$

b. The CO_3^{2-} ion in $BaCO_3$ is converted to an H_2CO_3 molecule; two H^+ ions are required:

$$BaCO_3(s) + 2 H^+(aq) \rightarrow Ba^{2+}(aq) + H_2CO_3(aq)$$

c. $MnS(s) + 2 H^+(aq) \rightarrow Mn^{2+}(aq) + H_2S(aq)$

H⁺ ion is not a universal solvent, but it is a good one

EXERCISE Would you expect AgCl to dissolve in strong acid? Would $PbSO_4$ react with H^+ ions? Answer: No; neither Cl^- nor SO_4^{2-} is a basic anion.

2. **Precipitates containing a transition metal cation (or Al^{3+}, Sn^{4+}, or Pb^{2+}) can often be brought into solution by adding a complexing agent.** The reagent, most often NH_3 or OH^-, forms a stable complex ion with the cation. This type of reaction is used often in qualitative analysis. For example, as we saw earlier, silver chloride can be brought into solution by treatment with ammonia, forming the $Ag(NH_3)_2^+$ complex (Equation 22.5). In another case, $Zn(OH)_2$ is dissolved by adding 6 M NaOH. The stable complex ion $Zn(OH)_4^{2-}$ is formed:

The complex ion is more stable than the solid

$$Zn(OH)_2(s) + 2 OH^-(aq) \rightarrow Zn(OH)_4^{2-}(aq) \tag{22.18}$$

EXAMPLE 22.9 When a precipitate of $Cd(OH)_2$ is treated with excess 6 M NH_3, it dissolves. Write a net ionic equation for the reaction involved.

Solution The reactants are $Cd(OH)_2(s)$ and $NH_3(aq)$. The products are the $Cd(NH_3)_4^{2+}$ ion (Table 22.2) and free OH^- ions, both in water solution. The net ionic equation for the reaction is

$$Cd(OH)_2(s) + 4 NH_3(aq) \rightarrow Cd(NH_3)_4^{2+}(aq) + 2 OH^-(aq)$$

EXERCISE Write a net ionic equation to explain why $Pb(OH)_2$ dissolves in excess NaOH, using Table 22.2. Answer: $Pb(OH)_2(s) + OH^-(aq) \rightarrow Pb(OH)_3^-(aq)$.

22.4
Solubility Equilibria
in Qualitative Analysis

Every reaction that we deal with in qualitative analysis can be associated with an equilibrium constant. Frequently, the constant is one of those discussed in Chapters 18, 20, or 21. Weak acid dissociation constants, K_a, and solubility product constants, K_{sp}, are perhaps used most often.

Qual analysis is just one equilibrium after another

K_a

Many of the solutions we work with in qualitative analysis are buffered. This keeps the concentration of H^+ or OH^- within narrow limits, allowing certain cations to precipitate and keeping others in solution. The NH_4^+–NH_3 buffer system is perhaps used more often than any other. It can be set up to maintain constant pH in the region 8 to 10 (K_a NH_4^+ = 5.6 × 10^{-10}; pK_a = 9.25).

EXAMPLE 22.10 In precipitating the alkaline earth carbonates in Group IV, it is important to keep the pH constant at about 9.50. An NH_4^+–NH_3 buffer is used to do this. Taking K_a NH_4^+ = 5.6 × 10^{-10}, calculate
a. the ratio $[NH_4^+]/[NH_3]$ required to give a pH of 9.50.
b. the volume of 1.0 M NH_4Cl that must be added to 10 mL of a solution 1.0 M in NH_3 to produce the ratio calculated in (a).

Solution
a. Let us first calculate $[H^+]$ and then the desired ratio:

$$[H^+] = 10^{-9.50} = 3.2 \times 10^{-10} \text{ M}$$

$$[H^+] = K_a \times \frac{[NH_4^+]}{[NH_3]}$$

$$\frac{[NH_4^+]}{[NH_3]} = \frac{[H^+]}{K_a} = \frac{3.2 \times 10^{-10}}{5.6 \times 10^{-10}} = 0.57$$

b. To solve this part of the problem, you must realize that since in the final solution both NH_4^+ and NH_3 are present in the same total volume,

$$\frac{[NH_4^+]}{[NH_3]} = \frac{\text{no. moles } NH_4^+ \text{ in buffer}}{\text{no. moles } NH_3 \text{ in buffer}}$$

The volume units cancel

no. moles NH_3 = 1.0 × 10^{-2} L × 1.0 mol/L = 1.0 × 10^{-2} mol

The number of moles of NH_4^+ ion is directly related to the volume, V, of NH_4Cl solution:

no. moles NH_4^+ = V(in L) × 1.0 mol/L = V mol

Hence,

$$\frac{\text{no. moles } NH_4^+ \text{ in buffer}}{\text{no. moles } NH_3 \text{ in buffer}} = \frac{V}{1.0 \times 10^{-2}} = 0.57$$

Solving for V,

$$V = (0.57)(1.0 \times 10^{-2}) = 5.7 \times 10^{-3} \text{ L} = 5.7 \text{ mL}$$

In other words, adding 5.7 mL of 1.0 M NH_4Cl to 10 mL of 1.0 M NH_3 will give a buffer with a pH of 9.50.

EXERCISE Suppose, by mistake, a student added 6.7 mL of 1.0 M NH_4Cl instead of 5.7 mL. What would be the pH of the buffer? Answer: 9.43.

K_{sp}

Precipitation reactions are more widely used than any other type in qualitative analysis. It is therefore hardly surprising that most equilibrium calculations in this area involve solubility product constants in one way or another. Example 22.11 illustrates one application of K_{sp} values, in this case in Group I analysis.

EXAMPLE 22.11 At 100°C, K_{sp} of $PbCl_2$ is 1.3×10^{-4}.
a. What is the solubility, in moles per liter, of $PbCl_2$ in hot water at 100°C?
b. Will a precipitate of $PbCrO_4$ ($K_{sp} = 1 \times 10^{-16}$) form in Group I analysis if enough K_2CrO_4 is added to a solution saturated with $PbCl_2$ at 100°C to make conc. $CrO_4^{2-} = 0.010$ M?

Solution
a. The solubility equation is

$$PbCl_2(s) \rightleftharpoons Pb^{2+}(aq) + 2\,Cl^-(aq)$$

If we let s be the solubility of $PbCl_2$, then

$$K_{sp}\ PbCl_2 = 1.3 \times 10^{-4} = [Pb^{2+}] \times [Cl^-]^2 = (s) \times (2s)^2 = 4s^3$$

Solving for s,

$$s = \left(\frac{1.3 \times 10^{-4}}{4}\right)^{1/3} = 0.032 \text{ mol/L}$$

This is roughly twice the calculated value at 25°C, which explains why $PbCl_2$ is heated with water to bring it into solution in Group I analysis.

b. (conc. Pb^{2+}) × (conc. CrO_4^{2-}) = (0.032) × (0.010) = 3.2×10^{-4}

Since this product is much larger than K_{sp} of $PbCrO_4$ (1×10^{-16}), we conclude that $PbCrO_4$ should precipitate from the hot saturated solution of $PbCl_2$, as indeed it does.

PbCl₂ is reasonably soluble, even at 25°C

EXERCISE Calculate the solubility of $PbCl_2$ at 25°C, where $K_{sp} = 1.7 \times 10^{-5}$. Answer: 0.016 M.

Relations Between K's

Often the equilibrium constant for a reaction in qualitative analysis is not, by itself, one of those that we have discussed: K_w, K_a, K_b, K_f, or K_{sp}. Instead, we have to derive the equilibrium constant for the reaction using the two general relations introduced in Chapter 20. These are

1. **The reciprocal rule:** $K = 1/K'$ **(22.19)**

where K and K' are the equilibrium constants for forward and reverse reactions.

2. **The rule of multiple equilibria:** $\qquad$ $K_3 = K_1 \times K_2$ $\qquad$ **(22.20)**

where Reaction 3 = Reaction 1 + Reaction 2.

You may recall from Chapter 20 that we used the rule of multiple equilibria to show that, for the system

$$H_2S(aq) \rightleftharpoons 2\,H^+(aq) + S^{2-}(aq)$$

$$K = K_a\,H_2S \times K_a\,HS^- = 1 \times 10^{-20}$$

Using this value for K, we can

—calculate the concentration of S^{2-} ions in Group II analysis (Example 22.12a).
—understand why certain cations precipitate in Group II analysis while others carry over into Group III (Example 22.12b and c).

EXAMPLE 22.12 Taking K for the system $H_2S(aq) \rightleftharpoons 2\,H^+(aq) + S^{2-}(aq)$ to be 1×10^{-20}, determine

a. $[S^{2-}]$ in Group II analysis, where $[H_2S] = 0.1$ M and $[H^+] = 0.3$ M.
b. whether CuS ($K_{sp} = 1 \times 10^{-35}$) will precipitate in Group II, taking conc. $Cu^{2+} = 0.02$ M.
c. whether NiS ($K_{sp} = 1 \times 10^{-19}$) will precipitate in Group II, taking conc. $Ni^{2+} = 0.02$ M.

Solution

a. $\quad K = \dfrac{[H^+]^2 \times [S^{2-}]}{[H_2S]} = 1 \times 10^{-20}$

Solving for $[S^{2-}]$,

$$[S^{2-}] = 1 \times 10^{-20} \times \frac{[H_2S]}{[H^+]^2} = 1 \times 10^{-20} \times \frac{0.1}{(0.3)^2}$$
$$= 1 \times 10^{-20}\ \text{M}$$

b. To determine whether a precipitate forms, we compare the concentration product, P, to K_{sp}:

$$P = (\text{conc. } Cu^{2+}) \times (\text{conc. } S^{2-}) = (2 \times 10^{-2}) \times (1 \times 10^{-20})$$
$$= 2 \times 10^{-22}$$

Since P is greater than K_{sp} of CuS (1×10^{-35}), we conclude that CuS should precipitate. It does; CuS comes down in Group II.

And all of it comes down, 99.9999% of it

c. $P = (\text{conc. } Ni^{2+}) \times (\text{conc. } S^{2-}) = (2 \times 10^{-2}) \times (1 \times 10^{-20})$
$\qquad = 2 \times 10^{-22}$

In this case, P is less than K_{sp} of NiS (1×10^{-19}). Hence, NiS should not precipitate. Fortunately, it doesn't, since Ni^{2+} is in Group III.

EXERCISE The K_{sp} value of a certain sulfide MS is 1×10^{-26}. Would M^{2+} fall in Group II or III? Answer: Group II.

The argument we have just gone through can be applied generally to the cations in Groups II and III. The sulfides of all the Group II cations are very insoluble (low K_{sp} values). This means that they are precipitated by H_2S in Group II, even though the concentration of S^{2-} is very low, about 10^{-20} M. In contrast, the Group III sulfides are more soluble (larger K_{sp} values). As a result, the Group III ions stay in solution when Group II is precipitated.

The rule of multiple equilibria and the reciprocal rule are perhaps most often used to find K for the dissolving of a precipitate. Table 22.3 illustrates this use in connection with the reactions

Qual procedures are actually more empirical than this discussion implies

$$MgCO_3(s) + 2 H^+(aq) \rightleftharpoons Mg^{2+}(aq) + H_2CO_3(aq)$$

$$Mg(OH)_2(s) + 2 H^+(aq) \rightleftharpoons Mg^{2+}(aq) + 2 H_2O$$

Notice that in both cases the calculated K is very large (1×10^9, 1×10^{17}). Small wonder that both $MgCO_3$ and $Mg(OH)_2$ dissolve readily in strong acid! Indeed, we find that this situation applies with virtually all metal hydroxides and carbonates. Almost without exception, the equilibrium constant for their reaction with strong acid is very large. This explains the observation made in Section 22.3 that hydroxides and carbonates can be depended upon to dissolve in 6 M HCl.

Table 22.3
Applications of the Rule of Multiple Equilibria
(and the Reciprocal Rule)

---

1. *Dissolving magnesium carbonate in strong acid (H^+)*

$$MgCO_3(s) \rightleftharpoons Mg^{2+}(aq) + CO_3^{2-}(aq) \qquad K_1 = K_{sp}\ MgCO_3$$

$$CO_3^{2-}(aq) + H^+(aq) \rightleftharpoons HCO_3^-(aq) \qquad K_2 = 1/K_a\ HCO_3^-$$

$$\underline{HCO_3^-(aq) + H^+(aq) \rightleftharpoons H_2CO_3(aq)} \qquad K_3 = 1/K_a\ H_2CO_3$$

$$MgCO_3(s) + 2 H^+(aq) \rightleftharpoons Mg^{2+}(aq) + H_2CO_3(aq) \quad K$$

This approach is useful with any fairly complex reaction

$$K = K_1 \times K_2 \times K_3 = \frac{K_{sp}\ MgCO_3}{K_a\ HCO_3^- \times K_a\ H_2CO_3} = \frac{2 \times 10^{-8}}{2 \times 10^{-17}} = 1 \times 10^9$$

2. *Dissolving magnesium hydroxide in strong acid (H^+)*

$$Mg(OH)_2(s) \rightleftharpoons Mg^{2+}(aq) + 2 OH^-(aq) \qquad K_1 = K_{sp}\ Mg(OH)_2$$

$$H^+(aq) + OH^-(aq) \rightleftharpoons H_2O \qquad K_2 = 1/K_w$$

$$\underline{H^+(aq) + OH^-(aq) \rightleftharpoons H_2O} \qquad K_3 = 1/K_w$$

$$Mg(OH)_2(s) + 2 H^+(aq) \rightleftharpoons Mg^{2+}(aq) + 2 H_2O \qquad K$$

$$K = K_1 \times K_2 \times K_3 = \frac{K_{sp}\ Mg(OH)_2}{(K_w)^2} = \frac{1 \times 10^{-11}}{1 \times 10^{-28}} = 1 \times 10^{17}$$

An approach similar to that shown in Table 22.3 can be used for a reaction in which a precipitate dissolves in a complexing agent. The calculation of K for one such reaction is shown in Example 22.13.

EXAMPLE 22.13 Calculate K for the reaction by which silver chloride dissolves in ammonia:

$$AgCl(s) + 2 NH_3(aq) \rightleftharpoons Ag(NH_3)_2{}^+(aq) + Cl^-(aq)$$

$$(K_{sp} \, AgCl = 1.6 \times 10^{-10}; \qquad K_f \, Ag(NH_3)_2{}^+ = 2 \times 10^7)$$

Solution First we break the reaction down into two steps:

$$AgCl(s) \rightleftharpoons Ag^+(aq) + Cl^-(aq) \qquad\qquad K_1$$

$$\underline{Ag^+(aq) + 2 NH_3(aq) \rightleftharpoons Ag(NH_3)_2{}^+(aq) \qquad\qquad K_2}$$

$$AgCl(s) + 2 NH_3(aq) \rightleftharpoons Ag(NH_3)_2{}^+(aq) + Cl^-(aq) \qquad K = K_1 \times K_2$$

Note that:

$$K_1 = K_{sp} \, AgCl = 1.6 \times 10^{-10}$$

$$K_2 = K_f \, Ag(NH_3)_2{}^+ = 2 \times 10^7$$

Hence,

$$K = (1.6 \times 10^{-10}) \times (2 \times 10^7) = 3 \times 10^{-3}$$

Since K is not big, [NH$_3$] needs to be high to dissolve AgCl (s = 0.3 moles/L in 6 M NH$_3$)

EXERCISE Calculate K for the reaction of AgI ($K_{sp} = 1 \times 10^{-16}$) with NH$_3$. Answer: 2×10^{-9}.

Summary

The qualitative analysis of cations is carried out in the chemistry laboratory by first separating them into groups based on selective precipitation (Table 22.1 and Examples 22.1 and 22.2). The Group I ions precipitate as chlorides ($AgCl$, $PbCl_2$, Hg_2Cl_2). These solids can be separated by taking advantage of the fact that AgCl dissolves in NH_3, whereas $PbCl_2$ dissolves in hot water (Example 22.3).

Many different kinds of reactions are involved in qualitative analysis. Most of these were discussed in Chapters 18 through 21. They include the

—formation of complex ions containing NH_3 or OH^- as ligands (Example 22.4).

—decomposition of complex ions; those containing NH_3 or OH^- are unstable in strong acid (Example 22.5).

—formation of precipitates. Cations can be precipitated by adding anions directly (Example 22.6) or by adding a molecule, such as H_2S or NH_3, that dissociates to form the appropriate anion (Example 22.7).

—dissolving of precipitates. Solids containing a basic anion can be dissolved by adding a strong acid (Example 22.8). Ammonia and sodium hydroxide often bring solids into solution by forming a complex with the cation (Example 22.9).

Equilibrium constants for solution reactions are very useful in qualitative analysis. This is particularly true for K_a (Example 22.10) and K_{sp} (Example 22.11). The reciprocal rule and the rule of multiple equilibria can also be applied (Examples 22.12 and 22.13).

Key Words and Concepts

complex ion	K_{sp}	qualitative analysis
flow sheet	net ionic equation	reciprocal rule
Groups I, II, III, IV	P (conc. product)	rule of multiple equilibria
K_a		

Questions and Problems

General

22.1 Complete the following table for cations.

CATION	ANALYTICAL GROUP	PRECIPITATING AGENT	PPT. FORMED
Pb^{2+}	——	——	——
Cu^{2+}	——	——	——
Cr^{3+}	——	——	——
Ca^{2+}	——	——	——

22.2 Identify the general characteristic of
  a. Group I cations that allows for their separation from cations of Groups II through IV.
  b. Group IV cations that separates them from those of Groups I through III.

22.3 Explain why tests for Na^+, K^+, and NH_4^+ are carried out individually and on the original solution rather than that remaining after testing for Group I to IV cations.

22.4 Explain why the concentration of S^{2-} in a solution of H_2S is pH-dependent. Why is $[H^+]$ kept rather high (0.3 M) in precipitating Group II?

22.5 Give the formula of
  a. an ion in Group I that forms a complex with NH_3.
  b. two Group II cations that form complexes with OH^-.
  c. a weak acid that forms a precipitate with Cu^{2+}.

22.6 What would happen if
  a. an unknown that had been tested for Groups I to III were tested for Na^+?
  b. a solution containing Cu^{2+} and Ni^{2+} were treated with H_2S at pH 9?
  c. acid were added to a solution containing the $Sn(OH)_6^{2-}$ ion?

22.31 Complete the following table for cations.

SPECIES	TEST REAGENT	RESPONSE TO TEST REAGENT
Na^+	flame test	——
Mg^{2+}	——	——
——	hot water	precipitate dissolves
——	H_2S	white ppt. forms

22.32 All Group II and most Group III cations precipitate as sulfides. Why not bring all these ions down as one group?

22.33 Sodium and potassium ions give flame tests but zinc does not. Explain.

22.34 Explain why CuS precipitates in Group II but NiS does not. What would happen if $[H^+]$ were too low in the Group II precipitation?

22.35 Give the formula of
  a. a Group I chloride that gives a precipitate with NH_3.
  b. two ions in Group II that form complexes with NH_3.
  c. two anions that form precipitates with Pb^{2+}.

22.36 Explain why
  a. Mg^{2+} does not precipitate in Groups I, II, or III.
  b. heating a solution of NH_4Cl with NaOH gives a gas with a pungent odor.
  c. gas bubbles form when $BaCO_3$ is treated with a strong acid.

22.7 Consider the Al(OH)$_4$$^-$ ion listed in Table 22.2.
   a. What is the coordination number of Al^{3+} in this complex?
   b. Write an electron configuration for Al^{3+}.
   c. Based on your answer to (b), what is the geometry of Al(OH)$_4$$^-$?

Group I Analysis

22.8 Suppose you are working with a Group I unknown that contains only Ag$^+$ and Pb^{2+}. State precisely what will be observed in each step of the analysis.

22.9 A chloride precipitate contains only one of the cations in Group I. Develop a one-step scheme that will allow you to identify the ion present.

22.10 A student, in analyzing a Group I unknown makes the following errors in procedure. What effect, if any, will they have on the results?
   a. To precipitate the ions, he adds HNO$_3$ instead of HCl.
   b. To the solution obtained by dissolving AgCl in NH$_3$, he adds HCl instead of HNO$_3$.
   c. To a mixed precipitate of AgCl and Hg$_2$Cl$_2$, he adds NaOH instead of NH$_3$.

22.11 Given the following observations on a Group I unknown, indicate which ions are definitely present, which are absent, and which are in doubt. (Note that in problems of this sort, confirmatory tests are usually omitted.) The precipitate formed when HCl is added to the unknown is unaffected by hot water but dissolves completely in 6 M NH$_3$.

22.12 Describe a simple test that would allow you to distinguish between the following:
   a. AgCl(s) and AgNO$_3$(s)
   b. AgCl(s) and PbCl$_2$(s)
   c. Ag$^+$(aq) and Hg$_2$$^{2+}$(aq)
   d. Hg$_2$$^{2+}$(aq) and Pb^{2+}(aq)

22.13 A solid unknown may contain one or more of the following:

AgCl, PbSO$_4$, Hg$_2$Cl$_2$, AgNO$_3$

The solid is stirred with water. A precipitate forms when the solution is treated with HCl. The solid remaining from the first step is treated with 6 M NH$_3$ and turns black; there is no precipitate when the NH$_3$ solution is acidified with HNO$_3$. State which solids are definitely present, which are absent, and which are in doubt.

22.37 In Group II analysis, tin is separated from antimony by the formation of an octahedral complex with the oxalate ion, C$_2$O$_4$$^{2-}$. Draw a sketch of the geometry of this complex.

22.38 Suppose you have a Group I unknown that contains only Pb^{2+} and Hg$_2$$^{2+}$. Describe your observations in each step of the analysis.

22.39 An unknown solution may contain only Ag$^+$ and Hg$_2$$^{2+}$ in Group I. Develop a simple scheme of analysis for these ions.

22.40 What effect, if any, will each of the following errors have on the analysis of a Group I unknown?
   a. The solution obtained by heating PbCl$_2$ with water is allowed to cool to room temperature before being centrifuged.
   b. The solution referred to in (a) is treated with HCl instead of K$_2$CrO$_4$.
   c. The solution obtained by dissolving AgCl in NH$_3$ is treated with NaOH instead of HNO$_3$.

22.41 On the basis of the following observations on a Group I unknown, indicate which ions are definitely present, which are absent, and which are in doubt. The precipitate formed when HCl is added to the unknown is partially soluble in hot water. The residue is partially soluble in 6 M NH$_3$.

22.42 Describe a simple test that would allow you to distinguish between the following:
   a. PbCrO$_4$(s) and PbSO$_4$(s)
   b. AgCl(s) and Hg$_2$Cl$_2$(s)
   c. Ag$^+$(aq) and Pb^{2+}(aq)
   d. Cl$^-$(aq) and OH$^-$(aq)

22.43 A solid unknown may contain any of the following:

PbCl$_2$, Hg$_2$(NO$_3$)$_2$, AgI, PbCO$_3$

The unknown is treated with cold water; none of it dissolves. The mixture is heated to 100°C, and the liquid is drawn off. Treatment of that liquid with K$_2$CrO$_4$ gives a yellow solution but no precipitate. When the original solid is treated with nitric acid, a gas is evolved. State which solids are present, absent, and in doubt.

Reactions and Equations

22.14 Write a net ionic equation for the formation of a complex ion when
a. Ag^+ reacts with ammonia.
b. NaOH is added to a solution of $Pb(NO_3)_2$.
c. NH_3 is added to a solution of $NiCl_2$.

22.15 Write a net ionic equation for the reaction, if any, that occurs when strong acid is added to a solution containing
a. $Al(OH)_4^-$ b. $Ag(NH_3)_2^+$
c. $AgCl_2^-$ d. $Zn(OH)_4^{2-}$

22.16 Write net ionic equations to explain why a precipitate forms when solutions of the following are mixed:
a. Bi^{3+} and OH^- b. Bi^{3+} and S^{2-}
c. Bi^{3+} and NH_3 d. Bi^{3+} and H_2S

22.17 Write net ionic equations for the reaction of H^+ with
a. $Ni(OH)_2$ b. $BaCO_3$ c. $Al(OH)_3$ d. $Sb(OH)_4^-$

22.18 Write a net ionic equation for the reaction with NH_3 by which
a. Zn^{2+} forms a complex ion.
b. $Zn(OH)_2$ dissolves.
c. $Al(OH)_3$ precipitates.

22.19 Write a net ionic equation for the reaction with OH^- by which
a. Al^{3+} forms a complex ion.
b. $Al(OH)_3$ dissolves.
c. Fe^{3+} forms a precipitate.

22.20 Write net ionic equations to explain why the following precipitates dissolve in 6 M HCl:
a. $CaCO_3$ b. $Mg(OH)_2$ c. ZnS

22.21 Write balanced net ionic equations for three reactions by which AgCl is brought into solution through complex ion formation (see Table 21.6).

22.22 Write balanced net ionic equations to explain the following observations:
a. A precipitate forms when solutions of $Pb(NO_3)_2$ and NaCl are mixed.
b. If $PbCl_2$ is washed with hot water, a yellow precipitate forms when K_2CrO_4 is added to the hot water.

22.44 Write a net ionic equation for the formation of a complex ion when
a. a solution of $SnCl_4$ is made strongly basic with NaOH.
b. a solution of $ZnCl_2$ is treated with NH_3.
c. NaOH is added to a solution of $Al(NO_3)_3$.

22.45 Write a net ionic equation for the reaction of each of the following complex ions with H^+:
a. $Zn(NH_3)_4^{2+}$ b. $Pb(OH)_3^-$
c. $Ag(CN)_2^-$ d. $Ni(NH_3)_6^{2+}$

22.46 Write net ionic equations to explain why a precipitate of $Mg(OH)_2$ forms when the following reagents are added to a solution containing Mg^{2+}:
a. NaOH   b. NH_3
c. Na_2S (note that the S^{2-} ion reacts with H_2O to form OH^-)

22.47 Write net ionic equations for the reactions of each of the following with strong acid:
a. $MgCO_3$ b. MnS c. $Fe(OH)_3$ d. $Cu(NH_3)_4^{2+}$

22.48 Write a net ionic equation for the reaction with NH_3 by which
a. AgCl dissolves.
b. Pb^{2+} forms a precipitate.
c. Ni^{2+} forms a complex ion.

22.49 Write a net ionic equation for the reaction with OH^- by which
a. $Mg(OH)_2$ precipitates.
b. Pb^{2+} first forms a precipitate that then dissolves as more OH^- is added.

22.50 Zinc(II) forms the complex ions $Zn(NH_3)_4^{2+}$, $Zn(OH)_4^{2-}$, and $Zn(CN)_4^{2-}$. Write net ionic equations to explain why $ZnCO_3$ dissolves in
a. NH_3 b. NaOH   c. NaCN   d. HCl

22.51 Lead carbonate, $PbCO_3$, can be dissolved by treatment with either strong acid (H^+) or strong base (OH^-). Write balanced equations for the reactions involved.

22.52 Write balanced net ionic equations to describe the following changes:
a. AgCl can be dissolved in 6 M NH_3.
b. When 6 M HNO_3 is added in excess to a solution of AgCl in 6 M NH_3, a white precipitate forms.

Solution Equilibria

22.23 An NH_4^+–NH_3 buffer has a pH of 9.00 (K_a NH_4^+ = 5.6×10^{-10}). Calculate
a. $[H^+]$
b. $[NH_4^+]/[NH_3]$
c. the volume of 0.10 M NH_4^+ that must be added to 5.0 mL of 0.10 M NH_3 to prepare the buffer.

22.24 Determine the pH of a solution made by adding 0.10 mol H_2S (K_a = 1×10^{-7}) to enough water to form a liter of solution.

22.25 A student adds enough HCl to a Group I unknown to make $[Cl^-]$ = 0.50 M. Some $PbCl_2$ precipitates.
a. What is the $[Pb^{2+}]$ remaining (K_{sp} $PbCl_2$ = 1.7×10^{-5})?
b. Under these conditions, will Pb^{2+} precipitate in Group II, where $[S^{2-}]$ is 1×10^{-20} M (K_{sp} PbS = 1×10^{-27})?

22.26 A solution of 10 mL of 0.10 M HCl is mixed with 20 mL of a solution 0.010 M in Pb^{2+}, giving a total volume of 30 mL.
a. What are the concentrations of Cl^- and Pb^{2+} after mixing?
b. Will a precipitate of $PbCl_2$ form (K_{sp} $PbCl_2$ = 1.7×10^{-5})?

22.27 A solution is 0.010 M in H^+ and 0.10 M in H_2S. Taking $[H^+]^2 \times [S^{2-}]/[H_2S]$ = 1×10^{-20}, K_{sp} CdS = 1×10^{-26}, and K_{sp} ZnS = 1×10^{-20}, determine
a. the concentration of S^{2-}.
b. whether CdS will precipitate, taking conc. Cd^{2+} = 0.10 M.
c. whether ZnS will precipitate, taking conc. Zn^{2+} = 0.10 M.

22.28 At what pH does $Mg(OH)_2$ start to precipitate from a solution in which the concentration of Mg^{2+} is 0.020 M?

22.29 Calculate K for the reaction

$$Zn(OH)_2(s) + 2\ OH^-(aq) \rightarrow Zn(OH)_4^{2-}(aq)$$

K_{sp} $Zn(OH)_2$ = 5×10^{-17}
K_f $Zn(OH)_4^{2-}$ = 3×10^{15}

22.53 Work Problem 22.23 for an NH_4^+–NH_3 buffer in which the pH is 10.00.

22.54 Determine the pH of a solution containing 0.10 mol H_2CO_3 (K_a = 4.2×10^{-7}) in 2.50 L of solution.

22.55 At 50°C, K_{sp} $PbCl_2$ is 3.7×10^{-5}.
a. What is the solubility, in moles per liter, of $PbCl_2$ at 50°C?
b. How many grams of $PbCl_2$ can be dissolved by heating with 2.0 g of water at 50°C?

22.56 A Group I unknown is prepared by mixing 2.0-mL portions of three solutions, 0.10 M $AgNO_3$, 0.10 M $Pb(NO_3)_2$, and 0.10 M $Hg_2(NO_3)_2$, to give 6.0 mL of solution.
a. What is the concentration of each cation in the unknown?
b. At what $[Cl^-]$ will $PbCl_2$ start to precipitate from this solution? (K_{sp} $PbCl_2$ = 1.7×10^{-5})

22.57 In Group II precipitation, $[H^+]$ = 0.3 M, $[H_2S]$ = 0.10 M. Under these conditions, show by calculation whether the following ions will precipitate, assuming their original concentrations are 0.10 M.
a. Hg^{2+} (K_{sp} HgS = 1×10^{-52})
b. Fe^{2+} (K_{sp} FeS = 1×10^{-17})

22.58 Sodium hydroxide is added to a solution originally 0.01 M in Mg^{2+} until the pH is 10.0. What fraction of the Mg^{2+} is left in solution?

22.59 Calculate K for the reaction

$$Cu(OH)_2(s) + 4\ NH_3(aq) \rightarrow Cu(NH_3)_4^{2+}(aq) + 2\ OH^-(aq)$$

K_{sp} $Cu(OH)_2$ = 2×10^{-19}
K_f $Cu(NH_3)_4^{2+}$ = 5×10^{12}

22.30 Given that K_{sp} for $Fe(OH)_2$ is 1×10^{-15} and that K_w is 1×10^{-14}, calculate K for the reaction

$$Fe(OH)_2(s) + 2 H^+(aq) \rightarrow Fe^{2+}(aq) + 2 H_2O$$

Would you expect $Fe(OH)_2$ to be more or less soluble than $Mg(OH)_2$ ($K_{sp} = 1 \times 10^{-11}$) in strong acid?

22.60 For $Al(OH)_3$, $K_{sp} = 5 \times 10^{-33}$. Calculate K for the reaction

$$Al(OH)_3(s) + 3 H^+(aq) \rightarrow Al^{3+}(aq) + 3 H_2O$$

Which would be the more soluble in strong acid, $Al(OH)_3$ or $Cr(OH)_3$ ($K_{sp} = 4 \times 10^{-38}$)?

*22.61 A Group III unknown contains only Ni^{2+} and Al^{3+}. It is treated with aqueous ammonia to give a colored precipitate. As more NH_3 is added, part of the precipitate dissolves to give a deep blue solution. The precipitate remaining goes into solution when treated with excess NaOH. If acid is slowly added to this solution, a white precipitate forms that dissolves as more acid is added. Write a balanced net ionic equation for each reaction that takes place.

*22.62 Calculate the solubility, in moles per liter, of AgBr in 6 M NH_3.

*22.63 If a metal hydroxide is to dissolve in strong acid (H^+), K for the reaction must be about 1×10^{-2}. On this basis, calculate the minimum value of K_{sp} for the hydroxide to dissolve if its general formula is
a. MOH   b. $M(OH)_2$ c. $M(OH)_3$

*22.64 $Al(OH)_3$ is precipitated from an NH_3–NH_4^+ buffer in which $[NH_3] = [NH_4^+]$ and orig. conc. $Al^{3+} = 0.1$ M. What fraction of the Al^{3+} originally present remains in solution after precipitation? (K_b $NH_3 = 1.8 \times 10^{-5}$; K_{sp} $Al(OH)_3 = 5 \times 10^{-33}$)

*22.65 Taking K_{sp} $Mn(OH)_2 = 2 \times 10^{-13}$, K_a $HC_2H_3O_2 = 2 \times 10^{-5}$, and $K_w = 1.0 \times 10^{-14}$, calculate K for the reaction

$$Mn(OH)_2(s) + 2 HC_2H_3O_2(aq) \rightleftharpoons Mn^{2+}(aq) + 2 H_2O + 2 C_2H_3O_2^-(aq)$$

Would you expect $Mn(OH)_2$ to be soluble in acetic acid?

In Chapter 4 we discussed briefly a type of reaction known as oxidation-reduction. Such a process, commonly called a "redox reaction," involves an exchange of electrons. The species that *loses electrons* is said to be *oxidized*. The other species, which *gains electrons,* is *reduced*. A simple example of a redox reaction is that which occurs between the elements sodium and chlorine. The product of the reaction is the ionic compound sodium chloride (Na^+, Cl^- ions):

$$2 \; Na(s) \; + \; Cl_2(g) \rightarrow 2 \; NaCl(s) \tag{23.1}$$

Sodium atoms lose electrons and are oxidized to Na^+ ions:

$$\text{oxidation:} \qquad 2 \; Na \rightarrow 2 \; Na^+ \; + \; 2 \; e^- \tag{23.1a}$$

At the same time, chlorine atoms in the Cl_2 molecule gain electrons and are reduced to Cl^- ions:

$$\text{reduction:} \qquad Cl_2 \; + \; 2 \; e^- \rightarrow 2 \; Cl^- \tag{23.1b}$$

In this chapter and in Chapter 24, our main interest will be in redox reactions that take place in water solution. A typical reaction of this type is that between zinc and hydrochloric acid:

$$Zn(s) \; + \; 2 \; H^+(aq) \rightarrow Zn^{2+}(aq) \; + \; H_2(g) \tag{23.2}$$

Here, zinc atoms are oxidized to Zn^{2+} ions:

$$\text{oxidation:} \qquad Zn(s) \rightarrow Zn^{2+}(aq) \; + \; 2 \; e^- \tag{23.2a}$$

while H^+ ions are reduced to H_2 molecules:

$$\text{reduction:} \qquad 2\,H^+(aq) \,+\, 2\,e^- \longrightarrow H_2(g) \qquad\qquad \textbf{(23.2b)}$$

From these two examples, it should be clear that

> — *oxidation and reduction occur together,* in the same reaction; you can't have one without the other.
> — *there is no net change in the number of electrons in a redox reaction.* Those given off in the oxidation half-reaction are taken on by another species in the reduction half-reaction.
> — *electrons appear on the right side of an oxidation half-equation* (Equations 23.1a and 23.2a); *in a reduction half-equation* (Equations 23.1b and 23.2b), *they appear on the left side.*

If something gains electrons, something else must lose them

The two redox reactions we have considered are simple ones; we can readily see what is happening to the electrons. Many redox reactions in water solution are more complex, involving polyatomic molecules or oxyanions. To simplify the electron bookkeeping in such cases, it is helpful to introduce a new concept, oxidation number (Section 23.1). Oxidation numbers can be used to develop a general method of balancing redox equations (Section 23.2). In the latter parts of this chapter, we will consider two different types of electrochemical cells in which redox reactions take place. In one of these, called an electrolytic cell (Section 23.3), electrical energy is used to bring about a nonspontaneous redox reaction. In a voltaic cell (Section 23.4), the reverse process takes place: a spontaneous redox reaction produces electrical energy.

23.1
Oxidation Number

The chemical equation for the reaction of hydrogen with chlorine

$$H_2(g) \,+\, Cl_2(g) \longrightarrow 2\,HCl(g) \qquad\qquad \textbf{(23.3)}$$

These are both redox reactions

resembles that for the reaction of sodium with chlorine

$$2\,Na(s) \,+\, Cl_2(g) \longrightarrow 2\,NaCl(s)$$

Indeed, the two reactions themselves have much in common. In both there is an exchange of electrons between atoms. The major difference lies in the extent to which electrons are transferred. In Reaction 23.3, the product is a gaseous molecule, HCl, rather than a pair of ions (Na^+, Cl^-) in a solid. The valence electron of hydrogen is shared with chlorine instead of being transferred to it. This distinction, however, is one of degree rather than kind. The electrons in the H—Cl covalent bond are displaced strongly toward the more electronegative chlorine. Insofar as "electron bookkeeping" is concerned, we might assign these electrons to the chlorine atom:

$$H \;\Big|\; :\!\overset{\cdot\cdot}{\underset{\cdot\cdot}{Cl}}\!:$$

By assigning electrons in this way we have, in a sense, given a -1 charge to chlorine. It now has one more valence electron (eight) than an isolated chlorine atom (which has seven valence electrons). The hydrogen atom, stripped of its valence electron by this assignment, has in effect acquired a +1 charge.

The accounting system we have just gone through is widely used in inorganic chemistry. The concept of oxidation number is used to refer to the charge an atom would have if the bonding electrons were assigned arbitrarily to the more electronegative element. In the HCl molecule, hydrogen is said to have an oxidation number of +1, and chlorine an oxidation number of −1. In water the bonding electrons are assigned to the more electronegative oxygen atom:

$$H \mid :\ddot{O}: \mid H$$

This gives oxygen an oxidation number of −2 (eight valence electrons vs. six in the neutral atom). A hydrogen atom in water has an oxidation number of +1 (zero valence electrons vs. one in the neutral atom). In a nonpolar covalent bond, the bonding electrons are split evenly between the two atoms:

$$:\ddot{Cl}\cdot \mid \cdot\ddot{Cl}: \qquad \text{oxidation number Cl} = 0$$

We should emphasize that the oxidation number of an atom in a molecule is an artificial concept. Unlike the charge of an ion, oxidation number cannot ordinarily be determined by experiment. The hydrogen atom in the HCl or H_2O molecule does not carry a full positive charge. We might regard its oxidation number of +1 in these molecules as a "pseudocharge."

Oxidation numbers make it easier to analyze redox reactions

Rules for Assigning Oxidation Numbers

In principle, oxidation numbers could be determined in any species by assigning bonding electrons in the manner just described. Such a method would require, however, that we know the Lewis structure of the species. In practice, oxidation numbers are ordinarily obtained in a much simpler way, applying certain arbitrary rules. There are four such rules:

1. **The oxidation number of an element in an elementary substance is 0.** For example, the oxidation number of chlorine in Cl_2 or of phosphorus in P_4 is 0.

2. **The oxidation number of an element in a monatomic ion is equal to the charge of that ion.** In the ionic compound NaCl, sodium has an oxidation number of +1, chlorine an oxidation number of −1. The oxidation numbers of aluminum and oxygen in Al_2O_3 (Al^{3+}, O^{2-} ions) are +3 and −2, respectively.

3. **Certain elements have the same oxidation number in all or almost all their compounds.** The Group 1 metals always exist as +1 ions in their compounds and, hence, are assigned an oxidation number of +1. By the same token, Group 2 elements always have oxidation numbers of +2 in their compounds. Fluorine, the most electronegative of all elements, has an oxidation number of −1 in all of its compounds.

An element with an oxidation no. of +1 is in the +1 oxidation state

Oxygen is ordinarily assigned an oxidation number of −2 in its compounds. (An exception arises in compounds containing the peroxide ion, O_2^{2-}, where the oxidation number of oxygen is −1.)

Hydrogen in its compounds ordinarily has an oxidation number of +1. (The only exception is in metal hydrides such as NaH and CaH_2, where hydrogen is present as the H^- ion, and hence is assigned an oxidation number of −1.)

4. **The sum of the oxidation numbers of all the atoms in a neutral species is 0; in an ion, it is equal to the charge of that ion.** The application of this very useful principle is illustrated in Example 23.1.

EXAMPLE 23.1 What is the oxidation number of sulfur in Na_2SO_4? of manganese in MnO_4^-?

Solution For Na_2SO_4, taking the oxidation number of sodium to be $+1$ and that of oxygen to be -2, we have

$$2(+1) + \text{oxid. no. S} + 4(-2) = 0; \qquad \text{oxid. no. S} = 8 - 2 = +6$$

In the MnO_4^- ion, taking the oxidation number of oxygen to be -2 and realizing that the sum must be -1:

$$\text{oxid. no. Mn} + 4(-2) = -1; \qquad \text{oxid. no. Mn} = +7$$

EXERCISE What is the oxidation number of Cr in Na_2CrO_4? in $Cr_2O_7^{2-}$?
Answer: $+6$

Oxidation and Reduction— A Working Definition

The concept of oxidation number leads directly to a working definition of the terms oxidation and reduction. **Oxidation** is defined as **an increase in oxidation number,** and **reduction** as a **decrease in oxidation number.** Examples include

Reduction reduces oxidation no.

$$2\ Al(s) + 3\ Cl_2(g) \rightarrow 2\ AlCl_3(s) \qquad \begin{array}{l}\text{Al oxidized (oxid. no. } 0 \rightarrow +3)\\ \text{Cl reduced (oxid. no. } 0 \rightarrow -1)\end{array} \qquad \textbf{(23.4)}$$

$$4\ As(s) + 5\ O_2(g) \rightarrow 2\ As_2O_5(s) \qquad \begin{array}{l}\text{As oxidized (oxid. no. } 0 \rightarrow +5)\\ \text{O reduced (oxid. no. } 0 \rightarrow -2)\end{array} \qquad \textbf{(23.5)}$$

These definitions are compatible with our earlier interpretation of oxidation and reduction in terms of loss and gain of electrons. An element that loses electrons must increase in oxidation number. The gain of electrons always results in a decrease in oxidation number. However, defining oxidation and reduction in terms of changes in oxidation number has one distinct advantage. It greatly simplifies the electron bookkeeping in redox reactions. Consider, for example, the reaction

$$HCl(g) + HNO_3(l) \rightarrow NO_2(g) + \tfrac{1}{2}\ Cl_2(g) + H_2O(l) \qquad \textbf{(23.6)}$$

We might also say that HCl is oxidized and HNO_3 is reduced

Analysis in terms of oxidation number reveals that chlorine is oxidized (oxid. no. $= -1$ in HCl, 0 in Cl_2). Nitrogen is reduced (oxid. no. $= +5$ in HNO_3, $+4$ in NO_2). It is much more difficult to decide precisely which atoms are "losing" or "gaining" electrons in this reaction.

Oxidizing and Reducing Agents

In a redox reaction, we usually have at least two reactants. One of these is referred to as the oxidizing agent, another as the reducing agent.
An oxidizing agent brings about the oxidation of another species. To do this,

it must accept electrons from that species. Hence, the oxidizing agent is itself reduced in the reaction.

A reducing agent brings about the reduction of another species. To do this, it must donate electrons to that species. Hence, the reducing agent is itself oxidized in the reaction.

To illustrate what these statements mean, consider Reaction 23.4. Here, Cl_2 is the oxidizing agent. It oxidizes Al $(0 \rightarrow +3)$. The Cl_2 is itself reduced to Cl^- ions $(0 \rightarrow -1)$. The reducing agent is aluminum metal, Al. It reduces Cl_2 molecules to Cl^- ions. In the process, Al is oxidized to Al^{3+} ions.

EXAMPLE 23.2 In Reaction 23.6, what is the oxidizing agent? the reducing agent?

Solution Nitric acid is the oxidizing agent: it oxidizes chlorine from -1 in HCl to 0 in Cl_2. The reducing agent is HCl: it reduces nitrogen from $+5$ in HNO_3 to $+4$ in NO_2. Note that HNO_3 is itself reduced (to NO_2), while HCl is oxidized (to Cl_2).

A good oxidizing agent is easily reduced

EXERCISE What is the oxidizing agent in Reaction 23.5? the reducing agent? Answer: O_2; As.

23.2
Balancing Redox Equations

Many of the equations written to represent redox reactions are simple enough to balance by inspection. This is the case, for example, with Equations 23.1 through 23.5. Frequently, however, you will be working with more complex redox reactions where the coefficients in the balanced equation are by no means obvious. In this section, we will discuss a general approach to balancing such equations. It is called the **half-equation** (or *ion-electron*) method and is perhaps the most straightforward way to balance a variety of redox equations for reactions in water solution.

The first step in this method involves breaking the overall equation down into two half-equations. One of these is an oxidation, the other is a reduction. The two half-equations are balanced separately. Finally, they are combined in such a way as to obtain an overall equation in which there is no net change in the number of electrons. We will now consider several examples of the half-equation method of balancing redox equations.

Tough problems are often best handled by breaking them up into parts

Reaction Between Cr^{3+} and Cl^- Ions

To illustrate the half-equation method, we start with a simple example. This involves the reaction that occurs when a direct electric current is passed through a water solution of chromium(III) chloride, $CrCl_3$, producing the two elements chromium and chlorine. The unbalanced equation for the reaction is

$$Cr^{3+}(aq) + Cl^-(aq) \rightarrow Cr(s) + Cl_2(g)$$

To balance this equation, we proceed as follows:

1. *Split the equation into two half-equations,* one oxidation and one reduction:

reduction: $Cr^{3+}(aq) \rightarrow Cr(s)$ **(1a)**

oxidation: $Cl^-(aq) \rightarrow Cl_2(g)$ **(1b)**

2. *Balance these half-equations, first with respect to atoms and then with respect to charge.* Equation 1a is balanced insofar as atoms are concerned, since there is one atom of Cr on both sides. The charges, however, are unbalanced; the Cr atom on the right has 0 charge while the Cr^{3+} ion on the left has a charge of $+3$. To correct this, we add three electrons to the left of 1a, arriving at

This is the reduction half-equation

$$Cr^{3+}(aq) + 3\ e^- \rightarrow Cr(s) \qquad \textbf{(2a)}$$

Equation 1b must first be balanced with respect to atoms, by providing two Cl^- ions to give one molecule of Cl_2:

$$2\ Cl^-(aq) \rightarrow Cl_2(g)$$

To balance charges, two electrons must be added to the right, giving a charge of -2 on both sides:

This is the oxidation half-equation

$$2\ Cl^-(aq) \rightarrow Cl_2(g) + 2\ e^- \qquad \textbf{(2b)}$$

3. Having arrived at two balanced half-equations, *combine them so as to make the number of electrons gained in reduction equal to the number lost in oxidation.* In Equation 2a, three electrons are gained; in 2b, two electrons are given off. To arrive at a final equation in which no electrons appear, multiply 2a by 2, 2b by 3, and add the resulting half-equations:

$2 \times 2a:$ $2\ Cr^{3+}(aq) + 6e^- \rightarrow 2\ Cr(s)$ **(3a)**

$3 \times 2b:$ $\underline{6\ Cl^-(aq) \qquad\qquad \rightarrow 3\ Cl_2(g) + 6\ e^-}$ **(3b)**

$\qquad\qquad\ 2\ Cr^{3+}(aq) + 6\ Cl^-(aq)\ \rightarrow 2\ Cr(s) + 3\ Cl_2(g)$ **(3)**

Reaction Between MnO_4^- and Fe^{2+} (Acidic Solution)

The equation just worked out was easy to balance because it involved only two species. One of these (Cr^{3+}) was reduced; the other (Cl^-) was oxidized. In many redox reactions, species other than those being reduced or oxidized take part in the reaction. Most commonly, such species contain hydrogen (oxid. no. $= +1$) or oxygen (oxid. no. $= -2$). The presence of ions or molecules of this type makes the redox equation more difficult to balance. However, the half-equation method can still be applied.

To illustrate the approach used, consider the reaction between Fe^{2+} ions and permanganate ions (MnO_4^-) in acidic solution (H^+):

$$MnO_4^-(aq) + Fe^{2+}(aq) \rightarrow Mn^{2+}(aq) + Fe^{3+}(aq) \qquad \text{(acidic solution)}$$

This reaction can be represented by the (unbalanced) equation:

$$MnO_4^-(aq) + H^+(aq) + Fe^{2+}(aq) \rightarrow Mn^{2+}(aq) + Fe^{3+}(aq) + H_2O$$

Note that the two elements that undergo a change in oxidation number are manganese ($+7 \rightarrow +2$) and iron ($+2 \rightarrow +3$). Neither hydrogen nor oxygen changes oxidation number, yet atoms of these elements participate in the reaction. The oxygen atoms originally in the MnO_4^- ion end up in H_2O molecules; the H^+ ions meet the same fate.

To balance this equation, we proceed as follows:

1. Recognize which species undergo oxidation and reduction. Represent these by appropriate half-equations:

oxidation: $Fe^{2+}(aq) \rightarrow Fe^{3+}(aq)$ (1a)

reduction: $MnO_4^-(aq) + H^+(aq) \rightarrow Mn^{2+}(aq) + H_2O$ (1b)

In acidic aqueous solution, H^+ and H_2O can be reactants or products

2. Balance the two half-equations with respect to atoms and charge. Half-equation 1a is balanced by writing

$$Fe^{2+}(aq) \rightarrow Fe^{3+}(aq) + e^-$$ (2a)

To balance half-equation 1b, we first make sure that there are the same number of Mn atoms on both sides—one. Next, oxygen is balanced by using H_2O. In this case a coefficient of 4 in front of H_2O on the right accounts for the four oxygens in an MnO_4^- ion:

$$MnO_4^-(aq) + H^+(aq) \rightarrow Mn^{2+}(aq) + 4\ H_2O$$

To complete the atom balance, the number of hydrogen atoms on the two sides must be equalized. The four H_2O molecules on the right contain eight hydrogen atoms; there must then be eight H^+ ions on the left:

$$MnO_4^-(aq) + 8\ H^+(aq) \rightarrow Mn^{2+}(aq) + 4\ H_2O$$

This half-equation now has a charge of $+2$ on the right and $+7$ on the left ($-1 + 8$). To balance each side with a $+2$ charge, five electrons are added to the left:

$$MnO_4^-(aq) + 8\ H^+(aq) + 5\ e^- \rightarrow Mn^{2+}(aq) + 4\ H_2O$$ (2b)

This half-equation is now completely balanced

Notice that the number of electrons required in this reduction, 5, corresponds to the difference in oxidation number of the manganese on the two sides of the equation. The oxidation number of manganese decreases by 5 units, from $+7$ in the MnO_4^- ion to $+2$ in Mn^{2+}. This relationship always holds for any half-equation. The number of electrons in the balanced half-equation is that required to take the species being oxidized or reduced from its initial to its final oxidation number.

3. Finally, half-equations 2a and 2b are combined so as to eliminate electrons from the final equation. To do this, we multiply 2a by 5 and add to 2b:

$$5 \times 2a: \qquad\qquad 5\ Fe^{2+}(aq) \rightarrow 5\ Fe^{3+}(aq) + 5\ e^- \qquad\qquad \textbf{(3a)}$$

$$2b: \quad MnO_4^-(aq) + 8\ H^+(aq) + 5\ e^- \rightarrow Mn^{2+}(aq) + 4\ H_2O \qquad \textbf{(3b)}$$

$$\overline{MnO_4^-(aq) + 8\ H^+(aq) + 5\ Fe^{2+}(aq) \rightarrow Mn^{2+}(aq) + 4\ H_2O + 5\ Fe^{3+}(aq)}\ \textbf{(3)}$$

It is helpful to compare the procedure we have just gone through for the MnO_4^-–Fe^{2+} reaction to that discussed earlier for the simpler Cr^{3+}–Cl^- reaction. They differ only in the balancing of the half-equations (step 2). This process is more tedious where MnO_4^- is involved, because of the extra elements present (H, O). In general, to balance a redox half-equation involving H^+ ions and H_2O molecules, you proceed as follows:

    a. Balance the atoms of the element being oxidized or reduced.
    b. Balance oxygen, using H_2O molecules.
    c. Balance hydrogen, using H^+ ions.
    d. Balance charge, using e^-.

Reaction Between MnO_4^- and I^- (Basic Solution)

Often we need to write balanced equations for redox reactions taking place in basic solution. Here, we should not have H^+ ions in the final equation; their concentration in basic solution is very small ($< 10^{-7}$ M). Instead, hydrogen in such equations should be in the form of OH^- ions or H_2O molecules. A simple way to accomplish this is to eliminate any H^+ ions appearing in the half-equations, "neutralizing" them by adding an equal number of OH^- ions to both sides. Consider, for example, the oxidation, in basic solution, of iodide by permanganate ions:

This is an easy way to deal with basic solutions

$$I^-(aq) + MnO_4^-(aq) \rightarrow I_2(aq) + MnO_2(s) \qquad \text{(basic solution)}$$

One can proceed exactly as in the foregoing example to obtain the half-equations

$$\text{oxidation:} \qquad 2\ I^-(aq) \rightarrow I_2(aq) + 2\ e^- \qquad\qquad \textbf{(2a)}$$

$$\text{reduction:} \qquad MnO_4^-(aq) + 4\ H^+(aq) + 3\ e^- \rightarrow MnO_2(s) + 2\ H_2O \qquad \textbf{(2b)}$$

The H^+ ions appearing in the reduction half-equation must now be removed to obtain an equation valid in basic solution. To do this, four OH^- ions are added to both sides, and water is formed on the left by combining H^+ with OH^- ions:

$$MnO_4^-(aq) + 4\ H^+(aq) + 3\ e^- \rightarrow MnO_2(s) + 2\ H_2O$$
$$\underline{\qquad\qquad + 4\ OH^-(aq) \qquad\qquad\qquad\qquad\qquad\qquad + 4\ OH^-(aq)}$$
$$MnO_4^-(aq) + 4\ H_2O + 3\ e^- \rightarrow MnO_2(s) + 2\ H_2O + 4\ OH^-(aq)$$

Eliminating two water molecules from each side, we arrive at

$$MnO_4^-(aq) + 2\ H_2O + 3\ e^- \rightarrow MnO_2(s) + 4\ OH^-(aq) \qquad \textbf{(2b')}$$

for the reduction half-reaction in basic solution. To obtain the overall equation, we proceed as before combining 2a and 2b' in such a way as to make the electron gain equal the electron loss:

3 × 2a: $\qquad\qquad\qquad 6\,I^-(aq) \rightarrow 3\,I_2(aq) + \cancel{6}\,e^-$ **(3a)**

2 × 2b': $\quad 2\,MnO_4^-(aq) + 4\,H_2O + \cancel{6}\,e^- \rightarrow 2\,MnO_2(s) + 8\,OH^-(aq)$ **(3b)**

$6\,I^-(aq) + 2\,MnO_4^-(aq) + 4\,H_2O \rightarrow 3\,I_2(aq) + 2\,MnO_2(s) + 8\,OH^-(aq)$ **(3)**

EXAMPLE 23.3   Balance the equation

$$Cl_2(g) \rightarrow Cl^-(aq) + ClO_3^-(aq) \qquad \text{(acidic solution)}$$

Solution   In this case, the same species, Cl_2, is being both oxidized and reduced. Part of the chlorine is reduced to Cl^- (oxid. no. Cl: $0 \rightarrow -1$), while part of it is oxidized to ClO_3^- (oxid. no. Cl: $0 \rightarrow +5$).

1.   reduction: $\quad Cl_2(g) \rightarrow Cl^-(aq)$

   oxidation: $\quad Cl_2(g) \rightarrow ClO_3^-(aq)$

   We now balance these equations in acidic solution, using H^+ ions and H_2O molecules. The reduction half-equation is readily balanced. We need two Cl^- ions to balance chlorine atoms and two electrons on the left to balance charge:

$$Cl_2(g) + 2\,e^- \rightarrow 2\,Cl^-(aq)$$

To balance the oxidation half-equation, we start by balancing Cl atoms:

$$Cl_2(g) \rightarrow 2\,ClO_3^-(aq)$$

To balance oxygen, we add 6 H_2O molecules to the left:

$$Cl_2(g) + 6\,H_2O \rightarrow 2\,ClO_3^-(aq)$$

To balance hydrogen, we add 12 H^+ ions to the right:

$$Cl_2(g) + 6\,H_2O \rightarrow 2\,ClO_3^-(aq) + 12\,H^+(aq)$$

To balance charge (0 on the left, $+10$ on the right at this point), we add 10 e^- to the right

$$Cl_2(g) + 6\,H_2O \rightarrow 2\,ClO_3^-(aq) + 12\,H^+(aq) + 10\,e^-$$

(Note that 10 e^- is the number that must be given off if two Cl atoms are to increase in oxidation number from 0 to $+5$).

2. $Cl_2(g) + 2\,e^- \rightarrow 2\,Cl^-(aq)$

   $Cl_2(g) + 6\,H_2O \rightarrow 2\,ClO_3^-(aq) + 12\,H^+(aq) + 10\,e^-$

   We now combine the half-equations in such a way as to eliminate electrons. To do this, we multiply the reduction half-equation by 5 and add it to the oxidation half-equation:

> Don't panic just because the same species is both oxidized and reduced

3.
$$5 \, Cl_2(g) \, + \, 10 \, e^- \rightarrow 10 \, Cl^-(aq)$$

$$\underline{Cl_2(g) \, + \, 6 \, H_2O \rightarrow 2 \, ClO_3{}^-(aq) \, + \, 12 \, H^+(aq) \, + \, 10 \, e^-}$$

$$6 \, Cl_2(g) \, + \, 6 \, H_2O \rightarrow 10 \, Cl^-(aq) \, + \, 2 \, ClO_3{}^-(aq) \, + \, 12 \, H^+(aq)$$

This equation can be simplified by dividing all the coefficients by two. Doing this, we obtain

$$3 \, Cl_2(g) \, + \, 3 \, H_2O \rightarrow 5 \, Cl^-(aq) \, + \, ClO_3{}^-(aq) \, + \, 6 \, H^+(aq)$$

When asked to balance a redox equation, check to make sure that your final equation is the one with the smallest whole-number coefficients. Sometimes, as here, a simplification is possible.

EXERCISE Write the balanced equation for this reaction in basic solution.
Answer: $3 \, Cl_2(g) \, + \, 6 \, OH^-(aq) \rightarrow 5 \, Cl^-(aq) \, + \, ClO_3{}^-(aq) \, + \, 3 \, H_2O$.

Reactions such as the one considered in Example 23.3, where the same species is both oxidized and reduced, are given a special name. They are called *disproportionation* reactions. In all disproportionations, part of the reactant increases in oxidation number while part decreases in oxidation number.

The procedure used to balance redox equations can be broken down into several simple steps.

1. Divide the overall equation into two half-equations (one involving the element that is oxidized, the other the element that is reduced).

2. Balance each half-equation separately, proceeding in the following order:
   a. Balance the number of atoms of the species whose oxidation number is changing.
   b. Balance oxygen by adding H_2O molecules to one side of the equation.
   c. Balance hydrogen by adding H^+ ions.
   d. Balance charge by adding electrons.

3. Multiply the two half-equations by numbers that will make the electron gain equal the electron loss. Then add the two half-equations, canceling electrons.

4. If asked to balance the equation for a basic solution, do so by adding OH^- ions. Add enough OH^- ions (to both sides) to "neutralize" the H^+ ions, converting them to H_2O molecules. This procedure can be used with the individual half-equations or with the overall equation.

With this kind of problem you can be sure you are right

5. Check your final equation to make sure that

—there are the same number of atoms of each element on both sides.
—the net charge is the same on both sides.
—the coefficients are the simplest whole numbers possible.

Mole Relations in Redox Reactions; Redox Titrations

A balanced redox equation, like any balanced equation, can be used to relate amounts of reactants and products. For reactions in solution, it can be used to relate volumes and concentrations (Example 23.4).

EXAMPLE 23.4 Consider the balanced equation for the reaction of $MnO_4{}^-$ with Fe^{2+} in acidic solution:

$$MnO_4^-(aq) + 5 Fe^{2+}(aq) + 8 H^+(aq) \rightarrow Mn^{2+}(aq) + 5 Fe^{3+}(aq) + 4 H_2O$$

Using this equation, calculate

a. the volume of 0.100 M $KMnO_4$ required to react with 25.0 mL of 0.100 M Fe^{2+}.

b. the concentration of Fe^{2+} in a solution if 20.0 mL of that solution is required to react with 18.0 mL of 0.100 M $KMnO_4$.

Solution

a. From the balanced equation, we see that

$$1 \text{ mol } MnO_4^- \simeq 5 \text{ mol } Fe^{2+}$$

We can readily obtain the number of moles of Fe^{2+}:

$$\text{moles } Fe^{2+} = 0.100 \frac{\text{mol}}{\text{L}} \times 0.0250 \text{ L} = 2.50 \times 10^{-3} \text{ mol}$$

Hence,

$$\text{moles of } MnO_4^- = 2.50 \times 10^{-3} \text{ mol } Fe^{2+} \times \frac{1 \text{ mol } MnO_4^-}{5 \text{ mol } Fe^{2+}}$$

$$= 5.00 \times 10^{-4} \text{ mol } MnO_4^-$$

$$\text{volume 0.100 M } KMnO_4 = 5.00 \times 10^{-4} \text{ mol} \times \frac{1 \text{ L}}{1.00 \times 10^{-1} \text{ mol}}$$

$$= 5.00 \times 10^{-3} \text{ L} = 5.00 \text{ mL}$$

A redox reaction typically goes essentially to completion

b. Here, we proceed in a similar way, except that the "unknown" is the concentration of Fe^{2+} rather than the volume of $KMnO_4$ solution:

$$\text{moles } MnO_4^- = 0.100 \frac{\text{mol}}{\text{L}} \times 0.0180 \text{ L} = 1.80 \times 10^{-3} \text{ mol}$$

$$\text{moles } Fe^{2+} = 1.80 \times 10^{-3} \text{ mol } MnO_4^- \times \frac{5 \text{ mol } Fe^{2+}}{1 \text{ mol } MnO_4^-}$$

$$= 9.00 \times 10^{-3} \text{ mol } Fe^{2+}$$

$$\text{conc. } Fe^{2+} = \frac{9.00 \times 10^{-3} \text{ mol}}{2.00 \times 10^{-2} \text{ L}} = 0.450 \text{ M}$$

EXERCISE   Suppose 20.0 mL of 0.100 M $KMnO_4$ were required to react with 18.0 mL of Fe^{2+}. What would be the concentration of Fe^{2+}? Answer: 0.556 M.

Example 23.4b (and the exercise that follows) shows how a redox reaction can be used to determine the concentration of a reactant, in this case Fe^{2+}. The experimental procedure involved is referred to as a *redox titration*. To carry out this titration, we start with a known volume (for example, 20.0 mL) of an acidified solution containing Fe^{2+} ions. A solution of $KMnO_4$ of known concentration (for example, 0.100 M) is added from a buret. The end point of the titration is easily detected. When all the Fe^{2+} ions are gone, the addition of one or two drops of excess MnO_4^- gives the pink or purple color of that ion.

The MnO_4^- ion is both a reactant and an indicator

23.3
Electrolytic Cells

In an electrolytic cell, a nonspontaneous redox reaction is made to occur by pumping electrical energy into the system. A generalized diagram for such a cell is shown in Figure 23.1. The storage battery at the left provides a source of direct electric current. From the terminals of the battery, two wires lead to the electrolytic cell. This consists of two electrodes, A and C, dipping into a liquid containing ions M^+ and X^-.

The battery acts as an electron pump, pushing electrons into C and removing them from A. To maintain electrical neutrality, some process within the cell must consume electrons at C and liberate them at A. This process is an oxidation-reduction reaction. At electrode C, known as the **cathode,** an ion or molecule undergoes **reduction** by accepting electrons. At the **anode,** A, electrons are produced by the **oxidation** of an ion or molecule. The overall cell reaction is the sum of the two half-reactions at the electrodes. While this reaction is taking place, there is a steady flow of ions to the two electrodes. Positive ions *(cations)* move toward the *cathode;* negative ions *(anions)* move toward the *anode.*

> The electrodes themselves may be oxidized or reduced

The chemical process that takes place when energy is supplied to an electrolytic cell is called *electrolysis.* In Chapter 4, we considered several examples of electrolysis. There, molten ionic compounds were involved. You may recall that when molten sodium chloride is electrolyzed the reactions are

$$
\begin{array}{ll}
\text{cathode:} & 2\,Na^+ + 2\,e^- \rightarrow 2\,Na(l) \\
\text{anode:} & \underline{\qquad 2\,Cl^- \rightarrow Cl_2(g) + 2\,e^-} \\
& 2\,Na^+ + 2\,Cl^- \rightarrow 2\,Na(l) + Cl_2(g)
\end{array}
\qquad \textbf{(23.7)}
$$

> An electrolytic cell converts electrical energy into chemical energy

The products are sodium metal, formed at the cathode, and chlorine gas, produced by oxidation at the anode. Electrolysis of molten compounds is of great industrial importance. Many metals are made that way, including sodium, magnesium, and aluminum.

In this chapter, we will consider electrolysis processes taking place in water solution. It is ordinarily less expensive and more convenient to use a water solution as the electrolyte rather than a molten salt. The products of electrolysis, however, may be quite different in the two cases.

> Here, a voltaic cell, the storage battery, is used to drive an electrolytic cell

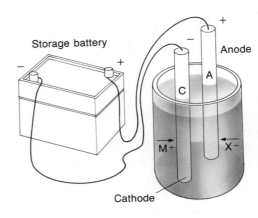

FIGURE 23.1 Schematic diagram of an electrolytic cell. Reduction occurs at the cathode (C). Oxidation occurs at the anode (A). During electrolysis the cations move toward the cathode, the anions toward the anode.

Products of Electrolysis in Water Solution

When a direct electric current is passed through a water solution of an electrolyte, several different half-reactions are possible. At the cathode, three different reductions are most common:

1. *A cation may be reduced to the corresponding metal atom.* This ordinarily occurs when the cation is derived from a transition or post-transition metal. Such cations are relatively easy to reduce. Electrolysis of a water solution of a copper(II) salt deposits copper metal at the cathode:

$$Cu^{2+}(aq) + 2\ e^- \rightarrow Cu(s)$$

Cu^{2+} ion is easily reduced

Similarly, electrolysis of a tin(II) salt forms tin at the cathode:

$$Sn^{2+}(aq) + 2\ e^- \rightarrow Sn(s)$$

2. *H^+ ions may be reduced to H_2 molecules:*

$$2\ H^+(aq) + 2\ e^- \rightarrow H_2(g)$$

This half-reaction occurs at the cathode when a water solution of a strong acid such as HCl or H_2SO_4 is electrolyzed.

3. *H_2O molecules may be reduced, forming H_2 molecules and OH^- ions:*

$$2\ H_2O + 2\ e^- \rightarrow H_2(g) + 2\ OH^-(aq)$$

This half-reaction occurs when the cation in solution is derived from a Group 1 metal (such as Na^+), a Group 2 metal (such as Ca^{2+}), or Al^{3+}. Such cations are very difficult to reduce to the metal. As a result, water molecules are reduced instead.

Some species must be reduced. The one most easily reduced is the logical candidate

At the anode of an electrolytic cell containing a water solution of an electrolyte, three oxidation half-reactions are most common.

1. *A monatomic anion may be oxidized to the corresponding nonmetal atom or molecule.* Such a half-reaction occurs when a water solution containing Cl^- ions is electrolyzed:

$$2\ Cl^-(aq) \rightarrow Cl_2(g) + 2\ e^-$$

A similar half-reaction occurs at the anode when the anion is Br^- or I^-; the products are $Br_2(l)$ or $I_2(s)$.

2. *OH^- ions may be oxidized, forming O_2 molecules and H_2O:*

$$2\ OH^-(aq) \rightarrow \tfrac{1}{2} O_2(g) + H_2O + 2\ e^-$$

This is the half-reaction at the anode when a water solution of a strong base, such as NaOH or $Ba(OH)_2$, is electrolyzed.

3. *H_2O molecules may be oxidized, forming O_2 molecules and H^+ ions:*

$$H_2O \rightarrow \tfrac{1}{2} O_2(g) + 2\ H^+(aq) + 2\ e^-$$

This half-reaction occurs when the anion in the solution is very difficult to oxidize. Among the common anions in this category are F^-, NO_3^-, and SO_4^{2-}. In the electrolysis of water solutions of fluoride, nitrate, or sulfate salts, the product at the anode is oxygen gas.

> **EXAMPLE 23.5** Predict the products that will be obtained and write a balanced equation for the redox reaction involved when water solutions of the following compounds are electrolyzed:
>
> a. HCl   b. $CuSO_4$ c. NaOH
>
> **Solution** In each case, we follow the rules described in the preceding discussion.
>
> a. The ions present in an HCl solution are H^+ and Cl^-. The H^+ ion is reduced while the Cl^- ion is oxidized:
>
> cathode: $\quad\quad\quad\quad\quad 2\,H^+(aq) + 2\,e^- \rightarrow H_2(g)$
> anode: $\quad\quad\quad\quad\quad\quad\quad 2\,Cl^-(aq) \rightarrow Cl_2(g) + 2\,e^-$
> $$\overline{2\,H^+(aq) + 2\,Cl^-(aq) \rightarrow H_2(g) + Cl_2(g)}$$
>
> b. Ions present: Cu^{2+} and SO_4^{2-}. The Cu^{2+} ion is reduced to copper metal. In contrast, SO_4^{2-} cannot be oxidized; H_2O molecules are oxidized instead:
>
> cathode: $\quad Cu^{2+}(aq) + 2\,e^- \rightarrow Cu(s)$
> anode: $\quad\quad\quad\quad\quad H_2O \rightarrow \tfrac{1}{2}\,O_2(g) + 2\,H^+(aq) + 2\,e^-$
> $$\overline{Cu^{2+}(aq) + H_2O \rightarrow Cu(s) + \tfrac{1}{2}\,O_2(g) + 2\,H^+(aq)}$$
>
> c. Ions present: Na^+ and OH^-. The Na^+ ion is very difficult to reduce; H_2O molecules are reduced instead:
>
> cathode: $\quad 2\,H_2O + 2\,e^- \rightarrow H_2(g) + 2\,OH^-(aq)$
> anode: $\quad\quad\quad 2\,OH^-(aq) \rightarrow \tfrac{1}{2}\,O_2(g) + H_2O + 2\,e^-$
> $$\overline{H_2O \rightarrow H_2(g) + \tfrac{1}{2}\,O_2(g)}$$
>
> Note that the overall reaction in this case is simply the electrolysis of water to form hydrogen and oxygen.
>
> **EXERCISE** Write a balanced equation for the overall reaction when a water solution of nickel(II) chloride is electrolyzed. Answer: $Ni^{2+}(aq) + 2\,Cl^-(aq) \rightarrow Ni(s) + Cl_2(g)$.

Ease of oxidation:

$I^- > Br^- > Cl^-$
$\quad > H_2O > SO_4^{2-}$

Ease of reduction:

$Ag^+ > Cu^{2+} > H^+$
$\quad > H_2O > Na^+$

From a commercial standpoint, the most important electrolysis carried out in water solution is that of sodium chloride (Fig. 23.2). Here, the reactions are

cathode: $\quad 2\,H_2O + 2\,e^- \rightarrow H_2(g) + 2\,OH^-(aq)$
anode: $\quad\quad\quad 2\,Cl^-(aq) \rightarrow Cl_2(g) + 2\,e^-$
$$\overline{2\,H_2O + 2\,Cl^-(aq) \rightarrow H_2(g) + Cl_2(g) + 2\,OH^-(aq)} \quad\quad \textbf{(23.8)}$$

Notice that one effect of the cell reaction is to replace Cl^- ions by an equal number of OH^- ions. Evaporation of the solution remaining after electrolysis yields solid sodium hydroxide, mixed with some unreacted sodium chloride. Much of the sodium hydroxide and virtually all the chlorine made in this country are prepared by this process; hydrogen is an important by-product.

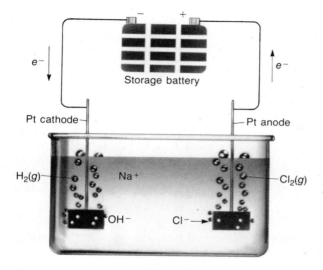

Cl_2 production: 10 million tons/yr

FIGURE 23.2 Electrolysis of an NaCl solution. At the anode, Cl^- ions are oxidized to $Cl_2(g)$. At the cathode, H_2O molecules are reduced to $H_2(g)$ and OH^- ions.

Electroplating

One of the most important applications of electrolytic cells is in the process of electroplating, in which a thin layer of metal (seldom as thick as 0.002 cm) is deposited on an electrically conducting surface. Electroplating is used for many different purposes. Sometimes it can increase the value or improve the appearance of an object, as is the case in gold and silver plating. Chromium plating is designed to provide an attractive shiny surface with improved wearing properties. Metals such as zinc or tin are plated over steel to protect it against corrosion. Some of the components of electrolytic cells used to plate various metals are listed in Table 23.1.

One of the simpler electroplating processes is that for copper (Fig. 23.3). Here, as in most electroplating cells, the metal to be plated is used as the anode (Cu)

Electroplating is an art as well as a science

Table 23.1
Electroplating Processes

METAL	ANODE	ELECTROLYTE	APPLICATION
Cu	Cu	20% $CuSO_4$, 7% H_2SO_4	electrotype
Ag	Ag	4% AgCN, 4% KCN, 4% K_2CO_3	tableware, jewelry
Au	Au, C, Ni-Cr	3% AuCN, 19% KCN, K_2HPO_4 buffer	jewelry
Cr	Pb	25% CrO_3, 0.25% H_2SO_4	automobile parts
Ni	Ni	30% $NiSO_4$, 2% $NiCl_2$, 1% H_3BO_3	base plate for Cr
Zn	Zn	4% $Zn(CN)_2$, 5% NaCN, 8% NaOH, 5% Na_2CO_3	galvanized steel
Sn	Sn	8% H_2SO_4, 7% $SnSO_4$	"tin" cans

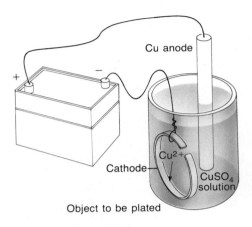

FIGURE 23.3 Electroplating of copper. The anode is made of pure copper to keep the concentration of Cu^{2+} constant. In this sort of cell there is no net reaction, simply the transfer of copper from one place to another.

and the electrolyte contains an ion (Cu^{2+}) derived from that metal. As copper is plated out at the cathode, it goes into solution at the anode

These reactions are also used to purify copper in industry

anode: $Cu(s) \rightarrow Cu^{2+}(aq) + 2\,e^-$

cathode: $Cu^{2+}(aq) + 2\,e^- \rightarrow Cu(s)$

thereby maintaining a constant concentration of Cu^{2+} in the electrolyte solution around the electrodes.

You will notice from Table 23.1 that metal cyanides are used in many electroplating processes. The CN^- ion acts as a complexing agent to lower the concentration of free metal ion. This tends to prevent the cation from plating too rapidly, which would give a rough or brittle coating. In silver plating the electrode reactions are

anode: $Ag(s) + 2\,CN^-(aq) \rightarrow Ag(CN)_2^-(aq) + e^-$

cathode: $Ag(CN)_2^-(aq) + e^- \rightarrow Ag(s) + 2\,CN^-(aq)$

Solutions containing cyanide ions are extremely toxic; many cases of water pollution have arisen from careless discharge of spent electrolyte by electroplating plants.

Quantitative Aspects of Electrolysis

There is a simple relationship between the amount of electricity passed through an electrolytic cell and the amounts of substances produced at the electrodes. The nature of that relationship should be clear from the half-equation for the electrode process:

HALF-EQUATION	QUANTITY OF CHARGE	AMOUNT OF PRODUCT
$Ag^+(aq) + e^- \rightarrow Ag(s)$	1 mol e^-	1 mol Ag = 107.9 g Ag
$Cu^{2+}(aq) + 2\,e^- \rightarrow Cu(s)$	2 mol e^-	1 mol Cu = 63.6 g Cu
$Au^{3+}(aq) + 3\,e^- \rightarrow Au(s)$	3 mol e^-	1 mol Au = 197.0 g Au

In other words, in these half-equations,

> 1 mol e^- ≏ 1 mol Ag (107.9 g Ag)
> 2 mol e^- ≏ 1 mol Cu (63.6 g Cu)
> 3 mol e^- ≏ 1 mol Au (197.0 g Au)

The quantity of electrical charge associated with one mole of electrons can be expressed in terms of the *coulomb* (abbreviated C):

> 1 mol electrons = 96,485 C $\qquad$ **(23.9)**

This means, for example, that

> 96,485 C are required to produce one mole of Ag from Ag^+.
> 2(96,485 C) are required to produce one mole of Cu from Cu^{2+}.
> 3(96,485 C) are required to produce one mole of Au from Au^{3+}.

The constant in Equation 23.9, 96,485 C/mol electrons, is referred to as the *Faraday* constant. This honors Michael Faraday, who first studied the quantitative aspects of electrochemistry over a century ago.

1 Faraday
= 96485 coulombs
= 1 mole e^-

In calculating amounts of substances produced in electrolytic cells, we frequently use a unit of current flow, the ampere (abbreviated A). When a current of one ampere flows through an electrical circuit, one coulomb passes a given point in the circuit in one second. The number of coulombs passing through a cell can be calculated by multiplying the rate of flow in amperes by the elapsed time in seconds:

> no. coulombs = (no. amperes) × (no. seconds) $\qquad$ **(23.10)**

The relations just discussed can be used in many practical calculations dealing with electrolytic cells (Example 23.6).

EXAMPLE 23.6 Chromium metal can be plated from an acidic solution containing chromium(VI) oxide, CrO_3.
a. Write a balanced half-equation for the reduction of CrO_3 to chromium.
b. How many grams of chromium will be plated by 1.00×10^4 C?
c. How long will it take to plate one gram of chromium using a current of 6.00 A?

Solution
a. The unbalanced half-equation is

> $CrO_3(aq) \rightarrow Cr(s)$

Proceeding as indicated in Section 23.2, we first add three H_2O molecules to the right to balance oxygen. This requires that we add 6 H^+ to the left side of the equation to balance hydrogen. Finally, to balance the charge, we add 6 e^- to the left:

> $CrO_3(aq) + 6 H^+(aq) + 6 e^- \rightarrow Cr(s) + 3 H_2O$

Here the electrons are reactants

b. From the coefficients of the balanced half-equation, we see that 6 mol of electrons are required for 1 mol of chromium (52.0 g Cr):

Yet another conversion factor problem

$$6 \text{ mol } e^- \simeq 52.0 \text{ g Cr}$$

To find the number of grams of chromium plated, we need only convert the 1.00×10^4 C given to moles of electrons (1 mol $e^- = 9.6485 \times 10^4$ C) and then to grams of chromium (52.0 g Cr $\simeq 6$ mol e^-):

$$\text{mass Cr} = 1.00 \times 10^4 \text{ C} \times \frac{1 \text{ mol } e^-}{9.6485 \times 10^4 \text{ C}} \times \frac{52.0 \text{ g Cr}}{6 \text{ mol } e^-}$$

$$= 0.898 \text{ g Cr}$$

c. The indicated procedure here is first to convert grams of chromium to moles of electrons and then to coulombs. From the number of coulombs and the number of amperes (6.00), we can readily calculate the time in seconds, using Equation 23.10:

$$\text{no. coulombs} = 1.00 \text{ g Cr} \times \frac{6 \text{ mol } e^-}{52.0 \text{ g Cr}} \times \frac{9.6485 \times 10^4 \text{ C}}{1 \text{ mol } e^-}$$

$$= 1.11 \times 10^4 \text{ C}$$

$$\text{time} = \frac{\text{coulombs}}{\text{amperes}} = \frac{1.11 \times 10^4}{6.00}$$

$$= 1.85 \times 10^3 \text{ s or about 30.8 min}$$

EXERCISE How long does it take to prepare 1.00 g Al by the electrolysis of Al_2O_3, using a current of 1.00 A? Answer: 1.07×10^4 s.

From an economic standpoint, the most important relationship in an electrolytic process is that between the amount of electrical energy absorbed and the amount of product formed. The electrical energy in joules is readily found if the voltage and number of coulombs are known. Since 1 joule (J) equals 1 volt (V) × 1 coulomb (C),

$$\text{no. joules} = (\text{no. volts}) \times (\text{no. coulombs}) \tag{23.11}$$

1 watt = 1 joule/sec

watts = volts
× amperes

The practical unit of electrical energy, the kilowatt-hour (kWh) can be related to the joule. By definition, 1 watt-second (W·s) is 1 joule (J). Hence,

$$1 \text{ kWh} = 1000 \text{ W} \times 3600 \text{ s} \times \frac{1 \text{ J}}{1 \text{ W·s}} = 3.6 \times 10^6 \text{ J} \tag{23.12}$$

EXAMPLE 23.7 Consider the electroplating of chromium, referred to in Example 23.6. If the applied voltage is 4.5 V, calculate the amount of electrical energy absorbed in plating 1.00 g Cr, first in joules and then in kilowatt-hours.

Solution In Example 23.6, we found that 1.11×10^4 C was required to plate one gram of chromium. If the voltage is 4.5 V,

$$\text{no. joules} = (4.5) \times (1.1 \times 10^4) = 5.0 \times 10^4 \text{ J}$$

To find the energy in kilowatt-hours, we use the conversion factor given by Equation 23.12:

$$\text{energy in kWh} = 5.0 \times 10^4 \text{ J} \times \frac{1 \text{ kWh}}{3.6 \times 10^6 \text{ J}} = 1.4 \times 10^{-2} \text{ kWh}$$

EXERCISE Suppose the cost of electrical energy is 6.0¢ per kilowatt-hour. How many grams of chromium can be plated for one dollar? Answer: 1.2×10^3 g.

23.4
Voltaic Cells

As we have noted, many redox reactions are spontaneous. Such reactions can be harnessed to produce electrical energy by carrying them out in a voltaic cell. With such a cell we can light a bulb, run an electric motor, or operate an electrolytic cell. All of us are familiar with certain types of voltaic cells. They include "dry cells" used in flashlights and calculators, and the lead storage battery in an automobile. To understand how a voltaic cell works, we start with some simple cells that are easily made in the general chemistry laboratory.

A voltaic cell converts chemical energy into electrical energy

The Zn–Cu^{2+} Cell
($Zn/Zn^{2+} \parallel Cu^{2+}/Cu$)

When a piece of zinc is added to a water solution containing Cu^{2+} ions, the following redox reaction takes place:

$$Zn(s) + Cu^{2+}(aq) \rightarrow Zn^{2+}(aq) + Cu(s) \tag{23.13}$$

In this reaction, copper metal plates out on the surface of the zinc. The blue color of the aqueous Cu^{2+} ion fades as it is replaced by the colorless aqueous Zn^{2+} ion (Color Plate 23.1). From Equation 23.13, we see that the reaction amounts to electron transfer from a zinc atom to a Cu^{2+} ion.

To design an electrical cell using Reaction 23.13 as a source of electrical energy, the electron transfer must occur indirectly; that is, the electrons given off by zinc atoms must be made to pass through an external electric circuit before they reduce Cu^{2+} ions to copper atoms. One way to do this is shown in Figure 23.4. Let us trace the flow of electric current through this cell.

Otherwise the cell is "shorted out"

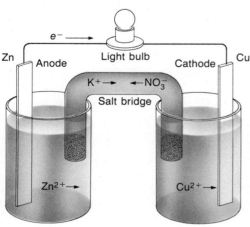

FIGURE 23.4 In this voltaic cell, the following spontaneous redox reaction takes place: $Zn(s) + Cu^{2+}(aq) \rightarrow Zn^{2+}(aq) + Cu(s)$. The salt bridge allows ions to pass from one solution to the other to complete the circuit. At the same time, it prevents direct contact between Zn atoms and Cu^{2+} ions.

1. At the zinc *anode,* electrons are produced by the *oxidation* half-reaction

$$Zn(s) \rightarrow Zn^{2+}(aq) + 2\ e^- \tag{23.13a}$$

Oxidation occurs at the anode. In a voltaic cell the anode is $(-)$

This electrode, which "pumps" electrons into the external circuit, is ordinarily marked as the negative pole of the cell.

2. Electrons generated by Reaction 23.13a move through the external circuit (left to right in Fig. 23.4). This part of the circuit may be a simple resistance wire, a light bulb, an electric motor, an electrolytic cell, or some other device that consumes electrical energy.

3. Electrons pass from the external circuit to the copper *cathode,* where they are used in the *reduction* of Cu^{2+} ions in the surrounding solution:

$$Cu^{2+}(aq) + 2\ e^- \rightarrow Cu(s) \tag{23.13b}$$

The copper electrode, which "pulls" electrons from the external circuit, is considered to be the positive pole of the cell.

4. To complete the circuit, ions must move through the aqueous solutions in the cell. As Reactions 23.13a and 23.13b proceed, a surplus of positive ions (Zn^{2+}) tends to build up around the zinc electrode. The region around the copper electrode tends to become deficient in positive ions as Cu^{2+} ions are consumed. To maintain electrical neutrality, cations must move toward the copper cathode or, alternatively, anions must move toward the zinc anode. In practice, both migrations occur.

It's not easy to make a cell without using a salt bridge

In the cell shown in Figure 23.4, movement of ions occurs through a *salt bridge.* In its simplest form a salt bridge may consist of an inverted U-tube, plugged with glass wool at each end. The tube is filled with a solution of a salt that takes no part in the electrode reactions; potassium nitrate, KNO_3, is frequently used. As current is drawn from the cell, K^+ ions move from the salt bridge into the copper half-cell to compensate for the Cu^{2+} ions consumed at the cathode. At the same time, NO_3^- ions move into the zinc half-cell to compensate for the charge on the Zn^{2+} ions formed at the anode.

The cell shown in Figure 23.4 is often abbreviated as

$$Zn/Zn^{2+} \parallel Cu^{2+}/Cu$$

In this notation,

Anode $\parallel$ Cathode

—the **anode** reaction (**oxidation**) is shown at the left. Zn atoms are oxidized to Zn^{2+} ions.
—the salt bridge (or other means of separating the half-cells) is indicated by the symbol $\parallel$.
—the **cathode** reaction (**reduction**) is shown at the right. Cu^{2+} ions are reduced to Cu atoms.

Other Salt Bridge Cells

Cells similar to that shown in Figure 23.4 can be set up for many different spontaneous redox reactions. Consider, for example, the reaction

$$Ni(s) + Cu^{2+}(aq) \rightarrow Ni^{2+}(aq) + Cu(s) \tag{23.14}$$

For this reaction the voltaic cell would closely resemble that in Figure 23.4. Indeed, the Cu^{2+}/Cu half-cell and the salt bridge would be identical to the one shown. The only difference would be in the oxidation half-cell. Here, we would use a nickel electrode, surrounded by a solution containing Ni^{2+} ions, such as $NiSO_4$ or $Ni(NO_3)_2$. The half-cell reactions would be

anode:	$Ni(s) \rightarrow Ni^{2+}(aq) + 2\ e^-$	(oxidation)
cathode:	$Cu^{2+}(aq) + 2\ e^- \rightarrow Cu(s)$	(reduction)

The anions do not participate in the reaction

The cell notation would be $Ni/Ni^{2+} \parallel Cu^{2+}/Cu$.

Another spontaneous redox reaction that can serve as a source of electrical energy is that between zinc metal and H^+ ions:

$$Zn(s) + 2\ H^+(aq) \rightarrow Zn^{2+}(aq) + H_2(g) \tag{23.15}$$

A voltaic cell using this reaction is shown in Figure 23.5. The Zn/Zn^{2+} half-cell and the salt bridge are the same as those in Figure 23.4. Since no metal is involved in the cathode half-reaction, we use an *inert* electrode; that is, the cathode is made of an unreactive material that conducts an electric current. In this case, it is convenient to use a special electrode made of platinum (graphite rods and Nichrome wires are often used as inert electrodes instead of platinum). Hydrogen gas is bubbled over the Pt electrode, which is surrounded by a solution containing H^+ ions (e.g., a solution of HCl).

The half-reactions occurring in the cell shown in Figure 23.5 are

anode:	$Zn(s) \rightarrow Zn^{2+}(aq) + 2\ e^-$	(oxidation)
cathode:	$2\ H^+(aq) + 2\ e^- \rightarrow H_2(g)$	(reduction)

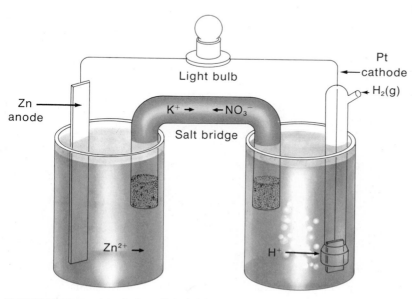

It's possible to have gases produced or consumed in voltaic cell reactions

FIGURE 23.5 A voltaic cell in which the reaction $Zn(s) + 2\ H^+\ (aq) \rightarrow Zn^{2+}\ (aq) + H_2(g)$ occurs. Hydrogen gass is bubbled over a specially prepared platinum electrode, which is surrounded by a solution containing H^+ ions.

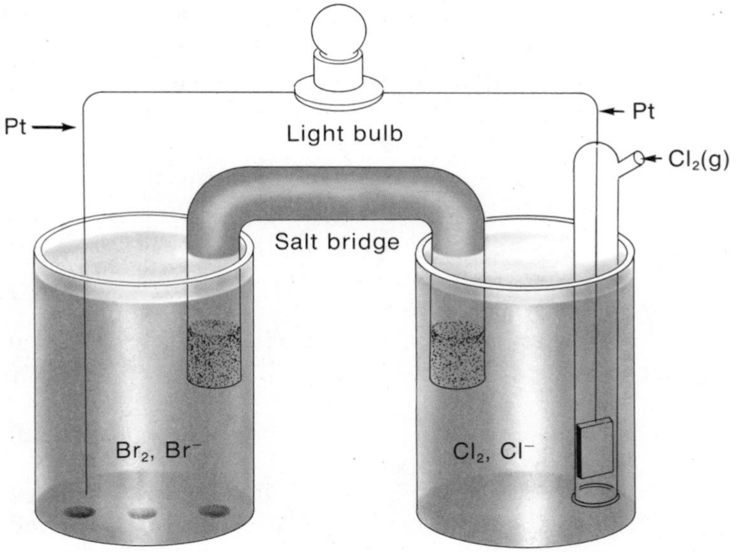

FIGURE 23.6 In this voltaic cell, the spontaneous redox reaction is $Cl_2(g) + 2\ Br^-(aq)$ $\rightarrow 2\ Cl^-(aq) + Br_2(l)$. Both electrodes are made of platinum.

The only problem with Pt electrodes is that they cost about $20/gram

The cell notation is $Zn/Zn^{2+} \parallel (Pt)H^+/H_2$. The symbol (Pt) is used to indicate the presence of an inert platinum cathode.

EXAMPLE 23.8 When chlorine gas is bubbled through a water solution of NaBr, a spontaneous redox reaction occurs:

$$Cl_2(g) + 2\ Br^-(aq) \rightarrow 2\ Cl^-(aq) + Br_2(l)$$

This reaction can serve as a source of electrical energy in the voltaic cell shown in Figure 23.6. In this cell,
a. what is the cathode reaction? the anode reaction?
b. which way do electrons move in the external circuit?
c. which way do anions move within the cell? cations?

Solution
a. cathode: $Cl_2(g) + 2\ e^- \rightarrow 2\ Cl^-(aq)$ (reduction)
   anode: $2\ Br^-(aq) \rightarrow Br_2(l) + 2\ e^-$ (oxidation)
b. From anode to cathode (left to right in Fig. 23.6).
c. Anions move to the anode (right to left); cations move to the cathode (left to right).

EXERCISE What is the notation for the cell shown in Figure 23.6? Answer: $(Pt)Br^-/Br_2 \parallel (Pt)Cl_2/Cl^-$.

Commercial Cells

As we will see in Chapter 24, salt bridge cells yield valuable information about the spontaneity of redox reactions. They have too high an internal resistance, however, to be used commercially. When an appreciable amount of current is drawn from a salt bridge cell, its voltage drops sharply. Commercial voltaic cells are designed to supply large amounts of current, at least for a short time. We will discuss two of the more familiar cells of this type.

DRY CELLS   The construction of the ordinary dry cell (Leclanché cell) used in flashlights is shown in Figure 23.7. The zinc wall of the cell is the anode. The graphite rod through the center of the cell is the cathode. The space between the electrodes is filled with a moist paste. This contains MnO_2, $ZnCl_2$, and NH_4Cl. When the cell operates, the half-reaction at the anode is

$$Zn(s) \rightarrow Zn^{2+}(aq) + 2\ e^- \tag{23.16a}$$

At the cathode, manganese dioxide is reduced to species in which Mn is in the $+3$ oxidation state, such as Mn_2O_3:

Dry cells are used to provide small amounts of electrical energy

$$2\ MnO_2(s) + 2\ NH_4^+(aq) + 2\ e^- \rightarrow Mn_2O_3(s) + 2\ NH_3(aq) + H_2O \tag{23.16b}$$

The overall reaction occurring in this voltaic cell is

$$Zn(s) + 2\ MnO_2(s) + 2\ NH_4^+(aq) \rightarrow Zn^{2+}(aq) + Mn_2O_3(s) + 2\ NH_3(aq) + H_2O \tag{23.16}$$

If too large a current is drawn from a Leclanché cell, the ammonia forms a gaseous insulating layer around the carbon cathode. When this happens the voltage drops sharply, and then returns slowly to its normal value of 1.5 V. This problem can be avoided by using an "alkaline" dry cell, in which the paste between the electrodes contains KOH rather than NH_4Cl. In this case the overall cell reaction is simply

$$Zn(s) + 2\ MnO_2(s) \rightarrow ZnO(s) + Mn_2O_3(s) \tag{23.17}$$

No gas is produced. The alkaline dry cell, although more expensive than the Leclanché cell, is widely used in calculators.

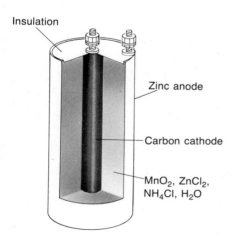

Insulation

Zinc anode

Carbon cathode

MnO_2, $ZnCl_2$, NH_4Cl, H_2O

FIGURE 23.7   Section of an ordinary Zn–MnO_2 dry cell. This cell produces 1.5 V and will deliver a current of about half an ampere for six hours.

STORAGE CELLS   A storage cell, unlike an ordinary dry cell, can be recharged repeatedly. This can be accomplished because the products of the reaction are deposited directly on the electrodes. By passing a current through a storage cell, it is possible to reverse the electrode reactions and restore the cell to its original condition.

The best known voltaic cell of this type is the lead storage battery. The 12-V battery used in automobiles consists of six voltaic cells of the type shown in Figure 23.8. A group of lead plates, the grills of which are filled with spongy gray lead, forms the anode of the cell. The multiple cathode consists of another group of plates of similar design filled with lead(IV) oxide, PbO_2. These two sets of plates alternate through the cell. They are immersed in a water solution of sulfuric acid, H_2SO_4, which acts as the electrolyte.

When a lead storage battery is supplying current, the lead in the anode grids is oxidized to Pb^{2+} ions. These immediately react with SO_4^{2-} ions in the electrolyte, precipitating $PbSO_4$ (lead sulfate) on the plates. At the cathode, lead dioxide is reduced to Pb^{2+} ions, which also precipitate as $PbSO_4$:

$$Pb(s) + SO_4^{2-}(aq) \rightarrow PbSO_4(s) + 2\ e^- \qquad \textbf{(23.18a)}$$
$$PbO_2(s) + 4\ H^+(aq) + SO_4^{2-}(aq) + 2\ e^- \rightarrow PbSO_4(s) + 2\ H_2O \qquad \textbf{(23.18b)}$$
$$\overline{Pb(s) + PbO_2(s) + 4\ H^+(aq) + 2\ SO_4^{2-}(aq) \rightarrow 2\ PbSO_4(s) + 2\ H_2O} \qquad \textbf{(23.18)}$$

Deposits of lead sulfate slowly build up on the plates, partially covering and replacing the lead and lead dioxide. As the cell discharges, the concentration of sulfuric acid decreases. For every mole of lead reacting, two moles of H_2SO_4

A 12-V lead storage battery can supply 300 amperes for a minute or so, about 5 horse-power

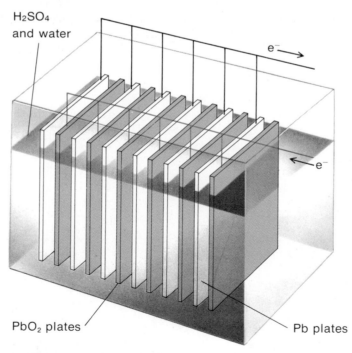

FIGURE 23.8   One cell of a lead storage battery. Two advantages of the lead storage battery are its ability to deliver large amounts of energy for a short time and its ability to be recharged. A disadvantage is its high mass/energy ratio.

(4 H$^+$, 2 SO$_4^{2-}$) are replaced by two moles of water. The state of charge of a storage battery can be checked by measuring the density of the electrolyte. When fully charged, the density is in the range of 1.25 to 1.30 g/cm^3. A density below 1.20 g/cm^3 indicates a low sulfuric acid concentration and hence a partially discharged cell.

A lead storage battery can be recharged and thus restored to its original condition. To do this, a direct current is passed through the cell in the reverse direction. While a storage battery is being charged, it acts as an electrolytic cell. The half-reactions 23.18a and 23.18b are reversed:

$$2 \; PbSO_4(s) + 2 \; H_2O \rightarrow Pb(s) + PbO_2(s) + 4 \; H^+(aq) + 2 \; SO_4^{2-}(aq)$$

The electrical energy required to bring about this nonspontaneous reaction in an automobile is furnished by an alternator equipped with a rectifier to convert alternating to direct current.

As you may have found from experience, lead storage batteries do not endure forever, particularly if they are allowed to stand for some time when discharged. Repeated quick-charging can cause Pb, PbO$_2$, and PbSO$_4$ to flake off the electrodes. This collects as a sludge at the bottom of the battery, often short-circuiting one or more cells. Discharged batteries are also susceptible to freezing, since the sulfuric acid concentration is low. If freezing occurs, the electrodes may warp and come in contact with one another.

Another type of rechargeable voltaic cell is the "nicad" storage battery, used for small appliances, tools, and calculators. The anode in this cell is made of cadmium metal and the cathode contains nickel(IV) oxide, NiO$_2$. The electrolyte is a concentrated solution of potassium hydroxide. The discharge reactions are

anode:	$Cd(s) + 2 \; OH^-(aq) \rightarrow Cd(OH)_2(s) + 2 \; e^-$ **(23.19a)**
cathode:	$NiO_2(s) + 2 \; H_2O + 2 \; e^- \rightarrow Ni(OH)_2(s) + 2 \; OH^-(aq)$ **(23.19b)**
	$\overline{Cd(s) + NiO_2(s) + 2 \; H_2O \rightarrow Cd(OH)_2(s) + Ni(OH)_2(s)}$ **(23.19)**

In the nicad battery, the Ni in NiO$_2$ is in the relatively rare +4 oxidation state

The insoluble hydroxides of cadmium and nickel deposit on the electrodes. Hence, the half-reactions are readily reversed during recharging. Nicad batteries are more expensive than lead storage batteries for a given amount of electrical energy delivered but also have a longer life.

23.5
Historical
Perspective

Michael Faraday
(1791–1867)

In the first half of the nineteenth century, the physical sciences were dominated by a series of remarkable Englishmen. In 1808, John Dalton (1766–1844) proposed his atomic theory of matter. At about the same time, Sir Humphry Davy (1778–1829) isolated six new chemical elements (Na, K, Mg, Ca, Sr, and Ba) by electrolysis of their molten carbonates or oxides. Later on, two

Perhaps the best of all time

English physicists, James Joule (1818–1889) and Lord Kelvin (1824–1907), laid the foundation for the laws of thermodynamics.

Probably the greatest experimental scientist of the nineteenth century and certainly the most prolific was Michael Faraday, who was born and lived all his life in what is now greater London. The son of a blacksmith, he had no formal education beyond the rudiments of reading, writing, and arithmetic. Apprenticed to a bookbinder at the age of 13, Faraday educated himself by reading virtually every book that came into the shop. One that particularly impressed him was a textbook, *Conversations in Chemistry,* written by Mrs. Jane Marcet. Within a few years he was carrying out simple experiments in his home laboratory and attending lectures given by Davy at the Royal Institution. Anxious to escape a life of drudgery as a tradesman, Faraday wrote out a copy of these lectures and submitted it to Davy with a request for employment. Shortly afterward, a vacancy arose and he was hired as a laboratory assistant.

Davy quickly recognized Faraday's talents and as time passed allowed him to work more and more independently. In his years with Davy, Faraday published papers covering almost every field of chemistry. They included studies on the condensation of gases (he was the first to liquefy ammonia), the reaction of silver compounds with ammonia, and the isolation of several organic compounds, the most important of which was benzene. In 1816, Faraday began a series of lectures at the Royal Institution that were brilliantly successful. In 1825 he succeeded Davy as director of the laboratory. As Faraday's reputation grew, it was said that "Humphry Davy's greatest discovery was Michael Faraday." Perhaps it was witticisms of this sort that led to an estrangement between master and protégé. Late in his life, Davy opposed Faraday's nomination as a Fellow of the Royal Society and is reputed to have cast the only vote against him.

In personality and temperament, Davy and Faraday were poles apart. To Sir Humphry Davy, science was a fascinating hobby that happened to bring him fame and fortune. He pursued it as his leisure, enjoying an active social life and writing mediocre poetry in his spare time. To Michael Faraday, science was an obsession; one of his biographers describes him as a "work maniac." An observer (Faraday had no students) said of him ". . . if he had to cross the laboratory for anything, he did not walk, he ran; the quickness of his perception was equalled by the calm rapidity of his movements." In 1839, he suffered a nervous breakdown, the result of overwork. For much of the rest of his life, Faraday was in poor health. He gradually gave up more and more of his social engagements but continued to do research at the same pace as before.

Faraday developed the laws of electrolysis between 1831 and 1834. In the summer of 1833, he showed that a given amount of electrical current passed through a solution of H_2SO_4 produces the same amount of H_2 regardless of the concentration of the acid, the size of the electrodes, or the voltage applied. He concluded that the amount of a substance produced by electrolysis must be directly proportional to the current passed and independent of other factors. In mid-December of the same year, he began a quantitative study of the electrolysis of several metal cations, including Sn^{2+}, Pb^{2+}, and Zn^{2+}. Despite taking a whole day off for Christmas, he managed to complete these experiments, write up the results of 3 years' work, and get his paper published in the Philosophic Transactions of the Royal Society on January 9, 1834. In this paper, Faraday introduced the basic vocabulary of electrochemistry, using for the first time the terms "anode," "cathode," "ion," "electrolyte," and "electrolysis."

Although we have emphasized Faraday's work in chemistry, his greatest contributions were to physics in the field of electricity and magnetism. In 1821, he discovered that a wire through which a current is flowing will rotate about a magnetic pole, and so produced the first electric motor. Ten years later, he showed that an electric potential could be created by rotating a copper disc between the poles of a permanent magnet, and thereby discovered the principle governing all modern electrical generators. These two contributions made possible the controlled use of electricity by mankind and were crucial steps in the development of modern society. No less a scientist than Albert Einstein ranked Faraday, along with Newton, Galileo, and Maxwell, as one of the greatest physicists of the ages.

The electric motor and generator are two of the greatest inventions of man. They don't hardly make them like Faraday any more

Summary

This chapter dealt with redox reactions in which

—One species is reduced (oxidation number decreases) and acts as an oxidizing agent.
—Another species is oxidized (oxidation number increases) and acts as a reducing agent.

These concepts are illustrated in Example 23.2. Oxidation numbers are assigned according to arbitrary rules (Example 23.1). Equations for redox reactions can be balanced by the half-equation method, either in acidic or basic solutions (Example 23.3). Balanced equations for such reactions are useful in calculations involving redox titrations (Example 23.4). They are also required to relate the amount of product formed by electrolysis to the quantity of electricity (Example 23.6) or the amount of electrical energy (Example 23.7) passed through a cell.

There are two kinds of electrochemical cells. In an electrolytic cell (Fig. 23.1), electrical energy is used to make a nonspontaneous redox reaction take place. The nature of that reaction depends upon which ions are present in the electrolyte (Example 23.5). In a voltaic cell (Fig. 23.4), a spontaneous redox reaction generates electrical energy. In both types of cells, oxidation occurs at the anode, reduction at the cathode. Anions move to the anode, cations to the cathode. Both types of cells are widely used commercially (Figs. 23.2, 23.3, 23.7, and 23.8).

Key Words and Concepts

ampere	Faraday constant	redox reaction
anode	half-equation	redox titration
cathode	joule	reducing agent
coulomb	kilowatt-hour	reduction
disproportionation	oxidation	volt
electrolytic cell	oxidation number	voltaic cell
electrolysis	oxidizing agent	

Questions and Problems

Oxidation Number; Oxidizing and Reducing Agents

23.1 Give the oxidation number of each atom in
   a. NO_2^- b. HIO_3 c. TeF_8^{2-}
   d. N_2O_3 e. Na_2MoO_4

23.2 Give the oxidation number of each atom in
   a. NOF b. C_2H_6O c. RuF_5
   d. N_2H_4 e. $HAsO_4^{2-}$

23.3 Classify each of the following half-equations as oxidation or reduction:
   a. $Ca(s) \rightarrow Ca^{2+}(aq)$
   b. $Fe^{3+}(aq) \rightarrow Fe^{2+}(aq)$
   c. $NO_3^-(aq) \rightarrow NO(g)$
   d. $Hg(l) \rightarrow Hg_2^{2+}(aq)$

23.4 Chlorine in its compounds shows oxidation numbers of -1, $+1$, $+3$, $+5$, and $+7$. Which of the following species can act as oxidizing agents? reducing agents?
   a. $HClO_2$ b. ClO_4^- c. Cl^-
   d. ClO^- e. Cl_2O_7

23.5 For each reaction (unbalanced equation) given below
   a. indicate the oxidation number of each atom.
   b. identify the oxidizing agent.
   c. identify the reducing agent.

$$Cr_2O_3(s) + Al(s) \rightarrow Cr(s) + Al_2O_3(s)$$
$$NO_3^-(aq) + Sn^{2+}(aq) + H^+(aq) \rightarrow$$
$$NO_2(g) + H_2O + Sn^{4+}(aq)$$

23.6 For each of the following reactions (unbalanced equations), identify the species oxidized, the species reduced, the oxidizing agent, and the reducing agent.
   a. $Fe^{3+}(aq) + I^-(aq) \rightarrow Fe^{2+}(aq) + I_2(s)$
   b. $ClO_3^-(aq) + H_2S(g) \rightarrow Cl^-(aq) + SO_4^{2-}(aq)$
$$+ H^+(aq)$$
   c. $ClO_3^-(aq) + S^{2-}(aq) + H_2O \rightarrow Cl^-(aq) + S(s)$
$$+ OH^-(aq)$$

Balancing Redox Equations

23.7 Balance the equations in Question 23.5.

23.8 Balance the equations in Question 23.6.

23.9 Write balanced equations for the following reactions in acid solution:
   a. $Se(s) + NO_3^-(aq) \rightarrow SeO_2(s) + NO(g)$
   b. $Sn^{2+}(aq) + O_2(g) \rightarrow Sn^{4+}(aq) + H_2O$
   c. $H_2O_2(aq) + Cr_2O_7^{2-}(aq) \rightarrow Cr^{3+}(aq) + O_2(g)$

23.31 Give the oxidation number of each atom in
   a. HCO_3^- b. $S_2O_3^{2-}$ c. $MgSO_4$
   d. ClO_4^- e. SF_6

23.32 Give the oxidation number of each atom in
   a. Sb_4O_{10} b. CaC_2O_4 c. HSO_4^-
   d. $Na_2Fe_2O_4$ e. HPO_3^{2-}

23.33 Classify each of the following half-equations as oxidation or reduction:
   a. $OH^-(aq) \rightarrow O_2(g)$
   b. $Co^{3+}(aq) \rightarrow Co^{2+}(aq)$
   c. $Cl_2(g) \rightarrow ClO_3^-(aq)$
   d. $AuCl_4^-(aq) \rightarrow AuCl_2^-(aq)$

23.34 Manganese in its compounds can have oxidation numbers of $+2$, $+3$, $+4$, $+6$, and $+7$. Which of the following species can act as oxidizing agents? reducing agents?
   a. Mn^{2+} b. MnO_4^- c. MnO_4^{2-}
   d. MnO_2 e. $MnO(OH)$

23.35 For the reaction (unbalanced equation) given below, identify
   a. the species oxidized.
   b. the species reduced.
   c. the oxidizing agent.
   d. the reducing agent.

$$CrO_4^{2-}(aq) + NO_2^-(aq) + H_2O \rightarrow$$
$$Cr(OH)_3(s) + NO_3^-(aq) + OH^-(aq)$$

23.36 Follow the directions for Question 23.6 for
   a. $Sn^{2+}(aq) + Fe^{3+}(aq) \rightarrow Sn^{4+}(aq) + Fe^{2+}(aq)$
   b. $NO_3^-(aq) + H^+(aq) + I_2(s) \rightarrow NO_2(g) + H_2O$
$$+ IO_3^-(aq)$$
   c. $Cl_2(g) + OH^-(aq) \rightarrow ClO^-(aq) + Cl^-(aq)$
$$+ H_2O$$

23.37 Balance the equation in Question 23.35.

23.38 Balance the equations in Question 23.36.

23.39 Balance the following equations (acidic solution):
   a. $Zn(s) + NO_3^-(aq) \rightarrow Zn^{2+}(aq) + NH_4^+(aq)$
   b. $Fe^{2+}(aq) + Cr_2O_7^{2-}(aq) \rightarrow Fe^{3+}(aq) + Cr^{3+}(aq)$
   c. $P_4(s) \rightarrow PH_3(g) + HPO_3^{2-}(aq)$

23.10 Write balanced equations for the following reactions in basic solution:
a. $NO_2^-(aq) + Br_2(l) \rightarrow NO_3^-(aq) + Br^-(aq)$
b. $MnO_4^-(aq) + Cl^-(aq) \rightarrow Mn^{2+}(aq) + ClO^-(aq)$
c. $Cl_2(g) \rightarrow Cl^-(aq) + ClO^-(aq)$

23.11 In gold plating, metallic gold is formed from the $Au(CN)_2^-$ ion. At the same time, $O_2(g)$ is formed. The plating solution is basic. Write equations for the two half-reactions and the overall reaction.

23.12 Consider the following half-equations for acidic solution:

$$Fe^{2+}(aq) \rightarrow Fe^{3+}(aq)$$

$$NO_2(g) \rightarrow NO_3^-(aq)$$

$$Al^{3+}(aq) \rightarrow Al(s)$$

$$ClO^-(aq) \rightarrow Cl_2(g)$$

a. Classify each as an oxidation or reduction.
b. Balance each half-equation.
c. Write as many balanced redox equations as possible by combining these half-equations.

23.13 Follow the directions for Question 23.12 for the following half-equations in basic solution:

$$S^{2-}(aq) \rightarrow S(s)$$

$$Cr(OH)_3(s) \rightarrow CrO_4^{2-}(aq)$$

$$Cl^-(aq) \rightarrow ClO_3^-(aq)$$

$$Fe(OH)_3(s) \rightarrow Fe(s)$$

Mole Relations in Redox Reactions; Titrations

23.14 An acidic solution of potassium permanganate reacts with $C_2O_4^{2-}$ ions to form $CO_2(g)$ and Mn^{2+}.
a. Write a balanced equation for the reaction.
b. If 21.6 mL of 0.100 M $KMnO_4$ are required to titrate 25.2 mL of a sodium oxalate solution, what was the concentration of $C_2O_4^{2-}$ in the original solution?

23.15 Consider the reaction of copper with nitric acid:

$$Cu(s) + 4\,H^+(aq) + 2\,NO_3^-(aq) \rightarrow$$
$$Cu^{2+}(aq) + 2\,H_2O + 2\,NO_2(g)$$

a. What volume of 16.0 M HNO_3 is needed to furnish the H^+ ions to react with 10.0 g of copper?
b. What mass of $NO_2(g)$ is formed when 10.0 g of copper reacts?

23.40 Balance the following equations (basic solution):
a. $Fe(OH)_2(s) + NO_3^-(aq) \rightarrow Fe(OH)_3(s) + NO_2^-(aq)$
b. $Cr(OH)_3(s) + ClO_3^-(aq) \rightarrow CrO_4^{2-}(aq) + Cl^-(aq)$
c. $Cl_2(g) \rightarrow ClO_4^-(aq) + Cl^-(aq)$

23.41 When moist air comes in contact with iron, corrosion takes place to form $Fe(OH)_3(s)$. Write balanced half-equations for the oxidation and reduction reactions (O_2 is reduced) and combine them to give the overall balanced equation.

23.42 Follow the directions for Question 23.12 for the following half-equations (acidic solution):

$$Cl_2(g) \rightarrow Cl^-(aq)$$

$$MnO_4^-(aq) \rightarrow Mn^{2+}(aq)$$

$$H_2S(g) \rightarrow S(s)$$

$$Cr^{3+}(aq) \rightarrow Cr_2O_7^{2-}(aq)$$

23.43 Follow the directions for Question 23.13 for the following half-equations in basic solution:

$$H_2(g) \rightarrow H_2O$$

$$OH^-(aq) \rightarrow O_2(g)$$

$$Ni(OH)_2(s) \rightarrow NiO_2(s)$$

$$MnO_4^-(aq) \rightarrow MnO(s)$$

23.44 It is possible to determine I^- in solution by titrating with MnO_4^- in acid; products include Mn^{2+} and I_2.
a. Write a balanced equation for this reaction.
b. If 27.6 mL of 0.0800 M $KMnO_4$ are required to titrate 18.4 mL NaI solution to the end point, what was the concentration of I^- in the original solution?

23.45 Consider the reaction of oxygen with iron(II) hydroxide:

$$4\,Fe(OH)_2(s) + O_2(g) + 2\,H_2O \rightarrow 4\,Fe(OH)_3(s)$$

a. What mass of $Fe(OH)_3$ is formed from 1.00 g $Fe(OH)_2$?
b. What volume of $O_2(g)$ at 25°C and 1.00 atm is required to react with 1.00 g $Fe(OH)_2$?

23.16 Consider the reaction

$$Cr_2O_7{}^{2-}(aq) + 14\ H^+(aq) + 6\ Fe^{2+}(aq) \rightarrow$$
$$2\ Cr^{3+}(aq) + 6\ Fe^{3+}(aq) + 7\ H_2O$$

It is found that 19.3 mL of a $K_2Cr_2O_7$ solution reacts with 25.0 mL of 0.0400 M Fe^{2+}.
a. What is the molarity of $K_2Cr_2O_7$?
b. How many moles of electrons are transferred when 25.0 mL of 0.0400 M Fe^{2+} reacts?

Electrolysis; Quantitative Relations

23.17 State what products will be formed at the anode and cathode when water solutions of the following compounds are electrolyzed:
a. HI   b. $CuCl_2$ c. KOH   d. $Ni(NO_3)_2$

23.18 Write balanced equations for the half-reactions at each electrode and for the overall reaction when solutions of the following compounds are electrolyzed:
a. $ZnBr_2$ b. NaF   c. $Ca(NO_3)_2$

23.19 Draw a diagram of an electrolytic cell in which the reaction is

$$2\ I^-(aq) + 2\ H_2O \rightarrow I_2(s) + H_2(g) + 2\ OH^-(aq)$$

Label anode and cathode and indicate the direction in which ions (I^-, Na^+) move within the cell.

23.20 Aluminum metal is formed by the electrolysis of Al_2O_3.
a. How many grams of Al are produced when 8.80×10^3 C pass through the cell?
b. How long does it take to form 0.500 g of Al using a current of 25.0 A?

23.21 A current of 2.50 A is drawn from an alkaline dry cell for eight minutes. Assuming that only Reaction 23.17 occurs,
a. how many grams of zinc are converted to ZnO?
b. how long can the cell operate at 2.50 A if it contains 18 g Zn and fails when 20% of the zinc is consumed?

23.22 Nickel metal is plated from a solution containing Ni^{2+} ions, using a voltage of 2.61 V. Calculate, per gram of Ni plated,
a. the number of coulombs required.
b. the amount of electrical energy required in joules and in kilowatt-hours.

23.46 Consider the reaction

$$2\ MnO_4{}^-(aq) + 3\ Mn^{2+}(aq) + 2\ H_2O \rightarrow$$
$$5\ MnO_2(s) + 4\ H^+(aq)$$

a. Calculate the volume of 0.0780 M $KMnO_4$ required to react with 30.0 mL of 0.110 M Mn^{2+}.
b. Find the mass of MnO_2 produced in (a).

23.47 What species will be formed at each electrode when water solutions of the following compounds are electrolyzed?
a. H_2SO_4 b. $CoCl_2$ c. NaI   d. $Cu(NO_3)_2$

23.48 Follow the directions of Question 23.18 for
a. $Ba(OH)_2$ b. $Pb(NO_3)_2$ c. HBr

23.49 Draw a diagram of an electrolytic cell in which the reaction is

$$Sn^{2+}(aq) + 2\ Cl^-(aq) \rightarrow Sn(s) + Cl_2(g)$$

Label anode and cathode and indicate the direction of movement of ions through the cell.

23.50 A solution of $ZnBr_2$ is electrolyzed to form zinc metal and liquid bromine. A total of 2.50×10^4 C passes through the cell.
a. How many grams of zinc are produced?
b. What volume of Br_2 (d = 3.12 g/cm³) is formed?
c. How long does the electrolysis take if the current is 6.12 A?

23.51 In the electrolysis of aqueous NaCl, 8.36×10^{21} electrons pass through the cell. Assuming 100% yield,
a. how many grams of OH^- are produced (see Equation 23.8)?
b. what volume of dry chlorine gas is produced at 25°C and 1.00 atm?

23.52 Silver metal is plated from a solution containing $Ag(CN)_2{}^-$ ions.
a. Write a balanced half-equation for the reduction.
b. How many coulombs are required to plate 1.00 g of silver?
c. How much electrical energy (joules and kilowatt-hours) is required if the voltage is 3.40 V?

23.23 A lead storage battery delivers a current of 2.00 A for one hour at a voltage of 12.0 V.
  a. How many grams of Pb are converted to $PbSO_4$?
  b. How much electrical energy is produced (kilowatt-hours)?

23.53 Aluminum is produced by the electrolysis of Al_2O_3, using a voltage of 6.0 V.
  a. How many joules of electrical energy are required to form 1.00 kg Al?
  b. What is the cost of the electrical energy in (a) at the rate of 6.0¢ per kilowatt-hour?

Voltaic Cells

23.24 Write a balanced chemical equation for the overall cell reaction represented as
  a. $Cr/Cr^{3+} \parallel Co^{2+}/Co$
  b. $Ni/Ni^{2+} \parallel Cu^{2+}/Cu$
  c. $(Pt)Br^-/Br_2 \parallel (Pt)I_2/I^-$

23.54 Write a balanced chemical equation for the overall cell reaction represented as
  a. $Mg/Mg^{2+} \parallel Fe^{2+}/Fe$
  b. $Al/Al^{3+} \parallel Zn^{2+}/Zn$
  c. $Sn/Sn^{2+} \parallel (Pt)O_2/H_2O$

23.25 Draw a diagram for a salt bridge cell for each of the following reactions. Label anode and cathode and indicate the direction of current flow throughout the circuit.
  a. $2 Fe^{3+}(aq) + Pb(s) \rightarrow 2 Fe^{2+}(aq) + Pb^{2+}(aq)$
  b. $Mg(s) + 2 H^+(aq) \rightarrow Mg^{2+}(aq) + H_2(g)$
  c. $Br_2(l) + 2 I^-(aq) \rightarrow 2 Br^-(aq) + I_2(s)$

23.55 Follow the directions for Question 23.25 for
  a. $Cd(s) + Co^{2+}(aq) \rightarrow Cd^{2+}(aq) + Co(s)$
  b. $Sn(s) + Cl_2(g) \rightarrow Sn^{2+}(aq) + 2 Cl^-(aq)$
  c. $Fe(s) + Cu(OH)_2(s) \rightarrow Cu(s) + Fe(OH)_2(s)$

23.26 Write balanced equations for
  a. the anode half-reaction in the ordinary dry cell.
  b. the anode half-reaction in the lead storage battery.
  c. the overall reaction in the alkaline dry cell.

23.56 Write balanced equations for
  a. the half-reaction at the cathode in the lead storage battery.
  b. the overall reaction in the lead storage battery.
  c. the overall reaction in a nickel-cadmium battery.

23.27 Suppose HCl rather than sulfuric acid were used in a lead storage battery. Write balanced equations for the half-reactions at each electrode and an equation for the overall reaction.

23.57 Suppose Mn^{2+} rather than Mn_2O_3 were produced in a dry Leclanché cell. Write balanced equations for the half-reactions and for the overall reaction.

General

23.28 In your own words explain why
  a. reduction causes a decrease in oxidation number.
  b. metallic sodium and I_2 are the only products when molten NaI is electrolyzed but electrolysis of aqueous NaI produces H_2 and I_2.
  c. an oxidizing agent becomes reduced.
  d. cations migrate to the cathode in an electrolytic cell.

23.58 Explain why
  a. sulfuric acid is used in a lead storage battery.
  b. a salt bridge is used in a voltaic cell.
  c. oxidation of Fe increases its oxidation number.
  d. complexing agents such as CN^- are used in electroplating.

23.29 Explain

a. why, on an electrochemical basis, it is likely that an electroplating process is less than 100% efficient.

b. how Co^{2+} can be a reducing agent or an oxidizing agent.

c. the essential difference between an electrolysis cell and a voltaic cell.

A lead storage battery is used to electrolyze a water solution of NaCl. Sketch the two cells, labeling anode and cathode, and indicate the direction of flow of current through all parts of the circuit.

23.59 Identify each of the following statements as true or false. If false, correct it to make it true.

a. Electrolysis of aqueous $FeCl_2$ produces iron metal and chlorine.

b. K^+ ions from a salt bridge move to the anode during operation of a voltaic cell.

c. When a lead storage battery is recharged, lead sulfate is deposited.

d. Vanadium is reduced in the conversion $VO^{2+} \rightarrow VO_2^+$.

23.60 An ordinary dry cell is used to electroplate copper, using Cu electrodes and a $CuSO_4$ solution. Sketch the two cells, labeling anode and cathode, and indicate the direction of flow of current through all parts of the circuit.

*23.61 Lead forms a series of oxides: PbO, PbO_2, Pb_2O_3, and Pb_3O_4.

a. Give the oxidation number of lead in each of these compounds.

b. Lead occurs as either $+2$ or $+4$ ions in these compounds. Rationalize this with the data from (a).

*23.62 Calcium in blood or urine can be determined by precipitation as CaC_2O_4. The precipitate is dissolved in strong acid and titrated with $KMnO_4$ (see Problem 23.14). A 24-h urine sample is collected from an adult patient, reduced to a small volume, and titrated with 26.2 cm^3 of 0.0946 M $KMnO_4$. How many grams of CaC_2O_4 are in the sample? Normal range for Ca^{2+} output for an adult is 100 to 300 mg per 24 h. Is the sample within the normal range?

*23.63 A nickel-cadmium battery is used to electrolyze a water solution of sodium chloride. How much cadmium is consumed to produce 1.00 L $Cl_2(g)$ at 24°C and 742 mm Hg?

*23.64 In Chapter 14, it was stated that the minimum amount of energy required to electrolyze one mole of $H_2O(l)$ is 237 kJ. How many coulombs are required for this electrolysis? What is the minimum voltage that can be used?

*23.65 Refer to Figure 13.9, Chapter 13, which gives the Lewis structures of the oxides of nitrogen. Assigning valence electrons to the more electronegative atom, determine the oxidation number of each atom in N_2O_3; in N_2O. (This question illustrates why oxidation number is ordinarily assigned according to arbitrary rules!)

In Section 23.4, we discussed the construction and operation of voltaic cells. These cells are of interest to chemists for reasons that go beyond their practical importance as a source of electrical energy. The property of a voltaic cell that is of particular concern to us is its **voltage,** which is a measure of the driving force of the cell reaction. By measuring cell voltages, we can decide whether a given redox reaction will occur.

The voltage of a cell depends upon two factors:

1. The nature of the cell reaction. Associated with every redox reaction is a standard voltage (Section 24.1), which can be calculated from a table of standard potentials (Table 24.1).

2. The concentrations of the species taking part in the reaction. The effect of concentration upon voltage can be calculated from the Nernst equation (Section 24.2).

The standard cell voltage, which we will refer to as E_{tot}^0, can be used for a variety of purposes. The sign of E_{tot}^0 describes the spontaneity of reaction. If the calculated value of E_{tot}^0 is positive, the redox reaction is spontaneous at standard concentrations. If E_{tot}^0 is negative, the reaction is nonspontaneous. This concept is a very useful one that helps us to decide whether a given redox reaction, such as the corrosion of a metal (Section 24.4), will occur in the laboratory or the world around us. We can also relate E_{tot}^0 to the standard free energy change for a reaction, ΔG^0, and the equilibrium constant for the reaction, K (Section 24.3).

24.1
Standard Voltages

The **standard voltage** for a given cell reaction is that measured when *all ions and molecules in solution are at a concentration of 1 M and all gases are at a pressure of 1 atm*. To illustrate, consider the $Zn/Zn^{2+} \parallel (Pt)H^+/H_2$ cell (recall Fig. 23.5, p.

707). We find that when the pressure of hydrogen gas is 1 atm and the concentrations of Zn^{2+} and H^+ are 1 M, the cell voltage is $+0.76$ V. This quantity is referred to as the standard voltage for the reaction and is given the symbol E^0_{tot}:

$$Zn(s) + 2\,H^+(aq,\ 1\ M) \to Zn^{2+}(aq,\ 1\ M) + H_2(g,\ 1\ atm);\ E^0_{tot} = +0.76\ V \qquad \textbf{(24.1)}$$

E^0_{ox} and E^0_{red}

You will recall that any redox reaction can be split into two half-reactions. One of these is an oxidation, and the other a reduction. For Reaction 24.1,

oxidation: $\qquad Zn(s) \to Zn^{2+}(aq,\ 1\ M) + 2\,e^-$ \qquad\qquad **(24.1a)**

reduction: $\qquad 2\,H^+(aq,\ 1\ M) + 2\,e^- \to H_2(g,\ 1\ atm)$ \qquad\qquad **(24.1b)**

It is possible to associate standard voltages with half-reactions such as these. Such a voltage is a measure of the driving force behind the half-reaction. The standard voltage for the oxidation half-reaction is given the symbol $\mathbf{E^0_{ox}}$. That for the reduction half-reaction is written as $\mathbf{E^0_{red}}$. The standard voltage for the cell reaction is the sum of these two quantities. Thus, we have

$$\begin{array}{ll} Zn(s) \to Zn^{2+}(1\ M) + 2\,e^- & E^0_{ox}\ (Zn \to Zn^{2+}) \\ \underline{2\,H^+(1\ M) + 2\,e^- \to H_2(1\ atm)} & \underline{E^0_{red}\ (H^+ \to H_2)} \\ Zn(s) + 2\,H^+(1\ M) \to Zn^{2+}(1\ M) + H_2(1\ atm) & E^0_{tot} = E^0_{ox} + E^0_{red} = +0.76\ V \end{array}$$

One equation with two unknowns cannot be solved

We would like to be able to calculate standard voltages for half-reactions such as those listed above. There is, however, a problem. As you can see from the equation just written, we have one known, E^0_{tot} ($+0.76$ V), and two unknowns, E^0_{ox} and E^0_{red}. No matter what we do, we cannot determine an individual half-reaction voltage experimentally. To resolve this dilemma, we make an arbitrary decision. We take the *standard voltage for the reduction of H^+ ions to H_2 gas to be zero*:

$$2\,H^+(aq,\ 1\ M) + 2\,e^- \to H_2(g,\ 1\ atm);\qquad E^0_{red}\ (H^+ \to H_2) = 0.00\ V$$

$E^0_{tot} = E^0_{ox} + E^0_{red}$
$\quad\ = 0.76\ V$

$E^0_{red} = 0.00\ V$

$E^0_{ox} = 0.76\ V$

Using this convention, and knowing that E^0_{tot} for the $Zn/Zn^{2+} \parallel (Pt)H^+/H_2$ cell is $+0.76$ V, it follows that the standard voltage for the oxidation of zinc must be $+0.76$ V; that is,

$$Zn(s) \to Zn^{2+}(aq,\ 1\ M) + 2\,e^-;\qquad E^0_{ox}\ (Zn \to Zn^{2+}) = +0.76\ V$$

As soon as one voltage is established, others can be determined from measurements on appropriate cells. Suppose, for example, we want to determine the standard voltage for the reduction of Cu^{2+} to Cu. One way to do this is to set up a $Zn/Zn^{2+} \parallel Cu^{2+}/Cu$ cell, using 1 M solutions of Zn^{2+} and Cu^{2+}, and measure the voltage. We find the standard voltage of this cell to be $+1.10$ V:

$$Zn(s) + Cu^{2+}(aq,\ 1\ M) \to Zn^{2+}(aq,\ 1\ M) + Cu(s);\qquad E^0_{tot} = +1.10\ V$$

In this cell, zinc is being oxidized and Cu^{2+} ions reduced. Hence,

Table 24.1
Standard Potentials in Water Solution at 25°C

Oxidizing Agent	Reducing Agent	E^0_{red} (V)
$Li^+(aq) + e^-$	$\rightarrow Li(s)$	-3.05
$K^+(aq) + e^-$	$\rightarrow K(s)$	-2.93
$Ba^{2+}(aq) + 2\,e^-$	$\rightarrow Ba(s)$	-2.90
$Ca^{2+}(aq) + 2\,e^-$	$\rightarrow Ca(s)$	-2.87
$Na^+(aq) + e^-$	$\rightarrow Na(s)$	-2.71
$Mg^{2+}(aq) + 2\,e^-$	$\rightarrow Mg(s)$	-2.37
$Al^{3+}(aq) + 3\,e^-$	$\rightarrow Al(s)$	-1.66
$Mn^{2+}(aq) + 2\,e^-$	$\rightarrow Mn(s)$	-1.18
$Zn^{2+}(aq) + 2\,e^-$	$\rightarrow Zn(s)$	-0.76
$Cr^{3+}(aq) + 3\,e^-$	$\rightarrow Cr(s)$	-0.74
$Fe^{2+}(aq) + 2\,e^-$	$\rightarrow Fe(s)$	-0.44
$Cr^{3+}(aq) + e^-$	$\rightarrow Cr^{2+}(aq)$	-0.41
$Cd^{2+}(aq) + 2\,e^-$	$\rightarrow Cd(s)$	-0.40
$PbSO_4(s) + 2\,e^-$	$\rightarrow Pb(s) + SO_4^{2-}(aq)$	-0.36
$Tl^+(aq) + e^-$	$\rightarrow Tl(s)$	-0.34
$Co^{2+}(aq) + 2\,e^-$	$\rightarrow Co(s)$	-0.28
$Ni^{2+}(aq) + 2\,e^-$	$\rightarrow Ni(s)$	-0.25
$AgI(s) + e^-$	$\rightarrow Ag(s) + I^-(aq)$	-0.15
$Sn^{2+}(aq) + 2\,e^-$	$\rightarrow Sn(s)$	-0.14
$Pb^{2+}(aq) + 2\,e^-$	$\rightarrow Pb(s)$	-0.13
$2\,H^+(aq) + 2\,e^-$	$\rightarrow H_2(g)$	0.00
$AgBr(s) + e^-$	$\rightarrow Ag(s) + Br^-(aq)$	0.07
$S(s) + 2\,H^+(aq) + 2\,e^-$	$\rightarrow H_2S(aq)$	0.14
$Sn^{4+}(aq) + 2\,e^-$	$\rightarrow Sn^{2+}(aq)$	0.15
$Cu^{2+}(aq) + e^-$	$\rightarrow Cu^+(aq)$	0.15
$SO_4^{2-}(aq) + 4\,H^+(aq) + 2\,e^-$	$\rightarrow SO_2(g) + 2\,H_2O$	0.20
$Cu^{2+}(aq) + 2\,e^-$	$\rightarrow Cu(s)$	0.34
$Cu^+(aq) + e^-$	$\rightarrow Cu(s)$	0.52
$I_2(s) + 2\,e^-$	$\rightarrow 2\,I^-(aq)$	0.53
$Fe^{3+}(aq) + e^-$	$\rightarrow Fe^{2+}(aq)$	0.77
$Hg_2^{2+}(aq) + 2\,e^-$	$\rightarrow 2\,Hg(l)$	0.79
$Ag^+(aq) + e^-$	$\rightarrow Ag(s)$	0.80
$2\,Hg^{2+}(aq) + 2\,e^-$	$\rightarrow Hg_2^{2+}(aq)$	0.92
$NO_3^-(aq) + 4\,H^+(aq) + 3\,e^-$	$\rightarrow NO(g) + 2\,H_2O$	0.96
$AuCl_4^-(aq) + 3\,e^-$	$\rightarrow Au(s) + 4\,Cl^-(aq)$	1.00
$Br_2(l) + 2\,e^-$	$\rightarrow 2\,Br^-(aq)$	1.07
$O_2(g) + 4\,H^+(aq) + 4\,e^-$	$\rightarrow 2\,H_2O$	1.23
$MnO_2(s) + 4\,H^+(aq) + 2\,e^-$	$\rightarrow Mn^{2+}(aq) + 2\,H_2O$	1.23
$Cr_2O_7^{2-}(aq) + 14\,H^+(aq) + 6\,e^-$	$\rightarrow 2\,Cr^{3+}(aq) + 7\,H_2O$	1.33
$Cl_2(g) + 2\,e^-$	$\rightarrow 2\,Cl^-(aq)$	1.36
$ClO_3^-(aq) + 6\,H^+(aq) + 5\,e^-$	$\rightarrow \frac{1}{2}\,Cl_2(g) + 3\,H_2O$	1.47
$Au^{3+}(aq) + 3\,e^-$	$\rightarrow Au(s)$	1.50
$MnO_4^-(aq) + 8\,H^+(aq) + 5\,e^-$	$\rightarrow Mn^{2+}(aq) + 4\,H_2O$	1.52
$PbO_2(s) + SO_4^{2-}(aq) + 4\,H^+(aq) + 2\,e^-$	$\rightarrow PbSO_4(s) + 2\,H_2O$	1.68
$H_2O_2(aq) + 2\,H^+(aq) + 2\,e^-$	$\rightarrow 2\,H_2O$	1.77
$Co^{3+}(aq) + e^-$	$\rightarrow Co^{2+}(aq)$	1.82
$F_2(g) + 2\,e^-$	$\rightarrow 2\,F^-(aq)$	2.87

Basic Solution

$Fe(OH)_2(s) + 2\,e^-$	$\rightarrow Fe(s) + 2\,OH^-(aq)$	-0.88
$2\,H_2O + 2\,e^-$	$\rightarrow H_2(g) + 2\,OH^-(aq)$	-0.83
$Fe(OH)_3(s) + e^-$	$\rightarrow Fe(OH)_2(s) + OH^-(aq)$	-0.56
$S(s) + 2\,e^-$	$\rightarrow S^{2-}(aq)$	-0.43
$Cu(OH)_2(s) + 2\,e^-$	$\rightarrow Cu(s) + 2\,OH^-(aq)$	-0.22
$CrO_4^{2-}(aq) + 4\,H_2O + 3\,e^-$	$\rightarrow Cr(OH)_3(s) + 5\,OH^-(aq)$	-0.12
$NO_3^-(aq) + H_2O + 2\,e^-$	$\rightarrow NO_2^-(aq) + 2\,OH^-(aq)$	0.01
$ClO_4^-(aq) + H_2O + 2\,e^-$	$\rightarrow ClO_3^-(aq) + 2\,OH^-(aq)$	0.36
$O_2(g) + 2\,H_2O + 4\,e^-$	$\rightarrow 4\,OH^-(aq)$	0.40
$ClO_3^-(aq) + 3\,H_2O + 6\,e^-$	$\rightarrow Cl^-(aq) + 6\,OH^-(aq)$	0.62
$ClO^-(aq) + H_2O + 2\,e^-$	$\rightarrow Cl^-(aq) + 2\,OH^-(aq)$	0.89

$$E^0_{ox} (Zn \rightarrow Zn^{2+}) + E^0_{red} (Cu^{2+} \rightarrow Cu) = +1.10 \text{ V}$$

Since the standard voltage for the oxidation of zinc must be the same here as in the Zn–H$^+$ cell, $+0.76$ V,

$$+0.76 \text{ V} + E^0_{red} (Cu^{2+} \rightarrow Cu) = +1.10 \text{ V}$$

$$E^0_{red} (Cu^{2+} \rightarrow Cu) = +1.10 \text{ V} - 0.76 \text{ V} = +0.34 \text{ V}$$

Standard voltages for half-reactions are constant at 25°C

Standard half-cell voltages are ordinarily obtained from a list of *standard potentials* such as that given in Table 24.1. **The potentials listed give us directly the standard voltages for reduction half-reactions.** For example, since the standard potentials listed in the table for $Zn^{2+} \rightarrow Zn$ and $Cu^{2+} \rightarrow Cu$ are -0.76 V and $+0.34$ V, respectively, we see immediately that

$$Zn^{2+}(aq) + 2 \ e^- \rightarrow Zn(s); \qquad E^0_{red} = -0.76 \text{ V}$$

$$Cu^{2+}(aq) + 2 \ e^- \rightarrow Cu(s); \qquad E^0_{red} = +0.34 \text{ V}$$

Standard voltages for oxidation half-reactions are obtained by changing the sign of the standard potential listed in Table 24.1. Thus, we have

$$Zn(s) \rightarrow Zn^{2+}(aq) + 2 \ e^-; \qquad E^0_{ox} = +0.76 \text{ V}$$

$$Cu(s) \rightarrow Cu^{2+}(aq) + 2 \ e^-; \qquad E^0_{ox} = -0.34 \text{ V}$$

$E^0_{ox} = -E^0_{red}$

In general, standard voltages for forward and reverse reactions (oxidation and reduction) are equal in magnitude but opposite in sign.

In the remainder of this section, we will consider some of the applications of standard voltages. We start by showing how they can be used to compare the strengths of different oxidizing and reducing agents. Later, we will consider their use in determining whether or not a given redox reaction will occur spontaneously in the laboratory.

Strength of Oxidizing and Reducing Agents

As pointed out earlier, an oxidizing agent gains electrons in a redox reaction. It follows that all the species listed in the column at the far left of Table 24.1 ($Li^+, \ldots, F_2$) are, at least in principle, oxidizing agents. A *"strong" oxidizing agent* is one that has a strong attraction for electrons and hence *can readily oxidize other species*. In contrast, a "weak" oxidizing agent does not gain electrons readily. It is capable of reacting only with those species that are very easily oxidized.

The strength of an oxidizing agent is directly related to its E^0_{red} value. **The more positive E^0_{red} is, the stronger the oxidizing agent.** Looking at Table 24.1, we see that oxidizing strength increases as we move down the table. The Li$^+$ ion, at the top of the left column, is a very weak oxidizing agent. E^0_{red} for Li$^+$ is a large *negative* number, which implies that it has little tendency to gain electrons:

$$Li^+(aq) + e^- \rightarrow Li(s); \qquad E^0_{red} = -3.05 \text{ V}$$

In practice, cations of the Group 1 metals (Li^+, Na^+, K^+, . . .) and the Group 2 metals (Mg^{2+}, Ca^{2+}, . . .) never act as oxidizing agents in water solution. Further down the list, the H^+ ion has a greater tendency to gain electrons:

$$2\,H^+(aq) + 2\,e^- \rightarrow H_2(g); \qquad E^0_{red} = 0.00\ V$$

It is capable of oxidizing metals such as Mg or Zn (to Mg^{2+} or Zn^{2+}). The strongest oxidizing agents are those at the bottom of the left column. Species such as $Cr_2O_7^{2-}$ ($E^0_{red} = +1.33$ V), Cl_2 ($E^0_{red} = +1.36$ V), and MnO_4^- ($E^0_{red} = +1.52$ V) are commonly used as oxidizing agents in redox reactions. The fluorine molecule, F_2, is in principle the strongest of all oxidizing agents:

$$F_2(g) + 2\,e^- \rightarrow 2\,F^-(aq); \qquad E^0_{red} = +2.87\ V$$

In practice, fluorine is seldom used as an oxidizing agent because it is too dangerous to work with. The F_2 molecule takes electrons away from just about anything, including water, often with explosive violence.

Most chemists prefer not to work with F_2

The argument we have just gone through can be applied, in reverse, to reducing agents. These species are listed in the column to the right of Table 24.1 (Li, . . . , F^-). In principle, at least, all of them can supply electrons to another species in a redox reaction. Their strength as reducing agents is directly related to their E^0_{ox} values. **The more positive E^0_{ox} is, the stronger the reducing agent.** Looking at the values, remembering that $E^0_{ox} = -E^0_{red}$,

$$Li(s) \rightarrow Li^+(aq) + e^-; \qquad E^0_{ox} = +3.05\ V$$

$$H_2(g) \rightarrow 2\,H^+(aq) + 2\,e^-; \qquad E^0_{ox} = 0.00\ V$$

$$2\,F^-(aq) \rightarrow F_2(g) + 2\,e^-; \qquad E^0_{ox} = -2.87\ V$$

Strongest reducing agent: Li

Weakest reducing agent: F^-

we conclude that reducing strength decreases as we move down the table. The strongest reducing agents are located at the upper right (Li, . . .), and the weakest at the lower right (. . . , F^-).

From this discussion, another principle emerges:

—*the species formed when a strong oxidizing agent gains electrons is itself a weak reducing agent.* For example, the F_2 molecule, a strong oxidizing agent, gains electrons to form the F^- ion, a weak reducing agent.

—*the species formed when a strong reducing agent gives off electrons is itself a weak oxidizing agent.* Thus an Li atom, which is a strong reducing agent, is converted to the Li^+ ion, which is a weak oxidizing agent.

EXAMPLE 24.1 Consider the following species: $Cr_2O_7^{2-}$, NO_3^-, Br^-, Mg, Sn^{2+}. Using Table 24.1, classify each of these as an oxidizing or reducing agent. Arrange the oxidizing agents in order of increasing strength; do the same with the reducing agents.

Solution We start by realizing that oxidizing agents (species that can be reduced) are located in the left column. Scanning that column, we find the following oxidizing agents:

Highest $E_{red}^0 \rightarrow$
   best oxidizing agent

Sn^{2+}:	E_{red}^0 ($Sn^{2+} \rightarrow Sn$) $= -0.14$ V
NO_3^-:	E_{red}^0 ($NO_3^- \rightarrow NO$) $= +0.96$ V
$Cr_2O_7^{2-}$:	E_{red}^0 ($Cr_2O_7^{2-} \rightarrow Cr^{3+}$) $= +1.33$ V

Reducing agents (species that can be oxidized) are found in the right column. Here we locate

Highest $E_{ox}^0 \rightarrow$
   best reducing agent

Mg:	E_{ox}^0 ($Mg \rightarrow Mg^{2+}$) $= +2.37$ V
Sn^{2+}:	E_{ox}^0 ($Sn^{2+} \rightarrow Sn^{4+}$) $= -0.15$ V
Br^-:	E_{ox}^0 ($Br^- \rightarrow Br_2$) $= -1.07$ V

(Note that Sn^{2+} can act as either an oxidizing agent, in which case it is reduced to Sn, or a reducing agent, in which case it is oxidized to Sn^{4+}).

The more positive the voltage, the stronger the oxidizing or reducing agent:

oxidizing agents: $Sn^{2+} < NO_3^- < Cr_2O_7^{2-}$
reducing agents: $Br^- < Sn^{2+} < Mg$

EXERCISE   Give the formula of a cation that is a stronger reducing agent than Sn^{2+}; a stronger oxidizing agent than Sn^{2+}. Answer: Cr^{2+}; Pb^{2+}, H^+, Sn^{4+}, ...

Calculation of E_{tot}^0 from E_{red}^0 and E_{ox}^0

As pointed out earlier, the standard voltage for a redox reaction is the sum of the standard voltages of the two half-reactions, reduction and oxidation; that is,

$$E_{tot}^0 = E_{red}^0 + E_{ox}^0 \tag{24.2}$$

This simple relation makes it possible, using Table 24.1, to calculate standard voltages for more than 3000 different redox reactions. Example 24.2 illustrates how this is done.

EXAMPLE 24.2   Using Table 24.1, calculate the standard voltage for the reaction

$$2 Fe^{3+}(aq) + 2 I^-(aq) \rightarrow 2 Fe^{2+}(aq) + I_2(s)$$

Solution   Splitting the reaction into two half-reactions, finding the appropriate values of E_{red}^0 and E_{ox}^0 from Table 24.1, and adding, we have

reduction:	$2 Fe^{3+}(aq) + 2 e^- \rightarrow 2 Fe^{2+}(aq)$	$E_{red}^0 = +0.77$ V
oxidation:	$2 I^-(aq) \rightarrow I_2(s) + 2 e^-$	$E_{ox}^0 = -0.53$ V
		$E_{tot}^0 = +0.24$ V

Notice that E_{red}^0 for Fe^{3+} is taken directly from Table 24.1, where we find

$$Fe^{3+}(aq) + e^- \rightarrow Fe^{2+}(aq); \qquad E_{red}^0 = +0.77 \text{ V}$$

We can't change a cell voltage by writing a number on a piece of paper

We do *not* multiply the voltage by two just because two Fe^{3+} ions appear in the balanced equation for the reaction. E_{tot}^0, E_{red}^0, or E_{ox}^0 for a given reaction is independent of the number of electrons transferred.

EXERCISE Calculate E_{tot}^0 for the cell $Cu/Cu^{2+} \parallel Ag^+/Ag$. Answer: $+0.46$ V.

You will notice that in all the examples we have worked, E_{tot}^0 is a positive quantity. This is generally true for reactions taking place in a voltaic cell. A spontaneous reaction taking place within the cell generates a positive voltage. If the calculated voltage is negative, the reaction as written cannot serve as a source of energy in a voltaic cell. It may, however, be possible to carry out the reaction in an electrolytic cell. Consider, for example,

$$
\begin{array}{ll}
2\ Cl^-(aq) \rightarrow Cl_2(g) + 2\ e^- & E_{ox}^0 = -1.36\ V \\
\underline{2\ H_2O + 2\ e^- \rightarrow H_2(g) + 2\ OH^-(aq)} & \underline{E_{red}^0 = -0.83\ V} \\
2\ Cl^-(aq) + 2\ H_2O \rightarrow Cl_2(g) + H_2(g) + 2\ OH^-(aq) & E_{tot}^0 = -2.19\ V
\end{array}
$$

You will recall (Chap. 23) that this reaction takes place when a water solution of sodium chloride is electrolyzed. A voltage of at least 2.19 V must be supplied, perhaps by a storage battery, to furnish the energy required for the electrolysis. A nonspontaneous reaction such as this one, with a negative E_{tot}^0 value, can be made to occur in an electrolytic cell. To do so, we must apply an external voltage at least equal in magnitude to E_{tot}^0.

2.19 V when $[OH^-] = $ 1 M

Spontaneity of Redox Reactions

As we have seen, the voltage of a cell in which a spontaneous redox reaction is taking place is always positive. Such a reaction will take place directly and spontaneously in the laboratory, perhaps in a test tube, beaker, or other container. This means that:

If the calculated voltage for a redox reaction is a positive quantity, the reaction will occur spontaneously in the laboratory. If the calculated voltage is negative, the reaction will not occur; instead, the reverse reaction will be spontaneous.

Ordinarily, we apply this principle at standard concentrations, where the calculated voltage is E_{tot}^0. Under these conditions, we can say that

—if $E_{tot}^0 > 0$, the reaction is spontaneous.
—if $E_{tot}^0 < 0$, the reaction is nonspontaneous; the reverse reaction will tend to occur.
—if, perchance, $E_{tot}^0 = 0$, the reaction is at equilibrium at standard concentrations; there is no tendency for reaction to occur in either direction.

To illustrate this principle, consider the problem of reducing Ni^{2+} ions to nickel ($E_{red}^0 = -0.25$ V). It should be possible to do this with zinc ($E_{ox}^0 = +0.76$ V) but not with copper ($E_{ox}^0 = -0.34$ V):

$$Zn(s) + Ni^{2+}(aq) \rightarrow Zn^{2+}(aq) + Ni(s)$$

$$E_{tot}^0 = E_{ox}^0 + E_{red}^0 = +0.76\ V - 0.25\ V = +0.51\ V \quad \text{(spontaneous)}$$

All ion concentrations are 1 M

$$Cu(s) + Ni^{2+}(aq) \rightarrow Cu^{2+}(aq) + Ni(s)$$

$$E_{tot}^0 = E_{ox}^0 + E_{red}^0 = -0.34\ V - 0.25\ V = -0.59\ V \quad \text{(nonspontaneous)}$$

Sure enough, if we add zinc metal to a solution of $NiCl_2$, a reaction occurs. Nickel metal plates out on the zinc and the green color of the Ni^{2+} ion fades. If, on the other hand, we add copper to a solution of $NiCl_2$, there is no evidence of reaction. In contrast, the reverse reaction

$$Ni(s) + Cu^{2+}(aq) \rightarrow Ni^{2+}(aq) + Cu(s); \qquad E^0_{tot} = +0.59 \text{ V}$$

takes place spontaneously. Nickel metal reacts with a solution of $CuCl_2$ to plate out copper and form Ni^{2+} ions in solution.

Example 24.3 illustrates a somewhat more subtle application of this principle.

EXAMPLE 24.3 Using the standard potentials listed in Table 24.1, decide whether
a. $Fe(s)$ will be oxidized to Fe^{2+} by treatment with 1 M hydrochloric acid (HCl).
b. $Cu(s)$ will be oxidized to Cu^{2+} by treatment with 1 M hydrochloric acid.
c. $Cu(s)$ will be oxidized to Cu^{2+} by treatment with 1 M nitric acid (HNO_3).

Solution
a. In order for iron to be oxidized, some species must be reduced. In hydrochloric acid, the only reducible species is the H^+ ion. Looking up the appropriate potentials,

$Fe(s) \rightarrow Fe^{2+}(aq) + 2\ e^-$	$E^0_{ox} = +0.44 \text{ V}$
$2\ H^+(aq) + 2\ e^- \rightarrow H_2(g)$	$E^0_{red} = 0.00 \text{ V}$
$Fe(s) + 2\ H^+(aq) \rightarrow Fe^{2+}(aq) + H_2(g)$	$E^0_{tot} = +0.44 \text{ V}$

The key to spontaneity in redox reactions is the sign of E^0_{tot}

Since the calculated voltage is positive, we deduce that the reaction should occur. In the laboratory, we find that it does. Iron filings dropped into hydrochloric acid dissolve, with the evolution of $H_2(g)$.
b. Proceeding in the same way,

$Cu(s) \rightarrow Cu^{2+}(aq) + 2\ e^-$	$E^0_{ox} = -0.34 \text{ V}$
$2\ H^+(aq) + 2\ e^- \rightarrow H_2(g)$	$E^0_{red} = 0.00 \text{ V}$
$Cu(s) + 2\ H^+(aq) \rightarrow Cu^{2+}(aq) + H_2(g)$	$E^0_{tot} = -0.34 \text{ V}$

We find, as predicted, that no reaction occurs when copper is added to 1 M hydrochloric acid.
c. In HNO_3, there is another possible oxidizing agent, the NO_3^- ion. Combining the proper half-equations,

$$\frac{3[Cu(s) \rightarrow Cu^{2+}(aq) + 2\ e^-]}{2[NO_3^-(aq) + 4\ H^+(aq) + 3\ e^- \rightarrow NO(g) + 2\ H_2O]}$$
$$3\ Cu(s) + 2\ NO_3^- + 8\ H^+(aq) \rightarrow 3\ Cu^{2+}(aq) + 2\ NO(g) + 4\ H_2O$$

$$E^0_{tot} = E^0_{ox} + E^0_{red}$$
$$= -0.34 \text{ V} + 0.96 \text{ V} = +0.62 \text{ V}$$

As predicted, nitric acid does indeed oxidize copper metal to Cu^{2+}; the reduction product is one of the lower oxidation states of nitrogen (such as NO or NO_2) rather than $H_2(g)$. The reaction is shown in Color Plate 24.1.

EXERCISE Which of the metals Zn, Cd, and Hg will react with HCl, based on E^0 values? Answer: $Zn(E^0_{ox} = +0.76 \text{ V})$ and $Cd\ (E^0_{ox} = +0.40 \text{ V})$.

Strictly speaking, the conclusions reached in Example 24.3 apply only at standard concentrations (1 M for species in solution, 1 atm for gases).

24.2
Effect of
Concentration upon
Voltage

To this point, we have dealt only with standard voltages: E_{tot}^0, E_{red}^0, E_{ox}^0. These, you will recall, apply at standard concentrations (1 M for species in solution, 1 atm for gases). As we have seen, standard voltages are useful for many purposes. However, we often carry out redox reactions where the concentrations of one or more species are far removed from 1 M. Under these conditions, we need to consider the effect of concentration upon voltage.

Qualitatively, we can readily predict the direction in which voltage will shift when concentrations are changed. The following relations are observed experimentally:

1. A reaction becomes more spontaneous if the concentration of a reactant increases or that of a product decreases. Under these conditions, the voltage increases (becomes more positive).

2. A reaction becomes less spontaneous if the concentration of a reactant decreases or that of a product increases. Under these conditions, the voltage decreases (becomes less positive).

These predictions are similar to those we get with Le Chatelier's Principle

Nernst Equation

Quantitatively, we use a relation called the Nernst equation to determine the effect of concentration upon voltage. Consider the general redox reaction at 25°C:

$$a A + b B \rightarrow c C + d D$$

where, A, B, C, D are species whose concentrations can be varied. The small letters a, b, c, d refer to the coefficients in the balanced equation. The Nernst equation can be written

$$E = E_{tot}^0 - \frac{RT}{nF} \ln \frac{(\text{conc. C})^c \times (\text{conc. D})^d}{(\text{conc. A})^a \times (\text{conc. B})^b}$$

We can derive this equation by using thermodynamics

In this equation,

E is the voltage at a given concentration.
E_{tot}^0 is the standard voltage.
n is the number of moles of electrons transferred for the equation as written.
R is the gas law constant, 8.31 J/mol·K.
T is the absolute temperature in K.
F is the Faraday constant, 96,485 coulombs/mol or, since 1 joule = 1 volt × 1 coulomb, F = 96,485 J/V·mol.

Substituting values for the constants R and F, taking T = 298 K, and converting to base 10 logarithms, the Nernst equation has the following form at 25°C:

$$E = E_{tot}^0 - \frac{(8.31)(298)(2.30)}{n(96,485)} \log_{10} \frac{(\text{conc. C})^c \times (\text{conc. D})^d}{(\text{conc. A})^a \times (\text{conc. B})^b}$$

We'll use this equation

$$E = E_{tot}^0 - \frac{0.0591}{n} \log_{10} \frac{(\text{conc. C})^c \times (\text{conc. D})^d}{(\text{conc. A})^a \times (\text{conc. B})^b} \qquad (24.3)$$

We should emphasize the following points about the Nernst equation (Equation 24.3):

Concentrations of species in aqueous solution are expressed as molarity, M.

Concentrations of gases are expressed as partial pressures (P).

Terms for solids (or pure liquids) do not appear.

To understand the use of the Nernst equation, consider the reaction

$$Zn(s) + Cu^{2+}(aq) \rightarrow Zn^{2+}(aq) + Cu(s); \qquad E_{tot}^0 = +1.10 \text{ V}$$

Get n from the half-cell reaction

Here, n = 2 (two moles of electrons are produced by the oxidation of one mole of Zn and are consumed by the reduction of one mole of Cu^{2+}). Hence,

$$E = +1.10 \text{ V} - \frac{0.0591}{2} \log_{10} \frac{(\text{conc. } Zn^{2+})}{(\text{conc. } Cu^{2+})}$$

In Table 24.2, we use this equation to calculate E for various values of the ratio (conc. Zn^{2+})/(conc. Cu^{2+}). Note that

—if (conc. Zn^{2+})/(conc. Cu^{2+}) is less than 1, E is larger than the standard voltage, + 1.10 V. This agrees with the qualitative statement made earlier. To make this ratio less than 1, we could decrease the concentration of the product, Zn^{2+}, or increase that of the reactant, Cu^{2+}. Either of these changes should make the reaction more spontaneous and hence increase E.

—if (conc. Zn^{2+})/(conc. Cu^{2+}) is greater than 1, E is smaller than the standard voltage, + 1.10 V. By increasing the concentration of the product, Zn^{2+}, relative to that of the reactant, Cu^{2+}, we make the reaction less spontaneous. Hence, the voltage decreases.

Voltage is not very sensitive to concentration

—only if the concentration ratio differs from 1 by several orders of magnitude does E differ appreciably from + 1.10 V. As an extreme case, consider what happens when (conc. Zn^{2+})/(conc. Cu^{2+}) becomes 10^{37}. E drops to zero. This is the condition in a completely discharged $Zn/Zn^{2+} \parallel Cu^{2+}/Cu$ cell. Virtually all the reactant, Cu^{2+}, has been consumed. The system is at equilibrium and there is no further driving force for the reaction. The cell is "dead." In general, *when a redox reaction taking place within a voltaic cell reaches equilibrium, the voltage becomes zero.*

Table 24.2
Voltage at 25°C for the Reaction
Zn(s) + Cu²⁺(aq) → Zn²⁺(aq) + Cu(s)

(conc. Zn^{2+})/(conc. Cu^{2+})	10^{-10}	10^{-5}	10^{-1}	1	10^1	10^5	10^{10}	10^{37}
E (V)	1.40	1.25	1.13	1.10	1.07	0.95	0.80	0.00

EXAMPLE 24.4 Consider a voltaic cell in which the following reaction occurs:

$$O_2(g) + 4 H^+(aq) + 4 Br^-(aq) \rightarrow 2 H_2O + 2 Br_2(l)$$

a. Set up the Nernst equation for the cell, relating E to E_{tot}^0.
b. Determine E_{tot}^0.
c. Calculate E when P O_2 = 1 atm, conc. H^+ = conc. Br^- = 0.1 M.

Solution

a. To find n, it is helpful to break the reaction down into two half-reactions:

$$O_2(g) + 4 H^+(aq) + 4 e^- \rightarrow 2 H_2O$$

$$4 Br^-(aq) \rightarrow 2 Br_2(l) + 4 e^-$$

Clearly, n = 4; four electrons are transferred from Br^- ions to O_2 molecules. The Nernst equation must then be

$$E = E_{tot}^0 - \frac{0.0591}{4} \log_{10} \frac{1}{(P\ O_2) \times (conc.\ H^+)^4 \times (conc.\ Br^-)^4}$$

b. $E_{tot}^0 = E_{red}^0\ O_2 + E_{ox}^0\ Br^- = +1.23\ V - 1.07\ V = +0.16\ V$

c. $E = +0.16\ V - \dfrac{0.0591}{4} \log_{10} \dfrac{1}{1(0.1)^4(0.1)^4}$

$= +0.16\ V - \dfrac{0.0591}{4} \log_{10} (1 \times 10^8)$

$= +0.16\ V - \dfrac{0.0591}{4} (8.0) = +0.04\ V$

In this equation, P_{gas} is in atm. The conc. term for pure liquids and solids is equal to 1

EXERCISE Calculate E for this cell under the same conditions except that P O_2 = 10 atm. Answer: +0.06 V.

The Nernst equation can also be used to determine the effect of changes in concentration upon the voltage of an individual half-cell, E_{red}^0 or E_{ox}^0. Consider, for example, the half-reaction

$$MnO_4^-(aq) + 8 H^+(aq) + 5 e^- \rightarrow Mn^{2+}(aq) + 4 H_2O; E_{red}^0 = +1.52\ V$$

Here the Nernst equation takes the form

$$E_{red} = +1.52\ V - \frac{0.0591}{5} \log_{10} \frac{(conc.\ Mn^{2+})}{(conc.\ MnO_4^-)(conc.\ H^+)^8}$$

This equation has the same form as with total reactions

where E_{red} is the observed reduction voltage corresponding to any given concentrations of Mn^{2+}, MnO_4^-, and H^+.

Use of the Nernst Equation to Determine Ion Concentrations

In chemistry, the most important use of the Nernst equation lies in the experimental determination of the concentration of ions in solution. Suppose we measure

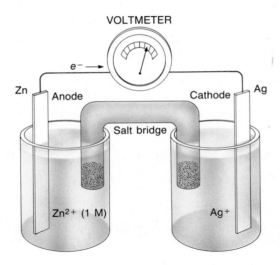

VOLTMETER

FIGURE 24.1 The reaction in this voltaic cell is

$$Zn(s) + 2\ Ag^+(aq) \rightarrow$$
$$Zn^{2+}(aq) + 2\ Ag(s)$$

From the measured cell voltage, it is possible to calculate the concentration of Ag^+ in the half-cell at the right, as shown in Example 24.5. If excess Cl^- ions are added to that half-cell, precipitating AgCl, the concentration of Ag^+ drops sharply, as does the voltage. The K_{sp} value of AgCl can be calculated from the measured voltage (Example 24.6).

the cell voltage E and know the concentrations of all but one species in the two half-cells. It should then be possible to calculate the concentration of that species by using the Nernst equation (Example 24.5).

EXAMPLE 24.5   Consider the cell shown in Figure 24.1, where the reaction is

$$Zn(s) + 2\ Ag^+(aq) \rightarrow Zn^{2+}(aq) + 2\ Ag(s)$$

a. Set up the Nernst equation for this cell reaction.
b. Determine E^0_{tot}.
c. Suppose the concentration of Zn^{2+} in the Zn/Zn^{2+} half-cell is maintained at 1 M. If the measured cell voltage is 1.21 V, what must be the concentration of Ag^+ in the Ag/Ag^+ half-cell?

Solution

a. Here, n = 2; two electrons are transferred from a Zn atom to two Ag^+ ions:

$$E = E^0_{tot} - \frac{0.0591}{2}\ log_{10}\ \frac{(conc.\ Zn^{2+})}{(conc.\ Ag^+)^2}$$

b. $E^0_{tot} = E^0_{red}\ Ag^+ + E^0_{ox}\ Zn = +0.80\ V + 0.76\ V = +1.56\ V$
c. The equation is

$$1.21 = 1.56 - \frac{0.0591}{2}\ log_{10}\ \frac{1}{(conc.\ Ag^+)^2}$$

Since $log\ 1/x^2 = -log\ x^2 = -2\ log\ x$, we can write

$$1.21 = 1.56 + \frac{0.0591}{2}\ (2\ log_{10}\ conc.\ Ag^+)$$

$$= 1.56 + 0.0591\ log_{10}\ (conc.\ Ag^+)$$

Solving,

$$log_{10}\ (conc.\ Ag^+) = \frac{1.21 - 1.56}{0.0591} = -6.0;$$

$$conc.\ Ag^+ = 1 \times 10^{-6}\ M$$

Zn   Anode

Cathode   Ag

e^-

Salt bridge

Zn^{2+} (1 M)

Ag^+

Voltaic cells are particularly useful when you want to determine small conc.

Like this one

EXERCISE Suppose conc. Zn^{2+} = 0.1 M. If E = +1.21 V, what is conc. Ag^+? Answer: 4×10^{-7} M.

The approach suggested by Example 24.5 is widely used to determine the concentration of ions in solution. It is very useful in cases where the concentration is low, perhaps 10^{-3} M or less. At such concentrations other approaches, such as titrations, are difficult to carry out accurately. To determine the concentration of an ion in solution using the Nernst equation, we need the following equipment:

1. A vacuum-tube voltmeter or potentiometer capable of measuring E to at least 0.01 V.

2. A reference half-cell of known voltage. In Example 24.5 and Figure 24.1, this is the Zn/Zn^{2+} half cell, E^0_{ox} = +0.76 V.

3. A half-cell whose voltage depends upon the concentration of the ion involved. In the simplest case (Fig. 24.1), this may consist of a metal wire or strip (Ag metal) dipping into a solution containing the cation (Ag^+) whose concentration is to be measured. More sophisticated half-cells are often used. Figure 24.2 shows a *glass* electrode used in the pH meter referred to in Chapter 19. This electrode is immersed in the solution whose [H^+] or pH is to be measured. The voltage across the thin, fragile glass membrane is a linear function of the pH of the solution outside the membrane.

No current flows, so E has its maximum value

A pH meter uses a special kind of voltaic cell

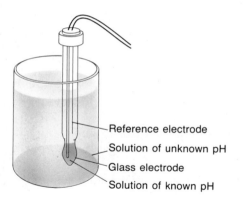

Reference electrode
Solution of unknown pH
Glass electrode
Solution of known pH

FIGURE 24.2 The pH of a solution can be determined with the aid of a "glass electrode." The voltage between the glass electrode and the reference electrode is directly related to pH. The leads from the electrodes are connected to a pH meter of the type referred to in Chapter 19.

Use of the Nernst Equation to Determine Solubility Product Constants

As we have just seen, the Nernst equation can be used to find the concentration of an ion in solution. With a properly designed cell, this concentration can in turn be used to calculate the value of a solubility product constant involving that ion. For example, we might determine [Ag^+] in a solution in which the concentration of Cl^- in equilibrium with AgCl(s) is known. From that information, we could calculate the solubility product constant of silver chloride (Example 24.6).

EXAMPLE 24.6 Consider again the voltaic cell referred to in Example 24.5 and shown in Figure 24.1, where conc. Zn^{2+} = 1 M. Suppose that, to the Ag/Ag^+ half-cell, we add an excess of hydrochloric acid, precipitating AgCl and making the concentration of Cl^- ion over the precipitate 0.1 M. Under these conditions, the cell voltage is found to be 1.04 V. Calculate

a. the concentration of Ag^+ in the Ag/Ag^+ half-cell.

b. K_{sp} of AgCl.

Solution

a. As before, we have

$$E = E_{tot}^0 - \frac{0.0591}{2} \log_{10} \frac{1}{(\text{conc. } Ag^+)^2}$$

Substituting $E = 1.04$ V, $E_{tot}^0 = 1.56$ V, and solving as in Example 24.5, we obtain

$$\log_{10} (\text{conc. } Ag^+) = \frac{1.04 - 1.56}{0.0591} = -8.8$$

$$\text{conc. } Ag^+ = 2 \times 10^{-9} \text{ M}$$

b. $K_{sp} = [Ag^+] \times [Cl^-]$. The concentration of Cl^- was stated to be 0.1 M; we have just calculated $[Ag^+] = 2 \times 10^{-9}$ M. Hence,

$$K_{sp} \text{ AgCl} = (2 \times 10^{-9}) \times (1 \times 10^{-1}) = 2 \times 10^{-10}$$

EXERCISE Suppose enough HI were added to the cell to precipitate AgI and make conc. $I^- = 0.1$ M. Would the cell voltage be greater or less than 1.04 V (K_{sp} AgI $= 1 \times 10^{-16}$)? Answer: Less (about 0.67 V).

Many equilibrium constants are best found by using voltaic cells

Most of the solubility product constants listed in Table 18.2, Chapter 18, were determined with voltaic cells using the approach illustrated in Example 24.6. In general, to calculate K_{sp} from data giving cell voltage as a function of concentration, you follow three steps:

1. Set up the Nernst equation for the cell. Substitute values for all known quantities including E, E_{tot}^0, and n.

2. Solve for the concentration of the ion in question (for example, Ag^+ in Example 24.6). This requires that you know how to handle logarithms (Appendix 4).

3. Using the result of (2) and other information given in the statement of the problem (such as conc. Cl^-), calculate K_{sp}.

24.3
Relations Between
E_{tot}^0, ΔG^0, and K

In this chapter, we have emphasized that E_{tot}^0 is a criterion of spontaneity of redox reactions. If E_{tot}^0 is positive, the reaction is spontaneous at standard concentrations (1 M for species in solution, 1 atm for gases). If E_{tot}^0 is negative, the reaction is nonspontaneous; energy must be supplied to make it take place.

In earlier chapters, we discussed two other indicators of spontaneity.

1. *The standard free energy change, ΔG^0.* If ΔG^0 is negative, the reaction is spontaneous at standard concentrations. If ΔG^0 is positive, the reaction is nonspontaneous.

2. *The equilibrium constant, K.* If K is greater than 1, reaction occurs spontaneously from left to right at standard concentrations (1 M, 1 atm). If K is less than 1, the reverse reaction is spontaneous at standard concentrations.

As you might expect, E_{tot}^0 is related to both the standard free energy change,

ΔG^0, and the equilibrium constant, K, for a redox reaction. In this section, we will discuss the nature of these relationships.

Relation Between Cell Voltage and Free Energy Change for a Redox Reaction

We pointed out in Chapter 14 that the free energy change, ΔG, is a measure of the maximum amount of useful work that can be obtained from a reaction. Specifically, at constant T and P:

$$\Delta G = -w_{max} \tag{24.4}$$

where w_{max} refers to the work *produced* by the reaction system. When a reaction is carried out in a voltaic cell, the "useful work" is the electrical energy that is generated. As noted in Chapter 23, electrical energy in joules is the product of coulombs times volts. This means that

$$w_{max} = nFE \tag{24.5}$$

where n is the number of moles of electrons that pass through the cell and F is the Faraday constant, which converts moles of electrons to coulombs. The symbol E stands for the maximum cell voltage, which is the quantity calculated from standard voltages such as those given in Table 24.1.

Putting these two relations together, we obtain

$$\Delta G = -nFE \tag{24.6}$$

At standard conditions, this equation becomes

$$\Delta G^0 = -nFE_{tot}^0 \tag{24.7}$$

Using Eq. 24.6 we can establish reaction spontaneity for any concentrations, not just 1 M

In this equation, ΔG^0 is the standard free energy change for the redox reaction carried out in the voltaic cell and E_{tot}^0 is the standard voltage for that reaction. Moreover, as indicated above:

n has the same meaning as in the Nernst equation; it is the number of moles of electrons transferred in the chemical equation written for the redox reaction.

F is the Faraday constant, 96,485 C/mol = 96,485 J/V·mol. If we use Equation 24.7 with F = 96,485 J/V·mol, we obtain ΔG^0 for the redox reaction in joules. Most often, we want ΔG^0 in kilojoules. Since 1 kJ = 10^3 J,

$$\Delta G^0 \text{ (kJ)} = -96.5 \, nE_{tot}^0 \tag{24.8}$$

From Equation 24.8, we see that ΔG^0 and E_{tot}^0 have opposite signs. This is reasonable; remember that a spontaneous reaction has a *negative* free energy change and a *positive* voltage. Equation 24.8 can readily be used to calculate ΔG^0 from E_{tot}^0 or vice versa (Example 24.7).

EXAMPLE 24.7 For the reaction $Cl_2(g) + 2 Br^-(aq) \rightarrow 2 Cl^-(aq) + Br_2(l)$, calculate

a. E_{tot}^0, using Table 24.1. b. ΔG^0, using Equation 24.8.

Solution

a. $E_{tot}^0 = E_{red}^0 \; Cl_2(g) + E_{ox}^0 \; Br^-(aq) = +1.36 \; V - 1.07 \; V = +0.29 \; V$

Since E_{tot}^0 is positive, we conclude that the reaction is spontaneous. Indeed it is; one way to make bromine in the laboratory is to shake a solution of NaBr with a saturated solution of chlorine gas.

b. In this equation, n = 2; that is,

$$Cl_2(g) + 2 \; e^- \rightarrow 2 \; Cl^-(aq); \qquad 2 \; Br^-(aq) \rightarrow Br_2(l) + 2 \; e^-$$

Hence, we have

$$\Delta G^0 \; (in \; kJ) = -2(96.5)(+0.29) = -56 \; kJ$$

As we would expect, ΔG^0 is negative for this spontaneous reaction.

EXERCISE Calculate E_{tot}^0 for a reaction for which ΔG^0 is $+10.0$ kJ and n = 2. Answer: -0.0518 V.

Relation Between Cell Voltage and the Equilibrium Constant for a Redox Reaction

To relate the standard cell voltage, E_{tot}^0, to the equilibrium constant, K, for the redox reaction taking place within the cell, it is convenient to start with the Nernst equation. For the general reaction

$$a \; A + b \; B \rightarrow c \; C + d \; D$$

$$E = E_{tot}^0 - \frac{0.0591}{n} \log_{10} \frac{(conc. \; C)^c \times (conc. \; D)^d}{(conc. \; A)^a \times (conc. \; B)^b}$$

Given the Nernst equation, we can derive the relation between K and E_{tot}^0, and between K and ΔG^0 for a reaction

As pointed out earlier, the cell voltage becomes zero when the redox reaction taking place within the cell is at equilibrium. In other words,

$$E = 0, \qquad when \; \frac{(conc. \; C)^c \times (conc. \; D)^d}{(conc. \; A)^a \times (conc. \; B)^b} = \frac{[C]^c \times [D]^d}{[A]^a \times [B]^b} = K$$

Substituting in the Nernst equation under these conditions, we obtain

$$0 = E_{tot}^0 - \frac{0.0591}{n} \log_{10} K$$

Solving for $\log_{10} K$,

$$\log_{10} K = \frac{n E_{tot}^0}{0.0591} \qquad (at \; 25°C) \tag{24.9}$$

Equation 24.9, which relates E_{tot}^0 to K, tells us that

if E_{tot}^0 is a positive quantity, $\log_{10} K$ is positive, and K is greater than 1.

if E_{tot}^0 is a negative quantity, $\log_{10} K$ is negative, and K is less than 1.

EXAMPLE 24.8 For the reaction $3 \text{ Ag}(s) + NO_3^-(aq) + 4 \text{ H}^+(aq) \rightarrow 3 \text{ Ag}^+(aq) + NO(g) + 2 H_2O$, calculate

a. E^0_{tot} using Table 24.1
b. K, using Equation 24.9.

Solution

a. $E^0_{tot} = E^0_{red} NO_3^- + E^0_{ox} Ag = +0.96 \text{ V} - 0.80 \text{ V} = +0.16 \text{ V}$

b. To find n, it is helpful to break the equation into half-equations:

$$3 \text{ Ag}(s) \rightarrow 3 \text{ Ag}^+(aq) + 3 e^-$$

$$NO_3^-(aq) + 4 \text{ H}^+(aq) + 3 e^- \rightarrow NO(g) + 2 H_2O$$

Clearly, n = 3. Substituting in Equation 24.9,

$$\log_{10} K = \frac{3(+0.16)}{0.0591} = 8.1; \quad K \approx 1 \times 10^8$$

EXERCISE Suppose n = 2 and $E^0_{tot} = 3.00$ V; calculate K. Answer: $K = 10^{102}$.

Table 24.3 is a list of values of K calculated from Equation 24.9, corresponding to various values of E^0_{tot} with n = 2. It turns out that if E^0_{tot} is greater than about 0.2 V, K is very large, of the order of 10^7 or greater. In such a case, the redox reaction goes virtually to completion under ordinary conditions. At the opposite extreme, note what happens when E^0_{tot} is less than about -0.2 V. Here, K is very small, 10^{-7} or less, and in essence the reaction does not occur at all. Only if E^0_{tot} falls within a quite narrow range, perhaps -0.2 to $+0.2$ V, will we get an equilibrium mixture containing appreciable amounts of both products and reactants.

This means that most redox reactions either go to completion or don't go at all

Table 24.3
Relation Between E^0_{tot} and K (n = 2)

E^0_{tot}	$\log_{10} K$	K	E^0_{tot}	$\log_{10} K$	K	E^0_{tot}	$\log_{10} K$	K
+1.00	+34	10^{34}	+0.10	+3.4	2500	−0.20	−6.8	2×10^{-7}
+0.80	+27	10^{27}	+0.05	+1.7	50	−0.40	−14	10^{-14}
+0.60	+20	10^{20}	0.00	0.0	1	−0.60	−20	10^{-20}
+0.40	+14	10^{14}	−0.05	−1.7	0.02	−0.80	−27	10^{-27}
+0.20	+6.8	6×10^6	−0.10	−3.4	0.0004	−1.00	−34	10^{-34}

24.4
Corrosion of Metals

Most metals corrode when exposed to the atmosphere; that is, they react with oxygen, water vapor, carbon dioxide, or some other component of air to form a compound. Gold and platinum are among the few metals that retain their shiny appearance indefinitely when exposed to air. These metals are very difficult to oxidize ($E^0_{ox} Au = -1.50$ V), which explains their resistance to corrosion. Silver is also difficult to oxidize ($E^0_{ox} = -0.80$ V); at ordinary temperatures, it does not react with oxygen or any of the other major components of the atmosphere. If sulfur compounds are present, however, silver slowly tarnishes, forming black silver sulfide, Ag_2S. With hydrogen sulfide, the reaction is

$$4\ Ag(s)\ +\ 2\ H_2S(g)\ +\ O_2(g) \rightarrow 2\ Ag_2S(s)\ +\ 2\ H_2O(l) \tag{24.10}$$

One way to remove tarnish from silverware is to wrap it loosely in aluminum foil, immerse in a dilute solution of $NaHCO_3$, and warm slightly. Aluminum, which is much more readily oxidized than silver (E^0_{ox} Al = +1.66 V), is oxidized while Ag^+ ions are reduced:

$$2\ Al(s)\ +\ 3\ Ag_2S(s) \rightarrow 2\ Al^{3+}(aq)\ +\ 3\ S^{2-}(aq)\ +\ 6\ Ag(s) \tag{24.11}$$

Since aluminum has such a large positive value of E^0_{ox}, we would expect it to react readily with oxygen of the air. Indeed it does; a piece of freshly cut aluminum quickly picks up a coating of aluminum oxide:

$$4\ Al(s)\ +\ 3\ O_2(g) \rightarrow 2\ Al_2O_3(s) \tag{24.12}$$

However, the Al_2O_3 coating, which is only about 5 nm (5×10^{-9} m) thick, adheres tightly to the surface of the aluminum. This prevents further corrosion and explains why aluminum pans do not disintegrate upon exposure to air.

Aluminum oxide is readily soluble in strong acid:

$$Al_2O_3(s)\ +\ 6\ H^+(aq) \rightarrow 2\ Al^{3+}(aq)\ +\ 3\ H_2O \tag{24.13}$$

It also dissolves in strong base because of complex ion formation:

Al will dissolve in a solution of drain cleaner

$$Al_2O_3(s)\ +\ 3\ H_2O\ +\ 2\ OH^-(aq) \rightarrow 2\ Al(OH)_4{}^-(aq) \tag{24.14}$$

Taking these reactions into account, you can see why it is not a good idea to expose aluminum cookware to acidic foods (cranberries, citrus fruits, vinegar) or strong bases such as lye (NaOH).

Copper in moist air slowly acquires a dull green coating (often seen on statuary). The green material is most often a 1:1 mole mixture of $Cu(OH)_2$ and $CuCO_3$. The corrosion reaction can be represented as

$$2\ Cu(s)\ +\ H_2O(g)\ +\ CO_2(g)\ +\ O_2(g) \rightarrow Cu(OH)_2 \cdot CuCO_3(s) \tag{24.15}$$

In the presence of sulfur dioxide, the solid formed contains copper(II) sulfate rather than the carbonate:

Cu corrodes very slowly

$$2\ Cu(s)\ +\ H_2O(g)\ +\ SO_2(g)\ +\ \tfrac{3}{2}\ O_2(g) \rightarrow Cu(OH)_2 \cdot CuSO_4(s) \tag{24.16}$$

The product, like that formed in Reaction 24.15, is green.

Several elements, including zinc and lead, undergo a reaction with air similar to that of copper (Reaction 24.15). The products $Zn(OH)_2 \cdot ZnCO_3$ and $Pb(OH)_2 \cdot PbCO_3$ are white and adhere tightly to the metal, preventing further corrosion. These compounds are, however, soluble in strong acid:

$$Zn(OH)_2 \cdot ZnCO_3(s)\ +\ 4\ H^+(aq) \rightarrow 2\ Zn^{2+}(aq)\ +\ 3\ H_2O\ +\ CO_2(g) \tag{24.17}$$

In the case of lead, the protective coating of $Pb(OH)_2 \cdot PbCO_3$ dissolves in the weak acid $HC_2H_3O_2$, acetic acid. This occurs because Pb^{2+} forms a very stable complex with the acetate ion, $C_2H_3O_2{}^-$. It has been suggested that the ancient Romans suffered from lead poisoning because they stored wine (containing some acetic acid) in pottery vessels glazed with lead compounds. Conceivably, this could have contributed to the decline and fall of the Roman Empire.

Maybe they just drank too much wine

Corrosion of Iron

From an economic standpoint, the most important corrosion reaction is that involving iron and steel. It is estimated that in the United States the annual cost of corrosion of ferrous metals exceeds 50 billion dollars. We see the results of corrosion all around us in junk piles and automobile graveyards. Perhaps as much as 20% of all the iron produced each year in this country goes to replace products whose usefulness has been destroyed by rust.

In Minnesota, cars rust out in about 10 years

To understand how iron corrodes, consider what happens when a sheet of iron is exposed to a water solution containing dissolved oxygen. The iron tends to oxidize according to the half-reaction

$$Fe(s) \rightarrow Fe^{2+}(aq) + 2\ e^- \qquad \textbf{(24.18a)}$$

At the same time, oxygen molecules in the solution are reduced:

$$\tfrac{1}{2} O_2(g) + H_2O + 2\ e^- \rightarrow 2\ OH^-(aq) \qquad \textbf{(24.18b)}$$

Adding these two half-equations, and noting that iron(II) hydroxide is insoluble, we obtain for the primary corrosion reaction

$$Fe(s) + \tfrac{1}{2} O_2(g) + H_2O \rightarrow Fe(OH)_2(s) \qquad \textbf{(24.18)}$$

Reaction 24.18 appears to be the reaction involved in the first step of the corrosion of iron or steel. Ordinarily, iron(II) hydroxide is further oxidized in a second step:

$$2\ Fe(OH)_2(s) + \tfrac{1}{2} O_2(g) + H_2O \rightarrow 2\ Fe(OH)_3(s) \qquad \textbf{(24.19)}$$

The final product is the loose, flaky deposit that we call rust. It has the reddish brown color of iron(III) hydroxide, $Fe(OH)_3$.

Experimentally, we find that the two half-reactions 24.18a and 24.18b do not occur at the same location. The rust on a nail extracted from an old building is concentrated near the head, which has been in contact with moist air (Fig. 24.3). The most serious pitting, caused by oxidation, is found along the shank of the nail, which is embedded in the wood. These observations suggest that oxidation is occurring along a surface some distance away from the point where oxygen is being reduced.

They can, but they usually don't

The fact that oxidation and reduction half-reactions take place at different locations suggest that corrosion occurs by an electrochemical mechanism. The surface of a piece of corroding iron may be visualized as consisting of a series of tiny voltaic cells. At *anodic areas,* iron is oxidized to Fe^{2+} ions; at *cathodic areas,* elementary oxygen is reduced to OH^- ions. Electrons are transferred through the iron, which acts like the external conductor of an ordinary voltaic cell. The

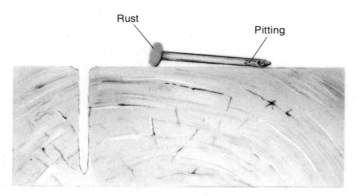

FIGURE 24.3 Corrosion of an iron nail driven into wood. Rust collects near the head of the nail, but pitting occurs along its length.

electrical circuit is completed by the flow of ions through the water solution or film covering the iron.

Many characteristics of corrosion are most readily explained in terms of an electrochemical mechanism. A perfectly dry metal surface is not attacked by oxygen; iron exposed to dry air does not corrode. This seems plausible if corrosion occurs through a voltaic cell, which requires a water solution through which ions can move to complete the circuit. The fact that corrosion occurs more readily in seawater than in fresh water has a similar explanation. The dissolved salts in seawater supply the ions necessary for the conduction of current.

The existence of discrete cathodic and anodic areas on a piece of corroding iron requires that adjacent surface areas differ from each other chemically. There are several ways in which one small area on a piece of iron or steel can become anodic or cathodic with respect to an adjacent area. Two of the most important are the following:

1. *The presence of impurities at scattered locations along the metal surface.* A tiny crystal of a less active metal such as copper or tin embedded in the surface of the iron acts as a cathode at which oxygen molecules are reduced. The iron atoms in the vicinity of these impurities are anodic and undergo oxidation to Fe^{2+} ions. This effect can be demonstrated on a large scale by immersing in water an iron plate that has been partially copper plated (Fig. 24.4). At the interface between the two metals, a voltaic cell is set up in which the iron is anodic and the copper cathodic. A thick deposit of rust forms at the interface. The formation of rust

> Ask any boat owner who runs his outboard motor in seawater

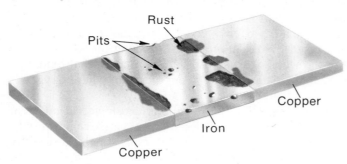

FIGURE 24.4 Corrosion of iron in contact with copper. Corrosion commonly occurs at points where dissimilar metals are connected.

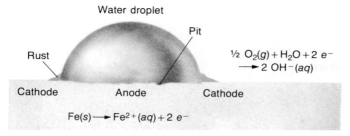

Water droplet
Pit
Rust
$\frac{1}{2} O_2(g) + H_2O + 2 e^-$
$\longrightarrow 2 OH^-(aq)$
Cathode   Anode   Cathode
$Fe(s) \longrightarrow Fe^{2+}(aq) + 2 e^-$

FIGURE 24.5 Corrosion of iron under a drop of water. The Fe^{2+} ions migrate toward the edge of the drop, where they precipitate as $Fe(OH)_2$, which is then further oxidized by air to $Fe(OH)_3$.

inside an automobile bumper where the chromium plate stops is another example of this phenomenon.

2. *Differences in oxygen concentration along the metal surface.* To illustrate this effect, consider what happens when a drop of water adheres to the surface of a piece of iron exposed to the air (Fig. 24.5). The metal around the edges of the drop is in contact with water containing a high concentration of dissolved oxygen. The water touching the metal beneath the center of the drop is depleted in oxygen, since it is cut off from contact with air. As a result, a small oxygen concentration cell is set up. The area around the edge of the drop, where the oxygen concentration is high, becomes cathodic; oxygen molecules are reduced there via Reaction 24.18b. Directly beneath the drop is an anodic area where the iron is oxidized. A particle of dirt on the surface of an iron object can act in much the same way as a drop of water to cut off the supply of oxygen to the area beneath it and thereby establish anodic and cathodic areas. This explains why garden tools left covered with soil are particularly susceptible to rusting.

After a while, some gardeners decide they don't mind a little rust

Iron or steel objects can be protected from corrosion in several different ways.

1. *Covering the surface with a protective coating.* This may be a layer of paint that cuts off access to moisture and oxygen. Under more severe conditions, it may be desirable to cover the surface of the iron or steel with a layer of another metal. Metallic plates, applied electrically (Cr, Ni, Cu, Ag, Zn, Sn) or by immersion at high temperatures (Zn, Sn), are ordinarily more resistant to heat and chemical attack than the organic coating left when paint dries. If the plating metal is more active than iron (for example, Zn), it, rather than iron, will be oxidized if the surface is broken. However, if the plating metal is less active than iron, there is a danger that cracks in its surface may enhance the corrosion of the iron or steel. This problem can arise with "tin cans," which are made by applying a layer of tin over a steel base. If the food in the can contains citric acid, some of the tin plate may dissolve, exposing the steel beneath.* When the can is opened, exposing the interior to the air, rust forms spontaneously on the iron surrounding the breaks in the tin surface. A thin coating of lacquer is ordinarily applied over the tin to prevent corrosive effects of this type.

The tin layer is very thin

2. *Bringing the object into electrical contact with a more active metal* such as magnesium or zinc. Under these conditions, the iron becomes cathodic and

---

*Tin forms an extremely stable complex with citrate ion and, hence, is attacked more readily by citric acid than by many stronger inorganic acids.

FIGURE 24.6 Cathodic protection of Alaska Oil Pipeline. (Photo, Alyeska Oil Pipeline Service Company)

In special cases, a small (−) DC voltage is applied to the metal, making it a cathode

hence, is protected against rusting; the more active metal serves as a sacrificial anode in a large-scale corrosion cell. This method of combating corrosion, known as **cathodic protection**, is particularly useful for steel objects such as cables or pipelines that are buried under soil or water (Fig. 24.6).

Summary

Voltaic cells can be used to measure standard voltages, E^0_{tot}. By arbitrarily setting the standard reduction voltage (E^0_{red}) of H^+ equal to zero, it is possible to obtain standard potentials for the reduction (E^0_{red}) of various species (Table 24.1). Standard voltages for oxidation (E^0_{ox}) can be obtained by changing the sign of the E^0_{red} value for the reverse half-reaction. Standard voltages for half-reactions can be used to do the following:

1. Compare the strengths of oxidizing agents and reducing agents (Example 24.1). A strong oxidizing agent has a large positive E^0_{red} value; a strong reducing agent has a large positive E^0_{ox} value.

2. Calculate the standard cell voltage (Example 24.2): $E^0_{tot} = E^0_{ox} + E^0_{red}$.

3. Determine whether or not a redox reaction is spontaneous (Example 24.3). The reaction is spontaneous if $E^0_{tot} > 0$.

The voltage of a cell depends upon the concentrations of reactants and products. The effect can be calculated from the Nernst equation (Example 24.4). The Nernst equation can also be used to obtain the concentration of a species in solution, if the cell voltage is known (Examples 24.5 and 24.6).

Knowing the standard voltage corresponding to a given redox reaction, we can calculate either the standard free energy change (Example 24.7) or the equilibrium constant (Example 24.8). The appropriate relations are

$$\Delta G^0 \text{ (in kJ)} = -96.5 \, nE^0_{tot}; \qquad \log_{10} K = \frac{nE^0_{tot}}{0.0591}$$

One of the most common redox reactions in the world around us is the corrosion of metals. Most metals are oxidized by exposure to air. In some cases (Al, Zn, and Pb), the product formed adheres tightly to the surface of the metal, preventing further corrosion. This is not the case with iron, where corrosion is a very serious problem. The corrosion of iron and steel occurs by an electrochemical mechanism (Figs. 24.3 through 24.6).

Key Words and Concepts

concentration cell	standard free energy change, ΔG^0	standard reduction voltage, E^0_{red}
corrosion	standard oxidation voltage, E^0_{ox}	standard voltage, E^0_{tot}
Faraday constant	standard potential, E^0_{red}	voltage (V)
Nernst equation		

Questions and Problems

Strength of Oxidizing and Reducing Agents

24.1 Using Table 24.1, arrange the following reducing agents in order of increasing strength:

SO_2, Au, K, F^-, Fe^{2+}

24.2 Consider the following species:

MnO_4^-, I^-, NO_3^-, Fe^{2+}, Ca

Using Table 24.1, classify each species as an oxidizing or reducing agent. Arrange the oxidizing agents in order of increasing strength; do the same with the reducing agents.

24.3 Use Table 24.1 to select
a. a reducing agent that will convert Pb^{2+} to Pb but not Cd^{2+} to Cd.
b. an oxidizing agent that converts I^- to I_2 but not Br^- to Br_2.
c. a reducing agent that converts Au^{3+} to Au but not $AuCl_4^-$ to Au.

24.31 Use Table 24.1 to classify each of the following as a reducing or oxidizing agent. Indicate the strongest one in each category.

Co^{2+}, Br_2, Cu^{2+}, Sn^{2+}, Al

24.32 Follow the directions of Question 24.2 for the following species:

Sn^{4+}, Ag^+, Br_2, Mn^{2+}, $Cr_2O_7^{2-}$, Cd

24.33 From Table 24.1 select a suitable species for each of the following (standard concentrations):
a. an oxidizing agent to convert Fe to Fe^{2+} but not Sn to Sn^{2+}.
b. a reducing agent capable of converting Co^{2+} to Co but not Mg^{2+} to Mg.
c. an oxidizing agent capable of converting Cl^- to Cl_2 but not Co^{2+} to Co^{3+}.

Calculation of E_{tot}^0

24.4 Calculate E_{tot}^0 for the following voltaic cells:
  a. $Al(s) + NO_3^-(aq) + 4 H^+(aq) \rightarrow$
     $NO(g) + 2 H_2O + Al^{3+}(aq)$
  b. $Cu(s) + NO_3^-(aq) + H_2O \rightarrow$
     $Cu(OH)_2(s) + NO_2^-(aq)$
  c. Cu-Cu^{2+} half-cell and I_2-I^- half-cell

24.5 Using Table 24.1, calculate E_{tot}^0 for
  a. $2 Cr^{2+}(aq) + Sn^{4+}(aq) \rightarrow 2 Cr^{3+}(aq) + Sn^{2+}(aq)$
  b. $Fe(s) + 2 H_2O \rightarrow Fe(OH)_2(s) + H_2(g)$
  c. iron(II) with oxygen in base.

24.6 Calculate E_{tot}^0 for the cells:
  a. $(Pt)H_2/H^+ \parallel (Pt)Cl_2/Cl^-$
  b. $Sn/Sn^{2+} \parallel Pb^{2+}/Pb$
  c. $(Pt)I^-/I_2 \parallel (Pt)Cl_2/Cl^-$

24.7 Suppose E_{red}^0 for $H^+ \rightarrow H_2$ were taken to be -1.00 V rather than zero. On this basis, what would be
  a. E_{ox}^0 $H_2 \rightarrow H^+$?
  b. E_{red}^0 $Cu^{2+} \rightarrow Cu$?
  c. E_{tot}^0 for the $Zn/Zn^{2+} \parallel Ag^+/Ag$ cell?

24.8 What is the minimum voltage that must be applied to achieve the following reactions in an electrolytic cell?
  a. $Ni^{2+}(aq) + 2 Cl^-(aq) \rightarrow Ni(s) + Cl_2(g)$
  b. electrolysis of a water solution of KI to give I_2, OH^-, and H_2.

24.34 Calculate E_{tot}^0 for the following voltaic cells:
  a. $Sn^{4+}(aq) + Pb(s) \rightarrow Sn^{2+}(aq) + Pb^{2+}(aq)$
  b. $O_2(g) + 4 H^+(aq) + 4 Cr^{2+}(aq) \rightarrow$
     $2 H_2O + 4 Cr^{3+}(aq)$
  c. Mn-Mn^{2+} half-cell and Ni-Ni^{2+} half-cell

24.35 Using Table 24.1, calculate E_{tot}^0 for
  a. $Fe(s) + 2 Fe^{3+}(aq) \rightarrow 3 Fe^{2+}(aq)$
  b. $Mn^{2+}(aq) + H_2O_2(aq) \rightarrow MnO_2(s) + 2H^+(aq)$
  c. $Cr(OH)_3$ reacting with O_2 in base.

24.36 Calculate standard voltages for the cells
  a. $Sn/Sn^{2+} \parallel Ag^+/Ag$
  b. $Zn/Zn^{2+} \parallel (Pt)Cl_2/Cl^-$
  c. $Ni/Ni^{2+} \parallel (Pt)Br_2/Br^-$

24.37 Suppose E_{red}^0 $Cu^{2+} \rightarrow Cu(s)$ were taken to be zero. On this basis, what would be
  a. E_{ox}^0 $Ag \rightarrow Ag^+$?
  b. E_{red}^0 $H^+ \rightarrow H_2$?
  c. E_{tot}^0 for the $Zn/Zn^{2+} \parallel Ag^+/Ag$ cell?

24.38 Which of the following electrolyses can be accomplished with a single dry cell (1.5 volts)?
  a. $Cd^{2+}(aq) + 2 Cl^-(aq) \rightarrow Cd(s) + Cl_2(g)$
  b. electrolysis of water to give H_2 and O_2.

Spontaneity and E_{tot}^0

24.9 Which of the following reactions are spontaneous at standard concentrations?
  a. $AuCl_4^-(aq) + 3 Fe^{2+}(aq) \rightarrow$
     $Au(s) + 4 Cl^-(aq) + 3 Fe^{3+}(aq)$
  b. $2 NO_3^-(aq) + 8 H^+(aq) + 6 Cl^-(aq) \rightarrow$
     $2 NO(g) + 4 H_2O + 3 Cl_2(g)$
  c. $Cu(s) + 2 H^+(aq) \rightarrow Cu^{2+}(aq) + H_2(g)$

24.10 Using Table 24.1, calculate E_{tot}^0 and decide whether each of the following reactions will occur at standard concentrations.
  a. $2 Al(s) + 3 Cu^{2+}(aq) \rightarrow 2 Al^{3+}(aq) + 3 Cu(s)$
  b. $3 Cd(s) + 8 H^+(aq) + 2 NO_3^-(aq) \rightarrow$
     $3 Cd^{2+}(aq) + 2 NO(g) + 4 H_2O$
  c. $Sn^{2+}(aq) + Pb(s) \rightarrow Sn(s) + Pb^{2+}(aq)$

24.11 Which of the following metals will react with 1 M HCl?
  a. Co   b. Au   c. Mg   d. Cu

24.39 Which of the following reactions are spontaneous at standard concentrations?
  a. $I_2(s) + 2 Br^-(aq) \rightarrow Br_2(l) + 2 I^-(aq)$
  b. $Zn(s) + 2 Fe^{3+}(aq) \rightarrow Zn^{2+}(aq) + 2 Fe^{2+}(aq)$
  c. $O_2(g) + 4 H^+(aq) + 4 Cl^-(aq) \rightarrow 2 H_2O +$
     $2 Cl_2(g)$

24.40 Using Table 24.1, calculate E_{tot}^0 and decide whether each of the following reactions will occur at standard concentrations.
  a. $Sn^{4+}(aq) + Sn(s) \rightarrow 2 Sn^{2+}(aq)$
  b. $2 Hg^{2+}(aq) + Sn^{2+}(aq) \rightarrow Hg_2^{2+}(aq) + Sn^{4+}(aq)$
  c. $Sn^{4+}(aq) + 2 Fe^{2+}(aq) \rightarrow Sn^{2+}(aq) + 2 Fe^{3+}(aq)$

24.41 Which of the following will be oxidized by 1 M HNO_3?
  a. F^- b. I^- c. Ag   d. Au

24.12 Predict what reaction, if any, will occur when liquid bromine is added to an acidic aqueous solution of each of the following (standard concentrations):
a. $Ca(NO_3)_2$
b. FeI_2
c. AgF

24.42 Using Table 24.1, decide what reaction, if any, will occur when the following are mixed (standard concentrations):
a. Co^{2+}, Co, Fe^{2+}
b. Fe^{2+}, Fe^{3+}, Ag^+
c. ClO_3^-, H^+, $Hg(l)$

Nernst Equation

24.13 Consider a voltaic cell in which the following reaction occurs:

$$Zn(s) + 2 H^+(aq) \rightarrow Zn^{2+}(aq) + H_2(g)$$

a. Set up the Nernst equation for this cell, relating E to E^0_{tot}.
b. Taking $E^0_{tot} = +0.76$ V, calculate E when conc. $Zn^{2+} = 1.0$ M, $P H_2 = 1.0$ atm, conc. $H^+ = 1.0 \times 10^{-3}$ M.

24.43 Calculate E for the cell in Problem 24.13 under the same conditions except that conc. $H^+ = 1.0 \times 10^{-4}$ M.

24.14 For the reduction of MnO_2 to Mn^{2+}, E^0_{red} is $+1.23$ V. Using the Nernst equation, calculate E_{red} when all species are at standard concentrations except H^+, which is 1.0×10^{-4} M.

24.44 Calculate E_{red} for the half-reaction referred to in Problem 24.14 under the same conditions except that the pH is 2.30.

24.15 Calculate the voltages of cells under the following conditions:
a. $Cd^{2+}(0.020 M) + Zn(s) \rightarrow Zn^{2+}(0.50 M) + Cd(s)$
b. $Sn^{2+}(0.10 M) + Cu^{2+}(1 M) \rightarrow Cu(s) + Sn^{4+}(0.010 M)$
Are the cell reactions spontaneous? Explain.

24.45 Calculate the voltages of cells under the following conditions:
a. $Fe(s) + Cu^{2+}(0.10 M) \rightarrow Fe^{2+}(0.010 M) + Cu(s)$
b. $Cu(s) + 2 H^+(0.10 M) \rightarrow Cu^{2+}(0.0010 M) + H_2(1 atm)$
Are the cell reactions spontaneous? Explain.

24.16 At what concentration of Cu^{2+} is the voltage zero for the cell

$$Ni(s) + Cu^{2+}(aq) \rightarrow Ni^{2+}(1 M) + Cu(s)?$$

24.46 At what concentration of H^+ is the voltage $+0.50$ V for the cell

$$Zn(s) + 2 H^+(aq) \rightarrow Zn^{2+}(1 M) + H_2(g, 1 atm)?$$

24.17 Consider the reaction:

$$O_2(g) + 4H^+(aq) + 4 Cl^-(aq) \rightarrow 2 Cl_2(g) + 2 H_2O$$

a. Calculate E^0_{tot}.
b. At what pressure of Cl_2 is the voltage zero, if all other species are at standard concentrations?

24.47 Consider the reaction:

$$2 Fe^{3+}(aq) + 2 I^-(aq) \rightarrow 2 Fe^{2+}(aq) + I_2(s)$$

a. Calculate E^0_{tot}.
b. At what concentration of I^- is the voltage zero, if all other species are at standard concentrations?

24.18 Consider a cell in which the reaction is:

$$Pb(s) + 2 H^+(aq) \rightarrow Pb^{2+}(aq) + H_2(g)$$

a. Calculate E^0_{tot}.
b. Chloride ions are added to the Pb/Pb^{2+} half-cell to precipitate $PbCl_2$. The voltage is measured to be $+0.21$ V. Taking conc. $H^+ = 1.0$ M and $P H_2 = 1.0$ atm, calculate conc. Pb^{2+}.
c. Taking $[Cl^-]$ in (b) to be 0.10 M, calculate K_{sp} for $PbCl_2$.

24.48 Consider a cell in which the reaction is:

$$2 Ag(s) + Cu^{2+}(aq) \rightarrow 2 Ag^+(aq) + Cu(s)$$

a. Calculate E^0_{tot} for this cell.
b. Chloride ions are added to the Ag/Ag^+ half-cell to precipitate $AgCl$. The measured voltage is $+0.06$ V. Taking conc. $Cu^{2+} = 1.0$ M, calculate conc. Ag^+.
c. Taking $[Cl^-]$ in (b) to be 0.10 M, calculate K_{sp} of $AgCl$.

24.19 A half-cell consisting of a silver electrode dipping into a solution of silver nitrate is connected to one in which a copper electrode is in contact with a 0.10 M solution of $Cu(NO_3)_2$. Excess HBr is added to the silver half-cell to give a precipitate of AgBr and make $Br^- = 0.10$ M. Under these conditions the voltage is 0.22 V and the Ag electrode is the anode. Calculate K_{sp} for AgBr.

24.49 The potential for the half-reaction

$$Au^{3+}(aq) + 3\ e^- \rightarrow Au(s)$$

is reduced from 1.50 V to 1.00 V when enough Cl^- is added to make conc. $AuCl_4^- = $ conc. $Cl^- = 1$ M. Calculate K for the reaction

$$AuCl_4^-(aq) \rightarrow Au^{3+}(aq) + 4\,Cl^-(aq)$$

E_{tot}^0, ΔG^0, and K

24.20 Calculate ΔG^0 and K at 25°C for cell reactions in which n = 2 and E_{tot}^0 is:
a. 0.00 V   b. +0.35 V   c. −0.35 V

24.50 Calculate ΔG^0 and K at 25°C for cell reactions in which n = 3 and E_{tot}^0 is
a. −1.00 V   b. −0.50 V   c. +0.50 V

24.21 For a certain cell, ΔG^0 is −21.6 kJ. Calculate E_{tot}^0 if n is
a. 1   b. 2   c. 3

24.51 Calculate E_{tot}^0 for cells in which n is 2 and ΔG^0 is
a. −20.0 kJ   b. +30.0 kJ

24.22 For a certain redox reaction at 25°C, the equilibrium constant is 1.0×10^{-6}. Calculate E_{tot}^0, taking n = 2.

24.52 For a certain redox reaction, K at 25°C is 2×10^{19}. Calculate E_{tot}^0, taking n = 2.

24.23 Using Table 24.1, calculate E_{tot}^0, ΔG^0, and K at 25°C for the reaction:

$$2\ Fe^{3+}(aq) + Sn^{2+}(aq) \rightarrow Sn^{4+}(aq) + 2\ Fe^{2+}(aq)$$

24.53 Using Table 24.1, calculate E_{tot}^0, ΔG^0, and K at 25°C for the reaction:

$$3\ Ag(s) + AuCl_4^-(aq) \rightarrow 3\ Ag^+(aq) + Au(s) + 4\ Cl^-(aq)$$

24.24 Calculate ΔG^0 for each of the reactions referred to in Problem 24.4.

24.54 Calculate ΔG^0 for each of the reactions referred to in Problem 24.34.

24.25 Calculate K at 25°C for each of the reactions referred to in Problem 24.5.

24.55 Calculate K at 25°C for each of the reactions referred to in Problem 24.35.

General

24.26 For a certain redox reaction, K is 2×10^{-18}. Is it possible to get electrical work out of this reaction at standard conditions? Explain your reasoning.

24.56 Can the reaction referred to in Problem 24.26 occur in a voltaic cell? an electrolytic cell?

24.27 Explain why corrosion
a. occurs when iron is in contact with both H_2O and O_2.
b. is prevented by washing garden tools thoroughly after use.

24.57 Explain why corrosion
a. usually occurs faster in seawater than in fresh water.
b. occurs most rapidly near the water line on a bridge support.

24.28 Explain why
a. silverware tarnishes even though E_{ox}^0 for Ag is −0.80 V.
b. aluminum objects do not corrode through when exposed to moist air.

24.58 Explain why
a. a sacrificial anode reduces corrosion.
b. it is unwise to store citrus fruit juices in a pewter container.

24.29 Explain
a. why acid rain samples should not be stored in aluminum containers.
b. why copper statuary acquires a green coating.

24.30 Consider the cell: $Ag/Ag^+ \parallel (Pt) Cl_2/Cl^-$. Will each of the following changes increase or decrease the voltage? Explain your answers.
a. Increasing the pressure of Cl_2.
b. Adding silver metal to the anode half-cell.
c. Adding Cl^- ions to the Cl_2/Cl^- half-cell.
d. Adding Cl^- ions to the Ag/Ag^+ half-cell.

24.59 Explain
a. how painting iron patio furniture protects it from rusting.
b. why we believe an electrochemical mechanism explains corrosion.

24.60 Consider the cell: $Cu/Cu^{2+} \parallel Al^{3+}/Al$. What effect will each of the following changes have on the voltage? Explain your answers.
a. Adding Cu^{2+} to the anode half-cell.
b. Adding OH^- ions to the Al/Al^{3+} half-cell.
c. Removing part of the Al electrode.
d. Adding Cu to the Cu/Cu^{2+} half-cell.

*24.61 Consider a voltaic cell in which the following reaction occurs:

$$Zn(s) + Sn^{2+}(aq) \rightarrow Zn^{2+}(aq) + Sn(s)$$

a. Calculate E^0_{tot} for the cell.
b. When the cell operates, what happens to the concentration of Zn^{2+}? the concentration of Sn^{2+}?
c. When the cell voltage drops to zero, what is the ratio of the concentration of Zn^{2+} to that of Sn^{2+}?
d. If the concentration of both cations is 1.0 M originally, what are their concentrations when the voltage drops to zero?

*24.62 In biological systems, acetate ion is converted to ethyl alcohol in a two-step process:

$$CH_3COO^-(aq) + 3 H^+(aq) + 2 e^- \rightarrow CH_3CHO(aq) + H_2O \; ; E^{0\prime} = -0.581 \text{ V}$$
$$\text{acetate ion} \qquad\qquad\qquad\qquad\qquad \text{acetaldehyde}$$

$$CH_3CHO(aq) \quad + 2 H^+(aq) + 2 e^- \rightarrow C_2H_5OH(aq) \; ; E^{0\prime} = -0.197 \text{ V}$$
$$\text{ethyl alcohol}$$

($E^{0\prime}$ is the standard reduction voltage at 25°C and a pH of 7.00).
a. Calculate $\Delta G^{0\prime}$ for each step and for the overall conversion.
b. Calculate $E^{0\prime}$ for the overall conversion.

*24.63 When 5.0 mL of 0.10 M Ce^{4+} is added to 5.0 mL of 0.20 M Fe^{2+}, the following reaction occurs:

$$Fe^{2+}(aq) + Ce^{4+}(aq) \rightarrow Fe^{3+}(aq) + Ce^{3+}(aq)$$

Taking the standard reduction voltage for Ce^{4+} to be +1.61 V, calculate:
a. K for the system.
b. the equilibrium concentration of each species.

*24.64 Given:

$$Ag^+(aq) + e^- \rightarrow Ag(s) \; ; E^0_{red} = +0.80 \text{ V}$$

$$AgCl(s) + e^- \rightarrow Ag(s) + Cl^-(aq) \; ; E^0_{red} = +0.22 \text{ V}$$

Use these E^0_{red} values and the Nernst equation to calculate the solubility product of AgCl.

*24.65 Consider the cell: $(Pt) H_2/H^+ \parallel (Pt) H^+/H_2$. In the anode half-cell, hydrogen gas at 1.0 atm is bubbled over a platinum electrode dipping into a solution which has a pH of 7.0. The other half-cell is identical to the first except that the solution around the platinum electrode has a pH of 0.0. What is the cell voltage?

Chapter 25
Chemistry
of the
Transition
Metals

In Chapter 8, we discussed the general properties of metals. We considered specifically the chemical properties of the metals in Groups 1 and 2. This chapter is devoted to the chemistry of some of the more important transition metals, the elements in the center of the Periodic Table (Fig. 25.1).

We will not attempt to discuss in detail the chemistry of these metals; to do so would require much more than one textbook chapter. Instead, we will look at some of the more important types of reactions involving the transition metals and the cations and oxyanions derived from them. Our goal will be to correlate their chemical behavior, attempting to draw general principles from what may seem to be a bewildering array of chemical facts. Our discussion will be organized around the chemistry of

— the metals themselves (Section 25.1).
— the cations, such as Cr^{3+} and Mn^{2+}, derived from these metals (Section 25.2).
— oxyanions of the transition metals, such as CrO_4^{2-} and MnO_4^- (Section 25.3).

Throughout this chapter, we will emphasize redox reactions of the type discussed in Chapters 23 and 24.

25.1
Reactions Involving
Transition Metals

In this section, we will examine some of the chemical properties of the transition metals in Figure 25.1. We will look first at the processes used to extract these metals from their ores. Then we will consider redox reactions involving these metals and their ions in water solution. Finally, we will describe some of the better known uses of these metals, as related to their properties.

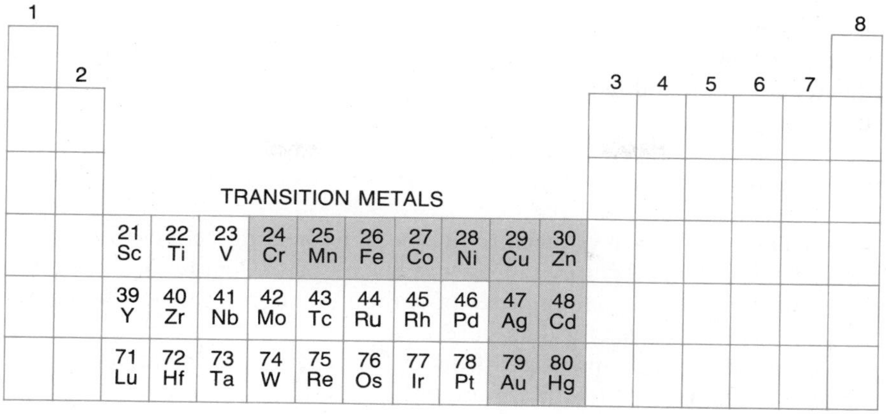

FIGURE 25.1 The transition elements are those in the centers of the fourth, fifth, and sixth periods. The more familiar transition metals, whose chemistry we will refer to in this chapter, are shown in color.

Metallurgy

Table 25.1 summarizes the methods used to extract the more common transition metals from their ores. Three metals (cobalt, silver, and cadmium) are not listed since they are formed mainly as by-products in the extraction of other metals (Co with Ni, Cd with Zn, Ag with Cu). Of the eight metals in Table 25.1, seven occur as either oxide or sulfide ores. In contrast, gold is found in elemental form, mixed with large amounts of rocky material.

Table 25.1
Metallurgy of the Transition Metals

METAL	PRINCIPAL ORE	EXTRACTION PROCESS
Cr	$Cr_2O_3 \cdot FeO$	heat with C to give Fe–Cr alloy; Cr_2O_3 reduced with Al to give pure Cr
Mn	$MnO_2 + Fe_2O_3$	heat with C to give Mn–Fe alloy; MnO_2 reduced with Al to give pure Mn
Fe	Fe_2O_3	see Chapter 4
Ni	NiS	flotation, roasting to NiO, reduction with C
Cu	Cu_2S, $CuFeS_2$	see Chapter 4
Zn	ZnS	flotation, roasting to ZnO, reduction with C
Au	Au	treatment with CN^- (see text discussion)
Hg	HgS	roasting in air to form $Hg(g)$

There are many processes used to produce metals

The forty-niners obtained gold by swirling gold-bearing rock and gravel with water in a pan. The less dense impurities were washed away, leaving gold nuggets and dust at the bottom of the pan. Nowadays, the gold content of available ores is much too low for panning to be effective. Instead, the ore is treated with sodium cyanide in the presence of air. The following redox reaction takes place:

$$4\ Au(s) + 8\ CN^-(aq) + O_2(g) + 2\ H_2O \rightarrow 4\ Au(CN)_2^-(aq) + 4\ OH^-(aq)$$

The reaction occurs in NaCN solution

(25.1)

The oxidizing agent is O_2, which takes gold to the $+1$ state. The cyanide ion acts as a complexing ligand, forming the stable $Au(CN)_2^-$ ion. Metallic gold is recovered from solution by adding zinc—the gold in the complex ion is reduced to the metal:

$$Zn(s) + 2\ Au(CN)_2^-(aq) \rightarrow Zn(CN)_4^{2-}(aq) + 2\ Au(s) \qquad (25.2)$$

Oxide ores are most often reduced to the metal by heating with carbon. This process as applied to the metallurgy of iron was discussed in Chapter 4. The metal formed frequently contains small amounts of carbon. With iron this is beneficial; steel (0 to 1.5% C) is more useful than pure iron.

A very pure metal can be produced from an oxide ore by using aluminum as a reducing agent. The reaction with chromium(III) oxide is typical:

$$Cr_2O_3(s) + 2\ Al(s) \rightarrow 2\ Cr(s) + Al_2O_3(s); \qquad \Delta H = -541\ kJ \qquad (25.3)$$

This type of reaction, known as the *thermite* process, is highly exothermic. With a mixture of Fe_2O_3 and powdered aluminum, a temperature of 3000°C can be reached. The molten iron formed this way was used at one time to weld steel rails and repair broken machinery.

Sulfide ores are reduced by a process described in Chapter 4. The purified ore is "roasted," i.e., heated in air. With the sulfide ore of copper, Cu_2S, the product is the free metal. A similar reaction occurs with cinnabar, HgS, the sulfide ore of mercury:

Because roasting can produce Cu, it was one of the first metals known

$$HgS(s) + O_2(g) \rightarrow Hg(g) + SO_2(g) \qquad (25.4)$$

With more active metals such as zinc, roasting the sulfide ore gives the oxide rather than the free metal:

$$2\ ZnS(s) + 3\ O_2(g) \rightarrow 2\ ZnO(s) + 2\ SO_2(g) \qquad (25.5)$$

Any zinc formed by a reaction analogous to 25.4 is immediately oxidized to ZnO. The oxide produced by roasting is then reduced to the metal by heating with carbon. The effective reducing agent here is carbon monoxide:

$$ZnO(s) + CO(g) \rightarrow Zn(s) + CO_2(g) \qquad (25.6)$$

Oxidation in Water Solution

Perhaps the best measure of the reactivity of a metal is its standard oxidation voltage, E_{ox}^0 (Table 25.2). The larger the value of E_{ox}^0 is, the more reactive the metal; i.e., the more readily it is oxidized. Looking at Table 25.2, we might expect that

—manganese ($E_{ox}^0 = +1.18$ V) is the most reactive transition metal. In many ways, this is true. Manganese is the only common transition metal that reacts (slowly) with cold water:

$$Mn(s) + 2\ H_2O(l) \rightarrow Mn(OH)_2(s) + H_2(g) \qquad (25.7)$$

—silver ($E^0_{ox} = -0.80$ V), mercury ($E^0_{ox} = -0.86$ V), and gold ($E^0_{ox} = -1.50$ V) are difficult to oxidize. Conversely, their cations (Ag^+, Hg^{2+}, Au^{3+}) are readily reduced to the free metal. You may have experienced this effect if you have ever spilled silver nitrate, $AgNO_3$, on your hands. Organic compounds in the skin reduce Ag^+ ions to give a black stain that consists of finely divided silver metal.

The stain is not attractive, but it does slough off with time

Table 25.2
Ease of Oxidation of Transition Metals

METAL		CATION	$E^0_{ox}(V)$	METAL		CATION	$E^0_{ox}(V)$
Mn	→	Mn^{2+}	$+1.18$	Cu	→	Cu^{2+}	-0.34
Cr	→	Cr^{2+}	$+0.90$	Ag	→	Ag^+	-0.80
Zn	→	Zn^{2+}	$+0.76$	Hg	→	Hg^{2+}	-0.86
Cd	→	Cd^{2+}	$+0.40$	Au	→	Au^{3+}	-1.50
Co	→	Co^{2+}	$+0.28$				
Ni	→	Ni^{2+}	$+0.25$				

All the metals listed in the left column of Table 25.2 react with strong, non-oxidizing acids such as HCl to evolve hydrogen gas. Zinc is commonly used for this purpose in the laboratory but manganese or nickel would serve as well:

$$Ni(s) + 2 H^+(aq) \rightarrow Ni^{2+}(aq) + H_2(g)$$

$$E^0_{tot} = E^0_{ox} \ Ni + E^0_{red} \ H^+ = +0.25 \ V + 0.00 \ V = +0.25 \ V$$

(25.8)

Zn reacts faster and is a lot cheaper

Metals with negative values of E^0_{ox}, listed at the right of Table 25.2, are too inactive to react with hydrochloric acid. The H^+ ion is not a strong enough oxidizing agent to convert a metal such as copper ($E^0_{ox} = -0.34$ V) to a cation. However, copper can be oxidized by nitric acid. Here, the oxidizing agent is the nitrate ion, NO_3^-, which may be reduced to NO_2 or NO:

$$3 \ Cu(s) + 8 \ H^+(aq) + 2 \ NO_3^-(aq) \rightarrow 3 \ Cu^{2+}(aq) + 2 \ NO(g) + 4 \ H_2O$$

(25.9)

$$E^0_{tot} = E^0_{ox} \ Cu + E^0_{red} \ NO_3^- = -0.34 \ V + 0.96 \ V = +0.62 \ V$$

Silver can also be oxidized by heating with nitric acid (Example 25.1).

EXAMPLE 25.1   Write a balanced redox equation, analogous to Equation 25.9, for the reaction of silver with nitric acid.

Solution   We proceed as described in Chapter 23. The half-equations are

oxidation: $Ag(s) \rightarrow Ag^+(aq)$

reduction: $NO_3^-(aq) \rightarrow NO(g)$

Balancing the half-equations, we obtain

oxidation: $Ag(s) \rightarrow Ag^+(aq) + e^-$

reduction: $NO_3^-(aq) + 4 H^+(aq) + 3 e^- \rightarrow NO(g) + 2 H_2O$

Multiplying the oxidation half-equation by three and adding to the reduction half-equation gives

$$3\ Ag(s) + NO_3^-(aq) + 4\ H^+(aq) \rightarrow 3\ Ag^+(aq) + NO(g) + 2\ H_2O$$

EXERCISE Write a balanced equation for the reaction of silver with nitric acid if the reduction product is NO_2 rather than NO. Answer: $Ag(s) + NO_3^-(aq) + 2\ H^+(aq) \rightarrow Ag^+(aq) + NO_2(g) + H_2O$

Gold ($E^0_{ox} = -1.50$ V) is too inactive to be oxidized by nitric acid alone ($E^0_{red} = +0.96$ V). It does, however, go into solution in *aqua regia*, a 3:1 mixture by volume of 12 M HCl and 16 M HNO_3:

$$Au(s) + 4\ H^+(aq) + 4\ Cl^-(aq) + NO_3^-(aq) \rightarrow AuCl_4^-(aq) + NO(g) + 2\ H_2O \qquad (25.10)$$

Formation of the $AuCl_4^-$ ion reduces $[Au^{3+}]$ to a very low value, which makes the oxidation of Au go more easily

This reaction is promoted by the formation of the very stable complex ion $AuCl_4^-$ ($K_f = 1 \times 10^{25}$). The function of the hydrochloric acid is to furnish the Cl^- ions needed to form this complex.

Uses

The major uses of the more common transition metals are listed in Table 25.3. Notice that many of the metals are most familiar as alloys with other metals. Often these alloys have properties that are more desirable than those of the pure metals. They may be harder (sterling silver: 92.5% Ag, 7.5% Cu), lower melting (Wood's metal), tougher (Mn steel), or more resistant to corrosion (stainless steel).

Table 25.3
Uses of the Transition Metals

Cr	chrome plate, stainless steel (Cr, Ni, Fe), nichrome (Ni, Cr)
Mn	alloy steels (rails, safes)
Fe	see Chapter 4
Co	alloy steels, alnico magnets (Al, Ni, Co, Fe), industrial catalyst
Ni	alloys such as alnico, monel (Ni, Cu, Fe), coinage (Ni, Cu)
Cu	see Chapter 4
Zn	galvanized iron, dry cells, brass (Cu, Zn), bronze (Cu, Zn, Pb)
Ag	tableware, jewelry, photography as AgBr
Cd	low-melting alloys such as Wood's metal (Bi, Pb, Sn, Cd), neutron absorption
Au	jewelry, gold leaf
Hg	thermometers, barometers, electrical contacts, lighting

Au is used a lot these days in contacts for integrated circuits

Some of the uses of cobalt and nickel depend upon a property that these metals share with iron: *ferromagnetism*. These three metals are strongly attracted into a magnetic field and retain their magnetism when removed from the field. Atoms of Fe, Co, and Ni, like those of most transition metals, have unpaired 3d electrons. This alone would make them paramagnetic, i.e., weakly attracted into a magnetic field. The much stronger ferromagnetism comes about because the

Unmagnetized (a)

Magnetized (b)

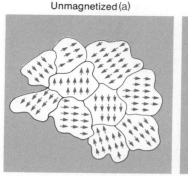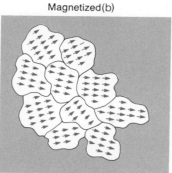

FIGURE 25.2 In a ferromagnetic substance, the magnetic moments of individual atoms (shown by arrows) are lined up parallel to one another through a large region (domain) of the solid. When the solid is magnetized, the individual domains line up parallel to one another as shown at the right.

magnetic moments of neighboring atoms line up so as to reinforce one another. As many as 10^{21} atoms with parallel moments may line up to form a "domain" (Fig. 25.2). In an unmagnetized sample, individual domains are oriented randomly (Fig. 25.2a). When the sample is brought into a magnetic field, the domains tend to line up parallel to one another (Fig. 25.2b). This creates the strong interaction that we call ferromagnetism. When the sample is removed from the field, the domains remain aligned, so the metal retains its magnetism.

The three metals in the copper subgroup (Cu, Ag, Au) have long been referred to as the "coinage" metals. Today this phrase is somewhat misleading, at least as far as United States coinage is concerned. Gold coins were taken out of circulation in 1934. The rising price of silver and the increasing demand for this metal in photography led to the elimination of silver coins in 1971. The present 10¢, 25¢, and 50¢ coins consist of an alloy of 75% Cu and 25% Ni, sandwiched around a copper core. The same Cu–Ni alloy is used to make the "nickel" coin. In 1982 the composition of the penny, which had been 95% copper, was changed. Pennies now being minted are 98% zinc, covered with a thin copper plate. Perhaps we should refer to Ni, Cu, and Zn as the "coinage" metals.

The silver was worth more than the coin

The largest use by far of silver is in photography. The thin, light-sensitive coating on photographic film contains a silver halide, most often AgBr, dispersed in gelatin. Upon exposure to light, a few of the Ag^+ ions in each grain of silver bromide are reduced to Ag atoms. Fewer than 1 in 10^6 Ag^+ ions are reduced, but enough silver is formed to make up what is known as the "latent image." After exposure, the film is treated in the dark with an organic reducing agent. The silver bromide grains that have been sensitized by exposure are reduced to silver metal. This process, known as development, can be represented by the half-equation

Color film also uses AgBr

$$AgBr(s) + e^- \rightarrow Ag(s) + Br^-$$

where the electron is furnished by the reducing agent. The small number of silver atoms making up the latent image act as a catalyst for this reaction.

After development, a film shows dark areas of silver metal where it was exposed and light areas of unchanged AgBr in the unexposed regions. This silver bromide must be removed so that the finished negative will not be light-sensitive.

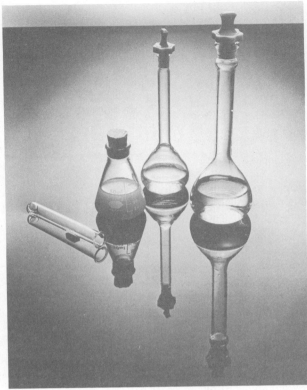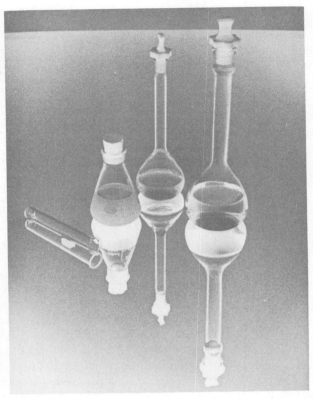

FIGURE 25.3 To prepare a positive print, the negative is placed over a piece of printing paper coated with AgBr. Light is passed through the negative to the printing paper. The amount of light reaching the print is inversely related to the thickness of the silver deposit on the negative. In this way, the light and dark areas of the negative are reversed. (Photo © Joel Gordon, 1979)

This is accomplished by dipping the negative into a water solution of sodium thiosulfate, $Na_2S_2O_3$. The reaction that occurs is referred to as fixing; it involves complex ion formation to bring silver bromide into solution:

$$AgBr(s) + 2\ S_2O_3{}^{2-}(aq) \rightarrow Ag(S_2O_3)_2{}^{3-}(aq) + Br^-(aq) \qquad \textbf{(25.11)}$$

The negative is thoroughly washed, dried, and used to prepare a positive (Fig. 25.3).

As Table 25.3 suggests, mercury is used for a variety of purposes. Since it expands linearly with temperature, mercury is the liquid of choice for accurate thermometers. Its high density (13.6 g/cm³) accounts for its use in barometers. As the only liquid metal, mercury has many applications as an electrical conductor. Mercury would have even more uses were it not for two factors. One is its relatively high cost ($12/kg). The other is the toxicity of mercury vapor and all of the soluble compounds of the element. The vapor attacks lung tissue and is a cumulative poison. As little as 5 mL of the liquid can saturate a laboratory with vapor, so you should be careful to clean up any spilled mercury.

Fortunately the vapor pressure is low, ~0.002 mm Hg at 25°C

21 Sc +3	22 Ti +4 +3 +2	23 V +5 +4 +3 +2	24 Cr +6 +3 +2	25 Mn +7 +6 +4 +3 +2	26 Fe +3 +2	27 Co +3 +2	28 Ni +2	29 Cu +2 +1	30 Zn +2
39 Y +3	40 Zr +4	41 Nb +5 +4	42 Mo +6 +5 +4 +3	43 Tc +7 +6 +4	44 Ru +8 +6 +4 +3	45 Rh +3 +2	46 Pd +4 +2	47 Ag +1	48 Cd +2
71 Lu +3	72 Hf +4	73 Ta +5	74 W +6 +5 +4	75 Re +7 +6 +5 +4	76 Os +8 +6 +4	77 Ir +4 +3	78 Pt +4 +2	79 Au +3 +1	80 Hg +2 +1

FIGURE 25.4 Oxidation states of the transition metals. The more common or stable states are shown in boldface.

25.2 Cations of the Transition Metals

Transition metals typically show a variety of oxidation numbers in their compounds (Fig. 25.4). In the lower oxidation states ($+1$, $+2$, $+3$), the metal is ordinarily in the form of a cation, such as Ag^+, Mn^{2+}, or Cr^{3+}. The reactions of transition metal cations will be considered in this section. Our discussion will be organized according to the types of reactions involved.

Complex Ion Formation

In water solution, transition metal cations are always present as complex ions of the type discussed in Chapter 21. The main species in dilute water solution is ordinarily a hydrated ion of the type listed in Table 25.4, p. 754. We find, for example, that

Actually, most cations in solution are hydrated

—solutions of $MnCl_2$, $MnSO_4$, or $Mn(NO_3)_2$ have the light pink color of the $Mn(H_2O)_6^{2+}$ ion.
—solutions of $NiCl_2$, $NiSO_4$, and $Ni(NO_3)_2$ have the deep green color of the $Ni(H_2O)_6^{2+}$ ion.

Solutions of $Zn(NO_3)_2$, $AgNO_3$, and $Cd(NO_3)_2$ are colorless, as are all the complex ions of Zn^{2+}, Ag^+, and Cd^{2+}. These cations, unlike those of most other transition metals, have completely filled d orbitals:

Electr. Conf.	Electr. Conf.	Electr. Conf.
$_{30}Zn^{2+}$ [$_{18}Ar$]$3d^{10}$	$_{47}Ag^+$ [$_{36}Kr$]$4d^{10}$	$_{48}Cd^{2+}$ [$_{36}Kr$]$4d^{10}$

Why does a d^{10} outer configuration produce a colorless cation?

A color change in a water solution of a transition metal salt ordinarily means that a new complex ion has been formed. When ammonia is added to a solution of $Ni(NO_3)_2$, the color changes from green to deep blue:

$$Ni(H_2O)_6^{2+}(aq) + 6\ NH_3(aq) \rightarrow Ni(NH_3)_6^{2+}(aq) + 6\ H_2O \qquad (25.12)$$

$$\underset{\text{green}}{\phantom{Ni(H_2O)_6^{2+}(aq)}} \qquad\qquad \underset{\text{blue}}{\phantom{Ni(NH_3)_6^{2+}(aq)}}$$

The color difference between these two complex ions can be explained by the crystal-field model of complex ion formation discussed in Chapter 21. In $Ni(NH_3)_6^{2+}$ there is a broad absorption band covering most of the visible spectrum, from about 500 to 700 nm. The light transmitted by solutions of $Ni(NH_3)_6^{2+}$ is rich in the blue (< 500 nm). In $Ni(H_2O)_6^{2+}$, the situation is somewhat different. The crystal-field splitting energy, Δ_0, is smaller with H_2O as a ligand. As a result, the absorption band shifts to longer wavelengths, starting at about 550 nm and extending beyond 700 nm. Hence, light transmitted by solutions containing the $Ni(H_2O)_6^{2+}$ ion is rich in the green (500 to 550 nm) region.

The colors of cobalt(II) complexes vary with coordination number. The octahedral complexes are ordinarily pink, whereas the tetrahedral complexes are blue. A common type of humidity indicator, containing cobalt(II) chloride, makes use of this color difference. At very low humidity, cobalt(II) is present as the $Co(H_2O)_4^{2+}$ complex. As the humidity increases, this species picks up water from the air to form the $Co(H_2O)_6^{2+}$ complex:

$$[Co(H_2O)_4]Cl_2(s) + 2\ H_2O(g) \rightarrow [Co(H_2O)_6]Cl_2(s) \qquad (25.13)$$

$$\underset{\text{blue}}{} \qquad\qquad \underset{\text{pink}}{}$$

Table 25.4
Complex Ions of the Transition Metals

CENTRAL CATION	COORDINATION NUMBER	EXAMPLES		
Cr^{3+}	6	$Cr(H_2O)_6^{3+}$, violet	$Cr(H_2O)_5Cl^{2+}$ green	
Mn^{2+}	6	$Mn(H_2O)_6^{2+}$ pink		
Fe^{2+}	6	$Fe(H_2O)_6^{2+}$, blue-green	$Fe(H_2O)_5NO^{2+}$ brown	
Fe^{3+}	6	$Fe(H_2O)_6^{3+}$, purple	$Fe(H_2O)_5(OH)^{2+}$, yellow	$Fe(H_2O)_5SCN^{2+}$ blood-red
Co^{2+}	4, 6	$Co(H_2O)_6^{2+}$, pink	$Co(H_2O)_4^{2+}$ blue	
Ni^{2+}	4, 6	$Ni(H_2O)_6^{2+}$, green	$Ni(NH_3)_6^{2+}$, blue	$Ni(CN)_4^{2-}$ yellow
Cu^{2+}	4, 6	$Cu(H_2O)_4^{2+}$, light blue	$Cu(NH_3)_4^{2+}$, deep blue	$CuCl_4^{2-}$ yellow
Zn^{2+}	4	$Zn(H_2O)_4^{2+}$, colorless	$Zn(NH_3)_4^{2+}$, colorless	$Zn(OH)_4^{2-}$ colorless
Ag^+	2	$Ag(H_2O)_2^+$, colorless	$Ag(NH_3)_2^+$, colorless	$Ag(S_2O_3)_2^{3-}$ colorless
Cd^{2+}	4	$Cd(H_2O)_4^{2+}$, colorless	$Cd(NH_3)_4^{2+}$ colorless	

If one of the complexes of a cation is colored, they all tend to be colored

Acid Dissociation

As pointed out in Chapter 19, hydrated transition metal cations behave as weak acids in water. The behavior of the $Cr(H_2O)_6^{3+}$ ion is typical:

$$Cr(H_2O)_6^{3+}(aq) \rightleftharpoons H^+(aq) + Cr(H_2O)_5(OH)^{2+}(aq) \qquad (25.14)$$

Dissociation constants for several ions of this type are listed in Table 25.5. Notice that the +3 ions as a group are considerably stronger acids (larger K_a values) than the +2 ions.

Table 25.5
Acid Dissociation Constants of Transition Metal Cations

+3 ION	K_a	+ 2 ION	K_a
$Co(H_2O)_6^{3+}$	2×10^{-2}	$Fe(H_2O)_6^{2+}$	2×10^{-7}
$Fe(H_2O)_6^{3+}$	2×10^{-3}	$Cu(H_2O)_4^{2+}$	3×10^{-8}
$Cr(H_2O)_6^{3+}$	1×10^{-4}	$Zn(H_2O)_4^{2+}$	1×10^{-9}
		$Cd(H_2O)_4^{2+}$	1×10^{-9}
		$Co(H_2O)_6^{2+}$	1×10^{-10}
		$Ni(H_2O)_6^{2+}$	1×10^{-10}
		$Mn(H_2O)_6^{2+}$	3×10^{-11}

The dissociation constants listed in Table 25.5 can be used to estimate the pH of a solution of a transition metal salt (Example 25.2).

EXAMPLE 25.2 Taking K_a of the $Cr(H_2O)_6^{3+}$ ion to be 1×10^{-4}, estimate $[H^+]$ and pH in a 0.1 M solution of $Cr(NO_3)_3$.

Solution The approach here is the same as that described in Chapter 20 with other weak acids:

$$Cr(H_2O)_6^{3+}(aq) \rightleftharpoons H^+(aq) + Cr(H_2O)_5(OH)^{2+}(aq)$$

$$K_a = \frac{[H^+] \times [Cr(H_2O)_5(OH)^{2+}]}{[Cr(H_2O)_6^{3+}]} = 1 \times 10^{-4}$$

We don't usually use metal salt solutions as acids

Taking our variable, x, to be the concentration of H^+ at equilibrium, we have

$$[H^+] = [Cr(H_2O)_5(OH)^{2+}] = x$$

$$[Cr(H_2O)_6^{3+}] = 0.1 - x \approx 0.1$$

$$\frac{x^2}{0.1} = 1 \times 10^{-4}$$

$$x = [H^+] = (1 \times 10^{-5})^{1/2} = 3 \times 10^{-3} \text{ M}; \quad pH = 2.5$$

We see that the solution is indeed acidic, somewhat more so than a 0.1 M solution of acetic acid (pH ≈ 2.9).

EXERCISE Calculate the pH of a 0.1 M solution of $Zn(NO_3)_2$. Answer: 5.0.

The composition of an equilibrium mixture such as that produced by Reaction 25.14 is sensitive to pH. Adding a strong acid to a solution of a transition metal salt shifts the equilibrium to the left, forming completely hydrated ions such as $Cr(H_2O)_6^{3+}$. Addition of base shifts the equilibrium in the other direction, favoring the formation of species such as $Cr(H_2O)_5(OH)^{2+}$. This effect can be observed with solutions of iron(III) salts, where the two complex ions have quite different colors:

$$Fe(H_2O)_6^{3+}(aq) \rightleftharpoons H^+(aq) + Fe(H_2O)_5(OH)^{2+}(aq) \qquad (25.15)$$

pale purple   yellow

Solutions of Fe(III) salts often contain added acid to prevent precipitation of species like $Fe(H_2O)_3(OH)_3$

If a salt such as $Fe(NO_3)_3$ is added to water, the concentration of $Fe(H_2O)_5(OH)^{2+}$ is high enough to give the solution a yellow color. Addition of 6 M HNO_3 shifts the equilibrium to the left, giving a nearly colorless solution. Conversely, if a solution of an iron(III) salt is made basic, the equilibrium shifts to the right. The yellow color becomes more intense and then changes to red as species such as $Fe(H_2O)_4(OH)_2^+$ and $Fe(H_2O)_3(OH)_3$ are formed.

Precipitation Reactions (Sulfides and Hydroxides)

A transition metal cation, M^{2+}, can be precipitated from solution as a sulfide by

—adding a solution of sodium sulfide

$$M^{2+}(aq) + S^{2-}(aq) \rightarrow MS(s) \qquad (25.16)$$

—saturating with hydrogen sulfide

$$M^{2+}(aq) + H_2S(aq) \rightleftharpoons MS(s) + 2 H^+(aq) \qquad (25.17)$$

As pointed out in Chapter 22, extremely insoluble sulfides such as CuS ($K_{sp} = 1 \times 10^{-35}$) can be precipitated by H_2S even in highly acidic solution, where the concentration of S^{2-} ions is very low. With more soluble sulfides such as ZnS ($K_{sp} = 1 \times 10^{-20}$), the solution must be made basic, which shifts the equilibrium system represented by Equation 25.17 to the right, forming a precipitate.

Sometimes when you add sulfide ions to a solution of a transition metal ion, you get a surprise. Consider, for example, what happens when the cation is Cr^{3+}. The precipitate formed is $Cr(OH)_3$ rather than Cr_2S_3. The OH^- ions in the precipitate come from the reaction of S^{2-} ions with water:

If you add solid Cr_2S_3 to water, you form $Cr(OH)_3$

$$\begin{array}{l} 3\ S^{2-}(aq) + 3\ H_2O \rightarrow 3\ HS^-(aq) + 3\ OH^-(aq) \\ \underline{Cr^{3+}(aq) + 3\ OH^-(aq) \rightarrow Cr(OH)_3(s)} \\ Cr^{3+}(aq) + 3\ S^{2-}(aq) + 3\ H_2O \rightarrow Cr(OH)_3(s) + 3\ HS^-(aq) \end{array} \qquad (25.18)$$

Transition metal hydroxides are ordinarily precipitated by adding either sodium hydroxide or ammonia. With Fe^{3+} the reactions are

NaOH: $Fe^{3+}(aq) + 3\ OH^-(aq) \rightarrow Fe(OH)_3(s)$ **(25.19)**

NH$_3$: $Fe^{3+}(aq) + 3\ NH_3(aq) + 3\ H_2O \rightarrow Fe(OH)_3(s) + 3\ NH_4^+(aq)$ **(25.20)**

All metal hydroxides dissolve in strong acid (for example, 6 M HCl). The reaction with iron(III) hydroxide is

$$Fe(OH)_3(s) + 3\ H^+(aq) \rightarrow Fe^{3+}(aq) + 3\ H_2O \qquad \textbf{(25.21)}$$

Water-insoluble hydroxides can be dissolved in 6 M NH$_3$ if the cation forms a stable complex with ammonia (Table 25.6). The behavior of copper(II) hydroxide is typical:

$$Cu(OH)_2(s) + 4\ NH_3(aq) \rightarrow Cu(NH_3)_4^{2+}(aq) + 2\ OH^-(aq) \qquad \textbf{(25.22)}$$

Complex ions of a cation are often more stable than its precipitates

If the cation forms a stable complex with OH$^-$ ions, 6 M NaOH can bring a metal hydroxide into solution:

$$Zn(OH)_2(s) + 2\ OH^-(aq) \rightarrow Zn(OH)_4^{2-}(aq) \qquad \textbf{(25.23)}$$

Hydroxides, such as $Zn(OH)_2$, that dissolve in both strong acid (6 M HCl) and strong base (6 M NaOH) are referred to as being *amphoteric*.

Table 25.6
Equilibrium Constants Involving Transition Metal Cations

SOLUBILITY PRODUCT CONSTANTS, K_{sp}			
SULFIDES		**HYDROXIDES**	
MnS	1×10^{-15}	$Cr(OH)_3$	4×10^{-38}
FeS	1×10^{-17}	$Mn(OH)_2$	4×10^{-14}
CoS	1×10^{-20}	$Fe(OH)_2$	1×10^{-15}
NiS	1×10^{-19}	$Fe(OH)_3$	5×10^{-38}
CuS	1×10^{-35}	$Co(OH)_2$	2×10^{-16}
ZnS	1×10^{-20}	$Ni(OH)_2$	1×10^{-16}
Ag$_2$S	1×10^{-49}	$Cu(OH)_2$	2×10^{-19}
CdS	1×10^{-26}	$Zn(OH)_2$	5×10^{-17}
HgS	1×10^{-52}	$Cd(OH)_2$	2×10^{-14}

FORMATION CONSTANTS, K_c					
$Co(NH_3)_6^{2+}$	1×10^5	$Cr(OH)_4^-$	1×10^{30}	$Fe(CN)_6^{4-}$	1×10^{35}
$Ni(NH_3)_6^{2+}$	5×10^8	$Cu(OH)_4^{2-}$	3×10^{18}	$Fe(CN)_6^{3-}$	1×10^{42}
$Cu(NH_3)_4^{2+}$	5×10^{12}	$Zn(OH)_4^{2-}$	3×10^{15}	$Ni(CN)_4^{2-}$	1×10^{31}
$Zn(NH_3)_4^{2+}$	1×10^9			$Cu(CN)_4^{2-}$	1×10^{30}
$Ag(NH_3)_2^+$	2×10^7			$Zn(CN)_4^{2-}$	1×10^{17}
$Cd(NH_3)_4^{2+}$	1×10^7			$Ag(CN)_2^-$	1×10^{21}
$Hg(NH_3)_4^{2+}$	2×10^{19}			$Cd(CN)_4^{2-}$	1×10^{19}
				$Au(CN)_2^-$	1×10^{38}
				$Hg(CN)_4^{2-}$	1×10^{41}

Cyanide complexes tend to be extremely stable, as indicated by their large K_f values

EXAMPLE 25.3 Write net ionic equations for the reactions by which
a. Ni^{2+} is precipitated by H_2S.
b. Ni^{2+} is precipitated by adding a solution of NaOH.
c. the precipitate in (b) dissolves in NH_3.

Solution
a. The precipitate is NiS. The equation is

$$Ni^{2+}(aq) + H_2S(aq) \rightarrow NiS(s) + 2 H^+(aq)$$

b. $Ni^{2+}(aq) + 2 OH^-(aq) \rightarrow Ni(OH)_2(s)$

c. The product is the complex ion $Ni(NH_3)_6^{2+}$. The equation is

$$Ni(OH)_2(s) + 6 NH_3(aq) \rightarrow Ni(NH_3)_6^{2+}(aq) + 2 OH^-(aq)$$

EXERCISE When ammonia is added to a solution of $Ni(NO_3)_2$, a precipitate first forms and then dissolves. Write net ionic equations for the two reactions involved. Answer:

$$Ni^{2+}(aq) + 2 NH_3(aq) + 2 H_2O \rightarrow Ni(OH)_2(s) + 2 NH_4^+(aq)$$

$$Ni(OH)_2(s) + 6 NH_3(aq) \rightarrow Ni(NH_3)_6^{2+}(aq) + 2 OH^-(aq)$$

Oxidation-Reduction Reactions

Several transition metals form more than one cation (for example, Fe^{2+}, Fe^{3+}). Table 25.7 lists standard reduction voltages (E^0_{red}) for several such systems. These voltages are shown above the arrows linking different oxidation states. For example,

This is how to interpret those arrows

$$Cr^{3+}(aq) + e^- \rightarrow Cr^{2+}(aq); \qquad E^0_{red} = -0.41 \text{ V}$$

$$Cr^{2+}(aq) + 2 e^- \rightarrow Cr(s); \qquad E^0_{red} = -0.90 \text{ V}$$

In general, we expect cations for which E^0_{red} is a large positive number to be easily reduced. These cations tend to be unstable in water solution. A case in point is the Mn^{3+} ion:

$$Mn^{3+}(aq) + e^- \rightarrow Mn^{2+}(aq); \qquad E^0_{red} = +1.51 \text{ V}$$

We would not expect Mn^{3+} ions to stay around very long in water solution, and indeed they don't. They react with H_2O molecules and are reduced to Mn^{2+} ions:

$$2 Mn^{3+}(aq) + H_2O \rightarrow 2 Mn^{2+}(aq) + \tfrac{1}{2} O_2(g) + 2 H^+(aq) \qquad \textbf{(25.24)}$$

$$E^0_{tot} = E^0_{red} \, Mn^{3+} + E^0_{ox} \, H_2O = +1.51 \text{ V} - 1.23 \text{ V} = +0.28 \text{ V}$$

And they're not common reagents

Manganese(III) is found only in insoluble oxides and hydroxides such as Mn_2O_3 and $MnO(OH)$.

From Table 25.7, we might predict that the Co^{3+} ion, like Mn^{3+}, would be unstable toward reduction:

$$Co^{3+}(aq) + e^- \rightarrow Co^{2+}(aq); \qquad E^0_{red} = +1.82 \text{ V}$$

Table 25.7
Ease of Reduction of Transition Metal Cations

Chromium Cr^{3+} $\xrightarrow{-0.41\ V}$ Cr^{2+} $\xrightarrow{-0.90\ V}$ Cr

Manganese Mn^{3+} $\xrightarrow{+1.51\ V}$ Mn^{2+} $\xrightarrow{-1.18\ V}$ Mn

Iron Fe^{3+} $\xrightarrow{+0.77\ V}$ Fe^{2+} $\xrightarrow{-0.44\ V}$ Fe

Cobalt Co^{3+} $\xrightarrow{+1.82\ V}$ Co^{2+} $\xrightarrow{-0.28\ V}$ Co

Copper Cu^{2+} $\xrightarrow{+0.15\ V}$ Cu^{+} $\xrightarrow{+0.52\ V}$ Cu

Gold Au^{3+} $\xrightarrow{+1.40\ V}$ Au^{+} $\xrightarrow{+1.69\ V}$ Au

Mercury Hg^{2+} $\xrightarrow{+0.92\ V}$ Hg_2^{2+} $\xrightarrow{+0.79\ V}$ Hg

Experimentally, we find that simple cobalt salts (such as $CoCl_2$, $CoSO_4$, . . .) always contain the Co^{2+} ion, never Co^{3+}. However, as you may recall from Chapter 21, complexes of cobalt(III), such as $Co(NH_3)_6^{3+}$, are quite common. The voltage for reduction of cobalt(III) changes drastically in the presence of a complexing agent such as ammonia:

$$Co(NH_3)_6^{3+}(aq) + e^- \rightarrow Co(NH_3)_6^{2+}(aq); \quad E^0_{red} = +0.1\ V$$

Water can't reduce $Co(NH_3)_6^{3+}$

The difference in E^0_{red} values reflects the much greater stability of cobalt(III) complexes as compared to cobalt(II). We find, for example, that

$$K_f\ Co(NH_3)_6^{3+} = 1 \times 10^{35}$$
$$K_f\ Co(NH_3)_6^{2+} = 1 \times 10^5$$

The Fe^{3+} ion ($E^0_{red} = +0.77\ V$) is less readily reduced than Mn^{3+} or Co^{3+}. In the presence of a suitable reducing agent, however, it can be converted to Fe^{2+} (Example 25.4).

EXAMPLE 25.4 When solutions of $FeCl_3$ and NaI are mixed, a redox reaction occurs and the solution takes on a reddish color. Suggest what this reaction might be and, using Table 24.1, p. 721, calculate E^0_{tot}.

Solution The Fe^{3+} ion is reduced to Fe^{2+}; the reddish color suggests that I_2 has been produced by the oxidation of I^- ions. The reaction is

$$\begin{array}{ll} 2\ Fe^{3+} + 2\ e^- \rightarrow 2\ Fe^{2+}(aq) & E^0_{red} = +0.77\ V \\ 2\ I^-(aq) \rightarrow I_2(s) + 2\ e^- & E^0_{ox} = -0.53\ V \\ \hline 2\ Fe^{3+}(aq) + 2\ I^-(aq) \rightarrow 2\ Fe^{2+}(aq) + I_2(s) & E^0_{tot} = +0.24\ V \end{array}$$

Since the overall voltage is positive, we expect the reaction to be spontaneous, as it is.

EXERCISE Would you expect Fe^{3+} ions to be reduced by Br^- ions? Answer: No, $E_{tot}^0 = -0.30$ V.

The cations in the center column of Table 25.7 (Cr^{2+}, Mn^{2+}, . . .) are in an intermediate oxidation state. They can either be oxidized to a cation of higher charge ($Cr^{2+} \rightarrow Cr^{3+}$) or reduced to the metal ($Cr^{2+} \rightarrow Cr$). With certain cations of this type, these two half-reactions occur simultaneously. Consider, for example, the Cu^+ ion. In water solution, copper(I) salts *disproportionate*, undergoing reduction to copper metal and oxidation to Cu^{2+}:

$$2\,Cu^+(aq) \rightarrow Cu(s) + Cu^{2+}(aq) \tag{25.25}$$

In solution, $[Cu^+]$ from these species is very low, which drives Eq. 25.25 to the left

As a result, the only stable copper(I) species are insoluble compounds such as CuCN ($K_{sp} = 1 \times 10^{-19}$) or complex ions such as $Cu(CN)_2^-$ ($K_f = 1 \times 10^{22}$).

EXAMPLE 25.5 For the disproportionation given by Equation 25.25, calculate
a. E_{tot}^0, using Table 25.7.
b. the equilibrium constant for the reaction.
c. the concentration of Cu^+ in equilibrium with 0.1 M Cu^{2+}.

Solution
a. $E_{tot}^0 = E_{red}^0\ Cu^+ \rightarrow Cu + E_{ox}^0\ Cu^+ \rightarrow Cu^{2+}$
$= +0.52$ V $- 0.15$ V $= +0.37$ V

b. $\log_{10} K = \dfrac{nE_{tot}^0}{0.0591}$

Since $Cu^+ \rightarrow Cu$ and $Cu^+ \rightarrow Cu^{2+}$ are one-electron changes, $n = 1$:

$$\log_{10} K = \dfrac{(1)(+0.37)}{0.0591} = +6.3; \qquad K = 2 \times 10^6$$

c. $K = \dfrac{[Cu^{2+}]}{[Cu^+]^2} = 2 \times 10^6;$ $[Cu^+]^2 = \dfrac{[Cu^{2+}]}{2 \times 10^6} = \dfrac{1 \times 10^{-1}}{2 \times 10^6}$
$= 5 \times 10^{-8}$

$[Cu^+] = (5 \times 10^{-8})^{1/2} = 2 \times 10^{-4}$ M

EXERCISE What is ΔG^0 for this reaction? Answer: -36 kJ.

The Cu^+ ion is one of the few species in the center column of Table 25.7 that disproportionates at standard concentrations. Ions of this type may be unstable in water solution for a quite different reason, however. Water ordinarily contains dissolved air; the O_2 in air may oxidize the cation. When a blue solution of a chromium(II) salt (Color Plate 25.1) is exposed to air, the color quickly changes to violet or green as the Cr^{3+} ion is formed by the reaction

$$2\,Cr^{2+}(aq) \rightarrow 2\,Cr^{3+}(aq) + 2\,e^- \qquad E_{ox}^0 = +0.41\ V$$
$$\dfrac{\tfrac{1}{2}\,O_2(g) + 2\,H^+(aq) + 2\,e^- \rightarrow H_2O}{2\,Cr^{2+}(aq) + \tfrac{1}{2}\,O_2(g) + 2\,H^+(aq) \rightarrow 2\,Cr^{3+}(aq) + H_2O} \qquad \begin{array}{l} E_{red}^0 = +1.23\ V \\ E_{tot}^0 = +1.64\ V \end{array} \tag{25.26}$$

As a result of this reaction, chromium(II) salts are difficult to prepare and even more difficult to store.

The Fe^{2+} ion ($E^0_{ox} = -0.77$ V) is much more stable toward oxidation than Cr^{2+}. However, iron(II) salts in water solution are slowly converted to iron(III) by dissolved oxygen. In acidic solution, the reaction is

$$2\ Fe^{2+}(aq) + \tfrac{1}{2}\ O_2(g) + 2\ H^+(aq) \rightarrow 2\ Fe^{3+}(aq) + H_2O \qquad \textbf{(25.27)}$$

$$E^0_{tot} = E^0_{ox}\ Fe^{2+} + E^0_{red}\ O_2 = -0.77\ V + 1.23\ V = +0.56\ V$$

A similar reaction takes place in basic solution. Iron(II) hydroxide is pure white when first precipitated but, in the presence of air, it turns first green and then brown as it is oxidized by O_2:

$$2\ Fe(OH)_2(s) + \tfrac{1}{2}\ O_2(g) + H_2O \rightarrow 2\ Fe(OH)_3(s) \qquad \textbf{(25.28)}$$

25.3
Oxyanions of the Transition Metals

When a metal is in an oxidation state higher than $+3$, it is a safe bet that the metal is covalently bonded to a highly electronegative nonmetal such as oxygen or fluorine. Consider, for example, the $+6$ state of chromium, found in the compounds

CrF_6	chromium(VI) fluoride	These substances do not form Cr^{6+} ions in solution
CrO_3	chromium(VI) oxide	
K_2CrO_4 (K^+, CrO_4^{2-} ions)	potassium chromate	
$K_2Cr_2O_7$ (K^+, $Cr_2O_7^{2-}$ ions)	potassium dichromate	

In each of these species, chromium atoms are joined by covalent bonds to oxygen or fluorine atoms.

In general chemistry, the only compounds of this type that you are likely to encounter are those, such as potassium chromate or potassium dichromate, that contain *oxyanions,* polyatomic negative ions containing oxygen. Transition metals form a wide variety of oxyanions. In this section we will focus attention upon those formed by chromium and manganese.

Chromium (CrO_4^{2-}, $Cr_2O_7^{2-}$)

Chromium in the $+6$ state forms two different oxyanions. These are the yellow chromate ion, CrO_4^{2-}, and the red dichromate ion, $Cr_2O_7^{2-}$ (Color Plate 25.1). The geometries of these ions are shown in Figure 25.5. Notice that, in both cases, chromium is bonded tetrahedrally to oxygen atoms. The CrO_4^{2-} ion is stable in basic or neutral solution; in acid it is converted to the $Cr_2O_7^{2-}$ ion:

$$2\ CrO_4^{2-}(aq) + 2\ H^+(aq) \rightleftharpoons Cr_2O_7^{2-}(aq) + H_2O; \quad K = 4 \times 10^{14} \quad \textbf{(25.29)}$$

yellow   red

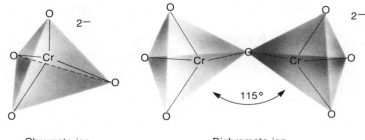

Chromate ion   Dichromate ion

FIGURE 25.5 Structures and geometries of the chromate and dichromate ions. In both ions the Cr atom is at the center of a tetrahedron, on the corners of which are O atoms.

The dichromate ion in acidic solution is a powerful oxidizing agent:

$$Cr_2O_7^{2-}(aq) + 14\ H^+(aq) + 6\ e^- \rightarrow 2\ Cr^{3+}(aq) + 7\ H_2O; \quad E_{red}^0 = +1.33\ V$$

In strong acid, $Cr_2O_7^{2-}$ oxidizes I^-, Br^-, or Fe^{2+} ions:

$$Cr_2O_7^{2-}(aq) + 14\ H^+(aq) + 6\ Fe^{2+}(aq) \rightarrow 2\ Cr^{3+}(aq) + 7\ H_2O + 6\ Fe^{3+}(aq) \quad \textbf{(25.30)}$$

In the laboratory, a solution of potassium dichromate in sulfuric acid is sometimes used to clean glassware. It is particularly effective in oxidizing greases and oils.

It will also oxidize skin very readily if you spill it

The oxidizing strength of $Cr_2O_7^{2-}$ decreases as pH increases (Example 25.6 and Fig. 25.6). The effect is a general one: *the oxidizing strength of an oxyanion, as measured by the value of E_{red}, is greatest in strongly acidic solution.*

EXAMPLE 25.6 Using the Nernst equation, calculate E_{red} for $Cr_2O_7^{2-}$ at pH 4.00, taking conc. Cr^{3+} = conc. $Cr_2O_7^{2-}$ = 1.0 M.

Solution Applying the Nernst equation to the half-reaction for the reduction of $Cr_2O_7^{2-}$,

Note the exponents from the balanced equation

$$E_{red} = E_{red}^0 - \frac{0.0591}{6} \log_{10} \frac{(conc.\ Cr^{3+})^2}{(conc.\ Cr_2O_7^{2-})(conc.\ H^+)^{14}}$$

Substituting numbers (conc. H^+ = 1.0×10^{-4} M, since pH = 4.00):

$$E_{red} = +1.33\ V - \frac{0.0591}{6} \log_{10} \frac{1}{(1 \times 10^{-4})^{14}}$$

$$= +1.33\ V - \frac{0.0591(56)}{6} = +0.78\ V$$

Note that the value of E_{red} decreases by 0.55 V (from 1.33 to 0.78 V) when the pH increases from 0 (conc. H^+ = 1.0 M) to 4. The oxidizing strength of the $Cr_2O_7^{2-}$ ion decreases accordingly.

EXERCISE At what pH does E_{red} = +1.00 V? Answer: 2.4.

The $Cr_2O_7^{2-}$ ion can act as an oxidizing agent in the solid state as well as in water solution. In particular, it can oxidize the NH_4^+ ion to molecular nitrogen.

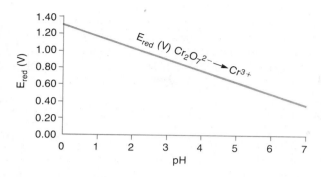

FIGURE 25.6 The oxidizing strength of the $Cr_2O_7^{2-}$ ion decreases with increasing pH (decreasing acidity). E_{red} drops by about 0.14 V when the pH increases by one unit.

When a pile of ammonium dichromate is ignited, a spectacular reaction occurs (Color Plate 25.2):

$$(NH_4)_2Cr_2O_7(s) \rightarrow N_2(g) + 4\ H_2O(g) + Cr_2O_3(s)$$

$$\text{red} \qquad\qquad\qquad\qquad \text{green}$$

(25.31)

The ammonium dichromate resembles a tiny volcano as it burns, emitting hot gases, sparks, and a voluminous green dust of chromium(III) oxide.

It's great for chem magic shows

Manganese (MnO_4^-, MnO_4^{2-})

The permanganate ion, MnO_4^-, like CrO_4^{2-}, has a tetrahedral geometry. It has an intense purple color, easily visible even in very dilute solution.* Crystals of solid potassium permanganate, $KMnO_4$, have a deep purple, almost black color. This compound is used to treat such diverse ailments as "athlete's foot" and rattlesnake bites. These applications depend upon the fact that the MnO_4^- ion is a very powerful oxidizing agent. This is especially true in acidic solution, where MnO_4^- is reduced to Mn^{2+}:

Also poison ivy, ringworm and, we're told, the seven-year itch

$$MnO_4^-(aq) + 8\ H^+(aq) + 5\ e^- \rightarrow Mn^{2+}(aq) + 4\ H_2O; \quad E^0_{red} = +1.52\ V$$

In basic solution, MnO_4^- is reduced to MnO_2, with a considerably smaller value of E^0_{red}:

$$MnO_4^-(aq) + 2\ H_2O + 3\ e^- \rightarrow MnO_2(s) + 4\ OH^-(aq); \quad E^0_{red} = +0.59\ V$$

However, even in basic solution, MnO_4^- can oxidize water:

$$4\ MnO_4^-(aq) + 2\ H_2O \rightarrow 4\ MnO_2(s) + 3\ O_2(g) + 4\ OH^-(aq) \qquad \textbf{(25.32)}$$

$$E^0_{tot} = E^0_{red}\ MnO_4^- + E^0_{ox}\ H_2O = +0.59\ V - 0.40\ V = +0.19\ V$$

This reaction accounts for the fact that laboratory solutions of $KMnO_4$ slowly decompose, producing a brownish solid (MnO_2) and gas bubbles (O_2).

---

*The deep purple color of glass that has been exposed to the sun for a long time is believed to be due to MnO_4^- ions. These are formed when ultraviolet light brings about the oxidation of manganese oxides in the glass.

The manganate ion, MnO_4^{2-}, in which manganese has an oxidation number of $+6$, has a deep green color. It can be prepared either by reducing MnO_4^- (oxid. no. Mn $= +7$) or oxidizing MnO_2 (oxid. no. Mn $= +4$). In either case, the reaction must be carried out in highly basic solution. The MnO_4^{2-} ion is unstable in acid, undergoing disproportionation:

$$3\ MnO_4^{2-}(aq) + 4\ H^+(aq) \rightarrow 2\ MnO_4^-(aq) + MnO_2(s) + 2\ H_2O \quad \textbf{(25.33)}$$

EXAMPLE 25.7 One way to prepare the MnO_4^{2-} ion is to oxidize MnO_2 in basic solution with molecular oxygen. Write a balanced equation for the redox reaction involved.

Solution The half-reactions are

oxidation: $\quad MnO_2(s) \rightarrow MnO_4^{2-}(aq)$
reduction: $\quad O_2(g) \rightarrow H_2O$

The balanced half-equations in *acidic* solution are

oxidation: $\quad MnO_2(s) + 2\ H_2O \rightarrow MnO_4^{2-}(aq) + 4\ H^+(aq) + 2\ e^-$
reduction: $\quad O_2(g) + 4\ H^+(aq) + 4\ e^- \rightarrow 2\ H_2O$

To obtain the balanced equation in acidic solution, we multiply the oxidation half-equation by two and add to the reduction half-equation. After simplification, we obtain

$$2\ MnO_2(s) + O_2(g) + 2\ H_2O \rightarrow 2\ MnO_4^{2-}(aq) + 4\ H^+(aq)$$

To find the equation in basic solution, we add 4 OH^- ions to each side, simplify, and obtain

$$2\ MnO_2(s) + O_2(g) + 4\ OH^-(aq) \rightarrow 2\ MnO_4^{2-}(aq) + 2\ H_2O$$

In practice, this reaction is ordinarily carried out by strongly heating a mixture of MnO_2 and KOH in the presence of air.

EXERCISE What volume of $O_2(g)$ at 27°C and 1.00 atm is required to react with 1.00 mol MnO_2? Answer: 12.3 L.

Summary

In this chapter, we have considered the chemistry of the more common transition metals and the cations and oxyanions derived from them. Most of these reactions were of the oxidation-reduction type. For example, we discussed

—reduction of metal ores (Section 25.1).
—oxidation of metals by acids (Section 25.1).
—redox equilibria between two different cations of the same metal (Section 25.2).
—reduction of oxyanions (Section 25.3).

We have also looked at other types of reactions involving transition metal cations (complex ion formation, acid dissociation, precipitation).

Most of the examples worked in this chapter involved redox reactions of the type discussed in Chapters 23 and 24. Principles reviewed include

—balancing redox equations (Examples 25.1 and 25.7).
—calculation and interpretation of E_{tot}^0 values (Examples 25.4 and 25.5).
—use of the Nernst equation (Example 25.6).
—calculation of ΔG^0 and K from E_{tot}^0 (Example 25.5).

Key Words and Concepts

acid dissociation constant, K_a	Nernst equation	standard oxidation voltage, E_{ox}^0
amphoteric	oxidizing agent	standard reduction voltage, E_{red}^0
aqua regia	oxyanion	standard voltage, E_{tot}^0
disproportionation	reducing agent	thermite reaction
ferromagnetism	roasting	

Questions and Problems

Reactions and Equations

25.1 Write a balanced equation to represent
  a. the roasting of Cu_2S to form copper metal.
  b. the reduction of NiO by carbon monoxide.
  c. the reduction of Fe_2O_3 with Al.

25.2 Write a balanced equation for the formation of a complex ion when
  a. a film covered with AgBr is treated with a solution of $Na_2S_2O_3$.
  b. the $Fe(H_2O)_6^{3+}$ ion acts as a weak acid.
  c. $Cu(OH)_2$ dissolves in 6 M NH_3.

25.3 Write a balanced equation for the formation of a precipitate when
  a. H_2S is added to a solution of nickel(II) nitrate.
  b. NH_3 is added to a solution of iron(III) chloride.
  c. a solution of cobalt(II) chloride is made basic with NaOH.

25.4 Write a balanced equation to show
  a. the acid behavior of $Co(H_2O)_6^{2+}$.
  b. why $Zn(OH)_2$ dissolves in strong acid.
  c. how manganese reacts with cold water.

25.5 When ammonia is added to aqueous $CuSO_4$, a precipitate first forms and then dissolves. Write a net ionic equation for
  a. precipitate formation.
  b. the dissolving of the precipitate.

25.31 Write a balanced equation to represent
  a. the roasting of Bi_2S_3 to form Bi_2O_3.
  b. the reduction of Fe_2O_3 by carbon monoxide.
  c. the reaction of MnO with Al.

25.32 Write a balanced equation for the formation of a complex ion when
  a. ammonia is added to a solution of nickel(II) nitrate.
  b. a solution containing the $Fe(H_2O)_5(OH)^{2+}$ ion is made strongly acidic.
  c. $Zn(OH)_2$ dissolves in 6 M NaOH.

25.33 Write a balanced equation for the formation of a precipitate when
  a. H_2S is added to a solution of cadmium sulfate.
  b. NH_3 is added to a solution of iron(II) chloride.
  c. a solution of $FeSO_4$ is made basic with NaOH.

25.34 Use a balanced equation to represent
  a. the roasting of ZnS in air.
  b. the evolution of hydrogen when manganese reacts with a strong acid.
  c. the dissolving of $Zn(OH)_2$ by a strong base.

25.35 Addition of concentrated NaOH solution to $Zn(NO_3)_2$ solution forms a precipitate that later dissolves as more NaOH is added. Write a net ionic equation for each of the two reactions involved.

25.6 Write a net ionic equation to account for
    a. the yellow color that appears when excess HCl is added to $Cu^{2+}(aq)$.
    b. the red color that appears when excess KSCN is added to $Fe^{3+}(aq)$.
    c. the release of hydrogen gas when hydrochloric acid reacts with cobalt metal.

25.36 Write a net ionic equation to account for
    a. the yellow precipitate that forms when $BaCl_2$ is added to a solution containing CrO_4^{2-}.
    b. the blue color that appears when excess NH_3 is added to $Ni^{2+}(aq)$.
    c. the fact that iron filings dissolve in HCl.

Balancing Redox Equations

25.7 Write a balanced redox equation to explain why
    a. nickel dissolves in a strong, nonoxidizing acid.
    b. copper dissolves in HNO_3 but not in HCl.

25.37 Write a balanced redox equation to explain why
    a. cadmium dissolves in dilute HCl.
    b. cadmium reacts with dilute HNO_3 to form $NO(g)$.

25.8 Write a balanced redox equation for the reaction of mercury with aqua regia, assuming the products include $HgCl_4^{2-}$ and $NO_2(g)$.

25.38 Write a balanced redox equation for the reaction of cadmium with aqua regia, assuming the products include $CdCl_4^{2-}$ and $NO(g)$.

25.9 Balance the following redox equations:
    a. $Cu(s) + NO_3^-(aq) \rightarrow Cu^{2+}(aq) + NO_2(g)$ (acidic solution)
    b. $Cr(OH)_3(s) + O_2(g) \rightarrow CrO_4^{2-}(aq)$ (basic solution)

25.39 Balance the following redox equations:
    a. $Fe(s) + NO_3^-(aq) \rightarrow Fe^{3+}(aq) + NO_2(g)$ (acidic solution)
    b. $Cr(OH)_3(s) + ClO^-(aq) \rightarrow CrO_4^{2-}(aq) + Cl^-(aq)$ (basic solution)

25.10 Balance the following redox equations:
    a. $Cd(s) + NO_3^-(aq) \rightarrow Cd^{2+}(aq) + NO(g)$ (acid solution)
    b. $Mn^{2+}(aq) + BiO_3^-(aq) \rightarrow MnO_4^-(aq) + Bi^{3+}(aq)$ (acid solution)

25.40 Balance the following redox equations:
    a. $Cr^{2+}(aq) + O_2(g) \rightarrow Cr^{3+}(aq)$ (acid solution)
    b. $MnO_2(s) + ClO^-(aq) \rightarrow MnO_4^-(aq) + Cl^-(aq)$ (basic solution)

E^0_{tot}; Calculation and Interpretation

25.11 Of the metals listed in Table 25.2, which would you expect to react with an oxidizing agent for which $E^0_{red} = -0.38$ V?

25.41 Of the +2 cations listed in Table 25.7, which would you expect to react with an oxidizing agent for which $E^0_{red} = +1.00$ V?

25.12 Of the cations listed in Table 25.7, show by calculation which ones should disproportionate at standard conditions.

25.42 Consider

$$M^{3+} \xrightarrow{+0.20V} M^{2+} \xrightarrow{+0.50V} M^+ \xrightarrow{+0.20V} M$$

Show by calculation which, if any, of these species should disproportionate.

25.13 Using data in Tables 25.2 and 25.7, decide which of the following reactions should react spontaneously at standard concentrations.
    a. $Ni + Zn^{2+}$
    b. $Cr^{2+} + Mn^{3+}$
    c. $Hg^{2+} + Hg$

25.43 Follow the directions for Question 25.13 for
    a. $Zn + Ni^{2+}$
    b. $Cr^{3+} + Mn^{2+}$
    c. $Co^{3+} + Co$

25.14 Using data in Tables 24.1 and 25.7, calculate E^0_{tot} for

a. $2 Co^{3+}(aq) + H_2O \rightarrow 2 Co^{2+}(aq) + \frac{1}{2} O_2(g) + 2 H^+(aq)$

b. $2 Cr^{2+}(s) + I_2(s) \rightarrow 2 Cr^{3+}(aq) + 2 I^-(aq)$

Which reactions are spontaneous at standard conditions?

25.15 Of the cations listed in the center column of Table 25.7, which one is the

a. strongest reducing agent?

b. strongest oxidizing agent?

25.16 Of the cations listed in Table 25.7, which ones would you expect to react with water by a reaction analogous to 25.24?

25.44 Follow the directions for Question 25.14 for

a. $Cr^{3+}(aq) + Co^{2+}(aq) \rightarrow Cr^{2+}(aq) + Co^{3+}(aq)$

b. $2 Fe(OH)_2(s) + \frac{1}{2} O_2(g) + H_2O \rightarrow 2 Fe(OH)_3(s)$

Which reactions are spontaneous at standard conditions?

25.45 Of the cations listed in the center column of Table 25.7, which one is the

a. weakest reducing agent?

b. weakest oxidizing agent?

25.46 Of the cations listed in Table 25.7, which ones would you expect to oxidize I^- ions to I_2 ($E^0_{ox} = -0.53$ V)?

Nernst Equation

25.17 Consider the reaction

$$2 Cu^+(aq) \rightarrow Cu^{2+}(aq) + Cu(s)$$

for which $E_{tot} = +0.37$ V. Using the Nernst equation, calculate

a. E when conc. Cu^{2+} = conc. $Cu^+ = 1 \times 10^{-4}$ M.

b. conc. Cu^+ when conc. $Cu^{2+} = 1$ M and E = 0.00 V.

25.18 Using the Nernst equation, calculate E_{red} for $Cr_2O_7^{2-}$ in 6.0 M H^+ when conc. Cr^{3+} = conc. $Cr_2O_7^{2-}$ = 0.10 M.

25.19 Consider

$$2 Co(NH_3)_6^{2+}(aq) + \frac{1}{2} O_2(g) + H_2O(l) \rightarrow$$
$$2 Co(NH_3)_6^{3+}(aq) + 2 OH^-(aq)$$

E^0_{red} $Co(NH_3)_6^{3+} = + 0.1$ V.

Use the Nernst equation to calculate E when the complex ions are 1 M, $P O_2$ = 1 atm, and the pH is

a. 4.00 b. 7.00

25.47 Consider the reaction

$$2 Ag^+(aq) \rightarrow Ag(s) + Ag^{2+}(aq)$$

for which $E_{tot} = -1.18$ V. Use the Nernst equation to calculate

a. E when conc. $Ag^+ = 1.0 \times 10^{-4}$ M = 5 × conc. Ag^{2+}.

b. Conc. Ag^{2+} when conc. $Ag^+ = 1.0$ M and E = 0.00 V.

25.48 Calculate the voltage for the reduction of MnO_4^- to Mn^{2+} ($E^0_{red} = +1.52$ V) at a pH of 7.00 when conc. $MnO_4^- = 0.20$ M and conc. $Mn^{2+} = 0.10$ M.

25.49 Repeat Problem 25.19 using $P O_2$ = 0.20 atm at a pH of

a. 6.00 b. 8.00

E^0_{tot}, K, and ΔG^0

25.20 Calculate ΔG^0 for the reactions in Problem 25.13.

25.21 Calculate K for the reactions in Problem 25.14.

25.50 Calculate ΔG^0 for the reactions in Problem 25.43.

25.51 Calculate K for the reactions in Problem 25.44.

25.22 Using Table 25.7, calculate, for the disproportion-ation of Fe^{2+},
  a. the equilibrium constant, K.
  b. the concentration of Fe^{3+} in equilibrium with 0.10 M Fe^{2+}.

25.23 Using Table 25.7, calculate ΔG^0 for
  a. $Fe(s) + 2\ Fe^{3+}(aq) \rightarrow 3\ Fe^{2+}(aq)$
  b. $Hg(l) + Hg^{2+}(aq) \rightarrow Hg_2^{2+}(aq)$

25.24 For the reaction

$$Zn(s) + 2\ Au(CN)_2^-(aq) \rightarrow$$
$$Zn(CN)_4^{2-}(aq) + 2\ Au(s)$$

ΔG^0 at 25°C is -124.7 kJ. Calculate E_{tot}^0 and K.

25.52 Using Table 25.7, calculate, for the disproportion-ation of Au^+,
  a. K
  b. the concentration of Au^+ in equilibrium with 0.10 M Au^{3+}.

25.53 Using Table 25.7, calculate ΔG^0 for the dispropor-tionation of
  a. 1 mol Fe^{2+} b. 1 mol Au^+

25.54 For the reaction

$$2\ Au(s) + AuCl_4^-(aq) + 2\ Cl^-(aq) \rightarrow$$
$$3\ AuCl_2^-(aq)$$

ΔG^0 at 25°C is $+44.2$ kJ. Calculate E_{tot}^0 and K.

General

25.25 Explain the difference between ferromagnetism and paramagnetism.

25.26 Describe briefly the various steps in the photo-graphic process.

25.27 Give the
  a. name of a transition metal that forms colorless aqueous ions.
  b. symbol of a transition metal found uncombined in nature.
  c. name of the common transition metal that reacts with cold water.
  d. formula of the oxyanion that imparts a deep pur-ple color to an aqueous solution.

25.28 Describe briefly how each of the following metals is obtained from its principal ore:
  a. Ni   b. Mn   c. Zn

25.29 Explain why
  a. a color change occurs when concentrated HCl is added to $Cr^{3+}(aq)$.
  b. aqua regia brings gold into solution, but either 16 M HNO_3 or 12 M HCl alone do not.
  c. a 0.10 M solution of $ZnCl_2$ is acidic.

25.30 Using Table 25.5, estimate the pH of 0.10 M $FeSO_4$.

25.55 What property of mercury is used in
  a. a barometer?
  b. a "silent" electrical switch?
  c. a thermometer?

25.56 In your own words, describe how hydrated co-balt(II) chloride indicates humidity changes.

25.57 What is
  a. a very reactive transition metal?
  b. the composition of aqua regia?
  c. a transition metal ion that disproportionates at standard conditions?
  d. the transition metal that is the only liquid metal at room temperature?

25.58 Describe the method used to obtain pure
  a. mercury   b. gold   c. chromium

25.59 Explain why
  a. zinc compounds are colorless.
  b. Mn^{3+} is seldom found in water solution.
  c. silver is not attacked by dilute sulfuric acid.

25.60 Using Table 25.5, estimate the pH of 0.10 M $CdCl_2$.

*25.61 For the reaction

$$2 Cu^+(aq) \rightleftharpoons Cu(s) + Cu^{2+}(aq); \quad K = 2 \times 10^6$$

The formation constants of $Cu(NH_3)_2^+$ and $Cu(NH_3)_4^{2+}$ are 5×10^{10} and 5×10^{12}, respectively. Calculate K for the reaction

$$2 Cu(NH_3)_2^+(aq) \rightleftharpoons Cu(s) + Cu(NH_3)_4^{2+}(aq)$$

Would you expect Cu^+ to disproportionate in a solution 1 M in NH_3?

*25.62 You want to oxidize 10.0 mL of 0.050 M $MnSO_4$ to MnO_4^-:

$$2 Mn^{2+}(aq) + 5 BiO_3^-(aq) + 14 H^+(aq) \rightarrow 2 MnO_4^-(aq) + 5 Bi^{3+}(aq) + 7 H_2O$$

a. How many grams of $NaBiO_3$ should you use?
b. What volume of 6.0 M HNO_3 is required?

*25.63 Taking K_{sp} $Fe(OH)_3 = 5 \times 10^{-38}$, K_a $HC_2H_3O_2 = 2 \times 10^{-5}$, and $K_w = 1.0 \times 10^{-14}$, calculate K for the reaction

$$Fe(OH)_3(s) + 3 HC_2H_3O_2(aq) \rightleftharpoons Fe^{3+}(aq) + 3 H_2O + 3 C_2H_3O_2^-(aq)$$

Would you expect $Fe(OH)_3$ to be soluble in acetic acid?

*25.64 Rust, which you can take to be $Fe(OH)_3$, can be dissolved by treating it with oxalic acid. An acid-base reaction occurs, and a complex ion is formed.
a. Write a balanced equation for the reaction.
b. What volume of 0.10 M $H_2C_2O_4$ would be required to remove a rust stain weighing 1.0 g?

*25.65 Consider the complex $Fe(NO)^{2+}$.
a. Assuming it consists of an Fe^{2+} ion and an NO molecule, calculate the number of unpaired electrons if it is a low-spin complex; a high-spin complex. (Remember that there is one unpaired electron in the NO molecule.)
b. Experimentally, it is found that the complex contains three unpaired electrons. Can you suggest an explanation for this number? (*Hint:* Consider the possibility of electron exchange between Fe^{2+} and NO.)

Chapter 26
Chemistry
of the
Nonmetals

In Chapter 13, we examined the molecular structures of the nonmetals and the compounds they form with one another. Our main concern in that chapter was with bonding and its effect upon physical properties. Here, the emphasis will be quite different. We will look at the chemical properties of selected nonmetals and some of the oxyacids and oxyanions derived from them. Virtually all of the reactions that we will discuss take place in water solution. Most of these are redox reactions of the type discussed in Chapters 23 and 24. A few are acid-base reactions (Chaps. 19 and 20).

Common oxidation numbers shown by the nonmetals are listed in Figure 26.1. As you can see, many nonmetals have a variety of different oxidation numbers. It is important to recognize two general points about these numbers:

1. In a compound where it has a *negative* oxidation number, a nonmetal is present as either

—a monatomic anion (for example, N^{3-}, O^{2-}, F^-).
—a species in which the nonmetal is covalently bonded to a less electronegative element such as hydrogen (for example, NH_3, H_2O, HF).

2. In a compound where it has a *positive* oxidation number, a nonmetal is bonded to a more electronegative element such as oxygen. Thus,

—nitrogen has an oxidation number of $+5$ in the HNO_3 molecule and the NO_3^- ion; in both of these species, the nitrogen atom is bonded to three oxygen atoms.
—sulfur has an oxidation number of $+6$ in the H_2SO_4 molecule and the SO_4^{2-} ion; in both cases, the sulfur atom is bonded to four oxygen atoms.

Most of our discussion in this chapter will focus upon species in which one of the nonmetals shown in color in Figure 26.1 has a positive oxidation number. In Section 26.2, we will describe and illustrate some of the general properties of ions and molecules of this type. Then we will consider the chemistry of

3	4	5	6	7	8
				1 H +1 −1	2 He
5 B +3	6 C +4 +2 −4	7 N +5 +4 +3 +2 +1 −3	8 O −1 −2	9 F −1	10 Ne
	14 Si +4 −4	15 P +5 +3 +1 −3	16 S +6 +4 +2 −2	17 Cl +7 +5 +3 +1 −1	18 Ar
	32 Ge +4 +2 −4	33 As +5 +3 −3	34 Se +6 +4 −2	35 Br +5 +1 −1	36 Kr +2
		51 Sb +5 +3 −3	52 Te +6 +4 −2	53 I +7 +5 +1 −1	54 Xe +6 +4 +2
			84 Po +2	85 At −1	86 Rn

FIGURE 26.1 Common oxidation states of the non-metals and metalloids. The elements whose redox chemistry will be emphasized in this chapter are shown in color. The nonmetals, though few in number, play an important role in chemistry, in part because they can exist in so many different oxidation states.

—chlorine, oxid. no. = +7, +5, +1 (Section 26.3).
—nitrogen, oxid. no. = +5, +3 (Section 26.4).
—sulfur, oxid. no. = +6, +4, +2 (Section 26.5).
—phosphorus, oxid. no. = +5 (Section 26.6).

We begin our study with the two most strongly electronegative nonmetals, oxygen and fluorine. These two nonmetals, in their compounds with other elements, never show positive oxidation numbers.

26.1
Oxygen and Fluorine

The redox chemistry of oxygen is relatively simple, at least in comparison with that of such nonmetals as chlorine, nitrogen, and sulfur. It shows only three common oxidation numbers: −2, −1, and 0. In most of its stable compounds, oxygen is present in the −2 oxidation state. This is the case, for example, in the

—water molecule (H_2O).
—oxide ion (O^{2-}) and hydroxide ion (OH^-).
—molecules of oxyacids ($HClO_4$, HNO_3, H_2SO_4, etc.).
—oxyanions (ClO_4^-, NO_3^-, HSO_4^-, SO_4^{2-}, CrO_4^{2-}, MnO_4^-, etc.).

Fluorine, the most electronegative of all elements, shows only one oxidation number in its compounds: -1. This is the case in

—molecules of nonmetal fluorides, such as HF, CF_4, and SF_6.
—the F^- ion, found in such compounds as NaF and SnF_2.

It doesn't take much F^- ion to prevent decay, and it is effective

As you may know, fluoride ions are commonly used in programs to prevent tooth decay. The outer enamel of teeth is mostly hydroxyapatite, $Ca_5(PO_4)_3(OH)$. The OH^- ion makes this compound susceptible to organic acids formed in the mouth by the fermentation of sugar. Exposure to F^- ions converts some of the hydroxyapatite to fluorapatite, $Ca_5(PO_4)_3F$. This is more resistant to acid, in part at least because the F^- ion is a weaker base than OH^-. The source of fluoride ions can be SnF_2 (stannous fluoride) in toothpaste or NaF, added to drinking water at the 1 ppm level.

The Elements: F_2, O_2, and O_3

Pure F_2 reacts with just about everything, including all organic substances

Fluorine ($E_{red}^0 = +2.87$ V) is such a powerful oxidizing agent that it cannot be prepared or used in water solution. Indeed, fluorine reacts violently with water:

$$F_2(g) + H_2O(l) \rightarrow 2\,F^-(aq) + \tfrac{1}{2}\,O_2(g) + 2\,H^+(aq) \qquad \textbf{(26.1)}$$
$$E_{tot}^0 = E_{red}^0\,F_2 + E_{ox}^0\,H_2O = +2.87\text{ V} - 1.23\text{ V} = +1.64\text{ V}$$

Other products, including ozone, O_3, are also produced.

Diatomic oxygen, O_2, is a strong oxidizing agent in acidic solution:

$$\tfrac{1}{2}\,O_2(g) + 2\,H^+(aq) + 2\,e^- \rightarrow H_2O; \qquad E_{red}^0 = +1.23\text{ V}$$

This half-reaction is especially important in living cells, where it is used to drive a variety of cellular processes. Since H^+ ions appear as a reactant in this half-equation, the oxidizing strength of O_2 decreases as pH increases. In basic solution (conc. $OH^- = 1$ M), the standard reduction voltage is much smaller:

$$\tfrac{1}{2}\,O_2(g) + H_2O + 2e^- \rightarrow 2\,OH^-(aq); \qquad E_{red}^0 = +0.40\text{ V}$$

EXAMPLE 26.1 Set up the Nernst equation for the half-reaction

$$\tfrac{1}{2}\,O_2(g) + 2\,H^+(aq) + 2\,e^- \rightarrow H_2O; \qquad E_{red}^0 = +1.23\text{ V}$$

and calculate E_{red} when $P\,O_2 = 1$ atm, conc. $H^+ = 1.0 \times 10^{-14}$ M.

Solution The Nernst equation has the form

$$E_{red} = E_{red}^0 - \frac{0.0591}{n} \log_{10} \frac{1}{(P\,O_2)^{1/2}\,(\text{conc. }H^+)^2}$$

Substituting numbers,

$$E_{red} = +1.23 \text{ V} - \frac{0.0591}{2} \log_{10} \frac{1}{(1)(1.0 \times 10^{-14})^2}$$

$$= +1.23 \text{ V} - \frac{0.0591}{2} (28.0) = +0.40 \text{ V}$$

Notice that this is the value of E_{red}^0 in basic solution:

$$\tfrac{1}{2} O_2(g) + H_2O + 2 e^- \rightarrow 2 OH^-(aq); \qquad E_{red}^0 = +0.40 \text{ V}$$

where conc. $OH^- = 1.0$ M and conc. $H^+ = 1.0 \times 10^{-14}$ M

EXERCISE Calculate E_{red} in neutral solution where conc. H^+ = conc. $OH^- = 1.0 \times 10^{-7}$ M. Answer: +0.82 V.

O_2 in water is still a good oxidizing agent

Of all oxidizing agents, elementary oxygen is the most abundant and, in many ways, the most important. Its presence in air ensures that all water supplies, including reagent solutions used in the laboratory, will ordinarily be saturated with atmospheric oxygen. We often forget this and are puzzled by such phenomena as

—the formation of white or yellow precipitates when hydrogen sulfide is used in qualitative analysis:

$$\tfrac{1}{2} O_2(g) + H_2S(g) \rightarrow S(s) + H_2O \qquad (26.2)$$

—the yellow color that solutions of NaI or KI acquire upon standing:

$$\tfrac{1}{2} O_2(g) + 2 I^-(aq) + 2 H^+(aq) \rightarrow I_2(aq) + H_2O \qquad (26.3)$$

—the cloudiness that develops in a solution of tin(II) chloride:

$$\tfrac{1}{2} O_2(g) + Sn^{2+}(aq) + H_2O \rightarrow SnO_2(s) + 2 H^+(aq) \qquad (26.4)$$

Next to fluorine, ozone, O_3, is perhaps the strongest of all oxidizing agents:

$$O_3(g) + 2 H^+(aq) + 2 e^- \rightarrow O_2(g) + H_2O; \qquad E_{red}^0 = +2.07 \text{ V}$$

Ozone is often used as a substitute for chlorine in treating municipal water supplies. A 2% mixture of O_3 in air is more effective than Cl_2 in oxidizing bacteria and other pollutants. Perhaps more important, the end products of ozone oxidation are less hazardous: chlorine can react with organic compounds in water to form suspected carcinogens such as $CHCl_3$. On the other hand, ozone is more expensive than chlorine. It also decomposes more rapidly, so that it offers little or no protection against bacteria that enter the water supply after treatment.

O_3 is made by passing O_2 between plates across which there is a high potential

$O_2 \longrightarrow$ 10000 V $\rightarrow$

$O_3 + O_2$

Hydrogen Peroxide (oxid. no. O = −1)

The molecular structure of hydrogen peroxide, H_2O_2, was discussed in Chapter 13. You may recall that its Lewis structure is

There is an O—O bond in the molecule.

Hydrogen peroxide can act as a strong oxidizing agent. In this case, it is reduced to H_2O (oxid. no. oxygen: $-1 \rightarrow -2$):

$$H_2O_2(aq) + 2\,H^+(aq) + 2\,e^- \rightarrow 2\,H_2O; \qquad E^0_{red} = +1.77\ V$$

Alternatively, H_2O_2 can act as weak reducing agent, being oxidized to O_2 (oxid. no. oxygen: $-1 \rightarrow 0$):

$$H_2O_2(aq) \rightarrow O_2(g) + 2\,H^+(aq) + 2\,e^-; \qquad E^0_{ox} = -0.68\ V$$

Hydrogen peroxide tends to decompose in water, which explains why its solutions soon lose their oxidizing power. The reaction involved is *disproportionation*, combining the two half-reactions referred to above:

$$
\begin{array}{ll}
H_2O_2(aq) + 2\,H^+(aq) + 2\,e^- \rightarrow & E^0_{red} = +1.77V \\
\underline{H_2O_2(aq) \rightarrow O_2(g) + 2\,H^+(aq) + 2\,e^-} & \underline{E^0_{ox} = -0.68\ V} \\
2\,H_2O_2(aq) \rightarrow O_2(g) + 2\,H_2O & E^0_{tot} = +1.09\ V \quad \textbf{(26.5)}
\end{array}
$$

EXAMPLE 26.2 Taking account of the fact that H_2O_2 can act as either an oxidizing agent ($E^0_{red} = +1.77\ V$) or a reducing agent ($E^0_{ox} = -0.68\ V$), determine whether the following reactions will occur (standard concentrations):

a. $H_2O_2(aq) + 2\,Fe^{2+}(aq) + 2\,H^+(aq) \rightarrow 2\,H_2O + 2\,Fe^{3+}(aq)$
b. $H_2O_2(aq) + 2\,Fe^{3+}(aq) \rightarrow O_2(g) + 2\,H^+(aq) + 2\,Fe^{2+}(aq)$
c. $H_2O_2(aq) + I_2(s) \rightarrow O_2(g) + 2\,H^+(aq) + 2\,I^-(aq)$
d. $H_2O_2(aq) + 2\,I^-(aq) + 2\,H^+(aq) \rightarrow 2\,H_2O + I_2(s)$

Use Table 24.1, p. 721, to find necessary values of E^0_{red} or E^0_{ox}.

Solution In each case, we calculate E^0_{tot} and note its sign.

a. H_2O_2 is reduced; Fe^{2+} ions are oxidized:

$$
\begin{aligned}
E^0_{tot} &= E^0_{red}\ H_2O_2 + E^0_{ox}\ Fe^{2+} \\
&= +1.77\ V - 0.77\ V = +1.00\ V \qquad \text{spontaneous}
\end{aligned}
$$

> What would happen if H_2O_2 were added in excess to 0.1 M $FeSO_4$?

b. H_2O_2 is oxidized; Fe^{3+} ions are reduced:

$$
\begin{aligned}
E^0_{tot} &= E^0_{ox}\ H_2O_2 + E^0_{red}\ Fe^{3+} \\
&= -0.68\ V + 0.77\ V = +0.09\ V \qquad \text{spontaneous}
\end{aligned}
$$

c. $E^0_{tot} = E^0_{ox}\ H_2O_2 + E^0_{red}\ I_2$
$\qquad = -0.68\ V + 0.53\ V = -0.15\ V \qquad$ nonspontaneous

d. $E^0_{tot} = E^0_{red}\ H_2O_2 + E^0_{ox}\ I^-$
$\qquad = +1.77\ V - 0.53\ V = +1.24\ V \qquad$ spontaneous

EXERCISE Will hydrogen peroxide oxidize the Mn^{2+} ion? Will it reduce Mn^{2+}? Answer: Yes; no.

26.2
Oxyacids and Oxyanions of the Nonmetals

Throughout the remainder of this chapter we will be dealing primarily with species in which a nonmetal (Cl, Br, I, N, S, or P) has a positive oxidation number. Many of these species are *oxyanions*, polyatomic anions in which a nonmetal atom is bonded to one or more oxygen atoms. Others are *oxyacids*, in which hydrogen atoms are bonded to one or more of the oxygen atoms of an oxyanion. Typical oxyanions are the perchlorate ion (ClO_4^-), the nitrate ion (NO_3^-), and the sulfate ion (SO_4^{2-}). The corresponding oxyacids are perchloric acid ($HClO_4$), nitric acid (HNO_3), and sulfuric acid (H_2SO_4).

In this section, we will look at some of the general properties of species of this type. We start by considering the system used to name oxyanions and oxyacids.

Nomenclature

Most nonmetals form at least two oxyanions and an equal number of oxyacids. To distinguish between these species, we assign names in a systematic way.

1. When a nonmetal forms two oxyanions, the suffix *-ate* is used for the anion in which the nonmetal is in the higher oxidation state. The suffix *-ite* is used for the anion containing the nonmetal in the lower oxidation state. Thus, we have

The *ate* anions are derived from *ic* acids; e.g., sulfate from sulfuric acid

SO_4^{2-} sulfate   oxid. no. S $= +6$

SO_3^{2-} sulfite   oxid. no. S $= +4$

2. When a nonmetal forms more than two oxyanions, the prefixes *per-* (highest oxidation state) and *hypo-* (lowest oxidation state) are used as well. This is necessary with the oxyanions of chlorine:

ClO_4^- perchlorate   oxid. no. Cl $= +7$

ClO_3^- chlorate   oxid. no. Cl $= +5$

ClO_2^- chlorite   oxid. no. Cl $= +3$

ClO^- hypochlorite   oxid. no. Cl $= +1$

3. The name of an oxyacid is directly related to that of the corresponding anion. The suffix *-ate* is replaced by *-ic; -ite* is replaced by *-ous*. Thus, we have

	$HClO_4$ perchloric acid
H_2SO_4 sulfuric acid	$HClO_3$ chloric acid
H_2SO_3 sulfurous acid	$HClO_2$ chlorous acid
	$HClO$ hypochlorous acid

Sulf*ite* ion is derived from sulfur*ous* acid

EXAMPLE 26.3   As we saw in Chapter 3, the compound Na_3PO_4 is called sodium phosphate. Name the following acids of phosphorous:

a. H_3PO_4 b. H_3PO_3 c. H_3PO_2

Solution

a. In H_3PO_4, as in Na_3PO_4, the oxidation number of phosphorus is $+5$. Since the name of the PO_4^{3-} ion is phosphate, that of H_3PO_4 must be *phosphoric acid*.

b. The oxidation number of phosphorus in H_3PO_3 is $+3$. This is lower than the $+5$ value in phosphoric acid. We call H_3PO_3 *phosphorous acid*.

c. Here, the oxidation number of phosphorus is still lower, $+1$. Following the rules described above, we refer to H_3PO_2 as *hypophosphorous acid*.

EXERCISE Nitrogen forms two oxyanions, NO_2^- and NO_3^-. What are their names? Answer: Nitrite, nitrate.

Acid Strength

Table 26.1 lists the dissociation constants of several oxyacids of the halogens. Generalizing from the values in the table, we can say that

1. *Acid strength increases with increasing electronegativity of the central nonmetal atom.* Of the three acids HClO, HBrO, and HIO, hypochlorous acid is the strongest (E.N. Cl = 3.0), while hypoiodous acid (E.N. I = 2.5) is the weakest. Hypobromous acid (E.N. Br = 2.8) is intermediate in strength between HClO and HIO.

2. *Acid strength increases with increasing oxidation number of the central nonmetal atom.* Of the four oxyacids of chlorine, $HClO_4$ (oxid. no. Cl = $+7$) is the strongest and HClO (oxid. no. Cl = $+1$) is the weakest.

Table 26.1
Dissociation Constants of Oxyacids of the Halogens

Oxyacid		Oxid. No. Halogen	Dissociation Constant, K_a
Perchloric acid	· $HClO_4$	$+7$	10^7
Chloric acid	$HClO_3$	$+5$	10^3
Chlorous acid	$HClO_2$	$+3$	1×10^{-2}
Hypochlorous acid	HClO	$+1$	3×10^{-8}
Hypobromous acid	HBrO	$+1$	2×10^{-9}
Hypoiodous acid	HIO	$+1$	2×10^{-11}

$HClO_4$ is a very strong acid, even stronger than HCl

These trends in acid strength can be related to molecular structure. In an oxyacid molecule, the hydrogen atom that dissociates is bonded to oxygen, which in turn is bonded to a nonmetal atom, X. We might represent the structure of an oxyacid as H—O—X and its dissociation in water as

$$H—O—X(aq) \rightleftharpoons H^+(aq) + XO^-(aq) \qquad (26.6)$$

For a proton, with its $+1$ charge, to separate from the molecule, the electron density around the oxygen should be as low as possible. This will weaken the O—H bond and favor dissociation. The electron density around the oxygen atom is decreased when:

1. X is a highly electronegative atom such as Cl. This draws electrons away from the oxygen atom and makes hypochlorous acid stronger than hypoiodous acid.

2. Additional, strongly electronegative oxygen atoms are bonded to X. These tend to draw electrons away from the oxygen atom bonded to H. Thus, we would predict that the ease of dissociation of a proton, and hence K_a, should increase in the following order; from left to right:

Seems reasonable, once you think about it

$$X-O-H \qquad O-X-O-H \qquad \underset{\underset{O}{|}}{O-X-O-H} \qquad \underset{\underset{O}{|}}{\overset{\overset{O}{|}}{O-X-O-H}}$$

oxid. no. X = +1 \qquad +3 \qquad +5 \qquad +7

Oxidizing and Reducing Properties

Most of the reactions of oxyanions and oxyacids that we will consider involve oxidation and reduction. It is helpful to keep in mind certain general principles that apply to all species of this type, regardless of the nonmetal involved.

1. *A species in which a nonmetal is in its highest oxidation state can act only as an oxidizing agent, never as a reducing agent.* Consider, for example, the ClO_4^- ion, in which chlorine is in its highest oxidation state, +7. In any redox reaction in which this ion takes part, chlorine must be reduced to a lower oxidation state. When that happens, the ClO_4^- ion acts as an oxidizing agent, taking electrons away from something else. The same argument applies to

—the NO_3^- ion (highest oxid. no. N = +5).
—the SO_4^{2-} ion (highest oxid. no. S = +6).

Both SO_4^{2-} and NO_3^- are very resistant to oxidation

By the same token, species in which a nonmetal is in its lowest oxidation state can act only as reducing agents. Such species include

—the Cl^- ion (lowest oxid. no. Cl = −1).
—the NH_3 molecule (lowest oxid. no. N = −3).

2. *A species in which a nonmetal is in an intermediate oxidation state can act as either an oxidizing agent or a reducing agent.* Consider, for example, the ClO_3^- ion (oxid. no. Cl = +5). It can be oxidized to the perchlorate ion, in which case ClO_3^- acts as a reducing agent:

$$ClO_3^-(aq) + H_2O \rightarrow ClO_4^-(aq) + 2 H^+(aq) + 2 e^-$$

Alternatively, the ClO_3^- ion can be reduced, perhaps to chloride ion. When that occurs, ClO_3^- acts as an oxidizing agent:

$$ClO_3^-(aq) + 6 H^+(aq) + 6 e^- \rightarrow Cl^-(aq) + 3 H_2O$$

Sometimes, with a species such as ClO_3^-, these two half-reactions occur together; the overall reaction is called disproportionation:

$$3 \ ClO_3^-(aq) + 3 \ H_2O \rightarrow 3 \ ClO_4^-(aq) + 6 \ H^+(aq) + 6 \ e^- \qquad E_{ox}^0 = -1.19 \ V$$
$$\underline{ClO_3^-(aq) + 6 \ H^+(aq) + 6 \ e^- \rightarrow Cl^-(aq) + 3 \ H_2O \qquad\qquad E_{red}^0 = +1.45 \ V}$$
$$4 \ ClO_3^-(aq) \rightarrow 3 \ ClO_4^-(aq) + Cl^-(aq) \qquad\qquad\qquad E_{tot}^0 = +0.26 \ V \qquad (26.7)$$

Three fourths of the ClO_3^- ions are oxidized to ClO_4^-, while one fourth of them are reduced to Cl^-. In general, *a species in an intermediate oxidation state is likely to disproportionate if the sum $E_{ox}^0 + E_{red}^0$ is a positive number ($E_{tot}^0 > 0$)*.

3. *The oxidizing strength of an oxyanion is directly related to the concentration of H^+ ion* (inversely related to pH). There is a simple reason for this. Consider, for example, the half-equation in which ClO_3^- acts as an oxidizing agent:

$$ClO_3^-(aq) + 6 \ H^+(aq) + 6 \ e^- \rightarrow Cl^-(aq) + 3 \ H_2O$$

Since H^+ is a reactant, a decrease in its concentration should make the half-reaction less spontaneous. Indeed it does: E_{red} decreases from $+1.45$ V when conc. $H^+ = 1.0$ M (pH = 0) to $+0.62$ V when conc. $H^+ = 1.0 \times 10^{-14}$ M (pH = 14).

4. *The reducing strength of an oxyanion is inversely related to the concentration of H^+ ion* (directly related to pH). To see why this should be true, consider the half-reaction in which ClO_3^- acts as a reducing agent:

$$ClO_3^-(aq) + H_2O \rightarrow ClO_4^-(aq) + 2 \ H^+(aq) + 2 \ e^-$$

Here, H^+ is a product. Decreasing its concentration should make the reaction more spontaneous and increase the voltage. Experimentally, we find that E_{ox} increases (algebraically) from -1.19 V when conc. $H^+ = 1.0$ M (pH = 0) to -0.36 V when conc. $H^+ = 1.0 \times 10^{-14}$ M (pH = 14).

EXAMPLE 26.4 Nitrogen can have oxidation numbers ranging from $+5$ to -3. Consider the NO_2^- ion.
a. Can the NO_2^- ion act as an oxidizing agent? a reducing agent?
b. How would you decide whether the NO_2^- ion would disproportionate in water solution?
c. If you wanted to oxidize NO_2^- to NO_3^-, would you work in acidic or basic solution?

Solution Note that the oxidation number of nitrogen in the NO_2^- ion is $+3$.
a. The NO_2^- ion can act as an oxidizing agent, in which case it is reduced to a species such as NO (oxid. no. N = $+2$) or N_2 (oxid. no. N = 0). It can act as a reducing agent, in which case it is oxidized, perhaps to NO_2 (oxid. no. N = $+4$) or NO_3^- (oxid. no. N = $+5$).
b. Add the standard voltages for the reduction and oxidation of the NO_2^- ion. Note that there will be several such voltages, one for each species formed. If any combination of $E_{ox}^0 + E_{red}^0$ is positive, the ion is likely to disproportionate.
c. When NO_2^- is oxidized to NO_3^-, it acts as a reducing agent. Applying the fourth principle related above, the reaction should proceed most readily in basic solution (low H^+ concentration). To confirm that this is the case, it is helpful to write the half-equation:

In most reactions, ClO_3^- ion is an oxidizing agent, and a strong one

The NO_2^- ion participates in many reactions. The NO_3^- ion does not, unless $[H^+]$ is high

$$NO_2^-(aq) + H_2O \rightarrow NO_3^-(aq) + 2\ H^+(aq) + 2\ e^-$$

Since H^+ ion is a product, the reaction should proceed more readily in basic solution.

EXERCISE If you wanted to reduce NO_2^- to N_2, would it be better to work in neutral solution (pH = 7) or acidic solution (pH = 0)? Answer: Acidic solution.

26.3
Chlorine
(Bromine, Iodine)

The elements chlorine, bromine, and iodine can act as oxidizing agents in water solution, in which case they are reduced to halide ions:

$$Cl_2(g) + 2\ e^- \rightarrow 2\ Cl^-(aq); \qquad E^0_{red} = +1.36\ V$$

$$Br_2(l) + 2\ e^- \rightarrow 2\ Br^-(aq); \qquad E^0_{red} = +1.07\ V$$

$$I_2(s) + 2\ e^- \rightarrow 2\ I^-(aq); \qquad E^0_{red} = +0.53\ V$$

Chlorination of water supplies takes advantage of the ability of chlorine to oxidize bacteria. Iodine, in its alcohol solution (tincture of iodine) is sometimes used as a mildly oxidizing antiseptic.

Since chlorine is a stronger oxidizing agent than bromine or iodine, it can be used to prepare these elements by oxidation of their anions, Br^- and I^-:

$$Cl_2(g) + 2\ Br^-(aq) \rightarrow 2\ Cl^-(aq) + Br_2(l)$$
$$E^0_{tot} = E^0_{red}\ Cl_2 + E^0_{ox}\ Br^- = +1.36\ V - 1.07\ V = +0.29\ V$$

(26.8)

$$Cl_2(g) + 2\ I^-(aq) \rightarrow 2\ Cl^-(aq) + I_2(s)$$
$$E^0_{tot} = E^0_{red}\ Cl_2 + E^0_{ox}\ I^- = +1.36\ V - 0.53\ V = +0.83\ V$$

(26.9)

These reactions are often used to test for the presence of Br^- or I^- ions (Color Plate 26.1)

Addition of chlorine to a solution containing either of these ions gives the free halogens, Br_2 or I_2. After addition of chlorine, the solution is often shaken with an organic solvent. The free halogen enters the organic layer, in which it is more soluble. It gives that layer its characteristic color, reddish brown (bromine) or violet (iodine).

The colors of I_2 and Br_2 in the solvent can be used to test for the presence of those halogens

Oxyanions and Oxyacids of the Halogens

Compounds of chlorine in the +7, +5, +3, and +1 states have been studied extensively and will occupy most of our attention here. The chemistry of bromine and iodine in positive oxidation states is less extensive. Perhaps the most stable state for these elements is the +5 ($NaBrO_3$, $NaIO_3$, HIO_3). In the +7 state, iodine forms two different oxyanions, IO_4^- and IO_6^{5-}, corresponding to the acids HIO_4 and H_5IO_6.

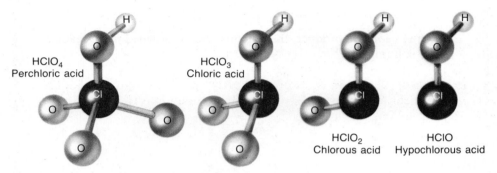

FIGURE 26.2 In perchloric acid, $HClO_4$, the Cl atom is at the center of a tetrahedron, bonded to four oxygen atoms; there is an H atom bonded to one of the oxygens. In $HClO_3$, $HClO_2$, and $HClO$, successive oxygen atoms are removed from corners of the tetrahedron.

The structures of the four oxyacids of chlorine are shown in Figure 26.2. Notice that, in each case, the hydrogen atom is bonded to oxygen rather than chlorine. This is generally true for oxyacids; *only those hydrogen atoms bonded to oxygen dissociate to form H^+ ions in water.*

The redox chemistry of the halogens and their compounds can be correlated by means of the *standard potential diagram* shown in Figure 26.3. We will use several such diagrams in this chapter. It is important to know how they are interpreted.

1. The numbers above the arrows correspond to E^0_{red} values. For example, for the reduction of ClO_4^- to ClO_3^- in acidic solution (1 M H^+),

It takes a while to get used to those diagrams

$$ClO_4^-(aq) + 2\ H^+(aq) + 2\ e^- \rightarrow ClO_3^-(aq) + H_2O; \qquad E^0_{red} = +1.19\ V$$

For the reduction of ClO_4^- to ClO_3^- in basic solution (1 M OH^-),

$$ClO_4^-(aq) + H_2O + 2\ e^- \rightarrow ClO_3^-(aq) + 2\ OH^-(aq); \qquad E^0_{red} = +0.36\ V$$

2. Values of E^0_{ox} can be obtained by changing the sign of E^0_{red}. For example,

$$ClO_3^-(aq) + H_2O \rightarrow ClO_4^-(aq) + 2\ H^+(aq) + 2\ e^-; \qquad E^0_{ox} = -1.19\ V$$

3. Species listed in the diagram are those that are present at high concentration in 1 M H^+ (acidic solution) or 1 M OH^- (basic solution). For +7 chlorine, the principal species, indeed the only species, in both acidic and basic solution is the ClO_4^- ion. Since $HClO_4$ is a strong acid, it is completely dissociated to ClO_4^- ions, regardless of the concentration of H^+. The situation is quite different with +1 chlorine, since $HClO$ is a weak acid. In acidic solution (1 M H^+), $HClO$ is the principal species; there are far more $HClO$ molecules than ClO^- ions. In basic solution (1 M OH^-), the principal species is ClO^-; most of the $HClO$ molecules have been converted to ClO^- ions.

Figure 26.3 confirms some of the general statements made earlier. Notice, for example, that E^0_{red} values are generally larger (more positive) in acidic than in basic solution. This means that reduction occurs more readily in acidic solution, so that species such as ClO_4^- and ClO_3^- are stronger oxidizing agents when the concentration of H^+ is high. Perhaps the most striking feature of Figure 26.3 is the fact that all the E^0_{red} values are positive. This means that:

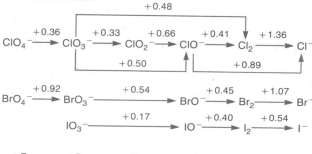

Oxidation number +7 +5 +3 +1 0 −1

Acidic Solution

Basic Solution

Oxidation number +7 +5 +3 +1 0 −1

The change in potential for a reaction as one goes from acidic to basic solution varies with the reaction

FIGURE 26.3 Values of E_{red}^0 for species of Cl, Br, and I in acidic solution (upper diagram and basic solution (lower diagram). Values of E_{ox}^0 can be obtained by changing the sign. Thus:

$$2\ ClO_3^-(aq) + 12H^+(aq) + 10e^- \rightarrow Cl_2(g) + 6\ H_2O; E_{red}^0 = +1.47\ V$$
$$Cl_2(g) + 6\ H_2O \rightarrow 2\ ClO_3^-(aq) + 12\ H^+(aq) + 10e^-; E_{ox}^0 = -1.47\ V$$

The halogens and all of their oxyacids and oxyanions are relatively strong oxidizing agents. For the oxyacids and oxyanions, this is particularly true in acidic solution.

This is not true for the halide ions

The data in Figure 26.3 can be used to calculate standard voltages for a large number of redox reactions. Several of these involve disproportionation, in which the same species is both oxidized and reduced (Example 26.5).

EXAMPLE 26.5 Consider the disproportionation of chlorous acid, $HClO_2$, to hypochlorous acid, $HClO$, and chlorate ion, ClO_3^-.
a. Write a balanced equation for this reaction.
b. Using Figure 26.3, calculate the standard voltage, E_{tot}^0, for the reaction.

Solution
a. The balanced half-equations, obtained in the usual way, are

reduction: $HClO_2(aq) + 2\ H^+(aq) + 2\ e^- \rightarrow HClO(aq) + H_2O$

oxidation: $HClO_2(aq) + H_2O \rightarrow ClO_3^-(aq) + 3\ H^+(aq) + 2\ e^-$

Adding the two half-equations and canceling out species that appear on both sides ($2\ e^-$, $2\ H^+$, H_2O) gives

$$2\ HClO_2(aq) \rightarrow HClO(aq) + ClO_3^-(aq) + H^+(aq)$$

b. $E^0_{tot} = E^0_{red} (HClO_2 \rightarrow HClO) + E^0_{ox} (HClO_2 \rightarrow ClO_3{}^-)$

From Figure 26.3, we read $E^0_{red} = +1.65$ V directly. To obtain E^0_{ox}, we note that E^0_{red} for the reverse reaction is $+1.21$ V. Changing the sign, we get $E^0_{ox} = -1.21$ V:

$$E^0_{tot} = +1.65 \text{ V} - 1.21 \text{ V} = +0.44 \text{ V}$$

The fact that E^0_{tot} is positive means that the $HClO_2$ molecule (oxid. no. Cl $= +3$) is unstable in water solution. $HClO_2$ disproportionates to $HClO$ (oxid. no. Cl $= +1$) and $ClO_3{}^-$ (oxid. no. Cl $= +5$).

EXERCISE Determine E^0_{tot} for the disproportion of the $ClO_2{}^-$ ion (in basic solution) to give ClO^- and $ClO_3{}^-$. Answer: $+0.33$ V.

As we have just seen, the $+3$ state of chlorine is unstable in water solution. Both the $HClO_2$ molecule and $ClO_2{}^-$ ion decompose spontaneously. Ordinarily, these reactions occur rather rapidly. This explains why you don't find compounds containing $+3$ chlorine in the general chemistry laboratory. Compounds containing $+3$ bromine or iodine are also unstable in water.

In the rest of this section, we will look at the properties of species in which the oxidation number of chlorine is $+1$, $+5$, or $+7$. Typically, such compounds as $NaClO$, $KClO_3$, and $KClO_4$ are found in reagent bottles in the laboratory or storeroom. The acids $HClO$, $HClO_3$, and $HClO_4$ are less stable. Indeed, perchloric acid, $HClO_4$, is the only oxyacid of chlorine that can be isolated in the pure state.

Hypochlorous Acid and the Hypochlorite Ion

When chlorine is added to water it undergoes the following reversible reaction:

$$Cl_2(g) + H_2O \rightleftharpoons HClO(aq) + H^+(aq) + Cl^-(aq) \tag{26.10}$$

The resulting solution is called "chlorine water." It contains equimolar amounts of the weak acid $HClO$ and the strong acid HCl. The concentrations of both acids in this solution are rather low, about 0.03 M.

When $P_{Cl_2} = 1$ atm

The position of the equilibrium in Reaction 26.10 is strongly affected by the concentration of H^+ ions. In basic solution, where $[H^+]$ is low, chlorine is much more soluble than in pure water. The overall reaction that occurs when chlorine is bubbled through a solution of sodium hydroxide at room temperature is

$$Cl_2(g) + 2 OH^-(aq) \rightarrow ClO^-(aq) + Cl^-(aq) + H_2O \tag{26.11}$$

The solution formed in Reaction 26.11 is sold under various trade names as a household bleach and disinfectant. It is prepared commercially by electrolyzing a stirred water solution of sodium chloride. Recall that the electrolysis of an $NaCl$ solution gives Cl_2 molecules and OH^- ions; stirring ensures that these species react with each other. The active ingredient of the resulting solution is the hypochlorite ion, which is a rather potent oxidizing agent:

$$ClO^-(aq) + H_2O + 2 e^- \rightarrow Cl^-(aq) + 2 OH^-(aq); \qquad E^0_{red} = +0.89 \text{ V}$$

EXAMPLE 26.6 For Reactions 26.10 and 26.11, calculate
a. E^0_{tot}, using Figure 26.3   b. the equilibrium constant, K.

Solution In both reactions, chlorine disproportionates, being reduced to the -1 state and oxidized to the $+1$ state.
a. Reaction 26.10 (acidic solution):

$$E^0_{tot} = E^0_{red} \; Cl_2 \rightarrow Cl^- + E^0_{ox} \; Cl_2 \rightarrow HClO$$
$$= +1.36 \; V - 1.63 \; V = -0.27 \; V$$

Reaction 26.11 (basic solution):

$$E^0_{tot} = E^0_{red} \; Cl_2 \rightarrow Cl^- + E^0_{ox} \; Cl_2 \rightarrow ClO^-$$
$$= +1.36 \; V - 0.41 \; V = +0.95 \; V$$

b. In general, $\log_{10} K = nE^0_{tot}/0.0591$. Here, $n = 1$ (why?), so that

Reaction 26.10: $\log_{10} K = (1)(-0.27)/0.0591 = -4.6; K = 3 \times 10^{-5}$

Reaction 26.11: $\log_{10} K = (1)(+0.95)/0.0591 = +16.0; K = 1 \times 10^{16}$

Clearly, the position of the equilibrium is much more favorable in basic solution, as we would predict from Le Châtelier's principle.

EXERCISE Calculate ΔG^0 at 25°C for the two reactions. Answer: $+26$ kJ, -92 kJ.

Chlorates and Perchlorates

In hot, concentrated solution, the reaction between chlorine and OH^- is quite different from the one that occurs at room temperature. Any ClO^- ions formed disproportionate to ClO_3^- and Cl^- ions ($E^0_{tot} = +0.39$ V). The net reaction is

$$3 \; Cl_2(g) + 6 \; OH^-(aq) \rightarrow ClO_3^-(aq) + 5 \; Cl^-(aq) + 3 \; H_2O \qquad (26.12)$$

Potassium chlorate is a powerful oxidizing agent in acidic solution (Fig. 26.3). It reacts violently with easily oxidized materials, including many organic compounds. It can be used as a laboratory source of oxygen if heated gently, with MnO_2 as a catalyst:

Many chemists tend to avoid chlorates and perchlorates

$$2 \; KClO_3(s) \rightarrow 2 \; KCl(s) + 3 \; O_2(g) \qquad (26.13)$$

Without a catalyst, at 350 to 400°C, $KClO_3$ disproportionates to potassium perchlorate, $KClO_3$, and potassium chloride, KCl:

$$4 \; KClO_4(s) \rightarrow 3 \; KClO_4(s) + KCl(s) \qquad (26.14)$$

Perchloric acid can be prepared by heating (very carefully!) a metal perchlorate with sulfuric acid:

$$KClO_4(s) + H_2SO_4(l) \rightarrow KHSO_4(s) + HClO_4(l) \qquad (26.15)$$

The pure acid and its concentrated water solution (above 60% $HClO_4$) are explosive and unsafe to work with. In cold, dilute solution, perchloric acid is a stable, very strong acid.

26.4
Nitrogen

Nitrogen can have all possible oxidation numbers between $+5$ and -3. Species stable in acidic and in basic solution for each of these oxidation states are listed in Table 26.2.

Table 26.2
Oxidation States of Nitrogen

OXID. NO.	ACIDIC SOLUTION	BASIC SOLUTION
$+5$	NO_3^-	NO_3^-
$+4$	$NO_2(g)$	$NO_2(g)$
$+3$	HNO_2*	NO_2^-
$+2$	$NO(g)$	$NO(g)$
$+1$	$N_2O(g)$	$N_2O(g)$
0	$N_2(g)$	$N_2(g)$
-1	NH_3OH^+	NH_2OH
-2	$N_2H_5^+$	N_2H_4
-3	NH_4^+	NH_3

*Slowly decomposes to NO and NO_3^-.

Nitrogen shows a remarkable array of oxidation states

A standard potential diagram for nitrogen-containing species is shown in Figure 26.4. Comparing this diagram to the analogous one for chlorine (Fig. 26.3), we note two differences:

—in several oxidation states, the stable nitrogen species is an oxide, such as NO or NO_2, rather than an oxyanion or oxyacid. There are only two common oxyanions of nitrogen, NO_3^- and NO_2^-.
—the E_{red}^0 values for nitrogen species are somewhat smaller than those for chlorine. For example, the NO_3^- ion has a somewhat lower E_{red}^0 value than does the ClO_3^- ion, which means that it is a somewhat weaker oxidizing agent. In practice, that turns out to be an advantage rather than a disadvantage. Nitrates are a great deal safer to work with than chlorates (or perchlorates or hypochlorites, for that matter).

Nitrates are not nearly so prone to explode

In the remainder of this section we will deal with the oxyacids and oxyanions of nitrogen in the $+5$ and $+3$ states. The Lewis structures of these species are:

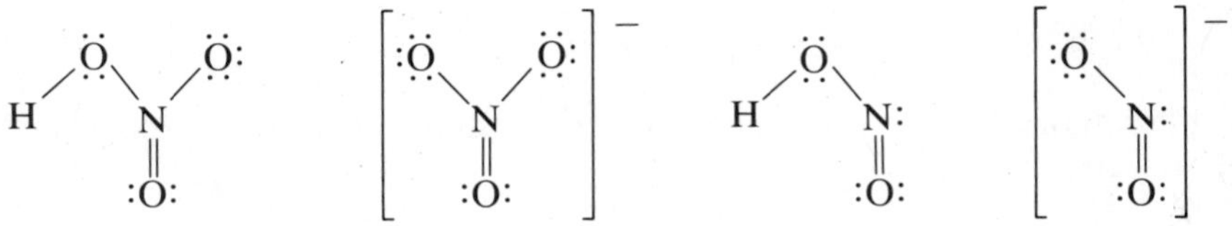

nitric acid HNO_3 — nitrate ion NO_3^- — nitrous acid HNO_2 — nitrite ion NO_2^-

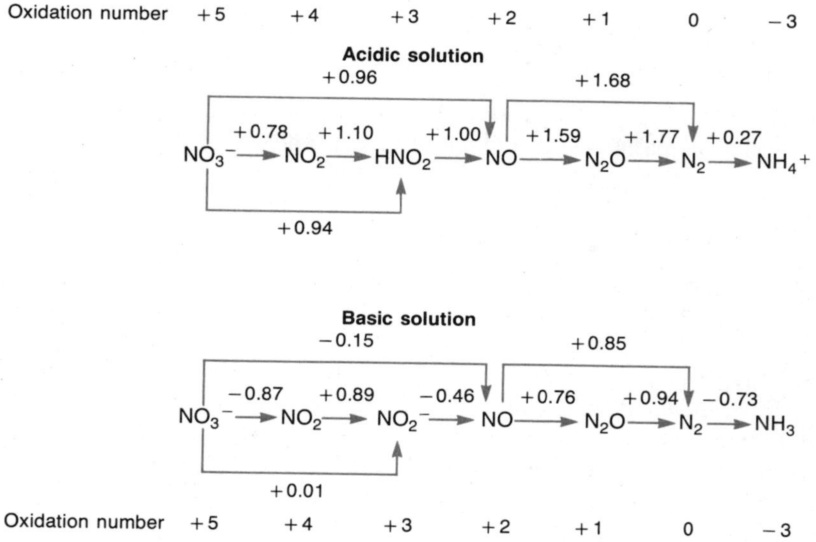

FIGURE 26.4 Values of E_{red}^0 for species containing nitrogen. These voltages are interpreted in the same way as those given in Figure 26.3.

+5 State (NO_3^-, HNO_3)

Nitric acid is a strong acid that is completely dissociated to H^+ and NO_3^- ions in dilute water solution:

$$HNO_3(aq) \rightarrow H^+(aq) + NO_3^-(aq)$$

Concentrated nitric acid (16 M) is colorless when pure. In sunlight, it turns yellow because it decomposes to NO_2:

Above 16 M HNO_3 you can always smell NO_2

$$4\ HNO_3(aq) \rightarrow 4\ NO_2(g) + 2\ H_2O + O_2(g) \qquad \textbf{(26.16)}$$

The concentrated acid (16 M) is a strong oxidizing agent. It can be reduced to any of the species listed below the NO_3^- ion in the "Acidic Solution" column of Table 26.2. Most often, the reduction product is nitrogen dioxide, NO_2. This is the case, for example, when 16 M HNO_3 reacts with the elements copper or sulfur:

$$Cu(s) + 2\ NO_3^-(aq) + 4\ H^+(aq) \rightarrow Cu^{2+}(aq) + 2\ NO_2(g) + 2\ H_2O \qquad \textbf{(26.17)}$$
$$S(s) + 6\ NO_3^-(aq) + 4\ H^+(aq) \rightarrow SO_4^{2-}(aq) + 6\ NO_2(g) + 2\ H_2O \qquad \textbf{(26.18)}$$

HNO_3 is a good solvent for many metals

Concentrated nitric acid can also be used to oxidize highly insoluble metal sulfides such as copper(II) sulfide, CuS (Example 26.7).

EXAMPLE 26.7 When 16 M HNO_3 reacts with CuS, the NO_3^- ion is reduced to $NO_2(g)$; sulfide ions in CuS are oxidized, mainly to elemental sulfur. Write a balanced equation for this redox reaction.

Solution The half-equations are

reduction: $NO_3^-(aq) \rightarrow NO_2(g)$

oxidation: $CuS(s) \rightarrow Cu^{2+}(aq) + S(s)$

(Note that the S^{2-} ions in CuS are oxidized; the Cu^{2+} ions pass into solution). The half-equations are balanced in the usual way:

reduction: $NO_3^-(aq) + 2 H^+(aq) + e^- \rightarrow NO_2(g) + H_2O$

oxidation: $CuS(s) \rightarrow Cu^{2+}(aq) + S(s) + 2 e^-$

We multiply the reduction half-equation by two and add to the oxidation half-equation, obtaining

$$CuS(s) + 2 NO_3^-(aq) + 4 H^+(aq) \rightarrow Cu^{2+}(aq) + S(s) + 2 NO_2(g) + 2 H_2O$$

Notice that in redox equations for reactions involving nitric acid, we always use NO_3^- ions and H^+ ions rather than HNO_3 molecules, since nitric acid is a strong acid.

EXERCISE What volume of 16 M HNO_3 is required to react with 0.10 g (1.0×10^{-3} mol) CuS? Answer: 0.25 mL.

Dilute nitric acid (6 M) is a weaker oxidizing agent than 16 M HNO_3. It also gives a wider variety of reduction products, depending upon the nature of the reducing agent. With inactive metals such as copper ($E_{ox}^0 = -0.34$ V), the major product is usually NO (oxid. no. N = +2):

$$3 Cu(s) + 2 NO_3^-(aq) + 8 H^+(aq) \rightarrow 3 Cu^{2+}(aq) + 2 NO(g) + 4 H_2O \qquad \textbf{(26.19)}$$

With very dilute acid and a strong reducing agent such as zinc ($E_{ox}^0 = +0.76$ V), reduction may go all the way to the NH_4^+ ion (oxid. no. N = −3):

This is one of the few reactions in which NO_3^- is reduced to NH_4^+

$$4 Zn(s) + NO_3^-(aq) + 10 H^+(aq) \rightarrow 4 Zn^{2+}(aq) + NH_4^+(aq) + 3 H_2O \qquad \textbf{(26.20)}$$

+3 State (NO_2^- and HNO_2)

Salts containing the nitrite ion are often made from the corresponding nitrates. Thus, sodium nitrite, $NaNO_2$, is prepared by heating sodium nitrate, $NaNO_3$:

$$2 NaNO_3(s) \rightarrow 2 NaNO_2(s) + O_2(g) \qquad \textbf{(26.21)}$$

Sodium nitrite is used as a food additive in hot dogs, ham, bacon, and cold cuts. It preserves the red color of the meat and also prevents the growth of an organism that causes botulism, an often fatal type of food poisoning. The use of $NaNO_2$ for this purpose has been questioned because it can produce compounds known as nitrosamines, which are carcinogenic (cancer-causing).

So far we have found no substitutes for nitrites as preservatives

A water solution of nitrous acid, HNO_2, is made by adding a strong acid to sodium nitrite, $NaNO_2$. Since HNO_2 is a weak acid ($K_a = 4.5 \times 10^{-4}$), the following acid-base reaction goes virtually to completion:

$$H^+(aq) + NO_2^-(aq) \rightarrow HNO_2(aq) \qquad \textbf{(26.22)}$$

At pH 0 (conc. H^+ = 1 M), about 99.95% of the NO_2^- ions are converted to HNO_2 molecules.

EXAMPLE 26.8 Taking K_a HNO_2 = 4.5×10^{-4}, calculate the ratio $[NO_2^-]/[HNO_2]$ at
a. pH 0.00 (conc. H^+ = 1.0 M).   b. pH 14.00 (conc. OH^- = 1.0 M).

Solution From the expression for K_a, we obtain a general relation to calculate the required ratio:

$$K_a = \frac{[H^+] \times [NO_2^-]}{[HNO_2]}; \qquad \frac{[NO_2^-]}{[HNO_2]} = \frac{K_a}{[H^+]} = \frac{4.5 \times 10^{-4}}{[H^+]}$$

a. At pH 0.00,

$$[H^+] = 1.0 \text{ M}; \qquad \frac{[NO_2^-]}{[HNO_2]} = 4.5 \times 10^{-4}$$

99.95% HNO_2
0.05% NO_2^-

Clearly, in 1 M H^+, the principal species is HNO_2. The fact that this ratio is so small means there are very few NO_2^- ions. This explains why, in Figure 26.4, the species listed in acidic solution for +3 nitrogen is the HNO_2 molecule rather than the NO_2^- ion.

b. At pH 14.00,

$$[H^+] = 1.0 \times 10^{-14} \text{ M}$$

$$\frac{[NO_2^-]}{[HNO_2]} = \frac{4.5 \times 10^{-4}}{1.0 \times 10^{-14}} = 4.5 \times 10^{10}$$

In this solution, there are 45 billion NO_2^- ions for every HNO_2 molecule. This explains why, in Figure 26.4, the species listed under +3 nitrogen in basic solution is the NO_2^- ion.

EXERCISE Which species, HNO_2 or NO_2^-, is present at the higher concentration in neutral solution (pH = 7)? Answer: NO_2^- ion.

In HNO_2, nitrogen is in an intermediate oxidation state, +3. Hence, at least in principle, nitrous acid can act as either an oxidizing or reducing agent. As you can see from Figure 26.4, nitrous acid is a strong oxidizing agent

At pH = 0, HNO_2 will oxidize Br^-, I^-, and SO_3^{2-}

$$HNO_2(aq) + H^+(aq) + e^- \rightarrow NO(g) + H_2O; \qquad E_{red}^0 = +1.00 \text{ V}$$

but a weak reducing agent

$$HNO_2(aq) + H_2O \rightarrow NO_3^-(aq) + 3 H^+(aq) + 2 e^-; \qquad E_{ox}^0 = -0.94 \text{ V}$$

and will reduce ClO_3^- and CrO_4^{2-}

Looking at the equations just written, you might guess that HNO_2 would disproportionate in water solution. Indeed it does; the overall reaction, obtained by combining these two half-equations, is

$$3 HNO_2(aq) \rightarrow 2 NO(g) + NO_3^-(aq) + H_2O + H^+(aq); \qquad \textbf{(26.23)}$$

HNO_2 in solution is pale blue

$$E_{tot}^0 = +0.06 \text{ V}$$

Ordinarily, this reaction occurs rather slowly, so that the HNO_2 stays around long enough for its properties to be studied. The pure acid cannot be isolated, however.

26.5
Sulfur

In its compounds, sulfur shows oxidation numbers of $+6$, $+4$, $+2$, and -2. The species found in acidic and basic solution in these oxidation states are listed in Table 26.3. Structures of the SO_4^{2-}, SO_3^{2-}, and $S_2O_3^{2-}$ anions are shown in Figure 26.5. In the protonated species (HSO_4^-, H_2SO_4; HSO_3^-, H_2SO_3), H atoms are bonded to oxygen atoms of the oxyanions.

Table 26.3
Oxidation States of Sulfur

OXID. NO. S	ACIDIC SOLUTION	BASIC SOLUTION
$+6$	HSO_4^-, SO_4^{2-}	SO_4^{2-}
$+4$	$SO_2(g)$, H_2SO_3, HSO_3^-	SO_3^{2-}
$+2$	$S_2O_3^{2-}*$	$S_2O_3^{2-}$
0	$S(s)$	$S(s)$
-2	$H_2S(g)$	HS^-, S^{2-}

*Decomposes to S and SO_2.

Figure 26.6 shows the standard potential diagram for sulfur species. As you can see, the oxyanions of sulfur are much weaker oxidizing agents than those of nitrogen (Fig. 26.4) or chlorine (Fig. 26.3). In basic solution, the E^0_{red} values are negative for all sulfur species, whereas they are usually positive for chlorine and nitrogen species. Compare, for example,

$$ClO_4^-(aq) + H_2O + 2\ e^- \rightarrow ClO_3^-(aq) + 2\ OH^-(aq); \qquad E^0_{red} = +0.36\ V$$

$$NO_3^-(aq) + H_2O + 2\ e^- \rightarrow NO_2^-(aq) + 2\ OH^-(aq); \qquad E^0_{red} = +0.01\ V$$

Sulfur oxyanions are poor oxidizing agents

$$SO_4^{2-}(aq) + H_2O + 2\ e^- \rightarrow SO_3^{2-}(aq) + 2\ OH^-(aq); \qquad E^0_{red} = -0.89\ V$$

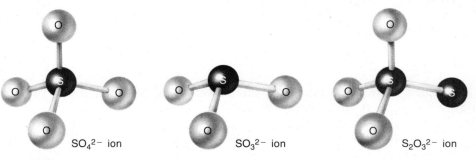

SO_4^{2-} ion SO_3^{2-} ion $S_2O_3^{2-}$ ion

FIGURE 26.5 In the SO_4^{2-} ion, the sulfur atom is at the center of a tetrahedron, bonded to four oxygen atoms. Removing one of these oxygens gives the sulfite ion, SO_3^{2-}. Replacing an oxygen by a sulfur atom gives the thiosulfate ion, $S_2O_3^{2-}$.

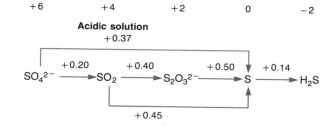

FIGURE 26.6 Values of E^0_{red} for species containing sulfur.

Note also that for sulfur:

1. Species in the $+6$ state (H_2SO_4, HSO_4^-, SO_4^{2-}) can act only as oxidizing agents, never as reducing agents, in redox reactions.

2. Species in the $+4$ state (SO_2, H_2SO_3, HSO_3^-, SO_3^{2-}), the $+2$ state ($S_2O_3^{2-}$), or the 0 state can act as either oxidizing or reducing agents.

3. Species in the -2 state (H_2S, HS^-, S^{2-}) can act only as reducing agents, never as oxidizing agents, in redox reactons.

SO_3^{2-} and S^{2-} tend to oxidize on exposure to air

$+6$ State (SO_4^{2-}, HSO_4^-, H_2SO_4)

You will recall from Chapter 19 that H_2SO_4 is a strong acid, completely dissociated to H^+ and HSO_4^- ions in dilute water solution. The HSO_4^- ion dissociates further to give H^+ and SO_4^{2-} ions:

$$H_2SO_4(aq) \rightarrow H^+(aq) + HSO_4^-(aq)$$

$$HSO_4^-(aq) \rightleftharpoons H^+(aq) + SO_4^{2-}(aq); \qquad K_a = 1.2 \times 10^{-2}$$

The dissociation constant of the HSO_4^- ion is relatively large. This explains why a solution made up by dissolving $NaHSO_4$ in water is acidic. Indeed, the pH of 0.10 M $NaHSO_4$ is about 1.5, only a little higher that that of 0.10 M HCl (pH = 1).

$NaHSO_4$ is a convenient solid source of H^+ ion

As you can see from Figure 26.6, sulfuric acid is not a very strong oxidizing agent. The concentrated acid (18 M), however, is capable of oxidizing copper metal:

$$Cu(s) + 4 H^+(aq) + SO_4^{2-}(aq) \rightarrow Cu^{2+}(aq) + SO_2(g) + 2 H_2O \qquad \textbf{(26.24)}$$

The SO_4^{2-} ion in neutral or basic solution is a very weak oxidizing agent (Example 26.9).

EXAMPLE 26.9 Consider the reduction of SO_4^{2-} to SO_3^{2-} in basic solution.
a. Write a balanced half-equation for the process.
b. Taking $E_{red}^0 = -0.89$ V, calculate E_{red} when conc. $OH^- = 10^{-7}$ M and all other species are at unit concentrations.

Solution
a. In acidic solution, the half-equation would be

$$SO_4^{2-}(aq) + 2\ H^+(aq) + 2\ e^- \rightarrow SO_3^{2-}(aq) + H_2O$$

Adding two OH^- ions to each side and simplifying,

$$SO_4^{2-}(aq) + H_2O + 2e^- \rightarrow SO_3^{2-}(aq) + 2\ OH^-(aq)$$

b. $E_{red} = E_{red}^0 - \dfrac{0.0591}{2} \log_{10} \dfrac{(\text{conc. } SO_3^{2-}) \times (\text{conc. } OH^-)^2}{(\text{conc. } SO_4^{2-})}$

$\qquad = -0.89\text{ V} - \dfrac{0.0591}{2} \log_{10} (10^{-7})^2$

$\qquad = -0.89\text{ V} - \dfrac{0.0591}{2} (-14) = -0.48\text{ V}$

EXERCISE What would be the value of E_{red} for this half-reaction at pH 10? Answer: -0.65 V.

It's not easy to reduce SO_3^{2-} to SO_3^{2-}

Concentrated sulfuric acid, in addition to being an acid and an oxidizing agent, is also a dehydrating agent. Small amounts of water can be removed from organic liquids such as gasoline by extraction with sulfuric acid. Sometimes it is even possible to remove the elements of water from a compound by treating it with 18 M H_2SO_4. This happens with table sugar, $C_{12}H_{22}O_{11}$; the product is a black char that is mostly carbon (Fig 26.7):

$$C_{12}H_{22}O_{11}(s) \rightarrow 12\ C(s) + 11\ H_2O(l) \tag{26.25}$$

FIGURE 26.7 Effect of adding concentrated H_2SO_4 to sugar. The black, gummy solid formed is mostly carbon.

When concentrated sulfuric acid dissolves in water, a great deal of heat is given off, nearly 100 kJ per mole of H_2SO_4. Sometimes enough heat is evolved to bring the solution to the boiling point. To prevent this and to avoid spattering, the acid should be added slowly to water, with constant stirring. If it comes in contact with the skin, concentrated sulfuric acid can cause painful chemical burns.

It treats skin much like sugar, so be careful with 18 M H_2SO_4

+4 State (SO_3^{2-}, HSO_3^-, H_2SO_3, SO_2)

Sulfur dioxide, SO_2, is very soluble in water. The concentration of its saturated solution at 25°C is about 1.3 mol/L. The high solubility is explained in part by a reversible reaction with H_2O to form sulfurous acid, H_2SO_3:

$$SO_2(g) + H_2O \rightleftharpoons H_2SO_3(aq) \tag{26.26}$$

The equilibrium constant for this reaction is not known, but it appears that much of the SO_2 remains unreacted.

Sulfurous acid is a weak acid that ionizes in two steps:

$$H_2SO_3(aq) \rightleftharpoons H^+(aq) + HSO_3^-(aq); \quad K_a = 1.7 \times 10^{-2}$$

$$HSO_3^-(aq) \rightleftharpoons H^+(aq) + SO_3^{2-}(aq); \quad K_a = 5.6 \times 10^{-8}$$

You can usually smell SO_2 above a solution of H_2SO_3

The species present in the +4 state in water solution depends upon the pH of the solution. As you can see from Figure 26.8, H_2SO_3 and SO_2 are the main species below pH 2. Between pH 2 and 7, the major species is the hydrogen sulfite ion, HSO_3^-. In basic solution, above pH 7, the sulfite ion, SO_3^{2-}, is predominant.

In basic solution, the SO_3^{2-} ion is readily oxidized to SO_4^{2-} by dissolved oxygen:

$$SO_3^{2-}(aq) + 2\ OH^-(aq) \rightarrow SO_4^{2-}(aq) + H_2O + 2\ e^- \qquad E_{ox}^0 = +0.89\ V$$

$$\frac{\frac{1}{2} O_2(g) + H_2O + 2\ e^- \rightarrow 2\ OH^-(aq)}{SO_3^{2-}(aq) + \frac{1}{2} O_2(g) \rightarrow SO_4^{2-}(aq)} \qquad \frac{E_{red}^0 = +0.40\ V}{E_{tot}^0 = +1.29\ V}$$

$$\tag{26.27}$$

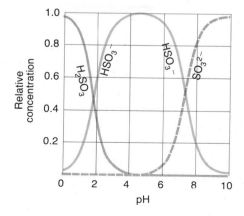

FIGURE 26.8 In the +4 state, sulfur can exist as H_2SO_3 (a water solution of SO_2), HSO_3^-, or SO_3^{2-}. The relative concentrations of these species change with pH. In strongly acidic solution, H_2SO_3 dominates. Close to pH 2, the concentration of HSO_3^- becomes equal to that of H_2SO_3; between pH 3 and pH 7, HSO_3^- is the major species present. In basic solution, above about pH 7, the SO_3^{2-} ion dominates.

This reaction accounts for the fact that solutions containing the sulfite ion, SO_3^{2-}, are almost always contaminated by the sulfate ion, SO_4^{2-}. Even in the solid state, salts such as Na_2SO_3 are somewhat unstable, slowly forming Na_2SO_4 on exposure to air.

+2 State ($S_2O_3^{2-}$)

The thiosulfate ion, $S_2O_3^{2-}$, is formed by heating a basic solution containing the sulfite ion, SO_3^{2-}, with elemental sulfur:

$$SO_3^{2-}(aq) + S(s) \rightarrow S_2O_3^{2-}(aq) \tag{26.28}$$

Sodium thiosulfate, $Na_2S_2O_3$, is made in this way from sodium sulfite, Na_2SO_3. The principal use of sodium thiosulfate ("hypo") is in photography (Chap. 25).

In acidic solution, the thiosulfate ion decomposes slowly as a result of the reaction

$$S_2O_3^{2-}(aq) + 2 H^+(aq) \rightarrow S(s) + SO_2(g) + H_2O \tag{26.29}$$

This is a disproportionation reaction: half the sulfur is reduced ($+2 \rightarrow 0$) and half is oxidized ($+2 \rightarrow +4$). From Figure 26.6, we see that E_{tot}^0 for this reaction in acidic solution is positive:

$$\begin{aligned} E_{tot}^0 &= E_{red}^0 (S_2O_3^{2-} \rightarrow S) + E_{ox}^0 (S_2O_3^{2-} \rightarrow SO_2) \\ &= +0.50 \text{ V} - 0.40 \text{ V} = +0.10 \text{ V} \end{aligned}$$

Thiosulfate ion is reasonably stable in neutral and basic solution

Thiosulfate ion is an important analytical reagent in redox titrations involving iodine. Iodine liberated by the oxidation of I^- ions can be titrated with $S_2O_3^{2-}$ ions using starch as an indicator. The redox reaction is

$$I_2(aq) + 2 S_2O_3^{2-}(aq) \rightarrow 2 I^-(aq) + S_4O_6^{2-}(aq) \tag{26.30}$$

Starch forms a blue-black complex with iodine. The disappearance of this color marks the equivalence point of the titration. Trace amounts of ozone in air can be measured by oxidizing I^- ions to I_2, which is then titrated by Reaction 26.30.

26.6 Phosphorus

Phosphorus, which lies below nitrogen in Group 5 of the Periodic Table, has many fewer oxidation states. The +5 state is by far the most important, although phosphorus can have oxidation numbers of +3, +1, and −3 (Table 26.4). The Lewis structures of the three oxyacids of phosphorus are

phosphoric acid, H_3PO_4 phosphorous acid, H_3PO_3 hypophosphorous acid, H_3PO_2

Table 26.4
Oxidation States of Phosphorus

OXID. NO. P	ACIDIC SOLUTION	BASIC SOLUTION
+5	H_3PO_4, $H_2PO_4^-$	HPO_4^{2-}, PO_4^{3-}
+3	H_3PO_3, $H_2PO_3^-$	HPO_3^{2-}
+1	H_3PO_2	$H_2PO_2^-$
0	P_4	P_4
−3	PH_3	PH_3

Notice that in the H_3PO_3 molecule one of the three hydrogen atoms is bonded to phosphorus. In H_3PO_2, two hydrogen atoms are bonded to phosphorus. These hydrogen atoms do not dissociate in water, which explains why the stable species in basic solution are the HPO_3^{2-} ion (+3 state) and the $H_2PO_2^-$ ion (+1 state).

The formula H_3PO_2 is somewhat misleading, since only one H^+ ion can dissociate

+5 State; Phosphoric Acid and Its Salts

The +5 acid of phosphorus, H_3PO_4, has properties quite different from those of the corresponding oxyacid of nitrogen, HNO_3. In particular:

1. Phosphoric acid is a weak acid. It dissociates in three steps:

$$H_3PO_4(aq) \rightleftharpoons H_2PO_4^-(aq) + H^+(aq); \quad K_1 = 7.5 \times 10^{-3}$$
$$H_2PO_4^-(aq) \rightleftharpoons HPO_4^{2-}(aq) + H^+(aq); \quad K_2 = 6.2 \times 10^{-8}$$
$$HPO_4^{2-}(aq) \rightleftharpoons PO_4^{3-}(aq) + H^+(aq); \quad K_3 = 1.7 \times 10^{-12}$$

H_3PO_4 is one of the stronger weak acids

2. Phosphoric acid is a very weak oxidizing agent; its E^0_{red} value is negative:

$$H_3PO_4(aq) + 2 H^+(aq) + 2 e^- \rightarrow H_3PO_3(aq) + H_2O; \quad E^0_{red} = -0.28 \text{ V}$$

In basic solution, reduction is still more difficult:

$$PO_4^{3-}(aq) + 2 H_2O + 2 e^- \rightarrow HPO_3^{2-}(aq) + 3 OH^-(aq); \quad E^0_{red} = -1.12 \text{ V}$$

Compounds of phosphorus in lower oxidation states are readily oxidized to the +5 state, particularly in basic solution.

Since phosphoric acid is tribasic, it forms three different types of salts. Salts containing the $H_2PO_4^-$ ion are slightly acidic; the acid dissociation

$$H_2PO_4^-(aq) \rightleftharpoons H^+(aq) + HPO_4^{2-}(aq); \quad K_a = 6.2 \times 10^{-8}$$

occurs to a greater extent than base dissociation

$$H_2PO_4^-(aq) + H_2O \rightleftharpoons OH^-(aq) + H_3PO_4(aq); \quad K_b = 1.3 \times 10^{-12}$$

In the stockroom we can find

H_3PO_4, 85%
$NaH_2PO_4 \cdot H_2O$
$Na_2HPO_4 \cdot 7 H_2O$
$Na_3PO_4 \cdot 12 H_2O$

As a result, salts such as NaH_2PO_4 and $Ca(H_2PO_4)_2$ have a pH close to 5. The situation is quite different with the HPO_4^{2-} ion, where K_b (1.6×10^{-7}) is larger than K_a (1.7×10^{-12}). Salts such as Na_2HPO_4 and $CaHPO_4$ are slightly basic, with a pH close to 9. The PO_4^{3-} ion is one of the strongest of the "weak bases":

$$PO_4^{3-}(aq) + H_2O \rightleftharpoons OH^-(aq) + HPO_4^{2-}(aq); \qquad K_b = 5.9 \times 10^{-3}$$

Salts containing this ion are highly basic; a 0.1 M solution of Na_3PO_4 has a pH of 12.3.

The salts of phosphoric acid have a variety of uses. Calcium dihydrogen phosphate, $Ca(H_2PO_4)_2$, is a major component of most fertilizers. It is water soluble and furnishes the phosphorus required for plant growth. Sodium phosphate, Na_3PO_4, is used in heavy-duty cleaning agents because its water solution is so highly basic. At one time phosphates were found in most detergents. That application has been curtailed because phosphates promote the growth of algae that can clog lakes and streams. The major use of phosphoric acid is in the manufacture of fertilizers such as $Ca(H_2PO_4)_2$. Smaller amounts of H_3PO_4 are used as a flavoring agent in cola drinks.

> Plants need K, N and P, usually in the form of K^+, NO_3^- or NH_4^+, and $H_2PO_4^-$

Summary

In this chapter, we have looked at some of the chemical properties of the nonmetals and the oxyanions and oxyacids derived from them. We have considered the nomenclature (Example 26.3), relative acid strengths (Table 26.1), and acid-base properties of these species (Example 26.8). Most of our discussion, however, has focused upon the redox chemistry of compounds in which a nonmetal is in a positive oxidation state.

All of the oxyanions and oxyacids of the nonmetals can, at least in principle, act as oxidizing agents. Oxidizing strength is measured by E^0_{red} values (Figs. 26.3, 26.4, and 26.6). Generally speaking, E^0_{red} values are highest for the oxyanions of chlorine and nitrogen (for example, ClO_4^-, ClO_3^-, ClO^-, NO_3^-). The oxyanions of sulfur, SO_4^{2-} and SO_3^{2-}, are relatively weak oxidizing agents; those of phosphorus are still weaker. For all oxyanions, oxidizing strength is greatest in acidic solution and falls off rapidly with increasing pH.

Any species containing a nonmetal atom in an intermediate oxidation state can act as either an oxidizing or reducing agent. It can be reduced to a lower state, acting as an oxidizing agent, or oxidized to a higher state, in which case it behaves as a reducing agent. Sometimes these two reactions occur together and the species disproportionates. We expect that to happen if the sum of $E^0_{red} + E^0_{ox}$ for a species is greater than zero.

In discussing the redox chemistry of the nonmetals, we have reviewed many of the principles introduced in Chapters 23 and 24. These include

—calculation and interpretation of E^0_{tot} values (Examples 26.2, 26.4, 26.5, and 26.6).
—relation between E^0_{tot}, K, and ΔG^0 (Example 26.6).
—balancing redox equations (Examples 26.5 and 26.7).
—the Nernst equation (Examples 26.1 and 26.9).

Key Words and Concepts

disproportionation	oxidizing agent	standard oxidation voltage, E^0_{ox}
electronegativity	oxyacid	standard reduction voltage, E^0_{red}
Nernst equation	oxyanion	standard voltage, E^0_{tot}
oxidation number (state)	reducing agent	

Questions and Problems

Oxyanions and Oxyacids

26.1 Name the following oxyacids and oxyanions of bromine:
a. $HBrO_2$ b. BrO_3^- c. BrO_4^- d. $HBrO$

26.2 Name the acid for which each of the following is the conjugate base:
a. NO_2^- b. ClO_2^- c. BrO_4^- d. $H_2PO_4^-$

26.3 The compound H_2SeO_3 is called selenous acid. Give the name of
a. H_2SeO_4 b. SeO_3^{2-}

26.4 Which would you expect to be the stronger acid:
a. $HBrO_3$ or $HBrO$? b. $HClO$ or $HBrO$?
c. H_2SO_3 or H_2SeO_3? d. HIO or HIO_2?

26.5 Classify each of the following as an oxidizing agent, a reducing agent, or a species capable of being either one:
a. BrO_3^- b. Br^- c. NO d. SO_3^{2-}

26.6 Of the species listed in the "Acidic Solution" column of Table 26.2, which ones are capable of acting as oxidizing agents? as reducing agents?

26.7 Taking K_a $HNO_2 = 4.5 \times 10^{-4}$, calculate the pH at which $[NO_2^-] = [HNO_2]$.

26.8 Taking K_a $H_2SO_3 = 1.7 \times 10^{-2}$ and K_a $HSO_3^- = 5.6 \times 10^{-8}$, estimate the pH at which
a. $[H_2SO_3] = [HSO_3^-]$
b. $[HSO_3^-] = 2.0 \times [SO_3^{2-}]$
c. $[H_2SO_3] = 0.50 \times [HSO_3^-]$
Check your answers by referring to Figure 26.8.

26.31 Name the following oxyacids and oxyanions:
a. HIO_2 b. PO_4^{3-} c. NO_2^- d. H_3PO_2

26.32 Name the acid for which each of the following is the conjugate base:
a. HSO_3^- b. IO_3^- c. SO_4^{2-} d. ClO_4^-

26.33 The compound Na_3AsO_4 is called sodium arsenate. Give the formula of
a. sodium arsenite b. arsenous acid

26.34 Which would you expect to be the stronger acid:
a. HNO_2 or HNO_3? b. H_3PO_4 or H_3AsO_4?
c. H_2SO_4 or H_2SO_3? d. $HClO_3$ or $HBrO_3$?

26.35 Follow instructions for Question 26.5 for
a. N^{3-} b. IO_2^- c. SO_4^{2-} d. Br_2

26.36 Of the species listed in the "Basic Solution" column of Table 26.2, which ones can act as oxidizing agents? as reducing agents?

26.37 Determine the pH at which $[HNO_2] = 2.0 \times [NO_2^-]$ (K_a $HNO_2 = 4.5 \times 10^{-4}$).

26.38 Follow the directions for Problem 26.8 to estimate the pH at which
a. $[HSO_3^-] = 7.0 \times [H_2SO_3]$
b. $[SO_3^{2-}] = 4.0 \times [HSO_3^-]$
c. $[HSO_3^-] = 3.0 \times [SO_3^{2-}]$

Balancing Redox Equations

26.9 Balance the following equations:
a. $Cl_2(g) + OH^-(aq) \rightarrow$
$$ClO_3^-(aq) + Cl^-(aq) + H_2O$$
b. $CuS(s) + NO_3^-(aq) + H^+(aq) \rightarrow$
$$Cu^{2+}(aq) + S(s) + NO_2(g) + H_2O$$
c. $Zn(s) + NO_3^-(aq) + H^+(aq) \rightarrow$
$$Zn^{2+}(aq) + NH_4^+(aq) + H_2O$$

26.10 Complete and balance the following redox equations:
a. $Br_2(l) + I^-(aq) \rightarrow$
b. $SCN^-(aq) + NO_3^-(aq) \rightarrow SO_4^{2-}(aq) + NO_2(g)$
$+ HCN(aq)$ (acidic solution)

26.39 Balance the following equations:
a. $Sn^{2+}(aq) + O_2(g) + H_2O \rightarrow SnO_2(s) + H^+(aq)$
b. $H_2O_2(aq) + I^-(aq) + H^+(aq) \rightarrow I_2(s) + H_2O$
c. $Fe(s) + O_2(g) + H_2O \rightarrow Fe(OH)_3(s)$ (basic solution)

26.40 Complete and balance the following redox equations:
a. I_2 and H_2O (basic solution) to give I^- and IO^-.
b. $I^-(aq) + SO_4^{2-}(aq)$ to yield, in acidic solution, $I_2(s)$ and $SO_2(g)$.

26.11 Write balanced equations for the disproportionation of
a. $HClO$ to $HClO_2$ and Cl_2 (acidic solution).
b. ClO^- to ClO_2^- and Cl_2 (basic solution).
c. N_2 to NO and NH_4^+ (acidic solution).

26.12 Write balanced equations for the oxidation of the following species by $Cr_2O_7^{2-}$, which is reduced to Cr^{3+}, in acidic solution:
a. I^- to I_2 b. Fe^{2+} to Fe^{3+}
c. HNO_2 to NO_3^- d. H_2S to S

E_{tot}^0; Calculation and Uses

26.13 Based on the potentials given in Figure 26.3, which bromine species is
a. the strongest oxidizing agent in acidic solution?
b. the strongest oxidizing agent in basic solution?
c. the strongest reducing agent in acidic solution?
d. the strongest reducing agent in basic solution?

26.14 Taking $E_{red}^0 H_2O_2 = +1.77$ V, determine which of the following species will be oxidized by hydrogen peroxide (use Table 24.1 to find E_{ox}^0 values).
a. Co^{2+} b. Cl^- c. Fe^{2+} d. Sn^{2+}

26.15 Using Figure 26.3, decide which of the following species should disproportionate in acidic solution (standard concentrations).
a. ClO_3^- b. $HClO$ c. Cl_2 d. Cl^-

26.16 Use Figure 26.4 to decide whether the following processes will occur at standard conditions:
a. $NO_3^-(aq) + HNO_2(aq) \rightarrow NO_2(g)$
b. $NO_3^-(aq) + NO_2^-(aq) \rightarrow NO_2(g)$
c. $NO_3^-(aq) + HNO_2(aq) \rightarrow NO(g)$
d. $NO(g) \rightarrow HNO_2(aq) + N_2O(g)$

26.17 Using appropriate data from this chapter, calculate E_{tot}^0 for each of the following reactions.
a. $ClO_3^-(aq) + NO_2^-(aq) \rightarrow$
$$ClO_2^-(aq) + NO_3^-(aq)$$
b. $SO_4^{2-}(aq) + 4 H^+(aq) + 2 Cl^-(aq) \rightarrow$
$$SO_2(g) + 2 H_2O + Cl_2(g)$$
c. $H_3PO_4(aq) + H^+(aq) + Cl^-(aq) \rightarrow$
$$H_3PO_3(aq) + HClO(aq)$$

Nernst Equation

26.18 Consider the half-reaction

$$NO_3^-(aq) + 2 H^+(aq) + e^- \rightarrow NO_2(g) + H_2O$$

$E_{red}^0 = +0.78$ V. Calculate E_{red} when NO_3^- is 1 M, P NO_2 is 1 atm, and the pH is
a. 0.0   b. 7.0   c. 14.0

26.41 Write balanced equations for the disproportionation of
a. ClO_3^- to ClO_4^- and ClO_2^-.
b. HNO_2 to NO_3^- and NO (acidic solution).
c. NO_2^- to NO_3^- and NO (basic solution).

26.42 Write balanced equations for the oxidation of the species in Question 26.12 by MnO_4^-, which is reduced to Mn^{2+}, in acidic solution.

26.43 Answer Question 26.13 for chlorine species.

26.44 Taking $E_{ox}^0 H_2O_2 = -0.68$ V, determine which of the following species will be reduced by hydrogen peroxide (use Table 24.1 to find E_{red}^0 values).
a. $Cr_2O_7^{2-}$ b. Fe^{2+} c. I_2 d. Br_2

26.45 Follow the directions of Question 26.15 for the following species in basic solution.
a. ClO_3^- b. ClO_2^- c. ClO^- d. Cl^-

26.46 Using Figure 26.6, decide whether each of the following processes will occur at standard conditions.
a. $SO_4^{2-}(aq) + S_2O_3^{2-}(aq) \rightarrow SO_3^{2-}(aq)$
b. $SO_4^{2-}(aq) + S_2O_3^{2-}(aq) \rightarrow SO_2(g)$
c. $S(s) + SO_2(g) \rightarrow S_2O_3^{2-}(aq)$

26.47 Follow the directions of Question 26.17 for the following reactions.
a. $2 NO_3^-(aq) + 4 H^+(aq) + 2 Br^-(aq) \rightarrow$
$$2 NO_2(g) + Br_2(l) + 2 H_2O$$
b. $3 Cl_2(g) + 6 OH^-(aq) \rightarrow$
$$ClO_3^-(aq) + 5 Cl^-(aq) + 3 H_2O$$
c. $3 NO(g) + 2 NH_3(aq) \rightarrow \frac{5}{2} N_2(g) + 3 H_2O$

26.48 Consider the half-reaction

$$ClO_4^-(aq) + 2 H^+(aq) + 2 e^- \rightarrow$$
$$ClO_3^-(aq) + H_2O$$

$E_{red}^0 = +1.19$ V. Calculate E_{red} when ClO_4^- and ClO_3^- are 1 M at pH
a. 0.0   b. 7.0   c. 14.0

26.19 For the reaction

$$2\,ClO_2^-(aq) \rightarrow ClO^-(aq) + ClO_3^-(aq)$$

$E_{tot}^0 = +0.33$ V. Calculate E when conc. $ClO_2^- =$ conc. $ClO^- =$ conc. $ClO_3^- = 0.10$ M.

26.20 For the reaction in Problem 26.19, taking all species to be 1 M except ClO_2^-, calculate the concentration of ClO_2^- at which E = 0.

E_{tot}^0, ΔG^0, and K

26.21 Calculate ΔG^0 and K for
  a. Equation 26.1   b. Equation 26.8
  c. Equation 26.27

26.22 Calculate ΔG^0 for each of the reactions in Problem 26.17.

26.23 For the reaction

$$3\,BrO_3^-(aq) + H^+(aq) \rightarrow 2\,BrO_4^-(aq) + HBrO(aq)$$

$\Delta G^0 = +98.0$ kJ. Calculate E_{tot}^0 and compare to the value obtained from Figure 26.3.

General

26.24 What is the oxidation number of Cl in
  a. ClO_3^- b. HClO   c. ClF_3 d. $HClO_4$

26.25 Give the formula of a compound or an oxyanion of phosphorus in which the oxidation number of phosphorus is
  a. +3   b. +5   c. −3

26.26 Describe how you would convert
  a. Br^- to Br_2.   b. $KClO_3$ to $KClO_4$.
  c. SO_3^{2-} to SO_4^{2-}.

26.27 Give the formula(s) of the product(s) formed when H^+ ions are added to a solution of
  a. NO_2^- b. S^{2-} c. $S_2O_3^{2-}$ d. ClO^- e. HSO_3^-

26.28 Give the formula of
  a. an anion in which S has an oxidation number of +2.
  b. two anions in which S has an oxidation number of +4.
  c. three different acids of sulfur.

26.29 Write the Lewis structure of
  a. $H_2PO_4^-$ b. HSO_4^- c. IO_4^-

26.49 For the reaction

$$2\,HClO_2(aq) \rightarrow HClO(aq) + ClO_3^-(aq) + H^+(aq)$$

$E_{tot}^0 = +0.44$ V. Calculate E when conc. $HClO_2 = 0.01$ M, conc. $HClO = 0.10$ M, conc. $ClO_3^- = 0.10$ M, pH = 5.

26.50 For the reaction in Problem 26.49, taking all species to be 1 M except $HClO_2$, calculate the concentration of $HClO_2$ at which the reaction would no longer be spontaneous, that is, E = 0.

26.51 Calculate ΔG^0 and K for
  a. Equation 26.5   b. Equation 26.7
  c. Equation 26.9

26.52 Calculate ΔG^0 for each of the reactions in Problem 26.47.

26.53 For the reaction

$$3\,BrO_3^-(aq) \rightarrow 2\,BrO_4^-(aq) + BrO^-(aq)$$

$\Delta G^0 = +147.0$ kJ. Calculate E_{tot}^0 and compare to the value obtained from Figure 26.3.

26.54 What is the oxidation number of N in
  a. NO_2^- b. NO_2 c. HNO_3 d. NH_4^+

26.55 Give the formula of a compound or oxyanion of bromine in which the oxidation number of bromine is
  a. +5   b. −1   c. +7

26.56 How would you prepare
  a. $HClO_4$ from $KClO_4$?   b. NO_2 from HNO_3?
  c. $S_2O_3^{2-}$ from S?

26.57 Give the formula(s) of the product(s) formed when OH^- ions are added to a solution of
  a. NH_4^+ b. $HClO_4$ c. HSO_3^- d. HClO

26.58 Give the formula of a compound of nitrogen that is
  a. a weak base.
  b. a strong acid.
  c. a weak acid.
  d. capable of oxidizing copper.
  e. formed by heating $NaNO_3$.

26.59 Write the Lewis structure of
  a. HSO_3^- b. SO_3^{2-} c. SO_3 d. H_2SO_3

26.30 Explain why

    a. fluorine can not be obtained by the electrolysis of NaF in water.

    b. O_3 is in some ways preferable to Cl_2 in water treatment.

    c. Concentrated H_2SO_4 converts table sugar to a black char.

26.60 Explain why

    a. in preparing dilute sulfuric acid, it is important to add the concentrated acid to water with constant stirring.

    b. HClO is a stronger acid than HBrO.

    c. nitric acid solutions often have a yellow or brown color.

*26.61 The compound hyponitrous acid has the molecular formula $H_2N_2O_2$. Draw a reasonable Lewis structure for this molecule. Are there isomers of $H_2N_2O_2$?

*26.62 Taking the Lewis structure of the SCN^- ion to be $:\ddot{S}{=}C{=}\ddot{N}:$, find the oxidation number of each atom (assign bonding electrons to the more electronegative atom).

*26.63 Using data from Chapter 5, calculate ΔH for the reaction

$$H_2SO_4(l) \rightarrow 2\,H^+(aq) + SO_4^{2-}(aq)$$

How many grams of water could be heated from 20.0 to 100.0°C when one mole of sulfuric acid undergoes this reaction?

*26.64 Of the 30 numbered equations in this chapter, how many involve oxidation and reduction?

*26.65 You may have noticed from Figures 26.3, 26.4, and 26.6 that E^0_{red} values are not additive. For example, $E^0_{red}\,(NO_3^- \rightarrow NO_2) + E^0_{red}\,(NO_2 \rightarrow HNO_2)$ does *not* equal $E^0_{red}\,(NO_3^- \rightarrow HNO_2)$. There is, however, a simple general relation that applies in situations such as this. Can you discover this relation from the data in these three figures?

The "ordinary chemical reactions" discussed to this point involve changes in the outer electronic structure of atoms or molecules. In contrast, nuclear reactions result from a change taking place within atomic nuclei. You will recall (Chap. 2) that atomic nuclei are represented by symbols such as

$$^{12}_{6}C \qquad ^{14}_{6}C$$

Here, the atomic number (number of protons in the nucleus) is shown as a subscript at the lower left. The mass number (number of protons + neutrons in the nucleus) appears as a superscript at the upper left. Nuclei with the same number of protons but different numbers of neutrons are called *isotopes*. The symbols written above represent two isotopes of the element carbon (at. no. = 6). One isotope has 6 neutrons in the nucleus and hence has a mass number of $6 + 6 = 12$. The heavier isotope has 8 neutrons and hence a mass number of 14.

At the present time, about 1500 different nuclei are known. Of these, somewhat less than 300 are stable; that is, they show no tendency to decompose or change in any way as time passes. The neutron-to-proton ratio required for stability varies with atomic number. For light elements, this ratio is close to one. For example, the isotopes $^{12}_{6}C$, $^{14}_{7}N$, and $^{16}_{8}O$ are stable. As atomic number increases, the ratio increases, reaching about 1.5 with heavy elements such as $^{206}_{82}Pb$. For any given element, only a few isotopes will have a neutron-to-proton ratio within the range required for stability (Fig. 27.1).

All isotopes of elements beyond Bi in the Periodic Table are radioactive

Isotopes that fall above or below the "belt of stability" shown in Figure 27.1 are unstable. They decompose spontaneously by a type of nuclear reaction called *radioactivity*. We will examine the nature of this process and some of its effects in Section 27.1. The rate at which unstable nuclei decompose ("decay") will be considered in Section 27.2. Two other important types of nuclear reactions are fission (Section 27.4) and fusion (Section 27.5).

Nuclear reactions are accompanied by energy changes that greatly exceed those associated with ordinary chemical reactions. The energy evolved when one

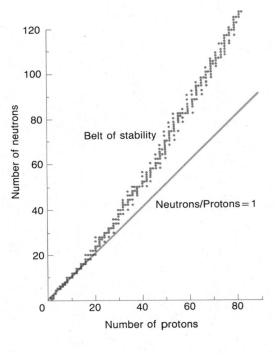

FIGURE 27.1 Stable isotopes (*colored dots*) have neutron-to-proton ratios that fall within a narrow range, referred to as a "belt of stability." For light isotopes of small atomic number, the stable ratio is 1.0; with heavier isotopes, it increases to about 1.5. There are no stable isotopes for elements of atomic number greater than 83 (Bi).

gram of radium undergoes radioactive decay is 500,000 times as great as that given off when the same amount of radium reacts with chlorine to form $RaCl_2$. Still larger amounts of energy are given off in nuclear fission and nuclear fusion. This energy is related to the change in mass that takes place in a nuclear reaction. We will examine the nature of this relationship in Section 27.3.

27.1 Radioactivity

In the reaction other nuclei are produced

An unstable nucleus undergoes a reaction called radioactive decomposition or decay. A few such nuclei occur in nature; their decomposition is referred to as *natural radioactivity*. Many more unstable nuclei have been made in the laboratory; the process by which such nuclei decompose is called *artificial radioactivity*. In this section, we will look at the characteristics of these reactions. We will be particularly interested in the nature of the radiation given off and the effect it has on human beings.

Natural Radioactivity

This process was discovered, almost accidentally, by Henri Becquerel, a French scientist, in 1896. While studying the fluorescence of uranium salts, he found that they gave off a new type of high-energy radiation, capable of blackening a photographic plate. This radiation seems never to have been detected before, even though the element uranium had been known for more than a century.

Becquerel showed that the rate of emission of radiation from a uranium salt was directly proportional to the amount of uranium present. There was one exception to this rule. A certain uranium ore called pitchblende gave off radiation

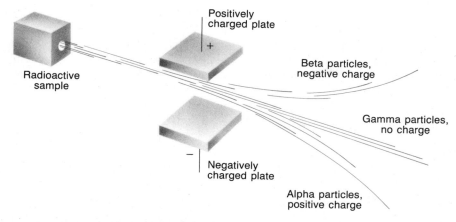

FIGURE 27.2 The direction in which beta particles are deflected shows that they are negatively charged. Alpha particles move so as to indicate that they carry a positive charge. Gamma rays are undeflected and so must be uncharged.

at a rate four times as great as one would calculate on the basis of its uranium content. In 1898 Marie and Pierre Curie, colleagues of Becquerel at the Sorbonne, searched for the active ingredient of pitchblende. They isolated a fraction of a gram of a new element from a ton of ore. This element was more intensely radioactive than uranium. They named it polonium, after Poland, Marie Curie's native country. Six months later the Curies isolated another, intensely radioactive, new element, radium. The Nobel Prize for physics in 1903 was awarded jointly to Becquerel and the Curies. Eight years later Marie Curie received an unprecedented second Nobel Prize, this time in chemistry.

> Her maiden name was Sklodowska
>
> EJS, CLS

The radiation given off in natural radioactivity can be separated by an electric or magnetic field into three distinct parts (Fig. 27.2):

1. **Alpha radiation** consists of a stream of positively charged particles (alpha particles) with a charge of $+2$ and a mass of 4 on the atomic mass scale. These particles are identical with the nuclei of ordinary helium atoms, ^4_2He.

When an alpha particle is ejected from the nucleus, the atomic number decreases by two units; the mass number decreases by four units. Consider, for example, the loss of an alpha particle by a uranium atom with atomic number 92 and mass number 238. This gives an isotope of thorium with atomic number 90 and mass number 234. The transmutation of elements, long sought in vain by the alchemists, occurs by reactions of this sort. The nuclear reaction is represented by the equation

$$^{238}_{92}\text{U} \rightarrow \, ^4_2\text{H} + \, ^{234}_{90}\text{Th} \tag{27.1}$$

> In a gram of $^{238}_{92}$U about 12000 atoms undergo this reaction every second

Here, as in all nuclear equations, there is a balance of both atomic number $(90 + 2 = 92)$ and mass number $(4 + 234 = 238)$ on the two sides.

2. **Beta radiation** is made up of a stream of negatively charged particles (beta particles) identical in their properties to electrons. The ejection of a beta particle (mass ≈ 0, charge $= -1$) converts a neutron (mass $= 1$, charge $= 0$) in the nucleus into a proton (mass $= 1$, charge $= +1$). Hence, beta emission leaves the mass number unchanged but increases the atomic number by one unit. An example of beta emission is the radioactive decay of thorium-234 (90 protons, 144 neutrons) to protactinium-234 (91 protons, 143 neutrons:

> A β particle is an electron

$$^{234}_{90}\text{Th} \rightarrow \,_{-1}^{0}e + \,^{234}_{91}\text{Pa} \qquad\qquad (27.2)$$

The symbol $_{-1}^{0}e$ is written to stand for a beta particle (electron).

3. **Gamma radiation** consists of high-energy photons of very short wavelength ($\lambda = 0.0005$ to 0.1 nm). The emission of gamma radiation accompanies most nuclear reactions. It results from an energy change within the nucleus. An excited nucleus resulting from alpha or beta emission gives off a photon and drops to a lower, more stable energy state. Gamma emission changes neither the atomic number nor the mass number. For that reason, we shall frequently omit it in writing nuclear equations.

In radioactive decay, the unstable nucleus is frequently called the *parent* nucleus. It decays to form a *daughter* nucleus, which may or may not be stable (Example 27.1).

> γ rays are short-wavelength x-rays

EXAMPLE 27.1 Thorium-232 undergoes radioactive decay in a three-step process:

a. $^{232}_{90}\text{Th} \rightarrow Q + \,^{4}_{2}\text{He}$ b. $Q \rightarrow R + \,_{-1}^{0}e$ c. $R \rightarrow T + \,^{228}_{90}\text{Th}$

Write a balanced nuclear equation for each step, identifying Q, R, and T by their nuclear symbols.

Solution The mass numbers and atomic numbers must balance on the two sides of the equation.

a. Alpha emission by $^{232}_{90}\text{Th}$ decreases the mass number by four and the atomic number by two. The mass number of Q is $232 - 4 = 228$; its atomic number is $90 - 2 = 88$. Locating element 88 in the Periodic Table, we find it to be radium, symbol Ra. Hence, the balanced nuclear equation is

$$^{232}_{90}\text{Th} \rightarrow \,^{4}_{2}\text{He} + \,^{228}_{88}\text{Ra}$$

b. When $^{228}_{88}\text{Ra}$ emits an electron, the mass number is unchanged but the atomic number increases by 1 to 89. The daughter product is an isotope of actinium, Ac:

$$^{228}_{88}\text{Ra} \rightarrow \,_{-1}^{0}e + \,^{228}_{89}\text{Ac}$$

c. The unbalanced equation in this case is

$$^{228}_{89}\text{Ac} \rightarrow T + \,^{228}_{90}\text{Th}$$

For the equation to balance, T must have a mass number of 0 and an atomic number of -1; T must be an electron. The balanced equation is

$$^{228}_{89}\text{Ac} \rightarrow \,_{-1}^{0}e + \,^{228}_{90}\text{Th}$$

> The radioactive decay series of $^{238}_{92}\text{U}$ includes 14 steps and ends at $^{206}_{82}\text{Pb}$

EXERCISE The first two steps in the conversion of ^{234}Pa to ^{206}Pb are beta emission followed by alpha emission. Write a balanced nuclear equation for each step. Answers:

$$^{234}_{91}\text{Pa} \rightarrow \,_{-1}^{0}e + \,^{234}_{92}\text{U}; \qquad ^{234}_{92}\text{U} \rightarrow \,^{4}_{2}\text{He} + \,^{230}_{90}\text{Th}$$

Induced Radioactivity; Bombardment Reactions

During the past 50 years, more than 1200 radioactive isotopes have been prepared in the laboratory. The number of such isotopes per element ranges from one (hydrogen and boron) to 34 (indium). They are all prepared by bombardment reactions in which a stable nucleus is converted to one that is radioactive. A typical reaction is that which occurs when the stable isotope of aluminum, $^{27}_{13}Al$, absorbs a neutron to form $^{28}_{13}Al$. The latter is unstable, decaying by electron emission to give a stable isotope of silicon, $^{28}_{14}Si$. The two steps involved in the process are

neutron bombardment: $\qquad ^{27}_{13}Al + ^{1}_{0}n \rightarrow ^{28}_{13}Al$ $\qquad\qquad$ **(27.3)**

radioactive decay: $\qquad\quad ^{28}_{13}Al \rightarrow ^{28}_{14}Si + ^{0}_{-1}e$ $\qquad\qquad$ **(27.4)**

The first radioactive isotopes to be made in the laboratory were prepared in 1934 by Irene (daughter of Marie and Pierre) Curie and her husband, Frederic Joliot. They achieved this by bombarding certain stable isotopes with high-energy alpha particles. One reaction was

$$^{27}_{13}Al + ^{4}_{2}He \rightarrow ^{30}_{15}P + ^{1}_{0}n \qquad\qquad \textbf{(27.5)}$$

The α particles were emitted by radium

The product, phosphorus-30, is radioactive. It decays by emitting a particle called a **positron,** which has the same mass as an electron, but a charge of $+1$ rather than -1:

$$^{30}_{15}P \rightarrow ^{30}_{14}Si + ^{0}_{1}e \qquad\qquad \textbf{(27.6)}$$

Positron emission is never observed in natural radioactivity. It is, however, a common mode of decay in induced radioactivity. Notice from Equation 27.6 that the result of positron emission is the conversion of a proton in the nucleus to a neutron (15 p, 15 n in $^{30}_{15}P$; 14 p, 16 n in $^{30}_{14}Si$). Positron emission occurs with "light" isotopes—that is, nuclei that have too few neutrons to be stable. An example is carbon-11 (6 p, 5 n), which decays by giving off a positron:

$$^{11}_{6}C \rightarrow ^{11}_{5}B + ^{0}_{1}e \qquad\qquad \textbf{(27.7)}$$

In contrast, the "heavy" isotope of the same element, carbon-14 (6 p, 8 n), decays by electron emission:

$$^{14}_{6}C \rightarrow ^{14}_{7}N + ^{0}_{-1}e \qquad\qquad \textbf{(27.8)}$$

An interesting application of bombardment reactions is in the preparation of very heavy elements. During the past forty years, 16 elements with atomic numbers greater than uranium have been prepared. Much of this work was done by a group at the University of California at Berkeley, under the direction first of Glenn Seaborg and then Albert Ghiorso. In the past fifteen years, a Russian group led by G. N. Flerov has made substantial contributions to the field.

Rather remarkably, man has created 16 elements

Some of the reactions used to prepare elements beyond uranium are listed in Table 27.1. Neutron bombardment is effective for the lower members of the series. However, the yield of product decreases rapidly with increasing atomic number. To form very heavy elements, it is necessary to bombard appropriate targets with high-energy positive ions, accelerated to very high velocities. With heavy bombarding particles such as carbon-12, it is possible to achieve a large increase in atomic number.

Table 27.1
Synthesis of Transuranium Elements

NEUTRON BOMBARDMENT						
Neptunium, plutonium	$^{238}_{92}U$ + $^{1}_{0}n$	$\rightarrow$ $^{239}_{92}U$	$\rightarrow$ $^{239}_{93}Np$ + $^{0}_{-1}e$			
	$^{239}_{93}Np$ $\rightarrow$ $^{239}_{94}Pu$ + $^{0}_{-1}e$					
Americium	$^{239}_{94}Pu$ + $2\,^{1}_{0}n$	$\rightarrow$ $^{241}_{94}Pu$	$\rightarrow$ $^{241}_{95}Am$ + $^{0}_{-1}e$			

POSITIVE ION BOMBARDMENT					
Curium		$^{239}_{94}Pu$ + $^{4}_{2}He$	$\rightarrow$	$^{242}_{96}Cm$ + $^{1}_{0}n$	
Californium		$^{242}_{96}Cm$ + $^{4}_{2}He$	$\rightarrow$	$^{245}_{98}Cf$ + $^{1}_{0}n$	
	or	$^{238}_{92}U$ + $^{12}_{6}C$	$\rightarrow$	$^{246}_{98}Cf$ + $4\,^{1}_{0}n$	
Element 104*		$^{249}_{98}Cf$ + $^{12}_{6}C$	$\rightarrow$	$^{257}_{104}Unq$ + $4\,^{1}_{0}n$	
Element 105*		$^{249}_{98}Cf$ + $^{15}_{7}N$	$\rightarrow$	$^{260}_{105}Unp$ + $4\,^{1}_{0}n$	
Element 106*		$^{249}_{98}Cf$ + $^{18}_{8}O$	$\rightarrow$	$^{263}_{106}Unh$ + $4\,^{1}_{0}n$	

*The equations given represent reactions used by the group at Berkeley. The names of elements 104 through 106 have not been established. For 104 and 105, the Berkeley group has suggested rutherfordium and hahnium, honoring Ernest Rutherford and Otto Hahn, discoverer of nuclear fission. The Russian group prefers the names bohrium and kurchatovium, after Niels Bohr and the Russian physicist I. V. Kurchatov. To resolve this impass, the International Union of Pure and Applied Chemistry has suggested that elements 104, 105, and 106 be called, at least temporarily, unnilquadium (symbol Unq), unnilpentium (symbol Unp) and unnilhexium (symbol Unh). This recommendation, which could only have been made by a committee, gives the scientific community a powerful incentive to decide between the rival claims of the American and Russian groups.

Big accelerators are used in these experiments

These names remind one of Gulliver's Travels

EXAMPLE 27.2 Dr. Seaborg has suggested that, by using very heavy nuclei as bombarding particles, it may be possible to synthesize elements of atomic numbers much higher than any now known. One reaction that has been considered is that of ^{40}Ar with ^{248}Cm. Assuming that the product nucleus contains 114 protons and 170 neutrons, write a balanced nuclear equation for this reaction.

Solution The unbalanced equation, from the information given, is

$$^{248}_{96}Cm + ^{40}_{18}Ar \rightarrow ^{284}_{114}X + \underline{\hspace{2cm}}$$

Atomic number balance is already present (114 on each side). The other product must be neutrons, $^{1}_{0}n$. Four neutrons must be formed to compensate for the deficiency of four in the mass number on the right side (288 vs. 284). The balanced equation is therefore

$$^{248}_{96}Cm + ^{40}_{18}Ar \rightarrow ^{284}_{114}X + 4\,^1_0n$$

EXERCISE   Suppose ^{244}Pu is bombarded by a heavy particle to produce four neutrons and an isotope of element 114 containing 174 neutrons. What is the bombarding nucleus? Answer: $^{48}_{20}$Ca.

Isotopes of the very heavy elements have very short half-lives. Moreover, most of them have been formed in very minute quantities, amounting in some cases to only a few atoms. One of the greatest achievements of scientists working in this field has been their ability to study the properties of these elements on submicrogram samples. Both chemical and physical evidence indicate that the elements of atomic numbers 89 through 102 are filling a second rare-earth series by completing the 5f sublevel. Element 103 is the first member of a new transition series, falling below lutetium in the Periodic Table.

Short half-lives go with "hot" isotopes and lots of energy radiated per gram per second

Biological Effects of Radiation

The harmful effect of radiation on human beings is caused by its ability to ionize and ultimately destroy the organic molecules of which body cells are composed. The extent of damage depends mainly upon two factors. These are the amount of radiation absorbed and the type of radiation. The former is commonly expressed in *rads*. A rad corresponds to the absorption of 10^{-2} J of energy per kilogram of tissue. The total biological effect of radiation is expressed in *rems*. This is found by multiplying the number of rads by an appropriate factor, n, for the particular type of radiation:

no. rems = n(no. rads)

Here, n = 1 for beta, gamma, and x-radiation and 10 for alpha radiation or high-energy neutrons. Table 27.2 lists some of the effects to be expected when a person is exposed to a single dose of radiation at various levels.

Small doses of radiation repeated over long periods of time can have very serious consequences. Many of the early workers in the field of radioactivity developed cancer in this way. Cases are known in which cancers developed as long as 40 years after initial exposure. Studies have shown an abnormally large

Table 27.2
Effect of Exposure to a Single Dose of Radiation

DOSE (REMS)	PROBABLE EFFECT
0 to 25	no observable effect
25 to 50	small decrease in white blood cell count
50 to 100	lesions, marked decrease in white blood cells
100 to 200	nausea, vomiting, loss of hair
200 to 500	hemorrhaging, ulcers, possible death
500+	fatal

number of cases of leukemia among the survivors of the atomic bombs dropped on Hiroshima and Nagasaki.

Radiation can also have a genetic effect; that is, it can produce mutations in plants and animals by bringing about changes in chromosomes. There is every reason to suppose that similar effects can arise in human beings. Surveys of the children of radiologists show an increased frequency of congenital defects. This is confirmed by studies of children born to the survivors of Nagasaki and Hiroshima. A disturbing aspect of this problem is that there seems to be no lower limit, or "threshold," below which the genetic effects of radiation become negligible. Even a small increase in background radiation can be expected to produce a proportional increase in undesirable mutations.

Table 27.3 lists average exposures to radiation of people living in the United States. Several features of the data in this table are worthy of comment.

1. About two thirds of radiation exposure comes from natural sources. The level depends upon the area in which a person lives. Cosmic radiation is much more intense at high elevations. A person living in Denver is exposed to about 100 millirems/year from this source. This is twice the national average.

2. The largest exposure to man-made sources of radiation is due to x-rays. The numbers shown in Table 27.3 assume only occasional x-ray examinations. A single dental x-ray corresponds to 20 millirems, and a chest x-ray to 50 to 200 millirems. These amounts can be much higher if the operator of the x-ray machine

A nuclear war would have many long-term effects, all bad

A fluoroscopy may involve large doses of radiation

Table 27.3
Typical Radiation Exposures in the United States
(1 Millirem = 10^{-3} Rem)

SOURCES	MILLIREMS/YR
I. Natural	
A. External to the body	
1. From cosmic radiation	50
2. From the earth	47
3. From building materials	3
B. Inside the body	
1. Inhalation of air	5
2. In human tissues (mostly $^{40}_{19}K$)	<u>21</u>
Total from natural sources	**126**
II. Man-made	
A. Medical procedures	
1. Diagnostic x-rays	50
2. Radiotherapy x-rays, radioisotopes	10
3. Internal diagnosis, therapy	1
B. Nuclear power industry	0.2
C. Luminous watch dials, TV tubes, industrial wastes	2
D. Radioactive fallout (nuclear tests)	<u>4</u>
Total from manmade sources	**67**
Total	**193**

is inexperienced or if the machine itself is defective. Clearly, excessive diagnostic use of x-rays should be avoided.

3. Exposure to radioactive fallout amounted to about 30 millirems/year prior to the nuclear test ban treaty in 1963. It has decreased steadily since then. Much of the present radiation from this source is coming from countries (France and China) that have continued atmospheric testing of nuclear weapons.

4. On the average, exposure to radiation from nuclear power plants is very low, about 0.2 millirem/year. This number is somewhat higher in the vicinity of the plant; nevertheless, a person sitting on a fence at the boundary of a plant all year long would receive only 5 millirems of radiation. The maximum possible exposure of a person standing at the gate of the Three Mile Island plant for the first 2 weeks of the March 1979 accident would have been 80 millirems. This is less than the radiation from a single chest x-ray.

Uses of Radioactive Isotopes

A large number of radioactive isotopes have been used in many areas of basic and applied research. A few of these are discussed below.

MEDICINE   The high-energy radiation given off by radium was used for many years in the treatment of cancer. Nowadays, cobalt-60, which is cheaper than radium and gives off even more powerful radiation, is used for this purpose. Certain types of cancer can be treated internally with radioactive isotopes. If a patient suffering from cancer of the thyroid drinks a solution of NaI containing radioactive iodide ions (^{131}I or ^{123}I), the iodine moves preferentially to the thyroid gland. There, the radiation destroys malignant cells without affecting the rest of the body.

This is called radiation therapy, to be distinguished from chemotherapy

Trace amounts of radioactive samples injected into the blood can be used to detect circulatory disorders. For example, a sodium chloride solution containing a small amount of radioactive sodium may be injected into the leg of a patient. By measuring the build-up of radiation in the foot, a physician can quickly find out whether the circulation in that area is abnormal.

The NaCl is said to be "tagged"

Positron emission tomography (PET) is a recently developed technique used to study brain disorders. The patient is given a dose of glucose ($C_6H_{12}O_6$) containing a small amount of carbon-11, a positron emitter. The brain is then scanned to detect positron emission from the radioactive, "labeled" glucose. In this way, differences in glucose uptake and metabolism in the brains of normal and abnormal patients are established. For example, PET scans have determined that the brain of a schizophrenic metabolizes only about 20% as much glucose as that of a normal individual.

CHEMISTRY   An early application of radioactive isotopes in chemistry was to study the nature of a dynamic equilibrium. In our discussion of chemical equilibria (Chap. 15), we defined the equilibrium state as one in which forward and reverse reactions are proceeding at the same rate. Consider, for example, a saturated solution of lead chloride:

$$PbCl_2(s) \rightleftharpoons Pb^{2+}(aq) + 2\ Cl^-(aq)$$

The fact that the concentration of Pb^{2+} does not change with time is explained by assuming that the two opposing processes, solution and precipitation, are occurring at the same rate. A radioactive isotope of lead, ^{212}Pb, can be used to check this model. A small amount of this isotope, in the form of $Pb(NO_3)_2$, is injected into a solution saturated with $PbCl_2$. Within a very short time, some of the radioactive lead appears in the lead chloride. This indicates that exchange between the solid and the solution is indeed occurring.

Radioactive isotopes are often used to trace the path of an element as it passes through various steps from reactant to final product. Organic chemists have learned a great deal about the mechanism of complex reactions by using carbon-14 as a tracer. One such reaction is the natural process of photosynthesis. The overall reaction can be represented as

$$6\ CO_2(g)\ +\ 6\ H_2O(l) \rightarrow C_6H_{12}O_6(s)\ +\ 6\ O_2(g) \tag{27.9}$$

The reactions are often fast, and there are many of them

This reaction proceeds through a series of steps in which successively more complex organic molecules are formed. To study the path of reaction, plants are exposed to CO_2 containing carbon-14. At various time intervals, the plants are analyzed to determine which organic compounds contain carbon-14 and hence are early products of photosynthesis. Research along these lines by Melvin Calvin at the University of California at Berkeley led to a Nobel Prize in chemistry in 1961.

27.2 Rate of Radioactive Decay

The rate at which a radioactive sample decays can be measured by counting the number of particles given off in unit time. Instruments for measuring radioactivity, such as the Geiger counter, do this automatically. In using such instruments, it is necessary to correct for the "background" radiation given off by natural sources.

First-Order Rate Law

The rate law follows from the fact that the odds for decay are the same for all nuclei of a given isotope

Radioactive decay is a first-order rate process; that is, the rate at which an isotope decays is directly proportional to amount:

rate of decay = k(amount of radioactive isotope)

You will recall (Chap. 16) that, for a first-order reaction, concentration, or in this case, amount, is related to time by the equation

$$\log_{10} \frac{X_0}{X} = \frac{kt}{2.30} \tag{27.10}$$

where X_0 is the amount of radioactive substance at zero time (i.e., when the counting process starts), and X is the amount remaining after time t. The first-order rate constant, k, is characteristic of the isotope undergoing radioactive decay.

Decay rates of radioactive isotopes are most often expressed in terms of their half-lives, $t_{1/2}$, rather than the first-order rate constant, k. As noted in Chapter 16, these two quantities are related by the equation

$$k = \frac{0.693}{t_{1/2}}$$ (27.11)

The application of these equations to radioactive decay processes is illustrated in Example 27.3.

EXAMPLE 27.3 Plutonium-240, produced in nuclear reactors, has a half-life of 6.58×10^3 years. Calculate
a. the first-order rate constant for the decay of plutonium-240.
b. the fraction of a sample that will remain after 100 (1.00×10^2) years.

Solution
a. $k = 0.693/(6.58 \times 10^3 \text{ yr}) = 1.05 \times 10^{-4}/\text{yr}$

b. $\log_{10} \frac{X_0}{X} = \frac{(1.05 \times 10^{-4}/\text{yr})(100 \text{ yr})}{2.30} = 4.57 \times 10^{-3}$

Taking antilogs, $X_0/X = 10^{0.00457} = 1.01$. The fraction remaining is X/X_0:

$X/X_0 = 1/1.01 = 0.99$; that is, 99% remains

This problem illustrates the difficulty inherent in storing or disposing of a long-lived radioactive species. Such isotopes cannot be released into the environment in the naive hope that they will decompose rapidly. In the case of plutonium-240, its radiation level would be virtually unchanged a century from now.

It takes a long time for some nuclei to "cool off"

EXERCISE How long does it take for three fourths of a ^{240}Pu sample to decay? Answer: 1.32×10^4 years.

Half-lives can be interpreted in terms of the level of radiation of the corresponding isotopes. Since uranium-238 has a very long half-life (4.5×10^9 years), it gives off radiation very slowly. At the opposite extreme is fermium-258, which decays with a half-life of 3.8×10^{-4} s. Within a second, virtually all the radiation from this isotope is gone. Species such as this produce a very high level of radiation during their brief existence.

Age of Rocks

Certain radioactive isotopes act as "natural clocks"; that is, they help us to determine the time at which rock deposits solidified. The time elapsed since then is referred to as the "age" of the rock. To see how this information is obtained, consider a uranium-bearing rock formed billions of years ago by solidification from a molten mass. Once the rock became solid, the products of radioactive decay of uranium could no longer diffuse away. Hence, they were incorporated into the rock. Over time these products, all of which have short half-lives, were converted to lead-206. The overall equation for the decay process can be written

There are several radioactive isotopes that can be used as clocks

$$^{238}_{92}\text{U} \rightarrow {}^{206}_{82}\text{Pb} + 8\,{}^{4}_{2}\text{He} + 6\,{}_{-1}^{0}e; \qquad t_{1/2} = 4.5 \times 10^9 \text{ yr} \qquad \textbf{(27.12)}$$

Knowing the half-life for this process, we should then be able, by measuring the ratio of lead-206 to uranium-238 in the rock today, to calculate the time that has elapsed since the rock solidified. If we should find, for example, that equal numbers of atoms of these two isotopes were present, we would infer that the rock must be about 4.5×10^9 (4.5 billion) years old.

Ages of rocks determined by this method range from 3 to 4.5×10^9 years. The larger number is often taken as an approximate value for the age of the earth. Analyses of rock samples from the moon indicate ages in the same range. This argues against the once prevalent idea that the moon was torn from the earth's surface by a violent event a long time after the earth solidified.

Age of Organic Material

During the 1950s Professor W. F. Libby of the University of Chicago and others worked out a method for determining the age of organic material. It is based upon the decay rate of carbon-14. The method can be applied to objects from a few hundred up to 50,000 years old. It has been used to determine the authenticity of canvases of Renaissance painters and to check the ages of relics left by prehistoric cavemen.

Carbon-14 is produced in the atmosphere by the interaction of neutrons from cosmic radiation with ordinary nitrogen atoms:

Even low-energy neutrons can penetrate nuclei

$$^{14}_{7}\text{N} + {}^{1}_{0}n \rightarrow {}^{14}_{6}\text{C} + {}^{1}_{1}\text{H} \qquad \textbf{(27.13)}$$

The carbon-14 formed by this nuclear reaction is eventually incorporated into the carbon dioxide of the air. A steady-state concentration, amounting to about one atom of carbon-14 for every 10^{12} atoms of carbon-12, is established in atmospheric CO_2. A living plant, taking in carbon dioxide, has this same $^{14}\text{C}/^{12}\text{C}$ ratio, as do plant-eating animals or human beings.

When a plant or animal dies, the intake of radioactive carbon stops. Consequently, the radioactive decay of carbon-14

$$^{14}_{6}\text{C} \rightarrow {}^{14}_{7}\text{N} + {}_{-1}^{0}e \qquad (t_{1/2} = 5720 \text{ yr})$$

takes over and the ratio $^{14}\text{C}/^{12}\text{C}$ drops. By measuring this ratio and comparing it to that in living plants, we can estimate the time at which the plant or animal died (Example 27.4).

EXAMPLE 27.4   A tiny piece of paper taken from the Dead Sea Scrolls, believed to date back to the first century AD, was found to have a $^{14}\text{C}/^{12}\text{C}$ ratio of 0.795 times that in a living plant. Estimate the age of the scrolls.

Solution   Knowing the half-life of carbon-14 ($t_{1/2} = 5720$ years), we can calculate the first-order rate constant from Equation 27.11. Then, using Equation 27.10, we can obtain the elapsed time:

$$k = \frac{0.693}{5720 \text{ yr}} = 1.21 \times 10^{-4}/\text{yr}$$

$$\log_{10} \frac{X_0}{X} = \frac{(1.21 \times 10^{-4}/\text{yr}) \times t}{2.30}$$

Since $X = 0.795 \, X_0$,

$$\log_{10} \frac{X_0}{X} = \log_{10} \frac{1.000}{0.795} = \log_{10} 1.26 = 0.100$$

Hence,

$$0.100 = \frac{(1.21 \times 10^{-4}/\text{yr}) \times t}{2.30}; \quad t = 1900 \text{ yr}$$

EXERCISE What is the age of a piece of charcoal in which the $^{14}C/^{12}C$ ratio is 0.400 times that in a living plant? Answer: 7560 years.

This method has been applied to date a wide variety of organic objects (Table 27.4). It depends upon an assumption that cannot be proved. For carbon-14 dating to be valid, the $^{14}C/^{12}C$ ratio in the atmosphere must have remained constant over the centuries. Ages estimated by this method agree within ± 10% with historical records, suggesting that the assumption is valid.

Tree-ring patterns are also useful for dating over the last 1000 years or so

Table 27.4
Representative Radiocarbon Dates*

SOURCE	AGE (YR)
Charcoal from the Lascaux cave in France, containing early cave paintings	15,500 ± 900
Bison bones from Lubbock, Texas, associated with prehistoric man	9,900 ± 350
Charcoal from a tree burned during the upheaval that formed Crater Lake, Oregon	6,500 ± 250
Wheat and barley from ancient Egypt	6,100 ± 250
Charcoal associated with early period from Stonehenge, England	3,800 ± 275
Wood from coffin from Ptolemaic period in Egypt	2,200 ± 450
Linen wrappings used for the Book of Isaiah in the Dead Sea Scrolls	1,900 ± 200
Ancient Manchurian lotus seeds, still fertile	1,000 ± 210

*Taken from Charles Compton, *Inside Chemistry*, New York, McGraw-Hill, 1979.

27.3 Mass-Energy Relations

We pointed out at the beginning of this chapter that the energy change in nuclear reactions is much greater than that for ordinary chemical reactions. The energy change can be calculated from Einstein's equation:

$$\Delta E = \Delta mc^2 \tag{27.14}$$

where Δm is the change in mass,* ΔE is the change in energy, and c is the speed of light. If we substitute for c the value 3.00×10^8 m/s, Equation 27.14 gives the relation between the energy change in *joules* and the mass change in *kilograms*:

SI units here

$$\Delta E \text{ (in joules)} = 9.00 \times 10^{16} \times \Delta m \text{ (in kilograms)}$$

In dealing with nuclear reactions, we usually want ΔE in *kilojoules* corresponding to a mass change in *grams*. Using the relations

$$1 \text{ kJ} = 10^3 \text{ J}; \qquad 1 \text{ kg} = 10^3 \text{ g}$$

we obtain

$$\Delta E \text{ (in kilojoules)} = 9.00 \times 10^{10} \, \Delta m \text{ (in grams)} \tag{27.15}$$

Using Equation 27.15 along with the appropriate nuclear masses (Table 27.5), we can calculate the energy change accompanying a nuclear reaction (Example 27.5).

EXAMPLE 27.5 For the radioactive decay of radium, $^{226}_{88}\text{Ra} \rightarrow ^{222}_{86}\text{Rn} + ^{4}_{2}\text{He}$, calculate ΔE in kilojoules when
a. one mole of radium decays. b. one gram of radium decays.

Solution
a. We first calculate Δm for the reaction and then obtain ΔE from Equation 27.15. For the decay of one mole of Ra (using Table 27.5),

$$\Delta m = \text{mass 1 mol } ^{4}_{2}\text{He} + \text{mass 1 mol } ^{222}_{86}\text{Rn} - \text{mass 1 mol } ^{226}_{88}\text{Ra}$$
$$= 4.0015 \text{ g} + 221.9703 \text{ g} - 225.9771 \text{ g}$$
$$= -0.0053 \text{ g}$$

(Note that since Δm is extremely small, it is necessary to know the masses of products and reactants very accurately to obtain the mass difference to two significant figures.)

$$\Delta E \text{ (in kilojoules)} = 9.00 \times 10^{10} \times (-0.0053) = -4.8 \times 10^8 \text{ kJ}$$

b. Since one mole of radium weighs 226 g, we have

In studying a reaction like this one, we work with only a few atoms, so the total amount of energy released is small

$$\Delta E = 1.00 \text{ g Ra} \times \frac{(-4.8 \times 10^8 \text{ kJ})}{226 \text{ g Ra}} = -2.1 \times 10^6 \text{ kJ}$$

EXERCISE Calculate ΔE for the decay of one mole of $^{239}_{94}\text{Pu}$: $^{239}_{94}\text{Pu} \rightarrow ^{4}_{2}\text{He} + ^{235}_{92}\text{U}$. Answer: -5.1×10^8 kJ.

Energy changes in ordinary chemical reactions are of the order of 50 kJ/g or less. For example, in the combustion of petroleum, about 46 kJ of heat is evolved

---

*Specifically, Δm = mass of products − mass of reactants; ΔE = energy of products − energy of reactants. In spontaneous nuclear reactions, the products weigh less than the reactants (Δm negative). In this case, the energy of the products is less than that of the reactants (ΔE negative), and energy is evolved to the surroundings.

Table 27.5
Nuclear Masses on the ^{12}C Scale*

	At. No.	Mass No.	Mass		At. No.	Mass No.	Mass
n	0	1	1.00867	Br	35	79	78.8992
H	1	1	1.00728		35	81	80.8971
	1	2	2.01355		35	87	86.9028
	1	3	3.01550	Rb	37	89	88.8913
He	2	3	3.01493	Sr	38	90	89.8869
	2	4	4.00150	Mo	42	99	98.8846
Li	3	6	6.01348	Ru	44	106	105.8832
	3	7	7.01436	Ag	47	109	108.8790
Be	4	9	9.00999	Cd	48	109	108.8786
	4	10	10.01134		48	115	114.8791
B	5	10	10.01019	Sn	50	120	119.8748
	5	11	11.00656	Ce	58	144	143.8817
C	6	11	11.00814		58	146	145.8868
	6	12	11.99671	Pr	59	144	143.8809
	6	13	13.00006	Sm	62	152	151.8857
	6	14	13.99995	Eu	63	157	156.8908
O	8	16	15.99052	Er	68	168	167.8951
	8	17	16.99474	Hf	72	179	178.9065
	8	18	17.99477	W	74	186	185.9138
F	9	18	17.99601	Os	76	192	191.9197
	9	19	18.99346	Au	79	196	195.9231
Na	11	23	22.98373	Hg	80	196	195.9219
Mg	12	24	23.97845	Pb	82	206	205.9295
	12	25	24.97925		82	207	206.9309
	12	26	25.97600		82	208	207.9316
Al	13	26	25.97977	Po	84	210	209.9368
	13	27	26.97439		84	218	217.9628
	13	28	27.97477	Rn	86	222	221.9703
Si	14	28	27.96924	Ra	88	226	225.9771
S	16	32	31.96329	Th	90	230	229.9837
Cl	17	35	34.95952	Pa	91	234	233.9934
	17	37	36.95657	U	92	233	232.9890
Ar	18	40	39.95250		92	235	234.9934
K	19	39	38.95328		92	238	238.0003
	19	40	39.95358		92	239	239.0038
Ca	20	40	39.95162	Np	93	239	239.0019
Ti	22	48	47.93588	Pu	94	239	239.0006
Cr	24	52	51.92734		94	241	241.0051
Fe	26	56	55.92066	Am	95	241	241.0045
Co	27	59	58.91837	Cm	96	242	242.0061
Ni	28	59	58.91897	Bk	97	245	245.0129
Zn	30	64	63.91268	Cf	98	248	248.0186
	30	72	71.91128	Es	99	251	251.0255
Ge	32	76	75.90380	Fm	100	252	252.0278
As	33	79	78.90288		100	254	254.0331

*Note that these are *nuclear masses*. The masses of the corresponding atoms can be calculated by adding the masses of each extranuclear electron (0.000549). For example, for an *atom* of $^4_2$He we have

$$4.00150 + 2(0.000549) = 4.00260$$

Similarly, for an atom of $^{12}_6$C,

$$11.99671 + 6(0.000549) = 12.00000$$

Nuclear masses can readily be measured to 7 significant figures, using mass spectroscopy

per gram of fuel burned. Looking at Example 27.5b, we see that ΔE for the radioactive decay of radium is about 50,000 times as great.

Using Table 27.5 it is possible to calculate, for various nuclei, what might be called an average mass per nuclear particle. To do this, we divide the mass of the nucleus by the mass number:

$$\text{average mass per nuclear particle} = \frac{\text{mass of nucleus on } ^{12}\text{C scale}}{\text{mass no.}} \quad \textbf{(27.16)}$$

Thus, for a deuteron, $^{2}_{1}\text{H}$, we have

$$\text{average mass per nuclear particle} = \frac{2.01355}{2} = 1.00678$$

Figure 27.3 shows a plot of this quantity vs. mass number. Notice that the curve has a broad minimum in the vicinity of mass numbers 50 to 100. Consider what would happen if a heavy nucleus such as $^{235}_{92}\text{U}$ were to split into smaller nuclei near this minimum. This process, referred to as *nuclear fission,* should result in a decrease in mass and hence an evolution of energy. The same effect would be obtained if very light nuclei such as $^{2}_{1}\text{H}$ were to combine with one another. Indeed, this process, called *nuclear fusion,* should evolve even more energy, since the average mass per nuclear particle drops off very sharply at the beginning of the curve.

This may take a while to understand, so take the time you need

27.4
Nuclear Fission

Shortly before World War II several groups of scientists were studying the products obtained by bombarding uranium with neutrons. They were looking for new elements with atomic numbers greater than 92. In 1938 Hahn and Strassman in Germany isolated a compound of a Group 2 element, which they originally believed

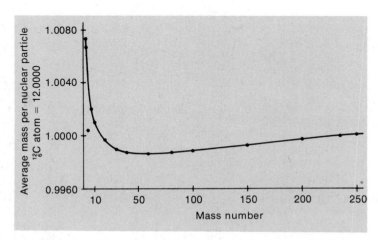

FIGURE 27.3   The average mass per nuclear particle goes through a minimum near mass number 50. This means that when very light nuclei combine (fusion) or very heavy nuclei split apart (fission) there is a decrease in mass and hence an evolution of energy.

to be radium (at. no. = 88). Later, they showed that this element was barium (at. no. = 56), indicating that a uranium atom had been split into fragments. Hahn's first reaction to this discovery was one of disbelief. He later stated, in January 1939, "As chemists, we should replace the symbol Ra . . . by Ba . . . [but], as nuclear chemists, closely associated with physics, we cannot decide to take this step in contradiction to all previous experience in nuclear physics."

If Hahn was reluctant to admit the possibility of an entirely new type of nuclear reaction, a former colleague of his, Lisa Meitner, was not. In a letter published with O. R. Frisch in January 1939, she stated: "At first sight, this result seems very hard to understand. . . . On the basis, however, of present ideas about the behavior of heavy nuclei, an entirely different picture of these new disintegration processes suggests itself. . . . It seems possible that the uranium nucleus . . . may, after neutron capture, divide itself into nuclei of roughly equal size." This revolutionary suggestion was confirmed by experiments carried out all over the world.

This observation had tremendous implications for mankind

With the outbreak of World War II, interest in nuclear fission focused on the enormous amount of energy released in the process. At Los Alamos, in the mountains of New Mexico, a group of scientists led by J. Robert Oppenheimer worked feverishly to produce the fission, or "atomic," bomb. Many of the members of this group were exiles from Nazi Germany. They were spurred on by the fear that Hitler would obtain the bomb first. Their work led to the explosion of the first atomic bomb in the New Mexico desert at 5:30 AM on July 16, 1945. Less than a month later (August 6, 1945) the world learned of this new weapon when another bomb was exploded over Hiroshima. This bomb killed 70,000 people and completely devastated an area of ten square kilometers. Three days later Nagasaki and its inhabitants met a similar fate. On August 14th, Japan surrendered and World War II was over.

Some scientists tried to prevent the use of these bombs, to no avail

The Fission Process ($^{235}_{92}U$)

Several isotopes of the heavy elements undergo fission if bombarded by neutrons of high enough energy. In practice, attention has centered upon two particular isotopes, $^{235}_{92}U$ and $^{239}_{94}Pu$. Both of these can be split into fragments by low-energy neutrons.

Our discussion will concentrate upon the uranium-235 isotope. It makes up only about 0.7% of naturally occurring uranium. The more abundant isotope, uranium-238, does not undergo the fission reaction. During World War II, several different processes were studied for the separation of these two isotopes. The most successful technique was that of gaseous effusion (Chap. 6), using the volatile compound UF_6 (bp = 56°C).

FISSION PRODUCTS When a uranium-235 atom undergoes fission, it splits into two unequal fragments and a number of neutrons and beta particles. The fission process is complicated by the fact that different uranium-235 atoms split up in many different ways. For example, while one atom of $^{235}_{92}U$ is splitting to give isotopes of rubidium (at. no. = 37) and cesium (at. no. = 55), another may break up to give isotopes of bromine (at. no. = 35) and lanthanum (at. no. = 57), while still another atom yields isotopes of zinc (at. no. = 30) and samarium (at. no. = 62):

$$\ce{^{90}_{37}Rb + ^{144}_{55}Cs + 2 ^1_0n} \tag{27.17}$$

$$\ce{^1_0n + ^{235}_{92}U \rightarrow} \; \nearrow \; \ce{^{87}_{35}Br + ^{146}_{57}La + 3 ^1_0n} \tag{27.18}$$

$$\searrow \; \ce{^{72}_{30}Zn + ^{160}_{62}Sm + 4 ^1_0n} \tag{27.19}$$

More than 200 isotopes of 35 different elements have been identified among the fission products of uranium-235.

The stable neutron-to-proton ratio near the middle of the Periodic Table, where the fission products are located, is considerably smaller (~1.2) than that of uranium-235 (1.5). Hence, the immediate products of the fission process contain too many neutrons for stability. In the case of rubidium-90, three steps are required to reach a stable nucleus:

The fission products are "hot"

$$\ce{^{90}_{37}Rb \rightarrow ^{90}_{38}Sr + ^0_{-1}e}; \quad t_{1/2} = 2.8 \text{ min}$$

$$\ce{^{90}_{38}Sr \rightarrow ^{90}_{39}Y + ^0_{-1}e}; \quad t_{1/2} = 29 \text{ yr}$$

$$\ce{^{90}_{39}Y \rightarrow ^{90}_{40}Zr + ^0_{-1}e}; \quad t_{1/2} = 64 \text{ h}$$

The radiation hazard associated with nuclear fallout arises from the formation of radioactive isotopes such as these. One of the most dangerous is strontium-90. In the form of strontium carbonate, $SrCO_3$, it is incorporated into the bones of animals and human beings.

You will notice from Equations 27.17 to 27.19 that two to four neutrons are produced by fission for every one consumed. Once a few atoms of uranium-235 split, the neutrons produced can bring about the fission of many more uranium-235 atoms. This creates the possibility of a chain reaction. This is precisely what happens in the atomic bomb. The energy evolved in successive fissions escalates to give a tremendous explosion within a few seconds.

For nuclear fission to result in a chain reaction, the sample must be large enough so that most of the neutrons are captured internally. If the sample is too small, most of the neutrons escape, breaking the chain. The *critical mass* of uranium-235 required to maintain a chain reaction in a bomb appears to be about 1 to 10 kg. In the bomb dropped on Hiroshima, the critical mass was achieved by using a conventional explosive to fire one piece of uranium-235 into another.

FISSION ENERGY   The evolution of energy in nuclear fission is directly related to the decrease in mass that takes place. About 80,000,000 kJ of energy is given off for every gram of $^{235}_{92}U$ that reacts. This is about 40 times as great as the energy change for simple nuclear reactions such as radioactive decay. The heat of combustion of coal is only about 30 kJ/g; the energy given off when TNT explodes is still smaller, about 2.8 kJ/g. Putting it another way, the fission of one gram of $^{235}_{92}U$ produces as much energy as the combustion of 2700 kg of coal or the explosion of 30 metric tons (3×10^4 kg) of TNT.

The Hiroshima bomb was equivalent to 20×10^3 tons of TNT, so about 7 kg of $^{235}_{92}U$ reacted

Nuclear Reactors

Even before the first atomic bomb exploded, scientists and political leaders began to speculate on the use of fission as a peacetime energy source. Nuclear reactors that convert the heat produced by the fission of uranium-235 into electrical energy

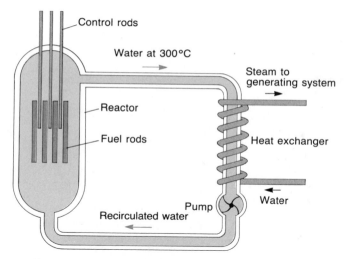

FIGURE 27.4 Nuclear reactor of the pressurized water type. The control rods are made of a material such as cadmium, which absorbs neutrons very effectively. The rate of fission is carefully monitored and controlled.

are now a reality. About 70 such reactors supply 10% of the electrical energy used in the United States. Without them, meeting energy needs in the 1980s would be extremely difficult.

The type of reactor that is most common in the United States today is shown in Figure 27.4. Water at a pressure of 140 atm is passed through the reactor to absorb the heat given off by fission. The water, coming out of the reactor core at 320°C, circulates through a closed loop containing a heat exchanger. A second stream of water at a lower pressure passes through the heat exchanger and is converted to steam at 270°C. This steam is used to drive a turbogenerator that produces electrical energy.

The most serious problem posed by nuclear reactors involves the radioactive products of fission. The amounts of such lethal isotopes as ^{90}Sr and ^{137}Cs that accumulate in a few months are equal to those produced by an atomic bomb of the size dropped on Hiroshima. If the reactor were broken open by a fire, explosion, or earthquake, the fallout could raise the radiation level in the vicinity to deadly concentrations. Despite all the precautions taken to prevent such an accident, there can be no guarantee that it will not happen. This was shown dramatically by the "near miss" at the Three Mile Island reactor in Pennsylvania in March 1979. A set of valves in the steam-generating system accidentally closed. Within seconds, the temperature and pressure within the reactor core increased to the danger point. The problem was compounded when the operators cut off the supply of emergency cooling water. For a considerable time, the fuel elements were uncovered and there was a very real danger that they might "melt down." Had that happened, the release of radiation from the plant could have been many times greater than the small amount that actually escaped.

At this point the construction of nuclear reactors is at a standstill in the U.S.

When a nuclear reactor is operating, the fuel rods undergo physical and chemical changes due to the enormous amount of radiation to which they are exposed. Each year, on the average, one fourth of these intensely radioactive rods must be replaced. Some of the spent fuel rods are reprocessed to recover uranium or plutonium; others are stored without treatment. Either approach gen-

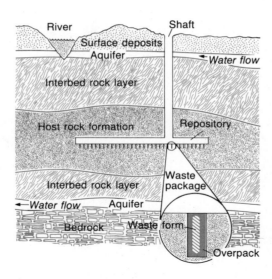

FIGURE 27.5 Proposed federal underground repository for nuclear waste. (Chemical and Engineering News, July 18, 1983, p. 30.)

erates a large amount of dangerous nuclear waste, of the order of 2×10^3 metric tons per year. In 1982, a federal program was established for the disposal of such waste. It is supposed to be buried in an underground site of the type shown in Figure 27.5.

This waste disposal problem has not been solved

Several factors enter into the selection and design of such a deep underground nuclear waste facility. To prevent radioactivity from reaching the surface, the site must be geologically stable, dry, and sufficiently deep to prevent inadvertent mining. Current proposals call for encasing the concentrated wastes in glass or ceramic containers for burial. The containers must withstand ground-water leaching and extensive radiation bombardment and extreme heat from the radioactive decay of the waste, especially if it is "new" (see Example 27.3). Recent calculations indicate that waste from reprocessed fuel requires 20,000 years to decay to a "safe" level. Unprocessed waste takes 100 times longer. Data like these do not tend to encourage people to accept the idea of having nuclear wastes stored in their backyard.

Breeder Reactors

From a long-range standpoint, nuclear reactors using uranium-235 will not be able to satisfy our future power needs. Sooner or later, our supplies of this relatively rare isotope will be depleted and we will have to turn to other sources of energy. One possibility is to convert the more abundant isotope, uranium-238, into plutonium-239:

$$^{238}_{92}\text{U} + ^{1}_{0}n \rightarrow ^{239}_{94}\text{Pu} + 2\ ^{0}_{-1}e \tag{27.20}$$

which can then undergo fission by reactions such as

$$^{239}_{94}\text{Pu} + ^{1}_{0}n \rightarrow ^{90}_{38}\text{Sr} + ^{147}_{56}\text{Ba} + 3\ ^{1}_{0}n \tag{27.21}$$

Notice that this two-step reaction sequence produces more neutrons than it consumes. Consequently, it leads to a chain reaction similar to the fission of uranium-235, producing comparable amounts of energy. By using this process, we could make use of nearly all the uranium found in nature rather than a tiny fraction of that element. So-called **breeder reactors** based on Reactions 27.20 and 27.21 could satisfy our energy needs for a century or more.

The scientific feasibility of the $^{238}_{92}U-^{239}_{94}Pu$ cycle was shown many years ago. The bomb exploded over Nagasaki contained plutonium-239 made from uranium-238. The first nuclear reactor to produce electrical energy in 1951 was of the breeder type. However, 35 years later, we do not yet have a breeder reactor in commercial operation. A combination of economic factors, technical problems, and safety hazards have held back the development of such reactors.

A major concern with breeder reactors is the behavior of plutonium-239. For one thing, it is highly radioactive. As little as ten micrograms (1×10^{-5} g) can cause cancer if it enters the body. Fears have also been expressed that plutonium from breeder reactors might be diverted to produce atomic bombs. Only about 5 kg of plutonium-239 would be required for such a bomb.

There are lots of pros and cons in connection with breeder reactors

27.5
Nuclear Fusion

Recall (Fig. 27.3) that very light isotopes, such as those of hydrogen, are unstable with respect to fusion into heavier isotopes. Indeed, the energy available from nuclear fusion is considerably greater than that given off in the fission of an equal mass of a heavy element (Example 27.6).

EXAMPLE 27.6 Calculate the amount of energy evolved, in kilojoules per gram of reactants, in
a. a fusion reaction, $^2_1H + ^2_1H \rightarrow ^4_2He$.
b. a fission reaction, $^{235}_{92}U \rightarrow ^{90}_{38}Sr + ^{144}_{58}Ce + ^1_0n + 4\ _{-1}^0e$.

Solution
a. We first calculate the change in mass per mole of product, using Table 27.5:

$$\Delta m = 4.00150\ g - 2(2.01355\ g) = -0.02560\ g$$

Converting to kilojoules,

$$\Delta E = -2.56 \times 10^{-2}\ g \times 9.00 \times 10^{10}\ kJ/g = -2.30 \times 10^9\ kJ$$

Since 4.03 g of deuterium is involved, ΔE per gram of reactant is

$$\Delta E = \frac{-2.30 \times 10^9\ kJ}{4.03} = -5.71 \times 10^8\ kJ$$

b. Proceeding as in (a), we find that, per mole of uranium reacting,

$$\Delta m = 89.8869\ g + 143.8817\ g + 1.0087\ g \\ + 4(0.00055\ g) - 234.9934\ g \\ = -0.2139\ g$$

Hence, for one mole of $^{235}_{92}U$,

$$\Delta E = -0.2139\ g \times 9.00 \times 10^{10}\ kJ/g = -1.93 \times 10^{10}\ kJ$$

For one gram of ^{235}U,

$$\Delta E = \frac{-1.93 \times 10^{10}\ kJ}{235} = -8.21 \times 10^7\ kJ$$

Comparing the answers to (a) and (b), we conclude that the fusion reaction produces about seven times as much energy per gram of starting material (57.1×10^7 vs. 8.21×10^7 kJ) as does the fission reaction. This factor varies from about three to ten, depending upon the particular reactions chosen to represent the fusion and fission processes.

EXERCISE Calculate ΔE per gram of reactant for $^2_1H + {}^3_1H \rightarrow {}^4_2He + {}^1_0n$. Answer: -3.38×10^8 kJ.

As an energy source, nuclear fusion possesses several additional advantages over nuclear fission. For one thing, fusion is a "clean" process in the sense that the products are stable isotopes such as ^4_2He, rather than the hazardous radioactive isotopes formed by fission. Equally important, light isotopes suitable for fusion are far more abundant than the heavy isotopes required for fission. We can calculate, for example (Problem 27.65), that the fusion of only 2×10^{-13} percent of the deuterium (^2_1H) in seawater would meet the total annual energy requirements of the world.

Unfortunately, fusion processes, unlike neutron-induced fission, have very high activation energies. In order to overcome the electrostatic repulsion between two deuterium nuclei and cause them to react, they have to be accelerated to velocities of about 10^6 m/s, about 10,000 times greater than ordinary molecular velocities at room temperature. The corresponding temperature for fusion, as calculated from kinetic theory (Problem 27.64), is of the order of 10^9 °C. In the hydrogen bomb, temperatures of this magnitude were achieved by using a fission reaction to trigger nuclear fusion. If fusion reactions are to be used to generate electricity, it will be necessary to develop equipment in which very high temperatures can be maintained long enough to allow fusion to occur and give off energy. In any conventional container, the reactant nuclei would quickly lose their high kinetic energies by collisions with the walls.

One fusion reaction currently under study is a two-step process involving deuterium and lithium as the basic starting materials:

$$\frac{\begin{aligned} {}^2_1H + {}^3_1H &\rightarrow {}^4_2He + {}^1_0n \\ {}^6_3Li + {}^1_0n &\rightarrow {}^4_2He + {}^3_1H \end{aligned}}{{}^2_1H + {}^6_3Li \rightarrow 2\,{}^4_2He} \qquad (27.22)$$

This process is attractive because it has a lower activation energy than other fusion reactions.

We're making progress with fusion, but there is a long way to go

Within the past few years, promising results have been obtained with Reaction 27.22 using "magnetic bottles" (Fig. 27.6) to confine the reactant nuclei. So far, prototype models using this principle have been able to sustain the reaction for a fraction of a second. To achieve a net evolution of energy, this time must be extended to at least one second. At least another 25 years will be required to develop commercial fusion reactors capable of making a significant contribution to our energy needs.

Another approach to nuclear fusion is shown in Figure 27.7. Tiny glass pellets (about 0.1 mm in diameter) filled with frozen deuterium and tritium are illuminated by a powerful laser beam. In principle at least, the neutrons produced should be able to react with lithium to complete Reaction 27.22 and give off energy. Unfortunately, the economic feasibility of this approach is extremely dubious. The

FIGURE 27.6 One way to carry out a controlled fusion reaction is to confine very light nuclei in a strong magnetic field. The problem is to extend the time during which the reaction occurs to one second or more.

glass pellets are expensive to produce and the laser beam has a very short life expectancy.

Perhaps the ultimate irony of our time is the fact that we have made so little use of the energy produced in a nuclear fusion process that has been going on since the universe was formed. The energy given off by the sun and other stars results from fusion reactions in which ordinary hydrogen is converted to helium. One mechanism that has been suggested for this process is

$$\begin{aligned}
{}_1^1H + {}_1^1H &\rightarrow {}_1^2H + {}_1^0e \\
{}_1^2H + {}_1^1H &\rightarrow {}_2^3He \\
\underline{{}_2^3He + {}_1^1H} &\rightarrow {}_2^4He + {}_1^0e \\
4\,{}_1^1H &\rightarrow {}_2^4He + 2\,{}_1^0e; \qquad \Delta E = -6.0 \times 10^8 \text{ kJ/g reactant} \qquad \textbf{(27.23)}
\end{aligned}$$

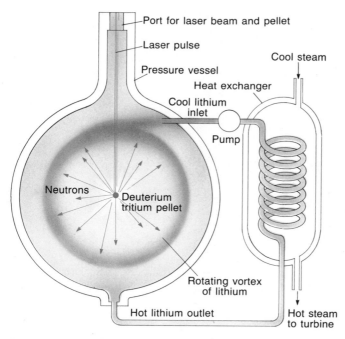

FIGURE 27.7 Schematic diagram of laser-induced fusion apparatus. A pellet of deuterium-tritium undergoes fusion on absorbing energy from a laser pulse. Neutrons from the fusion reaction react with molten lithium, liberating energy and regenerating tritium (Reaction 27.22).

Each day, processes such as this, occurring within the sun at a temperature of perhaps 10^9 °C, produce enormous quantities of energy, of which about 1×10^{19} kJ reaches the surface of the earth. This is roughly equivalent to the total amount of energy that mankind has consumed since the beginning of time.

Summary

Unstable radioactive nuclei decay by emitting

—alpha particles ($^4_2$He nuclei). Emission of an alpha particle decreases the atomic number by two units and the mass number by four units.
—beta particles (electrons). Emission of a beta particle increases the atomic number by one unit, leaving the mass number unchanged.
—gamma radiation.

Nuclear decay can be represented by a nuclear equation (Example 27.1). Stable nuclei can be made radioactive by bombardment with neutrons or high-energy, positively charged particles. Bombardment reactions are used to make isotopes of elements beyond uranium (Table 27.1 and Example 27.2).

Radioactive decay follows a first-order rate law (Example 27.3). Measurements of decay rate can be used to estimate the age of rocks or dead organic matter (Example 27.4).

Changes in mass and energy associated with a nuclear reaction can be calculated using the Einstein equation (Example 27.5). In nuclear fission, energy is released when a heavy nucleus splits into lighter ones. A critical mass of fissionable material is required for a self-sustaining chain reaction. Such reactions in nuclear reactors evolve heat to produce steam for generating electricity (Fig. 27.4). Nuclear fusion releases energy when light nuclei combine to form heavier nuclei (Example 27.6). Current research efforts to achieve controlled fusion use magnetic containment (Fig. 27.6) or laser bombardment (Fig. 27.7).

Key Words and Concepts

alpha (α) particle	fusion	positron
beta (β) particle	gamma (γ) radiation	rad
Einstein equation	half-life	radioactivity
fission	nuclear reactor	rem

Questions and Problems

Nuclear Equations

27.1 The blood volume of a patient can be measured using ^{51}Cr, a positron emitter administered as an Na_2CrO_4 solution. Write the nuclear equation for the decay of ^{51}Cr.

27.2 Write balanced nuclear equations for
a. alpha emission by ^{255}Lr.
b. positron emission by ^{77}Kr.
c. beta emission by ^{65}Ni.

27.31 Lead-210 is used to prepare eyes for corneal transplants. Its daughter product is ^{210}Bi. Identify the emission from ^{210}Pb.

27.32 Write balanced nuclear equations for
a. beta emission by ^{28}Mg.
b. alpha emission by ^{150}Nd.
c. positron emission by ^{72}As.

27.3 Thorium-231 is the product of alpha emission and is radioactive, emitting beta radiation. Determine
a. the parent nucleus of ^{231}Th.
b. the daughter product of ^{231}Th decay.

27.4 Balance the following nuclear equations by filling in the blanks:
a. $^{121}_{51}Sb + {}^{4}_{2}He \rightarrow$ ____ $+ {}^{1}_{1}H$
b. $^{238}_{92}U + {}^{1}_{1}H \rightarrow {}^{238}_{93}Np +$ ____
c. $^{27}_{13}Al + {}^{4}_{2}He \rightarrow$ ____ $+ {}^{1}_{0}n$

27.5 Balance the following nuclear equations by filling in the blanks:
a. $^{32}_{16}S +$ ____ $\rightarrow {}^{32}_{15}P + {}^{1}_{1}H$
b. $^{14}_{7}N +$ ____ $\rightarrow {}^{17}_{8}O + {}^{1}_{1}H$
c. $^{9}_{4}Be + {}^{4}_{2}He \rightarrow {}^{1}_{0}n +$ ____

27.6 Polonium-218 is formed by the nuclear decay of $^{238}_{92}U$ through a series of emissions in the order: α, β, β, α, α, α, α. Write a nuclear equation for each of these steps.

27.7 A certain natural radioactive series starts with ^{238}U and ends with ^{206}Pb. Each step in the series involves the loss of either an alpha or a beta particle. In the entire series, how many alpha particles are given off? how many beta particles?

Rate of Radioactive Decay

27.8 The half-life of barium-131 is 12.0 days. How many milligrams of ^{131}Ba remain after five half-lives, starting with 10.0 mg?

27.9 Polonium-210, an alpha emitter, has a half-life of 138 days.
a. What is the daughter product?
b. Determine the rate constant, k, for the nuclear decay of ^{210}Po.
c. How long will it take before only 10.0% of the original amount of ^{210}Po remains?

27.10 Phosphorus-32 is a beta emitter with a half-life of 14.3 days.
a. What is the rate constant, k, for this reaction?
b. How long will it be before only 5.0% of the original isotope remains?

27.11 The radioisotope ^{241}Am ($t_{1/2}$ = 433 yr) is used commercially in smoke detectors. How long will it take for the activity of ^{241}Am to drop to 90% of its original value?

27.33 Rubidium-87, a beta emitter, is the product of positron emission. Identify
a. the daughter product of ^{87}Rb decay.
b. the parent nucleus of ^{87}Rb.

27.34 Complete the following nuclear equations:
a. $^{65}_{30}Zn + {}^{1}_{1}H \rightarrow {}^{65}_{31}Ga +$ ____
b. $^{238}_{92}U + {}^{1}_{0}n \rightarrow {}^{0}_{-1}e +$ ____
c. $^{241}_{95}Am + {}^{4}_{2}He \rightarrow$ ____ $+ 2 {}^{1}_{0}n$

27.35 Complete the following nuclear equations:
a. $^{27}_{13}Al +$ ____ $\rightarrow {}^{24}_{11}Na + {}^{4}_{2}He$
b. $^{24}_{12}Mg +$ ____ $\rightarrow {}^{27}_{14}Si + {}^{1}_{0}n$
c. $^{10}_{5}B + {}^{1}_{1}H \rightarrow {}^{4}_{2}He +$ ____

27.36 The sequence of emissions that converts $^{238}_{92}U$ to $^{226}_{88}Ra$ is α, β, β, α, α. Write a nuclear equation for each step in the conversion.

27.37 A certain radioactive series starts with $^{237}_{93}Np$ and ends with $^{209}_{83}Bi$. How many alpha particles are emitted? how many beta particles?

27.38 How much ^{137}Cs ($t_{1/2}$ = 30.0 yr) was present originally if, after 120 years, 6.00 g remain?

27.39 Potassium-42, used to locate brain tumors, is a beta emitter with a rate constant, k, of 0.0561/h.
a. Write the nuclear equation for the decay of ^{42}K.
b. What is the half-life of ^{42}K?
c. What fraction of a ^{42}K sample remains after 62 h?

27.40 The radioactive isotope ^{24}Na has a rate constant, k, of 0.0462/h. What fraction of an ^{24}Na sample remains after two days?

27.41 Assuming that the earth is 4.5×10^9 yr old, what fraction of ^{235}U ($t_{1/2}$ = 7.1×10^8 yr) originally present remains today?

27.12 A scintillation counter is used to monitor the radioactivity level of a certain isotope. During a 24-h period the count rate drops from 600 counts/min to 320 counts/min.
   a. Calculate the rate constant for the decay of this isotope.
   b. What is its half-life?

27.13 The radioactive isotope tritium, ^3_1H, is produced in nature in much the same way as $^{14}_6C$. Its half-life is 12.3 yr. Estimate the age of a sample of Scotch whiskey that has a tritium content 0.59 times that of the water in the area where the whiskey was produced.

27.14 One way of dating rocks is to determine the relative amounts of ^{40}K and ^{40}Ar; the decay of ^{40}K has a half-life of 1.26×10^9 yr. Analysis of a certain lunar sample gives the following results in mole ratios:

$$^{40}Ar/^{40}K = 4.13 \qquad (t_{1/2}\ ^{40}K = 1.26 \times 10^9\ yr)$$
$$^{206}Pb/^{238}U = 0.66 \qquad (t_{1/2}\ ^{238}U = 4.5 \times 10^9\ yr)$$
$$^{87}Sr/^{87}Rb = 0.049 \qquad (t_{1/2}\ ^{87}Rb = 4.8 \times 10^{10}\ yr)$$

Using these data, obtain the best possible value for the age of the sample. Can you suggest why the K–Ar method gives a low result?

27.15 Why would the ^{238}U to ^{206}Pb method be inappropriate for determining the age of a sample thought to be about 500 yr old?

Mass-Energy Changes

27.16 For the reaction $^{230}_{90}Th \rightarrow\ ^{226}_{88}Ra\ +\ ^4_2He$,
   a. calculate Δm in grams when one mole of $^{230}_{90}Th$ decays.
   b. calculate ΔE in joules when one mole of $^{230}_{90}Th$ decays; one gram of $^{230}_{90}Th$ decays.

27.17 For the fission reaction

$$^1_0n\ +\ ^{235}_{92}U \rightarrow\ ^{89}_{37}Rb\ +\ ^{144}_{58}Ce\ +\ 3\ _{-1}^0e\ +\ 3\ ^1_0n$$

   a. how much energy (in kJ) is given off per gram of $^{235}_{92}U$?
   b. how many kilograms of TNT must be detonated to produce the same amount of energy? ($\Delta E = -2.76$ kJ/g)

27.18 Compare the energies given off per gram of reactant in the two fusion processes considered to occur during the formation of a star:
   a. $^2_1H\ +\ ^1_1H \rightarrow\ ^3_2He$
   b. $2\ ^3_2He \rightarrow\ ^4_2He\ +\ 2\ ^1_1H$

27.42 Iron-59 is a radioactive isotope ($t_{1/2} = 46$ days) used in anemia diagnosis. If the initial count rate of an ^{59}Fe sample is 560 counts/min, how long will it take for the count rate to drop to 400 counts/min?

27.43 An oil painting supposed to be by Rembrandt (1606–1669 AD) is checked by ^{14}C dating. The ^{14}C content ($t_{1/2} = 5720$ yr) of the canvas is 0.961 times that in a living plant. Could the painting have been by Rembrandt?

27.44 The ^{87}Rb to ^{87}Sr method of dating rocks was used to analyze lunar samples from the Apollo-15 mission. Estimate the age of the lunar sample in which
   a. the mole ratio of ^{87}Rb to ^{87}Sr is 25.0 (see Problem 27.14).
   b. the mole ratio of ^{87}Rb to ^{87}Sr is 20.0.

27.45 One objection to the ^{14}C method is that it assumes that the $^{14}C/^{12}C$ ratio in the atmosphere many years ago was the same as it is now. Suggest two ways in which the validity of this assumption might be checked.

27.46 Bk-245 decays by alpha emission. For one gram of ^{245}Bk, calculate Δm (grams) and ΔE (kilojoules).

27.47 Consider the fission reaction

$$^{239}_{94}Pu\ +\ ^1_0n \rightarrow\ ^{146}_{58}Ce\ +\ ^{90}_{38}Sr\ +\ 2\ _{-1}^0e\ +\ 4\ ^1_0n$$

   a. How many grams of $^{239}_{94}Pu$ would have to react to produce 1.00 kJ of energy?
   b. How many atoms of $^{239}_{94}Pu$ would have to react to produce 1.00 kJ of energy?

27.48 Taking Equation 27.23 to represent the reaction that produces the sun's energy, how many grams of hydrogen would have to be fused to provide the 1.0×10^{19} kJ that reaches the earth each day?

27.19 Determine, by calculation, whether the following nuclear reaction is spontaneous:

$$\ce{^{52}_{24}Cr -> ^{48}_{22}Ti + ^{4}_{2}He}$$

27.20 Using Table 27.5, calculate ΔE for the synthesis of 1.00×10^{-3} mol ^{239}Np from ^{238}U by the process listed in Table 27.1.

27.21 Arrange the following isotopes in order of increasing average mass per nuclear particle:
a. $^{1}_{1}$H   b. $^{4}_{2}$He   c. $^{59}_{27}$Co   d. $^{238}_{92}$U

27.22 The binding energy of a nucleus is the energy that would have to be absorbed to decompose the nucleus into neutrons and protons. Using Table 27.5, calculate the binding energy, in kilojoules per mole, of the $^{14}_{6}$C nucleus.

General

27.23 Which of the following isotopes of fluorine (at. mass = 19.00) would you expect to decompose by electron emission? positron emission?
a. ^{17}F   b. ^{18}F   c. ^{20}F   d. ^{21}F

27.24 Consider element 114, referred to in Example 27.2.
a. What element would it fall beneath in the Periodic Table?
b. Would it be a metal, nonmetal, or metalloid?
c. How would its atomic radius compare to that of bismuth?

27.25 Consider the interaction of beta and slow neutron radiation with human cells.
a. Calculate the number of rems absorbed by a human being exposed to 15 rads of slow neutrons (n = 3) and 10 rads of beta radiation (n = 1). Assume the dosage is additive.
b. Using Table 27.2, describe the effect of the absorption of the radiation in part a.

27.26 Using Table 27.3, estimate the percentage by which radiation exposure would be increased if atmospheric nuclear tests were still contributing 30 millirems/yr.

27.49 Will ^{18}F decay spontaneously by positron emission? ^{10}B? Show by calculation.

27.50 Curium-242 is produced by alpha bombardment of plutonium-239. Write a balanced nuclear equation for this process. What is ΔE for the formation of 1.00 mg of curium-242?

27.51 Arrange the following isotopes in order of decreasing average mass per nuclear particle:
a. $^{40}_{20}$Ca   b. $^{72}_{30}$Zn   c. $^{109}_{47}$Ag

27.52 Which has the larger binding energy, $^{4}_{2}$He or $^{6}_{3}$Li (see Problem 27.22)?

27.53 Answer Question 27.23 for the following isotopes of sodium (at. mass = 22.99):
a. ^{26}Na   b. ^{24}Na   c. ^{22}Na   d. ^{21}Na

27.54 If element 116 were synthesized,
a. what would be its outer electron configuration?
b. what Periodic Table group would it fall in?

27.55 How many rads of low-energy beta radiation (n = 1.7) would be equivalent to 15 rems of alpha radiation?

27.56 Suppose a person living in the vicinity of the Three Mile Island accident were exposed to radiation of 0.010 rem. Estimate the percentage by which this would increase his or her total radiation exposure for 1979.

27.27 Explain why
  a. ^{131}I and its -1 ion behave the same way in nuclear reactions.
  b. a nonradioactive $^{23}_{11}Na^+$ ion and a radioactive $^{24}_{11}Na^+$ ion behave chemically alike.
  c. alpha emission produces a nucleus having two less protons than the original nucleus.
  d. mass changes are not observed in ordinary chemical reactions.

27.28 Classify each of the following statements as true or false. If false, correct the statement to make it true.
  a. The mass number increases in beta emission.
  b. A radioactive species with a large rate constant, k, decays very slowly.
  c. A breeder reactor converts nonfissionable ^{238}U into a fissionable species.

27.29 Describe an experiment using a radioactive isotope to determine the cleaning effectiveness of a laundry detergent.

Describe advantages and disadvantages of nuclear reactors as compared to conventional energy sources.

27.57 Explain how
  a. alpha and beta radiation are separated by an electric field.
  b. radioactive ^{14}C can be used as a tracer to study steps in photosynthesis.
  c. elements of atomic number greater than 92 are prepared in the laboratory.
  d. a self-sustaining chain reaction occurs in nuclear fission.

27.58 In your own words explain
  a. how radioactive ^{32}P in a fertilizer can be used in plant growth studies.
  b. what is meant by a nuclear chain reaction.
  c. why radiation can cause mutations in plant or animal cells.

27.59 How could a radioactive isotope be used to determine the length of time it takes for water to flow underground from a chemical dump to a nearby stream?

27.60 Why will it be necessary to develop breeder reactors if we are to continue to produce electricity by nuclear fission in the next century?

*27.61 The activity (rate of decay) of a radioactive sample can be expressed in curies (Ci): 1 Ci = 3.7×10^{10} atoms disintegrating/second. A smoke detector contains a 0.90-millicurie source of ^{241}Am ($t_{1/2}$ = 433 yr). What mass of ^{241}Am is present?

*27.62 The sun has a mass of approximately 2×10^{33} g and emits about 4×10^{26} J/s due primarily to the process given in Equation 27.23. How long would it take for the sun to lose 90% of its mass of ^1_1H at this rate of energy emission?

*27.63 Plutonium-239 decays by the reaction $^{239}_{94}Pu \rightarrow ^{235}_{92}U + ^4_2He$, with a rate constant of 5.5×10^{-11}/min. In a one-gram sample of ^{239}Pu,
  a. how many grams decompose in 10 min?
  b. how much energy in kilojoules is given off in 10 min?
  c. what radiation dosage in rems is received by a 70-kg man exposed to a gram of ^{239}Pu for 10 min?

*27.64 It is possible to estimate the activation energy for fusion by calculating the energy required to bring two deuterons close enough to form an alpha particle. This energy can be obtained by using Coulomb's law: $E = \dfrac{q_1 \times q_2}{r}$, where q_1 and q_2 are the charges of the deutrons (4.8×10^{-10} esu), r is the radius of the helium nucleus ($\sim 1 \times 10^{-12}$ cm), and E is the energy in ergs.
  a. Estimate E in ergs per alpha particle.
  b. Using the equation $E = mv^2/2$, estimate the velocity (cm/s) that each deuteron must have if a collision between two of them is to supply the activation energy for fusion (m is the mass of the deuteron in grams).
  c. Using the equation $v = (3RT/MM)^{1/2}$, estimate the temperature that would have to be reached to achieve fusion $\left(R = 8.31 \times 10^7 \dfrac{erg}{mol \cdot K} \right)$.

*27.65 Consider the reaction $2 \, {}^2_1H \rightarrow {}^4_2He$.

a. Calculate ΔE in kilojoules per gram of deuterium fused.

b. How much energy is potentially available from the fusion of all the deuterium atoms in seawater? The percentage of deuterium in water is about 0.015%. The total mass of water in the oceans is 1.3×10^{27} g.

c. What fraction of the deuterium in the oceans would have to be consumed to supply the annual energy requirements of the world (2.3×10^{17} kJ)?

Chapter 28
Organic Molecules, Small and Large

Organic chemistry deals with the compounds of carbon, almost all of which are molecular. In Chapter 13 we looked at the simplest type of organic compound: hydrocarbons. These include *saturated* hydrocarbons such as methane, ethane, and propane

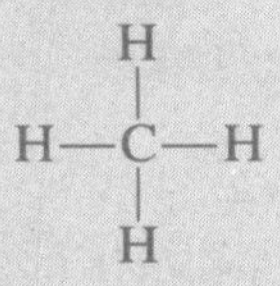

saturated: no multiple bonds

$$
\begin{array}{ccc}
& H & \\
& | & \\
H-\!\!&C&\!\!-H \\
& | & \\
& H &
\end{array}
\qquad
\begin{array}{cc}
H & H \\
| & | \\
H-\!C\!-\!C\!-H \\
| & | \\
H & H
\end{array}
\qquad
\begin{array}{ccc}
H & H & H \\
| & | & | \\
H-\!C\!-\!C\!-\!C\!-H \\
| & | & | \\
H & H & H
\end{array}
$$

methane, CH_4 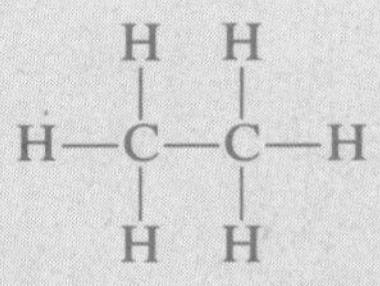   ethane, C_2H_6 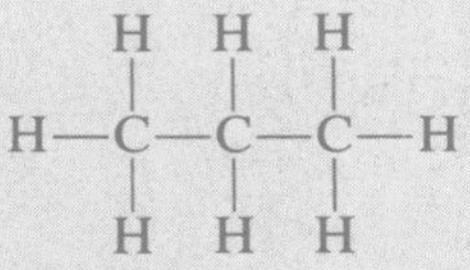   propane, C_3H_8

unsaturated: multiple (double or triple) bonds

unsaturated hydrocarbons such as ethylene and acetylene

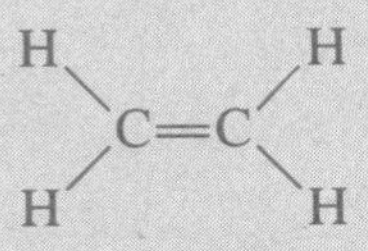

$$
\begin{array}{ccc}
H & & H \\
& \diagdown \!\! \diagup & \\
& C\!=\!C & \\
& \diagup \!\! \diagdown & \\
H & & H
\end{array}
\qquad\qquad\qquad
H-\!C\!\equiv\!C\!-H
$$

ethylene, C_2H_4 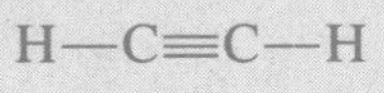   acetylene, C_2H_2

aromatic: one or more benzene rings

and *aromatic* hydrocarbons such as benzene, whose structure is commonly shown as

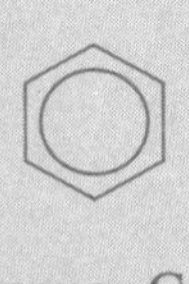

benzene, C_6H_6

In these molecules, and indeed in all the organic compounds we will consider

—**each carbon atom forms four covalent bonds.**
—**each hydrogen atom forms one covalent bond.**

Carbon forms four
bonds in essentially all
of its compounds

We start this chapter by considering the structures and properties of some simple organic molecules that contain Cl, O, or N atoms (Section 28.1) in addition to carbon and hydrogen. Typical compounds of this type are

$$
\begin{array}{ccc}
\begin{array}{c} H \\ | \\ H\!-\!C\!-\!Cl \\ | \\ H \end{array}
&
\begin{array}{c} H \\ | \\ H\!-\!C\!-\!O\!-\!H \\ | \\ H \end{array}
&
\begin{array}{c} H \\ | \\ H\!-\!C\!-\!N\!-\!H \\ | \quad | \\ H \quad H \end{array}
\end{array}
$$

methyl chloride,   methyl alcohol,   methyl amine,
CH_3Cl CH_3OH CH_3NH_2

In these molecules and in all the organic compounds we will consider,

—**each chlorine atom forms one covalent bond.**
—**each oxygen atom forms two covalent bonds.**
—**each nitrogen atom forms three covalent bonds.**

In Chapter 13, we pointed out that two quite different organic compounds can have the same molecular formula. Compounds of this type are called *isomers*. The phenomenon of isomerism helps to explain why there are so many organic compounds. *Structural isomers,* referred to in Chapter 13, differ in the order in which atoms are bonded to one another. In Section 28.2, we will extend the concept of isomerism to include two other types: geometric and optical isomers.

Two structural
isomers:

$$
\begin{array}{c} H \quad H \\ | \quad | \\ H\!-\!C\!-\!C\!-\!O\!-\!H \\ | \quad | \\ H \quad H \end{array} \text{(I)}
$$

The last two sections of this chapter are devoted to a survey of large organic molecules called *polymers*. These are built from small molecules called *monomers*. A polymer molecule may contain several thousand monomer units bonded to one another in a continuous chain. Man-made, synthetic polymers (Section 28.3) usually contain only one or two kinds of monomer units. This is also true of some natural polymers (Section 28.4), notably the carbohydrates starch and cellulose. In contrast, the natural polymers called *proteins* contain many different types of monomer units and for that reason have rather complex structures.

$$
\begin{array}{c} H \qquad H \\ | \qquad | \\ H\!-\!C\!-\!O\!-\!C\!-\!H \\ | \qquad | \\ H \qquad H \end{array} \text{(II)}
$$

28.1
Functional Groups in
Organic Molecules
Containing Cl, O,
and N Atoms

Many organic molecules can be considered to be derived from hydrocarbons by substituting a **functional group** for a hydrogen atom. The functional group can be a nonmetal atom or small group of atoms that is bonded to carbon. In this section we will consider a few of the more common functional groups. These are the

—Cl atom (organic chlorine compounds)

—OH group (alcohols)

—C—OH group (carboxylic acids)
$\parallel$
O

—C—O— group (esters)
$\parallel$
O

—NH$_2$ group (amines)

Organic Chlorine Compounds

The simplest compounds of this type are derived from methane, CH$_4$, by replacing successive hydrogen atoms with chlorine:

Almost no naturally occurring organic molecules contain chlorine

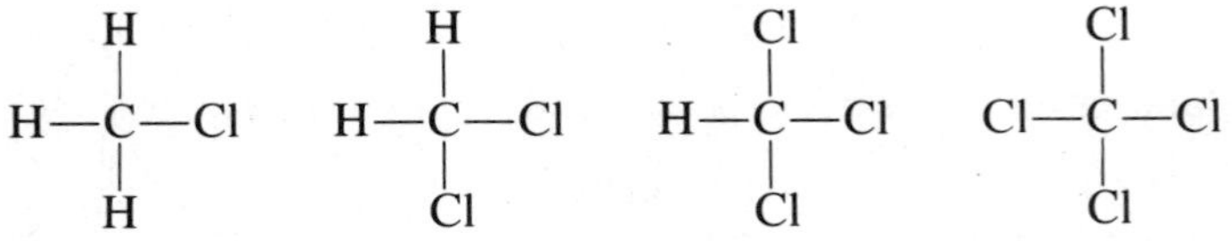

CH$_3$Cl　　CH$_2$Cl$_2$　　CHCl$_3$　　CCl$_4$

At one time chloroform, CHCl$_3$, and carbon tetrachloride, CCl$_4$, were widely used as solvents in the laboratory. They have now been replaced for that purpose by

Table 28.1
Some Organic Chlorine Compounds

NAME	FORMULA	USES
Dichlorodiphenyltrichloro-ethane (DDT)	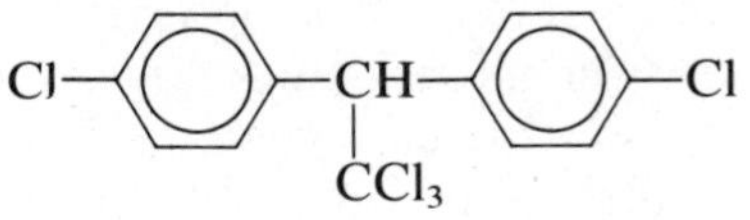	insecticide
2,4-dichlorophenoxyacetic acid (2,4-D)	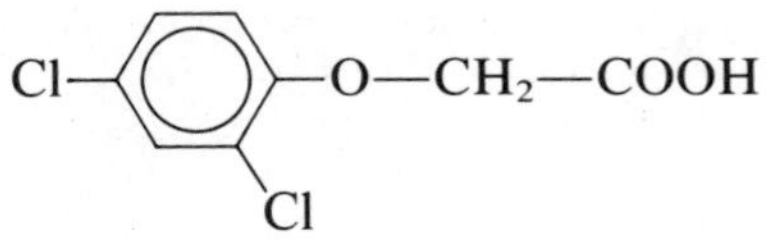	herbicide
2,3,3′,4′,5-pentachloro-biphenyl (a typical polychlorinated biphenyl—PCB)	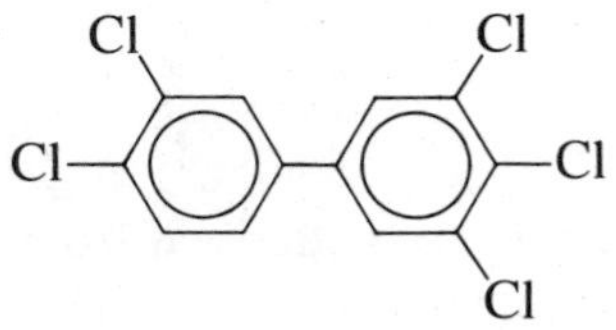	plasticizer, solvent, coolant
Vinyl chloride	CH$_2$=CHCl	plastics
2,3,7,8-tetrachloro-dibenzo-p-dioxin	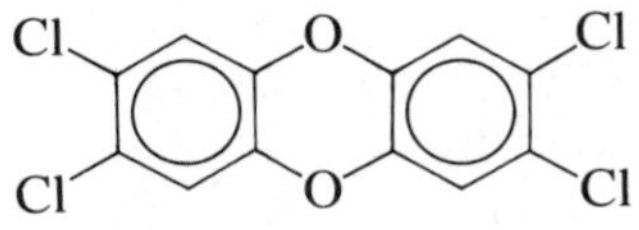	herbicide

dichloromethane, CH_2Cl_2, and 1,1,1-trichloroethane, CH_3CCl_3, which are much less toxic.

Organic chlorine compounds have a variety of uses (Table 28.1). Many of these compounds have been shown to be toxic or have other adverse effects upon the environment. Hardly a year passes without the use of an organic chlorine compound being banned or restricted. Recent attention has focused upon dioxin, a general term used to refer to a family of about 75 related chlorinated compounds (Table 28.1). In 1983, residents of Times Beach, Missouri, were evacuated permanently because of relatively high dioxin concentrations in the soil. Although no human deaths have been traced to dioxin, it is known to be lethal to guinea pigs at the 1 part per billion level.

Dioxin is often a trace contaminant in 2,4-D

Several organic compounds containing halogen atoms other than chlorine have been or are now used commercially. In Chapter 17, we mentioned freons (such as $CFCl_3$, CF_2Cl_2) as aerosol propellants and refrigerants. A compound that has been in the news lately is 1,2-dibromoethane:

$$
\begin{array}{c}
\quad H \quad\ H \\
\quad | \qquad | \\
H-C-C-H \\
\quad | \qquad | \\
\quad Br \quad Br
\end{array}
$$

This compound, often referred to as "EDB," was used as a fumigant in grain even though it was known to be carcinogenic in animals.

Alcohols

Alcohols can be considered to be derived from hydrocarbons by replacing one or more H atoms by —OH groups. The simplest, and most important, alcohols are methyl alcohol (methanol) and ethyl alcohol (ethanol)

$$
\begin{array}{c}
\quad H \\
\quad | \\
H-C-OH \\
\quad | \\
\quad H
\end{array}
\qquad
\begin{array}{c}
\quad H \quad\ H \\
\quad | \qquad | \\
H-C-C-OH \\
\quad | \qquad | \\
\quad H \quad\ H
\end{array}
$$

methyl alcohol, CH_3OH ethyl alcohol, C_2H_5OH

These are certainly the two most important alcohols

About 3×10^9 kg of methanol are produced annually in the United States from water gas, a mixture of carbon monoxide and hydrogen:

$$
CO(g) + 2\ H_2(g) \xrightarrow[\text{250 atm, 350°C}]{\text{ZnO, Cr}_2\text{O}_3} CH_3OH(g) \tag{28.1}
$$

It is also formed as a by-product when charcoal is made by heating wood in the absence of air. For this reason, methanol is sometimes called wood alcohol. Methanol is used in jet fuels and as a solvent, gasoline additive, and starting material for several industrial syntheses. It is poisonous, causing blindness or death. Although it is an intoxicant, it must not be used in alcoholic beverages.

It's also the main ingredient in windshield washer fluid

Ethanol, the most common alcohol, can be prepared by the fermentation of grains or sugar. A typical reaction is that of glucose, $C_6H_{12}O_6$:

$$C_6H_{12}O_6(aq) \rightarrow 2\ C_2H_5OH(aq)\ +\ 2\ CO_2(g) \tag{28.2}$$

95% ethanol is a strong dehydrating agent. Never drink it straight

A mixture of 95% ethanol and 5% water can be separated from the fermentation products by distillation. Ethanol, a colorless liquid, is the active ingredient of alcoholic beverages. There it is present in various concentrations (4% to 8% in beers, 12% to 15% in wine, and 40% or more in distilled spirits). The "proof" of an alcoholic beverage is twice the volume percent of ethyl alcohol. Thus, an 86 proof bourbon whiskey contains 43% ethanol.

Most of the 1.5×10^9 kg of industrial ethanol made annually in the United States is produced synthetically. The starting material is ethylene, a by-product of the distillation of crude oil. The reaction is

$$\tag{28.3}$$

This is an addition reaction. The water molecule is split and the fragments add to the ethylene double bond.

Certain alcohols contain two or more —OH groups per molecule. Perhaps the most familiar compounds of this type are ethylene glycol and glycerol:

ethylene glycol   glycerol

Glycerol is used in the antifreeze for R.V. water systems

Ethylene glycol is widely used as an antifreeze. Glycerol is formed as a by-product in making soaps and detergents. It is a viscous, sweet-tasting liquid, used in making drugs, antibiotics, plastics, and explosives (nitroglycerin).

Carboxylic Acids

Carboxylic acids can be considered to be derived from hydrocarbons by replacing one or more H atoms by a carboxyl group, $—\overset{\displaystyle O}{\underset{}{C}}—OH$, often abbreviated —COOH.

The most common carboxylic acid, at least in general chemistry, is acetic acid, whose formula may be written as

acetic acid

Table 28.2
Naturally Occurring Carboxylic Acids

NAME	STRUCTURE	NATURAL SOURCE
Acetic acid	CH_3-COOH	vinegar
Citric acid	$\begin{array}{c} OH \\ \vert \\ HOOC-CH_2-C-CH_2-COOH \\ \vert \\ COOH \end{array}$	citrus fruits
Lactic acid	$\begin{array}{c} CH_3-CH-COOH \\ \vert \\ OH \end{array}$	sour milk
Malic acid	$\begin{array}{c} HOOC-CH_2-CH-COOH \\ \vert \\ OH \end{array}$	apples
Oleic acid	$CH_3(CH_2)_7-CH=CH-(CH_2)_7-COOH$	vegetable oils
Oxalic acid	$HOOC-COOH$	rhubarb, spinach, cabbage, tomatoes
Stearic acid	$CH_3(CH_2)_{16}-COOH$	animal fats
Tartaric acid	$\begin{array}{c} HOOC-CH-CH-COOH \\ \vert \quad \vert \\ OH \quad OH \end{array}$	grape juice, wine

Acetic acid is the active ingredient of vinegar, responsible for its sour taste. A variety of other carboxylic acids are found in natural products (Table 28.2).

Treatment of a carboxylic acid with the strong base NaOH forms the sodium salt of the acid. With acetic acid, the acid-base reaction is

$$CH_3-\overset{\overset{\displaystyle O}{\|}}{C}-OH(aq) + OH^-(aq) \rightarrow CH_3-\overset{\overset{\displaystyle O}{\|}}{C}-O^-(aq) + H_2O \qquad \textbf{(28.4)}$$

Evaporation gives the salt sodium acetate, which contains Na^+ and CH_3COO^- ions. Many of the salts of carboxylic acids have important uses. Sodium and calcium propionate (Na^+ or Ca^{2+} ions, $CH_3CH_2COO^-$ ions) are added to bread, cake, and cheese to inhibit the growth of mold. *Soaps* are sodium salts of long-chain carboxylic acids such as stearic acid:

Sodium acetate is very soluble in water, ~20 moles/L at 100°C

$$CH_3(CH_2)_{16}-\overset{\overset{\displaystyle O}{\|}}{C}-OH \qquad Na^+, CH_3(CH_2)_{16}-\overset{\overset{\displaystyle O}{\|}}{C}-O^-$$

stearic acid   sodium stearate, a soap

Esters

The reaction between a carboxylic acid and an alcohol forms an ester, containing the functional group —C—O—, often abbreviated —COO—. The reaction
$$\overset{\|}{\underset{O}{}}$$
between acetic acid and methyl alcohol is typical:

$$CH_3-\underset{\underset{O}{\|}}{C}-OH(aq) + H\,O-CH_3(aq) \xrightarrow{\;H^+\;} CH_3-\underset{\underset{O}{\|}}{C}-O-CH_3(aq) + H_2O \qquad \textbf{(28.5)}$$

acetic acid   methyl alcohol   methyl acetate

The reaction is reversible

This reaction is ordinarily carried out in dilute H_2SO_4 solution; the H^+ ion acts as a catalyst. Evidence from tracer studies, using the $^{18}_{8}O$ isotope, indicates that the —OH group is removed from the acid rather than the alcohol.

Most esters have a pleasant odor. Butyl acetate gives bananas their odor. Ethyl acetate is one of at least five esters found in pineapples. In addition to being used in synthetic fragrances, esters are also industrial solvents and starting materials for plastics such as Plexiglas.

EXAMPLE 28.1   Give the structural formula of
a. the three-carbon alcohol with an —OH group at the end of the chain.
b. the three-carbon carboxylic acid.
c. the ester formed when these two compounds react.

Solution
a. $CH_3CH_2CH_2-OH$ b. $CH_3CH_2-\underset{\underset{O}{\|}}{C}-OH$

It's called propyl propionate

c. $CH_3CH_2-\underset{\underset{O}{\|}}{C}-O-CH_2CH_2CH_3$

EXERCISE   Which one of the following is *neither* an alcohol, a carboxylic acid, nor an ester? CH_3CH_2-OH; $H-\underset{\underset{O}{\|}}{C}-OH$; CH_3-O-CH_3;

$H-\underset{\underset{O}{\|}}{C}-O-CH_3$. Answer: CH_3-O-CH_3.

Fats are esters

Animal fats and vegetable oils are esters of long-chain carboxylic acids with glycerol. A typical fat molecule might have the structure

$$CH_3(CH_2)_{14}-COO-CH_2$$
$$CH_3(CH_2)_7CH{=}CH(CH_2)_7-COO-CH$$
$$CH_3(CH_2)_{16}-COO-CH_2$$

The three carboxylic acids from which this fat is derived are

  —palmitic acid, $CH_3(CH_2)_{14}COOH$.
  —oleic acid, $CH_3(CH_2)_7CH=CH(CH_2)_7COOH$.
  —stearic acid, $CH_3(CH_2)_{16}COOH$.

These acids are typical of those found in fats. Some "fatty acids" are *saturated,* such as palmitic and stearic acid; the hydrocarbon chain contains no multiple bonds. Others, such as oleic acid, are *unsaturated*; there are one or more carbon-carbon multiple bonds in the molecule.

So-called "saturated fats" contain relatively few carbon-carbon multiple bonds. They are solids at room temperature and are commonly found in animal products such as lard and butter. "Unsaturated fats" contain a higher proportion of multiple bonds. They are liquids at room temperature and are found in such vegetable products as corn oil and cottonseed oil. Recently, there has been a trend toward the use of unsaturated as opposed to saturated fats. It appears that saturated fats may raise the level of cholesterol in the blood, perhaps contributing to the risk of heart attacks and other circulatory problems.

Margarine usually contains unsaturated fats

Amines

Amines are perhaps visualized most simply as derivatives of ammonia, NH_3,

$$H-\overset{\displaystyle H}{\underset{\displaystyle |}{N}}-H$$

in which one or more of the H atoms are replaced by hydrocarbon groups. We distinguish between

These are all bases in water solution

  —*primary* amines, in which one H atom of NH_3 is replaced.
  —*secondary* amines, in which two H atoms of NH_3 are replaced
  —*tertiary* amines, in which all three H atoms of NH_3 are replaced.

Examples include

$CH_3-\overset{\displaystyle H}{\underset{\displaystyle |}{N}}-H$ or CH_3NH_2 $CH_3-\overset{\displaystyle CH_3}{\underset{\displaystyle |}{N}}-H$ or $(CH_3)_2NH$ $CH_3-\overset{\displaystyle CH_3}{\underset{\displaystyle |}{N}}-CH_3$ or $(CH_3)_3N$

  methylamine   dimethylamine   trimethylamine
  (primary)   (secondary)   (tertiary)

Alkaloids such as caffeine, nicotine, morphine, and coniine (Fig. 28.1) form an important class of naturally occurring amines. Caffeine occurs in tea leaves, coffee beans, and cola nuts used to make cola soft drinks. Nicotine is the most abundant alkaloid in tobacco. The painkillers morphine and codeine are obtained from unripe opium poppy seed pods. They are narcotics that, with repeated usage, cause addiction. Coniine, extracted from hemlock, is the alkaloid that killed Socrates. He was sentenced to drink a brew made from hemlock because of his unconventional teaching methods and religious practices.

Teaching was tough in those days

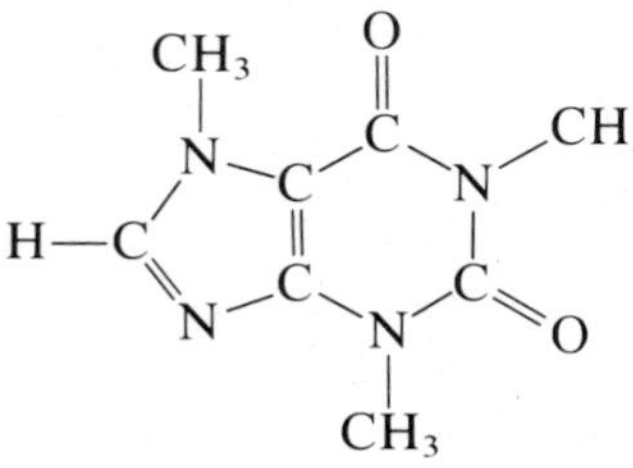

Caffeine

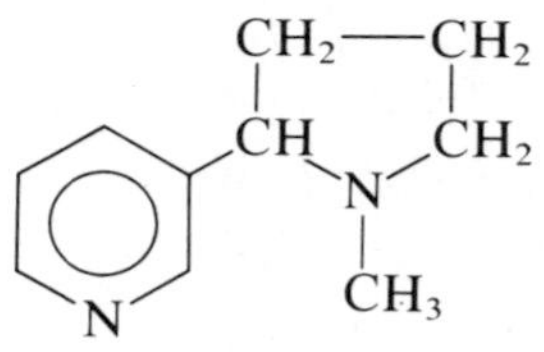

Nicotine

Alkaloids tend to be addictive, poisonous, or both

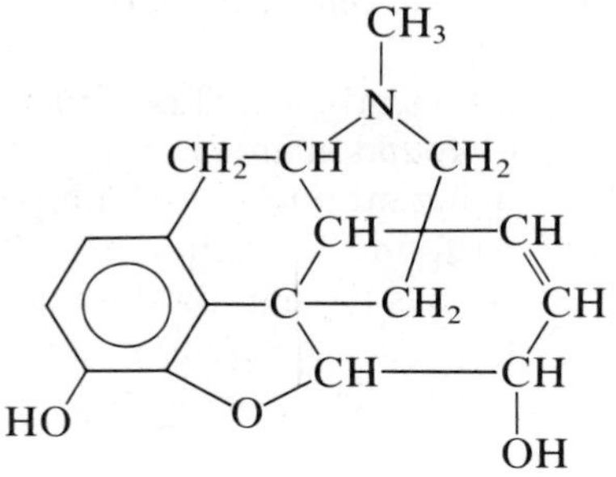

Morphine

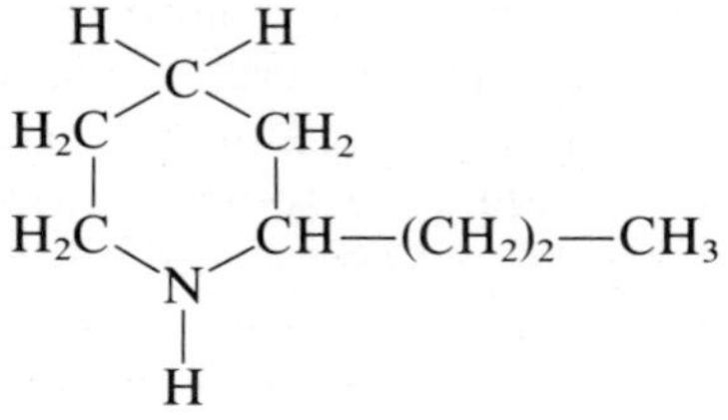

Coniine

FIGURE 28.1 Molecular structures of four alkaloids. Caffeine, nicotine, and morphine are tertiary amines; coniine is a secondary amine.

28.2 Isomerism in Organic Compounds

straight: unbranched

Isomers are distinctly different compounds, with different properties, that have the same molecular formula. In Chapter 13, we considered **structural isomers** of alkanes. You will recall that compounds such as butane and 2-methylpropane have the same molecular formula, C_4H_{10}, but different structural formulas. In these, as in all structural isomers, the order in which the atoms are bonded to each other differs. Specifically, butane is a ''straight-chain'' alkane, while 2-methylpropane has a branched carbon chain.

Structural isomerism is common among all types of organic compounds. Consider the following examples:

1. The three structural isomers of the alkene C_4H_8:

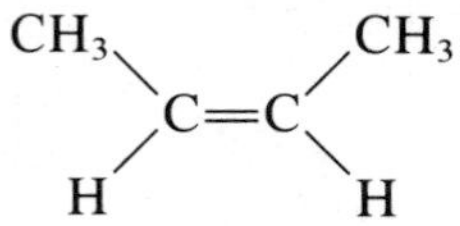

2-butene

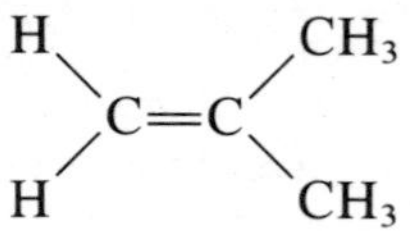

isobutylene

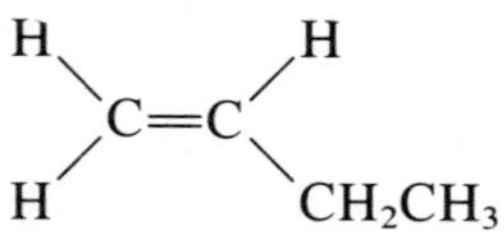

1-butene

2. The two structural isomers of the three-carbon alcohol C_3H_7OH:

As the molecules get bigger, the number of isomers goes up, fast

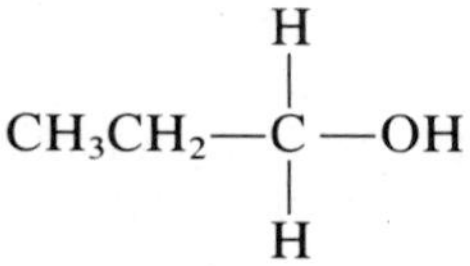

propyl alcohol

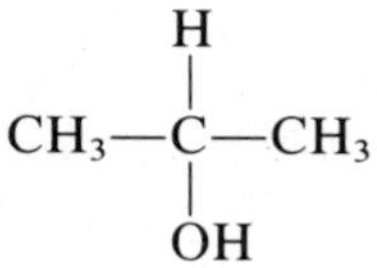

isopropyl alcohol

3. The two structural isomers dimethyl ether and ethyl alcohol, both of which have the molecular formula C_2H_6O:

$$CH_3-O-CH_3 \qquad CH_3-\underset{\underset{H}{|}}{\overset{\overset{H}{|}}{C}}-OH$$

dimethyl ether   ethyl alcohol

EXAMPLE 28.2 Consider the molecule $C_3H_6Cl_2$, which is derived from propane by substituting two Cl atoms for H atoms. Draw the structural isomers of $C_3H_6Cl_2$.

Solution There are four isomers. In one, both Cl atoms are bonded to a carbon atom at the end of the chain. In another, both Cl atoms are bonded to the central carbon atom:

$$CH_3-CH_2-\underset{\underset{Cl}{|}}{\overset{\overset{Cl}{|}}{C}}-H \qquad CH_3-\underset{\underset{Cl}{|}}{\overset{\overset{Cl}{|}}{C}}-CH_3$$

It can be tricky to find all the isomers

In the other two isomers, the Cl atoms are bonded to different carbons:

$$H-\underset{\underset{Cl}{|}}{\overset{\overset{H}{|}}{C}}-CH_2-\underset{\underset{Cl}{|}}{\overset{\overset{H}{|}}{C}}-H \qquad CH_3-\underset{\underset{Cl}{|}}{\overset{\overset{H}{|}}{C}}-\underset{\underset{Cl}{|}}{\overset{\overset{H}{|}}{C}}-H$$

EXERCISE How many structural isomers are there of $C_3H_5Cl_3$? Answer: Five.

Geometric Isomers

As we have seen, there are three structural isomers of the alkene C_4H_8. You may be surprised to learn that there are actually *four* different alkenes with this molecular formula. The "extra" compound arises because of a phenomenon called **geometric isomerism**. There are two different 2-butenes:

cis-2-butene   *trans*-2-butene
bp = 4°C   bp = 1°C

In the *cis* isomer, the two CH_3 groups (or the two H atoms) are as close to one another as possible. In the *trans* isomer, the two identical groups are farther apart. The two forms exist because there is no free rotation about the carbon-carbon double bond. The situation here is analogous to that with *cis-trans* isomers of

There is free rotation around single bonds

square planar complexes (Chap. 21). In both cases, the difference in geometry is responsible for isomerism; the atoms are bonded to each other in the same way.

Geometric, or *cis-trans*, isomerism is common among alkenes. Indeed, it occurs with all alkenes *except those in which two identical atoms or groups are attached to one of the double-bonded carbons.* Thus, although 2-butene has *cis* and *trans* isomers, isobutylene and 1-butene (p. 836) do not.

Name the compound:

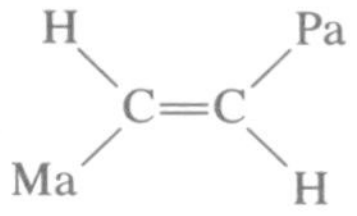

Ans. transparent

EXAMPLE 28.3 Draw all the isomers of the molecule $C_2H_2Cl_2$, in which two of the H atoms of ethylene are replaced by Cl atoms.

Solution

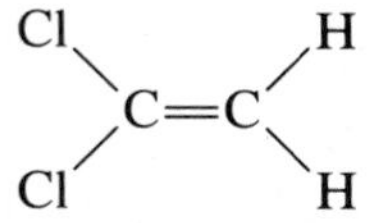

1,1-dichloroethylene

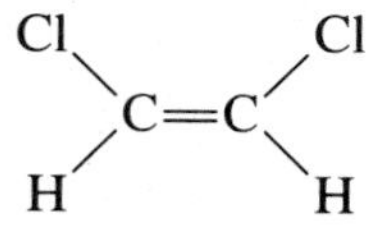

cis-1,2-dichloroethylene

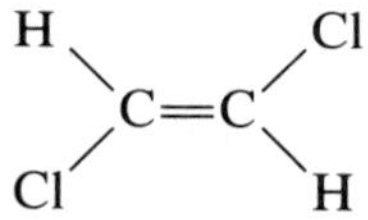

trans-1,2-dichloroethylene

Notice that 1,1-dichloroethylene, in which the two Cl atoms are bonded to the same carbon, does not show geometric isomerism.

EXERCISE Would you expect to find geometric isomers of

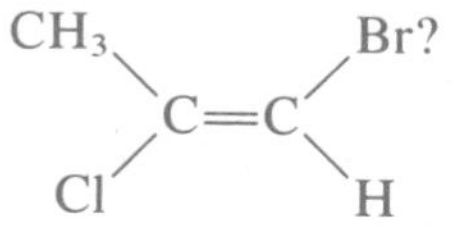

Answer: Yes. The other isomer is:

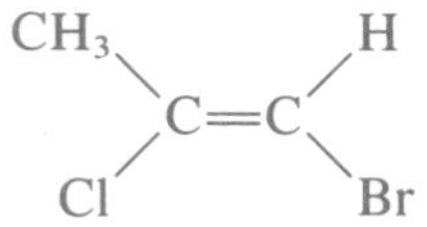

Optical Isomerism

Optical isomerism arises because of the tetrahedral nature of the bonding around a carbon atom. It occurs when at least one carbon atom in a molecule is bonded to four different atoms or groups. Consider, for example, the methane derivative

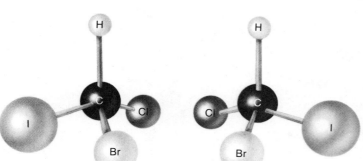

FIGURE 28.2 The two optical isomers of CHClBrI are mirror images of each other.

CHClBrI. As you can see from Figure 28.2, there are two different forms of this molecule, which are mirror images of one another. The mirror images are not superimposable; that is, you cannot place one molecule over the other so that identical groups are touching. In this sense, the two isomers resemble right- and left-hand gloves. Optical isomers differ from geometric isomers in that the latter are not mirror images of one another.

A carbon atom with four different atoms or groups attached to it is referred to as a **chiral center**. A molecule containing such a carbon atom shows optical isomerism. It exists in two different forms that are nonsuperimposable mirror images. These forms are referred to as optical isomers or *enantiomers*. Molecules may contain more than one chiral center, in which case there can be more than two enantiomers.

Such atoms used to be called asymmetric

EXAMPLE 28.4 In the following structural formulas, locate each chiral carbon atom:

a. CH_3—$\overset{\overset{\displaystyle Cl}{|}}{\underset{\underset{\displaystyle Cl}{|}}{C}}$—$\overset{\overset{\displaystyle H}{|}}{\underset{\underset{\displaystyle OH}{|}}{C}}$—$CH_3$ b. HO—$\overset{\overset{\displaystyle H}{|}}{\underset{\underset{\displaystyle O}{\|}}{C}}$—$\overset{\overset{\displaystyle H}{|}}{\underset{\underset{\displaystyle NH_2}{|}}{C}}$—$\overset{\overset{\displaystyle H}{|}}{\underset{\underset{\displaystyle H}{|}}{C}}$—$OH$

c. Cl—$\overset{\overset{\displaystyle H}{|}}{\underset{\underset{\displaystyle OH}{|}}{C}}$—$\overset{\overset{\displaystyle CH_3}{|}}{\underset{\underset{\displaystyle OH}{|}}{C}}$—$Cl$

Solution A chiral carbon has four different atoms or groups attached to it. In (a), the C atom bonded to OH is chiral. In (b), the C atom bonded to NH_2 is chiral. In (c), both C atoms are chiral.

EXERCISE Consider the glycerol molecule, whose structural formula is shown on p. 832. Would you expect glycerol to show optical isomerism? Answer: No.

The term "optical isomerism" comes from the effect that enantiomers have upon plane-polarized light, such as that produced by a Polaroid lens (Fig. 28.3). When this light is passed through a solution containing a single enantiomer, the

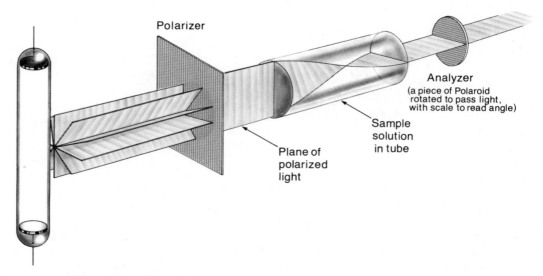

FIGURE 28.3 When ordinary light is passed through a polarizer, the light that emerges vibrates only in a single plane. If this "plane-polarized" light is passed through a solution containing an optically active compound, the plane is rotated, either to the right (clockwise) or to the left (counterclockwise).

plane is rotated from its original position. One isomer rotates it to the right (clockwise), and the other to the left (counterclockwise). If both isomers are present in equal amounts, we obtain what is known as a racemic mixture. In this case, the two rotations offset each other and there is no effect on plane-polarized light.

Enantiomers ordinarily resemble each other closely in their physical and chemical properties. For example, the two forms of lactic acid have the same melting point (52°C), density (1.25 g/cm³), and acid dissociation constant ($K_a = 1.4 \times 10^{-4}$):

The C atom in the middle is a chiral center

$$
\begin{array}{ccc}
\text{COOH} & & \text{COOH} \\
| & & | \\
\text{HO—C—H} & & \text{H—C—OH} \\
| & & | \\
\text{CH}_3 & & \text{CH}_3 \\
\text{I} & & \text{II}
\end{array}
$$

enantiomers of lactic acid

On the other hand, enantiomers frequently differ in their physiological activity. This was discovered by Louis Pasteur, the father of modern biochemistry. Working with a mixture of the optical isomers of lactic acid, he found that mold growth occurred only with enantiomer II. Apparently, the mold was unable to metabolize enantiomer I. A more modern example of this type involves amphetamine, often used illicitly as an "upper" or "pep pill." Amphetamine consists of two enantiomers:

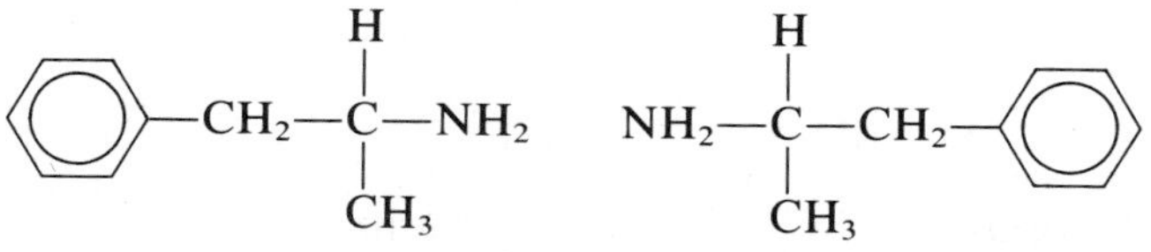

I   II

enantiomers of amphetamine

Where are the chiral carbon atoms?

Enantiomer I, called Dexedrine, is by far the stronger stimulant. It is about four times as active as Benzedrine, the racemic mixture of the two isomers.

28.3
Synthetic Polymers

Over the past 50 years, chemists have made a great many different polymers in the laboratory. These "synthetic" polymers usually contain only one or two different kinds of monomers, joined in chains that may be thousands of units long. Synthetic polymers can be divided into two general types, depending upon the way in which monomer units combine with one another:

    1. *Addition polymers,* in which monomer units add directly to one another.

    2. *Condensation polymers,* in which monomer units combine by splitting out a small molecule, most often water.

Addition Polymers

The monomer from which an addition polymer is made ordinarily contains a double bond. Upon polymerization, the double bond is converted to a single bond:

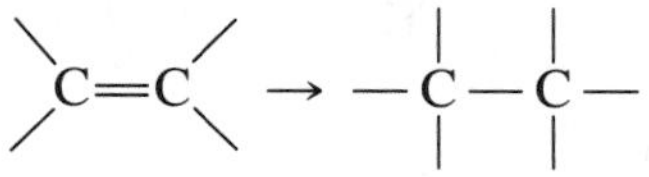

and successive monomer units add to one another. In the case of ethylene, we might represent the reaction as

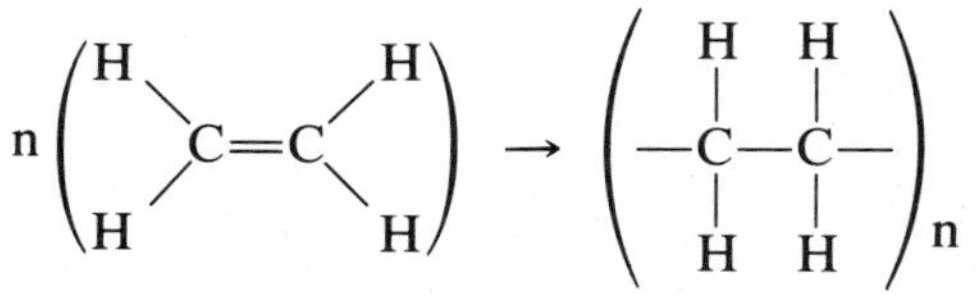

    ethylene   polyethylene

(28.6)

Here all the atoms in the monomer end up in the polymer

where n is a very large number, of the order of 2000.

    There are several different ways to initiate Reaction 28.6. One method uses a small amount of a very reactive species called a *free radical,* which contains an unpaired electron. This free radical, which we will represent simply as X·, reacts rapidly with ethylene to form a species that still contains an unpaired electron:

$$X\cdot + \underset{H}{\overset{H}{}}C=C\underset{H}{\overset{H}{}} \rightarrow X-\underset{H}{\overset{H}{\underset{|}{\overset{|}{C}}}}-\underset{H}{\overset{H}{\underset{|}{\overset{|}{C}}}}\cdot \qquad (28.7)$$

The product of this reaction is able to add another ethylene molecule, and then another, and then another, ... Eventually, very long chains are formed, each with an unpaired electron at one end. Two of these chains can combine to terminate the polymerization process. The product formed by free radical polymerization of ethylene is one in which most of the chains are branched. Its structure might be shown as

$$-\overset{H}{\underset{H}{C}}-\overset{H}{\underset{H}{C}}-\overset{H}{\underset{H}{C}}-\overset{H}{\underset{CH_2}{C}}-\overset{H}{\underset{H}{C}}-\overset{H}{\underset{H}{C}}-\overset{H}{\underset{H}{C}}-\overset{H}{\underset{H}{C}}-\overset{H}{\underset{H}{C}}-\overset{H}{\underset{H}{C}}-\overset{H}{\underset{CH_2}{C}}-\overset{H}{\underset{H}{C}}-$$
$$CH_3 CH_3$$

The properties of a polymer depend in part on the amount of branching present

In this polymer, known as *branched* polyethylene, neighboring chains are arranged in a somewhat random fashion, often overlapping each other. This produces a soft, flexible solid, used mostly in films and coatings where a pliable material is needed.

With a different type of catalyst, it is possible to produce *linear* polyethylene. This consists almost entirely of unbranched chains:

$$-\overset{H}{\underset{H}{C}}-\overset{H}{\underset{H}{C}}-\overset{H}{\underset{H}{C}}-\overset{H}{\underset{H}{C}}-\overset{H}{\underset{H}{C}}-\overset{H}{\underset{H}{C}}-\overset{H}{\underset{H}{C}}-\overset{H}{\underset{H}{C}}-\overset{H}{\underset{H}{C}}-\overset{H}{\underset{H}{C}}-\overset{H}{\underset{H}{C}}-\overset{H}{\underset{H}{C}}-\overset{H}{\underset{H}{C}}-\overset{H}{\underset{H}{C}}-\overset{H}{\underset{H}{C}}-\overset{H}{\underset{H}{C}}-$$

The early polyethylenes were produced at 100°C and 1000 atm pressure. They were highly branched

FIGURE 28.4 Two polyethylene bottles. The one on the left is made of semirigid, linear polyethylene. The other is made of more pliable, branched polyethylene. (From Chickos, J., et al.: *Chemistry; Its Role in Society*. New York, Heath, 1973, p. 94.)

Neighboring chains in linear polyethylene line up nearly parallel to each other. This gives a polymer that approaches a crystalline material. It is used for bottles, toys, and other semirigid objects (Fig. 28.4).

Table 28.3 lists several different addition polymers and the monomers from which they are derived. Several of the monomers are unsymmetrical in the sense that the two ends of the molecule are different. In vinyl chloride, for example, there is a CH_2 group at one end of the molecule and a CHCl group at the other end. When an unsymmetrical monomer polymerizes, the "head" of one molecule usually attaches itself to the "tail" of the next molecule. This gives a "head-to-tail" polymer in which different groups alternate along the polymer chain. The situation with vinyl chloride is typical. The structure of the polymer is:

Commercial production of these polymers began around 1940

Table 28.3
Some Common Addition Polymers

MONOMER	NAME	POLYMER	USES
$H_2C=CH_2$	ethylene	polyethylene	bags, coatings, toys
$H_2C=CH-CH_3$	propylene	polypropylene	beakers, milk cartons
$H_2C=CH-Cl$	vinyl chloride	polyvinyl chloride, PVC	floor tile, raincoats, pipe, phonograph records
$H_2C=CH-CN$	acrylonitrile	polyacrylonitrile, PAN	rugs; Orlon and Acrilan are copolymers with other monomers
$H_2C=CH-C_6H_5$	styrene	polystyrene	cast articles using a transparent plastic
$H_2C=C(CH_3)-C(=O)-OCH_3$	methyl methacrylate	Plexiglas, Lucite, acrylic resins	high-quality transparent objects, latex paints
$F_2C=CF_2$	tetrafluoroethylene	Teflon	gaskets, insulation, bearings, pan coatings

$$-\overset{\overset{\displaystyle H}{|}}{\underset{\underset{\displaystyle H}{|}}{C}}-\overset{\overset{\displaystyle H}{|}}{\underset{\underset{\displaystyle Cl}{|}}{C}}-\overset{\overset{\displaystyle H}{|}}{\underset{\underset{\displaystyle H}{|}}{C}}-\overset{\overset{\displaystyle H}{|}}{\underset{\underset{\displaystyle Cl}{|}}{C}}-\overset{\overset{\displaystyle H}{|}}{\underset{\underset{\displaystyle H}{|}}{C}}-\overset{\overset{\displaystyle H}{|}}{\underset{\underset{\displaystyle Cl}{|}}{C}}-\overset{\overset{\displaystyle H}{|}}{\underset{\underset{\displaystyle H}{|}}{C}}-\overset{\overset{\displaystyle H}{|}}{\underset{\underset{\displaystyle Cl}{|}}{C}}-\overset{\overset{\displaystyle H}{|}}{\underset{\underset{\displaystyle H}{|}}{C}}-\overset{\overset{\displaystyle H}{|}}{\underset{\underset{\displaystyle Cl}{|}}{C}}-$$

Note that CH_2 and $CHCl$ groups alternate along the chain.

EXAMPLE 28.5 Sketch the head-to-tail polymer derived from propylene (Table 28.3).

Solution

$$-\overset{\overset{\displaystyle H}{|}}{\underset{\underset{\displaystyle H}{|}}{C}}-\overset{\overset{\displaystyle H}{|}}{\underset{\underset{\displaystyle CH_3}{|}}{C}}-\overset{\overset{\displaystyle H}{|}}{\underset{\underset{\displaystyle H}{|}}{C}}-\overset{\overset{\displaystyle H}{|}}{\underset{\underset{\displaystyle CH_3}{|}}{C}}-\overset{\overset{\displaystyle H}{|}}{\underset{\underset{\displaystyle H}{|}}{C}}-\overset{\overset{\displaystyle H}{|}}{\underset{\underset{\displaystyle CH_3}{|}}{C}}-\overset{\overset{\displaystyle H}{|}}{\underset{\underset{\displaystyle H}{|}}{C}}-\overset{\overset{\displaystyle H}{|}}{\underset{\underset{\displaystyle CH_3}{|}}{C}}-\overset{\overset{\displaystyle H}{|}}{\underset{\underset{\displaystyle H}{|}}{C}}-\overset{\overset{\displaystyle H}{|}}{\underset{\underset{\displaystyle CH_3}{|}}{C}}-$$

EXERCISE Suppose the polymer were of the head-to-head type, with the CH_2 group of one molecule bonded to the CH_2 group of the next. Sketch the structure of this polymer. Answer:

$$-\overset{\overset{\displaystyle H}{|}}{\underset{\underset{\displaystyle H}{|}}{C}}-\overset{\overset{\displaystyle H}{|}}{\underset{\underset{\displaystyle CH_3}{|}}{C}}-\overset{\overset{\displaystyle H}{|}}{\underset{\underset{\displaystyle CH_3}{|}}{C}}-\overset{\overset{\displaystyle H}{|}}{\underset{\underset{\displaystyle H}{|}}{C}}-\overset{\overset{\displaystyle H}{|}}{\underset{\underset{\displaystyle H}{|}}{C}}-\overset{\overset{\displaystyle H}{|}}{\underset{\underset{\displaystyle CH_3}{|}}{C}}-\overset{\overset{\displaystyle H}{|}}{\underset{\underset{\displaystyle CH_3}{|}}{C}}-\overset{\overset{\displaystyle H}{|}}{\underset{\underset{\displaystyle H}{|}}{C}}-\overset{\overset{\displaystyle H}{|}}{\underset{\underset{\displaystyle H}{|}}{C}}-\overset{\overset{\displaystyle H}{|}}{\underset{\underset{\displaystyle CH_3}{|}}{C}}-$$

Most polymers decompose at high temperatures. For example, polyvinyl chloride starts to decompose at 100°C, liberating HCl. Teflon, a polymer of tetrafluoroethylene (Table 28.3) behaves quite differently. A hard waxy solid, Teflon is useful from about −70 to 250°C. Teflon is also extremely unreactive. It resists chemical attack by all known reagents except molten alkali metals. Its best known use is as a nonstick coating for frying pans. Teflon was discovered by accident at the Du Pont Company in 1938 during basic research on halogenated hydrocarbons. Louis Pasteur's maxim that "chance favors the prepared mind" was applied by Dr. Ray Plunkett and co-workers who discovered and developed Teflon.

Teflon is "tough" in just about all respects

Condensation Polymers

In forming a condensation polymer, monomer units combine by splitting out a small molecule, most often water. Usually, two different monomers are involved. Each of these monomers has a functional group, such as

$$-\overset{\displaystyle C}{\underset{\displaystyle \|}{\underset{\displaystyle O}{}}}-OH \qquad -OH \qquad -\overset{\displaystyle N}{\underset{\displaystyle |}{\underset{\displaystyle H}{}}}-H$$

at *both ends of the molecule*. We will consider two of the most common types of synthetic condensation polymers: polyesters and polyamides.

A **polyester** is formed when a dihydroxy alcohol, HO—R—OH, reacts with a dicarboxylic acid, HOOC—R′—COOH. We might represent the first step in the reaction as

$$HO-R-OH + HO-\underset{\underset{O}{\|}}{C}-R'-\underset{\underset{O}{\|}}{C}-OH \rightarrow HO-R-O-\underset{\underset{O}{\|}}{C}-R'-\underset{\underset{O}{\|}}{C}-OH + H_2O$$

| dihydroxy alcohol | dicarboxylic acid | ester with active end groups | (28.8) |

The COOH group at one end of the ester molecule can react with another alcohol molecule. The OH group at the other end can react with an acid molecule. This process can continue, leading eventually to a long-chain polymer containing 500 or more ester groups. The general structure of the polyester can be represented as

$$-\underset{\underset{O}{\|}}{C}-R'-\underset{\underset{O}{\|}}{C}-O-R-O-\underset{\underset{O}{\|}}{C}-R'-\underset{\underset{O}{\|}}{C}-O-R-O-$$

section of a polyester molecule

A thin polyester film was used to cover the wings and pilot compartment of the *Gossamer Albatross,* the first human-powered aircraft to cross the English Channel (Fig. 28.5).

One of the most familiar polyesters is Dacron, in which the monomers are ethylene glycol and terephthalic acid:

In developing polymers, chemists try all kinds of monomers. Some work well, most don't

$$HO-CH_2-CH_2-OH \qquad HO-\underset{\underset{O}{\|}}{C}-\bigcirc-\underset{\underset{O}{\|}}{C}-OH$$

ethylene glycol   terephthalic acid

FIGURE 28.5 The *Gossamer Albatross,* the first human-powered craft to fly across the English Channel (1979). Light-weight, durable polyesters cover the wings and body. (Photo courtesy of DuPont Corporation)

As a thin film the polymer is called Mylar, used in magnetic recording tape and weather balloons. The Dacron (Mylar) polymer has the structure

$$-\overset{O}{\underset{\|}{C}}-\overset{}{\text{⬡}}-\overset{O}{\underset{\|}{C}}-O-CH_2-CH_2-O-\overset{O}{\underset{\|}{C}}-\overset{}{\text{⬡}}-\overset{O}{\underset{\|}{C}}-O-CH_2-CH_2-O-$$

When a diamine (molecule containing two NH_2 groups) reacts with a dicarboxylic acid (two COOH groups), a **polyamide** is formed. This condensation polymerization is entirely analogous to that used to make polyesters. In this case, the NH_2 group of the diamine reacts with the COOH group of the dicarboxylic acid:

$$NH_2-R-\underset{\underset{H}{|}}{N}-H + HO-\overset{O}{\underset{\|}{C}}-R'-\overset{O}{\underset{\|}{C}}-OH \longrightarrow NH_2-R-\underset{\underset{H}{|}}{N}-\overset{O}{\underset{\|}{C}}-R'-\overset{O}{\underset{\|}{C}}-OH + H_2O$$

These chains are often hundreds of units long

Condensation can continue to form a long-chain polymer (Example 28.6).

EXAMPLE 28.6 Consider the diamine $H_2N-(CH_2)_6-NH_2$ and the dicarboxylic acid $HOOC-(CH_2)_4-COOH$. Give the
a. structural formula of the dimer formed when one molecule of diamine reacts with one molecule of dicarboxylic acid.
b. structure of a section of the polyamide formed by these two monomers.

Solution

a. $H_2N-(CH_2)_6-\underset{\underset{H}{|}}{N}-\overset{O}{\underset{\|}{C}}-(CH_2)_4-\overset{O}{\underset{\|}{C}}-OH$

b. $-\overset{O}{\underset{\|}{C}}-(CH_2)_4-\overset{O}{\underset{\|}{C}}-\underset{\underset{H}{|}}{N}-(CH_2)_6-\underset{\underset{H}{|}}{N}-\overset{O}{\underset{\|}{C}}-(CH_2)_4-\overset{O}{\underset{\|}{C}}-\underset{\underset{H}{|}}{N}-(CH_2)_6-\underset{\underset{H}{|}}{N}-$

EXERCISE Consider the dimer

$$H_2N-(CH_2)_5-\underset{\underset{H}{|}}{N}-\overset{O}{\underset{\|}{C}}-\text{⬡}-\overset{O}{\underset{\|}{C}}-OH$$

Write structural formulas for the amine and carboxylic acid used to make this amide. Answer:

$$H_2N-(CH_2)_5-NH_2 \text{ and } HOOC-\text{⬡}-COOH$$

The polyamide in Example 28.6 is Nylon 66 (the numbers indicate the number of carbon atoms in the amine and acid monomers). This polyamide was first made in 1935 by Wallace Carothers, the discoverer of neoprene. Since then, other nylons, all polyamides, have been synthesized. The earliest use of nylon was as a textile, especially for women's hosiery. Careful control of chain length during manufacturing gives nylon fibers of the desired sheerness and luster. The elasticity of nylon fibers is due in part to hydrogen bonds between adjacent polymer chains. These hydrogen bonds join carboxyl oxygen atoms on one chain to NH groups on adjacent chains. Bulk nylon can be molded or cut into a variety of shapes. Its high wear resistance and slight slipperiness make bulk nylon ideal for door latches and gears.

The discovery of nylon caused a fantastic increase in industrial chemical research

Polyurethanes are condensation polymers related to polyesters and polyamides. A typical polyurethane has the structure

$$-\text{O}-\text{CH}_2-\text{O}-\overset{\overset{\text{O}}{\|}}{\text{C}}-\underset{\underset{\text{H}}{|}}{\text{N}}-(\text{CH}_2)_4-\underset{\underset{\text{H}}{|}}{\text{N}}-\overset{\overset{\text{O}}{\|}}{\text{C}}-\text{O}-\text{CH}_2-\text{O}-\overset{\overset{\text{O}}{\|}}{\text{C}}-\underset{\underset{\text{H}}{|}}{\text{N}}-(\text{CH}_2)_4-\underset{\underset{\text{H}}{|}}{\text{N}}-\overset{\overset{\text{O}}{\|}}{\text{C}}-$$

Foam rubber for furniture, packaging materials, and many other products contain polyurethane. A polyurethane fabric (trademark Biomer) is used as the diaphragm in the Jarvik-7 heart. The first human recipient of this artificial heart was Dr. Barney Clark (Fig. 28.6).

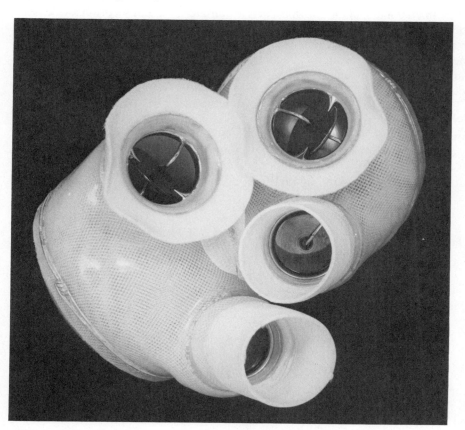

Synthetic polymers have had a large effect on our day-to-day lives, mostly a good effect

FIGURE 28.6 The Jarvik-7 artificial heart; the base is aluminum and the valves are pyrolytic graphite and titanium. The base, valves and valve-holding rings are coated with the polyurethane, Biomer®. Layers of this polymer serve as the diaphragm in the device. (Photo, University of Utah Medical Center)

28.4
Natural Polymers

Natural polymers produced by plants and animals are essential to all forms of life. The cell walls of plants and the woody structures of trees are composed largely of cellulose, a polymeric carbohydrate. Your body contains thousands of polymers, which make up tissues, blood, and skin. Many of these fall in the class of organic polymers known as proteins. Natural polymers are also useful in our daily lives. For centuries, clothes have been made from cotton (largely cellulose) and wool, silk, and leather (protein materials).

Carbohydrates

Carbohydrates, which comprise one of the three basic classes of foodstuffs, contain carbon, hydrogen, and oxygen atoms. Their general formula, $C_n(H_2O)_m$, is the basis for their name. One of the simplest and most important carbohydrates is **glucose,** molecular formula $C_6H_{12}O_6$. Glucose, in water solution, exists primarily as six-membered ring molecules, shown below. The rings are not planar but have a three-dimensional "chair" form. The carbon atoms at each corner of the ring are not shown in the structure; they are represented by the numbers 1 to 5. The sixth atom in the ring is oxygen. There is a CH_2OH group bonded to carbon atom 5. There is an H atom bonded to each of the ring carbons; an OH group is bonded to carbon atoms 1 through 4. The heavy lines indicate the front of the ring, projected toward you.

Glucose has many isomers and several chiral centers

The bonds to the H and OH groups from the ring carbons are oriented either in the plane of the ring (*equatorial*) or perpendicular to it (*axial*). In β-glucose, all four OH groups are equatorial. This keeps these rather bulky groups out of each other's way. In α-glucose, the OH group on carbon atom 1 is axial; all the others are equatorial. In water solution, glucose exists as an equilibrium mixture, about 37% in the alpha form and 63% in the beta form.

How in the world could we prove that? Ans. Optical rotation

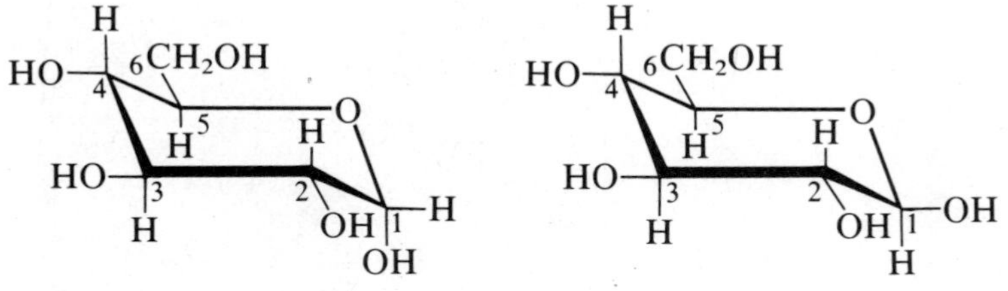

α-glucose β-glucose

Although glucose can exist as a simple sugar, it is most often found in nature in combined form, as part of a larger molecule. The simplest such molecule is that of maltose, a dimer formed from two α-glucose molecules. These are combined head to tail: carbon atom 1 of one molecule is joined through an oxygen atom to carbon atom 4 of the second molecule. To form maltose, the two OH groups on these carbon atoms react, condensing out H_2O and leaving the O atom bridge:

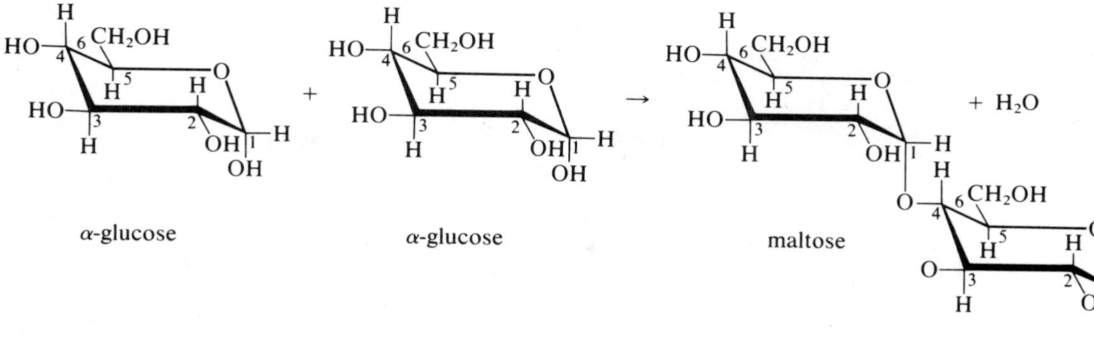

Sucrose, the compound we call "sugar" is, like maltose, a dimer. In this case, two monomers are involved. One is α-glucose; the other is another simple sugar called fructose (Fig. 28.7).

The polymerization process represented by Equation 28.9 can continue, since there are —OH groups at each end of the molecule. Eventually, a polymer is formed containing a large number of monomer units bridged by oxygen atoms. One such polymer is *starch,* a carbohydrate stored in the seeds and roots of many vegetables, including corn and potatoes. Another is *cellulose,* an indigestible carbohydrate that is the major component of wood and the stalks and leaves of plants.

Starch and cellulose are condensation polymers of glucose

Starch is actually a mixture of two types of α-glucose polymers. One of these, called amylose, is insoluble in water and comprises about 20% of natural starch. It consists of long single chains of 1000 or more α-glucose units joined head to tail, as in maltose (Fig. 28.8, p. 850). The other polymer found in starch is amylopectin, which is soluble in water. The linkage between α-glucose units is the same as in amylose. However, amylopectin has a highly branched structure. Short chains of 20 to 25 glucose units are linked through oxygen atom bridges. The oxygen atom joins a number 1 carbon atom at the end of one chain to a number 6 carbon atom on an adjacent chain.

Plants use starch as their main source of glucose. In animals, glucose is stored mainly in the form of glycogen. This polysaccharide has a structure similar to

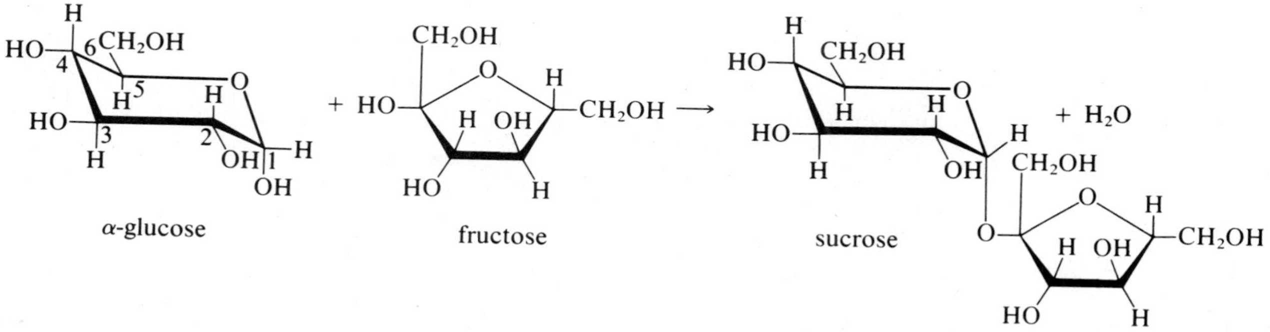

FIGURE 28.7 In sucrose, the monomer units are α-glucose and fructose. The reaction shown above can be reversed by acid or the enzyme sucrase. This occurs in digestion, which makes glucose and fructose available for absorption into the blood.

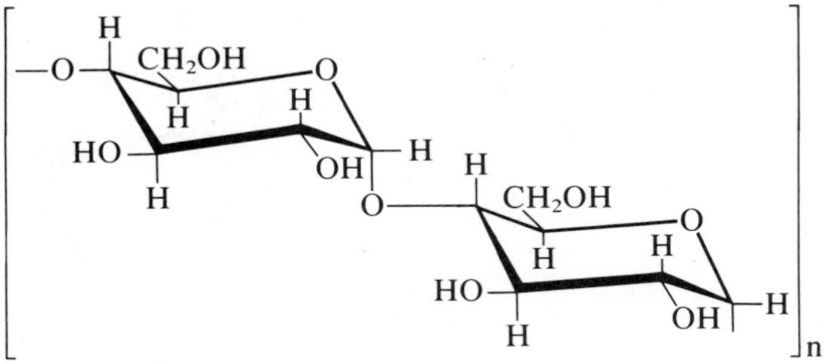

Amylose

FIGURE 28.8 In starch, glucose molecules are joined head-to-tail through oxygen atoms. A thousand or more glucose molecules may be linked in this way, either in long single chains (amylose) or branched chains (amylopectin).

that of amylopectin except that it is more highly branched. Glycogen is found principally in the liver and in muscle tissue.

EXAMPLE 28.7 What is the simplest formula of amylose?

Solution Amylose is a condensation polymer of glucose, whose formula is $C_6H_{12}O_6$. Every time a glucose unit is added to the polysaccharide chain a molecule of H_2O is eliminated. Hence, the unit added is not $C_6H_{12}O_6$, but rather $C_6H_{10}O_5$. This is the simplest formula of amylose.

This is also the formula for starch and cellulose

EXERCISE Amylose has a molar mass of about 3.00×10^5 g/mol. How many glucose units does an amylose molecule contain? Answer: 1850.

Cellulose contains long, unbranched chains of glucose units, about 10,000 per chain. It differs from starch in the way the glucose units are joined to each other. In starch, the oxygen bridge between the units is in the alpha position; in cellulose it is in the beta position (Fig. 28.9). Because humans lack the enzymes

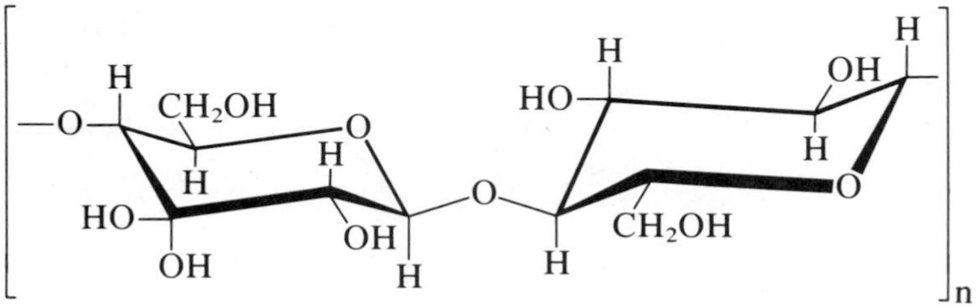

Cellulose

FIGURE 28.9 Cellulose structure. The bonding between glucose rings in cellulose is through oxygen bridges in the β position for each ring to the left of the bridge. This structure allows for ordered hydrogen bonding between chains and formation of long strong fibers. Cotton and wood fiber would have structures like that shown.

required to catalyze the hydrolysis of beta linkages, we cannot digest cellulose. The cellulose that we take in from fruits and vegetables remains undigested as "fiber."

The molecular structure of cellulose, unlike that of starch, allows for strong hydrogen bonding between polymer chains. This results in the formation of strong water-resistant fibers such as those found in cotton, which is 98% cellulose. Cotton actually has a tensile strength greater than that of steel. The major industrial source of cellulose is wood (~50% cellulose). Wood chips can be treated with hot NaOH solution, which dissolves some of the wood components and partially breaks down the polymer chains. The insoluble residue is impure cellulose in the form of wood pulp. It can be used directly to make paper or treated further to make various plastics. Although cellulose can be degraded completely to glucose, the process has not been practical economically. If it could be made so, the glucose could be used for food, or converted to ethanol by fermentation. Both such uses would be important, in view of world shortages of food and energy.

Wood contains cellulose and a lot of other substances

Proteins

The natural polymers known as proteins make up about 15% by mass of our bodies. They serve many functions. Fibrous proteins are the main components of hair, muscle, and skin. Other proteins found in body fluids transport oxygen, fats, and other substances needed for metabolism. Still others, such as insulin and vasopressin, are hormones. Enzymes, which catalyze reactions in the body, are chiefly protein.

The monomers from which proteins are derived have the general structure shown below. These compounds are called α-amino acids. They have an NH_2 group attached to the carbon atom (the alpha carbon) adjacent to a COOH group:

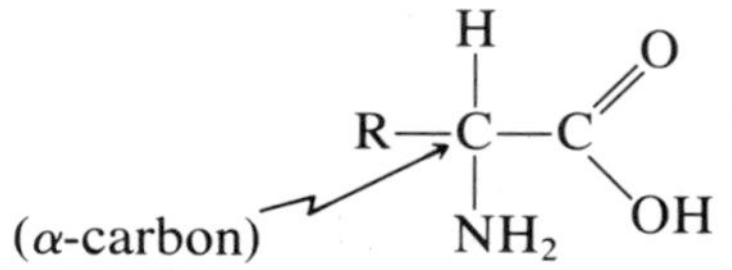

(α-carbon)

an α-amino acid

In our digestive system, proteins are broken down into α-amino acids, which are then reassembled in cells to form other proteins required by the body.

Natural proteins can be broken down into about 20 different α-amino acids (Table 28.4). These molecules differ in the nature of the R group attached to the alpha carbon. As you can see from the table, R can be

Some α-amino acids are quite simple, others are complex

— an H atom (glycine).
— a simple hydrocarbon group (alanine, valine, etc.).
— a more complex group containing one or more atoms of oxygen (serine, aspartic acid, etc.), nitrogen (lysine, etc.), or sulfur (methionine, etc.).

As noted in Section 28.3, molecules containing NH_2 and COOH groups can undergo condensation polymerization. Amino acids contain both groups in the

Table 28.4
The Common α-Amino Acids (names and abbreviations below structures)

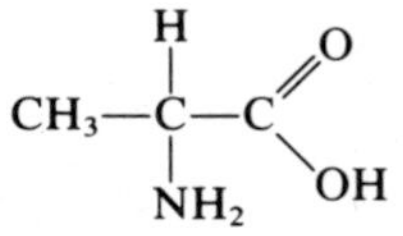

alanine   Ala

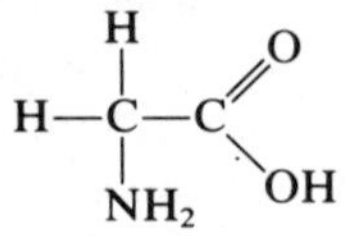

glycine   Gly

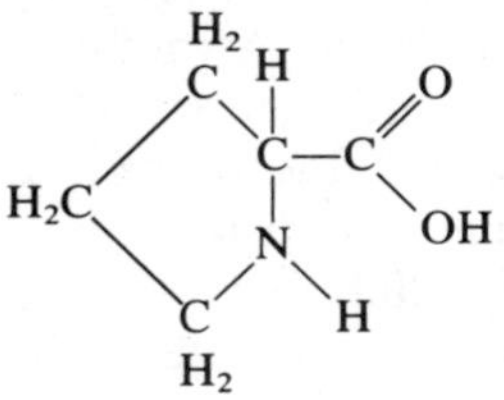

proline   Pro

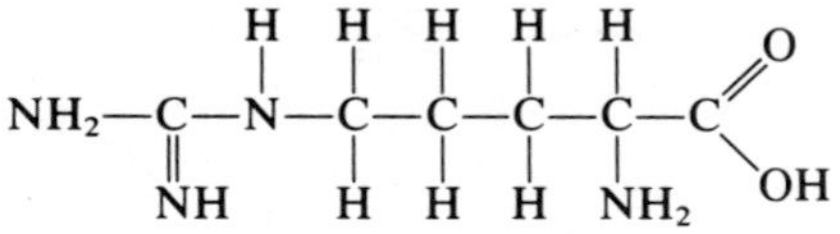

arginine*   Arg

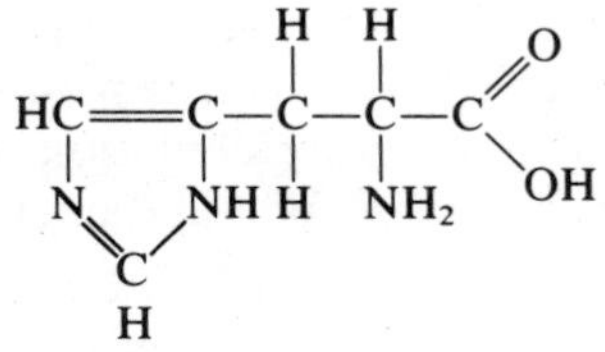

histidine*   His

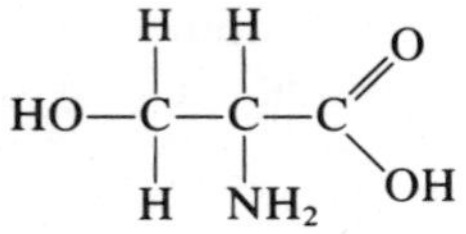

serine   Ser

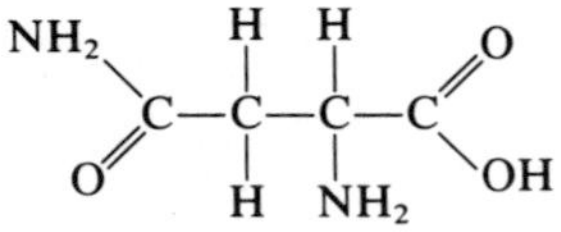

asparagine   Asn

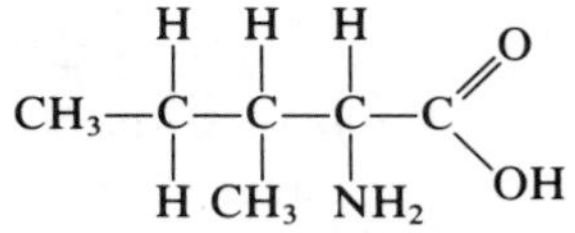

isoleucine*   Ile

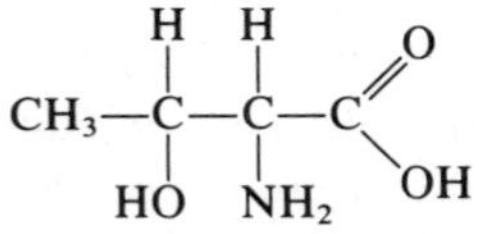

threonine*   Thr

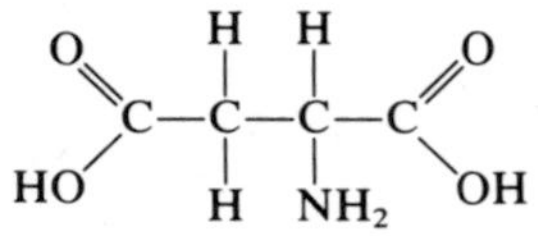

aspartic acid   Asp

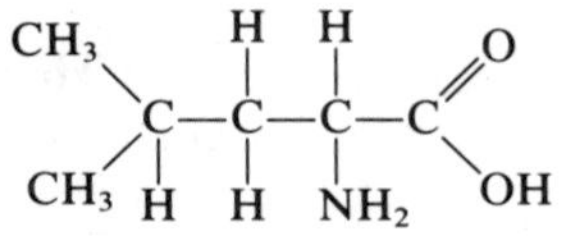

leucine*   Leu

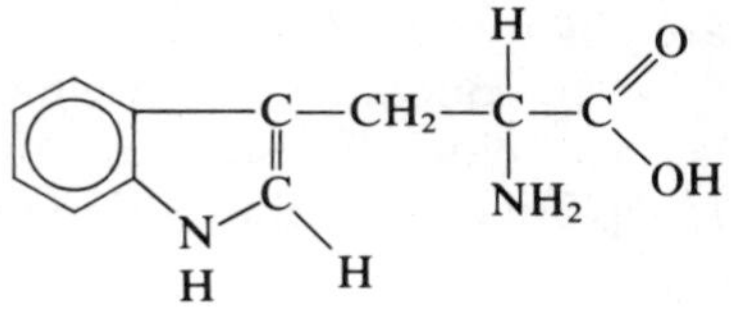

tryptophan*   Trp

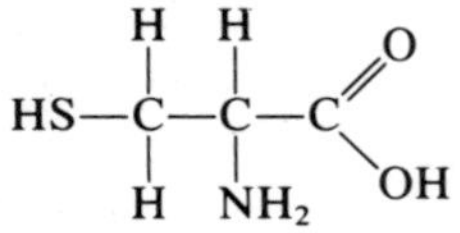

cysteine   Cys

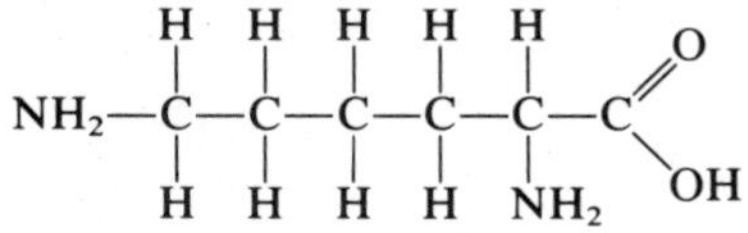

lysine*   Lys

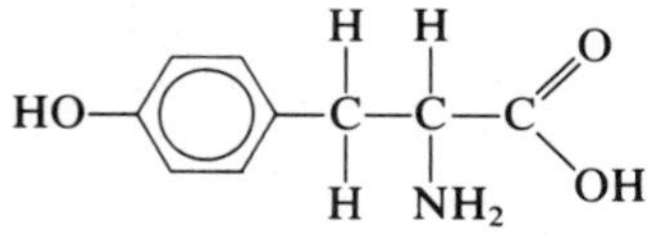

tyrosine   Tyr

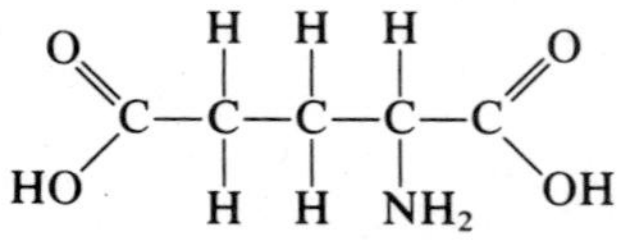

glutamic acid   Glu

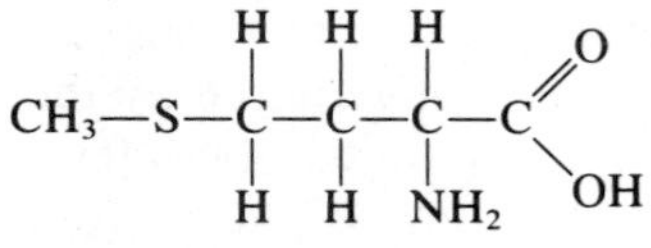

methionine*   Met

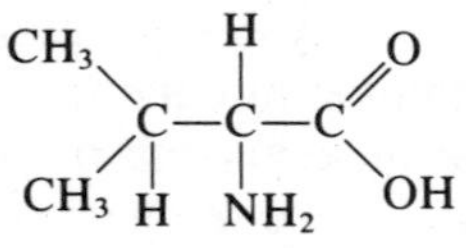

valine*   Val

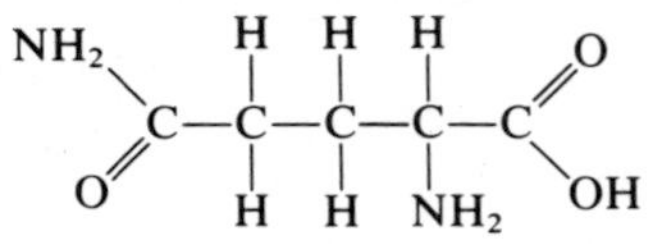

glutamine   Gln

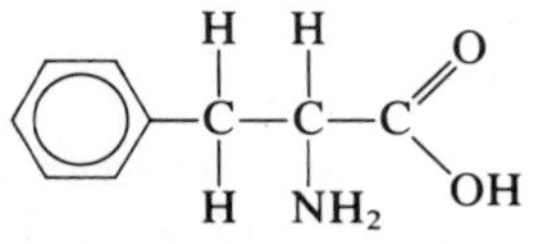

phenylalanine*   Phe

*These are essential amino acids that cannot be synthesized by the body and therefore must be obtained from the diet.

same molecule. Hence, two amino acid molecules can combine by the reaction of the COOH group in one molecule with the NH$_2$ group of the other molecule:

$$H-\underset{\underset{H}{|}}{\overset{\overset{H}{|}}{N}}-\underset{\underset{R_1}{|}}{\overset{\overset{O}{\|}}{C}}-\overset{\overset{O}{\|}}{C}-\boxed{OH + H}-\underset{\underset{H}{|}}{\overset{\overset{H}{|}}{N}}-\underset{\underset{R_2}{|}}{C}-\overset{\overset{O}{\|}}{C}-OH \rightarrow$$

The linkage is similar to that in nylon

$$H-\underset{\underset{H}{|}}{\overset{\overset{H}{|}}{N}}-\underset{\underset{R_1}{|}}{C}-\overset{\overset{O}{\|}}{C}-\underset{\underset{H}{|}}{\overset{\overset{H}{|}}{N}}-\underset{\underset{R_2}{|}}{C}-\overset{\overset{O}{\|}}{C}-OH + H_2O \qquad (28.10)$$

EXAMPLE 28.8 Draw the structure of the dimer formed when the COOH group of glycine reacts with the NH$_2$ group of serine.

Solution Here, R$_1$ is the group attached to the alpha carbon atom in glycine; R$_2$ is the corresponding group in serine. From Table 28.4,

$$R_1 = H; \qquad R_2 = CH_2OH$$

The structure of the dimer must be

$$H_2N-\underset{\underset{H}{|}}{\overset{\overset{H}{|}}{C}}-\overset{\overset{O}{\|}}{C}-N-\underset{\underset{CH_2OH}{|}}{\overset{\overset{H}{|}}{C}}-COOH$$

This molecule is called glycylserine, and is often abbreviated as Gly-Ser.

EXERCISE Draw the structure of the product, Ser-Gly, obtained by reacting the COOH group of serine with the NH$_2$ group of glycine. Answer:

$$H_2N-\underset{\underset{H}{|}}{\overset{\overset{CH_2OH}{|}}{C}}-\underset{\underset{O}{\|}}{C}-\underset{\underset{H}{|}}{N}-CH_2-COOH$$

The product of Reaction 28.10 has reactive groups at both ends of the molecule (NH$_2$ at one end, COOH at the other). Hence condensation can continue, giving the long-chain polymers called proteins. The general structure of a protein can be represented as

$$-\underset{\underset{H}{|}}{\overset{\overset{H}{|}}{N}}-\underset{\underset{R_1}{|}}{C}-\overset{\overset{O}{\|}}{C}-\underset{\underset{H}{|}}{\overset{\overset{H}{|}}{N}}-\underset{\underset{R_2}{|}}{C}-\overset{\overset{O}{\|}}{C}-\underset{\underset{H}{|}}{\overset{\overset{H}{|}}{N}}-\underset{\underset{R_3}{|}}{C}-\overset{\overset{O}{\|}}{C}-\underset{\underset{H}{|}}{\overset{\overset{H}{|}}{N}}-\underset{\underset{R_4}{|}}{C}-\overset{\overset{O}{\|}}{C}-$$

Proteins are condensation polymers of the 20 amino acids

The "peptide linkage" which is characteristic of proteins, is shown in color.

Proteins differ from the other polymers we have discussed in that they may contain up to 20 different monomers. This means that there is a huge number of possible proteins. Consider, for example, the proteins formed by linking together

50 amino acid units. Since there are 20 possibilities for each of these units, we have

$$20^{50} = 1 \times 10^{65}$$

possible proteins containing 50 monomer units.

As you can imagine, it is not an easy task to identify all the amino acid units present in a protein chain. To go further and determine the order in which these units are arranged might seem next to impossible. In recent years, however, this type of analysis has become possible. The detailed amino acid sequences in some very complex proteins have been determined. The first protein for which this was done was insulin, which contains two chains linked by disulfide (—S—S—) bridges. One chain contains 21 amino acid units; the other has 30. The amino acid sequence in insulin is

The amino acid sequence in hemoglobin, with 574 units, is known

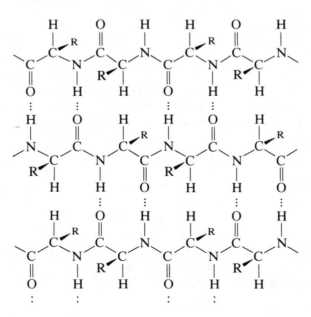

This structure was established by Frederick Sanger, who received the 1958 Nobel Prize in chemistry for his work.

In the body, proteins are built up by a series of reactions that in general produce a specific sequence of amino acids. Even tiny errors in this sequence may have serious effects. Among the genetic diseases known to be caused by improper sequencing are hemophilia, sickle cell anemia, and albinism. Sickle cell anemia is caused by the substitution of *one* valine unit for a glutamic acid unit in a chain containing 146 monomers.

FIGURE 28.10 β-keratin structure of silk fiber. When the R groups on the amino acid residues are small, the protein chains can hydrogen bond in the roughly planar structure shown. Since the bonding around the N and non-carbonyl carbon atoms is tetrahedral, the sheet formed has a pleated structure. In the drawing, the hydrogen bonds are shown by dots.

The way in which protein chains are oriented in three dimensions is determined in large part by hydrogen bonding. Oxygen atoms on C=O groups can interact with H atoms in nearby N—H groups to form these bonds. This can occur within a single protein chain or between neighboring chains.

There are two ways in which a protein chain can be oriented to give maximum hydrogen bonding. When amino acid units with small R groups are present, as in glycine or alanine, hydrogen bonding leads to a pleated-sheet structure (Fig. 28.10). The sheet shown lies in the plane of the paper. It consists of many parallel chains held together by hydrogen bonds between peptide linkages in adjacent chains. By following the middle chain in Figure 28.10, you can see that hydrogen bonding occurs at each amino acid unit. This pleated-sheet structure is found in muscles and silk fibers.

Most proteins have amino acid units with R groups that are too bulky for a pleated-sheet structure. These proteins form a coil called an *α-helix* to maximize hydrogen bonding between peptide linkages. At the left of Figure 28.11 is the

Detailed protein structure can be determined by x-ray crystallography

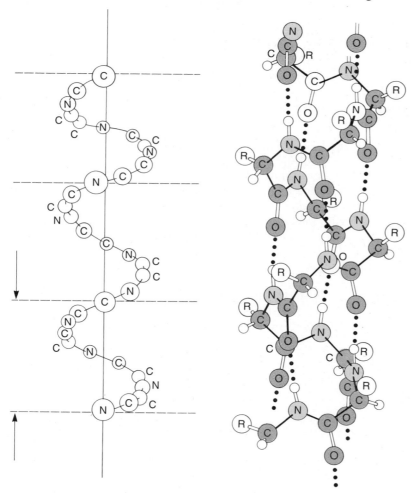

FIGURE 28.11 *α*-helix structure of proteins. The main atom chain in the helix is shown schematically on the left. The sketch on the right more nearly represents the actual positions of atoms and shows were intrachain hydrogen bonding occurs. Wool and many other fibrous proteins have the *α*-helix structure.

outline of the protein chain forming the helix. The more complete structure is given at the right. This shows where the hydrogen bonds form between amino acid units in the protein chain. Notice that the bulky R groups are all on the outside of the helix, where they have the most room. The actual structure of the helix is nearly independent of the nature of the R groups. The dimensions of the helix correspond closely to those observed in such fibrous proteins as wool, hair, skin, feathers, and fingernails.

Summary

Organic compounds may contain halogen, oxygen, and nitrogen atoms in addition to carbon and hydrogen. In such compounds, an atom or small group of atoms serves as a functional group (Example 28.1). Compounds containing the same functional group show similar chemical properties.

Isomerism is common among organic molecules. Isomers have the same molecular formula but differ in their structures and properties. Three types of isomerism were considered in this chapter:

TYPE	DIFFERENCE BETWEEN ISOMERS	EXAMPLE
Structural	bonding pattern of atoms	28.2
Geometric	distance between groups	28.3
Optical	molecules not superimposable	28.4

An organic polymer contains monomer units joined in an extended chain. Addition polymers are made by direct addition of monomer molecules containing double bonds (Table 28.3 and Example 28.5). A condensation polymer is made by splitting out a small molecule such as H_2O between two monomer molecules. Synthetic condensation polymers may be polyesters or polyamides (Example 28.6).

Natural polymers include

—the carbohydrates starch and cellulose, in which the monomer unit is derived from α-glucose (Example 28.7).
—proteins, which can be broken down to, or synthesized from, α-amino acids (Example 28.8).

Key Words and Concepts

addition polymer	condensation polymer	monomer
alcohol	enantiomer	optical isomerism
amine	equatorial	peptide linkage
α-amino acid	ester	polyamide
axial	fat	polyester
carbohydrate	functional group	polymer
carboxylic acid	geometric isomerism	protein
chiral center	isomer	structural isomerism

Questions and Problems

Functional Groups

28.1 Write the molecular formula of DDT (see Table 28.1).

28.2 Classify each of the following as a carboxylic acid, ester, and/or alcohol.
a. $HO—CH_2—CH_2—CH_2—OH$
b.

c.

28.3 Give the structural formula for
a. a four-carbon alcohol with an —OH group not at the end of the chain.
b. a five-carbon carboxylic acid.
c. the ester formed when these two compounds react.

28.4 Write the formula of the acid and the base that react to form calcium propionate.

28.5 Write the structural formula of the ester formed by methyl alcohol with
a. formic acid   b. acetic acid
c. $CH_3—(CH_2)_6—COOH$

28.6 Using three carbon atoms and the required number of hydrogen atoms, write the structural formula for
a. a primary amine.
b. a secondary amine.

28.7 What is the mass percent of nitrogen in caffeine (Fig. 28.1)?

28.31 Write the molecular formula of the dioxin given in Table 28.1.

28.32 Classify each of the following as a carboxylic acid, ester, and/or alcohol.
a. $CH_3—(CH_2)_3—OH$
b.

c.

28.33 Give the structural formula of
a. a three-carbon alcohol with the —OH group on the center carbon.
b. a four-carbon carboxylic acid.
c. the ester formed when these two compounds react.

28.34 Calcium stearate is a principal component of "bathtub ring." Write an acid-base reaction representing its formation.

28.35 Write the structural formulas of all the esters that can be formed by ethylene glycol with formic and acetic acids. (One or both of the —OH groups of ethylene glycol can react.)

28.36 Classify each of the following as a primary, secondary, or tertiary amine:

28.37 What is the molecular formula of nicotine (Fig. 28.1)?

Structural Isomerism

28.8 Draw structural formulas for all the isomers of $C_2H_3Cl_3$.

28.38 Draw structural formulas for all the isomers of $C_2H_4Br_2$.

28.9 How many dibromobenzenes are there?

28.10 Write structural formulas for all the structural isomers of double-bonded compounds with the molecular formula C_5H_{10}.

28.11 Draw Lewis structures for all the alcohols with molecular formula $C_5H_{12}O$.

Geometric and Optical Isomerism

28.12 Of the compounds in Question 28.10, which ones show geometric isomerism? Draw the *cis* and *trans* isomers.

28.13 Maleic acid and fumaric acid are the *cis* and *trans* isomers, respectively, of $C_2H_2(COOH)_2$, a dicarboxylic acid. Draw and label their structural formulas.

28.14 How many chiral carbon atoms are there in glucose? fructose (Fig. 28.7)?

28.15 Locate the chiral carbon(s), if any, in the following molecules:

a.
$$
\begin{array}{c}
\ \ \ \ \ \ \ \ \ \ \ \text{H} \ \ \text{H} \\
\ \ \ \ \ \ \ \ \ \ \ | \ \ \ | \\
\text{HO—C—C—C—C—H} \\
\ \ \ \ \ \ \ \ \| \ \ | \ \ | \ \ \| \\
\ \ \ \ \ \ \ \ \text{O} \ \ \text{OH} \ \text{OH} \ \text{O}
\end{array}
$$

b.
$$
\begin{array}{c}
\text{CH}_3\text{—C—C—OH} \\
\ \ \ \ \ \ \| \ \ \| \\
\ \ \ \ \ \ \text{O} \ \ \text{O}
\end{array}
$$

c.
$$
\begin{array}{c}
\ \ \ \ \ \ \ \ \ \ \ \ \ \text{H} \\
\ \ \ \ \ \ \ \ \ \ \ \ \ | \\
\text{CH}_3\text{—CH}_2\text{—C—COOH} \\
\ \ \ \ \ \ \ \ \ \ \ \ \ | \\
\ \ \ \ \ \ \ \ \ \ \ \ \text{NH}_2
\end{array}
$$

Addition Polymers

28.16 Consider a polymer made from tetrachloroethylene.
 a. Draw a portion of the polymer chain.
 b. What is the molar mass of the polymer if it contains 3.2×10^3 tetrachloroethylene molecules?
 c. What are the mass percents of C and Cl in the polymer?

28.39 How many tribromobenzenes are there?

28.40 Write structural formulas for all the structural isomers of compounds with the molecular formula C_4H_6ClBr in which Cl and Br are bonded to a double-bonded carbon.

28.41 Draw Lewis structures for all the saturated carboxylic acids with four carbon atoms per molecule.

28.42 Of the compounds in Question 28.40, which ones show geometric isomerism? Draw the geometric isomers.

28.43 For which of the following is geometric isomerism possible?
 a. $(CH_3)_2C{=}CCl_2$
 b. $CH_3ClC{=}CCH_3Cl$
 c. $CH_3BrC{=}CCH_3Cl$

28.44 Which of the following amino acids would you expect to show optical activity (Table 28.4)?
 a. glycine   b. alanine
 c. threonine   d. lysine

28.45 Locate the chiral carbon(s), if any, in the following molecules:

a.
$$
\begin{array}{c}
\ \ \ \ \ \ \ \ \ \ \ \text{H} \ \ \text{H} \\
\ \ \ \ \ \ \ \ \ \ \ | \ \ \ | \\
\text{CH}_3\text{—C—C—H} \\
\ \ \ \ \ \ \ \ \ \ \ | \ \ \ | \\
\ \ \ \ \ \ \ \ \ \ \ \text{OH} \ \text{OH}
\end{array}
$$

b.
$$
\begin{array}{c}
\text{H—C}{=}\text{C—CH}_2\text{—OH} \\
\ \ \ \ | \ \ \ | \\
\ \ \ \text{H} \ \ \text{H}
\end{array}
$$

c.
$$
\begin{array}{c}
\ \ \ \ \ \ \ \ \ \text{Cl} \ \ \text{F} \\
\ \ \ \ \ \ \ \ \ | \ \ \ | \\
\text{CH}_3\text{—C—C—Cl} \\
\ \ \ \ \ \ \ \ \ | \ \ \ | \\
\ \ \ \ \ \ \ \ \ \text{Cl} \ \ \text{H}
\end{array}
$$

28.46 Consider Teflon, the polymer made from tetrafluoroethylene.
 a. Draw a portion of the Teflon molecule.
 b. Calculate the molar mass of a Teflon molecule that contains 5.0×10^4 CF_2 units.
 c. What are the mass percents of C and F in Teflon?

28.17 Sketch a portion of the acrylonitrile polymer (Table 28.3) assuming it is a
a. head-to-tail polymer.
b. head-to-head, tail-to-tail polymer

28.18 The polymer whose structure is shown below is made from two different monomers. Identify the monomers.

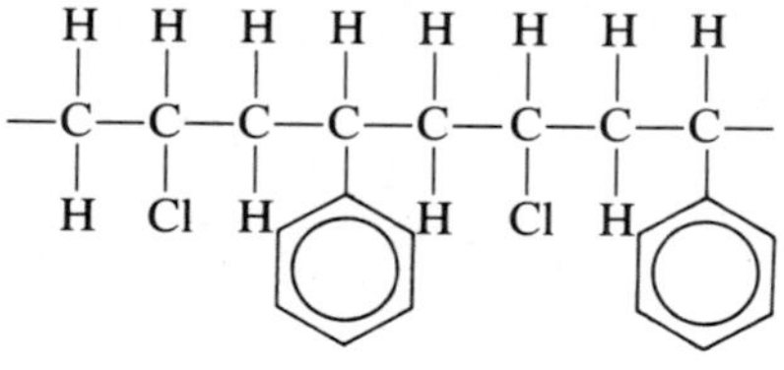

28.47 Styrene, $H_2C=\overset{\overset{\displaystyle H}{|}}{C}$—⬡ , forms a head-to-tail addition polymer. Sketch a portion of a polystryene molecule.

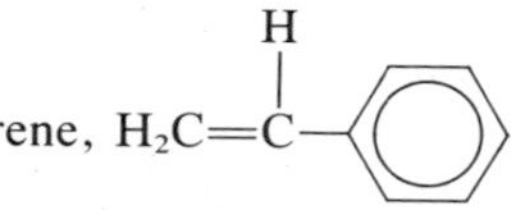

28.48 Show the structure of the monomer used to make the following addition polymers:

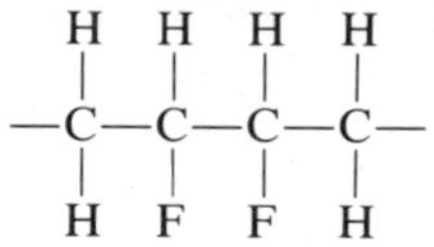

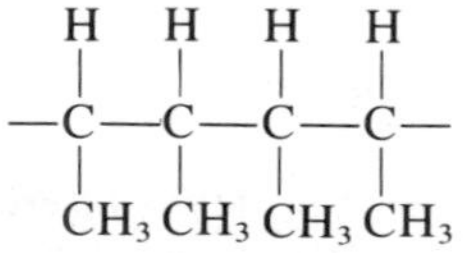

Condensation Polymers

28.19 A rather simple polymer can be made from ethylene glycol, $HO-CH_2-CH_2-OH$, and oxalic acid, $HO-\overset{\overset{\displaystyle O}{||}}{C}-\overset{\overset{\displaystyle O}{||}}{C}-OH$. Sketch a portion of the polymer chain obtained from these monomers.

28.20 *Para*-aminobenzoic acid is an "essential vitamin" for many bacteria:

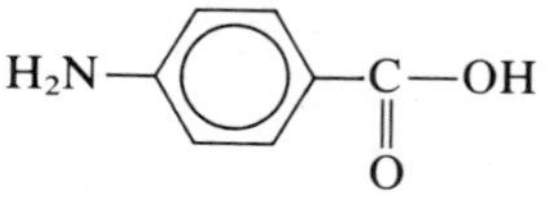

Sketch a portion of a polyamide polymer made from this monomer.

28.21 The following condensation polymer is made from a single monomer. Identify the monomer.

28.49 Lexan is a very rugged polyester in which the monomers can be taken to be carbonic acid, $HO-\overset{\overset{\displaystyle O}{||}}{C}-OH$, and

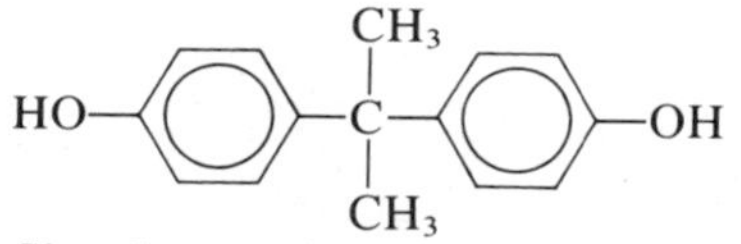

Sketch a section of the Lexan chain.

28.50 Nylon 6 is made from a single monomer:
$H_2N-(CH_2)_5-\overset{\overset{\displaystyle O}{||}}{C}-OH$

Sketch a section of the polymer chain in Nylon 6.

28.51 Identify the monomers from which the following condensation polymers are made:

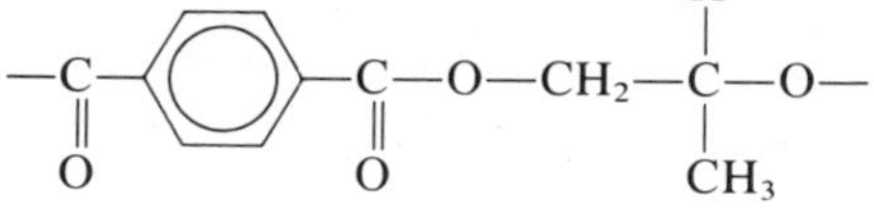

Carbohydrates

28.22 Write a chemical equation, using molecular formulas, for the reaction of sucrose with water to form glucose and fructose.

28.23 Cellulose consists of about 10,000 $C_6H_{10}O_5$ units linked together.
   a. What are the mass percents of C, H, and O in cellulose?
   b. What is the molar mass of cellulose?

28.52 Write a chemical equation, using molecular formulas, for the reaction of maltose with water to form glucose.

28.53 Starch has the same empirical formula as cellulose and a molar mass of about 1.0×10^5 g/mol.
   a. What are the mass percents of C, H, and O in starch?
   b. How many $C_6H_{10}O_5$ units are linked together in a starch molecule?

Amino Acids and Proteins

28.24 Give the structural formulas of two different dimers formed between arginine and serine.

28.25 On complete hydrolysis, a polypeptide gives two alanine, one leucine, one methionine, one phenylalanine, and one valine residue. Partial hydrolysis gives the following fragments: Ala-Phe, Leu-Met, Val-Ala, Phe-Leu. It is known that the first amino acid in the sequence is valine and the last one is methionine. What is the complete sequence of amino acids?

28.26 How would you explain the fact that the β-keratin structure present in silk is much less common in proteins than the α-helix structure?

28.54 Give the structural formulas of two different dimers formed between leucine and lysine.

28.55 Suppose that, in the polypeptide referred to in Question 28.25, the first amino acid is alanine and the last one is also alanine. What is the complete sequence of amino acids?

28.56 In β-keratin, how many hydrogen bonds are present per amino acid residue (Fig. 28.10)?

General

28.27 Give the structural formula of the compound containing the smallest number of carbon atoms that is an
   a. amine b. alcohol c. carboxylic acid d. ester

28.28 Which of the following monomers could form an addition polymer? a condensation polymer?
   a. C_2H_6
   b. C_2H_4
   c. $HO—CH_2—CH_2—OH$
   d. $HO—CH_2—CH_3$

28.29 Explain the difference between
   a. a synthetic and natural polymer.
   b. a polyester and polyamide.
   c. α- and β-glucose.

28.57 What are the mass percents of the elements in the alcohol, carboxylic acid, and ester that contain two carbon atoms per molecule?

28.58 How would you explain to a young science student how to decide whether a given compound might be useful as a monomer for addition polymerization? condensation polymerization?

28.59 Explain the difference between
   a. linear and branched polyethylene.
   b. glucose and fructose.
   c. maltose and sucrose.

28.30 Explain
a. why halogenated hydrocarbons are of environmental concern.
b. why wood alcohol is so named.

28.60 Name and give the formulas of four organic compounds identified as toxic.

***28.61** Heroin is a diester derivative of morphine (Fig. 28.1). It can be considered to be the reaction product of two moles of acetic acid per mole of morphine. What are the mass percents of the elements present in heroin?

***28.62** A certain PCB (Table 28.1) contains 49.4% C, 48.6% Cl and 2.07% H by mass. What is its molecular formula? Draw structural formulas for at least five isomers of this compound.

***28.63** Glycerol and orthophthalic acid, ⬡—COOH, form a cross-linked polymer called an alkyd resin used COOH
in floor coverings and dentures.
a. Write the structural formula for a portion of the polymer chain.
b. Use your answer in (a) to show how cross-linking can occur between the polymer chains to form a water-insoluble, macromolecular solid.

***28.64** Draw structures for all the alcohols with molecular formula $C_6H_{14}O$.

***28.65** Determine the mass percents of the elements in Nylon 66, whose structure is shown in Example 28.6.

Appendix 1
Constants, Reference Data, SI Units

Constants

Acceleration of gravity (standard)	9.8066 m/s^2
Avogadro's number	6.0220×10^{23}
Electronic charge	$1.6022 \times 10^{-19} \text{ C}$
Electronic mass	$9.1095 \times 10^{-28} \text{ g}$
Faraday constant	$9.6485 \times 10^4 \text{ J/V}$
Gas constant	$0.082057 \text{ L·atm/(mol·K)}$
	8.3144 J/(mol·K)
	$8.3144 \times 10^7 \text{ g·cm}^2/(\text{s}^2\text{·mol·K})$
Planck's constant	$6.6262 \times 10^{-34} \text{ J·s}$
Velocity of light	$2.9979 \times 10^8 \text{ m/s}$
π	3.1416
e	2.7183
ln x	$2.3026 \log_{10} x$
2.3026 R	19.145 J/(mol·K)
2.3026 RT (at 25°C)	5.7080 kJ/mol

Basic Equations

CHAPTER	RELATION	EQUATION
1	temperature conversions	$°F = 1.8(°C) + 32°; \ K = °C + 273$
5	heats of formation	$\Delta H = \Sigma \Delta H_f \text{ products} - \Sigma \Delta H_f \text{ reactants}$
	calorimetry	$q_{water} = 4.18 \dfrac{J}{g·°C} \times m \times \Delta t \text{ (coffee-cup)}$
		$q = q_{water} + C\Delta t \quad \text{(bomb)}$
	First Law of Thermodynamics	$\Delta E = q + w$
6	Ideal Gas Law	$PV = nRT$
	Dalton's Law	$P_{tot} = P_A + P_B + \cdots$
	Graham's Law	$\dfrac{time_B}{time_A} = \dfrac{rate_A}{rate_B} = \left(\dfrac{MM_B}{MM_A}\right)^{1/2}$

	molecular velocity	$u \text{ (in cm/s)} = \left(\dfrac{3 \times 8.31 \times 10^7 \times T}{MM} \right)^{1/2}$
7	energy vs. wavelength	$E_{hi} - E_{lo} = \dfrac{1.986 \times 10^{-25}}{\lambda} \dfrac{J \cdot m}{particle}$
		$= \dfrac{1.196 \times 10^5}{\lambda} \dfrac{kJ \cdot nm}{mol}$
	frequency vs. wavelength	$\nu = \dfrac{2.998 \times 10^8}{\lambda} \text{ m/s}$
	Bohr equation	$E_n = \dfrac{-2.179 \times 10^{-18}}{n^2} \text{ J/particle}$
		$= \dfrac{-1312}{n^2} \text{ kJ/mol}$
11	Vapor pressure vs. temperature	$\log_{10} \dfrac{P_2}{P_1} = \dfrac{\Delta H_{vap}}{2.30 \times 8.31} \left(\dfrac{T_2 - T_1}{T_2 T_1} \right)$
	Unit cells	$2r = s$ (simple cubic) $4r = s\sqrt{2}$ (face-centered cubic) $4r = s\sqrt{3}$ (body-centered cubic)
12	Henry's Law	$C_g = kP_g$
	Raoult's Law	$P_1 = X_1 P_1^\circ$
	freezing point lowering	$\Delta T_f = 1.86°C \times m$ (nonelectrolytes in water)
	boiling point elevation	$\Delta T_b = 0.52°C \times m$ (nonelectrolytes in water)
	osmotic pressure	$\pi = \dfrac{nRT}{V}$
14	change in entropy	$\Delta S^0 = \Sigma S^0_{products} - \Sigma S^0_{reactants}$
	change in free energy	$\Delta G^0 = \Delta H^0 - T \Delta S^0$
15	ΔG^0 vs. K	$\Delta G^0 \text{ (in kJ)} =$ $\qquad 2.30 \times 8.31 \times 10^{-3} \, T \log_{10} K$
	K_p vs. K_c	$K_p = K_c(RT)^{\Delta n_g}$
16	first-order rate law	$\log_{10} \dfrac{X_0}{X} = \dfrac{kt}{2.30}; t_{1/2} = 0.693/k$
	rate constant vs. temperature	$\log_{10} \dfrac{k_2}{k_1} = \dfrac{\Delta E_a}{2.30 \times 8.31} \left(\dfrac{T_2 - T_1}{T_2 T_1} \right)$
19	dissociation of water	$K_w = [H^+] \times [OH^-] = 1.0 \times 10^{-14}$
20	relation between K_a and K_b	$K_a \times K_b = K_w$
	reciprocal rule	$K = 1/K'$
	rule of multiple equilibria	$K = K_1 \times K_2 \times \cdots$
24	Nernst equation	$E = E^0_{tot} - \dfrac{0.0591}{n} \log_{10} Q$
	ΔG^0 vs. E^0_{tot}	$\Delta G^0 \text{ (in kJ)} = -96.5 \, n E^0_{tot}$

	K vs. E_{tot}^0	$\log_{10} K = \dfrac{nE_{tot}^0}{0.0591}$
27	rate of decay	$\log_{10} \dfrac{X_0}{X} = \dfrac{kt}{2.30}$; $k = 0.693/t_{1/2}$
	mass-energy	$\Delta E = 9.00 \times 10^{10} \dfrac{kJ}{g} \times \Delta m$

Vapor Pressure of Water (mm Hg)

T(°C)	vp	T(°C)	vp	T(°C)	vp	T(°C)	vp
0	4.58	21	18.65	35	42.2	92	567.0
5	6.54	22	19.83	40	55.3	94	610.9
10	9.21	23	21.07	45	71.9	96	657.6
12	10.52	24	22.38	50	92.5	98	707.3
14	11.99	25	23.76	55	118.0	100	760.0
16	13.63	26	25.21	60	149.4	102	815.9
17	14.53	27	26.74	65	187.5	104	875.1
18	15.48	28	28.35	70	233.7	106	937.9
19	16.48	29	30.04	80	355.1	108	1004.4
20	17.54	30	31.82	90	525.8	110	1074.6

Thermodynamic Data

	ΔH_f (kJ/mol)	S^0 (kJ/K·mol)	ΔG_f^0 (kJ/mol)		ΔH_f (kJ/mol)	S^0 (kJ/K·mol)	ΔG_f^0 (kJ/mol)
$Ag(s)$	0.0	+0.0427	0.0	$CH_3OH(l)$	−238.6	+0.1268	−166.2
$Ag^+(aq)$	+105.9	+0.0739	+77.1	$C_2H_2(g)$	+226.7	+0.2008	+209.2
$AgBr(s)$	−99.5	+0.1071	−95.9	$C_2H_4(g)$	+52.3	+0.2195	+68.1
$AgCl(s)$	−127.0	+0.0961	−109.7	$C_2H_6(g)$	−84.7	+0.2295	−32.9
$AgI(s)$	−62.4	+0.1142	−66.3	$C_3H_8(g)$	−103.8	+0.2699	−23.5
$Ag_2O(s)$	−30.6	+0.1217	−10.8	$n\text{-}C_4H_{10}(g)$	−124.7	+0.3100	−15.7
$Ag_2S(s)$	−31.8	+0.1456	−40.3	$n\text{-}C_5H_{12}(l)$	−173.1	+0.2628	−9.4
$Al(s)$	0.0	+0.0283	0.0	$C_2H_5OH(l)$	−277.6	+0.1607	−174.8
$Al^{3+}(aq)$	−524.7	−0.3134	−481.1	$CO(g)$	−110.5	+0.1979	−137.3
$Al_2O_3(s)$	−1669.8	+0.0510	−1576.4	$CO_2(g)$	−393.5	+0.2136	−394.4
$Ba(s)$	0.0	+0.067	0.0	$CO_3^{2-}(aq)$	−676.3	−0.0531	−528.1
$Ba^{2+}(aq)$	−538.4	+0.013	−560.7	$Ca(s)$	0.0	+0.0416	0.0
$BaCl_2(s)$	−860.1	+0.126	−810.9	$Ca^{2+}(aq)$	−543.0	−0.0552	−553.0
$BaCO_3(s)$	−1218.8	+0.1121	−1138.9	$CaCl_2(s)$	−795.0	+0.1138	−750.2
$BaO(s)$	−558.1	+0.0703	−528.4	$CaCO_3(s)$	−1207.0	+0.0929	−1128.8
$BaSO_4(s)$	−1465.2	+0.1322	−1353.1	$CaO(s)$	−635.5	+0.0397	−604.2
$Br_2(l)$	0.0	+0.1523	0.0	$Ca(OH)_2(s)$	−986.6	+0.0761	−896.8
$Br^-(aq)$	−120.9	+0.0807	−102.8	$CaSO_4(s)$	−1432.7	+0.1067	−1320.3
$C(s)$	0.0	+0.0057	0.0	$Cl_2(g)$	0.0	+0.2229	0.0
$CCl_4(l)$	−139.5	+0.2144	−68.6	$Cl^-(aq)$	−167.4	+0.0551	−131.2
$CH_4(g)$	−74.8	+0.1862	−50.8	$ClO_3^-(aq)$	−98.3	+0.1623	−2.6
$CHCl_3(l)$	−131.8	+0.2029	−71.5	$ClO_4^-(aq)$	−131.4	+0.1820	−8

	ΔH_f (kJ/mol)	S^0 (kJ/K·mol)	ΔG_f^0 (kJ/mol)		ΔH_f (kJ/mol)	S^0 (kJ/K·mol)	ΔG_f^0 (kJ/mol)
$Cr(s)$	0.0	+0.0238	0.0	$MgO(s)$	−601.8	+0.0268	−569.6
$CrO_4^{2-}(aq)$	−863.2	+0.0385	−736.8	$Mg(OH)_2(s)$	−924.7	+0.0631	−833.7
$Cr_2O_3(s)$	−1128.4	+0.0812	−1046.8	$MgSO_4(s)$	−1278.2	+0.0916	−1164.1
$Cu(s)$	0.0	+0.0333	0.0	$Mn(s)$	0.0	+0.0318	0.0
$Cu^{2+}(aq)$	+64.4	−0.0987	+65.0	$Mn^{2+}(aq)$	−218.8	−0.0736	−223.4
$CuO(s)$	−155.2	+0.0435	−127.7	$MnO(s)$	−384.9	+0.0602	−363.2
$Cu_2O(s)$	−166.7	+0.1008	−146.4	$MnO_2(s)$	−519.7	+0.0531	−466.1
$CuS(s)$	−48.5	+0.0665	−49.0	$MnO_4^-(aq)$	−518.4	+0.1900	−425.1
$CuSO_4(s)$	−769.9	+0.1134	−661.9	$N_2(g)$	0.0	+0.1915	0.0
$F_2(g)$	0.0	+0.2033	0.0	$NH_3(g)$	−46.2	+0.1925	−16.6
$F^-(aq)$	−329.1	−0.0096	−276.5	$NH_4^+(aq)$	−132.8	+0.1128	−79.5
$Fe(s)$	0.0	+0.0272	0.0	$NH_4Cl(s)$	−315.4	+0.0946	−203.9
$Fe^{2+}(aq)$	−87.9	−0.1134	−84.9	$NH_4NO_3(s)$	−365.1	+0.1510	−183.6
$Fe^{3+}(aq)$	−47.7	−0.2933	−10.6	$NO(g)$	+90.4	+0.2106	+86.7
$Fe_2O_3(s)$	−822.2	+0.0900	−741.0	$NO_2(g)$	+33.9	+0.2405	+51.8
$Fe_3O_4(s)$	−1120.9	+0.1464	−1018.0	$NO_3^-(aq)$	−206.6	+0.1464	−110.6
$H_2(g)$	0.0	+0.1306	0.0	$Na(s)$	0.0	+0.0510	0.0
$H^+(aq)$	0.0	0.0	0.0	$Na^+(aq)$	−239.7	+0.0602	−261.9
$HBr(g)$	−36.2	+0.1985	−53.2	$NaCl(s)$	−411.0	+0.0724	−384.0
$HCl(g)$	−92.3	+0.1867	−95.3	$NaF(s)$	−569.0	+0.0586	−541.0
$HCO_3^-(aq)$	−691.1	+0.0950	−587.1	$NaOH(s)$	−426.7	+0.0523	−377.1
$HF(g)$	−268.6	+0.1735	−270.7	$Ni(s)$	0.0	+0.0301	0.0
$HI(g)$	+25.9	+0.2063	+1.3	$Ni^{2+}(aq)$	−64.0	−0.1594	−48.2
$HNO_3(l)$	−173.2	+0.1556	−79.9	$NiO(s)$	−244.3	+0.0386	−216.3
$H_2O(g)$	−241.8	+0.1887	−228.6	$O_2(g)$	0.0	+0.2050	0.0
$H_2O(l)$	−285.8	+0.0699	−237.2	$OH^-(aq)$	−229.9	−0.0105	−157.3
$H_2O_2(l)$	−187.6	+0.0886	−114.0	$P_4(s)$	0.0	+0.1774	0.0
$HPO_4^{2-}(aq)$	−1298.7	−0.0360	−1094	$PCl_3(g)$	−306.4	+0.3117	−286.3
$H_2PO_4^-(aq)$	−1302.5	+0.0891	−1135	$PCl_5(g)$	−398.9	+0.3527	−324.6
$H_2S(g)$	−20.1	+0.2056	−33.0	$PO_4^{3-}(aq)$	−1284.1	−0.218	−1025
$H_2SO_4(l)$	−811.3	+0.1569	−687.5	$Pb(s)$	0.0	+0.0649	0.0
$Hg(l)$	0.0	+0.0774	0.0	$Pb^{2+}(aq)$	+1.6	+0.0213	−24.3
$HgO(s)$	−90.7	+0.0720	−58.5	$PbBr_2(s)$	−277.0	+0.1615	−260.4
$HgS(s)$	−58.2	+0.0778	−48.8	$PbCl_2(s)$	−359.2	+0.1364	−314.0
$I_2(s)$	0.0	+0.1167	0.0	$PbO(s)$	−217.9	+0.0695	−188.5
$I^-(aq)$	−55.9	+0.1094	−51.7	$PbO_2(s)$	−276.6	+0.0766	−219.0
$K(s)$	0.0	+0.0636	0.0	$Pb_3O_4(s)$	−734.7	+0.2113	−617.6
$K^+(aq)$	−251.2	+0.1025	−282.3	$S(s)$	0.0	+0.0319	0.0
$KBr(s)$	−392.2	+0.0964	−379.2	$S^{2-}(aq)$	+41.8	+0.0222	+83.7
$KCl(s)$	−435.9	+0.0827	−408.3	$SO_2(g)$	−296.1	+0.2485	−300.4
$KClO_3(s)$	−391.4	+0.1430	−289.9	$SO_3(g)$	−395.2	+0.2562	−370.4
$KF(s)$	−562.6	+0.0666	−533.1	$SO_4^{2-}(aq)$	−907.5	+0.0172	−742.0
$Mg(s)$	0.0	+0.0325	0.0	$Si(s)$	0.0	+0.0187	0.0
$Mg^{2+}(aq)$	−462.0	−0.1180	−456.0	$SiO_2(s)$	−859.4	+0.0418	−805.0
$MgCl_2(s)$	−641.8	+0.0895	−592.3	$Sn(s)$	0.0	+0.0515	0.0
$MgCO_3(s)$	−1113	+0.0657	−1029	$Sn^{2+}(aq)$	−10.0	−0.0247	−26.2

	ΔH_f (kJ/mol)	S^0 (kJ/K·mol)	ΔG_f^0 (kJ/mol)		ΔH_f (kJ/mol)	S^0 (kJ/K·mol)	ΔG_f^0 (kJ/mol)
$SnCl_2(s)$	−349.8	+0.1226	−304.6	$Zn(s)$	0.0	+0.0416	0.0
$SnCl_4(l)$	−545.2	+0.2586	−474.0	$Zn^{2+}(aq)$	−152.4	−0.1065	−147.2
$SnO(s)$	−286.2	+0.0565	−257.3	$ZnO(s)$	−348.0	+0.0439	−318.2
$SnO_2(s)$	−580.7	+0.0523	−519.7	$ZnS(s)$	−202.9	+0.0577	−198.3

Bond Energies (kJ/mol)

Br—Br	193	C—H	414	Cl—N	201	H—S	339
Br—C	276	C—I	218	Cl—O	205	I—I	151
Br—Cl	218	C—N	293	Cl—S	255	I—O	201
Br—F	255	C≡N	615	F—F	153	N—N	159
Br—H	368	C≡N	890	F—H	565	N═N	418
Br—I	180	C—O	351	F—I	277	N≡N	941
Br—N	243	C═O	715	F—N	272	N—O	222
Br—O	201	C≡O	1075	F—O	184	N═O	607
Br—S	213	C—S	259	F—S	285	O—O	138
C—C	347	C═S	477	H—H	436	O═O	498
C═C	612	Cl—Cl	243	H—I	297	O—S	347
C≡C	820	Cl—F	255	H—N	389	O═S	498
C—Cl	331	Cl—H	431	H—O	464	S—S	226
C—F	485	Cl—I	209				

Equilibrium Constants

SOLUBILITY CONSTANTS, K_{sp}

AgBr	5×10^{-13}	$CaSO_4$	3×10^{-5}	MnS	1×10^{-15}
Ag_2CO_3	1×10^{-11}	$CdCO_3$	5×10^{-12}	$NiCO_3$	1.4×10^{-7}
$AgC_2H_3O_2$	2×10^{-3}	$Cd(OH)_2$	2×10^{-14}	$Ni(OH)_2$	1×10^{-16}
AgCl	1.6×10^{-10}	CdS	1×10^{-26}	NiS	1×10^{-19}
Ag_2CrO_4	2×10^{-12}	$Co(OH)_2$	2×10^{-16}	$PbBr_2$	5×10^{-6}
AgI	1×10^{-16}	CoS	1×10^{-20}	$PbCO_3$	1×10^{-13}
Ag_3PO_4	1×10^{-15}	$Cr(OH)_3$	4×10^{-38}	$PbCl_2$	1.7×10^{-5}
Ag_2S	1×10^{-49}	$CuCO_3$	2×10^{-10}	$PbCrO_4$	1×10^{-16}
$Al(OH)_3$	5×10^{-33}	$Cu(OH)_2$	2×10^{-19}	PbF_2	4×10^{-8}
$BaCO_3$	2×10^{-9}	CuS	1×10^{-35}	PbI_2	1×10^{-8}
BaC_2O_4	1.5×10^{-8}	$Fe(OH)_2$	1×10^{-15}	$Pb(OH)_2$	4×10^{-15}
$BaCrO_4$	2×10^{-10}	$Fe(OH)_3$	5×10^{-38}	$Pb_3(PO_4)_2$	1×10^{-54}
BaF_2	2×10^{-6}	FeS	1×10^{-17}	PbS	1×10^{-27}
$BaSO_4$	1.4×10^{-9}	Hg_2Cl_2	1×10^{-18}	$PbSO_4$	1×10^{-8}
$CaCO_3$	5×10^{-9}	HgS	1×10^{-52}	$SrCO_3$	7×10^{-10}
CaC_2O_4	1.3×10^{-9}	$MgCO_3$	2×10^{-8}	$SrCrO_4$	4×10^{-5}
CaF_2	2×10^{-10}	MgC_2O_4	8.6×10^{-5}	$ZnCO_3$	2×10^{-10}
$Ca(OH)_2$	1.3×10^{-6}	$Mg(OH)_2$	1×10^{-11}	$Zn(OH)_2$	5×10^{-17}
$Ca_3(PO_4)_2$	1×10^{-33}	$Mn(OH)_2$	4×10^{-14}	ZnS	1×10^{-20}

DISSOCIATION CONSTANTS, WEAK ACIDS, K_a

$Co(H_2O)_6^{3+}$	2×10^{-2}	$HC_3H_5O_2$	1.4×10^{-5}	H_3BO_3	5.8×10^{-10}
H_2SO_3	1.7×10^{-2}	H_2CO_3	4.2×10^{-7}	NH_4^+	5.6×10^{-10}
HSO_4^-	1.2×10^{-2}	$Fe(H_2O)_6^{3+}$	2×10^{-7}	HCN	4.0×10^{-10}
H_3PO_4	7.5×10^{-3}	H_2S	1×10^{-7}	$Co(H_2O)_6^{2+}$	1×10^{-10}
$Fe(H_2O)_6^{3+}$	2×10^{-3}	$H_2PO_4^-$	6.2×10^{-8}	$Ni(H_2O)_6^{2+}$	1×10^{-10}
HF	7.0×10^{-4}	HSO_3^-	5.6×10^{-8}	HCO_3^-	4.8×10^{-11}
HNO_2	4.5×10^{-4}	$HClO$	3.2×10^{-8}	$Mn(H_2O)_6^{2+}$	3×10^{-11}
$HCHO_2$	1.8×10^{-4}	$Cu(H_2O)_4^{2+}$	3×10^{-8}	HIO	2×10^{-11}
$Cr(H_2O)_6^{3+}$	1×10^{-4}	$HBrO$	2×10^{-9}	HPO_4^{2-}	1.7×10^{-12}
$HC_7H_5O_2$	6.6×10^{-5}	$Zn(H_2O)_4^{2+}$	1×10^{-9}	HS^-	1×10^{-13}
$HC_2H_3O_2$	1.8×10^{-5}	$Cd(H_2O)_4^{2+}$	1×10^{-9}		

DISSOCIATION CONSTANTS, WEAK BASES, K_b

S^{2-}	1×10^{-1}	ClO^-	3.1×10^{-7}	$C_7H_5O_2^-$	1.5×10^{-10}
PO_4^{3-}	5.9×10^{-3}	SO_3^{2-}	1.8×10^{-7}	CHO_2^-	5.6×10^{-11}
IO^-	5×10^{-4}	HPO_4^{2-}	1.6×10^{-7}	NO_2^-	2.2×10^{-11}
CO_3^{2-}	2.1×10^{-4}	HS^-	1×10^{-7}	F^-	1.4×10^{-11}
CN^-	2.5×10^{-5}	HCO_3^-	2.4×10^{-8}	$H_2PO_4^-$	1.3×10^{-12}
NH_3	1.8×10^{-5}	$C_3H_5O_2^-$	7.1×10^{-10}	SO_4^{2-}	8.3×10^{-13}
$H_2BO_3^-$	1.7×10^{-5}	$C_2H_3O_2^-$	5.6×10^{-10}	HSO_3^-	5.9×10^{-13}
BrO^-	5×10^{-6}				

FORMATION CONSTANTS, COMPLEX IONS, K_f

$Ag(CN)_2^-$	1×10^{21}	$Co(NH_3)_6^{2+}$	1×10^5	$Fe(CN)_6^{3-}$	1×10^{42}
$AgCl_2^-$	1×10^5	$Co(NH_3)_6^{3+}$	1×10^{35}	$Hg(CN)_4^{2-}$	1×10^{41}
$Ag(NH_3)_2^+$	2×10^7	$Cr(OH)_4^-$	1×10^{30}	$Hg(NH_3)_4^{2+}$	2×10^{19}
$Ag(S_2O_3)_2^{3-}$	1×10^{13}	$Cu(CN)_4^{2-}$	1×10^{30}	$Ni(CN)_4^{2-}$	1×10^{31}
$Al(OH)_4^-$	1×10^{33}	$CuCl_2^-$	1×10^5	$Ni(NH_3)_6^{2+}$	5×10^8
$Au(CN)_2^-$	1×10^{38}	$Cu(NH_3)_2^+$	5×10^{10}	$Pb(OH)_3^-$	1×10^{14}
$Cd(CN)_4^{2-}$	1×10^{19}	$Cu(NH_3)_4^{2+}$	5×10^{12}	$Zn(CN)_4^{2-}$	1×10^{17}
$CdCl_4^{2-}$	1×10^3	$Cu(OH)_4^{2-}$	3×10^{18}	$Zn(NH_3)_4^{2+}$	1×10^9
$Cd(NH_3)_4^{2+}$	1×10^7	$Fe(CN)_6^{4-}$	1×10^{35}	$Zn(OH)_4^{2-}$	3×10^{15}

SI Units

Base Units

The International System of Units or *Système International* (SI), which represents an extension of the metric system, was adopted by the 11th General Conference of Weights and Measures in 1960. It is constructed from seven base units, each of which represents a particular physical quantity (Table I).

Table I
SI Base Units

PHYSICAL QUANTITY	NAME OF UNIT	SYMBOL
Length	metre	m
Mass	kilogram	kg
Time	second	s
Temperature	kelvin	K
Amount of substance	mole	mol
Electric current	ampere	A
Luminous intensity	candela	cd

Of the seven units listed in Table I, the first five are particularly useful in general chemistry. They are defined as follows:

1. The *metre* was redefined in 1983 to be equal to the distance light travels in a vacuum in 1/299 792 458 second.

2. The *kilogram* represents the mass of a platinum-iridium block kept at the International Bureau of Weights and Measures at Sevres, France.

3. The *second* was redefined in 1967 as the duration of 9 192 631 770 periods of a certain line in the microwave spectrum of cesium-133.

4. The *kelvin* is 1/273.16 of the temperature interval between the absolute zero and the triple point of water (0.01°C = 273.16 K).

5. The *mole* is the amount of substance that contains as many entities as there are atoms in exactly 0.012 kg of carbon-12.

Prefixes Used with SI Units

Decimal fractions and multiples of SI units are designated by using the prefixes listed in Table II. Those most commonly used in general chemistry are underlined.

Table II
SI Prefixes

FACTOR	PREFIX	SYMBOL	FACTOR	PREFIX	SYMBOL
10^{12}	tera	T	10^{-1}	deci	d
10^{9}	giga	G	10^{-2}	centi	c
10^{6}	mega	M	10^{-3}	milli	m
10^{3}	kilo	k	10^{-6}	micro	μ
10^{2}	hecto	h	10^{-9}	nano	n
10^{1}	deca	da	10^{-12}	pico	p
			10^{-15}	femto	f
			10^{-18}	atto	a

Derived Units

In the International System of Units, all physical quantities are represented by appropriate combinations of the base units listed in Table I. To choose a particularly simple example, the SI unit for volume, the cubic metre, represents the volume of a cube one metre on an edge. Again, in SI, the density of a substance can be expressed by dividing its mass in kilograms by its volume in cubic metres. A list of the derived units most frequently used in general chemistry is given in Table III.

Table III
SI Derived Units

Physical Quantity	Name of Unit	Symbol	Definition
Area	square metre	m^2	
Volume	cubic metre	m^3	
Density	kilogram per cubic metre	kg/m^3	
Force	newton	N	$kg \cdot m/s^2$
Pressure	pascal	Pa	N/m^2
Energy	joule	J	$kg \cdot m^2/s^2$
Electric charge	coulomb	C	$A \cdot s$
Electric potential difference	volt	V	$J/(A \cdot s)$

Perhaps the least familiar of these units to the beginning chemistry student are the ones used to represent force, pressure, and energy.

The *newton* is defined as the force required to impart an acceleration of one metre per second squared to a mass of one kilogram (recall that Newton's second law can be stated as force = mass × acceleration).

The *pascal* is defined as the pressure exerted by a force of one newton acting upon an area of one square metre (recall that pressure = force/area). Commonly, pressures are expressed in kilopascals:

$$1 \text{ kPa} = 10^3 \text{ Pa}$$

Typically, atmospheric pressure near sea level is in the vicinity of 100 kPa.

The **joule** is defined as the work done when a force of one newton ($kg \cdot m/s^2$) acts through a distance of one metre (recall that work = force × distance). Commonly, energies are expressed in kilojoules:

$$1 \text{ kJ} = 10^3 \text{ J}$$

In terms of more familiar units, a kilowatt-hour is 3600 kJ; a kilocalorie is 4.184 kJ.

Conversions Between SI and Other Units

Table IV lists conversion factors for translating units from other systems to the International System.

Table IV
Conversion Factors

QUANTITY	SI UNIT	OTHER UNIT	
Area	m^2	ft^2	$1 \; ft^2 = 0.929 \; 030 \; 4 \; m^2$
		acre	$1 \; acre = 4.046 \; 856 \times 10^3 \; m^2$
		cm^2	$1 \; cm^2 = 10^{-4} \; m^2$
		hectare	$1 \; hectare = 10^4 \; m^2$
Density	kg/m^3	g/cm^3	$1 \; g/cm^3 = 10^3 \; kg/m^3$
		lb/ft^3	$1 \; lb/ft^3 = 16.018 \; 46 \; kg/m^3$
Electric charge	coulomb (C)	mole electrons	$1 \; mol \; e^- = 9.6485 \times 10^4 \; C$
Electric potential	volt (V)	joule/coulomb	$1 \; V = 1 \; J/C$
Energy	joule (J)	calorie	$1 \; cal = 4.184 \; J$
		L·atm	$1 \; L{\cdot}atm = 101.3 \; J$
		erg	$1 \; erg = 10^{-7} \; J$
		kilowatt-hour	$1 \; kWh = 3.6 \times 10^6 \; J$
		BTU	$1 \; BTU = 1.055 \times 10^3 \; J$
Entropy	J/K	cal/K	$1 \; cal/K = 4.184 \; J/K$
Force	newton (N)	dyne	$1 \; dyn = 10^{-5} \; N$
Frequency	hertz (Hz)	cycle/second	$1 \; Hz = 1 \; cycle/s$
Length	metre (m)	inch	$1 \; in = 0.0254 \; m$
		mile	$1 \; mile = 1.609 \; 344 \; km$
		angstrom	$1 \; Å = 10^{-10} \; m = 10^{-1} \; nm$
		micron	$1 \; micron = 10^{-6} \; m$
Mass	kilogram (kg)	pound	$1 \; lb = 0.453 \; 592 \; 37 \; kg$
		metric ton (t)	$1 \; t = 10^3 \; kg$
		ton (short)	$1 \; ton = 2000 \; lb = 9.071 \; 847 \; 4 \times 10^2 \; kg$
Power	watt (w)	joule/second	$1 \; W = 1 \; J/s$
Pressure	pascal (Pa)	atmosphere	$1 \; atm = 101.325 \; kPa$
		bar	$1 \; bar = 10^5 \; Pa$
		mm Hg	$1 \; mm \; Hg = 133.322 \; Pa$
		lb/in^2	$1 \; lb/in^2 = 6.894 \; 757 \; kPa$
		torr	$1 \; torr = 133.322 \; Pa$
Temperature*	kelvin (K)	Celsius degree	$1°C = 1 \; K$
		Fahrenheit degree	$1°F = 5/9 \; K$
Surface tension	N/m	dynes/cm	$1 \; dyn/cm = 10^{-3} \; N/m$
Volume	m^3	litre	$1 \; L = 1 \; dm^3 = 10^{-3} \; m^3$
		cm^3	$1 \; cm^3 = 1 \; mL = 10^{-6} \; m^3$
		ft^3	$1 \; ft^3 = 28.316 \; 85 \; dm^3$
		gallon (US)	$1 \; gal = 4 \; qt = 3.785 \; 412 \; dm^3$

*Temperatures in degrees Celsius (t_c) or degrees Fahrenheit (t_f) can be converted to temperatures in Kelvin (T) by using the equations

$$t_c = T - 273.15; \qquad t_f = \frac{9}{5}T - 459.67$$

Appendix 2
Properties of
The Elements

ELEMENT	AT. NO.	mp(°C)	bp(°C)	E.N.	ION. ENER. (kJ/mol)	AT. RAD. (nm)	ION. RAD. (nm)
H	1	−259	−253	2.1	1312	0.037	(−1)0.208
He	2	−272	−269		2372	0.05	
Li	3	186	1326	1.0	520	0.152	(+1)0.060
Be	4	1283	2970	1.5	900	0.111	(+2)0.031
B	5	2300	2550	2.0	801	0.088	
C	6	3570	subl.	2.5	1086	0.077	
N	7	−210	−196	3.0	1402	0.070	
O	8	−218	−183	3.5	1314	0.066	(−2)0.140
F	9	−220	−188	4.0	1681	0.064	(−1)0.136
Ne	10	−249	−246		2081	0.070	
Na	11	98	889	0.9	496	0.186	(+1)0.095
Mg	12	650	1120	1.2	738	0.160	(+2)0.065
Al	13	660	2327	1.5	578	0.143	(+3)0.050
Si	14	1414	2355	1.8	786	0.117	
P	15	44	280	2.1	1012	0.110	
S	16	119	444	2.5	1000	0.104	(−2)0.184
Cl	17	−101	−34	3.0	1251	0.099	(−1)0.181
Ar	18	−189	−186		1520	0.094	
K	19	64	774	0.8	419	0.231	(+1)0.133
Ca	20	845	1420	1.0	590	0.197	(+2)0.099
Sc	21	1541	2831	1.3	631	0.160	(+3)0.081
Ti	22	1660	3287	1.5	658	0.146	
V	23	1890	3380	1.6	650	0.131	
Cr	24	1857	2672	1.6	653	0.125	(+3)0.064
Mn	25	1244	1962	1.5	717	0.129	(+2)0.080
Fe	26	1535	2750	1.8	759	0.126	(+2)0.075
Co	27	1495	2870	1.9	758	0.125	(+2)0.072
Ni	28	1453	2732	1.9	737	0.124	(+2)0.069
Cu	29	1083	2567	1.9	746	0.128	(+1)0.096
Zn	30	420	907	1.6	906	0.133	(+2)0.074
Ga	31	30	2403	1.6	579	0.122	(+3)0.062
Ge	32	937	2830	1.8	762	0.122	
As	33	814	subl.	2.0	944	0.121	

Element	At. No.	mp(°C)	bp(°C)	E.N.	Ion. Ener. (kJ/mol)	At. Rad. (nm)	Ion. Rad. (nm)
Se	34	217	685	2.4	941	0.117	(−2)0.198
Br	35	−7	59	2.8	1140	0.114	(−1)0.195
Kr	36	−157	−152		1351	0.109	
Rb	37	39	688	0.8	403	0.244	(+1)0.148
Sr	38	770	1380	1.0	550	0.215	(+2)0.113
Y	39	1509	2930	1.2	616	0.180	(+3)0.093
Zr	40	1852	3580	1.4	660	0.157	
Nb	41	2468	5127	1.6	664	0.143	
Mo	42	2610	5560	1.8	685	0.136	
Tc	43	2200	4700	1.9	702	0.136	
Ru	44	2430	3700	2.2	711	0.133	
Rh	45	1966	3700	2.2	720	0.134	
Pd	46	1550	3170	2.2	805	0.138	
Ag	47	961	2210	1.9	731	0.144	(+1)0.126
Cd	48	321	767	1.7	868	0.149	(+2)0.097
In	49	157	2000	1.7	558	0.162	(+3)0.081
Sn	50	232	2270	1.8	709	0.140	
Sb	51	631	1380	1.9	832	0.141	
Te	52	450	990	2.1	869	0.137	(−2)0.221
I	53	114	184	2.5	1009	0.133	(−1)0.216
Xe	54	−112	−107		1170	0.130	
Cs	55	28	690	0.7	376	0.262	(+1)0.169
Ba	56	725	1640	0.9	503	0.217	(+2)0.135
La	57	920	3469	1.1	538	0.187	(+3)0.115
Ce	58	795	3468	1.1	528	0.182	(+3)0.101
Pr	59	935	3127	1.1	523	0.182	(+3)0.100
Nd	60	1024	3027	1.1	530	0.182	(+3)0.099
Pm	61	1027	2727	1.1	536	0.181	
Sm	62	1072	1900	1.1	543	0.180	
Eu	63	826	1439	1.1	547	0.204	(+2)0.097
Gd	64	1312	3000	1.1	592	0.179	(+3)0.096
Tb	65	1356	2800	1.1	564	0.177	(+3)0.095
Dy	66	1407	2600	1.1	572	0.177	(+3)0.094
Ho	67	1461	2600	1.1	581	0.176	(+3)0.093
Er	68	1497	2900	1.1	589	0.175	(+3)0.092
Tm	69	1356	2800	1.1	597	0.174	(+3)0.091
Yb	70	824	1427	1.1	603	0.193	(+3)0.089
Lu	71	1652	3327	1.1	524	0.174	(+3)0.089
Hf	72	2225	5200	1.3	654	0.157	
Ta	73	2980	5425	1.5	761	0.143	
W	74	3410	5930	1.7	770	0.137	
Re	75	3180	5885	1.9	760	0.137	
Os	76	2727	4100	2.2	840	0.134	
Ir	77	2448	4500	2.2	880	0.135	
Pt	78	1769	4530	2.2	870	0.138	
Au	79	1063	2966	2.4	890	0.144	(+1)0.137
Hg	80	−39	357	1.9	1007	0.155	(+2)0.110

Element	At. No.	mp(°C)	bp(°C)	E.N.	Ion. Ener. (kJ/mol)	At. Rad. (nm)	Ion. Rad. (nm)
Tl	81	304	1457	1.8	589	0.171	(+3)0.095
Pb	82	328	1750	1.9	716	0.175	
Bi	83	271	1560	1.9	703	0.146	
Po	84	254	962	2.0	812	0.165	
At	85	302	334	2.2			
Rn	86	−71	−62		1037	0.14	
Fr	87	27	677	0.7			
Ra	88	700	1140	0.9	509	0.220	
Ac	89	1050	3200	1.1	490	0.20	
Th	90	1750	4790	1.3	590	0.180	
Pa	91	1600	4200	1.4	570		
U	92	1132	3818	1.4	590	0.14	

The rules used to name simple inorganic compounds are discussed in Chapter 3 (ionic and binary molecular compounds) and Chapter 26 (oxyanions and oxyacids). Here we will consider the systems used to name

—coordination compounds (Chapter 21).
—hydrocarbons (Chapter 13).

Coordination Compounds

The nomenclature of compounds containing complex ions (Chap. 21) is perhaps more involved than that of any other type of inorganic compound. Several rules are required, the more pertinent of which are as follows:

1. As in simple ionic compounds, the cation is named first, followed by the anion.

2. If there is more than one ligand of a particular type attached to the central atom, Greek prefixes are used to indicate the number of these ligands. Where the name of the ligand itself is complex (e.g., ethylenediamine), the number of such ligands is indicated by the prefixes *bis-* or *tris-* instead of *di-* or *tri-* and the name of the ligand is enclosed in parentheses.

3. In naming a complex ion, the names of anionic ligands are written first, followed by those of neutral ligands, and finally that of the central metal atom. This is exactly the reverse of the order in which the groups are listed in the formula of the complex ion: the symbol of the central atom is written first, followed by the formulas of neutral ligands, and then those of negatively charged ligands. In the formula of a coordination compound, the complex ion is often set off by brackets.

4. The names of anionic ligands are modified by substituting the suffix *-o* for the usual ending. Thus, we have

Cl^-	chloro	CO_3^{2-}	carbonato
OH^-	hydroxo	CN^-	cyano

The names of neutral ligands are ordinarily not changed. Two important exceptions are

H_2O	aquo	NH_3	ammine

5. The charge of the metal ion is indicated by a Roman numeral following the name of the metal. If the complex is an anion, the suffix *-ate* is added, often to the Latin stem of the name of the metal. Applying these rules,

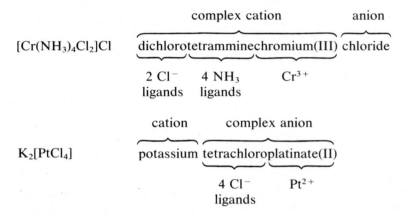

Other examples are

$[Co(NH_3)_6]Cl_3$ hexamminecobalt(III) chloride
$[Co(en)_3](NO_3)_3$ tris(ethylenediamine)cobalt(III) nitrate
$[Pt(H_2O)_3Cl]Br$ chlorotriaquoplatinum(II) bromide
$K_3[Fe(CN)_6]$ potassium hexacyanoferrate(III)
$K_4[Fe(CN)_6]$ potassium hexacyanoferrate(II)

Hydrocarbons

As organic chemistry developed, it became apparent that some systematic way of naming compounds was needed. About 50 years ago, the International Union of Pure and Applied Chemistry (IUPAC) devised a system that could be used for all organic compounds. To illustrate this system, we will show how it works with four different types of hydrocarbons considered in Chapter 13.

Alkanes

For straight-chain alkanes such as

$$CH_3—CH_2—CH_3 \qquad CH_3—CH_2—CH_2—CH_3$$
propane   butane

the IUPAC name consists of a single word. These names, for up to eight carbon atoms, are listed in Table I.

With alkanes containing a branched chain, such as

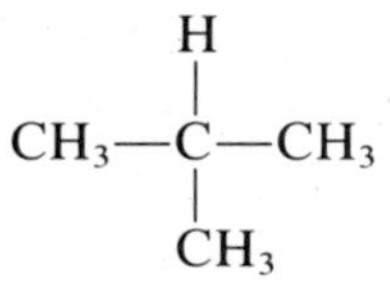

$$\begin{array}{c} H \\ | \\ CH_3—C—CH_3 \\ | \\ CH_3 \end{array}$$

2-methylpropane

Table 1
Nomenclature of Alkanes

STRAIGHT-CHAIN ALKANES		ALKYL GROUPS	
Methane	CH_4	Methyl	CH_3-
Ethane	CH_3CH_3	Ethyl	CH_3-CH_2-
Propane	$CH_3CH_2CH_3$	Propyl	$CH_3-CH_2-CH_2-$
Butane	$CH_3(CH_2)_2CH_3$		
Pentane	$CH_3(CH_2)_3CH_3$	Isopropyl	
Hexane	$CH_3(CH_2)_4CH_3$		
Heptane	$CH_3(CH_2)_5CH_3$		
Octane	$CH_3(CH_2)_6CH_3$		
		Butyl	$CH_3-CH_2-CH_2-CH_2-$

Isopropyl group:

$$CH_3-\overset{\displaystyle H}{\underset{\displaystyle CH_3}{C}}-$$

the name is more complex. A branched-chain alkane such as 2-methylpropane can be considered to be derived from a straight-chain alkane by replacing one or more hydrogen atoms by alkyl groups. The name consists of two parts:

— a **suffix** that identifies the parent straight-chain alkane. To find the suffix, count the number of carbon atoms in the longest chain. For a three-carbon chain, the suffix is *propane;* for a four-carbon chain it is *butane,* and so on.

— a **prefix** that identifies the branching alkyl group (Table I) and indicates by a number the carbon atom where branching occurs. In 2-methylpropane, referred to above, the methyl branch is located at the second carbon from the end of the chain:

$$C_1-\overset{\displaystyle }{\underset{\displaystyle |}{C_2}}-C_3$$

Following this system, the IUPAC names of the isomers of pentane are

$$CH_3-CH_2-CH_2-CH_2-CH_3$$
pentane

$$CH_3-\overset{\displaystyle H}{\underset{\displaystyle CH_3}{C}}-CH_2-CH_3$$
2-methylbutane

$$CH_3-\overset{\displaystyle CH_3}{\underset{\displaystyle CH_3}{C}}-CH_3$$
2,2-dimethylpropane

Notice that

— if the same alkyl group is at two branches, the prefix *di* is used (2,2-dimethylpropane). If there were three methyl branches, we would write trimethyl, and so on.

— the number in the name is made as small as possible. Thus, we write 2-methylbutane, numbering the chain from the left

$$C_1-\overset{\displaystyle }{\underset{\displaystyle |}{C_2}}-C_3-C_4$$

rather than from the right.

Alkenes

The names of straight-chain alkenes are derived from those of the corresponding alkanes containing the same number of carbon atoms. There are two modifications:

—the ending *-ane* is replaced by *-ene*

$$CH_3—CH_3 \qquad CH_2{=}CH_2$$

ethane   ethene

—with alkenes containing four or more carbon atoms, a number is used to designate the double-bonded carbon

$$CH_2{=}CH—CH_2—CH_3 \qquad CH_3—CH{=}CH—CH_3$$

1-butene   2-butene

Note that the number is made as small as possible. In 1-butene, the chain is numbered from the left

$$C_1{=}C_2—C_3—C_4$$

rather than from the right.

A branched-chain alkene can be considered to be derived from a straight-chain alkene by replacing one or more hydrogen atoms by alkyl groups. The name consists of two parts:

—a **suffix** that identifies the parent straight-chain alkene. To find the suffix, consider the longest chain *containing the double bond*. The compounds

$$CH_2{=}\underset{\underset{CH_3}{|}}{C}—CH_2—CH_3 \qquad CH_3—\underset{\underset{CH_3}{|}}{C}{=}\underset{\underset{H}{|}}{C}—CH_3$$

2-methyl-1-butene   2-methyl-2-butene

are considered to be derived from 1-butene and 2-butene, respectively; in both cases, the longest chain contains four carbon atoms.

—a **prefix** that identifies the branching alkyl group and indicates by a number where the branch occurs. In the two branched-chain alkenes just considered, there is a methyl group on the second carbon atom, so the prefix in each case is 2-methyl. The other structurally isomeric alkenes of molecular formula C_5H_{10} are

$$CH_2{=}CH—CH_2—CH_2—CH_3 \qquad CH_3—CH{=}CH—CH_2—CH_3 \qquad CH_2{=}CH—\underset{\underset{CH_3}{|}}{CH}—CH_3$$

1-pentene   2-pentene   3-methyl-1-butene

Alkynes

The IUPAC names of alkynes are identical to those of the corresponding alkenes except that the -*ene* ending is replaced by -*yne:*

$$CH{\equiv}C-CH_2-CH_3 \qquad CH{\equiv}C-\underset{\underset{CH_3}{|}}{CH}-CH_3$$

1-butyne

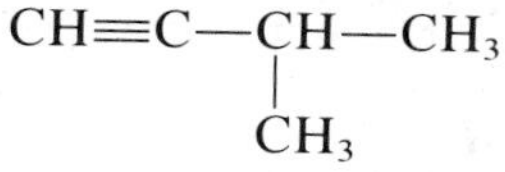

3-methyl-1-butyne

Derivatives of Benzene

In compounds derived from benzene by replacing hydrogen atoms with alkyl groups, the IUPAC name depends upon the number of groups present. With only one alkyl group, all we need do is to attach the name of that group as a prefix:

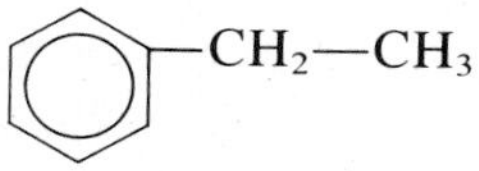

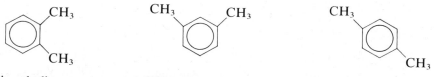

ethylbenzene 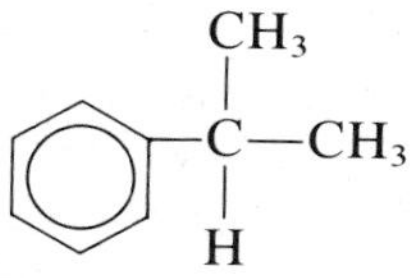   isopropylbenzene

If two groups are present, their relative positions on the ring must be indicated. To do this, we may use numbers to indicate the carbon atoms where the groups are located. More commonly, the prefixes *ortho-*, *meta-*, and *para-* are used:

1,2-dimethylbenzene   1,3-dimethylbenzene   1,4-dimethylbenzene
(*ortho*-dimethylbenzene)   (*meta*-dimethylbenzene)   (*para*-dimethylbenzene)

(Notice that the numbers are made as small as possible.)
    With more than two groups, numbers are always used. Thus, we have

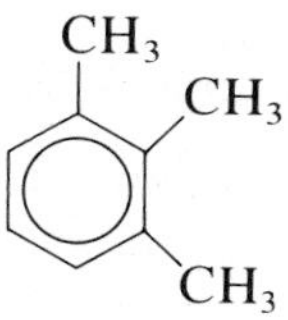

1,2,3-trimethylbenzene 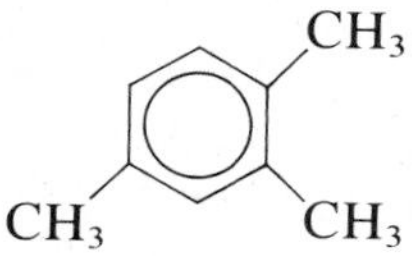   1,2,4-trimethylbenzene 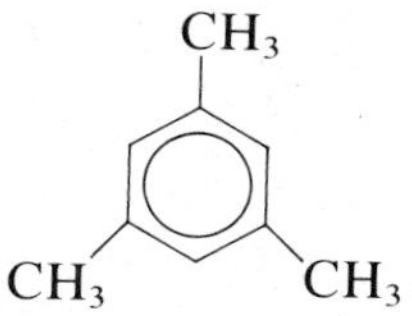   1,3,5-trimethylbenzene

Appendix 4
Review of Mathematics

The mathematics you will use in general chemistry is relatively simple. You will, however, be expected to

—make calculations involving exponential numbers, such as 6.022×10^{23} or 1.6×10^{-10}.

—work with logarithms or antilogarithms, particularly in problems involving pH:

$$\text{pH} = -\log_{10} (\text{conc. } H^+)$$

In this appendix, we will review each of these topics briefly. To start with, it will be helpful to comment on electronic calculators, which are very useful for all kinds of calculations in general chemistry.

Electronic Calculators

A so-called "scientific calculator," selling for between $20 and $50, is entirely adequate for general chemistry. Like any calculator, it will allow you to carry out such simple operations as addition, subtraction, multiplication, and division. Beyond that, make sure that the scientific calculator you buy can be used to

—enter and perform operations on numbers expressed in exponential (scientific) notation.

—find a common (base 10) logarithm or antilogarithm (number corresponding to a given logarithm).

—raise a number to any power, n, or extract the nth root of a number.

The first thing you should do after buying a calculator is to learn how to use it. Read the instruction manual and work with the calculator until you become familiar with it. To get started, try carrying out the following operations:

a. $2.2 \times 6.1 = 13.42$
b. $8.1/2.7 = 3$
c. $(64)^{1/2} = 8$
d. $(27)^{1/3} = 3$
e. $3^4 = 81$

(In d and e, you will need to use the **y**^{x} or **x**^{y} key. Refer to your instruction manual for the sequence of operations, which differs depending upon the brand of calculator.)

f. $\dfrac{16 \times 9}{3 \times 8} = 6$

(This is carried out as a single operation; you do *not* solve for intermediate answers.)

In working with a calculator, you should be aware of one of its limitations. It does not indicate the number of significant figures in the answer. Consider, for example, the operations in a and b above. Assume that 2.2, 6.1, 8.1, and 2.7 represent experimentally measured quantities. Following the rules for significant figures (Chap. 1), the answers should be 13 and 3.0, in that order, *not* 13.42 and 3, the numbers appearing on the calculator. In another case, if you are asked to obtain the reciprocal of 3.68, your answer should be

$1/3.68 = 0.272$

not 0.2717391 . . . , or whatever other number appears on your calculator.

Exponential Notation

In chemistry we frequently deal with very large or very small numbers. In one gram of the element carbon there are

50,150,000,000,000,000,000,000 atoms of carbon.

At the opposite extreme, the mass of a single atom is

0.00000000000000000000001994 g

Numbers such as these are very awkward to work with. For example, neither of the numbers just written could be entered directly on a calculator. To simplify operations involving very large or very small numbers, we use what is known as **exponential** or **scientific notation.** To express a number in exponential notation, we write it in the form

$C \times 10^n$

where C is a number between 1 and 10 (for example, 1, 2.62, 5.8) and n is a positive or negative integer such as 1, -1, -3. To find n, we count the number of places that the decimal point must be moved to give the coefficient, C. If the decimal point must be moved to the *left*, n is a *positive* integer; if it must be moved to the *right*, n is a *negative* integer. Thus, we have

$26.23 = 2.623 \times 10^1$	(decimal point moved 1 place to left)
$5609 = 5.609 \times 10^3$	(decimal point moved 3 places to left)
$0.0918 = 9.18 \times 10^{-2}$	(decimal point moved 2 places to right)

Numbers written in exponential notation can be given a very simple interpretation. Recognizing that $10^1 = 10$, $10^3 = 1000$, and $10^{-2} = 1/100 = 0.01$, we could express the three exponentials written above as

$$2.623 \times 10^1 \quad = 2.623 \times 10$$
$$5.609 \times 10^3 \quad = 5.609 \times 1000$$
$$9.18 \times 10^{-2} = 9.18 \times 0.01$$

The magnitude of a number written in exponential notation depends upon the values of both the coefficient, C, and the exponent, n. Suppose we compare two numbers that have the same value of n, such as 2.6×10^2 and 3.8×10^2. Here, the larger number is the one with the larger coefficient:

$$3.8 \times 10^2 > 2.6 \times 10^2; \quad 380 > 260$$

Suppose now that we compare two exponential numbers with different values of n. Here, *the larger number is the one that has the larger value of n:*

$$2.6 \times 10^2 > 4.8 \times 10^1 \qquad 260 \ > 48$$
$$3.2 \times 10^1 > 8.0 \times 10^{-1} \qquad 32 \quad > 0.80$$
$$2 \times 10^{-2} > 4 \times 10^{-3} \qquad 0.02 > 0.004$$

Multiplication and Division

A major advantage of exponential notation is that it simplifies the processes of multiplication and division. To *multiply,* we *add exponents:*

$$10^1 \times 10^2 = 10^{1+2} = 10^3; \qquad 10^6 \times 10^{-4} = 10^{6+(-4)} = 10^2$$

To *divide,* we *subtract* exponents:

$$10^3/10^2 = 10^{3-2} = 10^1; \qquad 10^{-3}/10^6 = 10^{-3-6} = 10^{-9}$$

To multiply one exponential number by another, we can first multiply the coefficients in the usual manner and then add exponents. To divide one exponential number by another, we can find the quotient of the coefficients and then subtract exponents. For example,

$$(5.00 \times 10^4) \times (1.60 \times 10^2) = (5.00 \times 1.60) \times (10^4 \times 10^2)$$
$$= 8.00 \times 10^6$$

$$(6.01 \times 10^{-3})/(5.23 \times 10^6) \quad = \frac{6.01}{5.23} \times \frac{10^{-3}}{10^6} = 1.15 \times 10^{-9}$$

It often happens that multiplication or division yields an answer that is not in standard exponential notation. Thus, we might have

$$(5.0 \times 10^4) \times (6.0 \times 10^3) = (5.0 \times 6.0) \times 10^4 \times 10^3 = 30 \times 10^7$$

The product is not in standard exponential notation since the coefficient, 30, does not lie between 1 and 10. To correct this situation, we could rewrite the coefficient as 3.0×10^1 and then add exponents:

$$30 \times 10^7 = (3.0 \times 10^1) \times 10^7 = 3.0 \times 10^8$$

In another case,

$$0.526 \times 10^3 = (5.26 \times 10^{-1}) \times 10^3 = 5.26 \times 10^2$$

Exponential Notation on the Calculator

On all scientific calculators, it is possible to enter numbers in exponential notation. The method used depends upon the brand of calculator. Most often, it involves using a key labeled **EXP, EE,** or **EEX.** Check your instruction manual for the procedure to be followed. To make sure you understand it, try entering the following numbers:

$$2.4 \times 10^6; \quad 3.16 \times 10^{-8}; \quad 6.2 \times 10^{-16}$$

Multiplication, division, and many other operations can be carried out directly on your calculator. Try the following exercises, using your calculator:

a. $(6.0 \times 10^2) \times (4.2 \times 10^{-4}) = ?$

b. $\dfrac{6.0 \times 10^2}{4.2 \times 10^{-4}} = ?$

c. $(2.50 \times 10^{-9})^{1/2} = ?$

d. $3.6 \times 10^{-4} + 4 \times 10^{-5} = ?$

The answers, expressed in exponential notation and following the rules of significant figures, are as follows: a. 2.5×10^{-1} b. 1.4×10^6 c. 5.00×10^{-5} d. 4.0×10^{-4}.

Logarithms and Antilogarithms

The common (base 10) logarithm of a number is the power to which 10 must be raised to give that number. For example,

since	$10^2 = 100,$	$\log 100 = 2$
since	$10^0 = 1,$	$\log 1 = 0$
since	$10^{-3} = 0.001,$	$\log 0.001 = -3$

Notice that numbers that are larger than 1 have a common logarithm greater than 0. Numbers smaller than 1 have a common logarithm which is less than 0.

An antilogarithm is, quite simply, the number corresponding to a given logarithm. Thus,

since	$\log 100 = 2,$	antilog $2 = 100$
since	$\log 1 = 0,$	antilog $0 = 1$
since	$\log 0.001 = -3,$	antilog $-3 = 0.001$

In other words, the numbers whose logarithms are 2, 0, and -3 are 100, 1, and

0.001, in that order. Notice that a positive antilogarithm signifies a number greater than 1, a negative antilogarithm a number less than 1.

The definitions we have just presented can be summarized in the following statement:

if $10^x = y$, log y = x, antilog x = y

Finding Logs and Antilogs on a Calculator

Most numbers do not have logarithms that are simple integers such as 2, 0, or −3. This is the case only for integral powers of 10 such as 10^2, 10^0, or 10^{-3}. Logarithms of numbers such as 4.14 or 0.526 cannot be found by inspection. Until quite recently, finding logarithms of such numbers involved using a table that took some considerable practice to interpret correctly. Now, with the advent of the low-priced scientific calculator, "log tables" are obsolete.

To obtain a base 10 logarithm using a calculator, all you need do is enter the number and press the **LOG** key. This way you should find that

$$\log 4.14 = 0.617 \ldots$$
$$\log 0.526 = -0.279 \ldots$$

The same process can be used to find the logarithm of an exponential number. Thus,

$$\log (4.14 \times 10^2) = 2.617 \ldots$$

Notice that $\log (4.14 \times 10^2) = \log 4.14 + \log 10^2 = 0.617 + 2 = 2.617$
In the general case,

$$\log (C \times 10^n) = \log C + n$$

This applies to negative as well as positive values of n. Thus,

$$\log (4.14 \times 10^{-1}) = 0.617 - 1 = -0.383$$

The process used to find an antilogarithm varies with the brand of calculator. Perhaps most commonly, you enter the number and press the **INV** and **LOG** keys to obtain the antilogarithm. If your calculator doesn't have an **INV** key, refer to the instruction manual to find out how to do this operation. Then, find the numbers whose logarithms are 0.616, 2.415, and −1.057. You should find that

antilog 0.616 = 4.13 . . . ,	which means that $10^{0.616} = 4.13 \ldots$
antilog 2.415 = 260 . . . ,	which means that $10^{2.415} = 260 \ldots$
antilog −1.057 = 0.0877 . . . ,	which means that $10^{-1.057} = 0.0877 \ldots$

Significant Figures in Logarithms and Antilogarithms

As with other operations, a calculator does not indicate the number of significant figures to be retained in a logarithm or antilogarithm. In the examples just worked, you probably found several more digits in the calculator display, beyond those listed. To decide how many figures to retain, use the following rules:

1. In taking the logarithm of a number, retain after the decimal point in the log as many digits as there are significant figures in the number. (This part of the logarithm is often referred to as the *mantissa;* digits that precede the decimal point comprise the *characteristic* of the logarithm.) To illustrate this rule, consider the following:

$$\log 2.00 = 0.301 \qquad \log (2.00 \times 10^3) = 3.301$$
$$\log 2.0 \;\; = 0.30 \qquad \log (2.0 \times 10^1) \;\; = 1.30$$
$$\log 2 \;\;\;\; = 0.3 \qquad \log (2 \times 10^{-3}) \;\; = 0.3 - 3 = -2.7$$

2. In taking the antilogarithm of a number, retain as many significant figures in the antilogarithm as there are after the decimal point in the number. Thus,

$$\text{antilog } 0.301 = 2.00 \qquad \text{antilog } 3.301 = 2.00 \times 10^3$$
$$\text{antilog } 0.30 \;\; = 2.0 \qquad \text{antilog } 1.30 \;\; = 2.0 \times 10^1$$
$$\text{antilog } 0.3 \;\; = 2 \qquad \text{antilog } -2.7 = 2 \times 10^{-3}$$

Operations Involving Logarithms

Since logarithms are exponents, the rules governing the use of exponents apply here as well.

Multiplication: $\log (xy) = \log x + \log y$

> Example: $\log (6.02 \times 2.00) = \log 6.02 + \log 2.00$
> $$= 0.780 + 0.301 = 1.081$$

Division: $\log (x/y) = \log x - \log y$

> Example: $\log (6.02/2.00) = \log 6.02 - \log 2.00$
> $$= 0.780 - 0.301 = 0.479$$

Raising to a power: $\log(x^n) = n \log x$

> Example: $\log (2.00^3) = 3 \log 2.00$
> $$= 3(0.301) = 0.903$$

Extracting a root: $\log (x^{1/n}) = \dfrac{1}{n} \log x$

> Example: $\log (4.00^{1/2}) = \tfrac{1}{2} \log 4.00$
> $$= \tfrac{1}{2}(0.602) = 0.301$$

Taking a reciprocal: $\log (1/x) = -\log x$

> Example: $\log (1/2.00) = -\log 2.00 = -0.301$

Natural Logarithms

For calculation purposes, common logarithms are most convenient since our number system is based upon multiples of 10. However, certain of the equations we use in general chemistry involve a different type of logarithm, taken to the base e, where, to four significant figures,

$$e = 2.718 \ldots$$

Logarithms to the base e are referred to as **natural** logarithms and are often written as "ln":

$$\log_e x \equiv \ln x; \qquad \log_{10} x \equiv \log x$$

To find the natural logarithm of a number on your calculator, enter the number and press the **LN** key. This way you should find that

$$\ln 10.0 = 2.303$$
$$\ln 2.00 = 0.693$$
$$\ln 1.00 = 0.000$$

There is a simple relationship between common and natural logarithms which you may be able to deduce from the examples just worked. Recalling that log 10.0 = 1.000,

$$\frac{\ln 10.0}{\log 10.0} = \frac{2.303}{1.000} = 2.303$$

In the general case,

$$\ln x = 2.303 \log x$$

For example, $\ln 2.00 = 2.303 \log 2.00 = 2.303(0.301) = 0.693$

Sometimes you will need to evaluate an expression such as $e^{0.250}$, where the base of natural logarithms, e, is raised to a power. This can be done by entering the exponent, 0.250, on your calculator and pressing the **e^x** key: (or **INV** and **LN x** keys):

$$e^{0.250} = 1.28$$

Chapter 1

1.31 a. mass   b. volume   c. volume   d. length   e. density   f. pressure
     g. energy

1.32 a. 502 m   b. 500 kg   c. 150 cm³   d. 12.0 g/cm³

1.33 a. −88.28°C   b. 184.87 K   **1.34** $\dfrac{cm^2 \cdot g}{s^2 \cdot mol \cdot K}$

1.35 a. 3   b. 3   c. 4   d. 3   e. 3   f. 1

1.36 0.0125 nm³   **1.37** a. 4   b. 1   c. 2   d. 2

1.38 a. 4.12 km²   b. 4.12 × 10⁶ m²   c. 4.43 × 10⁷ ft²

1.39 a. 0.2390 cal   b. 1.000 × 10⁻³ kJ   c. 9.869 × 10⁻³ L·atm

1.40 0.4046   **1.41** 5.43 m/s   **1.42** a. 256   b. 1.3; 42

1.43 84

1.44 a. Am, As, At, Es, In, I, Mn, Mo, Ne, Ni, N, No, Os, O, Sm, Se, Si, Na, S, Ta,
     Te, Tm, Sn, Ti   b. Ar, Ga, In, I, Ir, Ni, N, Ra, Rn, Ag, Na, V

1.45 a. 6   b. 5   c. 6   d. 4

1.46 a. K   b. Cu   c. Au   d. Sb   e. Pb

1.47 a. C   b. P   c. P   d. P

1.48 2.90 × 10⁻³ g/cm³   **1.49** 2.8 g/cm³

1.50 1.2 × 10³ kg   **1.51** a. 20.0   b. 17.9

1.52 a. 2.9 × 10⁻⁵ g   b. 3.4 × 10³ g

1.53 a. 35 g/100 g   b. 1.7 × 10³ g   c. 2.9 g

1.54 22.6°C   **1.55** 0.476 J/g·°C   **1.56** a. T   b. F   c. T

1.57 a. shake with water, filter
     b. dissolve in water; carry out fractional crystallization
     c. filter   d. distil

1.58 a. 82 g   b. 0 g   c. 11%

1.59 a. carry out reaction to determine chemical properties
     b. solute is one component of solution
     c. at normal bp, P = 1 atm
     d. mixture contains more than one substance

1.60 a, c   **1.61** −12.3°F   **1.62** 1.0 km²   **1.63** 34.5 cm

1.64 8.2 × 10⁻³ g   **1.65** 5.3°C

Chapter 2

2.31 See discussion, Section 2.1

2.32 a. Conservation of Mass   b. Constant Composition   c. Multiple Proportions

2.33 a. 0.4211 g O/g F; 0.8420 g O/g F   b. 0.8420/0.4211 = 2:1
c. F_2O and FO are one possible pair
2.34 94; 145   **2.35** $^{85}_{37}Rb$; $^{87}_{37}Rb$
2.36 a. 53   b. 78   c. 53   d. 78, 53
2.37 42, 56, 42; $^{80}_{34}Se^{2-}$, 36; $^{151}_{63}Eu^{3+}$, 60; 82, 125, 82
2.38 c < a < b < d
2.39 a. 14.0186   b. 72.65   c. 107.9596
2.40 24.31   **2.41** 92.5% Li-7
2.42 9.0% Ne-22, 90.7% Ne-20
2.43 a. 9, 10   b. 18, 18   c. 10, 10   d. 1, 0
2.44 27, 24; 14, 14; 34, 36; 18, 18   **2.45** a. K_2O, K_3N b. $FeCl_2$, $FeCl_3$
2.46 a. 1.6×10^{-2} g   b. 9.476×10^{21}
2.47 a. 7×10^{-7} g   b. 1.318×10^{24} **2.48** 1.47×10^{32}
2.49 a. 55   b. 3.312×10^{25} c. 7.02×10^{26} d. 5.28×10^{24}
2.50 a. 194.190 g/mol   b. 76.095 g/mol   c. 194.9103 g/mol
2.51 a. 0.242 mol   b. 0.104 mol   c. 0.0661 mol
2.52 a. 0.102 mol   b. 0.0509 mol   c. 0.0618 mol   d. 0.0325 mol
2.53 4.89×10^{-5}, 2.94×10^{19}, 8.83×10^{19}; 1.15, 7.53×10^{21}, 2.26×10^{22}
4.6×10^2, 5.0, 9.0×10^{24}; 6.1×10^{-8}, 6.6×10^{-10}, 4.0×10^{14}
2.54 a. 0.705 mol   b. 139 mL   c. 2.76×10^{-4} g
2.55 Dissolve to form 355 mL of solution:
a. 10.6 g   b. 20.2 g   c. 32.8 g
2.56 a. 0.22 g, 5.4 g   b. 0.50 L, 0.0083 L
2.57 0.750 mol/L, 9.61 g, 0.0240 L, 0.466 g
2.58 a. Anions heavier than corresponding atoms   b. Twice as many anions
c. $12.0 g/6.022 \times 10^{23}$ d. 6.02×10^{23} Na^+ ions, 6.02×10^{23} Cl^- ions
2.59 a. 183.207 g/mol   b. 0.346   c. 2.0×10^{14} **2.60** c < b < a < d
2.61 a. 39.0978 g   b. 18.9989 g   c. 58.0967 g   **2.62** 6.57×10^4 g/mol
2.63 6.03×10^{23} **2.64** 6.00×10^{23} **2.65** 2×10^{14}
2.66 a. 2.5×10^{24} b. 2.3×10^{-20} c. 3×10^2

Chapter 3

3.31 57.468% C, 5.425% H, 8.380% N, 19.14% O, 9.588% S
3.32 a. 51.2   b. 0.188 g   **3.33** 83.2% C, 16.76% H   **3.34** SnF_4
3.35 a. $C_{12}H_{22}O_{11}$ b. $NaSO_2$ c. $XeOF_4$
3.36 a. C_6H_6O b. 94.11, 188.22, 282.33 g/mol   **3.37** CHCl
3.38 Cu_2O **3.39** 46.0; 7   **3.40** a. C_2H_2 b. C_4H_4 c. C_6H_6
3.41 a. $Cr_2(SO_4)_3$ b. $LiNO_3$ c. PbI_2 d. $Sr(ClO_3)_2$ e. Cu_2Te f. $MnCO_3$
3.42 a. Li_3N b. Cu_2SO_4 c. silver selenide   d. iron(III) sulfate   e. $BaCrO_4$
f. ammonium perchlorate   g. $(NH_4)_3PO_4$
3.43 a. hydrazine   b. Se_2I_2 c. xenon trioxide   d. iodine pentafluoride
e. tetraphosphorus decoxide   f. N_2O
3.44 a. 0.200 mol   b. 1.50 mol   c. 2.92×10^{-3} mol
3.45 a. $2 H_2O_2(l) \rightarrow 2 H_2O(l) + O_2(g)$
b. $2 Al(s) + 3 MgO(s) \rightarrow 3 Mg(s) + Al_2O_3(s)$
c. $C_2H_5OH(l) + 3 O_2(g) \rightarrow 2 CO_2(g) + 3 H_2O(l)$
3.46 a. $6 Li(s) + N_2(g) \rightarrow 2 Li_3N(s)$
b. $4 Li(s) + O_2(g) \rightarrow 2 Li_2O(s)$
c. $2 Li(s) + S(s) \rightarrow Li_2S(s)$
d. $2 Li(s) + Cl_2(g) \rightarrow 2 LiCl(s)$
e. $2 Li(s) + I_2(s) \rightarrow 2 LiI(s)$
3.47 a. $2 Fe(s) + 3 Cl_2(g) \rightarrow 2 FeCl_3(s)$
b. $2 Sr(s) + O_2(g) \rightarrow 2 SrO(s)$

    c. $2 Mg(s) + CO_2(g) \rightarrow 2 MgO(s) + C(s)$

    d. $SiH_4(g) + 2 O_2(g) \rightarrow SiO_2(s) + 2 H_2O(g)$

    e. $Zn(s) + I_2(s) \rightarrow ZnI_2(s)$

3.48 a. 11.0   b. 7.75   c. 0.926   d. 22.4

3.49 a. 29.2 g   b. 36.3 g   c. 1.44 g   d. 17.0 g

3.50 a. $2 C_4H_{10}(g) + 13 O_2(g) \rightarrow 8 CO_2(g) + 10 H_2O(l)$

    b. 2.90 mol   c. 242 g

3.51 a. 230 g   b. 62 L   **3.52** 9.2 L   **3.53** no

3.54 a. $Cl_2(g) + 3 F_2(g) \rightarrow 2 ClF_3(g)$

    b. F_2 c. 4.77 mol   d. 1.01 mol

3.55 a. $2 C_2H_2(g) + 5 O_2(g) \rightarrow 4 CO_2(g) + 2 H_2O(l)$

    b. O_2 c. 28.2 g   d. 79.8

3.56 1.36×10^4 g   **3.57** 27 g $[Co(NH_3)_5Cl]Cl_2$, 16 g KSCN

3.58 a. only if anion and cation have equal charges

    b. Co is an element, CO is a compound

    c. Two reactants could be present in equivalent amounts

    d. reduction

    e. no molecules in $CaCl_2$

3.59 a. R   b. R   c. O   **3.60** 52.92% Al   **3.61** 194.20 g/mol

3.62 a. $2 Mg(s) + O_2(g) \rightarrow 2 MgO(s)$; $3 Mg(s) + N_2(g) \rightarrow Mg_3N_2(s)$

    $MgO(s) + H_2O(l) \rightarrow Mg(OH)_2(s)$

    b. 2.449 g   c. 0.170 g

3.63 80.3% NaCl   **3.64** a. V_2O_3, V_2O_5 b. 2.271 g

3.65 50   **3.66** 487.2 g/mol; Na, Fe

Chapter 4

4.31 Elemental: Ar, N_2, S   Electrolysis: Al, Mg, (Cu)

4.32 chloride, 2, oxides, sulfide

4.33 a. fract. dist. liquid air   b. See Figure 4.3   c. treatment of brine with Cl_2

    d. electrolysis NaCl   e. reduction of Fe_2O_3 with CO

4.34 a. more readily extracted   b. sulfur deposits too deep   c. F^- much more difficult to oxidize   d. would react to form NaCl

4.35 a. T   b. F   c. F   d. F

4.36 a. $Cl_2(g) + 2 I^-(aq) \rightarrow 2 Cl^-(aq) + I_2(s)$

    b. $2 Cl^-(aq) + 2 H_2O \rightarrow Cl_2(g) + H_2(g) + 2 OH^-(aq)$

    c. $CaF_2(s) + H_2SO_4(l) \rightarrow 2 HF(l) + CaSO_4(s)$

4.37 a. $Mg^{2+}(aq) + 2 OH^-(aq) \rightarrow Mg(OH)_2(s)$

    b. $Mg(OH)_2(s) + 2 H^+(aq) \rightarrow Mg^{2+}(aq) + 2 H_2O$

    c. $MgCl_2(l) \rightarrow Mg(l) + Cl_2(g)$

    d. $P_4(s) + 5 O_2(g) \rightarrow P_4O_{10}(s)$

4.38 a. $MgCO_3 \cdot CaCO_3(s) \rightarrow MgO(s) + CaO(s) + 2 CO_2(g)$

    b. $HgS(s) + O_2(g) \rightarrow Hg(g) + SO_2(g)$

4.39 a. $Cl_2(g) + 2 e^- \rightarrow 2 Cl^-(aq)$ reduction

    $2 I^-(aq) \rightarrow I_2(s) + 2 e^-$ oxidation

    b. $2 Al^{3+} + 6 e^- \rightarrow 2 Al(s)$ reduction

    $3 O^{2-} \rightarrow 3/2 O_2(g) + 6 e^-$ oxidation

    c. $2 Cl^-(aq) \rightarrow Cl_2(g) + 2 e^-$ oxidation

    $MnO_2(s) + 4 H^+(aq) + 2 e^- \rightarrow Mn^{2+}(aq) + 2 H_2O$ reduction

4.40 a. gain of e^- b. species consumed when product is formed in 100% yield

    c. member of Group 7   d. method of separating two liquids

4.41 a. steelmaking, breathing apparatus   b. synthesis of NH_3 c. light alloys

    d. structural metal (aircraft)   e. thermometers, barometers

4.42 Mn, Zn, Cr, Ni, Sn, Au   **4.43** a. 1.252   b. 1.962   c. 1.257   d. 1.711

4.44　a. 0.915 g　　b. 0.720 g　　c. 0.709 g　　d. 0.413 g　　e. 0.515 g
4.45　1.532 mol, 74.48 g; 490 g, 80.2 g; 0.510 mol, 151 g
4.46　a. 4.0×10^{-3}　b. ·0.32　　**4.47**　0.61 cm³　　**4.48**　a. 31.56
　　　b. 1.58×10^6
4.49　a. $Cu_3C_2O_8H_2$　　b. $Cu(OH)_2 \cdot 2\ CuCO_3$　　**4.50**　Bi_2S_3
4.51　a. 14.3 g　　b. 43.0 g　　c. 25.6 g　　d. 17.0 g
4.52　a. 12.6 mol　　b. 309 L　　c. 1.47×10^3 L
4.53　a. 3.04×10^3 g　　b. 1.62×10^3 g　　c. 3.23×10^2 kg
4.54　a. $MnO_2(s) + 4\ H^+(aq) + 2\ Br^-(aq) \rightarrow Mn^{2+}(aq) + 2\ H_2O + Br_2(l)$
　　　b. 0.0711 mol　　c. 0.143 g
4.55　a. 0.0282 mol　　b. 9.66 L　　**4.56**　33.9
4.57　a. $CaCO_3$　　b. 1.16 g　　**4.58**　a. 1.43 mol　　b. 0.941 g　　c. 47.0 g
4.59　a. $Mg(OH)_2$　　b. 2.5 mol　　c. 1.0 mol/L
4.60　a. Sb_4O_6　　b. 251 g　　c. 1.46×10^3 g
4.61　30 kg
4.62　2.45 mol/L　　**4.63**　12.52　　**4.64**　36　　**4.65**　1.9×10^{12}

Chapter 5

5.31　a. T　　b. F　　c. T　　d. T
5.32　a. $KBr(s) \rightarrow K^+(aq) + Br^-(aq)$　　b. +
5.33　a. endothermic　　b. products 160.7 kJ above reactants　　c. +0.941 kJ
　　　d. 1.06 g
5.34　a. +55.9 kJ　　b. +3.10 kJ
5.35　a. $2\ Mg(s) + CO_2(g) \rightarrow 2\ MgO(s) + C(s)$
　　　b. −812 kJ　　c. 8.77 kJ
5.36　6.3 kJ　　**5.37**　18.8 kJ; 12.6 kJ　　**5.38**　+71 kJ　　**5.39**　−580.7 kJ
5.40　a. −155.2 kJ　　b. −97.5 kJ　　**5.41**　+0.321 kJ
5.42　a. +446.6 kJ　　b. −16.0 kJ　　c. −363.2 kJ
5.43　a. −2219.9 kJ　　b. −1370.9 kJ　　**5.44**　−256.5 kJ　　**5.45**　−431.8 kJ
5.46　a. $C_2H_2(g) + 5/2\ O_2(g) \rightarrow 2\ CO_2(g) + H_2O(l)$
　　　　$C_4H_{10}(g) + 13/2\ O_2(g) \rightarrow 4\ CO_2(g) + 5\ H_2O\ (l)$
　　　b. 49.90 kJ; 49.52 kJ　　c. butane
5.47　a. +1.8 kJ　　b. +21 kJ　　**5.48**　+42.0 kJ　　**5.49**　-1.37×10^3 kJ
5.50　2.59×10^3 J/°C　　**5.51**　25.83°C　　**5.52**　a. −25 J　　b. +150 J
5.53　a. −1559.7 kJ　　b. 1553.5 kJ　　**5.54**　5.3×10^8　　**5.55**　58.4°C
5.56　a. 9.4×10^7 J　　b. $1.6　　**5.57**　See Section 5.6
5.58　a. independent of path　　b. endothermic　　c. increasing
5.59　0.139 J/g·°C
5.60　0.279 J/g·°C
5.61　a. 1.4×10^{16} kJ　　b. 8.7 km/L　　**5.62**　5.1×10^3 kJ
5.63　a. 1.71×10^5 J　　b. 513 g
5.64　25% empty　　**5.65**　a. −847.6 kJ　　b. 6300°C　　c. yes

Chapter 6

6.31　0.828 atm
6.32　0.0296 atm, 3.00 kPa; 632 mm Hg, 84.2 kPa; 758 mm Hg, 0.997 atm
6.33　a. 457 mm Hg　　b. 888 mm Hg　　**6.34**　383°C　　**6.35**　533 mL
6.36　1.13 atm　　**6.37**　0.07 m³　　**6.38**　188 atm　　**6.39**　3.13×10^{24}; 93.5 cm³
6.40　11.7 L, 0.496 mol; 73.3 atm, 0.768 mol; 0.0459 mol, 0.926 g; 75.3 K, 1.24 mol
6.41　a. 1.13 g/L　　b. 1.29 g/L　　c. 14.2 g/L
6.42　a. 133 g/mol　　b. $C_2H_3Cl_3$　　**6.43**　a. 9.20 g/mol　　b. 0.317
6.44　a. 53.1 g/mol　　b. B　　**6.45**　a. O_2　　b. 5.14 L

5

6.46 a. 0.0109 L   b. yes   **6.47** 4.34 g   **6.48** a. 0.0037   b. 0.031 g
6.49 P O_2 = P N_2 = 0.618 atm; P He = 4.94 atm
6.50 a. 730 mm Hg b. 0.161 g
6.51 a. 0.0391 mol   b. 0.00128 mol   c. 1.27 g
6.52 a. CO   b. increase T, decrease P   **6.53** 89 g/L
6.54 a. lighter   b. 1.10   **6.55** 749 mm Hg   **6.56** 2.6 K
6.57 a. 0°C   b. 2530 K   **6.58** 1 atm = 34 ft water
6.59 a. He   b. same   c. He   d. He
6.60 a.

6.61 0.0456 L·atm/mol·°R   **6.62** 62.8% Al   **6.63** 6.36 m
6.64 0.787 atm
6.65 $V_a = n_a RT/P$; $V = nRT/P$; $V_a/V = n_a/n = X_a$

Chapter 7

7.31 short λ : high energy, high frequency
7.32 a. 7.106×10^{-19} J/part.   b. 427.9 kJ/mol
7.33 a. 1.32×10^{-7} m   b. 132 nm   c. 2.27×10^{15}/s
7.34 a. 3.11 m   b. 3.11×10^9 nm   c. 3.11×10^{-3} km
7.35 a. 7.5a   b. 7.5a   c. 7.5d
7.36 Orbits consist of concentric circles with radii of 0.0529 nm, 0.212 nm, 0.476 nm, 0.846 nm; lines from **n** = 4, 3 or 2 to **n** = 1 are in Lyman series; **n** = 4 or 3 to **n** = 2 in Balmer series
7.37 1945 nm   **7.38** 5   **7.39** 36.44 kJ/mol   **7.40** 2.0×10^{-38} m
7.41 a. 0   b. 3, 2, 1, 0, −1, −2, −3   c. 0; 1, 0, −1; 2, 1, 0, −1, −2
7.42 a. d   b. 3   c. 10   **7.43** a. 16   b. 5   c. 7
7.44 a. ℓ = 0, 1, − − (**n** − 1)   b. $m_\ell = \ell$, − −, 0, − −, $-\ell$
7.45 a. $1s^2 2s^2 2p^6 3s^2 3p^2$ b. $1s^2 2s^2 2p^6 3s^2 3p^6 4s^2 3d^6$
    c. $1s^2 2s^2 2p^6 3s^1$ d. $1s^2 2s^2 2p^6 3s^2 3p^6 4s^2 3d^{10} 4p^1$
7.46 a. $[_{10}Ne] 3s^2 3p^4$ b. $[_{18}Ar] 4s^2$ c. $[_{18}Ar] 4s^2 3d^7$ d. $[_{18}Ar] 4s^2 3d^{10} 4p^4$
7.47 a. Cu   b. Ca   c. Ge   **7.48** a. He   b. Sc   c. Na
7.49 a. 0   b. 1/2   c. 12/26
7.50 a. G   b. E   c. I   d. I   e. G   f. E
7.51   1s   2s   2p   3s   3p   4s   3d
    a. (⇅) (⇅) (↑)(↑)( )
    b. (⇅) (⇅) (⇅)(⇅)(⇅) (⇅) (⇅)(⇅)(⇅) (⇅) (↑)(↑)(↑)( )( )
    c. (⇅) (⇅) (⇅)(⇅)(⇅) (⇅) (⇅)(⇅)(⇅) (⇅)
    d. all the above sublevels filled; three electrons with parallel spins in three 4p orbitals.
7.52 a. Na   b. Si   c. Oxygen   **7.53** a. 1   b. 4   c. 0
7.54 a. Zn   b. Sc, Cu   c. Ti, Ni   d. V, Co   e. Fe   f. Mn

7.55

n	1	1	2	2	2	2	2	2	2
ℓ	0	0	0	0	1	1	1	1	1
m_ℓ	0	0	0	0	1	1	0	0	−1
m_s	$+\frac{1}{2}$	$-\frac{1}{2}$	$+\frac{1}{2}$	$-\frac{1}{2}$	$+\frac{1}{2}$	$-\frac{1}{2}$	$+\frac{1}{2}$	$-\frac{1}{2}$	$+\frac{1}{2}$

7.56 a.

n	3	3
ℓ	0	0
m_ℓ	0	0
m_s	$+\frac{1}{2}$	$-\frac{1}{2}$

b.

n	3	3	3	3	3	3	3	3
ℓ	2	2	2	2	2	2	2	2
m_ℓ	2	2	1	1	0	0	−1	−2
m_s	$+\frac{1}{2}$	$-\frac{1}{2}$	$+\frac{1}{2}$	$-\frac{1}{2}$	$+\frac{1}{2}$	$-\frac{1}{2}$	$+\frac{1}{2}$	$+\frac{1}{2}$

c.

n	3	3	3	3	3
ℓ	1	1	1	1	1
m_ℓ	1	1	0	0	-1
m_s	$+\frac{1}{2}$	$-\frac{1}{2}$	$+\frac{1}{2}$	$-\frac{1}{2}$	$+\frac{1}{2}$

7.57 a. 1   b. 10   c. 2   **7.58** b < f < c < e < a = d

7.59 a. See Section 7.3   b. See Glossary   c. See Glossary   d. At right angles

7.60 a. false; evolves energy   b. false; longer wavelength   c. true   d. true

7.61 1312 kJ/mol

7.62 $E_{hi} - E_{lo} = -\dfrac{1312}{n^2} + \dfrac{1312}{4} = 328.0\,\dfrac{(n^2 - 4)}{n^2}$

$\lambda = \dfrac{1.196 \times 10^5}{E_{hi} - E_{lo}} = \dfrac{1.196 \times 10^5}{328.0}\,\dfrac{n^2}{(n^2 - 4)} = 364.6\,\dfrac{n^2}{n^2 - 4}$

7.63 $1s^4 1p^4$ **7.64** a. 3, 9, 15   b. 27   c. $1s^3 2s^3 2p^2$; $1s^3 2s^3 2p^9 3s^2$

7.65 a. 3.01×10^{-19} J   b. 660 nm

Chapter 8

8.31 a. horizontal row   b. Group 2 metal   c. element filling 4f sublevel
d. element filling d sublevel   e. intermediate between metal, nonmetal

8.32 a. probably 111 (compare Cu)   b. 115

8.33 a. 81   b. 86   c. 20   d. 102

8.34 a. $5s^2 5p^6$ b. $6s^1$ c. $6s^2 6p^4$ d. $5s^2 5p^3$

8.35 a. Cl < Si < Na   b. Na < Si < Cl   c. Na > Si > Cl

8.36 a. Cl   b. K   c. K

8.37 a. metal   b. nonmetal   c. metalloid   d. metal   e. metalloid

8.38 a. C, Si, Ge, Sn, Pb   b. Be, Mg, Ca   c. Ne, Ar, Kr, Xe, Rn
d. H, Li, Na, K, Rb, Cs, Fr

8.39 785°C   **8.40** a. CaO   b. $CaBr_2$ c. Ca_3N_2 d. $Ca(OH)_2$

8.41 a. $Ca(s) + Br_2(l) \rightarrow CaBr_2(s)$
b. $BaO_2(s) + 2\,H_2O \rightarrow H_2O_2(aq) + Ba^{2+}(aq) + 2\,OH^-(aq)$
c. $2\,K(s) + S(s) \rightarrow K_2S(s)$
d. $2\,K(s) + 2\,H_2O \rightarrow 2\,K^+(aq) + 2\,OH^-(aq) + H_2(g)$

8.42 $2\,Na(s) + O_2(g) \rightarrow Na_2O_2(s)$
$Na_2O_2(s) + 2\,H_2O \rightarrow H_2O_2(aq) + 2\,Na^+(aq) + 2\,OH^-(aq)$
$2\,Na(s) + 2\,H_2O \rightarrow H_2(g) + 2\,Na^+(aq) + 2\,OH^-(aq)$

8.43 a. Be, Mg   b. Ba   c. Be   d. Ca

8.44 a. $BaO(s) + H_2O \rightarrow Ba^{2+}(aq) + 2\,OH^-(aq)$
b. $HCO_3^-(aq) + H^+(aq) \rightarrow CO_2(g) + H_2O$
c. $Ca(OH)_2(s) + CO_2(g) \rightarrow CaCO_3(s) + H_2O(l)$
d. $CaSO_4 \cdot 1/2\,H_2O(s) + 3/2\,H_2O(l) \rightarrow CaSO_4 \cdot 2H_2O(s)$

8.45 117 kJ   **8.46** a. -65.3 kJ   b. -81.5 kJ

8.47 a. $+12.3$ kJ   b. -0.123 kJ   **8.48** a. -3.64 kJ   b. 174 g

8.49 0.548 L   **8.50** a. $2\,Na(s) + O_2(g) \rightarrow Na_2O_2(s)$ b. 1.60 L

8.51 a. 0.10 mol   b. 2.30 mol

8.52 $1s^2 2s^2 2p^6 3s^2 3p^6 4s^2 3d^{10} 4p^6 5s^2 4d^{10} 5p^6 6s^1$. A total of 27 orbitals are filled, each with two electrons of opposed spins; there is one electron in the 6s orbital.

8.53 a.

n	3	3	3	3	3
ℓ	2	2	2	2	2
m_ℓ	2	1	0	-1	-2
m_s	$+\frac{1}{2}$	$+\frac{1}{2}$	$+\frac{1}{2}$	$+\frac{1}{2}$	$+\frac{1}{2}$

b.

n	3	3	3	3	3	3	3	3	3	3	4	4	4	4	4
ℓ	2	2	2	2	2	2	2	2	2	2	2	2	2	2	2
m_ℓ	2	2	1	1	0	0	-1	-1	-2	-2	2	1	0	-1	-2
m_s	$+\frac{1}{2}$	$-\frac{1}{2}$	$+\frac{1}{2}$	$-\frac{1}{2}$	$+\frac{1}{2}$	$-\frac{1}{2}$	$+\frac{1}{2}$	$-\frac{1}{2}$	$+\frac{1}{2}$	$-\frac{1}{2}$	$+\frac{1}{2}$	$+\frac{1}{2}$	$+\frac{1}{2}$	$+\frac{1}{2}$	$+\frac{1}{2}$

c. **n**   3   3   3   3   3   3   3   3   3   3
  ℓ 2   2   2   2   2   2   2   2   2   2
  m$_\ell$ 2   2   1   1   0   0   −1   −1   −2   −2
  m$_s$ $+\frac{1}{2}$ $-\frac{1}{2}$ $+\frac{1}{2}$ $-\frac{1}{2}$ $+\frac{1}{2}$ $-\frac{1}{2}$ $+\frac{1}{2}$ $-\frac{1}{2}$ $+\frac{1}{2}$ $-\frac{1}{2}$

8.54 216 kJ/mol   **8.55** 285.2 nm; UV   **8.56** See Section 8.1

8.57 a. ΔE_4 and ΔE_5 b. ΔE_3 and ΔE_4 c. ΔE_7 and ΔE_8

8.58 hardness, melting point, chemical reactivity

8.59 a. several electrons of comparable energy
  b. effective nuclear charge increases
  c. outer s electrons readily removed

8.60 a. energy is absorbed   b. Ar before K   c. react violently with water

8.61 a. 0.15 nm   b. 300°C   c. 400°C   **8.62** 33% BaO

8.63 1.26 g/cm³, 2.08 g/cm³, 4.34 g/cm³   **8.64** 2.80×10^{23}

8.65 2704 kJ, −11474 kJ, 4986 kJ   **8.66** 0.26; 0.61 L

Chapter 9

9.31 a. O^{2-}, F^- b. Na^+, Ni^{2+}, Al^{3+} c. Na_2O, NaF, NiO, NiF_2, Al_2O_3, AlF_3

9.32 a. $CaCO_3$ b. $Fe_3(PO_4)_2$ c. K_2S d. MgI_2

9.33 a. bromine   b. sulfur   c. vanadium   d. oxygen

9.34 MgO, CaS, SrSe, BaTe

9.35 a. $1s^22s^22p^63s^23p^3$; $1s^22s^22p^63s^23p^6$
  b. $1s^22s^22p^63s^23p^64s^23d^{10}4p^65s^1$; $1s^22s^22p^63s^23p^64s^23d^{10}4p^6$
  c. $1s^22s^22p^63s^23p^64s^23d^1$; $1s^22s^22p^63s^23p^6$
  d. $1s^22s^22p^63s^23p^63d^7$; $1s^22s^22p^63s^23p^63d^6$

9.36 a. 1   b. 0   c. 1   d. 0   e. 0

9.37 a. P atom   b. Rb^+ ion   c. Sc^{3+} ion   d. Co^{3+} ion

9.38 a. Br < Br^- < I^- b. Cr^{3+} < Cr^{2+} < Cr   c. Se < Te < Te^{2-}
  d. F^- < Cl^- < I^-

9.39 −328 kJ

9.40 a. :Cl—C—Cl: b. :Cl—P—Cl: c. (:O—N=O:)⁻
        | |
        :Cl: :Cl:

  d. (H—N—H)⁺
        |
        H

9.41 a. (:Cl—P—Cl:)⁺   b. (:C≡N:)⁻   c. (:F—Cl—F:)⁺
        |
        :Cl:

  d. (:N=N=N:)⁻

9.42 a. H—N—N—F: b. (H—O—C—O:)⁻   c. :N≡N—O:
              | | |
              H :F: :O:

  d. :O=N—N=O:

9.43 (H—O≡C:)⁺

9.44 H—C—C—H, H—C—C—H
          | | | |
          :Cl: :Cl: H :Cl:

9.45 a. OH⁻   b. ClO⁻   c. CN⁻   d. PO₄³⁻

9.46 a. :C̈l—N—F̈:   b. (H—Ö—P—Ö—H)⁻   c. (:C̈l—B—C̈l:)⁻

with :C̈l: below N in (a), :Ö: above and below P in (b), :C̈l: above and below B in (c)

d. (H—Ö—S—Ö:)⁻ with :Ö: below S

9.47 a. ·N with :Ö: above and :Ö: below   b. ·C̈—H with H above and H below   c. H—Be—H   d. (·C̈=Ö:)⁻

9.48 a. :Se (with Ö double bond above, Ö below) ⟷ :Se (with Ö above, Ö double bond below)

b. N (with Ö, Ö above; :Ö: below) ⟷ N ⟷ N — three resonance structures

c. :O≡C—Ö: ⟷ :Ö=C=Ö: ⟷ :Ö—C≡O:

9.49 a. C—C (with :Ö:, :Ö. around)   b. C—C , C—C , C—C (three resonance structures)   c. no

9.50 N—F > N—O > N—S > N—N   **9.51** Si—As
9.52 a. N   b. Br   c. H   d. C
9.53 longest in H₃COH, shortest in CO   **9.54** CO strongest, H₃COH weakest
9.55 a. −941 kJ   b. −1543 kJ   c. −2268 kJ   **9.56** −85 kJ
9.57 −57 kJ

9.58 a. Mn²⁺: 1s²2s²2p⁶3s²3p⁶3d⁵   NO₃⁻: [:Ö N Ö:]⁻ with N=O and :Ö: below

b. Mg²⁺: 1s²2s²2p⁶   CO₃²⁻: [:Ö C Ö:]²⁻ with C=O and :Ö: below

c. K⁺: 1s²2s²2p⁶3s²3p⁶   SO₄²⁻: (:Ö—S—Ö:)²⁻ with :Ö: above and :Ö: below S

9.59 a. cation smaller   b. metals lose electrons
     c. can form atom from anion   d. less electronegative atom
9.60 a. octet rule requires electron pairs   b. loses 3 e⁻
     c. electrostatic attraction decreases as ions move farther apart

9.61

H—C C≡C (ring structure) H—C, H—C, C—H, C—H with C—H at bottom, forming a six-membered ring

9.62 :F̈—N̈=N̈—F̈:   :N̈=N̈—F̈—F̈:   :F̈=N̈—F̈—N̈:,   :N̈—F̈—F̈=N̈:,

$$:\ddot{N}=N\begin{matrix}\ddot{F}:\\\\\ddot{F}:\end{matrix}\quad,\quad :\ddot{F}=F\begin{matrix}\ddot{N}:\\\\\ddot{N}:\end{matrix}\quad,\quad \begin{matrix}:\ddot{N}-N:\\|\quad|\\:\ddot{F}-\ddot{F}:\end{matrix}\quad,\quad \begin{matrix}:\ddot{N}-\ddot{F}:\\|\quad|\\:\ddot{F}-N:\end{matrix}\quad,\text{ and many others}$$

9.63　a. 137 g/mol　　b. P　　c. $:\ddot{Cl}-\overset{\displaystyle|}{\underset{\displaystyle\ddot{Cl}:}{P}}-\ddot{Cl}:$

9.64　+2217 kJ　　　**9.65**　52%

Chapter 10

10.31　a. linear　　b. bent　　c. bent　　d. linear
10.32　a. tetrahedral　　b. triangular　　c. bent　　d. tetrahedral
10.33　a. 120°　　b. 120°　　c. 109.5°
10.34　a. $H-\ddot{O}-\ddot{Cl}:$　　b. $:\ddot{Cl}-C\equiv C-H$　　c. $(:\ddot{O}=C=\ddot{N}:)^-$

　　　　　bent　　　　　　　　　linear　　　　　　　　linear

　　　　d. $(H-\overset{\displaystyle|}{\underset{\displaystyle\ddot{O}:}{C}}-\ddot{O}:)^-$ with $||$

　　　　triangular

10.35　a. $:\ddot{F}-\overset{\displaystyle:\ddot{F}:}{\underset{\displaystyle:\ddot{F}:}{Si}}-\ddot{F}:$　　b. $:\ddot{Cl}-\overset{\displaystyle|}{\underset{\displaystyle\ddot{O}:}{C}}-\ddot{Cl}:$　　c. $(:\ddot{O}-\overset{\displaystyle|}{\underset{\displaystyle\ddot{O}:}{Br}}-\ddot{O}:)^-$　　d. $:\ddot{I}-\overset{\displaystyle|}{\underset{\displaystyle\ddot{I}:}{N}}-\ddot{I}:$

　　　　tetrahedral　　　　triangular　　　　pyramidal　　　　pyramidal
10.36　a. 0, 0, 2, 0　　b. 4, 3, 2, 4　　c. 109.5°, 120°, 109.5°, 109.5°
10.37　$H-\overset{\displaystyle H}{\underset{\displaystyle H}{C}}-\overset{\displaystyle H}{\underset{\displaystyle H}{C}}-\overset{\displaystyle|}{\underset{\displaystyle\ddot{O}:}{C}}-\ddot{O}-\ddot{O}-\overset{\displaystyle|}{\underset{\displaystyle\ddot{O}:}{N}}-\ddot{O}:$　109.5°, except around double-bonded C and N, where bond angle is 120°
10.38　a. 109.5°　　b. 120°　　c. 109.5° in CH₃ group, 180° elsewhere
10.39　$SnCl_2$, SO_2; unshared pairs　　**10.40**　b, c, d　　**10.41**　a, b, d nonpolar
10.42　$\begin{matrix}Cl\\ \\H\end{matrix}\!\!>\!\!C\!\!=\!\!C\!\!<\!\!\begin{matrix}H\\ \\Cl\end{matrix}$　is nonpolar
10.43　a. sp　　b. sp²　　c. sp²　　d. sp
10.44　a. sp³ around O atom at left; sp² around N and other O atom
　　　　b. sp² around C and double-bonded O atom; sp³ around other oxygen atoms
　　　　c. sp³
10.45　sp² around all C atoms and N; sp² around double-bonded O; sp³ around other O atom
10.46　a. sp²　　b. sp³　　c. sp　　d. sp　　**10.47**　a. sp³　　b. sp²　　c. sp³
10.48　a. 3 sigma, 1 pi　　b. 3 sigma　　c. 1 sigma, 2 pi　　d. 2 sigma, 2 pi
10.49　a. BF_3　　b. BCl_4^-　　c. H_2CO　　d. CO_2, C_2H_2
10.50　a. 6, sp³d²　　b. 5, sp³d　　c. 5, sp³d　　d. 6, sp³d²
10.51　a. square planar　　b. distorted tetrahedron　　c. trigonal bipyramid
　　　　d. octahedral
10.52　a. trigonal bipyramid　　b. distorted tetrahedron　　c. T-shaped
10.53　a. sp³d　　b. sp³d²　　c. sp³d

10.54

	σ_{2s}^b	σ_{2s}^*	π_{2p}^b	π_{2p}^b	σ_{2p}^b	π_{2p}^*	π_{2p}^*	σ_{2p}^*	Bonds	Unpaired e^-
a.	2	2	2	2	2	1	1		2	2
b.	2	2	2	2	2	2	1		$\frac{3}{2}$	1
c.	2	2	2	2	2	2	2		1	0

10.55

	σ_{2s}^b	σ_{2s}^*	σ_{2p}^b	π_{2p}^b	π_{2p}^b	π_{2p}^*	π_{2p}^*	σ_{2p}^*	Bonds	Unpaired e^-
Li_2	2								1	0
Be_2	2	2							0	0
B_2	2	2	2						1	0
C_2	2	2	2	1	1				2	2
N_2	2	2	2	2	2				3	0
O_2	2	2	2	2	2	1	1		2	2
F_2	2	2	2	2	2	2	2		1	0
Ne_2	2	2	2	2	2	2	2	2	0	0

10.56 a. 2   b. 1   c. $\frac{1}{2}$ d. $\frac{1}{2}$

10.57 a. Apply principles in Table 10.1
  b. pi bonds are the "extra" electron pairs in multiple bonds; others are sigma bonds
  c. See Figure 10.10

10.58 a.

$$\ddot{\underset{\ddot{F}}{\overset{:\ddot{F}:}{S}}}\!\!-\!\ddot{F}:\qquad :\ddot{F}:$$

  b. distorted tetrahedron; 90°, 120°
  c. yes   d. sp³d for S, sp³ for F

10.59 a.

$$\underset{:\ddot{F}:}{\overset{:\ddot{F}:}{\underset{:\ddot{F}}{\overset{:\ddot{F}}{S}}}}\!\!\!\ddot{F}:$$

  b. octahedral, 90°   c. no   d. sp³d² for S

10.60

XY_2E_2	sp³	yes
XY_3	sp²	no
XY_4E_2	sp³d²	no
XY_5	sp³d	no

10.61 H—N̈—N̈—H   109.5° bond angles, polar
              | |
              H   H

10.62 ClF_3 could be triangular, XeF_2 bent. Unshared pairs occupy more space than bonds

10.63 6, sp³d², octahedral

10.64 a. H—C—C=C—H and
             with H H H around first, and the second a triangular C—C—C ring with H's

  b. In 1st molecule; sp³, sp², sp². In 2nd molecule: all sp³
  c. In 1st molecule, 109.5° around CH₃ group, otherwise 120°. In 2nd molecule, all 60°.

10.65 14 electrons around Xe; pentagonal bipyramid

Chapter 11

11.31 a. straight line: 250 mm Hg at 80°C to 208 mm Hg at 20°C   b. 45°C
11.32 a. 4.84 L   b. 298 mm Hg   c. 240 mm Hg

11.33 a. 86°C   b. 74°C   c. 44°C

11.34 a. 2.90×10^4 J   b. 420 mm Hg

11.35 a. $-33°C$ b. 0.27   **11.36** 3.3×10^4 J   **11.37** a, b, c

11.38 a. solid   b. vapor   c. liquid, vapor   **11.39** a

11.40 b. 145°C   c. gas to liquid

11.41 a. molecular   b. metallic   c. ionic

11.42 a. molecular   b. metallic   c. metallic, ionic, network covalent

11.43 a. metallic   b. network covalent   c. molecular   d. ionic   e. ionic

11.44 a. metallic   b. molecular   c. network covalent   d. ionic   e. molecular

11.45 a. ions   b. molecules   c. ions, electrons   d. atoms

11.46 He < Ne < Ar < Xe   **11.47** a, d   **11.48** b, c, d

11.49 a. hydrogen bonding in H_2O b. hydrogen bonding in H_2O_2 c. molecular vs ionic   d. dipole forces in ICl

11.50 b, c

11.51 a. H_2; lower dispersion forces   b. PH_3; no H bonds
    c. PH_3; weaker dispersion forces   d. SO_2; molecular vs. network covalent

11.52 a. dispersion   b. dispersion   c. ionic bonds   d. dispersion

11.53 0.244 nm   **11.54** 0.0582 nm³   **11.55** a. 0.698 nm   b. 0.987 nm

11.56 a. 0.700 nm   b. 0.404 nm

11.57 a. 7.10 cm³   b. 5.05 cm³   c. 0.289

11.58 a. independent of volume   b. relation is more complex
    c. true for nonpolar molecular substances
    d. covalent bonds need not be broken

11.59 a. See Figure 11.16   b. $s \rightarrow l$ vs. $s \rightarrow g$ c. normal bp at 1 atm
    d. phase diagram contains two other curves

11.60 a. evaporation absorbs heat   b. vp drops as T decreases
    c. bp increases with P   d. road tar is molecular; SiO_2 is network covalent

11.61 Each Si bonded to 4 oxygen atoms, each O atom to 2 silicons

11.62 0.6°C; transfer of heat through skate   **11.63** 0.414

11.64 fraction occupied = V atom/V cell = $\dfrac{4\pi\, r^3/3}{8\, r^3}$

11.65 a. 3.20×10^{22} b. 1.97×10^{19} **11.66** MM = 56.8 g/mol

Chapter 12

12.31 Bubble CO_2 through water; heat carefully

12.32 a. $CaCl_2(s) \rightarrow Ca^{2+}(aq) + 2\ Cl^-(aq)$
    b. $LiNO_3(s) \rightarrow Li^+(aq) + NO_3^-(aq)$
    c. $Ni(ClO_3)_2(s) \rightarrow Ni^{2+}(aq) + 2\ ClO_3^-(aq)$
    d. $Ca_3(PO_4)_2(s) \rightarrow 3\ Ca^{2+}(aq) + 2\ PO_4^{3-}(aq)$

12.33 a. 0.040   b. 0.030   c. 0.060   d. 0.040

12.34 a. 9.73   b. 90.3   c. 2.34   **12.35** 26

12.36 X CH_3OH = 0.264

12.37 18.8 g   0.132 mol   0.352 L   0.375
     4.13 g   0.0291 mol   0.291 L   0.100
   102 g   0.718 mol   2.92 L   0.246

12.38 a. dilute 3.21 L of 0.200 M solution to 6.42 L
    b. dilute 1.18 L of 0.543 M solution to 6.42 L

12.39 a. 0.0564 M, 0.0564 M, 0.169 M
    b. 0.120 M, 0.360 M

12.40 a. 0.480 M   b. 0.080 M

12.41 a. 0.9468   b. 3.12   c. 12.6% alcohol, 87.4% water

12.42 2.56, 5.31

12.43 a. NaCl; ionic vs. molecular   b. CH_3OH; H bonding
    c. HOOH; H bonding   d. CO_2; molecular vs. network covalent

12.44 a. +15.2 kJ   b. decrease   **12.45** a. 0.55   b. 0.9955

12.46 a. 1.88, 186 mm Hg   b. 18.8, 169 mm Hg   c. 37.6, 150 mm Hg

12.47 76 mm Hg   **12.48** a. 1.36 atm   b. 3.26 atm   c. 4.07 atm

12.49 a. 12.9 g, 100.39°C   b. 3.47 g, 100.39°C   **12.50** 1.37×10^3 cm^3

12.51 -4.05°C, 84.7°C   **12.52** $C_{10}H_{14}N_2$ **12.53** 1.7×10^2 g/mol

12.54 2.60×10^4 g/mol   **12.55** a > d > b > c   **12.56** b

12.57 a. 2.71   b. 2.27   c. -5.04°C   d. 54.6 atm

12.58 a. i = 3 vs. i = 2   b. more ions   c. solution process endothermic
d. solution process exothermic

12.59 a. osmotic pressures must be equal   b. sugar will crystallize out
c. higher conc. of O_2 at low T   d. air dissolves under pressure
e. CO_2 comes out of solution as pressure drops

12.60 a. measure conductivity   b. CO_2 comes out of solution as T rises
c. moles per liter solution vs. moles per kilogram solvent
d. vapor pressure lowered

12.61 49%   **12.62** 0.0018 g/cm^3

12.63 a. 2.08 M   b. 1.87 mol   c. 47.3 L

12.64 $V_{gas} = \dfrac{n_{gas}RT}{P_{gas}}$; $n_{gas} = kP_{gas}$; $V_{gas} = kRT$

12.65 27 m

Chapter 13

13.31 a. long chain of S atoms   b. long chain molecules in random pattern
c. similar to graphite   d. silicon with small amount of B impurity

13.32 a. $N_2H_4(l) \rightarrow N_2(g) + 2\ H_2(g)$
b. $H_2(g) + Br_2(l) \rightarrow 2\ HBr(g)$
c. $NaCl(s) + H_2SO_4(l) \rightarrow HCl(g) + NaHSO_4(s)$
d. $NaBr(s) + H_3PO_4(l) \rightarrow HBr(g) + NaH_2PO_4(s)$

13.33 a. $2\ ZnS(s) + 3\ O_2(g) \rightarrow 2\ ZnO(s) + 2\ SO_2(g)$
b. $2\ CO(g) + O_2(g) \rightarrow 2\ CO_2(g)$
c. $P_4O_6(s) + 2\ O_2(g) \rightarrow P_4O_{10}(s)$
d. $2\ SO_2(g) + O_2(g) \rightarrow 2\ SO_3(g)$
e. $N_2(g) + O_2(g) \rightarrow 2\ NO(g)$

13.34 a. XeO_3, XeO_4 b. PF_3, PF_5 c. ClF, ClF_3 d. ClF_3, ClF_5, IF_7

13.35 a. C   b. N   c. P

13.36 a. Allotropes must be in same physical state
b. Oxide formulas quite different
c. High T favors lower oxide
d. Resonance forms are not distinct species
e. 6 electron pairs

13.37 a. In alkanes, all carbon-carbon bonds are single; double bond in alkene
b. Resonance forms differ in arrangement of electrons, isomers in arrangement of atoms
c. In straight-chain, carbon atoms are joined in a single, continuous chain
d. C_2H_4 vs. C_3H_6

13.38

13.39

13.40

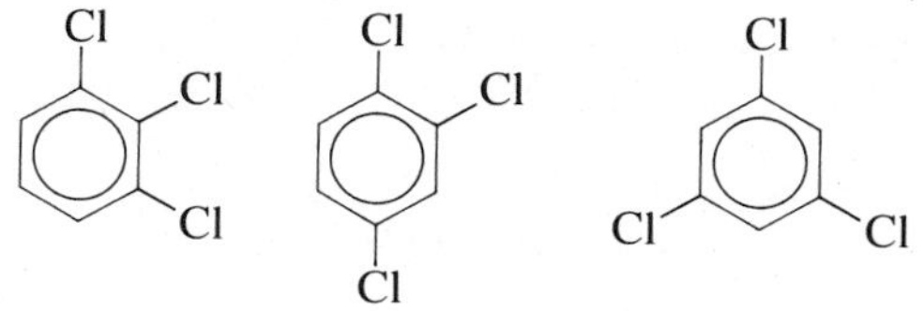

13.41　a. H—N̈—N̈—H　b. H—S̈e—H　c. ·N̈（:Ö above, :Ö below）　d. :F̈—P̈—F̈:（:F̈: below）

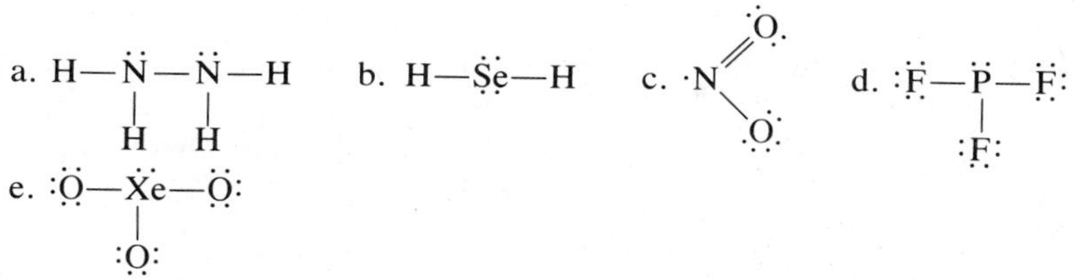

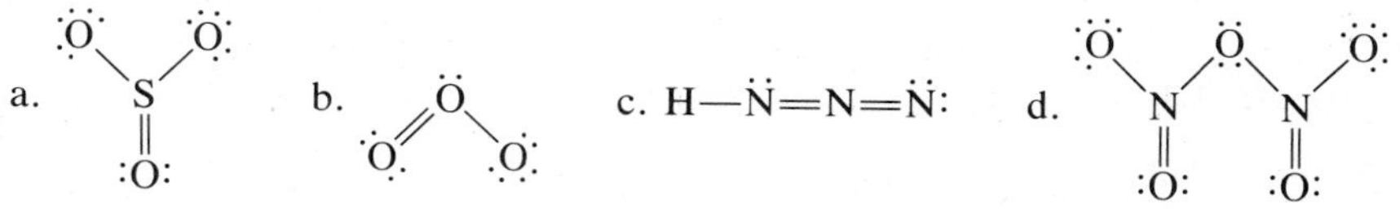

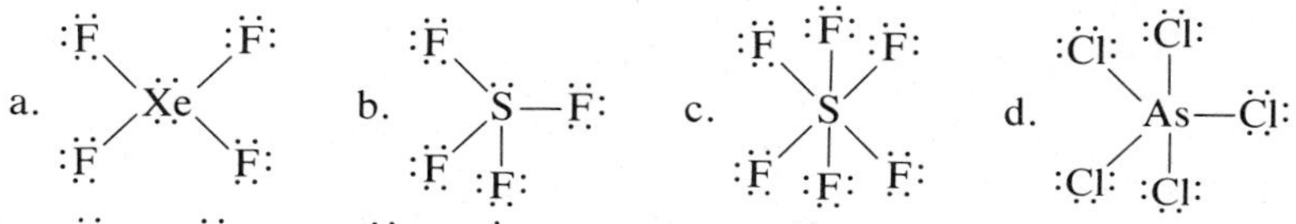

e. :Ö—Ẍe—Ö:（:Ö: below）

13.42　a. （S with :Ö:, :Ö:, and :Ö: below）　b. （O—O—O structure）　c. H—N̈=N̈=N̈:　d. （N—O—N bridged structure）

13.43　a. Xe with four F　b. S—F with F's　c. S with F's　d. As—Cl with Cl's

13.44　a. （resonance forms of S with O's）　b. （O—O—O resonance forms）

c. H—N̈=N̈=N̈: ⟷ H—N̈—N≡N: ⟷ H—N≡N—N̈:

d. （N—O—N resonance structures）, six other resonance forms

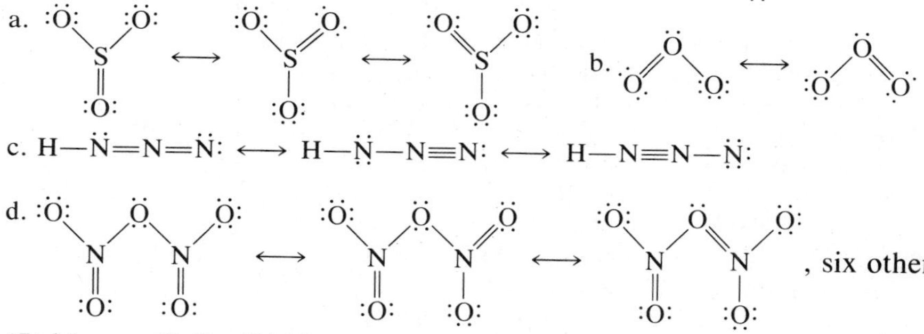

13.45　171 kJ　　**13.46**　464 kJ

13.47　a. 109°　b. 109°　c. 120°　d. 109°　e. 109°

13.48　a. square planar　b. distorted tetrahedron　c. octahedral
d. trigonal bypyramid

13.49　a. nonpolar　b. polar　c. polar　d. nonpolar

13.50　a. sp^3d^2　b. sp^3d　c. sp^3d^2　d. sp^3d

13.51　N_2O_5: 6 sigma, 2 pi　N_2O_2: 3 sigma, 2 pi　N_2O: 2 sigma, 2 pi
N_2O_4: 5 sigma, 2 pi　NO_2: 2 sigma, 1 pi
N_2O_3: 4 sigma, 2 pi　NO: 1 sigma, 1 pi

13.52　120°; polar

13.53　:Ï—C̈l: linear, polar, sp^3

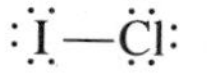

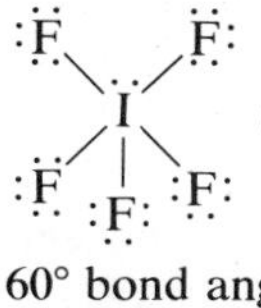

（IF5 structure） square pyramid, polar, sp^3d^2

13.54　60° bond angle, sp^3

13.55　a. less dense　b. absorbs heat　c. rhombic, monoclinic, vapor;
vapor; rhombic

13.56　5.2 atm

13.57　a. dispersion　b. covalent bond　c. covalent bond　d. H bond

13.58 $CH_3CH_2CH_2CH_2Cl$; less compact molecule has stronger dispersion forces

13.59 10.0, 3.47 **13.60** 0.454 mL

13.61

13.62 and others

13.63 a. S_2Cl_2 b. $S_2Cl_2(l) + 3 Cl_2(g) \rightarrow 2 SCl_4(s)$

13.64 pentagonal bipyramid **13.65** CI_4

Chapter 14

14.31 a. -1127.4 kJ b. -555.7 kJ c. -597.4 kJ

14.32 a. $+402.0$ kJ b. -93.4 kJ c. -26.2 kJ

14.33 a. -17.8 kJ b. $+25$ kJ c. -638.5 kJ

14.34 a. $+17.5$ kJ b. -94.3 kJ/mol **14.35** a. $-$ b. $+$ c. $-$ d. $-$

14.36 a. $-$ b. $-$ c. $-$

14.37 a. -198.3 J/K b. -164.0 J/K c. $+541.0$ J/K

14.38 a. -210.5 J/K b. -32.9 J/K c. -9.2 J/K

14.39 a. -255.4 J/K b. -131.5 J/K c. -267.7 J/K

14.40 a. -144.7 J/K b. $+236.8$ J/K c. $+156.7$ J/K

14.41 a. $+6.6$ kJ b. -180.4 kJ c. -10.0 kJ

14.42 a. -52.3 kJ b. -239.3 kJ c. $+38.3$ kJ

14.43 a. -1025.2 kJ b. -503.1 kJ c. -490.3 kJ

14.44 a. $+21.7$ kJ b. -40 kJ c. -681.3 kJ

14.45 a. -180.0 kJ b. -274.4 kJ c. -128.5 kJ

14.46 a. $+0.321$ kJ/K b. $+128.9$ kJ, -192 kJ

14.47 a. $+32.4$ kJ b. -653.0 kJ/mol c. 59.0 J/mol·K

14.48 a. reaction becomes spontaneous at high T
b. nonspontaneous at all T
c. reaction becomes spontaneous at high T

14.49 a. 2.06×10^4 K c. 1050 K

14.50 a. -1035.6 kJ b. -151.8 J/K c. 6820 K

14.51 a. $+353.4$ kJ, $+242.9$ kJ, $+132.4$ kJ, $+21.9$ kJ, -88.6 kJ
b. 840 K

14.52 330 K **14.53** 282 K **14.54** 291 K

14.55 a. 393.5 kJ b. 394.9 kJ

14.56 a. spontaneous below 23.8 mm Hg, nonspontaneous above
b. no; raise T

14.57 ΔH, ΔS

14.58 a. generally true only at low T b. only if ΔH can be ignored
c. number of moles of gas d. true

14.59 a. products more ordered than reactants
b. ΔS surroundings must be considered
c. ΔH, unlike ΔS and ΔG, is essentially independent of pressure, concentration

14.60 a. negative b. less

14.61 a. $+6.00$ kJ b. 0 c. $+0.0220$ kJ/K d. $+0.21$ kJ e. -0.23 kJ

14.62 1.247×10^4 kJ; 21¢ **14.63** a. 5.1 kJ b. 1.8×10^2 g

14.64 1176°C **14.65** 970°C, 2170°C

Chapter 15

15.31 a. 0.100 M   b. forward faster at 60 s; same at 120 s

15.32 a. $A(g) \rightleftharpoons 2\,B(g)$ b. no

15.33 a. $\dfrac{[N_2O] \times [SO_3]}{[NO]^2 \times [SO_2]}$ b. $\dfrac{[SO_3]}{[SO_2] \times [O_2]^{1/2}}$ c. $\dfrac{[H_2O]^2 \times [Cl_2]^2}{[O_2] \times [HCl]^4}$

15.34 a. $[CO_2]$ b. $\dfrac{[CH_4] \times [H_2O]}{[CO] \times [H_2]^3}$ c. $\dfrac{[SnCl_4]}{[Cl_2]^2}$ d. $\dfrac{[CH_4]}{[CO] \times [H_2]^3}$

15.35 a. $2\,BrCl(g) \rightleftharpoons Br_2(g) + Cl_2(g)$
b. $NO_2(g) \rightleftharpoons NO(g) + \frac{1}{2}\,O_2(g)$
c. $2\,NOBr(g) \rightleftharpoons N_2(g) + O_2(g) + Br_2(g)$
d. $4\,HCl(g) + O_2(g) \rightleftharpoons 2\,H_2O(g) + 2\,Cl_2(g)$

15.36 a. 0.50   b. 1×10^3 **15.37** 0.22   **15.38** 0.35

15.39 a. 0.0400 M   b. 0.348 M   c. 2.64×10^{-4}

15.40 a. no   b. $\leftarrow$ **15.41** a. $\rightarrow$ b. $\leftarrow$

15.42 both   **15.43** 1.0 M   **15.44** 0.025 M

15.45 $[H_2] = [I_2] = 0.026$ M; $[HI] = 0.198$ M

15.46

[I₂]	[Br₂]	[IBr]
a. 0.0068 M	0.0068 M	0.14 M
b. 0.007 M	0.007 M	0.14 M
c. 0.010 M	0.010 M	0.205 M

15.47 $[CO] = 0.22$ M, $[Cl_2] = 0.12$ M, $[COCl_2] = 0.080$ M

15.48 $[PBr_5] = 0.39$ M, $[Br_2] = 2.61$ M, $[PBr_3] = 0.61$ M

15.49 a. 0.0046 M   b. 2×10^{-17} M

15.50 a. 0.72   b. $[CO_2] = [H_2] = 0.014$ M; $[CO] = 0.016$ M; $[H_2O] = 0.018$ M

15.51 a. $\leftarrow$ b. $\leftarrow$ c. $\leftarrow$ d. no effect

15.52 a. $\rightarrow$ b. $\leftarrow$ c. no effect

15.53 decrease   **15.54** c or d   **15.55** 0.43, 1.6×10^{-4}, 2.7×10^{-6}

15.56 a. 0.0313, 1.00, 1.41   b. 0.00128, 0.0327, 0.0448

15.57 a. $+54.0$ kJ   b. 2.2×10^{-4} **15.58** 211 kJ, 106 kJ

15.59 a. $[H_2S] = 2x$ b. $[H_2] = 2.00 - 4x$ c. didn't use coefficients as exponents   d. errors in (a) and (b)

15.60 b, c   **15.61** 0.0122 mol/L, 0.0064 mol/L   **15.62** 0.50 mol

15.63 0.11   **15.64** 0.25, $+3.5$ kJ   **15.65** 3.7

Chapter 16

16.31 a. 1.5×10^{-4} mol/L·s   b. 0.0044 mol/L·min

16.32 a. $\dfrac{-1}{4}\dfrac{\Delta\text{conc. HBr}}{\Delta t}$ b. $\dfrac{1}{2}\dfrac{\Delta\text{conc. Br}_2}{\Delta t}$ c. $\dfrac{1}{2}\dfrac{\Delta\text{conc. H}_2\text{O}}{\Delta t}$

16.33 a. 0.0095 mol/L·min   b. 0.0093 mol/L·min

16.34 0.54 mol/L·s; 1.1 mol/L; 0.017 L/mol·s

16.35 a. rate = k   b. 2.5×10^{-4} mol/L·min   c. mol/L·min
d. 2.5×10^{-4} mol/L·min

16.36 a. 2.6×10^{-6} M   b. 2.1×10^{-9} M

16.37 a. rate = k(conc. J)²   b. 1.1×10^2 L/mol·min   c. 0.21 M

16.38 d   **16.39** a. 1   b. 2   c. 3   **16.40** zero

16.41 a. 2, 1   b. rate = k(conc. NO)² × (conc. Cl₂)   c. 8.0 L²/mol²·s
d. 0.26 mol/L·s

16.42 a. 3   b. rate = k(conc. X) × (conc. Z)²; k = 0.46 L²/mol²·h
c. 3.1×10^{-3} mol/L·h

16.43 1st order

16.44 a. 3×10^{-6} mol/L·s   b. 3.1×10^{-6} mol/L·s   c. (b) is more accurate

16.45 a. 0.12/s   b. 19 s   c. 2.3 mol/L
16.46 a. 0.310 M   b. 81.4 min   c. 40.8 min
16.47 a. 0.087 M   b. 23 min   **16.48** C fastest, B slowest
16.49 $E_a' = 130$ kJ; 105 kJ for catalyzed reaction   **16.50** $E_a' = 254$ kJ
16.51 2.1×10^2 kJ   **16.52** 1.2×10^2 kJ   **16.53** 280 K
16.54 30 kJ   **16.55** a. 1.9×10^{-9} b. 2.9×10^{-7} c. 5.9×10^{-6}
16.56 no
16.57 (1) rate $= k_2K_c(\text{conc. NO})^2 \times (\text{conc. } O_2)$
    (2) rate $= k_2K_c(\text{conc. NO})^2 \times (\text{conc. } O_2)$
16.58 2nd
16.59 a. increases f   b. increases collision rate   c. acts as catalyst, lowers E_a
16.60 a. reaction very slow at ordinary T   b. rate $= k(\text{conc.})^0 = k$
    c. See Section 16.5   d. See Section 16.7

16.61 a. integrate: $-dx/x^2 = kdt$ b. integrate: $-dx/x^3 = kdt$ to obtain
$$\frac{1}{2X^2} - \frac{1}{2X_o^2} = kt$$
16.62 rate $= k(\text{conc. A})^2 \times (\text{conc. B}) \times (\text{conc. C})$
16.63 280 g
16.64 $Cl_2(g) \rightleftharpoons 2\,Cl(g)$ initiating
$Cl(g) + CH_4(g) \rightarrow HCl(g) + CH_3(g)$ propagating
$CH_3(g) + Cl_2(g) \rightarrow CH_3Cl(g) + Cl(g)$ propagating
$CH_3(g) + Cl(g) \rightarrow CH_3Cl(g)$ terminating
16.65 rate $= k_3(\text{conc. } NO_2) \times (\text{conc. } NO_3) = k_3K_2(\text{conc. } N_2O_5^*) = k_3K_2K_1(\text{conc. } N_2O_5)$

Chapter 17

17.31 a. $4\,NH_3(g) + 5\,O_2(g) \rightarrow 4\,NO(g) + 6\,H_2O(g)$
    $2\,NO(g) + O_2(g) \rightarrow 2\,NO_2(g)$
    $3\,NO_2(g) + H_2O(l) \rightarrow NO(g) + 2\,HNO_3(l)$
    b. $S(s) + O_2(g) \rightarrow SO_2(g)$
    $SO_2(g) + \frac{1}{2}O_2(g) \rightarrow SO_3(g)$
    c. $NH_3(g) + HNO_3(l) \rightarrow NH_4NO_3(s)$
17.32 increase in P increases rate and yield; increase in T increases rate, decreases yield
17.33 a. 35%   b. 13%   **17.34** 5°C   **17.35** 1136 nm
17.36 623 kJ/mol
17.37 a. also increases yield   b. T must drop below -15°C   c. O_3 present in smog   d. SO_3 forms H_2SO_4
17.38 a. increase in earth's temperature   b. increased exposure to UV
    c. increased acid rain
17.39 a. removal of CO, unburned HC, oxides of N
    b. contains rare metals
    c. converts hemoglobin to more stable CO complex
17.40 a. 1.4×10^4 kg CO   b. no
17.41 a. -425.1 kJ   b. -373.9 kJ   c. -254.3 kJ
17.42 a. -169.2 J/K   b. $+100.7$ J/K   c. -374.5 J/K   d. $+361.1$ J/K
17.43 a. -133.1 kJ   b. -424.1 J/K   c. -6.7 kJ
17.44 a. $\Delta G° = -756.8 + 0.3118\,T$ b. spontaneous below 2150°C
17.45 1490°C   **17.46** a. lower T and P   b, c, d. lower T, increase P
17.47 2.5×10^{-4}
17.48 a. 11% NH_3, 22% N_2, 67% H_2 b. 15.8 mol
    c. $[NH_3] = 0.17$ mol/L, $[N_2] = 0.35$ mol/L, $[H_2] = 1.1$ mol/L
    d. 0.062

17.49 a. 0.45 b. $+7.4$ kJ c. -94.8 J/K
17.50 a. 1.0×10^5 b. 1.0×10^5
17.51 a. 5.0×10^{-5} mol/L·s b. 7.1×10^{-3} M
17.52 1×10^{-12} mol/L·s; 1×10^4 s **17.53** 3.8×10^2 L/mol·s
17.54 3.6×10^{-3} **17.55** 4×10^{24} s
17.56 1st order in H_2, 2nd order in NO; 3.8×10^{11} L²/mol²·min
17.57 a. $+73$ kJ b. $+40$ kJ **17.58** 20°C
17.59 rate constant $= k \times K_c$; K_c decreases as T increases
17.60 rate $= \dfrac{kK_c \times (\text{conc. } O_3)^2}{(\text{conc. } O_2)}$ **17.61** 1.2 M
17.62 a. 1.0×10^{22} L b. 6.3×10^{18} mol c. 2×10^{15} g
17.63 a. 1.8×10^{-15} mol/L b. 1.8×10^{-11} mol/L·s c. 2300 s
17.64 1.1×10^3 K **17.65** a. 4.0 g b. 1.5 L c. 1500

Chapter 18

18.31 $CaCl_2$, $Ce(NO_3)_3$, and $(NH_4)_2S$ are water soluble
18.32 a. NaOH b. Na_2S c. Na_2CO_3
18.33 a. $Ba^{2+}(aq) + CrO_4^{2-}(aq) \rightarrow BaCrO_4(s)$
 b. $Al^{3+}(aq) + 3\ OH^-(aq) \rightarrow Al(OH)_3(s)$
18.34 a. $Ba^{2+}(aq) + SO_4^{2-}(aq) \rightarrow BaSO_4(s)$
 b. $Fe^{3+}(aq) + 3\ OH^-(aq) \rightarrow Fe(OH)_3(s)$
 c. no reaction
 d. $Cu^{2+}(aq) + S^{2-}(aq) \rightarrow CuS(s)$
 e. $Pb^{2+}(aq) + 2\ Cl^-(aq) \rightarrow PbCl_2(s)$
18.35 a. $Ba^{2+}(aq) + CO_3^{2-}(aq) \rightarrow BaCO_3(s)$
 b. no reaction
 c. $Zn^{2+}(aq) + S^{2-}(aq) \rightarrow ZnS(s)$
 d. $Ca^{2+}(aq) + CO_3^{2-}(aq) \rightarrow CaCO_3(s)$
18.36 a. 0.0106 mol/L b. 0.0576 g
18.37 a. 15.4 mL b. 113 mL c. 119 mL
18.38 a. 0.00165, 0.00750 b. 0.00165 c. 2.0 mL
18.39 27.2 **18.40** 62.0 **18.41** 13.2
18.42 a. $Ca(IO_3)_2(s) \rightleftharpoons Ca^{2+}(aq) + 2\ IO_3^-(aq); K_{sp} = [Ca^{2+}] \times [IO_3^-]^2$
 b. $Ag_2SO_3(s) \rightleftharpoons 2\ Ag^+(aq) + SO_3^{2-}(aq); K_{sp} = [Ag^+]^2 \times [SO_3^{2-}]$
 c. $Mn_3(AsO_4)_2(s) \rightleftharpoons 3\ Mn^{2+}(aq) + 2\ AsO_4^{3-}(aq); K_{sp} = [Mn^{2+}]^3 \times [AsO_4^{3-}]^2$
 d. $PbC_2O_4(s) \rightleftharpoons Pb^{2+}(aq) + C_2O_4^{2-}(aq); K_{sp} = [Pb^{2+}] \times [C_2O_4^{2-}]$
18.43 a. $Zr(OH)_4(s) \rightleftharpoons Zr^{4+}(aq) + 4\ OH^-(aq)$
 b. $Pb(OH)Br(s) \rightleftharpoons Pb^{2+}(aq) + OH^-(aq) + Br^-(aq)$
 c. $K_2SiF_6(s) \rightleftharpoons 2\ K^+(aq) + SiF_6^{2-}(aq)$
 d. $Bi_2S_3(s) \rightleftharpoons 2\ Bi^{3+}(aq) + 3\ S^{2-}(aq)$
18.44 2×10^{-4} M, 2×10^{-5} M, 1×10^{-3} M, 1×10^{-5} M
18.45 a. 5×10^{-21} M b. 1×10^{-7} M c. 5×10^{-26} M
18.46 a. 2×10^{-8} M b. 0.02% **18.47** a. 1.3×10^{-4} M b. no
18.48 $MgCO_3$ precipitates **18.49** a. 5.0×10^{-3} M b. 1.0×10^{-2} M
 c. no
18.50 no **18.51** a. 7.87×10^{-3} M b. 1.95×10^{-6}
18.52 a. 1.9×10^{-3} g/L b. 2×10^{-3} M **18.53** 0.22 L
18.54 a. $256\ s^5$ b. s^3 c. $4s^3$ d. $108s^5$
18.55 a. $\leftarrow$ b. $\rightarrow$ c. $\rightarrow$
18.56 a. 6×10^{-7} M b. 7×10^{-9} M c. 1×10^{-16} M
18.57 less soluble at high T **18.58** 0.032 kg
18.59 a. Ca^{2+} b. greater c. $\frac{1}{4}$ d. $Mg(OH)_2$
18.60 a. Na^+ instead of Al^{3+} b. decreases, not increases c. nearly all

18.61 7×10^{-4}; 7×10^2

18.62 a. 1×10^{-5} M   b. Al^{3+}, Fe^{3+} c. $\approx 100\%$ d. 1.5 g

18.63 a. CoS   b. 1×10^{-4} M   **18.64** 12   **18.65** 0.013

Chapter 19

19.31 a. 1.0×10^{-12} M   b. 3.6×10^{-12} M   c. 1.4×10^{-4} M

      d. 3.3×10^{-15} M

19.32 a. 1.2   b. -0.18 c. 7.92   d. 3.36

19.33 a. 1×10^{-9} M, 1×10^{-5} M   b. 13 M, 7.7×10^{-16} M

      c. 6×10^{-6} M, 2×10^{-9} M   d. 1.3×10^{-7} M, 7.7×10^{-8} M

19.34 A, B   **19.35** 3.8, 5.2   **19.36** a. 4×10^{-3} M   b. 2×10^9

19.37 a. $HClO_4(aq) \rightarrow H^+(aq) + ClO_4^-(aq)$; strong

      b. $H_2SO_4(aq) \rightarrow H^+(aq) + HSO_4^-(aq)$; strong

      c. $HClO(aq) \rightleftharpoons H^+(aq) + ClO^-(aq)$; weak

      d. $HI(aq) \rightarrow H^+(aq) + I^-(aq)$; strong

      e. $HF(aq) \rightleftharpoons H^+(aq) + F^-(aq)$; weak

19.38 a. $Ni(H_2O)_5(OH)^+(aq) \rightleftharpoons H^+(aq) + Ni(H_2O)_4(OH)_2(aq)$

      b. $Fe(H_2O)_6^{2+}(aq) \rightleftharpoons Fe(H_2O)_5(OH)^+(aq) + H^+(aq)$

      c. $HSO_4^-(aq) \rightleftharpoons H^+(aq) + SO_4^{2-}(aq)$

      d. $Cr(H_2O)_5(OH)^{2+}(aq) \rightleftharpoons H^+(aq) + Cr(H_2O)_4(OH)_2^+(aq)$

19.39

$[H^+]$	$[OH^-]$	pH
a. 2.0×10^{-13} M	0.050 M	12.70
b. 0.12 M	8.3×10^{-14} M	0.92
c. 1.2×10^{-14} M	0.852 M	13.92
d. 2.1×10^{-13} M	0.048 M	12.68

19.40 a, c, d   **19.41** a. W   b. S   c. S   d. W

19.42 a. $CO_3^{2-}(aq) + H_2O \rightleftharpoons HCO_3^-(aq) + OH^-(aq)$

      b. $CsOH(s) \rightarrow Cs^+(aq) + OH^-(aq)$

      c. $Sr(OH)_2(s) \rightarrow Sr^{2+}(aq) + 2\ OH^-(aq)$

      d. $NO_2^-(aq) + H_2O \rightleftharpoons HNO_2(aq) + OH^-(aq)$

19.43 a. $HCO_3^-(aq) + H_2O \rightleftharpoons H_2CO_3(aq) + OH^-(aq)$

      b. $SO_3^{2-}(aq) + H_2O \rightleftharpoons HSO_3^-(aq) + OH^-(aq)$

      c. $HS^-(aq) + H_2O \rightleftharpoons H_2S(aq) + OH^-(aq)$

19.44 a. 7   b. < 7 c. > 7 d. 7   e. > 7

19.45 $Zn(H_2O)_4^{2+}(aq) \rightleftharpoons H^+(aq) + Zn(H_2O)_3(OH)^+(aq)$

      $C_2H_3O_2^-(aq) + H_2O \rightleftharpoons HC_2H_3O_2(aq) + OH^-(aq)$

      $S^{2-}(aq) + H_2O \rightleftharpoons HS^-(aq) + OH^-(aq)$

19.46 a. neutral

      b. acidic: $NH_4^+(aq) \rightleftharpoons H^+(aq) + NH_3(aq)$

      c. acidic: $Zn(H_2O)_4^{2+}(aq) \rightleftharpoons H^+(aq) + Zn(H_2O)_3(OH)^+(aq)$

      d. basic: $C_2H_3O_2^-(aq) + H_2O \rightleftharpoons HC_2H_3O_2(aq) + OH^-(aq)$

19.47 $FeCl_3$, $FeBr_3$, $Fe(NO_3)_3$, $Fe_2(SO_4)_3$; Na_3PO_4, K_3PO_4, $Ca_3(PO_4)_2$, $Ba_3(PO_4)_2$

19.48 a. $H^+(aq) + S^{2-}(aq) \rightarrow HS^-(aq)$

      b. $HCO_3^-(aq) + OH^-(aq) \rightarrow CO_3^{2-}(aq) + H_2O$

      c. $H^+(aq) + OH^-(aq) \rightarrow H_2O$

19.49 a. $H^+(aq) + OH^-(aq) \rightarrow H_2O$

      b. $H^+(aq) + PO_4^{3-}(aq) \rightarrow HPO_4^{2-}(aq)$

      c. $H^+(aq) + CH_3NH_2(aq) \rightarrow CH_3NH_3^+(aq)$

19.50 pH < 7; 19.48a, 19.49b, 19.49c

      pH $= 7$; 19.48c, 19.49a

      pH > 7; 19.48b

19.51 11.8 mL   **19.52** a. 37.6 mL   b. 589 mL   c. 1.10 L   **19.53** 44.0

19.54 a. any indicator   b. MR   c. PP   d. MR

19.55 a. 1.00   b. 1.48   c. 4.00   d. 7.00   e. 9.00   f. 12.30
19.56 a. $Ni(H_2O)_5(OH)^+$ b. $Zn(H_2O)_4^{2+}$ c. $C_2H_3O_2^-$ d. $H_2PO_4^-$
19.57

Acid	Base	Pairs
a. HF	H_2O	$HF-F^-$; $H_3O^+-H_2O$
b. H_2O	NH_3	H_2O-OH^-; $NH_4^+-NH_3$
c. NH_4^+	OH^-	$NH_4^+-NH_3$; H_2O-OH^-

19.58 H_3O^+: BL acid; BL and Lewis base (in principle)
Ni^{2+}: Lewis acid
NH_3: BL acid; BL and Lewis base
19.59 a. 2.80×10^{-13} M   b. 12.553   c. 0.0742 mol/L
19.60 a. H_2SO_4 b. bromthymol blue   c. $AgNO_3$ d. H_2O, NH_3, ——
19.61 no; yes   **19.62** Test pH of $AgNO_3$ solution   **19.63** 3.0×10^2 mL
19.64 6.4×10^7 g   **19.65** $-0.23°C$ **19.66** a. -5.59 kJ   b. -6.49 kJ

Chapter 20

20.31 a. $HCN(aq) \rightleftharpoons H^+(aq) + CN^-(aq)$; $K_a = \dfrac{[H^+] \times [CN^-]}{[HCN]}$

b. $PH_4^+(aq) \rightleftharpoons H^+(aq) + PH_3(aq)$; $K_a = \dfrac{[H^+] \times [PH_3]}{[PH_4^+]}$

c. $HS^-(aq) \rightleftharpoons H^+(aq) + S^{2-}(aq)$; $K_a = \dfrac{[H^+] \times [S^{2-}]}{[HS^-]}$

20.32 a. 1×10^{-2} b. 3×10^{-7} c. 1.5×10^{-4}
20.33 a. D < A < B < C   b. C   **20.34** 3.2×10^{-8}
20.35 6.6×10^{-5} **20.36** 8.1×10^{-4}
20.37 a. 1.8×10^{-4} M   b. 8.0×10^{-5} M   **20.38** a. 3.52   b. 3.89
20.39 a. 2.1×10^{-5} M   b. 4.8×10^{-10} M   c. 4.68   d. 0.0026
20.40 a. 1.7×10^{-3} M   b. 5.9×10^{-12} M   c. 2.77   d. 0.85
20.41 0.18 M   **20.42** a. 2.85   b. 3.15   c. 3.46   d. 4.15
20.43 3.1×10^{-8} M; 7.51
20.44 a. HSO_3^-, SO_3^{2-} b. HF, F^- c. HSO_4^-, SO_4^{2-}
20.45 a. 1.8   b. 1.8 L   **20.46** a. 4.44   b. 5.66   c. 5.29
20.47 a. 9.72   b. 8.3   **20.48** b   **20.49** 12.21
20.50 a. 1.4×10^{-4} b. 3.3×10^{-6} M, 8.52
20.51 a. formic acid   b. 5.5×10^{-11}, 5.5×10^{-10} c. acetate ion
20.52 10.25   **20.53** 11.17
20.54 a. 1.6×10^{-7} b. 1.7×10^2 c. 1.6×10^7
20.55 a. 5.6×10^4 b. 0.026   c. 1.8×10^9
20.56 a. 5.1×10^{-15} b. 0.51 M
20.57 a. false; much lower in 0.10 M $HC_2H_3O_2$ b. true
c. false; $K_b = 3.2 \times 10^{-10}$
20.58 Na_2CO_3 > NaCN > NaCl > HCN > H_2CO_3 > HCl
20.59 a. NH_3; 11.28   b. H^+, Cl^-; 0.52   c. NH_4^+, Cl^-; 4.82
20.60 a. 2.26   b. 9.04   c. 4.30   **20.61** % diss. = $K_a^{1/2}/[HB]^{1/2}$
20.62 0.96
20.63 HZ < HY < HX; K_a HZ = 2×10^{-7}; K_a HY = 2×10^{-4}
20.64 a. 2.37   b. 4.74   c. 7.44   d. 9.23   e. 11.00   f. 13.52
20.65 5×10^4
20.66 a. H_2CO_3 b. H_2CO_3 c. HCO_3^- d. HCO_3^- e. CO_3^{2-}

Chapter 21

21.31 a. 2 $C_2O_4^{2-}$ ions, 2 Cl^- ions   b. +3   c. $Co(NH_3)_4Cl_2^+$
21.32 a. +2   b. +3   c. +2   d. +2

21.33 a. urea   b. KCl   c. K_3PO_4 d. K_2SO_4

21.34 a. 6   b. 6   c. 4   d. 6

21.35 a. $Co(NH_3)_6^{3+}$ b. $Ag(CN)_2^-$ c. $Zn(OH)_4^{2-}$ d. $Pt(en)_3^{4+}$

21.36 b, c

21.37 a. CN—Ag—CN   b. 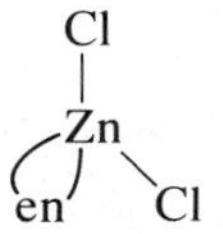   c. 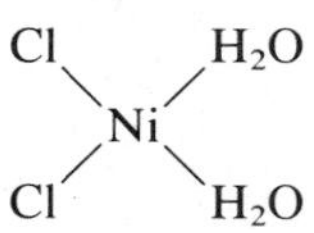   d. 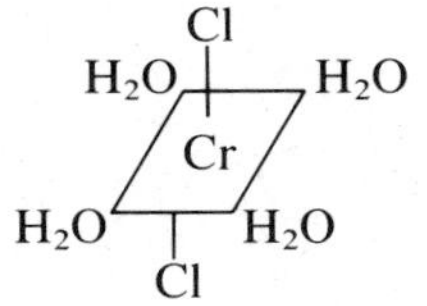   e.

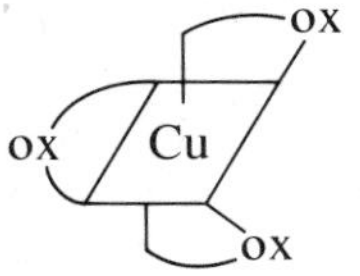

21.38

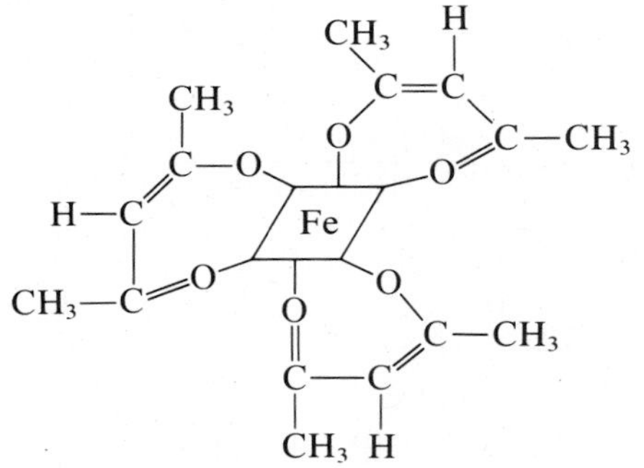

21.39 b. 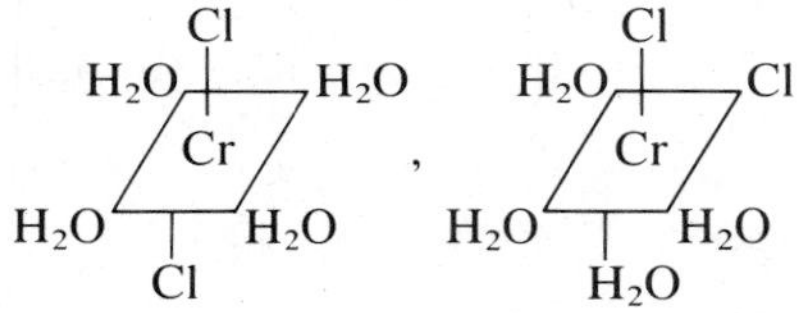   c.

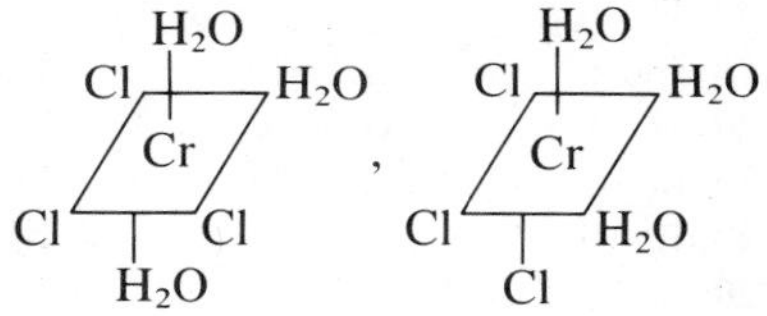

d.

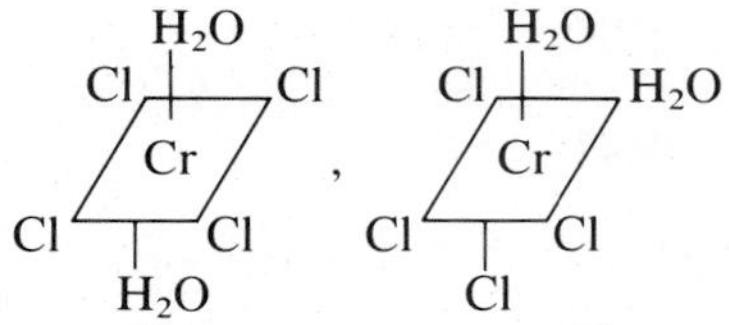

21.40

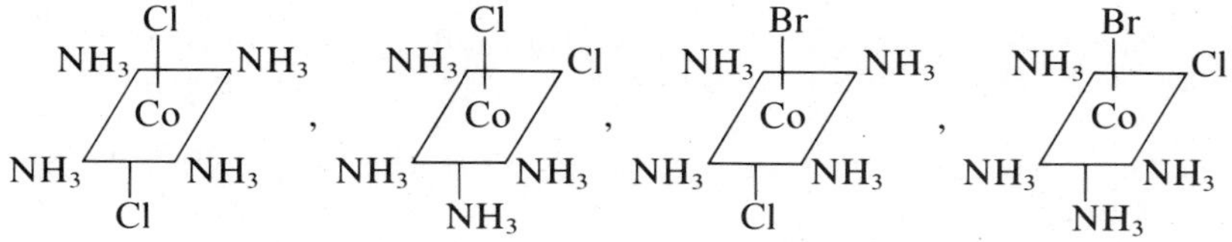

$[Co(NH_3)_4Cl_2]Br$ $[Co(NH_3)_4ClBr]Cl$

21.41 a. $1s^2 2s^2 2p^6 3s^2 3p^6 3d^6$
     b. $1s^2 2s^2 2p^6 3s^2 3p^6 3d^4$
     c. $1s^2 2s^2 2p^6 3s^2 3p^6 3d^{10}$
     d. $1s^2 2s^2 2p^6 3s^2 3p^6 3d^7$
     e. $1s^2 2s^2 2p^6 3s^2 3p^6 4s^2 3d^{10} 4p^6 4d^3$

21.42 (only the d sublevels are shown)
         3d
     a. (↑↓)(↑)(↑)(↑)(↑)   4
     b. (↑)(↑)(↑)(↑)( )   4
     c. (↑↓)(↑↓)(↑↓)(↑↓)(↑↓)   0

3d
d. $(↿⇂)(↿⇂)(↿)(↿)(↿)$ 3
4d
e. $(↿)(↿)(↿)( )( )$ 3

21.43
      3d   4s   4p   4d

a. $(↿⇂)(↿⇂)(↿)(↿)(↿)$ $\underline{(↿⇂)}$ $(↿⇂)(↿⇂)(↿⇂)$
b. $(↿⇂)(↿⇂)(↿)\underline{(↿⇂)(↿⇂)}$ $(↿⇂)$ $(↿⇂)(↿⇂)(↿⇂)$
c. $(↿⇂)(↿⇂)(↿⇂)\underline{(↿⇂)(↿⇂)}$ $(↿⇂)$ $(↿⇂)(↿⇂)(↿⇂)$ $(↿)(↿)( )( )( )$
d. $(↿⇂)(↿)(↿)(↿)(↿)\underline{(↿⇂)}$ $(↿⇂)$ $(↿⇂)(↿⇂)( )$

21.44
                        3d   4s   4p

square planar $(↿⇂)(↿⇂)(↿⇂)(↿⇂)\underline{(↿⇂)}$ $(↿⇂)$ $(↿⇂)(↿⇂)( )$
tetrahedral $(↿⇂)(↿⇂)(↿⇂)(↿)(↿)$ $\underline{(↿⇂)}$ $(↿⇂)(↿⇂)(↿⇂)$

21.45 a. 0   b. 3   c. 1   d. 0

21.46
        3s   3p
Al^{3+} $\underline{(↿⇂)}$ $(↿⇂)(↿⇂)(↿⇂)$

        5d   6s   6p
Au^+ $(↿⇂)(↿⇂)(↿⇂)(↿⇂)(↿⇂)$ $\underline{(↿⇂)}$ $(↿⇂)(↿⇂)(↿⇂)$

        4d   5s   5p
Cd^{2+} $(↿⇂)(↿⇂)(↿⇂)(↿⇂)(↿⇂)$ $\underline{(↿⇂)}$ $(↿⇂)(↿⇂)(↿⇂)$

        3d   4s   4p
Co^{2+} $(↿⇂)(↿⇂)(↿)(↿)(↿)$ $\underline{(↿⇂)}$ $(↿⇂)(↿⇂)(↿⇂)$

        3d   4s   4p
Cu^+ $(↿⇂)(↿⇂)(↿⇂)(↿⇂)(↿⇂)$ $\underline{(↿⇂)}$ $(↿⇂)(↿⇂)(↿⇂)$

        3d   4s   4p
Ni^{2+} $(↿⇂)(↿⇂)(↿⇂)(↿)(↿)$ $\underline{(↿⇂)}$ $(↿⇂)(↿⇂)(↿⇂)$

        3d   4s   4p
Zn^{2+} $(↿⇂)(↿⇂)(↿⇂)(↿⇂)(↿⇂)$ $\underline{(↿⇂)}$ $(↿⇂)(↿⇂)(↿⇂)$

21.47 a. $(↿⇂)(↿⇂)(↿⇂)$ $(↿⇂)(↿)(↿)$ b. $(↿⇂)(↿)(↿)$ $(↿)(↿)(↿)$
          $( )( )$ $(↿)(↿)$ $( )( )$ $(↿)( )$
      low-spin   high-spin   low-spin   high-spin

21.48 Mn^{3+}: $(↿⇂)(↿)(↿)$ and $(↿)(↿)(↿)$ Mn^{4+}: $(↿)(↿)(↿)$
           $( )( )$ $(↿)( )$ $( )( )$
                           $(↿)( )$

21.49 $Cr(CN)_6^{4-}$: $(↿⇂)(↿)(↿)$ $Cr(H_2O)_6^{2+}$: $(↿)(↿)(↿)$

21.50 a. 4   b. 4   c. 4   d. 2   e. 4   **21.51** 260 kJ/mol
21.52 add NH_3 and observe difference in rate of color change
21.53 1.3×10^5 s   **21.54** a. 4.0×10^{-4} M   b. 8.9×10^{-3} M
21.55 1×10^{-14} M   **21.56** 1×10^8
21.57 a. 6, not 5   b. no unpaired electrons   c. shorter wavelength
21.58 a. $Ni(H_2O)_6^{2+}$, $Cu(H_2O)_4^{2+}$ b. $Ag(NH_3)_2^+$, $Ag(CN)_2^-$ c. $Cu(H_2O)_4^{2+}$,
    $Ni(CN)_4^{2-}$ d. $Fe(H_2O)_6^{2+}$, $Co(H_2O)_6^{2+}$ e. $Fe(CN)_6^{4-}$, $Co(CN)_6^{4-}$
21.59 a. forms stable complex with Fe^{3+}
    b. forms complexes with metal ions in soil
    c. $Ni(H_2O)_6^{2+}$ converted to $Ni(NH_3)_6^{2+}$
    d. Co^{3+} complexes nonlabile

21.60 a. $PtN_2H_6Cl_4$ b.

21.61 $[Pt(NH_3)_4][PtCl_4]$ or $[Pt(NH_3)_3Cl][Pt(NH_3)Cl_3]$ **21.62** 0.92 g

21.63 a. $CuC_4H_{22}N_6SO_4$ b. $[Cu(en)_2(NH_3)_2]SO_4$ en, en

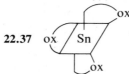

21.64 -37 kJ   **21.65** a. $Al(OH)_3(s) + OH^-(aq) \rightarrow Al(OH)_4^-(aq)$ b. 5

Chapter 22

22.31 yellow color; CO_3^{2-}, ppt. of $MgCO_3$; $PbCl_2$; Zn^{2+}

22.32 separation more difficult with increased number of ions

22.33 Zinc does not have energy levels with separation corresponding to visible λ

22.34 CuS less soluble than NiS; Group III might precipitate

22.35 a. Hg_2Cl_2 b. Cu^{2+} Cd^{2+} c. SO_4^{2-}, Cl^-, CrO_4^{2-}

22.36 a. does not form insoluble chloride or sulfide

b. $NH_4^+(aq) + OH^-(aq) \rightarrow NH_3(g) + H_2O$

c. $BaCO_3(s) + 2 H^+(aq) \rightarrow Ba^{2+}(aq) + CO_2(g) + H_2O$

22.37   ox / Sn — ox, ox, ox

22.38 White ppt. with HCl, partially soluble in hot water. Solution gives yellow ppt. with K_2CrO_4. Precipitate remaining turns grey with NH_3. Solution with NH_3 gives no ppt. on acidification.

22.39 Precipitate with HCl, then treat with NH_3. Grey ppt. indicates Hg_2^{2+}. Test NH_3 solution with H^+; white ppt. indicates Ag^+.

22.40 a. $PbCl_2$ will reprecipitate; $PbCrO_4$ may not form

b. White ppt. of $PbCl_2$ instead of yellow ppt. of $PbCrO_4$

c. No precipitate

22.41 All ions present

22.42 a. observe color   b. treat with NH_3 c. add SO_4^{2-} d. test pH

22.43 $PbCO_3$ present; $PbCl_2$, $Hg_2(NO_3)_2$ absent; AgI in doubt

22.44 a. $Sn^{4+}(aq) + 6 OH^-(aq) \rightarrow Sn(OH)_6^{2-}(aq)$

b. $Zn^{2+}(aq) + 4 NH_3(aq) \rightarrow Zn(NH_3)_4^{2+}(aq)$

c. $Al^{3+}(aq) + 4 OH^-(aq) \rightarrow Al(OH)_4^-(aq)$

22.45 a. $Zn(NH_3)_4^{2+} aq) + 4 H^+(aq) \rightarrow Zn^{2+}(aq) + 4 NH_4^+(aq)$

b. $Pb(OH)_3^-(aq) + 3 H^+(aq) \rightarrow Pb^{2+}(aq) + 3 H_2O$

c. $Ag(CN)_2^-(aq) + 2 H^+(aq) \rightarrow Ag^+(aq) + 2 HCN(aq)$

d. $Ni(NH_3)_6^{2+}(aq) + 6 H^+(aq) \rightarrow Ni^{2+}(aq) + 6 NH_4^+(aq)$

22.46 a. $Mg^{2+}(aq) + 2 OH^-(aq) \rightarrow Mg(OH)_2(s)$

b. $Mg^{2+}(aq) + 2 NH_3(aq) + 2 H_2O \rightarrow Mg(OH)_2(s) + 2 NH_4^+(aq)$

c. $Mg^{2+}(aq) + S^{2-}(aq) + 2 H_2O \rightarrow Mg(OH)_2(s) + H_2S(aq)$

22.47 a. $MgCO_3(s) + 2 H^+(aq) \rightarrow Mg^{2+}(aq) + H_2CO_3(aq)$

b. $MnS(s) + 2 H^+(aq) \rightarrow Mn^{2+}(aq) + H_2S(aq)$

c. $Fe(OH)_3(s) + 3 H^+(aq) \rightarrow Fe^{3+}(aq) + 3 H_2O$

d. $Cu(NH_3)_4^{2+}(aq) + 4 H^+(aq) \rightarrow Cu^{2+}(aq) + 4 NH_4^+(aq)$

22.48 a. $AgCl(s) + 2 NH_3(aq) \rightarrow Ag(NH_3)_2^+(aq) + Cl^-(aq)$

b. $Pb^{2+}(aq) + 2 NH_3(aq) + 2 H_2O \rightarrow Pb(OH)_2(s) + 2 NH_4^+(aq)$

c. $Ni^{2+}(aq) + 6 NH_3(aq) \rightarrow Ni(NH_3)_6^{2+}(aq)$

22.49 a. $Mg^{2+}(aq) + 2 OH^-(aq) \rightarrow Mg(OH)_2(s)$

b. $Pb^{2+}(aq) + 2 OH^-(aq) \rightarrow Pb(OH)_2(s)$

$Pb(OH)_2(s) + OH^-(aq) \rightarrow Pb(OH)_3^-(aq)$

22.50 a. $ZnCO_3(s) + 4 NH_3(aq) \rightarrow Zn(NH_3)_4^{2+}(aq) + CO_3^{2-}(aq)$
 b. $ZnCO_3(s) + 4 OH^-(aq) \rightarrow Zn(OH)_4^{2-}(aq) + CO_3^{2-}(aq)$
 c. $ZnCO_3(s) + 4 CN^-(aq) \rightarrow Zn(CN)_4^{2-}(aq) + CO_3^{2-}(aq)$
 d. $ZnCO_3(s) + 2 H^+(aq) \rightarrow Zn^{2+}(aq) + H_2CO_3(aq)$
22.51 $PbCO_3(s) + 2 H^+(aq) \rightarrow Pb^{2+}(aq) + H_2CO_3(aq)$
 $PbCO_3(s) + 3 OH^-(aq) \rightarrow Pb(OH)_3^-(aq) + CO_3^{2-}(aq)$
22.52 a. $AgCl(s) + 2 NH_3(aq) \rightarrow Ag(NH_3)_2^+(aq) + Cl^-(aq)$
 b. $Ag(NH_3)_2^+(aq) + Cl^-(aq) + 2 H^+(aq) \rightarrow AgCl(s) + 2 NH_4^+(aq)$
22.53 a. 1.0×10^{-10} b. 0.18 c. 0.90 mL   **22.54** 3.89
22.55 a. 2.1×10^{-2} b. 0.012 g   **22.56** a. 0.033 M   b. 2.3×10^{-2} M
22.57 HgS precipitates, FeS does not   **22.58** 10% **22.59** 1×10^{-6}
22.60 5×10^9; $Al(OH)_3$
22.61 $Ni^{2+}(aq) + 2 NH_3(aq) + 2 H_2O \rightarrow Ni(OH)_2(s) + 2 NH_4^+(aq)$
 $Al^{3+}(aq) + 3 NH_3(aq) + 3 H_2O \rightarrow Al(OH)_3(s) + 3 NH_4^+(aq)$
 $Ni(OH)_2(s) + 6 NH_3(aq) \rightarrow Ni(NH_3)_6^{2+}(aq) + 2 OH^-(aq)$
 $Al(OH)_3(s) + OH^-(aq) \rightarrow Al(OH)_4^-(aq)$
 $Al(OH)_4^-(aq) + H^+(aq) \rightarrow Al(OH)_3(s) + H_2O$
 $Al(OH)_3(s) + 3 H^+(aq) \rightarrow Al^{3+}(aq) + 3 H_2O$
22.62 0.02 M   **22.63** a. 1×10^{-16} b. 1×10^{-30} c. 1×10^{-44}
22.64 9×10^{-18} **22.65** 8×10^5

Chapter 23

23.31 a. $H = +1$, $O = -2$, $C = +4$ b. $O = -2$, $S = +2$ c. $Mg = +2$,
 $O = -2$, $S = +6$ d. $O = -2$, $Cl = +7$ e. $F = -1$, $S = +6$
23.32 a. $O = -2$, $Sb = +5$ b. $Ca = +2$, $O = -2$, $C = +3$ c. $H = +1$,
 $O = -2$, $S = +6$ d. $Na = +1$, $O = -2$, $Fe = +3$ e. $H = +1$,
 $O = -2$, $P = +3$
23.33 a. O   b. R   c. O   d. R
23.34 a. O, R agent   b. O agent   c. O, R agent   d. O, R agent
 e. O, R agent
23.35 a. NO_2^- b. CrO_4^{2-} c. CrO_4^{2-} d. NO_2^-
23.36 a. Sn^{2+} oxidized, Fe^{3+} reduced; Sn^{2+} reducing agent, Fe^{3+} oxidizing agent
 b. I_2 oxidized, NO_3^- reduced; I_2 reducing agent, NO_3^- oxidizing agent
23.37 $3 NO_2^-(aq) + 2 CrO_4^{2-}(aq) + 5 H_2O \rightarrow 3 NO_3^-(aq) + 2 Cr(OH)_3(s) + 4 OH^-(aq)$
23.38 a. $Sn^{2+}(aq) + 2 Fe^{3+}(aq) \rightarrow Sn^{4+}(aq) + 2 Fe^{2+}(aq)$
 b. $10 NO_3^-(aq) + I_2(s) + 8 H^+(aq) \rightarrow 10 NO_2(g) + 2 IO_3^-(aq) + 4 H_2O$
 c. $Cl_2(g) + 2 OH^-(aq) \rightarrow Cl^-(aq) + ClO^-(aq) + H_2O$
23.39 a. $4 Zn(s) + NO_3^-(aq) + 10 H^+(aq) \rightarrow 4 Zn^{2+}(aq) + NH_4^+(aq) + 3 H_2O$
 b. $6 Fe^{2+}(aq) + Cr_2O_7^{2-}(aq) + 14 H^+(aq) \rightarrow 6 Fe^{3+}(aq) + 2 Cr^{3+}(aq) + 7 H_2O$
 c. $P_4(s) + 6 H_2O \rightarrow 2 PH_3(g) + 2 HPO_3^{2-}(aq) + 4 H^+(aq)$
23.40 a. $2 Fe(OH)_2(s) + H_2O + NO_3^-(aq) \rightarrow 2 Fe(OH)_3(s) + NO_2^-(aq)$
 b. $2 Cr(OH)_3(s) + ClO_3^-(aq) + 4 OH^-(aq) \rightarrow 2 CrO_4^{2-}(aq) + Cl^-(aq)- + 5 H_2O$
 c. $4 Cl_2(g) + 8 OH^-(aq) \rightarrow ClO_4^-(aq) + 4 H_2O + 7 Cl^-(aq)$
23.41 $4 Fe(s) + 6 H_2O + 3 O_2(g) \rightarrow 4 Fe(OH)_3(s)$
23.42 a. R, R, O, O
 b. $Cl_2(g) + 2 e^- \rightarrow 2 Cl^-(aq)$
 $MnO_4^-(aq) + 8 H^+(aq) + 5 e^- \rightarrow Mn^{2+}(aq) + 4 H_2O$
 $H_2S(g) \rightarrow S(s) + 2 H^+(aq) + 2 e^-$
 $2 Cr^{3+}(aq) + 7 H_2O \rightarrow Cr_2O_7^{2-}(aq) + 14 H^+(aq) + 6 e^-$

c. $Cl_2(g) + H_2S(g) \rightarrow 2\ Cl^-(aq) + S(s) + 2\ H^+(aq)$
$3\ Cl_2(g) + 2\ Cr^{3+}(aq) + 7\ H_2O \rightarrow 6\ Cl^-(aq) + Cr_2O_7^{2-}(aq) + 14\ H^+(aq)$
$2\ MnO_4^-(aq) + 6\ H^+(aq) + 5\ H_2S(g) \rightarrow 2\ Mn^{2+}(aq) + 8\ H_2O + 5\ S(s)$
$6\ MnO_4^-(aq) + 10\ Cr^{3+}(aq) + 11\ H_2O \rightarrow 6\ Mn^{2+}(aq) + 22\ H^+(aq) +$
$\quad 5\ Cr_2O_7^{2-}(aq)$

23.43 a. O, O, O, R
b. $H_2(g) + 2\ OH^-(aq) \rightarrow 2\ H_2O + 2\ e^-$
$4\ OH^-(aq) \rightarrow O_2(g) + 2\ H_2O + 4\ e^-$
$Ni(OH)_2(s) + 2\ OH^-(aq) \rightarrow NiO_2(s) + 2\ H_2O + 2\ e^-$
$MnO_4^-(aq) + 3\ H_2O + 5\ e^- \rightarrow MnO(s) + 6\ OH^-(aq)$
c. $5\ H_2(g) + 2\ MnO_4^-(aq) \rightarrow 4\ H_2O + 2\ MnO(s) + 2\ OH^-(aq)$
$4\ MnO_4^-(aq) + 2\ H_2O \rightarrow 5\ O_2(g) + 4\ MnO(s) + 4\ OH^-(aq)$
$5\ Ni(OH)_2(s) + 2\ MnO_4^-(aq) \rightarrow 5\ NiO_2(s) + 4\ H_2O + 2\ MnO(s) +$
$\quad 2\ OH^-(aq)$

23.44 a. $2\ MnO_4^-(aq) + 10\ I^-(aq) + 16\ H^+(aq) \rightarrow 2\ Mn^{2+}(aq) + 8\ H_2O + 5\ I_2(s)$
b. 0.598 M
23.45 a. 1.19 g　　b. 0.0680 L　　**23.46**　a. 28.2 mL　　b. 0.478 g
23.47 a. H_2, O_2　　b. Co, Cl_2　　c. H_2, I_2　　d. Cu, O_2
23.48 a. $2\ H_2O + 2\ e^- \rightarrow H_2(g) + 2\ OH^-(aq)$
$\underline{2\ OH^-(aq) \rightarrow \frac{1}{2}\ O_2(g) + H_2O + 2\ e^-}$
$\quad H_2O \rightarrow H_2(g) + \frac{1}{2}\ O_2(g)$

b. $Pb^{2+}(aq) + 2\ e^- \rightarrow Pb(s)$
$\underline{H_2O \rightarrow \frac{1}{2}\ O_2(g) + 2\ H^+(aq) + 2\ e^-}$
$Pb^{2+}(aq) + H_2O \rightarrow Pb(s) + \frac{1}{2}\ O_2(g) + 2\ H^+(aq)$

c. $2\ H^+(aq) + 2\ e^- \rightarrow H_2(g)$
$\underline{2\ Br^-(aq) \rightarrow Br_2(l) + 2\ e^-}$
$2\ H^+(aq) + 2\ Br^-(aq) \rightarrow H_2(g) + Br_2(l)$

23.49

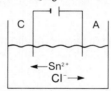

23.50 a. 8.47 g Zn　　b. 6.64 cm³　　c. 4.08×10^3 s
23.51 a. 0.236 g OH⁻　　b. 0.170 L
23.52 a. $Ag(CN)_2^-(aq) + e^- \rightarrow Ag(s) + 2\ CN^-(aq)$
b. 894 C
c. 3.04×10^3 J, 8.44×10^{-4} kWh
23.53 a. 6.4×10^7 J　　b. \$1.1
23.54 a. $Mg(s) + Fe^{2+}(aq) \rightarrow Mg^{2+}(aq) + Fe(s)$
b. $2\ Al(s) + 3\ Zn^{2+}(aq) \rightarrow 2\ Al^{3+}(aq) + 3\ Zn(s)$
c. $Sn(s) + \frac{1}{2}\ O_2(g) + 2H^+(aq) \rightarrow Sn^{2+}(aq) + H_2O$
23.55 a.　　　　　　　　　　　　　　b.

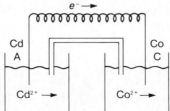
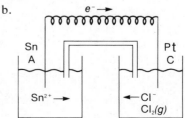

c.

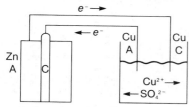

23.56 a. $PbO_2(s) + 4 H^+(aq) + SO_4^{2-}(aq) + 2 e^- \rightarrow PbSO_4(s) + 2 H_2O$
b. $PbO_2(s) + Pb(s) + 4 H^+(aq) + 2 SO_4^{2-}(aq) \rightarrow 2 PbSO_4(s) + 2 H_2O$
c. $Cd(s) + NiO_2(s) + 2 H_2O \rightarrow Cd(OH)_2(s) + Ni(OH)_2(s)$

23.57 $Zn(s) \rightarrow Zn^{2+}(aq) + 2 e^-$
$MnO_2(s) + 4 NH_4^+(aq) + 2 e^- \rightarrow Mn^{2+}(aq) + 4 NH_3(aq) + 2 H_2O$

$Zn(s) + MnO_2(s) + 4 NH_4^+(aq) \rightarrow Zn^{2+}(aq) + Mn^{2+}(aq) + 4 NH_3(aq) + 2 H_2O$

23.58 a. H^+ required for reaction, SO_4^{2-} ions plate $PbSO_4$
b. Allows current flow but prevents direct contact between reactants
c. Removal of electrons increases positive charge
d. Slower deposition of metal gives more adherent coat

23.59 a. T   b. F; move to cathode   c. F; $PbSO_4$ is removed   d. F; oxidized

23.60

23.61 a. $+2, +4, +3, +\frac{8}{3}$
b. Pb_2O_3: one Pb^{2+} per Pb^{4+}
c. Pb_3O_4: two Pb^{2+} per Pb^{4+}

23.62 0.794 g CaC_2O_4; yes   **23.63** 4.50 g   **23.64** 1.23 V

23.65 N_2O_3: N = $+4$ and $+2$; O = -2
N_2O: N = $+2$ and 0; O = -2

Chapter 24

24.31 oxidizing agents: $Co^{2+}, Br_2, Cu^{2+}, Sn^{2+}$; Br_2 strongest
reducing agents: Co^{2+}, Sn^{2+}, Al; Al strongest

24.32 oxidizing agents: $Mn^{2+} < Sn^{4+} < Ag^+ < Br_2 < Cr_2O_7^{2-}$
reducing agents: $Mn^{2+} < Cd$

24.33 a. any species between Fe^{2+} and Sn^{2+} in the left column of Table 24.1
b. any species between Mg and Co in the right column of Table 24.1
c. any species between Cl_2 and Co^{3+} in the left column of Table 24.1

24.34 a. $+0.28$ V   b. $+1.64$ V   c. $+0.93$ V
24.35 a. $+1.21$ V   b. $+0.54$ V   c. $+0.52$ V
24.36 a. $+0.94$ V   b. $+2.12$ V   c. $+1.32$ V
24.37 a. -0.46 V   b. -0.34 V   c. $+1.56$ V   **24.38** b   **24.39** b
24.40 a. yes   b. yes   c. no   **24.41** I^-, Ag
24.42 b. $Fe^{2+}(aq) + Ag^+(aq) \rightarrow Fe^{3+}(aq) + Ag(s)$
c. $2 ClO_3^-(aq) + 12 H^+(aq) + 10 Hg(l) \rightarrow Cl_2(g) + 6 H_2O + 5 Hg_2^{2+}(aq)$
24.43 $+0.52$ V   **24.44** $+0.96$ V
24.45 a. $+0.81$ V; spon.   b. -0.31 V; nonspon.   **24.46** 4×10^{-5} M
24.47 a. $+0.24$ V   b. 9×10^{-5} M

24.48 a. -0.46 V   b. 2×10^{-9} M   c. 2×10^{-10}

24.49 10^{-25}

24.50 a. $+290$ kJ, 2×10^{-51} b. $+150$ kJ, 10^{-25} c. -150 kJ, 10^{25}

24.51 a. $+0.104$ V   b. -0.155 V   **24.52** $+0.570$ V

24.53 $+0.20$ V, -58 kJ, 10^{10}

24.54 a. -54 kJ   b. -633 kJ

c. -180 kJ for $Mn(s) + Ni^{2+}(aq) \rightarrow Mn^{2+}(aq) + Ni(s)$

24.55 a. 8×10^{40} b. 10^{18}

c. 10^{106} for $4\,Cr(OH)_3(s) + 3\,O_2(g) + 8\,OH^-(aq) \rightarrow 4\,CrO_4^{2-}(aq) + 10\,H_2O$

24.56 electrolytic, not voltaic (at standard concentrations)

24.57 a. ions carry current   b. O_2 concentration cell

24.58 a. metal oxidized instead of Fe   b. tin forms complex with citrate ion

24.59 a. prevents access to O_2 and H_2O

b. existence of anodic and cathodic areas

24.60 a. decrease   b. decreases   c. no effect   d. no effect

24.61 a. $+0.62$ V   b. conc. Zn^{2+} increases, Sn^{2+} decreases   c. 10^{21}

d. conc. $Zn^{2+} = 2.0$ M, $Sn^{2+} = 2 \times 10^{-21}$ M

24.62 a. $+112$ kJ, 38.0 kJ, 150 kJ   b. -0.389 V

24.63 a. 10^{14} b. conc. Ce^{3+} = conc. Fe^{2+} = conc. Fe^{3+} = 0.050 M;

conc. $Ce^{4+} = 5 \times 10^{-16}$ M

24.64 2×10^{-10} **24.65** $+0.41$ V

Chapter 25

25.31 a. $2\,Bi_2S_3(s) + 9\,O_2(g) \rightarrow 2\,Bi_2O_3(s) + 6\,SO_2(g)$

b. $Fe_2O_3(s) + 3\,CO(g) \rightarrow 2\,Fe(s) + 3\,CO_2(g)$

c. $3\,MnO(s) + 2\,Al(s) \rightarrow 3\,Mn(s) + Al_2O_3(s)$

25.32 a. $Ni^{2+}(aq) + 6\,NH_3(aq) \rightarrow Ni(NH_3)_6^{2+}(aq)$

b. $Fe(H_2O)_5(OH)^{2+}(aq) + H^+(aq) \rightarrow Fe(H_2O)_6^{3+}(aq)$

c. $Zn(OH)_2(s) + 2\,OH^-(aq) \rightarrow Zn(OH)_4^{2-}(aq)$

25.33 a. $Cd^{2+}(aq) + H_2S(aq) \rightarrow CdS(s) + 2\,H^+(aq)$

b. $Fe^{2+}(aq) + 2\,NH_3(aq) + 2\,H_2O \rightarrow Fe(OH)_2(s) + 2\,NH_4^+(aq)$

c. $Fe^{2+}(aq) + 2\,OH^-(aq) \rightarrow Fe(OH)_2(s)$

25.34 a. $2\,ZnS(s) + 3\,O_2(g) \rightarrow 2\,ZnO(s) + 2\,SO_2(g)$

b. $Mn(s) + 2\,H^+(aq) \rightarrow Mn^{2+}(aq) + H_2(g)$

c. $Zn(OH)_2(s) + 2\,OH^-(aq) \rightarrow Zn(OH)_4^{2-}(aq)$

25.35 $Zn^{2+}(aq) + 2\,OH^-(aq) \rightarrow Zn(OH)_2(s)$

$Zn(OH)_2(s) + 2\,OH^-(aq) \rightarrow Zn(OH)_4^{2-}(aq)$

25.36 a. $Ba^{2+}(aq) + CrO_4^{2-}(aq) \rightarrow BaCrO_4(s)$

b. $Ni^{2+}(aq) + 6\,NH_3(aq) \rightarrow Ni(NH_3)_6^{2+}(aq)$

c. $Fe(s) + 2\,H^+(aq) \rightarrow Fe^{2+}(aq) + H_2(g)$

25.37 a. $Cd(s) + 2\,H^+(aq) \rightarrow Cd^{2+}(aq) + H_2(g)$

b. $3\,Cd(s) + 2\,NO_3^-(aq) + 8\,H^+(aq) \rightarrow 3\,Cd^{2+}(aq) + 2\,NO(g) + 4\,H_2O$

25.38 $3\,Cd(s) + 12\,Cl^-(aq) + 2\,NO_3^-(aq) + 8\,H^+(aq) \rightarrow 3\,CdCl_4^{2-}(aq) + 2\,NO(g)$

$+ 4\,H_2O$

25.39 a. $Fe(s) + 3\,NO_3^-(aq) + 6\,H^+(aq) \rightarrow Fe^{3+}(aq) + 3\,NO_2(g) + 3\,H_2O$

b. $2\,Cr(OH)_3(s) + 3\,ClO^-(aq) + 4\,OH^-(aq) \rightarrow 2\,CrO_4^{2-}(aq) + 3\,Cl^-(aq) +$

$5\,H_2O$

25.40 a. $4\,Cr^{2+}(aq) + O_2(g) + 4\,H^+(aq) \rightarrow 4\,Cr^{3+}(aq) + 2\,H_2O$

b. $2\,MnO_2(s) + 3\,ClO^-(aq) + 2\,OH^-(aq) \rightarrow 2\,MnO_4^-(aq) + 3\,Cl^-(aq) + H_2O$

25.41 Cr^{2+}, Fe^{2+}, Hg_2^{2+} **25.42** M^{2+} **25.43** a, c

25.44 a. -2.23 V; nonspon.   b. $+0.96$ V; spon.

25.45 a. Co^{2+} b. Mn^{2+} **25.46** Mn^{3+}, Fe^{3+}, Co^{3+}, Au^{3+}, Au^+, Hg^{2+}, Hg_2^{2+}

25.47 a. -1.38 V   b. 1×10^{-20} M   **25.48** $+0.86$ V

25.49 a. $+0.8$ V   b. $+0.6$ V

25.50 a. -98 kJ   b. $+185$ kJ   c. -405 kJ
25.51 a. 2×10^{-38} b. 3×10^{32}
25.52 a. 7×10^9 b. 2×10^{-4} M
25.53 a. $+78.0$ kJ   b. -19 kJ   **25.54** -0.229 V, 1.8×10^{-8}
25.55 a. high density   b. electr. cond.   c. uniform coefficient of thermal expansion
25.56 See Equation 25.13
25.57 a. Mn   b. 3 volumes 12 M HCl, 1 volume 16 M HNO_3
c. Cu^+, Au^+ d. Hg
25.58 a. heat HgS in air   b. See Equation 25.1
c. reduction of Cr_2O_3 with C or Al
25.59 a. d orbitals filled in Zn^{2+} b. Mn^{3+} oxidizes water
c. Ag too difficult to oxidize
25.60 5.0   **25.61** 0.004   **25.62** a. 0.35g   b. 0.58 mL
25.63 4×10^{-10}
25.64 a. $Fe(OH)_3(s) + 3 H_2C_2O_4(aq) \rightarrow Fe(C_3O_4)_3{}^{3-}(aq) + 3 H_2O + 3 H^+(aq)$
b. 0.28 L
25.65 a. 1, 5   b. NO^+ and high spin Fe^+

Chapter 26

26.31 a. iodous acid   b. phosphate   c. nitrite   d. hypohosphorous acid
26.32 a. sulfurous acid   b. iodic acid   c. hydrogen sulfate ion
d. perchloric acid
26.33 a. Na_3AsO_3 b. H_3AsO_3
26.34 a. HNO_3 b. H_3PO_4 c. H_2SO_4 d. $HClO_3$
26.35 a. red. agent   b. either   c. oxid. agent   d. either
26.36 All except NH_3 can act as oxidizing agents; all except $NO_3{}^-$ can be reducing agents
26.37 3.05   **26.38** a. 2.62   b. 7.85   c. 6.77
26.39 a. $2 Sn^{2+}(aq) + O_2(g) + 2 H_2O \rightarrow 2 SnO_2(s) + 4 H^+(aq)$
b. $H_2O_2(aq) + 2 H^+(aq) + 2 I^-(aq) \rightarrow I_2(s) + 2 H_2O$
c. $4 Fe(s) + 3 O_2(g) + 6 H_2O \rightarrow 4 Fe(OH)_3(s)$
26.40 a. $I_2(s) + 2 OH^-(aq) \rightarrow I^-(aq) + IO^-(aq) + H_2O$
b. $2 I^-(aq) + SO_4{}^{2-}(aq) + 4 H^+(aq) \rightarrow I_2(s) + SO_2(g) + 2 H_2O$
26.41 a. $2 ClO_3{}^-(aq) \rightarrow ClO_4{}^-(aq) + ClO_2{}^-(aq)$
b. $3 HNO_2(aq) \rightarrow NO_3{}^-(aq) + 2 NO(g) + H^+(aq) + H_2O$
c. $3 NO_2{}^-(aq) + H_2O \rightarrow NO_3{}^-(aq) + 2 NO(g) + 2 OH^-(aq)$
26.42 a. $2 MnO_4{}^-(aq) + 16 H^+(aq) + 10 I^-(aq) \rightarrow 2 Mn^{2+}(aq) + 8 H_2O + 5 I_2(s)$
b. $MnO_4{}^-(aq) + 8 H^+(aq) + 5 Fe^{2+}(aq) \rightarrow Mn^{2+}(aq) + 4 H_2O + 5 Fe^{3+}(aq)$
c. $2 MnO_4{}^-(aq) + H^+(aq) + 5 HNO_2(aq) \rightarrow 2 Mn^{2+}(aq) + 3 H_2O + 5 NO_3{}^-(aq)$
d. $2 MnO_4{}^-(aq) + 6 H^+(aq) + 5 H_2S(aq) \rightarrow 2 Mn^{2+}(aq) + 8 H_2O + 5 S(s)$
26.43 a. $HClO_2$ b. Cl_2 c. $ClO_3{}^-$ d. $ClO_2{}^-$
26.44 $Cr_2O_7{}^{2-}$, Br_2 **26.45** a, b, c   **26.46** none
26.47 a. -0.29 V   b. $+0.88$ V   c. $+1.58$ V
26.48 a. $+1.19$ V   b. $+0.78$ V   c. $+0.36$ V
26.49 $+0.53$ V   **26.50** 4×10^{-8} M
26.51 a. -210 kJ, 8×10^{36} b. -150 kJ, 10^{26} c. -160 kJ, 10^{28}
26.52 a. $+56$ kJ   b. -420 kJ   c. -915 kJ   **26.53** -0.381 V
26.54 a. $+3$ b. $+4$ c. $+5$ d. -3
26.55 a. $HBrO_3$ b. HBr   c. $HBrO_4$
26.56 a. See Eqn. 26.15   b. See Eqn. 26.17, 26.18   c. See Eqn. 26.28
26.57 a. NH_3 b. $ClO_4{}^-$ c. $SO_3{}^{2-}$ d. ClO^-

26.58 a. NH_3 b. HNO_3 c. HNO_2 d. HNO_3 e. $NaNO_2$

26.59 a. H—$\ddot{O}$—$\ddot{S}$—$\ddot{O}$:   b. :$\ddot{O}$—$\ddot{S}$—$\ddot{O}$:   c. :$\ddot{O}$—S—$\ddot{O}$:

             | | ||

       :$\ddot{O}$:   :$\ddot{O}$:   :$\ddot{O}$:

    d. H—$\ddot{O}$—$\ddot{S}$—$\ddot{O}$—H

              |

            :$\ddot{O}$:

26.60 a. to prevent vaporization of water
    b. Cl more electronegative than Br   c. $NO_2(g)$

26.61 H—$\ddot{O}$—N = N—$\ddot{O}$—H ; yes

26.62 S = 0, C = +2, N = −3   **26.63** 288 g

26.64 25   **26.65** $E_3 = (n_1E_1 + n_2E_2)/n_3$

Chapter 27

27.31 beta

27.32 a. $^{28}_{12}Mg \rightarrow {}^{0}_{-1}e + {}^{28}_{13}Al$
    b. $^{150}_{60}Nd \rightarrow {}^{4}_{2}He + {}^{146}_{58}Ce$
    c. $^{72}_{33}As \rightarrow {}^{0}_{-1}e + {}^{72}_{32}Ge$

27.33 a. $^{87}_{38}Sr$ b. $^{87}_{38}Sr$ **27.34** a. $^{1}_{0}n$ b. $^{239}_{93}Np$ c. $^{243}_{97}Bk$

27.35 a. $^{1}_{0}n$ b. $^{4}_{2}He$ c. $^{7}_{4}Be$

27.36 $^{238}_{92}U \rightarrow {}^{4}_{2}He + {}^{234}_{90}Th$; $^{234}_{90}Th \rightarrow {}^{0}_{-1}e + {}^{234}_{91}Pa$; $^{234}_{91}Pa \rightarrow {}^{0}_{-1}e + {}^{234}_{92}U$
    $^{234}_{92}U \rightarrow {}^{4}_{2}He + {}^{230}_{90}Th$; $^{230}_{90}Th \rightarrow {}^{4}_{2}He + {}^{226}_{88}Ra$

27.37 7 alpha, 4 beta   **27.38** 96.0 g

27.39 a. $^{42}_{19}K \rightarrow {}^{42}_{20}Ca + {}^{0}_{-1}e$ b. 12.4 h   c. 0.031

27.40 0.109   **27.41** 0.013   **27.42** 22 d

27.43 t = 329 yr   **27.44** a. 2.7×10^9 yr   b. 3.4×10^9 yr

27.48 Use a different radioactive method; check historical records

27.46 -2.8×10^{-5} g, -2.5×10^6 kJ

27.47 a. 1.33×10^{-8} g   b. 3.35×10^{13}

27.48 1.7×10^{10} g   **27.49** yes; no   **27.50** 4.71×10^3 kJ

27.51 c > a > b   **27.52** $^{6}_{3}Li$

27.53 a. electron   b. electron   c. positron   d. positron

27.54 a. $7s^2 7p^4$ b. 6   **27.55** 8.8   **27.56** 5.2%

27.57 a. different charges   b. follow path of C compound   c. See Table
    27.1   d. more neutrons produced than consumed

27.58 a. can see where P-32 goes
    b. reaction whose rate steadily increases
    c. has a genetic effect

27.59 Inject isotope into water and measure time to appear

27.60 U-235 will be consumed   **27.61** 2.6×10^{-4} g

27.62 $\sim 1 \times 10^{11}$ yr

27.63 a. 5.5×10^{-10} g   b. 1.2×10^{-3} kJ   c. 17 rems

27.64 a. 2×10^{-7} erg   b. 2×10^8 cm/s   c. 4×10^8 K

27.65 a. -5.71×10^8 kJ   b. 1.1×10^{32} kJ   c. 2.1×10^{-15}

Chapter 28

28.31 $C_{12}H_4O_2Cl_4$

28.32 a. alcohol   b. ester   c. ester, carboxylic acid

28.33 a. $CH_3-\overset{\displaystyle H}{\underset{\displaystyle OH}{C}}-CH_3$

b. $CH_3-CH_2-CH_2-\overset{\displaystyle }{\underset{\displaystyle \|}{\underset{\displaystyle O}{C}}}-OH$ (or isomer)

c. $CH_3-\overset{\displaystyle H}{\underset{\displaystyle CH_3}{C}}-O-\overset{\displaystyle }{\underset{\displaystyle \|}{\underset{\displaystyle O}{C}}}-CH_2-CH_2-CH_3$

28.34 $Ca^{2+}(aq) + 2\,OH^-(aq) + 2\,C_{17}H_{35}\,COOH(aq) \rightarrow$
$Ca\,(C_{17}H_{35}COO)_2(s) + 2\,H_2O$

28.35 $HO-CH_2-CH_2-O-\overset{\displaystyle }{\underset{\displaystyle \|}{\underset{\displaystyle O}{C}}}-H$, $HO-CH_2-CH_2-O-\overset{\displaystyle }{\underset{\displaystyle \|}{\underset{\displaystyle O}{C}}}-CH_3$,

$H-\overset{\displaystyle }{\underset{\displaystyle \|}{\underset{\displaystyle O}{C}}}-O-CH_2-CH_2-O-\overset{\displaystyle }{\underset{\displaystyle \|}{\underset{\displaystyle O}{C}}}-H$, $H-\overset{\displaystyle }{\underset{\displaystyle \|}{\underset{\displaystyle O}{C}}}-O-CH_2-CH_2-O-\overset{\displaystyle }{\underset{\displaystyle \|}{\underset{\displaystyle O}{C}}}-CH_3$,

$CH_3-\overset{\displaystyle }{\underset{\displaystyle \|}{\underset{\displaystyle O}{C}}}-O-CH_2-CH_2-O-\overset{\displaystyle }{\underset{\displaystyle \|}{\underset{\displaystyle O}{C}}}-CH_3$

28.36 a. secondary　b. tertiary　c. primary　d. secondary

28.37 $C_{10}H_{14}N_2$

28.38 $H-\overset{\displaystyle H}{\underset{\displaystyle H}{C}}-\overset{\displaystyle H}{\underset{\displaystyle Br}{C}}-Br$, $Br-\overset{\displaystyle H}{\underset{\displaystyle H}{C}}-\overset{\displaystyle H}{\underset{\displaystyle H}{C}}-Br$

28.39 3

28.40

$\overset{\displaystyle CH_3}{\underset{\displaystyle CH_3}{}}C=C\overset{\displaystyle Cl}{\underset{\displaystyle Br}{}}$, $\overset{\displaystyle C_2H_5}{\underset{\displaystyle H}{}}C=C\overset{\displaystyle Cl}{\underset{\displaystyle Br}{}}$, $\overset{\displaystyle CH_3}{\underset{\displaystyle Br}{}}C=C\overset{\displaystyle Cl}{\underset{\displaystyle CH_3}{}}$,

$\overset{\displaystyle C_2H_5}{\underset{\displaystyle Br}{}}C=C\overset{\displaystyle Cl}{\underset{\displaystyle H}{}}$, $\overset{\displaystyle H}{\underset{\displaystyle Br}{}}C=C\overset{\displaystyle Cl}{\underset{\displaystyle C_2H_5}{}}$

28.41 $H-\overset{\displaystyle H}{\underset{\displaystyle H}{C}}-\overset{\displaystyle H}{\underset{\displaystyle H}{C}}-\overset{\displaystyle H}{\underset{\displaystyle \ddot{O}:}{C}}-\overset{\displaystyle }{C}-\ddot{O}-H$, $H-\overset{\displaystyle H}{\underset{\displaystyle H}{C}}-\overset{\displaystyle H}{\underset{\displaystyle C}{C}}-\overset{\displaystyle H}{\underset{\displaystyle H}{C}}-H$

$\ddot{O}:\quad\ddot{O}\cdot\overset{\displaystyle }{\underset{\displaystyle H}{}}$

28.42 all except $\overset{\displaystyle CH_3}{\underset{\displaystyle CH_3}{}}C=C\overset{\displaystyle Cl}{\underset{\displaystyle Br}{}}$ show geometric isomerism

28.43 b, c　　**28.44** b, c, d

28.45 a. 2nd carbon from left
c. 3rd carbon from left

28.46 a. $-\overset{\displaystyle F}{\underset{\displaystyle F}{C}}-\overset{\displaystyle F}{\underset{\displaystyle F}{C}}-\overset{\displaystyle F}{\underset{\displaystyle F}{C}}-\overset{\displaystyle F}{\underset{\displaystyle F}{C}}-$　b. 2.5×10^6 g/mol　c. 24.02% C

28.47

$$-\overset{\underset{|}{H}}{\underset{H}{C}}-\overset{\underset{|}{H}}{\underset{C_6H_5}{C}}-\overset{\underset{|}{H}}{\underset{H}{C}}-\overset{\underset{|}{H}}{\underset{C_6H_5}{C}}-$$

28.48 a. $\overset{H}{\underset{H}{}}C=C\overset{H}{\underset{F}{}}$ b. $\overset{H}{\underset{CH_3}{}}C=C\overset{H}{\underset{CH_3}{}}$

28.49 $-O-\overset{\underset{\|}{}}{\underset{O}{C}}-O-\bigcirc-\overset{\underset{|}{CH_3}}{\underset{CH_3}{C}}-\bigcirc-$

28.50 $-\overset{\underset{|}{}}{\underset{H}{N}}-(CH_2)_5-\overset{\underset{\|}{}}{\underset{O}{C}}-\overset{\underset{|}{}}{\underset{H}{N}}-(CH_2)_5-\overset{\underset{\|}{}}{\underset{O}{C}}-$

28.51 a. $NH_2-CH_2-CH_2-NH_2$ and $HOOC-CH_2-COOH$

b. $HOOC-\bigcirc-COOH$ and $HO-CH_2-\overset{\underset{|}{H}}{\underset{CH_3}{C}}-OH$

28.52 $C_{12}H_{22}O_{11}(aq) + H_2O \rightarrow 2\ C_6H_{12}O_6(aq)$

28.53 a. 44.44% C, 6.217% H, 49.34% O b. 6.2×10^2

28.54 $CH_3-\overset{\underset{|}{H}}{\underset{CH_3}{C}}-CH_2-\overset{\underset{|}{H}}{\underset{NH_2}{C}}-\overset{\underset{\|}{}}{\underset{O}{C}}-\overset{\underset{|}{H}}{\underset{H}{N}}-\overset{\underset{|}{H}}{\underset{COOH}{C}}-(CH_2)_4-NH_2$ and

$NH_2-(CH_2)_4-\overset{\underset{|}{H}}{\underset{NH_2}{C}}-\overset{\underset{\|}{}}{\underset{O}{C}}-\overset{\underset{|}{H}}{\underset{H}{N}}-\overset{\underset{|}{H}}{\underset{COOH}{C}}-CH_2-\overset{\underset{|}{H}}{\underset{CH_3}{C}}-CH_3$

28.55 Ala-Phe-Leu-Met-Val-Ala   **28.56** 1

28.57 alcohol: 52.14% C, 13.13% H, 34.73% O
acid and ester: 40.00% C, 6.714% H, 53.29% O

28.58 addition: must have double bond
condensation: must have reactive groups at both ends of molecule

28.59 a. linear has unbranched chains   b. See Figure 28.7
c. maltose is a dimer of glucose; sucrose built from glucose + fructose

28.60 Chlorine compounds in Table 28.1; $CHCl_3$, CCl_4, CH_3OH

28.61 68.28% C, 6.276% H, 21.66% O, 3.793% N

28.62 $C_{12}Cl_4H_6$

28.63 a. $-O-CH_2-\overset{\underset{|}{H}}{\underset{OH}{C}}-CH_2-O-\overset{\underset{\|}{}}{\underset{O}{C}}-\bigcirc-\overset{\underset{\|}{}}{\underset{O}{C}}-$

b. orthophthalic acid condenses with OH groups in two adjacent chains

28.64 17 isomers   **28.65** 63.68% C, 9.80% H, 14.14% O, 12.38% N

Terms in this glossary are part of the basic chemical vocabulary; they are defined in the context used in this text. If you do not find the term you seek here, look for it in the Index, which will refer you to pages where the term is used.

A

abbreviated electron configuration brief notation in which only those electrons beyond the nearest noble gas are shown. The abbreviated electron configuration of the Fe atom is $[_{18}Ar] 4s^2 3d^6$.

absolute zero the lowest possible temperature at which matter might exist; 0 K, $-273.15°C$.

absorption spectrum a graph showing the absorption of radiation by a substance over a range of wavelengths.

acid a substance that on being dissolved in water produces a solution in which $[H^+]$ is greater than 10^{-7} M. Examples: HCl, HNO_3, H_2CO_3, CH_3COOH.

acid anhydride a nonmetal oxide that reacts with water to form an acidic solution.

acid-base indicator see *indicator, acid-base*.

acid-base titration a procedure used to determine the concentration of an acid or base. The volume of a solution of an acid (or base) of known concentration required to react with a known volume of base (or acid) is measured.

acid-dissociation constant (K_a) the equilibrium constant for the following reaction of an acid HB:

$$HB(aq) \rightleftarrows H^+(aq) + B^-(aq)$$

$$K_a = \frac{[H^+] \times [B^-]}{[HB]}$$

acidic solution an aqueous solution with pH less than 7.0 ($[H^+] > 1.0 \times 10^{-7}$ M).

acid rain rainfall that has a pH less than about 5.6, the value observed when pure water is saturated with atmospheric CO_2. Acid rain typically contains H_2SO_4 and/or HNO_3.

actinides elements 89 (Ac) through 102 (No).

activated complex a species, formed by collision of energetic particles, that can react to form products or other intermediates.

activation energy (E_a) the minimum energy that must be possessed by a pair of molecules if collision is to result in reaction.

actual yield the amount of product obtained from reaction.

addition polymer a polymer produced by reaction of a monomer, usually a derivative of ethylene, adding to itself; no other product is formed.

alcohol a substance containing an OH group attached to a hydrocarbon chain. Examples: C_2H_5OH, ethyl alcohol; C_4H_9OH, butyl alcohol.

alkali metal a metal in Group 1. Examples: Li, Na, K.

alkaline (basic) having an $[OH^-]$ that is greater than 1×10^{-7} M.

alkaline earth metal a member of Group 2. Examples: Be, Mg, Ca, Ba.

alkane a hydrocarbon containing only single carbon-carbon bonds. Examples: C_2H_6, C_6H_{14}.

alkene a hydrocarbon containing one carbon-carbon double bond. Examples: $CH_3CH=CH_2$, $H_2C=CH_2$.

alkyl group a hydrocarbon group, such as $CH_3—$, $C_2H_5—$, $C_3H_7—$.

alkyne a hydrocarbon containing one carbon-carbon triple bond. Example: $HC≡CH$.

allotrope one of two or more forms of an elementary substance. Examples: O_2 and O_3 are allotropic forms of oxygen; graphite and diamond are allotropes of carbon.

alloy a material made by melting together two or more elements, at least one of which is a metal. Upon cooling, the alloy crystallizes as a crude mixture, a solid solution, or an intermetallic compound.

alpha (α) particle a helium nucleus; He^{2+} ion.

amalgam a solution of a metal in mercury.

amine organic compound containing the $—\overset{\mid}{N}—$ functional group. Examples: Methylamine, CH_3NH_2; dimethylamine, $(CH_3)_2NH$; and trimethylamine, $(CH_3)_3N$.

α-amino acid the monomer units of proteins. The amino acid contains an acid group ($—COOH$) and an amine group ($—NH_2$). The amine group is attached to the alpha carbon, the one adjacent to the $—COOH$ group.

ampere (A) rate of flow of electric current such that one coulomb passes a given point in one second.

amphoteric capable of reacting with both H^+ and OH^- ions; usually an insoluble hydroxide. Examples: $Al(OH)_3$, $Zn(OH)_2$.

amplitude (ψ) height of a standing wave.

anhydride a substance derived from another by removal of water. Examples: SO_3 is the anhydride of H_2SO_4; CaO is the anhydride of $Ca(OH)_2$.

anion a species carrying a negative charge. Examples: Cl^-, CO_3^{2-}, $H_2PO_4^-$.

anode an electrode at which oxidation occurs. Example: If, at a copper electrode, the reaction that occurs is $Cu(s) \rightarrow Cu^{2+}(aq) + 2\ e^-$, then the copper metal is behaving as an anode.

antibonding orbital a molecular orbital that has decreased density between two proximate atoms. The energy of its two electrons is greater than that of those electrons in the separated atoms.

aqua regia a mixture of 3 volumes of 12 M HCl with 1 volume of 16 M HNO_3.

aromatic substance an organic compound containing a benzene ring. Examples: Benzene, C_6H_6, ; toluene, C_7H_8, $—CH_3$; naphthalene, $C_{10}H_8$, .

Arrhenius acid a species which, upon addition to water, increases the concentration of H^+.

Arrhenius base a species, which upon addition to water, increases the concentration of OH^-.

Arrhenius equation the equation that expresses the temperature dependence of the rate constant: $\log_{10} k = A - E_a/2.30\ RT$. The "two-point" form of this equation is

$$\log_{10} \frac{k_2}{k_1} = \frac{E_a}{2.30\ R} \left(\frac{T_2 - T_1}{T_2 T_1} \right)$$

atmosphere (atm) standard unit of pressure; equal to 101.325 kPa; equivalent to the pressure exerted by a mercury column 760 mm high.

atom smallest particle of an element; matter is composed of atoms in various chemical combinations. Example: The N atom is the smallest particle of the element nitrogen. Two nitrogen atoms combine to form an N_2 molecule, the smallest particle which has the properties of nitrogen as it is ordinarily found.

atomic mass an average mass of atoms of one element relative to that of another element; based upon the atomic mass of a ^{12}C- isotope taken to be exactly 12.

atomic number the number of protons in the nucleus of an atom; each element has a unique atomic number.

atomic orbital model see *valence bond model*.

atomic radius the radius of an atom, taken to be one half the distance between two nuclei in the ordinary form of the elementary substance. Example: The radius of the Cl atom is 0.099 nm, since the internuclear distance in the Cl_2 molecule is 0.198 nm.

atomic spectrum a diagram showing the wavelengths at which light is emitted by excited electrons of an atom.

aufbau principle the rule stating that electrons enter energy levels in an atom in order of increasing energy, filling one sublevel before moving into the next.

Avogadro's Law a principle stating that equal volumes of gases at the same temperature and pressure contain equal numbers of molecules.

Avogadro's number 6.022×10^{23}; the number of units in a mole.

axial adjective used to describe an atom or group that is perpendicular to the plane of a ring molecule.

B

balanced equation an equation for a chemical reaction in which the reactants and products contain equal numbers of each kind of atom participating in the reaction. Example: The equation $CH_4(g) + 2\ O_2(g) \rightarrow CO_2(g) + 2\ H_2O(l)$ is balanced since both reactants and products contain one C, four H, and four O atoms.

Balmer series a series of "lines" in the atomic spectrum of hydrogen resulting from the transition of an electron from a higher energy level to the level n = 2.

base a substance that on dissolving in water produces a solution in which $[OH^-]$ is greater than 10^{-7} M. Examples: $NaOH$, Na_2CO_3, NH_3.

base anhydride a metal oxide that reacts with water to form a basic solution.

base dissociation constant (K_b) the equilibrium constant for the following reaction of the base B^-:

$$B^-(aq) + H_2O \rightleftharpoons HB(aq) + OH^-(aq)$$

$$K_b = \frac{[HB] \times [OH^-]}{[B^-]}$$

basic solution an aqueous solution with pH > 7.0 ($[H^+] < 1.0 \times 10^{-7}$ M).

bent adjective used to describe a molecule containing three atoms in which the bond angle is less than 180°. Examples: H_2O and SO_2.

beta radiation ($_{-1}^{0}e$) one of the types of radiation emitted by unstable nuclei. Beta particles have properties identical to those of electrons.

bidentate ligand a ligand that forms two coordinate covalent bonds with a metal atom. Example: Ethylenediamine. $H_2N—CH_2—CH_2—NH_2$, which bonds through both nitrogen atoms.

binary compound a compound containing only two kinds of atoms. Examples: HCl, H_2O, C_2H_6.

body-centered cubic a crystalline structure in which the unit cell is a cube with one atom at each of its corners and an atom at its center.

Bohr model a model of the hydrogen atom derived by Niels Bohr. The model predicts that the electronic energy (kJ/mol) is $-1312/n^2$, where **n** is the principal quantum number.

boiling point (bp) that temperature of a liquid at which its vapor pressure equals the applied pressure; a liquid will tend to form bubbles and vaporize at its boiling point; usually reported at one atmosphere pressure.

boiling point elevation (ΔT_b) increase in the boiling point of a liquid caused by addition of a nonvolatile solute. For a nonelectrolyte, the boiling point elevation is given by the equation $\Delta T_b = k_b \times m$, where m is the molality and k_b is a constant for a given liquid (0.52°C for water).

bond a linkage between two atoms.

bond angle the angle between two covalent bonds, Example: The H—O—H angle in the water molecule is 105°.

bond distance the distance, usually measured in nanometers, between the nuclei of two atoms joined by a chemical bond.

bond energy enthalpy change ΔH associated with a reaction in which a bond is broken. Example: For the reaction $HCl(g) \rightarrow H(g) + Cl(g)$, $\Delta H = 431$ kJ; the bond energy, B.E., of the H—Cl bond is 431 kJ/mol of bonds.

bonding orbital an orbital associated with two atoms in which the energy of its two electrons is less than the energies of those electrons in the separated atoms. The presence of a populated bonding orbital between two atoms stabilizes the bond between them.

bond polarity see *polar bond, nonpolar bond*.

Boyle's Law a relation stating that when a gas sample is compressed at a constant temperature, the product of the pressure and the volume remains constant.

branched-chain alkane a saturated hydrocarbon in which not all of the carbon atoms are located in a single, continuous chain. The simplest branched-chain alkane is 2-methylpropane, which has the structure

$$CH_3—CH—CH_3.$$
$$|$$
$$CH_3$$

Brönsted-Lowry acid a species which donates a proton to another species. Example: In the reaction $HB(aq) + H_2O \rightleftharpoons H_3O^+(aq) + B^-(aq)$, HB behaves as a Brönsted-Lowry acid in that it donates a proton to H_2O.

Brönsted-Lowry base a species that accepts a proton from another species. Example: In the reaction just given, H_2O behaves as a Brönsted-Lowry base, since it accepts a proton from HB.

buffer a system whose pH changes only slightly when strong acid or base is added. A buffer ordinarily contains roughly equal amounts of a weak acid such as $HC_2H_3O_2$ and its conjugate base $(C_2H_3O_2^-)$.

C

calorie (cal) a unit of thermal energy equal to 4.184 joules.

calorimeter a device used to measure heat flow. Heat may be absorbed by water, q (in J) = $4.18 \times m_{water} \times \Delta t$, and by metal parts of the calorimeter, $q = C \times \Delta t$.

calorimeter constant (C) the product of the mass times the specific heat of a bomb calorimeter.

carbohydrate a class of organic compounds in which the general formula is $C_m(H_2O)_n$. Examples: glucose, sucrose, starch, and cellulose.

carbonate ion CO_3^{2-}.

carboxylic acid an organic compound containing the functional
group —C—OH. Example: acetic acid,
$$‖$$
$$O$$
$$CH_3—C—OH.$$
$$‖$$
$$O$$

catalyst a substance which affects the rate of a reaction without being used up itself. Example: A piece of platinum foil can act as a catalyst for the combustion of methane in air.

catalytic converter a device inserted into the exhaust system of an automobile, containing finely divided Pt, Rh, and Pd. This metal catalyst converts CO to CO_2, unburned hydrocarbons to CO_2 and H_2O, and NO to N_2.

cathode an electrode at which reduction occurs. Example: If, at a silver electrode, the reaction that occurs is $Ag^+(aq) + e^- \rightarrow Ag(s)$, then the silver metal is serving as a cathode.

cation an ion having a positive charge. Examples: Fe^{2+}, K^+, NH_4^+.

cation exchange a process by which a cation in water solution is "exchanged" for a different cation, originally present in a solid resin. Used to soften water by exchanging Ca^{2+} for Na^+ ions.

Celsius degree (°C) a unit of temperature based on there being 100° between the freezing and boiling points of water.

centi- (c) a prefix on a metric unit indicating a multiple of 10^{-2}. Example: 1 centimeter = 10^{-2} meter.

chain reaction a type of chemical reaction occurring in steps in which the product of a late step serves as a reactant in an earlier step, thereby allowing a reaction, once begun, to continue.

Charles' Law a relation stating that the volume of a gas sample at constant pressure is directly proportional to its absolute temperature.

chelating agent a complexing ligand that can form more than one bond with a central ion. Example: Ethylenediamine, $H_2N—CH_2—CH_2—NH_2$ (en), is a chelating agent that can form two bonds with a metal ion; its complex ion with Cu^{2+}, coordination number 4, has the formula $Cu(en)_2^{2+}$.

chemical equation an expression which qualitatively and quantitatively describes the reactants and products of a chemical reaction as to their nature and amount. Example: $N_2(g) + 3 H_2(g) \rightarrow 2 NH_3(g)$, a chemical equation, tells us that one mole of nitrogen gas reacts with three moles of hydrogen gas to form two moles of ammonia gas.

chemical property property of a substance related to its chemical changes.

chemical thermodynamics the use of thermodynamic principles to predict whether a chemical reaction will be spontaneous; ordinarily involves calculation of ΔH, ΔS, and/or ΔG for the reaction.

chiral center atom in a molecule that is bonded to four different groups; a source of optical isomerism.

chromatography   a separation method in which the components of a solution are adsorbed at different locations on a solid surface.

cis **isomer**   a geometric isomer in which two identical bonded atoms or groups are relatively close to one another. Example: In square planar $Pt(NH_3)_2Cl_2$, the *cis* isomer has the structure

$$Cl \diagdown \quad NH_3$$
$$Pt$$
$$Cl \diagup \quad NH_3$$

Clausius-Clapeyron equation   an equation relating the vapor pressure of a liquid, P, to the absolute temperature T:

$$\log_{10} \frac{P_2}{P_1} = \frac{\Delta H_{vap}}{2.30\ R} \left(\frac{T_2 - T_1}{T_2 T_1} \right)$$

where ΔH_{vap} is the heat of vaporization per mole; R is the gas constant.

coefficient   a number preceding a symbol or formula in a chemical equation.

coke   solid material, mostly carbon, formed by the destructive distillation of coal. Coke is widely used as a reducing agent in metallurgy.

colligative property   a physical property of a solution which depends on the concentration, but not the kind, of solute particles. Example: The vapor pressure depression of a solution depends on the mole fraction of the solute but not on the nature of the solute, and thus is a colligative property.

common ion effect   a decrease in solubility of an ionic solute brought about by adding a solution containing one of the ions present in the solute.

complex ion   an ion containing a central metallic cation to which two or more groups are attached by coordinate covalent bonds. Example: In the $Ag(NH_3)_2{}^+$ complex ion the electrons in the coordinate covalent bonds between Ag^+ and NH_3 are furnished by the NH_3 molecules.

complexing agent   anion or molecule that bonds to a metal atom to form a complex ion. Examples of complexing agents (ligands) are OH and NH_3.

compound   a chemical substance containing more than one element.

concentrated   adjective used to describe a solution which contains a relatively high concentration of solute.

concentration   refers to relative amounts of solute and solvent in a solution; may be stated in many ways, such as per cent solute by mass, or mole fraction, but very often is given in terms of molarity, which is the number of moles of solute per liter of solution. Example: In 6 molar NaOH, 6 M NaOH, there are 6 mol of NaOH in a liter of solution.

concentration cell   a voltaic cell in which the driving force is a difference in concentration of a species in the two half-cells.

concentration quotient   a ratio of concentrations having the same form as that of the equilibrium constant K.

condensation   conversion of a gas to a liquid or solid.

condensation polymer   a polymer formed from monomer units joined by splitting out a small molecule, usually water.

conductivity   the relative ease with which a sample will transmit electricity or heat (should specify which). Example: Since a much larger electrical current will flow through an aluminum rod at a given voltage than through a glass rod of the same shape, the electrical conductivity of aluminum is much greater than that of glass.

conjugate acid   the acid formed by adding an H^+ ion to a base; the $NH_4{}^+$ ion is the conjugate acid of NH_3.

conjugate base   the base formed by removing an H^+ ion from an acid; the F^- ion is the conjugate base of HF.

Conservation of Energy   the law which states that energy can neither be created nor destroyed.

contact process   a process used in the industrial preparation of sulfuric acid. SO_2 and O_2 are converted to SO_3 by bringing them into contact with a solid catalyst, which may be V_2O_5 or Pt.

conversion factor   a ratio, numerically equal to 1, by which a quantity can be converted to another, equivalent quantity. Example: To convert 0.202 g H_2O to moles, we multiply by the conversion factor 1 mol/18.0 g: 0.202 g $H_2O \times$ 1 mol/18.0 g = 0.0112 mol H_2O.

coordinate covalent bond   a covalent bond in which the electrons are furnished by only one of the bonded atoms; most commonly encountered in complex ions. Example: In the $Zn(OH)_4{}^{2-}$ ion, the electrons in the bonds between Zn^{2+} and the OH^- ions are all furnished by the hydroxide ions; these bonds are therefore coordinate covalent bonds.

coordination compound a compound in which either the cation or the anion is a complex ion. Examples: $[Cu(NH_3)_4]Cl_2$ and $K_4[Fe(CN)_6]$.

coordination number the number of bonds formed from the central metal to the ligands in a coordination complex.

corrosion a destructive chemical process, most often applied to the conversion of a metal to one of its compounds. Example: The corrosion of iron in contact with O_2 and H_2O to form first $Fe(OH)_2$ and eventually hydrated $Fe(OH)_3$.

coulomb (C) a unit of quantity of electricity; 96,485 C = 1 mol e^-

covalent bond a chemical link between two atoms, produced by shared electrons in the region between the atoms. Example: In the H_2O molecule there is a covalent bond between the O atom and each H atom; each bond contains two electrons, one furnished by the H atom and one by the O atom; both atoms share the electrons in the bond.

critical pressure the pressure at the critical temperature.

critical temperature the highest temperature at which a substance can exhibit liquid-vapor equilibrium. Equilibrium pressure at that point is called the critical pressure. Above that temperature, liquid cannot be condensed from the vapor at any pressure. Example: Water has a critical temperature of 374°C; above 374°C one cannot have liquid water in equilibrium with its vapor.

crystal-field model model of the bonding in complex ions. The bonding is considered to be essentially ionic; the only effect of the ligands is to change the relative energies of the d orbitals of the central metal ion.

crystal-field splitting energy the difference in energy between the two sets of d orbitals in a complex ion.

crystallize to separate from solution or melt as a solid.

cubic centimeter (cm³) a volume unit equal to the volume of a cube 1 cm on each edge; a milliliter.

D

Dalton's Law a relation stating that the total pressure of a gas mixture is equal to the sum of the partial pressures of its components.

deBroglie relation an equation used to describe the wave properties of matter: $\lambda = h/mv$.

deliquescence a process in which a soluble substance picks up water vapor from the air to form a solution. For deliquescence to occur, the vapor pressure of water in the air must be greater than that of the saturated solution.

delocalized orbital a molecular orbital in which the electron density is spread over the entire molecule instead of being concentrated between two atoms.

density (d) a property of a sample equal to its mass per unit volume. Example: The density of mercury is 13.5 g/cm³; that is, 1.00 cm³ Hg weighs 13.5 g.

deuterium a heavy isotope of hydrogen, $_1^2H$.

diamagnetic a descriptive term indicating that a substance does not contain unpaired electrons and so is not attracted into a magnetic field. Example: Since all of the electrons in NH_3 molecules are paired, NH_3 is diamagnetic.

diffusion a process by which one substance, by virtue of the kinetic properties of its particles, gradually mixes with another. Example: $H_2S(g)$ prepared in a test tube slowly diffuses into the surrounding air.

dilute adjective used to describe a solution containing a relatively small amount of solute; opposite of concentrated.

dipole a species in which there is a separation of charge, i.e., a positive charge at one point and a negative charge at a different point. Example: HF, H_2O.

dipole force an attractive force between molecules possessing separate positive and negative poles. Example: Since the HCl molecule has positive and negative ends, there are dipole forces between neighboring HCl molecules.

dispersion force an attractive force between molecules that arises because of the presence of temporary dipoles. Usually increases with molar mass.

disproportionation a reaction in which a species undergoes oxidation and reduction simultaneously. Example: $2\ Cu^+(aq) \rightarrow Cu(s) + Cu^{2+}(aq)$.

dissociation separation into two or more species; usually applied to weak acids or bases. Example: The dissociation of acetic acid in water to form H^+ ions and acetate ions only occurs to a small extent.

distillation a procedure in which a liquid is vaporized under conditions where the evolved vapor is later condensed and collected.

double bond two shared electron pairs between two bonded atoms.

ductility the ability of a solid to retain strength on being forced through an orifice; characteristic of metals.

E

E, E^0_{ox}, E^0_{red}, E^0_{tot} see *energy, standard oxidation voltage, standard reduction voltage, standard voltage,* respectively.

effective nuclear charge positive charge felt by the outermost electrons in an atom; approximately equal to the number of nuclear protons minus the number of electrons in inner, complete levels.

efflorescence the loss of water of hydration from a hydrate.

effusion movement of gas molecules through a pinhole or capillary.

Einstein's equation the relation $\Delta E = \Delta mc^2$ relating mass and energy changes.

electrode a general name for anode or cathode.

electrolysis the passage of a direct electric current through a solution containing ions, producing chemical changes at the electrodes.

electrolyte a substance that exists as ions in water solution. Example: NaCl (Na^+ and Cl^- in water solution).

electrolytic cell a cell in which the flow of electrical energy from an external source causes a redox reaction to occur.

electron the negatively charged component of atoms; exists in a roughly spherical cloud around atomic nucleus; carries 1 unit of negative charge and has a very low mass.

electron cloud a region of negative charge around an atomic nucleus; associated with an atomic orbital.

electron configuration a statement of the populations of the electronic energy sublevels in an atom. Example: The electron configuration of the Li atom is $1s^2 2s^1$; this means there are two electrons in the 1s sublevel and one electron in the 2s sublevel.

electronegativity a property of an atom that increases with its tendency to attract the electrons in a bond. Example: Since the Cl atom is more electronegative than the H atom, in the HCl molecule the bonding electrons are closer to Cl.

electron pair repulsion a principle used to predict the geometry of a molecule or polyatomic ion. Electron pairs around a central atom tend to orient themselves so as to be as far apart as possible.

electron-sea model a model of metallic bonding in which cations are considered as fixed points in a mobile "sea" of electrons.

electron spin a property of an electron loosely related to its spin around an axis. Only two spin states are allowed, usually described by quantum number m_s, which can assume the values $+\frac{1}{2}$ and $-\frac{1}{2}$.

electrostatic forces the forces between particles caused by their electric charges.

element a substance whose atoms are all chemically the same. The 108 known elements differ from one another in the number of nuclear protons.

empirical formula an expression that furnishes relative numbers of atoms of the elements in a chemical substance; expressed as the lowest possible set of integers. Often called the simplest formula. Examples: NaCl, H_2SO_4, CH_2, Fe, HO (hydrogen peroxide).

enantiomer one of a pair of optical iosmers.

endothermic a process in which heat is absorbed from the surroundings; ΔH is positive for an endothermic reaction.

end point the point during a reaction, usually in the course of a titration, at which a chemical indicator changes color. Example: The end point in a titration using phenolphthalein indicator occurs at pH 9.

energy (E) a property of a system which is related to its capacity to cause change; can be altered only by exchanging heat or work with the surroundings.

energy level the value of the **n** quantum number.

enthalpy (H) a property of a system which reflects its capacity to exchange heat, q, with its surroundings; defined so that $\Delta H = q$ for changes in the system that occur at constant pressure.

enthalpy change (ΔH) difference in enthalpy between products and reactants.

enthalpy of formation (ΔH_f) heat flow for the reaction in which a species is formed at constant pressure from elementary substances.

entropy (S) a property of a system related to its degree of organization; highly ordered systems have low entropy.

entropy change (ΔS) difference in entropy between products and reactants.

enzyme a protein catalyst in biological systems.

equatorial adjective used to describe an atom or group that is parallel to the plane of a ring molecule.

equilibrium a state of dynamic balance in which rates of forward and reverse reactions are equal so system does not change with time. Example: At 100°C, liquid water is in equilibrium with its vapor when the vapor is at a pressure of 1 atm.

equilibrium concentration concentration, in moles per liter, of a species at equilibrium. Represented by the symbol [ ].

equilibrium constant (K_c) a number characteristic of an equilibrium system at a particular temperature. Example: For the reaction

$$PCl_3(g) + Cl_2(g) \rightleftharpoons PCl_5(g) \text{ at } 240°C$$

$$K_c = \frac{[PCl_5]}{[PCl_3] \times [Cl_2]}$$

where the quantities in square brackets represent equilibrium concentrations in moles per liter (molarity).

equilibrium constant (K_p) a quantity similar to K_c except that partial pressure replaces molarity. Example: For the reaction

$$PCl_3(g) + Cl_2(g) \rightleftharpoons PCl_5(g) \text{ at } 240°C$$

$$K_p = \frac{(P_{PCl_5})}{(P_{PCl_3}) \times (P_{Cl_2})} = 0.82$$

where P represents partial pressure in atmospheres.

equivalence point the point during a reaction between A and B, usually during a titration, when an amount of B has been added that is required to react exactly with the amount of A present. Example: The equivalence point in the reaction $H^+(aq) + OH^-(aq) \rightarrow H_2O(l)$ occurs when the number of moles of OH^- ion added to an acid solution equals the number of moles of H^+ ion in the solution.

ester the product of the reaction between an alcohol and an acid. Example: When methanol, CH_3OH, reacts with acetic acid, CH_3COOH, the ester called methyl acetate, CH_3—O—CO—CH_3, is formed.

excited state an electronic state of a higher energy than the ground state.

exclusion principle the rule stating that in an atom no two electrons can have the same set of four quantum numbers.

exothermic a process in which heat is evolved to the surroundings; ΔH is negative for an exothermic reaction.

expanded octet more than four electron pairs about a central atom.

F

face-centered cubic a type of crystal structure in which the unit cell is a cube with identical atoms at each corner and at the center of each face.

Fahrenheit degree (°F) a degree based on the temperature scale on which water freezes at 32° and boils at 1 atm at 212°.

Faraday constant the constant that gives the number of coulombs equivalent to one mole of electrons: 96485 C/mol e^-.

fat an ester made from glycerol and a long-chain carboxylic acid; found in seeds and in fatty tissue of animals.

fatty acid a long-chain carboxylic acid. Example: stearic acid, $CH_3(CH_2)_{16}COOH$.

ferromagnetism a property, shown by iron and certain other substances, of being strongly attracted into a magnetic field.

filtration a process for separating a solid-liquid mixture by passing it through a barrier with fine pores, such as filter paper.

first ionization energy the energy that must be absorbed to remove the outermost electron from a species, forming a +1 ion.

First Law of Thermodynamics the statement that the change in energy, ΔE, of a system is the sum of the heat flow into the system, q, and the work done upon the system, w: $\Delta E = q + w$.

first-order reaction a reaction whose rate depends on reactant concentration raised to the first power. Example: Since the rate of the reaction $2 N_2O_5(g) \rightarrow 2 N_2O_4(g) + O_2(g)$ is given by the equation rate = k(conc. N_2O_5), the reaction is first order.

fission see *nuclear fission*.

five percent rule empirical rule that the approximation $a - x \approx a$ is valid if $x \leq 0.05 \, a$. The rule depends upon the generalization that the value of the constant in the equation in which x appears is seldom known to better than ±5%.

fixation of nitrogen any process which converts $N_2(g)$ into a nitrogen-containing compound. Example: The fixation of nitrogen by the Haber process occurs via the reaction of N_2 with H_2 to make ammonia, NH_3.

flame test a test carried out by observing the color imparted to a Bunsen burner flame by a sample. Sodium compounds give a yellow flame test; potassium compounds give a violet flame test, and so on.

flotation a separation process used to free finely divided ore from rocky impurities. With a soapy emulsion of oil and water, the ore concentrates at the surface and can be skimmed off.

flow sheet a diagram used to summarize a separation scheme in qualitative analysis.

formation constant (K_f) the equilibrium constant for the formation of a complex ion. Example:

$$Cu^{2+}(aq) + 4 \, NH_3(aq) \rightleftharpoons Cu(NH_3)_4^{2+}(aq)$$

$$K_f = \frac{[Cu(NH_3)_4^{2+}]}{[Cu^{2+}] \times [NH_3]^4} = 5 \times 10^{12}$$

formula the expression used to describe the relative number of atoms of the different elements present in a substance; molecular formula is used with substances having molecules; empirical formula is used with nonmolecular substances.

formula mass the sum of the atomic masses of the atoms in a formula.

fractional crystallization a process used to separate a pure solid from a mixture with another solid. The mixture is dissolved in the minimum amount of hot solvent. Upon cooling, one solid should crystallize from solution while the other remains in solution.

fractional distillation a procedure used to separate components with different boiling points from a solution. It is based on passing vapors from a boiling solution up a column along which the temperature gradually decreases: higher boiling components condense on the column and return to solution, lowest boiling component goes out of the top of the column, where it is condensed and collected.

free energy (G) a quantity defined as $H - TS$ which decreases in a reaction that is spontaneous at constant T and P.

free energy change (ΔG) the difference in free energy between products and reactants.

free energy of formation (ΔG_f) ΔG for the formation of a species from the elements.

free radical a species having an unpaired electron. Examples: The H atom, the NO molecule, and the CH_3 group.

freezing point (fp) the temperature at which a solid and liquid phase can coexist at equilibrium.

freezing point depression (ΔT_f) the decrease in the freezing point of a liquid caused by addition of a solute. For a nonelectrolyte, the freezing point depression ΔT_f is given by the equation $\Delta T_f = k_f \times m$, where m is the molality and k_f is a constant for a given liquid (1.86°C for water).

frequency (ν) for a wave, the number of complete cycles per unit time.

functional group a small group of atoms in an organic molecule which give the molecule its distinctive chemical behavior.

fusion the melting of a solid to a liquid; also, the reaction between small atomic nuclei to form a larger one. Example: The fusion reaction $2 \, {}^2_1H \rightarrow {}^4_2He$ would produce a large amount of energy.

G

G see *free energy*.

gamma radiation (γ) high-energy photons emitted by radioactive nuclei.

gas constant (R) the constant which appears in the

Ideal Gas Law equation, PV = nRT; depends upon units of P, V, and T; equals 0.0821 $\dfrac{\text{L} \cdot \text{atm}}{\text{mol} \cdot \text{K}}$ in the units listed.

Gas Law see *Ideal Gas Law*.

geometric isomer a species having the same kind and number of atoms as another species, but in which the geometric structure is different. Example: There are two geometric isomers with the molecular formula $Pt(NH_3)_2Cl_2$ (structures are both planar):

Gibbs-Helmholtz equation a relationship among ΔG, ΔH, and ΔS; $\Delta G = \Delta H - T\Delta S$.

Graham's Law a relation stating that the rate of effusion of a gas is inversely proportional to the square root of its molar mass.

gram (g) a unit of mass in the metric system; equal to mass of one cubic centimeter of water at 4°C.

greenhouse effect the effect of water and carbon dioxide in absorbing outgoing IR radiation, thereby raising the earth's temperature.

ground state the lowest allowed energy state of an atom, ion, or molecule.

group a vertical column of the Periodic Table.

Groups I, II, III, IV cation groups in qualitative analysis. Roman numerals are used to distinguish from groups in the Periodic Table.

H

H see *enthalpy*.

Haber process an industrial process used to make ammonia from nitrogen and hydrogen.

half-cell half of a voltaic or electrolytic cell, at which either oxidation or reduction occurs. Example: The half-cell reaction at the anode is one of oxidation.

half-equation an equation written to describe a half-reaction of oxidation or reduction. Example: $Zn(s) \rightarrow Zn^{2+}(aq) + 2\,e^-$ is an oxidation half-equation.

half-life ($t_{1/2}$) the time required for a reaction to convert half of the initial reactant to product(s).

halide ion F^-, Cl^-, Br^-, or I^-.

halogen an elementary substance in Group 7. Examples: F_2, Cl_2, Br_2.

hard water water containing excessive Ca^{2+} or Mg^{2+}.

head-to-tail polymerization process leading to the formation of a polymer of the type

heat that form of energy which flows between two samples of matter because of their difference in temperature.

heat flow the amount of heat, q, passing into or out of a system; q is positive if flow is into system, negative if out of system.

heat of formation see *enthalpy of formation*.

heat of fusion ΔH for the conversion of unit amount (one gram or one mole) of a solid to a liquid at constant P and T.

heat of sublimation ΔH for the conversion of unit amount (one gram or one mole) of a solid to a vapor at constant P and T.

heat of vaporization ΔH for the conversion of unit amount (one gram or one mole) of a liquid to a vapor at constant P and T.

Henry's Law a relation stating that the solubility of a gas in a liquid is directly proportional to its partial pressure.

hertz (Hz) a unit of frequency: 1 cycle per second.

Hess's Law a relation stating that the heat flow in a reaction which is the sum of two other reactions is equal to the sum of the two heat flows in those reactions.

heterogeneous having nonuniform composition.

heterogeneous catalysis catalysis that occurs upon a solid surface.

high-spin complex a complex which, for a particular metal ion, has the largest possible number of unpaired electrons.

homogeneous having uniform composition.

homogeneous catalysis catalysis that occurs within a solution.

Hund's rule a relation stating that, ordinarily, electrons will not pair in an orbital until all orbitals of equal energy contain one electron.

hybrid atomic orbital an orbital made from a mixture of s, p, d, or f orbitals. Example: An sp^2

hybrid orbital is derived from an s and two p orbitals.

hydrate a substance containing bound water. Example: $BaCl_2 \cdot 2 H_2O$ is a common hydrate.

hydrocarbon a substance containing only hydrogen and carbon atoms.

hydrogen bonds attractive forces between molecules arising from interaction between a hydrogen atom in one molecule and a strongly electronegative atom (N, O, F) in a neighboring molecule. Example: Hydrogen bonding in water is due to interaction between the H atoms and O atoms on different H_2O molecules.

hydronium ion the H_3O^+ ion characteristic of acidic water solutions.

hydroxide ion OH^-.

I

Ideal Gas Law the relationship between pressure, volume, temperature, and amount for any gas at moderate pressures: $PV = nRT$.

indicator, acid-base a chemical substance that changes color with pH change; usually color change occurs over about two pH units.

inert complex a complex ion which exchanges ligands very slowly.

inert gas noble gas.

infrared light having a wavelength greater than about 700 nm.

initial rate rate at the beginning of a reaction, before reactant concentrations have decreased appreciably from their original values.

interhalogen a compound containing two different kinds of halogen atoms, such as ICl.

intermolecular force the force between adjacent molecules (dipole, dispersion, or hydrogen bond).

internal energy see *energy*.

intramolecular force the force between atoms within a molecule (covalent bond).

ion a charged species.

ionic bond the electrostatic force between oppositely charged ions in an ionic compound.

ionic compound a substance in which component species are cations and anions. Examples: NaCl, CaO, NH_4NO_3.

ionic radius the radius of an ion, based on the assumption that ions in a crystal are in contact with nearest neighbors.

ionization constant a general term for dissociation constant of a weak acid or base; see *acid dissociation constant*.

ionization energy the energy required to remove an electron from a species.

ion product (P) the product of the actual concentrations of cation and anion in equilibrium with a slightly soluble electrolyte, each raised to the appropriate power. If P, upon mixing, is greater than K_{sp}, a precipitate forms.

isoelectronic (with) having the same number of electrons as.

isomer a species having the same number and kind of atoms as another species, but having different properties; structural, geometric, optical, and stereoisomers may occur. Example: Dimethyl ether, CH_3-O-CH_3 is a structural isomer of ethyl alcohol, CH_3-CH_2-OH.

isotope an atom having the same number of nuclear protons as another, but with a different number of neutrons. Example: Ordinary oxygen has three isotopes, all with eight protons in the nucleus, but with eight, nine, and ten neutrons, respectively.

J

joule (J) the base SI unit of energy; equal to kinetic energy of a two-kilogram mass moving at a speed of one meter per second.

K

K_a, K_b, K_c, K_f, K_{sp}, K_w see *acid dissociation constant, base dissociation constant, equilibrium constant, formation constant, solubility product constant,* and *water dissociation constant*, respectively.

Kelvin temperature scale (K) scale obtained by taking lowest attainable temperature to be 0 K, normal freezing point of water to be 273.15 K.

kilo- (k) a prefix on metric units indicating multiple of 1000. Example: One kilojoule equals 1000 J.

kilogram (kg) basic unit of mass in SI: 1000 g.

kilojoule (kJ) a unit of energy: 1000 J.

kilopascal (kPa) a pressure unit: 1 kPa is approximately the pressure exerted by a 10-g mass resting on a 1-cm^2 area; 101.3 kPa = 1 atm.

kilowatt-hour (kWh) a unit of energy: 1 kWh = 3.6×10^6 J.

kinetic associated with motion. Example: The kinetic energy of a particle of mass m at speed v is equal to $\frac{1}{2}mv^2$.

kinetics the study of rates of chemical reactions.

kinetic theory a model of molecular motion used to explain many of the properties of gases.

L

labile complex a complex ion which rapidly reaches equilibrium with ligands in surrounding solution.

lanthanides elements 57 (La) through 70 (Yb).

lattice energy ΔH for the process in which oppositely charged ions in the gas phase combine to form an ionic lattice in the solid phase.

Law of Charles and Gay-Lussac see *Charles' Law*.

Law of Combining Volumes a relation stating that relative volumes of different gases (at same T and P) involved in a reaction are in the same ratio as their coefficients in the balanced equation.

Law of Conservation of Mass a relation stating that in a chemical reaction the mass of the products equals the mass of the reactants.

Law of Constant Composition a relation stating that the relative masses of the elements in a given chemical compound are fixed.

Law of Dulong and Petit a relation stating that the heat capacity of one mole of any metal is about 25 J/°C.

Law of Multiple Proportions a relation stating that when two elements, A and B, form two compounds, the relative amounts of B which combine with a fixed amount of A are in a ratio of small integers. Example: In water and hydrogen peroxide, both of which contain hydrogen and oxygen, there are 8 and 16 g of oxygen, respectively, for each gram of hydrogen.

Le Châtelier's Principle a relation stating that, when a system at equilibrium is disturbed, it responds in such a way as to counteract the change.

Lewis acid a species which can accept a pair of electrons. Example: In the reaction $BF_3 + NH_3 \rightarrow BF_3NH_3$, the BF_3 accepts a pair of electrons from NH_3 and so behaves as a Lewis acid.

Lewis base a species which can donate a pair of electrons. Example: NH_3 in the above reaction.

Lewis structure the electronic structure of a molecule or ion in which electrons are shown by dots or dashes (electron pairs).

ligand a molecule or anion bonded directly to the central metal in a complex ion.

limiting reactant the least abundant reactant, based on the equation, in a chemical reaction; dictates the maximum amount of product which can be formed.

linear molecule a molecule containing three atoms in which the bond angle is 180°. Example: BeF_2, CO_2.

liter (L) a unit of volume: 1 L = 1000 cm³.

logarithm of a number the exponent to which another number, usually 10, must be raised to give the number. Examples: The logarithm of 100 to base 10 is 2, since $10^2 = 100$; the logarithm of 3.00 is 0.477, since $10^{0.477} = 3.00$.

low-spin complex a complex, which for a particular metal ion, has the smallest possible number of unpaired electrons.

luster the characteristic shiny appearance of a metal surface.

Lyman series a series of "lines" in the atomic spectrum of hydrogen resulting from the transition of an electron from a higher energy level to the level n = 1.

M

macromolecular network covalent.

main group a numbered group of the Periodic Table.

malleable capable of being shaped, as by pounding with a hammer.

manometer a U-tube containing a liquid, usually mercury; used to measure the pressure of a gas.

mass a property reflecting the amount of matter in a sample.

mass number an integer equal to the sum of the number of protons and neutrons in an atomic nucleus. Example: The mass number of a $^{37}_{17}Cl$ isotope is 37; the nucleus of that isotope contains 17 protons and 20 neutrons.

mass percent see *percent*.

matter a general term for any kind of material; the stuff of which pure substances are made.

Maxwellian distribution a relation describing the way in which molecular speeds, or energies, are shared among the molecules in a gas.

mean free path the average distance traveled by an atom or molecule between collisions.

mechanism a sequence of steps that occurs during the course of a chemical reaction.

melting point (mp) same as *freezing point*.

metal a substance having characteristic luster, malleability, and high electrical conductivity; readily loses electrons to form positive ions.

metallic character the extent to which a substance has properties typical of a metal.

metalloid an element with properties intermediate between those usually associated with metals and nonmetals. Examples: B, Si, Ge, As, Sb, Te.

metallurgy the science and processes of extracting metals from their ores.

meter (m) a unit of length in the metric system.

milli- (m) a prefix on a metric unit indicating multiple of 1×10^{-3}. Example: One millimeter equals one one thousandth of a meter, 0.001 m.

mixture two or more substances combined so that each substance retains its chemical identity.

mm Hg a unit of pressure: 1 atm = 760 mm Hg.

molality (m) a concentration unit, defined as equal to the number of moles of solute divided by number of kilograms of solvent. Example: A solution made by dissolving 0.10 mol of KNO_3 in 200 g of water would be 0.50 molal in KNO_3 (0.50 m KNO_3).

molarity (M) a concentration unit, defined equal to number of moles of solute divided by number of liters of solution. Example: In 6 molar HCl, 6 M HCl, there is 6 mol of HCl in one liter of solution.

molar mass (MM) the mass of one mole of a substance. Example: MM of O_2 = 32.00 g/mol; MM NaCl = 58.44 g/mol.

mole (mol) a collection of 6.022×10^{23} items. The mass in grams of one mole of a substance is numerically equal to its formula mass. Examples: A mole of O_2 weighs 32.00 g; a mole of NaCl weighs 58.44 g.

mole fraction a concentration unit, defined as equal to the number of moles of component divided by the total number of moles in solution. Example: In a solution in which there is 1 mol benzene, 2 mol CCl_4, and 7 mol acetone, the mole fraction of the acetone is 0.7.

molecular formula an expression stating the number and kind of each atom present in the molecule of a substance. Example: The molecular formula of hexane is C_6H_{14}, which means there are 6 C atoms and 14 H atoms in a hexane molecule.

molecular geometry the shape of a molecule describing the relative positions of atomic nuclei.

molecular orbital an orbital involved in the chemical bond between two atoms, and taken to be a linear combination of the orbitals on the two bonded atoms.

molecule an aggregate of atoms, which is the characteristic component particle in all gases, many pure liquids, and some solids; often contains only a few atoms; has relatively little physical interaction with other molecules. Examples: In nitrogen gas, liquid benzene, and solid glucose, there are N_2 molecules, C_6H_6 molecules, and $C_6H_{12}O_6$ molecules, respectively.

monodentate ligand (also called unindentate) a ligand that forms only one coordinate covalent bond with a metal atom. Examples: H_2O, NH_3, Cl^-.

monomer a small molecule that joins with other monomers to form a polymer.

multiple bond a double or triple bond.

multiple equilibria, rule of a rule stating that if Equation 1 + Equation 2 = Equation 3, then $K_1 \times K_2 = K_3$.

N

n + ℓ rule the rule that orbitals usually fill in order of increasing value of **n + ℓ**. (For example, 4s fills before 3d). For two orbitals with the same value of **n + ℓ**, the one with the lower **n** value fills first (3p before 4s).

nano- (n) a prefix on a metric unit indicating multiple of 1×10^{-9}. Example: One nanometer equals 10^{-9} m.

nanometer (nm) a unit of length equal to 10^{-9} m.

natural logarithm a logarithm based upon the number e, 2.7182818 . . .; if $\log_e X = Y$, then $e^Y = X$, $\log_{10} X = \log_e X/2.303$; base e comes from the calculus, where certain derivatives and integrals are most easily expressed in terms of e.

Nernst equation an equation relating cell voltage E to the standard voltage E_{tot}^0 and the concentrations of reactants and products: $E = E_{tot}^0 - \dfrac{0.0591}{2} \log_{10} Q$.

net ionic equation a chemical equation for a reaction, in which only those species that actually react are listed. Example: When 1 M HCl and 1 M NaOH solutions are mixed, the net ionic equation for the reaction is $H^+(aq) + OH^-(aq) \rightarrow H_2O$; the Cl^- and Na^+ ions in the solution do not react and therefore do not appear in the equation.

network covalent having a structure in which all the atoms in a crystal are linked by a network of covalent bonds. Examples: Found in C, SiO_2.

neutralization a reaction of an acid and a base to produce a neutral (pH 7) solution.

neutral solution an aqueous solution with pH 7.0 ($[H^+] = 1.0 \times 10^{-7}$ M).

neutron one of the particles in an atomic nucleus; mass = 1, charge = 0.

nitrogen fixation see *fixation of nitrogen*.

noble gas an element in Group 8, at the far right of the Periodic Table.

noble gas structure the ns^2np^6 outer electron structure in an atom or ion; a particularly stable structure attained by atoms obeying the octet rule; sometimes called an inert gas structure.

nonelectrolyte a substance that does not exist as ions in water solution. Example: Since ethyl alcohol does not ionize when dissolved in water, it is a nonelectrolyte.

nonmetal one of the elements in the upper right corner of the Periodic Table that does not show metallic properties. Example: Nitrogen gas, N_2.

nonpolar bond a chemical bond in which there are no positive and negative ends; found in homonuclear diatomic molecules. Examples: H_2, O_2.

nonpolar molecule a molecule in which there is no separation of charge and hence no negative and positive poles. Nonpolar molecules include H_2, and CO_2.

nonspontaneous reaction a reaction which cannot occur by itself without input of work from an external source; $\Delta G > 0$ for nonspontaneous reactions at T and P.

normal boiling point the boiling point at one atmosphere pressure.

n-type semiconductor a semiconductor in which current is carried through a solid by "extra" electrons introduced by electron-rich impurity atoms.

nuclear fission the splitting of a heavy nucleus by a neutron into two lighter nuclei, accompanied by the release of energy.

nuclear fusion the combining of light nuclei to form a heavier nucleus, accompanied by the release of energy.

nuclear reactor a device used to generate electrical energy using the heat given off by nuclear fission.

nucleus the small, dense, positively charged region at the center of an atom.

O

octahedral having the symmetry of a regular octahedron. In an octahedral species, a central atom is surrounded by six other atoms. Four of these are at the corners of a square; the other two are directly above and directly below the central atom.

octet a group of eight valence electrons surrounding an atom. All the noble-gas atoms except He have an octet of valence electrons.

octet rule the principle that bonded atoms tend to have a share in eight outermost electrons.

octet structure see *noble gas structure*.

opposed spins a term that refers to electrons with different values of m_s; two electrons in a single orbital must have opposed spins.

optical activity the ability to rotate the plane of a beam of transmitted polarized light; a property possessed by substances having a chiral center.

optical isomerism the phenomenon in which each member of a pair of molecules having the same molecular formula rotates a beam of plane polarized light in opposite directions. Such molecules have at least one chiral center.

orbital an electron cloud with an energy state characterized by given values of n, l, and m_ℓ quantum numbers; has a capacity for two electrons having paired spins; often associated with a particular region in the atom. Example: In an atom the electrons in the $2p_x$ orbital are in a dumbbell-shaped cloud concentrated along the x axis.

orbital diagram a sketch showing electron populations of atomic orbitals, including electron spins.

order of reaction an exponent to which the concentration of a reactant needs to be raised to give observed dependence of reaction rate on concentration. Example: If, for the reaction A → products, rate = k(conc. A)^2, the reaction is second order.

ore a natural mineral deposit from which a metal can be extracted profitably.

organic used to characterize any compound containing carbon, hydrogen, and possibly other elements. Example: Propionic acid, CH_3CH_2COOH, is an organic compound; SiO_2 is not.

osmosis the process by which a solvent moves from a region where its vapor pressure is high (dilute solution) to a region where its vapor pressure is low (concentrated solution); movement occurs through a semipermeable membrane.

osmotic pressure the excess pressure that must be applied to a solution to prevent the pure solvent from diffusing into the solution through a semipermeable membrane.

Ostwald process the industrial process used to make nitric acid from ammonia.

outer electron configuration a statement of the population of sublevels in outer principal levels; electrons in inner, filled levels are not included. Example: The outer electron configuration of the iodine atom is $5s^2 5p^5$.

overall order the number obtained by summing all the exponents in the rate expression. If rate = k(conc. A)^m × (conc. B)^n, then overall order = m + n.

oxidation a half-reaction involving a loss of electrons or, more generally, an increase in oxidation number. Example: If, at an electrode, the reaction is $Ag(s) \rightarrow Ag^+(aq) + e^-$, silver is undergoing oxidation, since its oxidation number is increasing from 0 to +1.

oxidation number a number which can be assigned to an atom in a molecule or ion which reflects, qualitatively, its state of oxidation; the number is determined by applying a set of rules. Examples: In the NO_3^- ion the oxidation numbers of the N and O atoms are +5 and −2, respectively.

oxide a compound containing oxygen. The *oxide ion,* found in metal oxides, has the formula O^{2-}.

oxidizing agent a species which accepts electrons from another. Example: In the reaction $Cl_2(aq)$ + $2\,Br^-(aq) \rightarrow 2\,Cl^-(aq) + Br_2(aq)$, Cl_2 serves as an oxidizing agent.

oxyacid an acid containing oxygen. Examples: HNO_3, H_2SO_4, and $HClO$ are oxyacids.

oxyanion an anion containing oxygen. Examples: NO_3^-, SO_4^{2-}, and ClO^- are oxyanions.

ozone an allotropic form of oxygen in which the molecule is O_3.

P

P see *ion product.*

paired electrons two electrons in the same orbital with spins equal to $+\frac{1}{2}$ and $-\frac{1}{2}$; an electron pair.

parallel spins a term that refers to electrons with the same m_s values; single electrons in different orbitals of the same energy have parallel spins.

paramagnetic having magnetic properties caused by unpaired electrons. Examples: The NO molecule, H atom, and CH_3 radical are paramagnetic.

partial pressure of A (P_A) that part of the total pressure in a gaseous mixture that can be attributed to component A. The partial pressure of A is equal to the pressure A would exert in the container if A were there by itself. Example: In a mixture of 3 mol N_2 and 1 mol O_2 at a total pressure of 2 atm, the partial pressure of N_2 is 1.5 atm and that of O_2 is 0.5 atm.

parts per million (ppm) for gases, the number of moles of solute per million moles of gas; for liquids and solids, the number of grams of solute per million grams of sample. Example: If a gas sample contains 6 parts per million CO, then in one mole of gas there would be 6 × 10^{-6} mol CO.

Pauli exclusion principle see *exclusion principle.*

peptide linkage the —N—C— group in proteins;
also called peptide bond.

percent A parts A in 100 parts of a sample; usually in mass percent but may be in mole percent or volume percent. Example: A 5% NaOH solution contains 5 g NaOH in 100 g of solution.

percent composition percentages by mass of the elements in a compound.

percent dissociation for a weak acid HB; percent diss. = 100 × [H⁺]/orig. conc. HB.

percent yield a quantity equal to 100 × actual yield/theoretical yield.

period a horizontal row of elements in the Periodic Table.

Periodic Law statement that the properties of elements are a periodic (cyclical) function of atomic number.

Periodic Table an arrangement of the elements into rows and columns in which those elements with similar properties occur in the same column.

peroxide a binary compound containing an oxygen-oxygen single bond or the peroxide ion, O_2^{2-}.

pH an alternative way to express H⁺ ion concentration: $pH = -\log_{10}[H^+]$.

phase diagram for one-component systems, a graph of pressure vs. temperature, showing conditions under which the pure substance will exist as a liquid, solid, or gas, and also the conditions under which two-phase and three-phase equilibria can exist.

photon an individual quantum of radiant energy of wavelength lambda (λ).

physical property a property of a substance related to its physical characteristics. Examples: density, melting point.

pi (π) bond a bond in which electrons are concentrated in orbitals that are located off the internuclear axis; one bond in a double bond is a pi bond, and there are two pi bonds in a triple bond.

pKₐ an expression of strength of a weak acid: $pK_a = -\log_{10} K_a$.

Planck's constant (h) the constant in the equation: $E = h\nu = hc/\lambda$. $h = 6.626 \times 10^{-34}$ J·s.

polar bond a chemical bond which has positive and negative ends; characteristic of all bonds between nonidentical atoms. Example: In CCl_4 the C—Cl bonds are polar, since the Cl atom tends to attract electrons more than the C atom does.

polarization a distortion of the electron distribution in a molecule, tending to produce positive and negative poles.

polar molecule a molecule in which there is a separation of charge and hence a positive and a negative pole. Examples: HF, H_2O.

pollutant a contaminant, or foreign species, present in a sample; usually has a deleterious effect on quality of sample as far as living things are concerned.

polyamide a condensation polymer formed from carboxylic acid and amine units.

polyatomic ion a charged species containing more than one atom.

polydentate ligand a ligand which forms two or more bonds to a central metal.

polyester a condensation polymer made up of ester units. Examples: Dacron, Kodel.

polymer a molecule made up from many units which are linked together chemically. Example: In the polymer called polyethylene, many $H_2C{=}CH_2$ units become linked together by a chemical reaction to form chains which have the structure

$$-\underset{\underset{H}{|}}{\overset{\overset{H}{|}}{C}}-\underset{\underset{H}{|}}{\overset{\overset{H}{|}}{C}}-\underset{\underset{H}{|}}{\overset{\overset{H}{|}}{C}}-\underset{\underset{H}{|}}{\overset{\overset{H}{|}}{C}}-\underset{\underset{H}{|}}{\overset{\overset{H}{|}}{C}}-\underset{\underset{H}{|}}{\overset{\overset{H}{|}}{C}}-$$

polypeptide a condensation polymer in which the monomer units are α-amino acids; in the polymer the amino acid residues are linked by peptide bonds; another name for a protein.

polyprotic acid an acid containing more than one ionizable hydrogen atom. Examples: H_2SO_4, H_3PO_4.

positron (^0_1e) a particle having the mass of an electron but a +1 charge.

post-transition metals the lower members of Periodic Table Groups 3, 4 and 5.

precipitate a solid that forms when two solutions are mixed.

pressure the force per unit area; often expressed in mm Hg, atmospheres, or kilopascals.

principal energy level the energy level designated by the principal quantum number **n.** The first element in each period of the Periodic Table introduces a new principal energy level.

principal quantum number (n) the most important quantum number, since it has greatest effect on energy of electron; cited first in the set of four quantum numbers associated with an electron.

product a substance formed as a result of a chemical reaction. Example: In the reaction $Ag^+(aq) + Cl^-(aq) \rightarrow AgCl(s)$, AgCl is the product.

property a characteristic of a sample of matter that is fixed by its state. Example: The density and energy of a mole of H_2 at 100°C and 1 atm are properties of that sample of hydrogen gas.

protein a polypeptide.

proton the nucleus of a hydrogen atom, the H^+ ion; a component of atomic nuclei with mass = 1, charge = +1.

p-type semiconductor a semiconductor in which current is carried through a solid by electron flow into "positive holes" in the crystal, introduced by electron-deficient impurity atoms.

pyramidal the geometry of a molecule in which one atom lies directly above the center of an equilateral triangle formed by three other atoms. Example: NH_3.

Q

Q see *heat flow*.

quadratic formula the formula used to obtain the two roots of the general quadratic equation: $ax^2 + bx + c = 0$. The formula is $x = \dfrac{-b \pm \sqrt{b^2 - 4ac}}{2a}$.

qualitative analysis the determination of the nature of the species present in a sample. Example: By qualitative analysis she found that the solution contained Cu^{2+}, Sn^{4+}, and Cl^- ions.

quantitative analysis the determination of how much of a given component is present in a sample. Example: The students discovered, by quantitative analysis, that the ore contained 42.45% iron by mass.

quantum mechanics approach used to calculate the energies and spatial distributions of small particles confined to very small regions of space.

quantum number a number used in the description of the energy levels available to atoms and molecules; an electron in an atom or ion has four quantum numbers to describe its state.

quantum theory a general theory which describes the allowed energies of atoms and molecules.

R

R see *gas constant*.

rad a unit of absorbed radiation equal to 10^{-2} J absorbed per kilogram of tissue.

radical see *free radical*.

radioactivity the ability possessed by some natural and synthetic isotopes (induced radioactivity) to undergo reactions involving nuclear transformations to other isotopes.

Raoult's Law relation between the vapor pressure (P) of a component of a solution and that of the pure component (P^0) at the same temperature: $P = XP^0$, where X is the mole fraction.

rare earth the name sometimes given to members of the lanthanide series.

rate constant the proportionality constant in the rate equation for a reaction. Example: If the rate equation is rate = k(conc. A)^n, then k is the rate constant.

rate-determining step the slowest step in a multistep reaction.

rate expression (law) a mathematical relationship describing the dependence of the reaction rate upon the concentration(s) of reactant(s).

rate of a reaction the magnitude of the change in concentration of a reactant or product divided by the time required for the change to occur (with both quantities relatively small). Example: For the reaction $A \rightarrow B$,

$$\text{rate} = \frac{\Delta(\text{conc. B})}{\Delta t} = \frac{-\Delta(\text{conc. A})}{\Delta t}.$$

reactant the starting material in a chemical reaction. Example: In the reaction $H_2(g) + \frac{1}{2} O_2(g) \rightarrow H_2O(l)$, H_2 and O_2 are both reactants.

reaction a chemical change in which new substances are formed. Example: When aluminum burns in air, the chemical reaction that occurs is described by the equation $2\,Al(s) + \frac{3}{2} O_2(g) \rightarrow Al_2O_3(s)$.

reaction mechanism see *mechanism*.

reciprocal rule the relation between equilibrium constants for forward (K_f) and reverse (K_r) reactions: $K_r = 1/K_f$.

redox reaction a reaction involving oxidation and reduction.

redox titration the titration of an oxidizing agent by a reducing agent, or vice versa.

reducing agent a species which furnishes electrons to another. Example: In the reaction $Zn(s) + 2\,H^+(aq) \rightarrow Zn^{2+}(aq) + H_2(g)$, the $Zn(s)$, metallic zinc, in the reducing agent.

reduction a half-reaction in which a species gains electrons or, more generally, decreases in oxidation number. Example: In the reaction $Zn(s)$

+ 2 H$^+$(aq) → Zn^{2+}(aq) + H$_2$(g), the H$^+$ ions (oxid. no. = +1) are reduced to H$_2$ (oxid. no. = 0).

relative humidity 100 × P/P$_0$, where P = pressure of water vapor in air, P$_0$ = equilibrium vapor pressure of water at same T.

rem a unit of absorbed radiation equal to n times the number of rads. The factor n depends upon the type of radiation absorbed.

resonance a model used to rationalize properties of octet rule species for which one Lewis structure is inadequate; resonance structure is taken to be an average of two or more Lewis structures which differ only in positions of electrons; species are said to exhibit resonance.

reverse osmosis a process by which pure water is obtained from a salt solution. Under pressure, the water passes out of the salt solution through a semipermeable membrane.

roasting a metallurgical process in which a sulfide ore is heated in air. Roasting may convert a metal sulfide (ZnS, HgS) to a metal oxide (ZnO) or to the free metal (Hg).

rule of multiple equilibria see *multiple equilibria*.

S

S see *entropy*.

salt a solid ionic compound made up from a cation other than H$^+$ and an anion other than OH$^-$ or O^{2-}. Examples: NaCl, CuSO$_4$, NH$_4$NO$_3$.

saturated hydrocarbon an alkane, a hydrocarbon in which all carbon-carbon bonds are single.

saturated solution a solution containing as much solute as the amount of solvent can dissolve at a specific temperature.

Schrödinger equation a wave equation that relates mass, potential energy, kinetic energy, and coordinates of a particle.

Second Law of Thermodynamics a basic law used to describe spontaneous processes; one statement says that all spontaneous processes occur with an overall increase in entropy.

second-order reaction a reaction whose rate depends on second power of reactant concentration; may be sum of exponents of two reactant concentrations. Examples: The two expressions, rate = k(conc. A)^2 and rate = k(conc. A) × (conc. B), are both associated with second-order reactions.

semiconductor a substance used in transistors and thermistors whose electrical conductivity depends on the presence of tiny amounts of impurities (such as As or B) in a very pure crystal of an element (such as Si or Ge).

semipermeable membrane a film which allows passage of solvent molecules such as H$_2$O but does not pass solute molecules such as proteins or, in some cases, ions.

shielding a term used to describe effect of inner electrons in decreasing the attraction of an atomic nucleus on outermost electrons.

SI unit a unit associated with the International System of Units; see Appendix 1.

sigma (σ) bond a chemical bond in which electron density on the internuclear axis is high, which is the case with all single bonds; double bonds contain one sigma bond and one pi bond; triple bonds contain one sigma and two pi bonds.

significant figures meaningful digits in a measured quantity; number of digits in a number when expressed in exponential notation. Example: 1.035 × 10^3 has four significant figures (exponential doesn't count).

simple cubic cell a unit cell containing atoms at each corner of a cube.

simplest formula see *empirical formula*.

single bond a pair of electrons shared between two bonded atoms.

slag a by-product, formed in a metallurgical process, that floats on the surface of a molten metal. Slags typically contain metal silicates such as CaSiO$_3$.

smog smoky fog containing harmful species such as SO$_2$, SO$_3$, NO$_2$, and O$_3$.

soap sodium salt of a fatty acid.

solar cell a voltaic cell in which the sun is the direct source of energy.

solubility the amount of a solute that dissolves in a given amount of solvent at a specified temperature. May be stated in various ways; moles solute per liter solution is common.

solubility product constant (K$_{sp}$) the equilibrium constant for the solution reaction of a relatively insoluble ionic compound. Example: For the reaction Ca(OH)$_2$(s) ⇌ Ca^2(aq) + 2 OH$^-$(aq), K$_{sp}$ = [Ca^{2+}] × [OH$^-$]^2.

solubility rules the rules used to classify ionic compounds as to their solubility in water.

solute the solution component present in smaller amount than the solvent.

solution a liquid, gas, or solid phase containing two or more components dispersed uniformly throughout the phase.

solvent a substance, usually a liquid, in which another subtance, called the solute, is dissolved.

species a general term referring to a molecule, ion, or atom.

specific heat (S.H.) the amount of heat required to raise the temperature of one gram of a substance by one degree Celsius.

spectator ion an ion which, though present, takes no part in a reaction, and hence is not included in the equation used to represent the reaction.

spectrochemical series an arrangement of ligands in order of decreasing tendency to split d orbitals of transition metal cation. Substitution of a ligand for another higher in the series gives a smaller splitting and hence a longer wavelength of light absorbed.

spectrum a pattern of characteristic wavelengths associated with excitation of an atom, molecule, or ion; also used as name of a pattern having a similar appearance obtained as a result of chromatographic or mass spectroscopic experiments.

spontaneous reaction a reaction which can occur by itself, without input of work from outside; $\Delta G < 0$ for spontaneous reactions at T and P.

square planar the geometry of a complex ion in which four ligands are located at the corners of a square, bonded to a metal atom at the center of the square.

square pyramid a pyramid which has a square as its base. Example: $XeOF_4$ has such a structure, with the Xe atom at the center of the square.

stable will not change spontaneously. Nature of change should be specified. Example: Water is stable at 25°C with respect to thermal decomposition to hydrogen and oxygen.

standard free energy change (ΔG^0) ΔG when reactants and products are in their standard state (1 atm for a gas, effectively 1 M for a species in aqueous solution).

standard molar entropy (S^0) S of a substance in its standard state (1 atm for a gas, 1 M for an ion in water solution).

standard oxidation voltage (E^0_{ox}) the voltage associated with an oxidation reaction at an electrode, when all solutes are 1 M (strictly speaking, unit activity) and all gases are at 1 atm.

standard potential identical with the standard reduction voltage (described below).

standard reduction voltage (E^0_{red}) the voltage associated with a reduction reaction at an electrode, when all solutes are 1 M and all gases are at 1 atm.

standard voltage (E^0_{tot}) the voltage of a cell in which all species are in their standard states (solids and liquids are pure, solutes are at unit activity, often taken to be 1 M, and gases are at 1 atm).

state the condition of a system when its properties are fixed. Example: A mole of H_2O at 25°C and 1 atm is in a definite state in that all of its properties have values that are fixed.

state property a property of a system which is fixed when the temperature, pressure, and composition are specified. Example: One mole of water at 25°C and 1 atm has a fixed volume, enthalpy, and entropy; V, H, and S are state properties.

stoichiometric having to do with masses (grams, moles) of reactants and products in a chemical equation.

straight-chain alkane a saturated hydrocarbon in which all the carbon atoms are arranged in a single, continuous chain.

strong as applied to acids, bases, and electrolytes, indicates complete dissociation into ions when in water solution. Example: HCl is a strong acid since it exists as H^+ and Cl^- ions in aqueous solution.

structural formula a formula showing the arrangement of atoms in a molecule or polyatomic ion.

structural isomers two or more species having the same molecular formula but different molecular structures.

sublevel a subdivision of an energy level as designated by the quantum number ℓ.

sublimation a change in state from solid to gas. Example: Iodine slowly sublimes in an open container; the sublimation is endothermic.

successive approximations a technique used to solve quadratic or higher degree equations. On the basis of a reasonable assumption or an educated guess, a first, approximate answer is obtained; that answer is used, with the original equation, to obtain a more nearly exact solution. The process can be repeated until an answer with the desired accuracy is obtained.

superoxide an ionic oxygen compound containing the superoxide ion, O_2^-.

supersaturated   containing more solute than equilibrium conditions would allow; unstable to addition of solute crystal. Example: It is easy to make a supersaturated solution of sodium acetate by cooling a hot concentrated solution of the salt carefully to 25°C.

surroundings   everything outside the system being studied.

symbol   a one- or two-letter abbreviation for the name of an element.

system   the sample of matter under consideration.

T

tetrahedral   having the symmetry of a regular tetrahedron. In a tetrahedral species, a central atom is surrounded by four other atoms directed so as to make all the bond angles 109.5°.

theoretical yield   the amount of product obtained from the complete conversion of the limiting reactant.

thermal   having to do with heat.

thermite reaction   the exothermic reaction of Al with a metal oxide; the products are Al_2O_3 and the free metal.

thermochemical equation   a chemical equation in which the value of ΔH is specified.

thermodynamics   the study of heat, work, and the related properties of mechanical and chemical systems.

titration   a process in which a solution is added to another solution with which it reacts under conditions such that the volume of added solution can be accurately measured.

torr   a unit of pressure equal to 1 mm Hg.

trans isomer   a geometric isomer in which two identical groups are as far apart as possible, as opposed to cis, where they are as close as possible. Example: The trans isomer of the square planar $Pt(NH_3)_2Cl_2$ molecule has the structure

transition metal   any one of the metals in the central groups in the fourth, fifth, and sixth periods in the Periodic Table. Examples: Fe, Zr, W.

translational energy   the energy of motion through space. Example: A falling raindrop has translational energy.

triangular bipyramid   a solid with five vertices and six sides; may be regarded as two pyramids fused through a base that is an equilateral triangle.

triple bond   three electron pairs shared between two bonded atoms.

triple point   that temperature and pressure at which the solid, liquid, and vapor of a pure substance can coexist in equilibrium.

U

ultraviolet radiation   light having a wavelength less than about 400 nm but greater than about 10 nm.

unit cell   the smallest unit of a crystal that, if repeated indefinitely, could generate the whole crystal.

unpaired electron   a single electron occupying an orbital by itself.

unsaturated hydrocarbon   a hydrocarbon containing double or triple carbon-carbon bonds.

unsaturated solution   a solution that is able to dissolve more solute.

unshared pair   a pair of electrons that "belongs" to a single atom and is not involved in bonding.

useful work   any work other than expansion work associated with a process; frequently refers to the electrical work obtained from a reaction.

V

valence bond model   the theory that atoms tend to become bonded by pairing and sharing their outer, or valence, electrons; also referred to as the atomic orbital model.

valence electrons   those electrons in the outermost shell. Example: In the carbon atom, with electron configuration $1s^2 2s^2 2p^2$, there are four valence electrons, those in the 2s and 2p orbitals.

van der Waals equation   an equation used to express the physical behavior of a real gas: $(P + n^2a/V^2)(V - nb) = nRT$.

van der Waals forces a general name sometimes given to intermolecular forces.

vapor a condensable gas.

vapor pressure the pressure exerted by a vapor when it is in equilibrium with the liquid from which it is derived. Example: If liquid water is admitted to an evacuated container at 60°C, the pressure in the container when the liquid and vapor reach equilibrium becomes 149.4 mm Hg; therefore, the vapor pressure of water at 60°C is 149.4 mm Hg.

vapor pressure lowering the decrease in the vapor pressure of a liquid caused by addition of a nonvolatile solute.

volatile easily evaporated.

volt (V) a unit of electrical potential: $1 \text{ V} = 1 \text{ J/C}$.

voltage electric potential; a measure of the tendency of a cell or other device to force electrons through an external circuit.

voltaic cell a device in which a spontaneous chemical reaction is used to produce electrical work.

VSEPR Valence Shell Electron Pair Repulsion model, used to predict molecular geometry; states that electron pairs tend to be as far apart as possible.

W

water dissociation constant (K_w) equal to $[H^+] \times [OH^-] = 1 \times 10^{-14}$ at 25°C.

water softening the removal of ions, particularly Ca^{2+} and Mg^{2+}, from water.

wavelength a characteristic property of light, similar to its color, and equal to the length of a full wave; often expressed in nanometers; can be measured with a spectroscope.

weak as applied to acids and bases, being partially ionized in water solution. Example: Acetic acid, $HC_2H_3O_2$, is a weak acid because in water solution it is only slightly ionized to H^+ and $C_2H_3O_2^-$ ions.

weak electrolytes a species that, in water solution, forms an equilibrium mixture of molecules and ions. Examples: HF, NH_3.

work (w) one of the effects that may be associated with an energy change; during a change a system may do work on its surroundings, equivalent to the raising of a mass. Work may be electrical, mechanical, or due to expansion or compression, and may be done by, or on, the system.

X

X an unknown quantity.

x-rays light rays having a wavelength of 0.001 to 1.0 nm.

Y

yield the amount of product obtained from a reaction.

Z

zeolite a type of silicate mineral used to soften water by cation exchange.

zero-order reaction a reaction whose rate is independent of reactant concentration.

Index

Kathy Coglon

Kathy Ogden